CHAPTER 3 Formulas, Applications, and Linear Inequalities

Geometric Formulas:

SQUARE

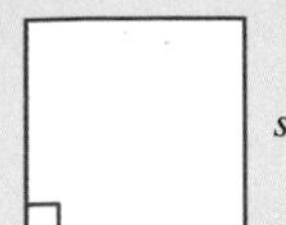

Perimeter: $P = 4s$

Area: $A = s^2$

RECTANGLE

Perimeter: $P = 2l + 2w$

Area: $A = lw$

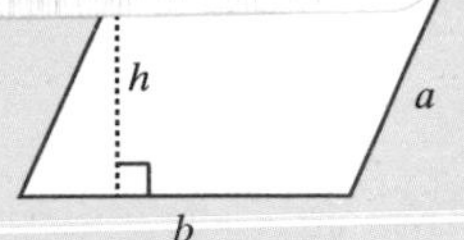

Perimeter: $P = 2a + 2b$

Area: $A = bh$

TRIANGLE

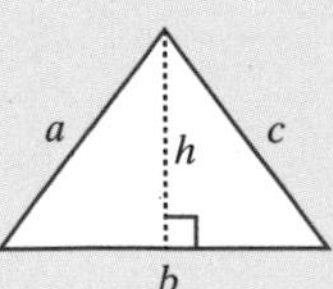

Perimeter: $P = a + b + c$

Area: $A = \frac{1}{2}bh$

CIRCLE

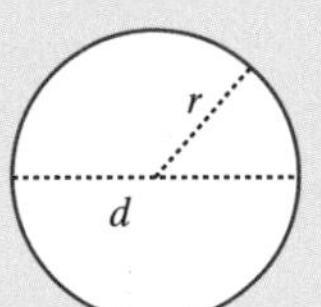

Circumference: $C = 2\pi r = \pi d$

Area: $A = \pi r^2$

TRAPEZOID

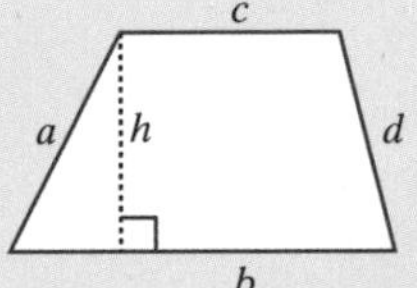

Perimeter: $P = a + b + c + d$

Area: $A = \frac{1}{2}h(b + c)$

Application Formulas:

Simple Interest: $I = Prt$

Temperature: $C = \frac{5}{9}(F - 32)$

Distance: $d = rt$

Perimeter: $P = 2l + 2w$

Lateral Surface Area: $L = 2\pi rh$

Force: $F = ma$

Sum of the Angles of a Triangle:

$\alpha + \beta + \gamma = 180°$

Ratio:

$\frac{a}{b}$ or $a : b$ or a to b

Volume Formulas:

Rectangular Solid: $V = lwh$

Rectangular Pyramid: $V = \frac{1}{3}lwh$

Right Circular Cylinder: $V = \pi r^2 h$

Right Circular Cone: $V = \frac{1}{3}\pi r^2 h$

Sphere: $V = \frac{4}{3}\pi r^3$

Proportion:

$\frac{a}{b} = \frac{c}{d}$, where $b \neq 0$

Similar Triangles:

$\Delta ABC \sim \Delta DEF$, if $\frac{AB}{DE} = \frac{BC}{EF}$ and $\frac{AB}{DE} = \frac{AC}{DF}$ and $\frac{BC}{EF} = \frac{AC}{DF}$

Addition Property of Inequality:

If $A < B$, then $A + C < B + C$.

If $A < B$, then $A - C < B - C$.

Multiplication Property of Inequality:

If $A < B$ and $C > 0$, then $AC < BC$.

If $A < B$ and $C > 0$, then $\frac{A}{C} < \frac{B}{C}$.

If $A < B$ and $C < 0$, then $AC > BC$.

If $A < B$ and $C < 0$, then $\frac{A}{C} > \frac{B}{C}$.

CHAPTER 4 Straight Lines and Functions

Summary of Formulas and Properties of Straight Lines:

1. $Ax + By = C$, where A and B do not equal 0. — Standard form
2. $m = \frac{y_2 - y_1}{x_2 - x_1}$, where $x_1 \neq x_2$. — Slope of a line
3. $y = mx + b$ — Slope-intercept form
4. $y - y_1 = m(x - x_1)$ — Point-slope form
5. $y = b$ — Horizontal line, $m = 0$
6. $x = a$ — Vertical line, m is undefined
7. Parallel lines have the same slope ($m_1 = m_2$).
8. Perpendicular lines have slopes that are negative reciprocals of each other $\left(m_2 = \frac{-1}{m_1} \text{ or } m_1 m_2 = -1 \right)$.

Relation, Domain, and Range:

A **relation** is a set of ordered pairs of real numbers.

The **domain**, ***D***, of a relation is the set of all first coordinates in the relation.

The **range**, ***R***, of a relation is the set of all second coordinates in the relation.

Function:

A **function** is a relation in which each domain element has a unique range element.

Vertical Line Test:

If **any** vertical line intersects the graph of a relation at more than one point, then the relation graphed is **not** a function.

CHAPTER 5 Exponents and Polynomials

Properties of Exponents:

For nonzero real numbers a and b and integers m and n,

The Exponent 1: $a = a^1$ (a is any real number.)

The Exponent 0: $a^0 = 1 \quad (a \neq 0)$

Product Rule: $a^m \cdot a^n = a^{m+n}$

Quotient Rule: $\dfrac{a^m}{a^n} = a^{m-n}$

Power Rule: $\left(a^m\right)^n = a^{mn}$

Negative Exponents Rule: $a^{-n} = \dfrac{1}{a^n}$ and $\dfrac{1}{a^{-n}} = a^n$

Power Rule for Products: $(ab)^n = a^n b^n$

Power Rule for Fractions: $\left(\dfrac{a}{b}\right)^n = \dfrac{a^n}{b^n}$

Classification of Polynomials:

Monomial: polynomial with one term

Binomial: polynomial with two terms

Trinomial: polynomial with three terms

Division Algorithm:

$$\frac{P(x)}{D(x)} = Q(x) + \frac{R(x)}{D(x)}, \quad (D(x) \neq 0)$$

FOIL Method:

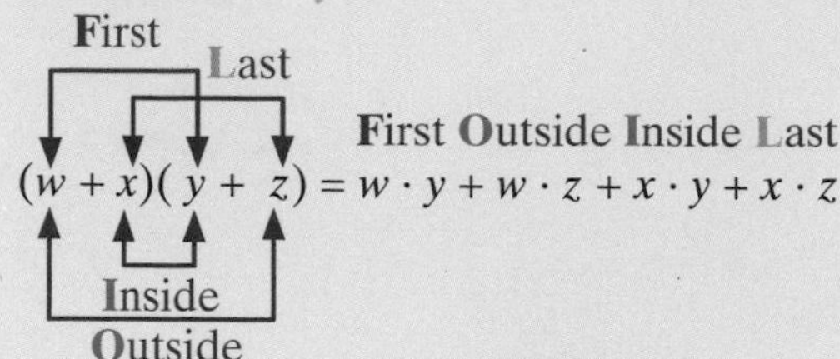

Special Products of Polynomials:

1. $(X+A)(X-A) = X^2 - A^2$: Difference of two squares
2. $(X+A)^2 = X^2 + 2AX + A^2$: Perfect square trinomial
3. $(X-A)^2 = X^2 - 2AX + A^2$: Perfect square trinomial
4. $(X-A)(X^2 + AX + A^2) = X^3 - A^3$: Difference of two cubes
5. $(X+A)(X^2 - AX + A^2) = X^3 + A^3$: Sum of two cubes

CHAPTER 6 Factoring Polynomials and Solving Quadratic Equations

Zero Factor Property:

If a and b are real numbers, and $a \cdot b = 0$, then $a = 0$ or $b = 0$ or both.

Quadratic Equation:

An equation that can be written in the form $ax^2 + bx + c = 0$ where a, b, and c are real numbers and $a \neq 0$ is called a **quadratic equation**.

Factor Theorem:

If $x = c$ is a root of a polynomial equation in the form $P(x) = 0$, then $x - c$ is a factor of the polynomial $P(x)$.

The Pythagorean Theorem:

In a right triangle, the square of the hypotenuse is equal to the sum of the squares of the legs.

$$c^2 = a^2 + b^2$$

CHAPTER 7 Rational Expressions

Rational Expression:

A **rational expression** is an expression of the form $\dfrac{P}{Q}$ (or in function notation, $\dfrac{P(x)}{Q(x)}$) where P and Q are polynomials and $Q \neq 0$.

Fundamental Principle of Fractions:

If $\dfrac{P}{Q}$ is a rational expression and K is a polynomial and $K \neq 0$, then

$$\frac{P}{Q} = \frac{P}{Q} \cdot \frac{K}{K} = \frac{P \cdot K}{Q \cdot K}.$$

Negative Signs in Rational Expressions:

$$-\frac{P}{Q} = \frac{P}{-Q} = \frac{-P}{Q} \quad \text{and} \quad \frac{P}{Q} = \frac{-P}{-Q} = -\frac{-P}{Q} = -\frac{P}{-Q}$$

Opposites in Rational Expressions:

In general, $\dfrac{-P}{P} = -1$ if $P \neq 0$. In particular, $\dfrac{a-x}{x-a} = -1$ if $x \neq a$.

Multiplication of Rational Expressions:

$\dfrac{P}{Q} \cdot \dfrac{R}{S} = \dfrac{P \cdot R}{Q \cdot S}$ where $Q, S \neq 0$.

Division of Rational Expressions:

$\dfrac{P}{Q} \div \dfrac{R}{S} = \dfrac{P}{Q} \cdot \dfrac{S}{R}$ where $Q, R, S \neq 0$.

Addition and Subtraction of Rational Expressions:

$\dfrac{P}{Q} + \dfrac{R}{Q} = \dfrac{P+R}{Q}$ and $\dfrac{P}{Q} - \dfrac{R}{Q} = \dfrac{P-R}{Q}$ where $Q \neq 0$.

Complex Fraction:

A **complex fraction** is a fraction in which the numerator or denominator is a fraction or the sum or difference of fractions.

Instructor's Annotated Edition

INTRODUCTORY &INTERMEDIATE ALGEBRA

INTRODUCTORY &INTERMEDIATE ALGEBRA

D. FRANKLIN WRIGHT
CERRITOS COLLEGE

Editor: Jennifer Knowles Butler
Developmental Editor: Marcel Prevuznak
Production Editors: Mary Janelle Cady, Mandy Glover
Answer Key Editors: Priyanka Bihani, Robin Lawson, Nina Miller,
Susan Rackley, Ashley Rankin, Kim Scott, Lindsay Stevens
Editorial Assistants: Kelly Epperson, K. V. Jagannadham, Lori Layne, P. Srikanth,
James Stambaugh, John Thomas, Ryan Wincey, Stephen Yackey
Layout: QSI (Pvt.) Ltd.: U. Nagesh, E. Jeevan Kumar
Art: Ayvin Samonte
Cover Art and Design: Johnson Design

A division of Quant Systems, Inc.

Library of Congress Control Number: 2004108127

Printed in the United States of America

ISBN:
Student: 0-918091-90-X
Student Solution Manual: 1-932628-03-7

CONTENTS

CHAPTER 3 Formulas, Applications, and Linear Inequalities 159

CHAPTER 4 Straight Lines and Functions 233

CHAPTER 5 Exponents and Polynomials 327

CHAPTER 6 Factoring Polynomials and Solving Quadratic Equations 403

CHAPTER 7 Rational Expressions 475

CHAPTER 8 Review of Chapters 1 – 7 545

CHAPTER 9 Systems of Linear Equations I 625

CHAPTER 10 Roots, Radicals, and Complex Numbers 679

CHAPTER 11 Quadratic Equations 747

CHAPTER 12 Quadratic Functions and Conic Sections 793

CHAPTER 13 Exponential and Logarithmic Functions 885

CHAPTER 14 Systems of Linear Equations II 971

PREFACE

Purpose and Style

Introductory & Intermediate Algebra is a comprehensive and versatile teaching tool. For the beginning student, this book develops the more abstract skills and reasoning abilities needed to master algebra. For the more experienced student, it provides a solid base for further studies in mathematics.

With feedback from users, insightful comments from reviewers, and skillful editing and design by the editorial staff at Hawkes Learning Systems, we have confidence that students and instructors alike will find that this text is indeed a superior teaching and learning tool. The text may be used independently or in conjunction with the software package ***Hawkes Learning Systems: Introductory & Intermediate Algebra*** developed by Hawkes Learning Systems.

Chapter 1 begins with the algebraic concepts of integers, real number lines, absolute value, and operations with integers. Geometric figures and related formulas are introduced to provide practice with algebraic concepts and as reference for word problems throughout the text. The transition from introductory algebra to intermediate algebra occurs midway through the text. We have included a chapter (Chapter 8) specifically designed as a comprehensive review of introductory algebra. This chapter is an excellent resource tool for both the beginning student and the experienced student.

The introductory portion of the text is written assuming that students have basic arithmetic knowledge and no previous experience with algebra; however, in the second half of the text, students will find that the review is comprehensive and that the pace of coverage is somewhat faster and in more depth than they have seen in previous courses. As with any text in mathematics, students should read the text carefully and thoroughly.

The style of the text is informal and nontechnical while maintaining mathematical accuracy. Each topic is developed in a straightforward step-by-step manner. Each section contains many carefully developed and worked out examples to lead the students successfully through the exercises and prepare them for examinations. Whenever appropriate, information is presented in list form for organized learning and easy reference. Common errors are highlighted and explained so that students can avoid such pitfalls and better understand the correct corresponding techniques. Practice problems with answers are provided in nearly every section to allow the students to "warm up" and to provide the instructor with immediate classroom feedback.

The NCTM and AMATYC curriculum standards have been taken into consideration in the development of the topics throughout the text. In particular:

- there is emphasis on reading and writing skills as they relate to mathematics
- techniques for using a graphing calculator are discussed early
- a special effort has been made to make the exercises motivating and interesting
- geometric concepts are integrated throughout
- statistical concepts, such as interpreting bar graphs and calculating elementary statistics, are included where appropriate

Real Numbers

CHAPTER 1

Did You Know?

Arithmetic operations defined on the set of positive integers, negative integers, and zero are studied in this chapter. The integer zero will be shown to have interesting properties under the operations of addition, subtraction, multiplication, and division.

Curiously, zero was not recognized as a number by early Greek mathematicians. When Hindu scientists developed the place-value numeration system we currently use, the zero symbol was initially a place holder but not a number. The spread of Islam transmitted the Hindu number system to Europe where it became known as the Hindu-Arabic system and replaced Roman numerals. The word zero comes from the Hindu word meaning "void," which was translated into Arabic as "sifr" and later into Latin as "zephirum," hence the derivation of our English words "zero" and "cipher."

Almost all of the operational properties of zero were known to the Hindus. However, the Hindu mathematician Bhaskara the Learned (1114 – 1185?) asserted that a number divided by zero was zero, or possibly infinite. Bhaskara did not seem to understand the role of zero as a divisor since division by zero is undefined and hence, an impossible operation in mathematics.

Albert Einstein, in his development of a proof that the universe was stable and unchangeable in time, divided both sides of one of his intermediate equations by a complicated expression that under certain circumstances could become zero. When the expression became zero, Einstein's proof did not hold and the possibility of a pulsating, expanding, or contracting universe had to be considered. This error was pointed out to Einstein and he was forced to withdraw his proof that the universe was stable. The

Einstein

"How can it b all a product of experience objects of rea

Einst

Did You Know?:

A feature at the beginning of every chapter presents some interesting math history related to the chapter at hand.

Introduction:

Presented before the first section of every chapter, this feature provides an introduction to the subject of the chapter and its purpose.

CHAPTER 1 Real Numbers

Numbers and number concepts form the foundation for the study of algebra. In this chapter, you will learn about positive and negative numbers and how to operate with these numbers. Believe it or not, the key number is 0. Pay particularly close attention to the idea of the magnitude of a number, called its absolute value, and the terminology used to represent different types of numbers.

1.1 The Real Number Line and Absolute Value

Objectives

After completing this section, you will be able to:

1. *Identify types of numbers.*
2. *Determine if given numbers are greater than, less than, or equal to other given numbers.*
3. *Determine absolute values.*

The study of algebra requires that we know a variety of types of numbers and their names. The set of numbers

$$N = \{1, 2, 3, 4, 5, 6, 7, 8, 9, 10, 11, \ldots\}$$

is called the **counting numbers** or **natural numbers**. The three dots indicate that the pattern is to continue without end. Putting 0 with the set of natural numbers gives the set of **whole numbers**.

$$W = \{0, 1, 2, 3, 4, 5, 6, \ldots\}$$

Thus, mathematicians make the distinction that 0 is a whole number but not a natural number.

To help in understanding different types of numbers and their relationships to each other, we begin with a "picture" called a **number line**. For example, choose some point on a horizontal line and label it with the number 0 (Figure 1.1).

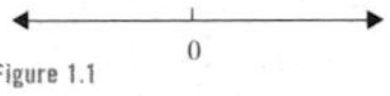

0

Figure 1.1

Now choose another point on the line to the right of 0 and label it with the number 1 (Figure 1.2).

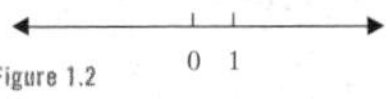

0 1

Figure 1.2

2

Objectives:

The objectives provide the students with a clear and concise list of skills presented in each section.

CHAPTER 1 Real Numbers

Example 1: Like Denominators

a. $\frac{3}{8}+\frac{4}{8}$

Solution: $\frac{3}{8}+\frac{4}{8}=\frac{3+4}{8}=\frac{7}{8}$

b. $\frac{9}{10}+\frac{3}{10}$

Solution: $\frac{9}{10}+\frac{3}{10}=\frac{9+3}{10}=\frac{12}{10}=\frac{\cancel{2}\cdot 6}{\cancel{2}\cdot 5}=\frac{6}{5}$

c. $\frac{2}{15}+\frac{3}{15}+\frac{1}{15}+\frac{6}{15}$

Solution: $\frac{2}{15}+\frac{3}{15}+\frac{1}{15}+\frac{6}{15}=\frac{2+3+1+6}{15}=\frac{12}{15}=\frac{\cancel{3}\cdot 4}{\cancel{3}\cdot 5}=\frac{4}{5}$

d. $\frac{3}{8x}+\frac{7}{8x}+\frac{5}{8x}$

Solution: $\frac{3}{8x}+\frac{7}{8x}+\frac{5}{8x}=\frac{3+7+5}{8x}=\frac{15}{8x}$

NOTES

In Example 1b above, the fraction $\frac{6}{5}$ is called an **improper fraction** because the numerator is larger than the denominator. Such unfortunate terminology implies that there is something wrong with improper fractions. This is not the case, and improper fractions are used throughout the study of mathematics. Improper fractions can be changed to mixed numbers, with a whole number and a fraction part. Divide the numerator by the denominator an numerator of the new fraction. For example,

$$\frac{6}{5}=1\frac{1}{5},\quad \frac{11}{7}=1\frac{4}{7},\quad \text{and}\quad \frac{3}{6}$$

The decision of whether to leave an answer in the for as a mixed number is optional. Generally, in algebra fraction form. However, in the case of an application are involved, a mixed number form may be preferred. write $\frac{3}{2}$ feet as $1\frac{1}{2}$ feet.

52

Examples:

Examples are denoted with titled headers indicating the problem solving skill being presented. Each section contains many carefully explained examples with lots of tables, diagrams, and graphs. Examples are presented in an easy to understand step-by-step fashion and annotated for additional clarification.

Notes:

Notes highlight common mistakes and give additional clarification to more subtle details.

SECTION 1.1 The Real Number Line and Absolute Value

The distance between a number and 0 on a number line is called its **absolute value** and is symbolized by two vertical bars, | |. Thus, | +7 | = 7 and | −7 | = 7. Similarly,

$|3| = 3$ (3 units, from 0 to 3)

$|-4| = 4$ (4 units, from −4 to 0)

$|-2| = 2$ (2 units, from −2 to 0)

$|0| = 0$

$\left|-\frac{4}{3}\right| = \frac{4}{3}$ ($\frac{4}{3}=1\frac{1}{3}$ units, from $-\frac{4}{3}$ to 0)

Since distance (similar to length) is never negative, the absolute value of a number is never negative. Therefore, the absolute value of a nonzero number is always positive.

Definition Boxes:

Definitions are presented in highly visible boxes for easy reference.

Absolute Value

*The **absolute value** of a real number is its distance from 0. Note that the absolute value of a real number is never negative.*

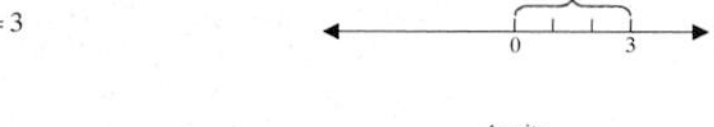

$|a| = a$ *if a is a positive number or 0.*

$|a| = -a$ *if a is a negative number.*

9

Calculator Instruction:

Step-by-step instructions are presented to introduce students to basic graphing skills with a TI-83 Plus calculator along with actual screen shots of a TI-83 Plus for visual reference.

The screen will display

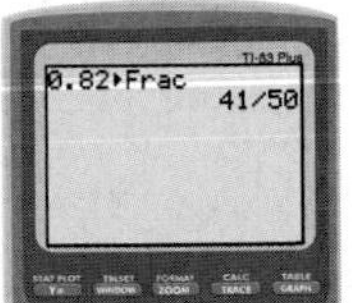

Note that the fraction is reduced.

b. Proceed as follows:

Enter the fraction $\frac{5}{8}$ by entering $5 \div 8$.

Press 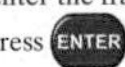.

The calculator will automatically give the answer in decimal form and the screen will appear as follows:

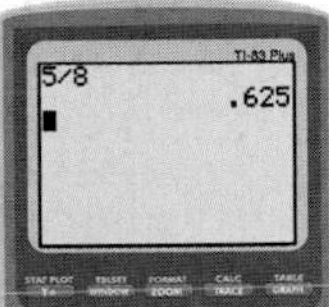

c. Proceed as follows:

Enter the sum of the fractions as $(3 \div 5) + (3 \div 20)$.

Press MATH.

Press ENTER.

Press ENTER.

The display on the screen will appear

Note t

Practice Problems:

Practice Problems are presented at the end of almost every section with answers giving the students an opportunity to practice their newly acquired skills.

b. $2(3^2 - 1) - 3 \cdot 2^3 = 2(9-1) - 3 \cdot 8$ Exponents.

$= 2(8) - 3 \cdot 8$ Subtract inside the parentheses.

$= 16 - 24$ Multiply.

$= -8$ Subtract (or add algebraically).

c. $9 - 2[(3 \cdot 5 - 7^2) \div 2 + 2^2] = 9 - 2[(3 \cdot 5 - 49) \div 2 + 4]$ Exponents.

$= 9 - 2[(15 - 49) \div 2 + 4]$ Multiply inside the parentheses.

$= 9 - 2[(-34) \div 2 + 4]$ Subtract inside the parentheses.

$= 9 - 2[-17 + 4]$ Divide inside the brackets.

$= 9 - 2[-13]$ Add inside the brackets.

$= 9 + 26$ Multiply.

$= 35$ Add.

Practice Problems

1. List all the prime numbers less than 12.

2. Find the prime factorization of 75.

3. Find the prime factorization of 117.

Find the value of each expression by using the Rules for Order of Operations.

4. $14 \div 2 + 2 \cdot 6 + 30 \div 3$

5. $3[8 + 2(1 - 10)] - 15 \div 5$

1.5 Exercises

In Exercises 1 – 16, rewrite each of the following integers in a base and exponent form without using an exponent of 1.

1. 25 **2.** 36 **3.** 49 **4.** 81 **5.** 121 **6.** 169 **7.** 64
8. 144 **9.** 8 **10.** 27 **11.** 125 **12.** 32 **13.** 243 **14.** 9
15. 100 **16.** 1000

Find the value of each of the following expressions in Exercises 17 – 35.

17. 8^2 **18.** 1^5 **19.** 10^2 **20.** 6^3 **21.** 7^3 **22.** 9^2
23. 10^4 **24.** 5^4 **25.** 13^0 **26.** $(-16)^0$ **27.** $(-15)^2$ **28.** 20^2
29. 30^3 **30.** 50^2 **31.** $(-10)^4$ **32.** $(-5)^3$ **33.** $(-6)^2$ **34.** -6^2
35. -2^6

Answers to Practice Problems: 1. 2, 3, 5, 7, 11 **2.** $3 \cdot 5^2$ **3.** $3^2 \cdot 13$ **4.** 29 **5.** -33

39

Exercises:

Each section includes a variety of paired and graded exercises to give the students much needed practice applying and reinforcing the skills learned in the section. More than 4600 carefully selected and graded exercises are provided in the sections. The exercises proceed from relatively easy to more difficult ones.

CHAPTER 1 Real Numbers

Chapter 1 Index of Key Ideas and Terms

Section 1.1 The Real Number Line and Absolute Value

82

Index of Key Ideas and Terms:

Each chapter contains an index highlighting the main concepts and skills presented in the chapter along with full definitions and page numbers for easy reference.

Writing and Thinking:

These exercises provide the student an opportunity to independently explore and expand on concepts presented in the chapter.

CHAPTER 2 Algebraic Expressions, Linear Equations, and Applications

Writing and Thinking About Mathematics

51. A man and his wife wanted to sell their house and contacted a realtor. The realtor stated that the price for the services (listing, advertising, selling, and paperwork) was 6% of the selling price. The realtor asked the couple how much cash they wanted after the realtor fee was deducted from the selling price. They said $141,000 and that therefore, the asking price should be 106% of $141,000 or a total of $149,460. The realtor declared that this was the wrong figure and that the percent used should be 94% and not 106%.

a. Explain why 106% is the wrong percent and $149,460 is not the right selling price.

b. Explain why and how 94% is the correct percent and just what the selling price should be for the couple to receive $141,000 after the realtor's fee is deducted.

Collaborative Learning Exercise

With the class separated into teams of two to four students, each team is to analyze the following problem and decide how to answer the related questions. Then each team leader is to present the team's answers and related ideas to the class for general discussion.

52. Sam works at a picture framing store and gets a salary of $500 per month plus a commission of 3% on sales he makes over $2000. Maria works at the same store, but she has decided to work on a straight commission of 8% of her sales.

a. At what amount of sales will Sam and Maria make the same amount of money?

b. Up to that point, who would be making more?

c. After that point, who would be making more? Explain briefly. (If you were offered a job at that store, which method of payment would you choose?)

Hawkes Learning Systems: Introductory & Intermediate Algebra

Percents and Applications

148

Hawkes Learning Systems:

Each section's exercises are followed by a feature which highlights corresponding lessons in *HLS: Introductory & Intermediate Algebra* software for ease of professor-assigned or student-motivated assignments, review, and instruction.

Additional Features

Calculator Problems: Each problem is designed to highlight the usefulness of a calculator in solving certain complex problems but maintain the necessity of understanding the concepts behind the problem.

Chapter Test: Provides an opportunity for the students to practice the skills presented in the chapter in a test format.

Cumulative Review: As new concepts build on previous concepts, the cumulative review provides the student with an opportunity to continually reinforce existing skills while practicing newer skills.

Answers: Answers are provided for odd numbered section exercises and for all even and odd numbered exercises in the Chapter Tests and Cumulative Reviews.

Teachers' Edition:

Answers: Answers to all the exercises are conveniently located in the margins next to the problems.

Teaching Notes: Suggestions for more in-depth classroom discussions and alternate methods and techniques are located in the margins.

Also included in this edition:

- New reader friendly layout
- Arrangement of chapters for better flow, continuity and progression
- Writing and Thinking About Mathematics
- Calculator Instructions (New emphasis on the graphing calculator)
- Calculator Problems
- Comprehensive review of introductory algebra topics
  (Chapter 8: Review of Chapters 1 – 7)

Content

The TI-83 Plus graphing calculator has been made an integral part of many of the presentations in this textbook. To get maximum benefits from the use of this text, the student must have one of these calculators (or a calculator with similar features). Directions are given for using the related calculator commands as they are needed throughout.

Chapter 1, Real Numbers, develops the algebraic concept of integers and the basic skills of operations with integers. Real, rational, and irrational numbers are discussed in conjuction with the real number line. Variables, absolute value, and exponents are defined and expressions are evaluated by using the rules for order of operations. Rational numbers (or fractions) are introduced. Two sections cover basic operations with fractions. A section on decimal numbers and change in value is also included. The chapter closes with a discussion of the properties of addition and multiplication with real numbers.

Chapter 2, Algebraic Expressions, Linear Equations, and Applications, shows how arithmetic concepts, through the use of variables and signed numbers, can be generalized with algebraic expressions. Algebraic expressions are simplified by combining like terms and evaluated by using the rules for order of operations. As a lead-in to interpreting and understanding word problems, a section involving translating English phrases and algebraic expressions is included. The chapter then goes on to develop the techniques for solving linear (or first-degree) equations in a step-by-step manner over two sections. Techniques include combining like terms and use of the distributive property. Applications relate to number problems, consecutive integers, and percent.

Chapter 3, Formulas, Applications, and Linear Inequalities, begins by applying the techniques acquired in the previous chapter to work with formulas. A variety of word problems as well as formulas related to the geometric concepts of perimeter, area, and volume are included. Other topics covered in this chapter are ratios and proportions, interval notation, and solving linear inequalities.

Chapter 4, Straight Lines and Functions, allows for the early introduction of a graphing calculator and the ideas and notation related to functions. Included are complete discussions on the three basic forms for equations of straight lines in a plane: the standard form, the slope-intercept form, and the point-slope form. Slope is discussed for parallel and perpendicular lines and treated as a rate of change. Functions are introduced and the vertical line test is used to tell whether or not a graph represents a function. Use of a TI-83 Plus graphing calculator is an integral part of this introduction to functions as well as part of graphing linear inequalities in the last section.

Chapter 5, Exponents and Polynomials, studies the properties of exponents in depth and shows how to read and write scientific notation. The remainder of the chapter is concerned with definitions and operations related to polynomials. Included are the FOIL method of multiplication with two binomials, special products of binomials, and the division algorithm.

Chapter 6, Factoring Polynomials and Solving Quadratic Equations, discusses methods of factoring polynomials, including finding common monomial factors, factoring by grouping, factoring trinomials by grouping and by trial-and-error, and factoring special products. A special subsection on factoring with negative exponents is included. The topic of solving quadratic equations is introduced, and quadratic equations are solved by factoring only. Applications with quadratic equations are included in two sections and involve topics such as the use of function notation to represent area, the Pythagorean Theorem, and consecutive integers. A section on using a graphing calculator to solve equations and inequalities provides the student with the opportunity to become more familiar with the calculator and to solve more difficult equations.

Chapter 7, Rational Expressions, provides still more practice with factoring and shows how to use factoring to operate with rational expressions. Included are the topics of multiplication, division, addition and subtraction with rational expressions, simplifying complex fractions, and solving equations and inequalities containing rational expressions. Applications are related to work, distance-rate-time, and variation.

Chapter 8, Review of Chapters 1 – 7, is a comprehensive review of topics from beginning algebra, which are covered in Chapters 1 through 7 in this text. Each section of Chapter 8 corresponds to an earlier chapter of the book. To aid the students in their review, various topics pertaining to the current text are listed in the left margin with page numbers so that the student can refer to previous, more detailed discussions.

Chapter 9, Systems of Linear Equations I, shows how to solve systems of linear equations three ways: by graphing (including the use of a graphing calculator), by substitution, and by addition. Applications are related to distance-rate-time, number problems, amounts and cost, interest, and mixture.

Chapter 10, Roots, Radicals, and Complex Numbers, introduces roots and fractional exponents and the use of a calculator to find estimated values. Arithmetic with radicals includes simplifying radical expressions, addition, subtraction, and rationalizing numerators and denominators. A new section on functions with radicals shows how to analyze the domain and range of radical functions and how to graph these functions by using a graphing calculator. Complex numbers are introduced along with the basic operations of addition, subtraction, multiplication, and division. These are skills needed for the work with quadratic equations and quadratic functions in Chapters 11 and 12.

Chapter 11, Quadratic Equations, reviews solving quadratic equations by factoring and introduces the methods of using the square root property and completing the square. The quadratic formula is developed by completing the square and students are encouraged to use the most efficient method for solving any particular quadratic equation. Applications are related to the Pythagorean Theorem, projectiles, geometry, and cost per person. The last two sections cover solving equations with radicals and solving equations in quadratic form.

Chapter 12, Quadratic Functions and Conic Sections, provides a basic understanding of conic sections (parabolas, circles, ellipses, and hyperbolas) and their graphs. The first section gives detailed analyses of parabolas as functions involving horizontal and vertical shifting and finding maximum and minimum values with applications. Quadratic inequalities are solved algebraically with factoring, the aid of number lines, and with techniques using the graphing calculator. Function notation is used in discussing reflections and translations of a variety of types of functions. Vertical and horizontal parabolas are then developed as conic sections. The distance formula and midpoint formula are included along with the discussion of circles. The thorough development of ellipses and hyperbolas includes graphs with centers not at the origin. Solving systems with nonlinear equations is the final topic of the chapter.

Chapter 13, Exponential and Logarithmic Functions, begins with a new section on the algebra of functions and leads to the development of the composition of functions and methods for finding the inverses of one-to-one functions. This introduction lays the groundwork for understanding the relationship between exponential functions and logarithmic functions. While the properties of real exponents and logarithms are presented completely, most numerical calculations are performed with the aid of a calculator. Special emphasis is placed on the number e and applications with natural logarithms. Students will find the applications with exponential and logarithmic functions among the most interesting and useful in their mathematical studies. Those students who plan to take a course in calculus should be aware that many of the applications found in calculus involve exponential and logarithmic expressions in some form.

Chapter 14, Systems of Linear Equations II, covers systems of three equations in three variables. The basic methods of graphing, substitution, and addition are included along with matrices and Gaussian elimination, determinants, and Cramer's Rule. Double subscript notation is now used with matrices for an easy transition to use of matrices in solving systems of equations with the TI-83 Plus calculator. Applications involve mixture, interest, work, algebra, and geometry. The last section discusses half-planes, graphing systems of linear inequalities, and linear programming, again including the use of a graphing calculator.

Chapter 15, Sequences, Series, and the Binomial Theorem, provides flexibility for the instructor and reference material for the students. The topics presented here, including permutations and combinations, are likely to appear in courses in probability and statistics, finite mathematics, and higher level courses in mathematics. Any of the topics covered in this chapter will give students additional mathematical experience and insight for future studies.

I recommend that the topics be covered in the order presented because most sections assume knowledge of the material in previous sections. This is particularly true of the cumulative review sections at the end of each chapter. Of course, time and other circumstances may dictate another sequence of topics. For example, in some programs, Chapters 1 and 2 might be considered review. In case of any changes, the instructor should be sure that the students are somewhat familiar with a graphing calculator.

About the annotations in the Instructor's Edition

The annotations in the margin entitled Teaching Notes: are ideas for on-the-spot helpful classroom suggestions. I am sure you are familiar with most of them; but many of us, myself included, have thought just after a class has been dismissed, "Oh, I wish I had mentioned _____ for interest or motivation or variety." These annotations, provided with a lot of help and insight from Professor Thomas Clark of Trident Technical College, include extra examples, historical comments, ideas for possible class discussions, and other comments that you might find interesting or helpful. Obviously, they are for use at your own discretion.

Thank you for using my text and I look forward to receiving any suggestions, comments, or errata that you might bring to my attention for future printing and editions.

Frank

Acknowledgements

I would like to thank Editor Jennifer Knowles Butler and Developmental Editor Marcel Prevuznak for their hard work and invaluable assistance in the development and production of this text.

Many thanks go to the following reviewers who offered their constructive and critical comments:

Russ Baker, *Howard Community College*
Linda Buchanan, *Howard College-Big Spring*
Connie Buller, *Metropolitan Community College-Fort Omaha*
Thom Clark, *Trident Technical College*
Elaine Elkind, *Kean University*
Terry Fung, *Kean University*
Theresa Hert, *Mount San Jacinto College*
Sandee House, *Georgia Perimeter College-Clarkston*
Linda Houston, *Tulsa Community College-Metro*
Laura Hoye, *Trident Technical College*
Charyl Link, *Kansas City Kansas Community College*
Lois Miller, *Golden West College*
Leah Pierce, *Crafton Hills College*
Virginia Puckett, *Miami-Dade Community College*
Michael Sanchez, *Sacramento City College*
Bill Schurter, *University of the Incarnate Word*
Peggie Smith, *Bowie State University*
Consuelo Stewart, *Howard Community College*
Rob Van Kirk, *Idaho State University*
Jack Wadhams, *Golden West College*
Karla Williams, *National Park Community College*

Finally, special thanks go to James Hawkes and Greg Hill for their faith in this edition and their willingness to commit so many resources to guarantee a top-quality product for students and teachers.

D. Franklin Wright

TO THE STUDENT

The goal of this text and of your instructor is for you to succeed in intermediate algebra. Certainly, you should make this your goal as well. What follows is a brief discussion about developing good work habits and using the features of this text to your best advantage. For you to achieve the greatest return on your investment of time and energy, you should practice the following three rules of learning.

1. Reserve a block of time for study every day.
2. Study what you don't know.
3. Don't be afraid to make mistakes.

How to use this book

The following seven-step guide will not only make using this book a more worthwhile and efficient task, but it will also help you benefit more from classroom lectures or the assistance that you receive in a math lab.

1. Try to look over the assigned section(s) before attending class or lab. In this way, new ideas may not sound so foreign when you hear them mentioned again. This will also help you see where you need to ask questions about material that seems difficult to you.

2. Read examples carefully. They have been chosen and written to show you all of the problem-solving steps that you need to be familiar with. You might even try to solve example problems on your own before studying the solutions that are given.

3. Work the section exercises faithfully as they are assigned. Problem-solving practice is the single most important element in achieving success in any math class, and there is no good substitute for actually doing this work yourself. Demonstrating that you can think independently through each step of each type of problem will also give you confidence in your ability to answer questions on quizzes and exams. Check the Answer Key periodically while working section exercises to be sure that you have the right ideas and are proceeding in the right manner.

4. Use the Writing and Thinking About Mathematics questions as an opportunity to explore the way that you think about math. A big part of learning and understanding mathematics is being able to talk about mathematical ideas and communicate the thinking that you do when you approach new concepts and problems. These questions can help you analyze your own approach to mathematics and, in class or group discussions, learn from ideas expressed by your fellow students.

5. Use the Chapter Index of Key Ideas and Terms as a recap when you begin to prepare for a Chapter Test. It will reference all the major ideas that you should be familiar with from that chapter and indicate where you can turn if review is needed. You can also use the Chapter Index as a final checklist once you feel you have completed your review and are prepared for the Chapter Test.
6. Chapter Tests are provided so that you can practice for the tests that are actually given in class or lab. To simulate a test situation, block out a one-hour, uninterrupted period in a quiet place where your only focus is on accurately completing the Chapter Test. Use the Answer Key at the back of the book as a self-check only after you have completed all of the questions on the test.
7. Cumulative Reviews will help you retain the skills that you acquired in studying earlier chapters. They appear after every chapter beginning with Chapter 2. Approach them in much the same manner as you would the Chapter Tests in order to keep all of your skills sharp throughout the entire course.

How to Prepare for an Exam

Gaining Skill and Confidence

The stress that many students feel while trying to succeed in mathematics is what you have probably heard called "math anxiety." It is a real-life phenomenon, and many students experience such a high level of anxiety during mathematics exams in particular that they simply cannot perform to the best of their abilities. It is possible to overcome this stress simply by building your confidence in your ability to do mathematics and by minimizing your fears of making mistakes.

No matter how much it may seem that in mathematics you must either be right or wrong, with no middle ground, you should realize that you can be learning just as much from the times that you make mistakes as you can from the times that your work is correct. Success will come. Don't think that making mistakes at first means that you'll never be any good at mathematics. Learning mathematics requires lots of practice. Most importantly, it requires a true confidence in yourself and in the fact that with practice and persistence the mistakes will become fewer, the successes will become greater, and you will be able to say, "I can do this."

Showing What You Know

If you have attended class or lab regularly, taken good notes, read your textbook, kept up with homework exercises, and asked for help when it was needed, then you have already made significant progress in preparing for an exam and conquering any anxiety. Here are a few other suggestions to maximize your preparedness and minimize your stress.

1. Give yourself enough time to review. You will generally have several days advance notice before an exam. Set aside a block of time each day with the goal of reviewing a manageable portion of the material that the test will cover. Don't cram!
2. Work lots of problems to refresh your memory and sharpen your skills. Go back to redo selected exercises from all of your homework assignments.
3. Reread your text and your notes, and use the Chapter Index of Key Ideas and Terms and the Chapter Test to recap major ideas and do a self-evaluated test simulation.
4. Be sure that you are well-rested so that you can be alert and focused during the exam.
5. Don't study up to the last minute. Give yourself some time to wind down before the exam. This will help you to organize your thoughts and feel more calm as the test begins.
6. As you take the test, realize that its purpose is not to trick you, but to give you and your instructor an accurate idea of what you have learned. Good study habits, a positive attitude, and confidence in your own ability will be reflected in your performance on any exam.
7. Finally, you should realize that your responsibility does not end with taking the exam. When your instructor returns your corrected exam, you should review your instructor's comments and any mistakes that you might have made. Take the opportunity to learn from this important feedback about what you have accomplished, where you could work harder, and how you can best prepare for future exams.

HAWKES LEARNING SYSTEMS: INTRODUCTORY & INTERMEDIATE ALGEBRA

Overview

This multimedia courseware allows students to become better problem-solvers by creating a mastery level of learning in the classroom. The software includes a "overview," "instruct," "practice," "tutor," and "certify" mode in each lesson, allowing students to learn through step-by-step interactions with the software. These automated homework system's tutorial and assessment modes extend instructional influence beyond the classroom. Intelligence is what makes the tutorials so unique. By offering intelligent tutoring and mastery level testing to measure what has been learned, the software extends the instructor's ability to influence students to solve problems. This courseware can be ordered either seperately or bundled together with this text.

Minimum Requirements

In order to run *HLS: Introductory & Intermediate Algebra*, you will need:

400 MHz or faster processor or equivalent
Windows® 98SE or later
64 MB RAM (128 MB recommended)
150 MB hard drive space
256 color display (800x600, 16-bit color recommended)
Internet Explorer 5.5 or later
CD-ROM drive

Getting Started

Before you can run *HLS: Introductory & Intermediate Algebra*, you will need an access code. This 30 character code is your personal access code. To obtain an access code, go to **http://www.hawkeslearning.com** and follow the links to the access code request page (unless directed otherwise by your instructor.)

Installation

Insert the *HLS: Introductory & Intermediate Algebra* Installation CD-ROM into the CD-ROM drive. Select the Start/Run command, type in the CD-ROM drive letter followed by \setup.exe. (For example, d:\setup.exe where d is the CD-ROM drive letter.)

The complete installation will use over 140 MB of hard drive space and will install the entire product, except the multimedia files, on your hard drive.

After selecting the desired installation option, follow the on-screen instructions to complete your installation of *HLS: Introductory & Intermediate Algebra.*

Starting the Courseware

After you have installed *HLS: Introductory & Intermediate Algebra* on your computer, to run the courseware, select Start/Programs/Hawkes Learning Systems/Introductory & Intermediate Algebra.

You will be prompted to enter your access code with a message box similar to the following:

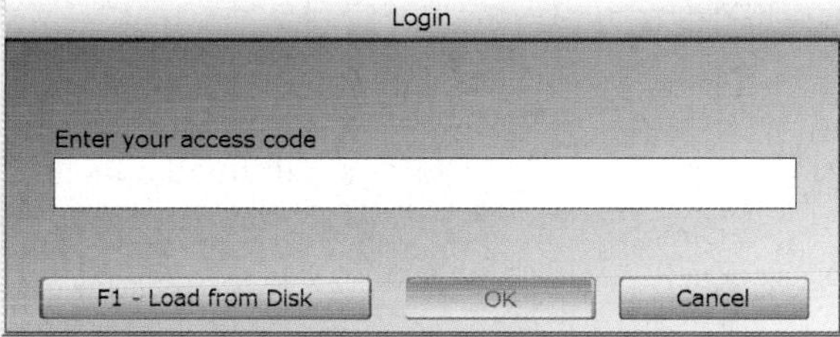

Type your access code into the box(es) provided. When you are finished, press OK.

If you typed in your access code correctly, you will be prompted to save the code to disk. If you choose to save your code to disk, typing in the access code each time you run *HLS: Introductory & Intermediate Algebra* will not be necessary. Instead, select the F1 - Load from Disk button when prompted to enter your access code and choose the path to your saved access code.

Now that you have entered your access code and saved it to diskette, you are ready to run a lesson. From the table of contents screen, choose the appropriate chapter and then choose the lesson you wish to run.

Features

Each lesson in *HLS: Introductory & Intermediate Algebra* has five modes: Overview, Instruct, Practice, Tutor, and Certify.

Overview: Overview provides you with a brief overview of the lesson. It presents an example of the type of question you will see in Practice and Certify, shows you how to input an answer, lists some of the specific features of the lesson, and tells you how many correct answers are needed to pass Certify.

Instruct: Instruct provides an expository on the material covered in the lesson in a multimedia environment. This same instruct mode can be accessed via the tutor mode.

Practice: Practice allows you to hone your problem-solving skills. It provides an unlimited number of randomly generated problems. Practice also provides access to the Tutor mode by selecting the Tutor button located by the Submit button.

Tutor: Tutor mode is broken up into several parts: Instruct, Explain Error, Step-by-Step, and Solution.

1. Instruct, which can also be selected directly from Practice mode, contains a multimedia lecture of the material covered in a lesson.
2. Explain Error is active whenever a problem is incorrectly answered. It will attempt to explain the error that caused you to incorrectly answer the problem.
3. Step-by-Step is an interactive "step through" of the problem. It breaks each problem into several steps, explains to you each step in solving the problem, and asks you a question about the step. After you answer the last step correctly, you have solved the problem.
4. Solution will provide you with a detailed "worked-out" solution to the problem.

Throughout the Tutor, you will see words or phrases colored green with a dashed underline. These are called Hot Words. Clicking on a Hot Word will provide you with more information on the word or phrase.

Certify: Certify is the testing mode. You are given a finite number of problems and a certain number of strikes (problems you can get wrong). If you answer the required number of questions, you will receive a certification code and a certificate. Write down your certification code and/or print out your certificate. The certification code will be used by your instructor to update your records. Note that the Tutor is not available in Certify.

Integration of Courseware and Textbook

Throughout this text, you will see an icon that helps to integrate the *Introductory & Intermediate Algebra* textbook and *HLS: Introductory & Intermediate Algebra* courseware.

This icon indicates which *HLS: Introductory & Intermediate Algebra* lessons you should run in order to test yourself on the subject material and to review the contents of a chapter.

Support

If you have questions about *HLS: Introductory & Intermediate Algebra* or are having technical difficulties, we can be contacted at the following:

Phone: (843) 571-2825
Email: techsupport@hawkeslearning.com
Web: www.hawkeslearning.com

Our support hours are 8:30 am to 5:30 pm, Eastern Time, Monday through Friday.

Real Numbers

CHAPTER 1

Did You Know?

Arithmetic operations defined on the set of positive integers, negative integers, and zero are studied in this chapter. The integer zero will be shown to have interesting properties under the operations of addition, subtraction, multiplication, and division.

Curiously, zero was not recognized as a number by early Greek mathematicians. When Hindu scientists developed the place-value numeration system we currently use, the zero symbol was initially a place holder but not a number. The spread of Islam transmitted the Hindu number system to Europe where it became known as the Hindu-Arabic system and replaced Roman numerals. The word zero comes from the Hindu word meaning "void," which was translated into Arabic as "sifr" and later into Latin as "zephirum," hence the derivation of our English words "zero" and "cipher."

Almost all of the operational properties of zero were known to the Hindus. However, the Hindu mathematician Bhaskara the Learned (1114 – 1185?) asserted that a number divided by zero was zero, or possibly infinite. Bhaskara did not seem to understand the role of zero as a divisor since division by zero is undefined and hence, an impossible operation in mathematics.

Einstein

Albert Einstein, in his development of a proof that the universe was stable and unchangeable in time, divided both sides of one of his intermediate equations by a complicated expression that under certain circumstances could become zero. When the expression became zero, Einstein's proof did not hold and the possibility of a pulsating, expanding, or contracting universe had to be considered. This error was pointed out to Einstein and he was forced to withdraw his proof that the universe was stable. The moral of this story is that although zero seems like a "harmless" number, its operational properties are different from those of the positive and negative integers.

"How can it be that mathematics, being after all a product of human thought independent of experience, is so admirably adapted to the objects of reality?"

Albert Einstein (1879 – 1955)

Numbers and number concepts form the foundation for the study of algebra. In this chapter, you will learn about positive and negative numbers and how to operate with these numbers. Believe it or not, the key number is 0. Pay particularly close attention to the idea of the magnitude of a number, called its absolute value, and the terminology used to represent different types of numbers.

1.1 The Real Number Line and Absolute Value

Objectives

After completing this section, you will be able to:

1. *Identify types of numbers.*
2. *Determine if given numbers are greater than, less than, or equal to other given numbers.*
3. *Determine absolute values.*

The study of algebra requires that we know a variety of types of numbers and their names. The set of numbers

$$N = \{1, 2, 3, 4, 5, 6, 7, 8, 9, 10, 11, \ldots\}$$

is called the **counting numbers** or **natural numbers**. The three dots indicate that the pattern is to continue without end. Putting 0 with the set of natural numbers gives the set of **whole numbers**.

$$W = \{0, 1, 2, 3, 4, 5, 6, \ldots\}$$

Thus, mathematicians make the distinction that 0 is a whole number but not a natural number.

Teaching Notes: Point out that this simple action represents the initial connection between numbers and geometry, and is the basis for graphing.

To help in understanding different types of numbers and their relationships to each other, we begin with a "picture" called a **number line**. For example, choose some point on a horizontal line and label it with the number 0 (Figure 1.1).

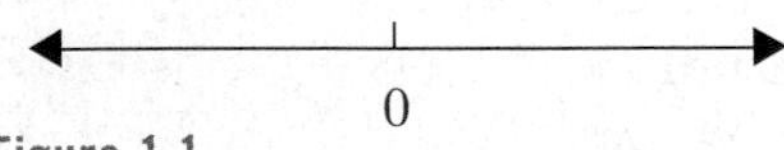

Figure 1.1

Now choose another point on the line to the right of 0 and label it with the number 1 (Figure 1.2).

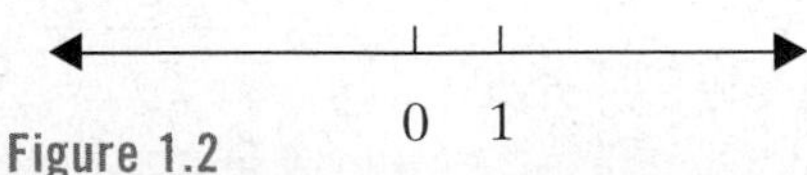

Figure 1.2

We now have a number line. Points corresponding to all the whole numbers are determined. The point corresponding to 2 is the same distance from 1 as 1 is from 0, 3 from 2, 4 from 3, and so on (Figure 1.3).

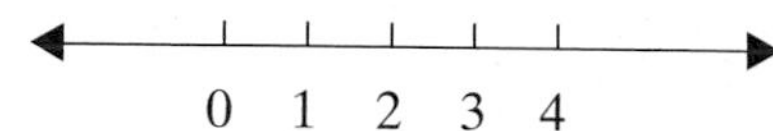

Figure 1.3

The **graph** of a number is the point that corresponds to the number and the number is called the **coordinate** of the point. We will follow the convention of using the terms "number" and "point" interchangeably. For example, a point can be called "seven" or "two". The graph of 7 is indicated by marking the point corresponding to 7 with a large dot (Figure 1.4).

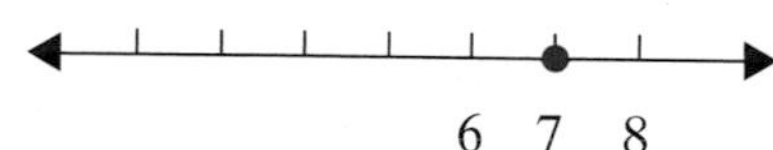

Figure 1.4

The graph of the set $A = \{ 2, 4, 6 \}$ is shown in Figure 1.5.

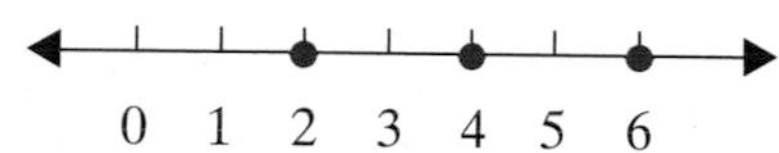

Figure 1.5

On a horizontal number line, the point one unit to the left of 0 is the **opposite** of 1. It is called **negative** 1 and is symbolized –1. Similarly, the point two units to the left of 0 is the opposite of 2, called negative 2, and symbolized –2, and so on (Figure 1.6).

The opposite of 1 is –1;
The opposite of 2 is –2;
The opposite of 3 is –3;
and so on.

The opposite of –1 is $-(-1) = +1$;
The opposite of –2 is $-(-2) = +2$;
The opposite of –3 is $-(-3) = +3$;
and so on.

NOTES The – sign indicates the opposite of a number as well as a negative number. It is also used, as we will see in Section 1.3, to indicate subtraction. To avoid confusion, you must learn (by practice) just how the – sign is used in each particular situation.

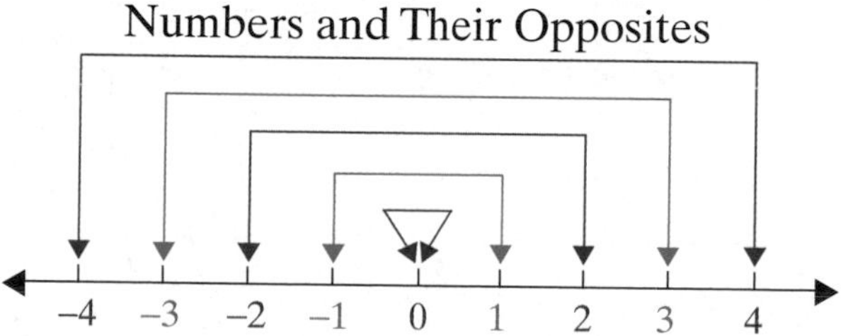

Figure 1.6

Integers

Teaching Notes:
Demonstrate $-(-a) = a$ and have students practice with examples.
As you know, the "double negative" comes up time and time again in algebra.

The set of numbers consisting of the whole numbers and their opposites is called the set of ***integers.***

The natural numbers are also called **positive integers**. Their opposites are called **negative integers**. Zero is its own opposite and is neither positive nor negative (Figure 1.7). Note that the opposite of a positive integer is a negative integer, and the opposite of a negative integer is a positive integer.

Integers:	$\{ \ldots, -4, -3, -2, -1, 0, 1, 2, 3, 4, \ldots \}$
Positive integers:	$\{ 1, 2, 3, 4, \ldots \}$
Negative integers:	$\{ \ldots, -4, -3, -2, -1 \}$

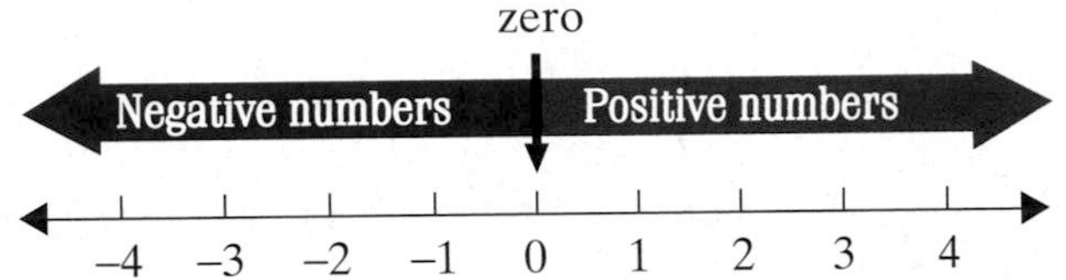

Figure 1.7

Example 1: Opposites

a. Find the opposite of 7.

Solution: -7

b. Find the opposite of -3.

Solution: $-(-3)$ or $+3$

In words, the opposite of -3 is $+3$.

Example 2: Number Line

a. Graph the set of integers $\{ -3, -1, 1, 3 \}$.

Solution:

–3 –2 –1 0 1 2 3

b. Graph the set of integers $\{ \ldots, -6, -5, -4, -3 \}$.

Solution:

...
–6 –5 –4 –3 –2 –1 0

The three dots above the number line indicate that the pattern in the graph continues without end.

The integers are not the only numbers that can be represented on a number line. Fractions and decimal numbers such as $\frac{1}{2}, -\frac{4}{3}, \frac{3}{4}$, and -2.3, as well as numbers such as π and $\sqrt{2}$, can also be represented (Figure 1.8).

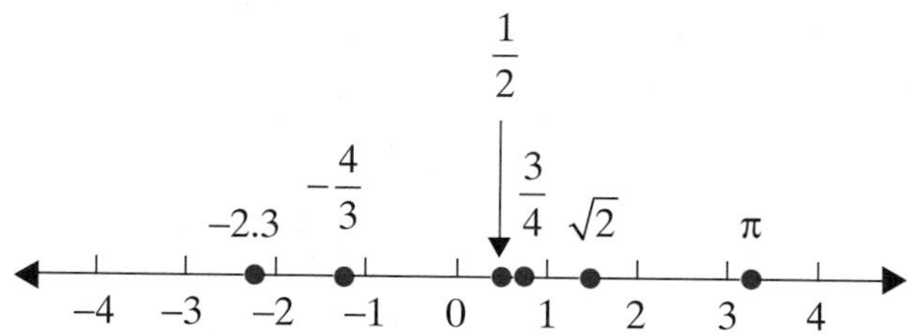

Figure 1.8

Teaching Notes:
You might mention that the word "rational", in this case, comes from "ratio."

Numbers that can be written as fractions and whose numerators and denominators are integers have the technical name **rational numbers**. Positive and negative decimal numbers that are terminating or infinitely repeating and the integers themselves can also be classified as rational numbers. For example, the following numbers are all rational numbers:

$$1.3 = \frac{13}{10},\quad 5 = \frac{5}{1},\quad -4 = \frac{-4}{1},\quad \frac{3}{8},\quad \text{and}\quad \frac{17}{6}.$$

To state general rules, to work with equations, and to form definitions we use variables. A **variable** is a letter (or other symbol) that can represent more than one number.

Rational Numbers

*A **rational number** is a number that can be written in the form of $\frac{a}{b}$ where a and b are integers and $b \neq 0$.*

OR

*A **rational number** is a number that can be written in decimal form as a terminating decimal or as an infinite repeating decimal.*

Other numbers on a number line, such as $\sqrt{2}, \sqrt{3}, \pi$, and $\sqrt[3]{5}$, are called **irrational numbers**. These numbers can be written as infinite nonrepeating decimal numbers. All rational numbers and irrational numbers are classified as **real numbers** and can be written in some decimal form. The number line is called the **real number line**.

We will discuss rational numbers (in both decimal form and fractional form) in detail later in Chapter 1 and irrational numbers in Chapter 10. For now, we are only interested in recognizing various types of numbers and locating their positions on the real number line.

With a calculator, you can find the following decimal values and approximations:

Examples of Rational Numbers:

$\frac{3}{4} = 0.75$ This decimal number is terminating.

Teaching Notes: Point out that any truncation of repeating decimals gives an approximation. Often, in algebra, we prefer the fractional form because it remains exact for rational numbers.

$\frac{1}{3} = 0.33333333...$ There is an infinite number of 3's in this repeating pattern.

$\frac{3}{11} = 0.27272727...$ This repeating pattern shows that there may be more than one digit in the pattern.

Examples of Irrational Numbers:

$\sqrt{2} = 1.414213562...$ There is an infinite number of digits with no repeating pattern.

$\pi = 3.141592653...$ There is an infinite number of digits with no repeating pattern. (The TI-83 Plus calculator will show a 4 in place of the ninth place digit 3 because it rounds off decimal numbers.)

1.41441444144441... There is an infinite number of digits and a pattern of sorts. However, the pattern is nonrepeating.

The following diagram (Figure 1.9) illustrates the relationships among the various categories of real numbers.

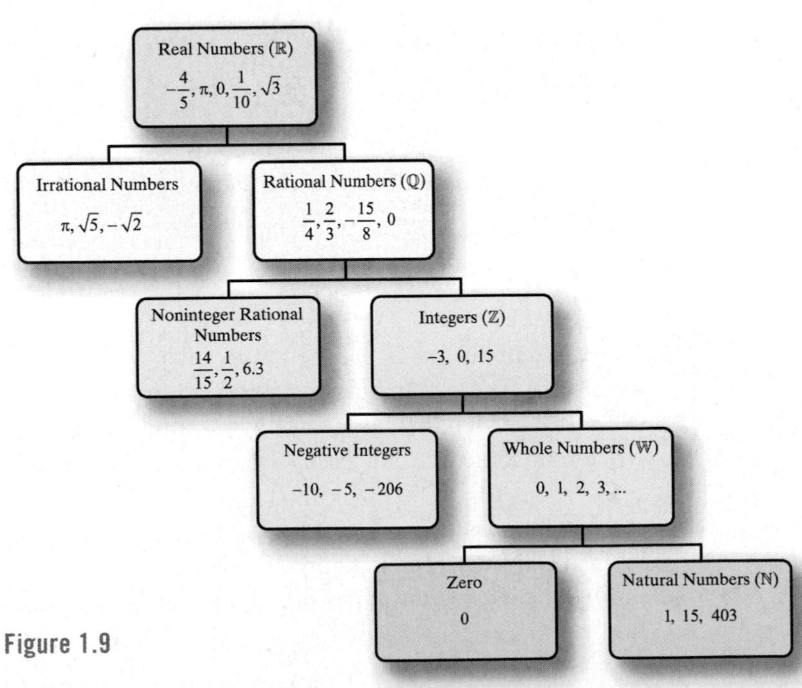

Figure 1.9

Inequality Symbols

On a horizontal number line, **smaller numbers are always to the left of larger numbers.** Each number is smaller than any number to its right and larger than any number to its left. Two symbols used to indicate order are

$<$, read "is less than"
and $>$, read "is greater than."

Using the real number line in Figure 1.10, you can see the following relationships:

Using $<$ or **Using** $>$

$0 < 3$	0 is less than 3
$-2 < 1$	-2 is less than 1
$-7 < -4$	-7 is less than -4
$\frac{1}{4} < \frac{9}{8}$	$\frac{1}{4}$ is less than $\frac{9}{8}$

$3 > 0$	3 is greater than 0
$1 > -2$	1 is greater than -2
$-4 > -7$	-4 is greater than -7
$\frac{9}{8} > \frac{1}{4}$	$\frac{9}{8}$ is greater than $\frac{1}{4}$

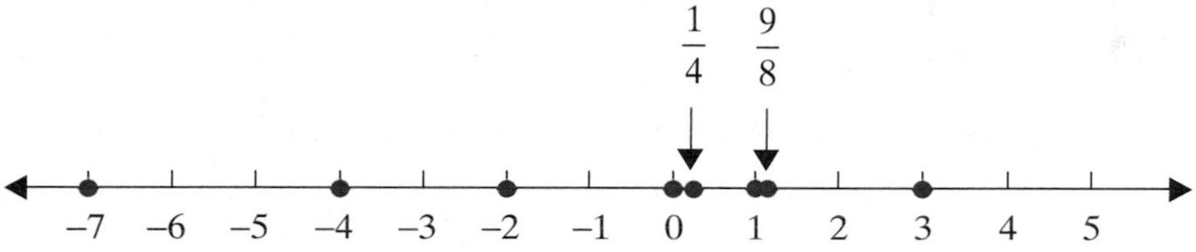

$\frac{1}{4}$ $\frac{9}{8}$

−7 −6 −5 −4 −3 −2 −1 0 1 2 3 4 5

Figure 1.10

Two other symbols commonly used are

$\leq$, read "is less than or equal to"
and $\geq$, read "is greater than or equal to."

For example, $5 \geq -10$ is true since 5 is greater than -10. Also, $5 \geq 5$ is true since 5 does equal 5.

Table of Symbols

$=$	*is equal to*	$\neq$	*is not equal to*
$<$	*is less than*	$>$	*is greater than*
$\leq$	*is less than or equal to*	$\geq$	*is greater than or equal to*

NOTES

Special Note About the Inequality Symbols.
Each symbol can be read from left to right as was just indicated in the Table of Symbols. However, each symbol can also be read from right to left. Thus, any inequality can be read in two ways. For example, $6 < 10$ can be read from left to right as "6 is less than 10", but also from right to left as "10 is greater than 6." We will see that this flexibility is particularly useful when reading expressions with variables in Appendix 1.

Example 3: Inequalities

a. Determine whether each of the following statements is true or false.

$7 < 15$	True, since 7 is less than 15.
$3 > -1$	True, since 3 is greater than –1.
$4 \geq -4$	True, since 4 is greater than –4.
$2.7 \geq 2.7$	True, since 2.7 is equal to 2.7.
$-5 < -6$	False, since –5 is greater than –6.

(**Note:** $7 < 15$ can be read as "7 is less than 15" or as "15 is greater than 7.")
(**Note:** $3 > -1$ can be read as "3 is greater than –1" or as "–1 is less than 3.")

b. Graph the set of **real numbers** $\left\{ -\frac{3}{4}, 0, 1, 1.5, 3 \right\}$
Solution:

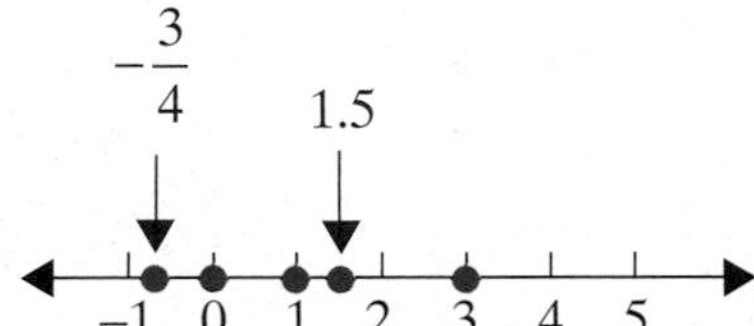

c. Graph all **natural numbers** less than or equal to 3.
Solution:

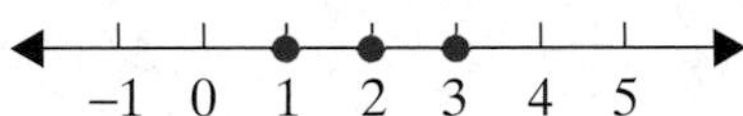

Remember that the natural numbers are 1, 2, 3, 4, ...

d. Graph all **integers** less than 0.
Solution:

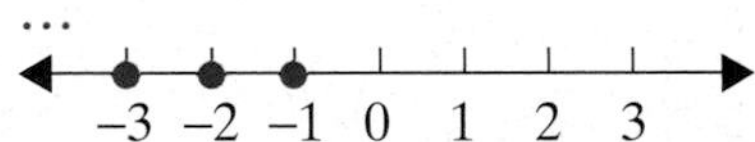

Absolute Value

In working with the real number line, you may have noticed that any integer and its opposite lie the same number of units from 0 on the number line. For example, both +7 and –7 are seven units from 0 (Figure 1.11). The + and – signs indicate direction and the 7 indicates distance.

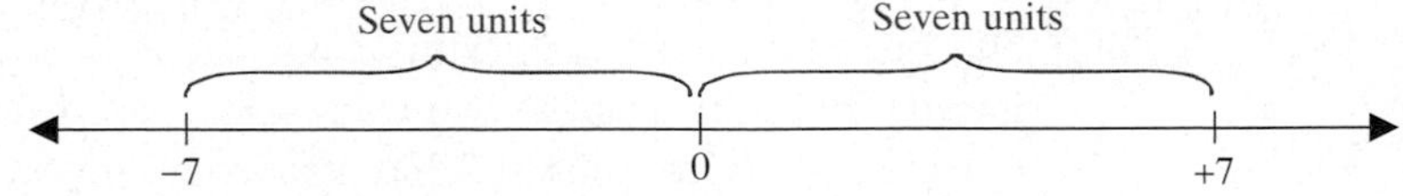

Figure 1.11

The distance between a number and 0 on a number line is called its **absolute value** and is symbolized by two vertical bars, $|\quad|$. Thus, $|+7| = 7$ and $|-7| = 7$. Similarly,

$|3| = 3$

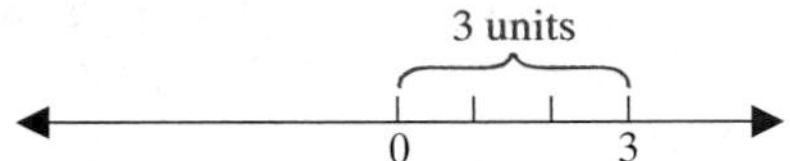

$|-4| = 4$

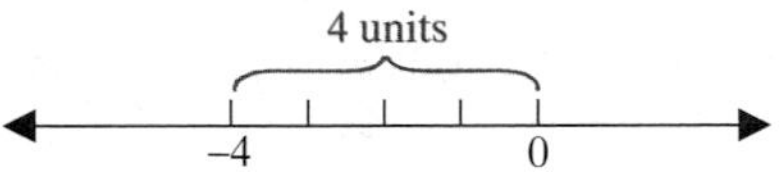

$|-2| = 2$

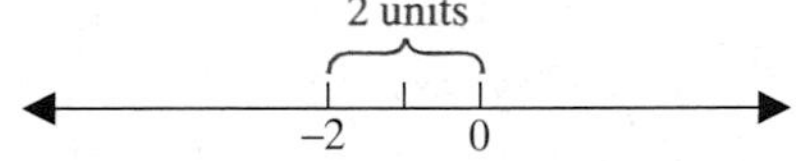

$|0| = 0$

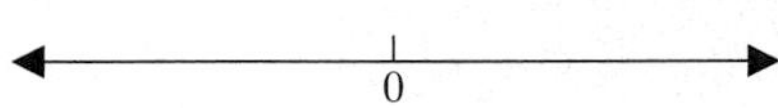

$\left|-\frac{4}{3}\right| = \frac{4}{3}$

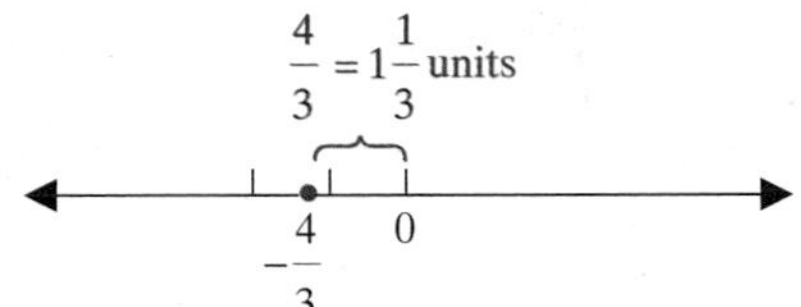

Since distance (similar to length) is never negative, the absolute value of a number is never negative. Therefore, the absolute value of a nonzero number is always positive.

Absolute Value

The ***absolute value*** *of a real number is its distance from 0. Note that the absolute value of a real number is never negative.*

$|a| = a$ *if a is a positive number or 0.*

$|a| = -a$ *if a is a negative number.*

NOTES

The symbol $-a$ should be thought of as the "opposite of a." Since a is a variable, a might represent a positive number, a negative number, or 0. This use of symbols can make the definition of absolute value difficult to understand at first. As an aid to understanding the use of the negative sign, consider the following examples.

If $a = -6$, then $-a = -(-6) = 6$.

Similarly,

If $x = -1$, then $-x = -(-1) = 1$.
If $y = -10$, then $-y = -(-10) = 10$.

Remember that $-a$ (the opposite of a) represents a positive number whenever a represents a negative number.

Example 4: Absolute Value

Teaching Notes:
You may want to show your students how to use the absolute value key and the negative key on the TI-83 Plus at this time. These ideas are included in the exercises.

a. $|6.3| = 6.3$

Solution: The number 6.3 is 6.3 units from 0. Also, 6.3 is positive so its absolute value is the same as the number itself.

b. $|-5.1| = -(-5.1) = 5.1$

Solution: The number –5.1 is 5.1 units from 0. Also, –5.1 is negative so its absolute value is its opposite.

c. If $|x| = 7$, what are the possible values for x?

Solution: $x = 7$ or $x = -7$ since $|7| = 7$ and $|-7| = 7$.

d. If $|x| = 1.35$, what are the possible values for x?

Solution: $x = 1.35$ or $x = -1.35$ since $|1.35| = 1.35$ and $|-1.35| = 1.35$.

e. True or False: $|-4| \geq 4$

Solution: True, since $|-4| = 4$ and $4 \geq 4$.

f. True or False: $\left|-5\frac{1}{2}\right| < 5\frac{1}{2}$

Solution: False, since $\left|-5\frac{1}{2}\right| = 5\frac{1}{2}$ and $5\frac{1}{2} \not< 5\frac{1}{2}$.

($\not<$ is read "is not less than")

g. If $|x| = -3$, what are the possible values for x?

Solution: There are no values of x for which $|x| = -3$. The absolute value can never be negative. There is **no solution**.

h. If $|x| < 3$, what are the possible integer values for x? Graph these numbers on a number line.

Solution: The integers are within 3 units of 0: $-2, -1, 0, 1, 2$.

i. If $|x| \geq 4$, what are the possible integer values for x? Graph these numbers on a number line.

Solution: The integers must be 4 or more units from 0: $\ldots, -7, -6, -5, -4, 4, 5, 6, 7, \ldots$

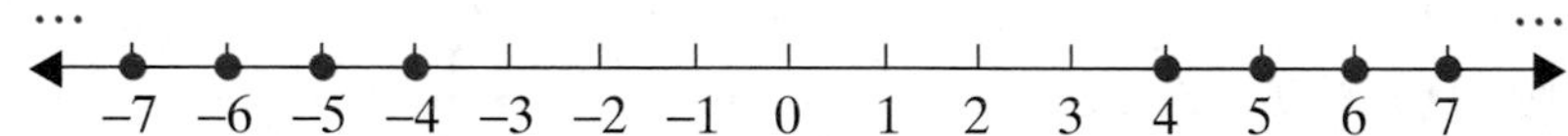

Practice Problems

Fill in the blank with the appropriate symbol: $<$, $>$, or $=$.

1. -2 ____ 1
2. $1\frac{6}{10}$ ____ 1.6
3. $-(-4.1)$ ____ -7.2
4. *Graph the set of all negative integers on a number line.*
5. *True or False:* $3.6 \leq |-3.6|$
6. *List the numbers that satisfy the equation* $|x| = 8$.
7. *List the numbers that satisfy the equation* $|x| = -6$.

Answers to Practice Problems: 1. $<$ **2.** $=$ **3.** $>$ **4.** ... –3 –2 –1 0 **5.** True **6.** 8, –8 **7.** No solution

1.1 Exercises

1.–5.
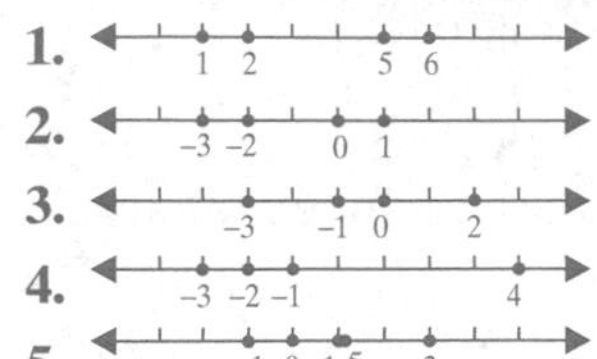

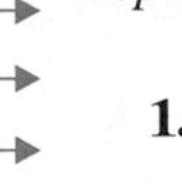

6. -2 -1 $-\frac{1}{3}$ 2
7. $-\frac{3}{4}$ 0 2 3.6
8. -3.4 -2 0.5 1 $\frac{5}{2}$
9. $-\frac{7}{2}$ -1.5 1 $\frac{4}{3}$ 2
10. -4 $-\frac{7}{3}$ -1 0.2 $\frac{5}{2}$
11. 1 2 3
12. 0 1 2 3 4 5 6
13.–15.
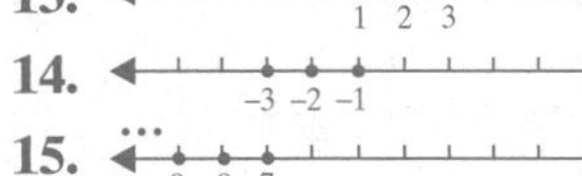

16. -1 0 1 2 3 4 5 ...
17. ... 3 4 5 6 7
18. 0 1 2 3 4
19. No Solution
20. -6 -5 -4 -3 -2 -1 0 1 2 3 ...
21. ... -9 -8 -7 7 8 9 ...
22. ... -5 -4 4 5 ...
23. -2 -1 0 1 2
24. -7 -6 -4 -2 0 2 4 6 7

25. < 26. <
27. > 28. =
29. < 30. <
31. = 32. <
33. > 34. >
35. < 36. <
37. = 38. <
39. <

In Exercises 1 – 10, graph each set of real numbers on a real number line. For decimal representations of fractions, use a calculator.

1. {1, 2, 5, 6} **2.** {–3, –2, 0, 1} **3.** {2, –3, 0, –1}

4. {–2, –1, 4, –3} **5.** $\left\{0, -1, \frac{5}{4}, 3, 1\right\}$ **6.** $\left\{-2, -1, -\frac{1}{3}, 2\right\}$

7. $\left\{-\frac{3}{4}, 0, 2, 3.6\right\}$ **8.** $\left\{-3.4, -2, 0.5, 1, \frac{5}{2}\right\}$ **9.** $\left\{-\frac{7}{2}, -1.5, 1, \frac{4}{3}, 2\right\}$

10. $\left\{-4, -\frac{7}{3}, -1, 0.2, \frac{5}{2}\right\}$

In Exercises 11 – 24, graph each set of integers on a real number line.

11. All positive integers less than 4
12. All whole numbers less than 7
13. All positive integers less than or equal to 3
14. All negative integers greater than or equal to –3
15. All integers less than –6
16. All integers greater than or equal to –1
17. All integers less than 8
18. All whole numbers less than or equal to 4
19. All negative integers greater than or equal to 2
20. All integers greater than or equal to –6
21. All integers more than 6 units from 0
22. All integers more than 3 units from 0
23. All integers less than 3 units from 0
24. All integers less than 8 units from 0

Fill in the blank in Exercises 25 – 39 with the appropriate symbol: <, >, or =.

25. 4 ____ 6 **26.** –3 ____ 1 **27.** –2 ____ –4 **28.** 5 ____ –(–5)

29. $\frac{1}{3}$ ____ $\frac{1}{2}$ **30.** $-\frac{2}{3}$ ____ $\frac{1}{8}$ **31.** $-\frac{2}{8}$ ____ $-\frac{1}{4}$ **32.** –8 ____ 0

33. 2.3 ____ 1.6 **34.** $-\frac{3}{4}$ ____ –1 **35.** $\frac{9}{16}$ ____ $\frac{3}{4}$ **36.** $-\frac{1}{2}$ ____ $-\frac{1}{3}$

37. –2.3 ____ $-2\frac{3}{10}$ **38.** 5.6 ____ –(–8.7) **39.** $-\frac{4}{3}$ ____ $-\left(-\frac{1}{3}\right)$

40. True **41.** True
42. False; $-9 < -8.5$
43. True **44.** True
45. False; $-6 > -8$
46. True **47.** True
48. True **49.** True **50.** True
51. True **52.** True
53. False; $|-7| = |7|$
54. True **55.** True
56. False; $\left|-\frac{5}{2}\right| > 2$
57. True **58.** True
59. False; $|-3.4| > 0$
60. False; $-|-3| > -|4|$
61. False; $-|5| < -|3.1|$
62. True **63.** True **64.** True
65. −4 4
66. −6 6
67. −9 9
68. −13 13
69. 0
70. No Solution
71. No Solution
72. −3.5 3.5
73. −4.7 4.7
74. $-\frac{5}{4}$ $\frac{5}{4}$
75. −4 −3 −2 −1 0 1 2 3 4
76. −5 −4 −3 −2 −1 0 1 2 3 4 5
77. ... −10 −8 −6 6 8 10 ...
78. ... −5 −4 −3 3 4 5 ...
79. −6 −5 −4 −3 −2 −1 0 1 2 3 4 5 6
80. −2 −1 0 1 2
81. 0 1 2 3 4 5 ...
82. ... −5 −4 −3 −2 −1 0
83. 0 1 2 3 4 5 ...
84. ... −5 −4 −3 −2 −1
85. Sometimes
86. Sometimes
87. Never
88. Sometimes
89. Sometimes

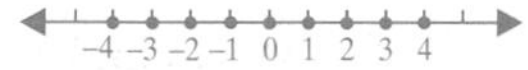

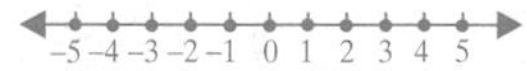

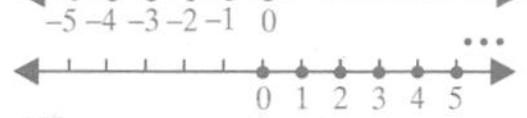

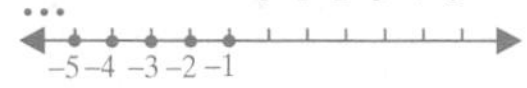

Determine whether each statement in Exercises 40 – 64 is true or false. If a statement is false, rewrite it in a form that is a true statement. (There may be more than one way to correct a statement.)

40. $0 = -0$ **41.** $-22 < -16$ **42.** $-9 > -8.5$ **43.** $11 = -(-11)$

44. $-17 \le 17$ **45.** $-6 < -8$ **46.** $4.7 \ge 3.5$ **47.** $-\frac{1}{3} \le 0$

48. $\frac{3}{5} > \frac{1}{4}$ **49.** $-2.3 < 1$ **50.** $|-5| = 5$ **51.** $|-8| \ge 4$

52. $|-6| \ge 6$ **53.** $|-7| < |7|$ **54.** $|-1.9| < 2$ **55.** $|-1.6| < |-2.1|$

56. $\left|-\frac{5}{2}\right| < 2$ **57.** $\frac{2}{3} < |-1|$ **58.** $3 > \left|-\frac{4}{3}\right|$ **59.** $|-3.4| < 0$

60. $-|-3| < -|4|$ **61.** $-|5| > -|3.1|$ **62.** $|-6.2| = 6.2$ **63.** $-|73| < |-73|$

64. $|2.5| = \left|-\frac{5}{2}\right|$

List the numbers, then graph the numbers on a real number line that satisfy the equations in Exercises 65 – 74.

65. $|x| = 4$ **66.** $|y| = 6$ **67.** $|x| = 9$ **68.** $13 = |x|$ **69.** $0 = |y|$

70. $-2 = |x|$ **71.** $|x| = -3$ **72.** $|x| = 3.5$ **73.** $|x| = 4.7$ **74.** $|x| = \left|-\frac{5}{4}\right|$

On a number line, graph the integers that satisfy the conditions stated in Exercises 75 – 84.

75. $|x| \le 4$ **76.** $|x| < 6$ **77.** $|x| > 5$ **78.** $|x| > 2$ **79.** $|x| < 7$

80. $|x| \le 2$ **81.** $|x| \le x$ **82.** $|x| = -x$ **83.** $|x| = x$ **84.** $|x| > x$

Choose the response that correctly completes each sentence in Exercises 85 – 89. Give one or more examples to illustrate your conclusion for each exercise.

85. $|x|$ is (never, sometimes, always) equal to x.

86. $|x|$ is (never, sometimes, always) equal to $-x$.

87. $|x|$ is (never, sometimes, always) a negative number.

88. $|x|$ is (never, sometimes, always) equal to 0.

89. $|x|$ is (never, sometimes, always) equal to a positive number.

90. 46 **91.** 61.4 **92.** –6

93. $\frac{1}{3}$ **94.** 1

95. If y is a negative number then $-y$ represents a positive number. For example, if $y = -2$, then $-y = -(-2) = 2$.

96. The magnitude of any number is called its absolute value.

Calculator Problems

A TI-83 Plus calculator has the absolute value command built in. To access the absolute value command press MATH*, go to* NUM *at the top of the screen and press* 1 *or* ENTER. *The command* ***abs(*** *will appear on the display screen. Then enter any arithmetic expression you wish and the calculator will print the absolute value of that expression.* (**Note:** *The negative sign is on the key marked* (–) *next to the* ENTER *key.)*

Follow the directions above and use your calculator to find the value of each of the following expressions in Exercises 90 – 94.

90. $|34 - 80|$

91. $|17.5 + 16.3 - 95.2|$

92. $-|10 - 16|$

93. $\dfrac{|-6|-|-3|}{|-6-3|}$

94. $\dfrac{|4|+|-4|}{|-8|}$

Writing and Thinking About Mathematics

95. Explain, in your own words, how a variable expression such as $-y$ might represent a positive number.

96. Explain, in your own words, the meaning of absolute value.

Hawkes Learning Systems: Introductory & Intermediate Algebra

The Real Number Line and Inequalities
Introduction to Absolute Values

1.2 Addition with Integers

Objectives

After completing this section, you will be able to:

1. *Add integers.*
2. *Determine if given integers are solutions for specified equations.*
3. *Complete statements about integers.*

Picture a straight line in an open field and numbers marked on a number line. An archer stands at 0 and shoots an arrow to +3, then stands at 3 and shoots the arrow 5 more units in the positive direction (to the right). Where will the arrow land? (Figure 1.12.)

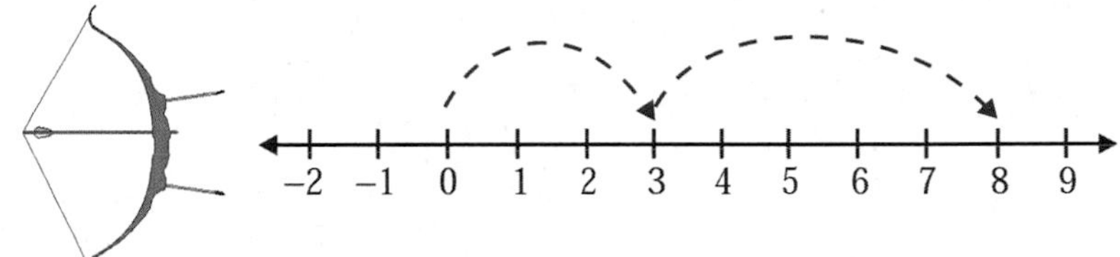

Figure 1.12

Naturally, you have figured out that the answer is +8. What you have done is add the two positive integers, +3 and +5.

$$(+3) + (+5) = +8 \quad \text{or} \quad 3 + 5 = 8$$

Suppose another archer shoots an arrow in the same manner as the first but in the opposite direction. Where would his arrow land? The arrow lands at –8. You have just added –3 and –5 (Figure 1.13).

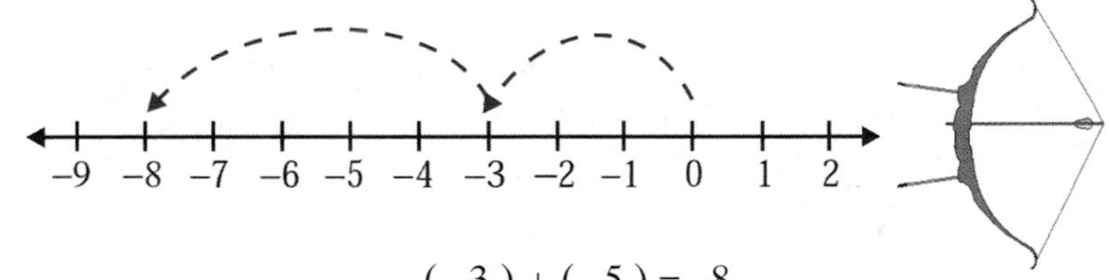

Figure 1.13

$$(-3) + (-5) = -8$$

If the archer stands at 0 and shoots an arrow to +3 and then the archer goes to +3 and turns around and shoots an arrow 5 units in the opposite direction, where will the arrow stick? Would you believe at –2? (Figure 1.14.)

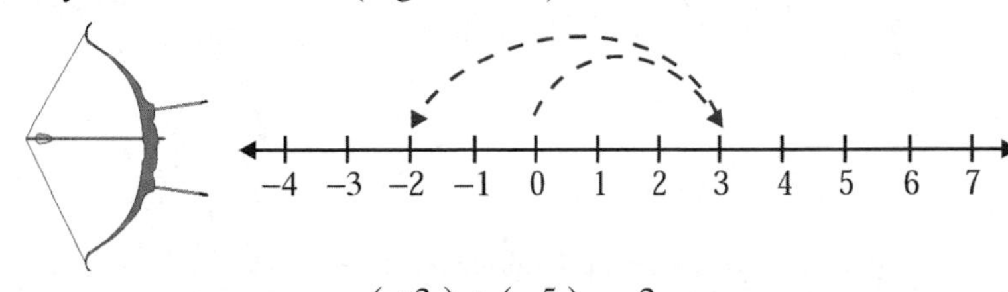

Figure 1.14

$$(+3) + (-5) = -2$$

For our final archer, the first shot is to –3. Then, after going to –3, he turns around and shoots 5 units in the opposite direction. Where is the arrow? It is at +2 (Figure 1.15).

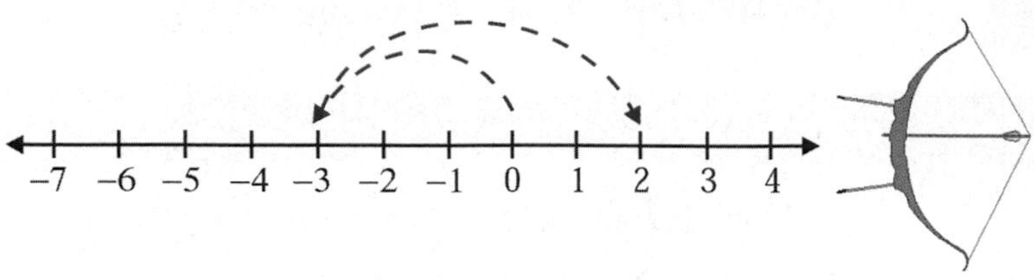

Figure 1.15

$$(-3) + (+5) = +2$$

In summary:

1. The sum of two positive integers is positive.

   (+3) + (+5) = +8

   positive plus positive is **positive**

2. The sum of two negative integers is negative.

   (–3) + (–5) = –8

   negative plus negative is **negative**

3. The sum of a positive integer and a negative integer may be negative or positive (or zero) depending on which number is further from 0.

Practice Problems

Find each sum. Add from left to right if there are more than two numbers.

1. $(-14) + (-6) =$	**2.** $(+16) + (-10) =$
3. $(-12) + (8) =$	**4.** $11 + 7 =$
5. $(-13) + (+13) =$	**6.** $(-11) + (-8) =$
7. $(+6) + (-7) + (-1) =$	**8.** $(+100) + (-100) + (+10) =$

Answers to Practice Problems: 1. –20 **2.** 6 **3.** –4 **4.** 18 **5.** 0 **6.** –19 **7.** –2 **8.** 10

You probably did quite well and understand how to add integers. The rules can be written out in the following rather formal manner.

Rules for Addition with Integers

Teaching Notes:
For many beginning students the absolute value notation is confusing. You might want to summarize the procedure as follows:
(1) When signs are the same, treat them first as positive and give their sum the common sign.
(2) When signs are different, first ignore the signs and subtract the smaller from the larger. Then give the difference the sign of the larger number.

1. *To add two integers with like signs, add their absolute values and use the common sign:*

$$(+7)+(+3)=+(|+7|+|+3|)=+(7+3)=+10$$
$$(-7)+(-3)=-(|-7|+|-3|)=-(7+3)=-10$$

2. *To add two integers with unlike signs, subtract their absolute values (the smaller from the larger) and use the sign of the number with the larger absolute value:*

$$(-12)+(+10)=-(|-12|-|+10|)=-(12-10)=-2$$
$$(+12)+(-10)=+(|+12|-|-10|)=+(12-10)=+2$$
$$(-15)+(+15)=(|-15|-|+15|)=(15-15)=0$$

An **equation** is a statement that two expressions are equal. Since equations in algebra are almost always written horizontally, you should become used to working with sums written horizontally. However, there are situations (as in long division) where sums (and differences) are written vertically with one number directly under another. We illustrate this technique in Example 1.

Example 1: Vertical

Find each sum.

a. $\begin{array}{r} -10 \\ \underline{7} \\ -3 \end{array}$ **b.** $\begin{array}{r} -4 \\ 6 \\ \underline{-15} \\ -13 \end{array}$ **c.** $\begin{array}{r} -5 \\ -8 \\ \underline{-9} \\ -22 \end{array}$ **d.** $\begin{array}{r} -10 \\ 3 \\ \underline{7} \\ 0 \end{array}$

Now that we know how to add positive and negative integers, we can determine whether or not a particular integer satisfies an equation that contains a variable. Recall, a **variable** is a letter (or symbol) that can represent one or more numbers. A number is said **to be a solution or to satisfy an equation** if it gives a true statement when substituted for the variable.

Example 2: Solutions

Determine whether or not the given integer is a solution to the given equation by substituting for the variable and adding.

a. $x + 5 = -2$; $x = -7$
Solution: $(-7) + 5 = -2$ is true, so -7 is a solution.

b. $y + (-4) = -6$; $y = -2$
Solution: $(-2) + (-4) = -6$ is true, so -2 is a solution.

c. $14 + z = -3$; $z = -11$
Solution: $14 + (-11) = -3$ is false since $14 + (-11) = +3$.
So, -11 is **not** a solution.

1.2 Exercises

1\. 13 2. 5 3. −4
4\. −10 5. 0 6. −6
7\. 5 8. −9 9. −13
10\. 0 11. −8 12. 9
13\. −10 14. 7 15. 0
16\. 13 17. 17 18. −18
19\. −29 20. 8 21. −9
22\. −14 23. −3 24. −3
25\. 22 26. −31 27. −54
28\. −3 29. −7 30. −10
31\. −16 32. 0 33. −26
34\. −25 35. 0 36. −2
37\. −32 38. −29
39\. 5 40. 3
41\. −83 42. 5
43\. 12 44. 7
45\. −32 46. −33
47\. −2 is a solution
48\. −3 is not a solution
49\. −4 is a solution
50\. −2 is a solution

Find the sum in Exercises 1 – 46.

1. $4 + 9$ **2.** $8 + (-3)$ **3.** $(-9) + 5$
4. $(-7) + (-3)$ **5.** $(-9) + 9$ **6.** $2 + (-8)$
7. $11 + (-6)$ **8.** $(-12) + 3$ **9.** $-18 + 5$
10. $26 + (-26)$ **11.** $-5 + (-3)$ **12.** $11 + (-2)$
13. $(-2) + (-8)$ **14.** $10 + (-3)$ **15.** $17 + (-17)$
16. $(-7) + 20$ **17.** $21 + (-4)$ **18.** $(-15) + (-3)$
19. $(-12) + (-17)$ **20.** $24 + (-16)$ **21.** $-4 + (-5)$
22. $(-6) + (-8)$ **23.** $9 + (-12)$ **24.** $-12 + 9$
25. $38 + (-16)$ **26.** $(-20) + (-11)$ **27.** $(-33) + (-21)$
28. $(-21) + 18$ **29.** $-3 + 4 + (-8)$ **30.** $(-9) + (-6) + 5$
31. $(-9) + (-2) + (-5)$ **32.** $(-21) + 6 + 15$ **33.** $-13 + (-1) + (-12)$
34. $-19 + (-2) + (-4)$ **35.** $27 + (-14) + (-13)$ **36.** $-33 + 29 + 2$
37. $-43 + (-16) + 27$ **38.** $-68 + (-3) + 42$ **39.** $-38 + 49 + (-6)$
40. $102 + (-93) + (-6)$

41. $\begin{array}{r} -21 \\ \underline{-62} \end{array}$ **42.** $\begin{array}{r} -12 \\ \underline{17} \end{array}$ **43.** $\begin{array}{r} -15 \\ 8 \\ \underline{19} \end{array}$

44. $\begin{array}{r} -7 \\ 23 \\ \underline{-9} \end{array}$ **45.** $\begin{array}{r} -163 \\ 204 \\ \underline{-73} \end{array}$ **46.** $\begin{array}{r} -93 \\ -87 \\ \underline{147} \end{array}$

In Exercises 47 – 60, determine whether or not the given number is a solution to the given equation by substituting and then evaluating.

47. $x + 4 = 2$; $x = -2$ **48.** $x + (-7) = 10$; $x = -3$
49. $-10 + x = -14$; $x = -4$ **50.** $y + 9 = 7$; $y = -2$

51. –6 is a solution
52. 8 is not a solution
53. 18 is a solution
54. –7 is not a solution
55. –10 is a solution
56. –2 is a solution
57. –2 is not a solution
58. –18 is a solution
59. –72 is not a solution
60. –4 is a solution
61. Sometimes
62. Sometimes
63. Never
64. Always
65. Never
66. Sometimes
67. Sometimes
68. Never
69. Always
70. Always
71. 84
72. –13
73. –97,714
74. 3807
75. –6143
76. $|0|+|0|=0$

51. $17 + y = 11;\ y = -6$ **52.** $z + (-1) = 9;\ z = 8$
53. $z + (-12) = 6;\ z = 18$ **54.** $x + 3 = -10;\ x = -7$
55. $x + (-5) = -15;\ x = -10$ **56.** $|x| + 15 = 17;\ x = -2$
57. $|y| + (-10) = -12;\ y = -2$ **58.** $-26 + |x| = -8;\ x = -18$
59. $42 + |z| = -30;\ z = -72$ **60.** $|x| + (-5) = -1;\ x = -4$

61. If x is a positive number and y is a negative number, then $x + y$ is (never, sometimes, always) a negative number.
62. If x and y are integers, then $x + y$ is (never, sometimes, always) equal to 0.
63. If x and y are positive numbers, then $x + y$ is (never, sometimes, always) equal to 0.
64. If y is an integer, then $y + (-y)$ is (never, sometimes, always) equal to 0.
65. If x and y are negative numbers, then $x + y$ is (never, sometimes, always) equal to 0.
66. If x and y are integers, then $x + y$ is (never, sometimes, always) negative.
67. If x and y are integers, then $x + y$ is (never, sometimes, always) positive.
68. If x represents a negative number, then $-x$ is (never, sometimes, always) negative.
69. If x represents a negative number, then $-x$ is (never, sometimes, always) positive.
70. If x represents a positive number, then $-x$ is (never, sometimes, always) negative.

Calculator Problems

Use your TI-83 Plus calculator to find the value of each of the following expressions. (Remember that the key marked (–) *next to the* ENTER *key is used to indicate negative numbers.)*

71. $47 + (-29) + 66$ **72.** $56 + (-41) + (-28)$
73. $(-16{,}945) + (-27{,}302) + (-53{,}467)$ **74.** $2{,}932 + 4{,}751 + (-3{,}876)$
75. $(-8{,}154) + 2{,}147 + (-136)$

Writing and Thinking About Mathematics

76. Describe, in your own words, how the sum of the absolute values of two numbers might be 0. (Is this even possible?)

Hawkes Learning Systems: Introductory & Intermediate Algebra

Addition with Integers

1.3 Subtraction with Integers

Objectives

After completing this section, you will be able to:

1. *Find the additive inverse of an integer.*
2. *Subtract integers.*
3. *Determine if given integers are solutions for specified equations.*

In basic arithmetic, subtraction is defined in terms of addition. For example, we know that the difference 32 – 25 is equal to 7 because 25 + 7 = 32. A beginning student in arithmetic does not know how to find a difference such as 15 – 20, where a larger number is subtracted from a smaller number, because negative numbers are not yet defined and there is no way to add a positive number to 20 and get 15. Now, with our knowledge of negative numbers, we will define subtraction in such a way that larger numbers may be subtracted from smaller numbers. We will still define subtraction in terms of addition, but we will apply our new rules of addition with integers.

Before we proceed to develop the techniques for subtraction with integers, we will state and illustrate an important relationship between any integer and its opposite.

Additive Inverse

The ***opposite*** *of an integer is called its* ***additive inverse****. The sum of a number and its additive inverse is zero. Symbolically, for any integer a,*

$$a + (-a) = 0$$

Example 1: Additive Inverse

a. Find the additive inverse (opposite) of 3.
Solution: The additive inverse of 3 is –3, 3 + (–3) = 0.

b. Find the additive inverse (opposite) of –7.
Solution: The additive inverse of –7 is –(–7) = +7, (–7) + (+7) = 0.

c. Find the additive inverse (opposite) of 0.
Solution: The additive inverse of 0 is –0 = 0.
That is, 0 is its own opposite, (0) + (–0) = 0 + 0 = 0.

In Section 1.2, we added integers such as

$$5 + (-2) = 3 \quad \text{and} \quad 26 + (-9) = 17.$$

Note that in each case, subtraction will give the same results; that is,

$$5 - 2 = 3 \quad \text{and} \quad 26 - 9 = 17.$$

Thus, it seems that subtraction and addition are closely related and, in fact, they are. From arithmetic, 5 – 2 is asking, "What number added to 2 gives 5?" In other words,

$$\begin{aligned} 5 - 2 = 3 &\quad \text{because} \quad 5 = 2 + 3 \\ 26 - 9 = 17 &\quad \text{because} \quad 26 = 9 + 17 \end{aligned}$$

What do we mean by 4 – (–1)? Do we mean, "What number added to –1 gives 4?" Precisely. Thus,

$$\begin{aligned} 4 - (-1) = 5 &\quad \text{because} \quad 4 = (-1) + 5 \\ 4 - (-2) = 6 &\quad \text{because} \quad 4 = (-2) + 6 \\ 4 - (-3) = 7 &\quad \text{because} \quad 4 = (-3) + 7 \end{aligned}$$

But note the following results:

$$\begin{aligned} 4 + (+1) &= 5 \\ 4 + (+2) &= 6 \\ 4 + (+3) &= 7 \end{aligned}$$

The following relationship between subtraction and addition becomes the basis for subtraction with all numbers:

$$\begin{aligned} 4 - (-1) = 4 + (+1) = 5 &\qquad (-4) - (-1) = (-4) + (+1) = -3 \\ 4 - (-2) = 4 + (+2) = 6 &\qquad (-4) - (-2) = (-4) + (+2) = -2 \\ 4 - (-3) = 4 + (+3) = 7 &\qquad (-4) - (-3) = (-4) + (+3) = -1 \end{aligned}$$

Example 2: Subtraction

a. $(-1) - (-4) = (-1) + (+4) = +3$
b. $(-1) - (-5) = (-1) + (+5) = +4$
c. $(-1) - (-8) = (-1) + (+8) = +7$
d. $(10) - (-2) = (10) + (+2) = +12$
e. $(-10) - (-5) = (-10) + (+5) = -5$

Practice Problems

Find each sum.

1. $(-4) - (-4) = (-4) + (+4) =$

2. $(-3) - (-8) = (-3) + (+8) =$

3. $(-3) - (+8) = (-3) + (-8) =$

4. $14 - (-5) = 14 + (5) =$

5. $14 - (5) = 14 + (-5) =$

Answers to Practice Problems: **1.** 0 **2.** 5 **3.** –11 **4.** 19 **5.** 9

You may have noticed that in subtraction, the **opposite** of the number being subtracted is **added**. For example, we could write

$$(-1)-(-4)=(-1)+[-(-4)]=(-1)+(+4)=+3$$

or add opposite

$$(-10)-(-3)=(-10)+[-(-3)]=(-10)+(+3)=-7$$

Subtraction

For any integers a and b,

$$a-b=a+(-b)$$

Teaching Notes: The following ideas might help students understand that the "–" is being used in two different ways. One way is as a verb ("subtract"), as in $a-b$. The other way is as an adjective ("negative"), as in $a+(-b)$. We tie the two uses together with $a+(-b)=a-b$.

This definition translates as, "To subtract b from a, **add** the **opposite** of b to a." In practice, the notation $a-b$ is thought of as addition of signed numbers. That is, since $a-b=a+(-b)$, we think of the plus sign, +, as being present in $a-b$. In fact, an expression such as 4 – 19 can be thought of as "four plus negative nineteen." We have

$$\begin{aligned} 4-19 &= 4+(-19)=-15 \\ -25-30 &= -25+(-30)=-55 \\ -3-(-17) &= -3+(+17)=14 \\ 24-11-6 &= 24+(-11)+(-6)=7 \end{aligned}$$

Generally, the second step is omitted and we go directly to the answer by computing the sum mentally.

$$\begin{gathered} 4-19=-15 \\ -25-30=-55 \\ -3-(-17)=14 \\ 24-11-6=7 \end{gathered}$$

The numbers may also be written vertically, that is, one underneath the other. In this case, the sign of the number being subtracted (the bottom number) is changed and addition is performed.

Example 3: Subtract and Add

a.

Subtract		**Add**
43		43
−(−25)	sign change →	+25
		68

b.

Subtract		**Add**
−38		−38
−(+11)	sign change →	−11
		−49

c. **Subtract** → **Add**

$$\begin{array}{r} -73 \\ -(-32) \\ \hline \end{array} \xrightarrow[\text{change}]{\text{sign}} \begin{array}{r} -73 \\ +32 \\ \hline -41 \end{array}$$

d. **Subtract** → **Add**

$$\begin{array}{r} 17 \\ -(+69) \\ \hline \end{array} \xrightarrow[\text{change}]{\text{sign}} \begin{array}{r} 17 \\ -69 \\ \hline -52 \end{array}$$

Now, using subtraction as well as addition, we can determine whether or not a number is a solution to an equation of a slightly more complex nature.

Example 4: Solutions

Determine whether or not the given number is a solution to the given equation by substituting and then evaluating.

a. $x - (-5) = 6;\ x = 1$
Solution: $1 - (-5) = 1 + (+5) = 6$ is true, so 1 is a solution.

b. $5 - y = 7;\ y = -2$
Solution: $5 - (-2) = 5 + (+2) = 7$ is true, so –2 is a solution.
Note that parentheses were used around –2. This should be done whenever substituting negative numbers for the variable.

c. $z - 14 = -3;\ z = 10$
Solution: $10 - 14 = -4$ and $-4 = -3$ is false, so 10 is **not** a solution.

Practice Problems

1. *What is the additive inverse of 85?*
2. *Find the difference:* $-6 - (-5)$
3. *Simplify:* $-6 - 4 - (-2)$
4. *True or false:* $-5 + (-3) < -5 - (-3)$
5. *Is* $x = 15$ *a solution to the equation* $x - 1 = -16$*?*

1.3 Exercises

1. –11 **2.** –17 **3.** 6 **4.** 23 **5.** –47 **6.** 34 **7.** 0 **8.** –100 **9.** 52 **10.** 257

Find the additive inverse for each integer given in Exercises 1 – 10.

1. 11	**2.** 17	**3.** –6	**4.** –23	**5.** 47
6. –34	**7.** 0	**8.** 100	**9.** –52	**10.** –257

Answers to Practice Problems: **1.** –85 **2.** –1 **3.** –8 **4.** True **5.** Not a solution

11. 5 **12.** −2 **13.** −10
14. 7 **15.** 12 **16.** −35
17. 3 **18.** 12 **19.** −16
20. −15 **21.** 24 **22.** −8

Simplify the expressions in Exercises 11 – 22.

11. 8 − 3 **12.** 5 − 7 **13.** −4 − 6 **14.** 3 − (−4)
15. 5 − (−7) **16.** −18 − 17 **17.** −8 − (−11) **18.** 0 − (−12)
19. −14 − 2 **20.** −8 − 7 **21.** 16 − (−8) **22.** 15 − 23

23. −15 **24.** −7
25. −16 **26.** −33
27. −57 **28.** −60
29. −54 **30.** 7

Subtract the bottom number from the top number in Exercises 23 – 30.

23.	**24.**	**25.**	**26.**
27	19	−23	−41
42	26	−7	−8

27.	**28.**	**29.**	**30.**
−21	−47	−27	−19
36	13	27	−26

31. 1 **32.** −7
33. 8 **34.** 42
35. −26

31. Find the difference between −5 and −6. (**Hint:** Subtract the numbers in the order given.)
32. Subtract −3 from −10.
33. Subtract −2 from 6.
34. Find the difference between 30 and −12. (**Hint:** Subtract the numbers in the order given.)
35. Subtract 13 from −13.

36. −15 **37.** 1
38. −1 **39.** −8
40. −3 **41.** −10
42. 7 **43.** −6
44. 0 **45.** −139

Perform the indicated operations in Exercises 36 – 45.

36. −6 + (−4) − 5 **37.** −7 − (−2) + 6 **38.** 6 + (−3) + (−4)
39. −3 + (−7) + 2 **40.** −5 − 2 − (−4) **41.** −8 − 5 − (−3)
42. −2 − 2 + 11 **43.** −3 − (−3) + (−6) **44.** 97 − 16 − (81)
45. −113 + 53 − 79

46. $-8 < -5$
47. $-1 > -7$
48. $-3 < 3$
49. $10 > -10$
50. $8 = 8$
51. $-6 < 6$
52. $0 > -27$
53. $-4 > -5$
54. $-237 > -248$
55. $-37 < -34$

Perform the operations on each side of the blank and then fill in the blank in Exercises 46 – 55 with the proper symbol: <, >, or =.

46. −6 + (−2) ________ 3 + (−8) **47.** −4 − (−3) ________ −4 + (−3)
48. 5 − 8 ________ 8 − 5 **49.** 7 − (−3) ________ −3 − 7
50. 11 + (−3) ________ 11 − 3 **51.** 0 − 6 ________ 0 − (−6)
52. −8 − (−8) ________ −14 − 13 **53.** −7 − (−3) ________ 4 − 9
54. −151 − 86 ________ −(107 + 141) **55.** 25 − 62 ________ −11 − 23

56. −8 is a solution
57. −3 is a solution
58. −2 is a solution
59. 3 is a solution
60. 4 is not a solution
61. 4 is not a solution
62. 5 is a solution
63. −10 is a solution
64. −1 is a solution
65. 12 is a solution
66. −4 is a solution
67. −9 is a solution
68. −5 is a solution
69. 16 is not a solution
70. −28 is a solution

In Exercises 56 – 70, determine whether or not the given number is a solution to the given equation by substituting and then evaluating.

56. $x + 5 = -3;\ x = -8$ **57.** $x - 6 = -9;\ x = -3$
58. $15 - y = 17;\ y = -2$ **59.** $11 - x = 8;\ x = 3$
60. $x - 3 = -7;\ x = 4$ **61.** $y - 2 = 6;\ y = 4$
62. $-9 - x = -14;\ x = 5$ **63.** $x + 13 = 3;\ x = -10$
64. $x - 2 = -3;\ x = -1$ **65.** $-18 - y = -30;\ y = 12$
66. $|x| - (-10) = 14;\ x = -4$ **67.** $|y| - 12 = -3;\ y = -9$
68. $|z| - 5 = 0;\ z = -5$ **69.** $-16 - |x| = 0;\ x = 16$
70. $|x| - |-3| = 25;\ x = -28$

71. 1044 > –39
72. –43,241 < 27,180
73. –15,254 > –35,090
74. Lost 7 pounds; 203 pounds
75. 2316 points
76. Gained 51 yards
77. 6° below 0 (or –6°)
78. 282 ft.

Calculator Problems

In Exercises 71 – 73, use your TI-83 Plus calculator to find the value indicated on each side of the blank and then fill in the blank with the proper symbol: <, >, or =.

71.	648 – (–396)	________	124 – 163
72.	–19,824 – 23,417	________	12,793 – (–14,387)
73.	–43,931 – (–28,677)	________	–(13,665 + 21,425)

Temperatures above 0 and below 0 as well as gains and losses can be thought of in terms of positive and negative numbers. Use positive and negative numbers to answer Exercises 74 – 78.

74. Harry and his wife went on a diet plan for 5 weeks. During those 5 weeks, Harry lost 5 pounds, gained 3 pounds, lost 2 pounds, lost 4 pounds, and gained 1 pound. What was his total loss (or gain) for the 5 weeks? If he weighed 210 pounds when he started the diet plan, what did he weigh at the end of the 5-week period? During the same time, his wife lost 10 pounds.

75. In a 5-day week the NASDAQ stock market posted a gain of 38 points, a loss of 65 points, a loss of 32 points, a gain of 10 points, and a gain of 15 points. If the NASDAQ started the week at 2350 points, what was the market at the end of the week?

76. In ten running plays in a football game the fullback gained 5 yards, lost 3 yards, gained 15 yards, gained 7 yards, gained 12 yards, lost 4 yards, lost 2 yards, gained 20 yards, lost 5 yards, and gained 6 yards. What was his net yardage for the game?

77. Beginning at a temperature of 10° above 0, the temperature in a scientific experiment was measured hourly for four hours. It dropped 5°, dropped 8°, dropped 6°, then rose 3°. What was the final temperature recorded?

78. A hiker, beginning at an altitude of 97 ft., ascends a peak 526 ft. Next, he descends 313 ft., and climbs another peak 157 ft. The hiker takes a rest and then continues to ascend the peak he is on for 219 ft. Finally he descends 404 ft. What is his final altitude?

Hawkes Learning Systems: Introductory & Intermediate Algebra

Subtraction with Integers

1.4 Multiplication and Division with Integers

Objectives

After completing this section, you will be able to:

1. *Multiply integers.*
2. *Divide integers.*
3. *Complete statements about the products and quotients of integers.*
4. *Determine if equations are true or false.*

Multiplication

Multiplication is shorthand for repeated addition. That is,

$$7+7+7+7+7=5\cdot 7=35$$

and

$$(-6)+(-6)+(-6)=3(-6)=-18.$$

Similarly,

$$(-2)+(-2)+(-2)+(-2)+(-2)=5(-2)=-10$$

Repeated addition with a negative integer results in a product of a positive integer and a negative integer. Since the sum of negative integers is negative, we have the following general rule.

The product of a positive integer and a negative integer is negative.

NOTES Multiplication can be represented by a raised dot, as in $5\cdot 7$, or by a number next to a parenthesis as in $3(-6)$ or $(3)(-6)$.

Example 1: Positive Times Negative

Teaching Notes: Remind students many times about the meaning of a numeral or variable next to a parenthesis. Some students will persist in translating $(3)(-6)$ to $3-6$ or -3.

a. $5(-3)=(-3)+(-3)+(-3)+(-3)+(-3)=-15$
b. $7(-10)=-70$
c. $42(-1)=-42$
d. $3(-5)=-15$

The product of two negative integers can be explained in terms of opposites. We also need the fact that for any integer a, we can think of the **opposite of** $\boldsymbol{a}$, $(-a)$, as the product of -1 and a. That is, we have

$$\boldsymbol{-a=-1\cdot a.}$$

Thus,

$$-4=-1\cdot 4=-1(4)$$

and in the product $-4(-7)$ we have,

$$-4(-7) = -1(4)(-7) = -1[(4)(-7)] = -1(-28) = -(-28) = 28.$$

Although one example does not prove a rule, this process can be used in general to arrive at the following correct conclusion:

The product of two negative integers is positive.

Example 2: Negative Times Negative

a. $(-4)(-9) = +36$
b. $-7(-5) = +35$
c. $-2(-6) = +12$
d. $(-1)(-5)(-3)(-2) = 5(-3)(-2) = -15(-2) = +30$

What happens if a number is multiplied by 0? For example, $3(0) = 0 + 0 + 0 = 0$. In fact,

Multiplication by 0 always gives a product of 0.

Example 3: Multiplication by 0

a. $6 \cdot 0 = 0$
b. $-13 \cdot 0 = 0$

The rules for multiplication can be summarized as follows.

Rules for Multiplication with Integers

If a and b are positive integers, then

1. *The product of two positive integers is positive:* $a \cdot b = ab$.
2. *The product of two negative integers is positive:* $(-a)(-b) = ab$.
3. *The product of a positive integer and a negative integer is negative:* $a(-b) = -ab$.
4. *The product of 0 and any integer is 0:* $a \cdot 0 = 0$ *and* $(-a) \cdot 0 = 0$.

Practice Problems

Find the following products.

1. $5(-3) =$ **2.** $-6(-4) =$
3. $-8(4) =$ **4.** $-12(0) =$
5. $-9(-2)(-1) =$ **6.** $3(-20)(5) =$

Answers to Practice Problems: 1. −15 **2.** 24 **3.** −32 **4.** 0 **5.** −18 **6.** −300

Division

The rules for multiplication lead directly to the rules for division since division is defined in terms of multiplication. For convenience, division is indicated in fraction form.

Division with Integers

Teaching Notes:
Remind students about the nature of defining – why it is needed. "Defining" often mystifies students. Point out that as our ideas grow in complexity, we sometimes use definitions to ensure that we avoid conflicts in notation and concepts.

For integers a, b, and x (where $b \neq 0$),

$$\frac{a}{b} = x \text{ means that } a = b \cdot x.$$

If a is any integer, then $\frac{a}{0}$ *is* ***undefined****, but* $\frac{0}{b} = 0$.

Now you may ask, "Why is division by 0 undefined?" The following explanation will answer that question.

Division by 0 is Undefined

1. *Suppose that $a \neq 0$ and $\frac{a}{0} = x$. Then, since division is related to multiplication, we must have $a = 0 \cdot x$. But this is not possible because $0 \cdot x = 0$ for any value of x and we stated that $a \neq 0$.*

2. *Suppose that $\frac{0}{0} = x$. Then, $0 = 0 \cdot x$ which is true for all values of x. But we must have a unique answer for x.*

   Therefore, in any case, we conclude that ***division by 0 is undefined****.*

Example 4: Division

a. $\frac{36}{9} = 4$ because $36 = 9 \cdot 4$.

b. $\frac{-36}{9} = -4$ because $-36 = 9(-4)$.

c. $\frac{36}{-9} = -4$ because $36 = -9(-4)$.

d. $\frac{-36}{-9} = +4$ because $-36 = -9(4)$.

The rules for division can be stated as follows.

Rules for Division with Integers

If a and b are positive integers,

1. *The quotient of two positive integers is positive:* $\frac{a}{b} = +\frac{a}{b}$.
2. *The quotient of two negative integers is positive:* $\frac{-a}{-b} = +\frac{a}{b}$.
3. *The quotient of a positive integer and a negative integer is negative:*

$$\frac{-a}{b} = -\frac{a}{b} \text{ and } \frac{a}{-b} = -\frac{a}{b}.$$

NOTES

The following common rules about multiplication and division with two non-zero integers are helpful in remembering the signs of answers.

1. If the integers have the same sign, both the product and quotient will be positive.
2. If the integers have different signs, both the product and quotient will be negative.

The quotient of two integers may not always be another integer. Just as with whole numbers, the quotient may be a fraction. For example, $\frac{-3}{6} = -\frac{1}{2}$ and $\frac{2}{-8} = -\frac{1}{4}$. We will discuss these ideas more thoroughly later in Chapter 1. In this section, the problems are set so that multiplication and division with integers will have an integer result. **Remember that division by 0 is not defined.**

Practice Problems

Find the quotients.

1. $\frac{-30}{10}$
2. $\frac{40}{-10}$
3. $\frac{-20}{-10}$
4. $\frac{-7}{0}$
5. $\frac{0}{13}$

Answers to Practice Problems: 1. –3 **2.** –4 **3.** 2 **4.** Undefined **5.** 0

Now that we have all the rules for addition, subtraction, multiplication, and division with integers, we can discuss the solutions to equations that involve any or all of these operations.

To indicate multiplication of a variable by a specific number, the times sign (or raised dot) is optional. For example,

$$23 \cdot x = 23x \quad \text{and} \quad -7 \cdot y = -7y.$$

In each case, the number is called the **coefficient** of the variable.

Example 5: Solutions

Determine whether or not the given integer is a solution to the given equation by substituting and then evaluating.

a. $7x = -21;\ x = -3$
Solution: $7(-3) = -21$ is true, so -3 is a solution.

b. $-8y = 56;\ y = -7$
Solution: $-8(-7) = 56$ is true, so -7 is a solution.

c. $\frac{y}{-4} = -10;\ y = -40$
Solution: $\frac{-40}{-4} = 10$ and $10 = -10$ is false, so -40 is not a solution.

d. $-5x + 7 = -3;\ x = 2$
Solution: $-5(2) + 7 = -10 + 7 = -3$ is true, so 2 is a solution.

1.4 Exercises

1. −12 **2.** −30 **3.** 56
4. 48 **5.** 57 **6.** 22
7. 56 **8.** −21 **9.** −30
10. 66 **11.** 26 **12.** −70
13. −60 **14.** 72 **15.** −24
16. −162 **17.** −288
18. 18 **19.** 0
20. −396

Find the product in Exercises 1 – 20.

1. $4 \cdot (-3)$	**2.** $(-5) \cdot 6$	**3.** $(-8)(-7)$	**4.** $12 \cdot 4$
5. $19 \cdot 3$	**6.** $(-11)(-2)$	**7.** $(-14)(-4)$	**8.** $(-3)(7)$
9. $(5)(-6)$	**10.** $(-11)(-6)$	**11.** $(-13)(-2)$	**12.** $10(-7)$
13. $(-5)(12)$	**14.** $(-8)(-9)$	**15.** $(-2)(-3)(-4)$	**16.** $(-6)(-3)(-9)$
17. $-8 \cdot 4 \cdot 9$	**18.** $-3 \cdot 2 \cdot (-3)$	**19.** $(-7)(-16) \cdot 0$	**20.** $(-9) \cdot 11 \cdot 4$

Find the quotient in Exercises 21 – 35.

21. 4 **22.** 2
23. –6 **24.** 2
25. –3 **26.** –17
27. 13 **28.** 0
29. 0
30. Undefined
31. Undefined
32. –17 **33.** –11
34. 5 **35.** –4

21. $\frac{-8}{-2}$ **22.** $\frac{-20}{-10}$ **23.** $\frac{-30}{5}$ **24.** $\frac{-26}{-13}$

25. $\frac{39}{-13}$ **26.** $\frac{-51}{3}$ **27.** $\frac{-91}{-7}$ **28.** $\frac{0}{6}$

29. $\frac{0}{-7}$ **30.** $\frac{-3}{0}$ **31.** $\frac{16}{0}$ **32.** $\frac{-34}{2}$

33. $\frac{44}{-4}$ **34.** $\frac{-60}{-12}$ **35.** $\frac{-36}{9}$

Correctly complete the sentences in Exercises 36 – 45 with positive, negative, 0, or undefined.

36. Negative
37. Negative
38. Positive
39. Negative
40. Positive
41. Negative
42. Positive
43. 0
44. 0
45. Undefined

36. If x is a positive integer, then $x(-x)$ is a __________ integer.
37. If x is a negative integer, then $x(-x)$ is a __________ integer.
38. If x is a negative integer and y is a negative integer, then xy is a __________ integer.
39. If x is a positive integer and y is a negative integer, then xy is a __________ integer.

40. If x and y are positive integers, then $\frac{x}{y}$ is a __________ number.

41. If x is a negative integer and y is a natural number then $\frac{x}{y}$ is a __________ number.

42. If x and y are negative numbers, then $\frac{x}{y}$ is a __________ number.

43. If x is any real number, then $x \cdot 0$ is ________.

44. If x is a nonzero real number, then $\frac{0}{x}$ is ________.

45. If x is any real number, then $\frac{x}{0}$ is ________.

Determine whether each statement in Exercises 46 – 55 is true or false. If a statement is false, rewrite it in a form that is true. (There may be more than one correct new form.)

46. True **47.** True
48. True **49.** True
50. True **51.** True
52. True **53.** False; $17+(-3) > (-14)+(-4)$
54. True **55.** True

46. $(-4)\cdot(6) < 3\cdot 8$
47. $(-7)\cdot(-9) = 3\cdot 21$
48. $(-12)\cdot(6) = 9(-8)$
49. $(-6)(9) = (18)(-3)$
50. $6(-3) \geq (-14)+(-4)$
51. $7+8 > (-10)+(-5)$
52. $-7+0 \leq (-7)\cdot(0)$
53. $17+(-3) < (-14)+(-4)$
54. $-4(9) = (-24)+(-12)$
55. $14+6 \leq -2(-10)$

56. 5 is a solution
57. –12 is a solution
58. 42 is not a solution
59. –72 is a solution
60. –90 is a solution
61. –8 is not a solution
62. 3 is a solution
63. 5 is a solution
64. –5 is a solution
65. –4 is a solution

In Exercises 56 – 65, determine whether or not the given number is a solution to the given equation by substituting and then evaluating.

56. $11x = 55;\ x = 5$

57. $-7x = 84;\ x = -12$

58. $\frac{x}{7} = -6;\ x = 42$

59. $\frac{y}{-6} = 12;\ y = -72$

60. $\frac{y}{10} = -9;\ y = -90$

61. $-9x = -72;\ x = -8$

62. $5x + 3 = 18;\ x = 3$

63. $-3x + 7 = -8;\ x = 5$

64. $4x - 3 = -23;\ x = -5$

65. $7x - 6 = -34;\ x = -4$

66. 1,179,360
67. –34,459,110
68. –22,032
69. –2671
70. 1682

Calculator Problems

Use a calculator to find the value of each expression in Exercises 66 – 70.

66. $(273)(-24)(-180)$

67. $(-4{,}613)(-45)(-166)$

68. $(54)(-17)(24)$

69. $(-77{,}459) \div 29$

70. $(-62{,}234) \div (-37)$

71. Dividing 0 by any number other than 0 gives a quotient of 0.
72. See page 28.

Writing and Thinking About Mathematics

71. Explain the conditions under which the quotient of two numbers is 0.

72. Explain, in your own words, why division by 0 is not a valid arithmetic operation.

Hawkes Learning Systems: Introductory & Intermediate Algebra

Multiplication and Division with Integers

1.5 Exponents, Prime Numbers, and Order of Operations

Objectives

After completing this section, you will be able to:

1. *Evaluate expressions with exponents.*
2. *Recognize prime numbers less than 50.*
3. *Determine the prime factorization of integers.*
4. *Follow the Rules for Order of Operations to evaluate expressions.*

Exponents

Repeated addition with the same number is indicated with multiplication. For example,

$$14 + 14 + 14 + 14 + 14 = 5 \cdot 14 = 70 \qquad \text{and} \qquad 6 + 6 + 6 = 3 \cdot 6 = 18$$

factors product　　　　factors product

The result of multiplication is called the **product**, and the numbers being multiplied are called **factors** of the product.

In a similar manner, multiplication by the same number can be indicated by using exponents. For example, if 2 is used as a factor 5 times, we can write

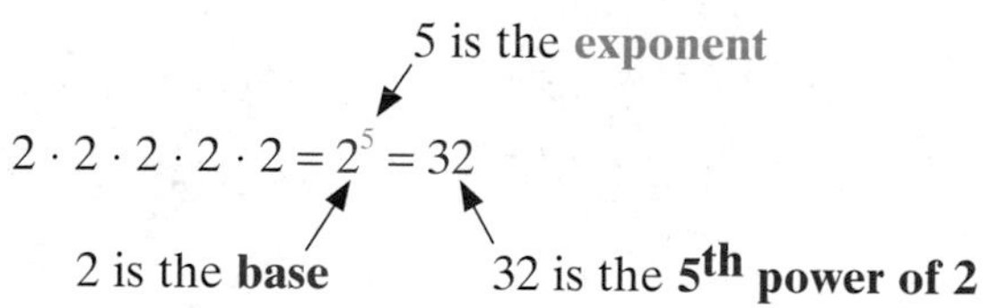

In the equation $2^5 = 32$, the number 32 is the **power**, 2 is the **base**, and 5 is the **exponent**. The expression 2^5 is read "two to the fifth power." We can also say that "32 is the fifth power of 2." Because of the wording in discussing exponents and powers, there can be some confusion as to just what number is the power. To help in understanding this rather technical distinction, think that the exponent describes the power.

With Repeated Multiplication	With Exponents
a. $4 \cdot 4 = 16$	$4^2 = 16$
b. $3 \cdot 3 \cdot 3 \cdot 3 = 81$	$3^4 = 81$

If an exponent is 2, then the base is said to be **squared**. If an exponent is 3, then the base is said to be **cubed**. For example, $5^2 = 25$ is read "five squared is equal to twenty-five"

Teaching Notes:
Draw the connection of "squared" and "cubed" to geometry. These are more than mere nicknames and students should know the connections.

and $7^3 = 343$ is read "seven cubed is equal to three hundred forty-three." If no exponent is written on a number or a variable, then the exponent is understood to be 1. Thus,

$$6 = 6^1 \text{ and } a = a^1.$$

The use of 0 as an exponent is a special case and, as we will see later in dealing with properties of exponents, 0 exponents are needed in simplifying algebraic expressions. To help in understanding the meaning of 0 as an exponent, consider the following patterns of powers.

$$\begin{array}{ll} 2^4 = 16 & 5^4 = 625 \\ 2^3 = 8 & 5^3 = 125 \\ 2^2 = 4 & 5^2 = 25 \\ 2^1 = 2 & 5^1 = 5 \\ 2^0 = ? & 5^0 = ? \end{array}$$

If you study these patterns of powers, you will see that each power of 2 is found by dividing the previous power by 2, and each power of 5 is found by dividing the previous power by 5. These results lead to the conclusion that $2^0 = 1$ and $5^0 = 1$. We have the following general properties of exponents.

General Properties of Exponents

In general, for any whole number n and any integer a,

1. $\underbrace{a \cdot a \cdot a \cdot a \cdot \ldots \cdot a}_{n \text{ factors}} = a^n$

2. $a^1 = a$

3. $a^0 = 1$ *(for* $a \neq 0$*)*
   (The expression 0^0 *is undefined.)*

Example 1: Calculator

Use a calculator to find the following powers.

a. 16^3 **b.** 52^4 **c.** 12^0 **d.** 85^1

Solutions:

With a Scientific Calculator

Your scientific calculator will have a key marked as $\mathbf{x^y}$ or $\mathbf{y^x}$.

a. To evaluate 16^3:
Enter the base **16**.
Press the key marked $\mathbf{x^y}$. (Note that nothing will appear on the display. The calculator is waiting for you to enter the exponent.)
Enter the exponent **3**.
Press = or **ENTER**.
The display should read **4096**.

b. To evaluate 52^4:
Enter the base **52**.
Press the key marked $\mathbf{x^y}$. (Note that nothing will appear on the display. The calculator is waiting for you to enter the exponent.)
Enter the exponent **4**.
Press = or **ENTER**.
The display should read **7311616**.

c. To evaluate 12^0:
Follow the steps outlined in Examples **a** and **b**.
The display should read **1**.

d. To evaluate 85^1:
Follow the steps outlined in Examples **a** and **b**.
The display should read **85**.

With a TI-83 Plus (or other graphing calculator)

Your TI-83 Plus calculator will have a key marked with a caret (^). This key indicates that an exponent is to follow. Each expression and its value are shown in the display screens.

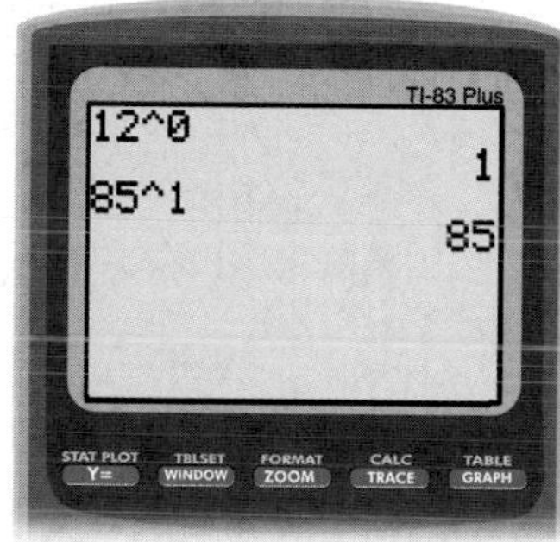

Prime Factors

In working with fractions and simplifying algebraic expressions, we sometimes want to see integers factored in their simplest form. That is, we want each number to be factored so that each factor is a **prime number**. Remember that **even integers** are divisible by 2 and **odd integers** are not divisible by 2.

Prime Number

*A **prime number** is a whole number (other than 0 or 1) that has exactly two different factors, itself and 1.*

A whole number (other than 0 or 1) that is not prime is a **composite number**.

You should memorize the prime numbers less than 50. Here is a list of the prime numbers less than 100:

2, 3, 5, 7, 11, 13, 17, 19, 23, 29, 31, 37, 41, 43, 47, 53, 59, 61, 67, 71, 73, 79, 83, 89, 97

NOTES

Note the following two facts:

1. 2 is the only even prime number.

2. While all other prime numbers are odd, not all odd numbers are prime. For example, 9 is an odd number that is not prime.

The following quick tests for divisibility can be helpful in finding beginning factors of composite numbers. The concepts of prime factors and divisibility can be applied to negative integers as well as positive integers. We simply treat a negative integer as –1 times a positive integer and factor the positive integer as before.

Tests for Divisibility

An integer is divisible

By 2: *if the units digit is 0, 2, 4, 6, or 8.*

By 3: *if the sum of the digits is divisible by 3.*

By 5: *if the units digit is 0 or 5.*

By 6: *if the number is divisible by both 2 and 3.*

By 9: *if the sum of the digits is divisible by 9.*

By 10: *if the units digit is 0.*

Example 2: Prime Factorization

Teaching Notes: You might want to show students the method of factoring using a factor tree. For example,

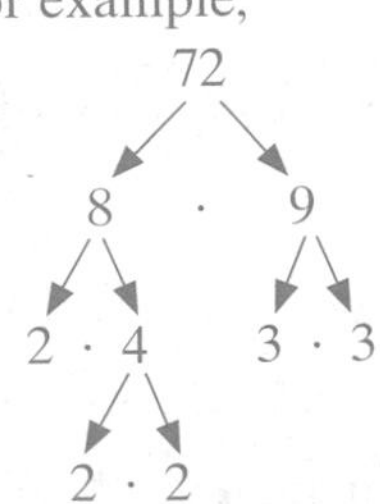

Find the prime factorization of each of the following composite numbers.

a. 72 **b.** 60 **c.** 165

Solution: By the tests for divisibility:

a. 72 is divisible by 2, 3, 6, and 9. $72 = 8 \cdot 9 = 2 \cdot 4 \cdot 3 \cdot 3 = 2^3 \cdot 3^2$.

b. 60 is divisible by 2, 3, 5, 6, and 10. $60 = 2 \cdot 3 \cdot 2 \cdot 5 = 2^2 \cdot 3 \cdot 5$.

c. 165 is divisible by 3 and 5. $165 = 5 \cdot 33 = 3 \cdot 5 \cdot 11$.

Therefore, the prime factorization of each number may be found by using any of these known factors.

You may have noticed that, regardless of how you begin, there is only one prime factorization for any composite number. This fact is known as the Fundamental Theorem of Arithmetic.

The Fundamental Theorem of Arithmetic

Every composite number has a unique prime factorization.

Teaching Notes: Some students will want to know "why do we care?". You might want to point out that one of the results of the Fundamental Theorem of Arithmetic is that renaming a fraction to its simplest form is unique.

Order of Operations

Mathematicians have agreed on a set of rules for the order of performing operations when evaluating numerical expressions that contain grouping symbols, exponents, and the operations of addition, subtraction, multiplication, and division. These rules are used throughout all levels of mathematics and science so that there is only one correct answer for the value of an expression, and so that everyone will get that answer. For example, evaluate the following expression:

$$36 \div 4 + 6 \cdot 2^2$$

Did you get 33 or 60? The correct answer is 33, which can be found by following a set of rules called the Rules for Order of Operations.

Rules for Order of Operations

Teaching Notes: You might demonstrate that calculators follow this convention of order of operations.

1. *Simplify within grouping symbols, such as parentheses ( ), brackets [ ], and braces { }. (Start with the innermost grouping symbol.)*
2. *Find any powers indicated by exponents.*
3. *Moving from* ***left to right****, perform any multiplications or divisions in the order they appear.*
4. *Moving from* ***left to right****, perform any additions or subtractions in the order they appear.*

NOTES Other grouping symbols are the absolute value bars, the fraction bar, and radicals such as the square root symbol.

A well known mnemonic device for remembering the Rules for Order of Operations is the following:

Please	Excuse	My	Dear	Aunt	Sally
Parentheses	**Exponents**	**Multiplication**	**Division**	**Addition**	**Subtraction**

NOTES

Remember that, while the mnemonic device is helpful, multiplication might not come before division and addition might not come before subtraction. The priority is from **left** to **right**.

For example:

$$12 \div 3 \cdot 4 = 4 \cdot 4 = 16$$

but,

$$12 \cdot 3 \div 4 = 36 \div 4 = 9$$

The Rules for Order of Operations can be particularly useful in determining the values of expressions involving negative numbers and exponents. For example, the two expressions

$$-7^2 \text{ and } (-7)^2$$

have two different values. By the order of operations, exponents come before multiplication. Thus,

$$-7^2 = -1 \cdot 7^2 = -1 \cdot 49 = -49 \quad \text{but} \quad (-7)^2 = (-7)(-7) = 49.$$

Remember that if the base is a negative number, then the negative number must be placed in parentheses. This distinction is particularly important when the exponent is an even integer. Thus,

$$-3^4 = -1 \cdot 3^4 = -1 \cdot 81 = -81 \text{ and } (-3)^4 = (-3)(-3)(-3)(-3) = 81$$

but

$$-3^3 = -1 \cdot 3^3 = -1 \cdot 27 = -27 \text{ and } (-3)^3 = (-3)(-3)(-3) = -27.$$

Example 3 shows several expressions evaluated in a step-by-step manner following the Rules for Order of Operations. Study these carefully.

Example 3: Order of Operations

Teaching Notes: You will probably need to emphasize the use of "PEMDAS" within a parenthetical expression.

Use the Rules for Order of Operations to evaluate each of the following expressions.

a. $36 \div 4 - 6 \cdot 2^2$ **b.** $2(3^2 - 1) - 3 \cdot 2^3$ **c.** $9 - 2[(3 \cdot 5 - 7^2) \div 2 + 2^2]$

Solution:

a. $36 \div 4 - 6 \cdot 2^2 = 36 \div 4 - 6 \cdot 4$ Exponents.

$= 9 - 24$ Divide and multiply, left to right.

$= -15$ Subtract.

b. $2(3^2-1)-3\cdot 2^3 = 2(9-1)-3\cdot 8$ Exponents.

$= 2(8)-3\cdot 8$ Subtract inside the parentheses.

$= 16-24$ Multiply.

$= -8$ Subtract (or add algebraically).

c. $9-2[(3\cdot 5-7^2)\div 2+2^2] = 9-2[(3\cdot 5-49)\div 2+4]$ Exponents.

$= 9-2[(15-49)\div 2+4]$ Multiply inside the parentheses.

$= 9-2[(-34)\div 2+4]$ Subtract inside the parentheses.

$= 9-2[-17+4]$ Divide inside the brackets.

$= 9-2[-13]$ Add inside the brackets.

$= 9+26$ Multiply.

$= 35$ Add.

Practice Problems

1. List all the prime numbers less than 12.

2. Find the prime factorization of 75.

3. Find the prime factorization of 117.

Find the value of each expression by using the Rules for Order of Operations.

4. $14 \div 2 + 2 \cdot 6 + 30 \div 3$

5. $3[8 + 2(1 - 10)] - 15 \div 5$

1. 5^2 **2.** 6^2 **3.** 7^2
4. 9^2 or 3^4 **5.** 11^2 **6.** 13^2
7. 8^2 or 4^3 or 2^6
8. 12^2 **9.** 2^3 **10.** 3^3
11. 5^3 **12.** 2^5 **13.** 3^5
14. 3^2 **15.** 10^2 **16.** 10^3

1.5 Exercises

17. 64 **18.** 1 **19.** 100
20. 216 **21.** 343 **22.** 81
23. 10,000 **24.** 625
25. 1 **26.** 1 **27.** 225
28. 400 **29.** 27,000
30. 2,500 **31.** 10,000
32. −125 **33.** 36
34. −36 **35.** −64

In Exercises 1 – 16, rewrite each of the following integers in a base and exponent form without using an exponent of 1.

1. 25 **2.** 36 **3.** 49 **4.** 81 **5.** 121 **6.** 169 **7.** 64
8. 144 **9.** 8 **10.** 27 **11.** 125 **12.** 32 **13.** 243 **14.** 9
15. 100 **16.** 1000

Find the value of each of the following expressions in Exercises 17 – 35.

17. 8^2 **18.** 1^5 **19.** 10^2 **20.** 6^3 **21.** 7^3 **22.** 9^2
23. 10^4 **24.** 5^4 **25.** 13^0 **26.** $(-16)^0$ **27.** $(-15)^2$ **28.** 20^2
29. 30^3 **30.** 50^2 **31.** $(-10)^4$ **32.** $(-5)^3$ **33.** $(-6)^2$ **34.** -6^2
35. -2^6

Answers to Practice Problems: **1.** 2, 3, 5, 7, 11 **2.** $3\cdot 5^2$ **3.** $3^2\cdot 13$ **4.** 29 **5.** −33

36. 194,481 **37.** 15,625
38. 147,008,443
39. 262,144 **40.** 1
41. –225 **42.** –248,832
43. –8 **44.** 10,000
45. 196

Use a calculator to find the value of each of the following expressions in Exercises 36 – 45.

36. 21^4 **37.** 5^6 **38.** 43^5 **39.** 2^{18} **40.** 63^0
41. -15^2 **42.** $(-12)^5$ **43.** $(6-8)^3$ **44.** $(20-30)^4$ **45.** $(18-32)^2$

46. List the prime numbers less than 50.

46. 2, 3, 5, 7, 11, 13, 17, 19, 23, 29, 31, 37, 41, 43 and 47

In Exercises 47 – 60, find the prime factorization of each integer.

47. 70 **48.** 80 **49.** 43 **50.** 59 **51.** 225 **52.** 65
53. 363 **54.** 120 **55.** 90 **56.** 128 **57.** 140 **58.** 340
59. 1000 **60.** 560

47. $2 \cdot 5 \cdot 7$
48. $2^4 \cdot 5$
49. 43 is prime
50. 59 is prime
51. $3^2 \cdot 5^2$ **52.** $5 \cdot 13$
53. $3 \cdot 11^2$ **54.** $2^3 \cdot 3 \cdot 5$
55. $2 \cdot 3^2 \cdot 5$ **56.** 2^7
57. $2^2 \cdot 5 \cdot 7$
58. $2^2 \cdot 5 \cdot 17$
59. $2^3 \cdot 5^3$ **60.** $2^4 \cdot 5 \cdot 7$

*Use the Rules for Order of Operations to evaluate each of the following expressions in Exercises 61 – 82. (**Note:** The fraction bar in Exercises 73 – 76 should be treated in the same manner as parentheses.)*

61. **a.** $24 \div 4 \cdot 6$ **b.** $24 \cdot 4 \div 6$
62. **a.** $20 \div 5 \cdot 2$ **b.** $20 \cdot 5 \div 2$
63. $15 \div (-3) \cdot 3 - 10$
64. $20 \cdot 2 \div 2^2 + 5(-2)$
65. $3^3 \div (-9) \cdot (4 - 2^2) + 5(-2)$
66. $4^2 \div (-8)(-2) + 3(2^2 - 5^2)$
67. $14 \cdot 3 \div (-2) - 6(4)$
68. $6(13 - 15)^2 \cdot 8 \div 2^2 + 3(-1)$
69. $-10 + 15 \div (-5) \cdot 3^2 - 10^2$
70. $16 \cdot 3 \div (2^2 - 5)$
71. $2 - 5[(-20) \div (-4) \cdot 2 - 40]$
72. $9 - 6[(-21) \div 7 \cdot 2 - (-8)]$

73. $\dfrac{14-56}{15-8}$ **74.** $\dfrac{13+2\cdot 6}{3\cdot 5-2\cdot 5}$ **75.** $\dfrac{6^2-2\cdot 7}{5^2-6^2}$ **76.** $\dfrac{8+2\cdot 5-3^3}{4^2-5^2}$

77. $(9-11)\left[(-10)^2\cdot 2+6(-5)^2-10^3+3\cdot 5\right]$ **78.** $6-20\left[(-15)\div 3\cdot 5+6\cdot 2\div 3\right]$

79. $8-9\left[(-39)\div(-13)+7(-2)-(-2)^3\right]$ **80.** $(7-10)\left[49\div(-7)+20\cdot 3-(-10)\right]$

81. $|10-30|\left[4^2\cdot|5-8|\div(-2)^3+|17-18|\right]$ **82.** $|16-20|\left[32\div|3-5|-5^2\right]$

61. **a.** 36 **b.** 16
62. **a.** 8 **b.** 50 **63.** –25
64. 0 **65.** –10 **66.** –59
67. –45 **68.** 45
69. –137 **70.** –48
71. 152 **72.** –3
73. –6 **74.** 5
75. –2 **76.** 1
77. 1270 **78.** 426
79. 35 **80.** –189
81. –100 **82.** –36

Writing and Thinking About Mathematics

83. Explain, in your own words, why the following expression cannot be evaluated.
$(24 - 2^4) + 6(3 - 5) \div (3^2 - 9)$

83. $(3^2 - 9) = 0$ and division by 0 is undefined.

Hawkes Learning Systems: Introductory & Intermediate Algebra

Factoring Positive Integers
Factoring Integers
Order of Operations

1.6 Multiplying and Dividing Fractions

Objectives

After completing this section, you will be able to:

1. *Reduce fractions to lowest terms.*
2. *Write fractions as equivalent fractions with specified denominators.*
3. *Multiply and divide fractions.*

We know, from Section 1.1, that fractions of the form

$$\frac{a}{b} \quad \frac{\leftarrow \text{numerator}}{\leftarrow \text{denominator}}$$

where a and b are integers and $b \neq 0$ are called **rational numbers**. For now we will use the terms **rational number** and **fraction** to mean the same thing. (Note that some fractions cannot be written with integers in the numerator and denominator. For example, $\frac{\pi}{3}$ is a fraction but also an infinite nonrepeating decimal and, therefore, an irrational number.)

Rational Number

A ***rational number*** *is a number that can be written in the form* $\frac{a}{b}$ *where a and b are integers and* $b \neq 0$.

OR

A ***rational number*** *is a number that can be written in decimal form as a terminating decimal or as an infinite repeating decimal.*

Examples of rational numbers (fractions):

$$\frac{1}{2}, \frac{3}{10}, \frac{-11}{7}, -\frac{13}{2}, -10, \text{ and } 15.$$

Note that every integer is also a rational number because integers can be written in fraction form with a denominator of 1. Thus,

$$0 = \frac{0}{1}, 1 = \frac{1}{1}, 2 = \frac{2}{1}, 3 = \frac{3}{1}, \text{ and so on.}$$

Also,

$$-1=\frac{-1}{1},\ -2=\frac{-2}{1},\ -3=\frac{-3}{1},\ -4=\frac{-4}{1},\ \text{and so on.}$$

In general, fractions can be used to indicate:

1. Equal parts of a whole, or
2. Division

Example 1: Fractions

Teaching Notes:
You might point out that the '÷' sign is a fraction bar with dots indicating numerator and denominator. Also, whether one actually interprets $\frac{a}{b}$ as an implied division depends on the context.

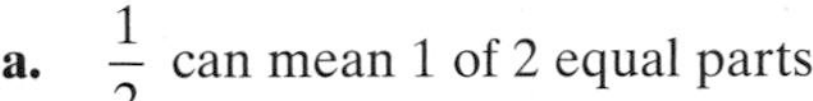

a. $\frac{1}{2}$ can mean 1 of 2 equal parts

If you read $\frac{1}{2}$ of a book, then you can view this as having read one of two equal parts of the book. If the book has 50 pages, then you read 25 pages, since, as we will see, $\frac{25}{50}=\frac{1}{2}$.

b. $\frac{-45}{9}$ can mean to divide: $(-45) \div 9$.

Before actually operating with fractions, we need to understand the use and placement of negative signs in fractions. Consider the following three results involving fractions, negative signs, and the meaning of division:

$$-\frac{16}{8}=-2,\ \frac{-16}{8}=-2,\ \text{and}\ \frac{16}{-8}=-2$$

Thus, the three different placements of a negative sign give the same results, as noted in the general statement or rule below.

Rules for the Placement of Negative Signs in Fractions

If a and b are real numbers, and $b \neq 0$, then

$$-\frac{a}{b}=\frac{-a}{b}=\frac{a}{-b}.$$

For example:

$$-\frac{3}{4b}=\frac{-3}{4b}=\frac{3}{-4b}$$

Example 2: Negative Signs

a. $-\frac{12}{4} = \frac{-12}{4} = \frac{12}{-4} = -3$

b. $-\frac{1}{5} = \frac{-1}{5} = \frac{1}{-5}$

Multiplication

Multiplication

To ***multiply*** *two fractions, multiply the numerators and multiply the denominators.*

$$\frac{a}{b} \cdot \frac{c}{d} = \frac{a \cdot c}{b \cdot d}$$

For example:

$$\frac{2}{3} \cdot \frac{7}{5} = \frac{2 \cdot 7}{3 \cdot 5} = \frac{14}{15}$$

Remember that the number 1 is called the **multiplicative identity** since the product of 1 with any number is that number. That is,

$$\frac{a}{b} \cdot 1 = \frac{a}{b}.$$

Thus, if $k \neq 0$, we have

$$\frac{a}{b} = \frac{a}{b} \cdot 1 = \frac{a}{b} \cdot \frac{\mathbf{k}}{\mathbf{k}} = \frac{a \cdot \mathbf{k}}{b \cdot \mathbf{k}}.$$

This relationship is called the **Fundamental Principle of Fractions**.

The Fundamental Principle of Fractions

$$\frac{a}{b} = \frac{a \cdot \mathbf{k}}{b \cdot \mathbf{k}}, \text{ where } \mathbf{k} \neq 0$$

Teaching Notes:
Remind students that this process of renaming does not alter the value.

We can use the Fundamental Principle to build a fraction to **higher terms** (find an equal fraction with a larger denominator) or reduce to **lower terms** (find an equal fraction with a smaller denominator).

To reduce a fraction, factor both the numerator and denominator, then use the Fundamental Principle to "divide out" any common factors. If the numerator and the denominator have no common prime factors, the fraction has been **reduced to lowest terms**. Finding the prime factorization of the numerator and denominator before reducing, while not necessary, will help guarantee a fraction is in lowest terms.

Example 3: Fundamental Principle of Fractions

a. Raise $\frac{3}{7}$ to higher terms with a denominator of 28.

Solution: Use $k = 4$ since $7 \cdot 4 = 28$.

$$\frac{3}{7} = \frac{3 \cdot 4}{7 \cdot 4} = \frac{12}{28}$$

b. Raise $-\frac{5}{8}$ to higher terms with a denominator of $16a$.

Solution: Use $k = 2a$ since $8 \cdot 2a = 16a$.

$$-\frac{5}{8} = -\frac{5 \cdot 2a}{8 \cdot 2a} = -\frac{10a}{16a}$$

c. Reduce $\frac{-12}{20}$ to lowest terms by using prime factorizations.

Solution: $$\frac{-12}{20} = \frac{-1 \cdot 2 \cdot 2 \cdot 3}{2 \cdot 2 \cdot 5} = -1 \cdot \frac{2}{2} \cdot \frac{2}{2} \cdot \frac{3}{5} = -1 \cdot 1 \cdot 1 \cdot \frac{3}{5} = -\frac{3}{5}$$

or $$\frac{-12}{20} = \frac{-1 \cdot 4 \cdot 3}{4 \cdot 5} = -1 \cdot \frac{4}{4} \cdot \frac{3}{5} = -1 \cdot 1 \cdot \frac{3}{5} = -\frac{3}{5}$$

or $$\frac{-12}{20} = \frac{-1 \cdot \overset{1}{\cancel{2}} \cdot \overset{1}{\cancel{2}} \cdot 3}{\underset{1}{\cancel{2}} \cdot \underset{1}{\cancel{2}} \cdot 5} = -\frac{3}{5}$$

d. Find the product $\frac{15ac}{28b^2} \cdot \frac{4bc}{9a^3}$ in lowest terms. (Do not find the product directly. Factor and reduce as you multiply.)

Solution: $$\frac{15ac}{28b^2} \cdot \frac{4bc}{9a^3} = \frac{\overset{1}{\cancel{3}} \cdot 5 \cdot \overset{1}{\cancel{a}} \cdot c \cdot \overset{1}{\cancel{4}} \cdot \overset{1}{\cancel{b}} \cdot c}{\underset{1}{\cancel{4}} \cdot 7 \cdot \underset{1}{\cancel{b}} \cdot b \cdot \underset{1}{\cancel{3}} \cdot 3 \cdot \underset{1}{\cancel{a}} \cdot a \cdot a} = \frac{5c^2}{21a^2b}$$

e. Find the product $\frac{4a}{12b} \cdot \frac{3b}{7a}$ in lowest terms. (Note that the number 1 is implied to be a factor even if it is not written.)

Solution: $\frac{4a}{12b} \cdot \frac{3b}{7a} = \frac{\overset{1}{\cancel{4}} \cdot \overset{1}{\cancel{a}} \cdot \overset{1}{\cancel{3}} \cdot \overset{1}{\cancel{b}} \cdot 1}{\underset{1}{\cancel{4}} \cdot \underset{1}{\cancel{3}} \cdot \underset{1}{\cancel{b}} \cdot 7 \cdot \underset{1}{\cancel{a}}} = \frac{1}{7}$

Here we write the factor 1 because all other factors have been "divided out."

We could write $\frac{4a}{12b} \cdot \frac{3b}{7a} = \frac{4 \cdot a \cdot 3 \cdot b}{4 \cdot 3 \cdot b \cdot 7 \cdot a}$

$$= \frac{4}{4} \cdot \frac{3}{3} \cdot \frac{a}{a} \cdot \frac{b}{b} \cdot \frac{1}{7}$$

$$= 1 \cdot 1 \cdot 1 \cdot 1 \cdot \frac{1}{7} = \frac{1}{7}$$

Finding the product of two fractions can be thought of as finding one fractional part **of** the other. Thus, to find a fraction **of** a number means to multiply the fraction and the number.

Example 4: Fraction of a Number

Find $\frac{2}{3}$ of $\frac{5}{7}$.

Solution: $\frac{2}{3} \cdot \frac{5}{7} = \frac{2 \cdot 5}{3 \cdot 7} = \frac{10}{21}$

Division

We will now see that division with fractions is accomplished by multiplication. That is, if we know how to multiply fractions, we automatically know how to divide them.

Reciprocal

If $a \neq 0$ and $b \neq 0$, the ***reciprocal*** *of $\frac{a}{b}$ is $\frac{b}{a}$, and $\frac{a}{b} \cdot \frac{b}{a} = 1$.*

Consider the division problem $\frac{2}{3} \div \frac{5}{6}$. The indicated division can be written in the form of a complex fraction as follows:

Teaching Notes:
You might want to show the students the form that they are familiar with:

$$\frac{2}{\cancel{3}_1} \cdot \frac{\cancel{6}^2}{5} = \frac{4}{5}.$$

However, I have found that the prime factorization technique works well with students who have trouble with fractions. See Example 5b and 5c on the next page.

$$\frac{2}{3} \div \frac{5}{6} = \frac{\frac{2}{3}}{\frac{5}{6}}$$ Write the division in fraction form.

$$= \frac{\frac{2}{3} \cdot \frac{6}{5}}{\frac{5}{6} \cdot \frac{6}{5}}$$ Use the reciprocal of the denominator and multiply by 1 in the form $\frac{\frac{6}{5}}{\frac{6}{5}}$.

$$= \frac{\frac{2}{3} \cdot \frac{6}{5}}{\frac{\cancel{5}^1}{\cancel{6}_1} \cdot \frac{\cancel{6}^1}{\cancel{5}_1}}$$ Simplify.

$$= \frac{\frac{2}{3} \cdot \frac{6}{5}}{1}$$ $\leftarrow \frac{5}{6} \cdot \frac{6}{5} = 1$

$$= \frac{2}{3} \cdot \frac{6}{5}$$

Thus, we have

$$\frac{2}{3} \div \frac{5}{6} = \frac{2}{3} \cdot \frac{6}{5} = \frac{2 \cdot 2 \cdot \cancel{3}^1}{\cancel{3}_1 \cdot 5} = \frac{4}{5}.$$

This example and the related discussion lead to the following definition.

Division

*To **divide** by a nonzero fraction, multiply by its reciprocal:*

$$\frac{\mathbf{a}}{\mathbf{b}} \div \frac{\mathbf{c}}{\mathbf{d}} = \frac{\mathbf{a}}{\mathbf{b}} \cdot \frac{\mathbf{d}}{\mathbf{c}}$$

Example 5: Division

a. $\left(-\frac{3}{4}\right) \div \left(-\frac{2}{5}\right)$

Solution: $\left(-\frac{3}{4}\right) \div \left(-\frac{2}{5}\right) = \left(-\frac{3}{4}\right) \cdot \left(-\frac{5}{2}\right) = \frac{15}{8}$

Note that, just as with integers, the product of two negative fractions is positive.

In algebra, $\frac{15}{8}$, an **improper fraction** (a fraction with the numerator greater than the denominator), is preferred to the mixed number $1\frac{7}{8}$.

Improper fractions are perfectly acceptable as long as they are reduced, meaning the numerator and denominator have no common prime factors. We will discuss mixed numbers in more detail in the next section.

b. $\frac{26}{35} \div \frac{39}{20}$

Solution: $\frac{26}{35} \div \frac{39}{20} = \frac{26}{35} \cdot \frac{20}{39} = \frac{2 \cdot \cancel{13}^{1} \cdot 2 \cdot 2 \cdot \cancel{5}^{1}}{\cancel{5}_{1} \cdot 7 \cdot 3 \cdot \cancel{13}_{1}} = \frac{8}{21}$

Note carefully that we factored and reduced before multiplying the numerator and denominator. It would not be wise to multiply first because we would then have to factor two large numbers. For example,

$$\frac{26}{35} \div \frac{39}{20} = \frac{26}{35} \cdot \frac{20}{39} = \frac{520}{1365}$$

and now we have to factor 520 and 1365. However, in the problem these were already factored since $520 = 26 \cdot 20$ and $1365 = 35 \cdot 39$.

c. $\frac{-21a^3}{5b^2} \div 3a^2b$

Solution: $\frac{-21a^3}{5b^2} \div 3a^2b = \frac{-21a^3}{5b^2} \cdot \frac{1}{3a^2b} = \frac{-7 \cdot \cancel{3}^{1} \cdot \cancel{a}^{1} \cdot \cancel{a}^{1} \cdot a \cdot 1}{5 \cdot b \cdot b \cdot \cancel{3}_{1} \cdot \cancel{a}_{1} \cdot \cancel{a}_{1} \cdot b} = \frac{-7a}{5b^3} \left(\text{or } -\frac{7a}{5b^3} \right)$

Note: The reciprocal of $3a^2b$ is $\frac{1}{3a^2b}$ since $3a^2b = \frac{3a^2b}{1}$. At this time we write $a^3 = a \cdot a \cdot a$, $a^2 = a \cdot a$, and $b^2 = b \cdot b$. We will learn more about exponents in Chapter 5.

If the product of two numbers is known and one of the numbers is also known, then the other number can be found by dividing the product by the known number. For example, with whole numbers, suppose the product of two numbers is 36 and one of the numbers is 9. What is the other number? Since $\frac{36}{9} = 4$, the other number is 4.

Example 6: Division

If the product of $\frac{3}{8}$ with another number is $-\frac{5}{16}$, what is the other number?

Solution: Divide the product by the given number.

$$-\frac{5}{16} \div \frac{3}{8} = -\frac{5}{16} \cdot \frac{8}{3} = -\frac{5 \cdot \overset{1}{\cancel{8}}}{2 \cdot \underset{1}{\cancel{8}} \cdot 3} = -\frac{5}{6}$$

The other number is $-\frac{5}{6}$.

Note that we could at least anticipate that the other number would be negative since the product is negative and the given number is positive.

1.6 Exercises

1. $\frac{5}{6} = \frac{5}{6} \cdot \frac{8}{8} = \frac{40}{48}$
2. $\frac{3}{13} = \frac{3}{13} \cdot \frac{4}{4} = \frac{12}{52}$
3. $\frac{0}{9} = \frac{0}{9} \cdot \frac{7b}{7b} = \frac{0}{63b}$
4. $\frac{-7}{24} = \frac{-7}{24} \cdot \frac{3x}{3x} = \frac{-21x}{72x}$
5. $\frac{2}{5}$ 6. $\frac{5}{9}$ 7. $\frac{3x}{7}$ 8. $\frac{4}{5}$
9. $\frac{3ab}{25}$ 10. $\frac{-2x}{17}$
11. $\frac{-2}{3y}$ 12. $\frac{3b}{4}$
13. $\frac{-1}{2x^2}$ 14. $\frac{-2x}{3}$
15. $\frac{-12y}{35}$ 16. $\frac{8x^3}{15y^3}$
17. $\frac{3}{8}$ 18. $\frac{10}{49}$ 19. 35
20. $\frac{1}{9}$ 21. $\frac{-1}{6}$
22. $\frac{-12}{35}$ 23. $\frac{81}{100}$

In Exercises 1 – 4, supply the missing numbers so that each fraction will be raised to higher terms as indicated.

1. $\frac{5}{6} = \frac{5}{6} \cdot \frac{?}{?} = \frac{?}{48}$ **2.** $\frac{3}{13} = \frac{3}{13} \cdot \frac{?}{?} = \frac{?}{52}$ **3.** $\frac{0}{9} = \frac{0}{9} \cdot \frac{?}{?} = \frac{?}{63b}$

4. $\frac{-7}{24} = \frac{-7}{24} \cdot \frac{?}{?} = \frac{?}{72x}$

Reduce each fraction to lowest terms in Exercises 5 – 16.

5. $\frac{18}{45}$ **6.** $\frac{35}{63}$ **7.** $\frac{150xy}{350y}$ **8.** $\frac{60a}{75a}$

9. $\frac{-12a^2b}{-100a}$ **10.** $\frac{6x^2}{-51x}$ **11.** $\frac{-30y}{45y^2}$ **12.** $\frac{66ab^2}{88ab}$

13. $\frac{-28}{56x^2}$ **14.** $\frac{34x^2}{-51x}$ **15.** $\frac{-12y^3}{35y^2}$ **16.** $\frac{8x^3}{15y^3}$

17. Find $\frac{1}{2}$ of $\frac{3}{4}$. **18.** Find $\frac{2}{7}$ of $\frac{5}{7}$. **19.** Find $\frac{7}{8}$ of 40. **20.** Find $\frac{1}{3}$ of $\frac{1}{3}$.

In Exercises 21 – 39, multiply or divide as indicated and reduce each answer to lowest terms.

21. $\frac{-3}{8} \cdot \frac{4}{9}$ **22.** $\frac{4}{5} \cdot \frac{-3}{7}$ **23.** $\frac{9}{10x} \div \frac{10}{9x}$

24. $\frac{4}{5a} \div \frac{1}{5a}$ **25.** $\frac{-16x}{7} \cdot \frac{49}{64x}$ **26.** $\frac{26b}{51a} \cdot \frac{4a}{-39}$

24. 4 25. $\frac{-7}{4}$

26. $\frac{-8b}{153}$ 27. $\frac{50}{9}$

28. $\frac{2}{5ab}$ 29. $\frac{4}{3y}$

30. $\frac{-273x^2}{80}$ 31. $\frac{-21}{8}$

32. $\frac{-1}{10}$ 33. Undefined

34. 0 35. $\frac{22}{a^2}$

36. $\frac{9ac}{10b^4}$ 37. $\frac{25}{98p^3q^2}$

38. $\frac{25}{8m^2}$ 39. $\frac{16}{5}$

40. 19 41. $\frac{25}{32}$

42. $\frac{12}{5}$

43. 2 inches

44. 350, 150

45. **a.** more than 60 **b.** less than 60 **c.** 72

46. 58,535,000 sq. mi.

47. $\frac{7}{20}$ and $\frac{13}{20}$

48. $\frac{35}{40}$ or $\frac{7}{8}$

27. $\frac{15}{4} \cdot \frac{5}{6} \cdot \frac{16}{9}$ **28.** $\frac{9a}{15} \cdot \frac{10}{3b} \cdot \frac{1}{5a^2}$ **29.** $\frac{-36x}{52xy} \cdot \frac{-26}{33x} \cdot \frac{22x}{9}$

30. $\frac{-9x}{40y} \cdot \frac{35x}{15} \cdot \frac{65y}{10}$ **31.** $\frac{-3}{2} \cdot \frac{7}{44} \cdot \frac{17}{1} \cdot \frac{22}{34}$ **32.** $\frac{21}{30} \div (-7)$

33. $\frac{4}{13} \div 0$ **34.** $0 \div \frac{7}{3x}$ **35.** $\frac{92}{7a} \div \frac{46a}{77}$

36. $\frac{45a}{30b^2} \cdot \frac{12ab}{18b^3} \div \frac{10ac}{9c^2}$ **37.** $\frac{15p}{3q^2} \cdot \frac{2pq}{28p^3} \div \frac{14p^2q^3}{10q^2}$ **38.** $\frac{35m}{21n^2} \div \frac{40mn^2}{27n^3} \cdot \frac{25m^2n}{9m^4}$

39. $\frac{72a^3}{4b^2} \div \frac{36ab^2}{16b^4} \div \frac{20a^2c^3}{8c^3}$

40. Multiply the quotient of $\frac{19}{2}$ and $\frac{13}{4}$ by $\frac{13}{2}$.

41. Divide the product of $\frac{5}{8}$ and $\frac{9}{10}$ by the product of $\frac{9}{10}$ and $\frac{4}{5}$.

42. Find the product of $\frac{11}{4}$ with the quotient of $\frac{8}{5}$ and $\frac{11}{6}$.

43. A glass is 6 inches tall. If the glass is $\frac{1}{3}$ full of milk, what is the height of the milk in the glass?

44. A study showed that $\frac{7}{10}$ of the students in an elementary school were over 4 feet tall. If the school had an enrollment of 500 students, how many were over 4 feet tall? How many were less than or equal to 4 feet tall?

45. A bus is carrying 60 passengers. This is $\frac{5}{6}$ of the capacity of the bus.

a. Is the capacity of the bus more or less than 60?

b. If you were to multiply 60 by $\frac{5}{6}$, would the product be more or less than 60?

c. What is the capacity of the bus?

46. The continent of Africa covers approximately 11,707,000 square miles. This is $\frac{1}{5}$ of the land area in the world. What is the approximate total land area in the world?

47. If you have \$20 and you spend \$7 on a glass of milk and a piece of pie, what fraction of your money did you spend? What fraction of your money do you still have?

48. In a class of 40 students, 5 received a grade of A. What fraction of the class did not receive an A?

49. $\frac{135}{32}$ ft. or $4\frac{7}{32}$ ft.

50. a. more than 180
b. 200 passengers

51. $\frac{-25}{24}$

52. 1050 students

53. Division by zero is undefined. For example we could write $0 = \frac{0}{1}$. Then the reciprocal would be $\frac{1}{0}$, but this reciprocal is undefined since division by zero is undefined. Thus, 0 does not have a reciprocal.

49. Suppose that a ball is dropped from a height of 30 ft. and that each bounce reaches to $\frac{3}{8}$ of the previous height. How high will the ball bounce on the second bounce?

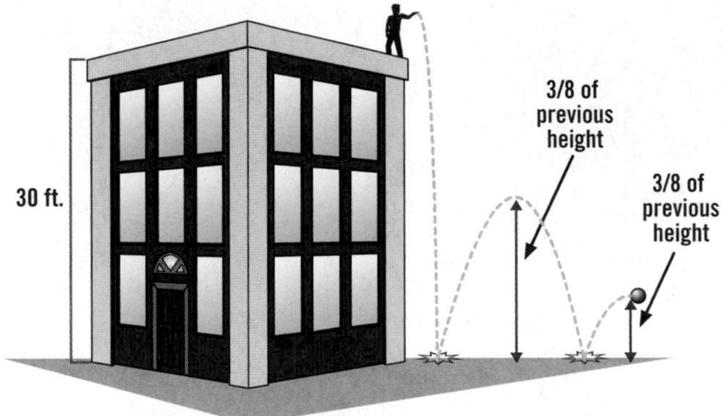

50. An airplane is carrying 180 passengers. This is $\frac{9}{10}$ of its capacity.

a. Is the capacity more or less than 180?
b. What is the capacity of the airplane?

51. The product of $\frac{3}{5}$ with another number is $\frac{-5}{8}$. What is the other number?

52. Valley Community College has 4000 students. Of these, $\frac{1}{8}$ are full-time students. Of the full-time students, $\frac{3}{10}$ are in favor of having a soccer team. Of the students that are not full-time, $\frac{4}{5}$ are in favor of having a soccer team. How many students in the school are not in favor of having a soccer team?

Writing and Thinking About Mathematics

53. Explain, in your own words, why 0 does not have a reciprocal.

Hawkes Learning Systems: Introductory & Intermediate Algebra

Reduction of Proper Fractions
Reduction of Improper Fractions
Reduction of Positive and Negative Fractions
Multiplying and Dividing Fractions

1.7 Adding and Subtracting Fractions

Objectives

After completing this section, you will be able to:

1. *Add and subtract fractions with like denominators.*
2. *Find the least common multiple (LCM) of two or more numbers.*
3. *Add and subtract fractions with unlike denominators.*
4. *Evaluate fractional expressions by using the Rules for Order of Operations.*

Addition and Subtraction

Finding the **sum** of two or more fractions with the same denominator is similar to adding whole numbers of some particular item. For example, the sum of 5 apples and 6 apples is 11 apples. Similarly, the sum of 5 seventeenths and 6 seventeenths is 11 seventeenths, or

$$\frac{5}{17}+\frac{6}{17}=\frac{11}{17}.$$

The following definition formally explains the above example.

Adding Two Fractions With Like Denominators

To **add** *two fractions* $\frac{a}{b}$ *and* $\frac{c}{b}$ *with common denominator b, add the numerators a and c and use the common denominator.*

$$\frac{a}{b}+\frac{c}{b}=\frac{a+c}{b}$$

A formal proof of this relationship involves the distributive property and the fact that

$$\frac{a}{b}=a\cdot\frac{1}{b}.$$

Proof:

$$\frac{a}{b}+\frac{c}{b}=a\cdot\frac{1}{b}+c\cdot\frac{1}{b}$$

$$=(a+c)\frac{1}{b} \qquad \text{Distributive property}$$

$$=\frac{a+c}{b}$$

Example 1: Like Denominators

a. $\frac{3}{8}+\frac{4}{8}$

Solution: $\frac{3}{8}+\frac{4}{8}=\frac{3+4}{8}=\frac{7}{8}$

b. $\frac{9}{10}+\frac{3}{10}$

Solution: $\frac{9}{10}+\frac{3}{10}=\frac{9+3}{10}=\frac{12}{10}=\frac{\cancel{2}\cdot 6}{\cancel{2}\cdot 5}=\frac{6}{5}$

c. $\frac{2}{15}+\frac{3}{15}+\frac{1}{15}+\frac{6}{15}$

Solution: $\frac{2}{15}+\frac{3}{15}+\frac{1}{15}+\frac{6}{15}=\frac{2+3+1+6}{15}=\frac{12}{15}=\frac{\cancel{3}\cdot 4}{\cancel{3}\cdot 5}=\frac{4}{5}$

d. $\frac{3}{8x}+\frac{7}{8x}+\frac{5}{8x}$

Solution: $\frac{3}{8x}+\frac{7}{8x}+\frac{5}{8x}=\frac{3+7+5}{8x}=\frac{15}{8x}$

Teaching Notes:
At some point, you might remind the student that mixed numbers have a "built-in" addition. For example, $3\frac{2}{5}=3+\frac{2}{5}$, and it shouldn't be confused with $3\cdot\frac{2}{5}$.

Teaching Notes:
You might show the students the benefits of using improper fractions to check a simple equation.

NOTES

In Example 1b above, the fraction $\frac{6}{5}$ is called an **improper fraction** because the numerator is larger than the denominator. Such unfortunate terminology implies that there is something wrong with improper fractions. This is not the case, and improper fractions are used throughout the study of mathematics. Improper fractions can be changed to mixed numbers, with a whole number and a fraction part. Divide the numerator by the denominator and use the remainder as the numerator of the new fraction. For example,

$$\frac{6}{5}=1\frac{1}{5},\quad \frac{11}{7}=1\frac{4}{7},\quad \text{and}\quad \frac{35}{6}=5\frac{5}{6}.$$

The decision of whether to leave an answer in the form of an improper fraction or as a mixed number is optional. Generally, in algebra we will keep the improper fraction form. However, in the case of an application where units of measurement are involved, a mixed number form may be preferred. That is, we would probably write $\frac{3}{2}$ feet as $1\frac{1}{2}$ feet.

To find the sum of fractions with different denominators, we need the concepts of **multiples** and **least common multiple**. **Multiples** of a number are the products of that number with the counting numbers.

Counting numbers:	**1,**	**2,**	**3,**	**4,**	**5,**	**6,**	**7,**	**8,**	**9,**	...
Multiples of 6:	6,	12,	18,	(24),	30,	36,	42,	(48),	54,	...
Multiples of 8:	8,	16,	(24),	32,	40,	(48),	56,	64,	(72),	...

For the multiples of 6 and 8, the common multiples are 24, 48, 72, 96, 120, . . .

The smallest of these, **24**, is called the **least common multiple (LCM)**. Note that the LCM is **not** $6 \cdot 8 = 48$. In this case, the LCM is 24 and it is smaller than the product of 6 and 8. We can use **prime factorizations** to find the LCM of two or more numbers using the following steps.

To Find the LCM

1. *List the prime factorization of each number.*
2. *List the prime factors that appear in any one of the prime factorizations.*
3. *Find the product of these primes using each prime the greatest number of times that it appears in any one prime factorization.*

   Note: *In terms of exponents, the LCM is the product of the highest power of each of the prime factors.*

Example 2: LCM

a. Find the LCM of the numbers 27, 15, and 60.

Solution:

$$\left.\begin{aligned} 27 &= 9 \cdot 3 = 3 \cdot 3 \cdot 3 = 3^3 \\ 15 &= 3 \cdot 5 \\ 60 &= 10 \cdot 6 = 2 \cdot 5 \cdot 2 \cdot 3 = 2^2 \cdot 3 \cdot 5 \end{aligned}\right\} \text{LCM} = 2^2 \cdot 3^3 \cdot 5 = 540$$

b. Find the LCM for $4x,\ x^2y,\ 6x^2$, and $18y^3$.
(**Hint:** Treat each variable as a prime factor.)

Solution:

$$\left.\begin{aligned} 4x &= 2^2 \cdot x \\ x^2y &= x^2 \cdot y \\ 6x^2 &= 2 \cdot 3 \cdot x^2 \\ 18y^3 &= 2 \cdot 3^2 \cdot y^3 \end{aligned}\right\} \text{LCM} = 2^2 \cdot 3^2 \cdot x^2 \cdot y^3 = 36x^2y^3$$

Adding Fractions with Different Denominators

1. *Find the LCM of the denominators.*
2. *Change each fraction to an equal fraction with the LCM as the denominator.*
3. *Add the new fractions.*

The LCM of the denominators of fractions is called the **least common denominator**, or **LCD**.

Example 3: Unlike Denominators

a. $\frac{1}{4}+\frac{3}{8}+\frac{3}{10}$

Solution:
$$\left.\begin{array}{r} 4=2^2 \\ 8=2^3 \\ 10=2\cdot 5 \end{array}\right\} \text{LCM= LCD=}2^3\cdot 5=40$$

To get the common denominator in each fraction, we see

$$40=4\cdot 10=8\cdot 5=10\cdot 4$$

$$\frac{1}{4}+\frac{3}{8}+\frac{3}{10}=\left(\frac{1}{4}\cdot\frac{10}{10}\right)+\left(\frac{3}{8}\cdot\frac{5}{5}\right)+\left(\frac{3}{10}\cdot\frac{4}{4}\right)$$

Multiply each fraction by 1 in the form $\frac{k}{k}$.

$$=\frac{10}{40}+\frac{15}{40}+\frac{12}{40}$$

Each fraction has the same denominator.

$$=\frac{37}{40}$$

Add the fractions.

b. $\frac{5}{21a}+\frac{5}{28a}$

Solution:
$$\left.\begin{array}{r} 21a=3\cdot 7\cdot a \\ 28a=2^2\cdot 7\cdot a \end{array}\right\} \text{LCM} = \text{LCD} = 2^2\cdot 3\cdot 7\cdot a=84a=21a\cdot 4=28a\cdot 3$$

In each fraction, the numerator and denominator are multiplied by the same number to get 84*a* as the denominator.

$$\frac{5}{21a}+\frac{5}{28a}=\left(\frac{5}{21a}\cdot\frac{4}{4}\right)+\left(\frac{5}{28a}\cdot\frac{3}{3}\right)$$
$$=\frac{20}{84a}+\frac{15}{84a}=\frac{35}{84a}$$
$$=\frac{\cancel{7}\cdot 5}{\cancel{7}\cdot 12a}=\frac{5}{12a}$$

Subtracting Fractions

*The **difference** of two fractions with a common denominator is found by subtracting the numerators and using the common denominator.*

$$\frac{a}{b}-\frac{c}{b}=\frac{a-c}{b}$$

Just as with addition, if the two fractions do not have the same denominator, find equal fractions with the least common denominator (LCD).

Example 4: Subtraction

Teaching Notes:
I have found that beginning students appreciate the technique of using factors or prime factors in reducing fractions. However, other students may use the GCF technique and you might want to discuss this with your class.

a. $\frac{1}{8a}-\frac{5}{8a}$

Solution: $\frac{1}{8a}-\frac{5}{8a}=\frac{1-5}{8a}=\frac{-4}{8a}=\frac{-1\cdot\cancel{4}}{\cancel{4}\cdot 2\cdot a}=\frac{-1}{2a}$ $\left(\text{or } -\frac{1}{2a}\right)$

b. $\frac{1}{45}-\frac{1}{72}$

Solution: $\left.\begin{array}{l}45=3^2\cdot 5\\ 72=2^3\cdot 3^2\end{array}\right\}$ $\text{LCM}=\text{LCD}=2^3\cdot 3^2\cdot 5=360=45\cdot 8=72\cdot 5$

$$\frac{1}{45}-\frac{1}{72}=\left(\frac{1}{45}\cdot\frac{8}{8}\right)-\left(\frac{1}{72}\cdot\frac{5}{5}\right)=\frac{8}{360}-\frac{5}{360}=\frac{3}{360}=\frac{\cancel{3}\cdot 1}{\cancel{3}\cdot 120}=\frac{1}{120}$$

Order of Operations

To evaluate an expression such as $\frac{1}{2}+\frac{3}{8}\div\frac{3}{4}$, we use the same rules for order of operations discussed in Section 1.5. These rules apply to all types of numbers. They are restated here for convenience and easy reference.

Rules for Order of Operations

1. *Simplify within grouping symbols, such as parentheses ( ), brackets [ ], and braces { }, working from the inner most grouping outward.*
2. *Find any powers indicated by exponents.*
3. *Moving from* ***left to right****, perform any multiplications or divisions in the order they appear.*
4. *Moving from* ***left to right****, perform any additions or subtractions in the order they appear.*

Example 5: Order of Operations

a. $\frac{1}{2}+\frac{3}{8}\div\frac{3}{4}$

Solution:

$$\frac{1}{2}+\frac{3}{8}\div\frac{3}{4}=\frac{1}{2}+\frac{3}{8}\cdot\frac{4}{3}$$ Divide (Multiply by the reciprocal).

$$=\frac{1}{2}+\frac{\cancel{3}\cdot\cancel{4}\cdot 1}{2\cdot\cancel{4}\cdot\cancel{3}}$$ Reduce.

$$=\frac{1}{2}+\frac{1}{2}$$ Add.

$$=\frac{1+1}{2}$$

$$=\frac{2}{2}$$ Reduce.

$$=1$$

Teaching Notes: Be careful to point out that in these examples the 'cancellation' is not across the '+' sign. For example,

$$\frac{2}{3}+\frac{1}{2} \neq \frac{\overset{1}{\cancel{2}}}{3}+\frac{1}{\underset{1}{\cancel{2}}}$$

Also, you may want to review the technique of reducing by cancellation:

$$\frac{2}{\underset{1}{\cancel{7}}}\cdot\frac{\overset{2}{\cancel{14}}}{3}+\frac{\cancel{5}}{\underset{4}{\cancel{8}}}\cdot\frac{\overset{1}{\cancel{2}}}{\cancel{5}}$$

$$=\frac{4}{3}+\frac{1}{4}$$

This is certainly acceptable for those students who are comfortable with fractions.

b. $\frac{2}{7}\cdot\frac{14}{3}+\frac{5}{8}\cdot\frac{2}{5}$

Solution: $\frac{2}{7}\cdot\frac{14}{3}+\frac{5}{8}\cdot\frac{2}{5}=\frac{2\cdot 14}{7\cdot 3}+\frac{5\cdot 2}{8\cdot 5}$ Multiply.

$=\frac{2\cdot\cancel{7}\cdot 2}{\cancel{7}\cdot 3}+\frac{\cancel{5}\cdot\cancel{2}\cdot 1}{\cancel{2}\cdot 4\cdot\cancel{5}}$ Reduce.

$=\frac{4}{3}+\frac{1}{4}$

$=\left(\frac{4}{3}\cdot\frac{4}{4}\right)+\left(\frac{1}{4}\cdot\frac{3}{3}\right)$ Common denominator is 12.

$=\frac{16}{12}+\frac{3}{12}=\frac{19}{12}$ Add.

c. Divide the sum of $\frac{5}{8}$ and $\frac{3}{4}$ by $\frac{3}{2}$.

Solution: First find the sum of $\frac{5}{8}$ and $\frac{3}{4}$.

$\frac{5}{8}+\frac{3}{4}=\frac{5}{8}+\left(\frac{3}{4}\cdot\frac{2}{2}\right)=\frac{5}{8}+\frac{6}{8}=\frac{11}{8}$ LCD is 8.

Now divide the sum by $\frac{3}{2}$.

$$\frac{11}{8}\div\frac{3}{2}=\frac{11}{8}\cdot\frac{2}{3}=\frac{11\cdot\cancel{2}}{\cancel{2}\cdot 4\cdot 3}=\frac{11}{12}$$

The answer is $\frac{11}{12}$.

d. $\frac{3}{x}+\frac{2}{5}$

Solution: In this case the LCD $= x\cdot 5 = 5x$

$\frac{3}{x}+\frac{2}{5}=\left(\frac{3}{x}\cdot\frac{5}{5}\right)+\left(\frac{2}{5}\cdot\frac{x}{x}\right)$

$=\frac{15}{5x}+\frac{2x}{5x}$

$=\frac{15+2x}{5x}$ This fraction cannot be reduced.

Practice Problems

Perform the indicated operations and reduce all answers to lowest terms.

1. $-5\cdot\left(\frac{3}{10}\right)^2$

2. $\frac{3}{4}\div\frac{4}{3}$

3. $\frac{7}{3}\div\frac{4}{5}\cdot\frac{3}{20}$

4. $\frac{1}{2a}-\frac{3}{2a}$

5. $\frac{5}{24}+\frac{7}{36}$

6. $\frac{1}{3}\div\frac{1}{2}+\frac{1}{5}\cdot\frac{5}{3}$

1.7 Exercises

1. a. 1, 2, 3, 6 **b.** 6, 12, 18, 24, 30, 36
2. a. 1, 2, 3, 4, 6, 12 **b.** 12, 24, 36, 48, 60, 72
3. a. 1, 3, 5, 15 **b.** 15, 30, 45, 60, 75, 90
4. a. 1, 2, 3, 6, 9, 18 **b.** 18, 36, 54, 72, 90, 108
5. 120 **6.** 1225 **7.** $40xy$ **8.** $480xyz$ **9.** $210x^2y^2$ **10.** $840x^2y$ **11.** $\frac{7}{9}$ **12.** 2 **13.** $\frac{5}{23}$ **14.** $\frac{1}{3}$ **15.** $-\frac{1}{6}$ **16.** $-\frac{1}{4}$ **17.** $\frac{4}{3}$ **18.** $\frac{2}{15a}$ **19.** $-\frac{28}{25x}$ **20.** $\frac{23}{24}$ **21.** $\frac{83}{60}$ **22.** $\frac{1}{20}$ **23.** $\frac{5}{42}$ **24.** $\frac{2}{9}$ **25.** $\frac{33}{70}$ **26.** $\frac{2}{9}$ **27.** $\frac{3}{4x}$ **28.** $\frac{5}{6y}$

Find (a) the factors of each of the numbers and (b) the first six multiples of each of the numbers in Exercises 1 – 4.

1. 6 **2.** 12 **3.** 15 **4.** 18

Find the LCM for each set of numbers or algebraic expressions in Exercises 5 – 10.

5. 24, 15, 10 **6.** 49, 25, 35 **7.** $8x$, $10y$, $20xy$
8. $20xz$, $24xy$, $32yz$ **9.** $14x^2$, $21xy$, $35x^2y^2$ **10.** $60x$, $105x^2y$, $120xy$

Perform the indicated operations in Exercises 11 – 40. Reduce each answer to lowest terms.

11. $\frac{2}{9}+\frac{5}{9}$ **12.** $\frac{2}{7}+\frac{8}{7}+\frac{4}{7}$ **13.** $\frac{14}{23}-\frac{9}{23}$ **14.** $\frac{7}{15}-\frac{2}{15}$

15. $\frac{11}{12}-\frac{13}{12}$ **16.** $\frac{5}{16}-\frac{9}{16}$ **17.** $\frac{5}{9}+\frac{7}{9}$ **18.** $\frac{11}{15a}-\frac{7}{15a}-\frac{2}{15a}$

19. $\frac{-19}{25x}-\frac{7}{25x}-\frac{2}{25x}$ **20.** $\frac{5}{6}+\frac{1}{8}$ **21.** $\frac{11}{12}+\frac{7}{15}$ **22.** $\frac{27}{40}-\frac{5}{8}$

23. $\frac{2}{7}-\frac{1}{6}$ **24.** $\frac{5}{24}+\frac{7}{18}-\frac{3}{8}$ **25.** $\frac{8}{35}+\frac{6}{10}-\frac{5}{14}$ **26.** $\frac{8}{9}-\frac{5}{12}-\frac{1}{4}$

27. $\frac{2}{3x}+\frac{1}{4x}-\frac{1}{6x}$ **28.** $\frac{7}{10y}-\frac{1}{5y}+\frac{1}{3y}$ **29.** $\frac{11}{24x}-\frac{7}{18x}+\frac{5}{36x}$ **30.** $\frac{7}{6b}-\frac{9}{10b}+\frac{4}{15b}$

31. $\frac{1}{28a}+\frac{8}{21a}-\frac{5}{12a}$ **32.** $\frac{5}{y}-\frac{1}{3}$ **33.** $\frac{6}{y}+\frac{1}{2}$ **34.** $\frac{x}{4}+\frac{1}{5}$

35. $\frac{x}{14}-\frac{1}{7}$ **36.** $\frac{y}{3}-\frac{1}{4}-\frac{1}{6}$ **37.** $\frac{a}{5}-\frac{1}{3}-\frac{1}{2}$ **38.** $\frac{a}{2}+\frac{1}{6}+\frac{2}{3}$

39. $\frac{b}{4}+\frac{1}{2}+\frac{3}{5}$ **40.** $\frac{1}{7}-\frac{10}{x}$

Answers to Practice Problems: 1. $-\frac{9}{20}$ **2.** $\frac{9}{16}$ **3.** $\frac{7}{16}$ **4.** $-\frac{1}{a}$ **5.** $\frac{29}{72}$ **6.** 1

29. $\frac{5}{24x}$ **30.** $\frac{8}{15b}$

31. 0 **32.** $\frac{15-y}{3y}$

33. $\frac{12+y}{2y}$

34. $\frac{5x+4}{20}$

35. $\frac{x-2}{14}$

36. $\frac{4y-5}{12}$

37. $\frac{6a-25}{30}$

38. $\frac{3a+5}{6}$

39. $\frac{5b+22}{20}$

40. $\frac{x-70}{7x}$

41. $\frac{11}{30}$ **42.** $\frac{1}{24}$

43. $\frac{31}{24}$ **44.** $\frac{21}{100}$

45. $\frac{7}{2}$ **46.** $\frac{-11}{12x}$

47. $\frac{-23}{21y}$ **48.** $\frac{-49y}{40x}$

49. $\frac{7}{5}$ **50.** $\frac{-7b}{10a}$

51. $\frac{1}{18009460}$

52. $\frac{991}{336}$ **53.** $\frac{92}{19}$

54. 60 ft. and 42 ft.

Simplify the expressions in Exercises 41 – 50 using the Rules for Order of Operations.

41. $\frac{3}{8}\cdot\frac{4}{5}+\frac{1}{15}$

42. $\frac{1}{3}\div\frac{1}{2}-\frac{5}{6}\cdot\frac{3}{4}$

43. $\left(\frac{5}{6}\right)^2\div\frac{5}{12}-\frac{3}{8}$

44. $\left(\frac{2}{5}\right)^2\cdot\frac{3}{8}+\frac{1}{5}\cdot\frac{3}{4}$

45. $\frac{3}{4a}\div\frac{3}{16a}-\frac{2b}{3}\cdot\frac{3}{4b}$

46. $\frac{3}{4x}-\frac{5}{6}\div\frac{5x}{8}-\frac{1}{3x}$

47. $\frac{-5}{7y}-\frac{1}{2y}\cdot\frac{2}{3}-\frac{1}{21y}$

48. $\left(\frac{1}{5x}-\frac{2}{3x}\right)\div\left(\frac{5}{7y}-\frac{1}{3y}\right)$

49. $\left(\frac{7}{10x}+\frac{3}{5x}\right)\div\left(\frac{1}{2x}+\frac{3}{7x}\right)$

50. $\left(\frac{1}{5a}-\frac{2}{3a}\right)\div\left(\frac{7}{7b}-\frac{1}{3b}\right)$

51. The California lottery is based on choosing any six of the integers from 1 to 51. The probability of winning the lottery can be found by multiplying the fractions:

$$\frac{6}{51}\cdot\frac{5}{50}\cdot\frac{4}{49}\cdot\frac{3}{48}\cdot\frac{2}{47}\cdot\frac{1}{46}$$

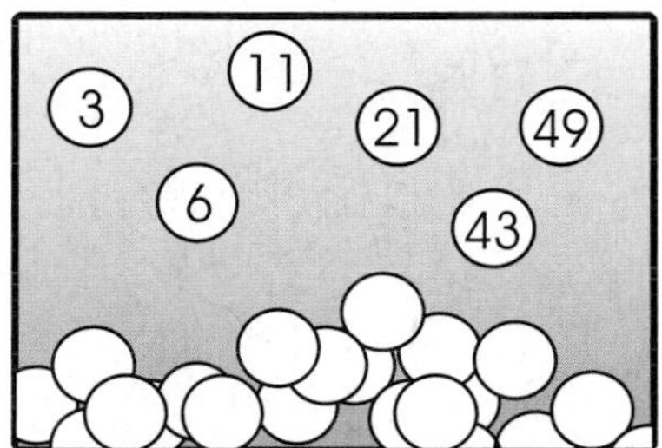

Multiply and reduce the product of these fractions to find the probability of winning the lottery in the form of a fraction with numerator 1.

52. The product of $\frac{9}{16}$ and $\frac{13}{7}$ is added to the quotient of $\frac{10}{3}$ and $\frac{7}{4}$. What is the sum?

53. Find the quotient if the sum of $\frac{4}{5}$ and $\frac{11}{15}$ is divided by the difference between $\frac{7}{12}$ and $\frac{4}{15}$.

54. In the scale for the blue print of the first floor of a house, 1 in. represents 48 ft. If on the drawing the first floor is $\frac{7}{8}$ in. wide by $\frac{5}{4}$ in. long, what will be the actual length and width of the first floor?

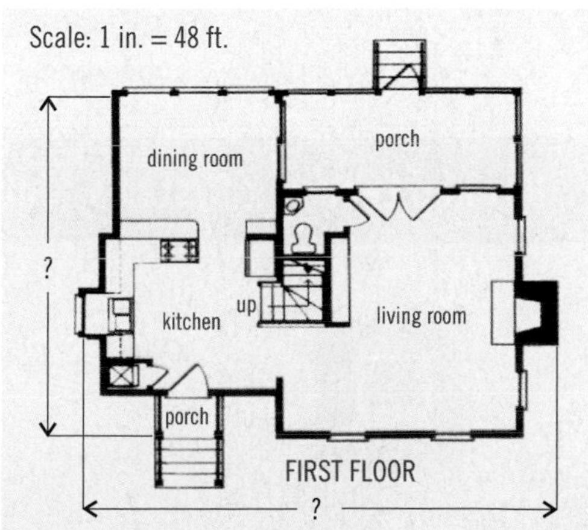

55. 344 yd.

56. $\frac{41}{40}$ in. by $\frac{9}{10}$ in.

57. a. $\frac{13}{30}$
b. \$1170

58. a. Vote for
b. Pass by 40 votes.

55. The seventeenth hole at the local golf course is a par 4 hole. Ralph drove his ball 258 yards. If this distance was $\frac{3}{4}$ the length of the hole, how long is the seventeenth hole?

56. A watch has a rectangular-shaped display screen that is $\frac{5}{8}$ inch by $\frac{1}{2}$ inch. The display screen has a border of metal that is $\frac{1}{5}$ inch thick. What are the dimensions (length and width) of the watch (including the metal border)?

57. Delia's income is \$2700 a month and she plans to budget $\frac{1}{3}$ of her income for rent and $\frac{1}{10}$ of her income for food.

a. What fraction of her income does she plan to spend each month on these two items?
b. What amount of money does she plan to spend each month on these two items?

58. The tennis club has 400 members, and they are considering putting in a new tennis court. The cost of the new court is going to involve an assessment of \$250 for each member. Of the six-tenths of the members who live near the club, $\frac{4}{5}$ are in favor of the assessment. However, of the other members who live further away, only $\frac{3}{10}$ are in favor of the assessment.

a. If a vote were taken today, would more than one-half of the members vote for or against the new court?
b. By how many votes would the question pass or fail if more than one-half of the members must vote in favor for the question to pass?

Hawkes Learning Systems: Introductory & Intermediate Algebra

Adding and Subtracting Fractions

1.8 Decimal Numbers and Change In Value

Objectives

After completing this section, you will be able to:

1. *Understand place value.*
2. *Read and write decimal numbers.*
3. *Round off decimal numbers.*
4. *Operate with decimal numbers.*
5. *Change fractions to decimal form.*
6. *Change decimals to fraction form.*
7. *Solve word problems that require the use of integers.*

Decimal numbers are used in most of the daily calculations we make, such as balancing a checkbook, measuring the distance we drive, and weighing a package to be mailed. Sometimes fractions are used and need to be changed to decimal form, as in the price of a stock of $\$3\frac{3}{4}$ to \$3.75 or $1\frac{1}{2}$ cups of flour to 1.5 cups. Even to be able to understand and use a calculator, we must be able to understand and operate with decimal numbers. In fact, in the year 2000, the stock markets did away with fractions in quoting the prices of stocks and began to use decimal numbers in stock pricing.

Place Value

A **decimal number** (or simply a **decimal**) is a fraction that has a power of ten in its denominator. Thus,

$$\frac{3}{10},\ \frac{51}{100},\ \frac{24}{10},\ \frac{17}{1},\ \text{and}\ \frac{3}{1000}$$

are all decimal numbers. Decimal numbers can be written with a decimal point and a place value system that indicates whole numbers to the left of the decimal point and fractions less than 1 to the right of the decimal point. The values of each place are indicated in Figure 1.16.

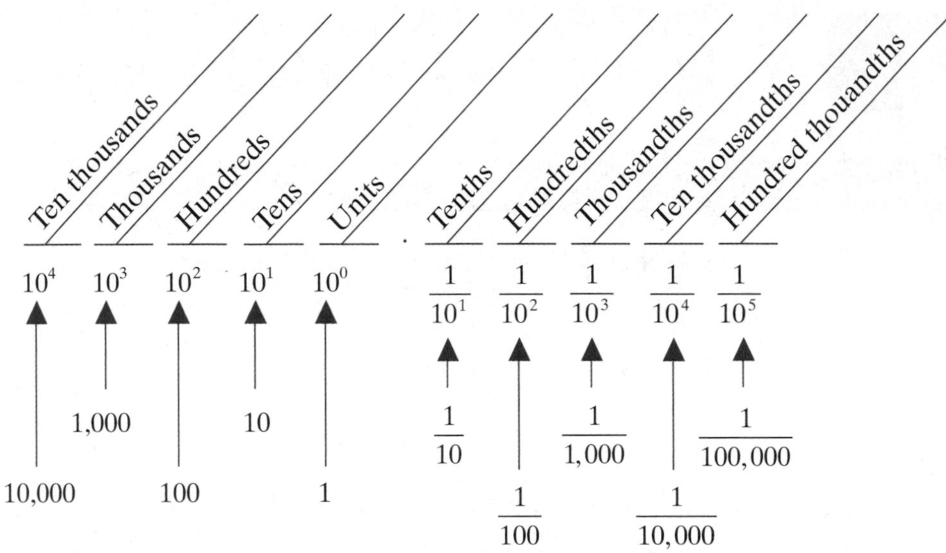

Figure 1.16

Reading and Writing Decimal Numbers

In reading a fraction such as $\frac{231}{1000}$, we read the numerator as a whole number ("two hundred thirty-one") and then attach the name of the denominator ("thousand**ths**"). Note that **ths** (or **th**) is used to indicate the fraction. **And** is used to indicate the decimal point. For example:

$\frac{231}{1000}$ is read "two hundred thirty-one thousandths."

$3\frac{17}{100}$ is read "three **and** seventeen hundredths."

Example 1: Writing Decimal Numbers

Write each of the following numbers in decimal notation and in words.

a. $64\frac{5}{10}$ **b.** $7\frac{9}{1000}$

Solution: **a.** 64.5 in decimal notation
"sixty-four and five tenths" in words

b. 7.009 in decimal notation
"seven and nine thousandths" in words

Rounding Off Decimal Numbers

Human made measuring devices such as rulers, yard sticks, meter sticks, clocks, and surveying instruments give only approximate measurements. For example, a carpenter may wish to cut a board of length 3.5 feet. As he cuts the board, the actual length cut will be approximately 3.5 feet, maybe 3.47 feet or 3.56 feet.

In this text we will use the following rules for rounding off decimal numbers. (Note that there are other rules for rounding off. For example, one rule is to round off to the nearest even digit.)

Rules for Rounding Off Decimal Numbers

1. *Look at the digit immediately to the right of the place of desired accuracy.*
2. *If this digit is 5 or greater, increase the digit in the desired place of accuracy by 1 and drop all the remaining digits to the right. (If a 9 is to be made larger, then the next digit to the left will also be increased by 1.)*
3. *If this digit is less than 5, leave the digit in the desired place of accuracy unchanged and drop all digits to the right.*

Example 2: Rounding

Round off each decimal number to the indicated place.

a. 43.8476 (nearest thousandth) **b.** 17.6035 (nearest hundredth)
c. 246.95 (nearest tenth)

Solution: **a.** 43.8476 7 is in the thousandths position. Look at the next digit to the right, 6. Since 6 is larger than 5, change 7 to 8 and drop the 6.

Thus, 43.8476 rounds off to 43.848 to the nearest thousandth.

b. 17.6035 0 is in the hundredths position. Look at the next digit to the right, 3. Since 3 is smaller than 5, leave the 0 and drop both the 3 and 5.

Thus, 17.6035 rounds off to 17.60 to the nearest hundredth. Note that the 0 is left in the answer because it indicates the place of rounding.

c. 246.95 9 is in the tenths position. Look at the next digit to the right, 5, and change the 9 to 10. But this also affects the 6.

Thus, 246.95 rounds off to 247.0 to the nearest tenth.

Operating with Decimal Numbers

To add or subtract decimal numbers, line up the decimal points, one under the other, then add or subtract as with whole numbers. Place the decimal point in the answer in line with the other decimal points. This technique guarantees that digits having the same place value will be added to or subtracted from each other.

The rules for operating with positive and negative numbers are the same regardless of the type of number. That is, with positive and negative decimal numbers or fractions, we follow the same rules we discussed earlier in this chapter for operating with positive and negative integers.

Example 3: Addition and Subtraction

Teaching Notes: Here might be the time to point out the distinction on the TI-83 Plus (and other calculators) between the '(–)' and the '–'. This gives students an alternative for Example 3b, without having to temporarily adjust the sign. For example, 13.99 – 26.872 = 13.99 + (–26.872), whereby students can use the '(–)' key.

a. Add: 37.498 + 5.63 + 42.781

Solution:

$$\begin{array}{r} 37.498 \\ 5.630 \\ +42.781 \\ \hline 85.909 \end{array}$$

b. Subtract: 13.99 – 26.872

Solution: Find the difference and then attach the negative sign.

$$\begin{array}{r} 26.872 \\ -13.990 \\ \hline 12.882 \end{array}$$

Thus, 13.99 – 26.872 = –12.882

To find the product of two decimal numbers, multiply as with whole numbers. Then place the decimal point so that the number of digits to its right is equal to the sum of the number of digits to the right of the decimal points in the numbers being multiplied.

Example 4: Multiplication

Find the products.

a. $\begin{array}{r} 4.78 \\ \times\ 0.3 \\ \hline \end{array}$

b. $\begin{array}{r} -16.4 \\ \times\ 0.517 \\ \hline \end{array}$

Solution:

$$\begin{array}{rl} 4.78 & \text{2 digits to the right} \\ \times\ 0.3 & \text{1 digit to the right} \\ \hline 1.434 & \text{3 digits to the right} \end{array}$$

Solution:

$$\begin{array}{r} -16.4 \\ \times\ 0.517 \\ \hline 1148 \\ 1640 \\ 82000 \\ \hline -8.4788 \end{array}$$

To find the quotient of two decimal numbers, move the decimal point in the divisor to the right to get a whole number. Move the decimal point in the dividend the same number of places. Divide as with whole numbers.

Example 5: Division

Find the quotient:

Solution: $3.2\overline{)51.52}$

$32.\overline{)515.2}$ Move decimal points one place to the right. This is effectively multiplying both numbers by 10. The divisor is now 32, a whole number.

$$\begin{array}{r} . \\ 32\overline{)515.2} \end{array}$$

Decimal point in quotient.

$$\begin{array}{r} 16.1 \\ 32\overline{)515.2} \\ \underline{32} \\ 195 \\ \underline{192} \\ 32 \\ \underline{32} \\ 0 \end{array}$$

Divide.

Fractions and Decimals

Fractions can be written in decimal form by dividing the numerator by the denominator. The result will be either

1. A terminating decimal, or
2. An infinite repeating decimal.

Examples of terminating decimals are:

$$\frac{1}{4}=0.25,\quad \frac{3}{8}=0.375,\quad \text{and}\quad 1\frac{4}{5}=\frac{9}{5}=1.8$$

Teaching Notes:
You might want to explain to your students that since in long division with integers the remainder must be less than the divisor b, there are only a finite number of possible remainders: namely, $0, 1, 2, \ldots, b-1$. If the remainder is 0, then the decimal is terminating. Otherwise, one of the remainders must eventually appear again. When this happens, the remainders will appear in the same order as before and the quotient will be a repeating pattern. This will help the students understand how the quotient is related to the remainders.

Examples of infinite repeating decimals are: $\frac{2}{3}, \frac{1}{7}$ and $\frac{4}{11}$. Long division shows the repeating decimal pattern for each:

$$\begin{array}{r} 0.6666\ldots \\ 3\overline{)2.0000\ldots} \\ \underline{18} \\ 20 \\ \underline{18} \\ 20 \\ \underline{18} \\ 20 \\ \underline{18} \\ 2 \end{array} \qquad \begin{array}{r} 0.14285714\ldots \\ 7\overline{)1.00000000\ldots} \\ \underline{7} \\ 30 \\ \underline{28} \\ 20 \\ \underline{14} \\ 60 \\ \underline{56} \\ 40 \\ \underline{35} \\ 50 \\ \underline{49} \\ 10 \\ \underline{7} \\ 30 \\ \underline{28} \\ 2 \end{array} \qquad \begin{array}{r} 0.3636\ldots \\ 11\overline{)4.0000\ldots} \\ \underline{33} \\ 70 \\ \underline{66} \\ 40 \\ \underline{33} \\ 70 \\ \underline{66} \\ 4 \end{array}$$

Thus,

$$\frac{2}{3} = 0.6666\ldots, \quad \frac{1}{7} = 0.14285714\ldots, \quad \text{and} \quad \frac{4}{11} = 0.3636\ldots$$

Or we can write a bar over the repeating pattern of digits as:

$$\frac{2}{3} = 0.\overline{6}, \quad \frac{1}{7} = 0.\overline{142857}, \quad \text{and} \quad \frac{4}{11} = 0.\overline{36}.$$

Example 6: Fraction to Decimal

Change $3\frac{7}{8}$ to decimal form.

Solution: $3\frac{7}{8} = \frac{31}{8}$ so we divide. With a calculator $\frac{31}{8} = 3.875$.

Or, we can write

$$3\frac{7}{8} = 3 + \frac{7}{8} = 3 + 0.875 = 3.875$$

Using long division, we get the same terminating decimal:

```
     3.875
  8)31.000
    24
     70
     64
      60
      56
       40
       40
        0
```

0 remainder means a terminating decimal.

To change a decimal to fraction form, just write the fraction part of the decimal in the numerator and use the position (place value) of the last digit on the right as the denominator. Reduce the fraction.

Example 7: Decimal to Fraction

Write the decimal number 0.56 in fraction form, reduced.

Solution: $0.56 = \frac{56}{100} = \frac{4 \cdot 14}{4 \cdot 25} = \frac{14}{25}$

Example 8: Calculator Examples

Use your calculator to perform the indicated operations.

a. $8.6321 + 7.5476 + 2.143 + 17.8293$

Solution: 36.1520

b. $(14.763)(0.47)(321.6)$

Solution: 2231.456976

continued on next page ...

Teaching Notes: You might want to discuss the approximation symbol, ≈, and the fact that $\frac{1}{7} \approx 0.1429$ is a more accurate statement than $\frac{1}{7} = 0.1429$.

c. Change the fraction $\frac{1}{7}$ to decimal form as an infinite repeating decimal. Then round off the decimal to ten thousandths (four decimal places).

Solution: $\frac{1}{7} = 1 \div 7 = 0.142857142857... = 0.\overline{142857}$ as an infinite repeating decimal.

(Remember that a bar can be placed over the repeating pattern of digits to indicate an infinite repeating decimal.)

$\frac{1}{7} \approx 0.1429$ Accurate to four decimal places.

read "approximately"

The TI-83 Plus calculator has commands that will change numbers in fraction form into decimal form and numbers in decimal form into fraction form. The two commands are found by pressing MATH and noting choices

1:>Frac and **2:>Dec**.

Choice **1:>Frac** will change a decimal number into fraction form. Note that if the decimal number is greater than 1, this form will be an improper fraction and not a mixed number. Choice **2:>Dec** will change a fraction into decimal form. Choice **2:>Dec** is not as useful as the first choice because the calculator automatically treats fractions as indicating division and gives the division in decimal form. Example 9 illustrates these two situations.

Example 9: Calculator Examples

Use a TI-83 Plus calculator (or other graphing calculator) to perform the following:

a. Change the decimal number 0.82 into fraction form.

b. Change the fraction $\frac{5}{8}$ into decimal form.

c. Find the sum $\frac{3}{5} + \frac{3}{20}$ in fraction form.

Solution: a. Proceed as follows:

Enter the numbers 0.82 (or just .82).

Press MATH.

Select **1:>Frac** by pressing ENTER.

Press ENTER again.

The screen will display

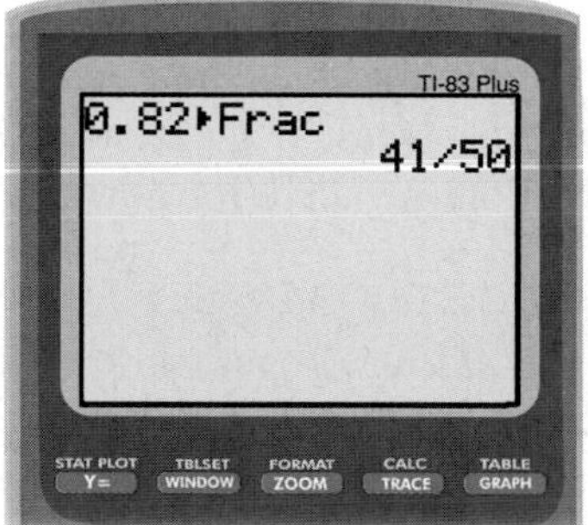

Note that the fraction is reduced.

b. Proceed as follows:

Enter the fraction $\frac{5}{8}$ by entering $5 \div 8$.

Press ENTER.

The calculator will automatically give the answer in decimal form and the screen will appear as follows:

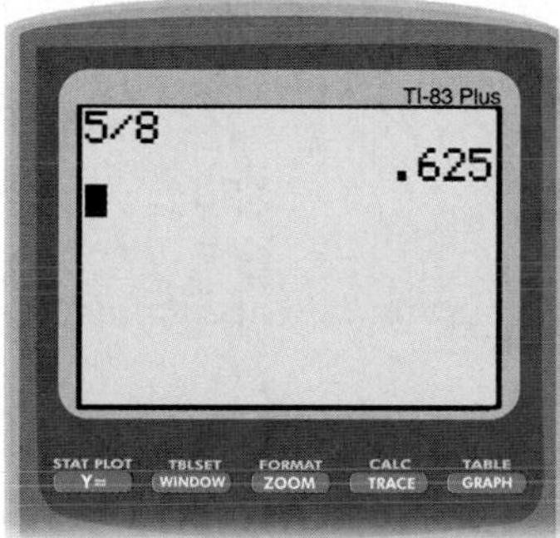

c. Proceed as follows:

Enter the sum of the fractions as $(3 \div 5) + (3 \div 20)$.

Press MATH.

Press ENTER.

Press ENTER.

The display on the screen will appear as follows:

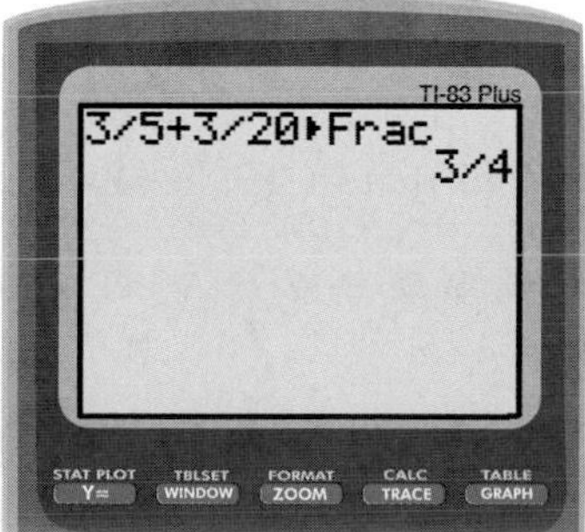

Note that the fraction is reduced.

Change In Value

Subtraction can be used to find the change in value between two readings of measures such as temperatures, distances, and altitudes. To calculate the change between two values, including direction (negative for down, positive for up) use the following rule:

Change In Value

First find the end value and then subtract the beginning value.

(Change In Value) = (End Value) – (Beginning Value)

Example 10: Change In Value

a. On a winter day, the temperature dropped from 35° F at noon to 6° below zero (–6° F) at 7 p.m. What was the change in temperature?

Solution: end temperature – beginning temperature = change in temperature

–6°	–	35°	=	
–6°	+	(–35°)	=	–41°

The change in temperature was –41° F. (This means that the temperature *dropped* by 41°.)

b. A jet pilot flew her plane from an altitude of 30,000 ft. to an altitude of 12,000 ft. What was the change in altitude?

Solution: end altitude – beginning altitude = change in altitude

12,000 – 30,000 = –18,000 ft.

(This means that the plane *descended* 18,000 feet.)

The **net change** in a measure is the algebraic sum of several signed numbers. Example 11 illustrates how positive and negative numbers can be used to find the net change of weight (gain or loss) over a period of time.

Example 11: Net Change

Sue weighed 130 lbs. when she started to diet. The first week she lost 7 lbs., the second week she gained 2 lbs., and the third week she lost 5 lbs. What was her weight after 3 weeks of dieting?

Solution:
$$\begin{aligned}130+(-7)+(+2)+(-5) &= 123+(+2)+(-5)\\ &= 125+(-5)\\ &= 120\text{ lbs.}\end{aligned}$$

1.8 Exercises

Write each of the numbers in Exercises 1 – 6 in decimal form and in words.

1. $\frac{86}{100}$ **2.** $\frac{75}{1000}$ **3.** $5\frac{1}{10}$ **4.** $7\frac{3}{100}$ **5.** $-18\frac{6}{100}$ **6.** $-21\frac{19}{1000}$

Write each of the numbers in Exercises 7 – 12 in decimal notation.

7. eighty-seven thousandths
8. two and thirty-five hundredths
9. negative five and fourteen hundredths
10. negative twenty and forty-five ten thousandths
11. seven and twenty-one ten thousandths
12. one hundred seventy-three and one hundred forty-six thousandths

In Exercises 13 – 18, round off each number to the place indicated.

13. 91.076 (nearest tenth)
14. 0.053 (nearest tenth)
15. 0.495 (nearest hundredth)
16. 6.0072 (nearest hundredth)
17. 67.0572 (nearest thousandth)
18. 75.4579 (nearest thousandth)

Add or subtract as indicated in Exercises 19 – 26.

19. 243.7 + 65.22 + 8.31
20. 29.51 + 17.2 + 72.4
21. 420.43 – 156.92
22. 87.0 – 66.31
23. 65.13 – 44.81 – 75.9
24. –84.1 + 17.63 – 12.98
25. –147.0 + 79.6 – 85.43
26. 6.49 + 103.81 – 59.62

Find the indicated products in Exercises 27 – 30.

27. $(60.4)(1.8)$ **28.** $(21.6)(0.83)$ **29.** $(1.23)(-6.2)$ **30.** $(-5.83)(-0.24)$

Find the indicated quotients in Exercises 31 and 32.

31. $5.1\overline{)77.01}$ **32.** $0.023\overline{)0.4922}$

Use long division to change the fractions to decimals in Exercises 33 – 36.

33. $\frac{3}{8}$ **34.** $\frac{4}{5}$ **35.** $\frac{1}{20}$ **36.** $\frac{3}{50}$

1. 0.86; eighty-six hundredths
2. 0.075; seventy-five thousandths
3. 5.1; five and one tenth
4. 7.03; seven and three hundredths
5. –18.06; negative eighteen and six hundredths
6. –21.019; negative twenty-one and nineteen thousandths
7. 0.087 **8.** 2.35
9. –5.14
10. –20.0045
11. 7.0021
12. 173.146 **13.** 91.1
14. 0.1 **15.** 0.50
16. 6.01 **17.** 67.057
18. 75.458 **19.** 317.23
20. 119.11 **21.** 263.51
22. 20.69 **23.** –55.58
24. –79.45 **25.** –152.83
26. 50.68 **27.** 108.72
28. 17.928 **29.** –7.626
30. 1.3992 **31.** 15.1
32. 21.4 **33.** 0.375
34. 0.8 **35.** 0.05
36. 0.06

37. −23 + 13 − 6; −16 lbs.
38. 24 − 17 + 2; $9
39. 4 + 3 − 9; −2°
40. −10 + 25 + 18 − 9; $24
41. 47 − 22 + 8 − 45; −$12
42. 20 − 42 + 58 − 11; $25
43. 14 − 6 + 11 − 15; 4°
44. −53 − 8 + 48 − 17; −$30
45. 187 − 241 + 82 + 26; $54
46. rose $2
47. 4 yd. gain
48. 3° above zero
49. 8th floor
50. 8 lbs. lost, 215 lbs.
51. $150
52. $220
53. $760
54. −18° F, −3° F
55. 35° F, 5° F

In Exercises 37 – 45, write each problem as a sum or difference, then simplify.

37. 23 lbs. lost, 13 lbs. gained, 6 lbs. lost
38. $24 earned, $17 spent, $2 earned
39. 4° rise, 3° rise, 9° drop
40. $10 withdrawal, $25 deposit, $18 deposit, $9 withdrawal
41. $47 earned, $22 spent, $8 earned, $45 spent
42. $20 won, $42 lost, $58 won, $11 lost
43. 14° rise, 6° drop, 11° rise, 15° drop
44. $53 withdrawal, $8 withdrawal, $48 deposit, $17 withdrawal
45. $187 profit, $241 loss, $82 profit, $26 profit

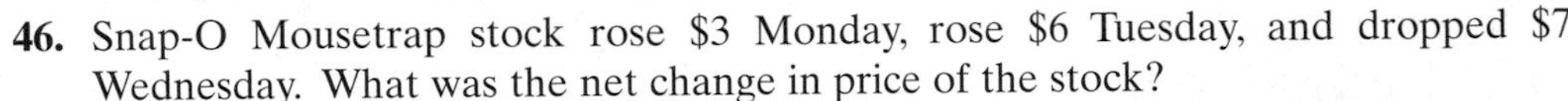

46. Snap-O Mousetrap stock rose $3 Monday, rose $6 Tuesday, and dropped $7 Wednesday. What was the net change in price of the stock?

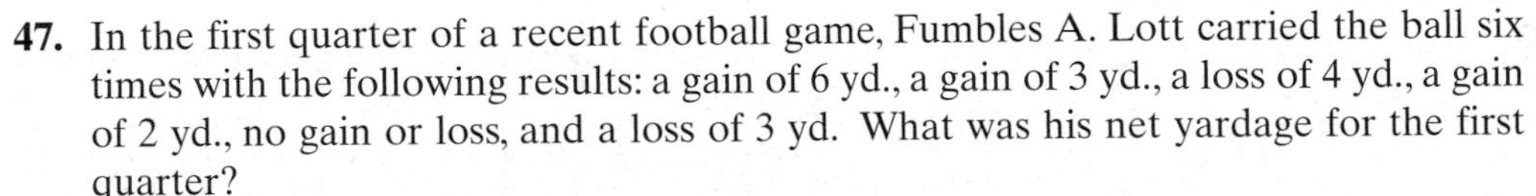

47. In the first quarter of a recent football game, Fumbles A. Lott carried the ball six times with the following results: a gain of 6 yd., a gain of 3 yd., a loss of 4 yd., a gain of 2 yd., no gain or loss, and a loss of 3 yd. What was his net yardage for the first quarter?

48. Beginning at 7° above zero, the temperature rose 4°, then dropped 2°, and then dropped 6°. Find the final temperature.

49. Starting at the third floor, an elevator went down 1 floor, up 3 floors, up 7 floors, and then down 4 floors. Find the final location of the elevator.

50. Bill lost 2 lbs. the first week of his diet, lost 6 lbs. the second week, gained 1 lb. the third week, lost 4 lbs. the fourth week, and gained 3 lbs. the fifth week. What was the total loss or gain? If Bill weighed 223 lbs. at the time he began his diet, what was his weight after 5 weeks of dieting?

51. Jeff works Friday, Saturday, and Sunday in a restaurant. His salary is $10 a night plus tips. Friday night he received $65 in tips but spent $9 for food. Saturday he bought a new shirt for $22, spent $19 for food and received $72 in tips. Sunday he spent $15 for food and received $48 in tips. How much money did he have left after three days of work?

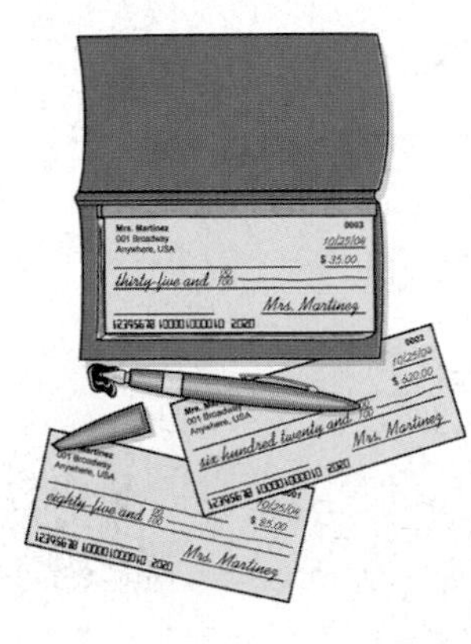

52. Mr. Chung received a bank statement indicating that he was overdrawn by $63. How much must he deposit to bring his balance to $157?

53. Mrs. Martinez knew that the balance in her checking account was $450. She made deposits of $300 and $750 and wrote checks for $85, $620, and $35. What was her new balance?

54. At 2 p.m. the temperature was 76° F. At 8 p.m. the temperature was 58° F. What was the change in temperature? What was the average change in temperature per hour?

55. The temperature at 5:00 a.m. was 8° below zero; at noon, the temperature was 27° F. What was the change in temperature? What was the average change in temperature per hour?

56. $-\$8$
57. 54 years
58. $-14{,}777$ ft.
59. 30,280 ft.
60. 85 years old
61. 0.4375
62. $0.2\overline{7}$ **63.** $-0.\overline{54}$
64. $-0.0\overline{45}$ **65.** $1.\overline{185}$
66. $46.\overline{6}$ **67.** $\frac{13}{10}$
68. $\frac{-39}{50}$ **69.** $\frac{1323}{250}$
70. $\frac{-23929}{1000}$ **71.** $\frac{69}{40}$
72. $\frac{-47}{150}$

73. a. For all positive numbers, the product will be less than the other number.
Ex: $\frac{1}{4}\cdot 2=\frac{1}{2}$

$\frac{1}{2}<2$

b. The product will equal the other number when the other number is 0.
Ex: $\frac{1}{2}\cdot 0=0$

$0=0$

56. Lotsa-Flavor Chewing Gum stock opened on Monday at \$47 per share and closed Friday at \$39 per share. Find the change in price of the stock.

57. The famous French mathematician René Descartes lived from 1596 to 1650. How long did he live?

58. If you travel from the top of Mt. Whitney, elevation 14,495 ft., to the floor of Death Valley, elevation 282 ft. below sea level, what is the change in elevation?

59. A submarine submerged 280 ft. below the surface of the sea fired a rocket that reached an altitude of 30,000 ft. What was the change in altitude of the rocket?

60. The great English mathematician and scientist Isaac Newton was born in 1642 and died in 1727. How old was he when he died?

In Exercises 61 – 66, use a calculator to find the decimal form of each fraction. If the decimal is non-terminating, write it using the bar notation over the repeating pattern of digits.

61. $\frac{7}{16}$ **62.** $\frac{5}{18}$ **63.** $-\frac{6}{11}$ **64.** $-\frac{1}{22}$ **65.** $\frac{32}{27}$ **66.** $\frac{140}{3}$

Use a calculator to find the answers to Exercises 67 – 72 and then to change each of the answers into fraction form.

67. $0.57+0.73$ **68.** $1.58-2.36$ **69.** $-2.78-1.93+10.002$

70. $13.175-16.32-20.784$ **71.** $\frac{7}{8}+\frac{3}{4}+\frac{1}{10}$ **72.** $\frac{9}{100}-\frac{17}{100}-\frac{9}{10}+\frac{2}{3}$

Writing and Thinking About Mathematics

73. Suppose that a fraction between 0 and 1, such as $\frac{1}{4}$ or $\frac{5}{6}$ is multiplied by some other number.

a. Discuss the situations (and give examples) in which the product will be less than the other number.

b. Discuss the situations (and give examples) in which the product will be equal to the other number.

c. Discuss the situations (and give examples) in which the product will be more than the other number.

Hawkes Learning Systems: Introductory & Intermediate Algebra

Decimals and Rounding
Decimals and Fractions
Applications: Change In Value

73. c. The product will be more than the other number when the other number is negative.
Ex: $\frac{1}{4}(-1)=-\frac{1}{4}$

$-\frac{1}{4}>-1$

1.9 Properties of Real Numbers

Objectives

After completing this section, you will be able to:

1. *Complete statements by using the properties of real numbers.*
2. *Name the real number properties that justify given statements.*

Recognizing Types of Numbers

In mathematics, it is important to be able to recognize and name various types of numbers. The conditions stated or implied in certain word problems may allow for only positive solutions or only integer solutions. For example, the height of a person or the length of a piece of string cannot be a negative number. The count of students in a class cannot be a negative integer. With these ideas in mind, we review some of the various types of numbers already discussed.

Whole Numbers

$W = \{0, 1, 2, 3, 4, 5, ...\}$

Integers

$Z = \{..., -5, -4, -3, -2, -1, 0, 1, 2, 3, 4, 5, ...\}$

The integers are the whole numbers and their opposites.

$0, 4, -6, 14,$ and -35 are all integers.

Rational Numbers

$$Q = \left\{\frac{a}{b} \text{ where } a \text{ and } b \text{ are integers and } b \neq 0\right\}$$

The rational numbers are numbers that can be written in fraction form with integers for the numerator and denominator. Rational numbers can also be written as terminating decimals or infinite repeating decimals.

$4, -7, 9.5, -16.8, \frac{3}{5}, 9\frac{3}{10},$ and $6.45454545\ldots$ are all rational numbers.

Irrational Numbers

In decimal form, irrational numbers are infinite nonrepeating decimals.

$\sqrt{3}, \sqrt{5}, -\sqrt{10}, \pi,$ and $\frac{2}{\sqrt{7}}$ are all irrational numbers.

Teaching Notes:
A brief discussion about how these numbers (the irrationals) might arise and how numerous they are might help the student relate to them. Approximating them with rationals can be instructive and can reinforce their meaning.

Real Numbers

$R = \{\text{all rational and irrational numbers}\}$

Any rational number or irrational number is also a real number. Thus all of the numbers listed so far in the text are real numbers.

Figure 1.9 from page 6 is shown again to emphasize the relationships among these various types of numbers.

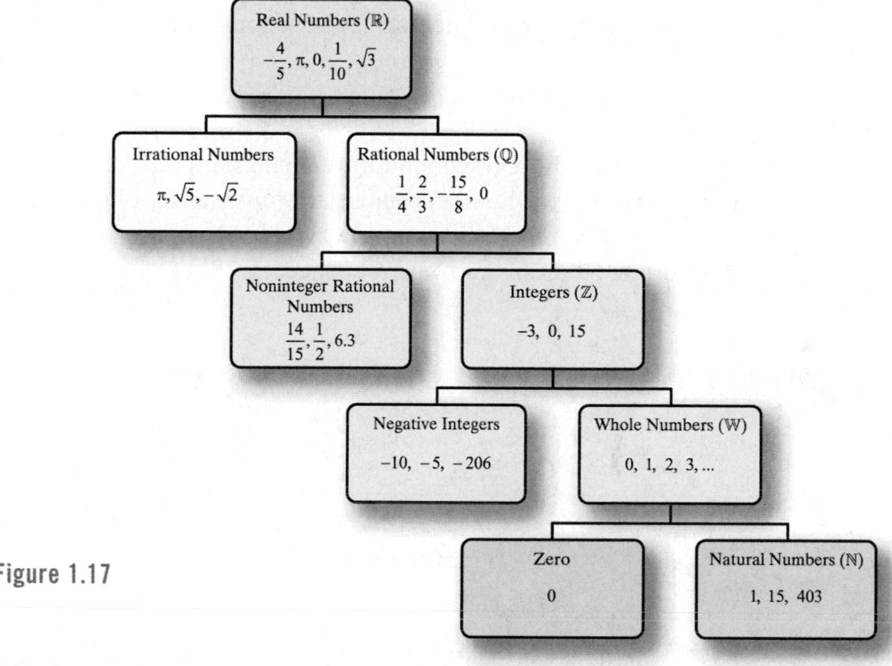

Figure 1.17

Example 1: Types of Numbers

Given the set of numbers $\{-43.2, -\sqrt{19}, -2, -\frac{3}{7}, 0, 0.58, \pi, \sqrt[3]{62}, 6, 13.5555\ldots\}$, tell which numbers are:

a. integers **b.** rational numbers **c.** irrational numbers **d.** real numbers

Solution:

a. The integers are $-2, 0,$ and 6.

b. The rational numbers are $-43.2, -2, -\frac{3}{7}, 0, 0.58, 6,$ and $13.5555\ldots$.

c. The irrational numbers are $-\sqrt{19}, \pi,$ and $\sqrt[3]{62}$.

d. All of the numbers are real numbers.

As we work with all types of real numbers (whole numbers, integers, rational numbers, and irrational numbers) and algebraic expressions, we will need to understand that certain properties are true for some operations and not for others. For example,

the order of the numbers in addition *does not* change the result:

$$17 + 5 = 5 + 17 = 22$$

but, the order of the numbers in subtraction *does* change the result:

$$17 - 5 = 12 \quad \text{but} \quad 5 - 17 = -12$$

Teaching Notes:
At some point, it may be important to show how this property of additon helps overcome the failure of subtraction to be commutative – e.g., that
$5 - 3 = 5 + (-3)$
$= -3 + 5.$

We say that addition is commutative while subtraction is not commutative. The various properties of real numbers under the operations of addition and multiplication are summarized here. These properties are used throughout algebra and mathematics in developing formulas and general concepts.

Properties of Addition

For any real numbers a, b, and c,

Examples

Commutative Property of Addition:

$a + b = b + a$ — $3 + 6 = 6 + 3$

Associative Property of Addition:

$(a + b) + c = a + (b + c)$ — $(2 + 5) + 4 = 2 + (5 + 4)$

Additive Identity:

$a + 0 = a$ — $20 + 0 = 20$

Additive Inverse:

$a + (-a) = 0$ — $10 + (-10) = 0$

$-8 + (+8) = 0$ *[Note that* $+8 = -(-8)$*.]*

Properties of Multiplication

For any real numbers a, b, and c,

	Examples
Commutative Property of Multiplication:	
$a \cdot b = b \cdot a$	$4 \cdot 9 = 9 \cdot 4$
Associative Property of Multiplication:	
$(a \cdot b) \cdot c = a \cdot (b \cdot c)$	$(6 \cdot 2) \cdot 7 = 6 \cdot (2 \cdot 7)$
Multiplicative Identity:	
$a \cdot 1 = a$	$-2 \cdot 1 = -2$
Zero Factor Property:	
$a \cdot 0 = 0 \cdot a = 0$	$-5 \cdot 0 = 0 \cdot (-5) = 0$
Multiplicative Inverse:	
$a \cdot \frac{1}{a} = 1 \quad (a \neq 0)$	$3 \cdot \frac{1}{3} = 1$

Distributive Property

Distributive Property (of Multiplication over Addition):

$a(b + c) = a \cdot b + a \cdot c$	$3(x + 5) = 3 \cdot x + 3 \cdot 5$
$(b + c)a = b \cdot a + c \cdot a$	$(y + 3)2 = y \cdot 2 + 3 \cdot 2$

NOTES The number 0 is called the **additive identity element** or simply the **additive identity**. Likewise, the number 1 is called the **multiplicative identity element** or simply the **multiplicative identity**.

As the following examples illustrate, subtraction and division are neither commutative nor associative.

Subtraction is not commutative: $a - b \neq b - a$

Example: $6 - 2 \neq 2 - 6$
because $6 - 2 = 4$
and $2 - 6 = -4$
and $4 \neq -4$.

continued on next page ...

Subtraction is not associative: $a-(b-c) \neq (a-b)-c$

Example: $10-(5-3) \neq (10-5)-3$
because $10-(5-3)=10-(2)=8$
and $(10-5)-3=5-3=2$
and $8 \neq 2$.

Division is not commutative: $a \div b \neq b \div a$

Example: $6 \div 2 \neq 2 \div 6$
because $6 \div 2 = 3$
and $2 \div 6 = \frac{1}{3}$
and $3 \neq \frac{1}{3}$.

Division is not associative: $a \div (b \div c) \neq (a \div b) \div c$

Example: $24 \div (4 \div 2) \neq (24 \div 4) \div 2$
because $24 \div (4 \div 2) = 24 \div (2) = 12$
and $(24 \div 4) \div 2 = (6) \div 2 = 3$
and $12 \neq 3$.

NOTES

The **multiplicative inverse** of a, $\frac{1}{a}$, is also called the **reciprocal** of a and is defined only for $a \neq 0$. For $a = 0$,

$\frac{1}{0}$ **is undefined.** (0 is the only real number that does not have a reciprocal.)

(That is, if $\frac{1}{0} = x$, then we must have $1 = 0 \cdot x$. But this is not possible because $0 \cdot x = 0$ for all x. Therefore, we say that $\frac{1}{0}$ is **undefined.**)

In illustrating the distributive property, we can write $5(x-6) = 5x - 30$ where -30 is the product of 5 and -6. Remember that the raised dot is optional when indicating multiplication between two variables or a number and a variable. As was stated in Section 1.4, the number is called the **coefficient** of the variable. Thus, 5 is the coefficient of x in the expression $5x$.

Example 2: State the Property

State the name of each property being illustrated.

a. $(-7) + 13 = 13 + (-7)$

Solution: Commutative Property of Addition

b. $8 + (9 + 1) = (8 + 9) + 1$

Solution: Associative Property of Addition

c. $(-25) \cdot 1 = -25$

Solution: Multiplicative Identity

d. $3(x + y) = 3x + 3y$

Solution: Distributive Property

e. $4(3 \cdot 2) = (4 \cdot 3) \cdot 2$

Solution: Associative Property of Multiplication

In each of the following equations, state the property illustrated and show that the statement is true for the value given for the variable by substituting the value in the equation and evaluating.

f. $x + 14 = 14 + x;\ x = -4$

Solution: The Commutative Property of Addition is illustrated.
$(-4) + 14 = 14 + (-4) = 10$

g. $2x + 5 = 5 + 2x;\ x = 10$

Solution: The Commutative Property of Addition is illustrated.
$2 \cdot 10 + 5 = 5 + 2 \cdot 10 = 25$

h. $-4(y + 5) = -4y - 20;\ y = -2$

Solution: The Distributive Property is illustrated.
$-4(-2 + 5) = -4(+3) = -12$
and $-4(-2 + 5) = -4(-2) + (-4)(5) = 8 - 20 = -12$

Practice Problems

Determine the property being illustrated.

1. $(-2 \cdot 5) \cdot 2 = -2 \cdot (5 \cdot 2)$
2. $15 \cdot 0 = 0 \cdot 15 = 0$
3. $2 + 7 = 7 + 2$
4. $2(y + 5) = 2 \cdot y + 2 \cdot 5$

1.9 Exercises

1. Given the set of numbers $\{-3.56, -\sqrt{8}, -\frac{5}{8}, -1, 0, 0.7, \pi, 4, \frac{13}{2}, \sqrt[3]{25}\}$, tell which numbers are
   a. integers **b.** rational numbers **c.** irrational numbers **d.** real numbers

2. Given the set of numbers $\{-123.21, -\sqrt{118}, -\frac{32}{11}, -2\frac{1}{3}, 0.9, \frac{\pi}{2}, \frac{17}{10}, 5, \frac{\sqrt{107}}{5}, 9.343434\ldots\}$, tell which numbers are
   a. integers **b.** rational numbers **c.** irrational numbers **d.** real numbers

Complete the expressions in Exercises 3 – 17 using the given property.

3. $7 + 3 =$ ____	Commutative property of addition
4. $(6 \cdot 9) \cdot 3 =$ ____	Associative property of multiplication
5. $19 \cdot 4 =$ ____	Commutative property of multiplication
6. $18 + 5 =$ ____	Commutative property of addition
7. $6(5 + 8) =$ ____	Distributive property
8. $16 + (9 + 11) =$ ____	Associative property of addition
9. $2 \cdot (3x) =$ ____	Associative property of multiplication
10. $3(x + 5) =$ ____	Distributive property
11. $3 + (x + 7) =$ ____	Associative property of addition
12. $9(x + 5) =$ ____	Distributive property
13. $6 \cdot 0 =$ ____	Zero factor property
14. $6 \cdot 1 =$ ____	Multiplicative identity
15. $0 + (x + 7) =$ ____	Additive identity
16. $0 \cdot (-13) =$ ____	Zero factor property
17. $2(x - 12) =$ ____	Distributive property

1. a. $-1, 0, 4$ **b.** $-3.56, -\frac{5}{8}, -1, 0, 0.7, 4, \frac{13}{2}$ **c.** $-\sqrt{8}, \pi, \sqrt[3]{25}$ **d.** All are real numbers
2. a. 5 **b.** $-123.21, \frac{-32}{11}, -2\frac{1}{3}, 0.9, \frac{17}{10}, 5,$ and $9.343434\ldots$ **c.** $-\sqrt{118}, \frac{\pi}{2},$ and $\frac{\sqrt{107}}{5}$ **d.** All are real numbers
3. $3 + 7$
4. $6 \cdot (9 \cdot 3)$
5. $4 \cdot 19$
6. $5 + 18$
7. $30 + 48$
8. $(16 + 9) + 11$
9. $(2 \cdot 3) \cdot x$
10. $3x + 15$
11. $(3 + x) + 7$
12. $9x + 45$
13. $0 \cdot 6 = 0$
14. 6
15. $x + 7$
16. $-13 \cdot 0 = 0$
17. $2x - 24$

Answers to Practice Problems: 1. Associative property of multiplication **2.** Zero factor property **3.** Commutative property of addition **4.** Distributive property

In Exercises 18 – 27, name the property of real numbers illustrated.

18. $5 + 16 = 16 + 5$ **19.** $5 \cdot 16 = 16 \cdot 5$ **20.** $32 \cdot 1 = 32$ **21.** $32 + 0 = 32$

22. $5 + (3 + 1) = (5 + 3) + 1$ **23.** $5 + (3 + 1) = (3 + 1) + 5$

24. $5(x + 7) = (x + 7) \cdot 5$ **25.** $13(y + 2) = (y + 2) \cdot 13$

26. $13(y + 2) = 13y + 26$ **27.** $6(2 \cdot 9) = (2 \cdot 9) \cdot 6$

In each of the following equations in Exercises 28 – 42, state the property illustrated and show that the statement is true for the value of $x = 4$, $y = -2$, or $z = 3$ by substituting the corresponding value in the equation and evaluating.

28. $6 \cdot x = x \cdot 6$ **29.** $19 + z = z + 19$

30. $8 + (5 + y) = (8 + 5) + y$ **31.** $(2 \cdot 7) \cdot x = 2 \cdot (7 \cdot x)$

32. $5(x + 18) = 5x + 90$ **33.** $(2z + 14) + 3 = 2z + (14 + 3)$

34. $(6 \cdot y) \cdot 9 = 6 \cdot (y \cdot 9)$ **35.** $11 \cdot x = x \cdot 11$

36. $z + (-34) = -34 + z$ **37.** $3(y + 15) = 3y + 45$

38. $2(3 + x) = 2(x + 3)$ **39.** $(y + 2)(y - 4) = (y - 4)(y + 2)$

40. $5 + (x - 15) = (x - 15) + 5$ **41.** $z + (4 + x) = (4 + x) + z$

42. $(x + y) + z = x + (y + z)$

In Exercises 43 – 47, first evaluate each expression by using the Rules for Order of Operations and then use the distributive property to evaluate the same expression. The value must be the same.

43. $6(3 + 8)$ **44.** $7(8 - 5)$ **45.** $10(2 - 9)$ **46.** $13(5 + 3)$

47. $5(14 - 16)$

18. Commutative property of addition
19. Commutative property of multiplication
20. Multiplicative identity
21. Additive identity
22. Associative property of addition
23. Commutative property of addition
24. Commutative property of multiplication
25. Commutative property of multiplication
26. Distributive property
27. Commutative property of multiplication
28. Commutative property of multiplication; $6 \cdot 4 = 4 \cdot 6 = 24$
29. Commutative property of addition; $19+3 = 3+19 = 22$
30. Associative property of addition; $8+(5+(-2)) = (8+5)+(-2) = 11$
31. Associative property of multiplication; $(2 \cdot 7) \cdot 4 = 2 \cdot (7 \cdot 4) = 56$
32. Distributive property; $5(4+18) = 5(4)+90 = 110$

Hawkes Learning Systems: Introductory & Intermediate Algebra

33. Associative property of addition; $(2(3)+14)+3 = 2(3)+(14+3) = 23$

Properties of Real Numbers

34. Associative property of multplication; $(6\cdot(-2))\cdot 9 = 6\cdot(-2\cdot 9) = -108$
35. Commutative property of multiplication; $11\cdot 4 = 4\cdot 11 = 44$
36. Commutative property of addition; $3+(-34) = (-34)+3 = -31$
37. Distributive property; $3(-2+15) = -6+45 = 39$
38. Commutative property of addition; $2(3+4) = 2(4+3) = 14$
39. Commutative property of multiplication; $(-2+2)(-2-4) = (-2-4)(-2+2) = 0$
40. Commutative property of addition; $5+(4-15) = (4-15)+5 = -6$
41. Commutative property of addition; $3+(4+4) = (4+4)+3 = 11$
42. Associative property of addition; $(4-2)+3 = 4+(-2+3) = 5$
43. $6(11) = 66$ and $6\cdot 3+6\cdot 8 = 66$
44. $7(3) = 21$ and $7\cdot 8-7\cdot 5 = 21$
45. $10(-7) = -70$ and $10\cdot 2-10\cdot 9 = -70$
46. $13(8) = 104$ and $13\cdot 5+13\cdot 3 = 104$
47. $5(-2) = -10$ and $5(14)-5(16) = -10$

Chapter 1 Index of Key Ideas and Terms

Section 1.1 The Real Number Line and Absolute Value

Types of Numbers

Counting numbers (or natural numbers) page 2

$\boldsymbol{N} = \{1, 2, 3, 4, 5, 6, 7, 8, 9, 10, 11, \ldots\}$

Whole numbers page 2

$\boldsymbol{W} = \{0, 1, 2, 3, 4, 5, 6, \ldots\}$

Integers page 4

Integers: $\{\ldots, -4, -3, -2, -1, 0, 1, 2, 3, 4, \ldots\}$

Positive integers: $\{1, 2, 3, 4, \ldots\}$

Negative integers: $\{\ldots, -4, -3, -2, -1\}$

The integer 0 is neither positive nor negative.

Rational numbers page 5

A **rational number** is a number that can be written in the form $\frac{a}{b}$, where a and b are integers and $b \neq 0$.

OR,

A **rational number** is a number that can be written in decimal form as a terminating decimal or as an infinite repeating decimal.

Irrational numbers page 5

Irrational numbers are numbers that can be written as infinite nonrepeating decimals.

Real numbers page 5

All rational and irrational numbers are classified as **real numbers**.

Diagram of Types of Numbers page 6

Inequality Symbols page 7

Read from left to right:

$<$ "is less than"

$>$ "is greater than"

$\leq$ "is less than or equal to"

$\geq$ "is greater than or equal to"

Note: Inequality symbols may be read from left to right or from right to left.

Absolute Value pages 8 - 10

The **absolute value** of a real number is its distance from 0. Note that the absolute value of a real number is never negative.

Symbolically,

$|a| = a$ if a is a positive number or 0.

$|a| = -a$ if a is a negative number.

Section 1.2 Addition with Integers

Addition with Integers page 17

1. To add two integers with like signs, add their absolute values and use the common sign.
2. To add two integers with unlike signs, subtract their absolute values (the smaller from the larger) and use the sign of the number with the larger absolute value.

Section 1.3 Subtraction with Integers

Additive Inverse page 20

The opposite of an integer is called its **additive inverse**.
Symbolically, for any integer a, $\boldsymbol{a + (-a) = 0}$.

Subtraction with Integers page 22

To subtract an integer, add its opposite.
Symbolically, for integers a and b, $\boldsymbol{a - b = a + (-b)}$.

Section 1.4 Multiplication and Division with Integers

Multiplication with Integers page 27

If a and b are positive integers, then

1. The product of two positive integers is positive: $\boldsymbol{a \cdot b = ab}$.
2. The product of two negative integers is positive: $\boldsymbol{(-a)(-b) = ab}$.
3. The product of a positive integer and a negative integer is negative: $\boldsymbol{a(-b) = -ab}$.
4. The product of 0 and any integer is 0: $\boldsymbol{a \cdot 0 = 0}$ and $\boldsymbol{(-a) \cdot 0 = 0}$.

Division with Integers pages 28 - 29

If a and b are positive integers,

1. The quotient of two positive integers is positive: $\boldsymbol{\frac{a}{b} = +\frac{a}{b}}$.
2. The quotient of two negative integers is positive: $\boldsymbol{\frac{-a}{-b} = +\frac{a}{b}}$.
3. The quotient of a positive integer and a negative integer is negative:

$$\boldsymbol{\frac{-a}{b} = -\frac{a}{b}} \text{ and } \boldsymbol{\frac{a}{-b} = -\frac{a}{b}}.$$

Division by 0 is undefined. page 28

Section 1.5 Exponents, Prime Numbers, and Order of Operations

Exponents page 34

In general, for any whole number n and integer a,

1. $\underbrace{a \cdot a \cdot a \cdot a \cdot \ldots \cdot a}_{n \text{ factors}} = a^n$
2. $a^1 = a$
3. $a^0 = 1$ (for $a \neq 0$)

(The expression 0^0 is undefined.)

Prime Numbers page 35

A **prime number** is a whole number (other than 0 or 1) that has exactly two different factors, itself and 1.

A whole number (other than 0 or 1) that is not prime is **composite**.

Tests for Divisibility page 36

An integer is divisible

By 2: if the units digit of an integer is 0, 2, 4, 6, or 8.
By 3: if the sum of the digits of an integer is divisible by 3.
By 5: if the units digit is 0 or 5.
By 6: if it is divisible by both 2 and 3.
By 9: if the sum of the digits is divisible by 9.
By 10: if the units digit is 0.

Rules for Order of Operations page 37

1. Simplify within grouping symbols, such as parentheses ( ), brackets [ ], and braces { }. (Starting with the innermost grouping and working out.)
2. Find any powers indicated by exponents.
3. Moving from **left to right**, perform any multiplications or divisions **in the order they appear**.
4. Moving from **left to right**, perform any additions or subtractions **in the order they appear**.

Section 1.6 Multiplying and Dividing Fractions

Rational Number page 41

A **rational number** is a number that can be written in the form $\frac{a}{b}$ where a and b are integers and $b \neq 0$.

OR,

A **rational number** is a number that can be written in decimal form as a terminating decimal or as an infinite repeating decimal.

Fraction pages 41 - 42

In general, fractions can be used to indicate:

1. Equal parts of a whole, or
2. Division

Placement of Negative Signs in Fractions page 42

If a and b are real numbers, and $b \neq 0$, then

$$-\frac{a}{b} = \frac{-a}{b} = \frac{a}{-b}.$$

Multiplication with Fractions page 43

$\frac{a}{b} \cdot \frac{c}{d} = \frac{a \cdot c}{b \cdot d}$ where $b \neq 0$ and $d \neq 0$.

The Fundamental Principle of Fractions page 43

$\frac{a}{b} = \frac{a \cdot \boldsymbol{k}}{b \cdot \boldsymbol{k}}$ where $k \neq 0$ and $b \neq 0$.

Reciprocal page 45

If $a \neq 0$ and $b \neq 0$, the **reciprocal** of $\frac{a}{b}$ is $\frac{b}{a}$, and $\frac{a}{b} \cdot \frac{b}{a} = 1$.

Division with Fractions page 46

To divide by a nonzero fraction, multiply by its reciprocal:

$\frac{a}{b} \div \frac{c}{d} = \frac{a}{b} \cdot \frac{d}{c}$ where $b \neq 0$, $c \neq 0$, and $d \neq 0$.

Section 1.7 Adding and Subtracting Fractions

Addition of Fractions with Like Denominators page 51

To add two fractions with a common denominator, add the numerators and use the common denominator.

$$\frac{a}{b}+\frac{c}{b}=\frac{a+c}{b}$$

Least Common Multiple page 53

To find the LCM of a set of integers

1. List the prime factorization of each number.
2. List the prime factors that appear in any one of the prime factorizations.
3. Find the product of these primes using each prime the greatest number of times that it appears in any one prime factorization.

Addition of Fractions with Different Denominators page 54

To add two fractions with different denominators

1. Find the LCM of the denominators (called the LCD).
2. Change each fraction to an equal fraction with the LCD.
3. Add the new fractions.

(**Note**: Treat variables as prime factors.)

Subtraction with Fractions page 55

To subtract two fractions with a common denominator, subtract the numerators and use the common denominator.

$$\frac{a}{b}-\frac{c}{b}=\frac{a-c}{b}$$

To subtract two fractions with different denominators, find the LCD, change each fraction to an equal fraction with the LCD, and subtract.

Order of Operations with Fractions page 56

Section 1.8 Decimal Numbers and Change In Value

Decimal Numbers

Place Value pages 61 - 62
Reading and Writing Decimal Numbers page 62
Rounding Off Decimal Numbers page 63
Operating with Decimal Numbers pages 64 - 65

continued on next page ...

Section 1.8 Decimal Numbers and Change In Value (continued)

Fractions and Decimals

Changing Fractions to Decimal Form page 66

Changing Decimals to Fraction Form page 67

Change In Value page 70

To find the **Change In Value**, subtract the Beginning Value from the End Value.

(Change In Value) = (End Value) – (Beginning Value)

Net Change is the algebraic sum of several signed numbers.

Section 1.9 Properties of Real Numbers

Properties of Real Numbers

Properties of Addition page 76

For any real numbers a, b, and c,

Commutative Property of Addition:

$$a + b = b + a$$

Associative Property of Addition:

$$(a + b) + c = a + (b + c)$$

Additive Identity:

$$a + 0 = a$$

Additive Inverse:

$$a + (-a) = 0$$

Properties of Multiplication page 77

For any real numbers a, b, and c,

Commutative Property of Multiplication:

$$a \cdot b = b \cdot a$$

Associative Property of Multiplication:

$$(a \cdot b) \cdot c = a \cdot (b \cdot c)$$

Multiplicative Identity:

$$a \cdot 1 = a$$

Zero Factor Property:

$$a \cdot 0 = 0 \cdot a = 0$$

Multiplicative Inverse:

$$a \cdot \frac{1}{a} = 1 \quad (a \neq 0)$$

Distributive Property (of Multiplication over Addition): page 77

$$a(b + c) = a \cdot b + a \cdot c$$

$$(b + c)a = b \cdot a + c \cdot a$$

Chapter 1 Review

For a review of the topics and problems from Chapter 1, look at the following lessons from *Hawkes Learning Systems: Introductory & Intermediate Algebra.*

The Real Number Line and Inequalities
Introduction to Absolute Values
Addition with Integers
Subtraction with Integers
Multiplication and Division with Integers
Factoring Positive Integers
Factoring Integers
Order of Operations
Reduction of Proper Fractions
Reduction of Improper Fractions
Reduction of Positive and Negative Fractions
Multiplying and Dividing Fractions
Adding and Subtracting Fractions
Decimals and Rounding
Decimals and Fractions
Applications: Change In Value
Properties of Real Numbers

Chapter 1 Test

1. Given the set of numbers $\{-10, -\pi, -\frac{3}{4}, -\sqrt{5}, 0, \frac{7}{6}, 3\frac{4}{9}, 7.121212\ldots\}$, tell which numbers are
   a. integers **b.** rational numbers **c.** irrational numbers **d.** real numbers

2. **a.** True or False; All integers are rational numbers. (Explain your answer.)
   b. True or False; All rational numbers are integers. (Explain your answer.)

3. Fill in the blanks with the proper symbol: <, >, or =.
   a. -4 ____ -2 **b.** $-(-2)$ ____ 0 **c.** $|-8|$ ____ $|8|$

4. Graph the following set of numbers on a real number line:
   $\left\{-2, -0.4, |-1|, \frac{7}{3}, 3.1\right\}$

5. On a real number line, graph the set of all integers less than or equal to 2.
6. What integers satisfy the equation $|y| = 7$?
7. List the prime numbers less than 50.

List the integers that satisfy the inequalities in Exercises 8 and 9, then graph the integers on a real number line.

8. $|x| < 3$
9. $|x| \geq 9$

In Exercises 10 – 15, perform the indicated operations.

10. $13 - 16 + 5 - 20$
11. $-28 - (-47)$
12. $9 - (-6) - 14$
13. $(-27)(-6)$
14. $\frac{-35}{-5}$
15. $(42)(-6)(3)(0)$

16. Use the Rules for Order of Operations to evaluate:
    a. $16 \div 4^2 + 2(12 - 5^2)$
    b. $15 - 60[7(3^2 - 3 \cdot 3) + (-2)(6)] - 10^3$
    c. $|16 - 20|[4^2 - (13 + 2^3) - |-6|]$

17. Find the prime factorization for
    a. 80 **b.** 175

18. Name each property illustrated.
    a. $7 \cdot 1 = 7$ **b.** $x + 3 = 3 + x$ **c.** $(9 + 1) + 2 = 9 + (1 + 2)$
    d. $-5 \cdot 0 = 0$ **e.** $6(x + 2) = 6x + 12$ **f.** $2(x + 5) = (x + 5) \cdot 2$

19. Use a calculator to find the value of each of the following expressions.
    a. $(-5 - 6)^7$ **b.** $32 \div 2^3 \cdot 5 - 16 + 4(7 - 8)$

---

1. a. The integers are -10 and 0
b. The rational numbers are $-10, \frac{-3}{4}, 0, \frac{7}{6}, 3\frac{4}{9}$ and $7.121212\ldots$
c. The irrational numbers are $-\pi$ and $-\sqrt{5}$
d. All are real numbers
2. a. True, since for any integer n, $n = \frac{n}{1}$ which is a rational number since the numerator and the denominator are both integers.
b. False, because rational numbers also include non-integers, such as $\frac{2}{5}$.
3. a. < **b.** > **c.** =
4.
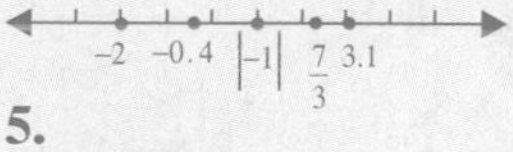

5.

6. $y = 7$ or $y = -7$
7. 2, 3, 5, 7, 11, 13, 17, 19, 23, 29, 31, 37, 41, 43 and 47
8. $\{-2, -1, 0, 1, 2\}$
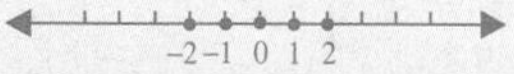

9. $\{\ldots, -10, -9, 9, 10, \ldots\}$

10. -18 **11.** 19 **12.** 1
13. 162 **14.** 7 **15.** 0
16. a. -25 **b.** -265 **c.** -44
17. a. $2^4 \cdot 5$ **b.** $5^2 \cdot 7$

18. a. Multiplicative identity
b. Commutative property of addition
c. Associative property of addition
d. Zero factor property
e. Distributive property
f. Commutative property of multiplication

19. a. −19,487,171 **b.** 0

20. 96 **21.** $90x^2y^2$

22. $\frac{-9}{2}$ **23.** $\frac{3}{5}$ **24.** $\frac{9}{10}$

25. $\frac{-2}{y}$ **26.** $\frac{29}{40}$

27. $\frac{15+n}{5n}$

28. Seventeen and three thousandths

29. 83.15 **30.** 36.53

31. −16.5 **32.** 17.952

33. 1.35

34. $\frac{-33}{16}$

35. a. −0.7 **b.** $15.80

36. a. more **b.** less **c.** $80

37. a. 15 gallons **b.** No, $1 short.

38. −33° F

In Exercises 20 and 21, reduce each fraction to lowest terms.

20. Find $\frac{3}{5}$ of 160.

21. Find the LCM of $10xy^2$, $18x^2y$, and $15xy$.

Perform the indicated operations and reduce each answer to lowest terms in Exercises 22 – 27.

22. $\frac{-15x}{7} \cdot \frac{42}{20x}$

23. $\frac{7}{12} \cdot \frac{9}{28} \div \frac{5}{16}$

24. $\frac{4}{15} + \frac{1}{3} + \frac{3}{10}$

25. $\frac{7}{2y} - \frac{8}{2y} - \frac{3}{2y}$

26. $\frac{7}{8} - \frac{1}{3} \div \frac{5}{6} + \frac{1}{4}$

27. $\frac{3}{n} + \frac{1}{5}$

28. Write the decimal number 17.003 in words.

29. Round off 83.1462 to the nearest hundredth.

Perform the indicated operations in Exercises 30 – 33.

30. 28.63 + 7.9

31. 19.1 − 35.6

32. (5.61)(3.2)

33. $27.27 \div 20.2$

34. Multiply the sum of $\frac{3}{4}$ and $\frac{9}{10}$ by the quotient of $\frac{2}{3}$ and $-\frac{8}{15}$. What is the product?

35. The Veri-Soft computer company stock opened on Monday at $16.50 per share and on the next five days rose $1.50, fell $2.10, rose $3.25, fell $1.75, and fell $1.60 per share. What was the net change in the price of the stock? What was the price of the stock at the closing on Friday?

36. The sale price of a jacket is $60. This is $\frac{3}{4}$ of the original price.

a. Is the original price more or less than $75?

b. If you multiply $60 by $\frac{3}{4}$, will the product be more or less than $60?

c. What was the original price?

37. You know that the gas tank in your car holds 20 gallons of gas and the gauge reads $\frac{1}{4}$ full.

a. How many gallons will be needed to fill the tank?

b. If the price of gas is $1.40 per gallon and you have $20, do you have enough cash to fill the tank? How much extra do you have or how much are you short?

38. At noon on a summer day in Los Angeles, the temperature was 98° F. By 6 p.m. the temperature had changed to 65° F. What was the change in temperature?

Algebraic Expressions, Linear Equations, and Applications

CHAPTER 2

Did You Know?

Traditionally, algebra has been defined to mean generalized arithmetic where letters represent numbers. For example, $3 + 3 + 3 + 3 = 4 \cdot 3$ is a special case of the more general algebraic statement that $x + x + x + x = 4 \cdot x$. The name algebra comes from an Arabic word, al-jabr, which means "to restore."

A notorious Moslem ruler Harun al-Rashid, the caliph made famous in *Tales of the Arabian Nights,* and his son Al-Mamun brought to their Baghdad court many renowned Moslem scholars. One of these scholars was the mathematician Mohammed ibn-Musa al-Khowarizmi, who wrote a text (c. A.D. 800) entitled *Ihm al-jabar w' al-muqabal.* The text included instructions for solving equations by adding terms to both sides of the equation, thus "restoring" equality. The abbreviated title of Al-Khowarizmi's text, *Al-jabar,* became our word for equation solving and operations on letters standing for numbers, **algebra**.

Al-Khowarizmi

Al-Khowarizmi's algebra was brought to western Europe through Moorish Spain in a Latin translation done by Robert of Chester (c. A.D. 1140). Al-Khowarizmi's name may have sounded familiar to you. It eventually was translated as "algorithm," which came to mean any series of steps used to solve a problem. Thus we speak of the division algorithm used to divide one number by another. Al-Khowarizmi is known as the father of algebra just as Euclid is known as the father of geometry.

One of the hallmarks of algebra is its use of specialized notation that enables complicated statements to be expressed using compact notation. Al-Khowarizmi's algebra used only a few symbols with most statements written out in words. He did not even use symbols for numbers! The development of algebraic symbols and notation occurred over the next 1000 years.

"Algebra is generous. She often gives more than is asked of her."

Jean le Rond d'Alembert (1717? – 1783)

Now you are ready to learn how to solve equations. You will also find a variety of useful formulas and use your equation-solving skills in manipulating these formulas. Of course, solving equations and working with formulas are useful skills for solving word problems. Remember that one of the major goals of mathematics is to develop the skills that will allow you, with reasoning, to solve a wide variety of problems you may see in your lifetime.

2.1 Simplifying and Evaluating Algebraic Expressions

Objectives

After completing this section, you will be able to:

1. *Simplify algebraic expressions by combining like terms.*
2. *Evaluate expressions for given values of the variables.*

Simplifying Algebraic Expressions

A single number is called a **constant**. Any constant, variable, the indicated product and/or quotient of constants, and powers of variables is called a **term**. Examples of terms are

$$16,\ 3x,\ -5.2,\ 1.3xy,\ -5x^2,\ 14a^2b^2,\ \text{and}\ -\frac{x}{y^2}.$$

A number written next to a variable (as in $10x$) or a variable written next to another variable (as in xy) indicates multiplication. In the term $5x^2$, the constant 5 is called the **numerical coefficient** of x^2 (or simply the **coefficient** of x^2).

If there is not a number written next to a variable, the coefficient is understood to be 1. For example,

$$x = 1 \cdot x,\ a^3 = 1 \cdot a^3,\ xy = 1 \cdot xy.$$

If a "−" sign is next to a variable, the coefficient is understood to be −1. For example,

$$-x = -1 \cdot x,\ -y^5 = -1 \cdot y^5,\ \text{and}\ -mn^2 = -1 \cdot mn^2.$$

Like Terms

Like terms *(or* ***similar terms****) are terms that are constants or terms that contain the same variables (if any) raised to the same powers.*

Like Terms

$-6,\ 1.84,\ 145,\ \frac{3}{4}$ are like terms because each term is a constant.

$-3a,\ 15a,\ 2.6a,\ \frac{2}{3}a$ are like terms because each term contains the same variable a, raised to the same power, 1. (Remember that $a = a^1$.)

$5xy^2$ and $-3.2xy^2$ are like terms because each term contains the same two variables, x and y, with x first-degree in both terms and y second-degree in both terms.

Unlike Terms

$8x$ and $-9x^2$ are unlike terms (**not** like terms) because the variable x is not of the same power in both terms.

Example 1: Like Terms

From the following list of terms, pick out the like terms:

$$-7,\ 2x,\ 4.1,\ -x,\ 3x^2y,\ 5x,\ -6x^2y,\ \text{and}\ 0$$

Solution: -7, 4.1, and 0 are like terms. All are constants.
$2x$, $-x$, and $5x$ are like terms.
$3x^2y$ and $-6x^2y$ are like terms.

Algebraic expressions (expressions that indicate operations with terms) such as

$$9x + 10y,\ 5n - n + 3m^2,\ 2x^2 - 8x + 10,\ \text{and}\ \frac{5x+3x}{4} - 8.2x^2$$

are not terms. However, we would like to simplify expressions that contain like terms. That is, we want to combine like terms. For example, by combining like terms we can write

$$9x + 6x = 15x \quad \text{and} \quad 6.3n + 2n - n = 7.3n$$

Like terms can be combined by applying the distributive property. Recall that the distributive property states that

$$a(b + c) = ab + ac$$
$$\text{or} \quad ab + ac = a(b + c)$$
$$\text{or} \quad ba + ca = (b + c)a$$

This last form is particularly useful when b and c are numerical coefficients. For example,

$$3x + 5x = (3 + 5)x \quad \text{By the distributive property}$$
$$= 8x \quad \text{Add the coefficients.}$$

$$3x^2 - 5x^2 = (3 - 5)x^2 \quad \text{By the distributive property}$$
$$= -2x^2 \quad \text{Add the coefficients algebraically.}$$

In the first expression, the like terms $3x$ and $5x$ are combined and the result is $8x$. In the second expression, the like terms $3x^2$ and $-5x^2$ are combined and the result is $-2x^2$.

Example 2: Combine Like Terms

Combine like terms whenever possible.

a. $8x + 10x$

Solution: $8x + 10x = (8 + 10)x = 18x$ By the distributive property

b. $6.5y - 2.3y$

Solution: $6.5y - 2.3y = (6.5 - 2.3)y = 4.2y$

c. $4(n - 7) + 5(n + 1)$

Solution: $4(n - 7) + 5(n + 1) = 4n - 28 + 5n + 5$ Use the distributive property twice.

$= 4n + 5n - 28 + 5$

$= 9n - 23$ Combine like terms.

Teaching Notes: Stressing the use of the distributive property here is extremely important. It is an opportunity to point out the use of the commutative and associative properties in these steps.

d. $2x^2 + 3a + x^2 - a$

Solution: $2x^2 + 3a + x^2 - a = 2x^2 + x^2 + 3a - a$

$= (2 + 1)x^2 + (3 - 1)a$ **Note:** $+x^2 = +1x^2$ and $-a = -1a$

$= 3x^2 + 2a$

e. $\frac{x+3x}{2}+5x$

Solution: A fraction bar is a symbol of inclusion, like parentheses. So combine like terms in the numerator first.

$$\begin{aligned}\frac{x+3x}{2}+5x &= \frac{4x}{2}+5x\\ &= \frac{4}{2}\cdot x+5x\\ &= 2x+5x\\ &= 7x\end{aligned}$$

Evaluating Algebraic Expressions

In most cases, if an expression is to be evaluated, like terms should be combined first and then the resulting expression evaluated by following the Rules for Order of Operations.

To Evaluate an Algebraic Expression

1. *Combine like terms, if possible.*
2. *Substitute the values given for any variables.*
3. *Follow the rules for order of operations.*

Remember to use parentheses around negative numbers when substituting. Without parentheses, an evaluation can be dramatically changed and lead to wrong answers, particularly when even exponents are involved. These ideas were discussed in Chapter 1 and we summarize with the exponent 2 as follows:

In general, except for $x = 0$,

1. $-x^2$ **is negative** $[\,-6^2 = -1 \cdot 6^2 = -36\,]$
2. $(-x)^2$ **is positive** $[\,(-6)^2 = (-6)(-6) = 36\,]$
3. $-x^2 \neq (-x)^2$ $[\,-36 \neq 36\,]$

Example 3: Evaluate Algebraic Expressions

a. Evaluate x^2 for $x = 3$ and for $x = -4$.

Solution: For $x = 3,\ x^2 = 3^2 = 9$
For $x = -4,\ x^2 = (-4)^2 = 16$

b. Evaluate $-x^2$ for $x = 3$ and for $x = -4$.

Solution: For $x = 3,\ -x^2 = -3^2 = -1 \cdot 9 = -9$
For $x = -4,\ -x^2 = -(-4)^2 = -1(16) = -16$

Simplify each expression below by combining like terms; then, evaluate the resulting expression using the given values for the variables. Remember to apply the Rules for Order of Operations.

c. $2x + 5 + 7x;\ x = -3$

Solution: Simplify first:

$$\begin{aligned} 2x + 5 + 7x &= 2x + 7x + 5 \\ &= 9x + 5 \end{aligned}$$

Now evaluate:

$$\begin{aligned} 9x + 5 &= 9(-3) + 5 \\ &= -27 + 5 \\ &= -22 \end{aligned}$$

d. $3ab - 4ab + 6a - a;\ a = 2, b = -1$

Solution: Simplify first:

$$3ab - 4ab + 6a - a = -ab + 5a$$

Now evaluate:

$$\begin{aligned} -ab + 5a &= -1(2)(-1) + 5(2) \\ &= 2 + 10 \\ &= 12 \end{aligned}$$

Note: $-ab = -1ab$

e. $\dfrac{5x + 3x}{4} + 2(x + 1);\ x = 5$

Solution: Simplify first:

$$\begin{aligned} \frac{5x + 3x}{4} + 2(x + 1) &= \frac{8x}{4} + 2x + 2 \\ &= 2x + 2x + 2 \\ &= 4x + 2 \end{aligned}$$

Now evaluate:

$$4x+2=4\cdot 5+2$$
$$=20+2$$
$$=22$$

f. $\frac{3}{4}x+\frac{3}{8}x;\ x=-4$

Solution: Simplify first: $\frac{3}{4}x+\frac{3}{8}x=\left(\frac{3}{4}+\frac{3}{8}\right)x=\left(\frac{6}{8}+\frac{3}{8}\right)x=\frac{9}{8}x$

Now evaluate:

$$\frac{9}{8}x=\frac{9}{\cancel{8}_{2}}\cdot(-\cancel{4}^{-1})=-\frac{9}{2}\quad\text{(or }-4.5\text{ in decimal form)}$$

NOTES

Important note about expressions involving fractions in algebra

An expression with a fraction as the coefficient can be written in two different forms. For example, for all values of x, $\frac{9}{8}x$ is the same as $\frac{9x}{8}$. To see this fact, note that we can write $\frac{9}{8}x=\frac{9}{8}\cdot\frac{x}{1}=\frac{9x}{8}$. However, algebraically, $\frac{9}{8x}$ (with the variable in the denominator) is *not* the same as $\frac{9}{8}x$.

Similarly,

$$\frac{5}{6}a=\frac{5a}{6}\quad\text{and}\quad\frac{5}{6a}\neq\frac{5}{6}a.$$

Practice Problems

Simplify the following expressions by combining like terms.

1. $-2x-5x$
2. $12y+6-y+10$
3. $5(x-1)+4x$
4. $2b^2-a+b^2+a$

Simplify the expression, then evaluate the resulting expression if $x=3$ and $y=-2$.

5. $2(x+3y)+4(x-y)$

Answers to Practice Problems: 1. $-7x$ **2.** $11y+16$ **3.** $9x-5$ **4.** $3b^2$ **5.** $6x+2y;\ 14$

2.1 Exercises

In Exercises 1 – 6, pick out the like terms in each list of terms.

1. $-5, \frac{1}{6}, 7x, 8, 9x, 3y$ **2.** $-2x^2, -13x^3, 5x^2, 14x^2, 10x^3$

3. $5xy, -x^2, -6xy, 3x^2y, 5x^2y, 2x^2$ **4.** $3ab^2, -ab^2, 8ab, 9a^2b, -10a^2b, ab, 12a^2$

5. $24, 8.3, 1.5xyz, -1.4xyz, -6, xyz, 5xy^2z, 2xyz^2$

6. $-35y, 1.62, -y^2, -y, 3y^2, \frac{1}{2}, 75y, 2.5y^2$

Find the value of each numerical expression in Exercises 7 – 10.

7. $(-8)^2$ **8.** -8^2 **9.** -11^2 **10.** $(-6)^2$

Simplify by combining like terms in each of the expressions in Exercises 11 – 42.

11. $8x + 7x$ **12.** $3y + 8y$ **13.** $5x - 2x$ **14.** $7x + (-3x)$

15. $-n - n$ **16.** $-x - x$ **17.** $6y^2 - y^2$ **18.** $16z^2 - 5z^2$

19. $23x^2 - 11x^2$ **20.** $18x^3 + 7x^3$ **21.** $4x + 2 + 3x$ **22.** $3x - 1 + x$

23. $2x - 3y - x$ **24.** $x + y + x - 2y$ **25.** $2x^2 + 5y + 6x^2 - 2y$

26. $4a + 2a - 3b - a$ **27.** $3(n + 1) - n$ **28.** $2(n - 4) + n + 1$

29. $5(a - b) + 2a - 3b$ **30.** $4a - 3b + 2(a + 2b)$

31. $3(2x + y) + 2(x - y)$ **32.** $4(x + 5y) + 3(2x - 7y)$

33. $2x + 3x^2 - 3x - x^2$ **34.** $2y^2 + 4y - y^2 - 3y$

35. $2(n^2 - 3n) + 4(-n^2 + 2n)$ **36.** $3n^2 + 2n - 5 - n^2 + n - 4$

37. $3x^2 + 4xy - 5xy + y^2$ **38.** $2x^2 - 5xy + 11xy + 3$ **39.** $\frac{x + 5x}{6} + x$

40. $2y - \frac{2y + 3y}{5}$ **41.** $y - \frac{2y + 4y}{3}$ **42.** $z - \frac{3z + 5z}{4}$

In Exercises 43 – 60, first (a) simplify the expressions and then (b) evaluate the simplified expression at x = 4, y = 3, a = –2, and b = –1.

43. $5x + 4 - 2x$ **44.** $7x - 17 - x$ **45.** $8.3x^2 - 5.7x^2 + x^2$

Answers:

1. $-5, \frac{1}{6}$ and 8 are like terms; $7x$ and $9x$ are like terms. **2.** $-2x^2$, $5x^2$ and $14x^2$ are like terms; $-13x^3$ and $10x^3$ are like terms. **3.** $-x^2$ and $2x^2$ are like terms; $5xy$ and $-6xy$ are like terms; $3x^2y$ and $5x^2y$ are like terms. **4.** $3ab^2$ and $-ab^2$ are like terms; $8ab$ and ab are like terms; $9a^2b$ and $-10a^2b$ are like terms. **5.** 24, 8.3 and –6 are like terms; $1.5xyz$, $-1.4xyz$ and xyz are like terms. **6.** $-35y$, $-y$ and $75y$ are like terms; 1.62 and $\frac{1}{2}$ are like terms. $-y^2$, $3y^2$ and $2.5y^2$ are like terms. **7.** 64 **8.** –64 **9.** –121 **10.** 36 **11.** $15x$ **12.** $11y$ **13.** $3x$ **14.** $4x$ **15.** $-2n$ **16.** $-2x$ **17.** $5y^2$ **18.** $11z^2$ **19.** $12x^2$ **20.** $25x^3$ **21.** $7x + 2$ **22.** $4x - 1$ **23.** $x - 3y$ **24.** $2x - y$ **25.** $8x^2 + 3y$ **26.** $5a - 3b$ **27.** $2n + 3$ **28.** $3n - 7$ **29.** $7a - 8b$ **30.** $6a+b$ **31.** $8x + y$ **32.** $10x - y$ **33.** $2x^2 - x$ **34.** $y^2 + y$ **35.** $-2n^2 + 2n$ **36.** $2n^2 + 3n - 9$ **37.** $3x^2 - xy + y^2$ **38.** $2x^2 + 6xy + 3$ **39.** $2x$ **40.** y **41.** $-y$ **42.** $-z$ **43. a.** $3x + 4$ **b.** 16 **44. a.** $6x - 17$ **b.** 7 **45. a.** $3.6x^2$ **b.** 57.6

46. $2.4(x+1)+1.3(x-1)$ 47. $\frac{7}{8}x+\frac{3}{4}x+\frac{1}{2}x$ 48. $\frac{7y}{10}-\frac{y}{5}$

49. $\frac{a}{6}+\frac{a}{2}-\frac{2a}{3}$ 50. $-\frac{3}{20}x-\frac{3}{10}x-\frac{2}{5}x$ 51. $x-10-3x+2$

52. $3(y-1)+2(y+2)$ 53. $4(y+3)+5(y-2)$ 54. $-5(x+y)+2(x-y)$

55. $6a+5a-a+13$ 56. $3a^2-a^2+4a-5$ 57. $5ab+b^2-2ab+b^3$

58. $5a+ab^2-2ab^2+3a$ 59. $\frac{3a+5a}{-2}+12a$ 60. $\frac{-4b-2b}{-3}+\frac{2b+5b}{7}$

Simplify each of the following expressions in Exercises 61 – 66.

61. a. $3a+3a$ b. $3a\cdot 3a$ c. $\frac{1}{3a}+\frac{1}{3a}$ d. $\frac{1}{3a}\cdot\frac{1}{3a}$ e. $\frac{1}{3a}\div\frac{1}{3a}$

62. a. $5x+5x$ b. $5x\cdot 5x$ c. $\frac{1}{5x}+\frac{1}{5x}$ d. $\frac{1}{5x}\cdot\frac{1}{5x}$ e. $\frac{1}{5x}\div\frac{1}{5x}$

63. a. $-7ab-7ab$ b. $(-7ab)\cdot(-7ab)$ c. $\frac{1}{7ab}+\frac{1}{7ab}$ d. $\frac{1}{7ab}\cdot\frac{1}{7ab}$ e. $\frac{1}{7ab}\div\frac{1}{7ab}$

64. a. $8y-8y$ b. $8y\cdot(-8y)$ c. $\frac{1}{8y}-\frac{1}{8y}$ d. $\frac{1}{8y}\cdot\left(-\frac{1}{8y}\right)$ e. $-\frac{1}{8y}\div\frac{1}{8y}$

65. a. $1.5a+1.5a$ b. $1.5a\cdot 1.5a$ c. $\frac{1}{1.5a}+\frac{1}{1.5a}$ d. $\frac{1}{1.5a}\cdot\frac{1}{1.5a}$ e. $\frac{1}{1.5a}\div\left(-\frac{1}{1.5a}\right)$

66. a. $\frac{4x}{3}+\frac{4x}{3}$ b. $\frac{4x}{3}\cdot\frac{4x}{3}$ c. $\frac{4x}{3}-\frac{4x}{3}$ d. $\frac{4x}{3}\cdot\frac{3}{4x}$ e. $\frac{4x}{3}\div\frac{3}{4x}$

Answers:

46. a. $3.7x+1.1$ b. 15.9
47. a. $\frac{17x}{8}$ b. 8.5
48. a. $\frac{y}{2}$ b. 1.5
49. a. 0 b. 0
50. a. $\frac{-17x}{20}$ b. -3.4
51. a. $-2x-8$ b. -16
52. a. $5y+1$ b. 16
53. a. $9y+2$ b. 29
54. a. $-3x-7y$ b. -33
55. a. $10a+13$ b. -7
56. a. $2a^2+4a-5$ b. -5
57. a. $3ab+b^2+b^3$ b. 6
58. a. $8a-ab^2$ b. -14
59. a. $8a$ b. -16
60. a. $3b$ b. -3
61. a. $6a$ b. $9a^2$ c. $\frac{2}{3a}$ d. $\frac{1}{9a^2}$ e. 1
62. a. $10x$ b. $25x^2$ c. $\frac{2}{5x}$ d. $\frac{1}{25x^2}$ e. 1
63. a. $-14ab$ b. $49a^2b^2$ c. $\frac{2}{7ab}$ d. $\frac{1}{49a^2b^2}$ e. 1
64. a. 0 b. $-64y^2$ c. 0 d. $\frac{-1}{64y^2}$ e. -1
65. a. $3a$ b. $2.25a^2$ c. $\frac{4}{3a}$ d. $\frac{4}{9a^2}$ e. -1
66. a. $\frac{8x}{3}$ b. $\frac{16x^2}{9}$ c. 0 d. 1 e. $\frac{16x^2}{9}$

Hawkes Learning Systems: Introductory & Intermediate Algebra

Variables and Algebraic Expressions
Simplifying Expressions
Simplifying Expressions with Parentheses
Evaluating Algebraic Expressions

2.2 Translating English Phrases and Algebraic Expressions

Objectives

After completing this section, you will be able to:

1. *Write the meanings of algebraic expressions in words.*
2. *Write algebraic expressions for word phrases.*

Translating English Phrases into Algebraic Expressions

Algebra is a language of mathematicians, and to understand mathematics, you must understand the language. We want to be able to change English phrases into their "algebraic" equivalents and vice versa. So, if a problem is stated in English, we can translate the phrases into algebraic symbols and proceed to solve the problem according to the rules developed for algebra.

The following examples illustrate how certain key words can be translated into algebraic symbols.

Example 1: Key Words

	English Phrase	Algebraic Expression
a.	3 **multiplied by** the number represented by x the **product** of 3 and x 3 **times** x	$3x$
b.	3 **added to** a number the **sum** of z and 3 z **plus** 3 3 **more than** z	$z+3$
c.	2 **times** the quantity found by **adding** a number to 1 **twice** the **sum** of x and 1 the **product** of 2 with the **sum** of x and 1	$2(x+1)$

d. **twice** x **plus** 1
the **sum** of **twice** x and 1
2 **times** x **increased** by 1
1 **more than** the **product** of 2 and a number

$2x+1$

e. the **difference** between 5 **times** a number and 3
5 **times** a number **minus** 3
5 **multiplied by** a number **less** 3
3 **less than** the **product** of a number and 5
3 **subtracted from** $5n$

$5n-3$

NOTES

In Example 1b, the phrase "the sum of z and 3" was translated as $z + 3$. If the expression had been translated as $3 + z$, there would have been no mathematical error because addition is commutative. That is, $z + 3 = 3 + z$. However, in 1e, the phrase "3 less than the product of a number and 5" must be translated as it was because subtraction is **not** commutative. Thus,

"3 less than 5 times a number" means $5n - 3$
while "5 times a number less than 3" means $3 - 5n$.

Therefore, be very careful when writing and/or interpreting expressions indicating subtraction. Be sure that the subtraction is in the order indicated by the wording in the problem. The same is true with expressions involving division.

Certain words, such as those in boldface in the previous examples, are the keys to the operations. Learn to look for these words and those from the following table.

Addition	Subtraction	Multiplication	Division
add	subtract (from)	multiply	divide
sum	difference	product	quotient
plus	minus	times	
more than	less than	twice	
increased by	decreased by	of (with fractions and percent)	
	less		

The words **quotient** and **difference** deserve special mention because their use implies that the numbers given are to be operated on in the order they are found. That is, division and subtraction are done with the values in the same order that they are given in the problem. For example:

the quotient of y and 5 ⟶ $\frac{y}{5}$

the quotient of 5 and y ⟶ $\frac{5}{y}$

the difference between 6 and x ⟶ $6 - x$

the difference between x and 6 ⟶ $x - 6$

If we did not have these agreements concerning subtraction and division, then the phrases just illustrated might have more than one interpretation and be considered **ambiguous**.

An **ambiguous phrase** is one whose meaning is not clear or for which there may be two or more interpretations. This is a common occurrence in ordinary everyday language, and misunderstandings occur frequently. Imagine the difficulties diplomats have in communicating ideas from one language to another trying to avoid ambiguities. Even the order of subjects, verbs, and adjectives may not be the same from one language to another. Translating grammatical phrases in any language into mathematical expressions is quite similar. To avoid ambiguous phrases in mathematics, we try to be precise in the use of terminology, to be careful with grammatical construction, and to follow the Rules for Order of Operations.

Translating Algebraic Expressions into English Phrases

Consider the three expressions to be translated into English:

$$7(n+1),\quad 6(n-3),\quad \text{and} \quad 7n+1.$$

In the first two expressions, we indicate the parentheses with a phrase such as "the quantity" or "the sum of" or "the difference between." Without the parentheses, we agree that the operations are to be indicated in the order given. Thus,

	$7(n+1)$	can be translated as "seven times the sum of a number and 1"
	$6(n-3)$	can be translated as "six times the difference between a number and 3"
while	$7n+1$	can be translated as "seven times a number plus 1."

Example 2: Algebraic Expression to Phrase

Write an English phrase that indicates the meaning of each algebraic expression.

a. $5x$ **b.** $2n + 8$ **c.** $3(a - 2)$

Solution:

Algebraic Expression	Possible English Phrase
a. $5x$	The product of a number and 5
b. $2n + 8$	Twice a number increased by 8
c. $3(a - 2)$	Three times the difference between a number and 2

Example 3: Phrase to Algebraic Expression

Change each phrase into an equivalent algebraic expression.

a. The quotient of a number and –4

b. 6 less than 5 times a number

c. Twice the sum of 3 and a number

Solution:

Phrase	Algebraic Expression
a. The quotient of a number and –4	$\frac{x}{-4}$
b. 6 less than 5 times a number	$5y - 6$
c. Twice the sum of 3 and a number	$2(3 + n)$

Practice Problems

Change the following phrases to algebraic expressions.

1. *7 less than a number*

2. *The quotient of y and 5*

3. *14 more than 3 times a number*

Change the following algebraic expressions into English phrases.

4. $10 - x$ **5.** $2(y - 3)$ **6.** $5n + 3n$

Answers to Practice Problems: **1.** $x - 7$ **2.** $\frac{y}{5}$ **3.** $3y + 14$ **4.** 10 decreased by a number **5.** twice the difference between a number and 3 **6.** 5 times a number plus 3 times the same number

2.2 Exercises

Translate each of the expressions in Exercises 1 – 12 into an equivalent English phrase. (There may be more than one correct translation.)

1. $4x$ **2.** $x + 6$ **3.** $2x + 1$ **4.** $4x - 7$
5. $7x - 5.3$ **6.** $3.2(x + 2.5)$ **7.** $-2(x - 8)$ **8.** $10(x + 4)$
9. $5(2x + 3)$ **10.** $3(4x - 5)$ **11.** $6(x - 1)$ **12.** $9(x + 3)$

Write each pair of expressions in words in Exercises 13 – 16. Notice the differences between the expressions and the corresponding English phrases.

13. $3x + 7$; $3(x + 7)$ **14.** $4x - 1$; $4(x - 1)$ **15.** $7x - 3$; $7(x - 3)$
16. $5(x + 6)$; $5x + 6$

Write the algebraic expression described by each of the word phrases in Exercises 17 – 40. Choose your own variable.

17. 6 added to a number
18. 7 more than a number
19. 4 less than a number
20. A number decreased by 13
21. 5 less than 3 times a number
22. The difference between twice a number and 10
23. The difference between x and 3, all divided by 7
24. 9 times the sum of a number and 2
25. 3 times the difference between a number and 8
26. 13 less than the product of 4 with the sum of a number and 1
27. 5 subtracted from three times a number
28. The sum of twice a number and four times the number
29. 8 minus twice a number
30. The sum of a number and 9 times the number
31. 4 more than the product of 8 with the difference between a number and 6
32. Twenty decreased by 4.8 times a number
33. The difference between three times a number and five times the same number
34. Eight more than the product of 3 times the sum of a number and 6
35. Six less than twice the difference between a number and 7
36. Four less than 3 times the difference between 7 and a number
37. Nine more than twice the sum of 17 and a number
38. **a.** 5 less than 3 times a number
b. 5 less 3 times a number
39. **a.** 6 less than a number
b. 6 less a number
40. **a.** 20 less than a number
b. A number less than 20

1. 4 times a number **2.** 6 more than a number **3.** 1 more than twice a number **4.** 7 less than 4 times a number **5.** 5.3 less than 7 times a number **6.** 3.2 times the sum of a number and 2.5 **7.** –2 times the difference between a number and 8 **8.** 10 times the sum of a number and 4 **9.** 5 times the sum of twice a number and 3 **10.** 3 times the difference of 4 times a number and 5 **11.** 6 times the difference between a number and 1 **12.** 9 times the sum of a number and 3 **13.** 3 times a number plus 7; 3 times the sum of a number and 7 **14.** 1 less than 4 times a number; 4 times the difference of a number and 1 **15.** The product of 7 and a number minus 3; 7 times the difference between a number and 3 **16.** 5 times the sum of a number and 6; the sum of 5 times a number and 6
17. $x + 6$ **18.** $x + 7$
19. $x - 4$ **20.** $x - 13$

21. $3x - 5$ **22.** $2x - 10$
23. $\frac{x-3}{7}$
24. $9(x + 2)$
25. $3(x - 8)$
26. $4(x + 1) - 13$
27. $3x - 5$
28. $2x + 4x$
29. $8 - 2x$
30. $x + 9x$
31. $8(x - 6) + 4$
32. $20 - 4.8x$
33. $3x - 5x$
34. $3(x + 6) + 8$
35. $2(x - 7) - 6$
36. $3(7 - x) - 4$
37. $2(17 + x) + 9$
38. **a.** $3x - 5$ **b.** $5 - 3x$
39. **a.** $x - 6$ **b.** $6 - x$
40. **a.** $x - 20$ **b.** $20 - x$
41. $\$4.95x$
42. $0.11x$
43. $7t + 3$
44. $60h + 20$
45. $6t + 3$
46. $20 + 0.15m$
47. $x + 8 + 2x = 3x + 8$
48. $0.09x + 250$
49. $c + 0.2c = 1.2c$
50. $2w + 2(2w - 3)$
$= 6w - 6$

Write the algebraic expression described by each of the word phrases in Exercises 41 – 50.

41. The cost of x pounds of candy at \$4.95 a pound

42. The annual interest on x dollars if the rate is 11% per year

43. The number of days in t weeks and 3 days

44. The number of minutes in h hours and 20 minutes

45. The points scored by a football team on t touchdowns (6 points each) and 1 field goal (3 points)

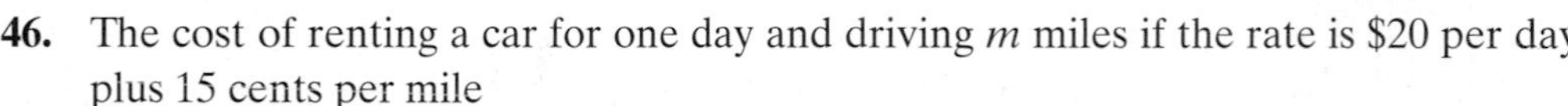

46. The cost of renting a car for one day and driving m miles if the rate is \$20 per day plus 15 cents per mile

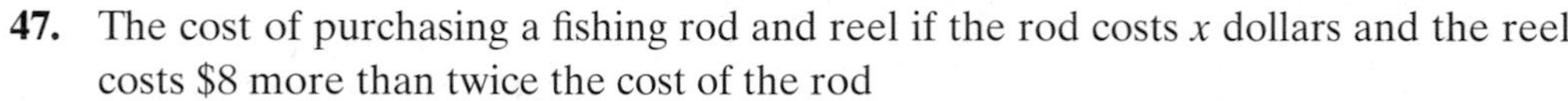

47. The cost of purchasing a fishing rod and reel if the rod costs x dollars and the reel costs \$8 more than twice the cost of the rod

48. A sales person's weekly salary if he receives \$250 as his base plus 9% of the weekly sales of x dollars

49. The selling price of an item that costs c dollars if the markup is 20% of the cost

50. The perimeter of a rectangle if the width is w centimeters and the length is 3 cm less than twice the width

Hawkes Learning Systems: Introductory & Intermediate Algebra

Translating Phrases into Algebraic Expressions

2.3 Solving Linear Equations: $x + b = c$ and $ax = c$

Objectives

After completing this section, you will be able to:

1. *Solve linear equations of the form* $x + b = c$.
2. *Solve linear equations of the form* $ax = c$.
3. *Solve applications involving linear equations.*

Many practical applications entail setting up an equation (or several equations) involving an unknown quantity. The objective is to find the value of this unknown so that the question asked in the application is answered. This means that before we can solve meaningful word problems, one of the basic skills needed is solving equations. In this section, we will discuss solving linear equations in the following two forms:

$$\boldsymbol{x + b = c} \qquad \text{and} \qquad \boldsymbol{ax = c}.$$

Then, in the following section, we will combine the techniques learned here and discuss solving linear equations in general of the form:

$$\boldsymbol{ax + b = c}.$$

An **equation** is a statement that two algebraic expressions are equal. That is, both expressions represent the same number. If an equation contains a variable, the value (or set of values) that gives a true statement when substituted for the variable is called the **solution** (or **solution set**) to the equation. The process of finding the solution is called **solving the equation**.

Linear Equation in x

If a, b, and c are constants and $a \neq 0$*, then a* ***linear equation in x*** *(or* ***first-degree equation in x****) is an equation that can be written in the form*

$$\boldsymbol{ax + b = c}.$$

Note: *The term* ***first-degree*** *is sometimes used because the variable x is understood to have an exponent of 1. That is,* $x = x^1$*. Also, variables other than x, such as y or z, may be used.*

All of the following equations are linear equations in one variable:

$$2x + 3 = 7, \qquad 5y + 6 = -13, \qquad z + 5 = 3z - 5.$$

Solving Equations of the Form $x + b = c$

To begin, we need the **Addition Property of Equality**.

Addition Property of Equality

If the same algebraic expression is added to both sides of an equation, the new equation has the same solutions as the original equation. Symbolically, if A, B and C are algebraic expressions, then the equations

$$A = B$$

and

$$A + C = B + C$$

have the same solutions.

Equations with the same solutions are said to be **equivalent**.

The objective of solving linear (or first-degree) equations is to get the variable by itself on one side of the equation and any constants on the other side. The following procedure will accomplish this.

Procedure for Solving an Equation that Simplifies to the Form $x + b = c$

1. *Combine any like terms on each side of the equation.*
2. *Use the Addition Property of Equality and add the opposite of a constant term or a variable term (or both) to both sides. The objective is to isolate the variable on one side of the equation with a coefficient of +1.*
3. *Check your answer by substituting it into the original equation.*

Teaching Notes: I have chosen to consider only equations that simplify to the form $ax + b = c$ to be linear equations. Equations that are identities (such as $x + 3 = x + 3$) or equations with no solution (such as $x + 3 = x + 2$) need to be considered separately.

Every linear equation has exactly one solution. This means that once we have found a solution, there is no need to search for another solution.

Example 1: Solving $x + b = c$

Solve each of the following linear equations:

a. $y - 8 = -2$

Solution: $y - 8 + 8 = -2 + 8$ Add 8, the opposite of -8, to both sides.

$y = 6$ Simplify.

Check: $y - 8 = -2$

$6 - 8 \stackrel{?}{=} -2$ Substitute $y = 6$.

$-2 = -2$ True statement.

continued on next page ...

d. $-x = 4$

Solution:

$-x = 4$

$-1x = 4$ — -1 is the coefficient of x.

$\frac{-1x}{-1} = \frac{4}{-1}$ — Divide by -1 so that the coefficient is $+1$.

$x = -4$

Check:

$-x = 4$

$-(-4) \stackrel{?}{=} 4$ — Substitute $x = -4$.

$4 = 4$ — True statement.

Problem Solving

George Pòlya (1877 – 1985), a famous professor at Stanford University, studied the process of discovery learning. Among his many accomplishments, he developed the following four-step process as an approach to problem solving:

1. Understand the problem.
2. Devise a plan.
3. Carry out the plan.
4. Look back over the results.

For a complete discussion of these ideas, see *How To Solve It* by Pòlya (Princeton University Press, 2nd edition, 1957). The following quote by Pòlya illustrates his sense of humor and his understanding of students' dilemmas: "The traditional mathematics professor of the popular legend is absent-minded. He usually appears in public with a lost umbrella in each hand. He prefers to face a blackboard and to turn his back on the class. He writes *a*, he says *b*, he means *c*, but it should be *d*. Some of his sayings are handed down from generation to generation."

There are a variety of types of applications discussed throughout this text and subsequent courses in mathematics, and you will find these four steps helpful as guidelines for understanding and solving all of them. Applying the necessary skills to solve exercises, such as combining like terms or solving equations, is not the same as accumulating the knowledge to solve problems. **Problem solving can involve careful reading, reflection, and some original or independent thought**.

Basic Steps for Solving Applications

1. *Understand the problem. For example,*

- ***a.*** *Read the problem carefully, maybe several times.*
- ***b.*** *Understand all the words.*
- ***c.*** *If it helps, restate the problem in your own words.*
- ***d.*** *Be sure that there is enough information.*

2. *Devise a plan. For example,*

- ***a.*** *Guess, estimate, or make a list of possibilities.*
- ***b.*** *Draw a picture or diagram.*
- ***c.*** *Represent the unknown quantity with a variable and form an equation.*

3. *Carry out the plan. For example,*

- ***a.*** *Try all the possibilities you have listed.*
- ***b.*** *Study your picture or diagram for insight into the solution.*
- ***c.*** *Solve any equation that you may have set up.*

4. *Look back over the results. For example,*

- ***a.*** *Can you see an easier way to solve the problem?*
- ***b.*** *Does your solution actually work? Does it make sense in terms of the wording of the problem? Is it reasonable?*
- ***c.*** *If there is an equation, check your answer in the equation.*

NOTES

You may find that many of the applications in this section can be solved by "reasoning," and there is nothing wrong with that approach. Reasoning is a fundamental part of all of mathematics. However, keep in mind that the algebraic techniques you are learning are important. They also involve reasoning and will prove very useful in solving more complicated problems in later sections and in later courses.

Example 3: Applications

a. When a \$2.13 tax was added to the price of a skirt, the total bill was \$37.63. What was the price of the skirt?

Solution: Let x = price of the skirt
Now, use the relationship: price of skirt + tax = total bill
Set up the equation:

$$x + 2.13 = 37.63$$

Solve the equation:

$$x + 2.13 - 2.13 = 37.63 - 2.13$$ Use the Addition Principle by adding –2.13 to both sides.

$$x = 35.50$$ Simplify both sides.

The price of the skirt was \$35.50.

b. The original price of a VCR was reduced by \$45.50. The sale price was \$215.90. What was the original price?

Solution: Let y = price of the VCR.
Now, use the relationship: price of VCR – reduction = sale price
Set up the equation:

$$y - 45.50 = 215.90$$

Solve the equation:

$$y - 45.50 + 45.50 = 215.90 + 45.50$$ Use the Addition Principle by adding 45.50 to both sides.

$$y = 261.40$$ Simplify both sides.

The original price of the VCR was \$261.40.

c. An exam is given with 15 problems and the students are allowed $1\frac{1}{2}$ hours to take it. How many minutes are allotted for each problem?

Solution: Here we let x = number of minutes allotted per problem. Then the product $15x$ will equal the total time for the exam. Since the time allowed is given in hours, we make the change

$$1\frac{1}{2} \text{ hours } = 90 \text{ minutes.}$$

Then the equation relating the problems and the time is

$$15x = 90$$

$$\frac{15x}{15} = \frac{90}{15}$$

$$x = 6$$

Each problem is allotted 6 minutes.

Practice Problems

Solve the following equations.

1. $-16 = x + 5$ **2.** $6y - 1.5 = 7y$ **3.** $4x = -20$

4. $\frac{3}{5}y = 33$ **5.** $1.7z + 2.4z = 8.2$ **6.** $-x = -8$

7. $3x = 10$ **8.** $5x - 4x + 1.6 = -2.7$

2.3 Exercises

1. $y = 6$
2. $x = 9$
3. $n = -10$
4. $x = -1.2$
5. $y = 5.2$
6. $x = -3$
7. $x = 3$
8. $x = -8$
9. $x = 25$
10. $x = -11$
11. $a = 0$
12. $n = 0$
13. $x = \frac{1}{2}$
14. $x = -\frac{2}{5}$
15. $x = 20$
16. $y = 10$
17. $x = -4$
18. $x = 12$
19. $y = -2$
20. $x = -113$
21. $a = 1$
22. $x = 5.1$
23. $x = -1$
24. $y = -2$
25. $a = 6.2$
26. 3.75 in.
27. 25 m, 34 m, 12 m

Solve each of the following equations in Exercises 1 – 25. (Remember that the objective is to isolate the variable on one side of the equation with a coefficient of 1.)

1. $y - 5 = 1$ **2.** $x + 14 = 23$ **3.** $n + 3 = -7$

4. $x + 3.6 = 2.4$ **5.** $2y = 10.4$ **6.** $-8x = 24$

7. $x + 8x = 12 + 15$ **8.** $10x - 2x = 36 - 100$ **9.** $13x - 12x - 4 = 21$

10. $-2x - 9 + 5x - 2 = 4x$ **11.** $7a - 6a + 17 = 17$ **12.** $3n - n + 14 - 2 - 12 = 3n$

13. $\frac{2}{3} = 5x - 4x + \frac{1}{6}$ **14.** $10x - 9x - \frac{1}{2} = -\frac{9}{10}$ **15.** $\frac{3x}{4} = 15$

16. $\frac{2y}{5} = 4$ **17.** $\frac{5}{2}x + 2x = 15 - 33$ **18.** $\frac{5}{3}x + \frac{2}{3}x = 21 + 7$

19. $6.4y + 8.8 = 2.0y$ **20.** $8x = 8.2x + 22.6$ **21.** $2.9a = 5a - 2.1$

22. $1.5x = 6.5x + 4.6 - 30.1$ **23.** $-\frac{1}{8}x - \frac{1}{4}x = \frac{5}{8} - \frac{1}{4}$ **24.** $\frac{5}{6}y + \frac{2}{3} = \frac{1}{2}y$

25. $5a + 12a - 6.2 = 8a + 8a$

Problems Related to Geometry

The perimeter, P, of a triangle is equal to the sum of the lengths of the sides, a, b, and c ($P = a + b + c$).

26. One side of a triangle is 11.5 inches long. A second side is 8.75 inches long. If the perimeter is 24 in., find the length of the third side.

27. The perimeter of a triangle is 71 meters. If the sides are related as shown in the following figure, find the length of each side.

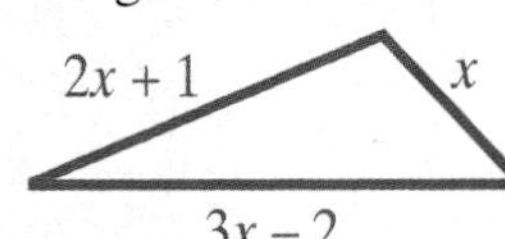

Answers to Practice Problems: **1.** $x = -21$ **2.** $y = -1.5$ **3.** $x = -5$ **4.** $y = 55$ **5.** $z = 2$ **6.** $x = 8$ **7.** $x = \frac{10}{3}$ **8.** $x = -4.3$

28. 28.7 cm

29. 16.125 m

30. 7.5 m

31. $x = 2$ in.

32. 4.5 cm

28. Two sides of a triangle measure 43.2 cm and 26.1 cm. Find the length of the third side if the perimeter is 98 cm.

29. The perimeter of a square is 4 times the length of a side ($P = 4s$). Find the length of a side if the perimeter of a square is $64\frac{1}{2}$ meters.

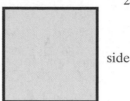

30. The area of a rectangle is found by multiplying the length times the width $(A = lw)$. Find the width of a rectangle with area 75 m^2 and length 10 m.

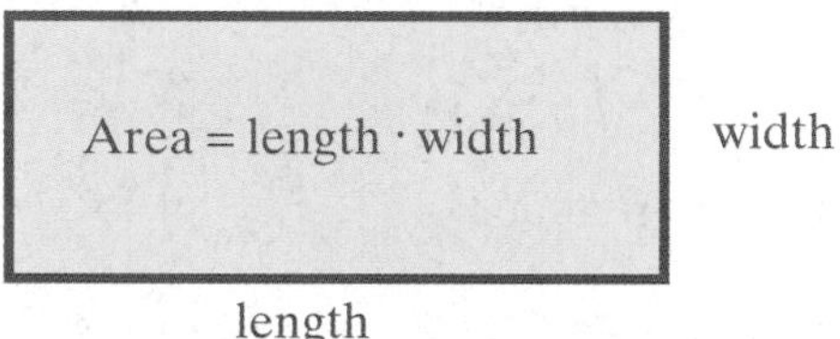

31. Find the value of x in the figure if the area of the shaded portion is 95 inches^2.

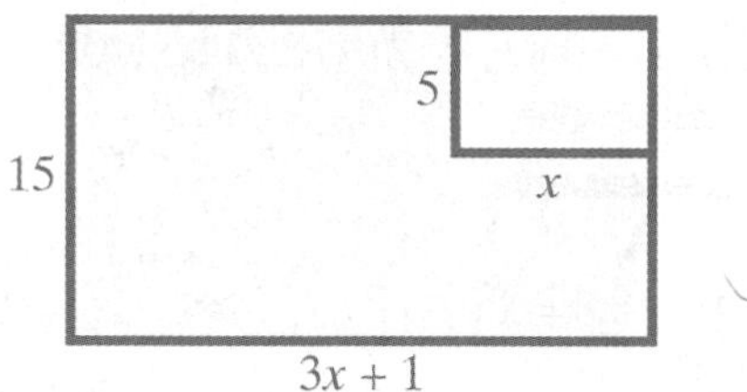

32. The volume of a rectangular box of cereal is the product of its length, width, and height $(V = lwh)$. Find the width of a cereal box with volume 1260 cm^3 if its length is 14 cm and its height is 20 cm.

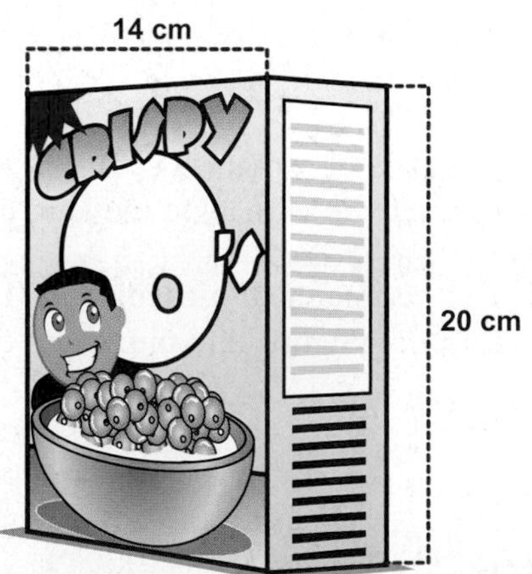

33. The area of a triangle is found by multiplying $\frac{1}{2}$ times the base times the height $(A = \frac{1}{2}bh)$. Find the length of the base of a triangle with area 180 in.^{2} and height 10 in.

$A = \frac{1}{2}bh$

height

base

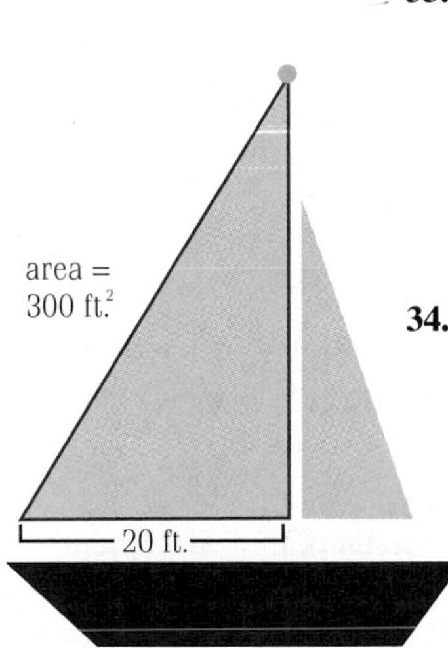

34. A sail is in the shape of a triangle. Find the height of the sail if its base is 20 ft. and its area is 300 ft^2.

The sum of the measures of the three angles of a triangle is 180° ($\alpha + \beta + \gamma = 180°$). (**Note**: α, β, and γ are the Greek lowercase letters alpha, beta, and gamma, respectively.)

35. A triangle with two equal angles is called an **isosceles** triangle. If two angles of an isosceles triangle both measure 45°, is the triangle a **right** triangle? (Does the third angle measure 90°?) Draw a sketch of such a triangle.

36. If all three angles of a triangle measure less than 90°, then the triangle is called an **acute** triangle. Is a triangle with two angles of measure 55° and 20° an acute triangle? What is the measure of the third angle? (**Note:** If a triangle is not acute and not right, then the measure of one of the angles is more than 90°, and it is called an **obtuse** triangle.)

33. 36 in.

34. 30 ft.

35. Yes

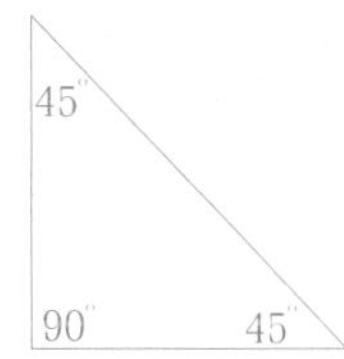

36. No, 105°

37. 0.25 years (or 3 months)

38. $10,666.67

39. 24% annual interest rate, $4,400

Problems Related to Business

Simple interest, I, is the product of the money invested or principal, P, the annual interest rate, r, and the time, t $(I = Prt)$. The time used in calculating simple interest is one year or less. Use 12 months in a year and 365 days in a year. (If the time is more than one year, then interest is calculated on the interest and a formula for compound interest is used.)

37. The interest earned on an investment of $3240 at 10% was $81. What was the time that the money was invested?

38. A savings account pays an annual interest rate of 5%. How much must be invested to make $400 in interest in 9 months?

39. If you have a credit card debt of $2000 and the simple interest for one month is $40, what is the interest rate that you are paying? (**Note:** Credit card companies usually charge compound interest, particularly on overdue payments.) If you continue to pay only the interest each month for 5 years and then pay off the $2000, how much will you have paid for the $2000 credit card debt?

40. \$5 million
41. \$855
42. \$2800
43. \$1690
44. 7 hrs.
45. 3.5 hrs.
46. $y = -50.753$
47. $x = -17.214$
48. $y = 26.087$
49. $x = 246$
50. $y = -1036$

The profit, P, is equal to the revenue, R, minus the cost, C ($P = R - C$).

40. Find the revenue (income) of a company that shows a profit of \$3.2 million and costs of \$1.8 million.

41. A suit cost the store \$675 and the owner said that he wants to make a profit of \$180. What price should he mark on the suit?

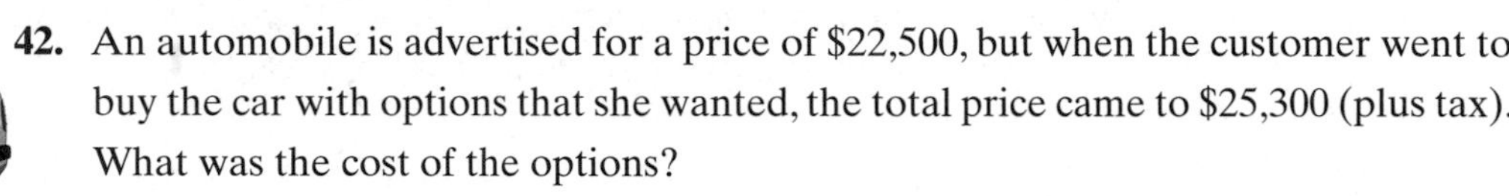

42. An automobile is advertised for a price of \$22,500, but when the customer went to buy the car with options that she wanted, the total price came to \$25,300 (plus tax). What was the cost of the options?

43. A computer was advertised on sale for \$1560. This was a reduction of \$130 from the original price. What was the original price?

The distance traveled, d, is equal to the product of the rate, r and the time, t ($d = rt$).

44. How long will a truck driver take to travel 350 miles if he averages 50 miles per hour?

45. How long will it take a train traveling at 40 miles per hour to go 140 miles?

Calculator Problems

Use a calculator to solve Exercises 46 – 50.

46. $y + 32.861 = -17.892$

47. $17.61x + 27.059 = 9.845 + 16.61x$

48. $14.38y - 8.65 + 9.73y = 17.437 + 23.11y$

49. $2.637x = 648.702$

50. $-0.3057y = 316.7052$

Hawkes Learning Systems: Introductory & Intermediate Algebra

Solving Linear Equations Using Addition and Subtraction
Applications of Linear Equations: Addition and Subtraction
Solving Linear Equations Using Multiplication and Division
Applications of Linear Equations: Multiplication and Division

2.4 Solving Linear Equations: $ax + b = c$

Objectives

After completing this section, you will be able to:

1. *Solve equations of the form* $ax + b = c$.
2. *Supply the reasons for each step in solving an equation.*
3. *Solve absolute value equations of the form* $|ax + b| = c$.

Now we are ready to solve equations of the general form $ax + b = c$. That is, we are ready to apply both the Addition Property and the Multiplication Property in solving one equation. The initial equation may be of any one of several forms with variables on both sides, parentheses, fractions, or decimals. Our goal is to make the process as simple as possible.

General Procedure for Solving Linear Equations

Teaching Notes:
This combination of principles is sometimes referred to as "The Golden Rule of Equation Solving" – "Do unto one side what you do unto the other."

1. *Simplify each side of the equation by removing any grouping symbols and combining like terms. (This may also involve multiplying both sides by the LCM of the denominators or by a power of 10 to remove decimals. In this manner we will be dealing only with integer constants and coefficients.)*
2. *Use the Addition Property of Equality to add the opposites of constants and/or variables so that variables are on one side and constants are on the other side.*
3. *Use the Multiplication Property of Equality to multiply both sides by the reciprocal of the coefficient (or to divide by the coefficient).*
4. *Check your answer by substituting it into the original equation.*

Study each of the following examples carefully. Remember that **the objective is to get the variable on one side by itself with a coefficient of 1**. Note that the equations are written one under the other and the equal signs are aligned vertically.

Example 1: $ax + b = c$

Solve each of the following equations.

a. $5x + 3 = 2x - 18$

Solution:

$5x+3=2x-18$	Write the equation.
$5x+3-3=2x-18-3$	Add -3 to both sides.
$5x=2x-21$	Simplify.
$5x-2x=2x-21-2x$	Add $-2x$ to both sides.
$3x=-21$	Simplify.
$\frac{\cancel{3}x}{\cancel{3}}=\frac{-21}{3}$	Divide both sides by 3. (Or, multiply by $\frac{1}{3}$.)
$x=-7$	Simplify.

Check:

$5(-7)+3 \stackrel{?}{=} 2(-7)-18$	Substitute $x=-7$.
$-35+3 \stackrel{?}{=} -14-18$	Simplify.
$-32=-32$	True statement.

b. $2(y-7) = 4(y+1) - 26$

Solution:

$2(y-7)=4(y+1)-26$	Write the equation.
$2y-14=4y+4-26$	Use the distributive property.
$2y-14=4y-22$	Combine like terms.
$2y-14+22=4y-22+22$	Add 22 to both sides. Here we will put the variables on the right side to get a positive coefficient of y.
$2y+8=4y$	Simplify.
$-2y+2y+8=-2y+4y$	Add $-2y$ to both sides.
$8=2y$	Simplify.
$\frac{8}{2}=\frac{2y}{2}$	Divide both sides by 2.
$4=y$	Simplify.

The number 4 does check, so the solution is 4.

As shown in Example 1c (on the next page), even if you use a calculator to do the arithmetic calculations, you should write the equations showing the steps used in the solving process.

c. $16.53 - 18.2z = 7.43$

Solution:

$16.53 - 18.2z = 7.43$	Write the equation.
$100(16.53 - 18.2z) = 100(7.43)$	Multiply both sides by 100. (This is not necessary; however, the resulting coefficients and constants will be integers and easier to work with than decimals.)
$1653 - 1820z = 743$	Simplify.
$1653 - 1820z - 1653 = 743 - 1653$	Add -1653 to both sides.
$-1820z = -910$	Simplify.
$\frac{-1820z}{-1820} = \frac{-910}{-1820}$	Divide both sides by -1820.
$z = \frac{1}{2}$ (or $z = 0.5$)	Simplify.

Check:

$16.53 - 18.2(0.5) = 7.43$	Substitute $z = 0.5$.
$16.53 - 9.1 \stackrel{?}{=} 7.43$	Simplify.
$7.43 = 7.43$	True statement.

d. $\frac{1}{2}x + \frac{3}{4}x + \frac{7}{2} = \frac{2}{3}x$

Solution:

$\frac{1}{2}x + \frac{3}{4}x + \frac{7}{2} = \frac{2}{3}x$	Write the equation.
$12\left(\frac{1}{2}x + \frac{3}{4}x + \frac{7}{2}\right) = 12\left(\frac{2}{3}x\right)$	Multiply both sides by 12, the LCM of the denominators.
$12\left(\frac{1}{2}x\right) + 12\left(\frac{3}{4}x\right) + 12\left(\frac{7}{2}\right) = 12\left(\frac{2}{3}x\right)$	Use the distributive property.
$6x + 9x + 42 = 8x$	Simplify. Now all constants and coefficients are integers.
$15x + 42 = 8x$	Combine like terms.
$15x + 42 - 15x = 8x - 15x$	Add $-15x$ to both sides.
$42 = -7x$	Simplify.
$\frac{42}{-7} = \frac{-7x}{-7}$	Divide both sides by -7.
$-6 = x$	Simplify.

Checking will show that -6 is the solution.

continued on next page ...

Note: The techniques illustrated in Examples 1c and 1d could have been used to solve equations with decimals and fractions in Section 2.3. Now you can see that there are two ways to handle these types of equations. Either work with the decimals or fractions as they are or multiply in such a way so that constants and coefficients are integers.

e. $-2(5x+13)-2=-6(3x-2)-41$

Solution:

$-2(5x+13)-2=-6(3x-2)-41$	
$-10x-26-2=-18x+12-41$	Use the distributive property. Be careful with the signs.
$-10x-28=-18x-29$	Simplify.
$-10x-28+18x=-18x-29+18x$	Add $18x$ to both sides.
$8x-28=-29$	Simplify.
$8x-28+28=-29+28$	Add 28 to both sides.
$8x=-1$	Simplify.
$\frac{8x}{8}=\frac{-1}{8}$	Divide both sides by 8.
$x=-\frac{1}{8}$	Simplify.

Checking will show that $-\frac{1}{8}$ is the solution.

ABOUT CHECKING: Checking can be quite time-consuming and need not be done for every problem. This is particularly important on exams. You should check only if you have time after the entire exam is completed.

Absolute Value Equations: $|ax+b|=c$

Absolute value was defined in Section 1.1 and that definition is restated here for convenience and emphasis.

Absolute Value

*The **absolute value** of a real number is its distance from 0. Note that the absolute value of a real number is never negative.*

$|a| = a$ *if a is a positive number or 0.*

$|a| = -a$ *if a is a negative number.*

In Section 1.1, we discussed a few simple equations of the form $|x| = c$ and the corresponding solutions resulting from the definition. For example:

If $|x| = 3$, then $x = 3$ or $x = -3$ because $|3| = 3$ and $|-3| = 3$.
If $|x| = -5$, then the equation has no solution because absolute value is never negative.

We now use the definition to solve absolute value equations of the form $|ax + b| = c$. The solutions depend on the number c.

Solutions of Absolute Value Equations of the Form $|ax + b| = c$

1. *If $c < 0$, then $|ax + b| = c$ has no solution.*
2. *If $c = 0$, then $|ax + b| = c$ has one solution. This solution can be found by solving the linear equation $ax + b = 0$.*
3. *If $c > 0$, then $|ax + b| = c$ has two solutions. These solutions can be found by solving the two linear equations $ax + b = c$ and $ax + b = -c$.*

Example 2: Absolute Value Equations

Solve each of the following absolute value equations.

a. $|3x + 1| = 7$

Solution: To solve $|3x + 1| = 7$, we solve the two linear equations
$3x + 1 = 7$ and $3x + 1 = -7$.

$$3x+1=7 \qquad\qquad 3x+1=-7$$
$$3x+1-1=7-1 \qquad\qquad 3x+1-1=-7-1$$
$$3x=6 \qquad\qquad 3x=-8$$
$$\frac{\cancel{3}x}{\cancel{3}}=\frac{6}{3} \qquad\qquad \frac{\cancel{3}x}{\cancel{3}}=\frac{-8}{3}$$
$$x=2 \qquad\qquad x=-\frac{8}{3}$$

Thus, the equation has two solutions: $x = 2$ and $x = -\frac{8}{3}$.

b. $|2n - 5| = 0$

Solution: To solve $|2n - 5| = 0$, we solve the linear equation $2n - 5 = 0$.

continued on next page ...

$$2n-5=0$$
$$2n-5+5=0+5$$
$$2n=5$$
$$\frac{\cancel{2}n}{\cancel{2}}=\frac{5}{2}$$
$$n=\frac{5}{2}$$

Thus, the equation has one solution: $n=\frac{5}{2}$.

c. $|5y+6|=-8$

Solution: The equation $|5y+6|=-8$ has no solution.

Practice Problems

Solve the following equations.

1. $x+14-6x=2x-7$

2. $6.4x+2.1=3.1x-1.2$

3. $5-(y-3)=14-4(y+2)$

4. $\frac{2n}{3}-\frac{1}{2}=n+\frac{1}{6}$

5. $\frac{3n}{14}+\frac{1}{4}=\frac{n}{7}-\frac{1}{4}$

6. $|5x-1|=4$

2.4 Exercises

For Exercises 1 – 4, supply the reasons for each step in solving the equation.

1. Use the distributive property.
Add 12 to both sides.
Simplify.
Add x to both sides.
Simplify.
Divide both sides by 4.
Simplify.

1.

$3(x-4)=-x+24$	Write the equation.
$3x-12=-x+24$	______________
$3x-12+12=-x+24+12$	______________
$3x=-x+36$	______________
$3x+x=-x+36+x$	______________
$4x=36$	______________
$\frac{\cancel{4}x}{\cancel{4}}=\frac{36}{4}$	______________
$x=9$	______________

Answers to Practice Problems: **1.** $x=3$ **2.** $x=-1$ **3.** $y=-\frac{2}{3}$ **4.** $n=-2$ **5.** $n=-7$ **6.** $x=1,-\frac{3}{5}$

2. Add –2 to both sides.
Simplify.
Add $-3y$ to both sides.
Simplify.
Divide both sides by 2.
Simplify.

3. Multiply both sides by 30.
Use the distributive property.
Simplify.
Add –5 to both sides.
Simplify.
Add $-12a$ to both sides.
Simplify.
Divide both sides by –2.
Simplify.

4. Multiply both sides by 10.
Simplify.
Add 1 to both sides.
Simplify.
Add $3x$ to both sides.
Simplify.
Divide both sides by 8.
Simplify.

5. $x=-3$ 6. $x=2$
7. $y=-4$ 8. $n=3$
9. $x=-5$ 10. $y=6$
11. $x=-0.12$
12. $x=-1.8$
13. $n=0$ 14. $x=0$
15. $x=0$ 16. $y=0$
17. $z=-1$ 18. $z=3$
19. $y=\frac{1}{5}$

2.

$$\begin{aligned} 5y+2&=3y+2 \\ 5y+2-2&=3y+2-2 \\ 5y&=3y \\ 5y-3y&=3y-3y \\ 2y&=0 \\ \frac{\cancel{2}y}{\cancel{2}}&=\frac{0}{2} \\ y&=0 \end{aligned}$$

Write the equation.
____
____
____
____
____
____

3.

$$\begin{aligned} \frac{1}{3}a+\frac{1}{6}&=\frac{2}{5}a-\frac{7}{10} \\ 30\left(\frac{1}{3}a+\frac{1}{6}\right)&=30\left(\frac{2}{5}a-\frac{7}{10}\right) \\ 30\left(\frac{1}{3}a\right)+30\left(\frac{1}{6}\right)&=30\left(\frac{2}{5}a\right)-30\left(\frac{7}{10}\right) \\ 10a+5&=12a-21 \\ 10a+5-5&=12a-21-5 \\ 10a&=12a-26 \\ 10a-12a&=12a-26-12a \\ -2a&=-26 \\ \frac{\cancel{-2}a}{\cancel{-2}}&=\frac{-26}{-2} \\ a&=13 \end{aligned}$$

Write the equation.
____
____
____
____
____
____
____
____
____

4.

$$\begin{aligned} 4.7-0.3x&=0.5x-0.1 \\ 10(4.7-0.3x)&=10(0.5x-0.1) \\ 47-3x&=5x-1 \\ 47-3x+1&=5x-1+1 \\ 48-3x&=5x \\ 48-3x+3x&=5x+3x \\ 48&=8x \\ \frac{48}{8}&=\frac{\cancel{8}x}{\cancel{8}} \\ 6&=x \end{aligned}$$

Write the equation.
____
____
____
____
____
____
____
____

In Exercises 5 – 40, solve each of the following linear (first-degree) equations.

5. $3x + 11 = 2$ **6.** $5x - 4 = 6$ **7.** $y - 7 = 5y + 9$

8. $5n - 3 = 2n + 6$ **9.** $3x + 2 = x - 8$ **10.** $3y + 18 = 7y - 6$

11. $-5x + 2.9 = 3.5$ **12.** $3x + 2.7 = -2.7$ **13.** $1.6n = 0.8n$

14. $5.3x = 0.2x$ **15.** $13x + 5 = 2x + 5$ **16.** $6y - 2.1 = y - 2.1$

17. $2(z + 1) = 3z + 3$ **18.** $6z - 3 = 3(z + 2)$ **19.** $16y + 23y - 5 = 14y$

20. $x = 5$ **21.** $x = -4$
22. $x = -1$ **23.** $x = -3$
24. $y = -3$ **25.** $x = -21$
26. $n = -15$ **27.** $y = 0$
28. $y = 1$ **29.** $x = -5$
30. $x = -\frac{3}{2}$ **31.** $n = \frac{1}{6}$
32. $x = -\frac{2}{3}$ **33.** $n = 2$
34. $x = \frac{1}{4}$ **35.** $x = \frac{7}{60}$
36. $x = \frac{8}{3}$ **37.** $x = \frac{8}{5}$
38. $y = -\frac{1}{2}$ **39.** $x = \frac{2}{3}$
40. $x = -\frac{7}{3}$
41. $n = \frac{2}{3}, -2$
42. $x = -\frac{5}{3}, 5$
43. $x = -22, 26$
44. $y = 0, \frac{20}{3}$
45. No solution
46. No solution
47. $x = 2$
48. $y = -\frac{2}{3}$
49. $x = -\frac{3}{2}, 2$
50. $x = -1, \frac{11}{3}$
51. $l = 64$ ft.

20. $5x - 2x + 4 = 3x + x - 1$

21. $6.5 + 1.2x = 0.5 - 0.3x$

22. $x - 0.1x + 0.9 = 0.2(x + 1)$

23. $0.25 + 3x + 6.5 = 0.75x$

24. $0.9y + 3 = 0.4y + 1.5$

25. $\frac{2}{3}x + 1 = \frac{1}{3}x - 6$

26. $\frac{4}{5}n + 2 = \frac{2}{5}n - 4$

27. $\frac{y}{5} + \frac{3}{4} = \frac{y}{2} + \frac{3}{4}$

28. $\frac{3}{8}\left(y - \frac{1}{2}\right) = \frac{1}{8}\left(y + \frac{1}{2}\right)$

29. $\frac{1}{2}(x + 1) = \frac{1}{3}(x - 1)$

30. $\frac{2x}{3} + \frac{x}{3} = -\frac{3}{4} + \frac{x}{2}$

31. $\frac{5n}{6} + \frac{1}{9} = \frac{3n}{2}$

32. $\frac{3}{4}x + \frac{1}{5}x = \frac{1}{2}x - \frac{3}{10}$

33. $\frac{1}{6}n + \frac{7}{15} - \frac{2}{5}n = 0$

34. $x + \frac{2}{3}x - 2x = \frac{x}{6} - \frac{1}{8}$

35. $3x + \frac{1}{2}x - \frac{2}{5}x = \frac{x}{10} + \frac{7}{20}$

36. $7(2x - 1) = 5(x + 6) - 13$

37. $5 - 3(2x + 1) = 4(x - 5) + 6$

38. $-2(y + 5) - 4 = 6(y - 2) + 2$

39. $8 + 4(2x - 3) = 5 - (x + 3)$

40. $8(3x + 5) - 9 = 9(x - 2) + 14$

Solve the absolute value equations in Exercises 41 – 50.

41. $|6n + 4| = 8$

42. $|3x - 5| = 10$

43. $\left|\frac{1}{4}x - \frac{1}{2}\right| = 6$

44. $\left|\frac{1}{5}y - \frac{2}{3}\right| = \frac{2}{3}$

45. $|3x + 4| = -9$

46. $|-2x + 1| = -3$

47. $|-5x + 10| = 0$

48. $|6y + 4| = 0$

49. $|-4x + 1| = 7$

50. $\left|-\frac{1}{2}x + \frac{2}{3}\right| = \frac{7}{6}$

The perimeter, P, of a rectangle is the sum of twice the length, l, and twice the width, w. ($P = 2l + 2w$)

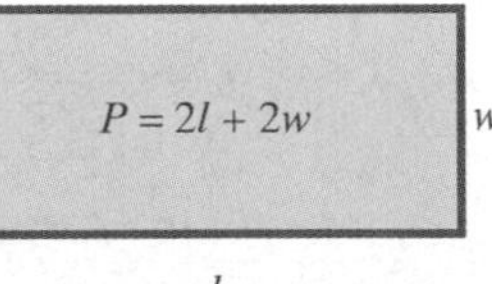

51. One hundred eighty-four feet of fencing is needed to enclose a rectangular-shaped garden plot. If the plot is 28 feet wide, what is the length?

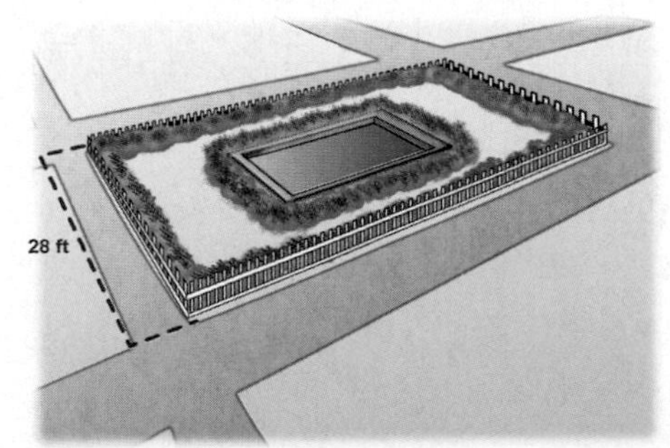

52. 36 ft.

53. 15 ft.

54. $l = 135$ yards

55. 10 cm

56. 18 in.

57. 12 ft., 38 ft.

52. Alan is installing a rectangular swimming pool in his yard. Because of the size of the yard, the pool can only be 18 ft. wide. Alan wants his pool to have a perimeter of 108 ft. What length should he make the pool?

53. Maria is laying down carpet in her home. She knows that her living room has a length of 25 ft. and a perimeter of 80 ft. How wide is her living room?

54. A rectangular shaped parking lot is to have a perimeter of 450 yards. If the width must be 90 yards because of a building code, what will the length need to be?

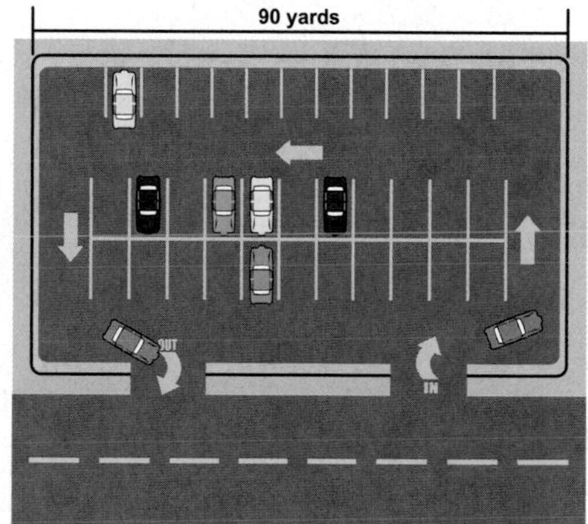

The area, A, of a trapezoid is $\frac{1}{2}$ of the height times the sum of the two parallel sides, b and c. (The parallel sides are sometimes called bases.) $\left[A = \frac{1}{2}h(b + c)\right]$

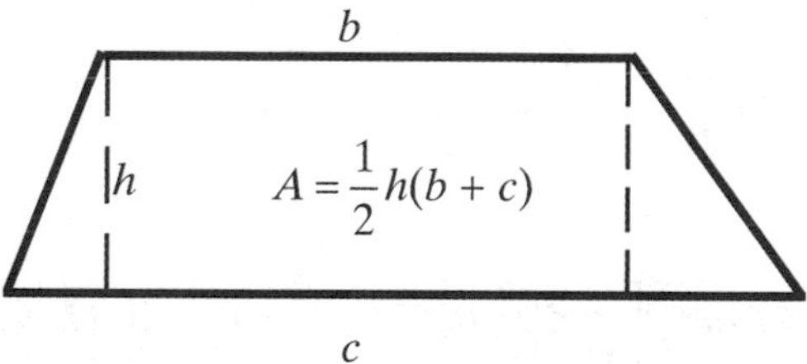

55. The area of a trapezoid is 108 cm^2. The height is 12 cm and the length of one of the parallel sides is 8 cm. Find the length of the second of the parallel sides.

56. The lengths of the bases of a trapezoid are 10 in. and 15 in. Find the height if the area is 225 square inches.

57. The height of a trapezoid is 10 ft. and its area is 250 ft^2. If one of the bases is 2 feet more than 3 times the other base, what are the lengths of the bases?

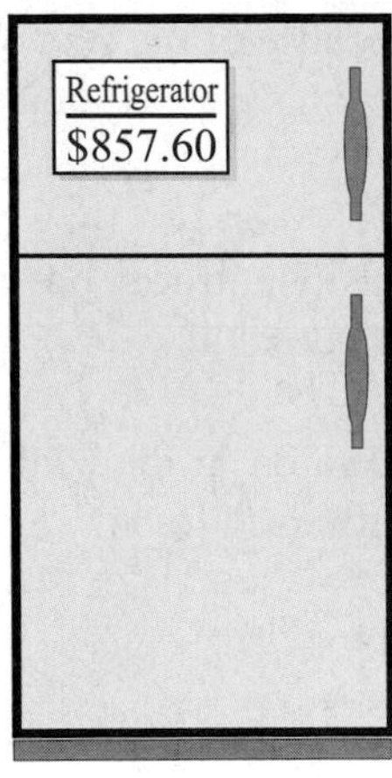

When purchasing an item on the installment plan, you find the total cost, C, by multiplying the monthly payment, p, by the number of months, t, and adding the product to the down payment, d. $(C = pt + d)$

58. A refrigerator costs \$857.60 if purchased on the installment plan. If the monthly payments are \$42.50 and the down payment is \$92.60, how long will it take to pay for the refrigerator?

59. A used automobile will cost \$3,250 if purchased on an installment plan. If the monthly payments are \$115 for 24 months, what will be the down payment?

Calculator Problems

Use a calculator to solve Exercises 60 – 63.

60. $0.17x - 23.0138 = 1.35x + 36.234$

61. $48.512 - 1.63x = 2.58x + 87.63553$

62. $0.32(x + 14.1) = 2.47x + 2.2188$

63. $1.6(9.3 + 2x) = 0.2(3x + 133.94)$

Writing and Thinking About Mathematics

64. Discuss, briefly, how you would apply Pòlya's four-step problem-solving process to a problem that you have faced today. (For example, what route to take to school, what time to spend studying algebra, what movie to see, etc.)

a. What was the problem?

b. What was the plan?

c. How did you carry out the plan?

d. Did your solution make sense? Could you have solved the problem in a different way?

58. 18 months

59. \$490

60. $x = -50.21$

61. $x = -9.293$

62. $x = 1.066604651$

63. $x = 4.58$

64. Answers will vary.

Hawkes Learning Systems: Introductory & Intermediate Algebra

Solving Linear Equations

2.5 Applications: Number Problems and Consecutive Integers

Objectives

After completing this section, you will be able to:

Solve word problems by writing and solving equations.

Pòlya's four-step process stated in Section 2.3 is outlined here for easy reference and to emphasize the importance of organization in problem solving.

Pòlya's Four-Step Process for Solving Problems

1. *Understand the problem. (Read the problem carefully and be sure that you understand all the terms used.)*
2. *Devise a plan. (Set up an equation, table, or chart relating the information.)*
3. *Carry out the plan. (Perform any operations indicated in Step 2.)*
4. *Look back over the results. (Ask yourself if the answer seems reasonable and if you could solve similar problems in the future.)*

In this section we will concentrate on two types of problems:

1. Number problems involving translating phrases, such as those we discussed in Section 2.2.
2. Problems related to the concept of consecutive integers.

Number Problems

In Section 2.2, we discussed translating English phrases into algebraic expressions. Now we will use those skills to read number problems and translate the sentences and phrases in the problem into a related equation. The solution of this equation will be the solution to the problem.

As you learn more and more abstract mathematical ideas, you will find that you will use these ideas and the related processes to solve a variety of everyday problems as well as problems in specialized fields of study. Generally, you may not even be aware of the fact that you are using your mathematical talents. However, these skills and ideas will be part of your thinking and problem solving techniques for the rest of your life.

Example 1: Number Problems

a. Three times the sum of a number and 5 is equal to twice the number plus 5. Find the number.

Solution: Let x = the unknown number.

$$\underbrace{\text{3 times the sum of a number and 5}}_{3(x+5)} \quad \underbrace{\text{is equal to}}_{=} \quad \underbrace{\text{twice the number plus 5}}_{2x+5}$$

$$\begin{aligned} 3x + 15 &= 2x + 5 \\ 3x + 15 - 2x &= 2x + 5 - 2x \\ x + 15 &= 5 \\ x + 15 - 15 &= 5 - 15 \\ x &= -10 \end{aligned}$$

The number is –10.

b. If a number is decreased by 36 and the result is 76 less than twice the number, what is the number?

Solution: Let n = the unknown number.

$$\underbrace{\text{a number decreased by 36}}_{n-36} \quad \underbrace{\text{the result is}}_{=} \quad \underbrace{\text{76 less than twice the number}}_{2n-76}$$

$$\begin{aligned} n - 36 - n &= 2n - 76 - n \\ -36 &= n - 76 \\ -36 + 76 &= n - 76 + 76 \\ 40 &= n \end{aligned}$$

The number is 40.

c. One integer is 4 more than three times a second integer. Their sum is 24. What are the two integers?

Solution: Let n = the second integer.
Then $3n + 4$ = the first integer.

$$\begin{aligned} (3n + 4) + n &= 24 \qquad \text{Their sum is 24.} \\ 4n + 4 &= 24 \\ 4n + 4 - 4 &= 24 - 4 \\ 4n &= 20 \\ \frac{4n}{4} &= \frac{20}{4} \\ n &= 5 \\ 3n + 4 &= 19 \end{aligned}$$

The two integers are 5 and 19.

d. Joe pays \$300 per month to rent an apartment. If this is $\frac{2}{5}$ of his monthly income, what is his monthly income?

Solution: Let x = Joe's monthly income.

$$\frac{2}{5}x = 300$$

$$\frac{5}{2} \cdot \frac{2}{5}x = \frac{5}{2} \cdot \frac{300}{1}$$

$$x = 750$$

Joe's monthly income is \$750.

Consecutive Integers

Remember that the set of **integers** consists of the whole numbers and their opposites:

$$\{\ \ldots, -4, -3, -2, -1, 0, 1, 2, 3, 4, \ldots\ \}$$

Even integers are integers that are divisible by 2. The even integers are

$$E = \{\ \ldots, -6, -4, -2, 0, 2, 4, 6, \ldots\ \}$$

Odd integers are integers that are not even. If an odd integer is divided by 2 the remainder will be 1. The odd integers are

$$O = \{\ \ldots, -5, -3, -1, 1, 3, 5, \ldots\ \}$$

In this discussion we will be dealing only with integers. Therefore, if you get a result that has a fraction or decimal number (not an integer), you will know that an error has been made and you should correct some part of your work.

The following terms and the ways of representing the integers must be understood before attempting the problems.

Consecutive Integers

Integers are **consecutive** *if each is 1 more than the previous integer. Three consecutive integers can be represented as*

$$n, \ n+1, \text{ and } \ n+2$$

Consecutive Odd Integers

Odd integers are **consecutive** *if each is 2 more than the previous odd integer. Three consecutive odd integers can be represented as*

$$\boldsymbol{n}, \; \boldsymbol{n+2}, \; \textit{and} \; \boldsymbol{n+4}$$

where n is an **odd** *integer.*

Consecutive Even Integers

Even integers are **consecutive** *if each is 2 more than the previous even integer. Three consecutive even integers can be represented as*

$$\boldsymbol{n}, \; \boldsymbol{n+2}, \; \textit{and} \; \boldsymbol{n+4}$$

where n is an **even** *integer.*

Note that consecutive even and consecutive odd integers are represented in the same way:

$$\boldsymbol{n}, \; \boldsymbol{n+2}, \; \text{and} \; \boldsymbol{n+4}$$

The value of the first integer, n, determines whether the remaining integers are odd or even. For example,

n is odd		**n is even**
If $n = 11$	**or**	If $n = 36$
then $n + 2 = 13$		then $n + 2 = 38$
and $n + 4 = 15$		and $n + 4 = 40$

Example 2: Consecutive Integers

a. Find three consecutive integers such that the sum of the first and third is 76 less than three times the second.

Solution: Let $n =$ the first integer
$n + 1 =$ the second integer
$n + 2 =$ the third integer

Set up and solve the related equation.

$$\begin{aligned} n + (n+2) &= 3(n+1) - 76 \\ 2n + 2 &= 3n + 3 - 76 \\ 2n + 2 &= 3n - 73 \\ 2n + 2 + 73 - 2n &= 3n - 73 + 73 - 2n \\ 75 &= n \\ 76 &= n + 1 \\ 77 &= n + 2 \end{aligned}$$

The three consecutive integers are 75, 76, and 77.

Check: $75 + 77 = 152$ and $3(76) - 76 = 228 - 76 = 152$

b. Three consecutive odd integers are such that their sum is –3. What are the integers?

Solution: Let n = the first odd integer
$n + 2$ = the second odd integer
$n + 4$ = the third odd integer

Set up and solve the related equation.

$$\begin{aligned} n + (n+2) + (n+4) &= -3 \\ 3n + 6 &= -3 \\ 3n &= -9 \\ n &= -3 \\ n + 2 &= -1 \\ n + 4 &= 1 \end{aligned}$$

The three consecutive odd integers are –3, –1, and 1.

Check: $(-3) + (-1) + (1) = -3$

c. Find three consecutive even integers such that three times the first is 10 more than the sum of the second and third.

Solution: Let n = the first even integer
$n + 2$ = the second even integer
$n + 4$ = the third even integer

Set up and solve the related equation.

$$\begin{aligned} 3n &= (n+2) + (n+4) + 10 \\ 3n &= 2n + 16 \\ 3n - 2n &= 2n + 16 - 2n \\ n &= 16 \\ n + 2 &= 18 \\ n + 4 &= 20 \end{aligned}$$

The three even integers are 16, 18, and 20.
(Checking shows that $3 \cdot 16$ is 10 more than $18 + 20$.)

2.5 Exercises

1. $x - 5 = 13 - x; 9$

2. $2x - 3 = x; 3$

3. $36 = 2x + 4; 16$

4. $15 - 2x = 27; -6$

5. $7x = 2x + 35; 7$

6. $2x - 3 = 6 - x; 3$

7. $3x + 14 = 6 - x; -2$

8. $\frac{x}{7} + 2 = -3; -35$

9. $\frac{2x}{5} = x + 6; -10$

10. $3(x + 4) = -9; -7$

11. $4(x - 5) = x + 4; 8$

12. $6x + 17 = 1 + 2x; -4$

13. $\frac{2x+5}{11} = 4 - x; 3$

14. $2x + 3x = 4(x + 3);$ 12

15. $x - 21 = 8x; -3$

16. $2(x - 10) = 6x + 14;$ $-\frac{17}{2}$

17. $x + x + 4 = 24;$ 10 cm

For Exercises 1 – 40, read each problem carefully, translate the various phrases into algebraic expressions, set up an equation, and solve the equation.

1. Five less than a number is equal to 13 decreased by the number. Find the number.

2. Three less than twice a number is equal to the number. What is the number?

3. Thirty-six is 4 more than twice a certain number. Find the number.

4. Fifteen decreased by twice a number is 27. Find the number.

5. Seven times a certain number is equal to the sum of twice the number and 35. What is the number?

6. The difference between twice a number and 3 is equal to 6 decreased by the number. Find the number.

7. Fourteen more than three times a number is equal to 6 decreased by the number. Find the number.

8. Two added to the quotient of a number and 7 is equal to –3. What is the number?

9. The quotient of twice a number and 5 is equal to the number increased by 6. What is the number?

10. Three times the sum of a number and 4 is equal to –9. Find the number.

11. Four times the difference between a number and 5 is equal to the number increased by 4. What is the number?

12. When 17 is added to six times a number, the result is equal to 1 plus twice the number. What is the number?

13. If the sum of twice a number and 5 is divided by 11, the result is equal to the difference between 4 and the number. Find the number.

14. Twice a number increased by three times the number is equal to 4 times the sum of the number and 3. Find the number.

15. If 21 is subtracted from a number and the result is 8 times the number, what is the number?

16. Twice the difference between a number and 10 is equal to 6 times the number plus 14. What is the number?

17. The perimeter (distance around) of a triangle is 24 centimeters. If two sides are equal and the length of the third side is 4 centimeters, what is the length of each of the other two sides?

18. $9 + 9 + x = 30$; 12 in.

18. The length of a wire is bent to form a triangle with two sides equal. If the wire is 30 inches long and the two equal sides are each 9 inches long, what is the length of the third side?

9" 9"

19. $x + x + 2x - 3 = 45$; 12 cm, 12 cm, 21 cm

20. $2(l - 16) + 2l = 172$; $l = 51$ m; $w = 35$ m

19. Two sides of a triangle have the same length and the third side is 3 cm less than the sum of the other two sides. If the perimeter of the triangle is 45 cm, what are the lengths of the three sides?

21. $3x + 1500 = 12{,}000$; \$3500

20. The perimeter of a rectangular shaped swimming pool is 172 meters. If the width is 16 meters less than the length, how long are the width and length of the pool?

22. $2x + 90{,}000 = 310{,}000$; \$110,000

21. A classic car is now selling for \$1500 more than three times its original price. If the selling price is now \$12,000, what was the car's original price?

23. $n + n + 2 = 60$; 29, 31

22. A real estate agent says that the current value of a home is \$90,000 more than twice its value when it was new. If the current value is \$310,000, what was the value of the home when it was new? The home is 25 years old.

24. $n + n + 1 + n + 2 = 69$; 22, 23, 24

23. The sum of two consecutive odd integers is 60. What are the integers?

25. $2n + 3(n + 1) = 83$; 16, 17

24. Find three consecutive integers whose sum is 69.

25. Find two consecutive integers such that twice the first plus three times the second equals 83.

26. $n + 2(n + 2) = 4(n + 4) - 54$; 42, 44, 46

26. Find three consecutive even integers such that the first plus twice the second is 54 less than four times the third.

27. $n + 2 + n + 4 - n = 66$; 60, 62, 64

27. Find three consecutive even integers such that if the first is subtracted from the sum of the second and third, the result is 66.

28. $4n = n + 2 + n + 4 + 44$; 25, 27, 29

28. Find three consecutive odd integers such that 4 times the first is 44 more than the sum of the second and third.

29. $n + n + 2 + n + 4 = n + 2 + 168$; 82, 84, 86

29. Find three consecutive even integers such that their sum is 168 more than the second.

30. $n + n + 1 + n + 2 + n + 3 = 90$; 21, 22, 23, 24

30. Find four consecutive integers whose sum is 90.

31. The perimeter of a rectangular parking lot is 410 yards. If the width is 75 yards, what is the length of the parking lot? The concrete in the parking lot is 4 inches thick.

31. $2l + 2(75) = 410$; 130 yards

32. Twenty-four plus a number is equal to twice the number plus three times the same number. What is the number?

32. $24 + x = 2x + 3x$; 6

33. $\frac{2l}{3} = 18$; 27 ft.

34. $x + x + 25{,}000 = 275{,}000$; lot: \$125,000, house: \$150,000

35. $x + x + 2 + x + 6 = 29$; 7 ft., 9 ft., 13 ft.

33. The width of a rectangular-shaped room is $\frac{2}{3}$ of the length. If the room is 18 ft. wide, find the length.

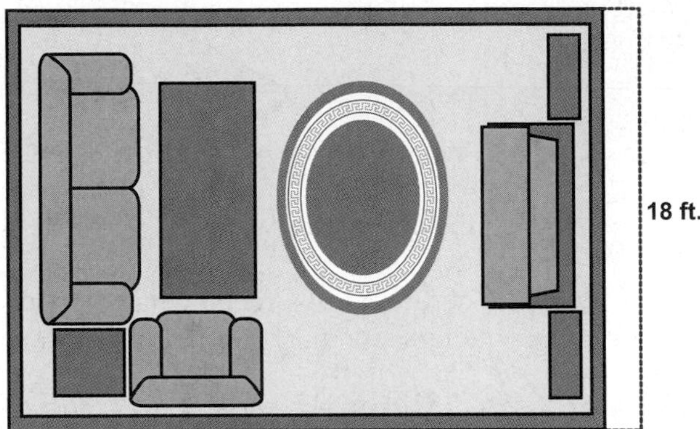

34. Joe Johnson decided to buy a lot and build a house on the lot. He knew that the cost of constructing the house was going to be \$25,000 more than the cost of the lot. He told a friend that the total cost was going to be \$275,000. As a test to see if his friend remembered the algebra they had together in school, he challenged his friend to calculate what he paid for the lot and what he was going to pay for the house. What was the cost of the lot and the cost of the house?

35. A 29 foot board is cut into three pieces at a sawmill. The second piece is 2 feet longer than the first and the third piece is 4 feet longer than the second. What is the length of each of the three pieces?

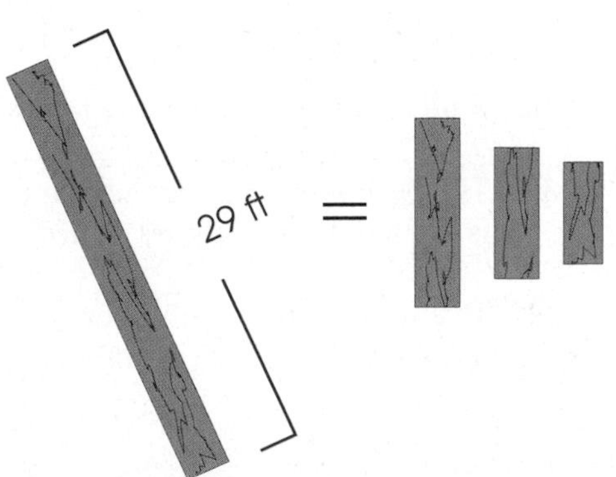

36. $2x = 50 - 10.50$; $19.75

37. $c + c + 49.50 = 125.74$; calculator: $38.12, textbook: $87.62

38. $x + 3x - 1 + 2x + 5 = 64$; 10 in., 29 in., 25 in.

39. $2n + 3(n + 2) = 2(n + 4) + 7$; 3, 5, 7

40. $3n + 2(n + 4) = 6(n + 2) - 20$; 16, 18, 20

Note: Answers for 41 – 45 may vary.

41. Find two consecutive integers whose sum is 33; 16, 17

42. Find two consecutive integers such that 3 times the the second is 53 more than the first; 25, 26

43. Find 3 consecutive even integers such that the sum of the first and the third is 3 times the second; –2, 0, 2

44. The difference between a number and $\frac{5}{6}$ times that number is equal to $\frac{5}{3}$; 10

45. The quotient of a number and 2 increased by $\frac{1}{3}$ is equal to 3 times the number divided by 4; $\frac{4}{3}$

36. Lucinda bought two boxes of golf balls. She gave the pro-shop clerk a 50-dollar bill and received $10.50 in change. What was the cost of one box of golf balls? (Tax was included.)

37. A mathematics student bought a graphing calculator and a textbook for a course in statistics. If the text cost $49.50 more than the calculator, and the total cost for both was $125.74, what was the cost of each item?

38. The three sides of a triangle are x, $3x - 1$, and $2x + 5$ (as shown in the figure). If the perimeter of the triangle is 64 inches, what is the length of each side?

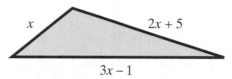

x $2x + 5$

$3x - 1$

39. Find three consecutive odd integers such that the sum of twice the first and three times the second is 7 more than twice the third.

40. Find three consecutive even integers such that the sum of three times the first and twice the third is twenty less than six times the second.

In Exercises 41 – 45, make up your own word problem that might use the given equation in its solution. Be creative! Then solve the equation and check to see that the answer is reasonable.

41. $n + (n + 1) = 33$

42. $3(n + 1) = n + 53$

43. $n + (n + 4) = 3(n + 2)$

44. $x - \frac{5}{6}x = \frac{5}{3}$

45. $\frac{x}{2} + \frac{1}{3} = \frac{3x}{4}$

Hawkes Learning Systems: Introductory & Intermediate Algebra

Applications: Number Problems and Consecutive Integers

2.6 Applications: Percent Problems (Discount, Taxes, Commission, Profit, and Others)

Objectives

After completing this section, you will be able to:

1. *Change decimals to percents.*
2. *Change percents to decimals.*
3. *Find percents of numbers.*
4. *Find fractional parts of numbers.*
5. *Solve word problems involving decimals, fractions, and percents.*

Our daily lives are filled with decimal numbers and percents: stock market reports, batting averages, won-lost records, salary raises, measures of pollution, taxes, interest on savings, home loans, discounts on clothes and cars, and on and on. Since decimal numbers and percents play such a prominent role in everyday life, we need to understand how to operate with them and apply them correctly in a variety of practical situations.

Review of Percent

The word **percent** comes from the Latin *per centum*, meaning "per hundred." So, **percent means hundredths**. Thus, 72%, $\frac{72}{100}$, and 0.72 all have the same meaning.

Similarly,

$$\frac{65}{100} = 0.65 = 65\% \quad \text{and} \quad \frac{124}{100} = 1.24 = 124\%$$

The following basic steps are used to change decimals and fractions to percents and vice versa. You may be familiar with these steps from your previous work in arithmetic, but some review should be helpful.

To Change a Decimal to a Percent

Step 1: *Move the decimal point two places to the right. (This is the same as multiplying by 100.)*

Step 2: *Add the % symbol.*

To Change a Fraction (or Mixed Number) to a Percent

Step 1: *Change the fraction to a decimal. (Divide the numerator by the denominator. If the number is a mixed number, first write it as an improper fraction then divide.)*

Step 2: *Change the decimal to a percent.*

Example 1: Decimals to Percents

Change each of the following numbers to percents.

a. 0.036 **b.** 0.25 **c.** $\frac{5}{8}$ **d.** $2\frac{1}{3}$

Solution: **a.** $0.036 = 3.6\%$ Move the decimal point two places to the right and add the % sign.

b. $0.25 = 25\%$ Move the decimal point two places to the right and add the % symbol.

c. $\frac{5}{8} = 0.625 = 62.5\%$ Change the fraction to decimal form by dividing 5 by 8, move the decimal point two places to the right, and add the % sign.

d. $2\frac{1}{3} = 2.33\frac{1}{3} = 233\frac{1}{3}\%$ The fraction $\frac{1}{3}$ is not a terminating decimal.

To Change a Percent to a Decimal

Step 1: *Move the decimal point two places to the left.*

Step 2: *Delete the % symbol.*

To Change a Percent to a Fraction (or Mixed Number)

Step 1: *Write the percent as a fraction with denominator 100 and delete the % symbol.*

Step 2: *Reduce the fraction (or change it to a mixed number if you prefer, with the fraction part reduced).*

Example 2: Percents to Decimals, Fractions, or Mixed Numbers

Change each of the following percents as indicated.

a. 47.5% to decimal form

b. 56% to fraction form reduced

c. 125% to mixed number form

Solution: **a.** $47.5\% = 0.475$

Move the decimal point two places to the left and delete the % sign.

b. $56\% = \frac{56}{100} = \frac{\cancel{4}\cdot 14}{\cancel{4}\cdot 25} = \frac{14}{25}$

Write the percent with denominator 100, delete the % symbol, and reduce.

c. $125\% = 1.25 = 1\frac{25}{100} = 1\frac{1}{4}$

Move the decimal point two places to the left, delete the % sign, change the decimal form to mixed number form and reduce.

The Basic Formula $R \cdot B = A$

Now, consider the statement

"15% of 80 is 12."

This statement has three numbers in it. In general, solving a percent problem involves knowing two of these numbers and trying to find the third. That is, **there are three basic types of percent problems**. We can break down the sentence in the following way.

Sentence →	15%	of	80	is	12
	↓	↓		↓	↓
Equation →	0.15	·	80	=	12
	↑	↑		↑	↑
	percent changed to decimal	"of" changed to times		"is" changed to =	
	↓	↓		↓	↓
Basic Formula →	Rate	·	Base	=	Amount

The terms that we have just discussed are explained in detail in the following box.

The Basic Formula $R \cdot B = A$

Teaching Notes:
Some students might find the following diagram helpful as a mnemonic device:

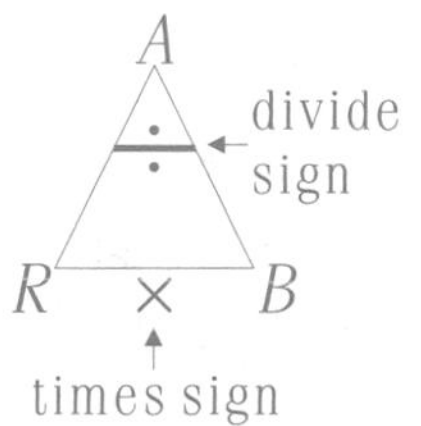

*R = **RATE** or percent (as a decimal or fraction)*

*B = **BASE** (number we are finding the percent of)*

*A = **AMOUNT** or percentage (a part of the base)*

"of" means to multiply.

"is" means equal (=).

The relationship among R, B, and A is given in the formula

$$\mathbf{R \cdot B = A \quad (or \quad A = R \cdot B)}$$

Even though there are just three basic types of percent problems, many people have difficulty deciding whether to multiply or divide in a particular problem. Using the formula $R \cdot B = A$ helps to avoid these difficulties. If the values of any two of the quantities in the formula are known, they can be substituted in the formula and then the missing number can be found by solving the equation. **The process of solving the equation for the unknown quantity determines whether multiplication or division is needed.**

The following examples illustrate how to substitute into the formula and how to solve the resulting equations.

Example 3: Percent of a Number

a. What is 72% of 800?

Solution: $R = 0.72$ and $B = 800$ and A is unknown.

$$R \cdot B = A$$
$$\downarrow \quad \downarrow \quad \downarrow$$
$$0.72 \cdot 800 = A \qquad \text{Here, simply multiply to find } A$$
$$576 = A$$

So, **576** is 72% of 800.

b. 57% of what number is 163.191?

Solution: $R = 0.57$ and B is unknown and A is 163.191.

$$R \cdot B = A$$
$$\downarrow \quad \downarrow \quad \downarrow$$
$$0.57 \cdot B = 163.191$$
$$\frac{0.57 \cdot B}{0.57} = \frac{163.191}{0.57} \qquad \text{Now, divide both sides by 0.57 to find } B.$$
$$B = 286.3$$

continued on next page ...

So, 57% of **286.3** is 163.191.

c. What percent of 180 is 45?

Solution: R is the unknown and B is 180 and A is 45.

$$R \cdot B = A$$
$$\downarrow \quad \downarrow \quad \downarrow$$
$$R \cdot 180 = 45$$
$$\frac{R \cdot 180}{180} = \frac{45}{180}$$ Now, divide both sides by 180 to find R.
$$R = 0.25$$
$$R = 25\%$$ Change the decimal to percent form.

So, **25%** of 180 is 45.

Applications with Percent

To sell goods that have been in stock for some time or simply to attract new customers, retailers and manufacturers sometimes offer a **discount**, a reduction in the selling price usually stated as a percent of the original price. The new, reduced price is called the **sale price**.

Sales tax is a tax charged on goods sold by retailers, and it is assessed by states and cities for income to operate various services. The **rate of sales tax** (a percent) varies by location.

Example 4: Applications

A bicycle was purchased at a discount of 25% of its original price of $1600. If 6% sales tax was added to the purchase price, what was the total paid for the bicycle?

Solution: There are two ways to approach this problem. One way is to find the discount and then subtract this amount from $1600. Another way is to subtract 25% from 100% to get 75% and then find 75% of $1600. In either case, the sales tax must be calculated on the sale price and added to the sale price.

Here, we will calculate the discount and then subtract from the original price.

$$R \cdot B = A$$
$$\downarrow \quad \downarrow \quad \downarrow$$
$$0.25 \cdot 1600 = 400$$ Discount

Original Price – Discount = $1600 – $400 = $1200   Sale Price

Sales Tax = 6% of Sale Price = $0.06 \cdot \$1200 = \$72$

Total Paid = Sale Price + Sales Tax = $1200 + $72 = $1272

The total paid for the bicycle, including sales tax, was $1272.

A **commission** is a fee paid to an agent or salesperson for a service. Commissions are usually a percent of a negotiated contract (as to a real estate agent) or a percent of sales.

Example 5: Commission

A saleswoman earns a salary of $1200 a month plus a commission of 8% on whatever she sells after she has sold $8000 in furniture. What did she earn the month she sold $25,000 worth of furniture?

Solution: First subtract $8000 from $25,000 to find the amount on which the commission is based.

$25,000 – $8000 = $17,000

Now, find the amount of the commission.

$$\begin{array}{ccccc} R & \cdot & B & = & A \\ \downarrow & & \downarrow & & \downarrow \\ 0.08 & \cdot & \$17{,}000 & = & \$1360 \text{ commission} \end{array}$$

Now, add the commission to her salary to find what she earned that month.

$1200 + $1360 = $2560 earned for the month

Teaching Notes:
Examples from real estate can be effective here. Some of your students may even be working in real estate and can relate their experiences in buying and selling real estate.

Percent of profit for an investment is the ratio (a fraction) that compares the money made to the money invested. If you make two investments of different amounts of money, then the amount of money you make on each investment is not a fair comparison. In comparing such investments, the investment with the greater percent of profit is considered the better investment. **To find the percent of profit, form the fraction (or ratio) of profit divided by the investment and change the fraction to a percent.**

Example 6: Percent of Profit

Calculate the percent of profit for both a and b and tell which is the better investment.
a. \$300 profit on an investment of \$2400
b. \$500 profit on an investment of \$5000

Solution: Set up ratios and find the corresponding percents.

a. $\frac{\$300 \text{ profit}}{\$2400 \text{ invested}} = \frac{300 \cdot 1}{300 \cdot 8} = \frac{1}{8} = 0.125 = 12.5\%$ percent of profit

b. $\frac{\$500 \text{ profit}}{\$5000 \text{ invested}} = \frac{500 \cdot 1}{500 \cdot 10} = \frac{1}{10} = 0.1 = 10\%$ percent of profit

Clearly, \$500 is more than \$300, but 12.5% is greater than 10%, so investment **a.** is the better investment.

Practice Problems

1. *Change 0.3 to a percent.*
2. *Change 6.4% to a decimal.*
3. *Change $\frac{1}{5}$ to a percent.*
4. *Find 12% of 200.*
5. *Find the percent of profit if \$400 is made on an investment of \$1500.*

2.6 Exercises

1. 91% **2.** 62.5%
3. 137% **4.** 0.75%
5. 37.5% **6.** 87%
7. 150% **8.** 66.67%
9. 0.69 **10.** 0.075
11. 0.113 **12.** 1.62
13. 0.005 **14.** 2.35
15. 0.82 **16.** 0.314
17. $\frac{7}{20}$ **18.** $\frac{18}{25}$
19. $\frac{13}{10}$ **20.** $\frac{2}{5}$

In Exercises 1 – 8, change each of the following numbers to percent form.

1. 0.91 **2.** 0.625 **3.** 1.37 **4.** 0.0075

5. $\frac{3}{8}$ **6.** $\frac{87}{100}$ **7.** $1\frac{1}{2}$ **8.** $\frac{2}{3}$

In Exercises 9 – 16, change each of the following percents to decimal form.

9. 69% **10.** 7.5% **11.** 11.3% **12.** 162%
13. 0.5% **14.** 235% **15.** 82% **16.** 31.4%

In Exercises 17 – 20, change each of the following percents to reduced fraction form.

17. 35% **18.** 72% **19.** 130% **20.** 40%

Answers to Practice Problems: **1.** 30% **2.** 0.064 **3.** 20% **4.** 24 **5.** 26.67%

21. a. 32% **b.** 28% **c.** 8%

22. a. 51.2% **b.** 21.4% **c.** 43.5%

23. a. 59.7% **b.** 16.7%

24. a. 15% **b.** 36%

21. Heather's monthly income is $2500. Using the given circle graph answer the following questions.

a. What percentage of her income does Heather spend on rent each month?

b. What percentage of her income is spent on food and entertaiment each month?

c. What percentage of her income does Heather save each month?

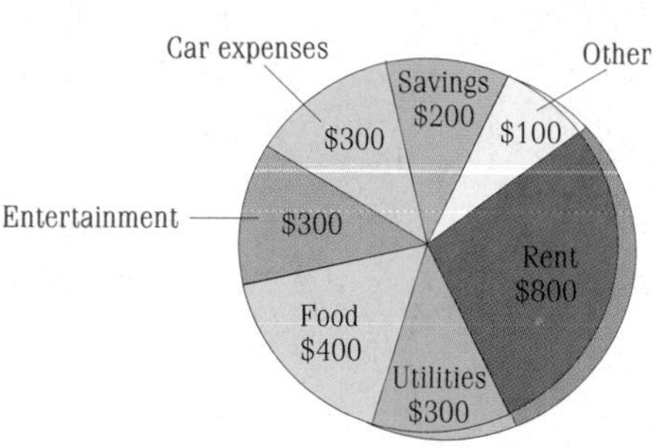

22. There were an estimated 20,832 earthquakes worldwide in 1999. Using the given graph answer the following questions. (Round answers to the nearest tenth of a percent.)

a. What percentage of earthquakes in 1999 were less than a magnitude of 4.0?

b. What percentage of earthquakes were of magnitude 3.0 – 3.9?

c. What percentage of earthquakes were 4.0 – 5.9?

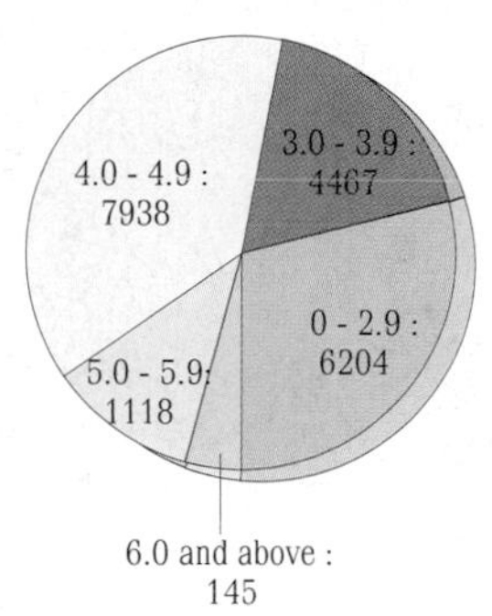

23. In 1998, the US adult population was estimated as 197.4 million. (Round each answer to the nearest tenth.)

a. What percentage of the US adult population was married in 1998?

b. What percentage of the adult population was either widowed or divorced?

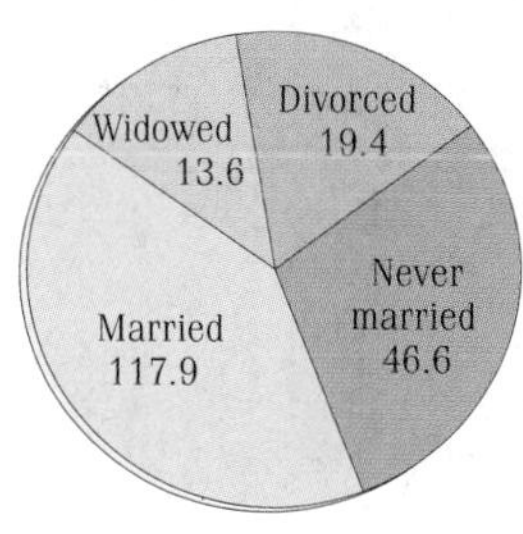

24. In July, Daddy Fat's Record Shop sold 3900 CD's.

a. R&B and Rap made up what percentage of their sales?

b. What percentage of their sales was not Rock or Country CD's?

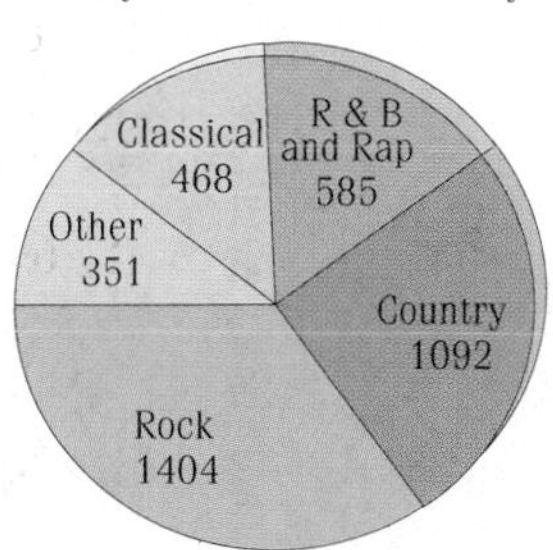

25. 61.56
26. 88.74
27. 40
28. 150
29. 2180
30. 225
31. 125%
32. 2500
33. 80
34. 58.56
35. a. 8% **b.** 10%
c. b
36. a. 20% **b.** 17%
c. a
37. a. \$990 **b.** 22%
38. \$85.33
39. \$1952.90
40. \$20,000
41. 1.5%
42. 129
43. 40

Find the missing rate, base, or amount in Exercises 25 – 34.

25. 81% of 76 is ______.
26. 102% of 87 is _______.
27. _______% of 150 is 60.
28. _______ % of 160 is 240.
29. 3% of _______ is 65.4
30. 20% of _______ is 45.
31. What percent of 32 is 40?
32. 1250 is 50% of what number?
33. 100 is 125% of what number?
34. Find 24% of 244.

35. What is the percent of profit if
a. \$400 is made on an investment of \$5000?
b. \$350 is made on an investment of \$3500?
c. Which was the better investment?

36. What is the percent of profit if
a. \$400 is made on an investment of \$2000?
b. \$510 is made on an investment of \$3000?
c. Which was the better investment?

37. Carlos bought 100 shares of CISCO stock at \$45 per share. One year later he sold the stock for \$5490.
a. What was the amount of his profit?
b. What was his percent of profit?

38. Patrick took his parents to dinner and the bill was \$70 plus 6% tax. If he left a tip of 15% of the total (food plus tax), what was the total cost of the dinner?

39. A calculator salesman's monthly income is \$1500 salary plus a commission of 7% of his sales over \$5000. What was his income the month that he sold \$11,470 in calculators?

40. A computer programmer was told that he would be given a bonus of 5% of any money his programs could save the company. How much would he have to save the company to earn a bonus of \$1000?

41. The property tax on a home was \$2550. What was the tax rate if the home was valued at \$170,000?

42. In one season a basketball player missed 14% of her free throws. How many free throws did she make if she attempted 150 free throws?

43. A student missed 6 problems on a statistics exam and received a grade of 85%. If all the problems were of equal value, how many problems were on the test?

44. The Golf Pro Shop had a set of golf clubs that were marked on sale for \$560. This was a discount of 20% off the original selling price.

a. What was the original selling price?

b. If the clubs cost the shop \$420, what was the percent of profit?

45. At a department store white sale, sheets and pillowcases were discounted 25%. Sheets were originally marked \$12.00 and pillowcases were originally marked \$5.00.

a. What was the sale price of the sheets?

b. What was the sale price of the pillowcases?

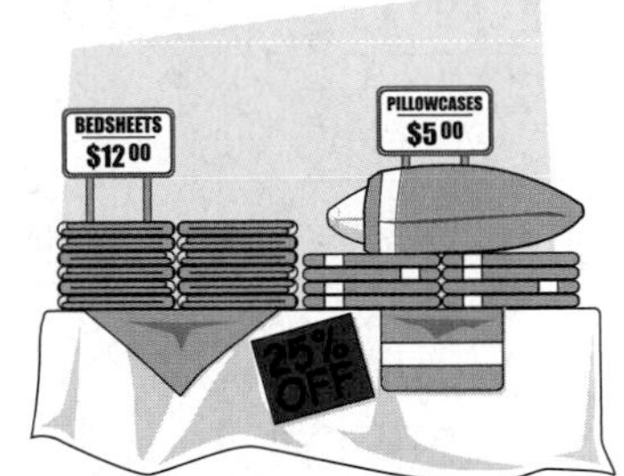

44. a. \$700 **b.** 33.33%

45. a. \$9 **b.** \$3.75

46. \$11.60, \$171.60

47. a. \$7134.29 **b.** \$6609.23

48. a. more **b.** \$30

49. a. 10% **b.** 11.11% **c.** The first percentage is a percentage of his original weight while the second percentage is a percentage of his weight after he lost weight.

50. a. \$3125 **b.** \$3375

46. If sales tax is figured at 7.25%, how much tax will be added to the total purchase price of three textbooks, priced at \$35.00, \$55.00, and \$70.00? What will be the total amount paid for the books?

47. The dealer's discount to the buyer of a new motorcycle was \$499.40. The manufacturer gave a rebate of \$1000 on this particular model.

a. What was the original price if the dealer discount was 7% of the original price?

b. What would a customer pay for the motorcycle if taxes were 5% of the final selling price and license fees were \$642.60?

48. Data cartridges for computer backup were on sale for \$27.00 (a package of two cartridges). This price was a discount of 10% off the original price.

a. Was the original price more than \$27 or less than \$27?

b. What was the original price?

49. A man weighed 200 pounds. He lost 20 pounds in 3 months. Then he gained back 20 pounds 2 months later.

a. What percent of his weight did he lose in the first 3 months?

b. What percent of his weight did he gain back?

c. The loss and the gain are the same amount, but the two percents are different. Explain.

50. A car dealer bought a used car for \$2500. He marked up the price so that he would make a profit of 25% of his cost.

a. What was the selling price?

b. If the customer paid 8% of the selling price in taxes and fees, what was the total cost of the car to the customer? The car was 6 years old.

51. a. Because the 6% comission is against the selling price, not the $141,000 the couple wanted.
b. The selling price is 100%. Therefore, $141,000 is 94% of the selling price (selling price(100%) – realtor fee (6%) = amount for couple(94%)). The selling price should be $150,000.

52. a. $8800 **b.** Sam **c.** Maria

Writing and Thinking About Mathematics

51. A man and his wife wanted to sell their house and contacted a realtor. The realtor stated that the price for the services (listing, advertising, selling, and paperwork) was 6% of the selling price. The realtor asked the couple how much cash they wanted after the realtor fee was deducted from the selling price. They said $141,000 and that therefore, the asking price should be 106% of $141,000 or a total of $149,460. The realtor declared that this was the wrong figure and that the percent used should be 94% and not 106%.

a. Explain why 106% is the wrong percent and $149,460 is not the right selling price.

b. Explain why and how 94% is the correct percent and just what the selling price should be for the couple to receive $141,000 after the realtor's fee is deducted.

Collaborative Learning Exercise

With the class separated into teams of two to four students, each team is to analyze the following problem and decide how to answer the related questions. Then each team leader is to present the team's answers and related ideas to the class for general discussion.

52. Sam works at a picture framing store and gets a salary of $500 per month plus a commission of 3% on sales he makes over $2000. Maria works at the same store, but she has decided to work on a straight commission of 8% of her sales.

a. At what amount of sales will Sam and Maria make the same amount of money?

b. Up to that point, who would be making more?

c. After that point, who would be making more? Explain briefly. (If you were offered a job at that store, which method of payment would you choose?)

Hawkes Learning Systems: Introductory & Intermediate Algebra

Percents and Applications

Chapter 2 Index of Key Ideas and Terms

Section 2.1 Simplifying and Evaluating Algebraic Expressions

Algebraic Terms page 92

Any constant or variable or the indicated product and/or quotient of constants and powers of variables is called a **term**.

Like Terms page 92

Like terms (or **similar terms**) are terms that are constants or terms that contain the same variables that are of the same power.

Combining Like Terms page 94

Evaluating Algebraic Expressions page 95

Section 2.2 Translating English Phrases and Algebraic Expressions

Translating English Phrases into Algebraic Expressions page 100

Translating Algebraic Expressions into English Phrases page 102

Section 2.3 Solving Linear Equations: $x + b = c$ and $ax = c$

Linear Equations page 106

If a, b, and c are constants and $a \neq 0$, then a **linear equation in *x*** (or **first-degree equation in *x***) is an equation that can be written in the form $ax + b = c$.

Addition Property of Equality page 107

If the same algebraic expression is added to both sides of an equation, the new equation has the same solutions as the original equation. Symbolically, if A, B, and C are algebraic expressions, then the equations

$$A = B$$
$$\text{and } A + C = B + C$$

have the same solutions.

continued on next page ...

Section 2.3 Solving Linear Equations: $x + b = c$ and $ax = c$ (continued)

Equivalent Equations page 107

Equations with the same solutions are said to be **equivalent**.

Multiplication Property of Equality page 109

If both sides of an equation are multiplied by (or divided by) the same nonzero constant, the new equation has the same solutions as the original equation. Symbolically, if A and B are algebraic expressions and C is any nonzero constant, then the equations

$$A = B$$

and $AC = BC$ where $C \neq 0$

and $\dfrac{A}{C} = \dfrac{B}{C}$ where $C \neq 0$

have the same solutions.

Procedure for Solving Applications pages 112 - 113

George Pòlya's four-step process as an approach to problem solving:

1. Understand the problem
2. Devise a plan.
3. Carry out the plan.
4. Look back over the results.

Section 2.4 Solving Linear Equations: $ax + b = c$

General Procedure for Solving Linear Equations page 119

1. Simplify each side of the equation by removing any grouping symbols and combining like terms. (This may also involve multiplying both sides by the LCM of the denominators or by a power of 10 to remove decimals. In this manner we will be dealing only with integer constants and coefficients.)
2. Use the Addition Property of Equality to add the opposites of constants and/or variables so that variables are on one side and constants are on the other side.
3. Use the Multiplication Property of Equality to multiply both sides by the reciprocal of the coefficient (or to divide by the coefficient).
4. Check your answer by substituting it into the original equation.

continued on next page ...

Section 2.4 Solving Linear Equations: $ax + b = c$ (continued)

Solutions of Absolute Value Equations page 123

1. If $c < 0$, then $|ax + b| = c$ has no solution.
2. If $c = 0$, then $|ax + b| = c$ has one solution. This solution can be found by solving the linear equation $ax + b = 0$.
3. If $c > 0$, then $|ax + b| = c$ has two solutions. These solutions can be found by solving the two linear equations $ax + b = c$ and $ax + b = -c$.

Section 2.5 Applications: Number Problems and Consecutive Integers

Number Problems pages 129 - 131

Consecutive Integers page 131

Consecutive integers are two integers that differ by 1.
Two consecutive integers can be represented as n and $n + 1$.

Consecutive Odd Integers page 132

Consecutive odd integers are two odd integers that differ by 2.
Two consecutive odd integers can be represented as n and $n + 2$ where n is odd.

Consecutive Even Integers page 132

Consecutive even integers are two even integers that differ by 2.
Two consecutive even integers can be represented as n and $n + 2$ where n is even.

Section 2.6 Applications: Percent Problems (Discount, Taxes, Commission, Profit, and Others)

Percent

To Change a Decimal to a Percent: page 138

Step 1: Move the decimal point two places to the right. (This is the same as multiplying by 100.)
Step 2: Add the % symbol.

To Change a Fraction (or Mixed Number) to a Percent: page 139

Step 1: Change the fraction to a decimal. (Divide the numerator by the denominator. If the number is a mixed number, first write it as an improper fraction then divide.)
Step 2: Change the decimal to a percent.

continued on next page ...

Section 2.6 Applications: Percent Problems (Discount, Taxes, Commission, Profit, and Others) (continued)

To Change a Percent to a Decimal: page 139
Step 1: Move the decimal point two places to the left.
Step 2: Delete the % symbol.

To Change a Percent to a Fraction (or Mixed Number): page 139
Step 1: Write the percent as a fraction with denominator 100 and delete the % symbol.
Step 2: Reduce the fraction (or change it to a mixed number if you prefer with the fraction part reduced).

The Formula for Solving Percent Problems page 141
$R \cdot B = A$ (or $A = R \cdot B$)
R = **RATE** or percent (as a decimal or fraction)
B = **BASE** (number we are finding the percent of)
A = **AMOUNT** or percentage (a part of the base)
"of " means to multiply.
"is" means equal (=).

Percent of Profit page 143
To find the **percent of profit**, form the fraction (or ratio) of profit divided by the investment and change the fraction to a percent.

Chapter 2 Review

For a review of the topics and problems from Chapter 2, look at the following lessons from *Hawkes Learning Systems: Introductory & Intermediate Algebra.*

Variables and Algebraic Expressions
Simplifying Expressions
Simplifying Expressions with Parentheses
Evaluating Algebraic Expressions
Translating Phrases into Algebraic Expressions
Solving Linear Equations Using Addition and Subtraction
Applications of Linear Equations: Addition and Subtraction
Solving Linear Equations Using Multiplication and Division
Applications of Linear Equations: Multiplication and Division
Solving Linear Equations
Applications: Number Problems and Consecutive Integers
Percents and Applications

Chapter 2 Test

1. $7-5x$
2. $4.4y$
3. $7a^2-5a+10$
4. **a.** $\frac{19y}{24}$ **b.** $\frac{19}{8}$
5. **a.** $2x-16$ **b.** -20
6. **a.** $3x+5$ **b.** 17
7. **a.** y^3+y^2-y-2 **b.** -17
8. $6x-3$
9. $2(x+5)$
10. $2(x+5)-4$
11. $3-2x$
12. $3(x+5)$
13. $\frac{x}{10}+x$
14. the quotient of a number and 6 decreased by twice the number
15. 9 less than twice a number decreased by 7
16. $-8.2y$
17. $\frac{2x}{3}$
18. $12x$
19. $36x^2$
20. $\frac{1}{36x^2}$
21. 1
22. $x=-3$
23. $x=5$
24. $x=0$
25. $x=-\frac{9}{8}$

Perform the indicated operations in Exercises 1 – 3.

1. $2x+3-7x+4$ **2.** $5.6y-y-2.3y+2.1y$ **3.** $2a^2+3a+5a^2-8a+10$

In Exercises 4 and 5, first (a) simplify each of the following expressions by combining like terms, and then (b) evaluate each simplified expression for $x=-2$ and $y=3$.

4. $\frac{2}{3}y+\frac{1}{4}y-\frac{1}{8}y$ **5.** $2(x-8)$

In Exercises 6 and 7, first (a) simplify each of the following expressions by combining like terms, and then (b) evaluate each simplified expression for $x=4$ and $y=-3$.

6. $4x+2-x+3$ **7.** $2y^2-3y+6-y^2+2y-8+y^3$

In Exercises 8 – 13, translate each English phrase into an equivalent algebraic expression.

8. 3 less than the product of a number and 6
9. Twice the sum of the number and 5
10. 4 less than twice the sum of a number and 5
11. Three decreased by twice a number
12. Three times the sum of a number and 5
13. The quotient of a number and 10 increased by the number

In Exercises 14 and 15, translate each algebraic expression into an English phrase.

14. $\frac{n}{6}-2n$ **15.** $2(n-7)-9$

In Exercises 16 and 17, perform the indicated operations.

16. $7.5y+2.3y-18y$ **17.** $\frac{3x}{7}+\frac{2x}{7}-\frac{x}{21}$

Simplify each of the following in Exercises 18 – 21.

18. $6x+6x$ **19.** $6x\cdot 6x$

20. $\frac{1}{6x}\cdot\frac{1}{6x}$ **21.** $\frac{1}{6x}\div\frac{1}{6x}$

Solve the equations in Exercises 22 – 29.

22. $\frac{5}{3}x+1=-4$ **23.** $4x-5-x=2x+5-x$

24. $8(3+x)=-4(2x-6)$ **25.** $\frac{3}{2}x+\frac{1}{2}x=-3+\frac{3}{4}$

26. $0.7x + 2 = 0.4x + 8$ **27.** $4(2x - 1) + 3 = 2(x - 4) - 5$

28. $|8x - 2| = 14$ **29.** $|3x + 1| = -17$

Find the missing number or percent in Exercises 30 and 31.

30. 62% of 180 is _____. **31.** 48 is _____% of 150.

32. What simple interest would be earned if \$5000 is invested at 9% for 9 months?

33. Which is a better investment, an investment of \$6000 that earns a profit of \$360 or an investment of \$10,000 that earns a profit of \$500? Explain your answer in terms of percent.

34. The triangle shown here indicates that the sides can be represented as $x, x + 1,$ and $2x - 1$. What is the length of each side if the perimeter is 12 meters?

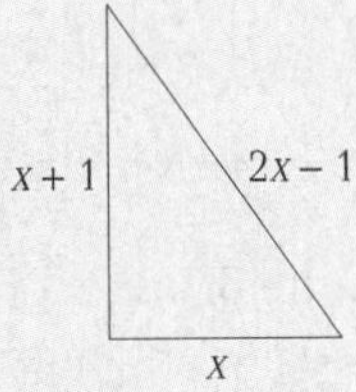

In Exercises 35 – 39, set up an equation for each word problem and solve.

35. One number is 5 more than twice another. Their sum is –22. Find the numbers.

36. Find two consecutive integers such that twice the first added to three times the second is equal to 83.

37. A man's suit was on sale for \$547.50.
 a. If this price was a discount of 25% from the original price, what was the original price?
 b. What would the suit cost a buyer if sales tax was 6% and the alterations were \$25?

38. Find three consecutive odd integers such that three times the second is equal to 27 more than the sum of the first and the third.

39. Cheryl decided to buy her algebra book, a solutions manual, and the required calculator. The book cost \$10 more than the calculator, and the solutions manual cost \$45.50 less than the calculator. If the total cost of all three items was \$156.50, what was the cost of each item?

Answers:

26. $x = 20$

27. $x = -2$

28. $x = -\frac{3}{2}, 2$

29. No solution

30. 111.6

31. 32%

32. \$337.50

33. The \$6000 investment is the better investment since it has a higher percent of profit, 6%, than the \$10,000 investment (5%).

34. 3 m, 4 m, 5 m

35. $(2y + 5) + y = -22$; $-9, -13$

36. $2n + 3(n + 1) = 83$; 16, 17

37. a. $(.75)x = 547.50$; \$730
b. $(0.06)547.50 + 547.50 + 25 = x$; \$605.35

38. $3(n + 2) = n + (n + 4) + 27$; $n = 25, 27, 29$

39. $(10 + c) + (c - 45.50) + c = 156.50$; book = \$74.00, manual = \$18.50, calculator = \$64.00

Cumulative Review: Chapters 1 – 2

1. 20
2. –14
3. 9
4. –15
5. –126
6. 459
7. 39
8. –124
9. True
10. True
11. False; $\frac{3}{4} \le |-1|$
12. [number line: integers –6 to 6 marked]
13. [number line: 0, 1, 2, 3, 4, 5, 6, ...]
14. [number line: –2, –1, 0, 1, 2]
15. Refer to discussion on "Division by 0 is Undefined" on page 28
16. 2, 3, 5, 7, 11, 13, 17, 19, 23, 29, 31, 37, 41, 43, 47
17. a. $2 \cdot 3 \cdot 5^2$ **b.** 5^3 **c.** $2^4 \cdot 3 \cdot 5$
18. a. 108 **b.** $72xy^2$
19. Associative property of multiplication
20. Commutative property of multiplication
21. Distributive property
22. Commutative property of addition
23. Associative property of addition
24. Multiplicative identity
25. $\frac{43}{40}$

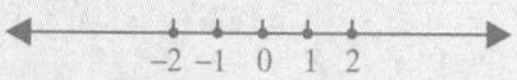

Perform the indicated operations in Exercises 1 – 8.

1. $18 + 7 - 8 + 3$ **2.** $9 - 16 - 11 + 4$ **3.** $26 - 17$ **4.** $18 - 33$
5. $(14)(-9)$ **6.** $(-27)(-17)$ **7.** $273 \div 7$ **8.** $744 \div (-6)$

Determine whether each expression in Exercises 9 – 11 is true or false. If a statement is false, rewrite it in a form that is a true statement. (There may be more than one way to correct a statement.)

9. $|-5| \le 5$ **10.** $2^3 - 8 > 8 - 3 \cdot 4$ **11.** $\frac{3}{4} \ge |-1|$

In Exercises 12 – 14, on a real number line, graph the integers that satisfy the stated inequality.

12. $|x| \le 6$ **13.** $|x| = x$ **14.** $|x| < 2.3$

15. Explain, in your own words, why division by 0 is undefined.

16. List the prime numbers less than 50.

17. Find the prime factorization of each integer:
a. 150 **b.** 125 **c.** 240

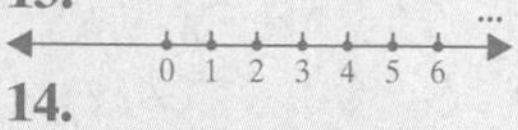

18. Find the LCM for each set of numbers or expressions:
a. 18, 27, 36 **b.** $12x, 9xy, 24xy^2$

Name the property of real numbers illustrated in Exercises 19 – 24.

19. $(3x)y = 3(xy)$ **20.** $x \cdot 5 = 5 \cdot x$ **21.** $4(3x + 1) = 12x + 4$
22. $x + 17 = 17 + x$ **23.** $(y + 11) + 4 = y + (11 + 4)$ **24.** $6 \cdot 1 = 6$

Perform the indicated operations in Exercises 25 – 32. Reduce each answer to lowest terms.

25. $\frac{3}{8} + \frac{7}{10}$ **26.** $\frac{5}{12} - \frac{4}{9}$ **27.** $\frac{4}{5y} + \frac{2}{3y}$ **28.** $\frac{7}{4a} - \frac{9}{6a}$

29. $\frac{3}{x} + \frac{1}{6}$ **30.** $\frac{11}{9} \div \frac{22}{15}$ **31.** $\frac{7b}{15} \div \frac{2b}{6}$ **32.** $\frac{8x}{21} \cdot \frac{7}{12x}$

26. $-\frac{1}{36}$ **27.** $\frac{22}{15y}$ **28.** $\frac{1}{4a}$ **29.** $\frac{18+x}{6x}$ **30.** $\frac{5}{6}$ **31.** $\frac{7}{5}$ **32.** $\frac{2}{9}$

33. 17.27
34. −15.31
35. 14.6
36. 58.19
37. 5.292
38. 19
39. 18
40. 90
41. 12
42. 164
43. 13
44. 37
45. $\frac{49}{40}$ **46.** $-\frac{31}{5}$
47. $\frac{1}{6}$ **48.** $\frac{-7}{45}$
49. $\frac{263}{576}$ **50.** $\frac{3}{5}$
51. 0 **52.** $\frac{-7}{16}$
53. $-2y-12$; -18
54. $4x+7$; -1
55. x^2+13x; -22
56. $x-7$; -9
57. $9-2x$
58. $3(x+10)$
59. $24x+5$
60. $6T+E+3F$

Perform the indicated operations in Exercises 33 – 37.

33. $15.8 + 9.1 - 7.63$ **34.** $19.31 - 34.62$ **35.** $24.3 + 6.81 - 16.51$

36. $(25.3)(2.3)$ **37.** $(14.7)(0.36)$

Simplify the expressions in Exercises 38 – 52 by using the rules for order of operations. Reduce all answers to lowest terms.

38. $3 \cdot 2^3 - 5$ **39.** $(18 + 2 \cdot 3) \div 4 \cdot 3$

40. $(13 \cdot 5 - 5) \div 2 \cdot 3$ **41.** $4[6 - (2 \cdot 5 - 2 \cdot 9) - 11]$

42. $5^2 \cdot 2^3 - 24 \div 2 \cdot 3$ **43.** $-36 \div (-2)^2 + 20 - 2(16 - 17)$

44. $(7 - 10)[49 \div (-7) + 20 \cdot 3 - 5 \cdot 15 - (-10)] + (-1)^0$

45. $\left(\frac{2}{3}\right)^2 \div \frac{5}{18} - \frac{3}{8}$ **46.** $\left(\frac{9}{10} + \frac{2}{15}\right) \div \left(\frac{1}{2} - \frac{2}{3}\right)$

47. $\frac{1}{2} + \frac{3}{8} \div \frac{3}{4} - \frac{5}{6}$ **48.** $\left(\frac{3}{4} - \frac{5}{6}\right) \cdot \left(\frac{6}{5} + \frac{2}{3}\right)$

49. $\frac{7}{18} + \frac{5}{24} - \left(\frac{3}{8}\right)^2$ **50.** $\left(\frac{3}{5}\right)^2 \div \frac{8}{5} \cdot \frac{8}{3}$

51. $\frac{1}{3} + \frac{2}{5} \cdot \frac{5}{8} \div \frac{3}{2} - \frac{1}{2}$ **52.** $\left(\frac{1}{3} - \frac{3}{4}\right) \div \left(\frac{2}{3} + \frac{2}{7}\right)$

Exercises 53 – 56: (a) Simplify each expression by combining like terms. (b) Evaluate the simplified form of each expression for x = −2 and y = 3.

53. $-4(y + 3) + 2y$ **54.** $3(x + 4) - 5 + x$ **55.** $2(x^2 + 4x) - (x^2 - 5x)$

56. $\frac{3(5x - x)}{4} - 2x - 7$

Exercises 57 – 60, write an algebraic expression described by each of the phrases.

57. The difference between 9 and twice a number

58. Three times the sum of a number and 10

59. The number of hours in x days and 5 hours

60. The number of points scored by a football team on T touchdowns (6 points each), E extra points (1 point each), and F field goals (3 points each).

61. $x = -1$

62. $x = 3$

63. $x = -11$

64. $x = -4$

65. $x = \frac{31}{8}$

66. $y = \frac{24}{5}$

67. $y = \frac{15}{2}$

68. $y = 0$

69. $-\frac{5}{51}$

70. -5

71. $\frac{10}{3}$ cups (or $3\frac{1}{3}$ cups)

72. 8 lollipops

73. **a.** 1000 **b.** 2500

74. \$80

75. \$506.94

76. $x = 11$

77. 18, 20, 22

Solve each of the equations in Exercises 61 – 68.

61. $9x - 8 = 4x - 13$ **62.** $6 - (x - 2) = 5$ **63.** $10 + 4x = 5x - 3(x + 4)$

64. $4.4 + 0.6x = 1.2 - 0.2x$ **65.** $-2(5x + 12) - 5 = -6(3x - 2) - 10$

66. $\frac{3}{4}y - \frac{1}{2} = \frac{5}{8}y + \frac{1}{10}$ **67.** $\frac{1}{2} + \frac{1}{5}y = \frac{2}{15}y + 1$ **68.** $-\frac{3}{4}y + \frac{3}{8}y = \frac{1}{6}y$

In Exercises 69 – 84, set up an equation or inequality for each problem and solve for the unknown quantity.

69. If the difference between $\frac{1}{2}$ and $\frac{11}{18}$ is divided by the sum of $\frac{2}{3}$ and $\frac{7}{15}$, what is the quotient?

70. From the sum of –2 and –11, subtract the product of –4 and 2.

71. Lucia is making punch for a party. She is making 5 gallons. The recipe calls for $\frac{2}{3}$ cup of sugar per gallon of punch. How many cups of sugar does she need?

72. Jonathan has 64 lollipops. He gives $\frac{3}{8}$ of them to Jennifer. She then gives $\frac{2}{3}$ of her lollipops to Allison. How many lollipops does Jennifer have left?

73. There are 4000 registered voters in Greenville, and $\frac{3}{8}$ of these voters are registered Republicans. A survey indicates that $\frac{2}{3}$ of the registered Republicans are in favor of Bond Measure *XX* and $\frac{3}{5}$ of the other registered voters are in favor of this measure.

a. How many of the registered Republicans are in favor of Measure *XX*?
b. How many of the registered voters are in favor of Measure *XX*?

74. Your neighbor has a used (red) pickup for sale and it is perfect for you. However, you have no money. With your new job you would be able to buy it in 6 months, but you know that the truck will certainly be sold by then. Your uncle has agreed to loan you \$2000 for 6 months at 8% interest and you have agreed to the terms. How much will you pay your uncle in interest for the 6 months loan?

75. Computers are on sale at a 30% discount. If you know that you will pay 6.5% in sales taxes, what will you pay for a computer that is priced at \$680?

76. If twice a certain number is increased by 3, the result is 8 less than three times the number. Find the number.

77. Find three consecutive even integers such that the sum of the first and twice the second is equal to 14 more than twice the third.

78. $-2, -1, 0, 1$

79. < 33.33 miles

80. The value of furniture sold was $12,500 each, Jay's and Kay's salary was $1250.

81. $10,000

82. These are equally good investments since the percent of profit, 8%, is the same for each investment.

83. $50,000

84. < 100 balloon animals

78. Find four consecutive integers such that the sum of the first, second, and fourth integers is 2 less than the third.

79. ALPHA Truck Rental charges $35 a day plus $0.55 per mile and BETA Truck Rental charges $45 a day but only $0.25 per mile. For how many miles can you drive an ALPHA truck in one day and keep the cost below the cost of a BETA truck?

80. Jay works at a furniture store for a salary of $800 per month plus a commission of 6% on his sales over $5000. Kay works at the same store but works only on a straight commission of 10%. In one particular month, Jay and Kay sold the same value in furniture and made the same amount of money. What was the dollar value of furniture sold by each? What amount of money did each make for that month?

81. What principal would you need to invest to earn $450 in interest in 6 months if the rate of interest was 9%?

82. Which is the better investment: (a) a profit of $800 on an investment of $10,000 or (b) a profit of $1200 on an investment of $15,000? Explain in terms of percents.

83. Agatha wants to make a profit of $5000 from two accounts over the next year. If she puts $10,000 into an account that earns 10% interest, how much money does she need to invest in an account that makes 8% interest in order to fulfill her goal?

84. Kidz Clownz charges $55 to hire a birthday clown for a day plus $0.12 per balloon animal. Clown Craze charges only $30 a day, but $0.37 per balloon animal. If you decide to hire from Clown Craze, how many balloon animals can you have created and still keep the cost below that of hiring from Kidz Clownz?

Formulas, Applications, and Linear Inequalities

CHAPTER 3

Did You Know?

In Chapter 3, you will find a great many symbols defined, as well as rules of manipulation for these symbolic expressions. Most people think that algebra has always existed, complete with all the common symbols in use today. That is not the case, since modern symbols did not appear consistently until the beginning of the sixteenth century. Prior to that, algebra was rhetorical. That is, all problems were written out in words using either Latin, Arabic, or Greek, and some nonstandard abbreviations. Numbers were written out. The common use of Hindu-Arabic numerals did not begin until the sixteenth century, although these numerals had been introduced into Europe in the twelfth century.

The sign for addition, +, was a contraction of the Latin *et*, which means "and." Gradually, the *e* was contracted and the crossed *t* became the plus sign. The minus sign or bar, –, is thought to be derived from the habit of early scribes of using a bar to represent the letter *m*. Thus the word *summa* was often written *sum̄a*. The bar came to represent the missing *m*, the first letter of the word *minus*. The radical symbol, $\sqrt{\ }$, is derived from a small printed *r*, which stood for the Latin word *radix*, or root. The symbol for times, a cross, ×, was developed from cross multiplication or for the purpose of indicating products in proportions. Thus

$$\begin{matrix}2 & & 6\\ & \times & \\ 3 & & 9\end{matrix} \text{ stood for } \frac{2}{3}=\frac{6}{9}.$$

The cross is not well suited for algebra, since it resembles the symbol *x*, which is used for variables. Therefore, a dot is usually used to indicate multiplication in algebra. The dot seemed to have developed from an Italian practice of separating columns in multiplication tables with a dot. Exponents were used as early as the fourteenth century by the mathematician Oresme (1320? – 1382), who gave the first known use of the rules for fractional exponents in a text book he wrote. The equal sign is attributed to Robert Recorde (1510? – 1558), who wrote, "I will sette as I doe often in woorke use, a paire of paralleles, or Gemowe [twin] lines of one lengthe, thus: = , because noe .2. thynges, can be moare equalle." As you can tell, the development of algebraic symbols occurred over a long period of time, and symbols became standardized through usage and convenience. If you are interested in the history of numerical symbolism, you will find more information in D.E. Smith's *History of Mathematics*, Volume II.

Recorde

"The Mathematician, carried along on his flood of symbols, dealing apparently with purely formal truths, may still reach results of endless importance for our description of the physical universe."

Karl Pearson (1857 – 1936)

3.1 Working with Formulas

Objectives

After completing this section, you will be able to:

1. *Evaluate formulas for given values of the variables.*
2. *Solve formulas for specified variables in terms of the other variables.*
3. *Use formulas to solve a variety of applications.*

Formulas are general rules or principles stated mathematically. There are many formulas in such fields of study as business, economics, medicine, physics, and chemistry as well as mathematics. Some of these formulas and their meanings are shown in the table below and on the next page.

SPECIAL COMMENT: Be sure to use the letters just as they are given in the formulas. In mathematics, there is little or no flexibility between capital and small letters as they are used in formulas. In general capital letters have special meanings that are different from corresponding small letters. For example, capital ***A*** may mean the area of a triangle and small ***a*** may mean the length of one side, two completely different ideas.

Teaching Notes:
The formulas shown in this section may represent for students, in familiar contexts, the notation of "modeling". Modeling with mathematics is a central theme and formulas are a primary example of modeling.

Formula	Meaning
1. $I = Prt$	The simple interest (I) earned by investing money is equal to the product of the principal (P) times the rate of interest (r) times the time (t) in one year or less. (**Note:** If more than one year is involved then the interest is compounded and another formula is used.)
2. $C = \frac{5}{9}(F-32)$	Temperature in degrees Celsius (C) equals $\frac{5}{9}$ times the difference between the Fahrenheit temperature (F) and 32.
3. $d = rt$	The distance traveled (d) equals the product of the rate of speed (r) and the time (t).
4. $P = 2l + 2w$	The perimeter (P) of a rectangle is equal to twice the length (l) plus twice the width (w).
5. $L = 2\pi rh$	The lateral surface area, L, (top and bottom not included) of a cylinder is equal to 2π times the radius (r) of the base times the height (h).

continued on next page ...

Formula	Meaning
6. $F = ma$	In physics, the force, F, acting on an object is equal to its mass (m) times its acceleration (a).
7. $\alpha + \beta + \gamma = 180$	The sum of the angles (α, β, and γ) of a triangle is 180°. (**Note:** α, β, and γ are the Greek lowercase letters alpha, beta, and gamma, respectively.)

Evaluating Formulas

If you know values for all but one variable in a formula, you can substitute those values and find the value of the unknown variable by using the techniques for solving equations discussed in this chapter. This section presents a variety of formulas from real-life situations, with complete descriptions of the meanings of these formulas. Working with these formulas will help you become familiar with a wide range of applications of algebra and provide practice in solving equations.

Example 1: Evaluating Formulas

a. A **note** is a loan for a period of 1 year or less, and the interest earned (or paid) is called **simple interest**. A note involves only one payment at the end of the term of the note and includes both principal and interest. The formula for calculating simple interest is:

$$I = Prt$$

where I = **Interest** (earned or paid)
P = **Principal** (the amount invested or borrowed)
r = **rate** of interest (stated as an annual or yearly rate)
t = **time** (one year or part of a year)

Note: The rate of interest is usually given in percent form and converted to decimal or fraction form for calculations. For the purpose of calculations, we will use 360 days in one year and 30 days in a month. Before the use of computers, this was common practice in business and banking.

Maribel loaned \$5000 to a friend for 6 months at an interest rate of 8%. How much will her friend pay her at the end of the 6 months?

Solution: Here, $P = \$5000$,
$r = 8\% = 0.08$,
$t = 6 \text{ months} = \frac{6}{12} \text{ year} = \frac{1}{2} \text{ year}$

Find the interest by substituting in the formula $I = Prt$ and evaluating.

continued on next page ...

$$I = 5000 \cdot \overset{0.04}{\cancel{0.08}} \cdot \frac{1}{\cancel{2}}$$
$$= 5000 \cdot 0.04$$
$$= \$200.00$$

The interest is \$200 and the amount to be paid at the end of 6 months is
Principal + Interest = \$5000 + \$200 = \$5200

b. Given the formula $C = \frac{5}{9}(F - 32)$, first find C if $F = 212°$ and then find F if $C = 20°$.

Solution: $F = 212°$, so substitute 212 for F in the formula.

$$C = \frac{5}{9}(212 - 32)$$
$$= \frac{5}{9}(180)$$
$$= 100$$

That is, 212° F is the same as 100° C. Water will boil at 212° F at sea level. This means that if the temperature is measured in degrees Celsius instead of degrees Fahrenheit, water will boil at 100° C at sea level.

$C = 20°$, so substitute 20 for C in the formula.

$$20 = \frac{5}{9}(F - 32) \qquad \text{Now solve for } F.$$
$$\frac{9}{5} \cdot 20 = \frac{9}{5} \cdot \frac{5}{9}(F - 32) \qquad \text{Multiply both sides by } \frac{9}{5}.$$
$$36 = F - 32 \qquad \text{Simplify.}$$
$$68 = F \qquad \text{Add 32 to both sides.}$$

That is, a temperature of 20° C is the same as a comfortable spring day temperature of 68° F.

c. The lifting force, F, exerted on an airplane wing is found by multiplying some constant, k, by the area, A, of the wing's surface and by the square of the plane's velocity, v. The formula is $F = kAv^2$. Find the force on a plane's wing of area 120 ft.^{2} if k is $\frac{4}{3}$ and the plane is traveling 80 miles per hour as it takes off.

Solution: We know that $k = \frac{4}{3}$, $A = 120$, and $v = 80$. Substitution gives

$$F = \frac{4}{3} \cdot 120 \cdot 80^2$$
$$= \frac{4}{3} \cdot 120 \cdot 6400$$
$$= 160 \cdot 6400$$
$$F = 1{,}024{,}000 \text{ lbs.} \quad \text{(The force is measured in pounds.)}$$

d. The perimeter of a triangle is 38 feet. One side is 5 feet long and a second side is 18 feet long. How long is the third side?

Solution 1: Using the formula $P = a + b + c$, substitute $P = 38$, $a = 5$, and $b = 18$. Then solve for the third side.

$$38 = 5 + 18 + c$$
$$38 = 23 + c$$
$$15 = c$$

The third side is 15 feet long.

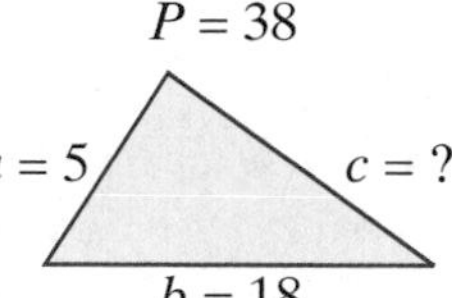

Solution 2: First solve for c in terms of P, a, and b. Then substitute for P, a, and b.

$$P = a + b + c$$
$$P - a - b = c$$ Treat a and b as constants. Add $-a - b$ to both sides.

or $$c = P - a - b$$

Substituting gives $c = 38 - 5 - 18$
$= 33 - 18 = 15$

Solving Formulas for Different Variables

We say that the formula $d = rt$ is "solved for" d in terms of r and t. Similarly, the formula $A = \frac{1}{2}bh$ is solved for A in terms of b and h, and the formula $P = S - C$ (profit is equal to selling price minus cost) is solved for P in terms of S and C. Many times we want to use a certain formula in another form. We want the formula "solved for" some variable other than the one given in terms of the remaining variables. **Treat the variables just as you would constants in solving linear equations**. Study the following examples carefully.

Example 2: Solving for Different Variables

a. Given $d = rt$, solve for t in terms of d and r. We want to represent the time in terms of distance and rate. We will use this concept later in word problems.

Solution: $d = rt$ Treat r and d as if they were constants.

$\frac{d}{r} = \frac{rt}{r}$ Divide both sides by r.

$\frac{d}{r} = t$ Simplify.

b. Given $P = a + b + c$, solve for a in terms of P, b, and c. This would be a convenient form for the case in which we know the perimeter and two sides of a triangle and want to find the third side.

continued on next page ...

Solution: $P = a + b + c$ Treat P, b, and c as if they were constants.

$P - b - c = a + b + c - b - c$ Add $-b-c$ to both sides.

$P - b - c = a$ Simplify.

c. Given $C = \frac{5}{9}(F - 32)$ as in Example 1b, solve for F in terms of C. This would give a formula for finding Fahrenheit temperature given a Celsius temperature value.

Solution: $C = \frac{5}{9}(F - 32)$ Treat C as a constant.

$\frac{9}{5} \cdot C = \frac{9}{5} \cdot \frac{5}{9}(F - 32)$ Multiply both sides by $\frac{9}{5}$.

$\frac{9}{5}C = F - 32$ Simplify.

$\frac{9}{5}C + 32 = F$ Add 32 to both sides.

Thus,

$$F = \frac{9}{5}C + 32 \text{ is solved for } F \text{ and } C = \frac{5}{9}(F - 32) \text{ is solved for } C.$$

These are two forms of the same formula.

Teaching Notes: You might ask the students if anyone observed that they could divide by 2 first. Explain that looking for a common factor or divisor can make the problem easier to solve later.

d. Given the equation $2x + 4y = 10$, (**i.**) solve first for x in terms of y, and then (**ii.**) solve for y in terms of x. This equation is typical of the algebraic equations that we will discuss in Chapter 6.

Solution: i. Solving for x yields

$2x + 4y = 10$ Treat $4y$ as a constant.

$2x + 4y - 4y = 10 - 4y$ Subtract $4y$ from both sides. (This is the same as adding $-4y$.)

$2x = 10 - 4y$

$\frac{2x}{2} = \frac{10 - 4y}{2}$ Divide both sides by 2.

$x = \frac{10}{2} - \frac{4y}{2}$ Simplify.

$x = 5 - 2y$

ii. Solving for y yields

$2x + 4y = 10$ Treat $2x$ as a constant.

$2x + 4y - 2x = 10 - 2x$ Subtract $2x$ from both sides.

$4y = 10 - 2x$ Simplify.

$\frac{4y}{4} = \frac{10 - 2x}{4}$ Divide both sides by 4.

$y = \frac{10}{4} - \frac{2x}{4}$ Simplify.

$y = \frac{5}{2} - \frac{x}{2}$

or we can write

$y = \frac{5 - x}{2}$ or $y = -\frac{1}{2}x + \frac{5}{2}$ All forms are correct.

e. Given $3x - y = 15$, solve for y in terms of x.

Solution: Solving for y gives

$3x - y = 15$

$-y = 15 - 3x$ Subtract $3x$ from both sides.

$-1(-y) = -1(15 - 3x)$ Multiply both sides by -1 (or divide both sides by -1).

$y = -15 + 3x$ Simplify using the distributive property.

f. Given $V = \frac{k}{P}$, solve for P in terms of V and k.

Solution: $V = \frac{k}{P}$

$V \cdot P = k$ Multiply both sides by P.

$\frac{V \cdot P}{V} = \frac{k}{V}$ Divide both sides by V.

$P = \frac{k}{V}$

g. Solve for R given the formula $F = \frac{1}{R + r}$.

Solution: $F = \frac{1}{R + r}$

$(R + r)F = (R + r)\frac{1}{R + r}$ Multiply both sides by the denominator, $R + r$.

$RF + rF = 1$ Use the distributive property.

continued on next page ...

$$RF = 1 - rF \quad \text{Add } -rF \text{ to both sides.}$$

$$\frac{R\cancel{F}}{\cancel{F}} = \frac{1 - rF}{F} \quad \text{Divide both sides by } F.$$

$$R = \frac{1 - rF}{F}$$

This problem illustrates the importance of writing the correct form of a variable in a formula. Note that the uppercase R and the lowercase r represent completely different quantities.

Practice Problems

1. $2x - y = 5$; *solve for y.*
2. $2x - y = 5$; *solve for x.*
3. $A = \frac{1}{2}bh$; *solve for h.*
4. $L = 2\pi rh$; *solve for r.*
5. $P = 2l + 2w$; *solve for w.*
6. $y = mx + b$; *solve for m.*

3.1 Exercises

For Exercises 1 – 5, refer to Example 1a for information concerning simple interest and the related formula $\boldsymbol{I = Prt}$.

Simple Interest

1. You want to borrow \$4000 at 12% for only 90 days. How much interest would you pay?
2. A savings account of \$3500 is left for 9 months and draws simple interest at a rate of 7%.
   a. How much interest is earned?
   b. What is the balance in the account at the end of the 9 months?
3. For how many days must you leave \$1000 in a savings account at 5.5% to earn \$11.00 in interest?
4. What is the rate of interest charged if a 3-month loan of \$2500 is paid off with \$2562.50?
5. What principal would you need to invest to earn \$450 in simple interest in 6 months if the rate of interest was 9%?

1. \$120
2. a. \$183.75
b. \$3683.75
3. 72 days
4. 10%
5. \$10,000

Answers to Practice Problems: **1.** $y = 2x - 5$ **2.** $x = \frac{y+5}{2}$ **3.** $h = \frac{2A}{b}$ **4.** $r = \frac{L}{2\pi h}$ **5.** $w = \frac{P-2l}{2}$ **6.** $m = \frac{y-b}{x}$

6. 48 ft./sec.

7. 176 ft./sec.

8. 2 seconds

In Exercises 6 – 20, read the descriptive information carefully and then substitute the values given in the problem for the corresponding variables in the formulas. Evaluate the resulting expression for the unknown variable.

Velocity

If an object is shot upward with an initial velocity v_0 feet per second, the velocity, v, in feet per second is given by the formula $v = v_0 - 32t$, where t is time in seconds.

6. Find the velocity at the end of 3 seconds if the initial velocity is 144 feet per second.

7. Find the initial velocity of an object if the velocity after 4 seconds is 48 feet per second.

8. An object projected upward with an initial velocity of 106 feet per second has a velocity of 42 feet per second. How many seconds have passed?

9. 4 milliliters

10. 100 milligrams

11. $1120

12. 5 years

13. 14

Medicine

In nursing, one procedure for determining the dosage for a child is

$$\text{child's dosage} = \frac{\text{age of child in years}}{\text{age of child} + 12} \cdot \text{adult dosage}$$

9. If the adult dosage of a drug is 20 milliliters, how much should a 3-year-old child receive?

10. If the adult dosage of a drug is 340 milligrams, how much should a 5-year-old child receive?

Investment

The amount of money due from investing P dollars is given by the formula $A = P + Prt$, where r is the rate expressed as a decimal and t is the time in years.

11. Find the amount due if $1000 is invested at 6% for 2 years.

12. How long will it take an investment of $600 at an annual rate of 5% to be worth $750?

Carpentry

The number, N, of rafters in a roof or studs in a wall can be found by the formula $N = \frac{l}{d} + 1$, where l is the length of the roof or wall and d is the center-to-center distance from one rafter or stud to the next. Note that l and d must be in the same units.

26 ft

13. How many rafters will be needed to build a roof 26 ft. long if they are placed 2 ft. on center?

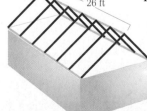

14. 16

15. 336 in. or 28 ft.

16. \$1030

17. \$337.50

18. 230

19. \$2400

20. 5%

21. $b = P - a - c$

22. $s = \frac{P}{3}$ **23.** $m = \frac{F}{a}$

24. $d = \frac{C}{\pi}$ **25.** $w = \frac{A}{l}$

26. $C = R - P$

27. $n = \frac{R}{p}$

28. $k = v - gt$

29. $P = A - I$

30. $h = \frac{L}{2\pi r}$

31. $m = 2A - n$

32. $a = P - 2b$

33. $t = \frac{I}{Pr}$ **34.** $E = RI$

35. $b = \frac{P-a}{2}$

36. $b^2 = c^2 - a^2$

37. $\beta = 180 - \alpha - \gamma$

14. A wall has studs placed 16 in. on center. If the wall is 20 ft. long, how many studs are in the wall?

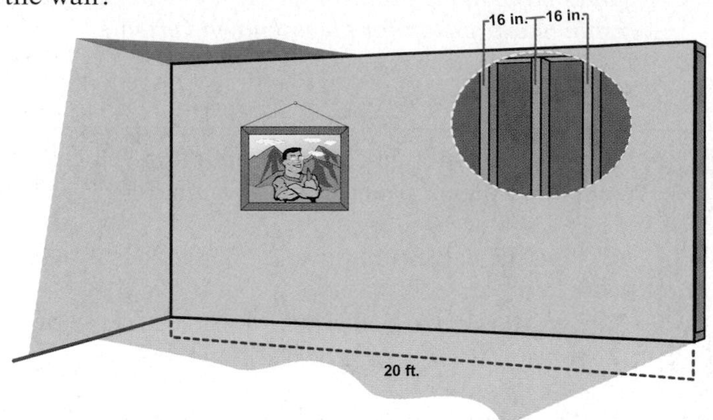

15. How long is a wall if it requires 22 studs placed 16 in. on center?

Cost

The total cost, C, of producing x items can be found by the formula $C = ax + k$, where a is the cost per item and k is the fixed costs (rent, utilities, and so on).

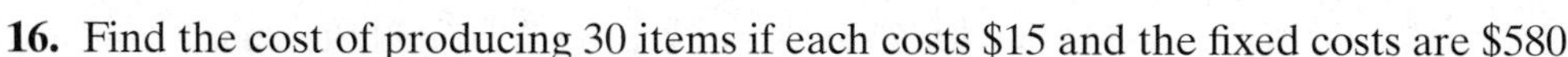

16. Find the cost of producing 30 items if each costs \$15 and the fixed costs are \$580.

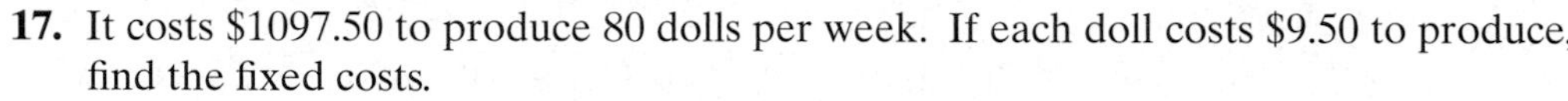

17. It costs \$1097.50 to produce 80 dolls per week. If each doll costs \$9.50 to produce, find the fixed costs.

18. It costs a company \$3.60 to produce a calculator. Last week the total costs were \$1308. If the fixed costs are \$480 weekly, how many calculators were produced last week?

Depreciation

Many items decrease in value as time passes. This decrease in value is called **depreciation**. One type of depreciation is called **linear depreciation**. The value, V, of an item after t years is given by $V = C - Crt$, where C is the original cost and r is the rate of depreciation expressed as a decimal.

19. If you buy a car for \$6000 and depreciate it linearly at a rate of 10% per year, what will be its value after 6 years?

20. A contractor buys a 4 year-old piece of heavy equipment valued at \$20,000. If the original cost of this equipment was \$25,000, find the rate of depreciation.

Solve for the indicated variable in Exercises 21 – 64.

21. $P = a + b + c$; solve for b.

22. $P = 3s$; solve for s.

23. $F = ma$; solve for m.

24. $C = \pi d$; solve for d.

25. $A = lw$; solve for w.

26. $P = R - C$; solve for C.

38. $x = \dfrac{y-b}{m}$

39. $h = \dfrac{V}{lw}$

40. $t = \dfrac{v_0 - v}{g}$

41. $b = \dfrac{2A}{h}$ 42. $I = \dfrac{E}{R}$

43. $\pi = \dfrac{A}{r^2}$

44. $L = \dfrac{R}{2A}$

45. $g = \dfrac{mv^2}{2K}$

46. $y = \dfrac{4-x}{4}$

47. $y = \dfrac{6-2x}{3}$

48. $y = 3x - 14$

49. $x = \dfrac{11-2y}{5}$

50. $x = \dfrac{2y-5}{2}$

51. $b = \dfrac{2A-hc}{h}$ or $b = \dfrac{2A}{h} - c$

52. $h = \dfrac{2A}{b+c}$

53. $x = \dfrac{8R+36}{3}$

54. $x = 5 - 3y$

55. $y = -x - 12$

27. $R = np$; solve for n.

28. $v = k + gt$; solve for k.

29. $I = A - P$; solve for P.

30. $L = 2\pi rh$; solve for h.

31. $A = \dfrac{m+n}{2}$; solve for m.

32. $P = a + 2b$; solve for a.

33. $I = Prt$; solve for t.

34. $R = \dfrac{E}{I}$; solve for E.

35. $P = a + 2b$; solve for b.

36. $c^2 = a^2 + b^2$; solve for b^2.

37. $\alpha + \beta + \gamma = 180$; solve for β.

38. $y = mx + b$; solve for x.

39. $V = lwh$; solve for h.

40. $v = -gt + v_0$; solve for t.

41. $A = \dfrac{1}{2}bh$; solve for b.

42. $R = \dfrac{E}{I}$; solve for I.

43. $A = \pi r^2$; solve for π.

44. $A = \dfrac{R}{2L}$; solve for L.

45. $K = \dfrac{mv^2}{2g}$; solve for g.

46. $x + 4y = 4$; solve for y.

47. $2x + 3y = 6$; solve for y.

48. $3x - y = 14$; solve for y.

49. $5x + 2y = 11$; solve for x.

50. $-2x + 2y = 5$; solve for x.

51. $A = \dfrac{1}{2}h(b+c)$; solve for b.

52. $A = \dfrac{1}{2}h(b+c)$; solve for h.

53. $R = \dfrac{3(x-12)}{8}$; solve for x.

54. $-2x - 5 = -3(x + y)$; solve for x.

55. $3y - 2 = x + 4y + 10$; solve for y.

56. $A = P + Prt$; solve for t.

57. $V = \dfrac{1}{3}\pi r^2 h$; solve for h.

58. $L = a + (n - 1)d$; solve for d.

59. $V^2 = v^2 + 2gh$; solve for g.

60. $S = 2\pi rh + 2\pi r^2$; solve for h.

61. $S = \dfrac{a}{1-r}$; solve for a.

62. $W = \dfrac{2PR}{R-r}$; solve for P.

63. $I = \dfrac{nE}{R+nr}$; solve for R.

64. $P = \dfrac{A}{1+ni}$; solve for A.

Make up a formula for each of the following situations in Exercises 65 – 68.

65. Each ticket for a concert costs $\$t$ per person and parking costs \$9.00. What is the total cost per car, C, if there are n people in a car?

66. ABC Car Rental charges \$25 per day plus \$0.12 per mile. What would you pay per day for renting a car from ABC if you were to drive the car x miles in one day?

67. Top-Of-The-Line computer company knows that the cost (labor and materials) of producing a computer is \$325 per computer per week and the fixed overhead costs (lighting, rent, etc.) are \$5400 per week. What are the company's weekly costs of producing n computers per week?

68. If T-O-T-L (see Exercise 67) sells its computers for \$683 each, what is its profit per week if it sells the same number, n, that it produces? (Remember that profit is equal to revenue minus costs, or $P = R - C$.)

56. $t = \dfrac{A - P}{Pr}$

57. $h = \dfrac{3V}{\pi r^2}$

58. $d = \dfrac{L - a}{n - 1}$

59. $g = \dfrac{V^2 - v^2}{2h}$

60. $h = \dfrac{S - 2\pi r^2}{2\pi r}$

61. $a = S(1 - r)$
$= S - Sr$

62. $P = \dfrac{W(R - r)}{2R}$
$= \dfrac{WR - Wr}{2R}$

63. $R = \dfrac{nE}{I} - nr$

64. $A = P(1 - ni)$
$= P - Pni$

65. $C = nt + 9$

66. $C = 0.12x + 25$

67. $C = 325n + 5400$

68. $C = 358n - 5400$

69. a. 0; No, because the numerator will be zero and thus the whole fraction will be equal to zero for all values of s.
b. $x < 70$
c. Answers will vary.

70. a. 1 **b.** –1
c. 1.5 **d.** –2

Writing and Thinking About Mathematics

69. The formula $z = \dfrac{x - \overline{x}}{s}$ is used extensively in statistics. In this formula, x represents one of a set of numbers, $\overline{x}$ represents the average (or mean) of those numbers in the set, and s represents a value called the standard deviation of the numbers. (The standard deviation is a positive number and is a measure of how "spread out" the numbers are.) The values for z are called z-scores, and they measure the number of standard deviation units a number x is from the mean $\overline{x}$.

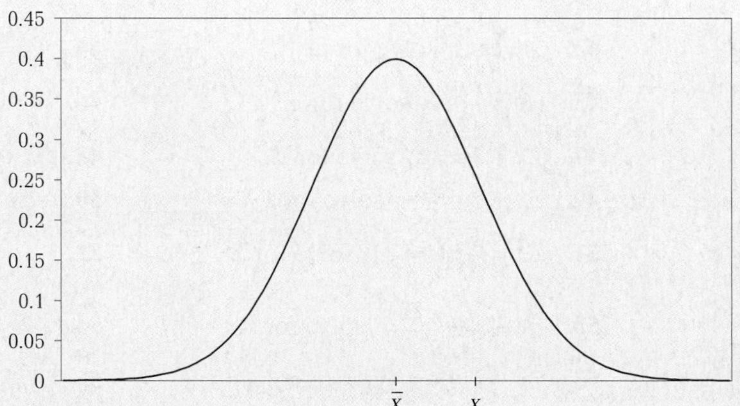

a. If $\overline{x} = 70$, what will be the z-score for $x = 70$? Does this z-score depend on the value of s? Explain.
b. For what values of x will the corresponding z-scores be negative?
c. Calculate your z-score on each of the last two scores in this class. (Your instructor will give you the mean and standard deviation for each test.) What do these scores tell you about your performance on the two exams?

70. Suppose that, for a particular set of exam scores, $\overline{x} = 72$ and $s = 6$. Find the z-score that corresponds to a score of
a. 78 **b.** 66 **c.** 81 **d.** 60

Hawkes Learning Systems: Introductory & Intermediate Algebra

Working with Formulas

3.2 Formulas in Geometry

Objectives

After completing this section, you will be able to:

Recognize and use appropriate geometric formulas for computations.

A **formula** is an equation that represents a general relationship between two or more quantities or measurements. Several variables may appear in a formula. And, as was illustrated by the variety of formulas presented in Section 3.1, formulas are useful in mathematics, economics, chemistry, medicine, physics, and many other fields of study. In this section, we will discuss some formulas related to simple geometric figures such as rectangles, circles, and triangles.

Perimeter

The **perimeter** of a geometric figure is the total distance around the figure. Perimeters are measured in units of length such as inches, feet, yards, miles, centimeters, and meters. The formulas for the perimeters of various geometric figures are shown here.

Formula	Figure Name	Figure
1. $P = 4s$	SQUARE	s
2. $P = 2l + 2w$	RECTANGLE	w, l
3. $P = 2a + 2b$	PARALLELOGRAM	h, a, b

continued on next page ...

Formula	Figure Name	Figure
4. $C = 2\pi r$ $C = \pi d$	CIRCLE	
5. $P = a + b + c + d$	TRAPEZOID	
6. $P = a + b + c$	TRIANGLE	

Terminology and Notation Related to Circles

Term	*Notation*	*Definition*
1. Circumference	$\boldsymbol{C}$	*The perimeter of a circle is called its* ***circumference****.*
2. Radius	$\boldsymbol{r}$	*The distance from the center of a circle to any point on the circle is called its* ***radius****.*
3. Diameter	$\boldsymbol{d}$	*The distance from one point on a circle to another point on the circle measured through its center is called its* ***diameter****.* *(**Note**: The diameter is twice the radius:* $\mathbf{d = 2r}$*.)*
4. Pi	π	*π is a Greek letter used for the constant 3.14159265358979…* *(π is an irrational number.)*

For the calculations involving π in this text we will round off to two decimal places and use $\pi \approx 3.14$. For more accuracy in any calculation with π, use the π key on the TI-83 Plus calculator (to do this first press the 2nd key and then the ^ key.) This key will give π accurate to 9 decimals. Remember, however, that π is an irrational number and is an infinite nonrepeating decimal.

Historically, π was discovered by mathematicians trying to find a relationship between the circumference and diameter of any circle. The result is that π is the ratio of the circumference to the diameter of any circle, a rather amazing fact.

Example 1: Perimeter

a. Find the perimeter of a rectangle with a length of 10 feet and a width of 8 feet.

Solution: **i.** Sketch the figure.

ii. $P = 2l + 2w$

$P = 2 \cdot 10 + 2 \cdot 8$

$P = 20 + 16 = 36$

The perimeter is 36 feet.

$w = 8$ ft.

$l = 10$ ft.

b. Find the circumference of a circle with a diameter of 3 centimeters.

Teaching Notes:
You might mention that 3π is also an acceptable answer. You also might want to encourage your students to use their calculator π key for more accuracy.

Solution 1: **i.** Sketch the figure.

ii. $C = \pi d$

$C \approx 3.14(3)$

$= 9.42$ cm

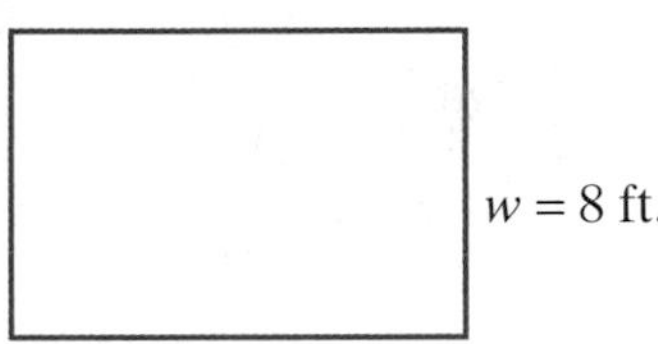

Solution 2: **i.** Sketch the figure.

ii. $C = 2\pi r$

$C \approx 2(3.14)(1.5)$

$= 9.42$ cm

Note that the result is the same with either formula.
The circumference is approximately 9.42 centimeters.

c. Find the perimeter of the triangle with sides labeled as in the figure.

Solution: **i.** Sketch the figure

ii. $P = a + b + c$

$P = 3 + 6.2 + 8.1$

$= 17.3$ in.

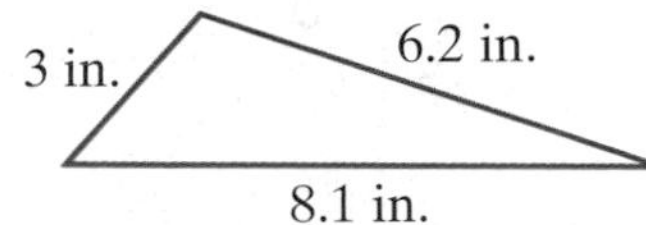

The perimeter is 17.3 inches.

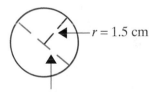

Area

Area is a measure of the interior, or enclosure, of a surface. Area is measured in square units such as square feet, square inches, square meters, or square miles. For example, the area enclosed by a square of 1 inch on each side is 1 sq. in. (or 1 in.^{2}) [Figure 3.1(a)], while the area enclosed by a square of 1 centimeter on each side is 1 sq. cm (or 1 cm^2) [Figure 3.1(b)].

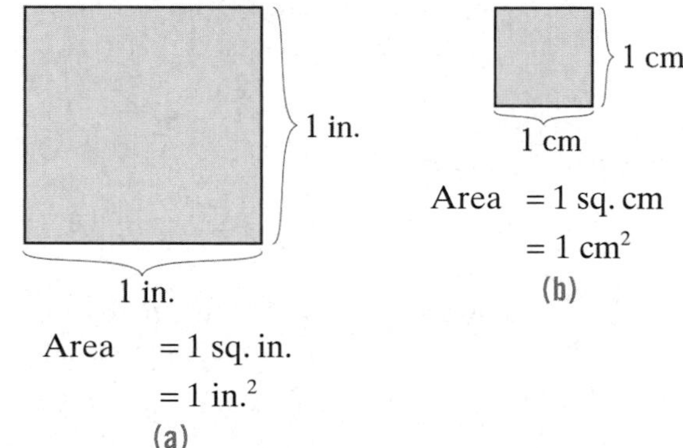

Figure 3.1

The formulas for finding the areas of several geometric figures are shown below and on the following page.

Formula	Figure Name	Figure
1. $A = s^2$	SQUARE	s, s
2. $A = lw$	RECTANGLE	w, l
3. $A = bh$	PARALLELOGRAM	h, 90° right angle, a, b

continued on next page ...

Formula	Figure Name	Figure
4. $A = \pi r^2$	CIRCLE	
5. $A = \frac{1}{2}h(b + c)$	TRAPEZOID	
6. $A = \frac{1}{2}bh$	TRIANGLE	

Example 2: Area

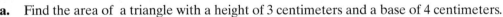

a. Find the area of a triangle with a height of 3 centimeters and a base of 4 centimeters.

Solution: **i.** Sketch the figure.

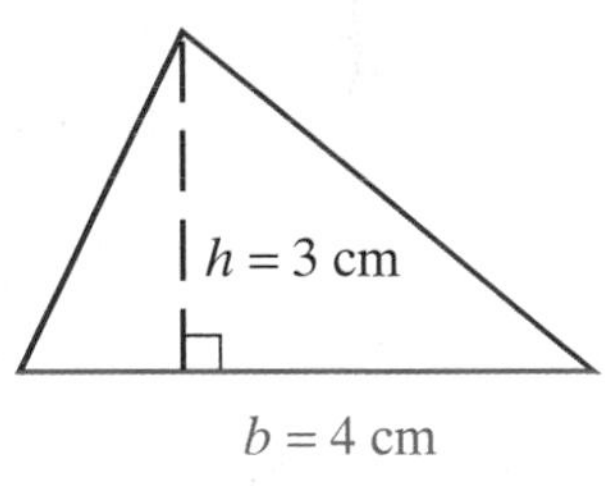

ii. $A = \frac{1}{2} \cdot b \cdot h$

$A = \frac{1}{2} \cdot 4 \cdot 3$

$= \frac{1}{2} \cdot 12 = 6$

The area is 6 square centimeters.

b. Find the area of a circle with a radius of 6 inches. (Use $\pi \approx 3.14$.)

Solution: **i.** Sketch the figure.

ii. $A = \pi r^2$

$A \approx 3.14(6)^2$

$= 3.14(36)$

$= 113.04$

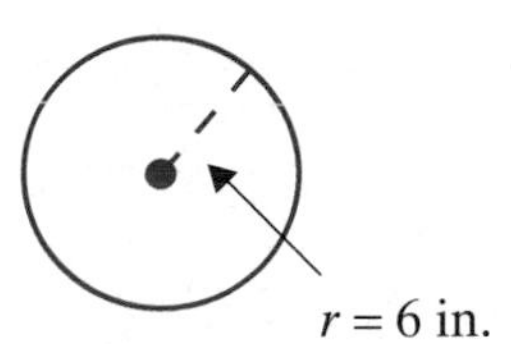

The area is approximately 113.04 square inches.

Teaching Notes:
For meaningful applications, relate to the student that the economics of package design and the analysis of material requirements can be determined from the formulas for area and volume.

Volume

Volume is a measure of the space enclosed by a three-dimensional figure. Volume is measured in cubic units, such as cubic inches, cubic centimeters, and cubic feet. For example, the volume enclosed by a cube of 1 inch on each edge is 1 cu. in. (or 1 in.^{3}) [Figure 3.2(a)], while the volume enclosed by a cube of 1 centimeter on each edge is 1 cu. cm (or 1 cm^3) [Figure 3.2(b)].

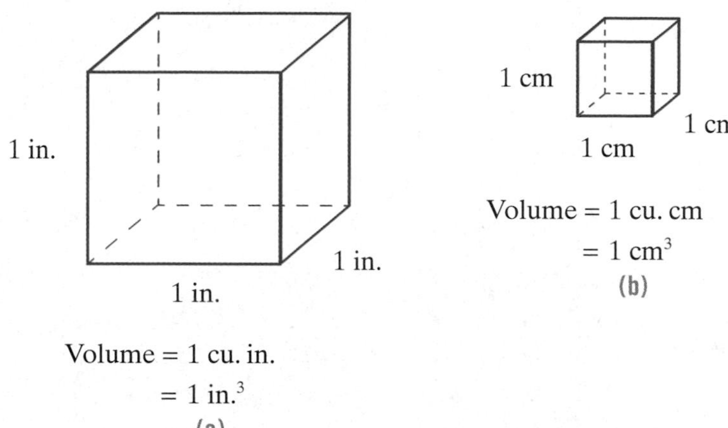

Figure 3.2

The formulas for finding the volume of several geometric figures are shown in Figure 3.3.

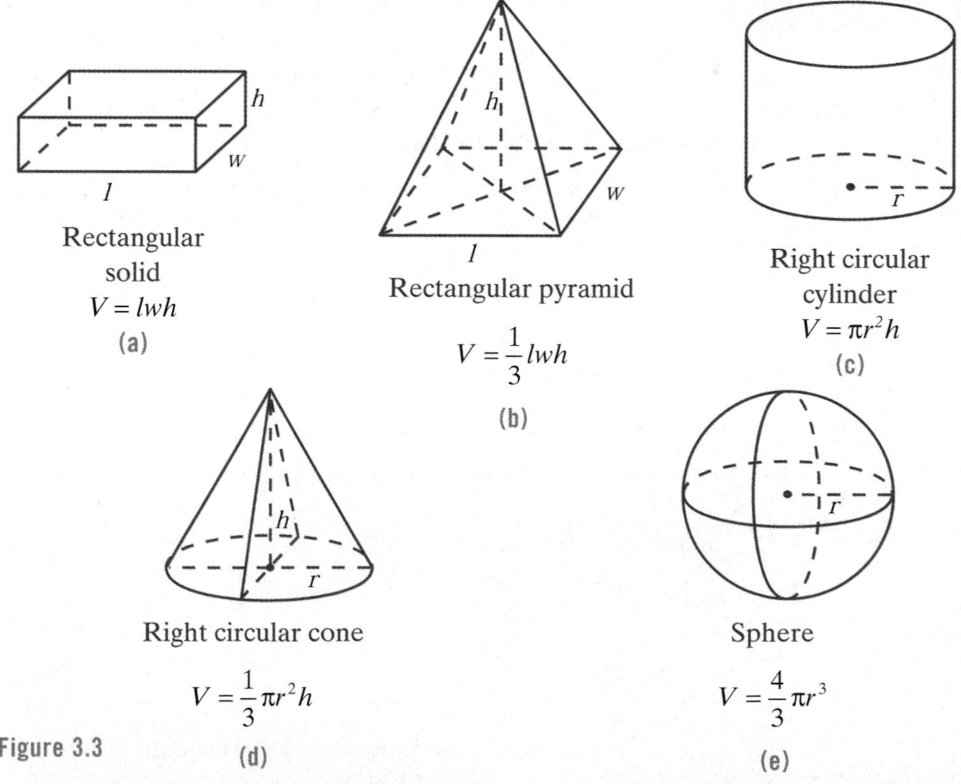

Figure 3.3

Example 3: Volume

a. Find the volume of a right circular cylinder with a radius of 2 centimeters and a height of 5 centimeters.

Solution: **i.** Sketch the figure.

ii. $V = \pi r^2 h$

$$V \approx 3.14\left(2^2\right)\cdot 5$$
$$= 3.14(4)\cdot 5$$
$$= 3.14(20)$$
$$= 62.80 \text{ cm}^3$$

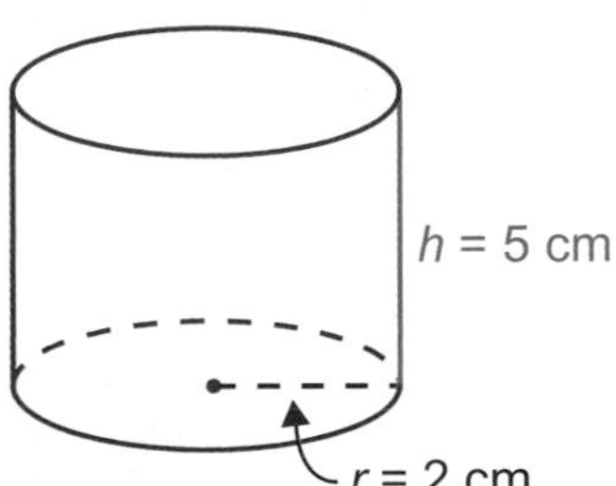

The volume is approximately 62.80 cubic centimeters.

b. Find the volume of a sphere with a radius of 4 feet.

Solution: **i.** Sketch the figure.

ii. $V = \frac{4}{3}\pi r^3$

$$V \approx \frac{4}{3}(3.14)\cdot 4^3$$
$$= \frac{4(3.14)\cdot 64}{3}$$
$$= \frac{803.84}{3} \text{ ft.}^3 \text{ or about } 267.95 \text{ ft.}^3$$

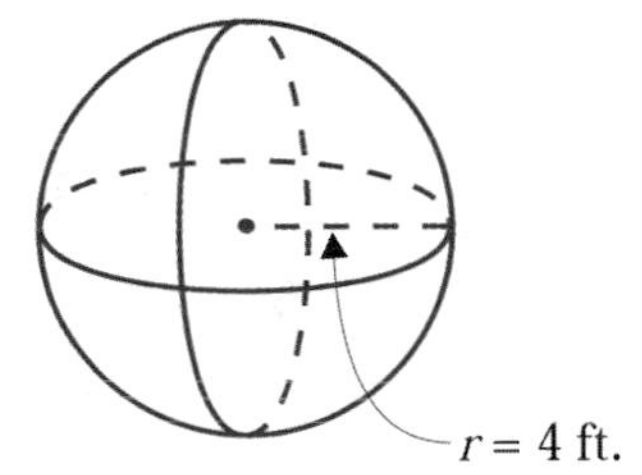

The volume is approximately 267.95 cubic feet.

Practice Problems

1. *Find the area of a square with sides 4 cm long.*
2. *Find the area of a circle with a diameter 6 in.*
3. *Find the perimeter of a rectangle 3.5 m long and 1.6 m wide.*

Answers to Practice Problems: **1.** 16 cm^2 **2.** 28.26 in.^{2} **3.** 10.2 m

3.2 Exercises

In Exercises 1 – 15, select the answer from the right-hand column that correctly matches the statement.

1. e

1. The formula for the perimeter of a square ______ **a.** $V=\frac{1}{3}lwh$

2. j

2. The formula for the circumference of a circle ______ **b.** $A=\frac{1}{2}h(b+c)$

3. k

3. The formula for the perimeter of a triangle ______ **c.** $A=s^2$

4. f

4. The formula for the area of a rectangle ______ **d.** $A=\frac{1}{2}bh$

5. c

5. The formula for the area of a square ______ **e.** $P=4s$

6. b

6. The formula for the area of a trapezoid ______ **f.** $A=lw$

7. d

7. The formula for the area of a triangle ______ **g.** $P=2l+2w$

8. o

8. The formula for the area of a parallelogram ______ **h.** $V=\frac{1}{3}\pi r^2h$

9. g

9. The formula for the perimeter of a rectangle ______ **i.** $V=\frac{4}{3}\pi r^3$

10. a

10. The formula for the volume of a rectangular pyramid ______ **j.** $C=2\pi r$

11. l

11. The formula for the volume of a rectangular solid ______ **k.** $P=a+b+c$

12. n

12. The formula for the area of a circle ______ **l.** $V=lwh$

13. m

13. The formula for the volume of a right circular cylinder ______ **m.** $V=\pi r^2h$

14. i

14. The formula for the volume of a sphere ______ **n.** $A=\pi r^2$

15. h

15. The formula for the volume of a right circular cone ______ **o.** $A=bh$

16. 72 mm; 324 mm^2

Find (a) the perimeter and (b) the area of each figure in Exercises 16 – 21.

17. 70 cm; 300 cm^2

16. $P=$ ______ $A=$ ______

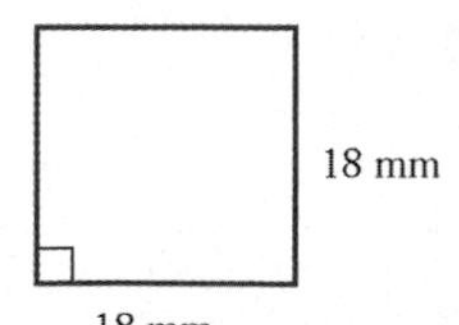

17. $P=$ ______ $A=$ ______

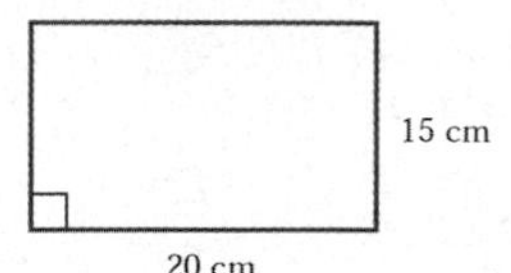

18. 38 in.; 72 in.^{2}

19. 54 mm; 126 mm^2

20. 32 cm; 44 cm^2

21. 31.4 in.; 78.5 in.^{2}

22. 108.85 cm^3

23. 1436.03 in.^{3}

24. 470.4 ft.^{3}

25. 729 m^3

26. 203.47 in.^{3}

18. $P =$ _____ $A =$ _____

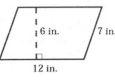

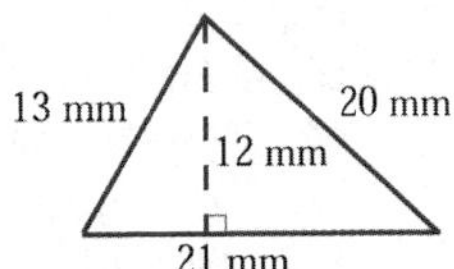

19. $P =$ _____ $A =$ _____

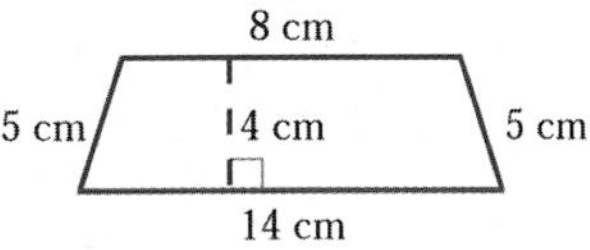

20. $P =$ _____ $A =$ _____

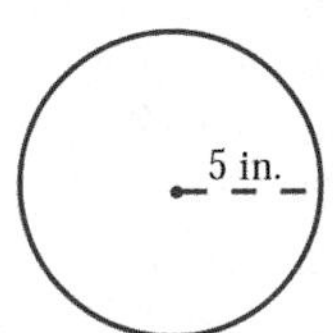

21. $C =$ _____ $A =$ _____

Find the volume of each figure in Exercises 22 – 26. (Use $\pi \approx 3.14$ and round to the nearest hundredth.)

22. $V =$ _____

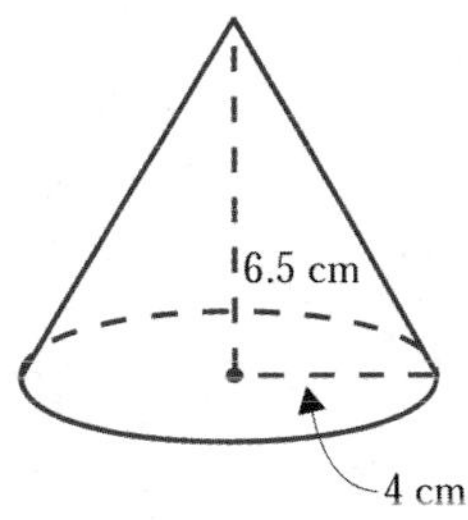

23. $V =$ _____

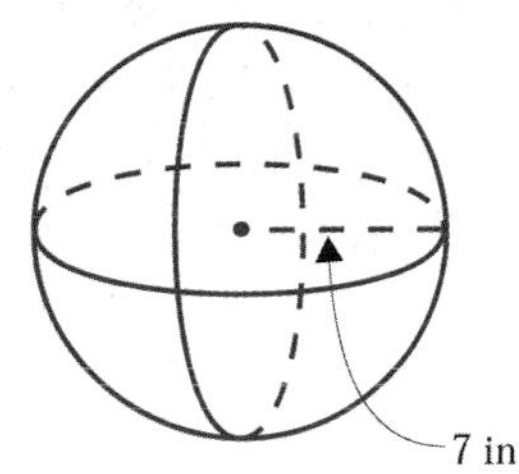

24. $V =$ _____

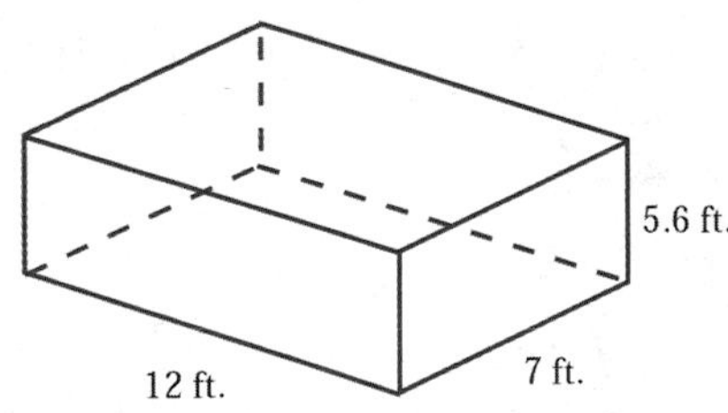

25. $V =$ _____

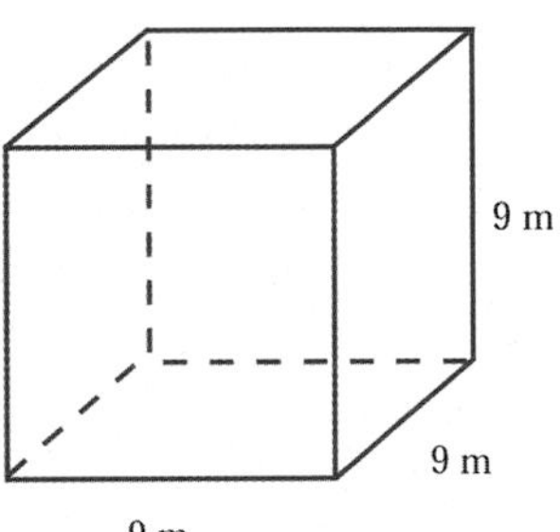

26. $V =$ _____

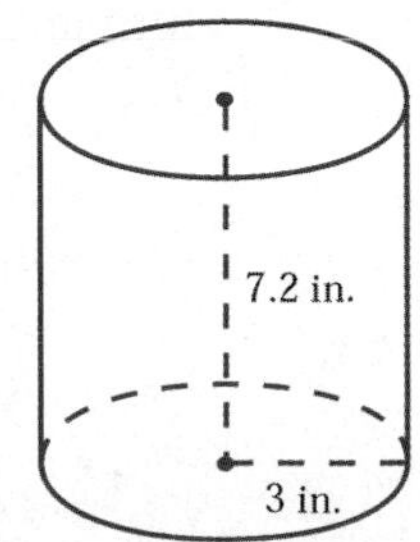

27. 56 in.

28. 81 ft.^{2}

29. 63 cm^2

30. 50.24 m

31. 6154.4 ft.^{3}

32. 24.2 m

33. 125 ft.^{3}

34. 2520 ft.^{2}

35. 272 cm^2

36. 1483.2 in.^{3}

37. 529.875 in.^{3}

38. 4186.67 cm^3

39. $\alpha = 180 - \beta - \gamma$, 43°

40. $s = \frac{P}{4}$, $2\frac{2}{3}$ m

41. $w = \frac{P - 2l}{2}$, 13 ft.

42. $b = \frac{A}{h}$, 47 in.

43. $a = P - b - c$, 61 in.

44. $r = \frac{C}{2\pi}$, 13 cm

Solve the problems in Exercises 27 – 38. (Use $\pi \approx 3.14$)

27. What is the perimeter of a rectangle with length of 17 in. and width of 11 in.?

28. Find the area of a square with sides that are 9 ft. long.

29. The base of a triangle is 14 cm and the height is 9 cm. Find the area.

30. Find the circumference of a circle with a radius of 8 m.

31. The radius of the base of a cylindrical tank is 14 ft. If the tank is 10 ft. high, find the volume.

32. The sides of a triangle are 6.2 m, 8.6 m, and 9.4 m. Find the perimeter.

33. What is the volume of a cube with edges of 5 ft.?

34. A rectangular garden plot is 60 ft. long and 42 ft. wide. Find the area.

35. A parallelogram has a base of 20 cm and a height of 13.6 cm. Find the area.

36. A rectangular box is 18 in. long, 10.3 in. wide, and 8 in. high. Find the volume.

37. The diameter of the base of a right circular cone is 15 in. If the height of the cone is 9 in., find the volume.

38. Find the volume of a sphere whose radius is 10 cm. (Round to the nearest hundredth.)

For Exercises 39 – 46, (a) state the formula that relates to the given information, (b) solve the formula for the unknown quantity, and (c) substitute the given values in the formula to determine the value of the unknown quantity. (Use $\pi \approx 3.14$.)

39. Two angles of a triangle measure 72° and 65°. Find the measure of the third angle.

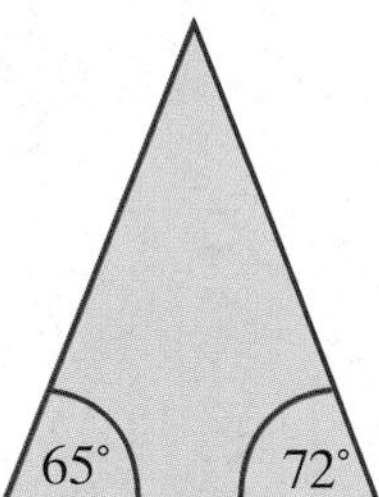

40. The perimeter of a square is $10\frac{2}{3}$ meters. Find the length of the sides.

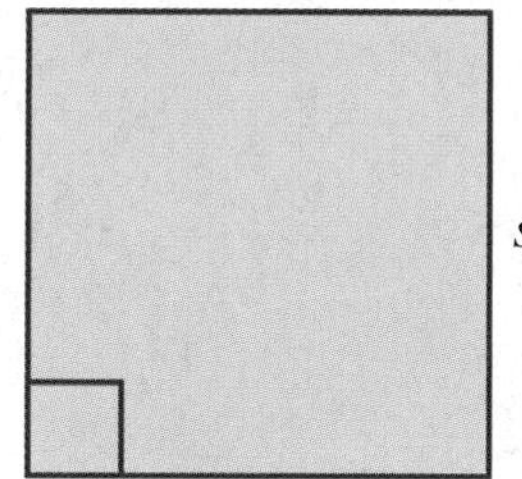

41. The perimeter of a rectangle is 88 feet. If the length is 31 feet, find the width.

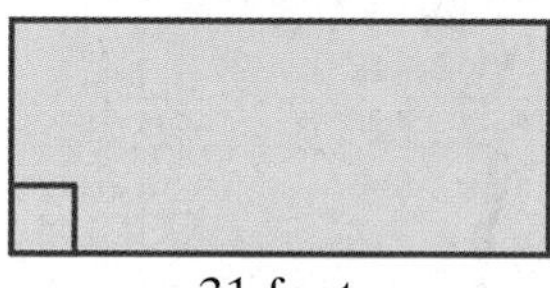

42. The area of a parallelogram is 1081 square inches. If the height is 23 inches, find the length of the base.

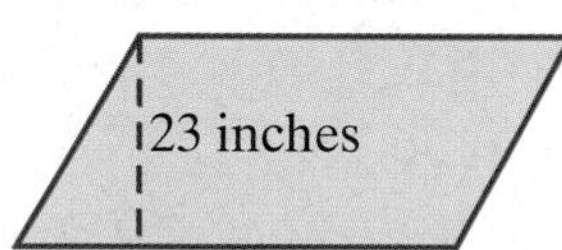

43. The perimeter of a triangle is 147 inches. Two of the sides measure 38 inches and 48 inches. Find the length of the third side.

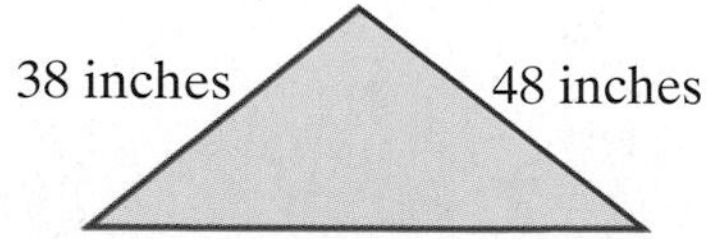

44. The circumference of a circle is 26π centimeters. Find the radius.

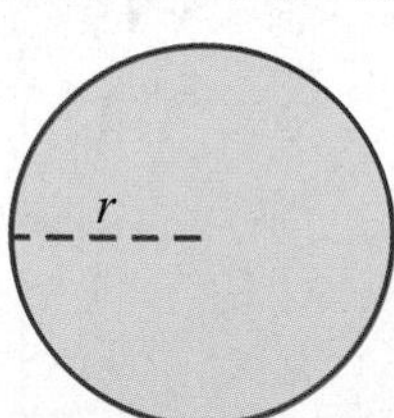

45. $A = \pi r^2$, 196π sq. ft. or 615.4 sq. ft.

46. $h = \frac{2A}{b+c}$, 6 m

47. 28 cm; 48 cm^2

48. 54.84 in., 200.52 in.^{2}

49. 212.04 m^2

50. 52.25 in.

51. 130.08 in.^{2}

52. 1326.93 cm^2

53. 6,334,233.21 cm^3

45. The radius of a circle is 14 feet. Find the area.

46. The area of a trapezoid is 51 square meters. One base is 7 meters long, and the other is 10 meters long. Find the height of the trapezoid.

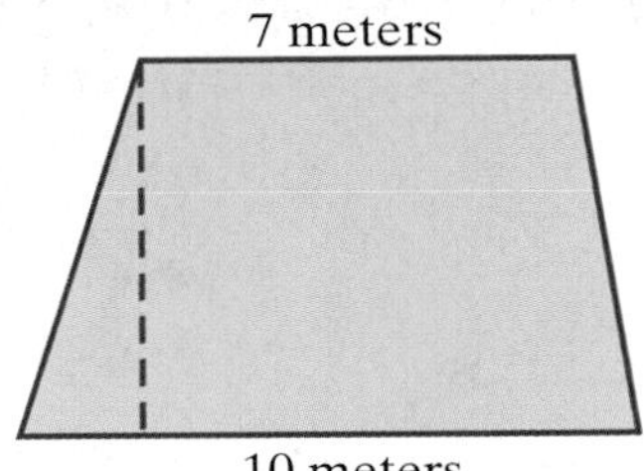

Find the perimeter and area for each figure in Exercises 47 and 48. (Use $\pi \approx 3.14$.)

47. $P =$ _____ $A =$ _____

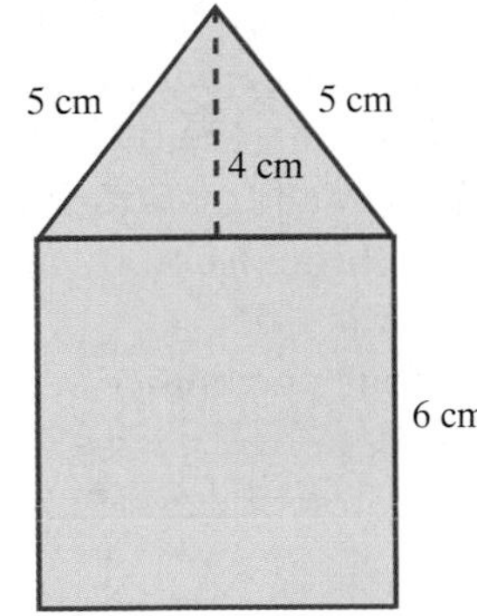

48. $P =$ _____ $A =$ _____

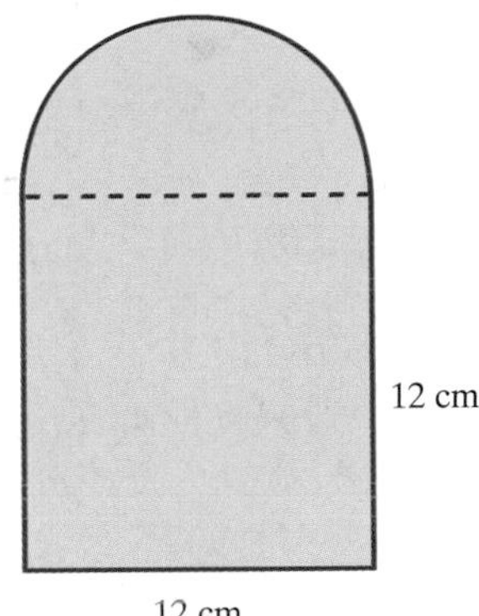

Calculator Problems

In Exercises 49 – 53, find the answers accurate to two decimal places.

49. Find the area of a rectangle that is 16.54 m long and 12.82 m wide.
50. The radius of a circle is 8.32 in. Find the circumference. (Use $\pi \approx 3.14$.)
51. Find the area of a trapezoid with bases of 22.36 in. and 17.48 in. and a height of 6.53 in.
52. Find the area of a triangle with a base of 63.52 cm and a height of 41.78 cm.
53. The radius of a sphere is 114.8 cm. Find the volume. (Use $\pi \approx 3.14$.)

Hawkes Learning Systems: Introductory & Intermediate Algebra

Formulas in Geometry

3.3 Applications

Objectives

After completing this section, you will be able to:

Solve the following by using first degree equations:

1. *Number problems,*
2. *Distance-rate-time problems,*
3. *Cost-profit problems,*
4. *Simple interest problems, and*
5. *Average problems.*

Teaching Notes:
A good practice for Step 4 is to include a conclusion that is a declarative statement that answers the question in the problem and *in the appropriate units of measurement*. This inclusion supports applications in science, business, and other applied areas.

Word problems (or applications) are designed to teach you to read carefully, to organize, and to think clearly. Whether or not a particular problem is easy for you depends a great deal on your personal experiences and general reasoning abilities. The problems generally do not give specific directions to add, subtract, multiply, or divide. You must decide what relationships are indicated through careful analysis of the problem.

As we mentioned in Section 2.3, among the many accomplishments of the famous Stanford University professor, George Pòlya, is the four-step process approach to problem solving. For emphasis and easy reference, we list the steps again.

1. Understand the problem.
2. Devise a plan.
3. Carry out the plan.
4. Look back over the results.

This strategy works for solving word problems at all levels of mathematics.

Problems involving numerical expressions will usually contain key words indicating the operations to be performed. These are basically the same words as listed in Section 2.2 to look for in translating phrases into algebraic expressions. We repeat a similar list here because those same words help in setting up equations when solving word problems.

Addition	Subtraction	Multiplication	Division	Equality
add	subtract	multiply	divide	gives
sum	difference	product	quotient	represents
plus	minus	times	ratio	amounts to
more than	less than	twice		is / was
increased by	decreased by	of (with fractions and percents)		is the same as

Example 1: Number Problem

The sum of two numbers is 36. If $\frac{1}{2}$ of the smaller number is equal to $\frac{1}{4}$ of the larger number, find the two numbers.

Solution: Analyze the problem and identify the key words.

The key words are **sum** (indicating addition) and **of** (indicating multiplication when used with fractions).

Assign variables to the unknown quantities.

Let x = smaller number
Since x + (larger number) = 36,
$36 - x$ = larger number.

Write an equation relating the given information.

$\frac{1}{2}$ of the smaller number	is equal to	$\frac{1}{4}$ of the larger number
$\frac{1}{2}x$	$=$	$\frac{1}{4}(36-x)$

Solve the equation.

$$\frac{1}{2}x = \frac{1}{4}(36-x)$$

$$4\cdot\frac{1}{2}x = 4\cdot\frac{1}{4}(36-x)$$ Multiplying both sides of the equation by 4 yields integer coefficients.

$$2x = 1(36-x)$$

$$2x = 36-x$$

$$3x = 36$$

$$x = 12$$ Smaller number

$$36-x = 24$$ Larger number

Check:

$$12+24 \stackrel{?}{=} 36$$

$$\frac{1}{2}(12) \stackrel{?}{=} \frac{1}{4}(24)$$

$$6 = 6$$

The two numbers are 12 and 24.

Problems involving distance usually make use of the relationship indicated by the formula $d = rt$, where r = rate, t = time, and d = distance. A chart or table showing the known and unknown values is quite helpful and is illustrated in the next example.

Example 2: Distance-Rate-Time

A motorist averaged 45 mph for the first part of a trip and 54 mph for the last part of the trip. If the total trip of 303 miles took 6 hours, what was the time for each part of the trip?

Solution:

Let t = time for 1st part of trip
$6 - t$ = time for 2nd part of trip

Analysis of Strategy
What is being asked for?
Total time minus time for 1st part of trip gives time for 2nd part of trip.

	rate	·	time	=	distance
1st Part	45		t		$45 \cdot t$
2nd Part	54		$6 - t$		$54(6 - t)$

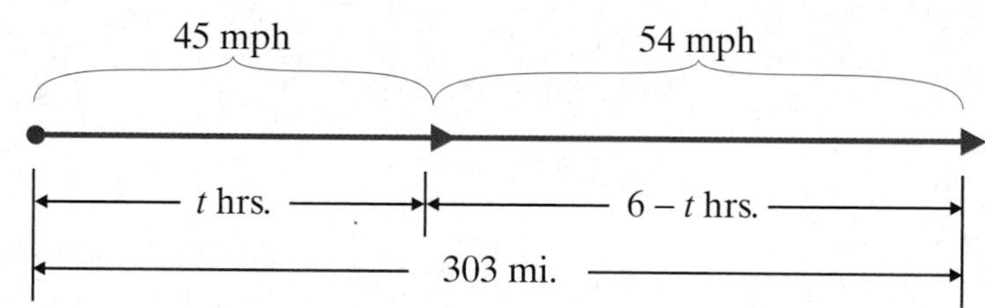

$$\underbrace{\text{1st part distance}} + \underbrace{\text{2nd part distance}} = \underbrace{\text{total distance}}$$

Form the equation relating the given information.

$$\begin{aligned} 45t + 54(6 - t) &= 303 \\ 45t + 324 - 54t &= 303 \\ 324 - 9t &= 303 \\ -9t &= -21 \end{aligned}$$

Solve the equation.

$$t = \frac{21}{9} = \frac{7}{3} \qquad \text{1st part of the trip}$$

$$6 - t = 6 - \frac{7}{3} = \frac{11}{3} \qquad \text{2nd part of the trip}$$

Check:

$$45 \cdot \frac{7}{3} = 15 \cdot 7 = 105 \text{ miles (1st part)}$$

$$54 \cdot \frac{11}{3} = 18 \cdot 11 = 198 \text{ miles (2nd part)}$$

$105 + 198 = 303$ miles total

The first part took $\frac{7}{3}$ hrs. or $2\frac{1}{3}$ hrs. The second part took $\frac{11}{3}$ hrs. or $3\frac{2}{3}$ hrs.

Problems involving cost come in a variety of forms. The next two examples illustrate the types of problems you will find in the exercises.

Example 3: Cost

a. The Berrys sold their house. After paying the real estate agent a commission of 6% of the selling price and then paying \$1,486 in other costs and \$90,000 on the mortgage, they received \$49,514. What was the selling price of the house?

Solution: Use the relationship $SP - C = P$,
that is, selling price − cost = profit.
Let s = selling price.
cost $= 0.06s + 1{,}486 + 90{,}000$

$$\underbrace{\text{selling price}} - \underbrace{\text{cost}} = \underbrace{\text{profit}}$$

$$s - (0.06s + 1{,}486 + 90{,}000) = 49{,}514$$
$$s - 0.06s - 1{,}486 - 90{,}000 = 49{,}514$$
$$0.94s = 141{,}000$$
$$s = 150{,}000$$

Check:

\$150,000	selling price	\$9,000	commission	\$150,000	selling price
0.06	commission %	1,486	costs	−100,486	expenses
\$9,000	commission	+ 90,000	mortgage	\$49,514	profit received
		\$100,486	total expenses		

The selling price was \$150,000.

b. A jeweler paid \$350 for a ring. He wants to price the ring for sale so that he can give a 30% discount on the selling price (or marked price) and still make a profit of 20% on his cost. What selling price should he mark for the ring?

Solution: Again, we make use of the relationship $SP - C = P$.
Let x = the selling price,
then $0.30x + 350$ = the cost.

continued on next page ...

$$\underbrace{\text{selling price}} - \underbrace{\text{cost}} = \underbrace{\text{profit}}$$

$$\begin{aligned} x - (0.30x + 350) &= 0.20(350) \\ x - 0.30x - 350 &= 70 \\ 0.70x &= 420 \\ x &= 600 \end{aligned}$$

His total cost is the discount plus what he paid for the ring.

The profit is 20% of what he paid originally.

Check:

\$600 selling price	\$180 discount	\$600 selling price
× 0.30 discount %	+350 cost	−530 total cost
\$180 discount	\$530 total cost	\$70 profit

As a double check,

\$350 original cost
× 0.20 profit %
\$70 profit

The jeweler should set the selling price at \$600.

To work problems related to interest on money invested for one year, you need to know the basic relationship between the principal P (amount invested), the annual rate of interest r, and the amount of interest I (money earned). This relationship is described by the formula $P \cdot r = I$. (This is the formula for simple interest, $I = Prt$, with $t = 1$.) We use this relationship in Example 4.

Example 4: Interest

A woman has had \$40,000 invested for one year, some with a savings and loan which paid 7%, the rest in a high-risk stock which yielded 12% for the year. If her interest income last year was \$3,550, how much did she have in the savings and loan and how much did she invest in the stock?

Solution: Let

$$\begin{aligned} x &= \text{amount invested at 7\%} \\ 40{,}000 - x &= \text{amount invested at 12\%} \end{aligned}$$

Total amount invested minus amount invested at 7% represents amount invested at 12%.

	principal	·	rate	=	interest
Savings and loan	x		0.07		$0.07(x)$
Stock	$40{,}000 - x$		0.12		$0.12(40{,}000 - x)$

$$\underbrace{\text{interest at }7\%}+\underbrace{\text{interest at }12\%}=\underbrace{\text{total interest}}$$

$$\begin{aligned} 0.07(x) + 0.12(40{,}000 - x) &= 3{,}550 \\ 7x + 12(40{,}000 - x) &= 355{,}000 && \text{Multiply both sides of the} \\ 7x + 480{,}000 - 12x &= 355{,}000 && \text{equation by 100.} \\ -5x &= -125{,}000 \\ x &= 25{,}000 \\ 40{,}000 - x &= 15{,}000 \end{aligned}$$

Check: $25{,}000(0.07) = 1{,}750$ and $15{,}000(0.12) = 1{,}800$ and $\$1{,}750 + \$1{,}800 = \$3550$.
The woman had \$25,000 in the savings and loan at 7% interest and invested \$15,000 in the stock at 12% interest.

Average (or Mean)

You are probably already familiar with the concept of **average** of a set of numbers. The average is also called the **arithmetic average** or **mean**. Your grade in most courses is related to an "average" of your exam scores. Magazines and newspapers report average income, average price of homes, average sales, batting averages, and so on. The mean is particularly important in the study of statistics. For example, traffic studies are interested in average speeds, census studies are concerned with the mean number of people living in a house, and universities study the mean test scores of incoming freshmen students.

Average

The ***average*** *(or* ***mean****) of a set of numbers is the value found by adding the numbers and then dividing the sum by the number of numbers in the set.*

Example 5: Average (or Mean)

a. At noon on five consecutive days in Aspen, Colorado the temperatures were –5°, 7°, 6°, –7°, and 14° (in degrees Fahrenheit). (Negative numbers represent temperatures below zero.) Find the average of these noon-day temperatures.

Solution: First, add the five temperatures.

$$-5 + 7 + 6 + (-7) + 14 = 15$$

continued on next page ...

Now, divide the sum, 15, by the number of temperatures, 5.

$$\frac{15}{5} = 3$$

The average noon temperature was 3° F.

b. In a placement exam for mathematics, a group of ten students had the following scores: 3 students scored 75, 2 students scored 80, 1 student scored 82, 3 students scored 85, and 1 student scored 88. What was the mean score for this group of students?

Solution: To find the total of all the scores, we multiply and then add. This is more efficient than adding all ten scores.

$75 \cdot 3 = 225$

$80 \cdot 2 = 160$

$82 \cdot 1 = 82$

$85 \cdot 3 = 255$ Multiply.

$88 \cdot 1 = 88$

$225 + 160 + 82 + 255 + 88 = 810$ Add.

$810 \div 10 = 81$ Divide by the number of scores.

The mean score on the placement test for this group of students was 81.

c. The following speeds (in miles per hour) of fifteen cars were recorded at a certain point on a freeway.

70	75	65	60	61
64	68	72	59	68
82	76	70	68	50

Find the average speed of these cars. (One car received a speeding ticket.)

Solution: Using a calculator, the sum of the speeds is 1008 mph. Dividing by 15 gives the average speed:

$1008 \div 15 = 67.2$ mph

d. Suppose that you have scores of 85, 92, 82 and 88 on four exams in your English class. What score will you need on the fifth exam to have an average of 90?

Solution: Let x = your score on the fifth exam

The sum of all the scores, including the unknown fifth exam, divided by 5 must equal 90.

$$\frac{85+92+82+88+x}{5}=90$$

$$\frac{347+x}{5}=90$$

$$5\cdot\frac{347+x}{5}=5\cdot 90$$

$$347+x=450$$

$$x=103$$

Assuming that each exam is worth 100 points, you cannot attain an average of 90 on the five exams.

3.3 Exercises

Refer to the formulas listed in Section 3.2 as necessary.

1. If 15 is added to a number, the result is 56 less than twice the number. Find the number.
2. A number subtracted from 20 is equal to three times the number. Find the number.
3. Nine less than twice a number is equal to the number. What is the number?
4. Find a number such that −64 is 12 more than four times the number.
5. What number gives a result of −2 when 5 is subtracted from the quotient of the number and 4?
6. If 6 is added to the quotient of a number and 3, the result is 1. What is the number?
7. Seven times a certain number is equal to the sum of three times the number and 28. What is the number?
8. Twelve more than five times a number is equal to the difference between 5 and twice the number. Find the number.
9. Four added to the quotient of a number and 6 is equal to 11 less than the number. What is the number?
10. The quotient of twice a number and 8 is equal to 3 more than the number. What is the number?
11. One number is 6 more than three times another. If their sum is 38, what are the two numbers?
12. The sum of two numbers is 98 and their difference is 20. Find the two numbers.

1. 71
2. 5
3. 9
4. −19
5. 12
6. −15
7. 7
8. −1
9. 18
10. −4
11. 8, 30
12. 39, 59

13. 46 ft. by 84 ft.

14. \$47

15. \$1500

16. 210 mi.

17. 78 min.

18. shirt \$50; tie \$35

19. 7 hrs

20. 375 mph, 450 mph

21. 3 hrs

22. 60 mph, 300 mi.

13. The length of a rectangular-shaped backyard is 8 feet less than twice the width. If 260 feet of fencing is needed to enclose the yard, find the dimensions of the yard.

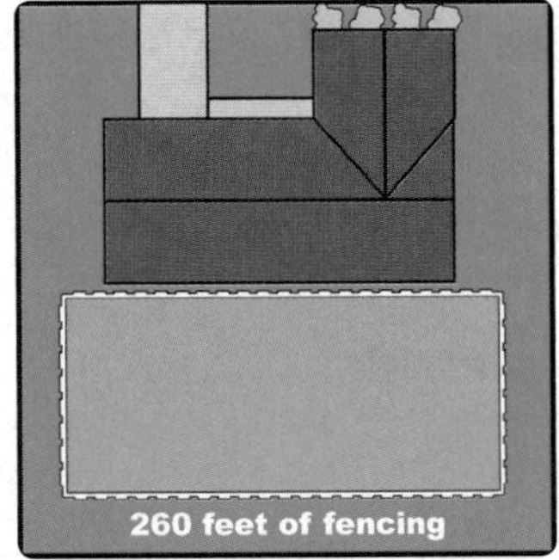

14. The price of a pair of trousers is reduced 15%. The sale price is \$39.95. Find the original price.

15. After a raise of 8%, Juan's salary is \$1620 per month. What was his salary before the increase?

16. The U-Drive Company charges \$20 per day plus 22¢ per mile driven. For a one-day trip, Louis paid a rent charge of \$66.20. How many miles did he drive?

17. For a long-distance call, the telephone company charges 35¢ for each of the first three minutes and 15¢ for each additional minute. If the cost of a call was \$12.30, how many minutes did the call last?

18. Willis bought a shirt and necktie for \$85. The shirt cost \$15 more than the tie. Find the cost of each.

19. The Reeds are moving across the state. Mr. Reed leaves $3\frac{1}{2}$ hours before Mrs. Reed. If he averages 40 mph and she averages 60 mph, how long will it take Mrs. Reed to overtake Mr. Reed?

20. Two planes, which are 2475 miles apart, fly toward each other. Their speeds differ by 75 mph. If they pass each other in 3 hours, what is the speed of each?

	rate	· time	= distance
1st plane	r	3	
2nd plane	$r + 75$	3	

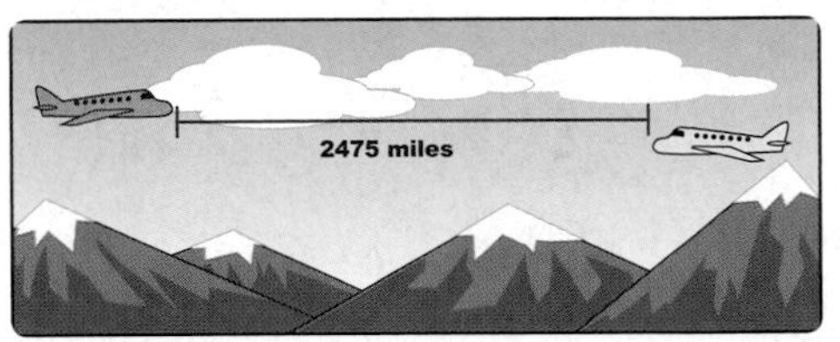

21. Jane rides her bike to Blue Lake. Going to the lake, she averages 12 mph. On the return trip, she averages 10 mph. If the round trip takes a total of 5.5 hours, how long does the return trip take?

	rate	· time	= distance
Going	12	$5.5 - t$	
Returning	10	t	

22. A car travels from one town to another in 6 hours. On the return trip, the speed is increased by 10 mph and the trip takes 5 hours. Find the rate on the return trip. How far apart are the towns?

	rate	· time	= distance
Going	r	6	
Returning	$r + 10$	5	

23. $4\frac{4}{5}$ hrs

24. 54 mph

25. 36 mph, 60 mph

26. 12 mi.

27. $112.50

28. 90 half gallons

29. 2,000 baskets

30. 320 sq. ft.

31. 62,500 pounds

32. 50 at $3.00, 400 at $4.50

23. Carol has 8 hours to spend on a mountain hike. She can walk up the trail at an average of 2 mph and can walk down at an average of 3 mph. How long should she plan to spend on the uphill part of the hike?

24. After traveling for 40 minutes, Mr. Koole had to slow to $\frac{2}{3}$ of his original speed for the rest of the trip due to heavy traffic. The total trip of 84 miles took 2 hours. Find his original speed.

25. A train leaves Los Angeles at 2:00 PM. A second train leaves the same station in the same direction at 4:00 PM. The second train travels 24 mph faster than the first. If the second train overtakes the first at 7:00 PM, what is the speed of each of the two trains?

26. Maria jogs to the country at a rate of 10 mph. She returns along the same route at 6 mph. If the total trip took 1 hour 36 minutes, how far did she jog?

27. A particular style of shoe costs the dealer $81 per pair. At what price should the dealer mark them so he can sell them at a 10% discount off the selling price and still make a 25% profit?

28. A grocery store bought ice cream for $2.60 a half gallon and stored it in two freezers. During the night, one freezer malfunctioned and ruined 15 half gallons. If the remaining ice cream is sold for $3.98 per half gallon, how many half gallons did the store buy if it made a profit of $64.50?

29. A farmer raises strawberries. They cost him $0.80 a basket to produce. He is able to sell only 85% of those he produces. If he sells his strawberries at $2.40 a basket, how many must he produce to make a profit of $2480?

30. Mary builds cabinets in her spare time. Good quality cabinet plywood costs $4.00 per square foot. There is approximately a 10% waste of material due to cutting and fitting. She also figures $240 per month for finishing material, glue, tools, etc. If she charges $9.20 per square foot of finished cabinet, how many square feet of plywood would she use if her profit is $1129.60 in one month?

31. A citrus farmer figures that his fruit costs 96¢ a pound to grow. If he lost 20% of the crop he produced due to a frost and he sold the remaining 80% at $1.80 a pound, how many pounds did he produce to make a profit of $30,000?

32. Mr. Wise bought $1950 worth of stock, some at $3.00 per share and some at $4.50 per share. If he bought a total of 450 shares of stock, how many of each did he buy?

33. 20 at $300; 24 at $250

34. 15 dozen

35. $14,000 at 5%; $11,000 at 6%

36. 4.5% on $10,000; 5.5% on $6000

37. 6.5% on $4,000; 6% on $3,000

38. $7400 at 5.5%; $2600 at 6%

39. $7,000 at 6%; $9,000 at 8%

40. improved $140,000; unimproved $80,000

41. 5°

42. 78.57

33. Last summer, Ernie sold surfboards. One style sold for $300 and the other sold for $250. He sold a total of 44 surfboards. How many of each style did he sell if the receipts from each style were equal?

34. The pro shop at the Divots Country Club ordered two brands of golf balls. Titleless balls cost $1.80 each and the Done Lob balls cost $1.50 each. The total cost of Titleless balls exceeded the total cost of the Done Lob balls by $108. If an equal number of each brand was ordered, how many dozen of each brand were ordered?

35. Amanda invests $25,000, part at 5% and the rest at 6%. The annual return on the 5% investment exceeds the annual return on the 6% investment by $40. How much did she invest at each rate?

36. The annual interest earned on a $6000 investment was $120 less than the interest earned on $10,000 invested at 1% less interest per year. What was the rate of interest on each amount?

37. The annual interest on a $4000 investment exceeds the interest earned on a $3000 investment by $80. The $4000 is invested at a 0.5% higher rate of interest than the $3000. What is the interest rate of each investment?

38. Mr. Hill invests ten thousand dollars, part at 5.5% and part at 6%. The interest from the 5.5% investment exceeds the interest from the 6% investment by $251. How much did he invest at each rate?

39. Two investments totaling $16,000 produce an annual income of $1140. One investment yields 6% a year, while the other yields 8% per year. How much is invested at each rate?

40. Sellit Realty Company gets a 6% fee for selling improved properties and 10% for selling unimproved land. Last week, the total sales were $220,000 and their total fees were $16,400. What were the sales from each of the two types of properties?

41. The temperature readings for 20 days at 3 PM at a local ski resort were recorded as follows:

24°	11°	−5°	14°	15°	5°	−6°
13°	−2°	−8°	−10°	32°	31°	−7°
−9°	4°	−18°	−9°	5°	20°	

What was the average of the recorded temperatures for these 20 days?

42. On an exam in history, a class of twenty-one students had the following test scores: 4 scored 65, 3 scored 70, 6 scored 78, 2 scored 82, 1 scored 85, 3 scored 91, and 2 scored 95. What was the mean score (to the nearest hundredth) on this test for the class?

43. 72.0 in.

44. 2.2 books

45. 31.5 hrs

The frequency of a number is simply a count of how many times that number appears. In statistics, data is commonly given in the table form of a frequency distribution as illustrated in Exercises 43 and 44. To find the mean, multiply each number by its frequency, add these products, and divide the sum by the sum of the frequencies.

43. The heights of twenty-two men were recorded in the following frequency distribution. Find the mean height (to the nearest tenth of an inch) for these men.

Height (in inches)	Frequency
68	2
69	3
72	8
73	5
74	2
75	1
78	1

44. The students in a psychology class were asked the number of books that they had read in the last month. The following frequency distribution indicates the results. Find the mean number of books read (to the nearest tenth) by these students.

Number of Books	Frequency
0	3
1	2
2	6
3	4
4	2
5	1

45. The bar graph shows the approximate amounts of time per week spent watching TV for six groups (by age and sex) of people 18 years of age and older. What is the average amount of time per week people over the age of 18 spend watching TV? (Assume each group has the same number of people.)

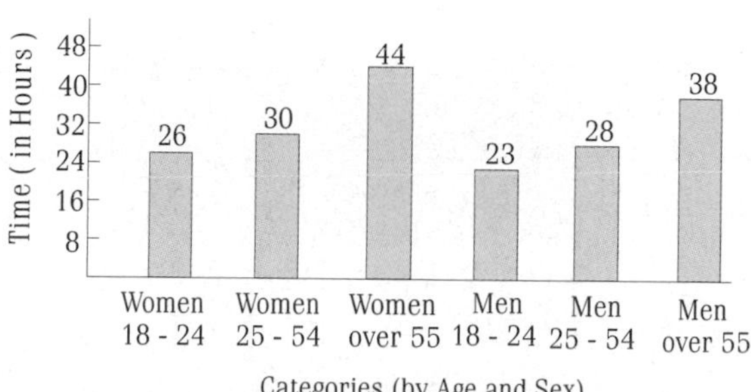

46. Approx. 22,000 sq. miles.

47. 94 or more

48. 62 min. or less

49. a. 53.25° F
b. 44° F
c. 16° F

50. a. 36,167
b. Northwestern
c. 8,000

46. The following bar graph shows the area of each of the five Great Lakes. Lake Superior with an area of about 32,000 square miles is the world's largest fresh water lake. What is the mean size of these lakes?

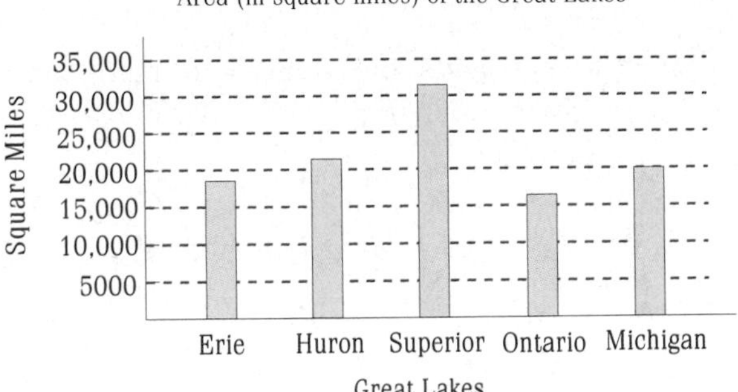

47. Gerald had scores of 80, 92, 89, and 95 on four exams in his algebra class. What score will he need on his fifth exam to have an average of 90 or better?

48. While riding her bike to the ocean and back home five times, Stacey timed herself at 60 min., 62 min., 55min. (the wind was helping), 58 min., and 63 min. She had set a goal of averaging 60 minutes for her rides. How many minutes will she need on her sixth ride to attain her goal?

49. Given the monthly temperatures over a year for Christchurch, New Zealand:

a. Find the average temperature for the year.

b. Find the minimum temperature for the year.

c. Find the difference in temperature between the months of June and December.

Source: http://www.weather.com

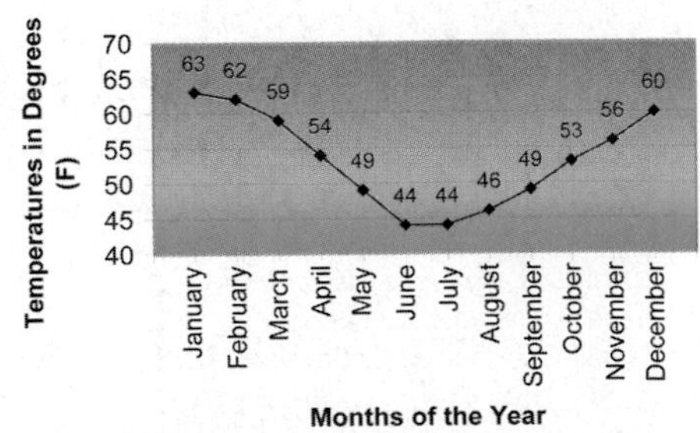

50. Given the enrollment at the Main Campuses of the following Big Ten Universities:

a. Find the average enrollment over the six schools. (Round to the nearest whole number.)

b. Find the University with the lowest enrollment.

c. Find the difference in enrollment between Ohio State and Penn State.

Source: Peterson's, 2002

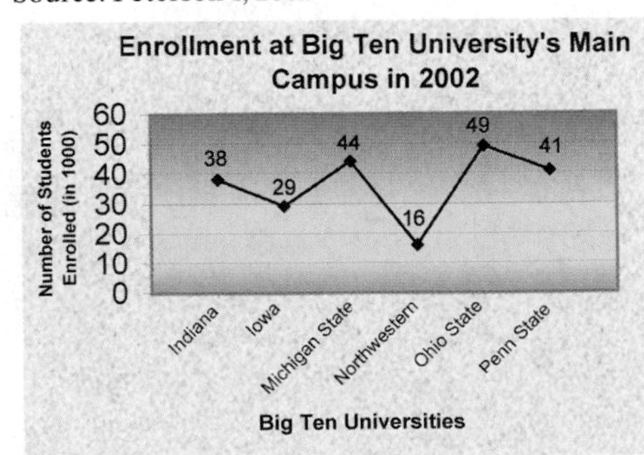

51. a. 3.2 in.
b. 7.1 in.
c. 6.9 in.

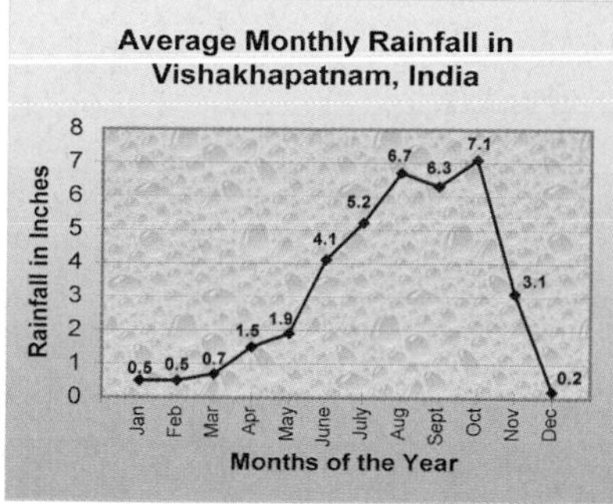

51. Given the monthly rainfall averages over a year for Vishakhapatnam, India:

a. Find the average rainfall for the year. (Round to the nearest tenth.)

b. Find the maximum rainfall for the year. (Round to the nearest tenth.)

c. Find the difference in rainfall between the months of October and December. (Round to the nearest tenth.)

Source: http://www.weather.com

52. a. 51,600,000
b. 33,000,000
c. 76,000,000

Number of Passengers at US Airports in 2001

Number of Passengers (in Millions)

Airports: ATL 76, ORD 67, LAX 62, JFK 29, BOS 24

52. Given the number of passengers at the following airports:

a. Find the average number of passengers.

b. Find the difference in passengers between JFK and LAX.

c. What was the total number of passengers to go through ATL?

Source: Airports Council International – North America

53. a. 11.3 days
b. 17 days
c. Feb. 3 and Nov. 12

53. Given the length of each Space Shuttle flight:

a. Find the average duration of the flights in the year 1995. (Round to the nearest tenth.)

b. Find the length of the flight on March 2nd.

c. Which two flights had the same duration?

Source: NASA

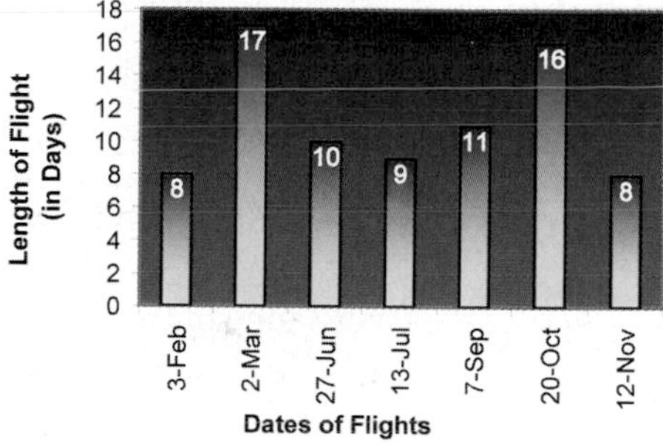

54. see page 182

Writing and Thinking About Mathematics

54. List the four steps in Polya's approach to problem solving. Then, in your own words, discuss how you used these steps in solving a "problem" you have had recently. Did you have trouble finding your car keys this morning? How did you decide what movie to see last weekend?

Hawkes Learning Systems: Introductory & Intermediate Algebra

Applications of Linear Equations
Applications: Average

3.4 Ratios and Proportions

Objectives

After completing this section, you will be able to:

1. *Understand the meaning of a ratio.*
2. *Understand that a proportion is an equation.*
3. *Be familiar with the terms means and extremes.*
4. *Be able to solve proportions.*
5. *Be able to use proportions to solve applications.*

Understanding Ratios

One of the uses of fractions (other than meaning to divide and to indicate part of a whole) is to indicate a comparison of two quantities. Such a comparison is called a **ratio**. For example,

$$\text{the } \textbf{ratio } \frac{\mathbf{5}}{\mathbf{6}} \text{ might mean } \frac{5\text{ feet}}{6\text{ feet}} \text{ or } \frac{5\text{ days}}{6\text{ days}}.$$

Note that in these examples, both the numerator and denominator are in the same units. If the units are not the same, then we must either change the units if possible or leave the units labeled in both numerator and denominator. Thus,

$$\text{by changing to a common unit, a ratio of } \frac{3\text{ dimes}}{4\text{ nickels}} = \frac{3\text{ dimes}}{2\text{ dimes}} = \frac{3}{2}.$$

$$\text{But, a ratio of } \frac{2\text{ cups flour}}{1\text{ cup water}} \text{ must be left as is with the units in place.}$$

Ratio

*A **ratio** is a comparison of two quantities by division. The ratio of a to b $(b \neq 0)$ can be written as*

$$\frac{a}{b} \quad or \quad a:b \quad or \quad a \text{ to } b.$$

Remember the following characteristics of ratios:

1. Ratios can be reduced just as fractions can be reduced.
2. Whenever the units of the numbers in a ratio are the same, the ratio has no units. We say that the ratio is an **abstract number**.

Teaching Notes:
You might mention that "per" means to divide. For example, 50 miles per hour can be thought of as $\frac{50 \text{ miles}}{1 \text{ hour}}$ or 3 cents per ounce can be thought of as $\frac{3 \text{ cents}}{1 \text{ ounce}}$.
The idea of an abstract number will probably be new to most students. However, an awareness of this concept would help them understand the need to label answers to word problems. It would also help them in solving word problems. They should also understand that labeling is absolutely necessary in setting up correct proportions for solving problems with proportions.

3. When the numbers in a ratio have different units, then the numbers must be labeled to clarify what is being compared. Such ratio is called a **rate**. For example, the ratio of 45 miles : 1 hour $\left(\text{or } \frac{45 \text{ miles}}{1 \text{ hour}}\right)$ is a rate of 45 miles per hour (or 45 mph).

Example 1: Ratio

a. During baseball season, major league players' batting averages are published in the newspapers. Suppose a player has a batting average of .320. What does this indicate?

Solution: A batting average is a ratio (or rate) of hits to times at bat. Thus, a batting average of .320 means

$$.320 = \frac{320 \text{ hits}}{1000 \text{ times at bat}}$$

Reducing gives

$$.320 = \frac{320}{1000} = \frac{\cancel{40} \cdot 8}{\cancel{40} \cdot 25} = \frac{8 \text{ hits}}{25 \text{ times at bat}}$$

This means that we can expect this player to hit safely at a rate of 8 hits for every 25 times he comes to bat.

Whenever possible, a ratio should be an abstract number. That is, we would like a ratio to represent a ratio of two quantities with the **same units of measure**. For example, to find the ratio of 4 nickels to 2 quarters, we can, and would prefer, to write the ratio with the same units (such as pennies or nickels). Thus, the ratio is

$\frac{4 \text{ nickels}}{2 \text{ quarters}} = \frac{4 \text{ nickels}}{10 \text{ nickels}} = \frac{2}{5}$ and the ratio of 4 nickels to 2 quarters is 2 to 5.

b. What is the reduced ratio of 250 centimeters (cm) to 2 meters (m)? Centimeters and meters are units of measure in the metric system.

Solution: Here we change the units so that they are the same. There are 100 centimeters in 1 meter. Therefore, 2 m = 200 cm, and the ratio is

$$\frac{250 \text{ cm}}{2 \text{ m}} = \frac{250 \text{ cm}}{200 \text{ cm}} = \frac{\cancel{50} \cdot 5}{\cancel{50} \cdot 4} = \frac{5}{4}$$

The reduced ratio can also be written as 5 : 4 or 5 to 4.

c. Inventory shows 5000 washers and 400 bolts. What is the ratio of washers to bolts? Write the solution as a ratio reduced to lowest terms.

Solution:

$$\frac{5000 \text{ washers}}{400 \text{ bolts}} = \frac{\cancel{200} \cdot 25}{\cancel{200} \cdot 2} = \frac{25 \text{ washers}}{2 \text{ bolts}}$$

Understanding Proportions

A **proportion** is a special type of equation that says two ratios are equal. For example,

$$\frac{3}{4} = \frac{6}{8} \text{ and } \frac{3.5}{7} = \frac{3.75}{7.5} \text{ are proportions.}$$

Proportion

*A **proportion** is a statement that two ratios are equal. In symbols,*

$$\frac{\mathbf{a}}{\mathbf{b}} = \frac{\mathbf{c}}{\mathbf{d}} \text{ is a proportion.}$$

A proportion has four **terms**. The first and fourth terms are called the **extremes**. The second and third terms are called the **means**.

To better understand the terms **means** and **extremes**, we can write a proportion in the following form:

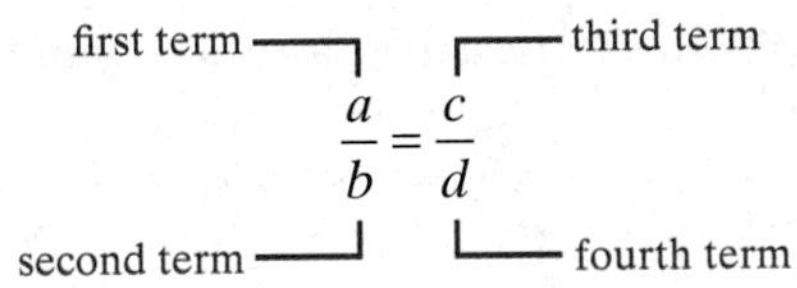

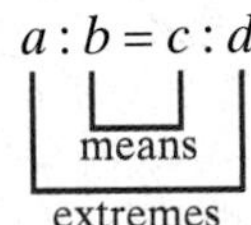

Now consider the following analysis of a proportion.

$\frac{a}{b} = \frac{c}{d}$	Write the proportion.
$bd\left(\frac{a}{b}\right) = \left(\frac{c}{d}\right)bd$	Multiply both sides by bd.
$a \cdot d = b \cdot c$	Simplify.

The last equation states that (assuming the original proportion is true) the product of the means is equal to the product of the extremes. This fact provides a useful technique for solving problems involving proportions, and we state it formally.

True Proportion

*In a **true proportion**, the product of the **extremes** is equal to the product of the **means**.*

In symbols,

$$\frac{a}{b}=\frac{c}{d} \text{ if and only if } a\times d=b\times c \text{ (where } b\neq 0 \text{ and } d\neq 0).$$

Example 2: Proportions

Determine whether each of the following proportions is true or false.

a. $\frac{6}{8}=\frac{11}{13}$ **b.** $\frac{2\frac{1}{3}}{7}=\frac{3\frac{1}{4}}{9\frac{3}{4}}$

Solution:

a. For the proportion $\frac{6}{8}=\frac{11}{13}$,
the product of the extremes is $6\times 13=78$, and
the product of the means is $8\times 11=88$.
Because the product of the extremes is not equal to the product of the means $(78\neq 88)$,
the proportion $\frac{6}{8}=\frac{11}{13}$ is false.

b. For the proportion $\frac{2\frac{1}{3}}{7}=\frac{3\frac{1}{4}}{9\frac{3}{4}}$,
change the means and extremes to improper fraction form and multiply:
The product of the means is $2\frac{1}{3}\cdot 9\frac{3}{4}=\frac{7}{3}\cdot\frac{39}{4}=\frac{7}{\cancel{3}_1}\cdot\frac{\cancel{39}^{13}}{4}=\frac{91}{4}$, and
the product of the extremes is $7\cdot 3\frac{1}{4}=\frac{7}{1}\cdot\frac{13}{4}=\frac{91}{4}$.
Since the product of the extremes is equal to the product of the means, the proportion is true.

Applications

Proportions can be used in solving certain types of word problems. In these applications there is some unknown quantity that can be represented as one term in a proportion. The problem is solved by finding the value of this unknown term that will make the proportion a true statement.

Solving for the Unknown Term in a Proportion

a. ***If the unknown term is in a denominator***, *solve by setting the product of the means equal to the product of the extremes and then solving the resulting equation.*

b. ***If the unknown term is in a numerator***, *solve the proportion directly by multiplying both sides by the reciprocal of the denominator as in Section 1.6.*

Example 3: Unknown Term

Find the unknown term in the following proportion: $\frac{6}{124}=\frac{3}{A}$ (In this case, the unknown term is in the denominator.)

Solution:

$\frac{6}{124}=\frac{3}{A}$ Write the proportion.

$6\times A=3(124)$ Write the product of the extremes equal to the product of the means.

$\frac{\cancel{6}A}{\cancel{6}}=\frac{372}{6}$ Solve the equation.

$A=62$

In using proportions to solve word problems, we must set up the proportion properly so that the ratios that are represented compare the units in the same order. Either Pattern A or Pattern B, described in the following example, must be followed.

Example 4: Setting up Proportions

Suppose that a car will travel 572 miles on 26 gallons of gas. How far would you expect to travel on 30 gallons of gas?

Solution:

Pattern A: Each ratio has different units, but they are in the same order. For example,

$$\frac{572\ \textbf{miles}}{26\ \textit{gallons}}=\frac{x\ \textbf{miles}}{30\ \textit{gallons}} \quad \text{or} \quad \frac{26\ \textit{gallons}}{572\ \textbf{miles}}=\frac{30\ \textit{gallons}}{x\ \textbf{miles}}$$

Pattern B: Each ratio has the same units, the numerators correspond, and the denominators correspond. For example,

$$\frac{572\ \textbf{miles}}{x\ \textbf{miles}} = \frac{26\ \textit{gallons}}{30\ \textit{gallons}} \quad \text{or} \quad \frac{x\ \textbf{miles}}{572\ \textbf{miles}} = \frac{30\ \textit{gallons}}{26\ \textit{gallons}}$$

With every one of the four equations illustrated in Pattern A and Pattern B, setting the product of the extremes equal to the product of the means will give the same equation to be solved, namely,

$$26x = 572 \times 30, \text{ where } x = 660.$$

So, you may use the form that occurs to you first. The solution will be the same regardless of the form chosen.

Example 5: Applying Proportions

An architect draws the plans for a building using a scale of $\frac{3}{4}$ inch to represent 10 feet. How many feet would 6 inches represent?

Solution:

Step 1: Let y represent the number of unknown feet.

Step 2: Set up a proportion and label the numerators and denominators to be sure that the pattern is correct.

Step 3: One such proportion (following Pattern B) is

$$\frac{\frac{3}{4}\ \textbf{inches}}{6\ \textbf{inches}} = \frac{10\ \textit{feet}}{y\ \textit{feet}}$$

Any of the proportions following Pattern A or Pattern B will give the same solution.

Step 4: Solve the proportion:

$$\frac{\frac{3}{4}\ \text{inches}}{6\ \text{inches}} = \frac{10\ \text{feet}}{y\ \text{feet}}$$

$$\frac{3}{4}y = 6 \cdot 10$$

$$\frac{\cancel{4}}{\cancel{3}} \cdot \frac{\cancel{3}}{\cancel{4}}y = \frac{4}{3} \cdot 60 \qquad \text{Multiply both sides by } \frac{4}{3}, \text{ the reciprocal of } \frac{3}{4}.$$

$$y = 80$$

Therefore, on the architect's drawing, 6 inches represents 80 feet.

Applications in Geometry (Similar Triangles)

We need the following facts concerning angles and triangles:

1. An angle can be measured in degrees. (A protractor, as shown in Figure 3.4, can be used to find the measure of an angle.)
2. Every triangle has six parts: three sides and three angles.
3. The sum of the measures of the angles of every triangle is 180°.
4. Each endpoint of the sides of a triangle is called a **vertex** of the triangle. (Capital letters are used to label the vertices, and these letters can be used to name the angles and the sides.)

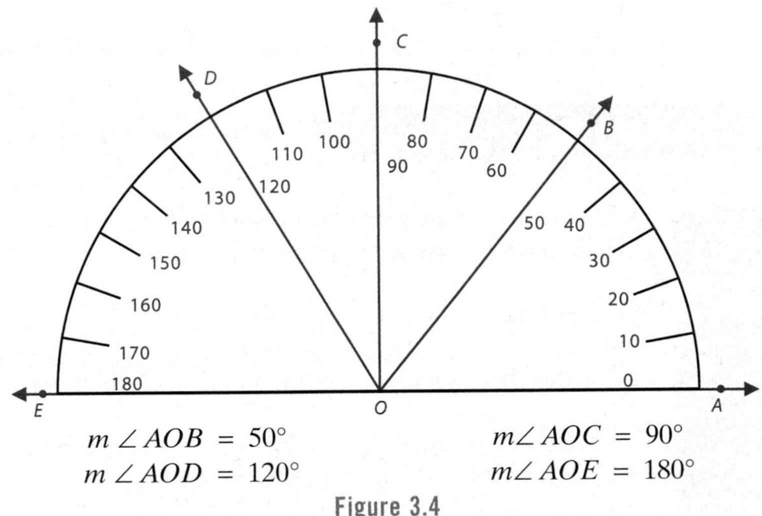

Figure 3.4

Teaching Notes:
You might want to show the students other types of similar geometric figures such as squares and rectangles. Related questions might be, "Are all squares similar?" or "Why are all squares similar, yet all rectangles are not similar?" You could also show them similar non-rectangular quadrilaterals and similar hexagons and develop a general discussion of the ideas related to the corresponding angles and proportionality.

To measure an angle with a protractor, lay the bottom edge of the protractor along one side of the angle with the vertex at the marked centerpoint on the protractor. Then, read the measure from the protractor where the other side crosses the arch part of the protractor.

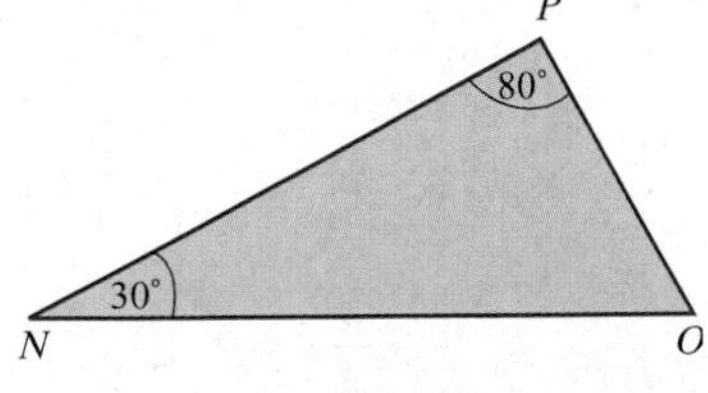

Figure 3.5

To illustrate the notation and information about triangles, consider triangle ***NOP*** (symbolized ΔNOP) in Figure 3.5. In this triangle,

$$m\angle N = 30° \quad \text{and} \quad m\angle P = 80°$$

The sum of the measures of the three angles must be 180°. Therefore, we can set up and solve an equation for the unknown $m\angle O$ as follows:

$$\begin{aligned} m\angle N + m\angle P + m\angle O &= 180^\circ \\ 30^\circ + 80^\circ + m\angle O &= 180^\circ \\ 110^\circ + m\angle O &= 180^\circ \\ m\angle O &= 180^\circ - 110^\circ \\ m\angle O &= 70^\circ \end{aligned}$$

In this manner, if we know the measures of two angles of a triangle, we can always find the measure of the third angle.

Two triangles are said to be **similar triangles** if they have the same "shape." They may or may not have the same "size." More formally, two triangles are similar if they have the following two properties:

Similar Triangles

In two similar triangles:

1. *Their* ***corresponding angles are equal****. (The corresponding angles have the same measure.)*
2. *Their* ***corresponding sides are proportional****.*

In similar triangles, **corresponding sides** are those sides opposite the equal angles in the respective triangles. (See Figure 3.6.)

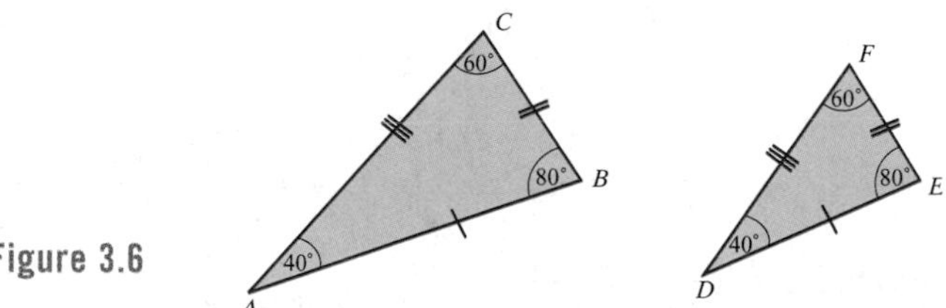

Figure 3.6

We write $\Delta ABC \sim \Delta DEF$. **(~ is read "is similar to.")** The corresponding sides are proportional, so the ratios of corresponding sides are equal and

$$\frac{AB}{DE} = \frac{BC}{EF} \quad \text{and} \quad \frac{AB}{DE} = \frac{AC}{DF} \quad \text{and} \quad \frac{BC}{EF} = \frac{AC}{DF}$$

To say that corresponding sides are proportional means that we can set up a proportion to solve for one of the unknown sides of two similar triangles. Example 6 illustrates how this can be done.

Example 6: Similar Triangles

Given that $\Delta ABC \sim \Delta PQR$, use the fact that corresponding sides are proportional and find the values of x and y.

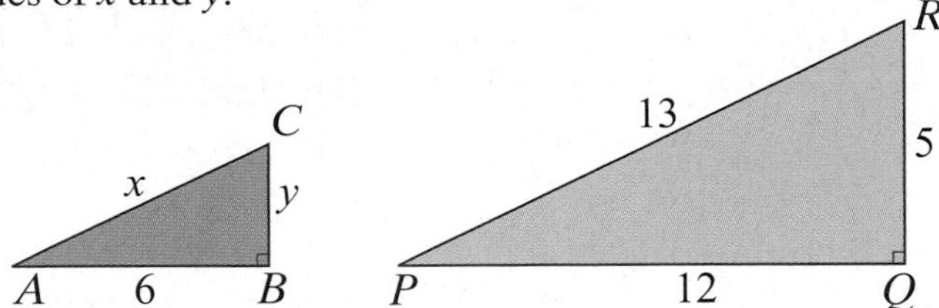

Solution: Set up two proportions and solve for the unknown terms.

$$\frac{x}{13} = \frac{6}{12} \qquad\qquad \frac{y}{5} = \frac{6}{12}$$

$$\frac{\cancel{13}}{1} \cdot \frac{x}{\cancel{13}} = \frac{13}{1} \cdot \frac{6}{12} \qquad\qquad \frac{\cancel{5}}{1} \cdot \frac{y}{\cancel{5}} = \frac{5}{1} \cdot \frac{6}{12}$$

$$x = \frac{13}{2}\left(\text{or } x = 6\frac{1}{2}\right) \qquad\qquad y = \frac{5}{2}\left(\text{or } y = 2\frac{1}{2}\right)$$

OR: Reducing first may lead to equations that are easier to solve. The answers will be the same.

$$\frac{x}{13} = \frac{6}{12} \qquad\qquad \frac{y}{5} = \frac{6}{12}$$

$$\frac{x}{13} = \frac{1}{2} \qquad\qquad \frac{y}{5} = \frac{1}{2}$$

$$\frac{\cancel{13}}{1} \cdot \frac{x}{\cancel{13}} = \frac{13}{1} \cdot \frac{1}{2} \qquad\qquad \frac{\cancel{5}}{1} \cdot \frac{y}{\cancel{5}} = \frac{5}{1} \cdot \frac{1}{2}$$

$$x = \frac{13}{2}\left(\text{or } x = 6\frac{1}{2}\right) \qquad\qquad y = \frac{5}{2}\left(\text{or } y = 2\frac{1}{2}\right)$$

Example 7: Corresponding Angles

Find the values of x and y in triangles ΔABC and ΔADE.

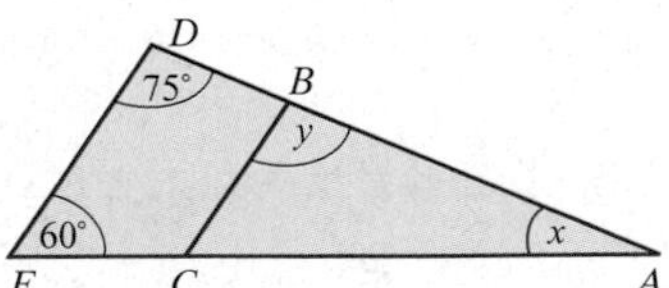

Solution: Let x be the measure of $\angle A$ and y be the measure of $\angle B$ in ΔABC as illustrated. Since y is in the same relative position as 75° in ΔADE ($\angle B$ and $\angle D$ are corresponding angles), we have $y = 75°$. In ΔADE, using the fact that the sum of the measures of the angles of a triangle must be 180° gives the equation $x + 75 + 60 = 180$. Solving for x gives $x = 45°$.

3.4 Exercises

1. 1
2. $\frac{2}{5}$
3. $\frac{1}{3}$
4. $\frac{1}{15}$
5. $\frac{1}{2}$
6. $\frac{6}{5}$
7. 50 miles / hr
8. $\frac{19\text{ mi}}{1\text{ gal}}$
9. 1
10. $\frac{1\text{ hit}}{4\text{ at-bats}}$
11. $\frac{\$7\text{ profit}}{\$100\text{ invested}}$
12. $\frac{\$3\text{ profit}}{\$10\text{ invested}}$
13. True
14. False
15. True
16. False
17. True
18. True
19. $x = 12$
20. $y = 60$
21. $x = 12\frac{1}{2}$
22. $A = 3$
23. $x = 156$
24. $x = 30$
25. $A = 180$
26. $B = 900$
27. $w = 6$
28. $x = 18$
29. $y = 72$
30. $x = -3$
31. $8.40
32. $24.84

In Exercises 1 – 12, write each of the comparisons as ratios reduced to lowest terms. Use common units in the numerator and denominator whenever possible.

1. 15 nickels to 3 quarters

2. 1 dime to 5 nickels

3. 8 hours to 1 day

4. 4 minutes to 1 hour

5. 12 inches to 2 feet

6. 2 yards to 5 feet

7. 250 miles to 5 hours

8. 38 miles to 2 gallons of gas

9. 100 centimeters to 1 meter

10. 125 hits to 500 times at bat

11. $70 profit to $1000 invested

12. $300 profit to $1000 invested

Determine whether each proportion is true or false in Exercises 13 – 18.

13. $\frac{5}{6} = \frac{10}{12}$

14. $\frac{2}{7} = \frac{5}{17}$

15. $\frac{-7}{21} = \frac{-4}{12}$

16. $\frac{6}{7} = \frac{8}{9}$

17. $\frac{-62}{31} = \frac{102}{-51}$

18. $\frac{8\frac{1}{2}}{2\frac{1}{3}} = \frac{4\frac{1}{4}}{1\frac{1}{6}}$

In Exercises 19 – 30, solve each proportion.

19. $\frac{3}{6} = \frac{6}{x}$

20. $\frac{3}{5} = \frac{y}{100}$

21. $\frac{x}{6} = \frac{25}{12}$

22. $\frac{A}{20} = \frac{15}{100}$

23. $\frac{78}{13} = \frac{x}{26}$

24. $\frac{3}{10} = \frac{x}{100}$

25. $\frac{A}{1000} = \frac{18}{100}$

26. $\frac{135}{B} = \frac{15}{100}$

27. $\frac{1}{4} = \frac{1\frac{1}{2}}{w}$

28. $\frac{18}{x+3} = \frac{6}{7}$

29. $\frac{2}{5} = \frac{26}{y-7}$

30. $\frac{x+11}{38} = \frac{4}{19}$

In Exercises 31 – 43, solve by using proportions.

31. Sales tax on a $16.00 CD is $1.40. What would be the sales tax on a CD player on sale for $96.00?

32. The last time you put gas in your car, 5 gallons of gasoline costs $10.35. You know your car has a 15 gallon tank and it has 3 gallons of gas in it. What would it cost to fill up the tank?

33. 5.5 in.

34. 60 ft.

35. $135

36. $500

37. Investor B, $100

38. 8 hrs

39. 259,200 revolutions

40. 60 ft.

41. 3 in. for the width, 7.5 in. for length

33. An architect plans to make a drawing of a house that uses the scale of 2.5 inches to 15 feet. If two points in the house are known to be 33 feet apart, how many inches apart on the drawing should these two points be?

34. Henry wanted to know the height of a tree in his yard. At a certain time of day he measured the shadow of the tree to be 21 feet long. At the same time of day a nearby lamppost that he knew was 10 feet tall cast a shadow 3.5 feet long. Henry said that he now knew the height of the tree. What was the height of the tree?

35. Investor A thinks that she should make $9 for every $100 she invests. How much does she expect to make on an investment of $1500?

36. Investor B thinks that she should make $15 for every $150 she invests. How much does she expect to make on an investment of $5,000?

37. Referring to Exercises 35 and 36, which investor would expect to make the most on an investment of $10,000? How much more?

38. You know that you drove to your grandmother's house in 6 hours. How long would you estimate to drive to your cousin's house if your grandmother lives 276 miles away and your cousin lives 368 miles away?

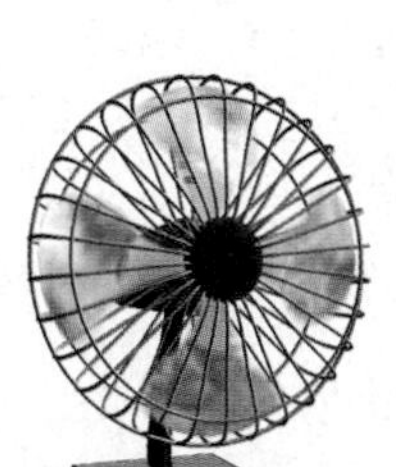

39. An electric fan makes 180 revolutions per minute. How many revolutions will the fan make if it runs for 24 hours?

40. An engineer would like to know the length of the shadow of a building that is 21 stories high at a particular time of day; however, this building is across town, and he needs the information now. So he simply goes outside and measures the length of the shadow of the building he is in, which is 14 stories tall. The shadow of his building is 40 feet long. With this information, he calculates the length of the shadow of the building across town. What is the length of that shadow?

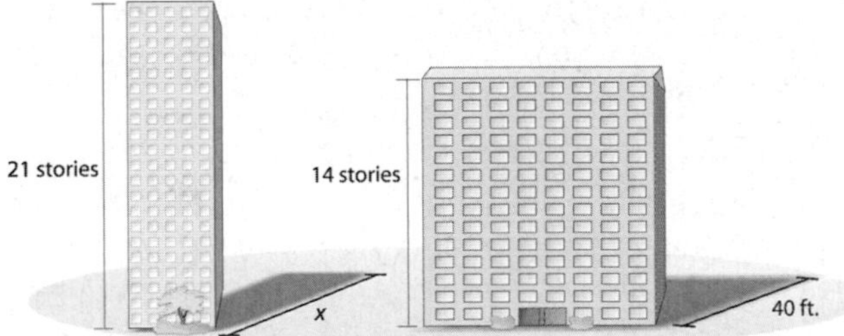

41. An architect is to draw plans for a city park. He intends to use a scale of $\frac{1}{2}$ inch to represent 25 feet. How many inches will be needed to use for the length and width of a rectangular playing field that is 50 yards by 125 yards? (1 yard = 3 feet)

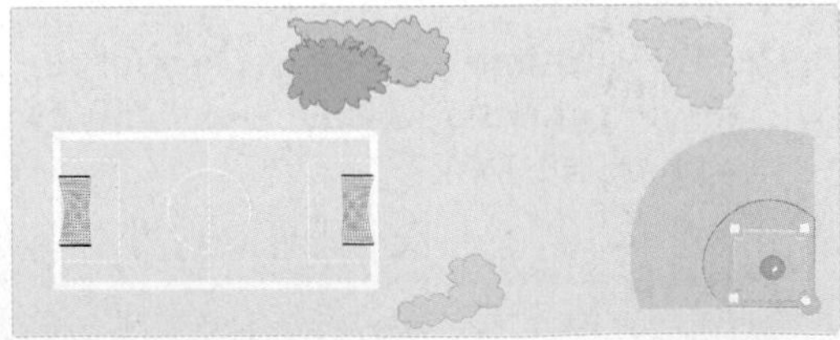

42. 7 in. by 10 in.

43. a. 7.5 mph
b. 55 mph

44. $x = 7\frac{1}{2}$, $y = 15$

45. $x = 3$, $y = 3$

46. $x = 7$, $y = 5$

47. $x = 6$, $y = 4$

48. $x = 50°$, $y = 60°$

49. $x = \frac{5}{2}$, $y = 2$

42. A cartographer (mapmaker) plans to use a scale of 2 inches to represent 30 miles. What rectangular shape of paper will she need to draw on if she plans to map a region that is 75 miles wide and 120 miles long and she wants to leave a 1-inch margin around all four edges of the paper?

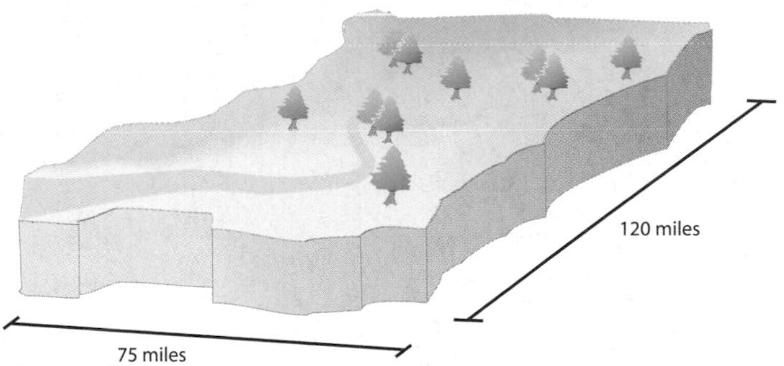

43. A test driver wants to increase the speed of the car he is driving by 3 miles per hour every 2 seconds. But he can only check his speed every 5 seconds because he is busy with other items during the test drive.

a. By how much should he increase his speed in 5 seconds?

b. If he starts checking his speed at 40 miles per hour, how fast should he be going in 10 seconds?

Exercises 44 – 51 each illustrate a pair of similar triangles. Find the values of x and y in each of these exercises.

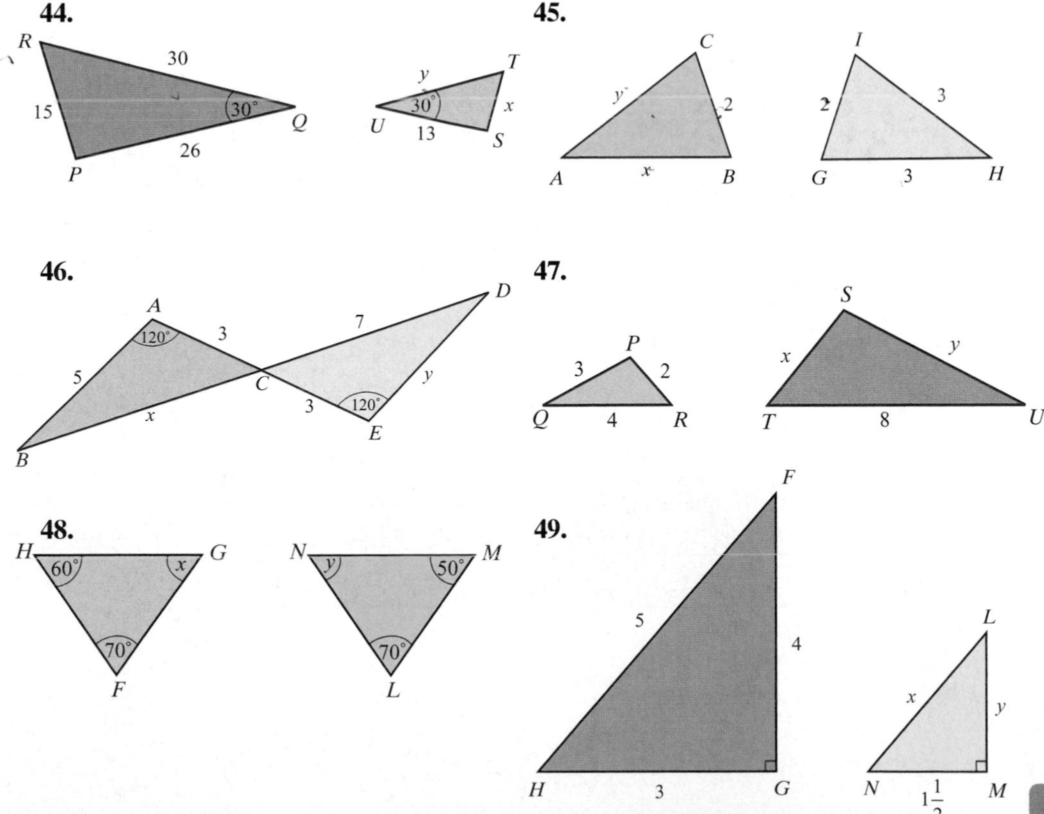

50. $x = 50°, y = 50°$

51. $x = 20°, y = 100°$

52. $a = 4, b = 5$

53. $a = 7, b = 5$

54. $a = 3, b = 7$

55. $x = 80°, y = 50°$

56. $x = 50°, y = 80°$

50.

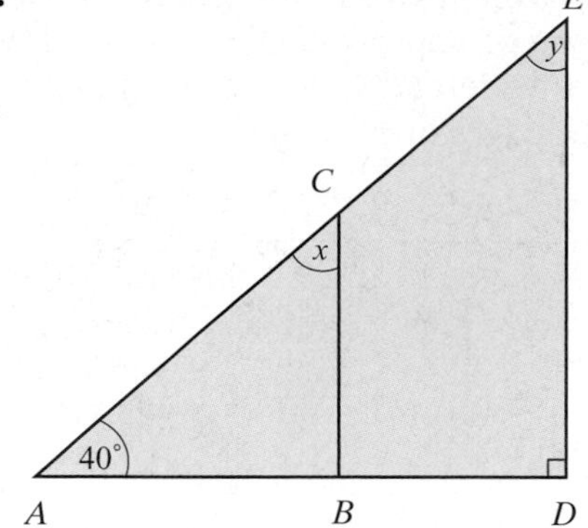

51.

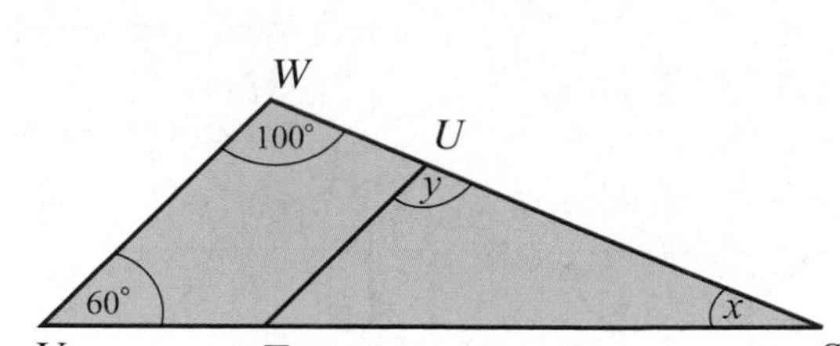

Exercises 52 – 58 each illustrate a pair of similar triangles. Find the values of the unknown variables in each of these exercises.

52.

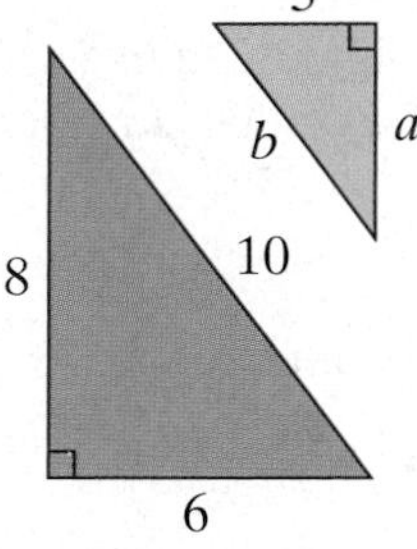

53.

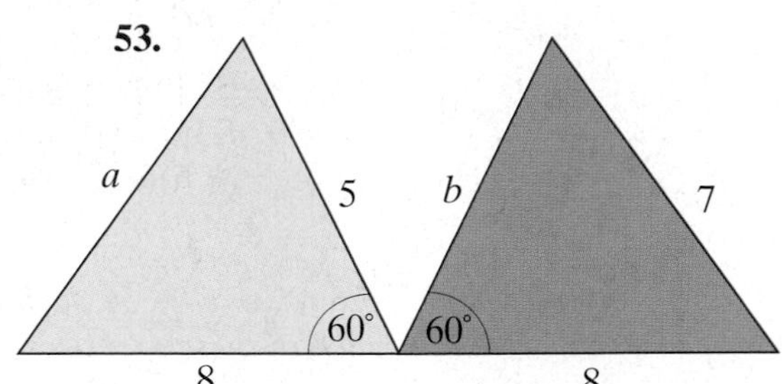

54.

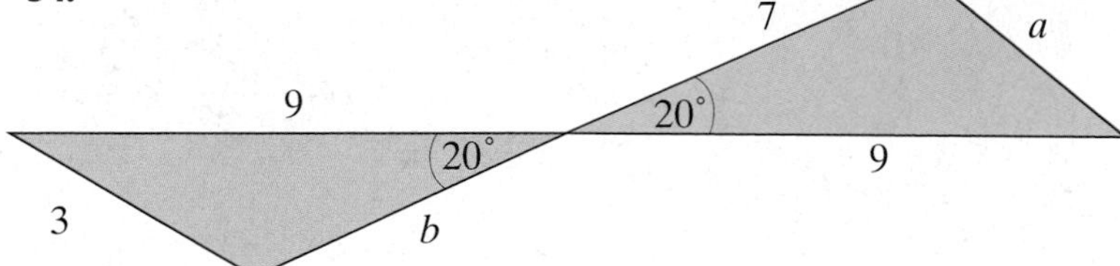

55.

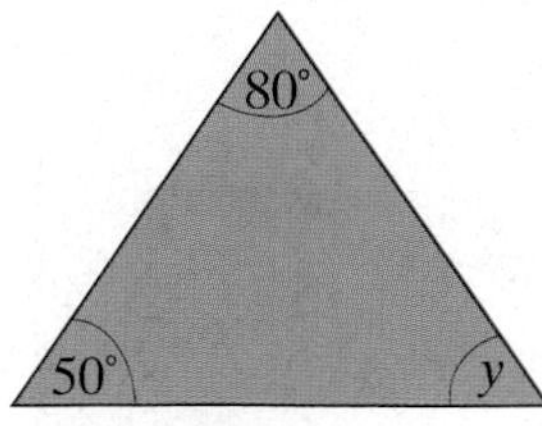

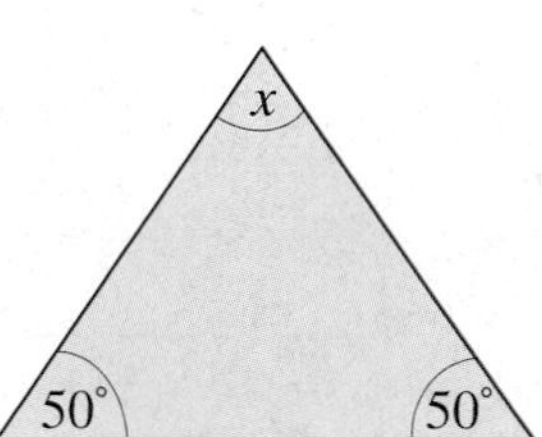

56.

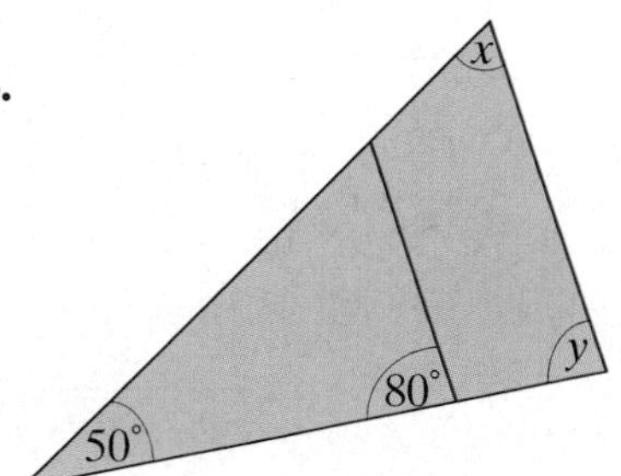

57. $a = 10, b = 8$
$x = 60°, y = 30°$

58. $x = 80°, y = 80°$

59. a. 55 lbs. **b.** 6 bags
c. $72

60. 400 ft.

61. 225, 100

57.

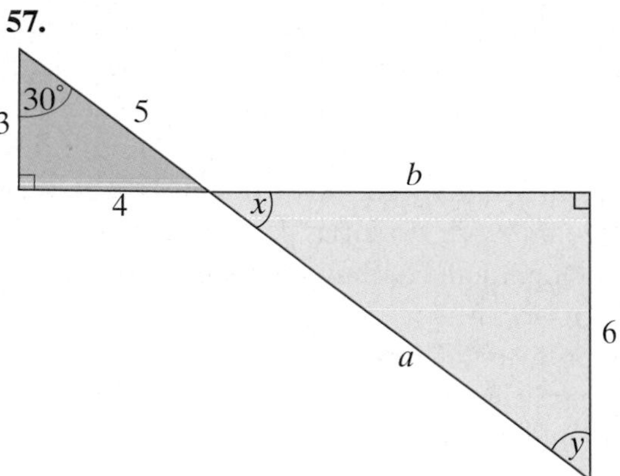

58.

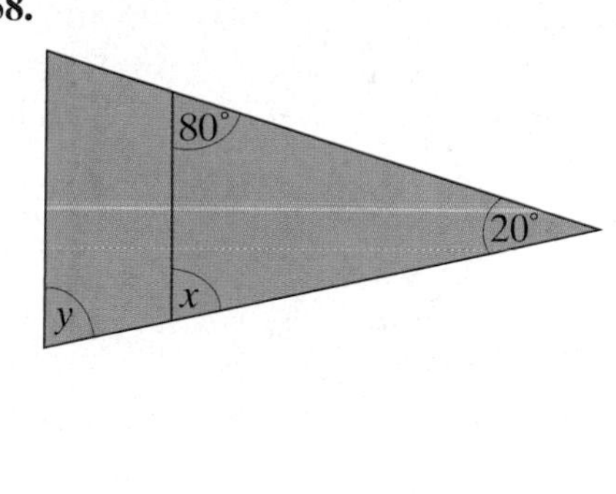

In Exercises 59 – 64, solve by using proportions.

59. One bag of Fertilizer & Weed Killer contains 10 pounds of fertilizer and weed treatment with a recommended coverage of 1200 square feet.

a. Your brother measured his lawn and asked you how many pounds of fertilizer he should buy. His lawn consists of two rectangular shapes, one 30 feet by 100 feet and the other 40 feet by 90 feet. How many pounds should he buy?

b. How many bags of fertilizer would you tell him to buy?

c. If each bag costs $12, how much would he have to pay?

60. A building casts a shadow 100 feet long and, at the same time of day, a man 6 feet tall casts a shadow $1\frac{1}{2}$ feet long. What is the height of the building?

61. A computer manufacturer is told to expect 3 defective microchips out of every 2000 produced by a particular machine. A second, older and slower, machine produces 1 defective chip in 1500 in the same amount of time. How many defective microchips should be expected from each machine in a production run of 150,000 chips?

62. 320 miles

63. 54 minutes

64. 8.75 hours

65. a. The statement is misleading because the numbers 4 and 5 are not in the same units.
b. The ratio of 4 quarters to 5 dollars is 1:5.

66. Figures drawn will vary. They are similar to the figures shown provided their corresponding angles are equal and their corresponding sides are proportional.

62. On a road map, two cities are in a straight line 5 inches apart. You know for a fact that these cities are 200 miles apart. On the same map, two other cities are in a straight line 8 inches apart. How many miles apart are these two cities?

63. Bruce can barbeque 9 hamburgers in 12 minutes. If he is having a large party for a total of 72 guests, and each person is going to eat one hamburger, how long will it take Bruce to cook enough hamburgers for everyone?

64. Eleanor's mother lives 78 miles away. Driving there takes Eleanor 1.5 hours. Eleanor's daughter lives 455 miles away. How long should Eleanor estimate a drive to her daughter's house takes?

Writing and Thinking About Mathematics

65. a. Explain why the following statement is misleading: The ratio of 4 quarters to 5 dollars is 4 : 5.
b. Rewrite the statement so that it is not misleading.

66. Many types of geometric figures are similar. For example, all circles are similar. For figures with line segments as sides (called polygons), corresponding sides are proportional and corresponding angles are equal (just as with triangles). Draw two figures similar to the figure shown here and explain why they are similar to this figure and to each other.

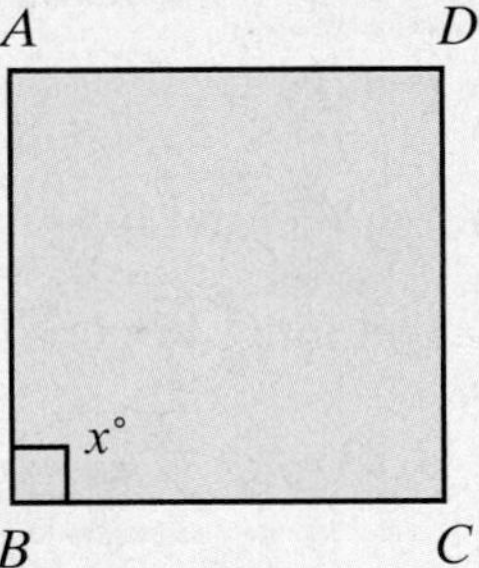

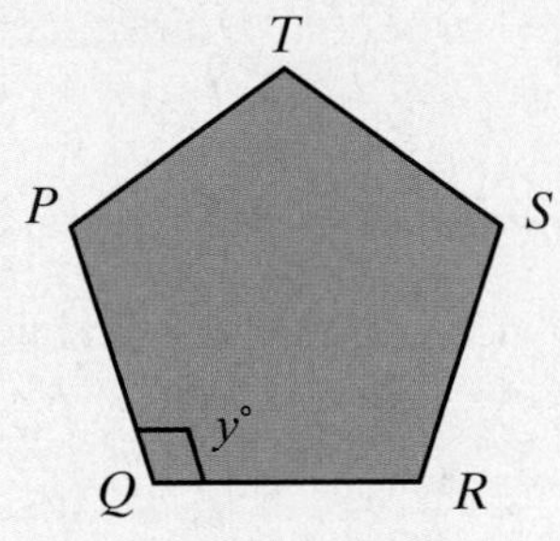

Hawkes Learning Systems: Introductory & Intermediate Algebra

Solving Equations: Ratios and Proportions

3.5 Linear Inequalities

Objectives

After completing this section, you will be able to:

1. *Understand intervals of real numbers.*
2. *Solve linear inequalities.*
3. *Write the solutions for inequalities using interval notation.*
4. *Graph the solutions for inequalities on real number lines.*
5. *Solve applications by using linear inequalities.*

Intervals of Real Numbers

Real numbers, graphs of real numbers on real number lines, and inequalities with the symbols <, >, ≤, and ≥ were discussed in Section 1.1. In this section, we will expand our work with inequalities to include the concept of **intervals of real numbers**. Remember that dots are used to indicate the graphs of individual real numbers on a real number line.

For example, the set of numbers $\left\{-\frac{3}{2}, -1, 0, 2, \sqrt{10}\right\}$ is graphed in Figure 3.7.

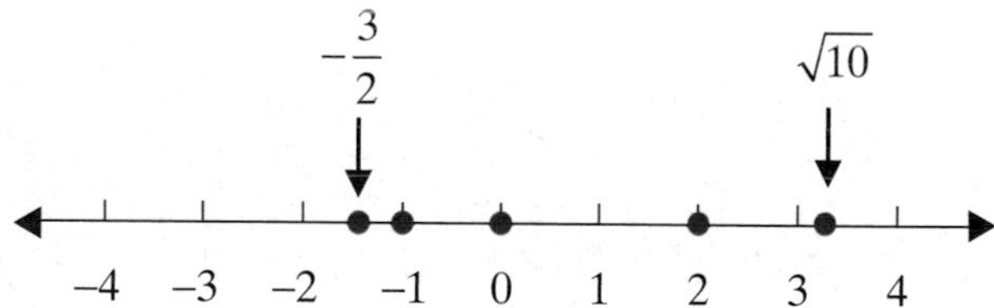

Figure 3.7

Irrational numbers, such as π, $\sqrt{2}$, $-\sqrt{7}$, and $\sqrt{10}$ have been discussed to some extent and will be discussed in more detail in Chapter 10. What is important here is that irrational numbers do indeed correspond to points on a real number line. In fact, we can make the following statement concerning real numbers and points on a line.

One-To-One Correspondence

There is a one-to-one correspondence between the real numbers and the points on a line. *That is, each point on a number line corresponds to one real number, and each real number corresponds to one point on a number line.*

Teaching Notes:
I have chosen to use the open and closed circles for the endpoints. You may want to introduce the other common techniques of parentheses and brackets.

Now, when an inequality such as $x < 8$ is given, the implication is that x can be **any real number less than 8**. That is, the **solution** to the inequality is the set of all real numbers less than 8. To indicate this solution graphically, an open circle is drawn around 8 on a number line and the line is shaded to the left of 8 as shown in Figure 3.8.

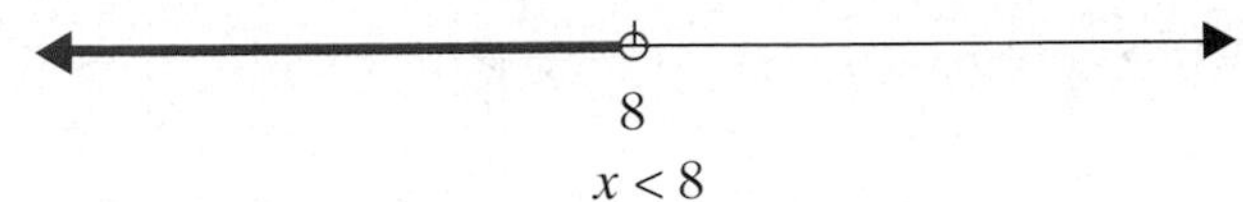

Figure 3.8

As an aid in reading inequalities and graphing inequalities correctly, note that an inequality may be read either from right to left or left to right. Also, read the variable first. For example,

Teaching Notes:
Inequalities seem to be underplayed in importance in the classroom. The "real world" uses inequality concepts at every turn. Cite several examples to illustrate common thinking in inequality terms: "...made to fit *within certain limits...*, ...manufactured to a specific *tolerance...*, ...must have an annual income *greater than...*, ...the ten year financial goal is to have grown *at least* by..., ...we need a *minimum* of...."

$x > 7$ is read from left to right as **"x is greater than 7."**

and $7 < x$ is read from right to left as **"x is greater than 7."**

A compound interval such as $1 < t \le 3$ is read

"t is greater than 1 and t is less than or equal to 3."

Intervals are classified and graphed as indicated in the Types of Intervals shown in Table 3.1 below. You should keep in mind the following facts as you study the tables:

1. x is understood to represent real numbers.
2. Open dots at endpoints a and b indicate that these points **are not** included in the graph.
3. Solid dots at endpoints a and b indicate that these points **are** included in the graph.

Types of Intervals

Name of Interval	Algebraic Notation	Interval Notation	Graph
Open Interval	$a < x < b$	(a, b)	a b
Closed Interval	$a \le x \le b$	$[a, b]$	a b
Half - open Interval	$a \le x < b$ $a < x \le b$	$[a, b)$ $(a, b]$	a b a b
Open Interval	$x > a$ $x < b$	(a, ∞) $(-\infty, b)$	a b
Half-open Interval	$x \ge a$ $x \le b$	$[a, \infty)$ $(-\infty, b]$	a b

Table 3.1

Example 1: Graph

Teaching Notes:
I tell my students that half-open intervals could also be called half-closed intervals. We generally agree to use just one term, namely, "half-open." Also, I tell them that there is no endpoint for an interval such as $x > 3$. We cannot go to some point called infinity.

a. Represent the following graph using algebraic notation, and tell what kind of interval it is.

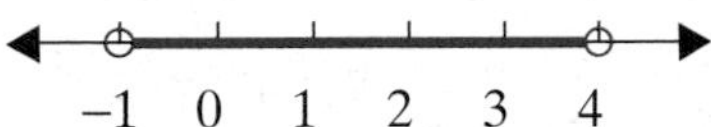

Solution: $-1 < x < 4$ is an open interval.

b. Represent the following graph using interval notation, and tell what kind of interval it is.

Solution: $[2, \infty)$ is a half-open interval.

c. Graph the half-open interval $-3 < x \leq 0$.

Solution:

Solving Linear Inequalities

In this section, we will solve **linear inequalities** such as $6x + 5 \leq -7$ and write the solution in interval notation as $(-\infty, -2]$. We say that "x is in $(-\infty, -2]$." (**Note:** The symbol for infinity, ∞ (or $-\infty$), is not a number. It is used to indicate that the interval is to continue without end.)

Linear Inequality

For real numbers a, b, and c ($a \neq 0$), $\mathbf{ax + b < c}$ *is called a* ***linear inequality*** *in x.*

(The definition also holds if $\leq$, $>$, or $\geq$ is used instead of $<$.)

Solving linear inequalities is similar to solving linear equations, with one important difference: **Multiplying or dividing both sides of an inequality by a negative number reverses the sense of the inequality**. "Is less than" becomes "is greater than," and vice versa.

For example,

Multiplying by –1	Multiplying by –4	Dividing by –6
a. $3 < 6$ $-1(3) ? -1(6)$ $-3 > -6$ ↑ reversed	**b.** $-2 \le 5$ $-4(-2) ? -4(5)$ $8 \ge -20$ ↑ reversed	**c.** $-6x > 12$ $\frac{-6x}{-6} ? \frac{12}{-6}$ $x < -2$ ↑ reversed

To solve first-degree inequalities, perform the following procedures and use the properties on these pages.

To Solve a Linear Inequality

1. *Simplify each side of the inequality by removing any grouping symbols and combining like terms.*
2. *Add the opposites of constants and/or variable expressions to both sides so that variables are on one side and constants are on the other.*
3. *Divide both sides by the coefficient of the variable and*
   a. *leave the direction of the inequality unchanged if the coefficient is positive; or*
   b. *reverse the direction of the inequality if the coefficient is negative.*

NOTES

Unless otherwise stated, we will assume that the set of values allowed for the variable (called the **replacement set**) in an inequality is the set of all real numbers. The **solution set** must be a part of the replacement set.

The procedure for solving linear inequalities uses the following procedures for finding equivalent inequalities just as we used the properties of equations to solve linear equations.

Addition Property of Inequality

If the same algebraic expression is added to both sides of an inequality, the new inequality is equivalent to the original inequality.

$$\text{If } A < B, \text{ then } A + C < B + C.$$

$$\text{If } A < B, \text{ then } A - C < B - C.$$

Multiplication Property of Inequality

If both sides of an inequality are multiplied (or divided) by the same positive expression, then the new inequality is equivalent to the original inequality.

$$\begin{cases} \text{If } A < B \text{ and } C > 0, \text{ then } AC < BC. \\ \text{If } A < B \text{ and } C > 0, \text{ then } \dfrac{A}{C} < \dfrac{B}{C}. \end{cases}$$

If both sides of an inequality are multiplied (or divided) by the same negative expression and the inequality is reversed, then the new inequality is equivalent to the original inequality.

$$\begin{cases} \text{If } A < B \text{ and } C < 0, \text{ then } AC > BC. \\ \text{If } A < B \text{ and } C < 0, \text{ then } \dfrac{A}{C} > \dfrac{B}{C}. \end{cases}$$

Example 2: Solving Linear Inequalities

Solve the following inequalities and graph their solution sets. Write the solutions in interval notation. As a check, you can substitute a number into the original inequality. (Any number will do. 0 is always easy to use.) If the result is a true statement, then this number should be in the solution interval. If the result is false, then this number should not be in the solution interval. **This check does not guarantee that the solution interval is correct**, but it can help determine whether the interval is headed in the right direction.

a. $6x + 5 \le -1$

Solution: $6x + 5 \le -1$

$6x \le -6$ — Add -5 to both sides.

$x \le -1$ — Divide both sides by 6.

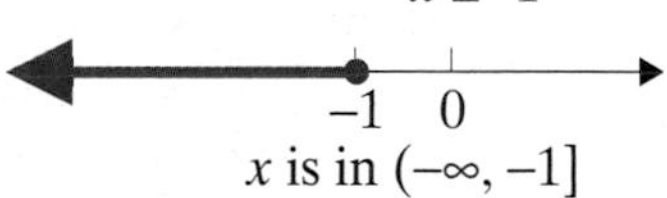

x is in $(-\infty, -1]$ — Use interval notation. Note that the interval $(-\infty, -1]$ is a half-open interval.

As a check, substituting 0 for x in the original inequality gives

$$6 \cdot 0 + 5 \le -1$$
$$0 + 5 \le -1$$
$$5 \le -1$$

This last statement is false and we see that 0 is not in the solution interval. This result indicates that the solution interval is going in the right direction.

continued on next page ...

b. $x - 3 > 3x + 4$

Solution:

$x - 3 > 3x + 4$

$-3 > 2x + 4$ Add $-x$ to both sides.

$-7 > 2x$ Add -4 to both sides.

$\frac{-7}{2} > x$ Divide both sides by 2.

or $x < -\frac{7}{2}$

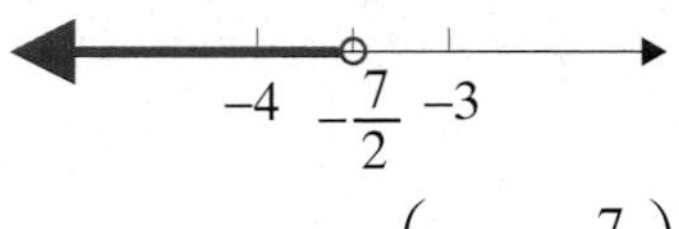

x is in $\left(-\infty, -\frac{7}{2}\right)$ Use interval notation. Note that the interval $\left(-\infty, -\frac{7}{2}\right)$ is an open interval.

c. $6 - 4x \le x + 1$

Solution:

$6 - 4x \le x + 1$

$6 - 5x \le 1$ Add $-x$ to both sides.

$-5x \le -5$ Add -6 to both sides.

$x \ge 1$ Divide both sides by -5. Note the reversal of the inequality sign!

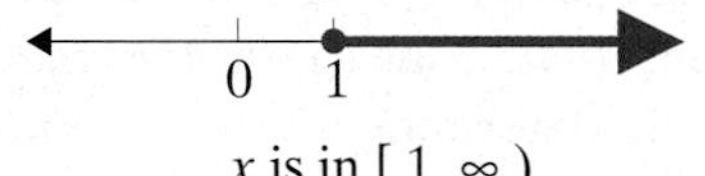

x is in $[1, \infty)$ Use interval notation. Note that the interval $[1, \infty)$ is a half-open interval.

d. $2x + 5 < 3x - (7 - x)$

Solution:

$2x + 5 < 3x - (7 - x)$

$2x + 5 < 3x - 7 + x$ Remove parentheses first.

$2x + 5 < 4x - 7$ Combine like terms.

$5 < 2x - 7$ Add $-2x$ to both sides.

$12 < 2x$ Add 7 to both sides.

$6 < x$ Divide both sides by 2.

5 6

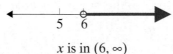

x is in $(6, \infty)$ Use interval notation. Note that the interval $(6, \infty)$ is an open interval.

Inequalities with Three Parts

Inequalities with three parts can arise when a variable or variable expression is to be between two numbers. (We will see this with some absolute value inequalities, too.) For example, the inequality

$$5 < x + 3 < 10$$

indicates that $x + 3$ is between 5 and 10. To solve this inequality (to isolate the variable x), subtract 3 from each part of the inequality as follows:

$$5 < x + 3 < 10$$
$$5 - 3 < x + 3 - 3 < 10 - 3$$
$$2 < x < 7$$

Thus, the values of x must be between 2 and 7. The solution set is the open interval (2, 7) and its graph is the following.

NOTES

Be careful when you write a three part inequality that the transitive property holds. For example writing $10 < x + 3 < 5$ would be wrong because the transitive property indicates $10 < 5$ which is false. With three part inequalities, both of the inequality symbols should point in the same direction and point toward the smaller number.

Example 3: Solving Inequalities with Three Parts

Solve the three part inequality $-3 \leq 4x - 1 \leq 11$ and graph the solution set.

Solution:

$-3 \leq 4x - 1 \leq 11$	Write the inequality.
$-3 + 1 \leq 4x - 1 + 1 \leq 11 + 1$	Add 1 to each part.
$-2 \leq 4x \leq 12$	Simplify.
$\frac{-2}{4} \leq \frac{4x}{4} \leq \frac{12}{4}$	Divide each part by 4.
$-\frac{1}{2} \leq x \leq 3$	Simplify.

The solution set is the closed interval $\left[-\frac{1}{2}, 3\right]$.

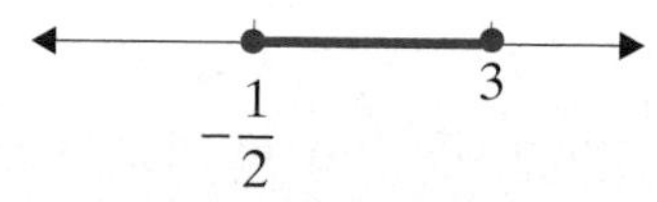

Applications of Linear Inequalities

We are generally familiar with the use of equations to solve word problems. In the following two examples, we show that solving inequalities can be related to real-world problems.

Example 4: Applications

a. A physics student has grades of 85, 98, 93, and 90 on four examinations. If he must average 90 or better to receive an A for the course, what score can he receive on the final exam and earn an A? (Assume that the final exam counts the same as the other exams.)

Solution: Let x = score on final exam. The average is found by adding the scores and dividing by 5.

$$\frac{85+98+93+90+x}{5} \geq 90$$

$$\frac{366+x}{5} \geq 90$$

$$\cancel{5}\left(\frac{366+x}{\cancel{5}}\right) \geq 5 \cdot 90$$

$$366+x \geq 450$$

$$x \geq 450-366$$

$$x \geq 84$$

If the student scores 84 or more on the final exam, he will average 90 or more and receive an A in physics.

b. Ellen is going to buy 30 stamps, some 23-cent and some 37-cent. If she has $8.02, what is the maximum number of 37-cent stamps she can buy?

Solution: Let x = number of 37-cent stamps, then $30 - x$ = number of 23-cent stamps. Ellen cannot spend more than $8.02.

$$0.37x+0.23(30-x) \leq 8.02$$

$$0.37x+6.90-0.23x \leq 8.02$$

$$0.14x+6.90 \leq 8.02$$

$$0.14x \leq 8.02-6.90$$

$$0.14x \leq 1.12$$

$$x \leq 8$$

Ellen can buy at most eight 37-cent stamps if she buys a total of 30 stamps.

Practice Problems

Solve the inequalities and graph the solutions.

1. $7 + x < 3$ **2.** $\frac{x}{2} + 1 \geq \frac{x}{3}$ **3.** $-5 \leq 2x + 1 < 9$

3.5 Exercises

1. False; $3 > -3$
2. False; $-14 < 4$
3. True **4.** True **5.** True
6. True **7.** True **8.** True
9. True **10.** True
11. half-open interval

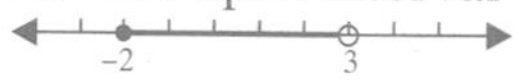

12. closed interval

13. half-open interval

14. open interval

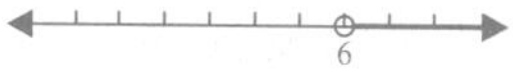

15. half-open interval

16. half-open interval

17. half-open interval

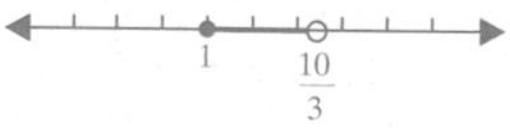

18. open interval

19. open interval

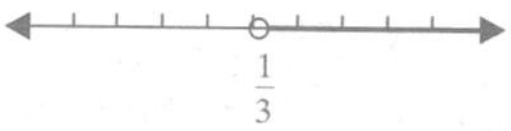

20. closed interval

In Exercises 1 – 10, determine whether each of the inequalities is true or false. If the inequality is false, rewrite it in a form that is true using the same numbers or expressions.

1. $3 < -3$ **2.** $-14 > 4$ **3.** $|-5| > -5$ **4.** $|-7| \leq 7$

5. $-3.4 > -3.5$ **6.** $\frac{1}{2} > \frac{1}{3}$ **7.** $-6 \leq 0$ **8.** $\frac{5}{8} \geq \frac{7}{16}$

9. $|-4| \leq |4|$ **10.** $-12.5 < -10.2$

In Exercises 11 – 20, graph the set of real numbers that satisfies each of the inequalities and tell what type of interval it represents.

11. $-2 \leq x < 3$ **12.** $-4 \leq x \leq -1$ **13.** $x \leq -4$ **14.** $x > 6$

15. $-2 < x \leq 1$ **16.** $x \geq -5$ **17.** $1 \leq x < \frac{10}{3}$ **18.** $x < \frac{13}{4}$

19. $x > \frac{1}{3}$ **20.** $3 \leq x \leq 10$

Solve the inequalities in Exercises 21 – 54 and graph the solutions.

21. $4x + 5 \geq -6$ **22.** $2x + 3 > -8$ **23.** $3x + 2 > 2x - 1$

24. $3y + 2 \leq y + 8$ **25.** $y - 6.1 \leq 4.2 - y$ **26.** $4x - 2.3 < 6x + 6.5$

27. $2.9x - 5.7 > -3 - 0.1x$ **28.** $3y - 21.9 \geq 11.04 - 3.1y$ **29.** $5y + 6 < 2y - 2$

30. $4 - 2x < 5 + x$ **31.** $4 + x > 1 - x$ **32.** $x - 6 > 3x + 5$

33. $\frac{x}{4} + 1 \leq 5 - \frac{x}{4}$ **34.** $\frac{x}{2} - 1 \leq \frac{5x}{2} - 3$ **35.** $\frac{x}{3} - 2 > 1 - \frac{x}{3}$

36. $\frac{5x}{3} + 2 > \frac{x}{3} - 1$ **37.** $-(x + 5) \leq 2x + 4$ **38.** $-3(2x - 5) \leq 3(x - 1)$

39. $x - (2x + 5) \geq 7 - (4 - x) + 10$ **40.** $x - 3(4 - x) + 5 \geq -2(3 - 2x) - x$

41. $-5 < 1 - x < 3$ **42.** $2 < 2x - 4 \leq 8$ **43.** $1 < 3x - 2 < 4$

Answers to Practice Problems: 1. $x < -4$

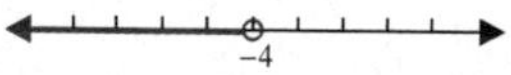

2. $x \geq -6$

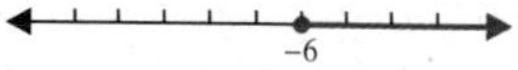

3. $-3 \leq x < 4$

21. $x \geq -\frac{11}{4}$

$-\frac{11}{4}$

22. $x > -\frac{11}{2}$

$-\frac{11}{2}$

23. $x > -3$

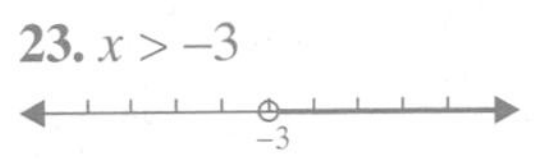

24. $y \leq 3$

25. $y \leq 5.15$

5.15

26. $x > -4.4$

−4.4

27. $x > 0.9$

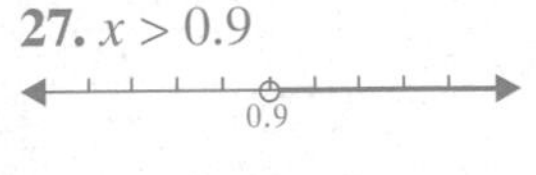

28. $y \geq 5.4$

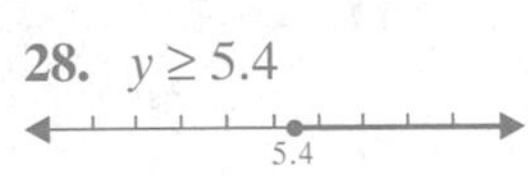

29. $y < -\frac{8}{3}$

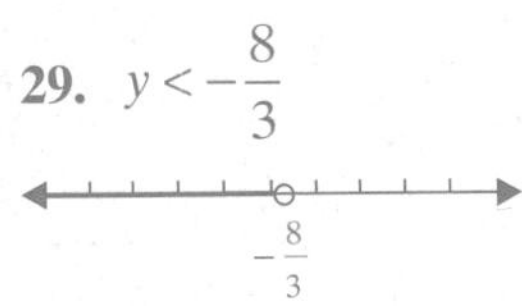

30. $x > -\frac{1}{3}$

$-\frac{1}{3}$

31. $x > -\frac{3}{2}$

$-\frac{3}{2}$

32. $x < -\frac{11}{2}$

$-\frac{11}{2}$

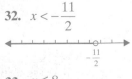

33. $x \leq 8$

8

44. $-5 < 4x + 1 < 7$

45. $-5 \leq 5x + 3 < 9$

46. $-1 < 3 - 2x < 6$

47. $\frac{3}{4} \leq \frac{1}{2}x + 6 \leq \frac{15}{2}$

48. $\frac{1}{3} \leq \frac{5}{6}x - 2 < \frac{14}{15}$

49. $-\frac{1}{2} < -\frac{2}{3}x + 5 \leq \frac{5}{9}$

50. $0 \leq -\frac{5}{8}x + 1 < \frac{9}{20}$

51. $1 \leq \frac{2}{3}x - 1 \leq 9$

52. $14 > -2x - 6 > 4$

53. $-11 \geq -3x + 2 > -20$

54. $-1.5 < 2x + 4.1 < 3.5$

55. To receive a B grade, a student must average 80 or more but less than 90. If John received a B in the course and had five grades of 94, 78, 91, 86, and 87 before taking the final exam, what were the possible grades for his final if there were 100 points possible?

1st Test	94
2nd Test	86
HW Avg.	78
Quiz Avg.	91
Last Test	87
Final Exam	??

56. The range for a C grade is 70 or more but less than 80. Before taking the final exam, Clyde had grades of 59, 68, 76, 84, and 69. If the final exam is counted as two tests, what is the minimum grade he could make on the final to receive a C? If there were 100 points possible, could he receive a B grade if an average of at least 80 points was required?

57. The temperature of a mixture in a chemistry experiment varied from 15° C to 65° C. What is the temperature expressed in degrees Fahrenheit?

$$\left[C = \frac{5}{9}(F - 32)\right]$$

58. The temperature at Braver Lake ranged from a low of 23° F to a high of 59° F. What is the equivalent range of temperatures in degrees Celsius?

$$\left[F = \frac{9}{5}C + 32\right]$$

59. The sum of the lengths of any two sides of a triangle must be greater than the third side. If a triangle has one side that is 17 cm and a second side that is 1 cm less than twice the third side, what are the possible lengths for the second and third sides?

60. The sum of four times a number and 21 is greater than 45 and less than 73. What are the possible values for the number?

34. $x \geq 1$

35. $x > \frac{9}{2}$

36. $x > -\frac{9}{4}$

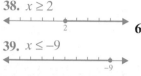

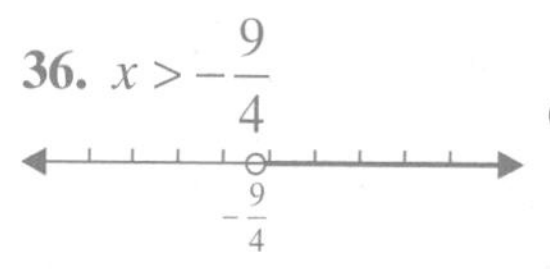

37. $x \geq -3$

38. $x \geq 2$

39. $x \leq -9$

40. $x \geq 1$

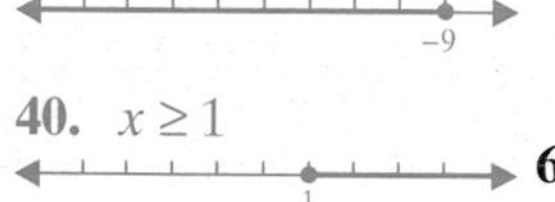

41. $-2 < x < 6$

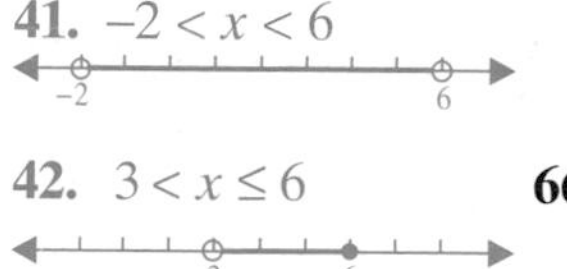

42. $3 < x \leq 6$

43. $1 < x < 2$

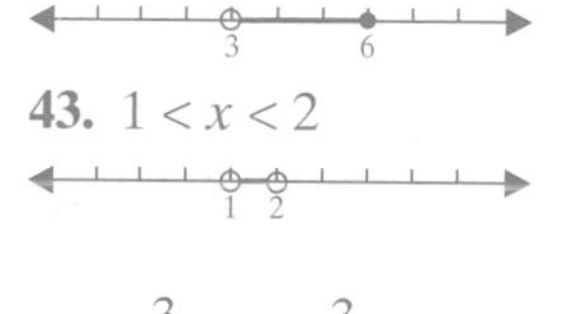

44. $-\frac{3}{2} < x < \frac{3}{2}$

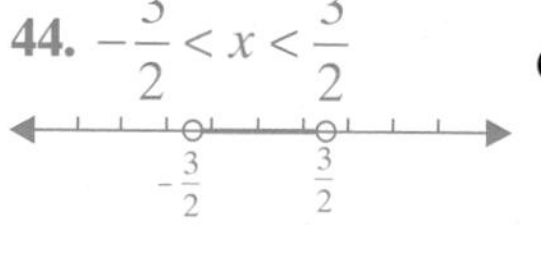

45. $-\frac{8}{5} \leq x < \frac{6}{5}$

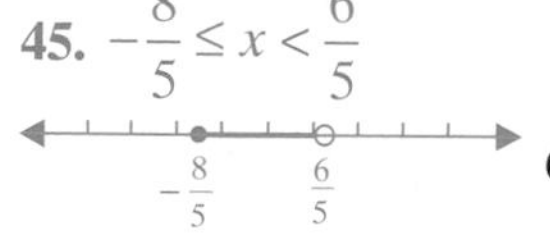

46. $-\frac{3}{2} < x < 2$

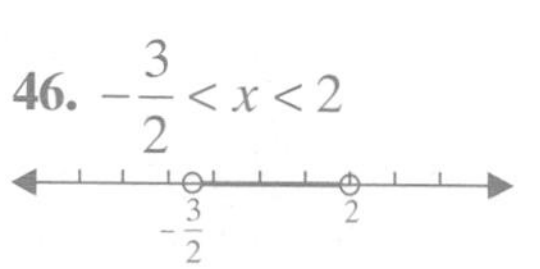

47. $-\frac{21}{2} \leq x \leq 3$

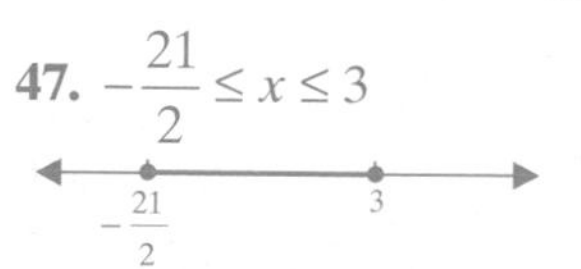

61. In order for Chuck to receive a B in his mathematics class, he must have a total of at least 400 points. If he has scores of 72, 68, 85, and 89, what scores can he make on the final and receive a B? (The maximum possible score on the final is 100.)

62. In Exercise 61, if the final exam counted twice, could Chuck receive a grade of A in the class if it takes 540 points for an A? Assume that the maximum possible score on the final is 100 points.

63. The Concert Hall has 400 seats. For a concert, the admission will be $6.00 for adults and $3.50 for students. If the expense of producing the concert is $1825, what is the least number of adult tickets that must be sold to realize a profit if all seats are taken?

64. The Pep Club is selling candied apples to raise money. The price per apple is $2.50 until Friday, when they will sell for $2.00. If the Pep Club sells 200 apples, what is the minimum number they must sell at $2.50 each in order to raise at least $475.00?

65. Better-Car Rental charges $15 per day plus $0.25 per mile and Best-Car Rental charges $25 per day plus $0.15 per mile. How many miles can be driven in one day with a Better car and still have a bill less than what a Best car would cost?

66. Gary and Susan want to have their wedding reception at a local restaurant and have found two that seem very nice. The Seaside Stop charges $500 plus $13.00 per plate for the guests and the Tall Shrimp charges $800 plus $10.50 per plate. How many guests can they have before the cost at Seaside Stop would become more than or equal to the cost at the Tall Shrimp?

67. A statistics student has grades of 82, 95, 93, and 78 on four hourly exams. He must average 90 or higher to receive an A for the course. What scores can he receive on the final exam and earn an A if:

a. the final is equivalent to a single hourly exam (100 points maximum)?

b. the final is equivalent to two hourly exams (200 points maximum)?

68. To receive a grade of B in a chemistry class, Melissa must average 80 or more but less than 90. If her five hourly exam scores were 75, 82, 90, 85, and 77, what score does she need on the final exam (100 points maximum) to earn a grade of B?

69. The sum of the lengths of any two sides of a triangle must be greater than the third side. A triangle has sides as follows: the first side is 18 mm, the second side is 3 mm more than twice the third side. What are the possible lengths of the second and third sides?

48. $\frac{14}{5} \le x < \frac{88}{25}$

49. $\frac{20}{3} \le x < \frac{33}{4}$

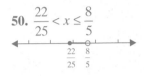

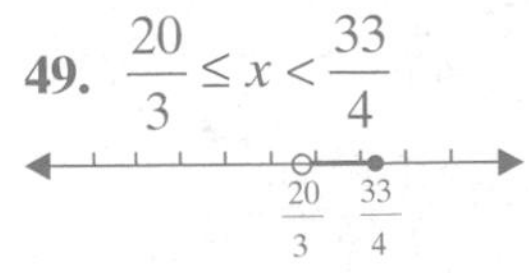

50. $\frac{22}{25} < x \le \frac{8}{5}$

51. $3 \le x \le 15$

70. Allison is going to the post office to buy 37 cent stamps and 23 cent postcards. She has $20 and wants to buy twice as many postcards as stamps. What is the largest number of 37 cent stamps she can buy?

71. You can rent a moped from Business X at $25 for the week with a free tank of gas. The same type of moped can be rented from Business Y at $15 for the week plus $1 for each mile you drive the moped. In one week, how many miles would you have to drive a moped for Business Y to be more expensive than Business X?

Hawkes Learning Systems: Introductory & Intermediate Algebra

Solving Linear Inequalities

52. $-10 < x < -5$

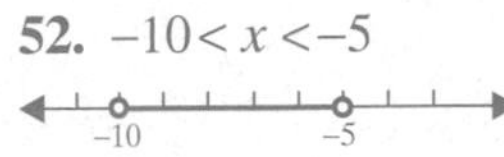

53. $\frac{13}{3} \le x < \frac{22}{3}$

54. $-2.8 < x < -0.3$

55. $44 \le x \le 100$

56. **a.** 67 **b.** No

57. 59° F to 149° F

58. –5° C to 15° C

59. The second side is between 11 cm and 35 cm and the third side is between 6 cm and 18 cm.

60. $6 < x < 13$

61. $86 \le x \le 100$

62. No

63. 171 adult tickets

64. 150 apples

65. less than 100 miles

66. less than 120 guests

67. **a.** The student cannot earn an A for the course.
b. The student must score at least 192 to earn an A for the course.

68. Melissa must score at least a 71 to earn a B for the course.

69. The second side must be more than 13 mm and less than 33 mm and the third more than 5 mm and less than 15 mm.

70. Allison can buy at most 24 of the 37 cent stamps.

71. You would have to drive more than 10 miles.

Chapter 3 Index of Key Ideas and Terms

Section 3.1 Working with Formulas

Section 3.2 Formulas in Geometry

Square: $P = 4s$
Rectangle: $P = 2l + 2w$
Parallelogram: $P = 2a + 2b$
Circle: $C = 2\pi r$ or $C = \pi d$
Trapezoid: $P = a + b + c + d$
Triangle: $P = a + b + c$

Square: $A = s^2$
Rectangle: $A = lw$
Parallelogram: $A = bh$
Circle: $A = \pi r^2$
Trapezoid: $A = \frac{1}{2}h(b+c)$
Triangle: $A = \frac{1}{2}bh$

Rectangular Solid: $V = lwh$
Rectangular Pyramid: $V = \frac{1}{3}lwh$
Right Circular Cylinder: $V = \pi r^2 h$
Right Circular Cone: $V = \frac{1}{3}\pi r^2 h$
Sphere: $V = \frac{4}{3}\pi r^3$

continued on next page ...

Section 3.2 Formulas in Geometry (continued)

Circles

Terminology and Notation Related to Circles: page 172

Term	Notation	Definition
1. Circumference	C	The perimeter of a circle is called its **circumference**.
2. Radius	r	The distance from the center of a circle to any point on the circle is called its **radius**.
3. Diameter	d	The distance from one point on a circle to another point on the circle measured through its center is called its **diameter**. (**Note:** The diameter is twice the radius: $d = 2r$.)
4. Pi	π	π is a Greek letter used for the constant 3.14159265358979… (π is an irrational number.)

Section 3.3 Applications

Applications pages 183 - 189

Number problems
Distance-rate-time
Cost
Interest
Average

Section 3.4 Ratios and Proportions

Ratio page 196

A **ratio** is a comparison of two quantities by division. The ratio of a to b $(b \neq 0)$ can be written as $\dfrac{a}{b}$ *or* $a:b$ *or* ***a to b***.

Proportions page 198

A **proportion** is a statement that two ratios are equal.
In symbols, $\dfrac{a}{b} = \dfrac{c}{d}$ is a proportion.

A proportion has four terms: The first and fourth terms are called the **extremes**. The second and third terms are called the **means**.

In a true proportion, the product of the extremes is equal to the product of the means. In symbols, $\dfrac{a}{b} = \dfrac{c}{d}$ if and only if $a \times d = b \times c$ (where $b \neq 0$ and $d \neq 0$).

continued on next page ...

Section 3.4 Ratios and Proportions (continued)

Applications page 200
Solving for the unknown term in a proportion.
Geometry (Similar Triangles)

Similar Triangles page 203
In two similar triangles:
1. Their **corresponding angles are equal.**
2. Their **corresponding sides are proportional.**

Section 3.5 Linear Inequalities

Intervals of Real Numbers pages 211 - 213
There is a one-to-one correspondence between the real numbers and the points on a line. That is, each point on a number line corresponds to one real number, and each real number corresponds to one point on a number line.

Types of Intervals page 212

Open Interval	$a < x < b,\ x > a,\ x < a$
Closed Interval	$a \le x \le b$
Half-Open Interval	$a < x \le b,\ a \le x < b, x \ge a,\ x \le a$

Interval Notation page 212
Types of intervals:

Open	$(a, b),\ (a, \infty),\ (-\infty, b)$
Closed	$[a, b]$
Half-open	$[a, \infty),\ (-\infty, b], [a, b),\ (a, b]$

Linear Inequalities page 213
For real numbers a, b, and c $(a \ne 0)$, $\boldsymbol{ax + b < c}$ is called a **linear inequality** in x. (The definition also holds if $\le$, $>$, or $\ge$ is used instead of $<$.)

To Solve a Linear Inequality page 214
1. Simplify each side of the inequality by removing any grouping symbols and combining like terms.
2. Add the opposites of constants and/or variable expressions to both sides so that variables are on one side and constants are on the other.
3. Divide both sides by the coefficient of the variable and
   a. leave the direction of the inequality unchanged if the coefficient is positive; or
   b. reverse the direction of the inequality if the coefficient is negative.

continued on next page ...

Section 3.5 Linear Inequalities (continued)

Addition Property of Inequality page 214

If the same algebraic expression is added to both sides of an inequality, the new inequality is equivalent to the original inequality.

If $A < B$, then $A + C < B + C$.

If $A < B$, then $A - C < B - C$.

Multiplication Property of Inequality page 215

If both sides of an inequality are multiplied (or divided) by the same positive expression, then the new inequality is equivalent to the original inequality.

$$\begin{cases} \text{If } A < B \text{ and } C > 0, \text{ then } AC < BC. \\ \text{If } A < B \text{ and } C > 0, \text{ then } \dfrac{A}{C} < \dfrac{B}{C}. \end{cases}$$

If both sides of an inequality are multiplied (or divided) by the same negative expression and the inequality is reversed, then the new inequality is equivalent to the original inequality.

$$\begin{cases} \text{If } A < B \text{ and } C < 0, \text{ then } AC > BC. \\ \text{If } A < B \text{ and } C < 0, \text{ then } \dfrac{A}{C} > \dfrac{B}{C}. \end{cases}$$

Chapter 3 Review

For a review of the topics and problems from Chapter 3, look at the following lessons from *Hawkes Learning Systems: Introductory & Intermediate Algebra.*

Working with Formulas
Formulas in Geometry
Applications of Linear Equations
Applications: Average
Solving Equations: Ratios and Proportions
Solving Linear Inequalities

Chapter 3 Test

In Exercises 1 and 2, solve each formula for the indicated variable.

1. $N = mrt + p$; solve for m.

2. $5x + 3y - 7 = 0$; solve for y.

In Exercises 3 – 5, solve the following proportions.

3. $\dfrac{\frac{x}{3}}{\frac{3}{8}} = \dfrac{\frac{1}{4}}{9}$

4. $\dfrac{21}{x+4} = \dfrac{7}{6}$

5. $\dfrac{6}{11} = \dfrac{18}{x+3}$

Solve each of the inequalities in Exercises 6 – 11 and graph each solution set on a real number line.

6. $3x - 8 \le -18$

7. $-\frac{3}{4}x - 6 > 9$

8. $2(x-7) < 4(2x+3)$

9. $-2 \le 4x + 7 \le 3$

10. $\frac{1}{2}x + \frac{3}{4} \le \frac{2}{3}x + \frac{1}{4}$

11. $-5.8 < 2x - 6 < 20.4$

12. Represent each interval by using inequality symbols and graph each interval on a real number line.

a. $(-4, 4)$ **b.** $[3, 7]$ **c.** $(-\infty, 0.75)$

13. Find (a) the perimeter and (b) the area of the quarter circle shown here. (Use $\pi \approx 3.14$.)

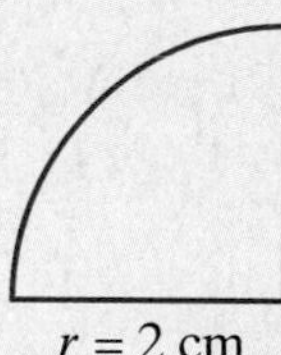

14. Find the height of a cone that has a volume of $24\pi\ \text{cm}^3$ and a diameter of 6 cm. (Use $\pi \approx 3.14$.)

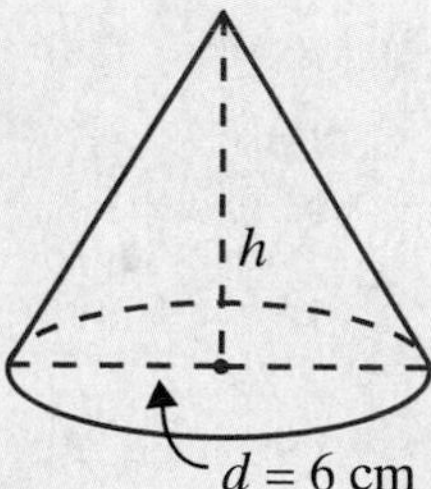

15. Find (a) the perimeter and (b) the area of the figure shown here. (Use $\pi \approx 3.14$.)

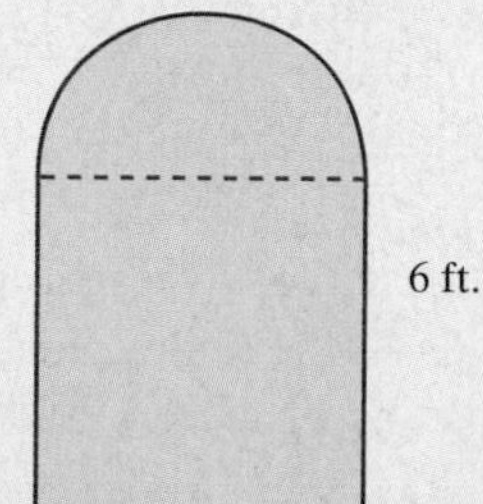

16. Find the volume of a sphere with diameter 20 cm. (Use $\pi \approx 3.14$.)

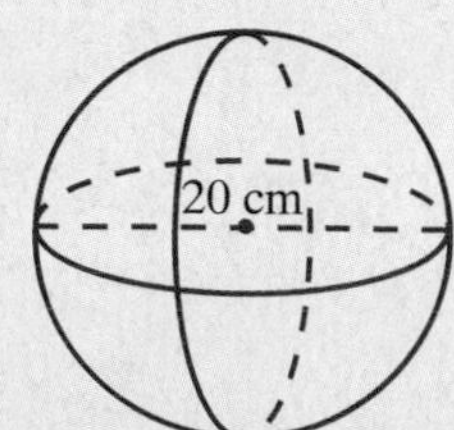

1. $m = \frac{N-p}{rt}$

2. $y = \frac{7-5x}{3}$

3. $x = \frac{1}{96}$

4. $x = 14$

5. $x = 30$

6. $x \le -\frac{10}{3}$

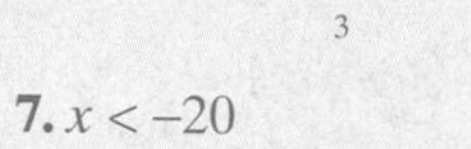

7. $x < -20$

8. $x > \frac{-13}{3}$

9. $\frac{-9}{4} \le x \le -1$

10. $x \ge 3$

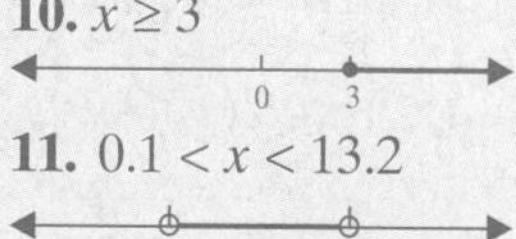

11. $0.1 < x < 13.2$

12. a. $-4 < x < 4$

b. $3 \le x \le 7$

c. $x < 0.75$

13. a. 3.14 cm

b. $3.14\ \text{cm}^2$

14. 8 cm

15. a. 27.42 ft.

b. $50.13\ \text{ft.}^2$

16. $4186.67\ \text{cm}^3$

17. $-20°$ C

18. $x = \frac{26}{3}$

19. **a.** 24 in. **b.** 888 in.^{2}

20. **a.** 200 ft.^{3}
b. 7.4 yd.^{3}

21. 100.48 in.^{3}

22. $11.25

23. 3.2 in.

24. $x = 3,\ y = 6.75$

25. $358.40 to $377.60

26. Between 9.1 and 18 gallons

17. Given the formula $C = \frac{5}{9}(F - 32)$ that relates Fahrenheit and Celsius temperatures, find the Celsius temperature equal to $-4°$ F.

18. Given the equation $3x - 2y = 18$, find the value of x that corresponds to the value of 4 for y.

In Exercises 19 – 26, set up an equation for each word problem and solve.

19. A rectangle is 37 inches long and has a perimeter of 122 inches.
a. Find the width of the rectangle.
b. Find the area of the rectangle.

20. **a.** Find the volume of concrete needed for a sidewalk that is 6 feet by 100 feet by 4 inches.
b. Find the number of cubic yards (to the nearest tenth) in the sidewalk. (**Note:** 12 in. = 1 ft. and 27 ft.^{3} = 1 yd.^{3})

21. If a cone has a height of 6 in. and a circular base with a radius of 4 in., what is its volume?

22. A manufacturing company expects to make a profit of $3 on a product that sells for $8. How much profit does the company expect to make on a product that it sells for $30?

23. An architect plans to make a drawing of a house that uses the scale of 2 inches to 15 feet. If two points in the house are known to be 24 feet apart, how many inches apart on the drawing should these two points be?

24. The two triangles shown in the figure below are similar triangles. Find the values of x and y.

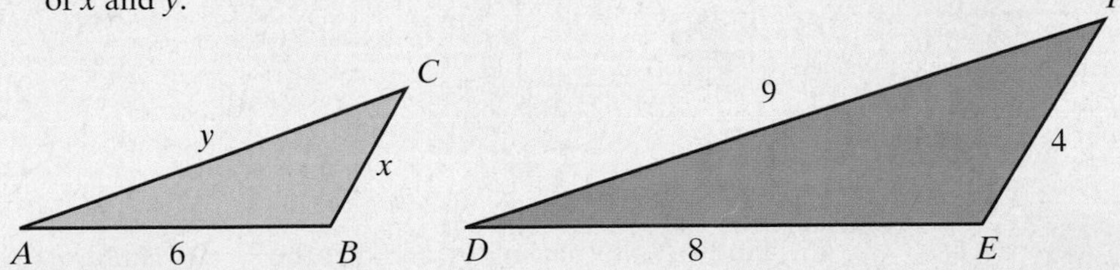

25. The regime fee of a condo includes the exterior maintenance, landscaping, and insurance of the condo. The regime fee this year is expected to increase from 12% to 18% of the previous year. If the fee last year was $320, what range of regime fee should the tenants expect this year?

26. On a recent trip to Virginia, a car that averages 33 miles per gallon ran out of gas after traveling at least 300 miles. If the tank holds 18 gallons of gas, what is the range of the number of gallons of gas that the tank had before driving to Virginia?

Cumulative Review: Chapters 1 – 3

1. $x + 15$
2. $3x + 8$
3. 540
4. $120a^2b^3$
5. 2, 16
6. –4, –9
7. –4, 14
8. 4, –12
9. False, $-15 < 5$
10. True
11. False, $\frac{7}{8} \geq \frac{7}{10}$
12. True
13. $x = \frac{8}{5}$
14. $x = -\frac{1}{2}$
15. $x = \frac{3}{8}$
16. $x = -3$
17. 133.33
18. 22
19. 16.38
20. $v = \frac{h + 16t^2}{t}$
21. $r = \frac{A - P}{Pt}$
22. $y = -\frac{14 - 5x}{3}$
23. $h = \frac{3V}{\pi r^2}$
24. $d = \frac{C}{\pi}$
25. $y = \frac{10 - 3x}{5}$

Use the distributive property to complete the expression in Exercises 1 and 2.

1. $3x + 45 = 3(\quad)$ **2.** $6x + 16 = 2(\quad)$

Find the LCM for each set of numbers or terms in Exercises 3 and 4.

3. 12, 15, 54 **4.** $6a^2, 24ab^3, 30ab, 40a^2b^2$

For each pair of numbers in Exercises 5 – 8, find two factors of the first number whose sum is the second number.

5. 32, 18 **6.** 36, –13 **7.** –56, 10 **8.** –48, –8

In Exercises 9 – 12, determine whether the inequality is true or false. If the inequality is false, rewrite it in a form that is true using the same numbers or expressions.

9. $-15 > 5$ **10.** $|-6| \geq 6$ **11.** $\frac{7}{8} \leq \frac{7}{10}$ **12.** $-10 < 0$

Solve the equations in Exercises 13 – 16.

13. $7(4 - x) = 3(x + 4)$ **14.** $-2(5x + 1) + 2x = 4(x + 1)$

15. $\frac{2}{3}x + \frac{1}{2} = \frac{3}{4}$ **16.** $1.5x - 3.7 = 3.6x + 2.6$

Find the missing number, to two decimal places, in Exercises 17 – 19.

17. 75% of ______ is 100. **18.** 20.24 is ______ % of 92.

19. 25.2% of 65 is ______ .

Solve for the indicated variable in Exercises 20 – 25.

20. $h = vt - 16t^2$; solve for v. **21.** $A = P + Prt$; solve for r.

22. $5x - 3y = 14$; solve for y. **23.** $V = \frac{1}{3}\pi r^2 h$; solve for h.

24. $C = \pi d$; solve for d. **25.** $3x + 5y = 10$; solve for y.

26. a. 17 b. −26
27. 21.7
28. $(-\infty, 2]$
29. $\left[\frac{13}{3}, \infty\right)$
30. $\left(\frac{9}{28}, \infty\right)$
31. $[-1.1, 7.4]$
32. $\left[-\frac{5}{2}, \frac{3}{2}\right]$
33. $(-6, 2)$
34. No solution
35. No solution
36. $a = -1.9, 0.9$
37. $x = \frac{-21}{2}, \frac{3}{2}$
38. $x = \frac{-8}{3}, 4$
39. $x = -30, 60$
40. a. $|x| \le 3$
b. $|x-7| < 1$
c. $|x-1.1| \le 0.2$
41. $x = 0$
42. $A = 100$
43. $x = \frac{1}{98}$
44. $x = 30$

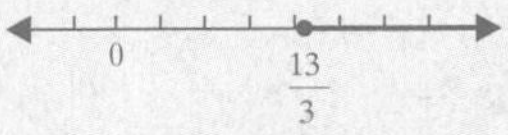

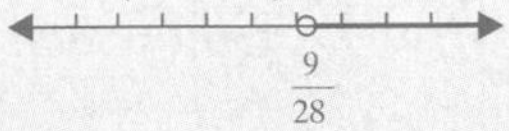

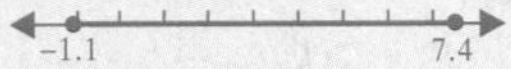

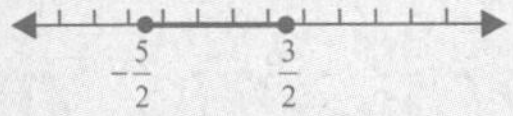

26. Find the mean of each set of numbers.

a. 17, 14, 20 **b.** −18, −22, −27, −37

27. Find the mean (to the nearest tenth) of the data in the following frequency distribution.

Clocked Speeds of Bicycles (to nearest mph)	Frequency
19	4
20	5
22	7
25	3
28	1

Solve each of the inequalities in Exercises 28 – 33. Graph each solution set on a real number line and write the answer in interval notation.

28. $2x + 5 - 3 \le 6$

29. $-3(7 - 2x) \ge 2 + 3x - 10$

30. $-\frac{1}{2}x + \frac{7}{8} < \frac{2}{3}x + \frac{1}{2}$

31. $-6.2 \le 2x - 4 \le 10.8$

32. $-11 \le 4x - 1 \le 5$

33. $0 < \frac{1}{2}x + 3 < 4$

Solve each of the absolute value equations in Exercises 34 – 39.

34. $|x| = -3$

35. $|y| = -1$

36. $|a + 0.5| = 1.4$

37. $|2x + 9| - 7 = 5$

38. $|3x - 2| + 3 = 13$

39. $\left|\frac{2x}{5} - 6\right| = 18$

40. Represent each interval by using an absolute value inequality.

a. $[-3, 3]$ **b.** $(6, 8)$ **c.** $[0.9, 1.3]$

Solve each of the proportions in Exercises 41 – 44.

41. $\frac{3}{x+4} = \frac{3}{4}$

42. $\frac{16}{A} = \frac{4}{25}$

43. $\frac{\frac{1}{3}}{14} = \frac{x}{\frac{3}{7}}$

44. $\frac{2.4}{x} = \frac{1.2}{15}$

45. $x = 3,\ y = 7$

46. $x = 9,\ y = 20°$

47. a. 45.7 cm
b. 139.25 cm^2

48. 1177.5 in.^{3}

49. 10.71 in.

The triangles shown in Exercises 45 and 46 are similar triangles. Solve for x and y.

45.

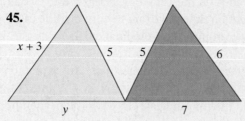

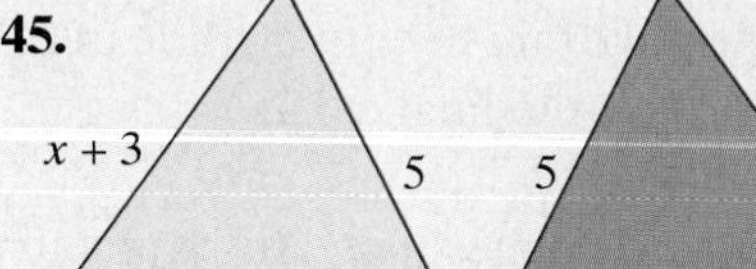

46.

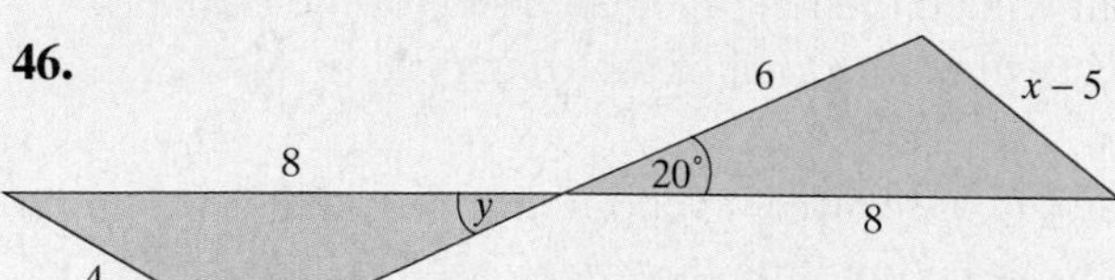

47. Find (a) the perimeter and (b) area of the figure shown here. (Use $\pi \approx 3.14$.)

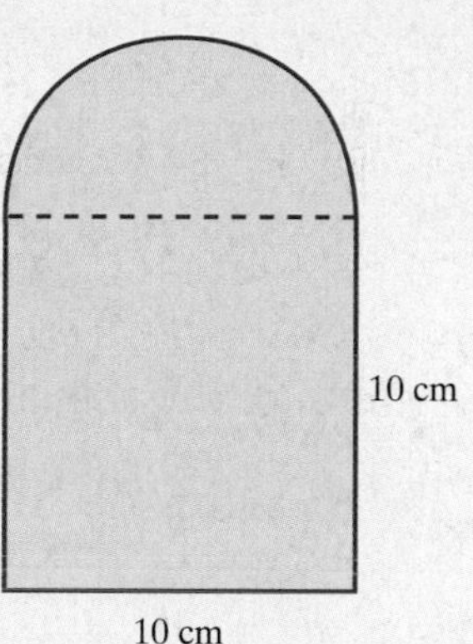

48. Find the volume of the circular cylinder with dimensions as shown in the figure. (Use $\pi \approx 3.14$.)

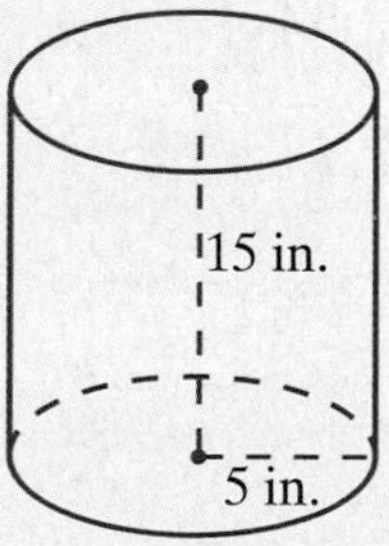

In Exercises 49 – 58, set up an equation or inequality for each problem and solve for the unknown quantity.

49. Find the height of a cone that has a volume of 175π in.^{3} and a radius of 7 in. (Use $\pi \approx 3.14$.)

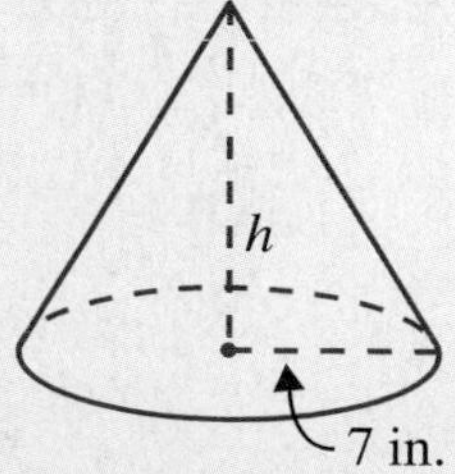

50. $86 \le x \le 100$

51. **a.** 5 in., 12 in., 13 in. **b.** 30 in.^{2}

52. 20, 22, 24

53. length = 25 cm, width = 16 cm

54. –19

55. $54 \le x \le 100$

56. less than 140 miles

57. 15, 17, 19

58. 87.25

50. The range for a grade of B is an average of 75 or more but less than 85 in an English class. If your grades on the first four exams were 73, 65, 77, and 74, what possible grades can you get on the final exam to earn a grade of B? (Assume that 100 is the maximum number of points on the final exam.)

51. An artist has a piece of wire 30 inches long and wants to bend it into the shape of a triangle so that the sides are of length x, $2x + 2$, and $2x + 3$. This triangle will be a right triangle and the two shortest sides will be perpendicular to each other.
a. What will be the length of each side of the triangle?
b. What will be the area of the triangle?

52. Find three consecutive even integers such that the sum of the second and third is equal to three times the first decreased by 14.

53. The length of a rectangle is 9 cm more than its width. The perimeter is 82 cm. Find the dimensions of the rectangle.

54. Twice the difference between a number and 16 is equal to four times the number increased by 6. What is the number?

55. To receive a B grade, a student must average 80 or more but less than 90. If Izumi received a B in the course and had four scores of 93, 77, 90, and 86 before taking the final exam, what were her possible scores for the final? (Assume that the maximum score on the final was 100.)

56. Acme Car Rental charges \$0.25 for each mile driven plus \$15 a day. Zenith Car Rental charges \$50 a day with no extra charge for mileage. How many miles a day can you drive an Acme car and still keep the cost less than a Zenith car?

57. Find three consecutive odd integers such that the sum of the first and three times the second is 28 more than twice the third.

58. From the syllabus in Alesha's English class, the final grade consists of 20% homework, 10% quiz average, 30% semester paper, and 40% final exam. Going into the final exam, Alesha has an 87 for her homework average, an 89 for her quiz average, and a 96 on her semester paper. What is the minimum grade Alesha can receive on the final exam to earn an A (90 to 100) in the class?

Straight Lines and Functions

CHAPTER 4

Did You Know?

In Chapter 4, you will be introduced to the idea of a graph of an algebraic equation. A graph is simply a picture of an algebraic relationship. This topic is more formally called **analytic geometry**. It is a combination of algebra (the equation) and geometry (the picture).

Descartes

The idea of combining algebra and geometry was not thought of until René Descartes wrote his famous *Discourse on the Method of Reasoning* in 1637. The third appendix in this book, "La Geometrie," made Descartes' system of analytic geometry known to the world. In fact, you will find that analytic geometry is sometimes called **Cartesian geometry**.

René Descartes is perhaps better known as a philosopher than as a mathematician; he is often referred to as the father of modern philosophy. His method of reasoning was to apply the same logical structure to philosophy that had been developed in mathematics, especially geometry.

In the Middle Ages, the highest forms of knowledge were believed to be mathematics, philosophy, and theology. Many famous people in history who have reputations as poets, artists, philosophers, and theologians were also creative mathematicians. Almost every royal court had mathematicians whose work reflected glory on the royal sponsor who paid the mathematician for his research and court presence.

Descartes, in fact, died in 1650 after accepting a position at the court of the young warrior-queen, Christina of Sweden. Apparently, the frail French philosopher-mathematician, who spent his mornings in bed doing mathematics, could not stand the climate of Sweden and the hardships imposed by Christina in her demand that Descartes tutor her in mathematics each morning at 5 o'clock in an unheated castle library.

"To divide each problem I examined it into as many parts as was feasible, and as was requisite for its better solution."

René Descartes (1596 – 1650)

Thanks to René Descartes (1596 – 1650), the ideas of algebra and geometry can be combined. He developed an entire theory connecting algebraic expressions and equations with points on geometric graphs. In particular, in this chapter, you will see that equations with two variables can be related to graphs or "pictures" of straight lines in a plane. By studying the equations in special forms, you will be able to tell the position and basic properties of the corresponding lines before you start to draw the graphs.

4.1 The Cartesian Coordinate System

Objectives

After completing this section, you will be able to:

1. *Name ordered pairs corresponding to points on graphs.*
2. *Graph ordered pairs in the Cartesian coordinate system.*
3. *Find ordered pairs that satisfy given equations.*

René Descartes (1596 – 1650), a famous French mathematician, developed a system for solving geometric problems using algebra. This system is called the **Cartesian coordinate system** in his honor. Descartes based his system on a relationship between points in a plane and **ordered pairs** of real numbers. This section begins by relating algebraic formulas with ordered pairs and then shows how these ideas can be related to geometry.

Equations in Two Variables

Equations such as $d = 60t$, $I = 0.05P$, and $y = 2x + 3$ represent relationships between pairs of variables. For example, in the first equation, if $t = 3$, then $d = 60 \cdot 3 = 180$. With the understanding that t is first and d is second, we can represent $t = 3$ and $d = 180$ in the form of an ordered pair (3, 180). In general, if t is the first number and d is the second number, then solutions to the equation $d = 60t$ can be written in the form of ordered pairs (t, d). Thus, we see that (180, 3) is different from (3, 180). **The order of the numbers in an ordered pair is critical.**

We say that (3, 180) **satisfies the equation** or **is a solution of the equation** $d = 60t$. Similarly, (100, 5) satisfies $I = 0.05P$ where $P = 100$ and $I = 0.05(100) = 5$. Also, (2, 7) satisfies $y = 2x + 3$, where $x = 2$ and $y = 2 \cdot 2 + 3 = 7$.

In an ordered pair such as (x, y), x is called the **first component** (or **first coordinate**) and y is called the **second component** (or **second coordinate**). To find ordered pairs that satisfy an equation such as $y = 2x + 3$, we can choose any value for one variable and then find the corresponding value for the other variable by substituting into the equation.

For example, for the equation $y = 2x + 3$:

Choices for x:	**Substitution:**	**Ordered Pairs:**
$x = 1$	$y = 2 \cdot 1 + 3 = 5$	$(1, 5)$
$x = -2$	$y = 2(-2) + 3 = -1$	$(-2, -1)$
$x = \frac{1}{2}$	$y = 2 \cdot \frac{1}{2} + 3 = 4$	$\left(\frac{1}{2}, 4\right)$

The ordered pairs (1, 5), (–2, –1), and $\left(\frac{1}{2}, 4\right)$ are just three ordered pairs that satisfy the equation $y = 2x + 3$. There are an infinite number of such ordered pairs. Any real number could have been chosen for x and the corresponding value for y calculated.

Since the equation $y = 2x + 3$ is solved for y, we say that the value of y "depends" on the choice of x. Thus, in an ordered pair of the form (x, y), the second coordinate, y, is called the **dependent variable** and the first coordinate, x, is called the **independent variable**.

In the following tables the first variable, in each case, is the independent variable and the second variable is the dependent variable. Corresponding ordered pairs would be of the form (t, d), (P, I), and (x, y). The choices for the values of the independent variables are arbitrary. There are an infinite number of other values that could have just as easily been chosen.

$d = 60t$	
t	d
5	$60(5) = 300$
10	$60(10) = 600$
12	$60(12) = 720$
15	$60(15) = 900$

$I = 0.05P$	
P	I
100	$0.05(100) = 5$
200	$0.05(200) = 10$
500	$0.05(500) = 25$
1000	$0.05(1000) = 50$

$y = 2x + 3$	
x	y
–2	$2(-2) + 3 = -1$
–1	$2(-1) + 3 = 1$
0	$2(0) + 3 = 3$
3	$2(3) + 3 = 9$

Graphing Ordered Pairs

The Cartesian coordinate system relates algebraic equations and ordered pairs to geometry. In this system, two number lines intersect at right angles and separate the plane into four **quadrants**. The **origin**, designated by the ordered pair (0, 0), is the point of intersection of the two lines. The horizontal number line is called the **horizontal axis** or ***x*-axis**. The vertical number line is called the **vertical axis** or ***y*-axis**. Points that lie on either axis are not in any quadrant. They are simply on an axis (Figure 4.1).

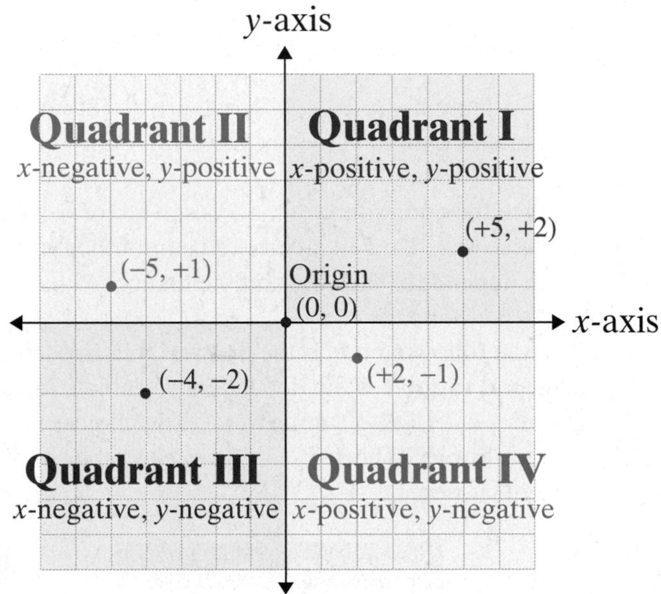

Figure 4.1

The following important relationship between ordered pairs of real numbers and points in a plane is the cornerstone of the Cartesian coordinate system.

One-to-One Correspondence

There is a ***one-to-one correspondence*** *between points in a plane and ordered pairs of real numbers.*

In other words, for each point there is one and only one corresponding ordered pair of real numbers, and for each ordered pair of real numbers there is one and only one corresponding point.

The **graphs of the points** A (2, 1), B (–2, 3), C (–3, –2), D (1, –2), and E (3, 0) are shown in Figure 4.2. [**Note:** An ordered pair of real numbers and the corresponding point on the graph are frequently used to refer to each other. Thus, the ordered pair (2, 1) and the point (2, 1) are interchangeable ideas.]

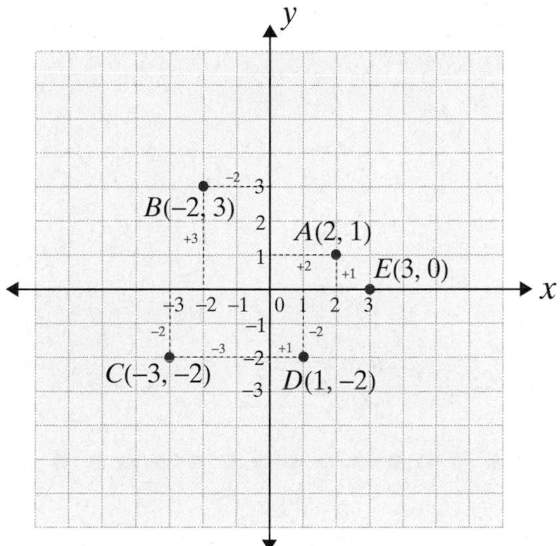

Figure 4.2

Example 1: Graph the Ordered Pairs

Graph the sets of ordered pairs.

a. {(–2, 1), (0, 2), (1, 3), (2, –3)}
(**Note:** The listing of ordered pairs in each set can be in any order.)

Solution:

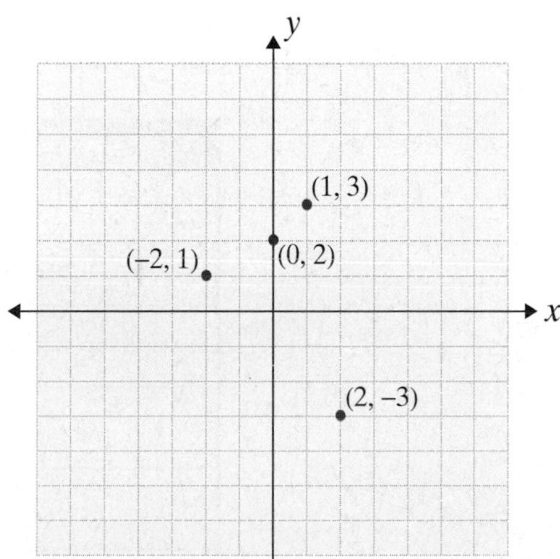

continued on next page ...

Teaching Notes:
You may need to remind your students that there is no significance in the order of the elements in a set. However, the order of each pair is important.

b. $\{(-1, 3), (0, 1), (1, -1), (2, -3), (3, -5)\}$

Solution:

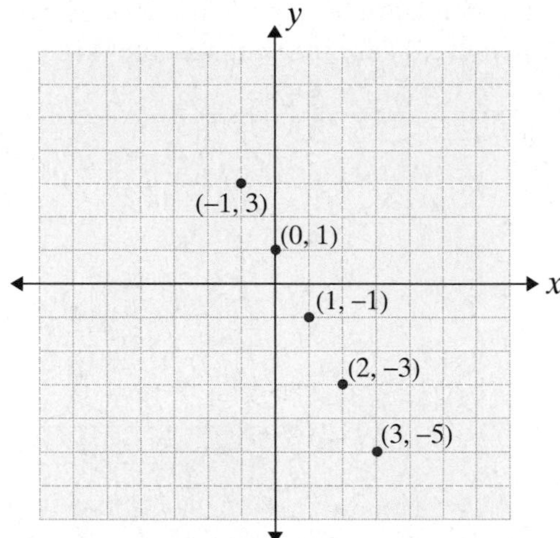

The points (ordered pairs) in Example 1b can be shown to satisfy the equation $y = -2x + 1$. For example, using $x = -1$ in the equation yields,

$$y = -2(\ -1\ ) + 1 = 2 + 1 = 3$$

and the ordered pair $(-1, 3)$ satisfies the equation. Similarly, letting $y = 1$ gives,

$$\begin{aligned} 1 &= -2x + 1 \\ 0 &= -2x \\ 0 &= x \end{aligned}$$

and the ordered pair $(0, 1)$ satisfies the equation.

We can write all the ordered pairs in Example 1b in table form.

$y = -2x + 3$	
x	y
-1	$-2(\ -1\ ) + 1 = 3$
0	$-2(\ 0\ ) + 1 = 1$
1	$-2(\ 1\ ) + 1 = -1$
2	$-2(\ 2\ ) + 1 = -3$
3	$-2(\ 3\ ) + 1 = -5$

Example 2: Ordered Pairs

a. Determine which, if any, of the ordered pairs $(0, -2)$, $\left(\frac{2}{3}, 0\right)$, and $(2, 5)$ satisfy the equation $y = 3x - 2$.

Solution: We will substitute 0, $\frac{2}{3}$, and 2 for x and see if the corresponding y-values match those in the given ordered pairs.

$x = 0$: $\quad y = 3(0) - 2 = -2 \quad$ so, $(0, -2)$ satisfies the equation.

$x = \frac{2}{3}$: $\quad y = 3\left(\frac{2}{3}\right) - 2 = 0 \quad$ so, $\left(\frac{2}{3}, 0\right)$ satisfies the equation.

$x = 2$: $\quad y = 3(2) - 2 = 4 \quad$ so, $(2, 4)$ satisfies the equation.

The point $(2, 5)$ does not satisfy the equation $y = 3x - 2$ since $y = 4$ when $x = 2$.

b. Determine the missing coordinate in each of the following ordered pairs so that the point will satisfy the equation $2x + 3y = 12$:

$$(0,\ \ ), (3,\ \ ), (\ \ , 0), (\ \ , -2).$$

Solution: For $(0,\ \ )$, let $x = 0$:

$$\begin{aligned} 2(0) + 3y &= 12 \\ 3y &= 12 \\ y &= 4 \end{aligned}$$

The ordered pair is $(0, 4)$.

For $(3,\ \ )$, let $x = 3$:

$$\begin{aligned} 2(3) + 3y &= 12 \\ 6 + 3y &= 12 \\ 3y &= 6 \\ y &= 2 \end{aligned}$$

The ordered pair is $(3, 2)$.

For $(\ \ , 0)$, let $y = 0$:

$$\begin{aligned} 2x + 3(0) &= 12 \\ 2x &= 12 \\ x &= 6 \end{aligned}$$

The ordered pair is $(6, 0)$.

For $(\ \ , -2)$, let $y = -2$:

$$\begin{aligned} 2x + 3(-2) &= 12 \\ 2x - 6 &= 12 \\ 2x &= 18 \\ x &= 9 \end{aligned}$$

The ordered pair is $(9, -2)$.

c. Complete the table below so that each ordered pair will satisfy the equation $y = 1 - 2x$.

x	y
0	
	3
$\frac{1}{2}$	
5	

continued on next page ...

Solution: Substituting each given value for x and y into the equation $y = 1 - 2x$ gives the following table of ordered pairs.

x	y
0	1
–1	3
$\frac{1}{2}$	0
5	–9

For $x = 0$:
$y = 1 - 2(0) = 1$

For $x = \frac{1}{2}$:
$y = 1 - 2\left(\frac{1}{2}\right) = 0$

For $y = 3$:
$3 = 1 - 2x$
$2 = -2x$
$-1 = x$

For $x = 5$:
$y = 1 - 2(5) = -9$

NOTES

Although this discussion is related to ordered pairs of real numbers, most of the examples use ordered pairs of **integers**. This is because ordered pairs of integers are relatively easy to locate on a graph and relatively easy to read from a graph. Ordered pairs with fractions, decimals, or radicals (irrational numbers) must be located by estimating the positions of the points. The precise coordinates intended for such points can be difficult or impossible to read because large dots must be used so the points can be seen. **Even with these difficulties, you should understand that we are discussing ordered pairs of real numbers and that points with fractions, decimals, and radicals as coordinates do exist and should be plotted by estimating their positions.**

Practice Problems

1. *Determine which ordered pairs satisfy the equation* $3x + y = 14$.

   a. $(5, -1)$ **b.** $(4, 2)$ **c.** $(-1, 17)$

2. *Given* $3x + y = 5$, *find the missing coordinate of each ordered pair so that it will satisfy the equation.*

   a. $(0,\ \ )$ **b.** $\left(\frac{1}{3},\ \ \right)$ **c.** $(\ \ , 2)$

3. *Complete the table so that each ordered pair will satisfy the equation* $y = \frac{2}{3}x + 1$.

x	y
0	
	–2
–3	
6	

4.1 Exercises

1. $\{(-5,1),(-3,3),(-1,1),(1,2),(2,-2)\}$

2. $\{(-3,1),(-1,5),(0,3),(1,5),(3,-2)\}$

3. $\{(-3,-2),(-1,-3),(-1,3),(0,0),(2,1)\}$

4. $\{(-3,2),(-1,-1),(1,5),(3,-2),(6,5)\}$

5. $\{(-4,4),(-3,-4),(0,-4),(0,3),(4,1)\}$

6. $\{(-4,-5),(-3,1),(0,-2),(1,1),(3,-1)\}$

*List the sets of ordered pairs corresponding to the graphs in Exercises 1 – 10. Assume that the grid lines are marked one unit apart. (**Note:** There is no particular order to be followed in listing ordered pairs.)*

1.

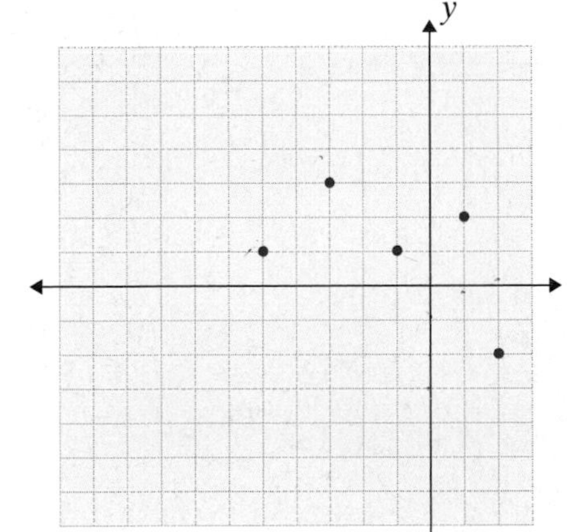

2.

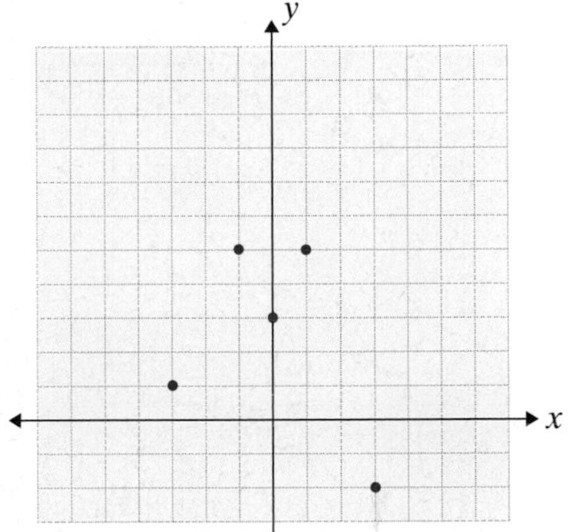

3.

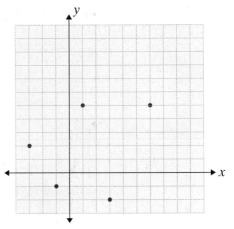

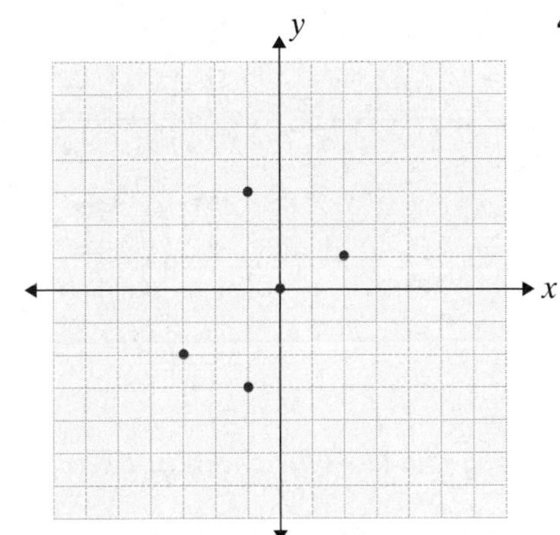

4.

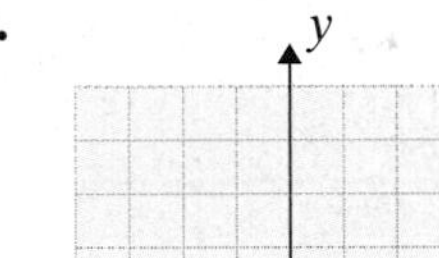

5.

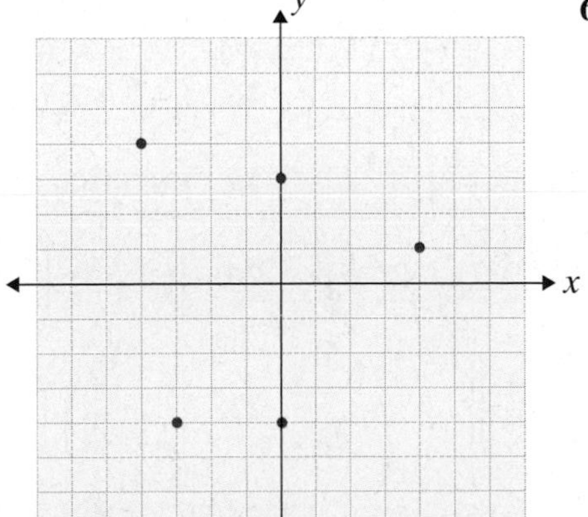

6.

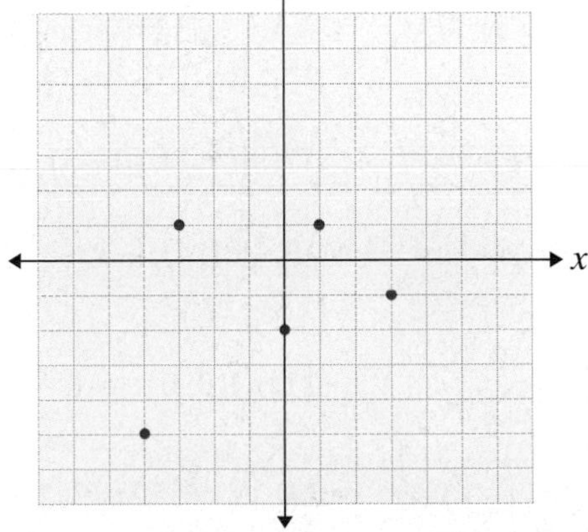

Answers to Practice Problems:

1. All satisfy the equation.

2. a. $(0, 5)$

b. $\left(\frac{1}{3}, 4\right)$

c. $(1, 2)$

3.

x	y
0	1
$-\frac{9}{2}$	−2
−3	−1
6	5

7. $\{(-6,2),(-1,6),(0,0),(1,-7),(6,3)\}$
8. $\{(-3,-5),(-1,4),(0,-1),(3,1),(6,0)\}$
9. $\{(-5,0),(-2,2),(-1,-4),(0,6),(2,0)\}$
10. $\{(-5,-2),(-3,4),(2,-2),(2,6),(4,3)\}$

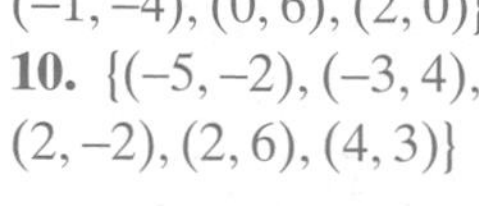

11.

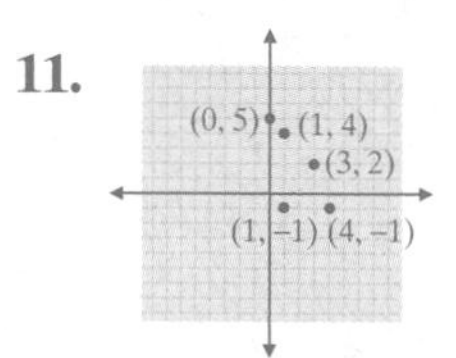

12.

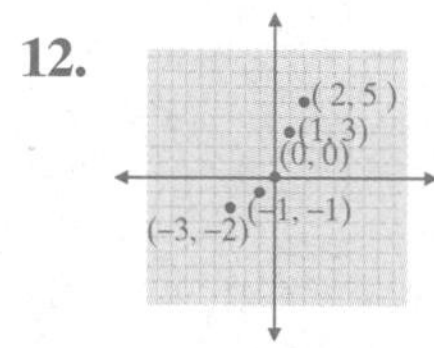

13.

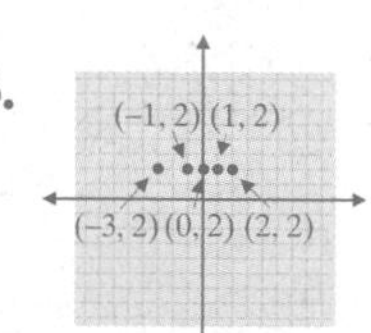

14.

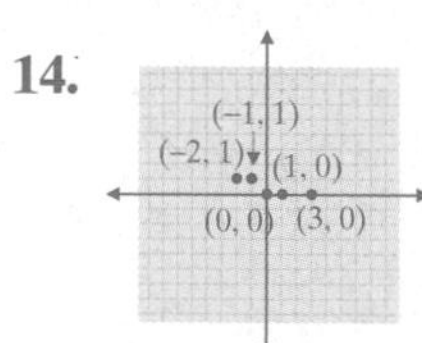

15.

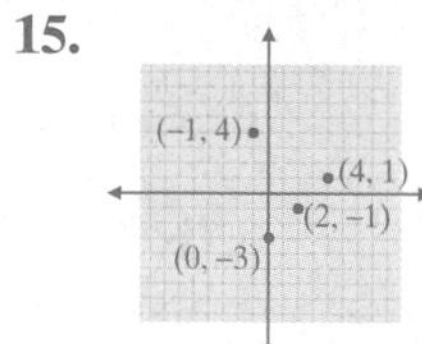

16. (3, 10), (2, 5), (1, 3), (0, 1), (−1, −1)

17. (4, 1), (1, −2), (2, −1), (0, −3)

7.

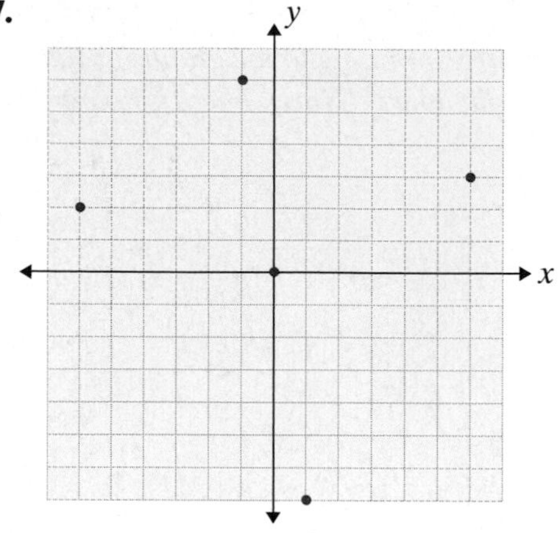

8.

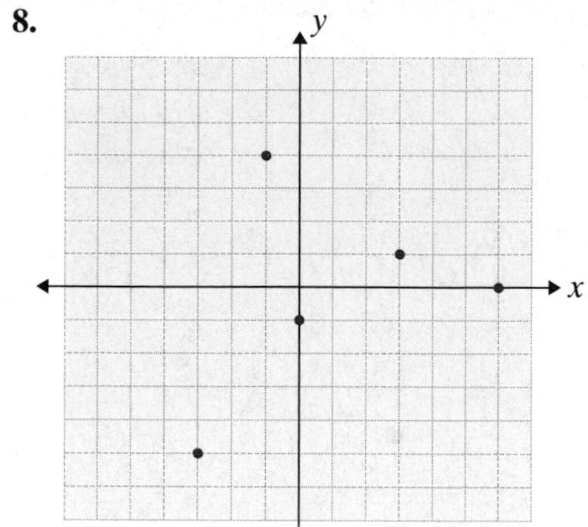

9.

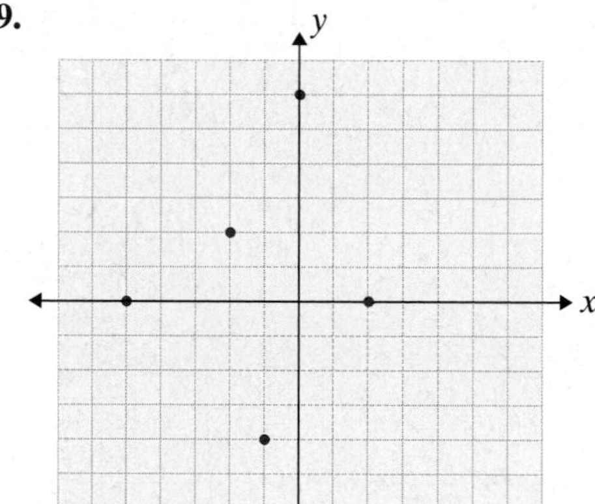

10.

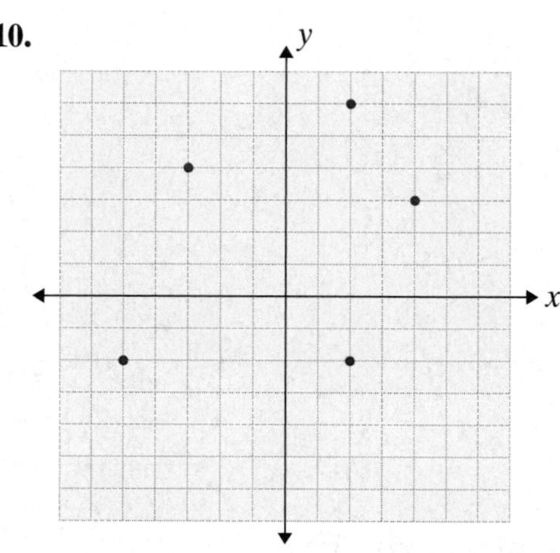

Graph the sets of ordered pairs and label the points in Exercises 11 – 24.

11. $\{(4,-1),(3,2),(0,5),(1,-1),(1,4)\}$

12. $\{(-1,-1),(-3,-2),(1,3),(0,0),(2,5)\}$

13. $\{(1,2),(0,2),(-1,2),(2,2),(-3,2)\}$

14. $\{(1,0),(3,0),(-2,1),(-1,1),(0,0)\}$

15. $\{(-1,4),(0,-3),(2,-1),(4,1)\}$

16. $\{(-1,-1),(0,1),(1,3),(2,5),(3,10)\}$

17. $\{(4,1),(0,-3),(1,-2),(2,-1)\}$

18. $\{(0,1),(1,0),(2,-1),(3,-2),(4,-3)\}$

19. $\{(1,4),(-1,-2),(0,1),(2,7),(-2,-5)\}$

20. $\left\{(1,-3),\left(-4,\frac{3}{4}\right),\left(2,-2\frac{1}{2}\right),\left(\frac{1}{2},4\right)\right\}$

21. $\left\{(0,0),\left(-1,\frac{7}{4}\right),\left(3,-\frac{1}{2}\right)\right\}$

22. $\left\{\left(\frac{3}{4},\frac{1}{2}\right),\left(2,-\frac{5}{4}\right),\left(\frac{1}{3},-2\right),\left(-\frac{5}{3},2\right)\right\}$

23. $\{(1.6,-2),(3,2.5),(-1,.5),(0,-2.3)\}$

24. $\{(-2,2),(-3,1.6),(3,0.5),(1.4,0)\}$

18. 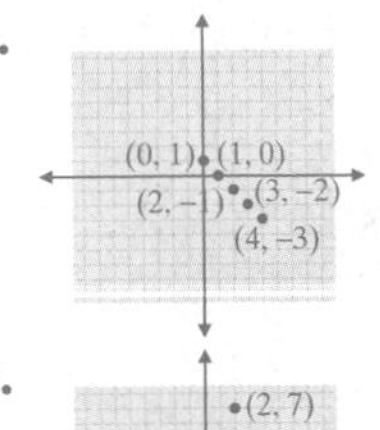

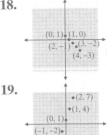

19.

20. 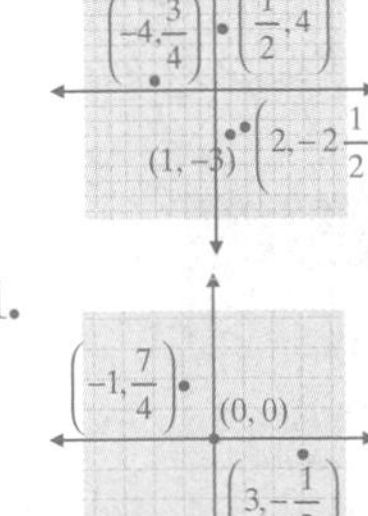

21.

22.

23. 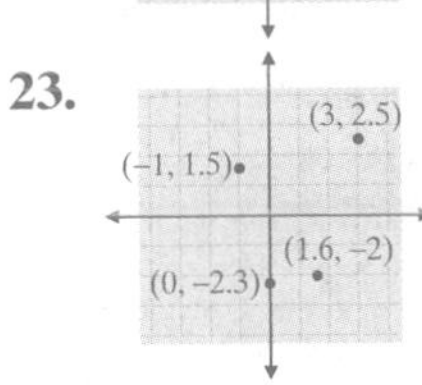

24. 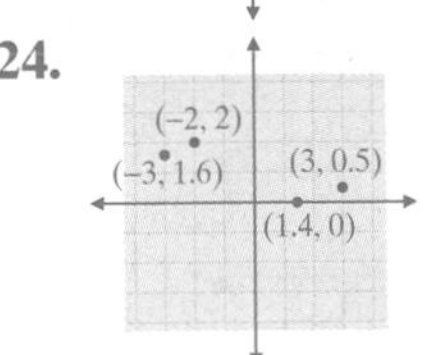

Determine which of the given ordered pairs satisfy the equation in Exercises 25 – 30.

25. $2x - y = 4$

a. $(1, 1)$
b. $(2, 0)$
c. $(1, -2)$
d. $(3, 2)$

26. $x + 2y = -1$

a. $(1, -1)$
b. $(1, 0)$
c. $(2, 1)$
d. $(3, -2)$

27. $4x + y = 5$

a. $\left(\frac{3}{4}, 2\right)$
b. $(4, 0)$
c. $(1, 1)$
d. $(0, 3)$

28. $2x - 3y = 7$

a. $(1, 3)$
b. $\left(\frac{1}{2}, -2\right)$
c. $\left(\frac{7}{2}, 0\right)$
d. $(2, 1)$

29. $2x + 5y = 8$

a. $(4, 0)$
b. $(2, 1)$
c. $(1, 1.2)$
d. $(1.5, 1)$

30. $3x + 4y = 10$

a. $(-2, 3)$
b. $(0, 2.5)$
c. $(4, -2)$
d. $(1.2, 1.6)$

Determine the missing coordinate in each of the ordered pairs so that it will satisfy the equation given in Exercises 31 – 40.

31. $x - y = 4$
$(0, \quad), (2, \quad), (\quad, 0), (\quad, -3)$

32. $x + y = 7$
$(0, \quad), (-1, \quad), (\quad, 0), (\quad, 3)$

33. $x + 2y = 6$
$(0, \quad), (2, \quad), (\quad, 0), (\quad, 4)$

34. $3x + y = 9$
$(0, \quad), (4, \quad), (\quad, 0), (\quad, 3)$

35. $4x - y = 8$
$(0, \quad), (1, \quad), (\quad, 0), (\quad, -4)$

36. $x - 2y = 2$
$(0, \quad), (4, \quad), (\quad, 0), (\quad, 3)$

37. $2x + 3y = 6$
$(0, \quad), (-1, \quad), (\quad, 0), (\quad, -2)$

38. $5x + 3y = 15$
$(0, \quad), (2, \quad), (\quad, 0), (\quad, 4)$

39. $3x - 4y = 7$
$(0, \quad), (1, \quad), (\quad, 0), \left(\quad, \frac{1}{2}\right)$

40. $2x + 5y = 6$
$(0, \quad), \left(\frac{1}{2}, \quad\right), (\quad, 0), (\quad, 2)$

25. b, c, d **26.** a, d
27. a, c **28.** b, c
29. a, c, d **30.** b, d

31. $(0, -4), (2, -2), (4, 0), (1, -3)$
32. $(0, 7), (-1, 8), (7, 0), (4, 3)$
33. $(0, 3), (2, 2), (6, 0), (-2, 4)$
34. $(0, 9), (4, -3), (3, 0), (2, 3)$
35. $(0, -8), (1, -4), (2, 0), (1, -4)$
36. $(0, -1), (4, 1), (2, 0), (8, 3)$
37. $(0, 2), \left(-1, \frac{8}{3}\right), (3, 0), (6, -2)$
38. $(0, 5), \left(2, \frac{5}{3}\right), (3, 0), \left(\frac{3}{5}, 4\right)$
39. $\left(0, -\frac{7}{4}\right), (1, -1), \left(\frac{7}{3}, 0\right), \left(3, \frac{1}{2}\right)$
40. $\left(0, \frac{6}{5}\right), \left(\frac{1}{2}, 1\right), (3, 0), (-2, 2)$

41. 0, −1, −6, 2

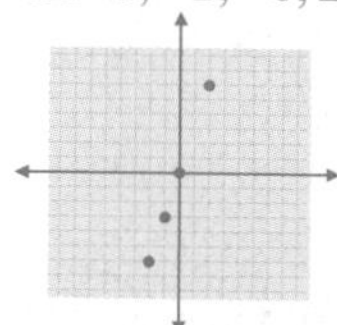

42. 0, −2, −6, 1

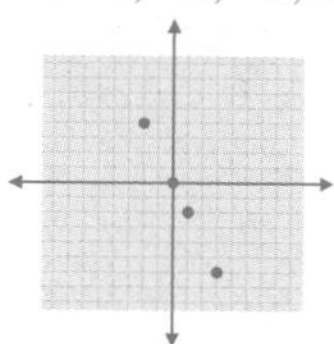

43. −3, 1, −7, $\frac{7}{4}$

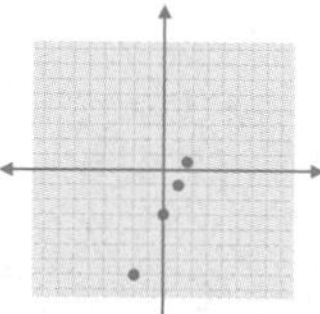

44. 5, −2, −1, $-\frac{3}{2}$

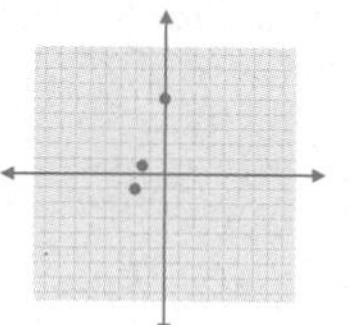

45. 7, $\frac{7}{3}$, 10, 6

46. 6, 3, 10, 5

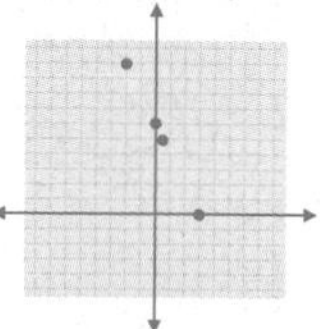

47. 2, 4, −1, −1

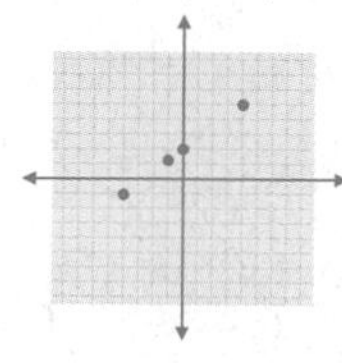

Complete the tables in Exercises 41 – 55 so that each ordered pair will satisfy the given equation. Graph the resulting sets of ordered pairs.

41. $y = 3x$

x	y
0	
	−3
−2	
	6

42. $y = -2x$

x	y
0	
	4
3	
	−2

43. $y = 2x - 3$

x	y
0	
	−1
−2	
	$\frac{1}{2}$

44. $y = 3x + 5$

x	y
0	
	−1
−2	
	$\frac{1}{2}$

45. $y = 7 - 3x$

x	y
0	
	0
−1	
$\frac{1}{3}$	

46. $y = 6 - 2x$

x	y
0	
	0
−2	
$\frac{1}{2}$	

47. $y = \frac{3}{4}x + 2$

x	y
0	
	5
−4	
	$\frac{5}{4}$

48. $y = \frac{3}{2}x - 1$

x	y
0	
	2
−2	
	$-\frac{5}{2}$

49. $3x - 5y = 9$

x	y
0	
	0
−2	
	−1

50. $4x + 3y = 6$

x	y
0	
	0
3	
	−1

51. $5x - 2y = 10$

x	y
0	
	0
−2	
	−1

52. $3x - 2y = 12$

x	y
	0
0	
	−3
2	

48. $-1, 2, -4, -1$

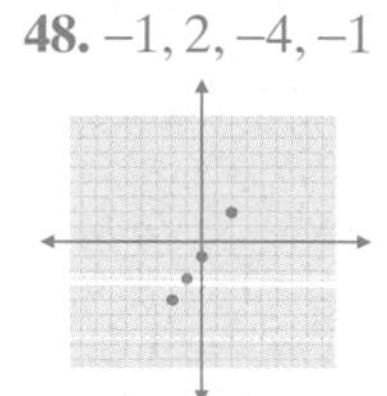

49. $-\frac{9}{5}, 3, -3, \frac{4}{3}$

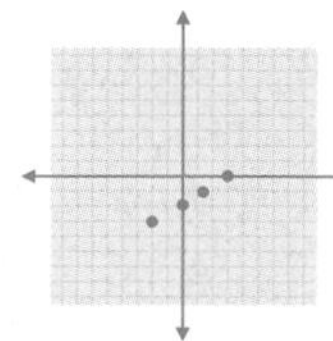

50. $2, \frac{3}{2}, -2, \frac{9}{4}$

51. $-5, 2, -10, \frac{8}{5}$

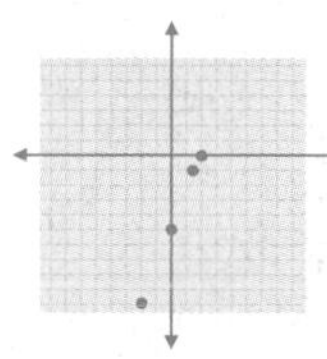

52. $4, -6, 2, -3$

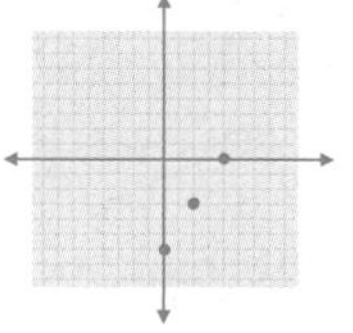

53. $-1.5, 0, 3.8, -1$

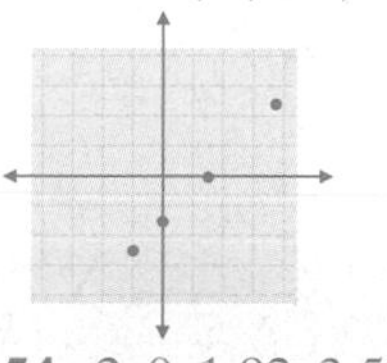

54. $2, 0, 1.92, 3.52$

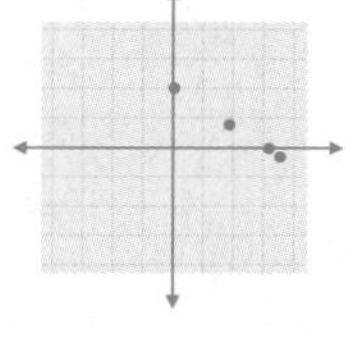

53.

$x - y = 1.5$	
x	y
0	
1.5	
	2.3
	−2.5

54.

$2x + 3.2y = 6.4$	
x	y
0	
3.2	
	0.8
	−0.2

55.

$3x + y = -2.4$	
x	y
	0
0	
	0.6
1.6	

56. Given the equation $I = 0.08P$, where I is the interest earned on a principal P at the rate of 8%:

a. Make a table of ordered pairs for the values of P and I if P has the values \$1000, \$2000, \$3000, \$4000, and \$5000.

b. Graph the points corresponding to the ordered pairs.

P	I
1000	
2000	
3000	
4000	
5000	

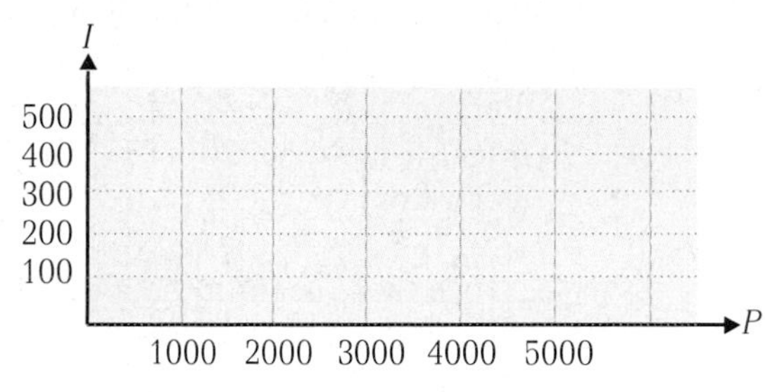

57. Given the equation $F = \frac{9}{5}C + 32$ where C is temperature in degrees Celsius and F is the corresponding temperature in degrees Fahrenheit:

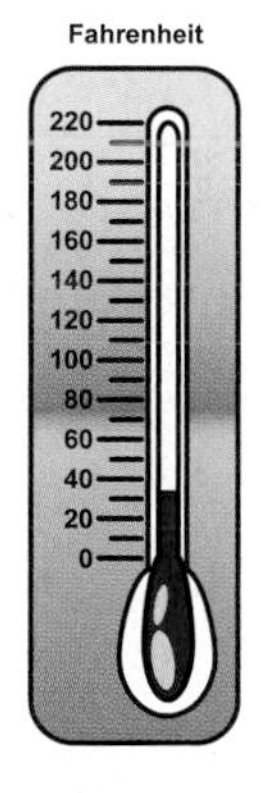

a. Make a table of ordered pairs for the values of C and F if C has the values −20°, −10°, −5°, 0°, 5°, 10°, and 15°.

b. Graph the points corresponding to the ordered pairs.

C	F
−20	
−10	
−5	
0	
5	
10	
15	

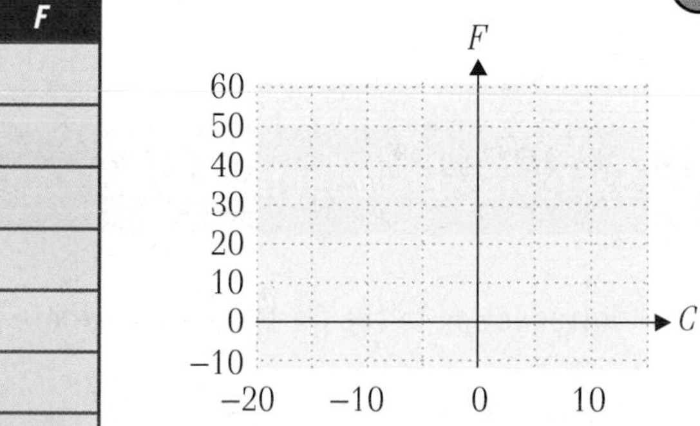

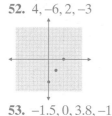

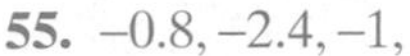

55. −0.8, −2.4, −1, −7.2

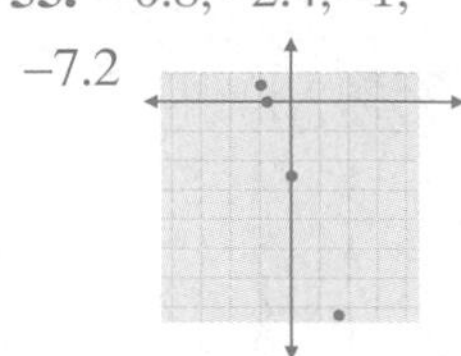

56. 80, 160, 240, 320, 400

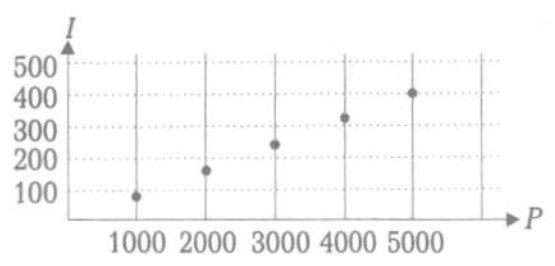

57. −4, 14, 23, 32,

41, 50, 59

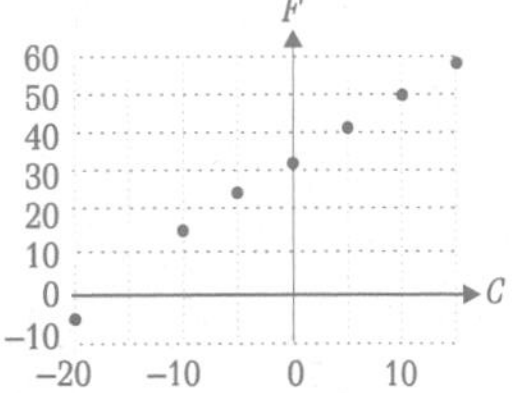

58. a.

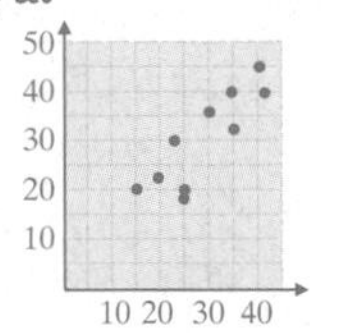

b. Yes, the more sit-ups a person can do it appears the more push-ups he can do.
c. Answers will vary.
59. Answers will vary.
60. Answers will vary. Not all scatterplots can be used to predict information related to the two variables graphed because not all variables are related.

Writing and Thinking About Mathematics

58. In statistics, data is sometimes given in the form of ordered pairs where each ordered pair represents two pieces of information about one person. For example, ordered pairs might represent the height and weight of a person or the person's number of years of education and that person's annual income. The ordered pairs are plotted on a graph. This type of graph is called a **scatter diagram** (or **scatter plot**). Such scatter diagrams are used to see if there is any pattern to the data and, if there is, they are used to predict the value for one of the variables if the value of the other is known. For example, if you know that a person's height is 5 ft. 6 in., then his or her weight might be predicted from information indicated in a scatter diagram that has several points of known information about height and weight.

a. The following table of values indicates the number of push-ups and the number of sit-ups that ten students did in a physical education class. Plot these points in a scatter diagram.

Person	#1	#2	#3	#4	#5	#6	#7	#8	#9	#10
x (push-ups)	20	15	25	23	35	30	42	40	25	35
y (sit-ups)	25	20	20	30	32	36	40	45	18	40

b. Does there seem to be a pattern in the relationship between push-ups and sit-ups? What is this pattern?

c. Using the scatter diagram in Part (a), predict the number of sit-ups that a student might be able to do if they have just done each of the following numbers of push-ups: 22, 32, 35, and 45. (**Note:** In each case, there is no one correct answer. The answers are only estimates based on the diagram.)

59. Ask ten friends or fellow students what their height and weight are. Organize the data in table form and then plot the corresponding scatter diagram. Knowing your own height, does the pattern indicated in the scatter diagram seem to predict your weight?

60. Do you think that all scatter diagrams can be used to predict information related to the two variables graphed? Explain.

Hawkes Learning Systems: Introductory & Intermediate Algebra

Introduction to the Cartesian Coordinate System

4.2 Graphing Linear Equations in Two Variables

Objectives

After completing this section, you will be able to:

1. *Find ordered pairs that satisfy given linear equations.*
2. *Graph lines in a Cartesian coordinate system by locating points that satisfy given linear equations.*
3. *Graph horizontal lines.*
4. *Graph vertical lines.*

The Standard Form: $Ax + By = C$

In Section 4.1, we discussed ordered pairs and graphed a few points (ordered pairs) that satisfied particular equations. Now we want to graph all the points that satisfy an equation. For example, consider the equation in two variables

$$2x + 3y = 6$$

or, solved for y, $$y = \frac{6-2x}{3}.$$

The solution to this equation consists of an infinite set of ordered pairs in the form (x, y) where the first coordinate represents the variable x (sometimes called the **independent variable**), and the second coordinate represents the variable y (sometimes called the **dependent variable**). To find some of these solutions, we form a table by choosing arbitrary values for x and finding the corresponding values for y by using the equation. We say that these ordered pairs **satisfy the equation**.

Teaching Notes:
Students can discover that solutions to linear equations are on the same line by having (guiding) them experiment with graphing. Then the claim that solutions to an equation of two variables, each of 1st degree, are on the same line will seem all the more reasonable.

$x = 0$: $$y = \frac{6-2\cdot 0}{3} = \frac{6}{3} = 2$$

$x = -3$: $$y = \frac{6-2\cdot(-3)}{3} = \frac{12}{3} = 4$$

$x = 3$: $$y = \frac{6-2\cdot 3}{3} = \frac{0}{3} = 0$$

$x = \frac{1}{2}$: $$y = \frac{6-2\cdot\frac{1}{2}}{3} = \frac{5}{3}$$

Choose x	Calculate y
0	2
–3	4
3	0
$\frac{1}{2}$	$\frac{5}{3}$

Thus, (0, 2), (–3, 4), (3, 0), and $\left(\frac{1}{2}, \frac{5}{3}\right)$ are ordered pairs that satisfy the equation $2x + 3y = 6$.

Solution of an Equation in Two Variables

*The **solution** (or **solution set**) of an equation in two variables, x and y, consists of all those ordered pairs of real numbers (x, y) that satisfy the equation.*

Example 1: Solution Set

a. Given the equation $3x + y = 9$, find the missing coordinate of each ordered pair so that the ordered pair belongs to the solution set of the equation.

i. (2,) **ii.** (0,) **iii.** (6,) **iv.** (, 0)

Solution: First solve the equation for y to make evaluations easier: $y = 9 - 3x$

i. For (2,), $x = 2$:
$y = 9 - 3 \cdot 2 = 9 - 6 = 3$
The ordered pair is (2, 3).

ii. For (0,), $x = 0$:
$y = 9 - 3 \cdot 0 = 9 - 0 = 9$
The ordered pair is (0, 9).

iii. For (6,), $x = 6$:
$y = 9 - 3 \cdot 6 = 9 - 18 = -9$
The ordered pair is (6, –9).

iv. For (, 0), $y = 0$:
$0 = 9 - 3x$
$3x = 9$
$x = 3$
The ordered pair is (3, 0).

b. Suppose that x belongs to the set $\left\{0, \frac{2}{3}, 1, 1.6\right\}$. Find the corresponding ordered pairs that satisfy the equation $x + y = 2$.

Solution: Solve for y to make evaluations easier: $y = 2 - x$
In table form:

x	$y = 2 - x$	(x, y)
0	$y = 2 - 0 = 2$	$(0, 2)$
$\frac{2}{3}$	$y = 2 - \frac{2}{3} = \frac{4}{3}$	$\left(\frac{2}{3}, \frac{4}{3}\right)$
1	$y = 2 - 1 = 1$	$(1, 1)$
1.6	$y = 2 - 1.6 = 0.4$	$(1.6, 0.4)$

Just as we use the terms **ordered pair** and **point** (graph of an ordered pair) interchangeably, we use the terms **equation** and **graph of an equation** interchangeably. The equations

$$2x + 3y = 4, \quad y = 7, \quad x = -1, \quad \text{and} \quad y = 3x + 5$$

are called **linear equations** (or first-degree equations in two variables) and their graphs will be straight lines.

Standard Form of a Linear Equation

Any equation of the form

$$\mathbf{Ax + By = C} \qquad \textit{where A and B are not both equal to 0}$$

is called the ***standard form*** *of a* ***linear equation****.*

Every straight line corresponds to some linear equation, and the graph of every linear equation is a straight line. Thus, because we know from geometry that two points determine a line, the graph of an equation can be found by locating any two points that satisfy the equation.

To Graph a Linear Equation in Two Variables

1. *Locate any two points that satisfy the equation. (Choose values for x and y that lead to simple solutions. Remember that there are an infinite number of choices for either x or y. But, once a value for x or y is chosen, the corresponding value for the other variable is found by substituting into the equation.)*
2. *Plot these two points on a Cartesian coordinate system.*
3. *Draw a straight line through these two points. (****Note:*** *Every point on that line will satisfy the equation.)*
4. *To check: Locate a third point that satisfies the equation and check to see that it does indeed lie on the line.*

Example 2: Equation in Two Variables

Graph each of the following linear equations.

a. $2x + 3y = 6$

Solution: Make a table with headings $\boldsymbol{x}$ and $\boldsymbol{y}$ and, whenever possible, choose values for x or y that lead to simple solutions. (In our previous discussion, we found four ordered pairs that satisfy this equation.)

continued on next page ...

x	y
0	2
–3	4
3	0
$\frac{1}{2}$	$\frac{5}{3}$

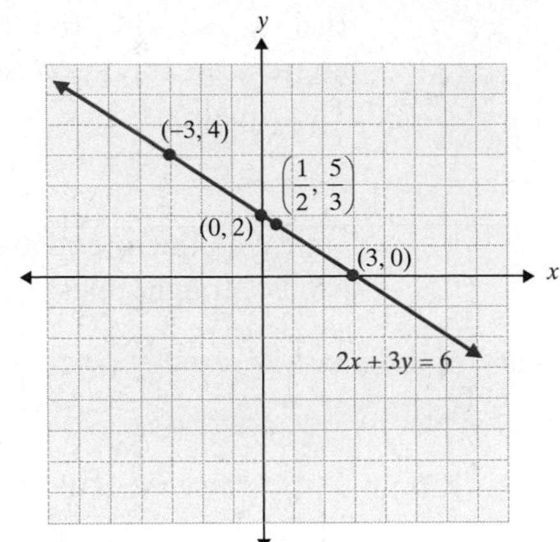

b. $x - 2y = 1$
Solution: Solve for x and substitute 0, 1, and 2 for y: $x = 1 + 2y$.

x	y
1	0
3	1
5	2

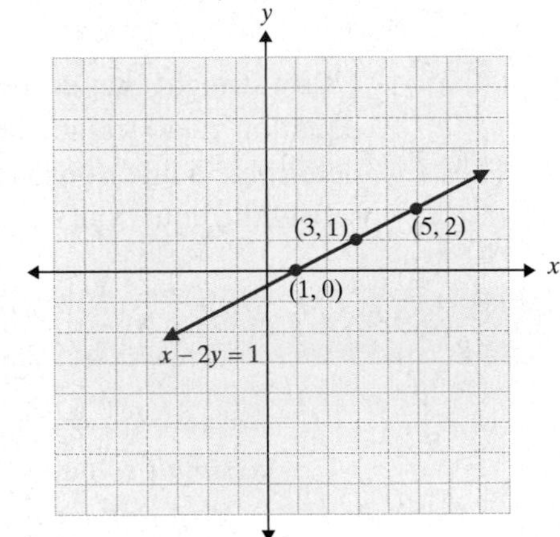

c. $y = 2x$
Solution: Substitute –1, 0, and 1 for x.

x	y
–1	–2
0	0
1	2

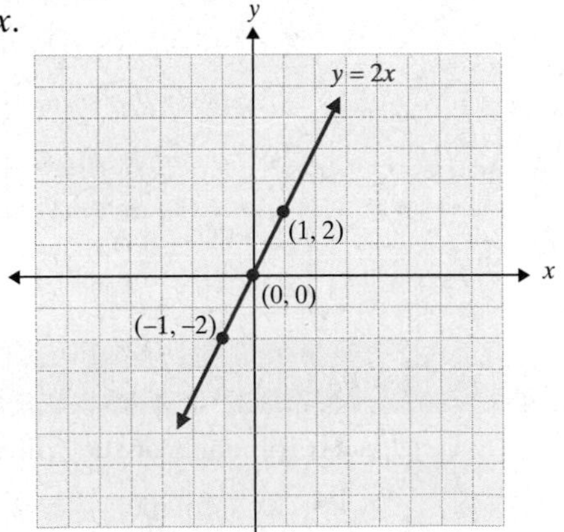

Locating the *y*-intercept and *x*-intercept

While the choice of the values for x or y can be arbitrary, letting $x = 0$ will locate the point on the graph where the line crosses (or intercepts) the y-axis. This point is called the ***y*-intercept**. The ***x*-intercept** is the point found by letting $y = 0$. This is the point where the line crosses (or intercepts) the x-axis. These two points are generally easy to locate and are frequently used as the two points for drawing the graph of a linear equation. If the line passes through the point $(0, 0)$, then the y-intercept and the x-intercept are the same point, namely the origin. In this case you will need to locate some other point to draw the graph. Example 3 illustrates this technique.

Example 3: *x*- and *y*-intercepts

Graph the following linear equations by locating the y-intercept and the x-intercept.

a. $x + 3y = 9$

Solution:

$$x = 0 \rightarrow 3y = 9$$
$$y = 3$$
$$y = 0 \rightarrow x = 9$$

$(0, 3)$ is the y-intercept.
$(9, 0)$ is the x-intercept.

Graph the two intercepts and draw the line that contains them.

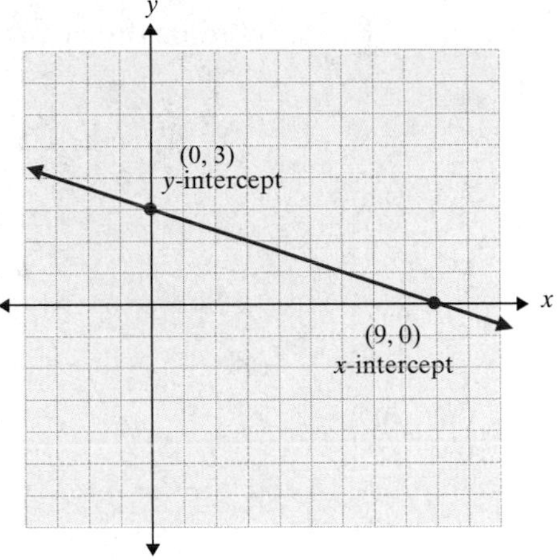

b. $3x - 2y = 12$

Solution:

$$x = 0 \rightarrow -2y = 12$$
$$y = -6$$

$$y = 0 \rightarrow 3x = 12$$
$$x = 4$$

$(0, -6)$ is the y-intercept.
$(4, 0)$ is the x-intercept.

Graph the two intercepts and draw the line that contains them.

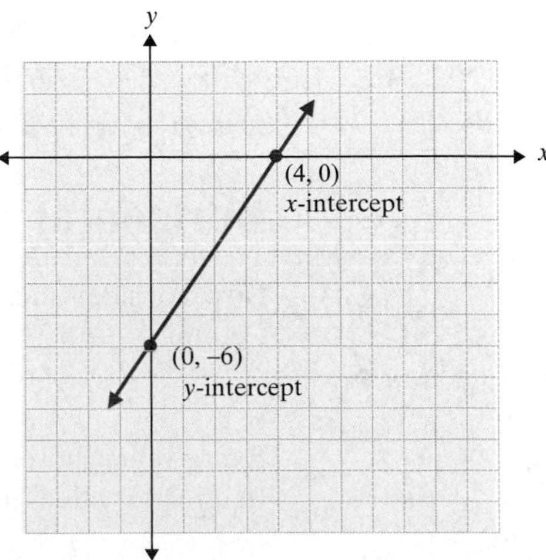

In general, the intercepts are easy to find because substituting 0 (for x or y) leads to an easy solution for the other variable. However, when the intercepts result in a point with fraction (or decimal) coordinates and estimation is involved, then a third point that satisfies the equation should be found to verify that the line is positioned correctly.

Lines that Contain the Origin

If the line goes through the origin, both the x-intercept and y-intercept will be 0. In this case, some other point must be used.

Example 4: Lines that Contain the Origin

Graph the linear equation $y = 3x$.

Solution: Locate two points on the graph.

$$x = 0 \rightarrow y = 0$$
$$x = 2 \rightarrow y = 6$$

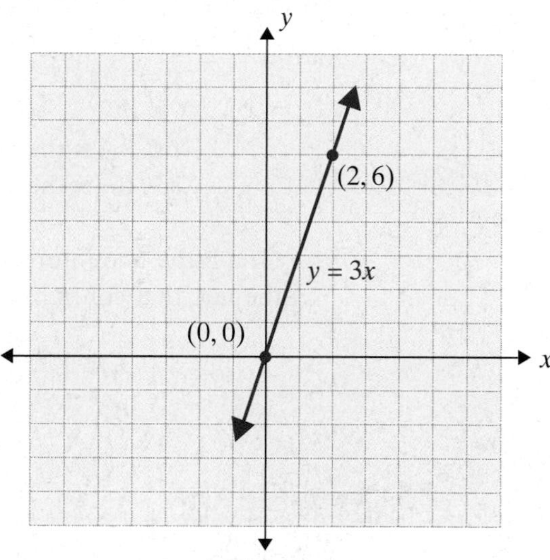

Horizontal and Vertical Lines

Now, consider an equation in the form $0x + By = C$ where the coefficient of x is 0. For example,

$$0x + y = 3 \qquad \text{or just} \qquad y = 3.$$

Every value chosen for x will be multiplied by 0 and the corresponding y-value will be 3. In effect, x-values have no influence on the y-value.

In table form three such points are:

x	y
–2	3
0	3
5	3

Regardless of what value is chosen for x, the corresponding y-value is 3. Thus, the graph of the equation $y = 3$ is a **horizontal line** (see Figure 4.3).

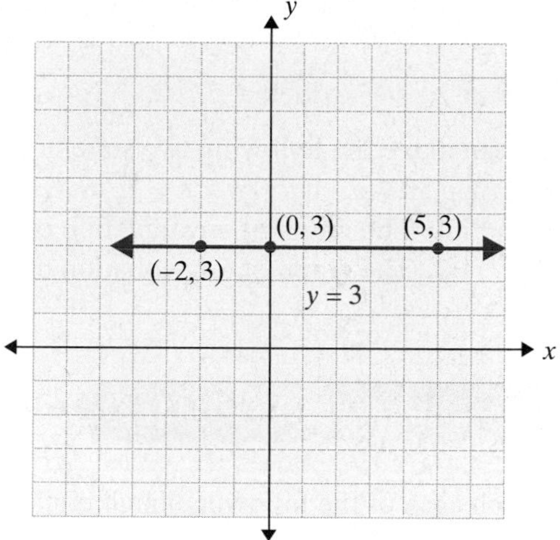

Figure 4.3

Next, consider an equation in the form $Ax + 0y = C$ where the coefficient of y is 0. For example,

$$x + 0y = 2 \qquad \text{or just} \qquad x = 2.$$

Every value chosen for y will be multiplied by 0 and the corresponding x-value will be 2. In effect the y-values have no influence on the value of x. In table form three such points are:

x	y
2	4
2	0
2	–3

Regardless of what is chosen for y, the corresponding x-value is 2. Thus, the graph of the equation $x = 2$ is a **vertical line** (see Figure 4.4).

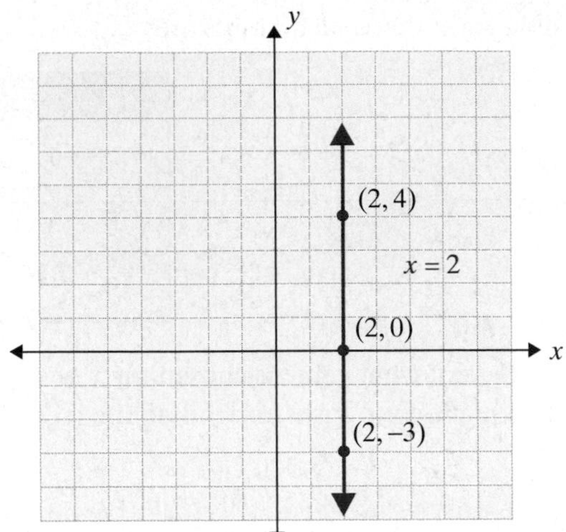

Figure 4.4

We can make the following two statements about horizontal and vertical lines:

1. The graph of any equation of the form $y = b$ is a horizontal line.
2. The graph of any equation of the form $x = a$ is a vertical line.

Example 5: Horizontal and Vertical Lines

Graph each of the following linear equations.

a. $2y = 5$

Solution: Solving for y:

$$2y = 5$$

$$y = \frac{5}{2}$$

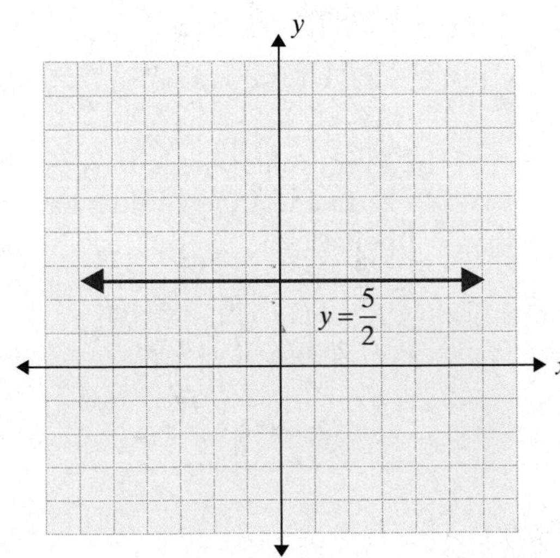

b. $x + 6 = 0$

Solution: Solving for x:

$$x + 6 = 0$$
$$x = -6$$

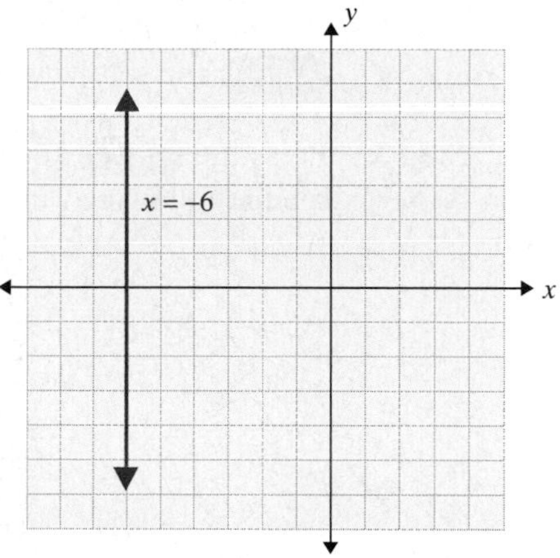

Using a TI-83 Plus Calculator to Graph Straight Lines

By following the steps outlined here you will be able to see the graphs of all straight lines that are not vertical. The calculator will not graph vertical lines using the methods described here.

To have the calculator graph a nonvertical straight line, you must first solve the equation for y. For example,

given the equation: $2x + y = 3$

solving for y gives: $y = -2x + 3$ (or $y = 3 - 2x$)

Step 1: Press the MODE key and set all the highlighted keys as shown in the diagram. If the mode is not as shown, press the down arrow until you reach the desired line and press ENTER.

It is particularly important that **Func** is highlighted. This stands for function. Function concepts and notation will be introduced later in this chapter.

Step 2: Press the ZOOM key and then press 6 to get the **ZStandard** window for your graphs. This will give scales from –10 to 10 for both the x-axis and the y-axis. If you later decide that you would like some other scales on either axis, press WINDOW and set the window values to the appropriate values.

Step 3: Press the Y= key in the upper left corner of the keyboard. Here you will see the following window.

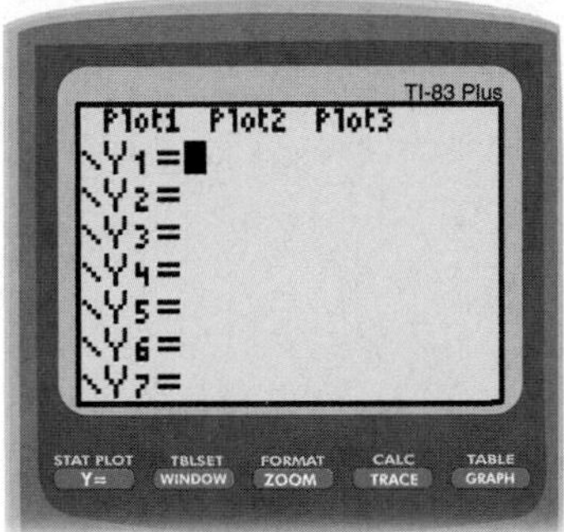

Step 4: Enter the expression for y here after $\mathbf{Y_1}$ **=**. The letter x is entered by pressing the X,T,θ,n key located just below the MODE key. Also note that the negative sign (–) is next to the ENTER key.

The screen should appear as follows:

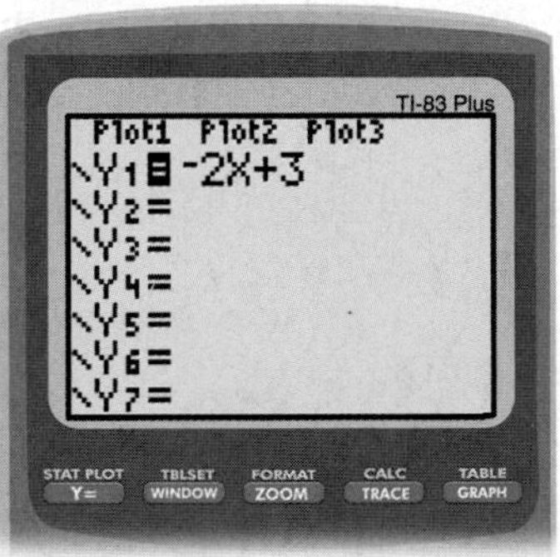

Note that the notation on the screen $\mathbf{Y_1}$ =, $\mathbf{Y_2}$ =, $\mathbf{Y_3}$ =, and so forth, allows you to graph several equations at once. We will discuss and use this feature later.

Step 5: Press the GRAPH key in the upper right corner of the keyboard. Your graph should appear as follows:

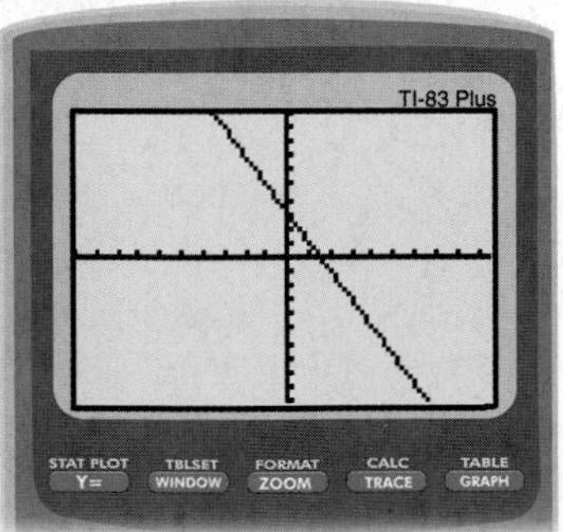

To clear the screen you may press CLEAR or 2nd QUIT.

Practice Problems

1. *For x in the set {−1, 2, 3}, find the corresponding ordered pairs that satisfy the equation $x - 2y = 3$.*
2. *Find the missing coordinate of each ordered pair so that it belongs to the solution set of the equation $2x + y = 4$:*
   (0,), (, 0), (, 8), (−1,).
3. *Does the ordered pair $\left(1, \frac{3}{2}\right)$ satisfy the equation $3x + 2y = 6$?*

4.2 Exercises

Find the missing coordinate of each ordered pair so that the ordered pair belongs to the solution set of the equation in Exercises 1 – 8.

1. $2x + y = 5$	**2.** $x + 2y = 6$	**3.** $3x - y = 4$	**4.** $x - 3y = 9$
a. (0,)	**a.** (0,)	**a.** (0,)	**a.** (0,)
b. (, 0)	**b.** (, 0)	**b.** (, 0)	**b.** (, 0)
c. (−2,)	**c.** (4,)	**c.** (2,)	**c.** (−3,)
d. (, 3)	**d.** (, −2)	**d.** (, 5)	**d.** (, −1)

5. $y = 5 - 2x$	**6.** $y = 5x - 3$	**7.** $3x - 2y = 6$	**8.** $5x + 2y = 10$
a. (0,)	**a.** (0,)	**a.** (0,)	**a.** (0,)
b. (, 0)	**b.** (, 0)	**b.** (, 0)	**b.** (, 0)
c. (2,)	**c.** (−1,)	**c.** (−2,)	**c.** (4,)
d. (, 7)	**d.** (, 7)	**d.** (, 3)	**d.** (, 10)

1. a. (0, 5) **b.** $\left(\frac{5}{2}, 0\right)$ **c.** (−2, 9) **d.** (1, 3)
2. a. (0, 3) **b.** (6, 0) **c.** (4, 1) **d.** (10, −2)
3. a. (0, −4) **b.** $\left(\frac{4}{3}, 0\right)$ **c.** (2, 2) **d.** (3, 5)
4. a. (0, −3) **b.** (9, 0) **c.** (−3, −4) **d.** (6, −1)
5. a. (0, 5) **b.** $\left(\frac{5}{2}, 0\right)$ **c.** (2, 1) **d.** (−1, 7)
6. a. (0, −3) **b.** $\left(\frac{3}{5}, 0\right)$ **c.** (−1, −8) **d.** (2, 7)
7. a. (0, −3) **b.** (2, 0) **c.** (−2, −6) **d.** (4, 3)
8. a. (0, 5) **b.** (2, 0) **c.** (4, −5) **d.** (−2, 10)
9. a **10.** e
11. d **12.** c
13. f **14.** b

15. $y = 2x$

16. $y = 3x$

17. $y = -x$

18.

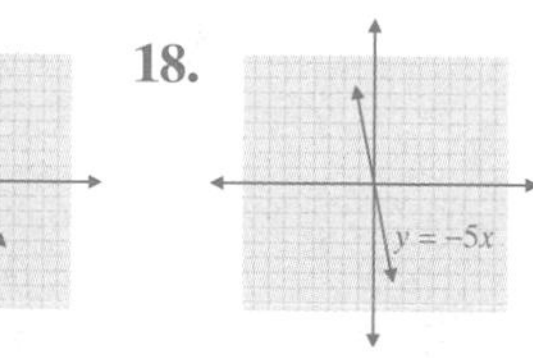

19. $y = x - 4$

20. $y = x + 3$

21. $y = x + 2$

22.

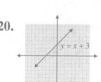

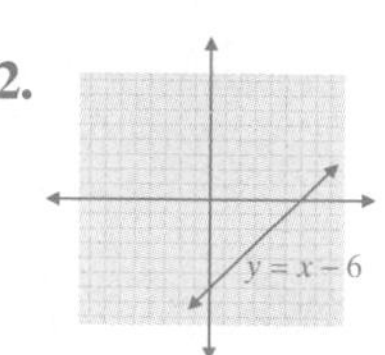

Answers to Practice Problems: 1. (−1, −2), $\left(2, -\frac{1}{2}\right)$, (3, 0) **2.** (0, 4), (2, 0), (−2, 8), (−1, 6) **3.** Yes

23. 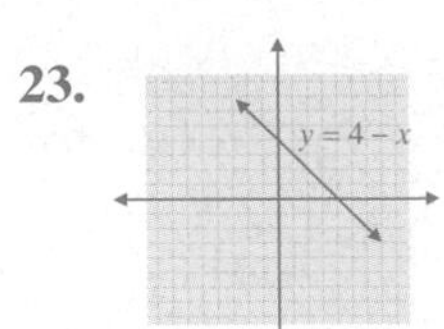

24. 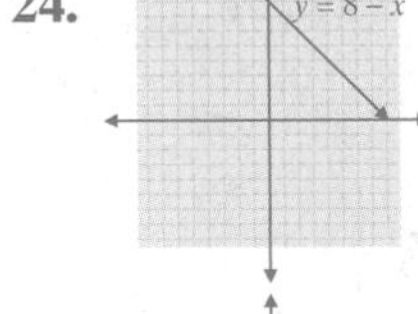

25. 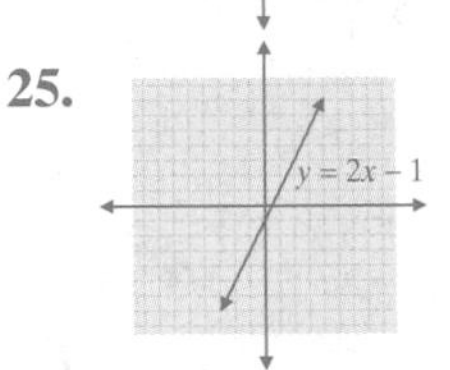

26. 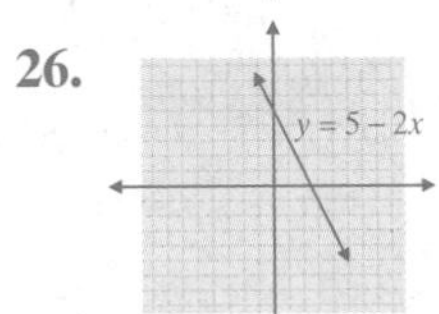

27. 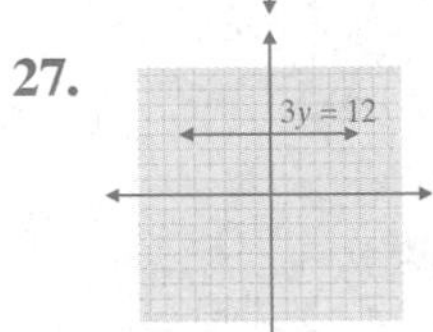

28. 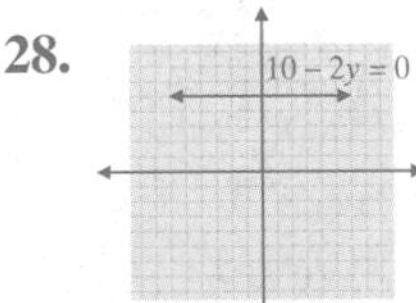

29. 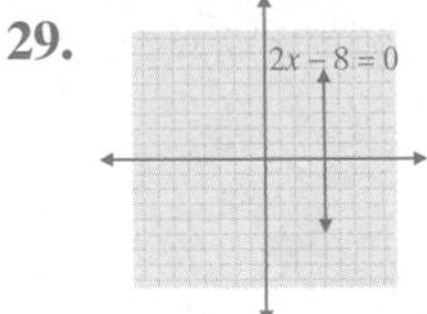

30. 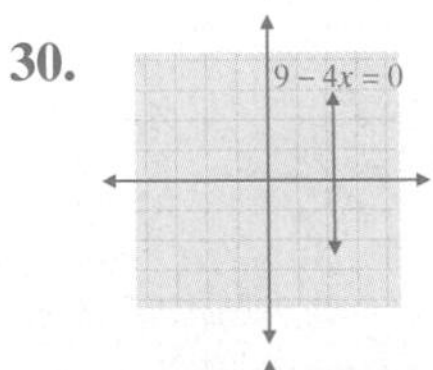

31. 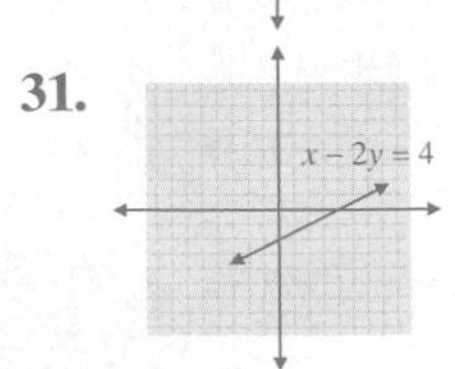

For Exercises 9 – 14, use your knowledge of x-intercepts and y-intercepts to match each of the following equations with its graph.

9. $4x + 3y = 12$

10. $4x - 3y = 12$

11. $x + 2y = 8$

12. $-x + 2y = 8$

13. $x + 4y = 0$

14. $5x - y = 10$

a. 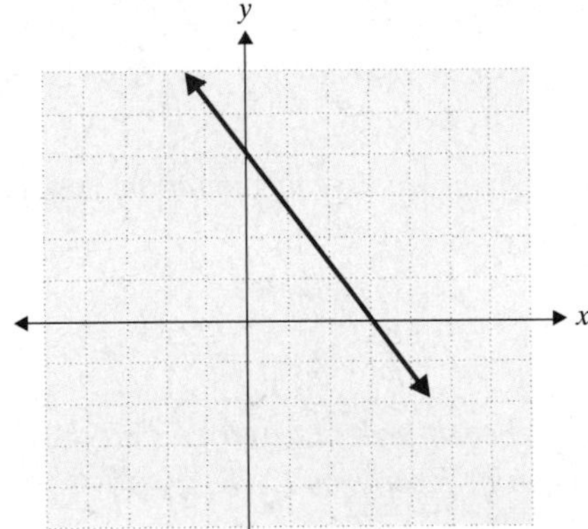

b. 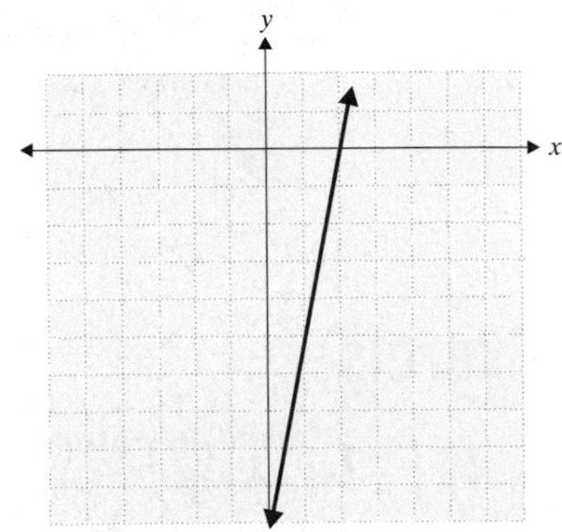

c. 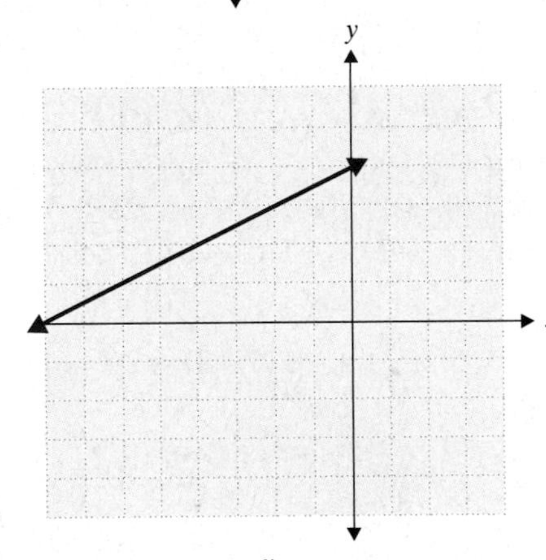

d. 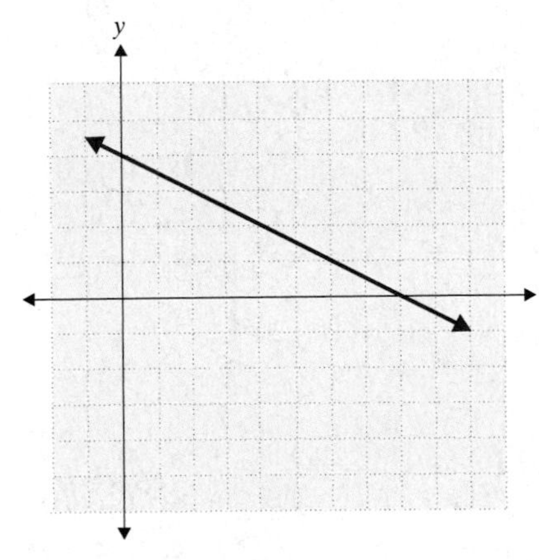

e. 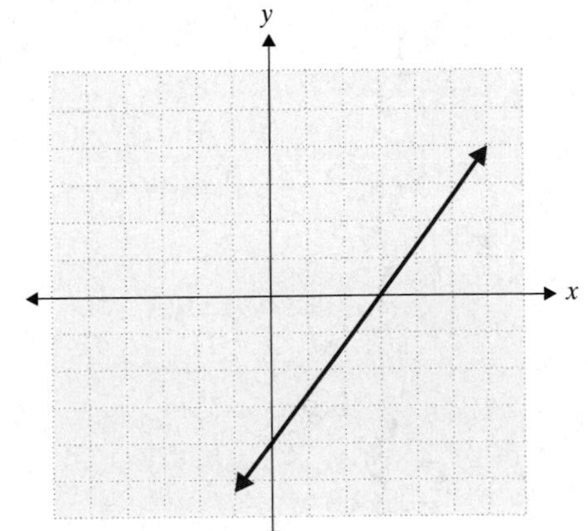

f. 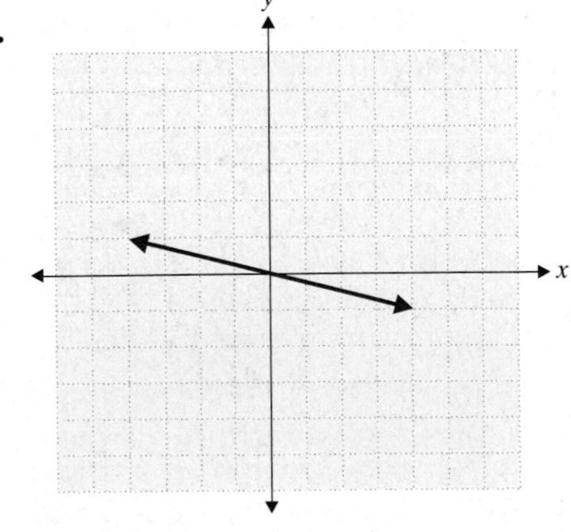

In Exercises 15 – 48, first tell (just by looking at the equation) whether the graph will be a vertical line, a horizontal line, or neither. Then, graph the line corresponding to each linear equation by locating any two (or three) ordered pairs (points) that satisfy the equation.

15. $y = 2x$ **16.** $y = 3x$ **17.** $y = -x$

18. $y = -5x$ **19.** $y = x - 4$ **20.** $y = x + 3$

21. $y = x + 2$ **22.** $y = x - 6$ **23.** $y = 4 - x$

24. $y = 8 - x$ **25.** $y = 2x - 1$ **26.** $y = 5 - 2x$

27. $3y = 12$ **28.** $10 - 2y = 0$ **29.** $2x - 8 = 0$

30. $9 - 4x = 0$ **31.** $x - 2y = 4$ **32.** $x + 3y = 5$

33. $2x + y = 0$ **34.** $2y = x$ **35.** $5y = 0$

36. $2y - 3 = 0$ **37.** $4x = 0$ **38.** $-\frac{3}{4}x - 1 = 0$

39. $2x + 3y = 7$ **40.** $4x + 3y = 11$ **41.** $3y - 2x = 4$

42. $3x - 2y = 6$ **43.** $5x + 2y = 9$ **44.** $2x - 7y = -14$

45. $4x + 2y = -10$ **46.** $y = \frac{1}{2}x + 1$ **47.** $y = \frac{1}{3}x - 3$

48. $\frac{2}{3}x + y = 4$

Graph the linear equations in Exercises 49 – 63 by locating the y-intercept and the x-intercept.

49. $x + y = 4$ **50.** $x - 2y = 6$ **51.** $3x - 2y = 6$

52. $3x - 4y = 12$ **53.** $5x + 2y = 10$ **54.** $3x + 7y = -21$

55. $2x - y = 9$ **56.** $4x + y = 7$ **57.** $x + 3y = 5$

58. $x - 6y = 3$ **59.** $\frac{1}{2}x - y = 4$ **60.** $\frac{2}{3}x - 3y = 4$

61. $\frac{1}{2}x - \frac{3}{4}y = 6$ **62.** $5x + 3y = 7$ **63.** $2x + 3y = 5$

32.
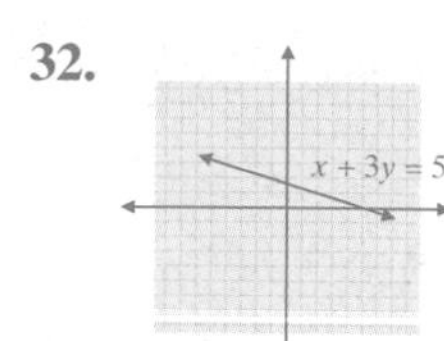

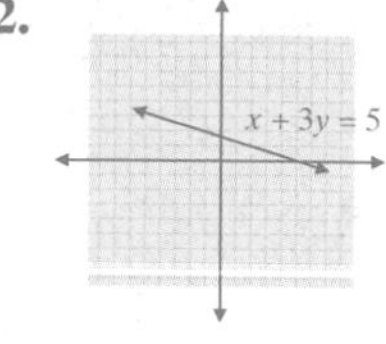

33. 2x + y = 0

34. 2y = x

35. 5y = 0

36. 2y − 3 = 0

37. 4x = 0

38.
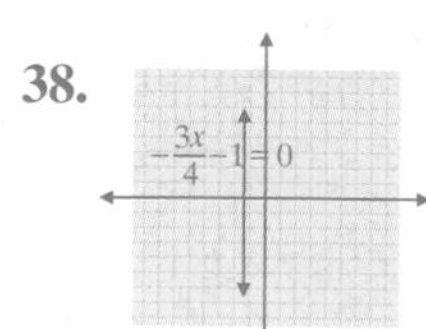

39.
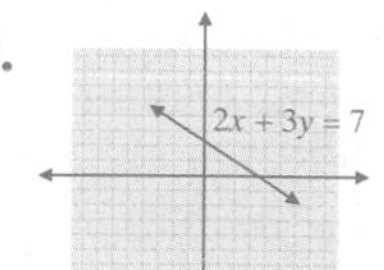

40.
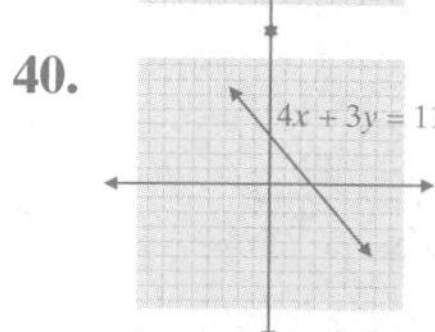

41.
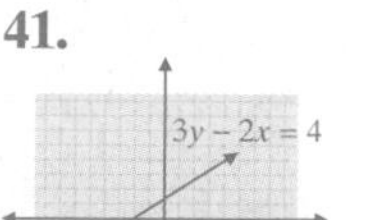

42.
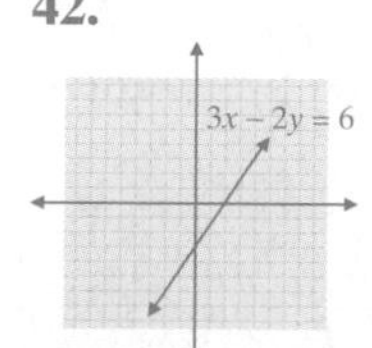

43.
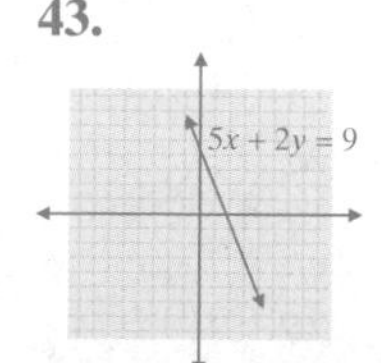

44.
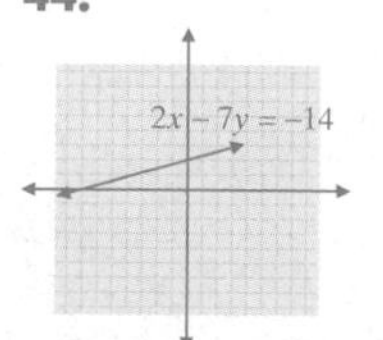

45.
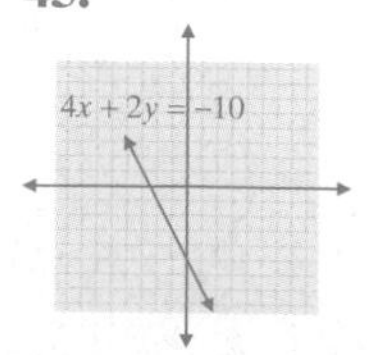

46.

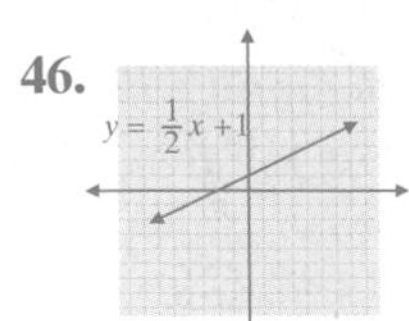

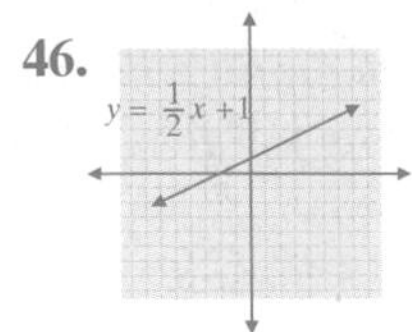

47.

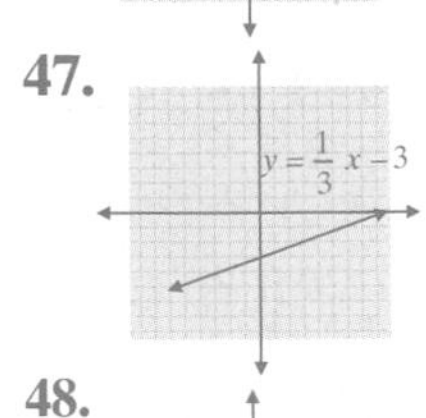

48.

$\frac{2}{3}x + y = 4$

49.

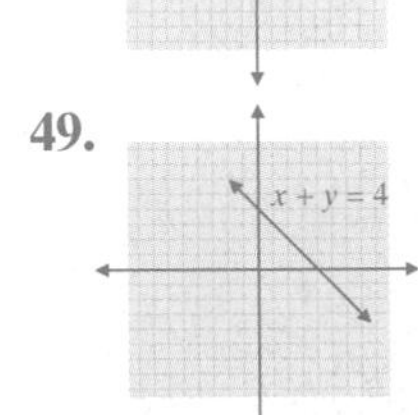

In Exercises 64 – 68, solve each of the equations for y and use your graphing calculator to graph each line.

64. $x + y = 5$ **65.** $2y + x = 0$ **66.** $3y - 15 = 0$

67. $3x - y = 4$ **68.** $5x - 2y = 10$

Writing and Thinking About Mathematics

Each of the following equations is not linear and the corresponding graph is not a straight line. Make a table and find several points, some where x is positive and some where x is negative, to determine the nature of the corresponding graph. After you have plotted your points and analyzed the nature of the graph, enter the expression for y in your graphing calculator to verify that you have a reasonably accurate graph.

69. $y = \dfrac{4}{x}$ **70.** $y = x^2$ **71.** $y = -x^2$

72. $y = x^2 - 5$ **73.** $y = x^3$

Hawkes Learning Systems: Introductory & Intermediate Algebra

Graphing Linear Equations by Plotting Points

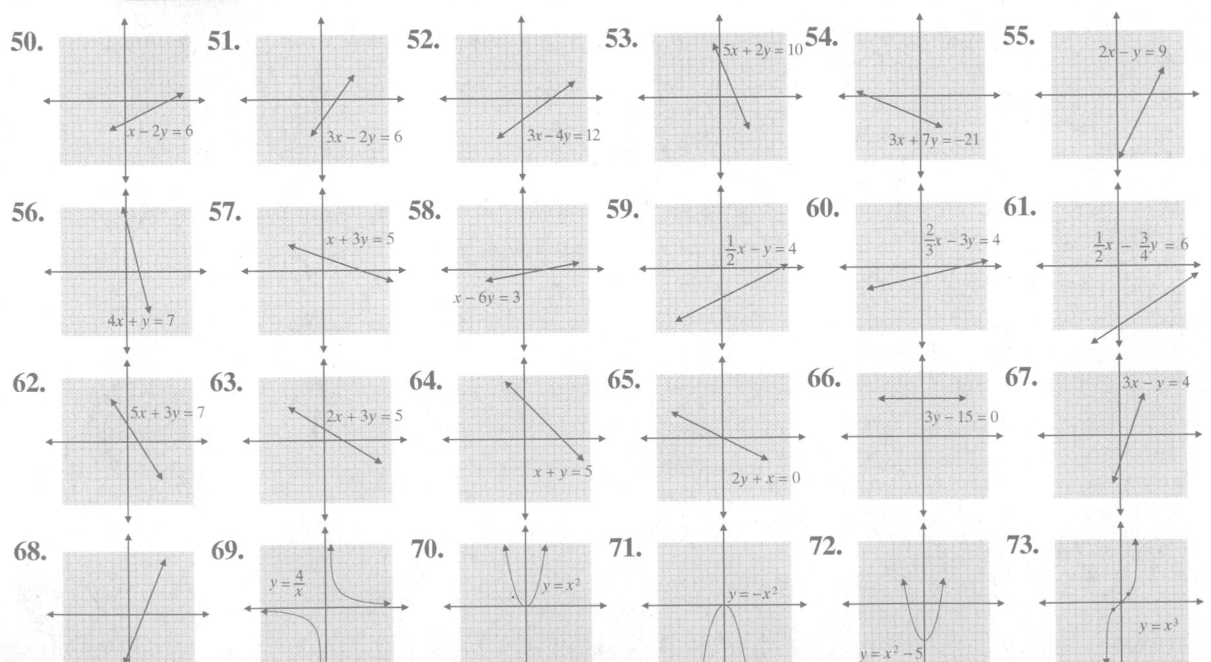

4.3 The Slope-Intercept Form: $y = mx + b$

Objectives

After completing this section, you will be able to:

1. *Find the slope of a line containing two given points.*
2. *Graph lines using the slope-intercept method.*
3. *Graph horizontal lines.*
4. *Graph vertical lines.*

Slope of a Line

In Section 4.2, we discussed the **standard form** of a linear equation in two variables:

$$Ax + By = C$$

In this section, we will analyze linear equations in another form using the concept of **slope**.

The term **slope** is common in phrases such as the slope of a roof, the slope of a road, or the slope of a mountain. If you ride a bicycle up a mountain road, you certainly know when the slope increases because you have to pedal harder. In construction, a roof that is to have a 7:12 pitch is constructed so that for every 7 inches of rise (vertical distance) there are 12 inches of run (horizontal distance). That is, the ratio of the rise to the run can be written as $\frac{\text{rise}}{\text{run}} = \frac{7}{12}$.

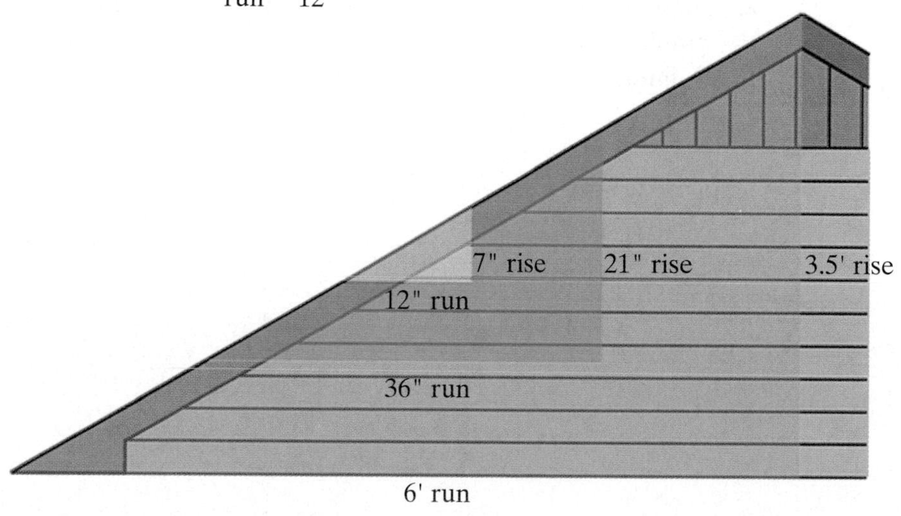

Figure 4.5

$$\frac{\text{rise}}{\text{run}} = \frac{7}{12} = \frac{21}{36} = \frac{3.5}{6}$$

If we know any two points on a line, say (x_1, y_1) and (x_2, y_2), we can calculate the **slope** using the following formula. (**Note:** The letter m is standard notation for representing the slope of a line.)

$$\textbf{slope} = m = \frac{\textbf{rise}}{\textbf{run}} = \frac{y_2 - y_1}{x_2 - x_1} = \frac{y_1 - y_2}{x_1 - x_2} \quad \textbf{where } x_1 \neq x_2$$

NOTES It does not matter which point is called (x_1, y_1) and which is called (x_2, y_2) as long as we are consistent. That is, be sure to subtract the coordinates in the same order in both the numerator and denominator.

The slope of a line is illustrated in Figure 4.6.

$$m = \frac{y_2 - y_1}{x_2 - x_1}$$

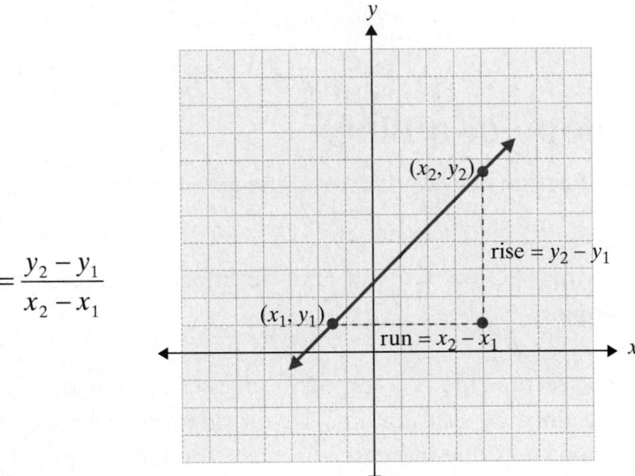

Figure 4.6

Example 1: Slope of a Line

a. Find the slope of the line that contains the two points (–1, 2) and (3, 5), and then graph the line.

Solution: Using $(x_1, y_1) = (-1, 2)$ and $(x_2, y_2) = (3, 5)$,

$$m = \frac{5-2}{3-(-1)}$$

$$= \frac{3}{4}$$

or, using $(x_1, y_1) = (3, 5)$ and $(x_2, y_2) = (-1, 2)$,

$$m = \frac{2-5}{-1-3}$$

$$= \frac{-3}{-4}$$

$$= \frac{3}{4}$$

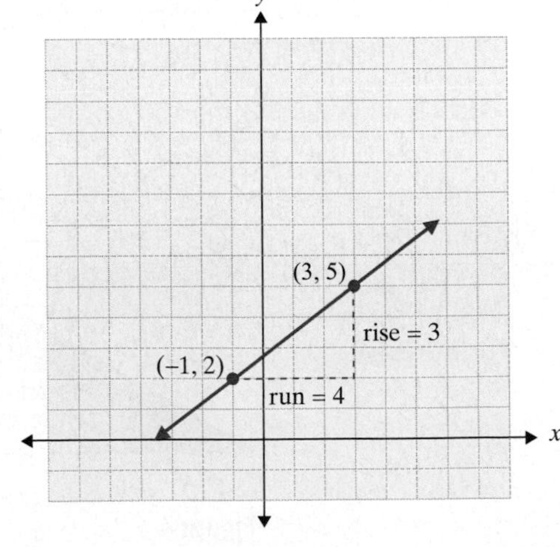

b. Find the slope of the line that contains the two points (0, 1) and (2, 6) and graph the line.

Solution: Using $(x_1, y_1) = (0, 1)$ and $(x_2, y_2) = (2, 6)$,

$$\text{slope} = m = \frac{1-6}{0-2} = \frac{-5}{-2} = \frac{5}{2}$$

$$\text{or } m = \frac{6-1}{2-0} = \frac{5}{2}$$

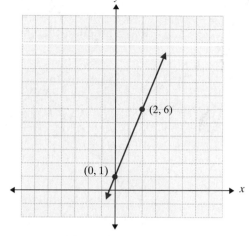

c. Find the slope of the line through the points (5, 1) and (1, 3) and graph the line.

Solution: Using $(x_1, y_1) = (5, 1)$ and $(x_2, y_2) = (1, 3)$,

$$\text{slope} = m = \frac{1-3}{5-1} = \frac{-2}{4} = -\frac{1}{2}$$

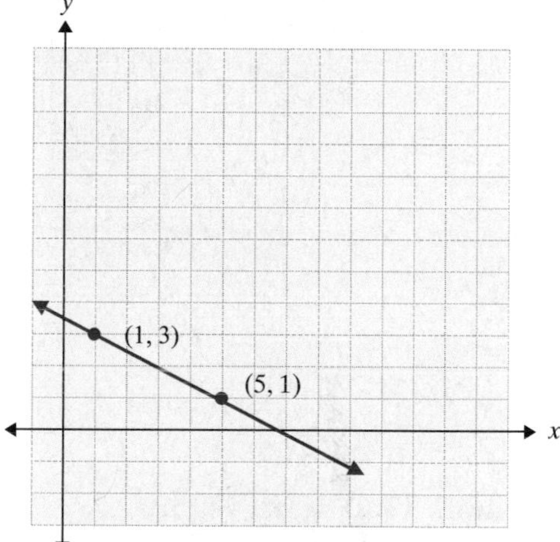

NOTES

After studying Examples 1a – 1c and the remaining examples in this section, you should be aware that lines with positive slope slant upward to the right and lines with negative slope slant downward to the right. This observation will be a great help in graphing straight lines and analyzing the graphs of many other types of curves.

Slope-Intercept Form: $y = mx + b$

Teaching Notes: A good idea is to have students practice (again) solving for y from the standard form of an equation. This skill needs frequent reinforcement.

There are certain relationships between the coefficients in the equation of a line and the graph of that line. For example, consider the equation

$$y = 5x - 7.$$

First, find two points on the line and calculate the slope.

(0, –7) and (2, 3) both satisfy the equation.

$$\text{slope} = m = \frac{-7-3}{0-2} = \frac{-10}{-2} = 5$$

$$\text{or } m = \frac{3-(-7)}{2-0} = \frac{10}{2} = 5$$

Observe that the slope m is the same as the coefficient of x in the equation $y = 5x - 7$. This is not just a coincidence. In fact, if a linear equation is solved for y, then the coefficient of x will always be the slope of the line. The proof of this statement follows:

Consider the equation

$$y = mx + b.$$

Suppose that (x_1, y_1) and (x_2, y_2) are any two points on the line. Then,

$$y_1 = mx_1 + b \qquad \text{and} \qquad y_2 = mx_2 + b$$

$$\begin{aligned} \text{slope} &= \frac{y_2 - y_1}{x_2 - x_1} \\ &= \frac{(mx_2 + b) - (mx_1 + b)}{x_2 - x_1} \\ &= \frac{mx_2 + b - mx_1 - b}{x_2 - x_1} \\ &= \frac{m\cancel{(x_2 - x_1)}}{\cancel{(x_2 - x_1)}} \\ &= m \end{aligned}$$

Therefore, the coefficient m in the equation $y = mx + b$ is the slope of the line.

As we discussed earlier, the **y-intercept** is the point where the graph of a line crosses the y-axis. The x-coordinate of this point will always be 0. If $x = 0$ in the general equation $y = mx + b$, then $y = m(0) + b = b$. Therefore, the y-intercept is the point $(0, b)$. The constant b is also called the y-intercept with the understanding that $x = 0$ when $y = b$.

Slope-Intercept Form

Any equation of the form

$$\mathbf{y = mx + b}$$

is called the ***slope-intercept*** *form for the equation of a line. The slope of the line is m and the y-intercept is b.*

An equation in the standard form

$$\boldsymbol{Ax + By = C} \qquad \text{with } B \neq 0$$

can be written in the slope-intercept form by solving for y.

$$Ax + By = C$$

$$By = -Ax + C$$

$$y = -\frac{A}{B}x + \frac{C}{B}$$

Thus, in general,

$$\boldsymbol{m} = -\frac{\boldsymbol{A}}{\boldsymbol{B}} \text{ and } \boldsymbol{b} = \frac{\boldsymbol{C}}{\boldsymbol{B}}.$$

Example 2: Slope-Intercept Form

a. Find the slope, m, and y-intercept, b, of the line $-2x + 3y = 6$ and graph the line.

Solution: Solving for y, $\quad -2x + 3y = 6$

$$3y = 2x + 6$$

$$y = \frac{2}{3}x + 2$$

Thus,

$$m = \frac{2}{3} \quad \text{and} \quad b = 2.$$

Now that the equation is in the slope-intercept form, the graph can be drawn by locating the y-intercept and using the slope as $\frac{\textbf{rise}}{\textbf{run}}$ to locate a second point on the line. **Note that when the rise is negative, move down, and when the run is negative, move left.**

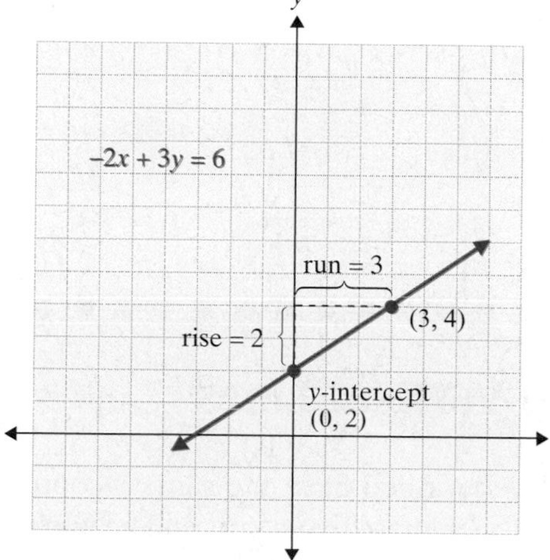

From the y-intercept, count 2 units up (rise) and 3 units right (run) to locate a second point on the line. This illustrates the slope

$$m = \frac{2}{3}.$$

Note: The same point can be located by moving 3 units right and then 2 units up.

continued on next page ...

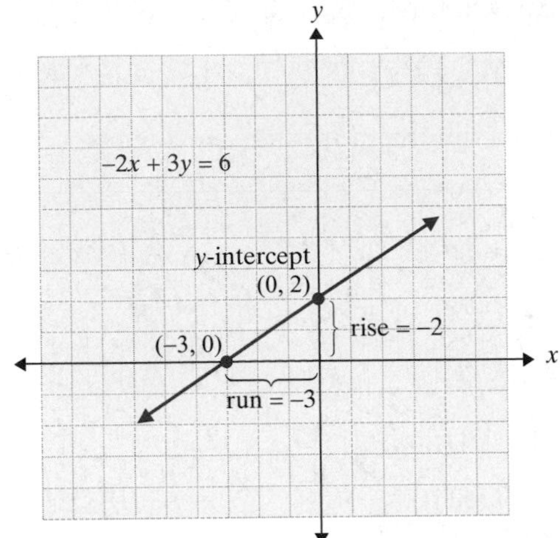

Or locate a second point by counting 2 units down (rise) and 3 units left (run). That is, interpret the slope m as

$$m = \frac{-2}{-3} = \frac{2}{3}.$$

b. Find the slope, m, and y-intercept, b, of the line $y = -3x + 2$ and graph the line.
Solution: The equation is already in the slope-intercept form with

$$m = -3 \quad \text{and} \quad b = 2.$$

To draw the graph, note that the slope $m = -3 = \frac{-3}{1} = \frac{\textbf{rise}}{\textbf{run}}$.

Treat the rise as –3 and the run as 1.

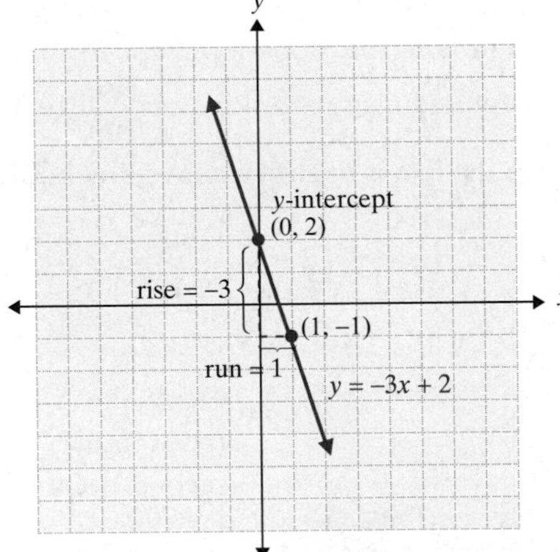

From the y-intercept, count 3 units down and 1 unit to the right to locate a second point.

Horizontal and Vertical Lines

Suppose that two points on a line have the same y-coordinate, such as $(-2, 3)$ and $(5, 3)$. Then the line through these two points will be **horizontal** as shown in Figure 4.7. The slope is

$$m = \frac{3-3}{5-(-2)} = \frac{0}{7} = 0$$

For any horizontal line, all of the y-values will be the same. Consequently, the formula for slope will always have 0 in the numerator. Therefore, **the slope of every horizontal line is 0.**

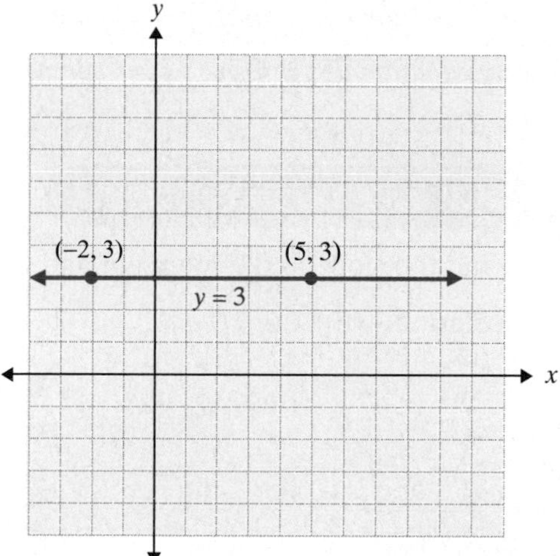

Figure 4.7

The equation of any horizontal line is in the form $y = 0x + b$ or $y = b$. In the case above, the y-coordinates are all 3, and the equation of the line is $y = 0x + 3$ or $y = 3$.

If two points have the same x-coordinates, such as (1, 3) and (1, −2), then the line through these two points will be **vertical** as in Figure 4.8. The slope is

$$m = \frac{-2-3}{1-1} = \frac{-5}{0}, \text{which is } \textbf{undefined}.$$

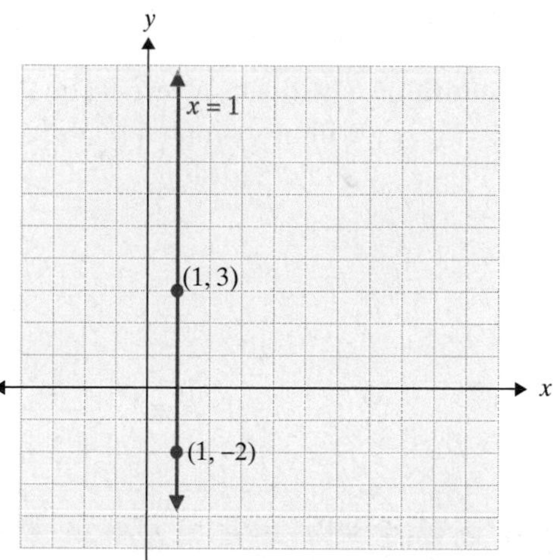

Figure 4.8

The equation of the vertical line in Figure 4.8 is $x = 1$ (in standard form, $x + 0y = 1$).

The x-coordinate is 1 for every point on the line.

In general, the equation $y = b$ restricts y-values to b with no restrictions on x, giving a horizontal line. Similarly, the equation $x = a$ restricts x-values to a with no restrictions on y, giving a vertical line.

Horizontal and Vertical Lines

Any equation of the form $\mathbf{y = b}$ *represents a* ***horizontal line*** *with* ***slope 0.***

Any equation of the form $\mathbf{x = a}$ *represents a* ***vertical line*** *with* ***undefined slope.***

Example 3: Horizontal and Vertical Lines

a. Find the slope and y-intercept and graph the line $3y + 6 = 0$.

Solution:

$$3y + 6 = 0$$
$$3y = -6$$
$$y = -2$$

The slope is 0, the y-intercept is -2 [or $(0, -2)$].

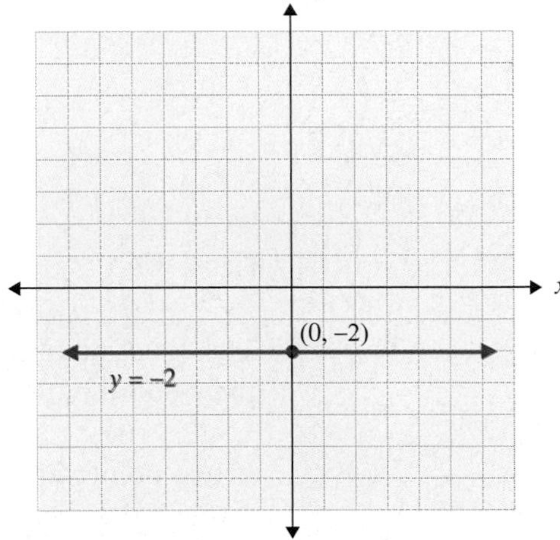

b. Graph the line $x = -2$.

Solution: The line is a vertical line with no y-intercept and an undefined slope.

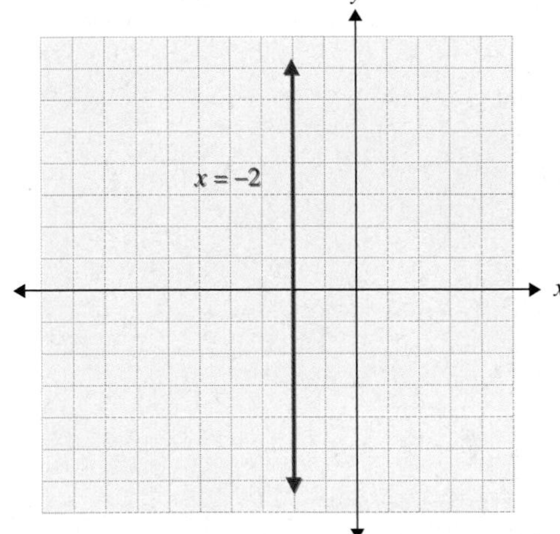

Characteristics of Slopes and Graphs of Lines

1. *Lines that have positive slope slant "upward" to the right.*
2. *Lines that have negative slope slant "downward" to the right.*
3. *Horizontal lines have slope 0.*
4. *Vertical lines have undefined slopes.*

Practice Problems

1. $m = 5$

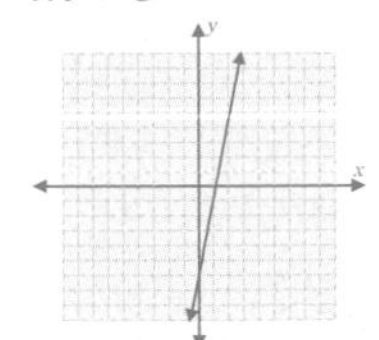

1. *Find the slope of the line through the two points (1, 3) and (4, 6).*
2. *What is the slope and the y-intercept of the line with the equation $2x + y = 5$?*
3. *Find the slope and y-intercept of the line with equation $5x - 4y = 20$.*

4.3 Exercises

2. $m = \frac{1}{2}$

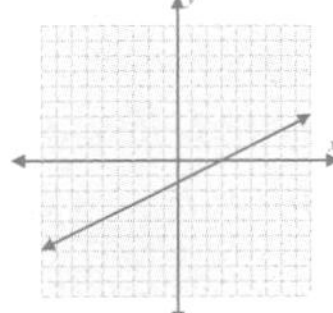

Graph the line determined by each pair of points in Exercises 1 – 12, and then find the slope of the line.

1. $(2, 4), (1, -1)$ **2.** $(5, 1), (3, 0)$ **3.** $(-3, 7), (4, -1)$

3. $m = -\frac{8}{7}$

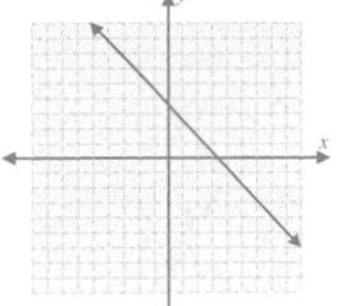

4. $(-6, 3), (1, 2)$ **5.** $(-5, 8), (3, 8)$ **6.** $(0, 0), (-2, -3)$

7. $\left(4, \frac{1}{2}\right), (-1, 2)$ **8.** $\left(\frac{3}{4}, \frac{3}{2}\right), (1, 2)$ **9.** $(-2, 3), (-2, -1)$

10. $(1, -2), (1, 4)$ **11.** $\left(\frac{3}{2}, \frac{4}{5}\right), \left(-2, \frac{1}{10}\right)$ **12.** $\left(\frac{7}{2}, \frac{3}{4}\right), \left(\frac{1}{2}, -3\right)$

4. $m = -\frac{1}{7}$

Find the equation and draw the graph of the line passing through the given y-intercept with the given slope in Exercises 13 – 24.

5. $m = 0$

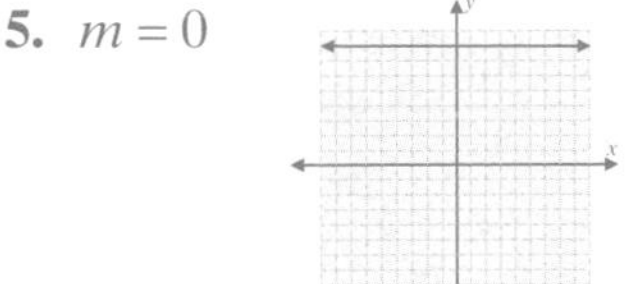

13. $(0, 0),\ m = \frac{2}{3}$ **14.** $(0, 1),\ m = \frac{1}{5}$ **15.** $(0, -3),\ m = -\frac{3}{4}$

16. $(0, -2),\ m = \frac{4}{3}$ **17.** $(0, 3),\ m = -\frac{5}{3}$ **18.** $(0, 2),\ m = 4$

19. $(0, -1),\ m = 2$ **20.** $(0, 4),\ m = -\frac{3}{5}$ **21.** $(0, -5),\ m = -\frac{1}{4}$

6. $m = \frac{3}{2}$

22. $(0, 5),\ m = -3$ **23.** $(0, -4),\ m = \frac{3}{2}$ **24.** $(0, 0),\ m = -\frac{1}{3}$

For Exercises 25 – 52, write the equation in slope-intercept form. Find the slope and the y-intercept, and then draw the graph.

7. $m = -\frac{3}{10}$

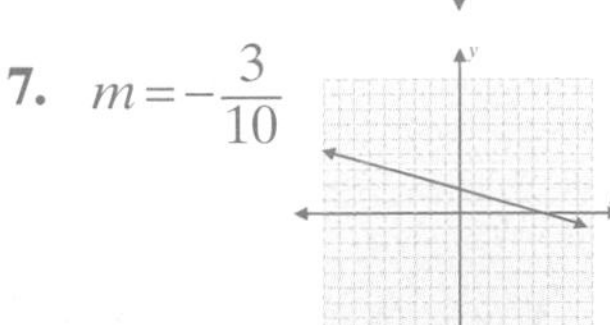

25. $y = 2x - 1$ **26.** $y = 3x - 4$ **27.** $y = 5 - 4x$ **28.** $y = 4 - x$

29. $y = \frac{2}{3}x + 2$ **30.** $y = \frac{2}{5}x + 2$ **31.** $x + y = 5$ **32.** $x - 2y = 6$

8. $m = 2$

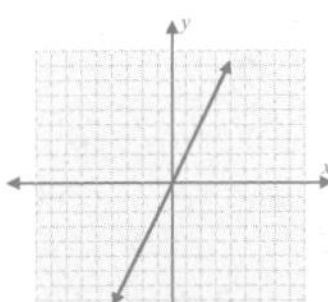

33. $x + 5y = 10$ **34.** $4x + y + 3 = 0$ **35.** $2y - 8 = 0$ **36.** $2x + 7y + 7 = 0$

9. Slope is undefined.

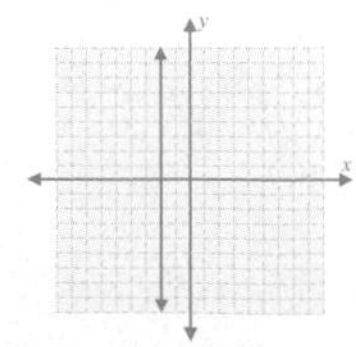

Answers to Practice Problems: **1.** $m = 1$ **2.** $m = -2, b = 5$ **3.** $m = \frac{5}{4}, b = -5$

10. Slope is undefined.

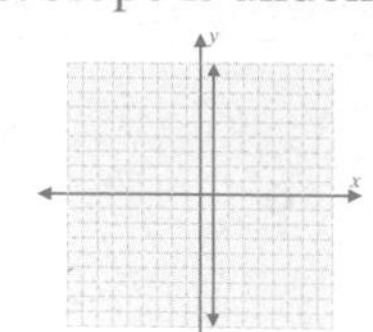

11. $m=\frac{1}{5}$

12. $m=\frac{5}{4}$

13. $y=\frac{2}{3}x$

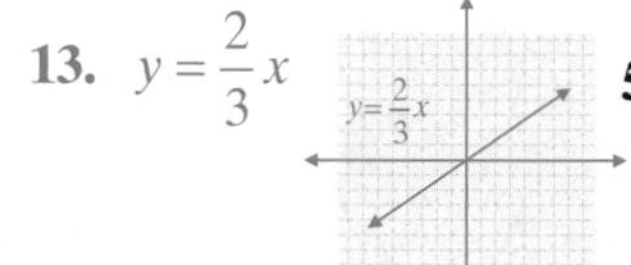

14. $y=\frac{1}{5}x+1$

15. $y=-\frac{3}{4}x-3$

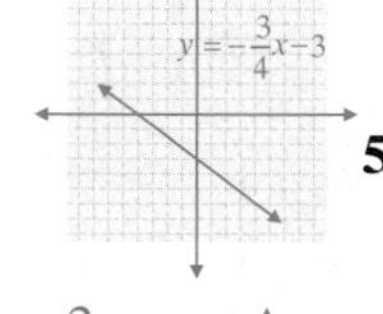

16. $y=\frac{4}{3}x-2$

17. $y=-\frac{5}{3}x+3$

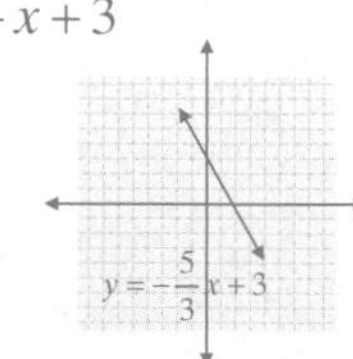

37. $4x+y=0$ **38.** $3y-9=0$ **39.** $2x=3y+6$ **40.** $4x=y+2$

41. $3x+9=0$ **42.** $3x+6=6y$ **43.** $5x-6y=10$ **44.** $4x+7=0$

45. $5-3x=4y$ **46.** $5x=11-2y$ **47.** $6x+4y=-7$ **48.** $7x+2y=4$

49. $6y=4+3x$ **50.** $6x+5y=-15$ **51.** $5x-2y+5=0$ **52.** $4x=3y-7$

53. In reference to the equation $y=mx+b$, sketch the graph of three lines for each of the two characteristics listed below.

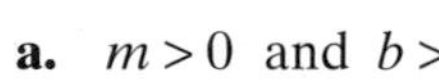

a. $m>0$ and $b>0$ **b.** $m<0$ and $b>0$

c. $m>0$ and $b<0$ **d.** $m<0$ and $b<0$

*In Exercises 54 – 61 the graph of a line is shown with two points highlighted. Find (**a**) the slope, (**b**) the y-intercept (if there is one), and (**c**) the equation of the line.*

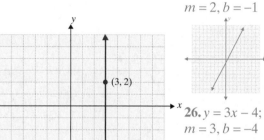

54.

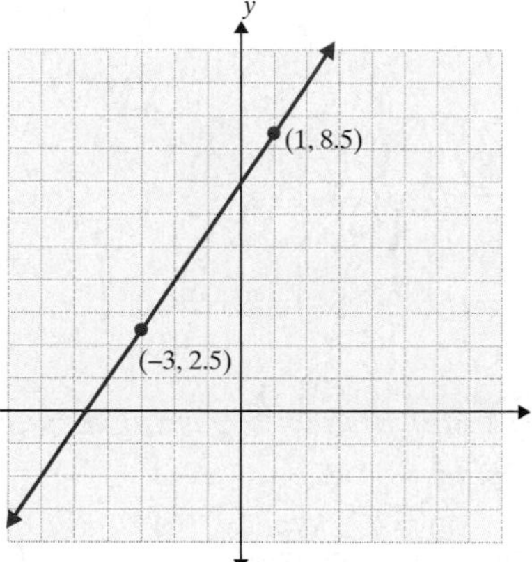

55.

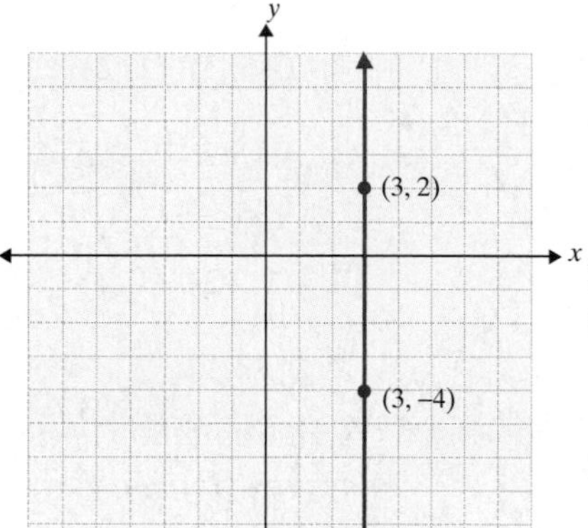

56.

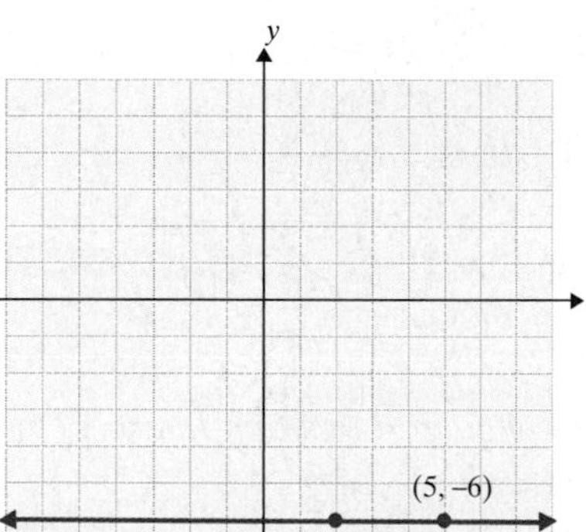

57.

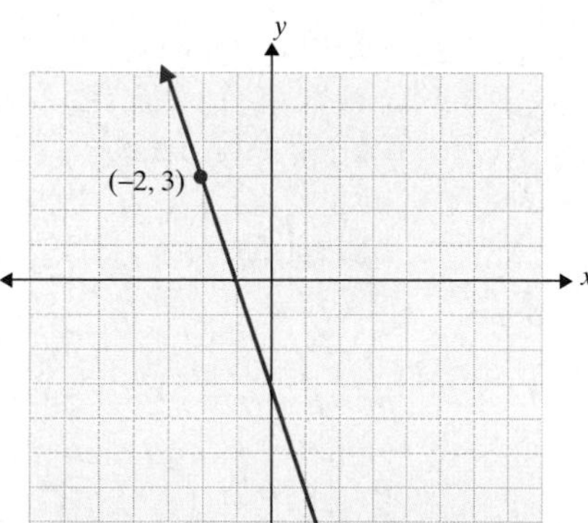

24. $y=-\frac{1}{3}x$

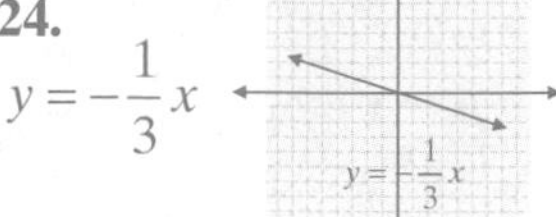

25. $y=2x-1$; $m=2, b=-1$

26. $y=3x-4$; $m=3, b=-4$

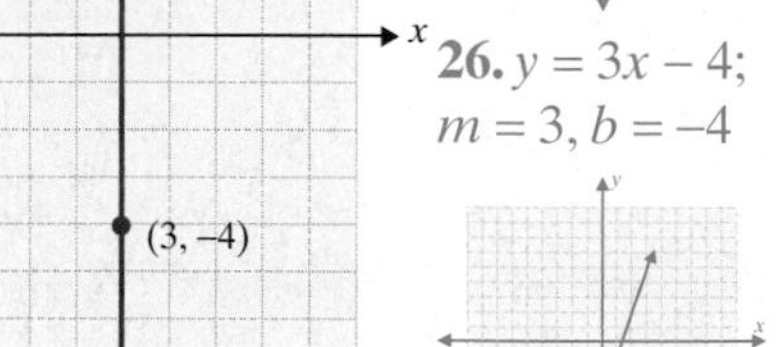

27. $y=-4x+5$; $m=-4, b=5$

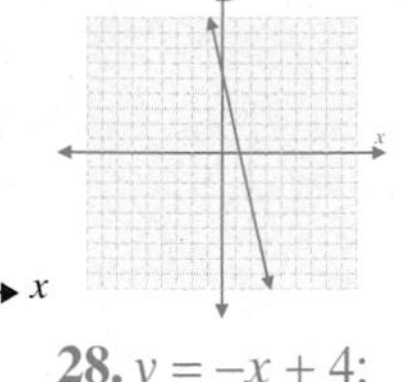

28. $y=-x+4$; $m=-1, b=4$

18. $y=4x+2$ **19.** $y=2x-1$ **20.** $y=-\frac{3}{5}x+4$ **21.** $y=-\frac{1}{4}x-5$ **22.** $y=-3x+5$ **23.** $y=\frac{3}{2}x-4$

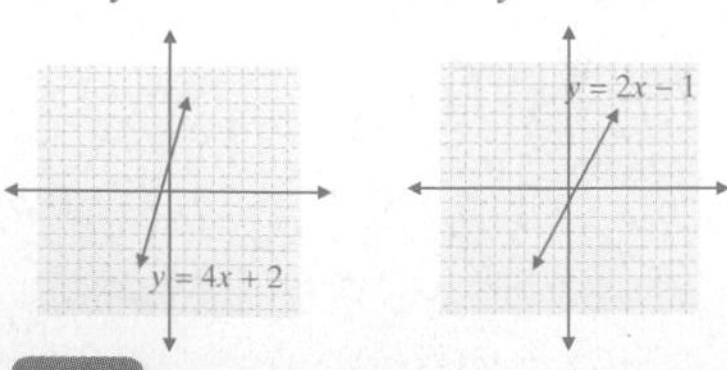

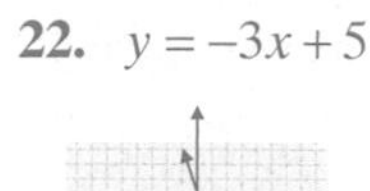

29. $y=\frac{2}{3}x+2; m=\frac{2}{3}, b=2$

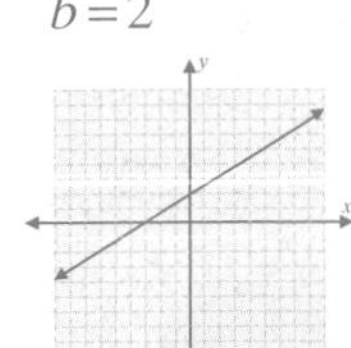

30. $y=\frac{2}{5}x+2; m=\frac{2}{5}, b=2$

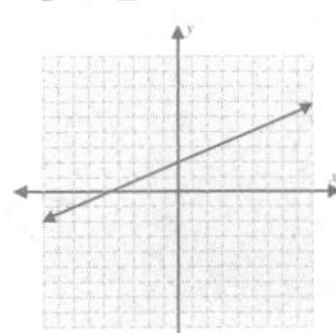

31. $y=-x+5; m=-1, b=5$

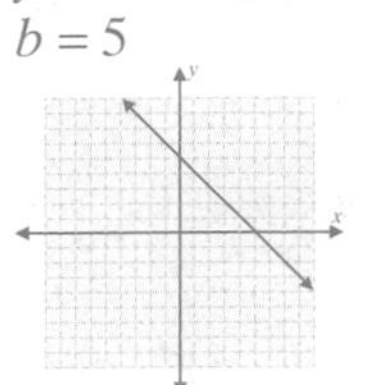

32. $y=\frac{1}{2}x-3; m=\frac{1}{2}, b=-3$

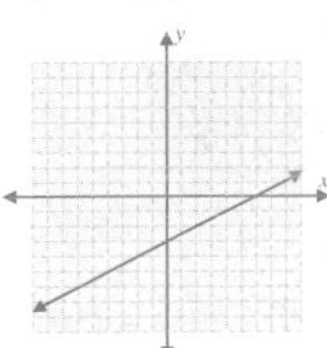

33. $y=-\frac{1}{5}x+2; m=-\frac{1}{5}, b=2$

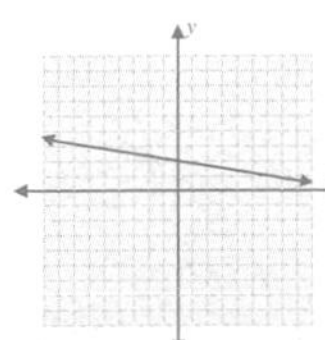

34. $y=-4x-3; m=-4, b=-3$

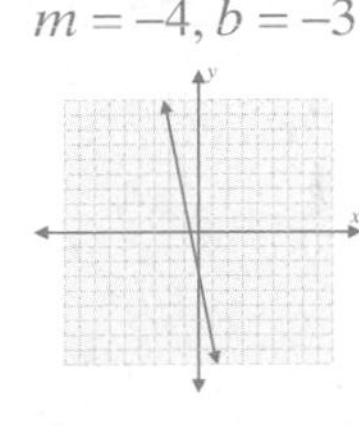

58.

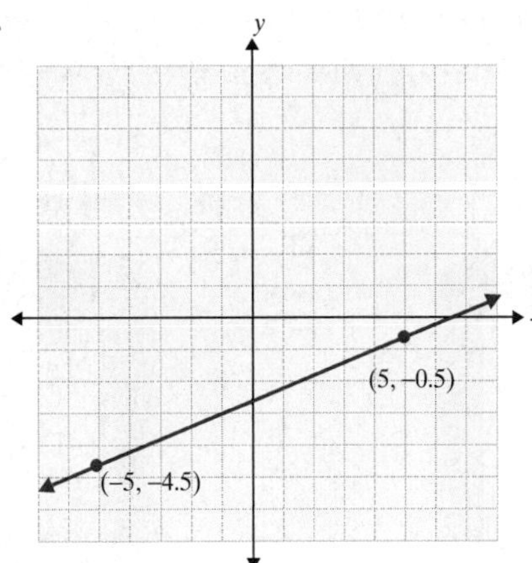

59.

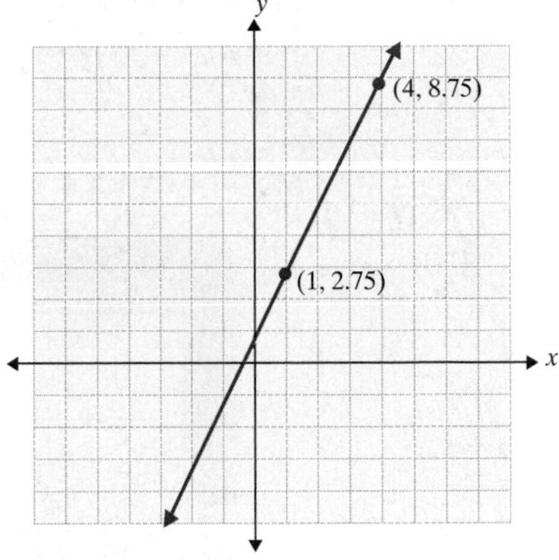

60.

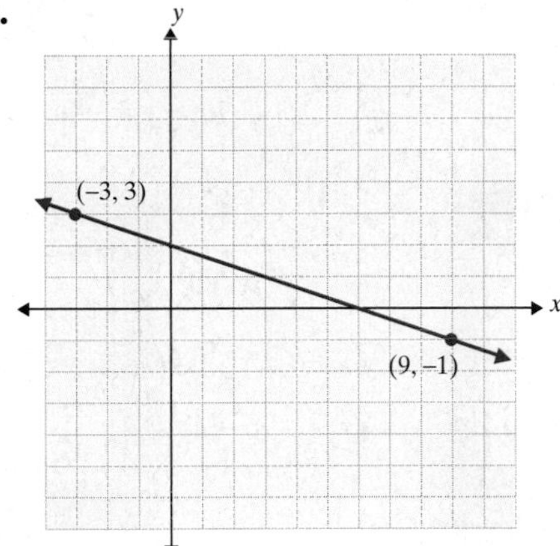

61.

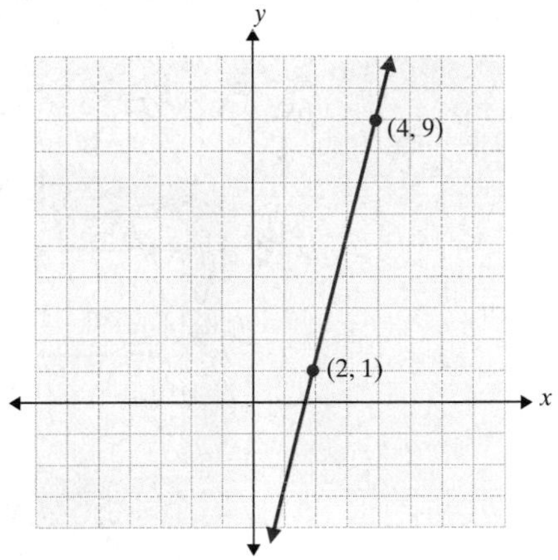

Points are said to be ***collinear*** *if they are on a straight line. If points are collinear, then the slope of the line through any two of them must be the same (because the line is the same line). Use this idea to determine whether or not the three points in each of the sets in Exercises 62 – 66 are collinear.*

62. $\{(-1, 3), (0, 1), (5, -9)\}$

63. $\{(-2, -4), (0, 2), (3, 11)\}$

64. $\{(-2, 0), (0, 30), (1.5, 5.25)\}$

65. $\left\{\left(\frac{2}{3}, \frac{1}{2}\right), \left(0, \frac{5}{6}\right), \left(-\frac{3}{4}, \frac{29}{24}\right)\right\}$

66. $\{(-1, -7), (1, 1), (4, 12)\}$

35. $y=4; m=0, b=4$

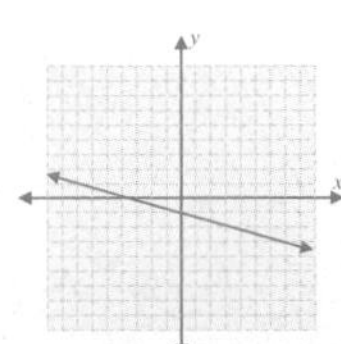

36. $y=-\frac{2}{7}x-1; m=-\frac{2}{7}, b=-1$

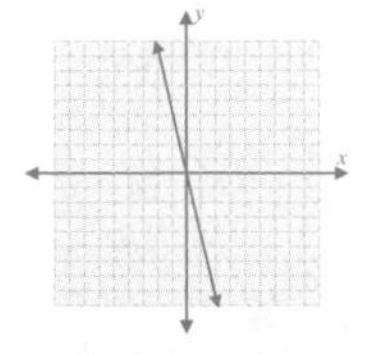

37. $y=-4x; m=-4, b=0$

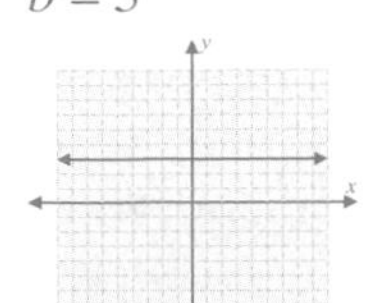

38. $y=3; m=0, b=3$

39. $y=\frac{2}{3}x-2;\ m=\frac{2}{3},$ $b=-2$

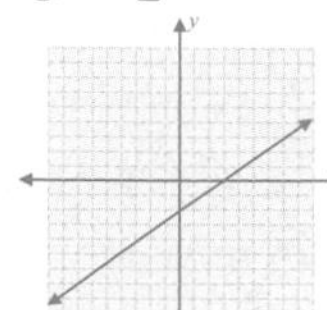

40. $y=4x-2;\ m=4,$ $b=-2$

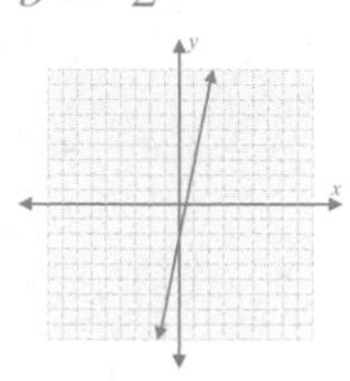

41. Cannot be written in slope intercept form; $x=-3$; slope is undefined, no y-intercept

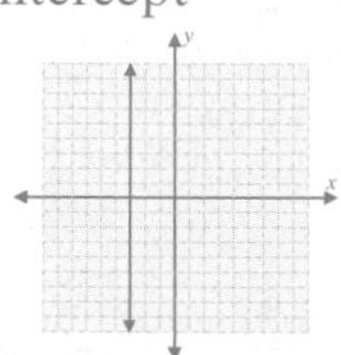

42. $y=\frac{1}{2}x+1;$ $m=\frac{1}{2},\ b=1$

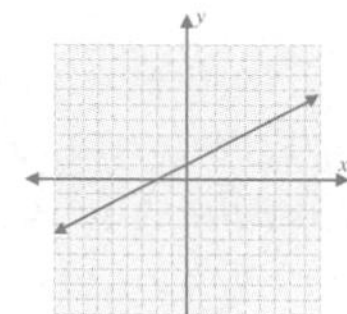

43. $y=\frac{5}{6}x-\frac{5}{3};$ $m=\frac{5}{6},\ b=-\frac{5}{3}$

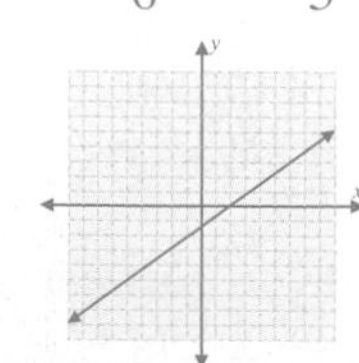

44. Cannot be written in slope intercept form; $x=-\frac{7}{4}$; slope is undefined, no y-intercept.

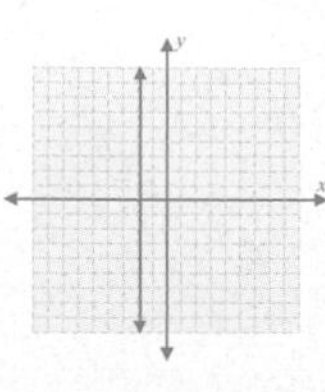

In Exercises 67 – 71, the calculator display shows an incorrect graph for the corresponding equation. Explain how you know, by just looking at the graph, that a mistake has been made.

67. $y=2x+5$

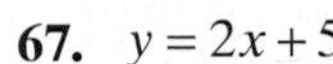

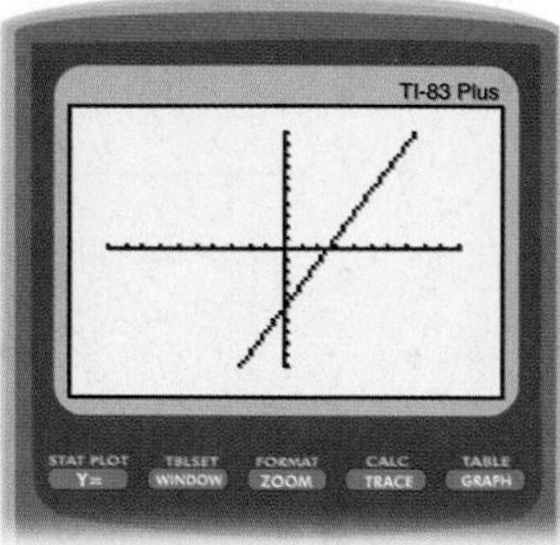

68. $y=-3x+4$

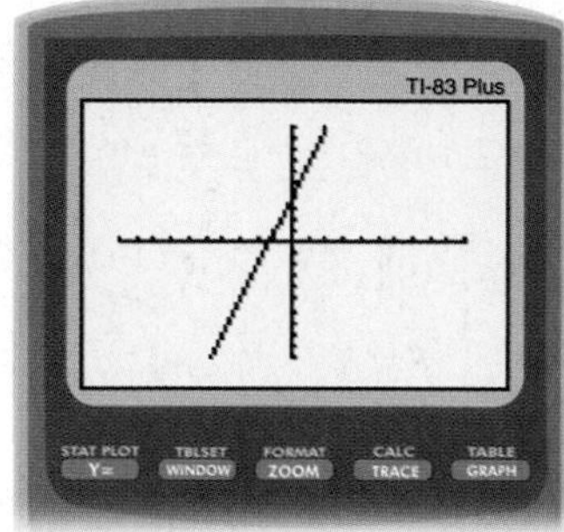

69. $y=\frac{2}{3}x-2$

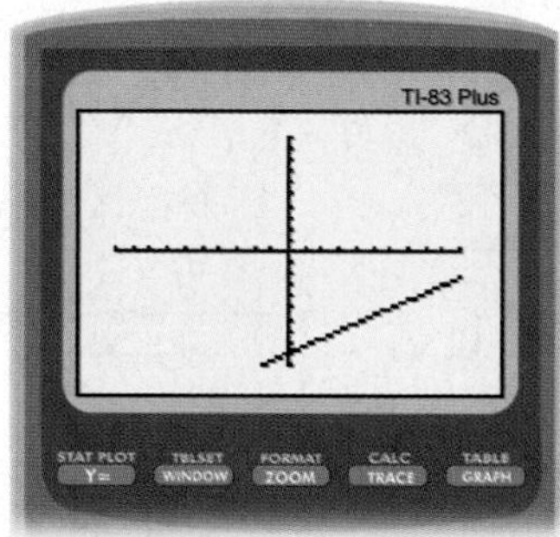

70. $y=-4x$

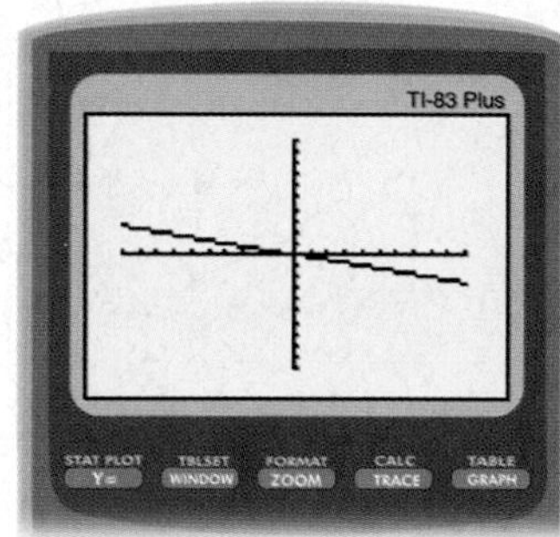

71. $y=-\frac{1}{3}x$

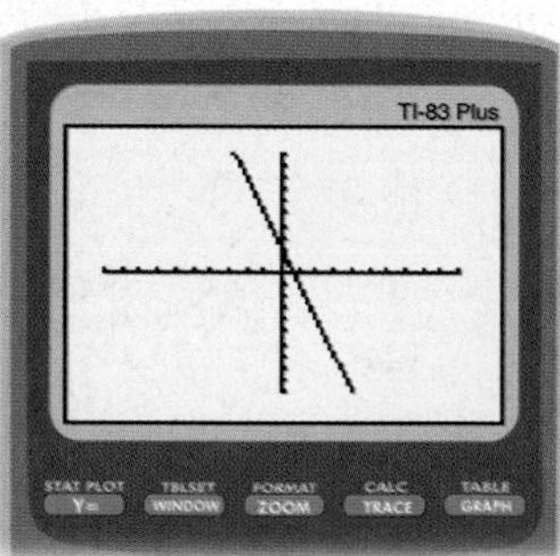

45. $y=-\frac{3}{4}x+\frac{5}{4};\ m=-\frac{3}{4},$ $b=\frac{5}{4}$

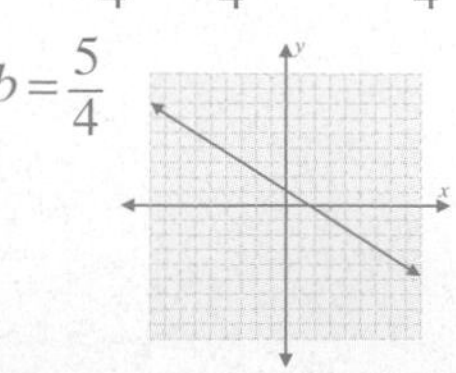

46. $y=-\frac{5}{2}x+\frac{11}{2};\ m=-\frac{5}{2},$ $b=\frac{11}{2}$

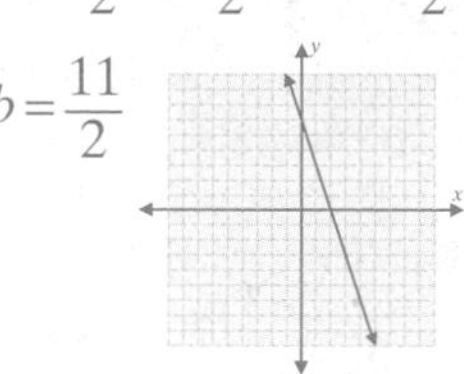

47. $y=-\frac{3}{2}x-\frac{7}{4};\ m=-\frac{3}{2},$ $b=-\frac{7}{4}$

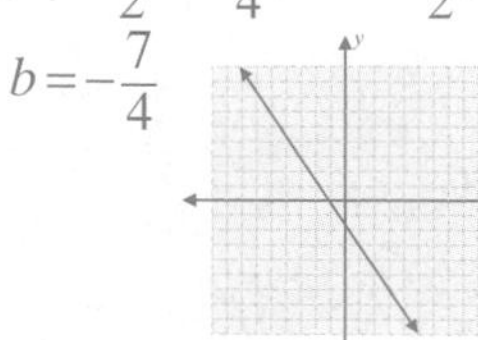

48. $y=-\frac{7}{2}x+2;\ m=-\frac{7}{2},$ $b=2$

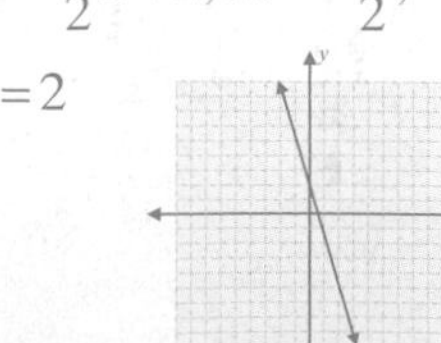

49. $y=\frac{1}{2}x+\frac{2}{3};\ m=\frac{1}{2},$ $b=\frac{2}{3}$

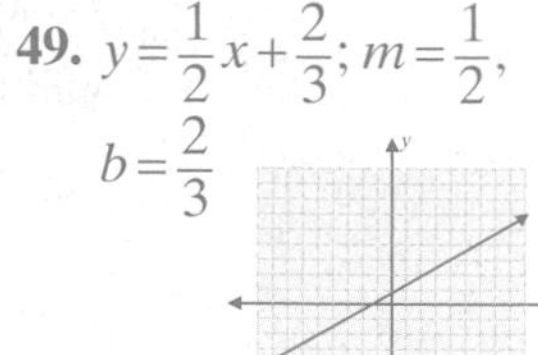

50. $y=-\frac{6}{5}x-3;\ m=-\frac{6}{5},$ $b=-3$

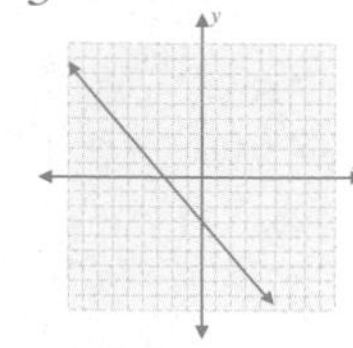

51. $y=\frac{5}{2}x+\frac{5}{2}$; $m=\frac{5}{2}$, $b=\frac{5}{2}$

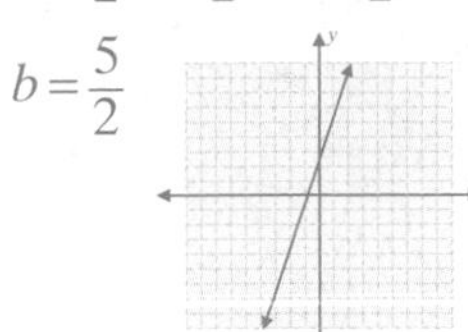

52. $y=\frac{4}{3}x+\frac{7}{3}$; $m=\frac{4}{3}$, $b=\frac{7}{3}$

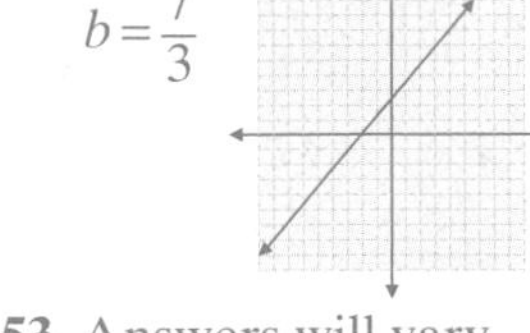

53. Answers will vary.
54. a. $\frac{3}{2}$ b. 7 c. $y=\frac{3}{2}x+7$
55. a. undefined
b. no y-intercept
c. $x=3$
56. a. 0 b. −6 c. $y=-6$
57. a. −3 b. −3
c. $y=-3x-3$
58. a. $\frac{2}{5}$ b. $-\frac{5}{2}$
c. $y=\frac{2}{5}x-\frac{5}{2}$
59. a. 2 b. $\frac{3}{4}$ c. $y=2x+\frac{3}{4}$

Writing and Thinking About Mathematics

72. a. Explain in your own words why the slope of a horizontal line must be 0.
b. Explain in your own words why the slope of a vertical line must be undefined.

73. a. Describe the graph of the line $y=0$.
b. Describe the graph of the line $x=0$.

74. Explain, in your own words, how the slope changes if the coordinates of the points (x_1, y_1) and (x_2, y_2) are interchanged in the formula for slope.

75. In the formula $y=mx+b$ explain the meaning of m and the meaning of b.

76. The slope of a road is called a "grade." A steep grade is cause for truck drivers to have slow speed limits in mountains. What do you think that a "grade of 12%" means? Draw a picture of a right triangle that would indicate a grade of 12%.

77. Ramps for persons in wheelchairs or otherwise handicapped are now built into most buildings and walkways. (If ramps are not present in a building, then there must be elevators.) What do you think that the slope of a ramp should be for handicapped access? Look in your library or contact your local building permit office to find the recommended slope for such ramps.

Hawkes Learning Systems: Introductory & Intermediate Algebra

Graphing Linear Equations in Slope-Intercept Form

60. a. $-\frac{1}{3}$ b. 2
c. $y=-\frac{1}{3}x+2$
61. a. 4 b. −7
c. $y=4x-7$
62. yes 63. yes
64. no 65. yes
66. no
67. y-intercept = −5 (should be 5)
68. slope = 3 (should be −3)
69. y-intercept = −9 (should be −2)
70. slope is reciprocal
71. slope is reciprocal and y-intercept = 2 (should be 0)
72. a. For any horizontal line, all of the y values will be the same. Consequently, the formula for slope will always have 0 in the numerator. Therefore the slope of every horizontal line is 0.
b. For any vertical line, all of the x values will be the same. Consequently, the formula for slope will always have 0 in the denominator. Therefore the slope of every vertical line is undefined.
73. a. The x-axis
b. The y-axis
74. (x_1, y_1) and (x_2, y_2),
Slope: $\frac{y_2-y_1}{x_2-x_1}=m.$.
After interchanging the coordinates of points the new points will be (y_1, x_1) and (y_2, x_2),
Slope:
$$\frac{x_2-x_1}{y_2-y_1}=\frac{1}{\frac{y_2-y_1}{x_2-x_1}}=\frac{1}{m}.$$
75. $y=mx+b$, $m=$ slope of the line, $b=$ y-intercept.
76. A grade of 12% the slope of the road is 12. For every 100 feet of horizontal distance (run) there is 12 feet of vertical distance (rise).

0.12
1

77. Answers will vary.

4.4 The Point-Slope Form: $y - y_1 = m(x - x_1)$

Objectives

After completing this section, you will be able to:

1. *Graph lines given the slope and one point on each line.*
2. *Write the equation of a line in standard form given either:*
   a. *the slope and one point, or*
   b. *two points.*
3. *Determining whether lines are parallel or perpendicular using slope.*
4. *Graph lines by finding the x- and y-intercepts.*

Graphing a Line Given a Point and the Slope

Lines represented by equations in the standard form $Ax + By = C$ and in the slope-intercept form $y = mx + b$ have been discussed in Sections 4.2 and 4.3. In Section 4.3, we discussed how to graph a line by using the y-intercept point $(0, b)$ and the slope by moving horizontally and vertically from the y-intercept.

Now, suppose that you are given the slope of a line and a point on the line. This point may or may not be the y-intercept. The graph of the line can be drawn by using the same technique discussed when using the y-intercept and slope. Consider the following example.

Example 1: Graph

Graph the line with slope $m = \frac{3}{4}$ which passes through the point $(4, 5)$.

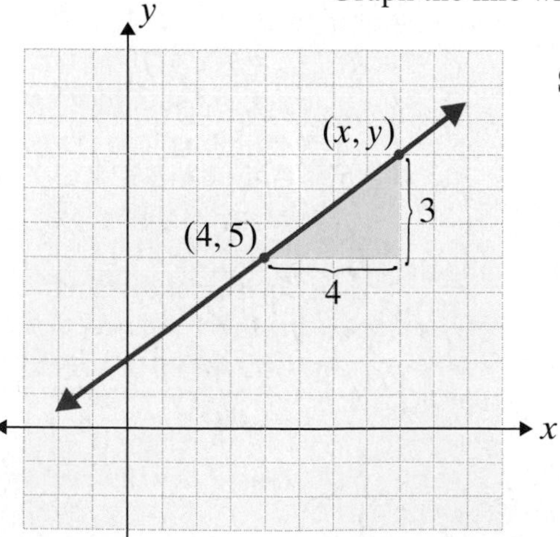

Solution: Start from the point $(4, 5)$ and locate another point on the line using the slope as $\frac{\text{rise}}{\text{run}} = \frac{3}{4}$. Moving 4 units to the right (run) and 3 units up (rise) from $(4, 5)$ will give another point on the line. From $(4, 5)$ you might have moved 4 units left and 3 units down, or 8 units right and then 6 units up. Just move so that the ratio of rise to run is 3 to 4, and you will locate a second point on the graph.

(For a negative slope, move either to the right and then down or to the left and then up.)

We have discussed and analyzed the **standard form** ($Ax + By = C$), **the slope-intercept form** ($y = mx + b$), horizontal lines ($y = b$), and vertical lines ($x = a$). In this section, we are going to develop another form for linear equations, and we will discuss parallel lines and perpendicular lines.

The Point-Slope Form: $y - y_1 = m(x - x_1)$

The objective here is to develop a formula for a line that passes through a given point, say (x_1, y_1), and has a given slope, say m. Now, if (x, y) is any point on the line, then the slope formula gives the equation

$$\frac{y - y_1}{x - x_1} = m$$

and multiplying both sides by the denominator (assuming the denominator is not 0 because m is defined) we have

$$y - y_1 = m(x - x_1) \qquad \text{The point-slope form}$$

For example, suppose that a point $(x_1, y_1) = (8, 3)$ and the slope $m = -\frac{3}{4}$ are given.

If (x, y) represents any point on the line other than (8, 3), then substituting into the formula for slope, gives

$$\frac{y - y_1}{x - x_1} = m \qquad \text{Formula for slope}$$

$$\frac{y - 3}{x - 8} = -\frac{3}{4} \qquad \text{Substituting given information}$$

$$y - 3 = -\frac{3}{4}(x - 8) \qquad \text{Point-slope form: } y - y_1 = m(x - x_1)$$

From this point-slope form, we can manipulate the equation to get the other two forms:

$$y - 3 = -\frac{3}{4}(x - 8)$$

$$y - 3 = -\frac{3}{4}x + 6$$

or $$y = -\frac{3}{4}x + 9 \qquad \text{Slope-intercept form: } y = mx + b$$

or $$3x + 4y = 36 \qquad \text{Standard form: } Ax + By = C$$

Point-Slope Form

An equation of the form

$$y - y_1 = m(x - x_1)$$

is called the **point-slope** *form for the equation of a line that contains the point* (x_1, y_1) *and has slope m.*

In Examples 2a and 2b, the equations of the lines are written in all three forms: point-slope form, slope-intercept form, and standard form. Generally any one of these forms is sufficient. However, there are situations in which one form is preferred over the others. Therefore, manipulation among the forms is an important skill. Also, if the answer in the text is in one form and your answer is in another form, you should be able to recognize that the answers are equivalent.

Example 2: Forms of Equations

a. Find the equation of the line containing the two points (–1, 2) and (4, –2).
Solution: First, find the slope.

$$m = \frac{y_2 - y_1}{x_2 - x_1}$$

$$m = \frac{-2-2}{4-(-1)}$$

$$= \frac{-4}{5}$$

$$= -\frac{4}{5}$$

Teaching Notes:
Emphasize, by using several examples, that either point may be used after the slope is determined. Have students demonstrate to their own satisfaction that one arrives at the same, or equivalent, equation for the line.

Now, use one of the given points and the point-slope form for the equation of a line. [(–1, 2) and (4, –2) are used here to illustrate that either point may be used.]

Using (–1, 2)		**Using (4, –2)**
$y - y_1 = m(x - x_1)$	Point-slope form	$y - y_2 = m(x - x_2)$
$y - 2 = -\frac{4}{5}[x - (-1)]$	Substitute	$y - (-2) = -\frac{4}{5}(x - 4)$
$y - 2 = -\frac{4}{5}x - \frac{4}{5}$		$y + 2 = -\frac{4}{5}x + \frac{16}{5}$
$y = -\frac{4}{5}x - \frac{4}{5} + 2$		$y = -\frac{4}{5}x + \frac{16}{5} - 2$
$y = -\frac{4}{5}x + \frac{6}{5}$	Slope-intercept form	$y = -\frac{4}{5}x + \frac{6}{5}$
or $4x + 5y = 6$	Standard form	$4x + 5y = 6$

b. Find the equation of the line with a slope of $-\frac{1}{2}$ and passing through the point (2, 3). Graph the line using the point and slope.

Solution: Substitute into the point-slope form:

$$y - y_1 = m(x - x_1) \qquad \text{Point-slope form}$$

$$y - 3 = -\frac{1}{2}(x - 2)$$

$$y = -\frac{1}{2}x + 4 \qquad \text{Slope-intercept form}$$

or $$x + 2y = 8 \qquad \text{Standard form}$$

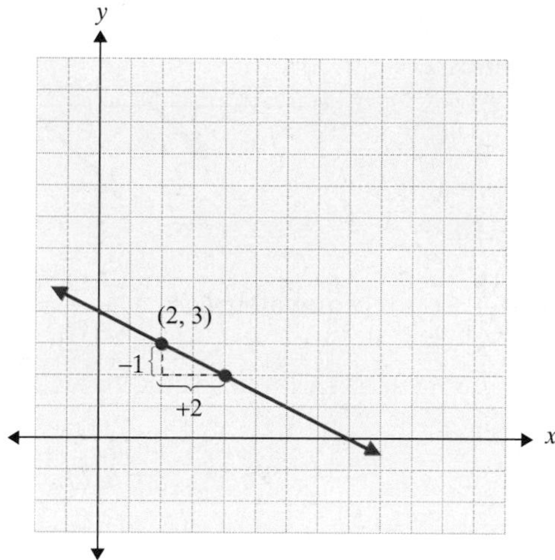

The point one unit down and two units right from (2, 3) will be on the line because the slope is $m = \frac{\text{rise}}{\text{run}} = \frac{-1}{2} = -\frac{1}{2}$.

With a negative slope, either the rise is negative and the run is positive, or the rise is positive and the run is negative. In either case, as the previous figure and the following figure illustrate, the line is the same.

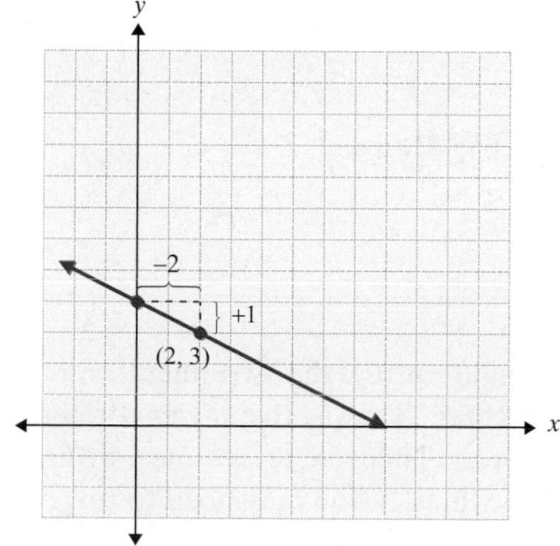

The point one unit up and two units to the left from (2, 3) is on the line because the slope is $m = \frac{\text{rise}}{\text{run}} = \frac{1}{-2} = -\frac{1}{2}$.

Parallel Lines and Perpendicular Lines

Consider two nonvertical parallel lines, L_1 and L_2, with slopes m_1 and m_2, respectively, as shown in Figure 4.9.

Teaching Notes:
Your students may have seen, in other courses, the ⊥ and || symbols to represent perpendicular and parallel lines, respectively. You may wish to describe these symbols to the entire class. However, they are not used in this text.

Since the lines are parallel, the two right triangles AOB and COD are similar triangles (from geometry). Therefore, the corresponding sides are proportional. Note that in Figure 4.9, AO and CO represent the run for lines L_1 and L_2, respectively. Similarly, BO and DO represent the rise for lines L_1 and L_2, respectively. That is,

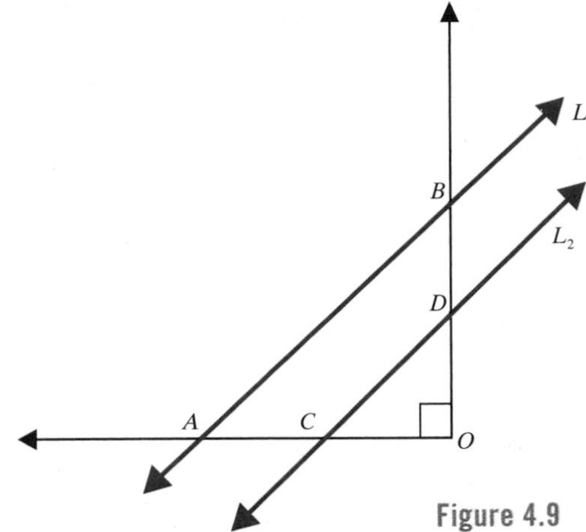

Figure 4.9

$$\frac{BO}{DO}=\frac{AO}{CO} \quad \text{or, equivalently,} \quad \frac{BO}{AO}=\frac{DO}{CO}.$$

But,

$$\frac{BO}{AO}=m_1 \quad \text{and} \quad \frac{DO}{CO}=m_2.$$

Thus, $m_1 = m_2$, and **parallel lines have the same slope**.

It can also be shown (using plane geometry) that, conversely, lines with the same slope are parallel. Therefore, we have the following theorem.

Parallel Lines

Two lines (neither vertical) are ***parallel*** *if and only if they have the same slope.*

All vertical lines are parallel.

In the following figure (Figure 4.10), the two lines L_1 and L_2 with slopes m_1 and m_2, respectively, are **perpendicular**. That is, the angles formed at the point of intersection are 90°. Using the Pythagorean Theorem three times, we find the following relationships between the slopes of these lines. (You may need to review some beginning algebra to follow this discussion in detail.)

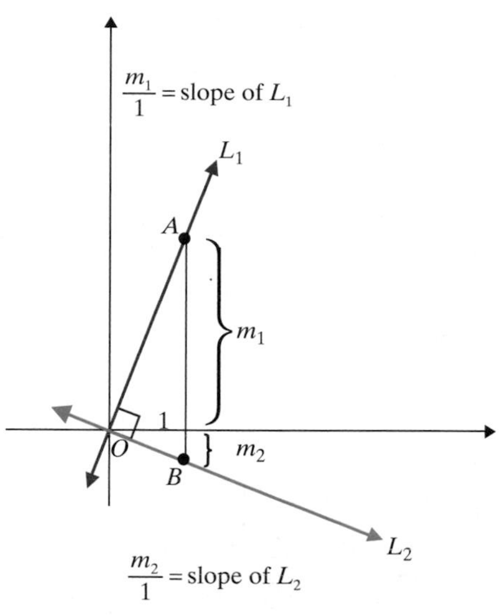

Figure 4.10

$$(AO)^2 = (m_1)^2 + 1^2$$

$$(BO)^2 = (m_2)^2 + 1^2$$

and $$(AB)^2 = (AO)^2 + (BO)^2$$

But, $AB = m_1 - m_2$ (since m_2 is negative). Substituting $m_1 - m_2$ for AB above gives

$$(m_1 - m_2)^2 = (AO)^2 + (BO)^2$$

$$(m_1 - m_2)^2 = (m_1)^2 + 1^2 + (m_2)^2 + 1^2$$

$$m_1^2 - 2m_1m_2 + m_2^2 = (m_1)^2 + (m_2)^2 + 2$$

$$-2m_1m_2 = 2$$

$$\mathbf{m_1m_2 = -1} \quad \text{or} \quad \mathbf{m_2 = \frac{-1}{m_1}}$$

This discussion constitutes proof of the following theorem.

Perpendicular Lines

Two lines (neither vertical) are ***perpendicular*** *if and only if their slopes are negative reciprocals of each other:*

$$m_2 = -\frac{1}{m_1} \qquad \text{or} \qquad m_1m_2 = -1.$$

Vertical lines are perpendicular to horizontal lines.

As illustrated below in Figure 4.11, the lines $y = 2x + 1$ and $y = 2x - 3$ are **parallel**. They have the same slope. The lines $y = \frac{2}{3}x + 1$ and $y = -\frac{3}{2}x - 2$ are **perpendicular**. Their slopes are negative reciprocals of each other. In other words, the product of their slopes is –1.

Figure 4.11

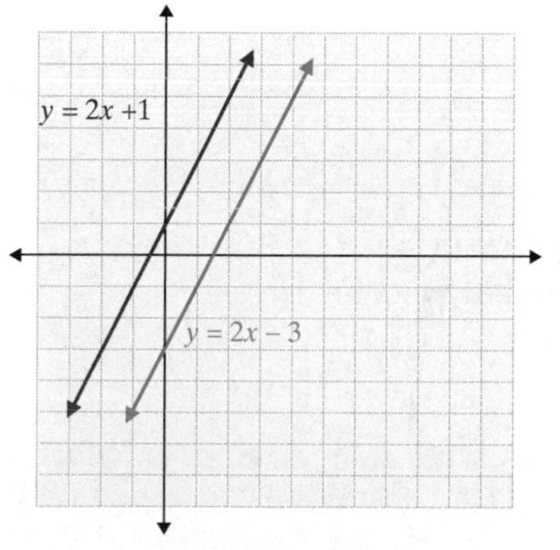

Parallel lines

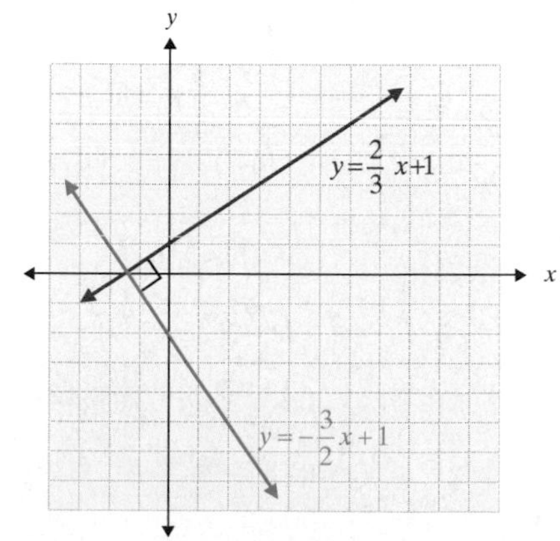

Perpendicular lines

Example 3: Slopes of Lines

Graph the following pairs of lines and state the slope of each line.

a. $y = -1$
$x = \sqrt{2}$

Solution:

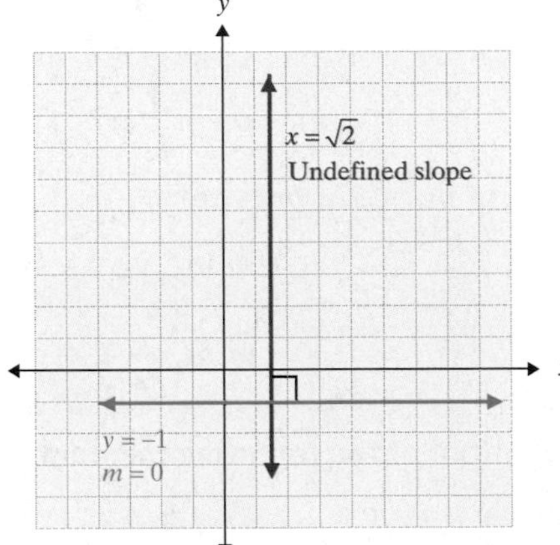

One line is vertical and the other is horizontal. The lines are perpendicular.

b.
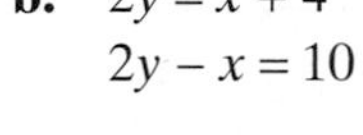
$2y = x + 4$
$2y - x = 10$

Solution:

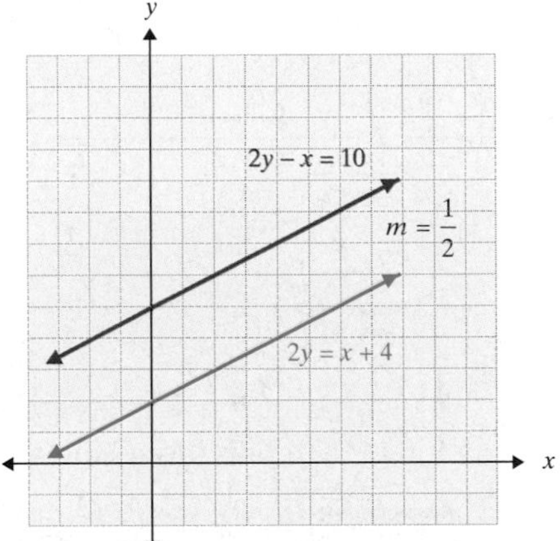

$m = \frac{1}{2}$ for both lines. The lines are parallel.

c. $3x - 4y = 8$
$3y = -4x + 3$

Solution:

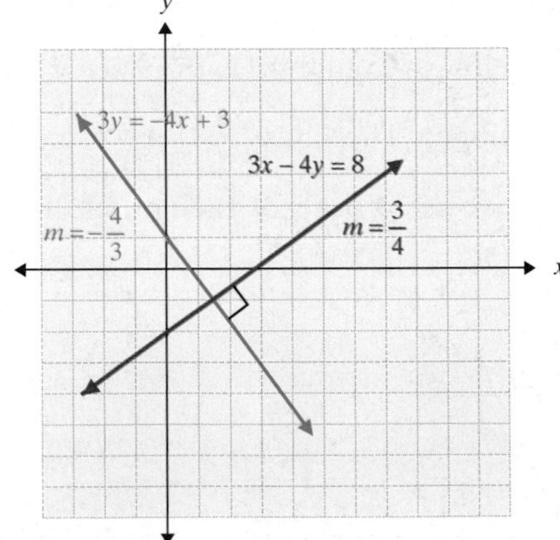

The lines are perpendicular.

$$-\frac{4}{3} \cdot \frac{3}{4} = -1$$

For easy reference, the following table summarizes what we know about straight lines.

Summary of Formulas and Properties of Straight Lines

1.	$Ax + By = C$	*Standard form*
2.	$m = \dfrac{y_2 - y_1}{x_2 - x_1}$	*Slope of a line*
3.	$y = mx + b$	*Slope-intercept form*
4.	$y - y_1 = m(x - x_1)$	*Point-slope form*
5.	$y = b$	*Horizontal line, slope 0*
6.	$x = a$	*Vertical line, undefined slope*

7. *Parallel lines have the same slope* ($m_1 = m_2$).

8. *Perpendicular lines have slopes that are negative reciprocals of each other:*

$$\left(m_2 = \frac{-1}{m_1} \quad \text{or} \quad m_1 m_2 = -1 \right).$$

Example 4 illustrates how to use information about slopes to find the equation of a line.

Example 4: How to Use Information About Slopes

Find the equation in standard form for the line parallel to the line $5x + 3y = 1$ and passing through the point $(2, 3)$.

Solution: First solve for y to find the slope m.

$$5x + 3y = 1$$

$$3y = -5x + 1$$

$$y = -\frac{5}{3}x + \frac{1}{3}$$

Thus, any line parallel to this line has slope $-\frac{5}{3}$.

Now use the point-slope form $y - y_1 = m(x - x_1)$ with $m = -\frac{5}{3}$ and $(x_1, y_1) = (2, 3)$.

$y - 3 = -\frac{5}{3}(x - 2)$	Point-slope form
$3(y - 3) = -5(x - 2)$	Multiply both sides by 3
$3y - 9 = -5x + 10$	Simplify
$5x + 3y = 19$	Standard form

Slope as a Rate of Change

Teaching Notes: You might consider using some "real-life" linear relationships such as the Celsius / Fahrenheit formulas.

The average speed that you ride your bicycle is the rate of change of distance with respect to time. For example, if you ride your bicycle 24 miles in 2 hours then your average rate is

$$r = \frac{d}{t} = \frac{24 \text{ miles}}{2 \text{ hours}} = \frac{12 \text{ mi.}}{1 \text{ hr.}} \left(\text{or } 12 \text{ mph}\right).$$

If time and distance are represented on the horizontal and vertical axes, respectively, then the slope of the line segment joining the two points (0, 0) and (2, 24) is average speed or average rate of change of distance with respect to time.

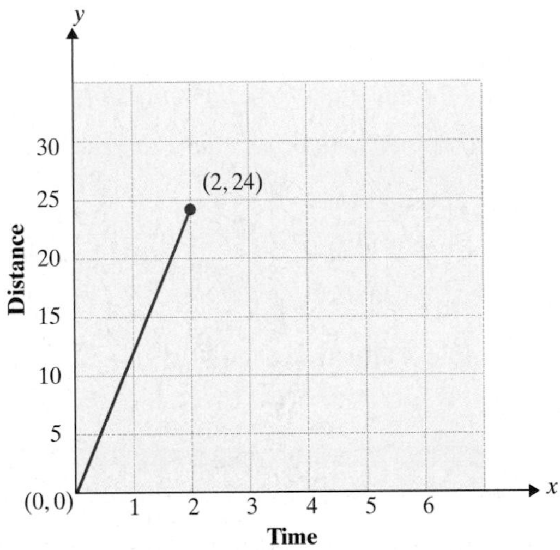

Figure 4.12

The rate of change of the average price of a computer rose at \$70/yr. from 1985 to 1990, then continued to rise at \$90/yr. until 1995, when it began to drop at a rate of \$92/yr. The number of households owning computers rose at 4.16 households per yr. from 1994 to 1997, and at 4.8 households per year from 1997 to 2000.

In general, the ratio of a change in one variable (say y) to a change in another variable (say x) is called the rate of change of y with respect to x. Graphically, this rate of change is the slope of the line segment joining the appropriate points. Figure 4.13 shows how two different rates of change can be interpreted as slope on a graph.

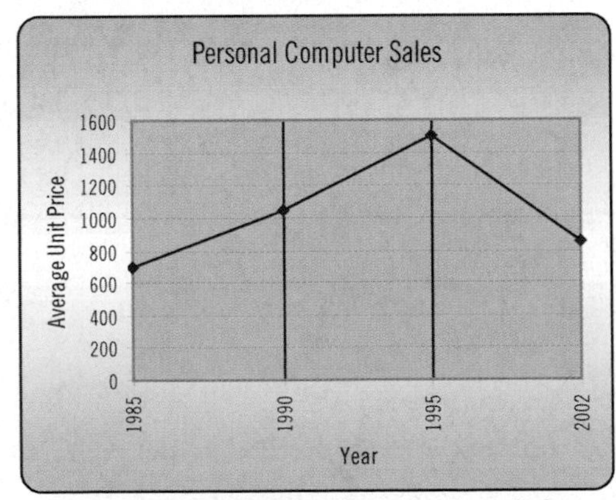

Source: Consumer Electonics Association

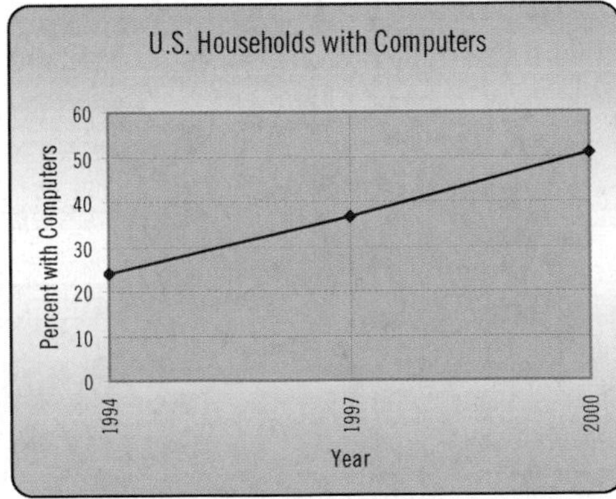

Source: U.S. Dept. of Commerce

Figure 4.13

Practice Problems

Find a linear equation in standard form that satisfies the given conditions.

1. *Passes through the point (4, –1) with $m = 2$*
2. *Parallel to $y = -3x + 4$ and contains the point (–1, 5)*
3. *Perpendicular to $2x + y = 1$ and passes through the origin (0, 0)*
4. *Contains the two points (6, –2) and (2, 0)*

1. $2x + y = -3$

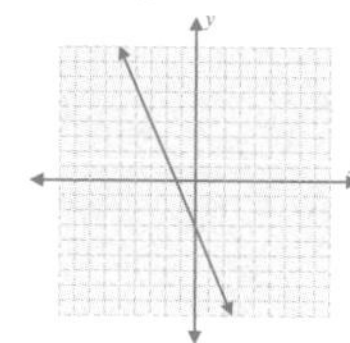

4.4 Exercises

Exercises 1 – 15 give either two points on a line or the slope and a point on a line. Find an equation in standard form that satisfies the given conditions and graph each line.

1. $m = -2$; $(-2, 1)$ **2.** $m = 3$; $(3, 4)$ **3.** $(-5, 2)$; $(3, 6)$ **4.** $(-3, 4)$; $(2, 1)$

5. $m = -\frac{1}{3}$; $(5, -1)$ **6.** $m = \frac{3}{4}$; $\left(0, \frac{1}{2}\right)$ **7.** $(4, 2)$; $(4, -3)$ **8.** $(5, 2)$; $(1, -3)$

9. $m = 0$; $(2, 3)$ **10.** $m = -\frac{5}{7}$; $\left(-2, \frac{1}{2}\right)$ **11.** $(-2, 7)$; $(3, 1)$

12. $(2, -5)$; $(4, -5)$ **13.** $\left(\frac{5}{2}, 0\right)$; $\left(-2, \frac{1}{3}\right)$ **14.** $m = 0$; $(-3, -1)$

15. $m = -\frac{4}{3}$; $\left(\frac{2}{3}, 1\right)$

Graph each line in Exercises 16 – 23 by finding the x- and y-intercepts.

16. $2x + y = 4$ **17.** $2x + y = 6$ **18.** $3x - 2y = 6$ **19.** $2x - 3y = 6$
20. $-2x + 5y = 10$ **21.** $-3x + 5y = 15$ **22.** $3x + 4y = 9$ **23.** $3x - 4y = 9$

Find an equation in standard form for each line that satisfies the given conditions in Exercises 24 – 36.

24. Parallel to $3x + y = 5$ and passes through (2, 1)

25. Parallel to $2x + 4y = 9$ and passes through (1, 6)

26. Parallel to $7x - 3y = 1$ and contains the point (1, 0)

27. Parallel to $5x = 7 + y$ and contains the point (–1, –3)

28. Parallel to the x-axis and passes through (–1, 3)

29. Parallel to the y-axis and contains the point (2, –4)

30. Perpendicular to $4x + 3y = 4$ and passes through the point (2, 2)

31. Perpendicular to $5x - 3y + 4 = 0$ and passes through the point (4, –1)

2. $3x - y = 5$

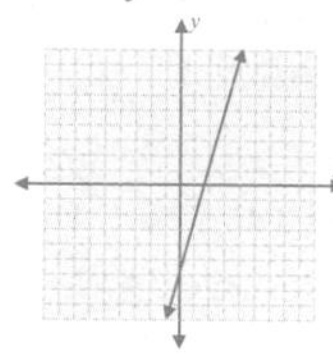

3. $x - 2y = -9$

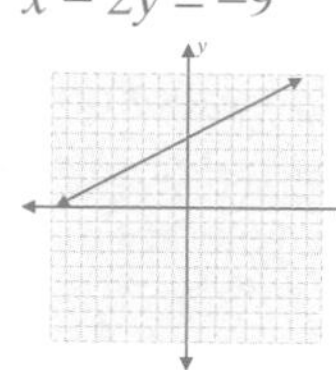

4. $3x + 5y = 11$

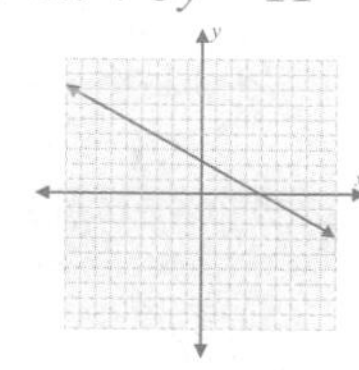

5. $x + 3y = 2$

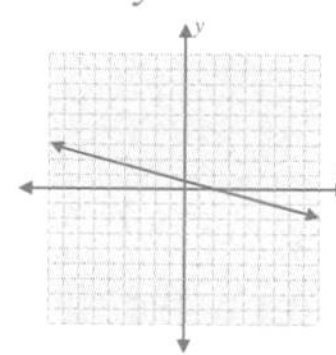

6. $3x - 4y = -2$

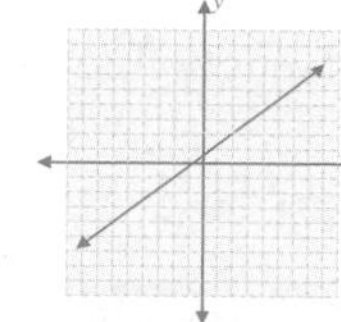

Answers to Practice Problems: 1. $2x - y = 9$ **2.** $3x + y = 2$ **3.** $x - 2y = 0$ **4.** $x + 2y = 2$

7. $x = 4$

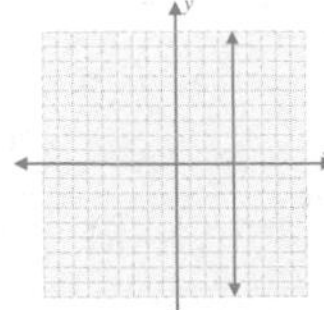

8. $5x - 4y = 17$

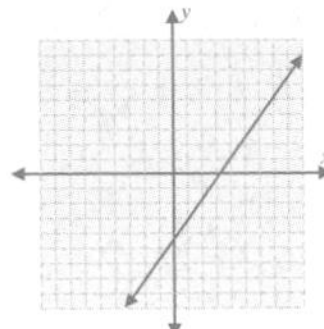

9. $y = 3$

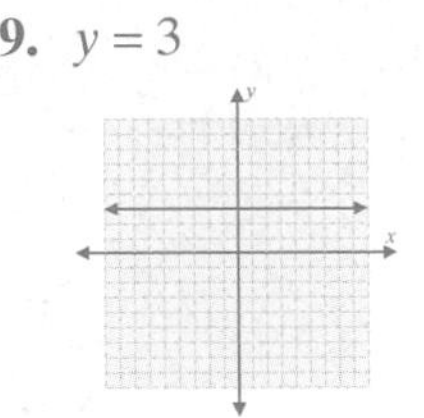

10. $10x + 14y = -13$

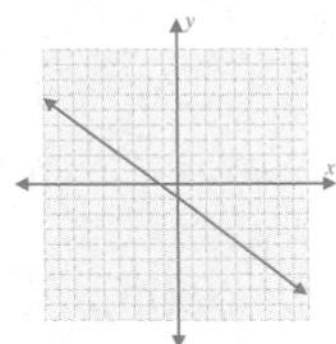

11. $6x + 5y = 23$

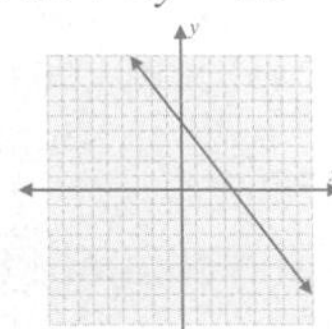

12. $y = -5$

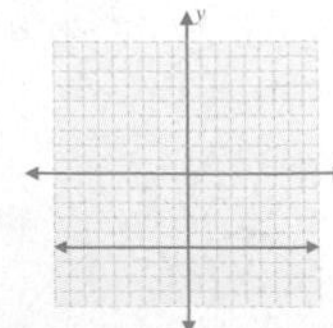

32. Perpendicular to $5x - 2y - 4 = 0$ and contains $(-3, 5)$

33. Perpendicular to $8 - 3x - 2y = 0$ and contains the point $(-4, -2)$

34. Perpendicular to $3x - y = 4$ and passes through the origin

35. Perpendicular to $2x - y = 7$ and has the same y-intercept as $x - 3y = 6$

36. Perpendicular to $3x - 2y = 4$ and has the same y-intercept as $5x + 4y = 12$

37. Show that the points $A(-2, 4)$, $B(0, 0)$, $C(6, 3)$, and $D(4, 7)$ are the vertices of a rectangle. (Plot the points and show that opposite sides are parallel and that adjacent sides are perpendicular.)

38. Show that the points $(0, -1)$, $(3, -4)$, $(6, 3)$, and $(9, 0)$ are the vertices of a parallelogram. (Plot the points and show that opposite sides are parallel.)

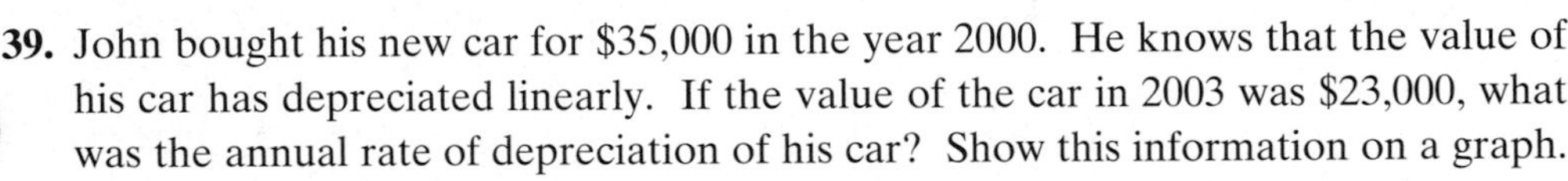

39. John bought his new car for \$35,000 in the year 2000. He knows that the value of his car has depreciated linearly. If the value of the car in 2003 was \$23,000, what was the annual rate of depreciation of his car? Show this information on a graph.

40. The number of homes in the United States with personal computers was about 33 million in 1995 and about 53 million in 2000. If the growth in personal computers per home was linear, what was the approximate rate of growth per year from 1995 to 2000? Show this information on a graph.

41. The following table shows the estimated number of internet users from 1998 – 2002. The number of users for each year is shown in millions.

Year	Internet Users
1998	73
1999	102
2000	124
2001	143
2002	164

Source: International Telecommunications Union Yearbook of Statistics

a. plot these points on a graph

b. connect the points with line segments

c. find the slope of each line segment

d. interpret the slope as a rate of change

42. The following table shows the urban growth from 1850 to 2000 in New York, NY.

Year	Population
1850	515,547
1900	3,437,202
1950	7,891,957
2000	8,008,278

Source: U.S. Census Bureau

a. plot these points on a graph

b. connect the points with line segments

c. find the slope of each line segment

d. interpret the slope as a rate of change

13. $2x + 27y = 5$

14. $y = -1$

15. $12x + 9y = 17$

16.

17.

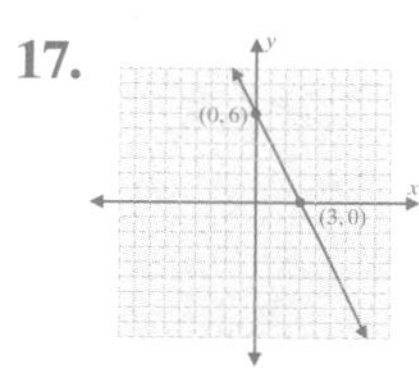

18.

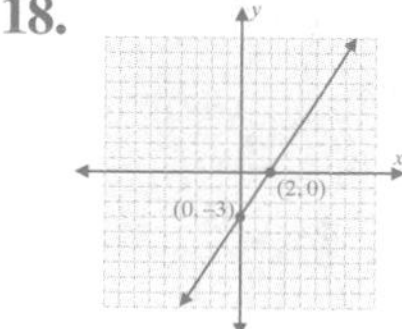

19.

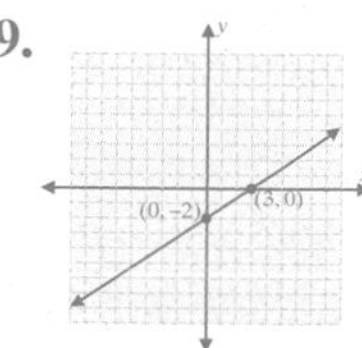

20.

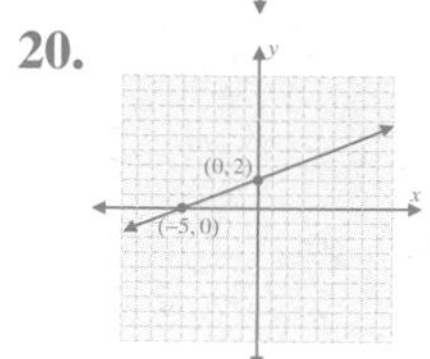

43. The following graph shows the number of female active duty military personnel over a span from 1945 to 2001. The number of women listed include both officers and enlisted personnel from the Army, the Navy, the Marine Corps, and the Air Force.

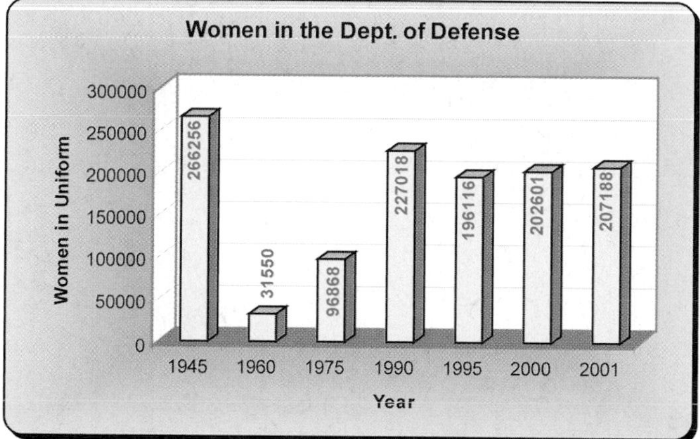

Source: U.S. Dept. of Defense

a. plot these points on a graph
b. connect the points with line segments
c. find the slope of each line segment
d. interpret the slope as a rate of change

44. The following graph shows the rates of marriage per 1,000 people in the U.S., over a span from 1920 to 2000.

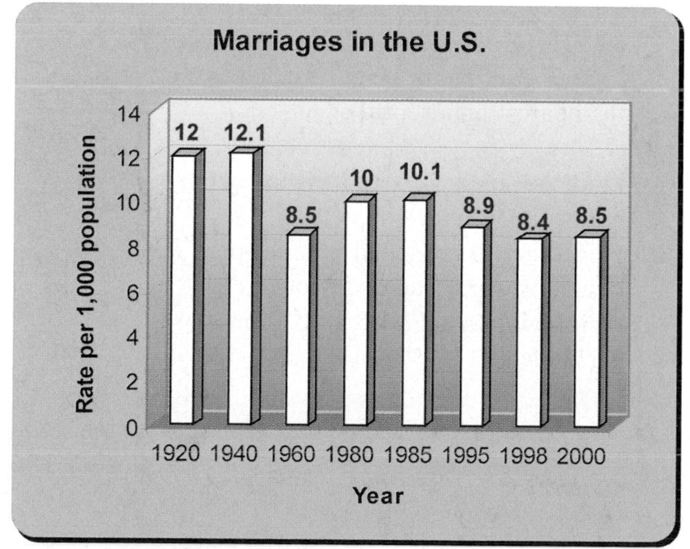

Source: U.S. National Center for Health Statistics

a. plot these points on a graph
b. connect the points with line segments
c. find the slope of each line segment
d. interpret the slope as a rate of change

21.

22.

23.

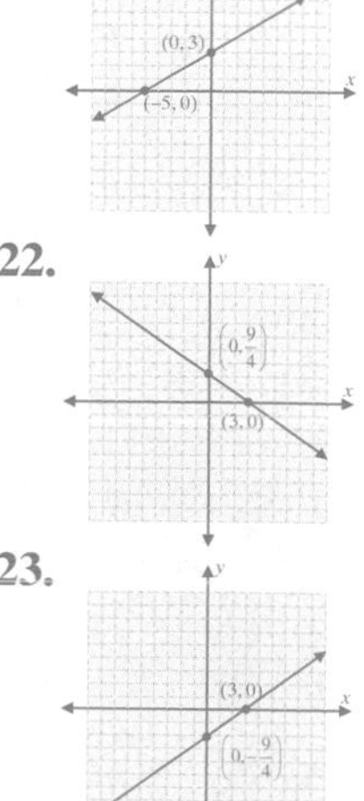

24. $3x + y = 7$
25. $x + 2y = 13$
26. $7x - 3y = 7$
27. $5x - y = -2$
28. $y = 3$
29. $x = 2$
30. $3x - 4y = -2$
31. $3x + 5y = 7$
32. $2x + 5y = 19$
33. $2x - 3y = -2$
34. $x + 3y = 0$
35. $x + 2y = -4$
36. $2x + 3y = 9$
37.

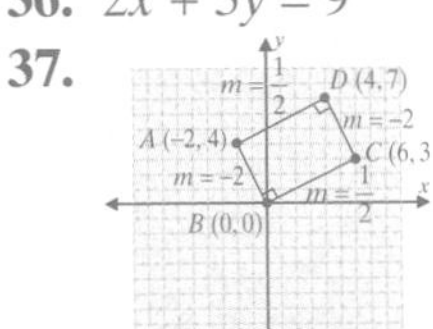

38.

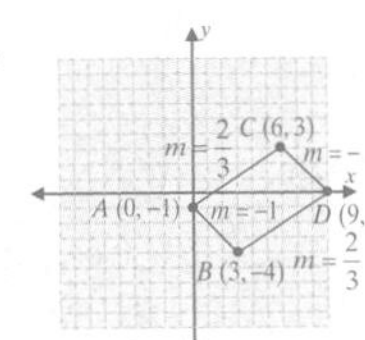

39. $ 4000/year

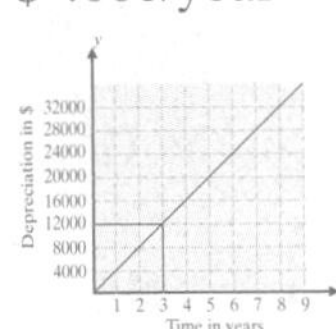

40. 4 million/year

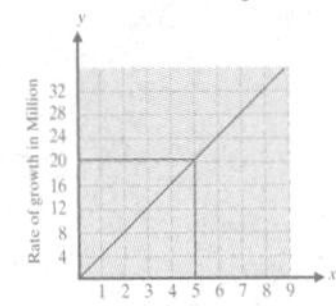

41. a. – b.

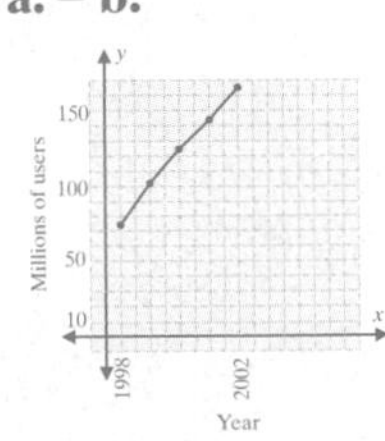

c. 29, 22, 19, 21
d. rate of change inc. 29 million people/year from 1998 - 1999, 22 million ppy from 1999 - 2000, 19 million ppy from 2000 - 2001, and 21 million ppy from 2001 - 2002

The graphs in exercises 45 and 46 show the relationship between time and distance in three different situations. Answer the following questions for each exercise by interpreting the slopes of the appropriate line segments:

a. *What is the average speed from point a to point b?*
b. *What is the average speed from point b to point c?*
c. *What is the average speed from point c to point d?*
d. *What is the average speed from point a to point d?*

45. Distance walking

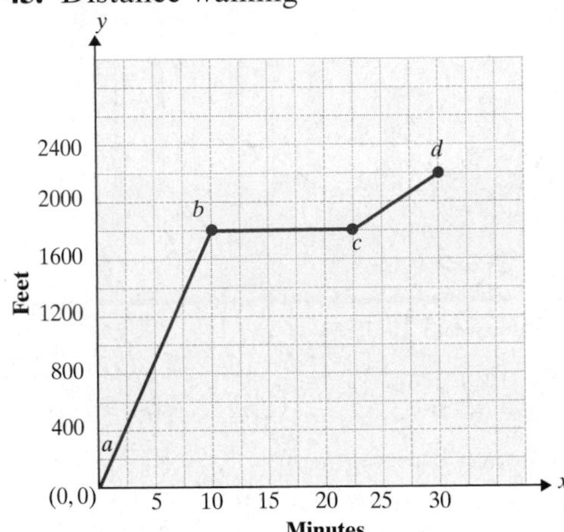

46. Distance driving a car

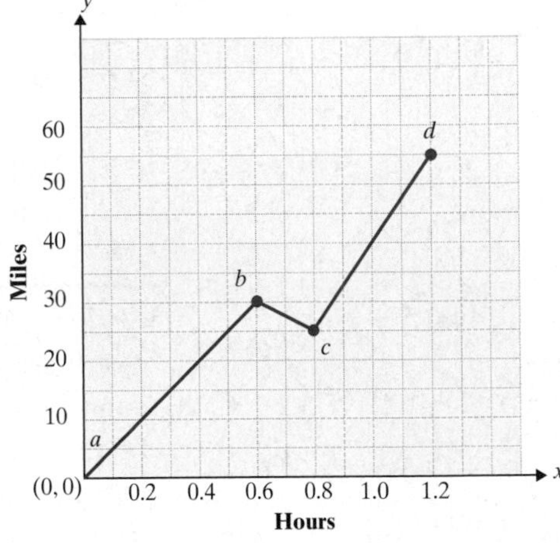

47. Distance of Lance Armstrong riding a bike in his winning stages of the 2002 *Tour de France*
(**Note:** Times have been rounded to the nearest hundredth)

a. Find the average speed in km per hour for each stage depicted.
b. Find the average speed in miles per hour for each stage depicted (by multiplying by 0.62, because 1 km = 0.62 miles).

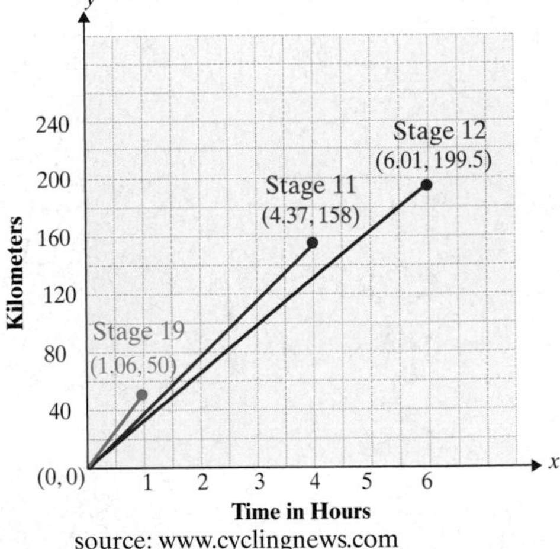

source: www.cyclingnews.com

*In Exercises 48 – 53, determine whether each pair of lines is (**a**) parallel, (**b**) perpendicular, or (**c**) neither. Graph both lines.*

48. $\begin{cases} y = -2x + 3 \\ y = -2x - 1 \end{cases}$

49. $\begin{cases} y = 3x + 2 \\ y = -\dfrac{1}{3}x + 6 \end{cases}$

50. $\begin{cases} 4x + y = 4 \\ x - 4y = 8 \end{cases}$

51. $\begin{cases} 2x + 3y = 5 \\ 3x + 2y = 10 \end{cases}$

52. $\begin{cases} 2x + 2y = 9 \\ 2x - y = 6 \end{cases}$

53. $\begin{cases} 3x - 4y = 16 \\ 4x + 3y = 15 \end{cases}$

42. a. – b.

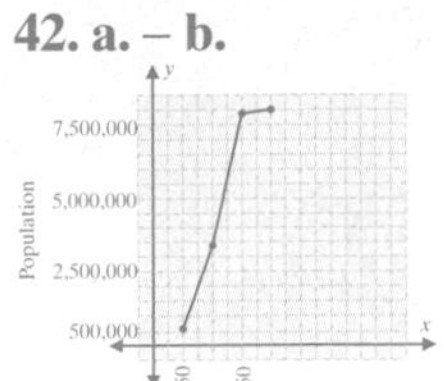

c. 58,433.1; 89,095.1; 2326.42

d. From 1850 - 1900, urban growth was 58,433.1 people per year (ppy); from 1900 - 1950 it was 89,095.1 ppy; from 1950 - 2000 it was 2326.42 ppy. Urban growth has slowed down in the last 50 years.

Writing and Thinking About Mathematics

54. Discuss the meaning of slope in five situations that you have observed related to daily life and why this is important. (For example, the slope or "pitch" of the roof of a house in the mountains is particularly important when there is a heavy snow.)

55. Discuss, in your own words, how to find the equation of a line if you are given two points on the line.

56. Discuss the difference between the concepts of a line having "slope 0" and a line having undefined slope.

Hawkes Learning Systems: Introductory & Intermediate Algebra

Graphing Linear Equations in Point-Slope Form
Finding the Equation of a Line

43. a. – b.

c. –15,647.067; 4354.53; 8676.67; –6180.4; 1297; 4587

d. rate of change dec. 15,647.067 women per year (wpy) from 1945 - 1960; inc. 4354.53 wpy from 1960 - 1975; inc. 8676.67 wpy from 1975 - 1990; dec. 6180.4 wpy from 1990 - 1995; inc. 1297 wpy from 1995 - 2000; and inc. 4587 wpy from 2000 - 2001

44. a. – b.

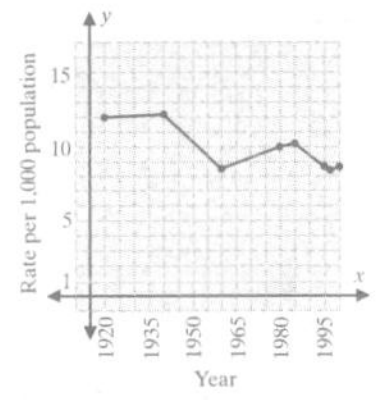

44. c. 0.005, –0.18, 0.075, 0.02, –0.12, –0.167, 0.05

d. rate of change inc. 0.005 marriages/1000 people from 1920 - 1940, dec. 0.18 from 1940 - 1960, inc. 0.075 from 1960 - 1980, inc 0.02 from 1980 - 1985, dec. 0.12 from 1985 - 1995, dec. 0.167 from 1995 - 1998, and inc. 0.05 from 1998 - 2000

45. a. 180 ft./min.
b. 0 ft./min.
c. 53.3 ft./min.
d. 73.3 ft./min.

46. a. 50 mph **b.** –25 mph
c. 75 mph **d.** 45.8 mph

47. a. stg 11: 36.16 kph
stg 12: 33.19 kph
stg 19: 47.17 kph
b. stg 11: 22.42 mph
stg 12: 20.58 mph
stg 19: 29.25 mph

48.

Parallel

49.

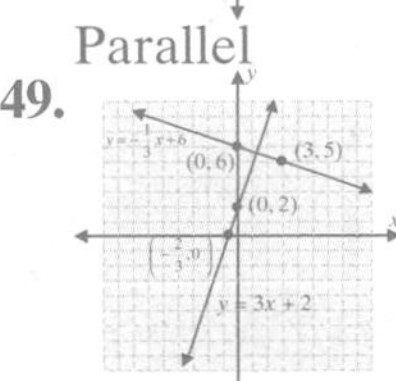

Perpendicular

50.

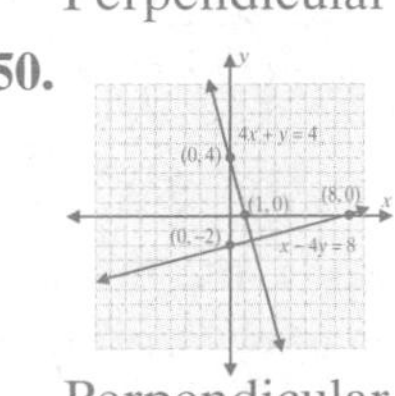

Perpendicular

51.

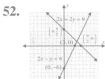

Neither

52.

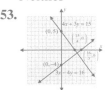

Neither

53.

Perpendicular

54. Answers will vary.

55. Use the two points to find the slope; then use the slope and one of the points in the point-slope form.

56. "slope 0" → 0 numerator → no rise → horizontal line
undefined slope → 0 denominator → no run → vertical line

4.5 Introduction to Functions and Function Notation

Objectives

After completing this section, you will be able to:

1. *State the domain and range of a relation and a function.*
2. *Use the vertical line test to determine whether or not a graph represents a function.*
3. *Write functions as sets of ordered pairs.*
4. *Use function notation.*
5. *Graph functions by using a graphing calculator.*

Everyday use of the term **function** is not far from the technical use in mathematics. For example, distance traveled is a function of time; profit is a function of sales; heart rate is a function of exertion; and interest earned is a function of principal invested. In this sense, one variable "depends on" (or "is a function of ") another.

Mathematicians distinguish between graphs of real numbers as those that represent **functions** and those that do not. Thus, the concept of a function is one of the most important concepts in mathematics. For example, every equation of the form $y = mx + b$ represents a function and we say that y "is a function of " x. Straight lines that are not vertical are the graphs of functions. As the following discussion indicates, vertical lines do not represent functions.

NOTES The ordered pairs discussed in this text will be ordered pairs of real numbers. However, more generally in some other course, the ordered pairs might be other types of entries such as (parent, child), (city, state), or (name, batting average).

Relations and Functions

Relation, Domain, and Range

*A **relation** is a set of ordered pairs of real numbers.*

*The **domain**, **D**, of a relation is the set of all first coordinates in the relation.*

*The **range**, **R**, of a relation is the set of all second coordinates in the relation.*

In graphing relations, the horizontal axis is called the **domain axis**, and the vertical axis is called the **range axis** (See Figure 4.1 on page 236).

Example 1: Relation, Domain, and Range

Find the domain and range for each of the following relations.

a. $r = \{(5,7),(\sqrt{6},2),(\sqrt{6},3),(-1,2)\}$

Solution: $D = \{5,\sqrt{6},-1\}$ All the first coordinates in r

$R = \{7,2,3\}$ All the second coordinates in r

Note that $\sqrt{6}$ is written only once in the domain and 2 is written only once in the range, even though each appears more than once in the relation.

b. $f = \{(-1,1),(1,5),(0,3)\}$

Solution: $D = \{-1,1,0\}$ All the first coordinates in f

$R = \{1,5,3\}$ All the second coordinates in f

The relation $f = \{(-1,1),(1,5),(0,3)\}$, used in Example 1b, meets a particular condition in that each first coordinate has a unique corresponding second coordinate. Such a relation is called a **function**. Notice that r in Example 1a is **not** a function because the first coordinate $\sqrt{6}$ has more than one corresponding second coordinate. Also, for ease in discussion and understanding, the relations illustrated in Examples 1 and 2 have only a finite number of ordered pairs. The graphs of these relations would be isolated dots or points. As we will see, the graphs of most relations and functions have an infinite number of points, and their graphs are smooth curves. (**Note:** Straight lines are also deemed to be curves in mathematics.)

Function

*A **function** is a relation in which each domain element has a exactly one corresponding range element.*

Functions have the following two characteristics:

1. A function is a relation in which each first coordinate appears only once.
2. A function is a relation in which no two ordered pairs have the same first coordinate.

Example 2: Functions

Determine whether or not each of the following relations is a function.

a. $r = \{(2,3),(1,6),(2,\sqrt{5}),(0,-1)\}$

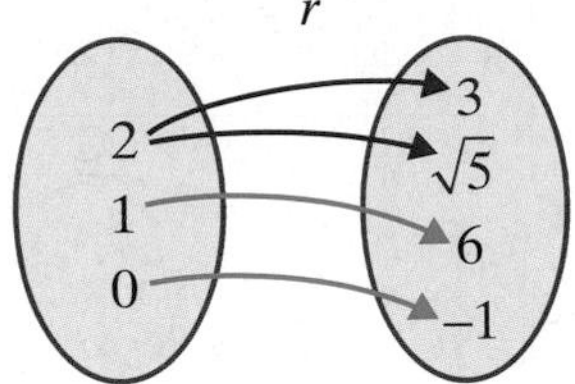

Solution: r is not a function. The number 2 appears as a first coordinate more than once.

b. $t = \{(1,5),(3,5),(\sqrt{2},5),(-1,5),(-4,5)\}$

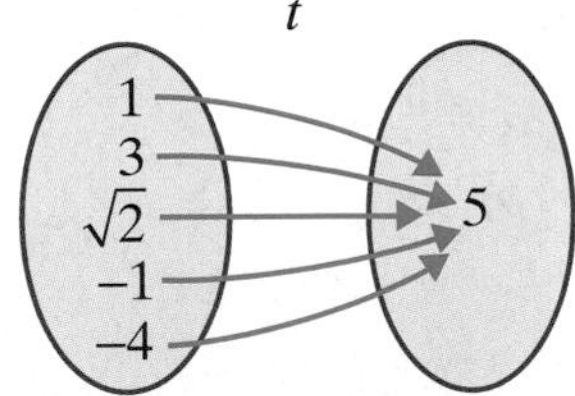

Solution: t is a function. Each first coordinate appears only once. The fact that the second coordinates are all the same has no effect on the definition of a function.

If one point on the graph of a relation is directly above or below another point on the graph, then these points have the same first coordinate (or x-coordinate). Such a relation is **not** a function. Therefore, the following **vertical line test** can be used to tell whether or not a graph represents a function (See Figure 4.14).

Vertical Line Test

*If **any** vertical line intersects the graph of a relation at more than one point, then the relation graphed is **not** a function.*

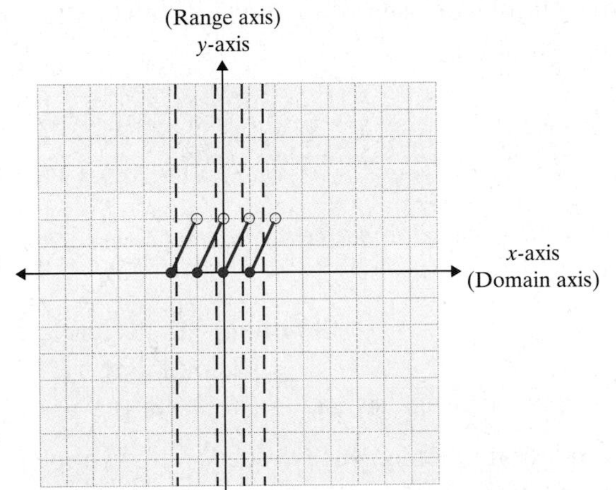

From the graph, we see that the domain of the function is the interval of real numbers $[-2, 2)$ and the range of the function is the interval of real numbers $[0, 2)$.

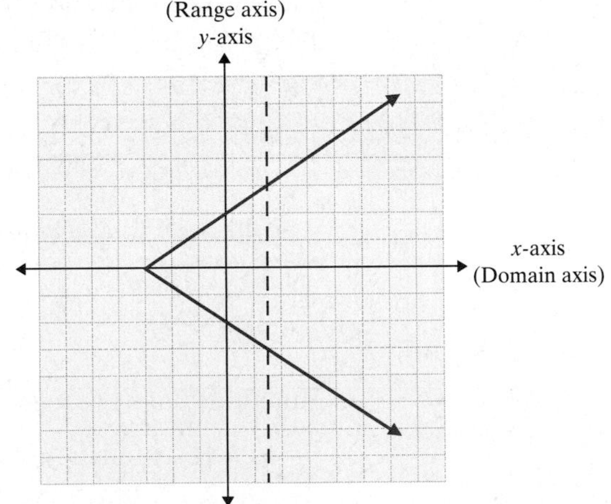

This graph is **not** a function because the vertical line drawn intersects the graph at more than one point. Thus, for that x-value, there is more than one corresponding y-value.

Figure 4.14

Example 3: Vertical Line Test

Use the vertical line test to determine whether or not each of the following graphs represents a function.

a.

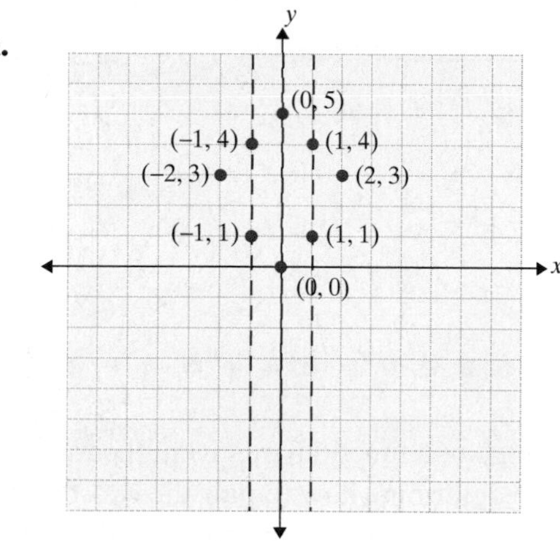

Solution: The relation is **not** a function since a vertical line can be drawn that intersects the graph at more than one point. Listing the ordered pairs shows that several x-coordinates appear more than once:

$$\begin{Bmatrix} (-2,3),(-1,1),(-1,4),(0,0), \\ (0,5),(1,1),(1,4),(2,3) \end{Bmatrix}$$

For this relation, we see from the graph that $D=\{-2,-1,0,1,2\}$ and $R=\{0,1,3,4,5\}$.

b.

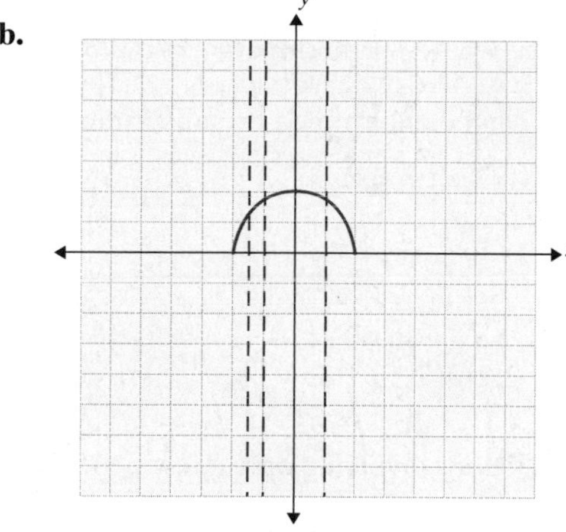

Solution: The relation is a function. No vertical line will intersect the graph at more than one point. Several vertical lines are drawn to illustrate this. For this function, we see from the graph that $D = [-2, 2]$ and $R=[0,2]$.

c.

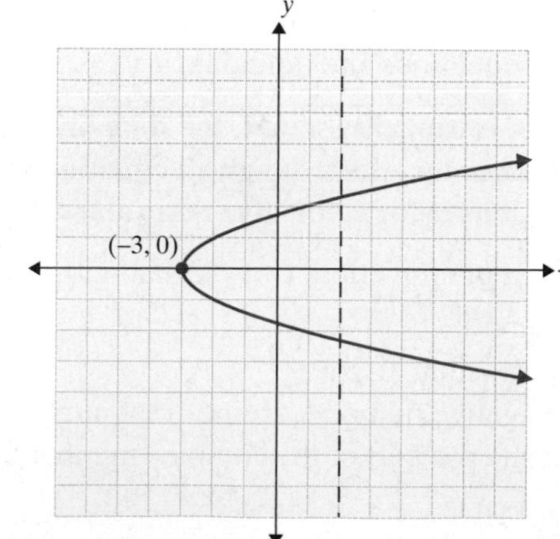

Solution: The relation is **not** a function. At least one vertical line (drawn) intersects the graph at more than one point.

continued on next page ...

Teaching Notes:
Point out to your students that just because a relation is not a function doesn't mean that it is not interesting or useful. Also, by imposing certain restrictions on domains or ranges, many relations can be made into functions.

d.

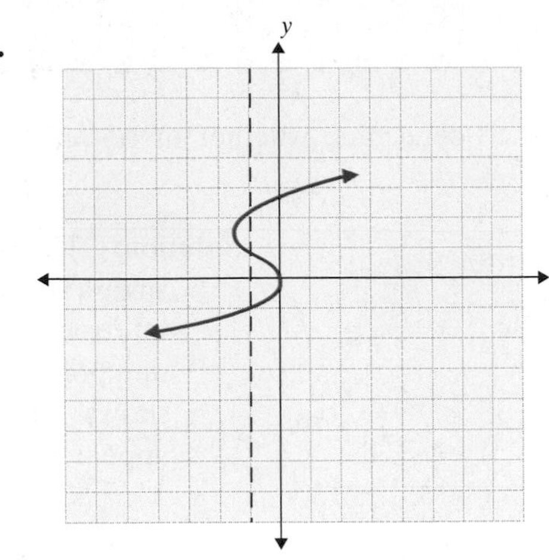

Solution: The relation is **not** a function. At least one vertical line (drawn) intersects the graph at more than one point.

If only the domain and range of a function are given, the function cannot be determined because there is no way to determine how to pair the coordinates. However, if a rule or equation is given that relates x and y, this rule can be used to determine how values of x and y are paired. **Every equation that can be solved for *y* represents a function, and we say that *y* is a function of *x*.** For example,

given the equation $y = \frac{2}{x-1}$ and the domain $\{2,3,5\}$,

the ordered pairs can be determined by substituting 2, 3, and 5 for x.

$x = 2:$ $y = \frac{2}{2-1} = 2$

$x = 3:$ $y = \frac{2}{3-1} = 1$

$x = 5:$ $y = \frac{2}{5-1} = \frac{1}{2}$

$$\text{The function is} \left\{ (2,2),(3,1),\left(5,\frac{1}{2}\right) \right\}.$$

In general, relations or functions involving only a few points are not of interest. In fact, if the domain had not been restricted in the previous discussion, the equation $y = \frac{2}{x-1}$ represents a function whose domain consists of all real numbers except 1. That is, $x \neq 1$ because the denominator cannot be 0 (division by 0 is undefined). We adopt the following rule concerning equations and domains:

Unless a finite domain is explicitly stated, the domain will be considered to be the set of all real *x*-values for which the given equation is defined. That is, the domain consists of all values for *x* that give real values for *y*.

In determining the domain of a function, two facts about real numbers are particularly important:

1. No denominator can equal 0, and
2. Square roots of negative numbers are not real numbers. (**Note:** Such non-real numbers do exist and are part of the complex number system that we will discuss in Chapter 10.)

Example 4: Ordered Pairs

Given the equation (or rule relating x and y)

$$y = x^2 + 1 \qquad \text{and} \qquad D = \{-1, 0, 1, 2, 3\},$$

find the function as a set of ordered pairs.

Solution: Find the ordered pairs by setting up a table and substituting for x in the equation.

x	$y = x^2 + 1$
–1	$y = (-1)^2 + 1 = 2$
0	$y = (0)^2 + 1 = 1$
1	$y = (1)^2 + 1 = 2$
2	$y = (2)^2 + 1 = 5$
3	$y = (3)^2 + 1 = 10$

The function is the following set of ordered pairs:

$$\{(-1,2),(0,1),(1,2),(2,5),(3,10)\}.$$

Example 5: Domain

Find the domain of the function represented by each of the following equations.

a. $y = \dfrac{2x+1}{x-5}$

Solution: The domain is all real numbers for which the expression $\dfrac{2x+1}{x-5}$ is defined.

Thus, $D = \{x \mid x \neq 5\}$.

Note: This mathematical notation means that x can be any real number except 5.

b. $y = \sqrt{x-2}$

Solution: For $\sqrt{x-2}$ to be a real number, the expression under the square root sign must be non-negative. So we have

$$x - 2 \geq 0 \qquad \text{or} \qquad x \geq 2.$$

Thus, $D = \{x \mid x \geq 2\}$.

Linear Functions

All non-vertical straight lines represent functions. If $B \neq 0$, then an equation in the standard form $Ax + By = C$ can be solved for y, and we have the form

$$y = -\frac{A}{B}x + \frac{C}{B} \qquad \text{or} \qquad y = mx + b.$$

Linear Function

*A **linear function** is a function represented by an equation of the form*

$$y = -\frac{A}{B}x + \frac{C}{B} \quad (\text{or} \quad y = mx + b) \quad \text{where } B \neq 0.$$

The domain of a linear function is the set of all real numbers.

If the graph of a linear function is not a horizontal line, then the range is also the set of all real numbers. If the line is horizontal, then the domain is still all real numbers; however, the range is just a single number. For example, the graph of the linear equation $y = 5$ is a horizontal line. The domain of the function is all real numbers and the range is the number 5. Figure 4.15 shows two linear functions and the domain and range of each function.

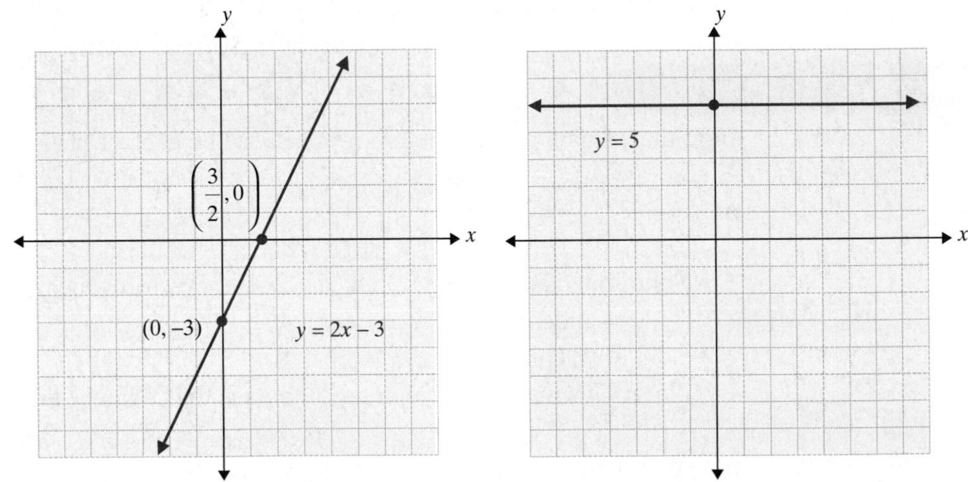

The graph of the linear function $y = 2x - 3$ is shown here.

$D = \{\text{all real numbers}\} = (-\infty, \infty)$

$R = \{\text{all real numbers}\} = (-\infty, \infty)$

The graph of the linear function $y = 5$ is shown here.

$D = \{\text{all real numbers}\} = (-\infty, \infty)$

$R = \{5\}$

Figure 4.15

Function Notation

We have used ordered pair notation (x, y) to represent points on the graphs of relations and functions. As the vertical line test will verify, equations of the form

$$y = mx + b$$

where the equation is solved for y, represent linear functions. The use of x and y in this manner is standard in mathematics and is particularly useful when graphing because we are familiar with the concepts of the x-axis and y-axis.

Another notation, called **function notation**, is more convenient for indicating calculations of values of a function and for indicating operations performed with functions. In function notation,

instead of writing y, write $f(x)$, read "f **of** x."

The letter f is the name of the function. Any letter will do. The letters f, g, h, F, G, and H are commonly used. We have used r, s, and t in previous examples.

Suppose that $y = 3x - 2$ is a given linear equation. Since the equation is solved for y it represents a linear function and we can replace y with the notation $f(x)$ as follows:

$$y = 3x - 2 \quad \text{can be written in the form} \quad f(x) = 3x - 2.$$

That is, the notation $f(x)$ represents the y-value for some corresponding x-value.

Now, in function notation, $f(5)$ means to replace x with 5 in the function:

$$f(5) = 3 \cdot 5 - 2 = 15 - 2 = 13.$$

Thus, the ordered pair $(5, 13)$ is the same as $(5, f(5))$.

Example 6: Function Evaluation

For the function $g(x) = 4x + 3$, find

i. $g(2)$ **ii.** $g(-3)$ **iii.** $g(0)$

Solution:
i. $g(2) = 4 \cdot 2 + 3 = 11$
ii. $g(-3) = 4(-3) + 3 = -9$
iii. $g(0) = 4 \cdot 0 + 3 = 3$

Not all functions are linear functions just as not all graphs are straight lines. The function notation is valid for a wide variety of types of functions. The following examples illustrate the use of function notation with functions other than linear functions.

Example 7: Nonlinear Functions

a. For the function $f(x)=x^2-2x+1$, find

i. $f(-2)$ **ii.** $f(0)$ **iii.** $f(4)$

Solution: **i.** $f(-2)=(-2)^2-2(-2)+1=4+4+1=9$

ii. $f(0)=0^2-2(0)+1=0-0+1=1$

iii. $f(4)=4^2-2\cdot 4+1=16-8+1=9$

b. For the function $h(x)=2x^3-5x$, find

i. $h(-1)$ **ii.** $h(3)$ **iii.** $h(-4)$

Solution: **i.** $h(-1)=2(-1)^3-5(-1)=2(-1)+5=-2+5=3$

ii. $h(3)=2(3)^3-5(3)=54-15=39$

iii. $h(-4)=2(-4)^3-5(-4)=2(-64)+20=-128+20=-108$

c. Given the function $f(x)=x^2-10$ with restricted domain $D=\{-1,0,2,3\}$, write the function as a set of ordered pairs.

Solution: For each x-value in D, find the corresponding y-value by substituting into $f(x)$.

$$f(-1)=(-1)^2-10=1-10=-9$$
$$f(0)=0^2-10=0-10=-10$$
$$f(2)=2^2-10=4-10=-6$$
$$f(3)=3^2-10=9-10=-1$$

So, the function can be written as the following set of ordered pairs:

$$\{(-1,-9),(0,-10),(2,-6),(3,-1)\}$$

Practice Problems

1. *State the domain and range of the relation* $\{(5,6),(7,8),(9,10),(10,11)\}$. *Is the relation a function? Explain briefly.*
2. *Write the function as a set of ordered pairs given* $y = 2x - 5$ *and* $D = \left\{-4, 0, \frac{1}{2}, 3\right\}$.
3. *State the domain of the function represented by the equation* $y = \sqrt{x+3}$.

Using a TI-83 Plus Calculator to Graph Functions

There are many types and brands of graphing calculators available. For convenience and so that directions can be specific, only the TI-83 Plus graphing calculator is used in the related discussions in this text. Other graphing calculators may be used, but the steps required may be different from those indicated in the text. If you do choose to use another calculator, be sure to read the manual for your calculator and follow the related directions.

Teaching Notes:
It is essential that students learn to check, and to be very adept at changing, the minimum and maximum values for the ranges of the x- and y-axes. A lack of good skills in setting these values may lead students to much frustration in this section.

In any case, remember that a calculator is just a tool to allow for fast calculations and to help in understanding some abstract concepts. A calculator does not replace the need for algebraic knowledge and skills.

Pressing certain keys will give a list of options called a **menu**. You may choose from the menu by pressing the corresponding numerical key or highlighting your choice by pressing an arrow key and pressing ENTER. Be aware that even for simple calculations, a calculator follows the rules for order of operations. For example,

$2+\frac{3}{4}$ would be entered as $2+\frac{3}{4}$ and will give the answer 2.75;

$\frac{2+3}{4}$ would be entered as $\frac{(2+3)}{4}$ and will give the answer 1.25.

Note that the parentheses are needed to indicate a numerator (or denominator) with more than one number.

You should practice and experiment with your calculator until you feel comfortable with the results. **Do not be afraid of making mistakes.**

Answers to Practice Problems: 1. $D=\{5,7,9,10\}; R=\{6,8,10,11\}$ Yes, the relation is a function because each x-coordinate appears only once. **2.** $\left\{(-4,-13),(0,-5),\left(\frac{1}{2},-4\right),(3,1)\right\}$ **3.** $D=\{x \mid x \geq -3\}$

The following six keys and the related menus are important for the exercises in this text. The CLEAR key or 2nd QUIT will get you out of most trouble and allow you to start over.

Some Basics about the TI-83 Plus

1. MODE Turn the calculator ON and press the MODE key. The screen should be highlighted as shown below. If it is not, use the arrow keys in the upper right corner of the keyboard to highlight the correct words and press ENTER. See the manual for the meanings of the terms not highlighted.

Note: A highlighted 4 in the `Float` setting indicates 4-digit accuracy. This may be changed at any time for more or fewer digits in the accuracy of calculations.

After these settings have been checked, press 2nd and QUIT.

2. WINDOW Press the WINDOW key and the standard window will be displayed:

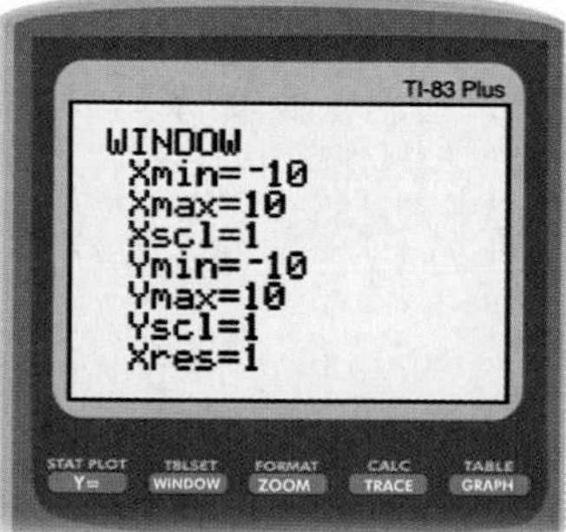

This window can be changed at any time by changing the individual numbers or pressing the ZOOM key and whatever number is needed in the menu displayed. Because of the shape of the display screen this standard screen is not a square screen. For example, be aware that the slopes of lines are not truly depicted unless the screen is in a scale of about 3 : 2. For example, a square screen can be attained by setting Xmin = –15 and Xmax = 15 to give the x-axis a length of 30 and the y-axis a length of 20, or a ratio of 3 : 2.

3. Y= The Y= key is in the upper left corner of the keyboard. This key will allow ten different functions to be entered. These functions are labeled as $Y_1 \ldots Y_{10}$. The variable x may be entered by using the X,T,θ,n key. The ^ key can be used to indicate exponents. For example, the equation $y = x^2 + 3x$ would be entered as:

$$Y_1 = X\text{^}2 + 3X$$

To change an entry, practice with the keys DEL, CLEAR, and 2nd INS.

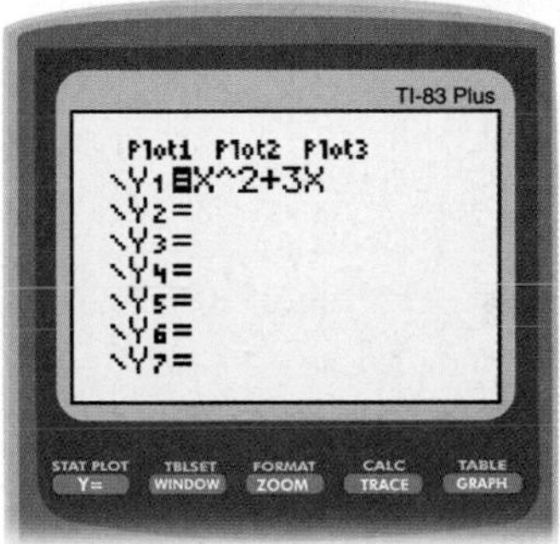

4. GRAPH If this key is pressed, then the screen will display the graph of whatever functions are indicated in the Y= list with the = sign highlighted by using whatever scales are indicated in the current WINDOW. In many cases the WINDOW must be changed to accommodate the domain and range of the function or to show a point where two functions intersect.

5. TRACE The TRACE key will display the current graph even if it is not already displayed and give the x- and y- coordinates of a point highlighted on the graph. The curve may be traced by pressing the left and right arrow keys. At each point on the graph, the corresponding x- and y-coordinates are indicated at the bottom of the screen. **(Remember that because of the limitations of the pixels (lighted dots) on the screen, these x- and y-coordinates are accurate only part of the time. Generally, they are only approximations.)**

6. CALC The CALC key (press 2nd TRACE) gives a menu with seven items. Items 2 – 5 are used with graphs.

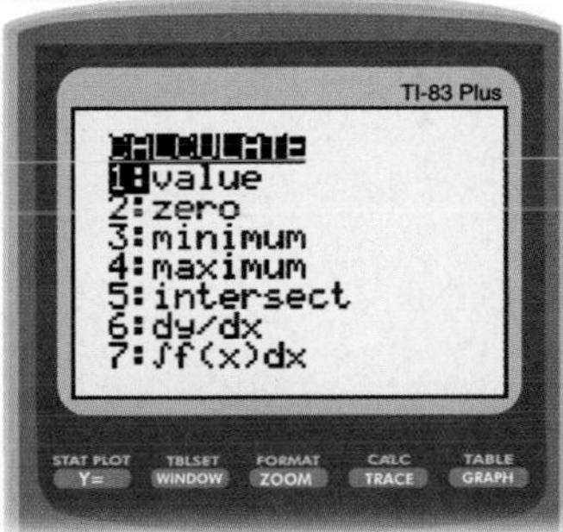

After displaying a graph, select CALC. Then press 2 and follow the steps outlined below to locate the point where the graph crosses the x-axis (the x-intercept). The graph must actually cross the axis. This point is called a **zero** of the function because the y-value will be 0.

Step 1: With the left arrow, move the cursor to the left of the x-intercept on the graph. Press ENTER in response to the question "`LeftBound?`".

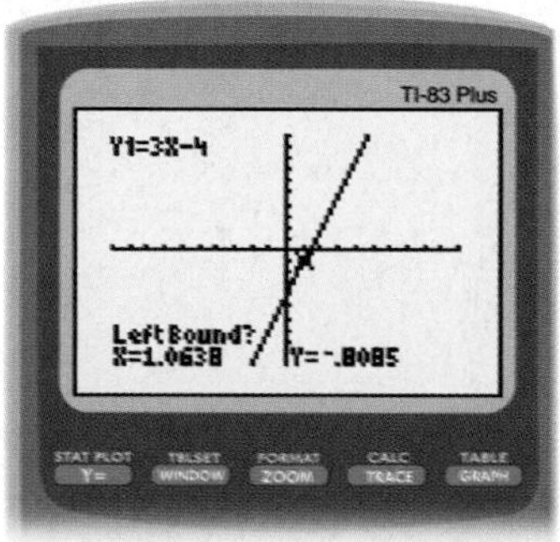

Step 2: With the right arrow, move the cursor to the right of the x-intercept on the graph. Press ENTER in response to the question "`RightBound?`".

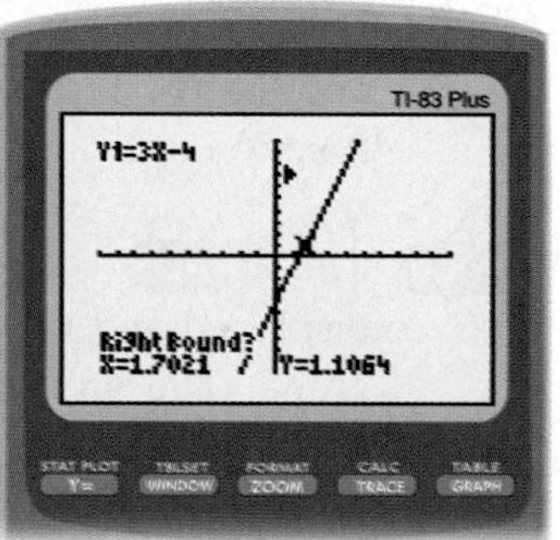

Step 3: With the left arrow, move the cursor near the x-intercept. Press ENTER in response to the question "`Guess?`". The calculator's estimate of the zero will appear at the bottom of the display.

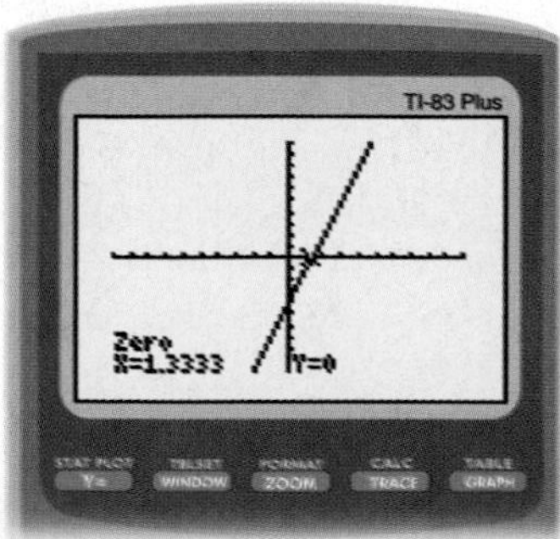

Example 8: Graphing Calculator

Use a graphing calculator to find the graphs of each of the following functions. Use the TRACE key to find the point where each graph intersects the x-axis. Sketch a copy of each graph on graph paper and label the x-intercepts.

Teaching Notes:
Tell your students that if the message
`ERR:SYNTAX`
`1: Quit`
`2: Goto`
appears on the display, the error will be highlighted if they press "`2: Goto`".

a. $y = -3x - 1$ (It is important that the (–) key be used to indicate the negative sign in front of $3x$. Use of the subtraction key is a common error.)

Solution:

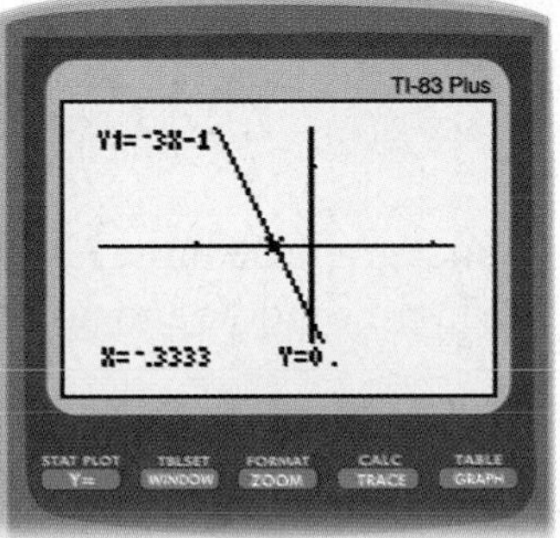

Note: Vertical lines are not functions and cannot be graphed by the calculator.

b. $y = \sqrt{x-4}$

Solution:

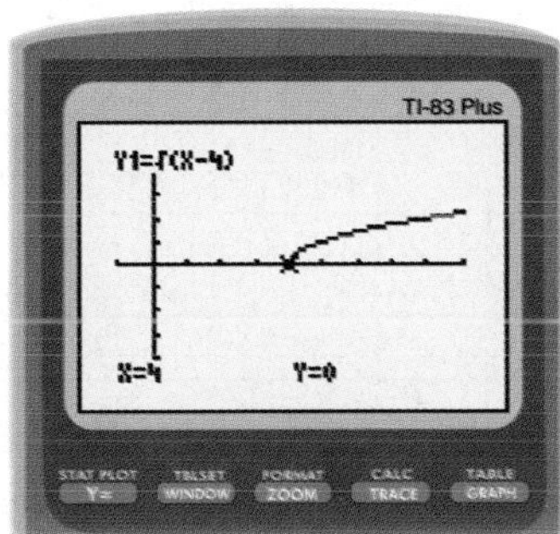

Note: Be sure to include the expression "$(x-4)$" in parentheses after the $\sqrt{\ }$ sign.

c. $y = x^2 + 3x$

Solution: Since the graph of this function has two x-intercepts, we have shown the graph twice. Each graph shows the coordinates of a distinct x- intercept.

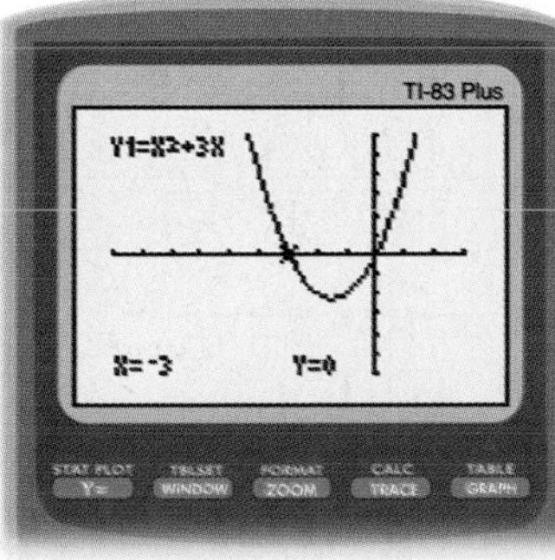

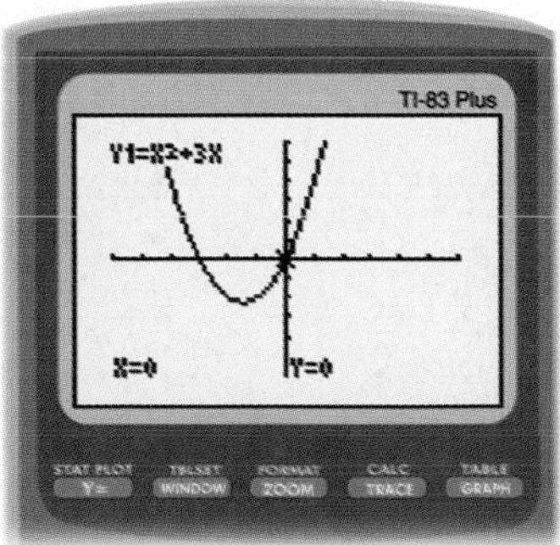

continued on next page ...

d. $y = 2x - 1$; $y = 2x + 1$; $y = 2x + 3$

Solution:

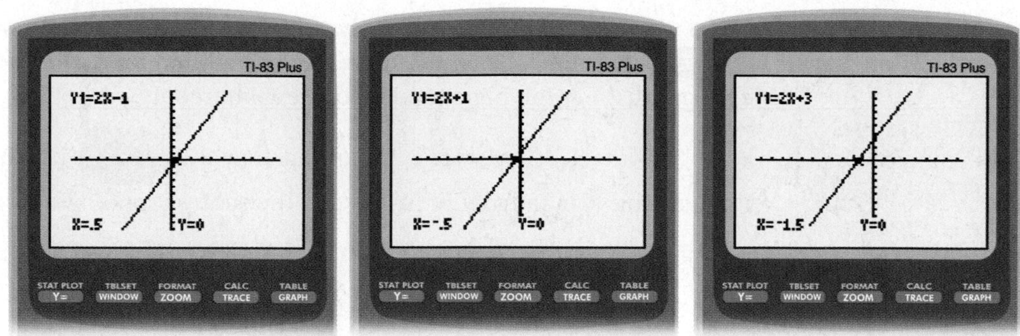

First graph all three functions in the standard window. Then change the window to a square window and notice the difference in the accuracy of the slopes of the lines.

4.5 Exercises

List the sets of ordered pairs corresponding to the points in Exercises 1 – 8. State the domain and range and indicate which of the relations are also functions.

1. $\left\{\begin{matrix}(-4,0),(-1,4),(1,2),\\(2,5),(6,-3)\end{matrix}\right\}$; $D=\{-4,-1,1,2,6\}$; $R=\{0,4,2,5,-3\}$; function

2. $\left\{\begin{matrix}(-4,-1),(0,0),(0,3),\\(2,-2),(3,1)\end{matrix}\right\}$; $D=\{-4,0,2,3\}$; $R=\{-1,0,3,-2,1\}$; not a function

3. $\left\{\begin{matrix}(-5,-4),(-4,-2),\\(-2,-2),(1,-2),(2,1)\end{matrix}\right\}$; $D=\{-5,-4,-2,1,2\}$; $R=\{-4,-2,1\}$; function

4. $\left\{\begin{matrix}(-2,-1),(0,3),\\(3,0),(4,-2),(6,2)\end{matrix}\right\}$; $D=\{-2,0,3,4,6\}$; $R=\{-1,3,0,-2,2\}$; function

1.

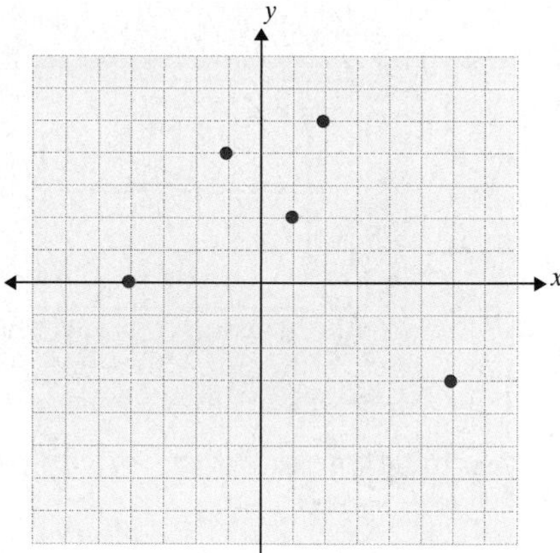

3.

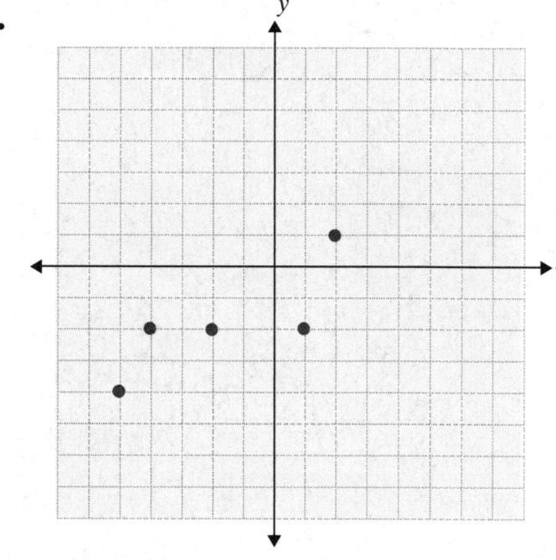

2.

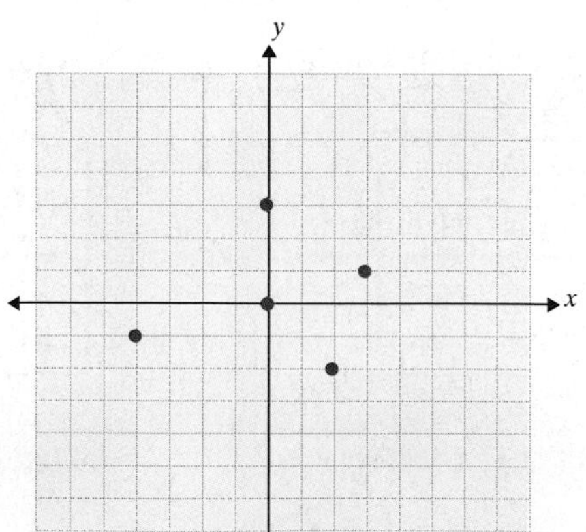

4.

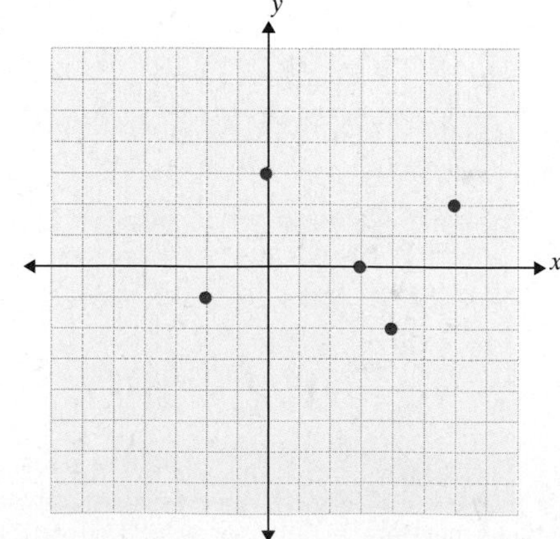

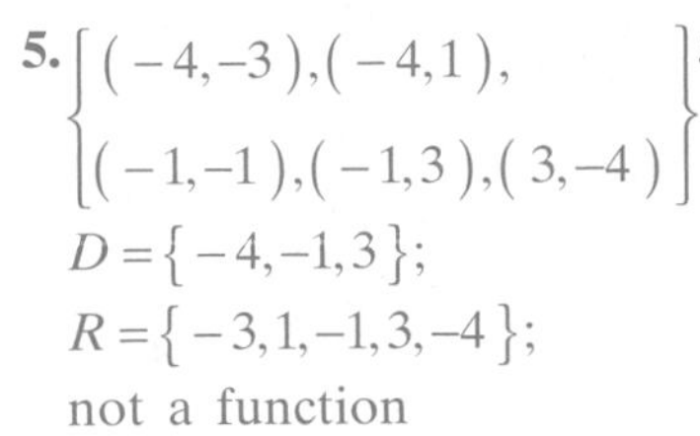

5. $\left\{\begin{matrix}(-4,-3),(-4,1),\\(-1,-1),(-1,3),(3,-4)\end{matrix}\right\}$

$D=\{-4,-1,3\};$

$R=\{-3,1,-1,3,-4\};$

not a function

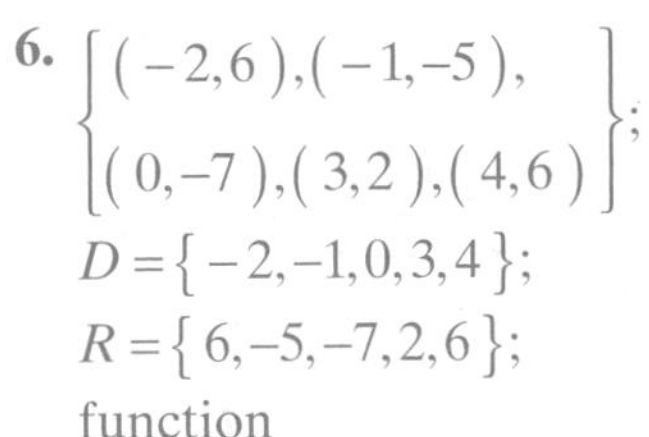

6. $\left\{\begin{matrix}(-2,6),(-1,-5),\\(0,-7),(3,2),(4,6)\end{matrix}\right\};$

$D=\{-2,-1,0,3,4\};$

$R=\{6,-5,-7,2,6\};$

function

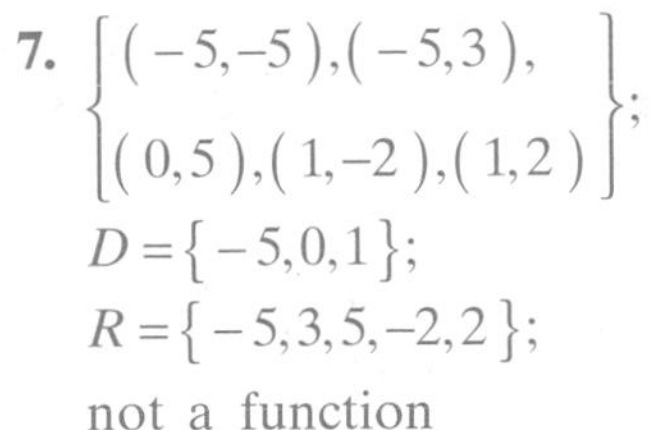

7. $\left\{\begin{matrix}(-5,-5),(-5,3),\\(0,5),(1,-2),(1,2)\end{matrix}\right\};$

$D=\{-5,0,1\};$

$R=\{-5,3,5,-2,2\};$

not a function

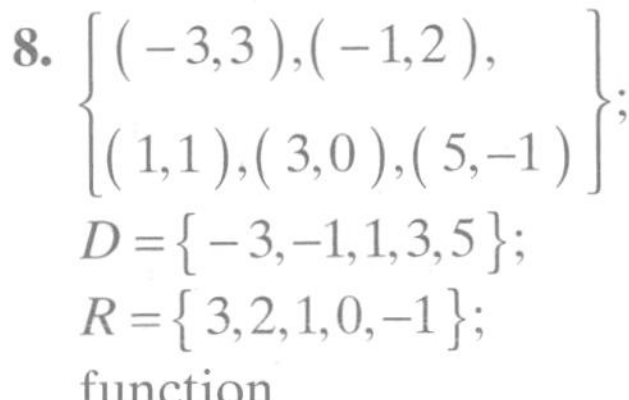

8. $\left\{\begin{matrix}(-3,3),(-1,2),\\(1,1),(3,0),(5,-1)\end{matrix}\right\};$

$D=\{-3,-1,1,3,5\};$

$R=\{3,2,1,0,-1\};$

function

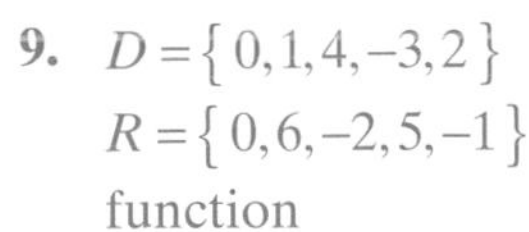

9. $D=\{0,1,4,-3,2\}$

$R=\{0,6,-2,5,-1\}$

function

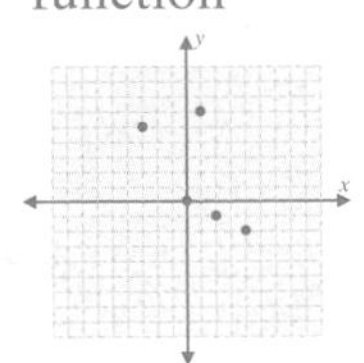

10. $D=\{1,2,-1,0,4\}$

$R=\{-5,-3,2,3\}$

function

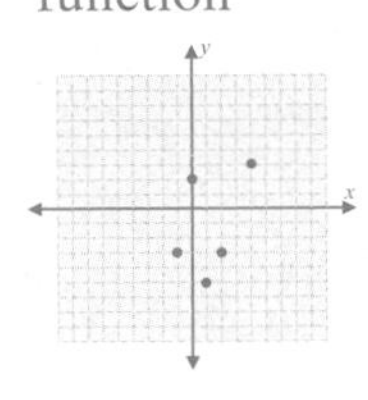

5.

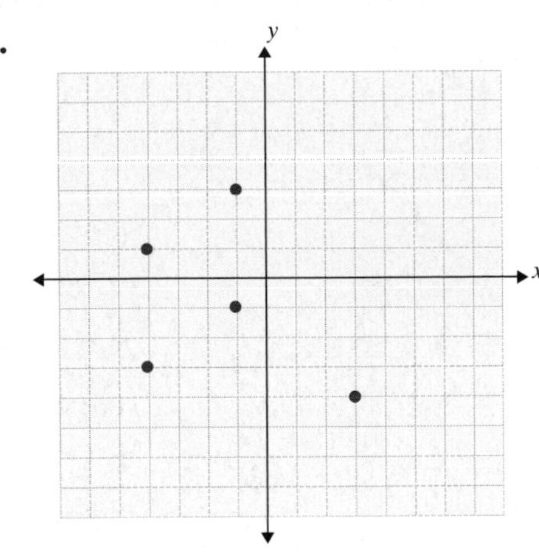

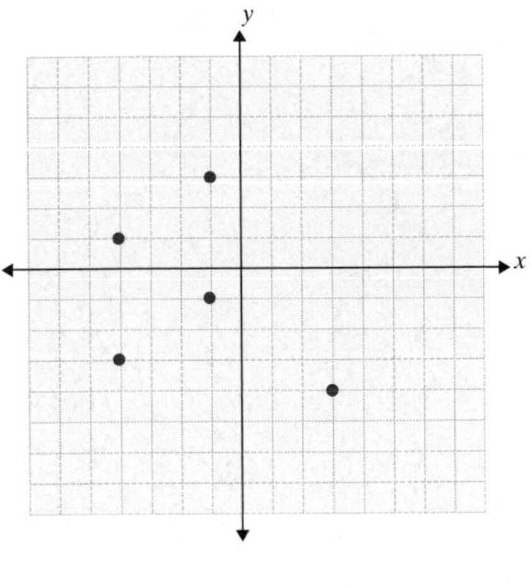

7.

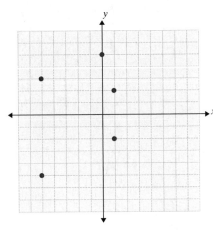

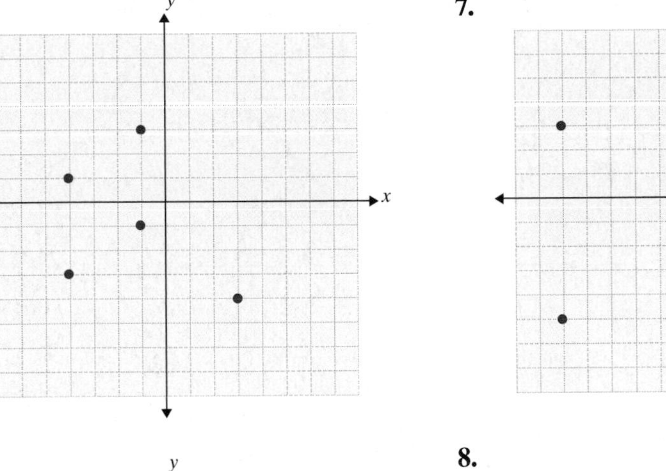

6.

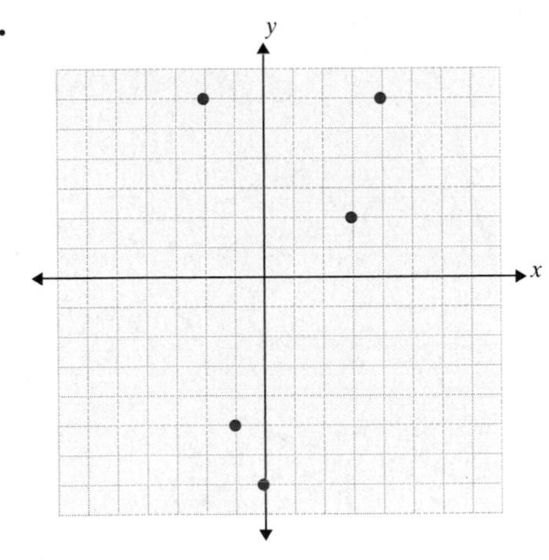

8.

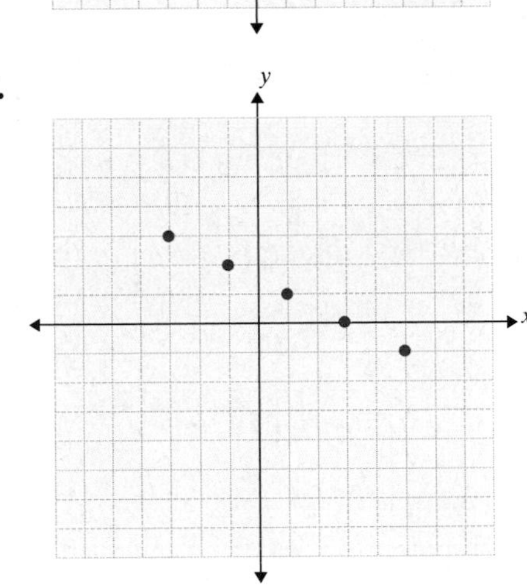

Graph the relations in Exercises 9 – 16. State the domain and range and indicate which of the relations are functions.

9. $\{(0,0),(1,6),(4,-2),(-3,5),(2,-1)\}$

10. $\{(1,-5),(2,-3),(-1,-3),(0,2),(4,3)\}$

11. $\{(-4,4),(-3,4),(1,4),(2,4),(3,4)\}$

12. $\{(-3,-3),(0,1),(-2,1),(3,1),(5,1)\}$

13. $\{(0,2),(-1,1),(2,4),(3,5),(-3,5)\}$

14. $\{(-1,-4),(0,-3),(2,-1),(4,1),(1,1)\}$

15. $\{(-1,4),(-1,2),(-1,0),(-1,6),(-1,-2)\}$

16. $\{(0,0),(-2,-5),(2,0),(4,-6),(5,2)\}$

11. $D=\{-4,-3,1,2,3\}$

$R=\{4\}$

function

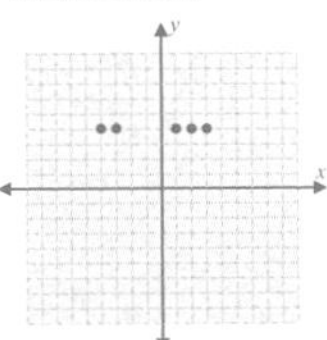

12. $D=\{-3,0,-2,3,5\}$

$R=\{-3,1\}$

function

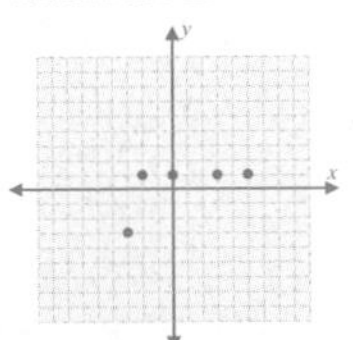

13. $D=\{0,-1,2,3,-3\}$
$R=\{2,1,4,5\}$
function

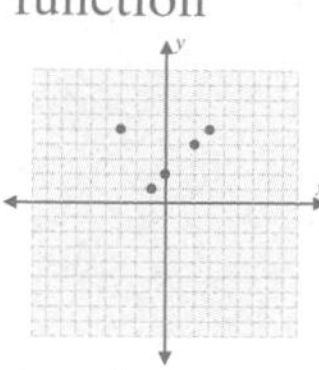

14. $D=\{-1,0,2,4,1\}$
$R=\{-4,-3,-1,1\}$
function

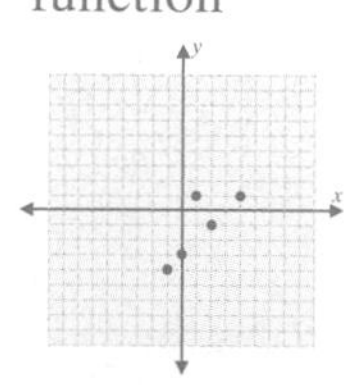

15. $D=\{-1\}$
$R=\{4,2,0,6,-2\}$
not a function

16. $D=\{0,-2,2,4,5\}$
$R=\{0,-5,-6,2\}$
function

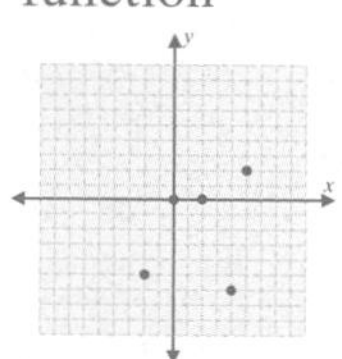

17. function
$D=(-\infty,\infty)$
$R=(0,\infty)$

18. function
$D=(-\infty,\infty)$
$R=(0,\infty)$

19. function
$D=(-6,6]$
$R=\{-6,-4,-2,0,2,4\}$

20. function
$D=(-\infty,\infty)$
$R=(-\infty,3)$

21. not a function

22. function
$D=(-\infty,\infty)$
$R=(-\infty,5)$

In Exercises 17 – 26, use the vertical line test to determine whether or not each graph is a function. If the graph represents a function, state the domain and range of the function.

17.

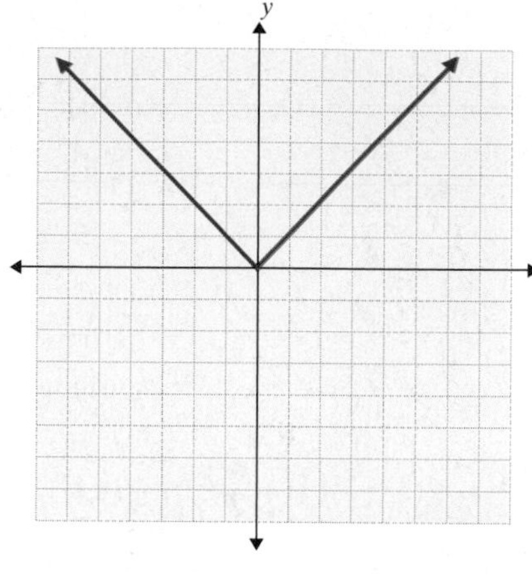

18.

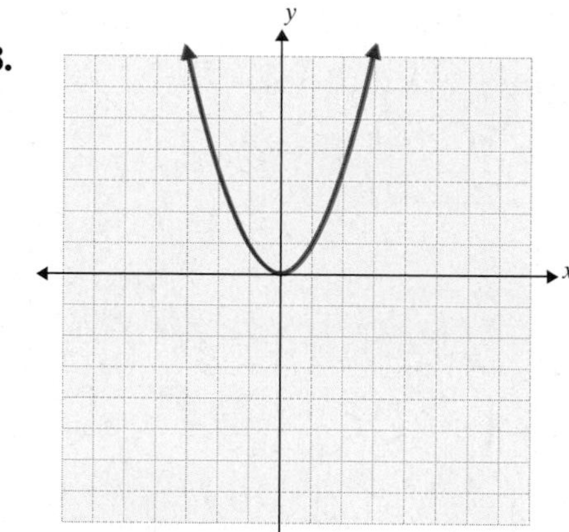

19.

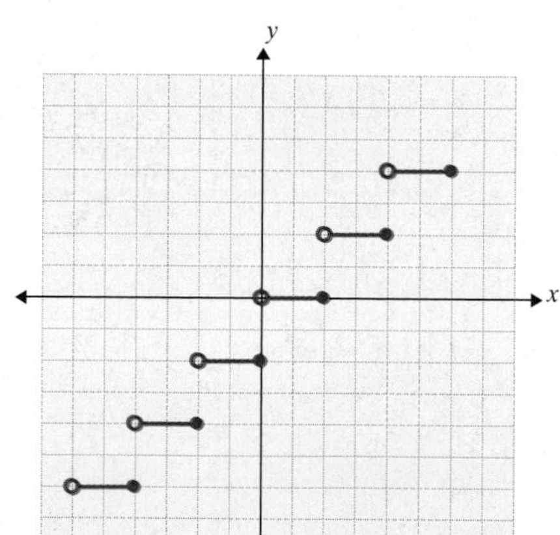

20.

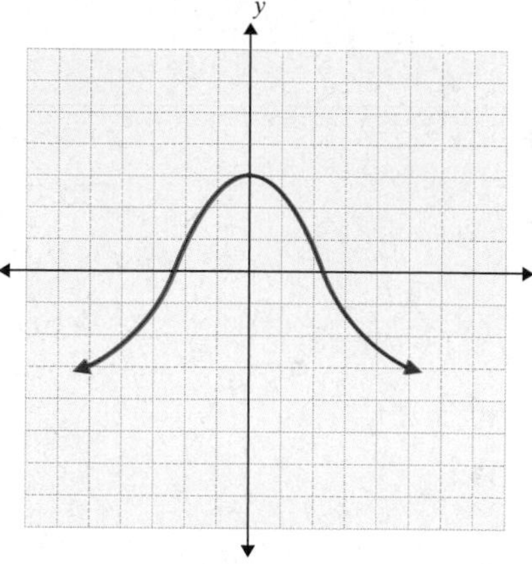

21.

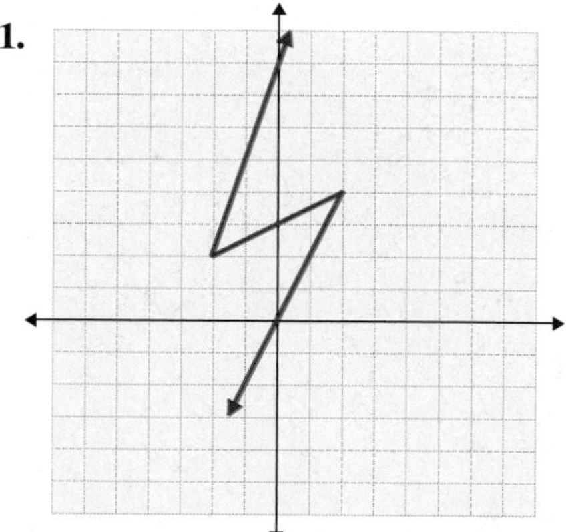

22.

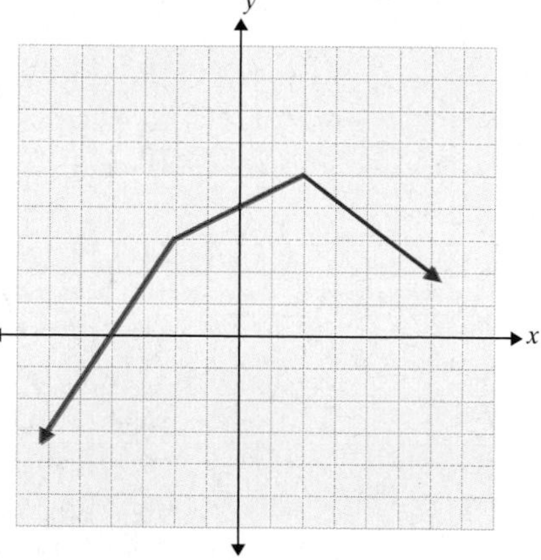

23. not a function

24. not a function

25. function
$D=(-\infty,\infty)$
$R=\left(-\frac{3}{2},\frac{3}{2}\right)$

26. not a function

27. $\left\{(-9,-26),\left(-\frac{1}{3},0\right),(0,1),\left(\frac{4}{3},5\right),(2,7)\right\}$

28. $\left\{(-4,5),\left(-2,\frac{7}{2}\right),(0,2),\left(3,-\frac{1}{4}\right),(4,-1)\right\}$

29. $\left\{\begin{array}{l}(-2,-11),(-1,-2),\\(0,1),(1,-2),\\(2,-11)\end{array}\right\}$

30. $\left\{\begin{array}{l}(-1,3),(0,0),\\\left(\frac{1}{2},-\frac{15}{8}\right),\\(1,-3),(2,0)\end{array}\right\}$

31. $f(-2)=-3$, $f(-1)=-1, f(0)=1$, $f(1)=3, f(5)=11$

32. $f(-2)=-11$, $f(-1)=-8, f(0)=-5$, $f(1)=-2, f(5)=10$

33. $g(-2)=10$, $g(-1)=6, g(0)=2$, $g(1)=-2, g(5)=-18$

34. $g(-2)=-5$, $g(-1)=-6, g(0)=-7$, $g(1)=-8, g(5)=-12$

35. $h(-2)=4$, $h(-1)=1, h(0)=0$, $h(1)=1, h(5)=25$

36. $h(-2)=-8$, $h(-1)=-1, h(0)=0$, $h(1)=1, h(5)=125$

37. $F(-2)=16$, $F(-1)=9, F(0)=4$, $F(1)=1, F(5)=9$

38. $G(-2)=0$, $G(-1)=2, G(0)=6$, $G(1)=12, G(5)=56$

39. $H(-2)=8$, $H(-1)=7, H(0)=0$, $H(1)=-7, H(5)=85$

40. $C(-2)=3$, $C(-1)=-6$, $C(0)=-7, C(1)=0$, $C(5)=108$

41. $P(-2)=9.5$, $P(-1)=6, P(0)=3.5$, $P(1)=2, P(5)=6$

23.

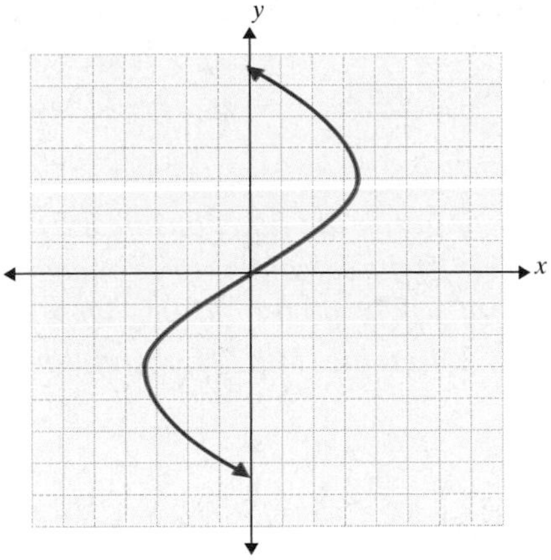

25.

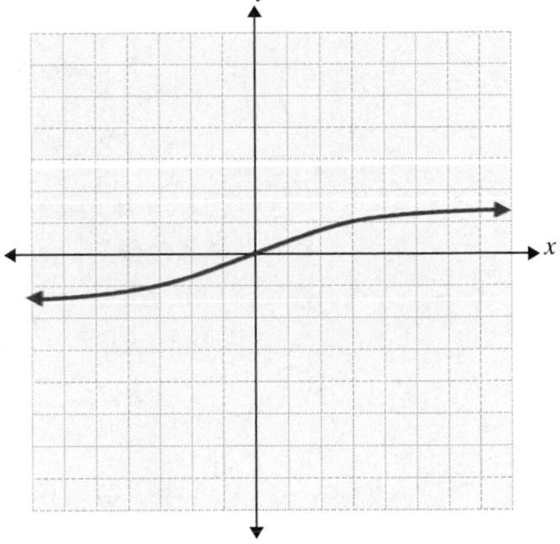

24.

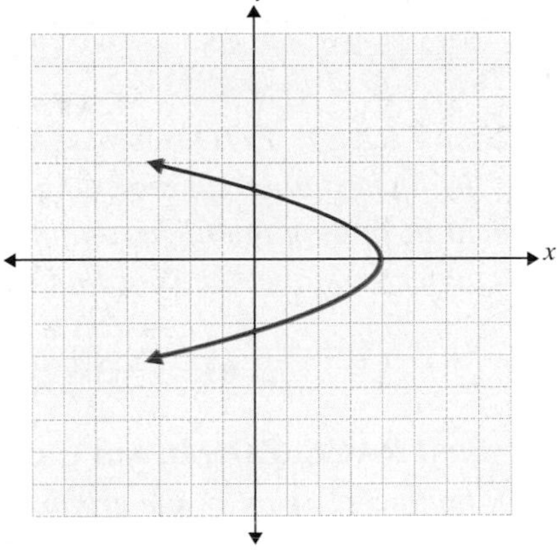

26.

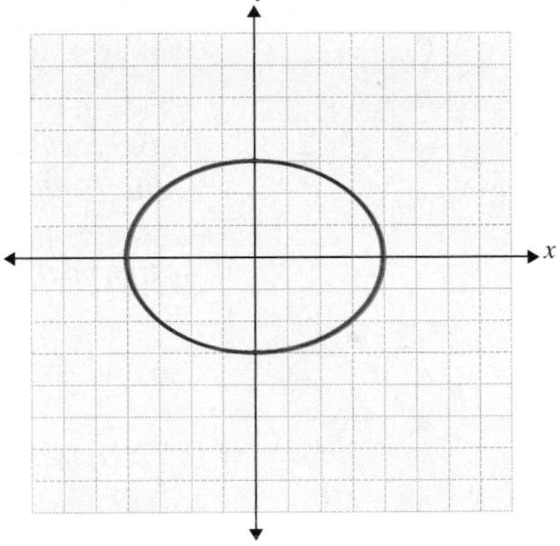

In Exercises 27 – 30, express the function as a set of ordered pairs for the given equation and domain.

27. $y=3x+1;\ D=\left\{-9,-\frac{1}{3},0,\frac{4}{3},2\right\}$

28. $y=-\frac{3}{4}x+2;\ D=\{-4,-2,0,3,4\}$

29. $y=1-3x^2;\ D=\{-2,-1,0,1,2\}$

30. $y=x^3-4x;\ D=\left\{-1,0,\frac{1}{2},1,2\right\}$

For Exercises 31 – 43, assume that each function has the same domain, $D=\{-2,-1,0,1,5\}$ and find the corresponding functional value (y-value) for each x-value in the domain.

31. $f(x)=2x+1$

32. $f(x)=3x-5$

33. $g(x)=-4x+2$

34. $g(x)=-x-7$

35. $h(x)=x^2$

36. $h(x)=x^3$

37. $F(x)=x^2-4x+4$

38. $G(x)=x^2+5x+6$

39. $H(x)=x^3-8x$

40. $C(x)=4x^2+3x-7$

41. $P(x)=0.5x^2-2x+3.5$

42. $R(x)=1.5x^2+4x-9$

43. $f(x)=x^3-5x^2+7x-2$

42. $R(-2)=-11, R(-1)=-11.5$, $R(0)=-9, R(1)=-3.5, R(5)=48.5$

43. $f(-2)=-44, f(-1)=-15$, $f(0)=-2, f(1)=1, f(5)=33$

44. $D = \{ x \mid x \neq -3 \}$

45. $D = \left\{ x \mid x \neq -\frac{1}{2} \right\}$

46. $D = \left\{ x \mid x \geq -\frac{5}{2} \right\}$

47. $D = \left\{ x \mid x \leq \frac{4}{3} \right\}$

48. y = 6

49. y = 4x

50. y = −2x + 3

51.

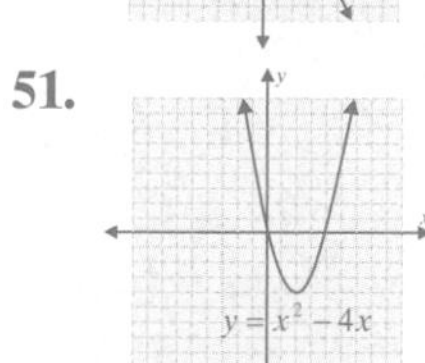

52.

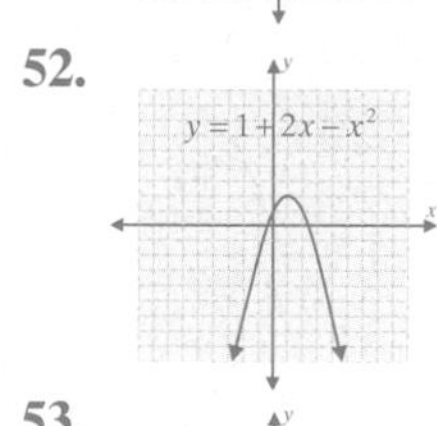

53. y = √(x + 5)

54.

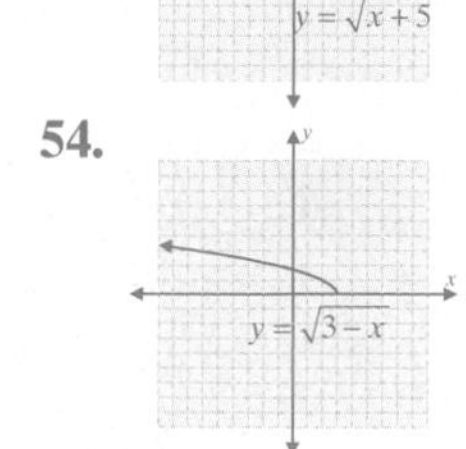

Find the domain of each function represented by the equations in Exercises 44 – 47.

44. $y = \dfrac{x+5}{x+3}$ **45.** $y = \dfrac{x-1}{2x+1}$ **46.** $y = \sqrt{2x+5}$ **47.** $y = \sqrt{4-3x}$

Use a graphing calculator to graph the functions in Exercises 48 – 59. Use the ***TRACE, ZOOM,*** *and* ***CALC*** *features of the calculator to estimate x-intercepts, if any. (Remember, at these points the value of y will be 0.) Sketch each function on graph paper. For absolute value functions, select the* ***MATH*** *menu, then the* ***NUM*** *menu, and then* ***1: abs(*** *.*

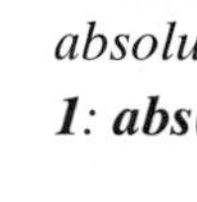

48. $y = 6$ **49.** $y = 4x$ **50.** $y = -2x + 3$ **51.** $y = x^2 - 4x$

52. $y = 1 + 2x - x^2$ **53.** $y = \sqrt{x+5}$ **54.** $y = \sqrt{3-x}$ **55.** $y = |x + 2|$

56. $y = |x^2 - 3x|$ **57.** $y = x^3 - 2x^2 + 1$ **58.** $y = -x^3 + 3x - 1$ **59.** $y = x^4 - 13x^2 + 36$

In Exercises 60 and 61, use the ***TRACE, ZOOM,*** *and* ***CALC*** *features of the calculator to estimate the coordinates of the highest point on the graph. (****HINT:*** *Item 4 on the* ***CALC*** *menu* ***4: maximum*** *will help in finding the highest point of a function, if there is one.) Sketch the graph on graph paper and label the coordinates of the highest point.*

60. $y = 4x - x^2$ **61.** $y = 3 - 2x - x^2$

In Exercises 62 and 63, use the ***TRACE, ZOOM,*** *and* ***CALC*** *features of the calculator to estimate the coordinates of the lowest point on the graph. (****HINT:*** *Item 3 on the* ***CALC*** *menu* ***3: minimum*** *will help in finding the lowest point of a function, if there is one.) Sketch the graph on graph paper and label the coordinates of the lowest point.*

62. $y = 2x^2 - x + 1$ **63.** $y = 3(x - 1)^2 + 2$

In Exercises 64 – 67, use the ***TRACE, ZOOM,*** *and* ***CALC*** *features of the calculator to estimate the coordinates of the point(s) of intersection on the graphs. (****HINT:*** *Item 5 on the* ***CALC*** *menu* ***5: intersect*** *will help in finding the point of intersection of two functions, if there is one.) In the* ***Y=*** *menu use both* ***Y1 =*** *and* ***Y2 =*** *to be able to graph both functions at the same time. Sketch these functions on graph paper and label the point(s) of intersection.*

64. $y = 3x + 2$, $y = 4 - x$ **65.** $y = 2 - x$, $y = x$ **66.** $y = 2x - 1$, $y = x^2$ **67.** $y = x + 3$, $y = -x^2 + x + 7$

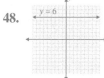

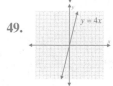

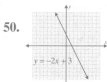

55.

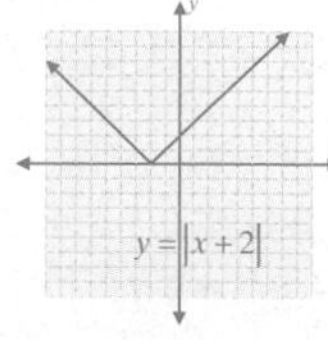

56.

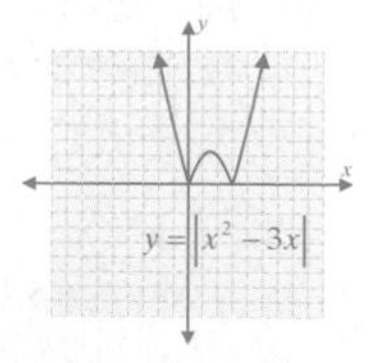

57.

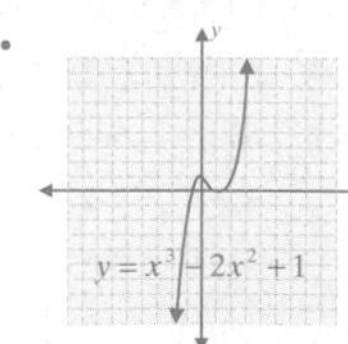

58.

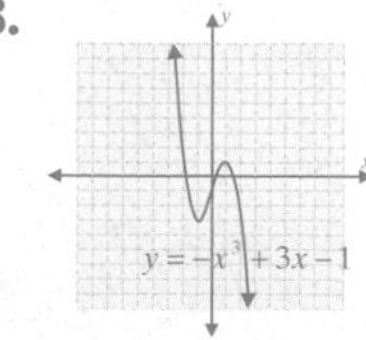

59. 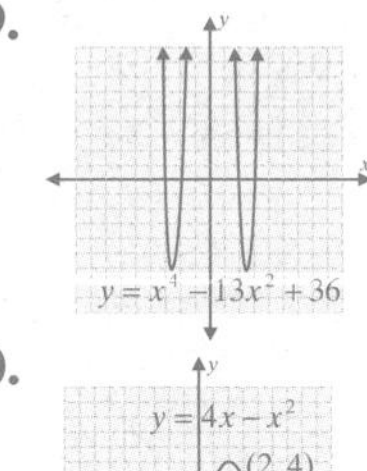

60.
$y = 4x - x^2$
(2, 4)

61. 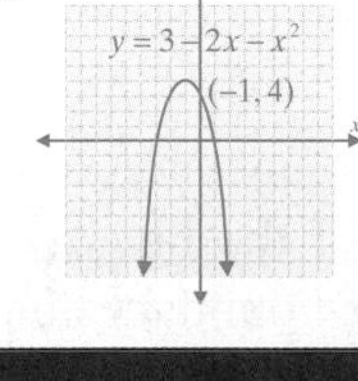

Writing and Thinking About Mathematics

68. Explain in your own words how you find the domain of a function.

a. graphically

b. algebraically

69. Which, if any, of the following functions has no restriction for the values in the domain? If the function's domain has restrictions, explain why.

a. $3x - 2y = -1$

b. $\sqrt{x+2} = y$

c. $\dfrac{1}{3x-1} = y$

Hawkes Learning Systems: Introductory & Intermediate Algebra

Introduction to Functions and Function Notation

62. 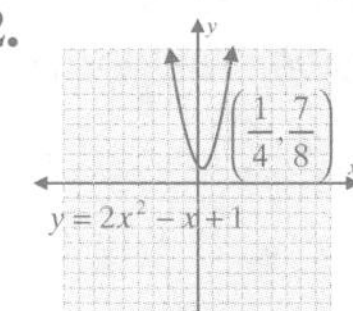

63. 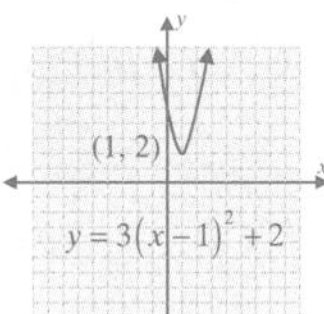

64. 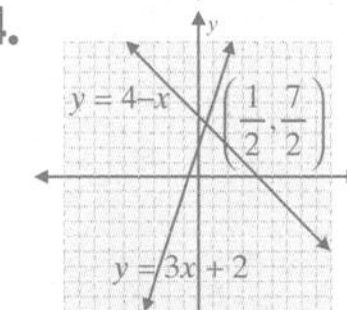

65. 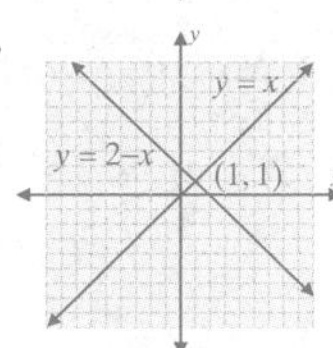

66. 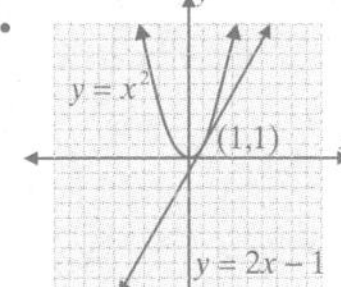

67. 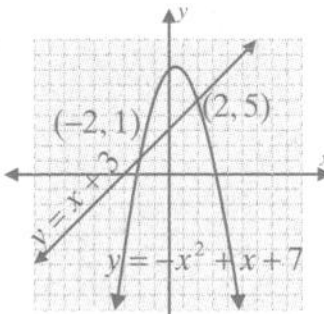

68. a. – b. Answers will vary.

69. a. No restrictions

b. $D = (-2, \infty)$ because the square root of a negative number does not exist.

c. $D = \left\{ x \mid x \neq \dfrac{1}{3} \right\}$ because a zero in the denominator makes the function undefined.

4.6 Graphing Linear Inequalities in Two Variables

Objectives

After completing this section, you will be able to:

Graph linear inequalities.

Graphing Linear Inequalities

A straight line separates a plane into two **half-planes**. The points on one side of the line are in one of the half-planes, and the points on the other side of the line are in the other half-plane. The line itself is called the **boundary line**. If the boundary line is included with a half-plane, then the half-plane is said to be **closed**. If the boundary line is not included, then the half-plane is said to be **open**. (Note the similarity between the terminology for open and closed intervals and half-planes.)

There are two basic methods for deciding which side of the line is the graph of the solution set of the inequality. In both methods, the boundary line must be graphed first.

Two Methods for Graphing Linear Inequalities

First, graph the boundary line (dashed if the inequality is $<$ or $>$, solid if the inequality is $\leq$ or $\geq$).

Method 1

a. *Test any one point obviously on one side of the line.*

b. *If the test-point satisfies the inequality, shade the half-plane on that side of the line. Otherwise, shade the other half-plane.*

*(**Note:** The point (0, 0), if it is not on the boundary line, is usually the easiest point to test.)*

Method 2

a. *Solve the inequality for y (assuming that the line is not vertical).*

b. *If the solution shows $y <$ or $y \leq$, then shade the half-plane below the line.*

c. *If the solution shows $y >$ or $y \geq$, then shade the half-plane above the line.*

*(**Note:** If the boundary line is vertical, then it is of the form $x = a$ and Method 1 should be used.)*

Figure 4.16 shows both (**a**) an open half-plane and (**b**) a closed half-plane with the line $5x - 3y = 15$ as the boundary line.

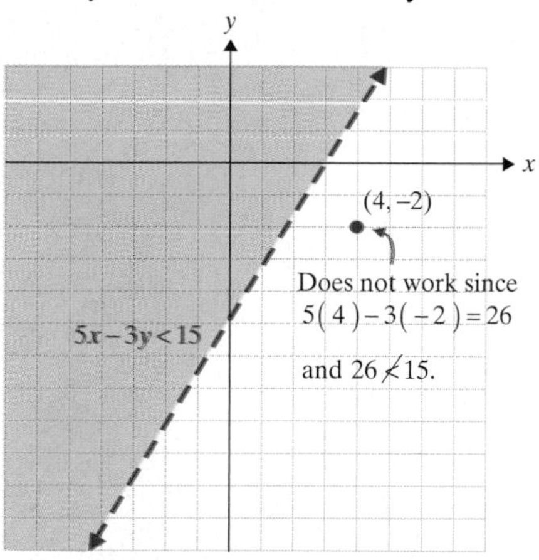

(a) The points on the line
$5x - 3y = 15$
are not included so the line is dashed.
The half-plane is open.

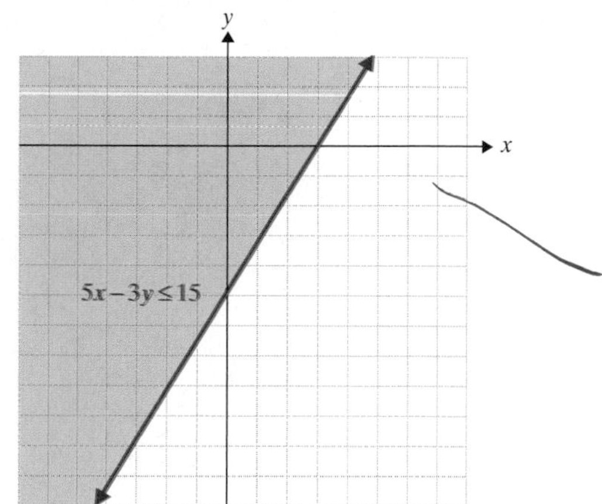

(b) The points on the line
$5x - 3y = 15$
are included so the line is solid.
The half-plane is closed.

Figure 4.16

Example 1: Graphing Linear Inequalities

Teaching Notes:
While some students will recognize that the origin (0, 0) is the easiest point to use as a test point, it is a good idea to have students "test" some points on both sides of the boundary line upon first introduction, so that they develop confidence in the test (and solution set).

Graph the following inequalities.

a. Graph the half-plane that satisfies the inequality $2x + y \leq 6$.

Solution: Method 1 is used in this example.

Step 1: Graph the line $2x + y = 6$ as a solid line.

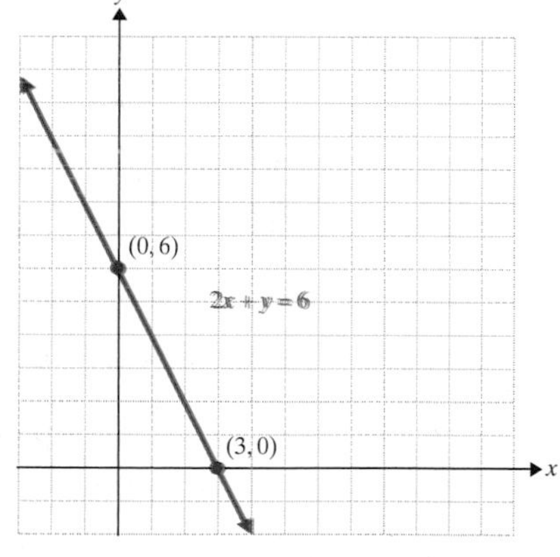

Step 2: Test any point on one side of the line. In this example, we have chosen (0, 0).

$$2 \cdot 0 + 0 \leq 6$$
$$0 \leq 6$$

This is a true statement.

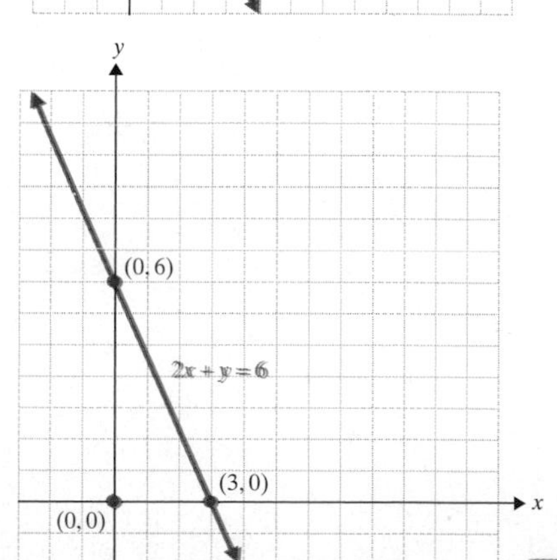

continued on next page ...

Step 3: Shade the points on the same side as the point (0, 0).

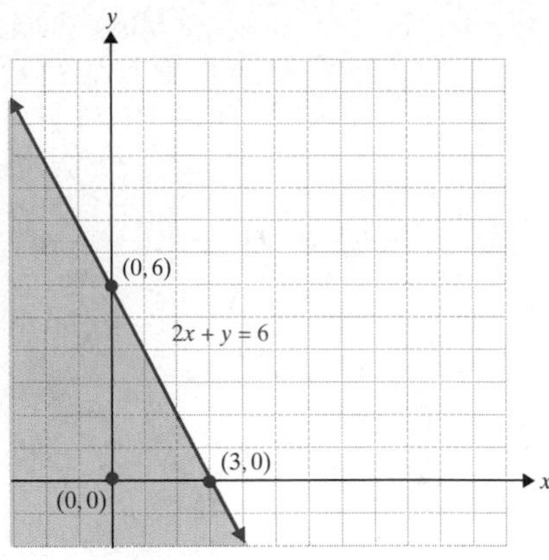

b. Graph the solution set to the inequality $y > 2x$.

Solution: Since the inequality is already solved for y, Method 2 is easy to apply.

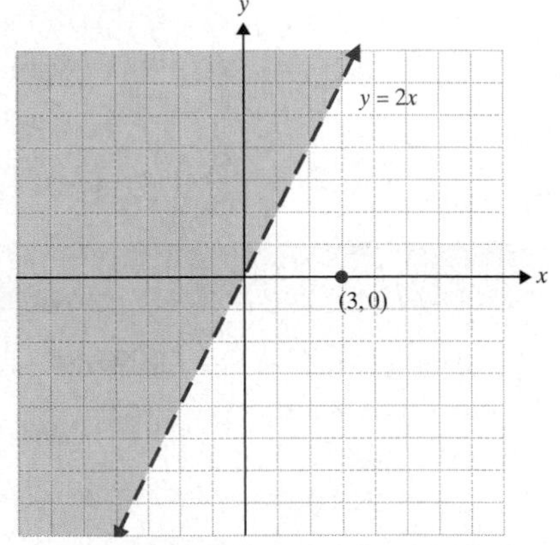

Step 1: Graph the line $y = 2x$ as a dashed line.

Step 2: By Method 2, the graph consists of those points above the line. Shade the half-plane above the line.

[**Note:** As a check, we see that the point (3, 0) gives $0 > 2 \cdot 3$, a false statement.]

c. Graph the half-plane that satisfies the inequality $y > 1$.

Solution: Again, the inequality is already solved for y and Method 2 is used.

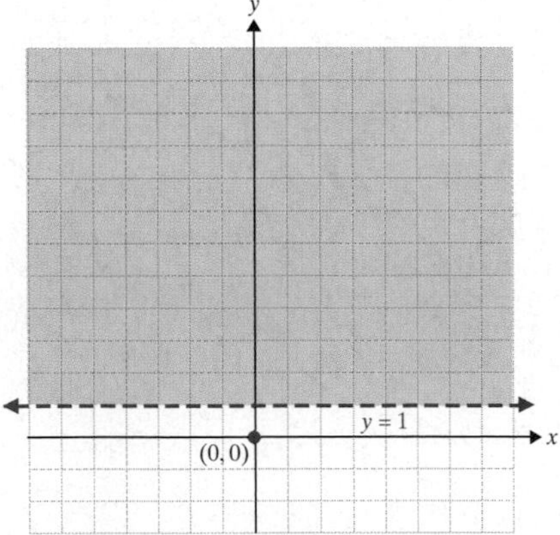

Step 1: Graph the horizontal line $y = 1$ as a dashed line.

Step 2: By Method 2, shade the half-plane above the line.

d. Graph the solution set to the inequality $x \le 0$.

Solution: The boundary line is a vertical line and Method 1 is used.

Step 1: Graph the line $x = 0$ as a solid line. Note that this is the y-axis.

Step 2: Test the point $(-2, 1)$.

$$-2 \le 0$$

This statement is true.

Step 3: Shade the half-plane on the same side of the line as $(-2, 1)$. This half-plane consists of the points with x-coordinate 0 or negative.

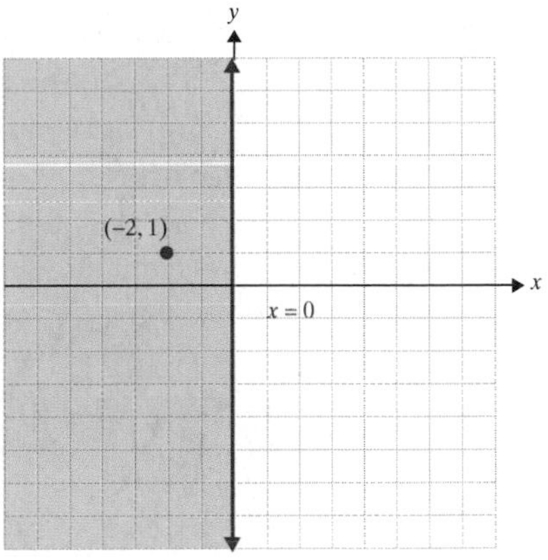

Using a TI-83 Plus Graphing Calculator to Graph Linear Inequalities

The first step in using the TI-83 Plus (or any other graphing calculator) to graph a linear inequality is to solve the inequality for y. This is necessary because this is the way that the boundary line equation can be graphed as a function. Thus, Method 2 for graphing the correct half-plane is appropriate.

Note that when you press the Y= key on the calculator, a slash (\ ) appears to the left of the Y expression as in \Y1 =. This slash is actually a command to the calculator to graph the corresponding function as a solid line or curve. If you move the cursor to position over the slash and hit ENTER, you will find the following options appearing:

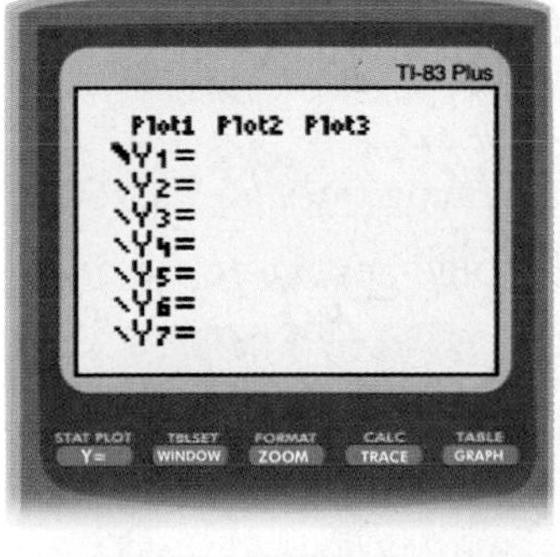
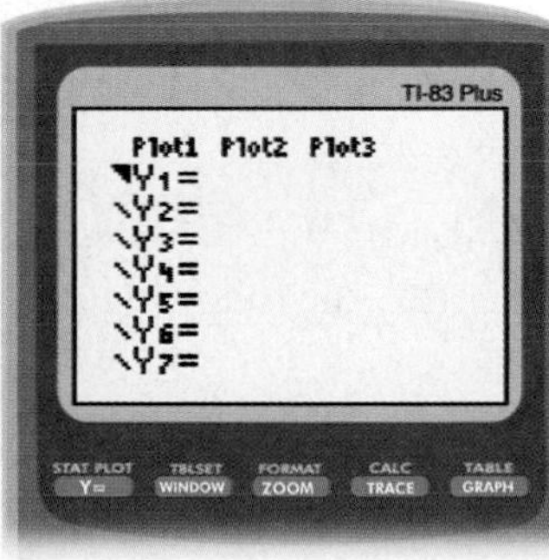
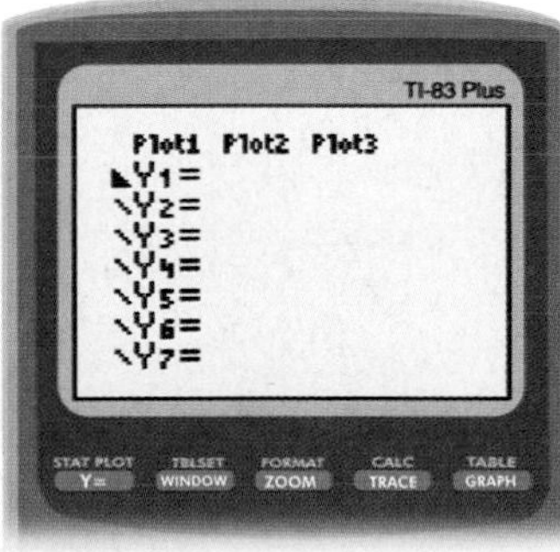
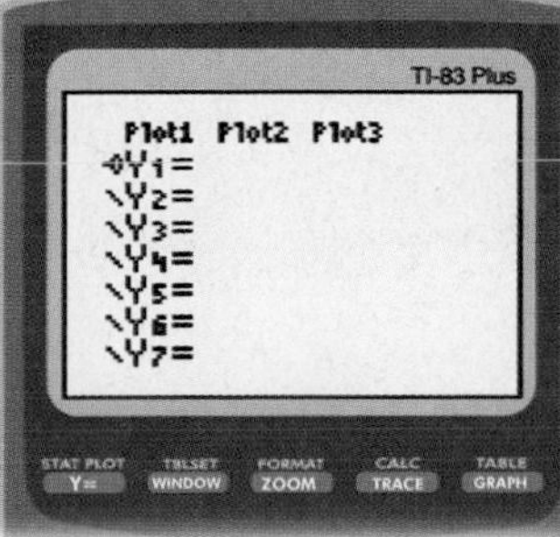
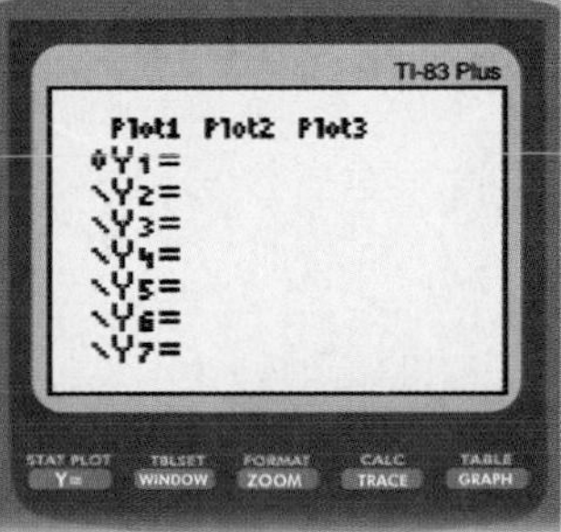
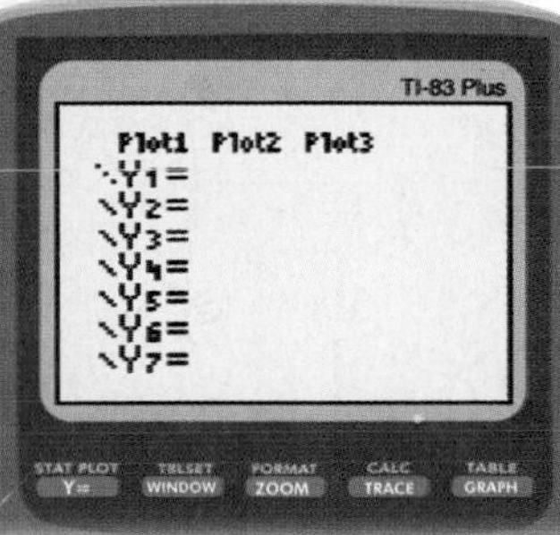

1.

If the slash (which is actually four dots if you look closely) becomes a set of three dots, then the corresponding graph of the function will be dotted. By setting the shading above the dots, the corresponding graph on the display will show shading above the line or curve. By setting the shading below the dots, the corresponding graph on the display will show shading below the line or curve. The actual shading occurs only when the slash is four dots. So, the calculator is not good for determining whether the boundary curve is included or not. The following example illustrates two situations.

2. 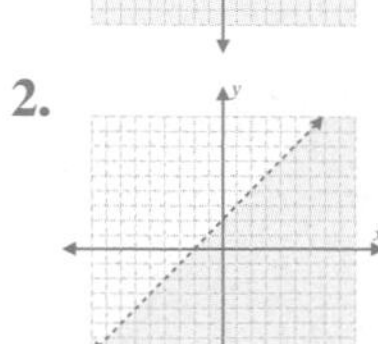

Example 2: Graphing Using a T1 - 83 Plus

3. 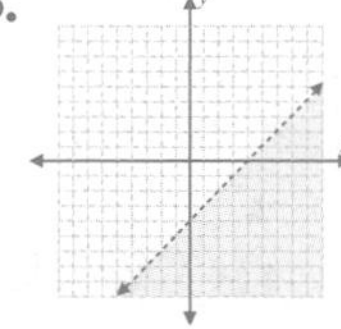

a. Graph the linear inequality $2x + y \leq 7$.

Solution:

Step 1: Solving the inequality for y gives: $y \leq -2x + 7$.

Step 2: Press the key Y= and enter the function: \Y1 = –2X + 7.

4. 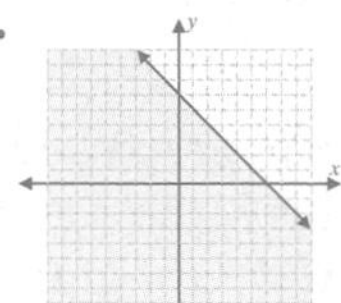

Step 3: Go to the \ and hit ENTER three times so that the display appears as follows:

5. 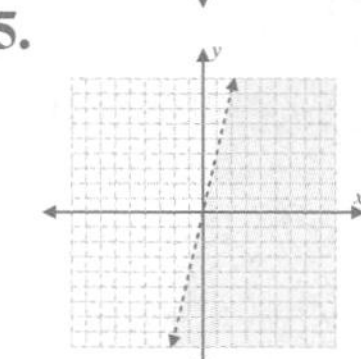

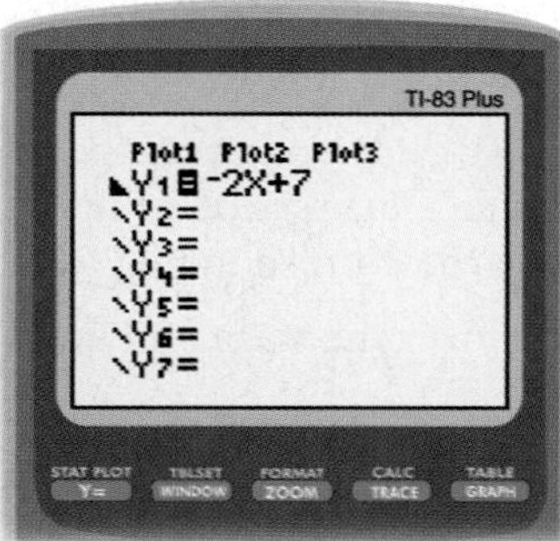

6. 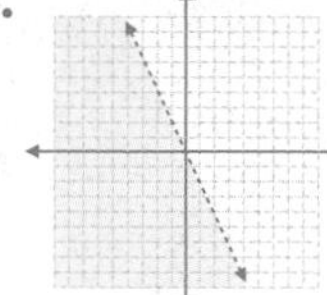

Step 4: Press GRAPH and (assuming that the WINDOW settings are the same) the following graph should appear on the display.

7. 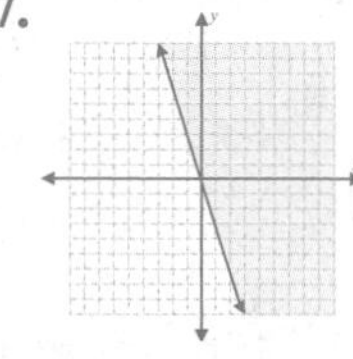

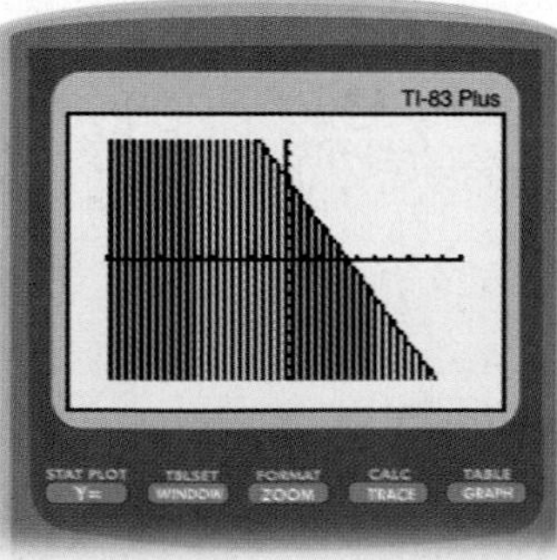

8. 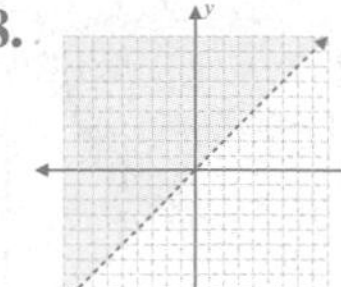

9. 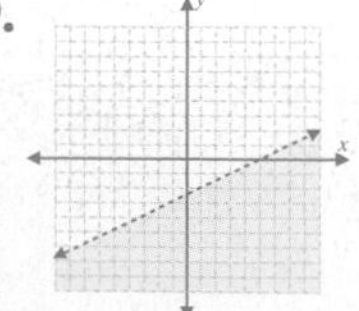

10. 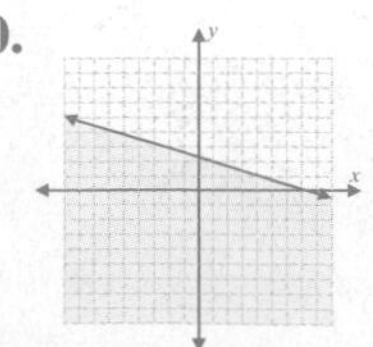

11. 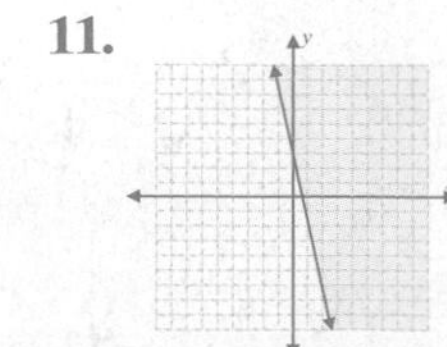

12. 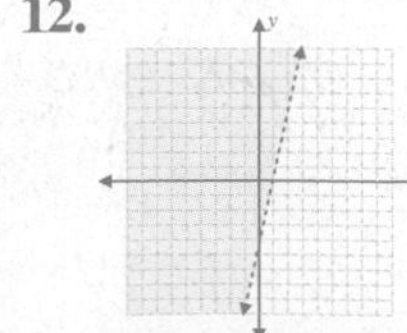

13. 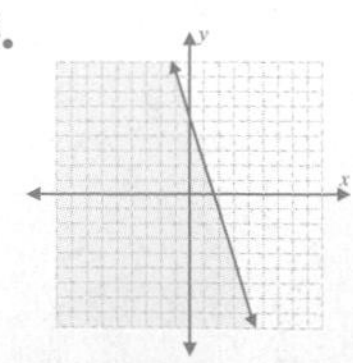

14. 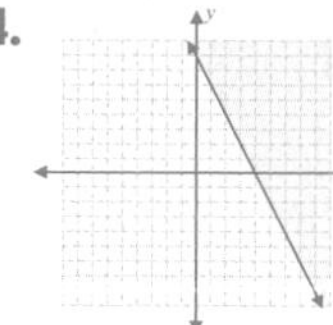

15.

16. 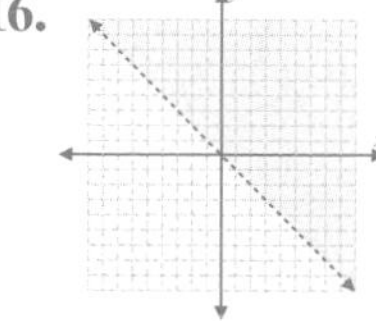

17.

18. 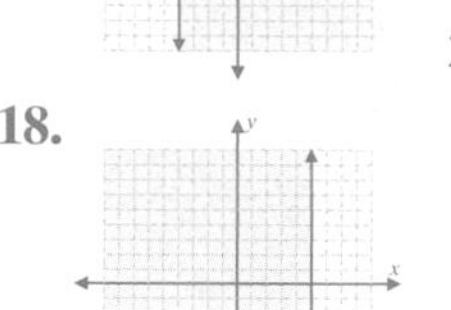

19.

22.

23. 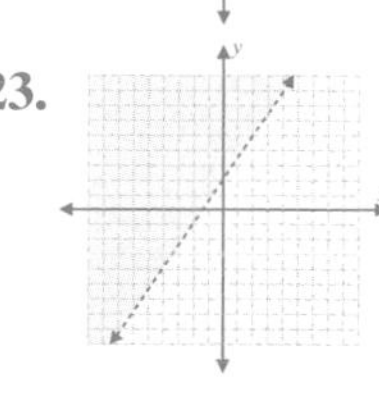

b. Graph the linear inequality $-5x + 4y > -8$.

Solution:

Step 1: Solving the inequality for y gives: $y > \frac{5}{4}x - 2$.

Step 2: Press the key Y= and enter the function: $\backslash Y_1 = (5/4)X - 2$.

Step 3: Go to the \ and hit ENTER two times so that the display appears as follows:

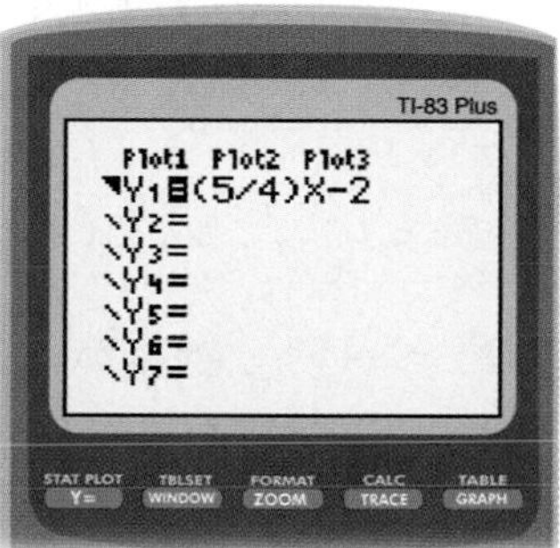

Step 4: Press GRAPH and (assuming that the WINDOW settings are the same) the following graph should appear on the display.

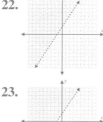

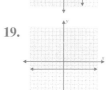

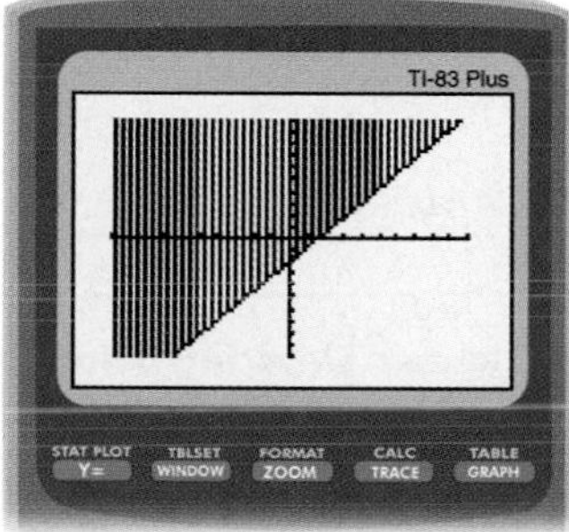

Practice Problems

20.

21.

1. *Which of the following points satisfy the inequality $x + y < 3$?*

   a. $(2, 1)$ **b.** $\left(\frac{1}{2}, 3\right)$ **c.** $(0, 5)$ **d.** $(-5, 2)$

2. *Which of the following points satisfy the inequality $x - 2y \geq 0$?*

   a. $(2, 1)$ **b.** $(1, 3)$ **c.** $(4, 2)$ **d.** $(3, 1)$

3. *Which of the following points satisfy the inequality $x < 3$?*

   a. $(1, 0)$ **b.** $(0, 1)$ **c.** $(4, -1)$ **d.** $(2, 3)$

Answers to Practice Problems: **1.** d. **2.** a., c., d. **3.** a., b., d.

24. 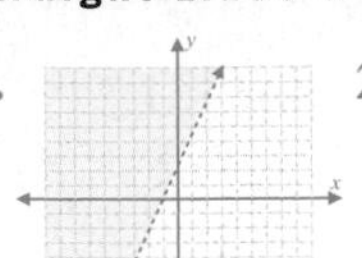
25.
26. 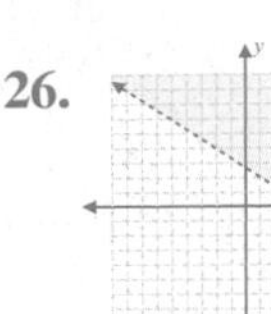
27. 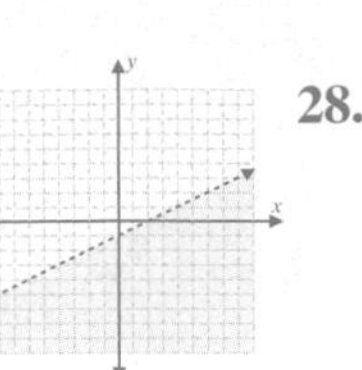
28.

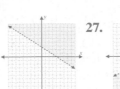

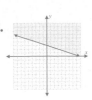

4.6 Exercises

29. 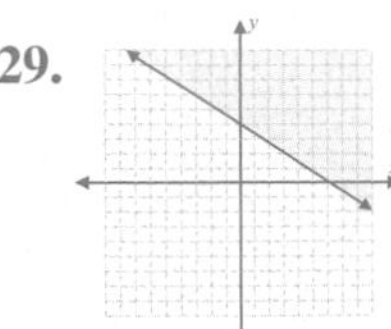

30.

31.

32.

33.

34.

Graph the solution set of each of the linear inequalities in Exercises 1 – 30.

1. $x + y \leq 7$ **2.** $x - y > -2$ **3.** $x - y > 4$ **4.** $x + y \leq 6$

5. $y < 4x$ **6.** $y < -2x$ **7.** $y \geq -3x$ **8.** $y > x$

9. $x - 2y > 5$ **10.** $x + 3y \leq 7$ **11.** $4x + y \geq 3$ **12.** $5x - y < 4$

13. $y \leq 5 - 3x$ **14.** $y \geq 8 - 2x$ **15.** $2y - x \leq 0$ **16.** $x + y > 0$

17. $x + 4 \geq 0$ **18.** $x - 5 \leq 0$ **19.** $y \geq -2$ **20.** $y + 3 < 0$

21. $4x + 3y < 8$ **22.** $3x < 2y - 4$ **23.** $3y > 4x + 6$ **24.** $5x < 2y - 5$

25. $x + 3y < 7$ **26.** $3x + 4y > 11$ **27.** $\frac{1}{2}x - y > 1$ **28.** $\frac{1}{3}x + y \geq 3$

29. $\frac{2}{3}x + y \geq 4$ **30.** $2x - \frac{4}{3}y > 8$

Use your graphing calculator to graph each of the linear inequalities in Exercises 31 – 40.

31. $y > \frac{1}{2}x$ **32.** $x - y \leq 5$ **33.** $x + 2y > 8$

34. $3x + 2y \geq 12$ **35.** $2x + y \leq 6$ **36.** $y \geq -3$

37. $x - 3y \geq 9$ **38.** $y \leq -4$ **39.** $2x + 6y \geq 0$

40. $3x - 4y > 15$

Writing and Thinking About Mathematics

41. Explain in your own words how to test to determine which side of the graph of an inequality should be shaded.

42. Describe the difference between a closed and open half-plane.

Hawkes Learning Systems: Introductory & Intermediate Algebra

Graphing Linear Inequalities

36.

37. 

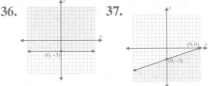

38. 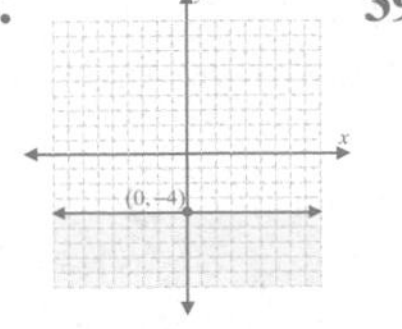

39. 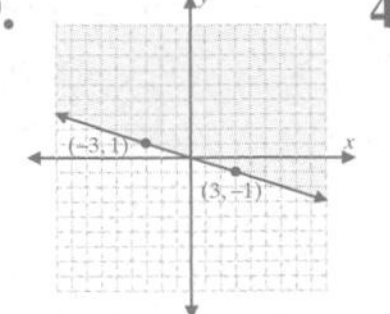

40. 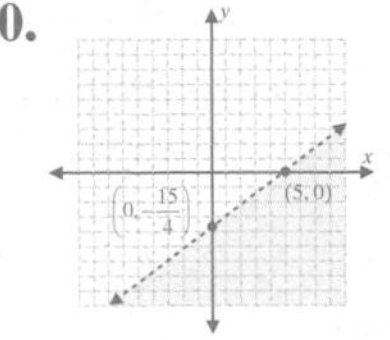

35. 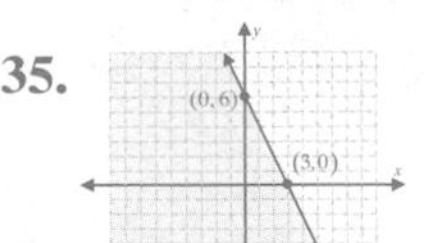

41. Answers will vary.

42. A closed interval includes points on the line and is symbolized by a solid boundary line. An open interval does not include points on the line and is symbolized by a dashed boundary line.

Chapter 4 Index of Key Ideas and Terms

Section 4.1 The Cartesian Coordinate System

Cartesian Coordinate System page 234

Ordered Pairs page 234

First coordinate (or first component)

Second coordinate (or second component)

Quadrants page 236

Origin page 236

Axes page 236

Horizontal Axis (or x-axis) page 236

Vertical Axis (or y-axis) page 236

One-To-One Correspondence page 236

There is a **one-to-one correspondence** between points in a plane and ordered pairs of real numbers.

Graphs of Points page 237

Section 4.2 Graphing Linear Equations in Two Variables

Solution (or Solution Set) of an Equation page 248

The **solution** (or **solution set**) of an equation in two variables, x and y, consists of all those ordered pairs of real numbers (x, y) that satisfy the equation.

Linear Equation page 249

Any equation of the form $Ax + By = C$, where A and B are not both 0, is called the **standard form** of a **linear equation**.

continued on next page ...

Section 4.2 Graphing Linear Equations in Two Variables (continued)

To Graph a Linear Equation page 249

1. Locate any two points that satisfy the equation. (Choose values for x and y that lead to simple solutions. Remember that there are an infinite number of choices for either x or y. But, once a value for x or y is chosen, the corresponding value for the other variable is found by substituting into the equation.)
2. Plot these two points on a Cartesian coordinate system.
3. Draw a straight line through these two points. (**Note:** Every point on that line will satisfy the equation.)
4. To check: Locate a third point that satisfies the equation and check to see that it does indeed lie on the line.

y-intercept page 251

The **y-intercept** is the point where the graph of a line crosses the y-axis. The x-coordinate will be 0.

x-intercept page 251

The **x-intercept** is the point where the graph of a line crosses the x-axis. The y-coordinate will be 0.

Horizontal and Vertical Lines pages 252 - 255

1. The graph of any equation of the form $y = b$ is a horizontal line.
2. The graph of any equation of the form $x = a$ is a vertical line.

Section 4.3 The Slope-Intercept Form: $y = mx + b$

Slope (of a line) pages 261 - 262

$$\text{slope} = m = \frac{rise}{run} = \frac{y_2 - y_1}{x_2 - x_1} = \frac{y_1 - y_2}{x_1 - x_2}$$

Slope-Intercept Form page 264

Any equation of the form $y = mx + b$ is called the **slope-intercept form** of the equation of a line. The slope is m and the y-intercept is b.

Horizontal Lines pages 266 - 268

Any equation of the form $y = b$ represents a **horizontal line** with **slope 0**.

continued on next page ...

Section 4.3 The Slope-Intercept Form: $y = mx + b$ (continued)

Vertical Lines pages 267 - 268

Any equation of the form $x = a$ represents a **vertical line** with **undefined slope**.

Section 4.4 The Point-Slope Form: $y - y_1 = m(x - x_1)$

Point-Slope Form page 276

An equation of the form $y - y_1 = m(x - x_1)$ is called the **point-slope form** for the equation of a line. The line contains the point (x_1, y_1) and has slope m.

Parallel Lines page 278

Two lines (neither vertical) are **parallel** if and only if they have the same slope. All vertical lines are parallel.

Perpendicular Lines page 279

Two lines (neither vertical) are **perpendicular** if and only if their slopes are negative reciprocals of each other:

$m_2 = -\frac{1}{m_1}$ *or* $m_1 \cdot m_2 = -1$.

Vertical lines are perpendicular to horizontal lines

Slope as a Rate of Change page 282

Section 4.5 Introduction to Functions and Function Notation

Relations page 288

A **relation** is a set of ordered pairs of real numbers. The **domain**, ***D***, of a relation is the set of all first coordinates in the relation. The **range**, ***R***, of a relation is the set of all second coordinates in the relation.

Functions page 289

A **function** is a relation in which each domain element has exactly one corresponding range element.

Functions have the following two characteristics:

1. A **function** is a relation in which each first coordinate appears only once.
2. A **function** is a relation in which no two ordered pairs have the same first coordinate.

continued on next page ...

Section 4.5 Introduction to Functions and Function Notation (continued)

Vertical Line Test page 290

If **any** vertical line intersects the graph of a relation at more than one point, then the relation graphed is **not** a function.

Function Notation: $f(x)$ page 295

The notation $f(x)$ is read "f of x."
The notation $f(x)$ represents the y-value for some corresponding x-value.

Using a Calculator to Graph Linear Functions pages 297 - 302

Section 4.6 Graphing Linear Inequalities in Two Variables

Two Methods for Graphing Linear Inequalities (without a calculator) pages 308 - 311

Using a Calculator to Graph Linear Inequalities pages 311 - 313

Chapter 4 Review

For a review of the topics and problems from Chapter 4, look at the following lessons from *Hawkes Learning Systems: Introductory & Intermediate Algebra.*

Introduction to the Cartesian Coordinate System
Graphing Linear Equations by Plotting Points
Graphing Linear Equations in Slope-Intercept Form
Graphing Linear Equations in Point-Slope Form
Finding the Equation of a Line
Introduction to Functions and Function Notation
Graphing Linear Inequalities

Chapter 4 Test

1. Which of the following points lie on the line determined by the equation $x - 4y = 7$?
   a. $(2, -2)$
   b. $(-1, -2)$
   c. $\left(5, -\frac{1}{2}\right)$
   d. $(0, 7)$

2. List the ordered pairs corresponding to the points on the graph.

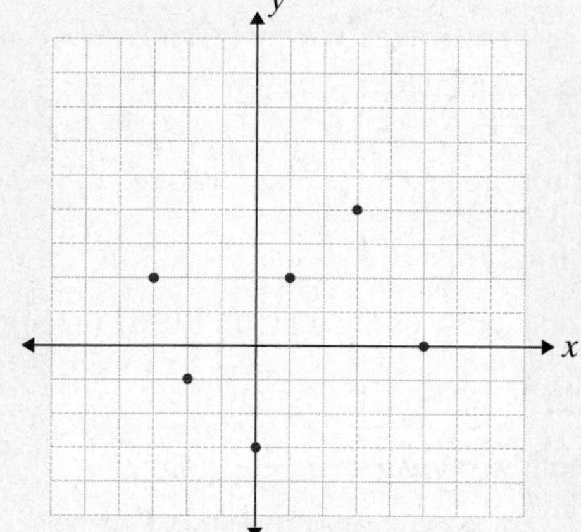

3. Graph the following set of ordered pairs:
$$\left\{(0, 2), (4, -1), (-3, 2), (-1, -5), \left(2, \frac{3}{2}\right)\right\}.$$

4. Graph the line $2x - y = 4$.

5. Graph the equation $3x + 4y = 9$ by locating and labeling the x-intercept and y-intercept.

6. Graph the line passing through the point $(1, -5)$ with the slope $m = \frac{5}{2}$.

7. For the line $3x + 5y = -15$, determine the slope, m, and the y-intercept, b. Then graph the line.

8. What is the slope of the line that contains the two points $(4, -3)$ and $\left(2, \frac{1}{2}\right)$?

9. Write an equation for the line passing through the point $(-3, 2)$ with the slope $m = \frac{1}{4}$.

Answers

1. b and c
2. $(-3, 2), (-2, -1), (0, -3), (1, 2), (3, 4), (5, 0)$
3.
4.

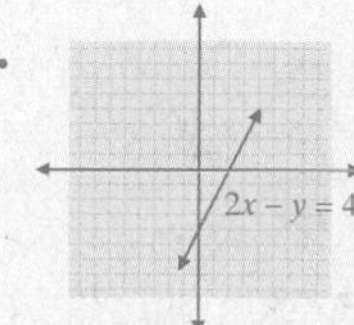

5. y-intercept $= \frac{9}{4}$
x-intercept $= 3$

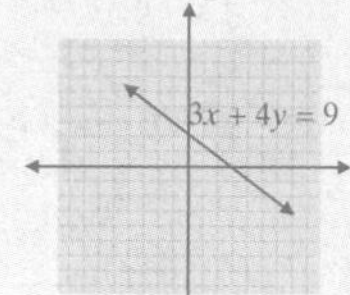

6.

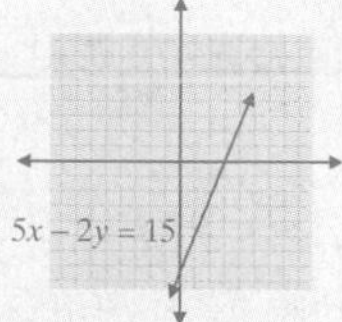

7. $m = -\frac{3}{5}, b = -3$

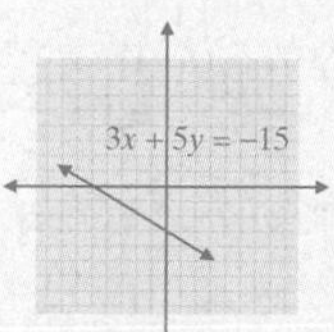

8. $m = -\frac{7}{4}$
9. $y - 2 = \frac{1}{4}(x + 3)$
or $y = \frac{1}{4}x + \frac{11}{4}$

10. $y = -x + 4$; $m = -1$
11. $y = -2, m = 0$
12. a. $C = 4t + 9$
b. Because t represents the number of hours worked and one cannot work a negative number of hours.
c. For every hour a person rents the carpet cleaning machine he will be charged \$4.
13. These lines are parallel since they have the same slope, but different y-intercepts.
14. $y = 2x - 8$
15. $D = \{-1, 0, 2, 5\}$
$R = \{-3, 4, 6\}$
16. It is not a function because the domain element, 2, has more than one range element, –5 and 0.
17. a. –7 **b.** 14
18. a. 22 **b.** 7
19. It is a function.
$D = \{-6, -4, -3, 3, 7\}$
$R = \{-4, -3, 0, 1, 3\}$
20. It is a function
$D: -2 \le x \le 2$
$R: 0 \le y \le 4$

10. Write an equation for the line passing through the points (3, 1) and (–2, 6). What is the slope of this line?

11. Write an equation for the line parallel to the x-axis through the point (3, –2). What is the slope of this line?

12. A carpet cleaning machine can be rented for a fixed cost of \$9.00 plus \$4.00 per hour.

a. Write a linear equation representing the cost, C, in terms of hours used, t.
b. Explain why this equation only makes sense if t is positve.
c. Explain the meaning of the slope of this line.

13. Determine whether the two lines $2x + 3y = 6$ and $y = -\frac{2}{3}x + 1$ are or are not parallel. Explain.

14. Write an equation for the line that passes through the point (1, –6) and is parallel to the line $y - 2x = 4$.

15. State the domain and range of the relation $\{(0, 6), (-1, 4), (2, -3), (5, 6)\}$.

16. Given the relation $r = \{(2, -5), (3, -4), (2, 0), (1, -5)\}$, determine whether or not this relation is a function. Explain.

17. For the function $f(x) = 3x - 4$, find
a. $f(-1)$ **b.** $f(6)$

18. For the function $h(x) = 3x^2 - 5$, find
a. $h(-3)$ **b.** $h(2)$

Use the vertical line test to determine whether each graph in Exercises 19 – 22 does or does not represent a function. If the graph represents a function, state its domain and range.

19.

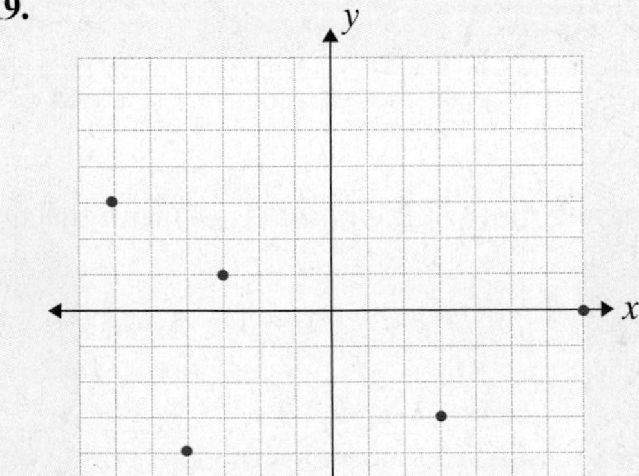

20.

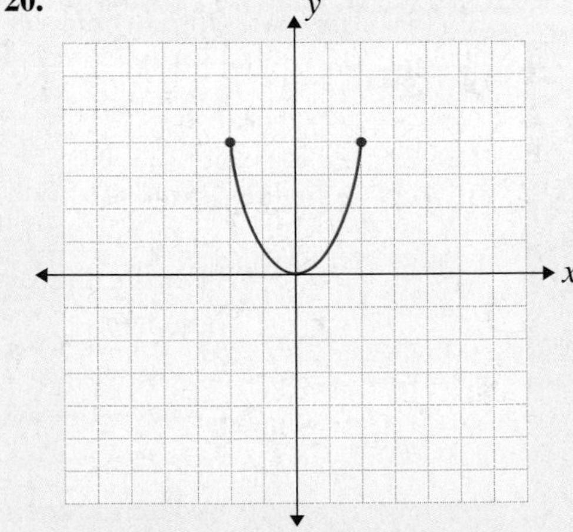

21. Not a function

22. It is a function
D: All real numbers
$R: y \geq 0$

23.

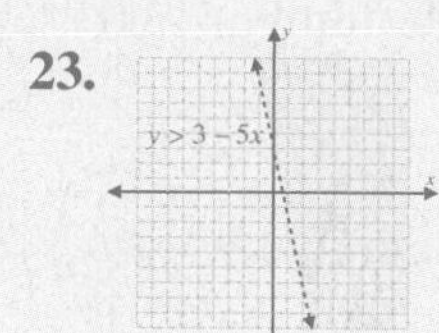

24.

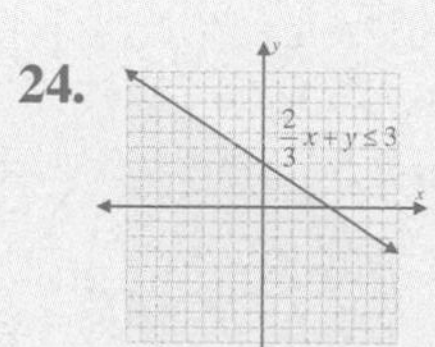

25.

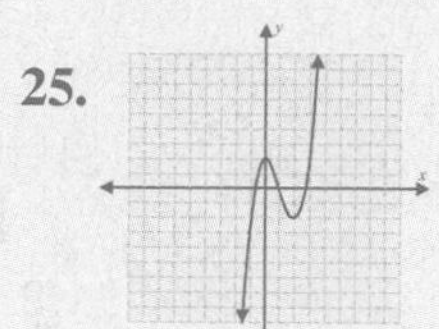

21.

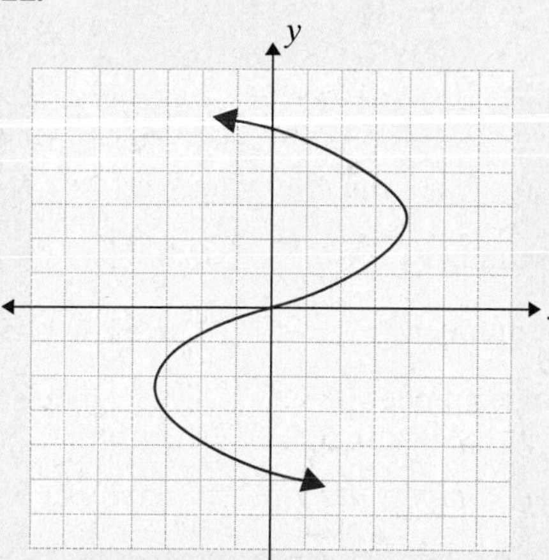

22.

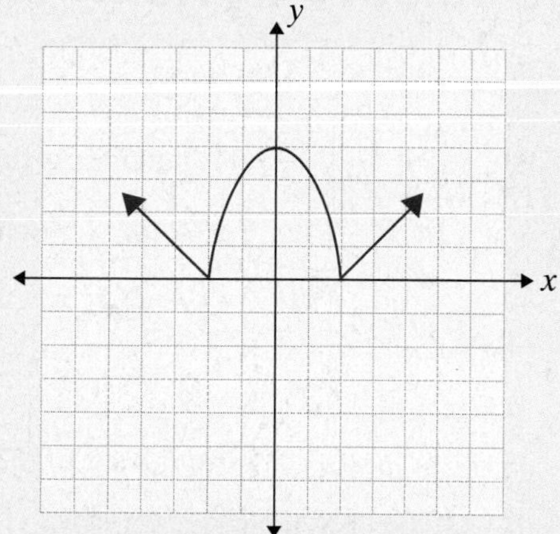

In Exercises 23 – 24, graph the half-plane that satisfies the inequality.

23. $y > 3 - 5x$

24. $\frac{2}{3}x + y \leq 3$

25. Use your graphing calculator to graph the function $y = x^3 - 3x^2 + 2$. Sketch the graph on your paper.

Cumulative Review: Chapters 1 – 4

1. a. $\{\sqrt{9}\}$ b. $\{0, \sqrt{9}\}$
c. $\left\{-10,-\sqrt{25},-1.6,0,\frac{1}{5},\sqrt{9}\right\}$
d. $\{-10,-\sqrt{25},0,\sqrt{9}\}$
e. $\{-\sqrt{7},\pi,\sqrt{12}\}$
f. $\left\{-10,-\sqrt{25},-1.6,-\sqrt{7},0,\frac{1}{5},\sqrt{9},\pi,\sqrt{12}\right\}$

2. 

3. 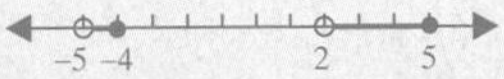

4. Associative property of addition
5. Distributive property
6. Inverse property of addition
7. Identity property of multiplication
8. Transitive property of order
9. Trichotomy property of order

10. −20 11. 11
12. 22 13. −1
14. −0.8 15. $-\frac{1}{12}$
16. 84 17. −40
18. −2 19. $\frac{2}{3}$
20. Undefined
21. 0 22. −2
23. −35 24. 26
25. −59 26. $-2x-12$
27. 0
28. $3x^3-x^2+4x-1$
29. $-10x+4$

1. Given the set of numbers $\left\{-10,-\sqrt{25},-1.6,-\sqrt{7},0,\frac{1}{5},\sqrt{9},\pi,\sqrt{12}\right\}$, list those numbers that belong to each of the following sets:

a. $\{x \mid x \text{ is a natural number}\}$ **b.** $\{x \mid x \text{ is a whole number}\}$

c. $\{x \mid x \text{ is a rational number}\}$ **d.** $\{x \mid x \text{ is an integer}\}$

e. $\{x \mid x \text{ is an irrational number}\}$ **f.** $\{x \mid x \text{ is a real number}\}$

Graph each of the sets of real numbers described in Exercises 2 and 3 on a real number line.

2. $\{x \mid -1.8 < x < 5 \text{ and } x \leq 3\}$ **3.** $\{x \mid -5 < x \leq -4 \text{ or } 2 < x \leq 5\}$

Name the property of real numbers that justifies each statement in Exercises 4 – 9. All variables represent real numbers.

4. $7+(x+3)=(7+x)+3$ **5.** $3(y+6)=3y+18$

6. $\frac{4}{5}+\left(-\frac{4}{5}\right)=0$ **7.** $x \cdot 1 = x$

8. If $x<10$ and $10<y$, then $x<y$. **9.** Either $x<-9$, $x=-9$, or $x>-9$.

Perform the indicated operations in Exercises 10 – 21.

10. $(-13)+(-7)$ **11.** $|-9|+2$ **12.** $17-(-5)$ **13.** $|9|-|-10|$

14. $6.5+(-4.2)-3.1$ **15.** $\frac{3}{4}-\frac{2}{3}+\left(-\frac{1}{6}\right)$ **16.** $(-7)\cdot(-12)$

17. $8(-5)$ **18.** $22 \div (-11)$ **19.** $(-4)\div(-6)$ **20.** $8 \div 0$

21. $0 \div \frac{3}{5}$

Find the value of each expression in Exercises 22 – 25 by using the rules for order of operations.

22. $12-6\div 2\cdot 3-5$ **23.** $5-(13\cdot 5-5)\div 3\cdot 2$

24. $3^2+5\cdot 4-10+|7|$ **25.** $6\cdot 4-2^3-(5\cdot 10)-5^2$

Simplify each expression in Exercises 26 – 29 by combining like terms.

26. $-4(x+3)+2x$ **27.** $x+\frac{x-5x}{4}$

28. $(x^3+4x-1)-(-2x^3+x^2)$ **29.** $-2[7x-(2x+5)+3]$

30. $x = 2$
31. $x = -4$
32. $x = -4$
33. $x = -8$
34. $x = 2.3$ or $x = -3.3$
35. $x = \frac{10}{3}$ or $x = 2$
36. $x = 62$
37. $x = 21$
38. $A = 2$
39. a. $n = 2A - m$
b. $w = \frac{P - 2l}{2}$
40. $-\frac{1}{3}$
41. 7 hrs
42. 10 miles
43. 81
44. 84
45. a. 5000 per hour
b. 600,000 per week
46. 4
$(4, \infty)$
47. −12.8
$[-12.8, \infty)$
48. 2
$(-\infty, 2]$
49. −5
$(-\infty, -5]$
50. No solution
51. $x = -\frac{8}{3}, \frac{4}{3}$

Solve each of the equations in Exercises 30 – 38.

30. $9x - 11 = x + 5$

31. $5(1 - 2x) = 3x + 57$

32. $5(2x + 3) = 3(x - 4) - 1$

33. $\frac{7x}{8} + 5 = \frac{x}{4}$

34. $|2x + 1| = 5.6$

35. $|2(x - 4) + x| = 2$

36. $\frac{5}{8} = \frac{35}{x - 6}$

37. $\frac{2}{x - 7} = \frac{3}{x}$

38. $\frac{A}{2.8} = \frac{3}{4.2}$

39. Solve each equation for the indicated variable.

a. Solve for n: $A = \frac{m + n}{2}$

b. Solve for w: $P = 2l + 2w$

40. The difference between twice a number and 3 is equal to the difference between five times the number and 2. Find the number.

41. The local supermarket charges a flat rate of $5, plus $3 per hour for rental of a carpet cleaner. If it cost Ron $26 to rent the machine, how many hours did he keep it?

42. Stephanie rode her new moped to Rod's house. Traveling the side streets, she averaged 20 mph. To save time on the return trip, they loaded the bike into Rod's truck and took the freeway, averaging 50 mph. The freeway distance is 2 miles less than the distance on the side streets and saves 24 minutes. Find the distance traveled on the return trip.

43. On a placement exam for English, a group of ten students had the following scores: 3 students scored 76, 2 students scored 79, 1 student scored 81, 3 students scored 85 and 1 student scored 88. What was the mean score for this group of students?

44. The mathematics component of the entrance exam at a certain Midwestern college consists of three parts: one part on geometry, one part on algebra, and one part on trigonometry. Prospective students must score at least 50 on each part and average at least 70 on the three parts. Beth learned that she had scored 60 and 66 on the first two parts of the exam. What minimum score did she need on the third part to pass this portion of the exam and gain entrance to the college?

45. You are in charge of setting a machine that produces hairpins. The machine is to run 24 hours per day, 5 days per week.

a. What setting of hairpins per hour should you set the machine to if it should produce 120,000 hairpins per day?

b. How many hairpins will the machine produce per 5-day week?

52. 1.05×10^{-2}

53. 9.0×10^{-4}

54. a. $(0,-4)$ **b.** $(2,0)$ **c.** $(1,-2)$ **d.** $(3,2)$

55. a. $(0,2)$ **b.** $(6,0)$ **c.** $\left(2,\frac{4}{3}\right)$ **d.** $(9,-1)$

56. $y=-\frac{1}{5}x+2;$ $m=-\frac{1}{5},\ b=2$

57. $y=-3x+1;$ $m=-3,\ b=1$

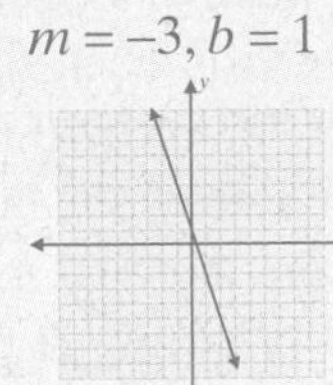

58. $y=\frac{3}{7}x-1;$ $m=\frac{3}{7},\ b=-1$

59. $y=-4x;$ $m=-4,\ b=0$

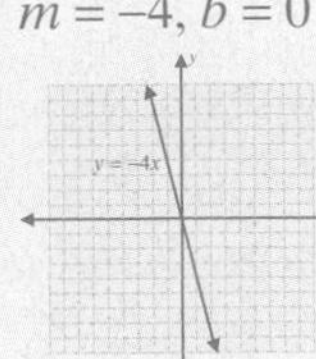

Solve the inequalities in Exercises 46 – 51 and graph the solutions. Write the solution set in interval notation. Assume that x is a real number.

46. $5x-7>x+9$ **47.** $5x+10\le6(x+3.8)$ **48.** $x+8-5x\ge2(x-2)$

49. $\frac{2x+1}{3}\le\frac{3x}{5}$

In Exercises 50 and 51, solve the equation.

50. $|5x+2|=-7$ **51.** $|3x+2|+4=10$

In Exercises 52 and 53, write each number in scientific notation and simplify.

52. $(2100)(0.000005)$ **53.** $\frac{(270,000)(0.00014)}{42,000}$

In Exercises 54 and 55, find the missing coordinate of each ordered pair so that the ordered pair belongs to the solution set of the given equation.

54. $2x-y=4$
a. (0,)
b. (, 0)
c. (1,)
d. (, 2)

55. $x+3y=6$
a. (0,)
b. (, 0)
c. (2,)
d. (, –1)

Write each equation in Exercises 56 – 61 in slope-intercept form. Find the slope and the y-intercept and draw the graph.

56. $x+5y=10$ **57.** $3x+y=1$ **58.** $3x-7y=7$

59. $y=-4x$ **60.** $x+2y=4$ **61.** $x+4=0$

Find the equation in standard form for the line determined by the given point and slope or two points in Exercises 62 – 67.

62. $(6,-1),\ m=\frac{2}{5}$ **63.** $(-1,2),\ m=-\frac{4}{3}$ **64.** $(0,0),\ m=2$

65. $(5,2)$, undefined slope **66.** $(0,3),(5,-1)$ **67.** $(5,-2),(1,6)$

Find the equation in standard form for the line satisfying the conditions in Exercises 68 – 71.

68. Parallel to $3x+2y-6=0$, passing through $(2,3)$
69. Parallel to the y-axis, passing through $(1,-7)$
70. Perpendicular to $4x+3y=5$, passing through $(4,0)$
71. Perpendicular to $3x-5y=1$, passing through $(6,-2)$

72. Write an equation for the line parallel to $x-2y=5$ having the same y-intercept as $5x+3y=9$.

73. For the function $f(x) = -3x + 4$, find **a.** $f(5)$ **b.** $f(-6)$

74. For the function $F(x) = 2x^2 - 5x + 1$, find **a.** $F(2)$ **b.** $F(-5)$

Graph the linear inequalities in Exercises 75 and 76.

75. $y \geq 4x$

76. $3x + y < 2$

77. Use the vertical line test to determine whether or not the given graphs represent a function.

a.

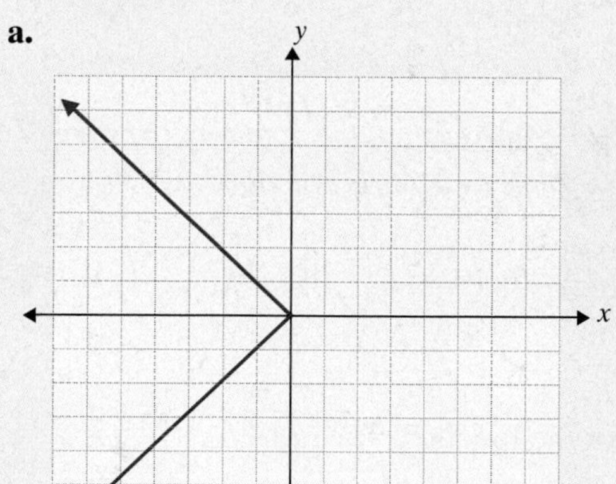

b.

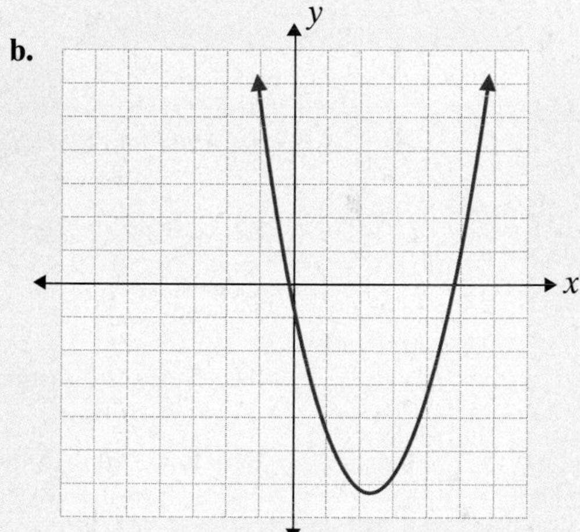

For each relation represented in Exercises 78 – 81, state the domain and range and state whether or not the relation represents a function.

78.

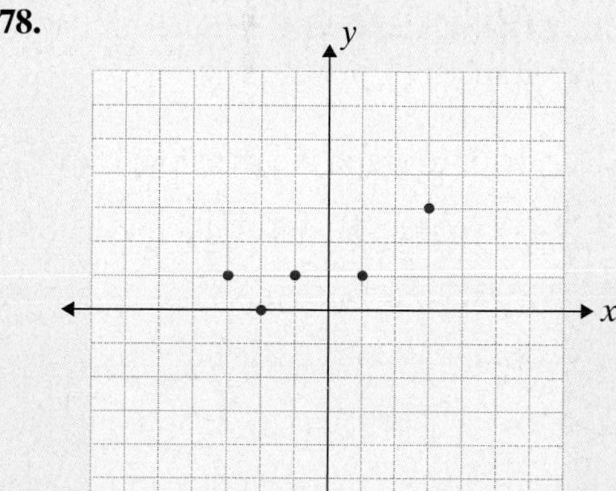

79.

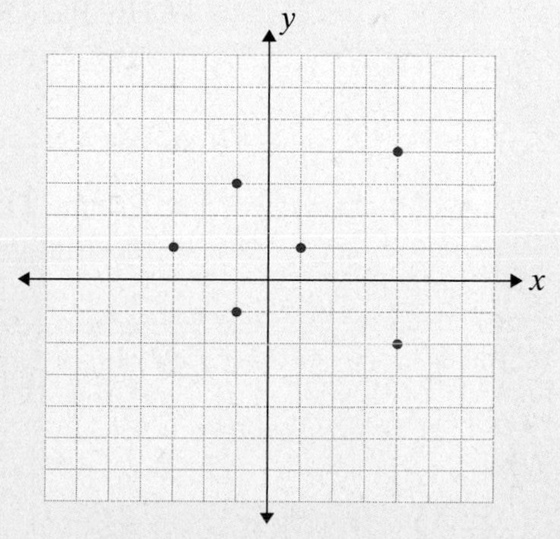

60. $y = -\frac{1}{2}x + 2$; $m = -\frac{1}{2}$, $b = 2$

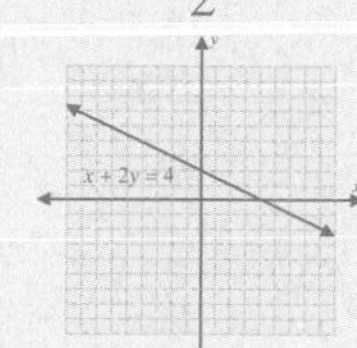

61. $x = -4$; m = undefined, no y-intercept

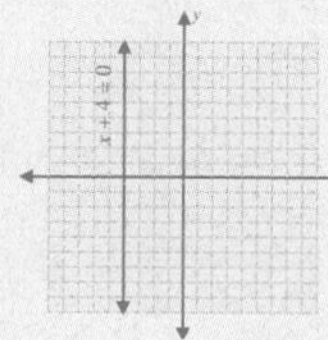

62. $2x - 5y = 17$
63. $4x - 3y = -10$
64. $2x - y = 0$
65. $x = 5$
66. $4x + 5y = 15$
67. $2x + y = 8$
68. $3x + 2y = 12$
69. $x = 1$
70. $3x - 4y = 12$
71. $5x + 3y = 24$
72. $x - 2y = -6$
73. a. -11 **b.** 22
74. a. -1 **b.** 76
75.

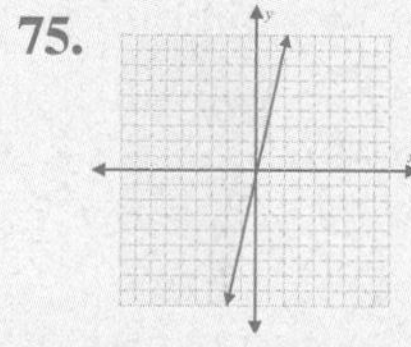

76.

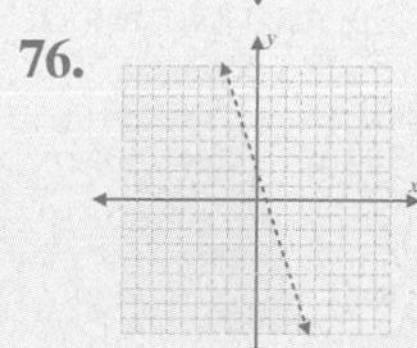

77. a. not a function **b.** it is a function
78. $D = \{-3, -2, -1, 1, 3\}$; $R = \{0, 1, 3\}$; It is a function
79. $D = \{-3, -1, 1, 4\}$; $R = \{-2, -1, 1, 3, 4\}$; It is not a function

80. $D = \{-2, 0, 2, 3, 4\}$; $R = \{-1, 0, 1, 2, 3, 4\}$; It is not a function

81. $D: -5 < x \leq -1$ and $0 < x \leq 6$, $R: \{-4, -2, 2, 4, 5\}$ It is a function.

82.

a.

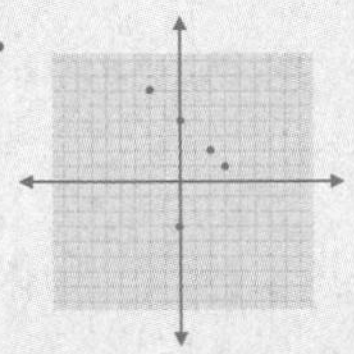

b. $D: \{-2, 0, 2, 3\}$; $R: \{-3, 1, 2, 4, 6\}$

c. It is not a function

83.

a.

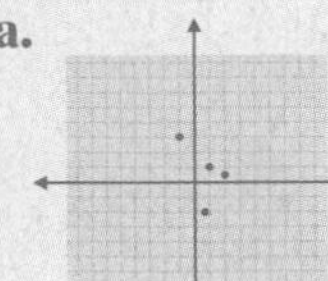

b. $D: \{-1, 1, \frac{2}{3}, 2\}$; $R: \{-2, 1, \frac{1}{2}, 3\}$

c. It is a function

84. $D = \{x \mid x \geq -2\}$

85. a.

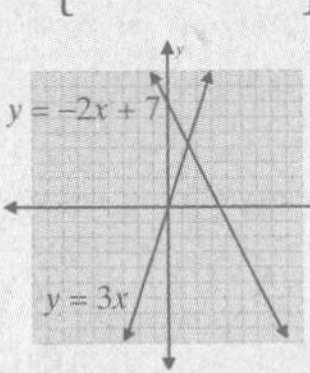

b. $\left(\frac{7}{5}, \frac{21}{5}\right)$

86. a.

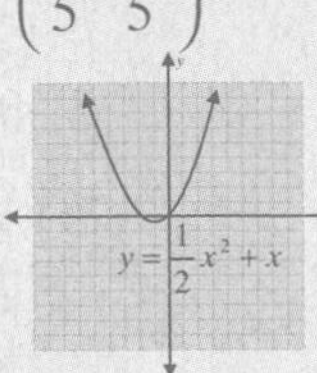

b. $\left(-1, -\frac{1}{2}\right)$

87. a.

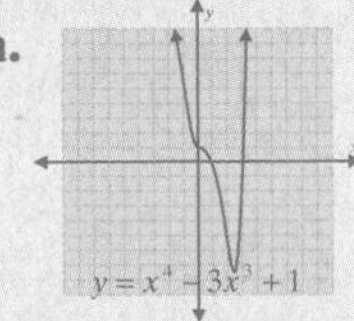

b. $(0.7648, 0)$, $(2.9615, 0)$

80.

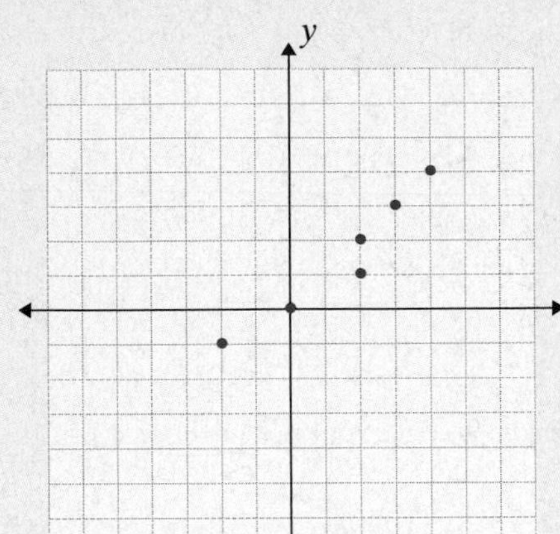

81.

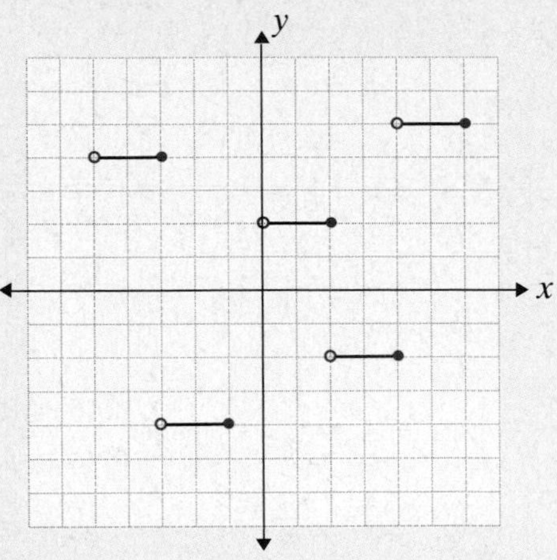

In Exercises 82 and 83, (a) graph each of the relations, (b) state the domain and range, and (c) state whether the relation is or is not a function.

82. $\{(0, 4), (2, 2), (3, 1), (-2, 6), (0, -3)\}$ **83.** $\left\{\left(2, \frac{1}{2}\right), (-1, 3), \left(\frac{2}{3}, -2\right), (1, 1)\right\}$

84. State the domain of the function $y = \sqrt{3x + 6}$.

85. **a.** Use a graphing calculator to graph the linear functions $y = -2x + 7$ and $y = 3x$, and then sketch the graphs on graph paper.

b. Use the **TRACE**, **ZOOM**, and **CALC** features of the calculator to estimate the point of intersection of these two linear functions.

86. **a.** Use a graphing calculator to graph the function $y = \frac{1}{2}x^2 + x$, and then sketch the graph on graph paper.

b. Use the **TRACE**, **ZOOM**, and **CALC** features of the calculator to estimate the lowest point on the graph.

87. **a.** Use a graphing calculator to graph the function $y = x^4 - 3x^3 + 1$.

b. Use the **TRACE** and **ZOOM** features of the calculator to estimate the values of the x-intercepts.

Exponents and Polynomials

CHAPTER 5

Did You Know?

One of the most difficult problems for students in beginning algebra is to become comfortable with the idea that letters or symbols can be manipulated just like numbers in arithmetic. These symbols may be the cause of "math anxiety." A great deal of publicity has recently been given to the concept that a large number of people suffer from math anxiety, a painful uneasiness caused by mathematical symbols or problem-solving situations. Persons affected by math anxiety find it almost impossible to learn mathematics, or they may be able to learn but be unable to apply their knowledge or do well on tests. Persons suffering from math anxiety often develop math avoidance, so they avoid careers, majors, or classes that will require mathematics courses or skills. The sociologist Lucy Sells has determined that mathematics is a critical filter in the job market. Persons who lack quantitative skills are channeled into high unemployment, low-paying, non-technical areas.

What causes math anxiety? Researchers are investigating the following hypotheses:

1. A lack of skills that leads to a lack of confidence and, therefore, to anxiety;
2. An attitude that mathematics is not useful or significant to society;
3. Career goals that seem to preclude mathematics;
4. A self-concept that differs radically from the stereotype of a mathematician;
5. Perceptions that parents, peers, or teachers have low expectations for the person in mathematics;
6. Social conditioning to avoid mathematics.

We hope that you are finding your present experience with algebra successful and that the skills you are acquiring now will enable you to approach mathematical problems with confidence.

"Mathematics is the queen of the sciences and arithmetic the queen of mathematics."

Karl F. Gauss (1777 – 1855)

In Chapter 5, you will learn the rules of exponents and how exponents can be used to simplify very large and very small numbers. Astronomers and chemists are very familiar with these ideas and the notation used is appropriately called scientific notation. This notation is also used in scientific calculators. (Multiply 5,000,000 by 5,000,000 on your calculator and read the results.)

Polynomials and operations with polynomials (the topics in Chapters 5, 6, and 7) appear at almost every level of mathematics from elementary and intermediate algebra through statistics, calculus, and beyond. Be aware that the knowledge and skills you learn about polynomials will be needed again and again in any mathematics courses you take in the future.

5.1 Exponents

Objectives

After completing this section, you will be able to:

1. *Simplify expressions with constant or single-variable bases using the properties of integer exponents.*
2. *Recognize which property of exponents is used to simplify an expression.*

The Product Rule

In Section 1.5, an **exponent** was defined as a number that tells how many times a number (called the **base**) is used in multiplication. This definition is limited because it is valid only if the exponents are positive integers. In this section, we will develop four properties of exponents that will help in simplifying algebraic expressions and expand your understanding of exponents to include variable bases, negative exponents, and the exponent 0. (In Chapter 10 you will study fractional exponents.)

From Section 1.5, we know that

Exponent

$$6^2 = 6 \cdot 6 = 36$$

Base

and

$$6^3 = 6 \cdot 6 \cdot 6 = 216.$$

Also, the base may be a variable so that

$$x^3 = x \cdot x \cdot x \quad \text{and} \quad x^5 = x \cdot x \cdot x \cdot x \cdot x.$$

Teaching Notes:
Be sure to emphasize that the exponent counts the number of factors in these expressions. This is especially important when students are dealing with expressions such as $2^3 \cdot 2^4$, because there is a tendency for them to multiply the base numbers and erroneously write 4^7.

Now, to find the products of expressions such as $6^2 \cdot 6^3$ or $x^3 \cdot x^5$ and to simplify these products, we can write down all the factors as follows:

$$6^2 \cdot 6^3 = (6 \cdot 6) \cdot (6 \cdot 6 \cdot 6) = 6^5$$

and

$$x^3 \cdot x^5 = (x \cdot x \cdot x) \cdot (x \cdot x \cdot x \cdot x \cdot x) = x^8.$$

With these examples in mind, what do you think would be a simplified form for the product $3^4 \cdot 3^3$? You were right if you thought 3^7. That is, $3^4 \cdot 3^3 = 3^7$. Notice that in each case, the base stays the same.

The preceding discussion, along with the basic concept of whole-number exponents, leads to the following **Product Rule for Exponents**.

Product Rule for Exponents

If a is a nonzero real number and m and n are integers, then

$$\boldsymbol{a^m \cdot a^n = a^{m+n}}.$$

In words, to multiply two powers with the same base, keep the base and add the exponents.

NOTES

Remember (see Section 1.5) about the **exponent 1**. If a variable or constant has no exponent written, the exponent is understood to be 1.

For example,

$$y = y^1$$

and

$$7 = 7^1.$$

In general, for any real number a,

$$\boldsymbol{a = a^1}.$$

Two different expressions with negative signs and exponents, such as $(-2)^4$ and -2^4, need special clarification. In the expression $(-2)^4$, -2 is the base and 4 is the exponent. Thus, we can write

$$(-2)^4 = (-2)(-2)(-2)(-2) = 16.$$

In contrast, in the expression -2^4, only the 2 is the base of the exponent 4. Thus, we have a different result:

$$-2^4 = (-1)\cdot 2^4 = (-1)(2)(2)(2)(2) = -16$$

This type of expression may be easier to understand with a variable. For example, $-x^2$ is understood to have the coefficient -1. Therefore,

$$-x^2 = -1\cdot x^2$$ Here x is the base and -1 is the coefficient.

and $$(-x)^2 = (-x)(-x) = x^2.$$ Here $(-x)$ is the base.

Example 1: The Product Rule

Use the Product Rule for exponents to simplify the following expressions.

a. $x^2 \cdot x^4$

Solution: $x^2 \cdot x^4 = x^{2+4} = x^6$

b. $y \cdot y^6$

Solution: $y \cdot y^6 = y^1 \cdot y^6 = y^{1+6} = y^7$

c. $4^2 \cdot 4$

Solution: $4^2 \cdot 4 = 4^{2+1} = 4^3 = 64$ Note that the base stays 4. That is, the bases are not multiplied.

d. $2^3 \cdot 2^2$

Solution: $2^3 \cdot 2^2 = 2^{3+2} = 2^5 = 32$ Note that the base stays 2. That is, the bases are not multiplied.

e. $(-2)^4(-2)^3$

Solution: $(-2)^4(-2)^3 = (-2)^{4+3} = (-2)^7 = -128$

f. $2y^2 \cdot 3y^9$

Solution: $2y^2 \cdot 3y^9 = 2 \cdot 3 \cdot y^2 \cdot y^9 = 6y^{2+9} = 6y^{11}$ Note that the coefficients are multiplied and the exponents are added.

g. $(-6xy^2)(-8xy^3)$

Solution: $(-6xy^2)(-8xy^3) = (-6)(-8)x^{1+1}y^{2+3} = 48x^2y^5$

The Exponent 0

The Product Rule is stated for m and n as **integers**. This means that the rule is valid if an exponent is 0 or a negative integer. Therefore, we need to develop an understanding of the meaning of 0 as an exponent and negative integers as exponents.

Study the following patterns of numbers. What do you think are the missing values for $2^0, 3^0$, and 5^0?

$2^5 = 32$	$3^5 = 243$	$5^5 = 3125$
$2^4 = 16$	$3^4 = 81$	$5^4 = 625$
$2^3 = 8$	$3^3 = 27$	$5^3 = 125$
$2^2 = 4$	$3^2 = 9$	$5^2 = 25$
$2^1 = 2$	$3^1 = 3$	$5^1 = 5$
$2^0 = ?$	$3^0 = ?$	$5^0 = ?$

Notice that in the column of powers of 2, each number is $\frac{1}{2}$ of the preceding number. Since $\frac{1}{2}\cdot 2 = 1$, a reasonable guess is that $2^0 = 1$. Similarly, in the column of powers of 3, $\frac{1}{3}\cdot 3 = 1$, so $3^0 = 1$ seems reasonable. Also, $\frac{1}{5}\cdot 5 = 1$, so $5^0 = 1$ fits the pattern. In fact, the missing values are $2^0 = 1, 3^0 = 1$, and $5^0 = 1$.

Another approach to understanding 0 as an exponent involves the Product Rule. Consider the following:

$$2^0 \cdot 2^3 = 2^{0+3} = 2^3 \qquad \text{Using the Product Rule.}$$

and

$$1 \cdot 2^3 = 2^3.$$

So,

$$2^0 \cdot 2^3 = 1 \cdot 2^3 \text{ and } 2^0 = 1.$$

Similarly,

$$7^0 \cdot 7^2 = 7^{0+2} = 7^2$$

and

$$1 \cdot 7^2 = 7^2.$$

So,

$$7^0 \cdot 7^2 = 1 \cdot 7^2 \text{ and } 7^0 = 1.$$

This discussion leads directly to the **rule for 0 as an exponent**.

The Exponent 0

If a is a nonzero real number, then

$$a^0 = 1.$$

The expression 0^0 is undefined.

NOTES Throughout this text, unless specifically stated otherwise, we will assume that the bases of exponents are nonzero.

Example 2: The Exponent 0

Simplify the following expressions using the rule for 0 as an exponent.

a. 10^0

Solution: $10^0 = 1$

b. $x^0 \cdot x^3$

Solution: $x^0 \cdot x^3 = x^{0+3} = x^3$ or $x^0 \cdot x^3 = 1 \cdot x^3 = x^3$

c. $(-6)^0$

Solution: $(-6)^0 = 1$

The Quotient Rule

Teaching Notes: While we are emphasizing the Quotient Rule for Exponents here, you might want to discuss the fact that $\frac{5^4}{5^2} = \frac{625}{25} = 25$; and, as long as the base is numerical, we can always check this way.

Now consider a fraction in which the numerator and denominator are powers with the same base, such as $\frac{5^4}{5^2}$ or $\frac{x^5}{x^2}$. We can write,

$$\frac{5^4}{5^2} = \frac{\cancel{5} \cdot \cancel{5} \cdot 5 \cdot 5}{\cancel{5} \cdot \cancel{5} \cdot 1} = \frac{5^2}{1} = 25 \text{ or } \frac{5^4}{5^2} = 5^{4-2} = 5^2 = 25$$

and

$$\frac{x^5}{x^2} = \frac{\cancel{x} \cdot \cancel{x} \cdot x \cdot x \cdot x}{\cancel{x} \cdot \cancel{x} \cdot 1} = \frac{x^3}{1} = x^3 \text{ or } \frac{x^5}{x^2} = x^{5-2} = x^3.$$

In fractions, as just illustrated, the exponents can be subtracted. Again, the base remains the same. We now have the following **Quotient Rule for Exponents**.

Quotient Rule for Exponents

If a is a nonzero real number and m and n are integers, then

$$\frac{a^m}{a^n} = a^{m-n}.$$

In words, to divide two powers with the same base, keep the base and subtract the exponents. (Subtract the denominator exponent from the numerator exponent.)

Example 3: The Quotient Rule

a. 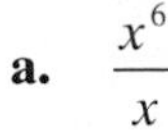 $\frac{x^6}{x}$

Solution: $\frac{x^6}{x} = x^{6-1} = x^5$

b. $\frac{y^8}{y^2}$

Solution: $\frac{y^8}{y^2} = y^{8-2} = y^6$

c. $\frac{x^2}{x^2}$

Solution: $\frac{x^2}{x^2} = x^{2-2} = x^0 = 1$

Note how this example shows another way to justify the idea that $a^0 = 1$. Since the numerator and denominator are the same and not 0, it makes sense that the fraction is equal to 1.

d. $\frac{15x^{15}}{3x^3}$

Solution: $\frac{15x^{15}}{3x^3} = 5x^{15-3} = 5x^{12}$

Note that the coefficients are divided and the exponents are subtracted.

e. $\frac{20x^{10}y^6}{4x^2y^3}$

Solution: $\frac{20x^{10}y^6}{4x^2y^3} = 5x^{10-2}y^{6-3} = 5x^8y^3$

Again, note that the coefficients are divided and the exponents are subtracted.

Negative Exponents

Now we will see that the Quotient Rule for exponents leads directly to an understanding of negative exponents. In the fractions in Example 3 (except for 3c), for each base the larger exponent was in the numerator. Therefore, when the exponents were subtracted (the exponent in the denominator was subtracted from the exponent in the numerator), the resulting expression had only positive exponents. But, if the larger exponent is in the denominator, the resulting expression will have a negative exponent. For example,

using the Quotient Rule: $\frac{3^3}{3^5} = 3^{3-5} = 3^{-2}$

and reducing: $\frac{3^3}{3^5} = \frac{\cancel{3} \cdot \cancel{3} \cdot \cancel{3} \cdot 1}{\cancel{3} \cdot \cancel{3} \cdot \cancel{3} \cdot 3 \cdot 3} = \frac{1}{3^2}$

Thus, $3^{-2} = \frac{1}{3^2}$, and the **Rule for Negative Exponents** follows.

Rule for Negative Exponents

If a is a nonzero real number and n is an integer, then

$$a^{-n} = \frac{1}{a^n}.$$

Remember that a negative exponent indicates a fraction, not a negative number.

Example 4: Negative Exponents

Use the Rule for Negative Exponents to simplify each expression so that it contains only positive exponents.

a. 5^{-1}

Solution: $5^{-1} = \frac{1}{5^1} = \frac{1}{5}$ Using the Rule for Negative Exponents.

b. x^{-3}

Solution: $x^{-3} = \frac{1}{x^3}$ Using the Rule for Negative Exponents.

c. $x^{-9} \cdot x^7$

Solution: Here we use the Product Rule first and then the Rule for Negative Exponents.

$$x^{-9} \cdot x^7 = x^{-9+7} = x^{-2} = \frac{1}{x^2}$$

Teaching Notes: Scientific notation (as discussed in Section 5.2) is an example of preference for the use of negative exponents.

Each of the expressions in Example 5 is simplified by using the appropriate rules for exponents. Study each example carefully. In each case, the expression is considered simplified if each base appears only once and each base has only positive exponents.

(**Note:** There is nothing wrong with negative exponents. In fact, negative exponents are preferred in later courses. However, so that all answers are the same, in this course we will consider expressions to be simplified if they have only positive exponents.)

Example 5: Combining the Rules for Exponents

a. 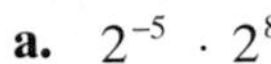 $2^{-5} \cdot 2^{8}$

Solution: $2^{-5} \cdot 2^{8} = 2^{-5+8} = 2^{3} = 8$ Using the Product Rule with positive and negative exponents.

b. $\dfrac{10^{-5}}{10^{-2}}$

Solution: $\dfrac{10^{-5}}{10^{-2}} = 10^{-5-(-2)} = 10^{-5+2}$ Using the Quotient Rule with positive and negative exponents.

$= 10^{-3} = \dfrac{1}{10^{3}}$ or $\dfrac{1}{1000}$ Using the Rule for Negative Exponents.

c. $\dfrac{15x^{10} \cdot 2x^{2}}{3x^{15}}$

Solution: $\dfrac{15x^{10} \cdot 2x^{2}}{3x^{15}} = \dfrac{30x^{10+2}}{3x^{15}}$ Using the Product Rule.

$= 10x^{12-15}$ Using the Quotient Rule.

$= 10x^{-3}$

$= \dfrac{10}{x^{3}}$ Using the Rule for Negative Exponents.

d. $\dfrac{x^{6}y^{3}}{x^{2}y^{5}}$

Solution: $\dfrac{x^{6}y^{3}}{x^{2}y^{5}} = x^{6-2}y^{3-5} = x^{4}y^{-2}$ Using the Quotient Rule with two variables.

$= \dfrac{x^{4}}{y^{2}}$ Using the Rule for Negative Exponents.

e. $\dfrac{24x^{6}}{16x^{-1}}$

Solution: $\dfrac{24x^{6}}{16x^{-1}} = \dfrac{(\cancel{8} \cdot 3)x^{6-(-1)}}{\cancel{8} \cdot 2} = \dfrac{3x^{6+1}}{2} = \dfrac{3x^{7}}{2}$ Using the Quotient Rule with positive and negative exponents.

Having negative exponents in the denominator can be confusing. Remember to subtract the exponent in the denominator even if it is negative.

f. $\dfrac{y^{-3}}{y^{-8}}$

Solution: $\dfrac{y^{-3}}{y^{-8}} = y^{-3-(-8)} = y^{-3+8} = y^{5}$

continued on next page ...

g. $\dfrac{a^{-10}}{a^{-6}}$

Solution: $\dfrac{a^{-10}}{a^{-6}} = a^{-10-(-6)} = a^{-10+6} = a^{-4} = \dfrac{1}{a^4}$

In Examples 5h and 5i, assume that k represents a nonzero integer. Follow the appropriate rules for exponents.

h. $a^k \cdot a$

Solution: $a^k \cdot a = a^k \cdot a^1 = a^{k+1}$

i. $\dfrac{x^{2k+2}}{x^k}$

Solution: $\dfrac{x^{2k+2}}{x^k} = x^{2k+2-k} = x^{k+2}$

NOTES

Special Note about Using the Quotient Rule:
Regardless of the size of the exponents or whether they are positive or negative, the following single subtraction rule can be used with the Quotient Rule.

(numerator exponent – denominator exponent)

This subtraction will always lead to the correct answer.

Summary of Properties of Exponents

If a and b are nonzero real numbers and m and n are integers,

1. The Exponent 1: $a = a^1$

2. The Exponent 0: $a^0 = 1 \quad (a \neq 0)$

3. Product Rule: $a^m \cdot a^n = a^{m+n}$

4. Quotient Rule: $\dfrac{a^m}{a^n} = a^{m-n}$

5. Negative Exponents: $a^{-n} = \dfrac{1}{a^n}, \quad \dfrac{1}{a^{-n}} = a^n$

Practice Problems

Simplify each expression.

1. $2^3 \cdot 2^4$ **2.** $\dfrac{2^3}{2^4}$ **3.** $\dfrac{x^7 \cdot x^{-3}}{x^{-2}}$

4. $\dfrac{10^{-8} \cdot 10^2}{10^{-7}}$ **5.** $\dfrac{14x^{-3}y^2}{2x^{-3}y^{-2}}$ **6.** $\left(9x^4\right)^0$

5.1 Exercises

Simplify each expression Exercises 1 – 50 and tell which rule (or rules) for exponents you used. The final form of the expressions with variables should contain only positive exponents. Assume that all variables represent nonzero numbers.

1. $3^2 \cdot 3$ **2.** $7^2 \cdot 7^3$ **3.** $8^3 \cdot 8^0$ **4.** 3^{-1}

5. 4^{-2} **6.** $(-5)^{-2}$ **7.** 6^{-3} **8.** $(-2)^4 \cdot (-2)^0$

9. $3 \cdot 2^3$ **10.** $(-4)^3 \cdot (-4)^0$ **11.** $6 \cdot 3^2$ **12.** $-4 \cdot 5^3$

13. $-2 \cdot 3^3$ **14.** $3 \cdot 2^{-3}$ **15.** $4 \cdot 3^{-2}$ **16.** $-3 \cdot 5^{-2}$

17. $-5 \cdot 2^{-2}$ **18.** $x^2 \cdot x^3$ **19.** $x^3 \cdot x$ **20.** $y^2 \cdot y^0$

21. $y^3 \cdot y^8$ **22.** x^{-3} **23.** y^{-2} **24.** $2x^{-1}$

25. $5y^{-4}$ **26.** $-8y^{-2}$ **27.** $-10x^{-3}$ **28.** $5x^6y^{-4}$

29. x^0y^{-2} **30.** $3x^0 + y^0$ **31.** $5y^0 - 3x^0$

32. $\dfrac{7^3}{7}$ **33.** $\dfrac{9^5}{9^2}$ **34.** $\dfrac{10^3}{10^4}$ **35.** $\dfrac{10}{10^5}$

36. $\dfrac{2^3}{2^6}$ **37.** $\dfrac{x^4}{x^2}$ **38.** $\dfrac{x^5}{x^3}$ **39.** $\dfrac{x^3}{x}$

40. $\dfrac{y^6}{y^4}$ **41.** $\dfrac{x^7}{x^3}$ **42.** $\dfrac{x^8}{x^3}$ **43.** $\dfrac{x^{-2}}{x^2}$

44. $\dfrac{x^{-3}}{x}$ **45.** $\dfrac{x^4}{x^{-2}}$ **46.** $\dfrac{x^5}{x^{-1}}$ **47.** $\dfrac{x^{-3}}{x^{-5}}$

48. $\dfrac{x^{-4}}{x^{-1}}$ **49.** $\dfrac{y^{-2}}{y^{-4}}$ **50.** $\dfrac{y^3}{y^{-3}}$

1. 27; product rule **2.** 16807; product rule **3.** 512; product rule, 0 exponent rule **4.** $\frac{1}{3}$; negative exponent rule **5.** $\frac{1}{16}$; negative exponent rule **6.** $\frac{1}{25}$; negative exponent rule **7.** $\frac{1}{216}$; negative exponent rule **8.** 16; product rule, 0 exponent rule **9.** 24 **10.** –64; product rule, 0 exponent rule **11.** 54 **12.** –500 **13.** –54 **14.** $\frac{3}{8}$; negative exponent rule **15.** $\frac{4}{9}$; negative exponent rule **16.** $\frac{-3}{25}$; negative exponent rule **17.** $\frac{-5}{4}$; negative exponent rule **18.** x^5; product rule **19.** x^4; product rule **20.** y^2; product rule, 0 exponent rule **21.** y^{11}; product rule **22.** $\frac{1}{x^3}$; negative exponent rule **23.** $\frac{1}{y^2}$; negative exponent rule **24.** $\frac{2}{x}$; negative exponent rule **25.** $\frac{5}{y^4}$; negative exponent rule **26.** $\frac{-8}{y^2}$; negative exponent rule **27.** $\frac{-10}{x^3}$; negative exponent rule

Answers to Practice Problems: 1. $2^7 = 128$ **2.** $\frac{1}{2}$ **3.** x^6 **4.** 10 **5.** $7y^4$ **6.** 1

28. $\frac{5x^6}{y^4}$; negative exponent rule **29.** $\frac{1}{y^2}$; negative exponent rule and 0 exponent rule
30. 4; 0 exponent rule
31. 2; 0 exponent rule
32. 49; quotient rule
33. 729; quotient rule
34. $\frac{1}{10}$; quotient rule and negative exponent rule **35.** $\frac{1}{10000}$; quotient rule and negative exponent rule
36. $\frac{1}{8}$; quotient rule and negative exponent rule
37. x^2; quotient rule
38. x^2; quotient rule
39. x^2; quotient rule
40. y^2; quotient rule
41. x^4; quotient rule
42. x^5; quotient rule
43. $\frac{1}{x^4}$; quotient rule and negative exponent rule **44.** $\frac{1}{x^4}$; quotient rule and negative exponent rule **45.** x^6; quotient rule **46.** x^6; quotient rule **47.** x^2; quotient rule **48.** $\frac{1}{x^3}$; quotient rule and negative exponent rule
49. y^2; quotient rule
50. y^6; quotient rule

Use any of the appropriate properties of exponents to simplify the expressions in Exercises 51 – 85 so that they contain only positive exponents.

51. -6^2 **52.** $3\cdot 2^2$ **53.** 5^{-1} **54.** -5^{-2}

55. $(-4)^{-3}$ **56.** $(-8)^{-2}$ **57.** $x^5\cdot x^7$ **58.** $x^3\cdot x^5$

59. $x^4\cdot x^0\cdot x$ **60.** $x^2\cdot x^{-1}$ **61.** $y^{-2}\cdot y^{-1}$ **62.** $x^{-2}\cdot x^3\cdot x^5$

63. $x^3\cdot x^{-7}\cdot x^2$ **64.** $y^{-3}\cdot y^{-2}\cdot y^0$ **65.** $\frac{x^3}{x^4}$ **66.** $\frac{x^{12}}{x^4}$

67. $\frac{y^5}{y^0}$ **68.** $\frac{x^2}{x^{-1}}$ **69.** $\frac{y^{-2}}{y^2}$ **70.** $\frac{y^2}{y^{-5}}$

71. $\frac{x^2x^4}{x^{-4}}$ **72.** $\frac{x^3x^5}{x^4}$ **73.** $\frac{x^0x^3}{x^6}$ **74.** $\frac{x\cdot x^3}{x^5}$

75. $\frac{x^2x^4}{x^{-2}}$ **76.** $\frac{x^{-1}x^3}{x^{-4}}$ **77.** $\frac{x\cdot x^{-2}}{x^2x^{-3}}$ **78.** $\frac{x^{16}}{x^{-2}x^{-8}}$

79. $x^k\cdot x$ **80.** $x^k\cdot x^3$ **81.** $x^k\cdot x^{2k}$ **82.** $x^{3k}\cdot x^4$

83. $\frac{x^k}{x^2}$ **84.** $\frac{x^{2k}}{x^k}$ **85.** $\frac{x^{k+1}}{x^3}$

Simplify each expression Exercises 86 – 115 and tell which rule (or rules) for exponents you used. The final form of the expressions with variables should contain only positive exponents. Assume that all variables represent nonzero numbers.

86. $3x^3\cdot x^0$ **87.** $3y\cdot y^4$ **88.** $(5x^2)(2x^2)$ **89.** $(3x^2)(3x)$

90. $(4x^3)(9x^0)$ **91.** $(5x^2)(3x^4)$ **92.** $(-2x^2)(7x^3)$ **93.** $(3y^3)(-6y^2)$

94. $(-4x^5)(3x)$ **95.** $(6y^4)(5y^5)$ **96.** $\frac{8y^3}{2y^2}$ **97.** $\frac{12x^4}{3x}$

98. $\frac{9y^5}{3y^3}$ **99.** $\frac{-10x^5}{2x}$ **100.** $\frac{-8y^4}{4y^2}$ **101.** $\frac{12x^6}{-3x^3}$

102. $\frac{21x^4}{-3x^2}$ **103.** $\frac{10\cdot 10^3}{10^{-3}}$ **104.** $\frac{10^4\cdot 10^{-3}}{10^{-2}}$ **105.** $\frac{10\cdot 10^{-1}}{10^2}$

106. $(9x^2)^0$ **107.** $(9x^2y^3)(-2x^3y^4)$ **108.** $(-3xy)(-5x^2y^{-3})$ **109.** $\frac{-8x^2y^4}{4x^3y^2}$

110. $\frac{-8x^{-2}y^4}{4x^2y^{-2}}$ **111.** $(-2x^{-3}y^5)^0$ **112.** $(3a^2b^4)(4ab^5c)$ **113.** $(-6a^3b^{-4})(4a^{-2}b^8)$

51. -36 **52.** 12
53. $\frac{1}{5}$ **54.** $-\frac{1}{25}$
55. $-\frac{1}{64}$ **56.** $\frac{1}{64}$
57. x^{12} **58.** x^8
59. x^5 **60.** x
61. $\frac{1}{y^3}$ **62.** x^6
63. $\frac{1}{x^2}$ **64.** $\frac{1}{y^5}$
65. $\frac{1}{x}$ **66.** x^8

114. $\frac{36a^5b^0c}{-9a^{-5}b^{-3}}$ **115.** $\frac{25y^6 \cdot 3y^{-2}}{15xy^4}$

Calculator Problems

Use a calculator to evaluate each expression.

116. $(2.16)^0$ **117.** $(-5.06)^2$ **118.** $(1.6)^{-2}$

119. $(6.4)^5 \cdot (2.3)^2$ **120.** $(-14.8)^2 \cdot (21.3)^2$

Writing and Thinking About Mathematics

121. Discuss, briefly, why each of the following statements is WRONG.

a. $3^2 \cdot 3^2 = 6^2$ **b.** $3^2 \cdot 2^2 = 6^4$ **c.** $3^2 \cdot 3^2 = 9^4$

Hawkes Learning Systems: Introductory & Intermediate Algebra

Simplifying Integer Exponents I

67. y^5 **68.** x^3
69. $\frac{1}{y^4}$ **70.** y^7
71. x^{10} **72.** x^4
73. $\frac{1}{x^3}$ **74.** $\frac{1}{x}$
75. x^8 **76.** x^6
77. 1 **78.** x^{26}
79. x^{k+1} **80.** x^{k+3}
81. x^{3k} **82.** x^{3k+4}
83. x^{k-2} **84.** x^k
85. x^{k-2}

86. $3x^3$; product rule or 0 exponent rule **87.** $3y^5$; product rule **88.** $10x^4$; product rule **89.** $9x^3$; product rule **90.** $36x^3$; product rule, 0 exponent rule **91.** $15x^6$; product rule **92.** $-14x^5$; product rule **93.** $-18y^5$; product rule **94.** $-12x^6$; product rule **95.** $30y^9$; product rule **96.** $4y$; quotient rule **97.** $4x^3$; quotient rule **98.** $3y^2$; quotient rule **99.** $-5x^4$; quotient rule **100.** $-2y^2$; quotient rule **101.** $-4x^3$; quotient rule **102.** $-7x^2$; quotient rule **103.** 10^7; product and quotient rules **104.** 10^3; product and quotient rules **105.** $\frac{1}{100}$; product, quotient, and negative exponent rules **106.** 1; 0 exponent rule **107.** $-18x^5y^7$; product rule **108.** $\frac{15x^3}{y^2}$; product and negative exponent rules **109.** $\frac{-2y^2}{x}$; quotient and negative exponent rules **110.** $\frac{-2y^6}{x^4}$; quotient and negative exponent rules **111.** 1; 0 exponent rule **112.** $12a^3b^9c$; product rule **113.** $-24ab^4$; product rule **114.** $-4a^{10}b^3c$; quotient rule **115.** $\frac{5}{x}$; quotient and negative exponent rules **116.** 1 **117.** 25.6036 **118.** .390625 **119.** 56,800.9424896 **120.** 99,376.2576 **121.** **a.** $a^m \cdot a^n = a^{m+n}$ **b.** $3^2 \cdot 2^2 = 3 \cdot 3 \cdot 2 \cdot 2 = 6^2$ **c.** $a^m \cdot a^n = a^{m+n}$

5.2 More on Exponents and Scientific Notation

Objectives

After completing this section, you will be able to:

1. *Simplify powers of expressions by using the properties of integer exponents.*
2. *Write a decimal number in scientific notation.*
3. *Operate with decimal numbers by using scientific notation.*

The summary of the rules for exponents given in Section 5.1 is repeated here for easy reference.

Summary of Properties of Exponents

If a and b are nonzero real numbers and m and n are integers,

1. The Exponent 1:	$a = a^1$
2. The Exponent 0:	$a^0 = 1 \quad (a \neq 0)$
3. Product Rule:	$a^m \cdot a^n = a^{m+n}$
4. Quotient Rule:	$\frac{a^m}{a^n} = a^{m-n}$
5. Negative Exponents:	$a^{-n} = \frac{1}{a^n}, \quad \frac{1}{a^{-n}} = a^n$

Power Rule

Now, consider what happens when a power is raised to a power. For example, to simplify the expressions $(x^2)^3$ and $(2^5)^2$, we can write

$$\left(x^2\right)^3 = x^2 \cdot x^2 \cdot x^2 = x^{2+2+2} = x^6$$

and

$$\left(2^5\right)^2 = 2^5 \cdot 2^5 = 2^{5+5} = 2^{10}.$$

However, this technique can be quite time-consuming when the exponent is large such as in $(3y^3)^{17}$. The **Power Rule for Exponents** gives a convenient way to handle powers raised to powers.

Power Rule for Exponents

If a is a nonzero real number and m and n are integers, then

$$\left(a^m\right)^n = a^{mn}.$$

In words, the value of a power raised to a power can be found by multiplying the exponents and keeping the base.

Example 1: Power Rule

Simplify each expression using the Power Rule for Exponents.

a. $(x^2)^4$

Solution: $(x^2)^4 = x^{2\cdot 4} = x^8$

b. $(x^5)^{-2}$

Solution: $(x^5)^{-2} = x^{5(-2)} = x^{-10} = \dfrac{1}{x^{10}}$

or $\left(x^5\right)^{-2} = \dfrac{1}{\left(x^5\right)^2} = \dfrac{1}{x^{5\cdot 2}} = \dfrac{1}{x^{10}}$

c. $(y^{-7})^2$

Solution: $(y^{-7})^2 = y^{(-7)2} = y^{-14} = \dfrac{1}{y^{14}}$

In Examples 1d – 1f, simplify each of the expressions using any of the five properties of exponents that apply. You might try each example on scratch paper first and then look at the solutions shown in the example. There may be more than one correct procedure, and **you should apply whichever property you "see" first**.

d. $\dfrac{x^{10}x^2}{x^3}$

Solution: $\dfrac{x^{10}x^2}{x^3} = \dfrac{x^{10+2}}{x^3} = \dfrac{x^{12}}{x^3} = x^{12-3} = x^9$ or

$\dfrac{x^{10}x^2}{x^3} = x^{10+2-3} = x^9$

e. $\dfrac{x^6x^{-2}}{x^7}$

Solution: $\dfrac{x^6x^{-2}}{x^7} = \dfrac{x^{6+(-2)}}{x^7} = \dfrac{x^{6-2}}{x^7} = \dfrac{x^4}{x^7} = x^{4-7} = x^{-3} = \dfrac{1}{x^3}$ or

$\dfrac{x^6x^{-2}}{x^7} = x^{6-2-7} = x^{-3} = \dfrac{1}{x^3}$

continued on next page ...

f. $\dfrac{3^{-5} \cdot 3^{9}}{3^{3} \cdot 3}$

Solution: $\dfrac{3^{-5} \cdot 3^{9}}{3^{3} \cdot 3} = \dfrac{3^{-5+9}}{3^{3+1}} = \dfrac{3^{4}}{3^{4}} = 3^{4-4} = 3^{0} = 1$

In Examples 1g and 1h, assume that k represents a nonzero integer. Follow the appropriate rules for exponents.

g. $x^{2k} \cdot x^{2k}$

Solution: $x^{2k} \cdot x^{2k} = x^{2k+2k} = x^{4k}$

h. $\dfrac{x^{3} \cdot x^{k}}{\left(x^{2}\right)^{k}}$

Solution: $\dfrac{x^{3} \cdot x^{k}}{\left(x^{2}\right)^{k}} = \dfrac{x^{k+3}}{x^{2k}} = x^{k+3-2k} = x^{3-k}$ or $\dfrac{1}{x^{k-3}}$

Power Rule for Products

If the base of an exponent is a product, we will see that each factor in the product can be raised to the power indicated by the exponent. For example, $(10x)^{3}$ indicates that the product of 10 and x is to be raised to the 3rd power and $(-2x^{2}y)^{5}$ indicates that the product of -2, x^{2}, and y is to be raised to the 5th power. We can simplify these expressions as follows:

$$(10x)^{3} = 10x \cdot 10x \cdot 10x = 10 \cdot 10 \cdot 10 \cdot x \cdot x \cdot x = 10^{3} \cdot x^{3} = 1000x^{3}$$

and

$$\begin{aligned}(-2x^{2}y)^{5} &= (-2x^{2}y) \cdot (-2x^{2}y) \cdot (-2x^{2}y) \cdot (-2x^{2}y) \cdot (-2x^{2}y) \\ &= (-2)(-2)(-2)(-2)(-2) \cdot x^{2} \cdot x^{2} \cdot x^{2} \cdot x^{2} \cdot x^{2} \cdot y \cdot y \cdot y \cdot y \cdot y \\ &= (-2)^{5} \cdot (x^{2})^{5} \cdot y^{5} \\ &= -32x^{10}y^{5}\end{aligned}$$

We can simplify expressions such as these in a much easier fashion by using the following **Power Rule for Products of Exponents.**

Power Rule for Products

If a and b are nonzero real numbers and n is an integer then

$$(ab)^{n} = a^{n}b^{n}.$$

In words, a power of a product is found by raising each factor to that power.

Example 2: Power Rule for Products

Simplify each expression using the Power Rule for Products.

a. $(5x)^2$

Solution: $(5x)^2 = 5^2 \cdot x^2 = 25x^2$

b. $(xy)^3$

Solution: $(xy)^3 = x^3 \cdot y^3 = x^3y^3$

c. $(-7ab)^2$

Solution: $(-7ab)^2 = (-7)^2a^2b^2 = 49a^2b^2$

d. $(ab)^{-5}$

Solution: $(ab)^{-5} = a^{-5} \cdot b^{-5} = \frac{1}{a^5} \cdot \frac{1}{b^5} = \frac{1}{a^5b^5}$

or, using the Rule for Negative Exponents first and then the Power Rule for Products,

$$(ab)^{-5} = \frac{1}{(ab)^5} = \frac{1}{a^5b^5}$$

e. $(x^2y^{-3})^4$

Solution: $\left(x^2y^{-3}\right)^4 = \left(x^2\right)^4 \cdot \left(y^{-3}\right)^4 = x^8 \cdot y^{-12} = x^8 \cdot \frac{1}{y^{12}} = \frac{x^8}{y^{12}}$

NOTES

Special Note about Negative Numbers and Exponents:

In an expression such as $-x^2$, we know that -1 is understood to be the coefficient of x^2. That is,

$$-x^2 = -1 \cdot x^2$$

The same is true for expressions with numbers only such as -7^2. That is,

$$-7^2 = -1 \cdot 7^2 = -1 \cdot 49 = -49.$$

We see that the exponent refers to 7 and **not** to -7. For the exponent to refer to -7 as the base, -7 **must be in parentheses** as follows:

$$(-7)^2 = (-7) \cdot (-7) = +49.$$

As another example,

$$-2^0 = -1 \cdot 2^0 = -1 \cdot 1 = -1 \text{ and } (-2)^0 = 1.$$

In general,

$$-x^2 \neq (-x)^2.$$

This distinction is critical only with even exponents. With odd exponents, the results are the same whether or not you apply the exponent to the coefficient -1. For example,

$$-2^3 = -1 \cdot 2^3 = -1 \cdot 8 = -8 \text{ and } (-2)^3 = -8.$$

Power Rule for Fractions

Expressions with fractions (or quotients) are treated much the same as expressions with products. For example,

$$\left(\frac{y}{3}\right)^4 = \frac{y}{3}\cdot\frac{y}{3}\cdot\frac{y}{3}\cdot\frac{y}{3} = \frac{y\cdot y\cdot y\cdot y}{3\cdot 3\cdot 3\cdot 3} = \frac{y^4}{3^4} = \frac{y^4}{81}$$

and

$$\left(\frac{a}{x}\right)^3 = \frac{a}{x}\cdot\frac{a}{x}\cdot\frac{a}{x} = \frac{a\cdot a\cdot a}{x\cdot x\cdot x} = \frac{a^3}{x^3}.$$

These examples illustrate the **Power Rule for Fractions**.

Power Rule for Fractions

If a and b are nonzero real numbers and n is an integer, then

$$\left(\frac{a}{b}\right)^n = \frac{a^n}{b^n}.$$

Example 3: Properties of Exponents

Simplify the following expressions by using any of the properties of exponents that apply.

a. $\left(\frac{5x}{3b}\right)^3$

Solution: $\left(\frac{5x}{3b}\right)^3 = \frac{(5x)^3}{(3b)^3} = \frac{5^3x^3}{3^3b^3} = \frac{125x^3}{27b^3}$

b. $\left(\frac{a^3b^{-3}}{4}\right)^2$

Solution: $\left(\frac{a^3b^{-3}}{4}\right)^2 = \frac{(a^3b^{-3})^2}{4^2} = \frac{(a^3)^2(b^{-3})^2}{4^2} = \frac{a^6b^{-6}}{16}$ **or** $\frac{a^6}{16b^6}$

c. $\frac{(3^{-2}x^{-3})^{-1}}{(x^{-2}y^3)^3(2x^{-1}y^2)^{-1}}$

Solution: $\frac{(3^{-2}x^{-3})^{-1}}{(x^{-2}y^3)^3(2x^{-1}y^2)^{-1}} = \frac{3^2x^3}{x^{-6}y^9\cdot 2^{-1}x^1y^{-2}} = \frac{2^1\cdot 9x^3}{x^{-5}y^7}$

$$= \frac{18x^{3-(-5)}}{y^7} = \frac{18x^8}{y^7} \text{ or } 18x^8y^{-7}$$

d. $\left(\dfrac{2x^2y^{-3}}{x^{-3}y}\right)^{-2}$

Solution: $\left(\dfrac{2x^2y^{-3}}{x^{-3}y}\right)^{-2} = \left(2x^{2-(-3)}y^{-3-1}\right)^{-2} = \left(2x^{2+3}y^{-3-1}\right)^{-2} = \left(2x^5y^{-4}\right)^{-2}$

$= 2^{-2}x^{-10}y^8$ **or** $\dfrac{1}{4}x^{-10}y^8$ **or** $\dfrac{y^8}{4x^{10}}$

e. $\left(\dfrac{x^{2k}y^k}{x^ky}\right)^2$

Solution: $\left(\dfrac{x^{2k}y^k}{x^ky}\right)^2 = \left(x^{2k-k}y^{k-1}\right)^2$

$= \left(x^ky^{k-1}\right)^2 = x^{2k}y^{2(k-1)} = x^{2k}y^{2k-2}$

f. $\dfrac{1}{5x^{-2}}$

Solution: $\dfrac{1}{5x^{-2}} = \dfrac{1}{5\cdot\dfrac{1}{x^2}} = \dfrac{1}{\dfrac{5}{x^2}} = 1\cdot\dfrac{x^2}{5} = \dfrac{x^2}{5}$

Notice that the exponent –2 applies only to the x and has no effect on the coefficient 5.

NOTES

Another general approach with fractions involving negative exponents is to note that

$$\left(\frac{a}{b}\right)^{-n} = \frac{a^{-n}}{b^{-n}} = \frac{b^n}{a^n} = \left(\frac{b}{a}\right)^n.$$

In effect, there are two basic shortcuts with negative exponents and fractions:

1. Taking the reciprocal of a fraction changes the sign of any exponent on the fraction.
2. Moving any factor from numerator to denominator or vice versa, changes the sign of the corresponding exponent.

Example 4: Two Approaches

a. Either of the following approaches can be used to simplify $\left(\dfrac{a^2}{b^3}\right)^{-3}$.

i. $\left(\dfrac{a^2}{b^3}\right)^{-3} = \left(\dfrac{b^3}{a^2}\right)^3 = \dfrac{\left(b^3\right)^3}{\left(a^2\right)^3} = \dfrac{b^9}{a^6}$

ii. $\left(\dfrac{a^2}{b^3}\right)^{-3} = \dfrac{\left(a^2\right)^{-3}}{\left(b^3\right)^{-3}} = \dfrac{a^{-6}}{b^{-9}} = \dfrac{b^9}{a^6}$

continued on next page ...

b. Either of the following approaches can be used to simplify $\left(\frac{x}{3y^4}\right)^{-2}$.

i. $\left(\frac{x}{3y^4}\right)^{-2} = \left(\frac{3y^4}{x}\right)^{2} = \frac{\left(3y^4\right)^2}{x^2} = \frac{9y^8}{x^2}$

ii. $\left(\frac{x}{3y^4}\right)^{-2} = \frac{x^{-2}}{\left(3y^4\right)^{-2}} = \frac{\left(3y^4\right)^2}{x^2} = \frac{9y^8}{x^2}$

In general,

$$\frac{1}{a^{-n}} = a^n \quad \text{and} \quad \frac{a^{-n}}{b^{-m}} = \frac{b^m}{a^n} \quad \text{and} \quad \left(\frac{a}{b}\right)^{-n} = \left(\frac{b}{a}\right)^n.$$

Remember that the choice of steps is yours and that, as long as you correctly apply the properties of exponents, the answer will be the same regardless of the order of the steps. The following table provides a summary of the properties of exponents for easy reference.

Summary of Properties of Exponents

If a and b are nonzero real numbers and m and n are integers,

1. The Exponent 1: $a = a^1$ ***(a is any real number.)***

2. The Exponent 0: $a^0 = 1 \quad (a \neq 0)$

3. Product Rule: $a^m \cdot a^n = a^{m+n}$

4. Quotient Rule: $\frac{a^m}{a^n} = a^{m-n}$

5. Power Rule: $\left(a^m\right)^n = a^{mn}$

6. Negative Exponents: $a^{-n} = \frac{1}{a^n}, \quad \frac{1}{a^{-n}} = a^n$

7. Power Rule for Products: $(ab)^n = a^n b^n$

8. Power Rule for Fractions: $\left(\frac{a}{b}\right)^n = \frac{a^n}{b^n}$

Scientific Notation

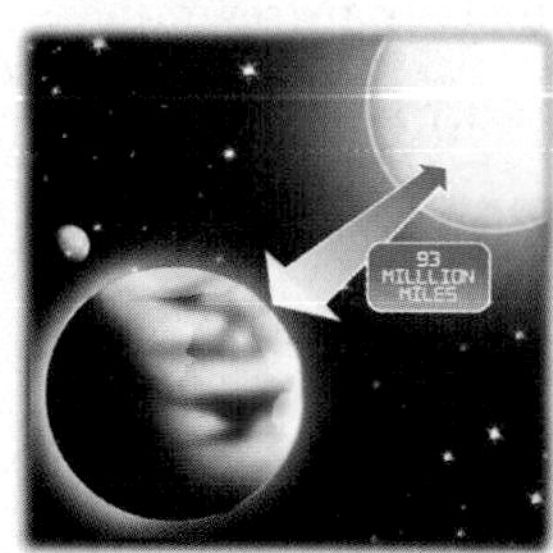

A basic application of integer exponents occurs in scientific disciplines, such as astronomy and biology, when very large and very small numbers are involved. For example, the distance from the earth to the sun is approximately 93,000,000 miles, and the approximate radius of a carbon atom is 0.000 000 007 7 centimeters.

In **scientific notation** (an option in all scientific and graphing calculators), **decimal numbers are written as the product of a number greater than or equal to 1 and less than 10, and an integer power of 10**. In scientific notation there is just one digit to the left of the decimal point. For example,

$$93{,}000{,}000 = 9.3\times10^7 \quad \text{and} \quad 0.0000000077 = 7.7\times10^{-9}.$$

The exponent tells how many places the decimal point is to be moved and in what direction. If the exponent is positive, the decimal point is moved to the right:

$$2.7\times 10^3 = 2700. \qquad \text{3 places right}$$

A negative exponent indicates that the decimal point should move to the left:

$$3.92\times10^{-6} = 0.00000392 \qquad \text{6 places left}$$

Scientific Notation

If N is a decimal number, then in ***scientific notation***

$$N = a \times 10^n \text{ where } 1 \le a < 10 \text{ and } n \text{ is an integer.}$$

Example 5: Decimals in Scientific Notation

Teaching Notes:
While mathematicians seem to prefer not to leave a final answer standing with negative exponents, scientific notation is an exception to this "rule" of simplification.

Write the following decimal numbers in scientific notation.

a. 867,000,000,000 **b.** 420,000
c. 0.0036 **d.** 0.000 000 025

Solution:

a. $867{,}000{,}000{,}000 = 8.67\times10^{11}$

b. $420{,}000 = 4.2\times10^5$

c. $0.0036 = 3.6\times10^{-3}$

d. $0.000000025 = 2.5\times10^{-8}$

Example 6: Properties of Exponents

Simplify the following expressions by first writing the decimal numbers in scientific notation and then using the properties of exponents.

a. $\dfrac{0.0023 \times 560,000}{0.00014}$

Solution: $\dfrac{0.0023 \times 560,000}{0.00014} = \dfrac{2.3 \times 10^{-3} \times 5.6 \times 10^{5}}{1.4 \times 10^{-4}}$

$$= \frac{2.3 \times \overset{4}{\cancel{5.6}}}{\cancel{1.4}} \times \frac{10^{-3} \times 10^{5}}{10^{-4}}$$

$$= 9.2 \times \frac{10^{2}}{10^{-4}} = 9.2 \times 10^{2-(-4)}$$

$$= 9.2 \times 10^{6}$$

b. $\dfrac{8.1 \times 8200}{9,000,000 \times 4.1}$

Solution: $\dfrac{8.1 \times 8200}{9,000,000 \times 4.1} = \dfrac{8.1 \times 8.2 \times 10^{3}}{9.0 \times 10^{6} \times 4.1}$

$$= \frac{\overset{0.9}{\cancel{8.1}} \times \overset{2}{\cancel{8.2}}}{\cancel{9.0} \times \cancel{4.1}} \times \frac{10^{3}}{10^{6}}$$

$$= 1.8 \times 10^{3-6} = 1.8 \times 10^{-3}$$

c. Light travels approximately 3×10^{8} meters per second. How many meters per minute does light travel?

Solution: Since there are 60 seconds in one minute, multiply by 60.

$$3 \times 10^{8} \times 60 = 180 \times 10^{8} = 1.8 \times 10^{2} \times 10^{8} = 1.8 \times 10^{10}$$

Thus, light travels 1.8×10^{10} meters per minute.

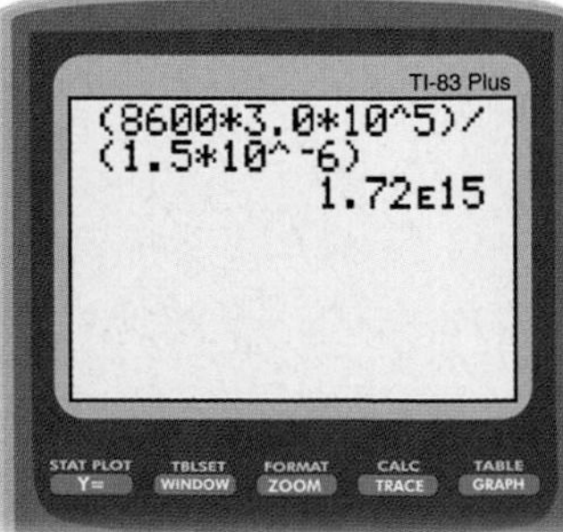

d. Use a calculator to evaluate the expression $\dfrac{8600(3.0 \times 10^{5})}{1.5 \times 10^{-6}}$. Leave the answer in scientific notation.

(**Note:** The caret key ^ is used to indicate an exponent.)

Solution: With a TI-83 Plus calculator (set in scientific notation mode) the display should appear as shown at left:

(**Note:** The numerator and denominator must be set in parentheses.)

e. Use a calculator to find the number of miles that light travels in one year (a light-year) if light travels 186,000 miles per second.

Solution: We know

$$60 \text{ sec} = 1 \text{ min}$$
$$60 \text{ min} = 1 \text{ hr}$$
$$24 \text{ hr} = 1 \text{ day}$$
$$365 \text{ days} = 1 \text{ yr}$$

Multiplication should give the following display on your calculator:

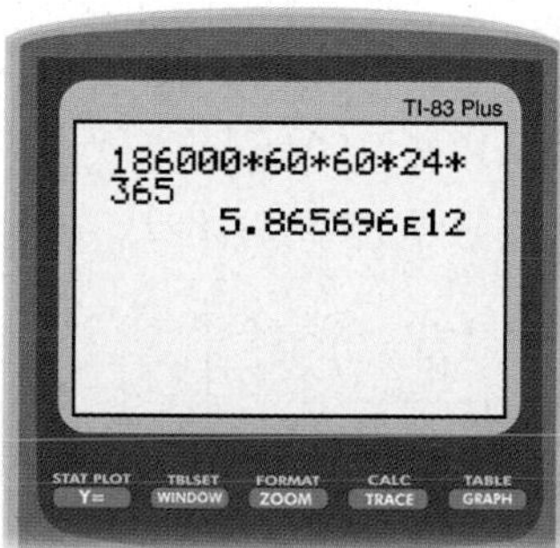

Thus, a light-year is 5,865,696,000,000 miles.

Practice Problems

Simplify each expression.

1. $\dfrac{x^{-2}x^5}{x^{-7}}$ **2.** $\left(\dfrac{a^2b^3}{4}\right)^0$ **3.** $\dfrac{-3^2\cdot 5}{2\cdot(-3)^2}$ **4.** $\left(\dfrac{5x}{3b}\right)^{-2}$

5. *Write the number in scientific notation: 186,000 miles per second (speed of light in miles per second).*

1. -81 **2.** -25 **3.** 81 **4.** -400 **5.** $1,000,000$ **6.** 4096 **7.** $\dfrac{1}{x^4}$ **8.** $\dfrac{1}{x^6}$ **9.** $4x^8$ **10.** $\dfrac{x^2}{3}$ **11.** $36x^6$ **12.** $9x^8$

5.2 Exercises

13. $\dfrac{x^4}{y^6}$ **14.** $\dfrac{1}{x^2y^6}$ **15.** $-108x^6$ **16.** $\dfrac{7}{y^8}$ **17.** -3 **18.** $\dfrac{5x^2}{y}$ **19.** $\dfrac{a^4}{b^8}$ **20.** $\dfrac{8a^3}{b^6}$ **21.** $\dfrac{27x^3}{y^3}$ **22.** $\dfrac{16x^2}{y^4}$

Simplify the expressions in Exercises 1 – 68 so that they contain only positive exponents.

1. -3^4 **2.** -5^2 **3.** $(-3)^4$ **4.** -20^2 **5.** $(-10)^6$

6. $(-4)^6$ **7.** $\left(x^{-2}\right)^2$ **8.** $\left(x^2\right)^{-3}$ **9.** $\left(-2x^4\right)^2$ **10.** $\left(3x^{-2}\right)^{-1}$

11. $(6x^3)^2$ **12.** $(-3x^4)^2$ **13.** $\left(x^2y^{-3}\right)^2$ **14.** $\left(xy^3\right)^{-2}$ **15.** $4(-3x^2)^3$

16. $7(y^{-2})^4$ **17.** $-3(7xy^2)^0$ **18.** $5(x^2y^{-1})$ **19.** $\left(\dfrac{a}{b^2}\right)^4$ **20.** $\left(\dfrac{2a}{b^2}\right)^3$

21. $\left(\dfrac{3x}{y}\right)^3$ **22.** $\left(\dfrac{-4x}{y^2}\right)^2$ **23.** $\left(\dfrac{6m^3}{n^5}\right)^0$ **24.** $\left(\dfrac{3x^2}{y^3}\right)^2$

Answers to Practice Problems: **1.** x^{10} **2.** 1 **3.** $\dfrac{-5}{2}$ **4.** $\dfrac{9b^2}{25x^2}$ **5.** 1.86×10^5

23. 1 **24.** $\dfrac{9x^4}{y^6}$

25. $4x^4y^4$ 26. $\dfrac{y^2}{x^2}$

27. $\dfrac{b}{2a}$ 28. $\dfrac{y^{10}}{4x^2}$

29. $\dfrac{1}{3xy^2}$ 30. $\dfrac{1}{64a^6b^9}$

31. $-\dfrac{x^3y^6}{27}$ 32. $25x^2y^4$

33. m^2n^4 34. $16a^4b^4$

35. $-\dfrac{y^2}{49x^2}$ 36. $\dfrac{1}{8a^3b^6}$

37. $\dfrac{25x^6}{y^2}$ 38. $\dfrac{y^8}{16x^8}$

39. $\dfrac{x^6}{y^6}$ 40. $\dfrac{8a^6}{b^9}$

41. $\dfrac{36y^{14}}{x^4}$ 42. $\dfrac{49y^4}{x^6}$

43. $\dfrac{24y^5}{x^7}$ 44. $\dfrac{4y^4}{25x^6}$

45. $\dfrac{36b^4}{a^2}$ 46. $-\dfrac{x^2y^3}{3}$

47. $\dfrac{x^2}{36y^6}$ 48. $\dfrac{27}{8a^3b^6}$

49. $\dfrac{x^4}{25y^2}$ 50. $\dfrac{3y^4}{x^3}$

51. $x^{2k}y^k$ 52. $x^{2k}y^{2m}$

53. $x^{2k+1}y^{2+n}$

54. $x^{5n}y^{3-k}$

55. x^ny^{k+1}

56. $x^{k+3}y^{2k}$ 57. $\dfrac{x^4}{y}$

58. $\dfrac{b^7}{a^2}$ 59. $\dfrac{9y^7}{x^6}$

60. $\dfrac{2x^3}{y^5}$ 61. $\dfrac{a^6}{b^3}$

62. $\dfrac{3y^7}{4x^2}$ 63. $\dfrac{7}{2x^3y}$

64. $\dfrac{80x^{10}}{y^9}$ 65. $\dfrac{x^{13}}{y^{20}}$

25. $\left(\dfrac{-2x^2}{y^{-2}}\right)^2$ **26.** $\left(\dfrac{x}{y}\right)^{-2}$ **27.** $\left(\dfrac{2a}{b}\right)^{-1}$ **28.** $\left(\dfrac{2x}{y^5}\right)^{-2}$

29. $\left(\dfrac{3x}{y^{-2}}\right)^{-1}$ **30.** $\left(\dfrac{4a^2}{b^{-3}}\right)^{-3}$ **31.** $\left(\dfrac{-3}{xy^2}\right)^{-3}$ **32.** $\left(\dfrac{5xy^3}{y}\right)^2$

33. $\left(\dfrac{m^2n^3}{mn}\right)^2$ **34.** $\left(\dfrac{2ab^3}{b^2}\right)^4$ **35.** $\left(\dfrac{-7^2x^2y}{y^3}\right)^{-1}$ **36.** $\left(\dfrac{2ab^4}{b^2}\right)^{-3}$

37. $\left(\dfrac{5x^3y}{y^2}\right)^2$ **38.** $\left(\dfrac{2x^2y}{y^3}\right)^{-4}$ **39.** $\left(\dfrac{x^3y^{-1}}{y^2}\right)^2$ **40.** $\left(\dfrac{2a^2b^{-1}}{b^2}\right)^3$

41. $\left(\dfrac{6y^5}{x^2y^{-2}}\right)^2$ **42.** $\left(\dfrac{7x^{-2}y}{xy^{-1}}\right)^2$ **43.** $\dfrac{\left(3x^2y^{-1}\right)^{-2}}{\left(6x^{-1}y\right)^{-3}}$ **44.** $\left(\dfrac{2x^{-3}}{5y^{-2}}\right)^2$

45. $\left(\dfrac{2^{-1}a}{3b^2}\right)^{-2}$ **46.** $\left(\dfrac{-3x^{-2}}{y^3}\right)^{-1}$ **47.** $\left(\dfrac{6x}{x^2y^{-3}}\right)^{-2}$ **48.** $\left(\dfrac{2ab^4}{3b^2}\right)^{-3}$

49. $\left(\dfrac{5^{-1}x^3y^{-2}}{xy^{-1}}\right)^2$ **50.** $\left(\dfrac{x^2y^{-3}}{3x^{-1}y}\right)^{-1}$ **51.** $\left(x^2y\right)^k$ **52.** $\left(x^ky^m\right)^2$

53. $\left(x^{2k}y^2\right)\left(xy^n\right)$ **54.** $\left(x^{4n}y^3\right)\left(x^ny^{-k}\right)$ **55.** $\left(x^{n+2}y^k\right)\left(x^{-2}y\right)$

56. $\left(x^{k+1}y^{3k}\right)\left(x^2y^{-k}\right)$ **57.** $\left(\dfrac{x^4y}{y^2}\right)\left(\dfrac{x^{-1}y^2}{y^{-1}}\right)^0$ **58.** $\left(\dfrac{a^2b}{ab^{-2}}\right)\left(\dfrac{a^{-3}b}{b^{-3}}\right)$

59. $\left(\dfrac{x^2y}{y^2}\right)^{-1}\left(\dfrac{3x^{-2}y}{y^{-2}}\right)^2$ **60.** $\left(\dfrac{x^2y^{-3}}{y^{-1}}\right)^2\left(\dfrac{xy^2}{2y}\right)^{-1}$ **61.** $\left(\dfrac{a^3b^{-1}}{ab^{-2}}\right)\left(\dfrac{a^0b^3}{a^4b^{-1}}\right)^{-1}$

62. $\left(\dfrac{5x^3y}{x^{-2}y^3}\right)^{-1}\left(\dfrac{4x^{-2}y^{-1}}{15xy^4}\right)^{-1}$ **63.** $\left(\dfrac{7x^{-2}y}{xy^4}\right)^2\left(\dfrac{14xy^{-3}}{x^4y^2}\right)^{-1}$ **64.** $\dfrac{\left(4^{-2}x^{-3}y\right)^{-1}}{\left(x^{-2}y^2\right)^3\left(5xy^{-2}\right)^{-1}}$

65. $\dfrac{\left(xy^{-2}\right)^4\left(x^{-2}y^3\right)^{-2}}{\left(xy^2\right)^{-1}\left(xy^{-2}\right)^{-4}}$ **66.** $\dfrac{\left(6x^2y\right)\left(x^{-1}y^3\right)^2}{\left(x^{-1}y\right)^2\left(3x^2y\right)^3}$ **67.** $\dfrac{\left(x^3y^4\right)^{-1}\left(x^{-2}y\right)^2}{\left(xy^2\right)^{-3}\left(xy^{-1}\right)^2}$

68. $\dfrac{\left(x^{-3}y^{-5}\right)^{-2}\left(x^2y^{-3}\right)^3}{\left(x^3y^{-4}\right)^2\left(x^{-1}y^{-2}\right)^{-2}}$

Write each expression in Exercises 69 – 80 in decimal notation.

69. 4.72×10^5 **70.** 6.91×10^{-4} **71.** 1.28×10^{-7}

72. 1.63×10^8 **73.** 9.23×10^{-3} **74.** 5.88×10^6

75. 4.2×10^{-2} **76.** 8.35×10^{-3} **77.** 7.56×10^6

78. 6.132×10^{-5} **79.** 8.515×10^8 **80.** 9.374×10^7

66. $\frac{2y^2}{9x^4}$ 67. $\frac{y^6}{x^6}$
68. x^4y^5
69. 472,000
70. 0.000691
71. 0.000000128
72. 163,000,000
73. 0.00923
74. 5,880,000
75. 0.042
76. 0.00835
77. 7,560,000
78. 0.00006132
79. 851,500,000
80. 93,740,000
81. 8.6×10^4
82. 9.27×10^5
83. 3.62×10^{-2}
84. 6.1×10^{-3}
85. 1.83×10^7
86. 3.76×10^8
87. 4.79×10^5
88. 3.67×10^{-4}
89. 8.71×10^{-7}
90. 5.28×10^7
91. 4.29×10
92. 8.4×10^{-7}
93. 1.44×10^5
94. 3.0×10^7
95. 6×10^{-4}
96. 4.0×10
97. 1.2×10^{-1}
98. 5.0×10^2
99. 5×10^3
100. 2.4×10^4
101. 5.6×10^{-2}
102. 8.0×10^2
103. 1.8×10^{12} cm/min.; 1.08×10^{14} cm/hr.
104. 9.75×10^{-19} g
105. 6×10^8 ounces

For Exercises 81 – 102, write each expression in scientific notation. Show the steps you use as in Examples 6a and 6b in the text. Do not use your calculator.

81. 86,000 **82.** 927,000 **83.** 0.0362

84. 0.0061 **85.** 18,300,000 **86.** 376,000,000

87. 479,000 **88.** 0.000367 **89.** 0.000000871

90. $52,800 \times 1,000$ **91.** $143,000 \times 0.0003$ **92.** 0.007×0.00012

93. $0.036 \times 4,000,000$ **94.** $\frac{27,000}{0.0009}$ **95.** $\frac{1800 \times 0.00045}{1350}$

96. $\frac{0.0032 \times 120}{0.0096}$ **97.** $\frac{0.084 \times 0.0093}{0.21 \times 0.031}$ **98.** $\frac{0.0070 \times 50 \times 0.55}{1.4 \times 0.0011 \times 0.25}$

99. $\frac{0.36 \times 5200}{0.00052 \times 720}$ **100.** $\frac{0.0016 \times 0.09 \times 460}{0.00012 \times 0.023}$ **101.** $\frac{760 \times 84 \times 0.063}{900 \times 0.38 \times 210}$

102. $\frac{420 \times 0.016 \times 80}{0.028 \times 120 \times 0.2}$

For Exercises 103 – 115, use your calculator and leave all answers in scientific notation.

103. Light travels approximately 3×10^{10} centimeters per second. How many centimeters would this be per minute? per hour?

104. An atom of gold weighs approximately 3.25×10^{-22} grams. What would be the weight of 3,000 atoms of gold?

105. An ounce of gold contains 5×10^{22} atoms. All the gold ever taken out of the Earth is estimated to be 3.0×10^{31} atoms. How many ounces of gold is this?

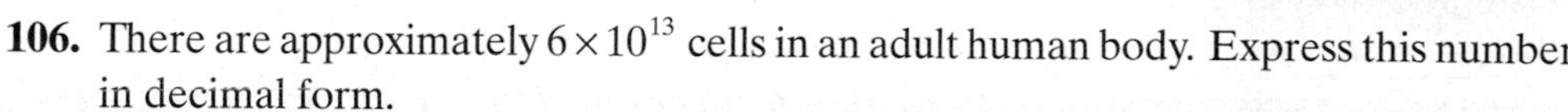

106. There are approximately 6×10^{13} cells in an adult human body. Express this number in decimal form.

107. Write the number 1.0×10^{10} in decimal form. About how many years are in 1.0×10^{10} minutes?

108. One light-year is approximately 9.46×10^{15} meters. The distance to a certain star is about 4.3 light years. How many meters is this?

109. One light-year is about 5.87×10^{12} miles. The mean distance from the sun to Pluto is 3.675×10^9 miles. How many light years is this?

106. 60,000,000,000,000
107. 10,000,000,000; 19,025.87519 years
108. 4.0678×10^{16} m
109. $\approx 6.26 \times 10^{-4}$ light years
110. 1.9926×10^{-26} kg
111. 6.642×10^{-26} kg
112. 1.04×10^{8}
113. 1.15×10^{7} sq. mi.
114. 6.214×10^{-3} mi.
115. 4.356×10^{8} sq. ft.
116. 1.0×10^{2}
117. 3.9
118. 1.844×10^{2}
119. 2.0×10^{-7}
120. 1.2×10^{6}
121. 1.0×10^{14}

110. The weight of an atom is measured in atomic mass units (amu), where 1 amu $= 1.6605 \times 10^{-27}$ kilograms. The atomic mass of carbon 12 is 12 amu. Express the atomic mass of carbon 12 in kilograms.

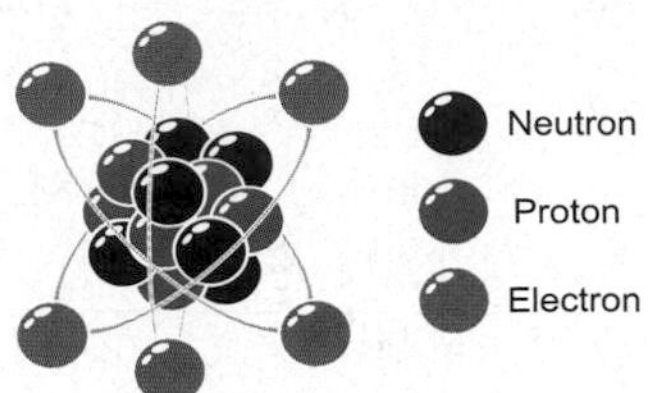

111. The atomic mass of argon is about 40 amu. Express this mass in kilograms. (See Exercise 110.)

112. A company can produce 2 million erasers in a week. How many erasers can the same company produce in a year?

113. The area of land that exists on Earth is approximately 57,500,000 square miles. What is the number of square miles where mountains exist if mountains cover 20% of the earth's area of land?

114. The number of miles in a centimeter is approximately 0.000006214. How many miles are in 1000 centimeters? (Hint: Multiply miles by centimeters.)

115. One acre of land is equal to 43,560 square feet. How many square feet are in 10,000 acres?

In Exercises 116 – 121, use your calculator (set in scientific notation mode) to evaluate each expression. Leave the answer in scientific notation.

116. $\dfrac{5.4 \cdot 0.003 \cdot 5000}{15 \cdot 0.0027 \cdot 20}$

117. $\dfrac{0.0005 \cdot 650 \cdot 3.3}{0.00011 \cdot 2500}$

118. $\dfrac{(1.4 \times 10^{-3})(922)}{(3.5 \times 10^{3})(2.0 \times 10^{-6})}$

119. $\dfrac{0.0084 \cdot 0.003}{0.21 \cdot 600}$

120. $\dfrac{0.02(3.9 \times 10^{3})}{0.013(5.0 \times 10^{-3})}$

121. $\dfrac{(43{,}000)(3.0 \times 10^{5})}{(8.6 \times 10^{-2})(1.5 \times 10^{-3})}$

Hawkes Learning Systems: Introductory & Intermediate Algebra

Simplifying Integer Exponents II
Scientific Notation

5.3 Identifying and Evaluating Polynomials

Objectives

After completing this section, you will be able to:

1. *Define a polynomial.*
2. *Classify a polynomial as a monomial, binomial, trinomial, or a polynomial with more than three terms.*
3. *Evaluate a polynomial for given values of the variable.*

Definition of a Polynomial

A **term** is an expression that involves only multiplication and/or division with constants and/or variables. Remember that a number written next to a variable indicates multiplication, and the number is called the **numerical coefficient** (or **coefficient**) of the variable. For example,

$$3x,\quad -5y^2,\quad \frac{x}{y},\quad \text{and}\quad 17$$

are all algebraic terms. In the term $3x$, 3 is the coefficient of x. In the term $-5y^2$, -5 is the coefficient of y^2. 1 is the coefficient of the term $\frac{x}{y}$. A term that consists of only a number, such as 17, is called a **constant** or a **constant term**.

Monomial

*A **monomial in x** is a term of the form*

$$kx^n$$

where k is a real number and n is a whole number.

***n** is called the **degree** of the term, and **k** is called the **coefficient**.*

A monomial may have more than one variable, and the degree of such a monomial is the sum of the degrees of its variables. For example,

$4x^2y^3$ is a 5th degree monomial in x and y.

However, in this chapter, only monomials of one variable (note that any variable may be used in place of x) will be discussed. Monomials may have fractional or negative coefficients; however, monomials may **not** have fractional or negative exponents. These facts are part of the definition since in the expression kx^n, k (the coefficient) can be any real number, but n (the exponent) must be a whole number.

Expressions that **are not** monomials: $3\sqrt{x},\ -15x^{\frac{2}{3}},\ 4a^{-2}$

Expressions that **are** monomials: $17,\ -3x,\ 5y^2,\ \frac{2}{7}a^4$

Since $x^0 = 1$, a nonzero constant can be multiplied by x^0 without changing its value. Thus, we say that a **nonzero constant is a monomial of degree 0**. For example,

$$17 = 17x^0 \qquad \text{and} \qquad -6 = -6x^0$$

which means that the constants 17 and –6 are monomials of degree 0. However, for the special number 0, we can write

$$0 = 0x^2 = 0x^5 = 0x^{13}$$

and we say that **the constant 0 is not a monomial.**

A **polynomial** is a monomial or the algebraic sum of monomials. Examples of polynomials are

$$3x,\ \ y+5,\ \ 4x^2-7x+1,\ \text{and}\ \ a^{10}+5a^3-2a^2+6.$$

In the formal definition of a polynomial, we make use of **subscript notation** for the coefficients. A subscript is a number written to the right and below a letter. Subscripts allow the use of the same letter while indicating different numbers. For example, a_1 (read "a sub 1") and a_2 (read "a sub 2") indicate two different numbers even though the same letter, a, is used.

Polynomial in *x*

*A **polynomial in x** is an expression of the form*

$$a_nx^n + a_{n-1}x^{n-1} + a_{n-2}x^{n-2} + \ldots + a_1x^1 + a_0$$

where $n, n-1, n-2, \ldots, 1, 0$ are whole numbers

and $a_n, a_{n-1}, \ldots, a_1, a_0$ are real numbers with $a_n \neq 0$.

n** is called **the degree of the polynomial.

a_n** is called **the leading coefficient of the polynomial.

As an example,

$$x^3 + 5x^2 - 8x + 14 \text{ is a third-degree polynomial in } x.$$

We can write

$$x^3 + 5x^2 - 8x + 14 = a_3x^3 + a_2x^2 + a_1x + a_0.$$

Since the corresponding coefficients must be equal, we have

$$a_3 = 1 \text{ (the leading coefficient is 1)}, a_2 = 5, a_1 = -8, \text{ and } a_0 = 14.$$

Similarly,

$$3y^4 - 2y^3 + 4y - 6 \text{ is a fourth-degree polynomial in } y$$

and writing

$$3y^4 - 2y^3 + 4y - 6 = a_4y^4 + a_3y^3 + a_2y^2 + a_1y + a_0$$

we see that

$$a_4 = 3 \text{ (the leading coefficient is 3)}, a_3 = -2, a_2 = 0, a_1 = 4, \text{ and } a_0 = -6.$$

In each of these examples, the terms have been written so that the exponents decrease in order from left to right. We say that the terms are written in **descending order**. If the terms increase in order from left to right, then we say that the terms are written in **ascending order**. As a general rule, for consistency and style in operating with polynomials, the polynomials in this chapter will be written in descending order.

Some forms of polynomials are used so frequently that they are given special names. These names are indicated in the following box with examples. Remember that all are also classified as polynomials.

Classification of Polynomials

		Example	
Monomial	*polynomial with one term*	$7a^3$	***Third-degree monomial***
Binomial	*polynomial with two terms*	$3x^2 + 5x$	***Second-degree binomial***
Trinomial	*polynomial with three terms*	$2x^5 + 3x^2 + 1$	***Fifth-degree trinomial***

Example 1: Simplify Polynomials

Simplify each of the following polynomials by combining like terms. Write the polynomial in descending order and state the degree and type of the polynomial.

a. $5x^3 + 7x^3$

Solution: $5x^3 + 7x^3 = (5 + 7)x^3 = 12x^3$ Third-degree monomial.

continued on next page ...

1. Monomial
2. Trinomial
3. Not a polynomial
4. Binomial
5. Trinomial
6. Not a polynomial
7. Not a polynomial
8. Trinomial
9. Binomial
10. Binomial
11. $4y$; first degree monomial. $a_1 = 4,\ a_0 = 0$
12. $5x^2 - x$; second degree binomial; $a_2 = 5,\ a_1 = -1,\ a_0 = 0$

b. $5x^3 + 7x^3 - 2x$

Solution: $5x^3 + 7x^3 - 2x = 12x^3 - 2x$ Third-degree binomial.

c. $\frac{1}{2}y + 3y - \frac{2}{3}y^2 - 7$

Solution: $\frac{1}{2}y + 3y - \frac{2}{3}y^2 - 7 = -\frac{2}{3}y^2 + \frac{7}{2}y - 7$ Second-degree trinomial.

d. $x^2 + 8x - 15 - x^2$

Solution: $x^2 + 8x - 15 - x^2 = 8x - 15$ First-degree binomial.

e. $-3y^4 + 2y^2 + y^{-1}$

Solution: This expression is not a polynomial since y has a negative exponent.

Evaluating a Polynomial

Teaching Notes: Some students may be perplexed by the notation "$p(x)$" and continue confusing it as meaning multiplication. Remind the students of the context for functions and be aware of the dual interpretations. Also, mention that the use of "p" is purely arbitrary and other letters may be used.

In evaluating any numerical expression, we follow the Rules for Order of Operations discussed in Section 1.5. Similarly, to evaluate a polynomial for a given value of the variable, we substitute the value for the variable wherever it occurs in the polynomial and follow the rules for order of operations.

A polynomial in x can be thought of as a function of x. Thus, using the notation for functions discussed in Section 4.5, we can write $p(x)$ = a polynomial in x. This notation is particularly appropriate when evaluating a polynomial, as illustrated in Example 2.

Example 2: Evaluating Polynomials

a. Evaluate $p(x) = 4x^2 + 5x - 15$ for $x = 3$.

Solution: For $p(x) = 4x^2 + 5x - 15$,

$$\begin{aligned} p(3) &= 4 \cdot 3^2 + 5 \cdot 3 - 15 \quad \text{Substitute 3 for } x. \\ &= 4 \cdot 9 + 15 - 15 \\ &= 36 + 15 - 15 \\ &= 36 \end{aligned}$$

13. $x^3 + 3x^2 - 2x$; third degree trinomial; $a_3 = 1,\ a_2 = 3,\ a_1 = -2,\ a_0 = 0$
14. $3x^2$; second degree monomial; $a_2 = 3,\ a_1 = 0,\ a_0 = 0$

b. Evaluate $p(y) = 5y^3 + y^2 - 3y + 8$ for $y = -2$.

Solution: For $p(y) = 5y^3 + y^2 - 3y + 8$,

$$\begin{aligned} p(-2) &= 5(-2)^3 + (-2)^2 - 3(-2) + 8 \quad \text{Note the use of parentheses around } -2. \\ &= 5(-8) + 4 - 3(-2) + 8 \\ &= -40 + 4 + 6 + 8 \\ &= -22 \end{aligned}$$

Practice Problems

15. $-2x^2$; second degree monomial; $a_2 = -2,\ a_1 = 0,\ a_0 = 0$

Combine like terms and state the degree and type of the polynomial.

1. $8x^3 - 3x^2 - x^3 + 5 + 3x^2$ **2.** $5y^4 + y^4 - 3y^2 + 2y^2 + 4$

3. *For the polynomial* $p(x) = x^2 - 5x - 5$, *find (a)* $p(3)$ *and (b)* $p(-1)$.

5.3 Exercises

16. $-y$; first degree monomial; $a_1 = -1,\ a_0 = 0$

17. 0; not a polynomial; $a_0 = 0$

18. $4x^2 - 3x + 2$; second degree trinomial; $a_2 = 4,\ a_1 = -3,\ a_0 = 2$

19. $6a^5 - 7a^3 - a^2$; fifth degree trinomial; $a_5 = 6,\ a_4 = 0,\ a_3 = -7;\ a_2 = -1,\ a_1 = 0,\ a_0 = 0$

20. $-x^2$; second degree monomial; $a_2 = -1,\ a_1 = 0,\ a_0 = 0$

21. $2y^3 + 4y$; third degree binomial; $a_3 = 2,\ a_2 = 0,\ a_1 = 4,\ a_0 = 0$

22. $-x + 10$; first degree binomial; $a_1 = -1,\ a_0 = 10$

23. 4; monomial of degree 0; $a_0 = 4$

24. $4x^2 - 10x$; second degree binomial; $a_2 = 4,\ a_1 = -10,\ a_0 = 0$

In Exercises 1 – 10, identify the expression as a monomial, binomial, trinomial, or not a polynomial.

1. $3x^4$ **2.** $5y^2 - 2y + 1$ **3.** $-2x^{-2}$ **4.** $8x^3 - 7$

5. $14a^7 - 2a - 6$ **6.** $17x^{\frac{2}{3}} + 5x^2$ **7.** $6a^3 + 5a^2 - a^{-3}$ **8.** $-3y^4 + 2y^2 - 9$

9. $\frac{1}{2}x^3 - \frac{2}{5}x$ **10.** $\frac{5}{8}x^5 + \frac{2}{3}x^4$

Simplify the polynomials in Exercises 11 – 30. Write the polynomial in descending order and state the degree and type of the simplified polynomial. For each polynomial, write the values for $a_n, a_{n-1}, a_{n-2}, \ldots, a_1$, *and* a_0.

11. $y + 3y$ **12.** $4x^2 - x + x^2$ **13.** $x^3 + 3x^2 - 2x$

14. $3x^2 - 8x + 8x$ **15.** $x^4 - 4x^2 + 2x^2 - x^4$ **16.** $2 - 6y + 5y - 2$

17. $-x^3 + 6x + x^3 - 6x$ **18.** $11x^2 - 3x + 2 - 7x^2$ **19.** $6a^5 + 2a^2 - 7a^3 - 3a^2$

20. $2x^2 - 3x^2 + 2 - 5x^2 - 2 + 5x^2$ **21.** $4y - 8y^2 + 2y^3 + 8y^2$

22. $2x + 9 - x + 1 - 2x$ **23.** $5y^2 + 3 - 2y^2 + 1 - 3y^2$

24. $13x^2 - 6x - 9x^2 - 4x$ **25.** $7x^3 + 3x^2 - 2x + x - 5x^3 + 1$

26. $-3y^5 + 7y - 2y^3 - 5 + 4y^2 + y^2$ **27.** $x^4 + 3x^4 - 2x + 5x - 10 - x^2 + x$

28. $a^3 + 2a^2 - 6a + 3a^3 + 2a^2 + 7a + 3$ **29.** $2x + 4x^2 + 6x + 9x^3$

30. $15y - y^3 + 2y^2 - 10y^2 + 2y - 16$

In Exercises 31 – 40, evaluate the polynomials for the given value of the variable.

31. $p(x) = x^2 + 14x - 3; x = -1$ **32.** $p(y) = y^3 - 5y^2 + 6y + 2; y = 2$

33. $p(x) = 3x^3 - 9x^2 - 10x - 11;\ x = 3$ **34.** $p(x) = -5x^2 - 8x + 7; x = -3$

35. $p(x) = 8x^4 + 2x^3 - 6x^2 - 7; x = -2$ **36.** $p(a) = a^3 + 4a^2 + a + 2; a = -5$

Answers to Practice Problems: 1. $7x^3 + 5$; third degree binomial **2.** $6y^4 - y^2 + 4$; fourth degree trinomial **3. a.** -11 **b.** 1

25. $2x^3+3x^2-x+1$; third degree polynomial; $a_3=2$, $a_2=3$, $a_1=-1$, $a_0=1$

26. $-3y^5-2y^3+5y^2+7y-5$; fifth degree polynomial; $a_5=-3, a_4=0, a_3=-2$, $a_2=5$, $a_1=7$, $a_0=-5$

27. $4x^4-x^2+4x-10$; fourth degree polynomial; $a_4=4$, $a_3=0$, $a_2=-1$, $a_1=4$, $a_0=-10$

28. $4a^3+4a^2+a+3$; third degree polynomial; $a_3=4$, $a_2=4$, $a_1=1$, $a_0=3$

29. $9x^3+4x^2+8x$; third degree trinomial; $a_3=9$, $a_2=4$, $a_1=8$, $a_0=0$

30. $-y^3-8y^2+17y-16$; third degree polynomial; $a_3=-1$, $a_2=-8$, $a_1=17$, $a_0=-16$

31. $p(-1)=-16$

32. $p(2)=2$

37. $p(a)=2a^4+3a^2-8a; a=-1$

38. $p(y)=-4y^3+5y^2+12y-1; y=-10$

39. $p(x)=x^5-x^3+x-2; x=2$

40. $p(x)=3x^6-2x^5+x^4-x^3-3x^2+2x-1; x=1$

33. $p(3)=-41$ **34.** $p(-3)=-14$
35. $p(-2)=81$ **36.** $p(-5)=-28$
37. $p(-1)=13$ **38.** $p(-10)=4379$
39. $p(2)=24$ **40.** $p(1)=-1$

Writing and Thinking About Mathematics

First-degree polynomials are also called **linear polynomials**, second-degree polynomials are called **quadratic polynomials**, and third-degree polynomials are called **cubic polynomials**. The related functions are called linear functions, quadratic functions, and cubic functions, respectively.

41. Use a graphing calculator to graph the following linear functions.
a. $p(x)=2x+3$ **b.** $p(x)=-3x+1$ **c.** $p(x)=\frac{1}{2}x$

42. Use a graphing calculator to graph the following quadratic functions.
a. $p(x)=x^2$ **b.** $p(x)=x^2+6x+9$ **c.** $p(x)=-x^2+2$

43. Use a graphing calculator to graph the following cubic functions.
a. $p(x)=x^3$ **b.** $p(x)=x^3-4x$ **c.** $p(x)=x^3+2x^2-5$

44. Make up a few of your own linear, quadratic, and cubic functions and graph these functions with your calculator. Using the results from Exercises 41, 42, and 43, and your own functions, describe in your own words:
a. the nature of graphs of linear functions.
b. the nature of graphs of quadratic functions.
c. the nature of graphs of cubic functions.

44. a. – c. Answers will vary.

Hawkes Learning Systems: Introductory & Intermediate Algebra

Identifying Polynomials
Evaluating Polynomials

41. a.
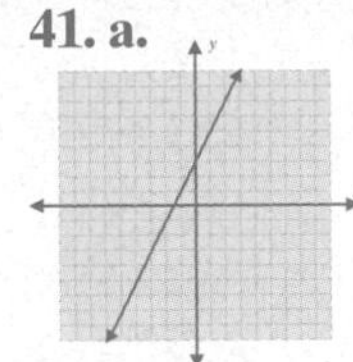
b.
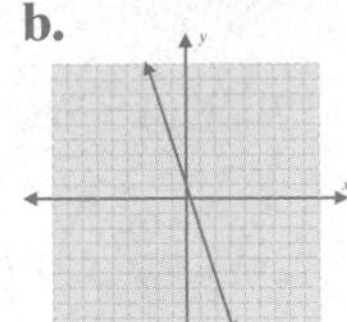
c.
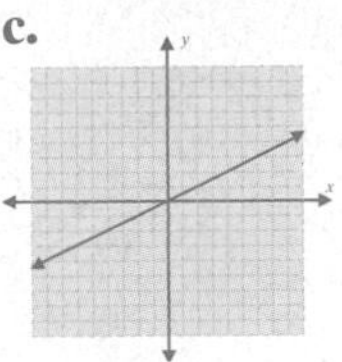

42. a.
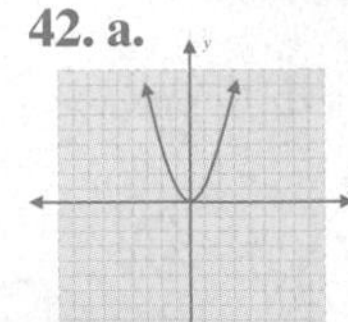
b.
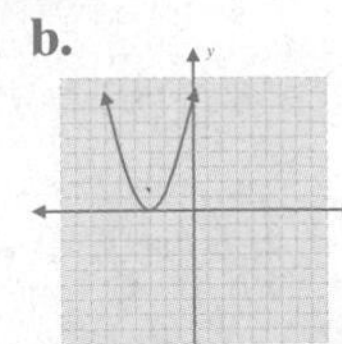
c.
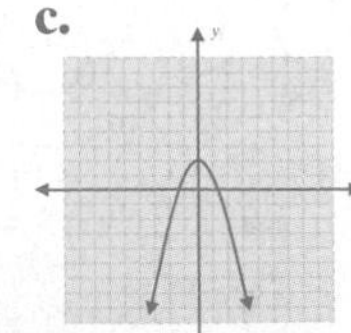

43. a.
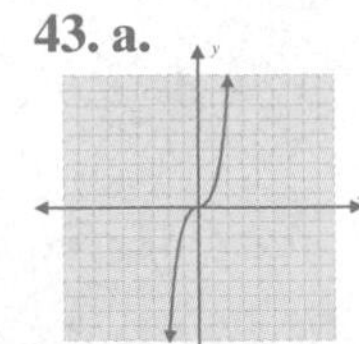
b.
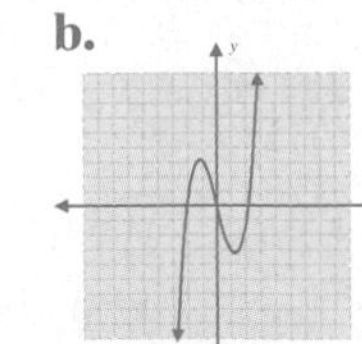
c.
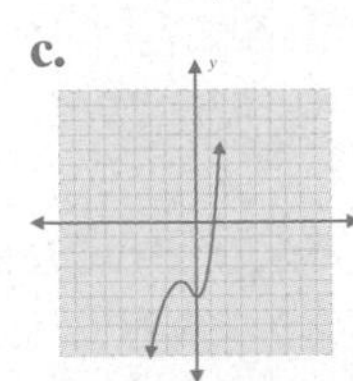

5.4 Adding and Subtracting Polynomials

Objectives

After completing this section, you will be able to:

1. *Add polynomials.*
2. *Subtract polynomials.*
3. *Simplify expressions by removing grouping symbols and combining like terms.*

We have discussed identifying polynomials in terms of degree and type. In the remainder of this chapter, we will discuss algebraic operations with polynomials. That is, we will learn to add, subtract, multiply, and divide with polynomials.

Addition with Polynomials

The **sum** of two or more polynomials is found by combining like terms. The polynomials may be written horizontally or vertically. For example,

Teaching Notes: For simplifying expressions like these, the roles of the commutative and associative properties are important and students need to be reminded that these properties are being used frequently.

$$\begin{aligned}&(x^2 - 5x + 3) + (2x^2 - 8x - 4) + (3x^3 + x^2 - 5)\\&= 3x^3 + (x^2 + 2x^2 + x^2) + (-5x - 8x) + (3 - 4 - 5)\\&= 3x^3 + 4x^2 - 13x - 6\end{aligned}$$

If the polynomials are written vertically, we write like terms one beneath the other in a column format and add the like terms in each column.

$$\begin{array}{rrrr} & x^2 & -\ 5x & +\ 3 \\ & 2x^2 & -\ 8x & -\ 4 \\ 3x^3 & +\ x^2 & & -\ 5 \\ \hline 3x^3 & +4x^2 & -13x & -\ 6 \end{array}$$

Example 1: Addition with Polynomials

a. Add as indicated: $(5x^3 - 8x^2 + 12x + 13) + (-2x^2 - 8) + (4x^3 - 5x + 14)$

Solution: $(5x^3 - 8x^2 + 12x + 13) + (-2x^2 - 8) + (4x^3 - 5x + 14)$

$= (5x^3 + 4x^3) + (-8x^2 - 2x^2) + (12x - 5x) + (13 - 8 + 14)$

$= 9x^3 - 10x^2 + 7x + 19$

b. Find the sum: $(x^3 - x^2 + 5x) + (4x^3 + 5x^2 - 8x + 9)$

Solution:

$$\begin{array}{r} x^3 - x^2 + 5x \phantom{{}+9} \\ 4x^3 + 5x^2 - 8x + 9 \\ \hline 5x^3 + 4x^2 - 3x + 9 \end{array}$$

Subtraction with Polynomials

If a negative sign is written in front of a polynomial in parentheses, the meaning is the opposite of the entire polynomial. The opposite can be found by changing the sign of every term in the polynomial.

$$-(2x^2 + 3x - 7) = -2x^2 - 3x + 7$$

We can also think of the opposite of a polynomial as –1 times the polynomial, then applying the distibutive property as follows:

$$\begin{aligned} -(2x^2 + 3x - 7) &= -1(2x^2 + 3x - 7) \\ &= -1(2x^2) - 1(3x) - 1(-7) \\ &= -2x^2 - 3x + 7 \end{aligned}$$

The result is the same with either approach. So the **difference** between two polynomials can be found by changing the sign of each term of the second polynomial and then combining like terms.

Teaching Notes:
Some students may need to be reminded that "combining" like terms is just a use of the distributive property.

$$\begin{aligned} (5x^2 - 3x - 7) - (2x^2 + 5x - 8) &= 5x^2 - 3x - 7 - 2x^2 - 5x + 8 \\ &= 5x^2 - 2x^2 - 3x - 5x - 7 + 8 \\ &= 3x^2 - 8x + 1 \end{aligned}$$

If the polynomials are written in a vertical format, one beneath the other, we change the signs of the terms of the polynomial being subtracted and then combine like terms.

Subtract:

$$\begin{array}{r} 5x^2 - 3x - 7 \\ -(2x^2 + 5x - 8) \\ \hline \end{array} \longrightarrow \begin{array}{r} 5x^2 - 3x - 7 \\ -2x^2 - 5x + 8 \\ \hline 3x^2 - 8x + 1 \end{array}$$

Example 2: Subtraction with Polynomials

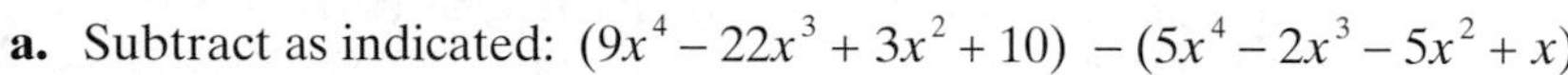

a. Subtract as indicated: $(9x^4 - 22x^3 + 3x^2 + 10) - (5x^4 - 2x^3 - 5x^2 + x)$

Solution:
$$\begin{aligned} &(9x^4 - 22x^3 + 3x^2 + 10) - (5x^4 - 2x^3 - 5x^2 + x) \\ &= 9x^4 - 22x^3 + 3x^2 + 10 - 5x^4 + 2x^3 + 5x^2 - x \\ &= 9x^4 - 5x^4 - 22x^3 + 2x^3 + 3x^2 + 5x^2 - x + 10 \\ &= 4x^4 - 20x^3 + 8x^2 - x + 10 \end{aligned}$$

b. Find the difference:
$$\begin{array}{r} 8x^3 + 5x^2 - 14 \\ -(-2x^3 + \ x^2 + 6x) \\ \hline \end{array}$$

Solution:
$$\begin{array}{r} 8x^3 + 5x^2 - 14 \\ -(-2x^3 + x^2 + 6x) \\ \hline \end{array} \longrightarrow \begin{array}{r} 8x^3 + 5x^2 + 0x - 14 \\ 2x^3 - x^2 - 6x + 0 \\ \hline 10x^3 + 4x^2 - 6x - 14 \end{array}$$

Write in 0's for missing powers to help with alignment of like terms.

If an expression contains more than one pair of grouping (or inclusion) symbols, such as parentheses ( ), brackets [ ], or braces { }, simplify by working to remove the innermost pair of symbols first.

Example 3: Simplify

Simplify each of the following expressions.

a. $5x - [\,2x + 3\,(4 - x) + 1\,] - 9$

Solution:
$$\begin{aligned} &5x - [\,2x + 3(4 - x) + 1\,] - 9 \\ &= 5x - [\,2x + 12 - 3x + 1\,] - 9 \\ &= 5x - [\,-x + 13\,] - 9 \\ &= 5x + x - 13 - 9 \\ &= 6x - 22 \end{aligned}$$

Work with the parentheses first since they are included inside the brackets.

b. $10 - x + 2\,[\,x + 3\,(x - 5) + 7\,]$

Solution:
$$\begin{aligned} &10 - x + 2\,[\,x + 3(x - 5) + 7\,] \\ &= 10 - x + 2\,[\,x + 3x - 15 + 7\,] \\ &= 10 - x + 2\,[\,4x - 8\,] \\ &= 10 - x + 8x - 16 \\ &= 7x - 6 \end{aligned}$$

Work with the parentheses first since they are included inside the brackets.

Practice Problems

Add or subtract as indicated and simplify the result.

1. Add: $(15x + 4) + (3x^2 - 9x - 5)$
2. Subtract: $(-5x^3 - 3x + 4) - (3x^3 - x^2 + 4x - 7)$
3. Simplify: $2 - [3a - (4 - 7a) + 2a]$
4. $(3x^2 - 2x + 5) + (2x^2 - x + 3)$
5. $(x^3 - 2x^2) - (x^2 - 1)$
6. $(5x^2 - 9x - 11) - (x^2 - 3x + 1)$

5.4 Exercises

Find the indicated sum in Exercises 1 – 29.

1. $(2x^2 + 5x - 1) + (x^2 + 2x + 3)$
2. $(x^2 + 2x - 3) + (x^2 + 5)$
3. $(x^2 + 7x - 7) + (x^2 + 4x)$
4. $(x^2 + 3x - 8) + (3x^2 - 2x + 4)$
5. $(2x^2 - x - 1) + (x^2 + x + 1)$
6. $(3x^2 + 5x - 4) + (2x^2 + x - 6)$
7. $(-2x^2 - 3x + 9) + (3x^2 - 2x + 8)$
8. $(x^2 + 6x - 7) + (3x^2 + x - 1)$
9. $\left(5x^2 + 8x - 3\right) + \left(-2x^2 + 6x - 4\right)$
10. $\left(x^2 - 9x + 2\right) + \left(-x^2 + 2x - 8\right)$
11. $\left(4x^2 - 8xy - 2y^2\right) + \left(-9x^2 + 5xy - 6y^2\right)$
12. $\left(x^2 + y^2\right) + \left(2x^2 - 5y^2\right)$
13. $(-4x^2 + 2x - 1) + (3x^2 - x + 2) + (x - 8)$
14. $(8x^2 + 5x + 2) + (-3x^2 + 9x - 4)$
15. $(x^2 - 3) + (-2x^2 + x + 4) + (x - 1)$
16. $(x^3 + 2x - 9) + (x^2 - 5x + 2) + (x^3 - 4x^2 + 1)$

17. $$\begin{array}{r} x^2 + 4x - 4 \\ \underline{-2x^2 + 3x + 1} \end{array}$$

18. $$\begin{array}{r} x^3 + 3x^2 + x \phantom{{}-4} \\ \underline{-2x^3 - x^2 + 2x - 4} \end{array}$$

19. $$\begin{array}{r} 7x^3 + 5x^2 + x - 6 \\ -3x^2 + 4x + 11 \\ \underline{-3x^3 - 2x^2 - 5x + 2} \end{array}$$

20. $$\begin{array}{r} 5x^2 - 3x + 11 \\ \underline{-2x^2 + x - 6} \end{array}$$

21. $$\begin{array}{r} 2x^2 - 5x - 6 \\ \underline{-3x^2 + 2x - 1} \end{array}$$

22. $$\begin{array}{r} 2x^2 + 4x - 3 \\ \underline{3x^2 - 9x + 2} \end{array}$$

23. $$\begin{array}{r} x^3 + 2x^2 + x - 2 \\ \underline{x^3 - 2x^2 - 3x - 1} \end{array}$$

24. $$\begin{array}{r} 5x^3 - 4x^2 \phantom{{}-6x} - 9 \\ \underline{2x^3 - 3x^2 - 6x + 5} \end{array}$$

Answers (margin):

1. $3x^2 + 7x + 2$
2. $2x^2 + 2x + 2$
3. $2x^2 + 11x - 7$
4. $4x^2 + x - 4$
5. $3x^2$
6. $5x^2 + 6x - 10$
7. $x^2 - 5x + 17$
8. $4x^2 + 7x - 8$
9. $3x^2 + 14x - 7$
10. $-7x - 6$
11. $-5x^2 - 3xy - 8y^2$
12. $3x^2 - 4y^2$
13. $-x^2 + 2x - 7$
14. $5x^2 + 14x - 2$
15. $-x^2 + 2x$
16. $2x^3 - 3x^2 - 3x - 6$
17. $-x^2 + 7x - 3$
18. $-x^3 + 2x^2 + 3x - 4$
19. $4x^3 + 7$

Answers to Practice Problems: **1.** $3x^2 + 6x - 1$ **2.** $-8x^3 + x^2 - 7x + 11$ **3.** $-12a + 6$ **4.** $5x^2 - 3x + 8$ **5.** $x^3 - 3x^2 + 1$ **6.** $4x^2 - 6x - 12$

20. $3x^2-2x+5$
21. $-x^2-3x-7$
22. $5x^2-5x-1$
23. $2x^3-2x-3$
24. $7x^3-7x^2-6x-4$
25. $10x^4+2x^3-4x^2$ $+2x+1$
26. $34x^3+13x^2-8x+12$
27. $2x^3+11x^2-4x-3$
28. x
29. $5x^3+11x^2+10x-14$
30. x^2+x+6
31. $2x^2+5x-11$
32. $-x^4-2x^3-16$
33. $-4x^4-7x^3-11x^2$ $+5x+13$
34. $-3x^2-6x-2$
35. $3x^2-4x-8$
36. $9x^2+3x$
37. $2x^2+13x+9$
38. $4x^3-8x+5$
39. $2x^4-5x^3-x-3$
40. $-x^4+2x^3-3x^2-x+5$
41. $3x^4-7x^3+3x^2-5x$ $+9$
42. $-x^5-2x^3-3x^2-5x$ -22
43. $-4x^2+6x-11$
44. $4x^2+3x+11$
45. $-2x^3+3x^2-2x-8$
46. $6x^2-7x+18$
47. $2x^3+4x^2+3x-10$
48. $5x^2+2x+3$
49. $8x^4+6x^2+15$
50. $14x^2-15$
51. $-5x^2+8x-12$
52. $4x^3+4x^2-7x+24$

25. $$\begin{array}{r} 3x^4+3x^3+x^2+x+2 \\ \underline{7x^4-x^3-5x^2+x-1} \end{array}$$

26. $$\begin{array}{r} 14x^3+13x^2+10x-13 \\ \underline{20x^3\qquad\qquad -18x+25} \end{array}$$

27. $$\begin{array}{r} x^3+3x^2\qquad -4 \\ 7x^2+2x+1 \\ \underline{x^3+x^2-6x\quad} \end{array}$$

28. $$\begin{array}{r} x^3+2x^2\qquad -5 \\ -2x^3\qquad +x-9 \\ \underline{x^3-2x^2\qquad +14} \end{array}$$

29. $$\begin{array}{r} x^3+5x^2+7x-3 \\ 4x^2+3x-9 \\ \underline{4x^3+2x^2\qquad -2} \end{array}$$

Find the indicated difference in Exercises 30 – 56.

30. $(2x^2+4x+8)-(x^2+3x+2)$

31. $(3x^2+7x-6)-(x^2+2x+5)$

32. $(x^4+8x^3-2x^2-5)-(2x^4+10x^3-2x^2+11)$

33. $(-3x^4+2x^3-7x^2+6x+12)-(x^4+9x^3+4x^2+x-1)$

34. $(x^2-9x+2)-(4x^2-3x+4)$

35. $(2x^2-x-10)-(-x^2+3x-2)$

36. $(7x^2+4x-9)-(-2x^2+x-9)$

37. $(6x^2+11x+2)-(4x^2-2x-7)$

38. $\left(6x^3-5x+1\right)-\left(2x^3+3x-4\right)$

39. $\left(2x^4+3x\right)-\left(5x^3+4x+3\right)$

40. $\left(2x^3-3x^2+6\right)-\left(x^4+x+1\right)$

41. $(3x^4-2x^3-8x-1)-(5x^3-3x^2-3x-10)$

42. $(x^5+6x^3-3x^2-5)-(2x^5+8x^3+5x+17)$

43. $(9x^2+6x-5)-(13x^2+6)$

44. $(8x^2+9)-(4x^2-3x-2)$

45. $(x^3+4x^2-7)-(3x^3+x^2+2x+1)$

46. $$\begin{array}{r} 14x^2-6x+9 \\ \underline{-\left(8x^2+x-9\right)} \end{array}$$

47. $$\begin{array}{r} x^3+6x^2\qquad -3 \\ \underline{-\left(-x^3+2x^2-3x+7\right)} \end{array}$$

48. $$\begin{array}{r} 9x^2-3x+2 \\ \underline{-\left(4x^2-5x-1\right)} \end{array}$$

49. $$\begin{array}{r} 5x^4+8x^2+11 \\ \underline{-\left(-3x^4+2x^2-4\right)} \end{array}$$

50. $$\begin{array}{r} 11x^2+5x-13 \\ \underline{-\left(-3x^2+5x+2\right)} \end{array}$$

51. $$\begin{array}{r} -3x^2+7x-6 \\ \underline{-\left(2x^2-x+6\right)} \end{array}$$

52. $$\begin{array}{r} 5x^3\qquad -10x+15 \\ \underline{-\left(x^3-4x^2-3x-9\right)} \end{array}$$

53. $$\begin{array}{r} x^3-8x^2+12x+5 \\ \underline{-\left(-3x^3+8x^2+2x+5\right)} \end{array}$$

54. $$\begin{array}{r} 3x^3\qquad +9x-17 \\ \underline{-\left(x^3+5x^2-2x-6\right)} \end{array}$$

55. $$\begin{array}{r} -3x^4+10x^3-8x^2-7x-6 \\ \underline{-\left(2x^4+x^3\qquad +5x+6\right)} \end{array}$$

56. $$\begin{array}{r} 2x^4-5x^3-6x^2+7x+7 \\ \underline{-\left(x^4\qquad +2x^2+4x+10\right)} \end{array}$$

53. $4x^3 - 16x^2 + 10x$
54. $2x^3 - 5x^2 + 11x - 11$
55. $-5x^4 + 9x^3 - 8x^2 - 12x - 12$
56. $x^4 - 5x^3 - 8x^2 + 3x - 3$
57. $4x - 13$
58. $-12x + 22$
59. $-7x + 9$
60. $-x - 15$
61. $2x + 17$
62. $-9x + 26$
63. $3x^3 + 13x^2 + 9$
64. $10x^3 - 3x^2 + 7$
65. $8x^2 - x - 2$
66. $-7x^2 + 6x - 2$
67. $3x - 19$
68. $5x - 7$ 69. $7x - 1$
70. $2x^2 - 3x$
71. $4x^2 - x$
72. $7x^3 - 2x^2 + 9x$
73. $x^2 + 11x - 8$
74. $5x^2 - 16x - 14$
75. $7x^3 - x^2 - 7$
76. $7x - 6$
77. Any monomial or algebraic sum of monomials.
78. Answers will vary.

Simplify each of the expressions in Exercises 57 – 71, and write the polynomials in order of descending powers.

57. $5x + 2(x - 3) - (3x + 7)$

58. $-4(x - 6) - (8x + 2)$

59. $11 + [\,3x - 2(1 + 5x)]$

60. $2x + [\,9x - 4(3x + 2) - 7\,]$

61. $8x - [\,2x + 4(x - 3) - 5\,]$

62. $17 - [\,-3x + 6(2x - 3) + 9\,]$

63. $3x^3 - [\,5 - 7(x^2 + 2) - 6x^2\,]$

64. $10x^3 - [\,8 - 5(3 - 2x^2) - 7x^2\,]$

65. $(2x^2 + 4) - [\,-8 + 2(7 - 3x^2) + x\,]$

66. $-[\,6x^2 - 3(4 + 2x) + 9\,] - (x^2 + 5)$

67. $2[\,3x + (x - 8) - (2x + 5)] - (x - 7)$

68. $-3[\,-x + (10 - 3x) - (8 - 3x)] + (2x - 1)$

69. $(x^2 - 1) + x[\,4 + (3 - x)]$

70. $x(x - 5) + [\,6x - x(4 - x)]$

71. $x(2x + 1) - [\,5x - x(2x + 3)]$

72. Subtract $2x^2 - 4x$ from $7x^3 + 5x$.

73. Subtract $3x^2 - 4x + 2$ from the sum of $4x^2 + x - 1$ and $6x - 5$.

74. Subtract $-2x^2 + 6x + 12$ from the sum of $2x^2 + 3x - 4$ and $x^2 - 13x + 2$.

75. Add $5x^3 - 8x + 1$ to the difference between $2x^3 + 14x - 3$ and $x^2 + 6x + 5$.

76. Find the sum of $10x - 2(3x + 5)$ and $3(x - 4) + 16$.

Writing and Thinking About Mathematics

77. Write the definition of a polynomial.

78. Explain, in your own words, how to subtract one polynomial from another.

79. Describe what is meant by the degree of a polynomial.

80. Give two examples that show how the sum of two binomials might not be a binomial.

81. Give two examples that show
   a. how the sum of two cubic polynomials might not be a cubic polynomial.
   b. how the difference of two quadratic polynomials might not be a quadratic polynomial.

Hawkes Learning Systems: Introductory & Intermediate Algebra

Adding and Subtracting Polynomials

79. The largest of the degrees of its terms after like terms have been combined.
80. Answers will vary. 81. a. Answers will vary. b. Answers will vary.

5.5 Multiplying Polynomials

Objectives

After completing this section, you will be able to:

1. *Multiply polynomials.*
2. *Multiply two binomials by using the FOIL method.*

Up to this point, we have multiplied terms such as $5x^2 \cdot 3x^4 = 15x^6$ by using the Product Rule for Exponents. Also, we have applied the distributive property to expressions such as $5(2x + 3) = 10x + 15$.

Now we will use both of these procedures to multiply polynomials. We will discuss first the product of a monomial with a polynomial of two or more terms; second, the product of two binomials; and third, the product of a binomial with a polynomial of more than two terms.

Teaching Notes: While we continue to emphasize the use of the distributive property in so many instances of simplifying, it is also important to point out our continued use of the commutative and associatve properties toward this end. For example, in this demonstration we reach a point of having "$5x \cdot 3$" and we simplify this as "$15x$" (by the commutative and associative properties of multiplication).

Multiplying a Polynomial by a Monomial

Using the distributive property $a(b + c) = ab + ac$ with multiplication indicated on the left, we can find the product of a monomial with a polynomial of two or more terms as follows:

$$5x(2x + 3) = 5x \cdot 2x + 5x \cdot 3 = 10x^2 + 15x$$
$$3x^2(4x - 1) = 3x^2 \cdot 4x + 3x^2(-1) = 12x^3 - 3x^2$$
$$-4a^5(a^2 - 8a + 5) = -4a^5 \cdot a^2 - 4a^5(-8a) - 4a^5(5) = -4a^7 + 32a^6 - 20a^5$$

Multiplying Two Polynomials

Now suppose that we want to multiply two binomials, say, $(x + 3)(x + 7)$. We will apply the distributive property in the following way with multiplication indicated on the right of the parentheses.

Compare $(x + 3)(x + 7)$ to

$$(a + b)c = ac + bc.$$

Think of $(x + 7)$ as taking the place of c. Thus,

$$\begin{array}{lcccccc} & (a + b)\,c & = & ac & + & bc \\ & \downarrow\ \downarrow\ \searrow & & \swarrow\downarrow & & \swarrow\downarrow \\ \text{takes the form} & (x + 3)(x + 7) & = & x(x + 7) & + & 3(x + 7) \end{array}$$

Completing the products on the right, using the distributive property twice again, gives

$$\begin{aligned}(x+3)(x+7) &= x(x+7)+3(x+7)\\ &= x\cdot x + x\cdot 7 + 3\cdot x + 3\cdot 7\\ &= x^2+7x+3x+21\\ &= x^2+10x+21\end{aligned}$$

In the same manner,

$$\begin{aligned}(x+2)(3x+4) &= x(3x+4)+2(3x+4)\\ &= x\cdot 3x + x\cdot 4 + 2\cdot 3x + 2\cdot 4\\ &= 3x^2+4x+6x+8\\ &= 3x^2+10x+8\end{aligned}$$

Similarly,

$$\begin{aligned}(2x-1)(x^2+x-5) &= 2x(x^2+x-5)-1(x^2+x-5)\\ &= 2x\cdot x^2 + 2x\cdot x + 2x(-5) - 1\cdot x^2 - 1\cdot x - 1(-5)\\ &= 2x^3+2x^2-10x-x^2-x+5\\ &= 2x^3+x^2-11x+5\end{aligned}$$

One quick way to check if your products are correct is to substitute some convenient number for x into the original two factors and into the product. Choose any nonzero number that you like. The values of both expressions should be the same. For example, let $x = 1$. Then,

$$(x+2)(3x+4) = (1+2)(3\cdot 1+4) = (3)(7) = 21$$

and

$$3x^2+10x+8 = 3\cdot 1^2+10\cdot 1+8 = 3+10+8 = 21$$

The product $(x+2)(3x+4) = 3x^2+10x+8$ seems to be correct. We could double check by letting $x = 5$.

$$(x+2)(3x+4) = (5+2)(3\cdot 5+4) = (7)(19) = 133$$

and

$$3x^2+10x+8 = 3\cdot 5^2+10\cdot 5+8 = 75+50+8 = 133$$

Convinced? This is just a quick check, however, and is not foolproof unless you try this process with more values of x than indicated by the degree of the product.

The product of two polynomials can also be found by writing one polynomial under the other. **The distributive property is applied by multiplying each term of one polynomial by each term of the other**. Consider the product $(2x^2+3x-4)(3x+7)$.

Now, writing one polynomial under the other and applying the distributive property, we obtain

Multiply by +7:

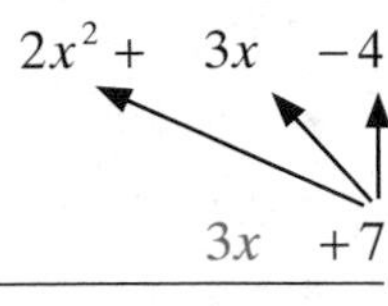

$$\begin{array}{rrr} 2x^2 + & 3x & -4 \\ & 3x & +7 \\ \hline 14x^2 + & 21x & -28 \end{array}$$

Multiply by $3x$:

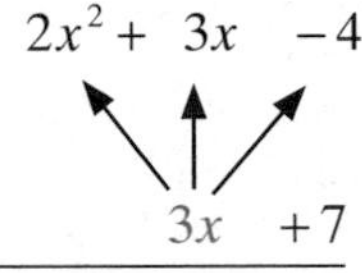

$$\begin{array}{rrrr} & 2x^2 + & 3x & -4 \\ & & 3x & +7 \\ \hline & 14x^2 + & 21x & -28 \\ 6x^3 + & 9x^2 - & 12x & \\ \hline \end{array}$$

Align the like terms so that they can be easily combined.

Finally, combine like terms:

$$\begin{array}{rrrr} & 2x^2 & +3x & -4 \\ & & 3x & +7 \\ \hline & 14x^2 & +21x & -28 \\ 6x^3 + & 9x^2 & -12x & \\ \hline 6x^3 & +23x^2 & +9x & -28 \end{array}$$

Combine like terms.

Example 1: Multiply

Find each product.

a. $-4x\left(x^2 - 3x + 12\right) = -4x \cdot x^2 - 4x(-3x) - 4x \cdot 12$

$= -4x^3 + 12x^2 - 48x$

b. $x^2y(x^2 + 3y^2) = x^2y \cdot x^2 + x^2y \cdot 3y^2$ Use the distributive property.

$= x^4y + 3x^2y^3$

c. $(2x+1)(x-5) = 2x(x-5) + 1(x-5)$ Use the distributive property.

$= 2x^2 - 10x + x - 5$

$= 2x^2 - 9x - 5$ Simplify.

d. $x^{3k}(x^k + x) = x^{3k} \cdot x^k + x^{3k} \cdot x$ Use the distributive property.

$= x^{3k+k} + x^{3k+1}$ Add the exponents.

$= x^{4k} + x^{3k+1}$ Simplify.

continued on next page ...

e. $(3a+4)(2a^2+a+5) = (3a+4)2a^2 + (3a+4)a + (3a+4)5$

$$= 6a^3 + 8a^2 + 3a^2 + 4a + 15a + 20$$

$$= 6a^3 + 11a^2 + 19a + 20$$

f. $$\begin{array}{r} 7y^2 - 3y + 2 \\ 2y + 3 \\ \hline \end{array}$$

Solution:

$$\begin{array}{r} 7y^2 - 3y + 2 \\ 2y + 3 \\ \hline 21y^2 - 9y + 6 \\ 14y^3 - 6y^2 + 4y \quad \\ \hline 14y^3 + 15y^2 - 5y + 6 \end{array}$$

Multiply by 3.

Multiply by $2y$.

Combine like terms.

g. $(x - 5)(x + 2)(x - 1)$

Solution: First multiply $(x - 5)(x + 2)$, then multiply this result by $(x - 1)$.

$$(x-5)(x+2) = x(x+2) - 5(x+2)$$

$$= x^2 + 2x - 5x - 10$$

$$= x^2 - 3x - 10$$

$$(x^2 - 3x - 10)(x-1) = (x^2 - 3x - 10)x + (x^2 - 3x - 10)(-1)$$

$$= x^3 - 3x^2 - 10x - x^2 + 3x + 10$$

$$= x^3 - 4x^2 - 7x + 10$$

The FOIL Method

In the case of the **product of two binomials** such as $(2x+5)(3x-7)$, the **FOIL** method is useful. **F-O-I-L** is a mnemonic device (memory aid) to help in remembering which terms of the binomials to multiply together. First, by using the distributive property we can see how the terms are multiplied:

$$(2x+5)(3x-7) = 2x(3x-7) + 5(3x-7)$$

$$= 2x \cdot 3x + 2x(-7) + 5 \cdot 3x + 5(-7)$$

↑	↑	↑	↑
First Terms	**O**utside Terms	**I**nside Terms	**L**ast Terms
F	**O**	**I**	**L**

Now, by using the FOIL method (mentally), except for combining like terms we can go directly to the answer:

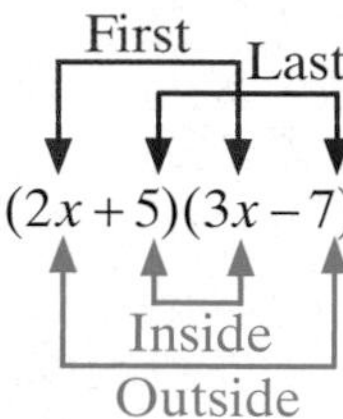

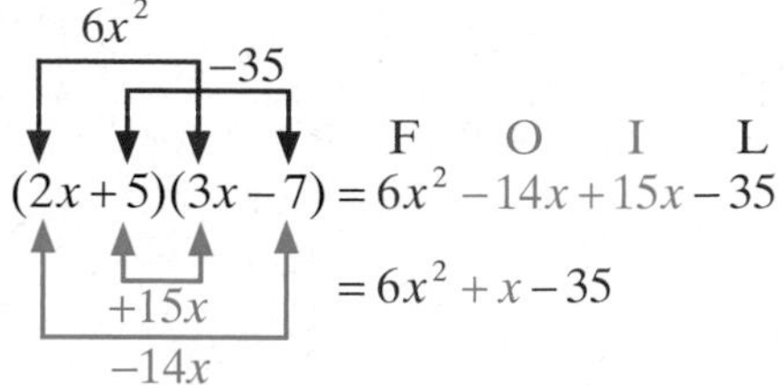

Example 2: FOIL Method

a. Find the product $(x+3)(2x+8)$ by using the **FOIL** method.

Solution:

First: $2x^2$ Last: 24

F O I L

$$(x+3)(2x+8)=2x^2+8x+6x+24$$
$$=2x^2+14x+24$$

Inside: $6x$ Outside: $8x$

b. Find the product $(2x-3)(3x-5)$ by using the **FOIL** method.

Solution:

First: $6x^2$ Last: 15

F O I L

$$(2x-3)(3x-5)=6x^2-10x-9x+15$$
$$=6x^2-19x+15$$

Inside: $-9x$ Outside: $-10x$

c. Find the product $(2x+1)(4x+3)$ by using the **FOIL** method.

Solution:

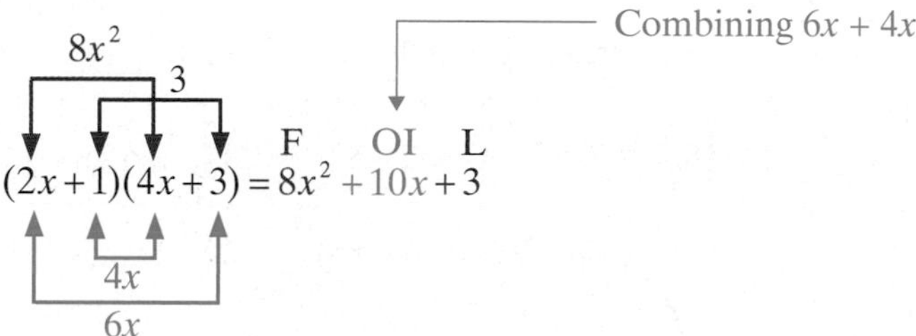

continued on next page ...

1. $5x^3-10x^2+15x$
2. $6x^4+10x^3-2x^2$
3. $x^3y^2+4xy^3$
4. $x^3z-4x^2yz+x^2z^2$
5. $-6x^5-15x^3$
6. $4x^7-12x^6+4x^5$
7. $-20x^4+30x^2$
8. $18x^6-9x^5+45x^4$
9. $-y^5+8y-2$
10. $-14y^4-21y^2-7$
11. $-4x^8+8x^7-12x^4$
12. $a^7+2a^6-5a^3+a^2$
13. $-35t^5+14t^4+7t^3$
14. $25x^5-5x^4+10x^3$
15. $-x^4-5x^2+4x$
16. $-2x^7+2x^6-4x^5$
17. $6x^2-x-2$
18. $3x^2+25x+28$
19. $9a^2-25$
20. $6x^2-x-5$
21. $-10x^2+39x-14$
22. y^3-y^2+y-1
23. x^3+5x^2+8x+4
24. $4x^3-x^2+x+3$
25. x^2+x-12
26. $x^2+2x-35$
27. $a^2-2a-48$
28. x^2-2x-8
29. x^2-3x+2
30. $x^2-15x+56$
31. $x^2-3x-18$
32. $x^2-7x+10$
33. x^2-9x+8
34. x^2+6x+8

d. Find the product $(x+6)(x-6)$ by using the **FOIL** method.

Solution:

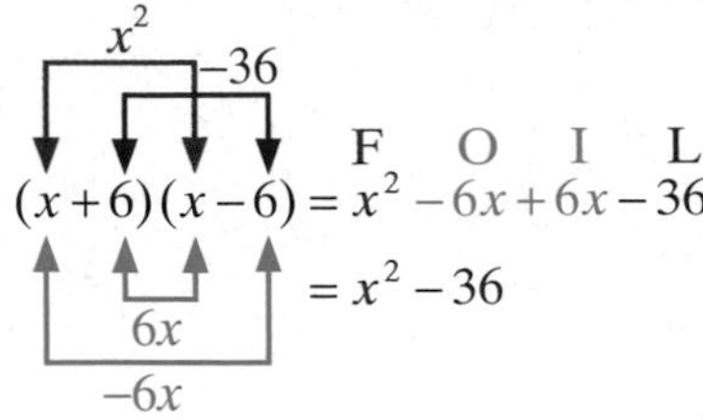

Practice Problems

Find each product.

1. $\begin{array}{r} 3x^2+x-1 \\ \underline{\quad 2x} \end{array}$

2. $\begin{array}{r} x-7 \\ \underline{x+3} \end{array}$

3. $\begin{array}{r} x^4+x-4 \\ \underline{x-1} \end{array}$

4. $\begin{array}{r} x^2+3x-2 \\ \underline{5x-1} \end{array}$

5. $\begin{array}{r} x+2 \\ x-2 \\ \underline{x+1} \end{array}$

5.5 Exercises

Find the indicated products in Exercises 1 – 65 and simplify if possible.

1. $5x(x^2-2x+3)$
2. $2x^2(3x^2+5x-1)$
3. $xy^2(x^2+4y)$
4. $x^2z(x-4y+z)$
5. $-3x^2(2x^3+5x)$
6. $4x^5(x^2-3x+1)$
7. $5x^2(-4x^2+6)$
8. $9x^3(2x^3-x^2+5x)$
9. $-1(y^5-8y+2)$
10. $-7(2y^4+3y^2+1)$
11. $-4x^3(x^5-2x^4+3x)$
12. $a^2(a^5+2a^4-5a+1)$
13. $7t^3(-5t^2+2t+1)$
14. $5x^3(5x^2-x+2)$
15. $-x(x^3+5x-4)$
16. $-2x^4(x^3-x^2+2x)$
17. $3x(2x+1)-2(2x+1)$
18. $x(3x+4)+7(3x+4)$
19. $3a(3a-5)+5(3a-5)$
20. $6x(x-1)+5(x-1)$
21. $5x(-2x+7)-2(-2x+7)$
22. $y(y^2+1)-1(y^2+1)$
23. $x(x^2+3x+2)+2(x^2+3x+2)$

Answers to Practice Problems: **1.** $6x^3+2x^2-2x$ **2.** $x^2-4x-21$ **3.** $x^5-x^4+x^2-5x+4$ **4.** $5x^3+14x^2-13x+2$ **5.** x^3+x^2-4x-4

35. $3t^2 - 3t - 60$
36. $-4x^2 + 4x + 168$
37. $x^3 + 11x^2 + 24x$
38. $t^3 - 11t^2 + 28t$
39. $2x^2 - 7x - 4$
40. $3x^2 + 11x - 4$
41. $6x^2 + 17x - 3$
42. $9t^2 - 25$
43. $4x^2 - 9$
44. $8x^2 + 23x + 15$
45. $16x^2 + 8x + 1$
46. $25x^2 - 20x + 4$
47. $2y^2 - 11y - 6$
48. $3y^2 + 17y + 10$
49. $3x^2 - 19x + 20$
50. $2x^2 - 5x + 2$
51. $6y^2 + 13y + 6$
52. $15y^2 - y - 2$
53. $8x^2 - 37x - 15$
54. $14x^2 - 9x - 18$
55. $27x^2 - 15x - 2$
56. $15x^2 - 13x - 44$
57. $y^3 + 2y^2 + y + 12$
58. $2x^3 - 13x^2 - 3x + 2$
59. $x^{k+2} + 3x^2$
60. $x^{2k+3} + x^4$
61. $x^{2k} - 2x^k - 15$
62. $x^{2k} - 36$
63. $x^{2k} + 5x^k + 4$
64. $2x^{2k} + x^k - 6$
65. $3x^{2k} + 17x^k + 10$
66. $3x^2 - 8x - 35$
67. $5x^3 + 6x^2 - 22x - 9$
68. $-16x^3 + 50x^2 + 25x - 14$

24. $4x(x^2 - x + 1) + 3(x^2 - x + 1)$ **25.** $(x + 4)(x - 3)$

26. $(x + 7)(x - 5)$ **27.** $(a + 6)(a - 8)$ **28.** $(x + 2)(x - 4)$

29. $(x - 2)(x - 1)$ **30.** $(x - 7)(x - 8)$ **31.** $(x + 3)(x - 6)$

32. $(x - 2)(x - 5)$ **33.** $(x - 8)(x - 1)$ **34.** $(x + 2)(x + 4)$

35. $3(t + 4)(t - 5)$ **36.** $-4(x + 6)(x - 7)$ **37.** $x(x + 3)(x + 8)$

38. $t(t - 4)(t - 7)$ **39.** $(2x + 1)(x - 4)$ **40.** $(3x - 1)(x + 4)$

41. $(6x - 1)(x + 3)$ **42.** $(3t + 5)(3t - 5)$ **43.** $(2x + 3)(2x - 3)$

44. $(8x + 15)(x + 1)$ **45.** $(4x + 1)(4x + 1)$ **46.** $(5x - 2)(5x - 2)$

47. $(2y + 1)(y - 6)$ **48.** $(y + 5)(3y + 2)$ **49.** $(3x - 4)(x - 5)$

50. $(2x - 1)(x - 2)$ **51.** $(2y + 3)(3y + 2)$ **52.** $(5y - 2)(3y + 1)$

53. $(8x + 3)(x - 5)$ **54.** $(7x + 6)(2x - 3)$ **55.** $(9x + 1)(3x - 2)$

56. $(5x - 11)(3x + 4)$ **57.** $(y + 3)(y^2 - y + 4)$ **58.** $(2x + 1)(x^2 - 7x + 2)$

59. $x^2(x^k + 3)$ **60.** $x^3(x^{2k} + x)$ **61.** $(x^k + 3)(x^k - 5)$

62. $(x^k + 6)(x^k - 6)$ **63.** $(x^k + 1)(x^k + 4)$ **64.** $(2x^k - 3)(x^k + 2)$

65. $(3x^k + 2)(x^k + 5)$

In Exercises 66 – 76, find the indicated products and simplify.

66. $\begin{array}{r} 3x + 7 \\ \underline{x - 5} \end{array}$ **67.** $\begin{array}{r} x^2 + 3x + 1 \\ \underline{5x - 9} \end{array}$ **68.** $\begin{array}{r} 8x^2 + 3x - 2 \\ \underline{-2x + 7} \end{array}$

69. $\begin{array}{r} 2x^2 + 3x + 5 \\ \underline{x^2 + 2x - 3} \end{array}$ **70.** $\begin{array}{r} 6x^2 - x + 8 \\ \underline{2x^2 + 5x + 6} \end{array}$ **71.** $\begin{array}{r} 2x^2 - 5x - 6 \\ \underline{3x + 1} \end{array}$

72. $\begin{array}{r} x^2 + 2x + 1 \\ \underline{x^2 + 2x + 1} \end{array}$ **73.** $\begin{array}{r} x^3 - 3x + 4 \\ \underline{2x - 3} \end{array}$ **74.** $\begin{array}{r} 2x^3 + 6x^2 + 5 \\ \underline{x^2 + 5} \end{array}$

75. $\begin{array}{r} x^3 - 7x - 4 \\ \underline{4x - 6} \end{array}$ **76.** $\begin{array}{r} x^3 - 5x + 14 \\ \underline{2x - 3} \end{array}$

69. $2x^4 + 7x^3 + 5x^2 + x - 15$

70. $12x^4 + 28x^3 + 47x^2 + 34x + 48$

71. $6x^3 - 13x^2 - 23x - 6$

72. $x^4 + 4x^3 + 6x^2 + 4x + 1$

73. $2x^4 - 3x^3 - 6x^2 + 17x - 12$

74. $2x^5 + 6x^4 + 10x^3 + 35x^2 + 25$

75. $4x^4 - 6x^3 - 28x^2 + 26x + 24$

76. $2x^4 - 3x^3 - 10x^2 + 43x - 42$

77. a. $3x^2 + 2x - 8$ b. 8

78. a. $4t^2 + 17t - 42$ b. 8

79. a. $2x^2 + 3x - 5$ b. 9

80. a. $5a^2 + 17a - 12$ b. 42

81. a. $6x^2 - 13x - 8$ b. -10

82. a. $3x^2 + 2x - 16$ b. 0

Find the product in Exercises 77 – 101 and simplify if possible. Check by letting the variable equal 2.

77. $(3x - 4)(x + 2)$
78. $(t + 6)(4t - 7)$
79. $(2x + 5)(x - 1)$
80. $(5a - 3)(a + 4)$
81. $(2x + 1)(3x - 8)$
82. $(x - 2)(3x + 8)$
83. $(7x + 1)(x - 2)$
84. $(3x + 7)(2x - 5)$
85. $(2x + 3)(2x + 3)$
86. $(5y + 2)(5y + 2)$
87. $(x + 3)(x^2 - 4)$
88. $(y^2 + 2)(y - 4)$
89. $(2x + 7)(2x - 7)$
90. $(3x - 4)(3x + 4)$
91. $(x + 1)(x^2 - x + 1)$
92. $(x - 2)(x^2 + 2x + 4)$
93. $(7a - 2)(7a - 2)$
94. $(5a - 6)(5a - 6)$
95. $(2x + 3)(x^2 - x - 1)$
96. $(3x + 1)(x^2 - x + 9)$
97. $(x + 1)(x + 2)(x + 3)$
98. $(t - 1)(t - 2)\ t - 3)$
99. $(a^2 + a - 1)(a^2 - a + 1)$
100. $(y^2 + y + 2)(y^2 + y - 2)$
101. $(t^2 + 3t + 2)^2$

Simplify the expressions in Exercises 102 – 106.

102. $(x - 3)(x + 5) - (x + 3)(x + 2)$
103. $(y - 2)(y - 4) + (y - 1)(y + 1)$
104. $(2a + 1)(a - 5) + (a - 4)(a - 4)$
105. $(2t + 3)(2t + 3) - (t - 2)(t - 2)$
106. $(y + 6)(y - 6) + (y + 5)(y - 5)$

Hawkes Learning Systems: Introductory & Intermediate Algebra

Multiplying a Polynomial by a Monomial
Multiplying Two Polynomials
The FOIL Method

83. a. $7x^2 - 13x - 2$ b. 0

84. a. $6x^2 - x - 35$ b. -13

85. a. $4x^2 + 12x + 9$ b. 49

86. a. $25y^2 + 20y + 4$ b. 144

87. a. $x^3 + 3x^2 - 4x - 12$ b. 0

88. a. $y^3 - 4y^2 + 2y - 8$ b. -12

89. a. $4x^2 - 49$ b. -33

90. a. $9x^2 - 16$ b. 20

91. a. $x^3 + 1$ b. 9

92. a. $x^3 - 8$ b. 0

93. a. $49a^2 - 28a + 4$ b. 144

94. a. $25a^2 - 60a + 36$ b. 16

95. a. $2x^3 + x^2 - 5x - 3$ b. 7

96. a. $3x^3 - 2x^2 + 26x + 9$ b. 77

97. a. $x^3 + 6x^2 + 11x + 6$ b. 60

98. a. $t^3 - 6t^2 + 11t - 6$ b. 0

99. a. $a^4 - a^2 + 2a - 1$ b. 15

100. a. $y^4 + 2y^3 + y^2 - 4$ b. 32

101. a. $t^4 + 6t^3 + 13t^2 + 12t + 4$ b. 144

102. $-3x - 21$

103. $2y^2 - 6y + 7$

104. $3a^2 - 17a + 11$

105. $3t^2 + 16t + 5$

106. $2y^2 - 61$

5.6 Special Products of Polynomials

Objectives

After completing this section, you will be able to:

1. *Multiply binomials that result in the difference of two squares.*
2. *Square binomials with a result of a perfect square trinomial.*
3. *Multiply polynomials that result in the difference of two cubes.*
4. *Multiply polynomials that result in the sum of two cubes.*

The Difference of Two Squares: $(X + A)(X - A) = X^2 - A^2$

Using the FOIL method to multiply $(x + 6)(x - 6)$, we find

$$(x+6)(x-6) = x^2 - 6x + 6x - 36 = x^2 - 36$$

with the result that the two middle terms are opposites of each other ($-6x$ and $+6x$) and with the resulting product having only two terms. This simplified product is in the form of a difference and both terms are squares. Thus, we have the following special case called the difference of two squares.

Difference of Two Squares

$$(X + A)(X - A) = X^2 - A^2$$

When the two binomials are in the form of the sum and difference of the same two terms, the product will always be the difference of the squares of the terms. In such a case, we can write the answer directly with no calculations. You should memorize the following squares of the positive integers from 1 to 20. The squares of integers are called **perfect squares**.

Perfect Squares

1, 4, 9, 16, 25, 36, 49, 64, 81, 100, 121, 144, 169, 196, 225, 256, 289, 324, 361, 400

Example 1: Difference of Two Squares

Teaching Notes:
You may want to use the FOIL method several times on examples such as $(x + 5)(x - 5)$ to help convince the students that the middle term is 0 and that the FOIL method always works.

Find each product.

a. $(x + 5)(x - 5)$

Solution: The two binomials represent the sum and difference of x and 5. So, the product is the difference of their squares.

$$(x + 5)(x - 5) = x^2 - 5^2 = x^2 - 25$$

b. $(8x + 3)(8x - 3)$

Solution: $(8x + 3)(8x - 3) = (\ 8x\ )^2 - (\ 3\ )^2 = 64x^2 - 9$

c. $(x^3 + 2)(x^3 - 2)$

Solution: $(x^3 + 2)(x^3 - 2) = (\ x^3\ )^2 - 2^2 = x^6 - 4$

Perfect Square Trinomials: $(X + A)^2 = X^2 + 2AX + A^2$ $(X - A)^2 = X^2 - 2AX + A^2$

We now want to consider the case where the two binomials being multiplied are the same. That is, we want to consider the **square of a binomial**. As the following discussion shows, there is a pattern that, after some practice, allows us to go directly to the product,

$$\begin{aligned}(x + 5)^2 = (x + 5)(x + 5) &= x^2 + 5x + 5x + 25 \\ &= x^2 + 2 \cdot 5x + 25 \\ &= x^2 + 10x + 25\end{aligned}$$

$$\begin{aligned}(x + 7)^2 = (x + 7)(x + 7) &= x^2 + 2 \cdot 7x + 49 \\ &= x^2 + 14x + 49\end{aligned}$$

$$(x + 10)^2 = (x + 10)(x + 10) = x^2 + 20x + 100$$

and the basic pattern is,

$$(X + A)^2 = (X + A)(X + A) = X^2 + 2AX + A^2$$

The result, $X^2 + 2AX + A^2$, is called a **perfect square trinomial** because it is a trinomial that results from squaring a binomial.

One interesting device for remembering the result of squaring a binomial is the square shown in Figure 5.1 where the total area is the sum of the shaded areas.

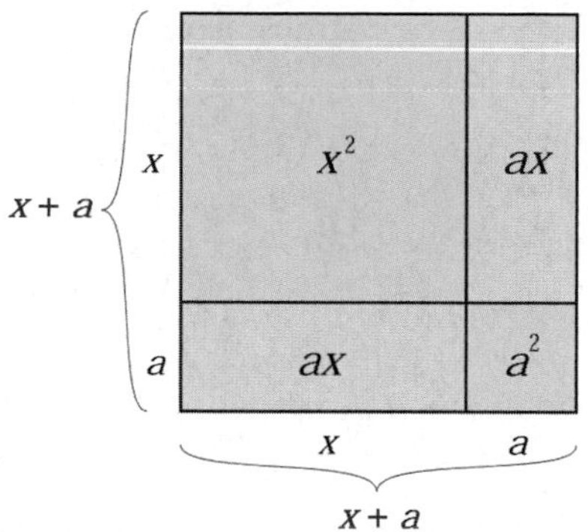

For the area of the square:

$$(x+a)^2 = x^2 + ax + ax + a^2$$
$$= x^2 + 2ax + a^2$$

Figure 5.1

Another perfect square trinomial results if an expression of the form $(x - a)$ is squared. The sign between x and a is $-$ instead of $+$.

$$\begin{aligned}(x-5)^2 = (x-5)(x-5) &= x^2 - 5x - 5x + 25\\ &= x^2 - 2 \cdot 5x + 25\\ &= x^2 - 10x + 25\end{aligned}$$

$$(x-7)^2 = (x-7)(x-7) = x^2 - 2 \cdot 7x + 49 = x^2 - 14x + 49$$

$$(x-10)^2 = (x-10)(x-10) = x^2 - 20x + 100$$

We have the following two formulas for perfect square trinomials.

Perfect Square Trinomials

$$\mathbf{(X+A)^2 = X^2 + 2AX + A^2}$$
$$\mathbf{(X-A)^2 = X^2 - 2AX + A^2}$$

Example 2: Perfect Square Trinomials

Find the following products.

a. $(2x+3)^2$

Solution: The pattern for squaring a binomial gives

$$\begin{aligned}(2x+3)^2 &= (\ 2x\ )^2 + 2 \cdot 3 \cdot 2x + (\ 3\ )^2\\ &= 4x^2 + 12x + 9\end{aligned}$$

continued on next page ...

b. $(5x-1)^2$

Solution: $(5x-1)^2 = (5x)^2 - 2(1)(5x) + (1)^2$
$= 25x^2 - 10x + 1$

c. $(9-x)^2$

Solution: $(9-x)^2 = (9)^2 - 2(9)(x) + x^2$
$= 81 - 18x + x^2$

d. $(y^3+1)^2$

Solution: $(y^3+1)^2 = (y^3)^2 + 2(1)(y^3) + 1^2$
$= y^6 + 2y^3 + 1$

NOTES

Many beginning algebra students make the following error:

$(x+a)^2 = x^2 + a^2$ WRONG
$(x+6)^2 = x^2 + 36$ WRONG

Avoid this error by remembering that **the square of a binomial is a trinomial**.

$(x+6)^2 = x^2 + 2 \cdot 6x + 36$
$= x^2 + 12x + 36$ RIGHT

Example 3: Abstract Products

a. $[(x+3)-y]^2$

Solution: Treat $(x+3)$ as a single term.

$$(X-A)^2 = X^2 - 2AX + A^2$$
$$[(x+3)-y]^2 = (x+3)^2 - 2(x+3)y + y^2$$
$$= x^2 + 6x + 9 - 2xy - 6y + y^2$$

b. $(x^k+3)(x^k-3)$ (Assume that k represents a whole number.)

Solution: $(X+A)(X-A) = X^2 - A^2$

$$(x^k+3)(x^k-3) = (x^k)^2 - 3^2$$
$$= x^{2k} - 9$$ Difference of two squares

c. $(3y^{2k}-5)(3y^{2k}-5)$

Solution: Using the FOIL method,

$$\begin{aligned}(3y^{2k}-5)(3y^{2k}-5) &= 3y^{2k}\cdot 3y^{2k}-5\cdot 3y^{2k}-5\cdot 3y^{2k}-5(-5)\\ &= 9y^{4k}-30y^{2k}+25\end{aligned}$$

Difference of Two Cubes and Sum of Two Cubes: $(X-A)(X^2+AX+A^2)=X^3-A^3$ $(X+A)(X^2-AX+A^2)=X^3+A^3$

Two more special cases of multiplying polynomials result in the **difference of two cubes** and the **sum of two cubes**. For example, consider the following product:

$$\begin{aligned}(x-3)(x^2+3x+9) &= x(x^2+3x+9)-3(x^2+3x+9)\\ &= x^3+3x^2+9x-3x^2-9x-27\\ &= x^3-27\end{aligned}$$

All of the terms in the middle drop out and we are left with the difference of two cubes:

$$(x-3)(x^2+3x+9)=x^3-27=(x)^3-(3)^3 \quad \longleftarrow \text{ the difference of two cubes}$$

If we change just two signs, $(x-3)$ to $(x+3)$ and (x^2+3x+9) to (x^2-3x+9), then the product is the sum of two cubes:

$$\begin{aligned}(x+3)(x^2-3x+9) &= x(x^2-3x+9)+3(x^2-3x+9)\\ &= x^3-3x^2+9x+3x^2-9x+27\\ &= x^3+27 \quad \longleftarrow \text{ the sum of two cubes}\end{aligned}$$

Difference of Two Cubes and Sum of Two Cubes

$$(X-A)(X^2+AX+A^2)=X^3-A^3 \qquad \textit{Difference of two cubes}$$

$$(X+A)(X^2-AX+A^2)=X^3+A^3 \qquad \textit{Sum of two cubes}$$

Example 4: Difference and Sum of Two Cubes

Find the following products.

a. $(x-5y)(x^2+5xy+25y^2)$

Solution: $(x-5y)(x^2+5xy+25y^2)=x^3-125y^3$ the difference of two cubes

b. $(2x+3)(4x^2-6x+9)$

Solution: $(2x+3)(4x^2-6x+9)=8x^3+27$ the sum of two cubes

All five formulas discussed in this section are listed here for easy reference.

Special Products of Polynomials

Teaching Notes:
You might want to emphasize to students that they should become familiar with the structure of the special products because without this, the factoring sections will be more difficult.

Formula	*Classification*
I. $(X+A)(X-A)=X^2-A^2$	*Difference of two squares*
II. $(X+A)^2=X^2+2AX+A^2$	*Perfect square trinomial*
III. $(X-A)^2=X^2-2AX+A^2$	*Perfect square trinomial*
IV. $(X-A)(X^2+AX+A^2)=X^3-A^3$	*Difference of two cubes*
V. $(X+A)(X^2-AX+A^2)=X^3+A^3$	*Sum of two cubes*

NOTES

CAUTION!
Be careful to note that the two trinomials X^2+AX+A^2 and X^2-AX+A^2 in Formulas **IV** and **V** are not perfect square trinomials.

You should memorize these formulas. Along with the FOIL method of multiplication of two binomials, they are part of the foundation for our work with factoring and with algebraic fractions in the next two chapters.

Practice Problems

Find the indicated products.

1. $(x+10)(x-10)$ **2.** $(x+3)^2$ **3.** $(2x-1)(x+3)$

4. $(2x-5)^2$ **5.** $(x^2+4)(x^2-3)$ **6.** $(x+5)(x^2-5x+25)$

Answers to Practice Problems: **1.** x^2-100 **2.** x^2+6x+9 **3.** $2x^2+5x-3$ **4.** $4x^2-20x+25$ **5.** x^4+x^2-12 **6.** x^3+125

5.6 Exercises

Find the product in Exercises 1 – 30, and identify those that are the difference of two squares or perfect square trinomials.

1. $(x+3)(x-3)$ **2.** $(x-7)^2$ **3.** $(x-5)^2$

4. $(x+4)(x+4)$ **5.** $(x-6)(x+6)$ **6.** $(x+9)(x-9)$

7. $(x+8)(x+8)$ **8.** $(x+12)(x-12)$ **9.** $(2x+3)(x-1)$

10. $(3x+1)(2x+5)$ **11.** $(3x-4)^2$ **12.** $(5x+2)(5x-2)$

13. $(2x+1)(2x-1)$ **14.** $(3x+1)^2$ **15.** $(3x-2)(3x-2)$

16. $(4x+5)(4x-5)$ **17.** $(3+x)^2$ **18.** $(8-x)(8-x)$

19. $(5-x)(5-x)$ **20.** $(11-x)(11+x)$ **21.** $(5x-9)(5x+9)$

22. $(4-x)^2$ **23.** $(2x+7)(2x+7)$ **24.** $(3x+2)^2$

25. $(9x+2)(9x-2)$ **26.** $(6x+5)(6x-5)$ **27.** $(5x^2+2)(2x^2-3)$

28. $(4x^2+7)(2x^2+1)$ **29.** $(1+7x)^2$ **30.** $(2-5x)^2$

Find the indicated products and simplify in Exercises 31 – 88. If a product involves three polynomials, find the product of any two and multiply this product by the third polynomial.

31. $(x+2)(5x+1)$ **32.** $(7x-2)(x-3)$ **33.** $(4x-3)(x+4)$

34. $(x+11)(x-8)$ **35.** $(3x-7)(x-6)$ **36.** $(x+7)(2x+9)$

37. $(5+x)(5+x)$ **38.** $(3-x)(6-x)$ **39.** $(x^2+1)(x^2-1)$

40. $(x^2+5)(x^2-5)$ **41.** $(x^2+3)(x^2+3)$ **42.** $(x^2-4)^2$

43. $(x^3-2)^2$ **44.** $(x^3+8)(x^3-8)$ **45.** $(x^2-6)(x^2+9)$

46. $(x^2+3)(x^2-5)$ **47.** $\left(x+\frac{2}{3}\right)\left(x-\frac{2}{3}\right)$ **48.** $\left(x-\frac{1}{2}\right)\left(x+\frac{1}{2}\right)$

1. x^2-9, difference of two squares
2. $x^2-14x+49$, perfect square trinomial
3. $x^2-10x+25$, perfect square trinomial
4. $x^2+8x+16$, perfect square trinomial
5. x^2-36, difference of two squares
6. x^2-81, difference of two squares
7. $x^2+16x+64$, perfect square trinomial
8. x^2-144, difference of two squares
9. $2x^2+x-3$, neither
10. $6x^2+17x+5$, neither
11. $9x^2-24x+16$, perfect square trinomial
12. $25x^2-4$, difference of two squares
13. $4x^2-1$, difference of two squares
14. $9x^2+6x+1$, perfect square trinomial
15. $9x^2-12x+4$, perfect square trinomial
16. $16x^2-25$, difference of two squares
17. x^2+6x+9, perfect square trinomial
18. $x^2-16x+64$, perfect square trinomial

19. $x^2-10x+25$, perfect square trinomial
20. $121-x^2$, difference of two squares
21. $25x^2-81$, difference of two squares
22. $x^2-8x+16$, perfect square trinomial
23. $4x^2+28x+49$, perfect square trinomial
24. $9x^2+12x+4$, perfect square trinomial
25. $81x^2-4$, difference of two squares
26. $36x^2-25$, difference of two squares
27. $10x^4-11x^2-6$, neither
28. $8x^4+18x^2+7$, neither
29. $49x^2+14x+1$, perfect square trinomial
30. $25x^2-20x+4$, perfect square trinomial
31. $5x^2+11x+2$
32. $7x^2-23x+6$
33. $4x^2+13x-12$
34. $x^2+3x-88$
35. $3x^2-25x+42$
36. $2x^2+23x+63$
37. $x^2+10x+25$
38. $x^2-9x+18$
39. x^4-1
40. x^4-25
41. x^4+6x^2+9
42. x^4-8x^2+16
43. x^6-4x^3+4
44. x^6-64
45. x^4+3x^2-54
46. x^4-2x^2-15

49. $\left(x+\frac{3}{4}\right)\left(x-\frac{3}{4}\right)$ **50.** $\left(x+\frac{3}{8}\right)\left(x-\frac{3}{8}\right)$ **51.** $\left(x+\frac{3}{5}\right)\left(x+\frac{3}{5}\right)$

52. $\left(x+\frac{4}{3}\right)\left(x+\frac{4}{3}\right)$ **53.** $\left(x-\frac{5}{6}\right)^2$ **54.** $\left(x-\frac{2}{7}\right)^2$

55. $\left(x+\frac{1}{4}\right)\left(x-\frac{1}{2}\right)$ **56.** $\left(x-\frac{1}{5}\right)\left(x+\frac{2}{3}\right)$ **57.** $\left(x+\frac{1}{3}\right)\left(x+\frac{1}{2}\right)$

58. $\left(x-\frac{4}{5}\right)\left(x-\frac{3}{10}\right)$ **59.** $(3x+1)^2$ **60.** $(4x-3)^2$

61. $(5x-2y)^2$ **62.** $(7x+4y)^2$ **63.** $(4x+7)(4x-7)$

64. $(3x+5)(3x-5)$ **65.** $(2x-3y)(2x+3y)$ **66.** $(6x-y)(6x+y)$

67. $x(3x^2-4)(3x^2+4)$ **68.** $3x(7x^2+8)(7x^2-8)$ **69.** $(x-1)(x^2+x+1)$

70. $(y+4)(y^2-4y+16)$ **71.** $(x+3)(x^2+6x+9)$ **72.** $(y-5)(y^2+3y+2)$

73. $(x^3+2)^2$ **74.** $(2x^3-3)^2$ **75.** $(2x^3-7)(2x^3+7)$

76. $(x+2y)(x^2-2xy+4y^2)$ **77.** $(x-3y)(x^2+3xy+9y^2)$

78. $(8y^2-7)(3y^2+2)$ **79.** $4x^2y(x^2+6y^2)(x^2-6y^2)$

80. $3xy(x^2-6y^2)(x^2+3y^2)$ **81.** $x^2y(5x^2+y^2)(2x^2-3y^2)$

82. $x^3(x-2y)\left(x^2+2xy+4y^2\right)$ **83.** $[(x+y)+2][(x+y)-2]$

84. $[(x+1)+y][(x+1)-y]$ **85.** $[(5x-y)+3]^2$ **86.** $[(2x+1)-y]^2$

87. $[(x+4)-2y]^2$ **88.** $[(x-3y)+5]^2$

47. $x^2-\frac{4}{9}$
48. $x^2-\frac{1}{4}$
49. $x^2-\frac{9}{16}$
50. $x^2-\frac{9}{64}$
51. $x^2+\frac{6}{5}x+\frac{9}{25}$
52. $x^2+\frac{8}{3}x+\frac{16}{9}$
53. $x^2-\frac{5}{3}x+\frac{25}{36}$
54. $x^2-\frac{4}{7}x+\frac{4}{49}$
55. $x^2-\frac{1}{4}x-\frac{1}{8}$
56. $x^2+\frac{7x}{15}-\frac{2}{15}$
57. $x^2+\frac{5}{6}x+\frac{1}{6}$
58. $x^2-\frac{11}{10}x+\frac{6}{25}$

59. $9x^2+6x+1$

60. $16x^2-24x+9$

61. $25x^2-20xy+4y^2$

62. $49x^2+56xy+16y^2$

63. $16x^2-49$

64. $9x^2-25$

65. $4x^2-9y^2$

66. $36x^2-y^2$

67. $9x^5-16x$

68. $147x^5-192x$

69. x^3-1

70. y^3+64

71. $x^3+9x^2+27x+27$

72. $y^3-2y^2-13y-10$

73. x^6+4x^3+4

74. $4x^6-12x^3+9$

75. $4x^6-49$

76. x^3+8y^3

77. x^3-27y^3

78. $24y^4-5y^2-14$

79. $4x^6y-144x^2y^5$

80. $3x^5y-9x^3y^3-54xy^5$

81. $10x^6y-13x^4y^3$
$-3x^2y^5$

82. $x^6-8x^3y^3$

83. $x^2+2xy+y^2-4$

84. x^2+2x-y^2+1

85. $25x^2+y^2+30x$
$-6y-10xy+9$

86. $4x^2+y^2+4x$
$-2y-4xy+1$

87. $x^2+8x-4xy$
$-16y+4y^2+16$

88. $x^2+10x-6xy$
$-30y+9y^2+25$

89. A square is 20 inches on each side. A square x inches on each side is cut from each corner of the square.

a. Represent the area of the remaining portion of the square in the form of a polynomial function $A(x)$.

b. Represent the perimeter of the remaining portion of the square in the form of a polynomial function $P(x)$.

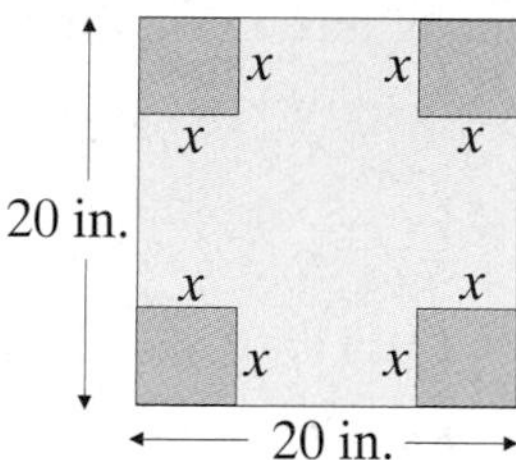

90. In the case of binomial probabilities, if x is the probability of success in one trial of an event, then the expression $f(x)=15x^4(1-x)^2$ is the probability of 4 successes in 6 trials where $0 \le x \le 1$.

a. Represent the expression $f(x)$ as a single polynomial.

b. If a fair coin is tossed, the probability of heads occurring is $\frac{1}{2}$. That is, $x=\frac{1}{2}$. Find the probability of 4 heads occurring in 6 tosses.

91. A rectangle has sides $(x+3)$ ft. and $(x+5)$ ft. If a square x feet on a side is cut from the rectangle, represent the remaining area (light shaded area in the figure shown) in the form of a polynomial function $A(x)$.

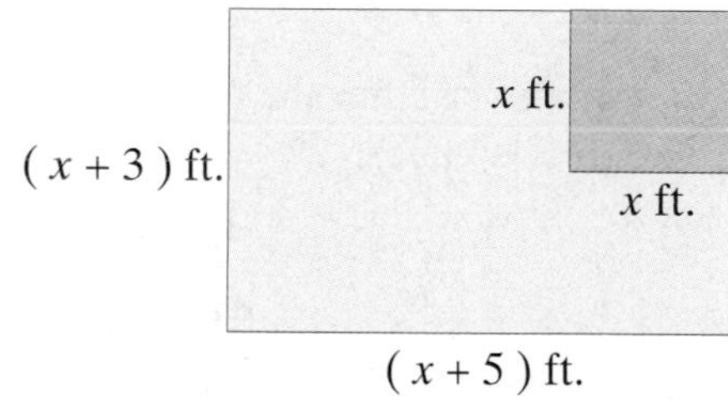

92. A pool, 20 meters by 50 meters, is surrounded by a concrete deck that is x meters wide.

a. Represent the area covered by the deck and the pool in the form of a polynomial function.

b. Represent the area covered by the deck only in the form of a polynomial function.

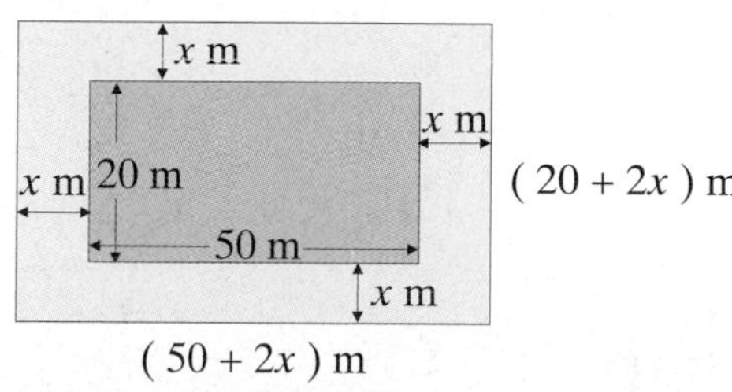

89. a. $A(x) = 400 - 4x^2$ **b.** $P(x) = 80$

90. a. $f(x) = 15x^4 - 30x^5 + 15x^6$ **b.** 0.234375

91. $A(x) = 8x + 15$

92. a. $A(x) = 4x^2 + 140x + 1000$ **b.** $A(x) = 140x + 4x^2$

93. a. $A(x) = 150 - 4x^2$ **b.** $P(x) = 50$ **c.** $V(x) = 150x - 50x^2 + 4x^3$

93. A rectangular piece of cardboard that is 10 inches by 15 inches has squares of length x inches on a side cut from each corner. (Assume that $0 < x < 5$.)

a. Represent the remaining area in the form of a polynomial function $A(x)$.

b. Represent the perimeter of the remaining figure in the form of a polynomial function $P(x)$.

c. If the flaps of the figure are folded up, an open box is formed. Represent the volume of this box in the form of a polynomial function $V(x)$.

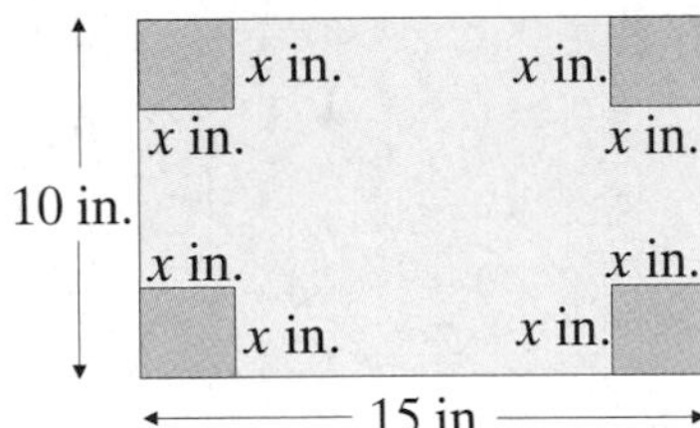

94. −12, 12

94. Find two values of b that will make $9x^2 + bx + 4$ a perfect square trinomial.

95. 1

95. Find one value of c that will make $9x^2 + x + c$ a perfect square trinomial.

96. factor: $x^2 + 5xy + 25y^2$; difference: $x^3 - 125y^3$

96. $x - 5y$ is one factor of the difference of two cubes. Find the remaining factor and find the difference of two cubes.

97. factor: $x^2 - 3x + 9$; sum: $x^3 + 27$

97. $x + 3$ is one factor of the sum of two cubes. Find the remaining factor and find the sum of two cubes.

98. Sum; factor: $4x^2 - 6x + 9$; sum: $8x^3 + 27$

98. Determine whether the following factor is from the difference of two cubes or the sum of two cubes: $2x + 3$. Find the remaining factor and find the sum or difference.

Hawkes Learning Systems: Introductory & Intermediate Algebra

Special Products

5.7 Dividing Polynomials

Objectives

After completing this section, you will be able to:

1. *Divide polynomials by monomials.*
2. *Divide polynomials by other polynomials using long division.*

Fractions, such as $\frac{135}{8}$ and $\frac{7}{8}$, in which the numerator and denominator are integers are called rational numbers. Fractions in which the numerator and denominator are polynomials are called **rational expressions**. (No denominator can be 0).

$$\text{Rational numbers: } \frac{2}{3}, \frac{-5}{16}, \frac{22}{7}, \frac{0}{6}, \frac{3}{4}, -\frac{3}{4}.$$

$$\text{Rational expressions: } \frac{x}{x^2+1}, \frac{x^2+5x+6}{x^2+7x+12}, \frac{x^2+2x+1}{x}, \frac{1}{5x}.$$

In this section, we want to treat a rational expression as an indicated division problem. From this basis, there are two situations to consider:

1. the denominator (divisor) is a monomial, or
2. the denominator (divisor) is not a monomial.

Division by a Monomial

We know from arithmetic that the sum of fractions with the same denominator can be written as a single fraction by adding the numerators and using the common denominator. For example,

$$\frac{3}{a}+\frac{2b}{a}+\frac{5c}{a}=\frac{3+2b+5c}{a}.$$

If, instead of adding the fractions, we want to divide the numerator by the denominator (with a monomial in the denominator), we divide each term in the numerator by the monomial denominator and simplify each fraction.

$$\frac{3x^3+6x^2+9x}{3x}=\frac{3x^3}{3x}+\frac{6x^2}{3x}+\frac{9x}{3x}=x^2+2x+3$$

Similarly,

$$\frac{x^2+2x+1}{x}=\frac{x^2}{x}+\frac{2x}{x}+\frac{1}{x}=x+2+\frac{1}{x}$$

and

$$\frac{3xy+6xy^2-18}{2y}=\frac{3xy}{2y}+\frac{6xy^2}{2y}-\frac{18}{2y}=\frac{3x}{2}+3xy-\frac{9}{y}$$

Example 1: Division by a Monomial

Divide each polynomial by the monomial denominator by writing a sum of fractions. Reduce each fraction, if possible.

a. $\frac{8x^2-14x+1}{2}$

Solution: $$\frac{8x^2-14x+1}{2}=\frac{8x^2}{2}-\frac{14x}{2}+\frac{1}{2}=4x^2-7x+\frac{1}{2}$$

b. $\frac{12x^3+3x^2-9x}{3x^2}$

Solution: $$\frac{12x^3+3x^2-9x}{3x^2}=\frac{12x^3}{3x^2}+\frac{3x^2}{3x^2}-\frac{9x}{3x^2}$$
$$=4x+1-\frac{3}{x}$$

c. $\frac{10x^2y+25xy+3y^2}{5xy^2}$

Solution: $$\frac{10x^2y+25xy+3y^2}{5xy^2}=\frac{10x^2y}{5xy^2}+\frac{25xy}{5xy^2}+\frac{3y^2}{5xy^2}$$
$$=\frac{2x}{y}+\frac{5}{y}+\frac{3}{5x}$$

The Division Algorithm

In arithmetic, the process (or series of steps) that we follow in dividing two numbers is called the **division algorithm** (or **long division**). By this division algorithm, we can find $135 \div 8$ as follows:

Teaching Notes:
It may be helpful to remind students that long division is just repeated subtraction carried out in a compressed (economical) fashion.

$$
\begin{array}{rl}
 & 16 \quad \longleftarrow \text{Quotient} \\
\text{Divisor} \longrightarrow & 8\overline{)135} \quad \longleftarrow \text{Dividend} \\
 & \underline{8} \quad \longleftarrow \text{Subtract} \\
 & 55 \\
 & \underline{48} \quad \longleftarrow \text{Subtract} \\
 & 7 \quad \longleftarrow \text{Remainder (The remainder is always smaller than the divisor.)}
\end{array}
$$

Check: $8 \cdot 16 + 7 = 128 + 7 = 135$ (Multiply the divisor times the quotient and add the remainder. The result should be the original dividend.)

We can also write the division in fraction form and the remainder over the divisor, giving a mixed number.

$$135 \div 8 = \frac{135}{8} = 16\frac{7}{8}$$

Long division with polynomials is a similar process. If the denominator is not a monomial, and the degree of the numerator is equal to or greater than the degree of the denominator, the division can be indicated in fraction form as $\frac{P}{D}$ where P and D are polynomials. This method of division is called the **division algorithm** or **long division**. (An algorithm is a process or series of steps for solving a problem.)

The Division Algorithm

For polynomials P and D, the ***division algorithm*** *gives*

$$\frac{P}{D} = Q + \frac{R}{D}, D \neq 0 \quad \left[\text{or, in function notation, } \frac{P(x)}{D(x)} = Q(x) + \frac{R(x)}{D(x)}, D(x) \neq 0\right]$$

where Q and R are polynomials and ***the degree of R < degree of D.***

The actual process of long division is not clear from this abstract definition. Although you are familiar with the process of long division with integers, this same procedure seems more complicated with polynomials. Long division with polynomials is illustrated in detail in the following two examples. Study them carefully.

Example 2: The Division Algorithm

Teaching Notes:
Another approach is to ask "What times the first term in the divisor will give us the first term in the dividend?" This is often easier to see mentally than to think "division". For example, "What times x will give us $3x^2$?" Students can use this at every division step.

a. $\dfrac{3x^2-5x+2}{x-4}$ or $(3x^2-5x+2)\div(x-4)$

Solution:

	Steps	Explanation
Step 1:	$x-4\overline{)3x^2-5x+2}$	Write both polynomials in order of descending powers. If any powers are missing, fill in with 0's.
Step 2:	$\begin{array}{r} 3x \\ x-4\overline{)3x^2-5x+2} \end{array}$	Mentally divide $3x^2$ by x. Since $\dfrac{3x^2}{x}=3x$, write $3x$ in the quotient.
Step 3:	$\begin{array}{r} 3x \\ x-4\overline{)3x^2-5x+2} \\ (3x^2-12x) \end{array}$ ← $3x(x-4)$	Multiply $3x$ times $(x-4)$ and write the product $3x^2-12x$ below the polynomial $3x^2-5x+2$.
Step 4:	$\begin{array}{r} 3x \\ x-4\overline{)3x^2-5x+2} \\ \underline{-3x^2+12x} \\ +7x+2 \end{array}$	Subtract $3x^2-12x$ by changing signs and adding. Bring down the next term, +2.
Step 5:	$\begin{array}{r} 3x+7 \\ x-4\overline{)3x^2-5x+2} \\ \underline{-3x^2+12x} \\ +7x+2 \end{array}$	Divide $+7x$ by x. Since $\dfrac{+7x}{x}=+7$, write +7 in the quotient.
Step 6:	$\begin{array}{r} 3x+7 \\ x-4\overline{)3x^2-5x+2} \\ \underline{-3x^2+12x} \\ +7x+2 \\ \underline{(+7x-28)} \end{array}$ ← $7(x-4)$	Multiply +7 times $(x-4)$. Write the product $+7x-28$ below $+7x+2$.

Teaching Notes:
Point out to students that as they work this method, the lead term will always be eliminated at each initial step until the remaining term is of smaller degree (the remainder).

Step 7:

$$\begin{array}{r} 3x + 7 \\ x-4\overline{\big)\,3x^2 - 5x + 2} \\ \underline{-3x^2 + 12x} \\ +\ 7x + 2 \\ \underline{-\ 7x + 28} \\ 30 \end{array}$$

Subtract $7x-28$ by changing signs and adding.

30 ⟵ Remainder

Step 8: $\dfrac{3x^2-5x+2}{x-4}=3x+7+\dfrac{30}{x-4}$ Write the remainder over the divisor and add this fraction to the quotient.

Step 9: **Check:** Multiply the divisor and quotient, and add the remainder. The result should be the dividend.

$$\begin{aligned} Q\cdot D+R &= (3x+7)(x-4)+30 \\ &= 3x^2-12x+7x-28+30 \\ &= 3x^2-5x+2 \end{aligned}$$

Thus, the quotient is $3x+7$ and the remainder is 30.

The answer can also be written in the form $Q+\dfrac{R}{D}$ as $3x+7+\dfrac{30}{x-4}$.

b. $\dfrac{4x^3-7x-5}{2x+1}$ or $\left(4x^3-7x-5\right)\div\left(2x+1\right)$

This example will be done in fewer steps. Note that $0x^2$ is inserted so that like terms will be aligned.

Solution:

	Steps	**Explanation**
Step 1:	$2x+1\overline{\big)4x^3+0x^2-7x-5}$	x^2 is a missing power, so $0x^2$ is supplied.

continued on next page ...

Step 2:

$$\begin{array}{r|l} & 2x^2 \\ 2x+1 & 4x^3+0x^2-7x-5 \\ & \underline{-(4x^3+2x^2)} \end{array}$$

$\frac{4x^3}{2x} = 2x^2$ and $2x^2(2x+1) = 4x^3+2x^2$.

Step 3:

$$\begin{array}{r|l} & 2x^2 \\ 2x+1 & 4x^3+0x^2-7x-5 \\ & \underline{-4x^3-2x^2} \\ & \quad -2x^2-7x \end{array}$$

Subtract $4x^3+2x^2$ and bring down $-7x$.

Step 4:

$$\begin{array}{r|l} & 2x^2-x \\ 2x+1 & 4x^3+0x^2-7x-5 \\ & \underline{-4x^3-2x^2} \\ & \quad -2x^2-7x \\ & \quad \underline{-(-2x^2-x)} \\ & \qquad -6x-5 \end{array}$$

$-\frac{2x^2}{2x} = -x$. Multiply $-x$ times $(2x+1)$. Subtract this term and bring down -5.

Step 5:

$$\begin{array}{r|l} & 2x^2-x-3 \\ 2x+1 & 4x^3+0x^2-7x-5 \\ & \underline{-4x^3-2x^2} \\ & \quad -2x^2-7x \\ & \quad \underline{+2x^2+x} \\ & \qquad -6x-5 \\ & \qquad \underline{+6x+3} \\ & \qquad\quad -2 \end{array}$$

$-2 \longleftarrow$ Remainder

Continue dividing, using the same procedure. The remainder is –5.

Step 6: **Check:** $(2x+1)(2x^2-x-3)-2 = 4x^3-2x^2-6x+2x^2-x-3-2$

$$= 4x^3-7x-5$$

The answer can also be written in the form

$$2x^2-x-3+\frac{-2}{2x+1} \text{ or } 2x^2-x-3-\frac{2}{2x+1}.$$

The division algorithm can be used to divide two polynomials whenever the degree of the dividend, P (the numerator), is greater than or equal to the degree of the divisor, D (the denominator). The remainder, R, must be of smaller degree than the divisor, D.

Example 3: Long Division

a. Use the division algorithm (or long division) to divide:

$$\left(x^3-3x^2-5x-8\right)\div\left(x^2+2x+3\right)$$

Solution:

$$\begin{array}{r} x-5 \\ x^2+2x+3\overline{)\,x^3-3x^2-5x-8} \\ \underline{-\left(x^3+2x^2+3x\right)} \\ -5x^2-8x-8 \\ \underline{-\left(-5x^2-10x-15\right)} \\ 2x+7 \end{array}$$

Dividend

Remainder is first-degree, which is smaller degree than the divisor.

$$Q+\frac{R}{D}=x-5+\frac{2x+7}{x^2+2x+3}$$

Check: $\left(x^2+2x+3\right)(x-5)+(2x+7)=x^3+2x^2+3x-5x^2-10x-15+2x+7$

$$=x^3-3x^2-5x-8$$

b. Simplify $\dfrac{25x^3+9x+2}{5x+1}$ by using long division.

Solution:

$$\begin{array}{r} 5x^2-x+2 \\ 5x+1\overline{)\,25x^3+0x^2+9x+2} \\ \underline{25x^3+5x^2} \\ -5x^2+9x \\ \underline{-5x^2-x} \\ 10x+2 \\ \underline{10x+2} \\ 0 \end{array}$$

Divide $25x^3\div5x=5x^2$. Write $5x^2$ in the quotient. Multiply $5x^2$ times $5x+1$ and **subtract**. Continue the process **until the degree of the remainder is smaller than the degree of the divisor.**

$$\frac{25x^3+9x+2}{5x+1}=5x^2-x+2$$

continued on next page ...

c. Divide $\frac{x^4 - 5x^3 + 2x^2 - 6x + 1}{x^2 + 1}$ by using the division algorithm.

Solution:

$$\begin{array}{r|l} & x^2 - 5x + 1 \\ \hline x^2 + 1 & x^4 - 5x^3 + 2x^2 - 6x + 1 \\ & \underline{x^4 \qquad\quad + x^2} \\ & \quad -5x^3 + x^2 - 6x \\ & \quad \underline{-5x^3 \qquad\quad -5x} \\ & \qquad\qquad x^2 - x + 1 \\ & \qquad\qquad \underline{x^2 \qquad +1} \\ & \qquad\qquad\quad -x \end{array}$$

So,

$$\frac{x^4 - 5x^3 + 2x^2 - 6x + 1}{x^2 + 1} = x^2 - 5x + 1 + \frac{-x}{x^2 + 1}$$

$$= x^2 - 5x + 1 - \frac{x}{x^2 + 1}$$

NOTES

In Example 3b, because the remainder is 0, both $5x + 1$ and $(5x^2 - x + 2)$ are **factors** of the polynomial $25x^3 + 9x + 2$. That is, the product of the divisor and the quotient is the dividend:

$$(5x + 1)(5x^2 - x + 2) = 25x^3 + 9x + 2$$

In general, if $\frac{P}{D} = Q$ or $D \cdot Q = P$, then D and Q are **factors** of P.

(We will discuss factors and factoring in detail in Chapter 6.)

Practice Problems

1. *Express the quotient as a sum of fractions:* $\frac{4x^2 + 6x + 1}{2x}$.

Use the division algorithm to divide.

2. $(3x^2 + 8x - 4) \div (x + 2)$
3. $(x^3 + 4x^2 - 5) \div (x^2 + x - 1)$

Answers to Practice Problems: 1. $2x + 3 + \frac{1}{2x}$ **2.** $3x + 2 - \frac{8}{x + 2}$ **3.** $x + 3 + \frac{-2x - 2}{x^2 + x - 1}$

5.7 Exercises

1. $x^2+2x+\frac{3}{4}$
2. $3x^2-5x+\frac{1}{2}$
3. $2x^2-3x-\frac{3}{5}$
4. $3x^2-4x+\frac{5}{3}$
5. $2x+5$
6. $8x-7$
7. $x+6-\frac{3}{x}$
8. $-2x-3+\frac{8}{x}$
9. $2x+3-\frac{3}{2x}$
10. $3x-2+\frac{1}{x}$
11. $2x-3-\frac{1}{x}$
12. $x-2y-\frac{3}{5x}$
13. $x+3y-\frac{11y}{7x}$
14. $2x+y-\frac{1}{2x}$
15. $\frac{3x^2}{4}-2y-\frac{y^2}{x}$
16. $\frac{2x^2}{3}-2xy+\frac{5}{y}$
17. $\frac{5x}{8}-1+\frac{2}{y}$
18. $\frac{3x}{7}-2-\frac{1}{x}$
19. $\frac{8x^2}{9}-xy+\frac{5}{9y}$
20. $8xy+2$

Express each quotient in Exercises 1 – 20 as a sum of fractions and simplify if possible.

1. $\dfrac{4x^2+8x+3}{4}$ **2.** $\dfrac{6x^2-10x+1}{2}$ **3.** $\dfrac{10x^2-15x-3}{5}$ **4.** $\dfrac{9x^2-12x+5}{3}$

5. $\dfrac{2x^2+5x}{x}$ **6.** $\dfrac{8x^2-7x}{x}$ **7.** $\dfrac{x^2+6x-3}{x}$ **8.** $\dfrac{-2x^2-3x+8}{x}$

9. $\dfrac{4x^2+6x-3}{2x}$ **10.** $\dfrac{3x^3-2x^2+x}{x^2}$ **11.** $\dfrac{6x^3-9x^2-3x}{3x^2}$

12. $\dfrac{5x^2y-10xy^2-3y}{5xy}$ **13.** $\dfrac{7x^2y^2+21xy^3-11y^3}{7xy^2}$ **14.** $\dfrac{12x^3y+6x^2y^2-3xy}{6x^2y}$

15. $\dfrac{3x^3y-8xy^2-4y^3}{4xy}$ **16.** $\dfrac{2x^3y^2-6x^2y^3+15xy}{3xy^2}$ **17.** $\dfrac{5x^2y^2-8xy^2+16xy}{8xy^2}$

18. $\dfrac{3x^3y-14x^2y-7xy}{7x^2y}$ **19.** $\dfrac{8x^3y^2-9x^2y^3+5xy}{9xy^2}$ **20.** $\dfrac{24x^2y^2+12xy-6xy}{3xy}$

Divide in Exercises 21 – 55 by using the long division procedure. Write your answers in the form $Q+\frac{R}{D}$. Check each answer by showing that $P=Q\cdot D+R$.

21. $278 \div 23$ **22.** $326 \div 64$

23. $437 \div 59$ **24.** $(x^2+3x+2)\div(x+2)$

25. $(x^2+x-6)\div(x+3)$ **26.** $(y^2+8y+15)\div(y+4)$

27. $(a^2-2a-15)\div(a-2)$ **28.** $(x^2-7x-18)\div(x+5)$

29. $(y^2-y-42)\div(y+4)$ **30.** $(4a^2-21a+2)\div(a-6)$

31. $(5y^2+14y-7)\div(y+5)$ **32.** $(8x^2+10x-4)\div(2x+3)$

33. $(8c^2+2c-14)\div(2c+3)$ **34.** $(6x^2+x-4)\div(2x-1)$

35. $(10m^2-m-6)\div(5m-3)$ **36.** $(x^2-6)\div(x+2)$

37. $(x^2+3x)\div(x+5)$ **38.** $(2x^3+4x^2-x+1)\div(x-3)$

39. $(y^3-9y^2+26y-24)\div(y-2)$ **40.** $(3t^3+10t^2+6t+3)\div(3t+1)$

21. $12+\frac{2}{23}$ **22.** $5+\frac{3}{32}$

23. $7+\frac{24}{59}$ **24.** $x+1$

25. $x-2$

26. $y+4-\frac{1}{y+4}$

27. $a-\frac{15}{a-2}$

28. $x-12+\frac{42}{x+5}$

29. $y-5-\frac{22}{y+4}$

30. $4a+3+\frac{20}{a-6}$

31. $5y-11+\frac{48}{y+5}$

32. $4x-1-\frac{1}{2x+3}$

33. $4c-5+\frac{1}{2c+3}$

34. $3x+2-\frac{2}{2x-1}$

35. $2m+1-\frac{3}{5m-3}$

36. $x-2-\frac{2}{x+2}$

37. $x-2+\frac{10}{x+5}$

38. $2x^2+10x+29+\frac{88}{x-3}$

39. $y^2-7y+12$

40. $t^2+3t+1+\frac{2}{3t+1}$

41. $3a^2+1+\frac{2}{4a-1}$

42. $2x^2-7x+28-\frac{118}{x+4}$

41. $(12a^3-3a^2+4a+1)\div(4a-1)$ **42.** $(2x^3+x^2-6)\div(x+4)$

43. $(x^3+2x^2-5)\div(x-5)$ **44.** $(x^3-8)\div(x-2)$

45. $(x^3+27)\div(x+3)$ **46.** $(x^3+7x^2+x-2)\div(x^2-x+1)$

47. $(2x^3-x+3)\div(x^2-2)$ **48.** $(x^3+3x^2+1)\div(x^2+2x+3)$

49. $\frac{8x^3-27}{2x-3}$ **50.** $\frac{27a^3-64}{3a-4}$ **51.** $\frac{4x^2-x+5}{x-1}$ **52.** $\frac{5x^3-4x^2+81}{x+3}$

53. $\frac{3x^3-10x^2-3x-20}{x-4}$ **54.** $\frac{x^4+2x^2-5}{x^2+1}$ **55.** $\frac{2a^3+3a^2+6}{a^2+2}$

Use the division algorithm to divide in Exercises 56 – 85 and write the answer in the form $Q+\frac{R}{D}$ where the degree of R is less than the degree of D. Assume that no divisor is 0.

56. $\frac{21x^2+25x-3}{7x-1}$ **57.** $\frac{15x^2-14x-11}{3x-4}$ **58.** $\frac{2x^3+7x^2+10x-6}{2x+3}$

59. $\frac{6x^3-7x^2+14x-8}{3x-2}$ **60.** $\frac{21x^3+41x^2+13x+5}{3x+5}$ **61.** $\frac{6x^3-4x^2+5x-7}{x-2}$

62. $\frac{x^3-x^2-10x-10}{x-4}$ **63.** $\frac{2x^3-3x^2+7x+4}{2x-1}$ **64.** $\frac{10x^3+11x^2-12x+9}{5x+3}$

65. $\frac{6x^3+19x^2-3x-7}{6x+1}$ **66.** $\frac{2x^3-7x+2}{x+4}$ **67.** $\frac{2x^3+4x^2-9}{x+3}$

68. $\frac{9x^3-19x+9}{3x-2}$ **69.** $\frac{16x^3+7x+12}{4x+3}$ **70.** $\frac{6x^3+11x^2+25}{2x+5}$

71. $\frac{4x^3-8x^2-9x}{2x-3}$ **72.** $\frac{3x^3+5x^2+7x+9}{x^2+2}$

73. $\frac{2x^4+2x^3+3x^2+6x-1}{2x^2+3}$ **74.** $\frac{x^4+x^3-4x+1}{x^2+4}$

75. $\frac{2x^4+x^3-8x^2+3x-2}{x^2-5}$ **76.** $\frac{6x^3+5x^2-8x+3}{3x^2-2x-1}$ **77.** $\frac{x^3-9x^2+20x-38}{x^2-3x+5}$

78. $\frac{3x^4-7x^3+5x^2+x-2}{x^2+x+1}$ **79.** $\frac{2x^4+9x^3-x^2+6x+9}{x^2-3x+1}$ **80.** $\frac{x^4+3x-7}{x^2+2x-3}$

43. $x^2+7x+35+\frac{170}{x-5}$

44. x^2+2x+4

45. x^2-3x+9

81. $\frac{3x^4-2x^3+4x^2-x+3}{3x^2+x-1}$ **82.** $\frac{x^3-27}{x-3}$ **83.** $\frac{x^3+125}{x+5}$

84. $\frac{x^5-1}{x^2+1}$ **85.** $\frac{x^6-1}{x^3-1}$

Hawkes Learning Systems: Introductory & Intermediate Algebra

Division by a Monomial
The Division Algorithm

46. $x+8+\frac{8x-10}{x^2-x+1}$

47. $2x+\frac{3x+3}{x^2-2}$

48. $x+1-\frac{5x+2}{x^2+2x+3}$

49. $4x^2+6x+9$

50. $9a^2+12a+16$

51. $4x+3+\frac{8}{x-1}$

52. $5x^2-19x+57-\frac{90}{x+3}$

53. $3x^2+2x+5$

54. $x^2+1-\frac{6}{x^2+1}$

55. $2a+3-\frac{4a}{a^2+2}$

56. $3x+4+\frac{1}{7x-1}$

57. $5x+2+\frac{-3}{3x-4}$

58. $x^2+2x+2+\frac{-12}{2x+3}$

59. $2x^2-x+4$

60. $7x^2+2x+1$

61. $6x^2+8x+21+\frac{35}{x-2}$

62. $x^2+3x+2+\frac{-2}{x-4}$

63. $x^2-x+3+\frac{7}{2x-1}$

64. $2x^2+x-3+\frac{18}{5x+3}$

65. $x^2+3x-1+\frac{-6}{6x+1}$

66. $2x^2-8x+25+\frac{-98}{x+4}$

67. $2x^2-2x+6+\frac{-27}{x+3}$

68. $3x^2+2x-5+\frac{-1}{3x-2}$

69. $4x^2-3x+4$

70. $3x^2-2x+5$

71. $2x^2-x-6+\frac{-18}{2x-3}$

72. $3x+5+\frac{x-1}{x^2+2}$

73. $x^2+x+\frac{3x-1}{2x^2+3}$

74. $x^2+x-4+\frac{-8x+17}{x^2+4}$

75. $2x^2+x+2+\frac{8x+8}{x^2-5}$

76. $2x+3+\frac{6}{3x^2-2x-1}$

77. $x-6+\frac{-3x-8}{x^2-3x+5}$

78. $3x^2-10x+12+\frac{-x-14}{x^2+x+1}$

79. $2x^2+15x+42+\frac{117x-33}{x^2-3x+1}$

80. $x^2-2x+7+\frac{-17x+14}{x^2+2x-3}$

81. $x^2-x+2+\frac{-4x+5}{3x^2+x-1}$

82. x^2+3x+9

83. $x^2-5x+25$

84. $x^3-x+\frac{x-1}{x^2+1}$

85. x^3+1

Chapter 5 Index of Key Ideas and Terms

Section 5.1 Exponents

Section 5.2 More on Exponents and Scientific Notation

If N is a decimal number, then in scientific notation

$N = a \times 10^n$ where $1 \leq a < 10$ and n is an integer.

Section 5.3 Identifying and Evaluating Polynomials

A **monomial in *x*** is a term of the form kx^n where k is a real number and n is a whole number. n is called the **degree** of the term, and k is called the **coefficient**.

A **polynomial in *x*** is an expression of the form

$$a_n x^n + a_{n-1}x^{n-1} + \ldots + a_1 x + a_0$$

where $n, n-1, n-2, \ldots, 1, 0$ are whole numbers and $a_n, a_{n-1}, \ldots, a_1, a_0$ are real numbers with $a_n \neq 0$.
n is called the **degree** of the polynomial.
a_n is called the **leading coefficient** of the polynomial.

continued on next page ...

Section 5.3 Identifying and Evaluating Polynomials (continued)

Classification of Polynomials page 355

Monomial: a polynomial with one term
Binomial: a polynomial with two terms
Trinomial: a polynomial with three terms

Evaluation of Polynomials page 356

A polynomial can be treated as a function and the notation $p(x)$ can be used.

Section 5.4 Adding and Subtracting Polynomials

Addition with Polynomials page 359
Subtraction with Polynomials page 360

Section 5.5 Multiplying Polynomials

Multiplication with Polynomials pages 365 - 367

The FOIL Method pages 368 - 369

Section 5.6 Special Products of Polynomials

Summary of Formulas of Special Products page 378

I. $(X+A)(X-A)=X^2-A^2$: Difference of two squares page 373

II. $(X+A)^2=X^2+2AX+A^2$: Perfect square trinomial page 375

III. $(X-A)^2=X^2-2AX+A^2$: Perfect square trinomial page 375

IV. $(X-A)(X^2+AX+A^2)=X^3-A^3$: Difference of two cubes page 377

V. $(X+A)(X^2-AX+A^2)=X^3+A^3$: Sum of two cubes page 377

80. Find three consecutive even integers such that the sum of the second and third is equal to three times the first decreased by 14.

81. The length of a rectangle is 9 centimeters more than its width. The perimeter is 72 centimeters. Find the dimensions of the rectangle.

82. Twice the difference between a number and [illegible] is equal to four times the number increased by [illegible]. What is the number?

84. A investor has two investments that total [illegible]. One investment earns interest at [illegible], and the other investment earns interest at [illegible]. If the total amount of interest from the two investments is $570 in one year, how much money did he have invested at each rate?

85. Stan has decided to fence in his backyard using his house as the fourth side of a rectangular shape. He wants to use 100 feet of fencing to enclose a rectangle. He wants the length to be [illegible] feet longer than the width of his fenced area. What will be the dimensions of his fenced area?

86. Tess went bike riding. She biked from her home to the beach at a rate of 10 mph and returned at a rate of [illegible] mph. She [illegible] and her return trip took 20 minutes less than the trip to the beach. How far is her home from the beach?

Factoring Polynomials and Solving Quadratic Equations

CHAPTER 6

Did You Know?

You have noticed by now that almost every algebraic skill somehow relates to equation solving and applied problems. The emphasis on equation solving has always been a part of classical algebra.

In Italy, during the Renaissance, it was the custom for one mathematician to challenge another mathematician to an equation-solving contest. A large amount of money, often in gold, was supplied by patrons or sponsoring cities as the prize. At that time, it was important not to publish equation-solving methods, since mathematicians could earn large amounts of money if they could solve problems that their competitors could not. Equation-solving techniques were passed down from a mathematician to an apprentice, but they were never shared.

A Venetian mathematician, Niccolo Fontana (1500? – 1557) known as Tartaglia, "the stammerer," discovered how to solve third-degree or cubic equations. At that time, everyone could solve first- and second-degree equations and special kinds of equations of higher degree. Tartaglia easily won equation-solving contests simply by giving his opponents third-degree equations to solve.

Tartaglia planned to keep his method secret, but after receiving a pledge of secrecy, he gave his method to Girolamo Cardano (1501 – 1576). Cardano broke his promise by publishing one of the first successful Latin algebra texts, *Ars Magna*, "The Great Art." In it, he included not only Tartaglia's solution to the third-degree equations but also a pupil's (Ferrari) discovery of the general solution to fourth-degree equations. Until recently, Cardano received credit for discovering both methods.

It was not until 300 years later that it was shown that there are no general algebraic methods for solving fifth- or higher-degree equations. As you can see, a great deal of time and energy has gone into developing the methods of equation solving that you are learning.

"Algebra is the intellectual instrument which has been created for rendering clear the quantitative aspects of the world."

Alfred North Whitehead (1861 – 1947)

ac-Method (Grouping)

Not all polynomials can be factored by using the known formulas. Two general methods for factoring trinomials in the form

$$ax^2 + bx + c \qquad \text{where } a, b, \text{ and } c \text{ are integers}$$

are the ***ac*-method** (or **grouping**) and the **FOIL method** (or **trial-and-error**).

The ***ac*-method** is very systematic. It also involves a technique called factoring by grouping that we will develop here on a limited basis. Factoring by grouping will be discussed in more detail in Section 6.3.

Consider the problem of factoring $2x^2 + 9x + 10$ in which $a = 2, b = 9,$ and $c = 10$.

Analysis of Factoring by the *ac*-Method

Teaching Notes:
While the *ac*-method is systematic, in practice students tend to obtain factoring results about as easily and quickly by using trial-and-error. It may be a good idea to emphasize that the trial-and-error method is not a random method, as one has to know the factors of the first and last terms. Furthermore, some instructors would argue that the trial-and-error method reinforces the concept of factoring more closely than the *ac*-method.

	General Method	***Example***
	$ax^2 + bx + c$	$2x^2 + 9x + 10$
Step 1:	*Multiply* $a \cdot c$	*Multiply* $2 \cdot 10 = 20$
Step 2:	*Find two integers whose product is ac and whose sum is b. If this is not possible, then the trinomial is* ***not factorable.***	*Find two integers whose product is 20 and whose sum is 9. (In this case,* $4 \cdot 5 = 20$ *and* $4 + 5 = 9$.)
Step 3:	*Rewrite the middle term* (bx) *using the two numbers found in Step 2 as coefficients.*	*Rewrite the middle term* $(9x)$ *using 4 and 5 as coefficients.* $2x^2 + 9x + 10$ $= 2x^2 + 4x + 5x + 10$
Step 4:	*Factor by grouping the first two terms and the last two terms.*	*Factor by grouping the first two terms and the last two terms.* $2x^2 + 4x + 5x + 10$ $= (2x^2 + 4x) + (5x + 10)$

continued on next page ...

Analysis of Factoring by the *ac*-Method (continued)

Step 5:	*Factor out the common binomial factor. This will give two binomial factors of the trinomial* $ax^2 + bx + c$.	*Factor out the common binomial factor* $(x+2)$. *Thus,* $2x^2 + 9x + 10$ $= 2x^2 + 4x + 5x + 10$ $= 2x(x+2) + 5(x+2)$ $= (x+2)(2x+5)$

Example 2: *ac*-Method

a. Factor $x^2 - 2x - 15$ using the *ac*-method.
Solution: $a = 1, b = -2$, and $c = -15$

Step 1: Find the product $a \cdot c$: $1(-15) = -15$.

Step 2: Find two integers whose product is -15 and whose sum is -2.
$(-5)(+3) = -15$ and $-5 + 3 = -2$

Step 3: Rewrite $-2x$ as $-5x + 3x$ to obtain
$x^2 - 2x - 15 = x^2 - 5x + 3x - 15$.

Step 4: Factor by grouping.
$x^2 - 2x - 15 = x^2 - 5x + 3x - 15 = x(x-5) + 3(x-5)$

Step 5: Factor out the common binomial factor $(x-5)$.
$x^2 - 2x - 15 = x(x-5) + 3(x-5) = (x-5)(x+3)$

b. Factor $18x^3 - 39x^2 + 18x$ using the *ac*–method.
Solution: First factor out the greatest common factor $3x$.

$18x^3 - 39x^2 + 18x = 3x(6x^2 - 13x + 6)$

Now factor the trinomial $6x^2 - 13x + 6$ with $a = 6, b = -13$, and $c = 6$.

Step 1: Find the product $a \cdot c$: $6(6) = 36$.

Step 2: Find two integers whose product is 36 and whose sum is –13.
Note: This may take some time and experimentation. We do know that both numbers must be negative since the product is positive and the sum is negative.

$(-9)(-4) = +36$ and $-9 + (-4) = -13$

continued on next page ...

Step 3: Rewrite $-13x$ as $-9x$ $-4x$ to obtain
$6x^2-13x+6=6x^2-9x-4x+6$.

Step 4: Factor by grouping.

$$\begin{aligned}6x^2-13x+6&=6x^2-9x-4x+6\\&=(6x^2-9x)+(-4x+6)\\&=3x(2x-3)-2(2x-3)\end{aligned}$$

Note: -2 is factored from the last two terms so that there will be a common binomial factor $(2x-3)$.

Step 5: Factor out the common binomial factor $(2x-3)$.

$$\begin{aligned}6x^2-13x+6&=6x^2-9x-4x+6\\&=3x(2x-3)-2(2x-3)\\&=(2x-3)(3x-2)\end{aligned}$$

Thus, for the original expression,

$$\begin{aligned}18x^3-39x^2+18x&=3x(6x^2-13x+6)\\&=3x(2x-3)(3x-2)\end{aligned}$$

Remember to include the original GCF in the final product.

FOIL Method (Trial-and-Error)

The **FOIL method** (or **Trial-and-Error method**) of factoring trinomials is actually applying the FOIL method of multiplication of two binomials in reverse. We consider two basic forms:

1. The leading coefficient is 1: ⟶ x^2+bx+c
2. The leading coefficient is not 1: ⟶ ax^2+bx+c

For example, consider factoring a trinomial with leading coefficient 1 such as

$$x^2+17x+30.$$

Because the leading coefficient is 1, we know that the first terms in the two binomial factors are x and x:

$\mathbf{F}=x^2$ $\mathbf{L}=30$

$$x^2+17x+30=(x\qquad)(x\qquad)$$

I

O

For **O** and **I**, we need two factors of 30 whose sum is 17.
Possible pairs are $30 \cdot 1$, $15 \cdot 2$, $10 \cdot 3$, and $6 \cdot 5$.

Because $15 \cdot 2 = 30$ and $15 + 2 = 17$, we have

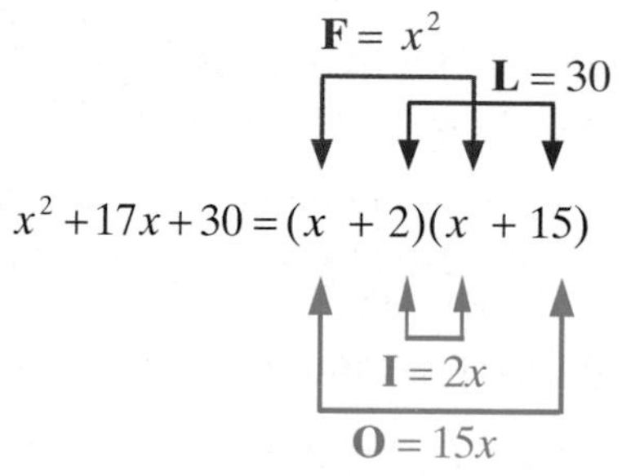

$$\overbrace{}^{\mathbf{F} = x^2} \quad \overbrace{}^{\mathbf{L} = 30}$$

$$x^2 + 17x + 30 = (x + 2)(x + 15)$$

$$\underbrace{}_{\mathbf{I} = 2x} \quad \underbrace{}_{\mathbf{O} = 15x}$$

Similarly, to factor $x^2 + 11x + 30$, we need the factors of 30 whose sum is 11.

$$6 \cdot 5 = 30 \text{ and } 6 + 5 = 11.$$

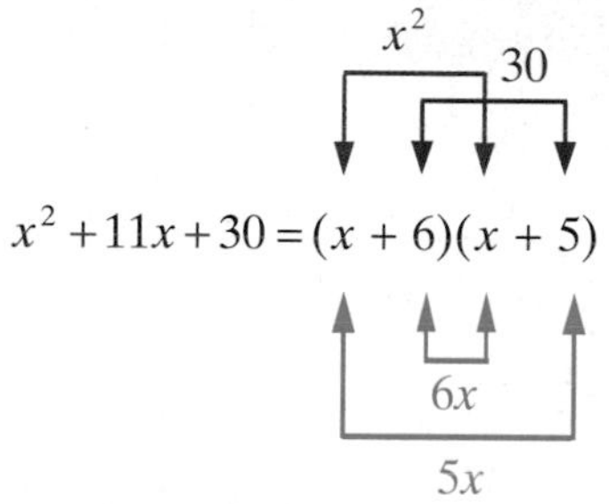

$$\overbrace{}^{x^2} \quad \overbrace{}^{30}$$

$$x^2 + 11x + 30 = (x + 6)(x + 5)$$

$$\underbrace{}_{6x} \quad \underbrace{}_{5x}$$

To factor a trinomial with leading coefficient other than 1, we use the FOIL method also, but with more of a trial and error approach. For example, to factor

$$4x^2 - x - 5,$$

the product of the first two terms of the binomial factors is to be $4x^2$, so we might have

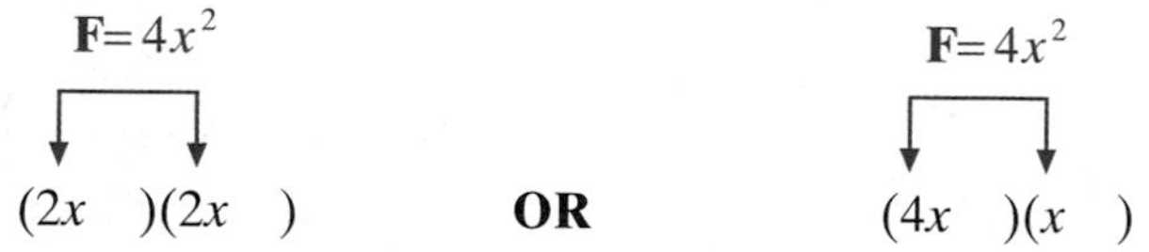

$$\overbrace{}^{\mathbf{F} = 4x^2} \qquad\qquad \overbrace{}^{\mathbf{F} = 4x^2}$$

$$(2x \quad)(2x \quad) \qquad \textbf{OR} \qquad (4x \quad)(x \quad)$$

The product of the last terms, **L**, is to be –5. The factors could be –5(1) or 5(–1). Try each of the possible combinations until the correct product is found. If all possible pairs are tried and none gives the correct product, then the trinomial is **not factorable using integer coefficients**.

1. $(2x+1)(2x-5) = 4x^2 - 10x + 2x - 5 = 4x^2 - 8x - 5 \neq 4x^2 - x - 5$

2. $(2x-1)(2x+5) = 4x^2 + 10x - 2x - 5 = 4x^2 + 8x - 5 \neq 4x^2 - x - 5$

3. $(4x+1)(x-5) = 4x^2 - 20x + x - 5 = 4x^2 - 19x - 5 \neq 4x^2 - x - 5$

continued on next page ...

4. $(4x-1)(x+5)=4x^2+20x-x-5=4x^2+19x-5\neq 4x^2-x-5$

5. $(4x+5)(x-1)=4x^2-4x+5x-5=4x^2+x-5\neq 4x^2-x-5$

6. $(4x-5)(x+1)=4x^2+4x-5x-5=4x^2-x-5$

We have found the factors on the last try. With practice, most of the steps can be done mentally, and the final form can be found more quickly.

a. For example, in the above discussion, since the first pair did not work, there is no point in trying the second pair because the only difference is the sign of the middle term. The same is true with the third and fourth pairs. Thus, two attempts could have been eliminated immediately.

b. Also, by looking at the fifth attempt, we can see that the middle term is only incorrect by the sign.

c. Therefore, we know the sixth pair must work before we even try it.

Example 3: FOIL Method (Trial-and-Error)

a. Factor $x^2-10x+16$.

Solution: The coefficient of x^2 is 1. We know $(-8)(-2)=16$ and $(-8)+(-2)=-10$. Thus, we have

F $=x^2$ **L** $=16$

$$x^2-10x+16=(x-8)(x-2)$$

I $=-8x$

O $=-2x$

b. Factor $-5x^2-5x+60$.

Solution:

$-5x^2-5x+60=-5(x^2+x-12)$ The greatest common factor is -5.
Note: If the leading coefficient is negative then factor out a negative common factor.

$=-5(x-3)(x+4)$ $(-3)(4)=-12$ and $-3x+4x=x$

c. Factor $5x^2+23x-10$.

Solution: $(5x-10)(x+1)$ $5x-10x=-5x\neq 23x$

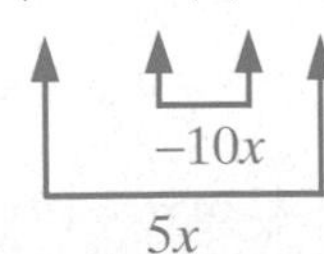

$(5x-1)(x+10)$ $\quad$ $50x - x = 49x \neq 23x$

$-x$

$50x$

$(5x+2)(x-5)$ $\quad$ $-25x + 2x = -23x \neq 23x$

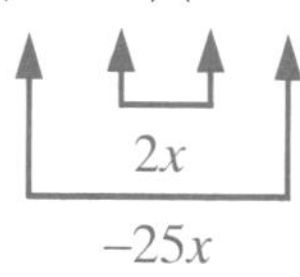

$(5x-2)(x+5)$ $\quad$ $25x - 2x = 23x$

$-2x$

$25x$

$$5x^2 + 23x - 10 = (5x-2)(x+5)$$

NOTES

Summary Note: When factoring polynomials, always look for a common monomial factor first. Then, if there is one, remember to include this common monomial factor as part of the answer. Not all polynomials are factorable. For example, no matter what combinations are tried, $x^2 + 3x + 5$ does not have two binomial factors with integer coefficients. (There are no factors of +5 that will add to +3.) We say that the polynomial is **irreducible** (or **not factorable** or **prime**). **An irreducible (or prime) polynomial is one that cannot be factored as the product of polynomials with integer coefficients.**

Practice Problems

Factor completely.

1. $x^2 - 12x + 36$	**2.** $x^2 + 2x - 35$
3. $-x^2 - 2x + 15$	**4.** $50y^6 - 2y^4$
5. $8x^2 + 13x - 6$	**6.** $2x^2(x+1) + 9x(x+1) - 18(x+1)$

Answers to Practice Problems: **1.** $(x-6)^2$ **2.** $(x+7)(x-5)$ **3.** $-1(x-3)(x+5)$ **4.** $2y^4(5y+1)(5y-1)$ **5.** $(8x-3)(x+2)$ **6.** $(x+1)(2x-3)(x+6)$

Teaching Notes:
In applying the zero factor property, it may be helpful to have students practice the process by writing a limited amount of verbiage to fill in the meaning of the several steps. After factoring, have students write out "Set:" (the factor = 0) and "Solve:" The words provide reminders of important steps in the process.

These solutions can be checked by substituting them, one at a time, into the original equation. Thus,

Check: $(5-5)(2\cdot5+7)\overset{?}{=}0$ $\qquad$ $\left(-\frac{7}{2}-5\right)\left(2\cdot\left(-\frac{7}{2}\right)+7\right)\overset{?}{=}0$

$(0)(17)\overset{?}{=}0$ $\qquad$ $\left(-\frac{17}{2}\right)(-7+7)\overset{?}{=}0$

$0=0$ $\qquad$ $\left(-\frac{17}{2}\right)(0)\overset{?}{=}0$

$0=0$

We say that the solutions are $x=5$ and $x=-\frac{7}{2}$ or that the solution set is $\left\{5,-\frac{7}{2}\right\}$.

In Example 1, the polynomial was given in factored form. However, multiplying the factors gives the product $(x-5)(2x+7)=2x^2-3x-35$, and the equation could have been given in the following form:

$$2x^2-3x-35=0$$

The expression $2x^2-3x-35$ is a quadratic expression and the equation $2x^2-3x-35=0$ is called a **quadratic equation**.

Quadratic Equation

An equation that can be written in the form

$\boldsymbol{ax^2+bx+c=0}$ $\qquad$ *where a, b, and c are real numbers and* $a\neq0$

is called a ***quadratic equation.***

Many polynomial equations, including quadratic equations, can be solved by factoring and using the zero factor property. The following list of steps outlines the procedure.

To Solve an Equation by Factoring

1. *Add or subtract terms so that one side of the equation is 0.*
2. *Factor the polynomial expression.*
3. *Set each factor equal to 0 and solve for the variable.*

Example 2: Quadratic Equations

Solve the following quadratic equations by factoring.

a. $y^2 - 6y = 27$

Solution:
$$y^2 - 6y = 27$$
$$y^2 - 6y - 27 = 0 \quad \text{Add } -27 \text{ to both sides. } \textbf{One side must be 0.}$$
$$(y-9)(y+3) = 0 \quad \text{Factor the left-hand side.}$$
$$y-9=0 \text{ or } y+3=0 \quad \text{Set each factor equal to 0.}$$
$$y=9 \qquad y=-3 \quad \text{Solve each linear equation.}$$

b. $4x^2 + 4x = 0$

Solution:
$$4x^2 + 4x = 0$$
$$4x(x+1) = 0 \quad \text{Factor out } 4x.$$
[Caution: Do not divide both sides by $4x$. You will lose a solution.]
$$4x = 0 \text{ or } x+1=0 \quad \text{Set each factor equal to 0.}$$
$$x=0 \qquad x=-1 \quad \text{Solve each linear equation.}$$

c. $(3z+6)(4z+12) = -3$

Solution:
$$(3z+6)(4z+12) = -3$$
$$12z^2 + 60z + 72 = -3 \quad \text{Multiply factors on left-hand side.}$$
$$12z^2 + 60z + 75 = 0 \quad \textbf{One side must be 0.}$$
$$3(4z^2 + 20z + 25) = 0 \quad \text{Factor out the GCF.}$$
$$4z^2 + 20z + 25 = 0 \quad \text{Divide both sides by the constant 3.}$$
$$(2z+5)^2 = 0 \quad \text{Factor the perfect square trinomial.}$$
$$2z+5 = 0 \quad \text{Since both factors are the same, there is only one equation to solve.}$$
$$z = -\frac{5}{2}$$

There is only one root because the factor $2z + 5$ is repeated. In this case, the root is called a **double root** or a **root of multiplicity two**.

continued on next page ...

Check: $\left[3\left(-\frac{5}{2}\right)+6\right]\left[4\left(-\frac{5}{2}\right)+12\right]\stackrel{?}{=}-3$

$$\left(-\frac{15}{2}+6\right)(-10+12)\stackrel{?}{=}-3$$

$$\left(-\frac{3}{2}\right)(2)\stackrel{?}{=}-3$$

$$-3=-3$$

Sometimes higher-degree polynomial equations can be solved by factoring. In particular, factoring is relatively easy if one of the factors is a monomial. This may give a second factor that is quadratic or some other familiar form.

Example 3: Cubic Equations

Solve the following **third-degree** (or **cubic**) equation by factoring: $100x = 4x^3$

Solution: $100x = 4x^3$ — Add $-100x$ to both sides. **Either the left side or the right side must be 0.**

$0 = 4x^3 - 100x$

$0 = 4x(x^2 - 25)$ — Factor out the monomial $4x$.

$0 = 4x(x+5)(x-5)$ — Factor the difference of two squares.

$4x = 0$ or $x+5=0$ or $x-5=0$ — Set each factor equal to 0 and solve.

$x = 0$ $\quad$ $x = -5$ $\quad$ $x = 5$

Checking will show that each of the numbers 0, –5, and 5 is a solution.

Finding an Equation Given the Roots

To help develop a complete understanding of the concepts of factors, factoring, and solutions to equations, we consider the problem of finding an equation that has certain given solutions (or roots). For example, to find an equation that has the roots

$$x = 4 \text{ and } x = -7$$

we proceed as follows:

1. Write the corresponding linear equations with 0 on one side.

$$x - 4 = 0 \text{ and } x + 7 = 0$$

2. Form the product of the factors $(x-4)$ and $(x+7)$ and set this product equal to 0.

$$(x-4)(x+7)=0$$

3. Multiply the factors. The resulting quadratic equation

$$x^2+3x-28=0$$

has the two roots 4 and –7.

The reasoning is based on the following theorem called the **Factor Theorem**.

Factor Theorem

If $x = c$ is a root of a polynomial equation in the form $P(x) = 0$, then $x - c$ is a factor of the polynomial $P(x)$.

Proof:

The Division Algorithm says that

$$P(x)=(x-c)\cdot Q(x)+r.$$

Now, by the Remainder Theorem (see Appendix 2), $r = P(c)$.
But, because c is a root of the equation, we know that $P(c) = 0$.
Thus, $P(x)=(x-c)\cdot Q(x)+0=(x-c)\cdot Q(x)$ and $(x - c)$ is a factor of $P(x)$.

Example 4: Factor Theorem

Find a polynomial equation that has the given roots: $x = 3$ and $x = -\frac{2}{3}$

Solution: $x = 3$ $\quad$ $x = -\frac{2}{3}$

$x-3=0$ $\quad$ $x+\frac{2}{3}=0$ $\quad$ Set each equation equal to zero.

$3x+2=0$ $\quad$ Multiplying both sides by 3 yields integer coefficients.

Form an equation by setting the product of the two factors equal to 0.

$$(x-3)(3x+2)=0$$

$3x^2-7x-6=0$ $\quad$ This equation has the given roots.

1. $x=2$ or $x=3$
2. $x=2$ or $x=-5$
3. $x=-2$ or $x=\frac{9}{2}$
4. $x=-7$ or $x=\frac{4}{3}$
5. $x=-3$
6. $x=\pm 10$
7. $x=-5$
8. $x=\pm 5$
9. $x=0$ or $x=2$
10. $x=0$ or $x=-3$
11. $y^2-y-6=0$
12. $x^2-12x+35=0$
13. $8x^2-10x+3=0$
14. $18y^2-15y+2=0$

Special Comment: All of the quadratic equations in this section can be solved by factoring. That is, all of the quadratic polynomials are factorable. However, as we have seen in some of the previous sections, not all polynomials are factorable. In Chapter 11, we will develop techniques (other than factoring) for solving quadratic equations whether the quadratic polynomial is factorable or not.

NOTES

COMMON ERROR

A **common error** is to divide both sides of an equation by the variable x. This error is illustrated below.

$$3x^2 = 6x$$

$$\frac{3x^2}{x} = \frac{6x}{x}$$ **WRONG:** DO NOT divide by x because you lose the solution $x = 0$.

$$3x = 6$$

$$x = 2$$

Factoring is the method to use. By factoring, you will find all solutions as shown in the previous examples.

Practice Problems

15. $p^3-p^2-6p=0$
16. $m^3+3m^2-4m=0$
17. $x^2+8x+15=0$
18. $4z^2+3z-1=0$
19. $y^3-4y^2-3y+18=0$
20. $x^3+3x^2+3x+1=0$

Solve the following equations by factoring.

1. $y^2-4y=21$
2. $3x^2-16x+5=0$
3. $z^2+6z=-9$
4. $x^3=25x$
5. $x^2-6x=0$
6. $6x^2-x-1=0$
7. $(x-2)^2-25=0$
8. $x^3-8x^2+16x=0$
9. *Find a quadratic equation with integer coefficients that has the roots* $\frac{2}{3}$ *and* $\frac{1}{2}$.

6.4 Exercises

21. $x=-9$ or $x=-4$
22. $x=-9$ or $x=-8$
23. $x=6$ or $x=8$
24. $y=7$ or $y=5$
25. $x=-5$ or $x=10$
26. $x=-6$ or $x=4$
27. $x=-7$ or $x=7$

Solve the equations in Exercises 1 – 10 by setting each factor equal to 0 and solving the resulting linear equations.

1. $(x-3)(x-2) = 0$
2. $(x+5)(x-2)=0$
3. $(2x-9)(x+2)=0$
4. $(x+7)(3x-4)=0$
5. $0=(x+3)(x+3)$
6. $0=(x+10)(x-10)$
7. $(x+5)(x+5)=0$
8. $(x+5)(x-5)=0$
9. $2x(x-2)=0$

Answers to Practice Problems: **1.** $y=7, y=-3$ **2.** $x=\frac{1}{3}, x=5$ **3.** $z=-3$ **4.** $x=0, x=5, x=-5$ **5.** $x=0, x=6$ **6.** $x=\frac{1}{2}, x=-\frac{1}{3}$ **7.** $x=7, x=-3$ **8.** $x=0, x=4$ **9.** $6x^2-7x+2=0$

10. $3x(x+3)=0$

For Exercises 11 – 20, write a polynomial equation with integer coefficients that has the given roots.

11. $y=3, y=-2$ **12.** $x=5, x=7$ **13.** $x=\frac{1}{2}, x=\frac{3}{4}$

14. $y=\frac{2}{3}, y=\frac{1}{6}$ **15.** $p=0, p=3, p=-2$ **16.** $m=0, m=-4, m=1$

17. $x=-5, x=-3$ **18.** $z=\frac{1}{4}, z=-1$

19. $y=-2, y=3, y=3$ (3 is a double root.)

20. $x=-1, x=-1, x=-1$ (–1 is a triple root.)

Solve the equations in Exercises 21 – 82 by factoring.

21. $x^2+13x+36=0$ **22.** $x^2+17x+72=0$ **23.** $5x^2-70x+240=0$

24. $2y^2-24y+70=0$ **25.** $4x^2=20x+200$ **26.** $7x^2+14x=168$

27. $3x^2=147$ **28.** $64-49x^2=0$ **29.** $3x^2+10=17x$

30. $2x^2=3x-1$ **31.** $6x^2-11x+4=0$ **32.** $4y^2=14y-6$

33. $2x^2-72=0$ **34.** $3x^2-27=0$ **35.** $4z^2-49=0$

36. $9x^2-16=0$ **37.** $2z^2+3=7z$ **38.** $34x+6=12x^2$

39. $12x^2=6-x$ **40.** $12x^2+5x=3$ **41.** $6x^2+x=35$

42. $50y^2-98=0$ **43.** $150y^2-96=0$ **44.** $8y^2+6y=35$

45. $(x+5)(x-7)=13$ **46.** $(2x+3)(x-1)=-2$

47. $x(x-5)+9=3(x-1)$ **48.** $x(2x+3)-2=2(x+4)$

49. $2x(x+3)-14=x(x-2)+19$ **50.** $x(3x+5)=x(x+2)+14$

51. $18y^2-15y+2=0$ **52.** $14+11y=15y^2$ **53.** $63x^2=40x+12$

54. $12z^2-47z+11=0$ **55.** $3x^3+15x^2+18x=0$ **56.** $x^3=4x^2+12x$

57. $16x^3-100x=0$ **58.** $112x-2x^2=2x^3$ **59.** $12x^3+2x^2=70x$

60. $21x^3=13x^2-2x$ **61.** $63x=3x^2+30x^3$ **62.** $14x^3+60x^2=50x$

63. $\frac{x^2}{3}-2x+3=0$ **64.** $\frac{x^2}{9}=1$ **65.** $\frac{x^2}{5}-x-10=0$

66. $\frac{2}{3}x^2+2x-\frac{20}{3}=0$ **67.** $\frac{x^2}{8}+x+\frac{3}{2}=0$ **68.** $\frac{x^2}{6}-\frac{1}{2}x-3=0$

69. $x^2-x+\frac{1}{4}=0$ **70.** $x^2-\frac{7}{6}x+\frac{1}{3}=0$ **71.** $x^3+8x=6x^2$

72. $x^3=x^2+30x$ **73.** $6x^3+7x^2=-2x$ **74.** $3x^3=8x-2x^2$

28. $x=-\frac{8}{7}$ or $x=\frac{8}{7}$

29. $x=\frac{2}{3}$ or $x=5$

30. $x=\frac{1}{2}$ or $x=1$

31. $x=\frac{4}{3}$ or $x=\frac{1}{2}$

32. $y=3$ or $y=\frac{1}{2}$

33. $x=-6$ or $x=6$

34. $x=-3$ or $x=3$

35. $z=-\frac{7}{2}$ or $z=\frac{7}{2}$

36. $x=-\frac{4}{3}$ or $x=\frac{4}{3}$

37. $z=\frac{1}{2}$ or $z=3$

38. $x=-\frac{1}{6}$ or $x=3$

39. $x=-\frac{3}{4}$ or $x=\frac{2}{3}$

40. $x=-\frac{3}{4}$ or $x=\frac{1}{3}$

41. $x=-\frac{5}{2}$ or $x=\frac{7}{3}$

42. $y=-\frac{7}{5}$ or $y=\frac{7}{5}$

43. $y=-\frac{4}{5}$ or $y=\frac{4}{5}$

44. $y=-\frac{5}{2}$ or $y=\frac{7}{4}$

45. $x=-6$ or $x=8$

46. $x=-1$ or $x=\frac{1}{2}$

47. $x=2$ or $x=6$

48. $x=-\frac{5}{2}$ or $x=2$

49. $x=-11$ or $x=3$

50. $x=-\frac{7}{2}$ or $x=2$

51. $y=\frac{1}{6}$ or $y=\frac{2}{3}$

52. $y=-\frac{2}{3}$ or $y=\frac{7}{5}$
53. $x=-\frac{2}{9}$ or $x=\frac{6}{7}$
54. $z=\frac{1}{4}$ or $z=\frac{11}{3}$
55. $x=-3$ or $x=-2$ or $x=0$
56. $x=-2$ or $x=6$ or $x=0$
57. $x=-\frac{5}{2}$ or $x=0$ or $x=\frac{5}{2}$

75. $0=x^2-100$ **76.** $0=x^2-121$ **77.** $3x^2-75=0$

78. $5x^2-45=0$ **79.** $x^2+8x+16=0$ **80.** $x^2+14x+49=0$

81. $3x^2=18x-27$ **82.** $5x^2=10x-5$

Find the solution sets for the polynomial equations in Exercises 83 – 88.

83. $x(2x-1)(3x+1)=0$ **84.** $2x(x-5)(2x+3)=0$

85. $(x+1)(x-2)(x-6)=0$ **86.** $(x-3)(2x-5)(x-4)(4x+9)=0$

87. $(4x+1)(3x-2)(x-8.5)(6x-1)=0$ **88.** $x^3(5x-1)(2x+3)=0$

Hawkes Learning Systems: Introductory & Intermediate Algebra

Solving Quadratic Equations by Factoring

58. $x=-8$ or $x=7$ or $x=0$
59. $x=-\frac{5}{2}$ or $x=0$ or $x=\frac{7}{3}$
60. $x=\frac{2}{7}$ or $x=\frac{1}{3}$ or $x=0$
61. $x=-\frac{3}{2}$ or $x=0$ or $x=\frac{7}{5}$
62. $x=0$ or $x=\frac{5}{7}$ or $x=-5$
63. $x=3$
64. $x=\pm 3$
65. $x=-5$ or $x=10$
66. $x=2$ or $x=-5$
67. $x=-2$ or $x=-6$
68. $x=-3$ or $x=6$
69. $x=\frac{1}{2}$
70. $x=\frac{1}{2}$ or $x=\frac{2}{3}$
71. $x=0$ or $x=2$ or $x=4$
72. $x=0$ or $x=-5$ or $x=6$
73. $x=0$ or $x=-\frac{1}{2}$ or $x=-\frac{2}{3}$
74. $x=0$ or $x=-2$ or $x=\frac{4}{3}$
75. $x=\pm 10$
76. $x=\pm 11$
77. $x=\pm 5$
78. $x=\pm 3$
79. $x=-4$
80. $x=-7$
81. $x=3$
82. $x=1$
83. $x=0$ or $x=\frac{1}{2}$ or $x=-\frac{1}{3}$
84. $x=0$ or $x=5$ or $x=-\frac{3}{2}$
85. $x=-1$ or $x=2$ or $x=6$
86. $x=3$ or $x=\frac{5}{2}$ $x=4$ or $x=-\frac{9}{4}$
87. $x=-\frac{1}{4}$ or $x=\frac{2}{3}$ $x=8.5$ or $x=\frac{1}{6}$
88. $x=0$ or $x=\frac{1}{5}$ or $x=-\frac{3}{2}$

6.5 Applications of Quadratic Equations

Objectives

After completing this section, you will be able to:

Solve word problems by writing quadratic equations that can be factored and solved.

Whether or not word problems cause you difficulty depends a great deal on your personal experiences and general reasoning abilities. These abilities are developed over a long period of time. A problem that is easy for you, possibly because you have had experience in a particular situation, might be quite difficult for a friend, and vice versa.

Most problems do not say specifically to add, subtract, multiply, or divide. You are to know from the nature of the problem what to do. You are to ask yourself, "What information is given? What am I trying to find? What tools, skills, and abilities do I need to use?"

Word problems should be approached in an orderly manner. Have an "attack plan."

Attack Plan for Word Problems

1. *Read the problem carefully at least twice.*
2. *Decide what is asked for and assign a variable or variable expression to the unknown quantities.*
3. *Organize a chart, or a table, or a diagram relating all the information provided.*
4. *Form an equation. (A formula of some type may be necessary.)*
5. *Solve the equation.*
6. *Check your solution with the wording of the problem to be sure it makes sense.*

Several types of problems lead to quadratic equations. The problems in this section are set up so that the equations can be solved by factoring. More general problems and approaches to solving quadratic equations are discussed in Chapter 11.

Example 1: Applications

a. One number is four more than another and the sum of their squares is 296. What are the numbers?

Solution: Let x = smaller number. Then, $x + 4$ = larger number.

continued on next page ...

Teaching Notes: You might explain this "disposal" of constant factors by dividing both sides by the factor to create an equivalent equation. As the constant factor contains no variable, it cannot contribute anything toward causing the left hand side to equal zero.

$x^2 + (x + 4)^2 = 296$ Add the squares.

$x^2 + x^2 + 8x + 16 = 296$

$2x^2 + 8x - 280 = 0$ Write the equation in standard form.

$2(x^2 + 4x - 140) = 0$ Factor out the GCF.

$(x + 14)(x - 10) = 0$ The constant factor 2 does not affect the solution.

$x + 14 = 0$ or $x - 10 = 0$

$x = -14$ $\quad$ $x = 10$

$x + 4 = -10$ $\quad$ $x + 4 = 14$

There are two sets of answers to the problem: 10 and 14 or –14 and –10.

Check: $10^2 + 14^2 = 100 + 196 = 296$

and $(-14)^2 + (-10)^2 = 196 + 100 = 296$

b. In an orange grove, there are 10 more trees in each row than there are rows. How many rows are there if there are 96 trees in the grove?

Solution: Let r = number of rows.
Then, $r + 10$ = number of trees per row.
Set up the equation and solve.

$$r(r+10) = 96$$

$$r^2 + 10r = 96$$

$$r^2 + 10r - 96 = 0$$

$$(r-6)(r+16) = 0$$

$r - 6 = 0$ or $r + 16 = 0$

$r = 6$ $\quad$ $r = -16$

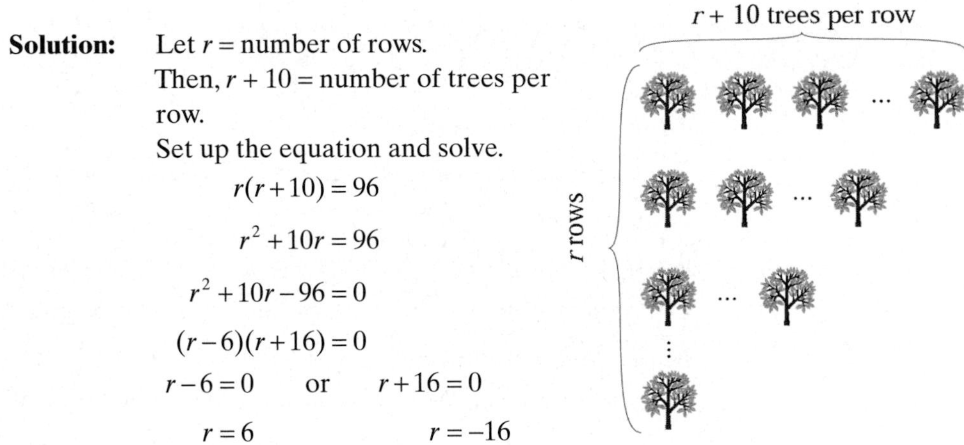

There are 6 rows in the grove ($6 \cdot 16 = 96$ trees).

Note: While –16 is a solution to the equation, –16 does not fit the conditions of the problem and is discarded. You cannot have –16 rows.

c. A rectangle has an area of 135 square meters and perimeter of 48 meters. What are the dimensions of the rectangle?

Solution: The area of a rectangle is the product of its length and width ($A = lw$).
The perimeter of a rectangle is given by $P = 2l + 2w$.
Since the perimeter is 48 meters, then the length plus the width must be 24 meters (one-half of the perimeter).

Let w = width. Then, $24 - w$ = length.
Set up the equation and solve.

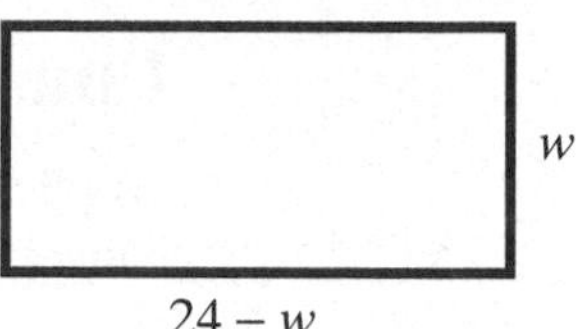

$$w(24-w)=135$$ Area = lw = length times width

$$24w-w^2=135$$

$$0=w^2-24w+135$$

$$0=(w-9)(w-15)$$

Note: 0 can be on either side of the equation.

$$w-9=0 \quad \text{or} \quad w-15=0$$

$$w=9 \qquad\qquad w=15$$

$$24-9=15 \qquad 24-15=9$$

The width is 9 meters and the length is 15 meters ($9 \cdot 15 = 135$).

d. A man wants to build a block wall shaped like a rectangle along three sides of his property. If 180 feet of fencing is needed and the area of the lot is 4000 square feet, what are the dimensions of the lot?

Solution: Let x = one of two equal sides. Then, $180 - 2x$ = third side.
Set up the equation and solve.

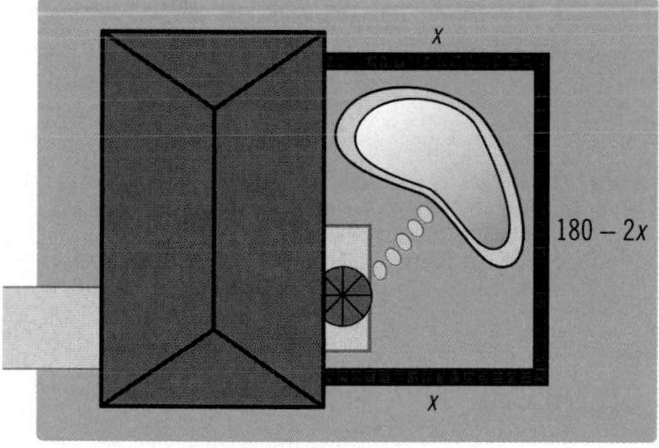

Area = 4000 sq. feet

$$x(180-2x)=4000$$

$$180x-2x^2=4000$$

$$0=2x^2-180x+4000$$

$$0=2(x^2-90x+2000)$$

$$0=2(x-50)(x-40)$$

$$x-50=0 \quad \text{or} \quad x-40=0$$

$$x=50 \qquad\qquad x=40$$

$$180-2(50)=80 \qquad 180-2(40)=100$$

From this information, there are two possible answers: the lot is 50 ft. by 80 ft., or the lot is 40 ft. by 100 ft.

Consecutive Integers

Applications related to integers often involve one of the following three categories: **consecutive integers**, **consecutive odd integers**, or **consecutive even integers**.

Consecutive Integers

Integers are consecutive if each is 1 more than the previous integer.

Three consecutive integers can be represented as $\boldsymbol{n}$, $\boldsymbol{n+1}$, *and* $\boldsymbol{n+2}$.

For example: 5, 6, 7

Consecutive Even Integers

Even integers are consecutive if each is 2 more than the previous even integer.

Three consecutive even integers can be represented as $\boldsymbol{n}$, $\boldsymbol{n+2}$, *and* $\boldsymbol{n+4}$.

For example: 24, 26, 28

Consecutive Odd Integers

Odd integers are consecutive if each is 2 more than the previous odd integer.

Three consecutive odd integers can be represented as $\boldsymbol{n}$, $\boldsymbol{n+2}$, *and* $\boldsymbol{n+4}$.

For example: 41, 43, 45

Note that consecutive even and consecutive odd integers are represented in the same way. The value of the first integer *n* determines whether the remaining integers are even or odd.

Example 2: Consecutive Integers

a. Find two consecutive positive integers such that the sum of their squares is 265.

Solution: Let n = first integer, and
$n + 1$ = next consecutive integer.
Set up and solve the related equation.

$$n^2 + (n+1)^2 = 265$$

$$n^2 + n^2 + 2n + 1 = 265$$

$$2n^2 + 2n - 264 = 0$$

$$n^2 + n - 132 = 0$$

$$(n+12)(n-11) = 0$$

$$n + 12 = 0 \quad \text{or} \quad n - 11 = 0$$

$$n = -12 \qquad\qquad n = 11$$

$$n + 1 = -11 \qquad\qquad n + 1 = 12$$

Teaching Notes:
You might want to show your students that another way to represent three consecutive even integers and three consecutive odd integers is $2n$, $2n + 2$, and $2n + 4$; and $2n + 1$, $2n + 3$, and $2n + 5$, respectively. In these representations, they need not be concerned about whether n is even or odd, but they do need to understand that n is not one of the numbers.

Consider the solution $n = -12$. The next consecutive integer, $n + 1$, is -11. While it is true that the sum of their squares is 265, we must remember that the problem calls for **positive** consecutive integers. Therefore, we can only consider positive solutions. Hence, the two integers are 11 and 12.

b. Find three consecutive odd integers such that the product of the first and second is 68 more than the third.

Solution: Let n = first integer
and $n + 2$ = second consecutive odd integer
and $n + 4$ = third consecutive odd integer.
Set up and solve the related equation.

$$n(n+2) = n+4+68$$

$$n^2 + 2n = n + 72$$

$$n^2 + n - 72 = 0$$

$$(n+9)(n-8) = 0$$

$$n+9=0 \quad \text{or} \quad n-8=0$$

$$n=-9 \qquad\qquad n=8$$

$$n+2=-7 \qquad\qquad n+2=10$$

$$n+4=-5 \qquad\qquad n+4=12$$

The three consecutive odd integers are -9, -7, and -5. Note that 8, 10 and 12 are even and therefore cannot be considered a solution to the problem.

The Pythagorean Theorem

Pythagoras

A geometric topic that often generates quadratic equations is right triangles. In a **right triangle**, one of the angles is a right angle (measures 90°), and the side opposite this angle (the longest side) is called the **hypotenuse**. The other two sides are called **legs**. Pythagoras (560 – 480 B.C.), a famous Greek mathematician, is given credit for proving the following very important and useful theorem (even though history indicates that the Chinese knew of this theorem centuries before Pythagoras). Now there are entire books written that contain only proofs of the Pythagorean Theorem developed by mathematicians since the time of Pythagoras. (You might want to visit the library!)

You will see the Pythagorean Theorem stated again in Chapter 11 and used throughout your studies in mathematics.

The Pythagorean Theorem

In a right triangle, the square of the hypotenuse is equal to the sum of the squares of the legs.

$$c^2 = a^2 + b^2$$

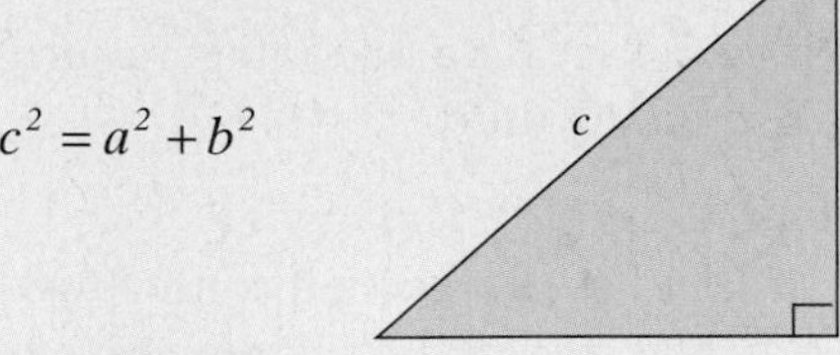

Example 3: The Pythagorean Theorem

A support wire is to be 25 feet long and stretched from a tree to a point on the ground. The point of attachment on the tree is to be 5 feet higher than the distance from the base of the tree to the point on the ground. How far up the tree is the point of attachment?

Solution:

Let x = distance from base of tree to point on ground (See diagram.)

then $x + 5$ = height of point of attachment

By the Pythagorean Theorem, we have

$$(x+5)^2 + x^2 = 25^2$$

$$x^2 + 10x + 25 + x^2 = 625$$

$$2x^2 + 10x - 600 = 0$$

$$2(x^2 + 5x - 300) = 0$$

$$2(x-15)(x+20) = 0$$

$$x = 15 \text{ or } x = -20$$

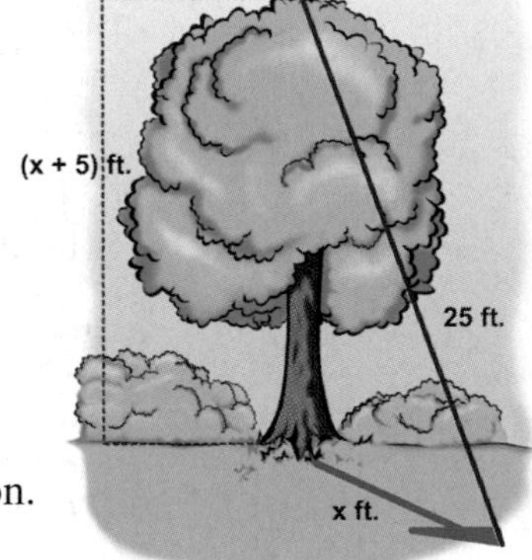

Because distance is positive, –20 is not a possible solution.

The solution is $x = 15$

and $x + 5 = 20$.

Thus, the point of attachment is to be 20 feet up the tree.

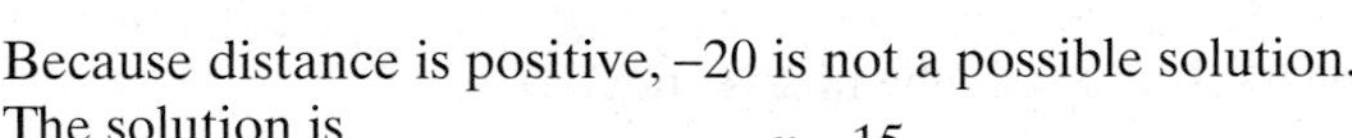

6.5 Exercises

Determine a quadratic equation for each of the following problems in Exercises 1 – 57. Then solve the equation.

1. $x^2 = 7x; x = 0, 7$
2. $x^2 = 2x; x = 0, 2$
3. $x^2 = x + 12; x = 4$
4. $x^2 + 3x = 28; x = 4$

1. The square of an integer is equal to seven times the integer. Find the integer.

2. The square of an integer is equal to twice the integer. Find the integer.

3. The square of a positive integer is equal to the sum of the integer and 12. Find the integer.

5. $x(x+7)=78$; $x=-13, 6$; so the numbers are –13 and –6 or 13 and 6.

6. $x(2x+3)=27$; $x=3$; so the numbers are 3 and 9.

7. $x^2+3x=54$; $x=6$; so the number is 6.

8. $(x+6)^2-x^2=132$; $x=8$; so the numbers are 8 and 14.

9. $x+(x+8)^2=124$; $x=3$; so the numbers are 3 and 11.

10. $x^2+(2x-3)^2=74$; $x=5$ or $-\frac{13}{5}$; so the numbers are 5 and 7 or $-\frac{13}{5}$ and $-\frac{41}{5}$

11. $x^2+(x-5)^2=97$; $x=-4, 9$; so the numbers are –9 and –4 or 9 and 4.

12. $x(x-3)=x+32$; $x=8$.

13. $x(x+1)=72$; $x=8$, so the numbers are 8 and 9.

14. $x(x+1)=110$; $x=10, -11$; so the two consecutive integers are 10 and 11 or –10 and –11.

4. If the square of a positive integer is added to three times the integer, the result is 28. Find the integer.

5. One number is seven more than another. Their product is 78. Find the numbers.

6. One positive number is three more than twice another. If the product is 27, find the numbers.

7. If the square of a positive integer is added to three times the number, the result is 54. Find the number.

8. One number is six more than another. The difference between their squares is 132. What are the numbers?

9. The difference between two positive integers is 8. If the smaller is added to the square of the larger, the sum is 124. Find the numbers.

10. One number is three less than twice another. The sum of their squares is 74. Find the numbers.

11. One number is five less than another. The sum of their squares is 97. Find the numbers.

12. Find a positive integer such that the product of the integer with a number three less than the integer is equal to the integer increased by 32.

13. The product of two consecutive positive integers is 72. Find the integers.

14. Find two consecutive integers whose product is 110.

15. Find two consecutive positive integers such that the sum of their squares is 85.

16. Find two consecutive positive integers such that the square of the second integer added to four times the first is equal to 41.

17. Find two consecutive positive integers such that the difference between their squares is 17.

18. The product of two consecutive odd integers is 63. Find the integers.

19. The product of two consecutive even integers is 120. Find the integers.

20. The product of two consecutive even integers is 168. Find the integers.

21. The length of a rectangle is twice the width. The area is 72 square inches. Find the length and width of the rectangle.

22. The length of a rectangle is three times the width. If the area is 147 square centimeters, find the length and width of the rectangle.

15. $x^2+(x+1)^2=85$; $x=6$; so the numbers are 6 and 7.
16. $4x+(x+1)^2=41$; $x=4$; so the two consecutive integers are 4 and 5.
17. $(x+1)^2-x^2=17$; $x=8$; so the numbers are 8 and 9.
18. $x(x+2)=63$; $x=7,-9$; so the numbers are –7 and –9 or 7 and 9.
19. $x(x+2)=120$; $x=-12, 10$; so the numbers are –12 and –10 or 12 and 10.
20. $x(x+2)=168$; $x=-14, 12$; so the numbers are –14 and –12 or 14 and 12.
21. $w(2w)=72$; width = 6 in. and length = 12 in.
22. $w(3w)=147$; $w=7$, width = 7 cm and length = 21 cm.
23. $w(w+12)=85$; $w=5$, width = 5 m and length = 17 m.
24. $w(w+3)=108$; $w=9$, width = 9 cm and length = 12 cm.
25. $l(l-4)=117$; $l=13$, width = 9 ft. and length = 13 ft.
26. $\frac{1}{2}b(b-4)=16$; $b=8$, base is 8 ft. and height is 4 ft.

23. The length of a rectangular yard is 12 meters greater than the width. If the area of the yard is 85 square meters, find the length and width of the yard.

24. The length of a rectangle is three centimeters greater than the width. The area is 108 square centimeters. Find the length and width of the rectangle.

25. The width of a rectangle is 4 feet less than the length. The area is 117 square feet. Find the length and width of the rectangle.

26. The height of a triangle is 4 feet less than the base. The area of the triangle is 16 square feet. Find the length of the base and height of the triangle.

27. The base of a triangle exceeds the height by 5 meters. If the area is 42 square meters, find the length of the base and height of the triangle.

28. The base of a triangle is 15 inches greater than the height. If the area is 63 square inches, find the length of the base.

29. The base of a triangle is 6 feet less than the height. The area is 56 square feet. Find the length of the height.

30. The perimeter of a rectangle is 32 inches. The area of the rectangle is 48 square inches. Find the dimensions of the rectangle.

31. The area of a rectangle is 24 square centimeters. If the perimeter is 20 centimeters, find the length and width of the rectangle.

32. The perimeter of a rectangle is 40 meters and the area is 96 square meters. Find the dimensions of the rectangle.

33. An orchard has 140 orange trees. The number of rows exceeds the number of trees per row by 13. How many trees are there in each row?

34. One formation for a drill team is rectangular. The number of members in each row exceeds the number of rows by 3. If there is a total of 108 members in the formation, how many rows are there?

35. A theater can seat 144 people. The number of rows is 7 less than the number of seats in each row. How many rows of seats are there?

36. The length of a rectangle is 7 centimeters greater than the width. If 4 centimeters are added to both the length and width, the new area would be 98 square centimeters. Find the dimensions of the original rectangle.

37. The width of a rectangle is 5 meters less than the length. If 6 meters are added to both the length and width, the new area will be 300 square meters. Find the dimensions of the original rectangle.

38. Susan is going to fence a rectangular flower garden in her back yard. She has 50 feet of fencing and she plans to use the house as the fence on one side of the garden. If the area is 300 square feet, what are the dimensions of the flower garden?

39. A rancher is going to build a corral with 52 yards of fencing. He is planning to use the barn as one side of the corral. If the area is 320 square yards, what are the dimensions?

40. The area of a square is 81 square centimeters. How long is each side of the square?

41. The length of a rectangle is three times the width. If the area is 48 square feet, find the length and width of the rectangle.

42. The length of a rectangle is five times the width. If the area is 180 square inches, find the length and width of the rectangle.

43. One number is eight more than another. Their product is –16. What are the numbers?

44. One number is 10 more than another. If their product is –25, find the numbers.

45. Find two consecutive positive integers such that the sum of their squares is 113.

46. The product of two consecutive even integers is 624. Find the integers.

27. $\frac{1}{2}(h+5)h=42$; $h=7$, height is 7 m and base is 12 m.
28. $\frac{1}{2}(h+15)h=63$; $h=6$; base is 21 in.
29. $\frac{1}{2}(h-6)h=56$; $h=14$, height is 14 ft.
30. $2(l+w)=32$; $lw=48$; length = 12 in. and width = 4 in.
31. $2(l+w)=20$; $lw=24$; width = 4 cm, length = 6 cm.
32. $2(l+w)=40$; $lw=96$; length = 12 m and width = 8 m.
33. $x(x+13)=140$; $x=7$; so there are 7 trees in each row.
34. $n(n+3)=108$; $n=9$; so there are 9 rows.
35. $x(x+7)=144$; $x=9$; so there are 9 rows.
36. $(b+11)(b+4)=98$; $b=3$ cm; so the rectangle is 3 cm by 10 cm.
37. $(l+6)(l+1)=300$; $l=14$ m, so the rectangle is 14 m by 9 m.
38. $2w+l=50$; $lw=300$, length = 20 ft. and width = 15 ft. or length = 30 ft. and width = 10 ft.
39. $w(52-2w)=320$; $w=10$ yd., 16 yd., so the corral is 10 yd. by 32 yd. or 16 yd. by 20 yd.

Chapter 6 Index of Key Ideas and Terms

Section 6.1 Greatest Common Factor and Factoring by Grouping

Greatest Common Factor (GCF) page 404

Procedure for Finding the GCF page 405

1. Find the prime factorization of all integers and integer coefficients.
2. List all factors common to all terms, including variables.
3. Choose the greatest power of each factor common to all terms.
4. Multiply these powers to find the GCF.
(**Note:** If there is no common prime factor or variable, then the GCF is 1.)

Factoring by Grouping page 409

Section 6.2 Special Factoring Techniques I

***ac*-Method of Factoring** page 416

FOIL Method of Factoring (Trial-and-Error) page 418

Section 6.3 Special Factoring Techniques II

Review of Factoring by Grouping page 427

Factoring with Negative Exponents page 428

Section 6.4 Solving Quadratic Equations by Factoring

Solving Equations by Factoring page 433

Zero Factor Property page 433

If the product of two factors is 0, then one or both of the factors must be 0. That is, if a and b are real numbers, if $a \cdot b = 0$, then $a = 0$ or $b = 0$.

continued on next page ...

Section 6.4 Solving Quadratic Equations by Factoring (continued)

Quadratic Equation page 434

An equation that can be written in the form $ax^2 + bx + c = 0$ where a, b, and c are real numbers and $a \neq 0$ is called a **quadratic equation**.

Factor Theorem page 437

If $x = c$ is a root of a polynomial equation in the form $P(x) = 0$, then $x - c$ is a factor of the polynomial $P(x)$.

Section 6.5 Applications of Quadratic Equations

Attack Plan for Word Problems page 441

1. Read the problem carefully at least twice.
2. Decide what is asked for and assign a variable or variable expression to the unknown quantities.
3. Organize a chart, or a table, or a diagram relating all the information provided.
4. Form an equation.
   (A formula of some type may be necessary.)
5. Solve the equation.
6. Check your solution with the wording of the problem to be sure it makes sense.

Consecutive Integers page 444

Integers are consecutive if each is 1 more than the previous integer.

Consecutive Even Integers page 444

Even integers are consecutive if each is 2 more than the previous even integer.

Consecutive Odd Integers page 444

Odd integers are consecutive if each is 2 more than the previous odd integer.

Pythagorean Theorem pages 445 - 446

In a right triangle, the square of the hypotenuse is equal to the sum of the squares of the legs.

Section 6.6 Additional Applications of Quadratic Equations

Section 6.7 Using a Graphing Calculator to Solve Equations and Absolute Values

Section 6.8 Additional Factoring Practice

1. Always look for a common monomial factor first. If the leading coefficient is negative, factor out a negative monomial even if it is just -1.

2. Check the number of terms.

- **a.** Two terms:
  - **(1)** difference of two squares? - factorable
  - **(2)** sum of two squares? - not factorable
  - **(3)** difference of two cubes? - factorable
  - **(4)** sum of two cubes? - factorable
- **b.** Three terms:
  - **(1)** perfect square trinomial?
  - **(2)** use *ac*-method?
  - **(3)** use trial-and-error method?
- **c.** Four terms:
  - **(1)** group terms with a common factor and factor out any common binomial factor.

3. Check the possibility of factoring any of the factors.

Chapter 6 Review

For a review of the topics and problems from Chapter 6, look at the following lessons from *Hawkes Learning Systems: Introductory & Intermediate Algebra.*

Greatest Common Factor of Two or More Terms
Greatest Common Factor of a Polynomial
Factoring Expressions by Grouping
Special Factorizations - Squares
Factoring Trinomials by Grouping
Factoring Trinomials by Trial and Error
Special Factorizations - Cubes
Solving Quadratic Equations by Factoring
Applications of Quadratic Equations
Using a Graphing Calculator to Solve Equations

Chapter 6 Test

Find the GCF for each of the sets of terms in Exercises 1 and 2.

1. 30, 75, 90 **2.** $40x^2y,\ 48x^2y^3,\ 56x^2y^2$

In Exercises 3 and 4, simplify each expression. Assume that no denominator is 0.

3. $\dfrac{16x^4}{8x}$ **4.** $\dfrac{42x^3y^3}{-6x^3y}$

Factor completely, if possible, each of the polynomials in Exercises 5 – 20.

5. $20x^3y^2 + 30x^3y + 10x^3$ **6.** $x^2 - 9x + 20$ **7.** $-x^2 - 14x - 49$

8. $6x^2 - 6$ **9.** $12x^2 + 2x - 10$ **10.** $3x^2 + x - 24$

11. $16x^2 - 25y^2$ **12.** $2x^3 - x^2 - 3x$ **13.** $6x^2 - 13x + 6$

14. $2xy - 3y + 14x - 21$ **15.** $4x^2 + 25$ **16.** $-3x^3 + 6x^2 - 6x$

17. $x^{3k} + 125$ **18.** $3x^{-3} + 15x^{-5}$ **19.** $(y^2-4y+4)-25x^2$

20. $y^3 - 64x^6$

Solve the equations in Exercises 21 – 27.

21. $(x + 2)(3x - 5) = 0$ **22.** $x^2 - 7x - 8 = 0$ **23.** $-3x^2 = 18x$

24. $\dfrac{2x^2}{5} - 6 = \dfrac{4x}{5}$ **25.** $0 = 4x^2 - 17x - 15$ **26.** $0 = 8x^2 - 2x - 15$

27. $(2x - 7)(x + 1) = 6x - 19$

28. One number is 10 less than five times another number. Their product is 120. Find the numbers.

29. The length of a rectangle is 7 centimeters less than twice the width. If the area of the rectangle is 165 square centimeters, find the length and width.

30. The product of two consecutive positive integers is 342. Find the two integers.

31. The difference between two positive numbers is 9. If the smaller is added to the square of the larger, the result is 147. Find the numbers.

1. 15
2. $8x^2y$
3. $2x^3$
4. $-7y^2$
5. $10x^3(2y+1)(y+1)$
6. $(x-5)(x-4)$
7. $-(x+7)(x+7)$
8. $6(x+1)(x-1)$
9. $2(6x-5)(x+1)$
10. $(x+3)(3x-8)$
11. $(4x-5y)(4x+5y)$
12. $x(x+1)(2x-3)$
13. $(2x-3)(3x-2)$
14. $(y+7)(2x-3)$
15. Not factorable
16. $-3x(x^2-2x+2)$
17. $(x^k+5)(x^{2k}-5x^k+25)$
18. $3x^{-5}(x^2+5)$
19. $(y-2+5x)(y-2-5x)$
20. $(y-4x^2)(y^2+4x^2y+16x^4)$
21. $x=-2,\ \frac{5}{3}$
22. $x=8,-1$
23. $x=0,-6$
24. $x=5,-3$
25. $x=5,\ -\frac{3}{4}$
26. $x=-\frac{5}{4},\ \frac{3}{2}$
27. $x=4,\ \frac{3}{2}$
28. $x=6,-4$; so the numbers are 6 and 20 or –4 and –30.
29. Length = 15 cm, Width = 11 cm
30. $n=18, 19$

31. 12, 3
32. a. $A(x) = 240 - (x^2 + 3x) = -(x^2 + 3x - 240)$
b. $P(x) = 64$
33. 10 meters and 24 meters
34. $x = -2.5$ and 3

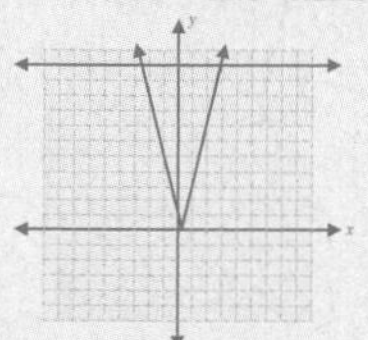

35. $x = 4.9$

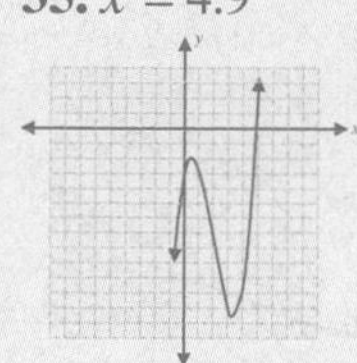

32. A sheet of metal is in the shape of a rectangle with width 12 inches and length 20 inches. A slot (see the figure) of width x inches and length $x + 3$ inches is cut from the top of the rectangle.

a. Write a polynomial function, $A(x)$, that represents the area of the remaining figure.

b. Write a polynomial function, $P(x)$, that represents the perimeter of the remaining figure.

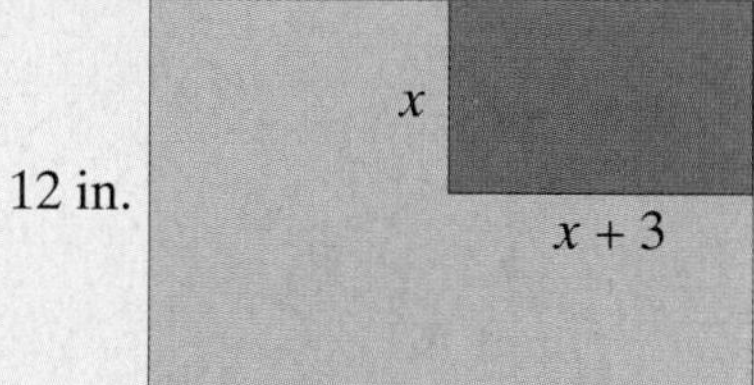

33. In a right triangle, the hypotenuse is 26 meters and one of the legs is 14 meters longer than the other leg. How long are the legs of the triangle?

Use a graphing calculator to solve the following equations in Exercises 34 and 35. Sketch the graphs of the functions you used and find the solutions accurate to 2 decimal places.

34. $|4x - 1| = 11$ **35.** $x^3 - 6x^2 + 6x = 3$

Cumulative Review: Chapter 1 – 6

1. 120 **2.** $168x^2y$
3. $\frac{55}{48}$ **4.** $\frac{19}{60a}$
5. $\frac{3}{10}$ **6.** $\frac{75x}{23}$
7. $-8x^2y$ **8.** $4x^2y$
9. $-\frac{xy^4}{3}$ **10.** $7x^2y$
11. a. $D = \{2, 3, 5, 7.1\}$
b. $R = \{-3, -2, 0, 3.2\}$
c. It is a function because each first coordinate (domain) has only one corresponding second coordinate (range).

Find the LCM for each set of terms in Exercises 1 and 2.

1. 20, 12, 24 **2.** $8x^2,\ 14x^2y,\ 21xy$

Perform the indicated operation in Exercises 3 – 6. Reduce all answers to lowest terms.

3. $\frac{7}{12} + \frac{9}{16}$ **4.** $\frac{11}{15a} - \frac{5}{12a}$ **5.** $\frac{6x}{25} \cdot \frac{5}{4x}$ **6.** $\frac{40}{92} \div \frac{2}{15x}$

Simplify each of the expressions in Exercises 7 – 10 so that it has only positive exponents.

7. $\frac{-24x^4y^2}{3x^2y}$ **8.** $\frac{36x^3y^5}{9xy^4}$ **9.** $\frac{-4x^2y}{12xy^{-3}}$ **10.** $\frac{21xy^2}{3x^{-1}y}$

11. Given the relation $r = \{(2, -3), (3, -2), (5, 0), (7.1, 3.2)\}$.

a. What is the domain of the relation?
b. What is the range of the relation?
c. Is the relation a function? Explain.

12. a. 58 **b.** 4 **c.** $\frac{11}{4}$

13. $x + y + 2 = 0$

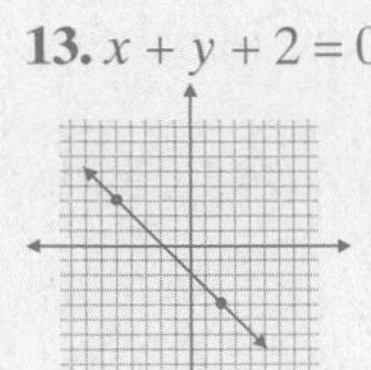

14. $2x - y + 8 = 0$

15. Function

16. Not a function

17. $(-2.076, -.692)$ and $(2.409, .803)$

18. $13x + 1$

19. $x^2 + 12x + 3$

20. $3x^2 - 5x + 3$

21. $4x^2 + 5x - 8$

22. $x^2 - 3x - 5$

23. $-4x^2 - 2x + 3$

24. $2x^2 + x - 28$

25. $-3x^2 - 17x + 6$

26. $x^2 + 12x + 36$

27. $4x^2 - 28x + 49$

28. $3x + 1 + \frac{22}{x-2}$

29. $x^2 - 8x + 38 - \frac{142}{x+4}$

30. $4x + 4 - \frac{11x+18}{x^2+2}$

31. $6x^4 + 9x^3 + 16x^2 + 24x + \frac{63}{2} + \frac{197}{2(2x-3)}$

32. $4(2x - 5)$

33. $6(x - 16)$

12. For the function $f(x) = x^2 - 3x + 4$, find

a. $f(-6)$ **b.** $f(0)$ **c.** $f\left(\frac{1}{2}\right)$

13. Find the equation of the line determined by the two points $(-5, 3)$ and $(2, -4)$. Graph the line.

14. Find the equation of the line parallel to the line $2x - y = 7$ and passing through the point $(-1, 6)$. Graph both lines.

In Exercises 15 and 16, use the vertical line test to determine whether each of the graphs does or does not represent a function.

15.

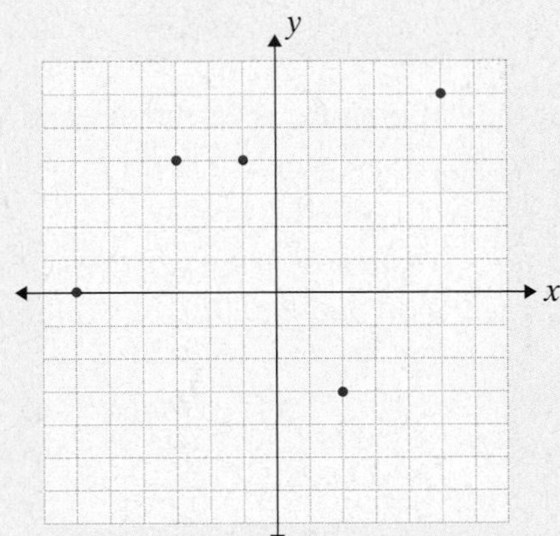

16.

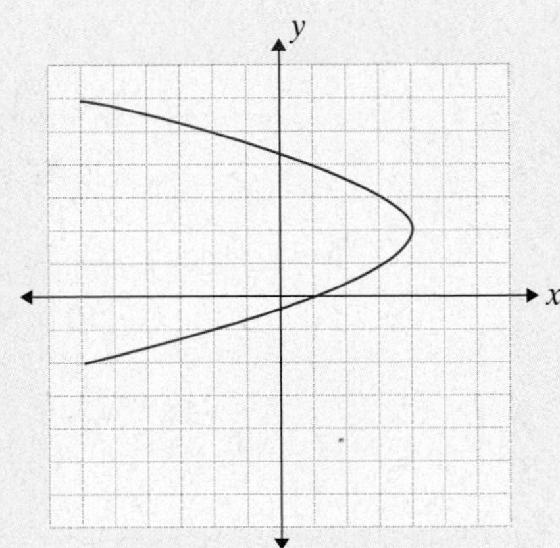

17. Use a graphing calculator to graph the two functions $f(x) = \frac{1}{3}x$ and $g(x) = x^2 - 5$. Estimate the points of intersection.

Perform the indicated operations in Exercises 18 – 27 and simplify by combining like terms.

18. $2(4x + 3) + 5(x - 1)$

19. $2x(x + 5) - (x + 1)(x - 3)$

20. $(x^2 + 3x - 1) + (2x^2 - 8x + 4)$

21. $(2x^2 + 6x - 7) + (2x^2 - x - 1)$

22. $(2x^2 - 5x + 3) - (x^2 - 2x + 8)$

23. $(x + 1) - (4x^2 + 3x - 2)$

24. $(2x - 7)(x + 4)$

25. $-(x + 6)(3x - 1)$

26. $(x + 6)^2$

27. $(2x - 7)^2$

34. $(x-6)(2x-3)$
35. $(2x-3)(3x+4)$
36. $5(x-2)$
37. $-4x(3x+4)$
38. $8xy(2x-3)$
39. $5x^2(2x^2-5x+1)$
40. $(2x+1)(2x-1)$
41. $(y-10)(y-10)$
42. $3(x+4y)(x-4y)$
43. $(x-9)(x+2)$
44. $5(x+4)(x+4)$
45. Not Factorable
46. $(5x+2)(5x+2)$
47. $(x+1)(3x+2)$
48. $2x(x-5)(x-5)$
49. $4x(x^2+25)$
50. $(x+2)(y+3)$
51. $(x-2)(a+b)$
52. $(2x+3y)(4x^2-6xy+9y^2)$
53. $(x^2-6)(x^4+6x^2+36)$
54. $3(x^k-2)(x^{2k}+2x^k+4)$
55. $x^{-4}(x^2+5x+1)$
56. $(x^k-5y)(x^k+5y)$
57. $(x+y^2+3)(x-y^2+3)$
58. $6x^{-4}(x^2-6)$
59. $(x^4-5)(x^8+5x^4+25)$
60. $x^2-3x-10$
61. $6y^2+y-1$
62. $4p^2-29p+7$

In Exercises 28 – 31, divide by using long division and write the answers in the form $Q+\frac{R}{D}$.

28. $(3x^2-5x+20)\div(x-2)$ **29.** $\dfrac{x^3-4x^2+6x+10}{x+4}$

30. $\dfrac{4x^3+4x^2-3x-10}{x^2+2}$ **31.** $\dfrac{12x^5+5x^3-9x+4}{2x-3}$

Factor each expression as completely as possible in Exercises 32 – 59.

32. $8x-20$ **33.** $6x-96$ **34.** $2x^2-15x+18$
35. $6x^2-x-12$ **36.** $5x-10$ **37.** $-12x^2-16x$
38. $16x^2y-24xy$ **39.** $10x^4-25x^3+5x^2$ **40.** $4x^2-1$
41. $y^2-20y+100$ **42.** $3x^2-48y^2$ **43.** $x^2-7x-18$
44. $5x^2+40x+80$ **45.** x^2+x+3 **46.** $25x^2+20x+4$
47. $3x^2+5x+2$ **48.** $2x^3-20x^2+50x$ **49.** $4x^3+100x$
50. $xy+3x+2y+6$ **51.** $ax-2a-2b+bx$ **52.** $8x^3+27y^3$
53. x^6-216 **54.** $3x^{3k}-24$ **55.** $x^{-2}+5x^{-3}+x^{-4}$
56. $x^{2k}-25y^2$ **57.** $(x^2+6x+9)-y^4$ **58.** $6x^{-2}-36x^{-4}$
59. $x^{12}-125$

In Exercises 60 – 62, find a polynomial equation with integer coefficients that has the given roots.

60. $x=-2,\ x=5$ **61.** $y=\frac{1}{3},\ y=-\frac{1}{2}$ **62.** $p=\frac{1}{4},\ p=7$

Solve the equations in Exercises 63 – 77.

63. $(x-7)(x+1)=0$ **64.** $x(3x+5)=0$ **65.** $4x^2+9x-9=0$
66. $21x-3x^2=0$ **67.** $0=x^2+8x+12$ **68.** $x^2=3x+28$
69. $x^3+5x^2-6x=0$ **70.** $\frac{1}{4}x^2+x-15=0$ **71.** $0=15-12x-3x^2$
72. $x^3+14x^2+49x=0$ **73.** $8x=12x+2x^2$ **74.** $6x^2=24x$
75. $2x(x-2)=x+25$ **76.** $(4x-3)(x+1)=2$ **77.** $2x(x+5)(x-2)=0$

78. A boat can travel 21 miles downstream in 3 hours. The return trip takes 7 hours. Find the speed of the boat in still water and the speed of the current.

79. Karl invested a total of $10,000 in two separate accounts. One account paid 6% interest and the other paid 8% interest. If the annual income from both accounts was $650, how much did he have invested in each account?

80. The difference between two positive numbers is 5. If the smaller is added to the square of the larger, the result is 177. Find the numbers.

81. Find two consecutive integers such that the sum of their squares is equal to 145.

82. A ball is thrown upward with an initial velocity of 88 feet per second $(v_0 = 88)$. When will the ball be 120 feet above the ground? $\left[h(t) = -16t^2 + v_0 t.\right]$

83. The demand for a certain commodity is given by the function $D = -25p^2 + 50p + 3900$. Find the selling price, p, if 2700 units are sold.

84. Find the height of a cylinder if the volume is 863.5 cm^3 and the radius is 5 cm (round to the nearest whole integer).

85. In a particular right triangle, the hypotenuse is 4 in. longer than three times the shorter leg and the other leg is 1 in. shorter than the hypotenuse. What are the lengths of the sides of the triangle (round to the nearest whole integer)?

86. Three consecutive even integers are such that the square of the largest integer is 48 less than the sum of the squares of the other two integers. What are the integers?

Use a graphing calculator to solve (or estimate the solutions of) the equations in Exercises 87 – 90. Sketch the graphs that you use and find any approximate solutions accurate to two decimal places.

87. $x^2 + 2x = -5$ **88.** $|3x - 5| = 4$ **89.** $|x| = \left|\frac{1}{3}x - 4\right|$ **90.** $x^3 = 2x - 5$

63. $x = -1, 7$
64. $x = 0, -\frac{5}{3}$
65. $x = -3, \frac{3}{4}$
66. $x = 0, 7$
67. $x = -2, -6$
68. $x = -4, 7$
69. $x = 0, 1, -6$
70. $x = -10, 6$
71. $x = -5, 1$
72. $x = 0, -7$
73. $x = 0, -2$
74. $x = 0, 4$
75. $x = 5, -\frac{5}{2}$
76. $x = 1, -\frac{5}{4}$
77. $x = 0, -5, 2$
78. boat = 5 mph, current = 2 mph
79. \$7500 at 6%, \$2500 at 8%.
80. 13 and 8
81. 8 and 9 or −8 and −9
82. $t = 2.5, 3$ seconds
83. $p = \$8$
84. 11 cm
85. 7, 24, 25
86. 10, 12, 14

87. No real solution

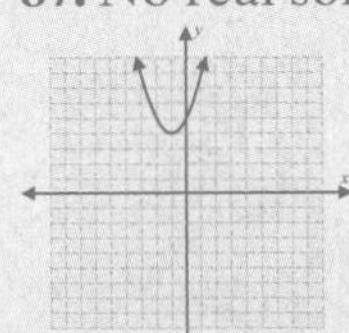

89. $x = -6, 3$

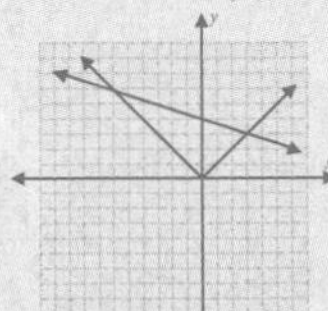

88. $x = 0.33, 3$

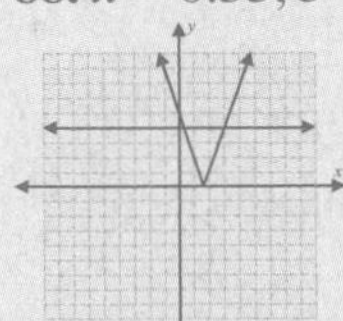

90. $x = -2.09$

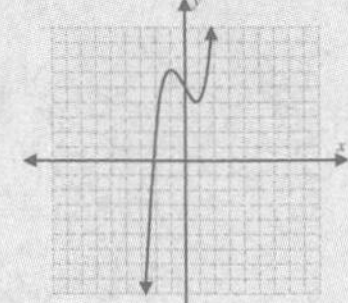

CHAPTER 7

Rational Expressions

Did You Know?

An important property related to rational expressions is the **cross-multiplication property**, which states that two rational expressions are equal if the cross products are equal. Symbolically, this property is stated as $\frac{a}{b}=\frac{c}{d}$ if and only if $a \cdot d = b \cdot c$. This property is the key to understanding how a missing term of a proportion can be found if the remaining three terms are known. For example, using the cross-multiplication property to solve the proportion $\frac{x}{6}=\frac{3}{4}$ yields $4x = 18$, and the solution of this simple linear equation is $x=\frac{9}{2}$, the missing term of the proportion. This type of solution illustrates the so-called Rule of Three, which has been known for over 3000 years. The Hindu mathematician Brahmagupta (c. 628) called the rule by that name in his writings, although problems of this type exist in ancient Egyptian and Chinese writings. Brahmagupta taught and wrote in the town of Ujjain in Central India, a center of Hindu science in the seventh century. Brahmagupta is one of the most famous Hindu mathematicians and he stated the Rule of Three as follows: "In the Rule of Three, argument, fruit and requisition are the names of the terms. Requisition multiplied by fruit and divided by argument is the produce;" or, as in our example, $x=3\cdot\left(\frac{6}{4}\right)$. In this case, the terms of the proportion have been given very fanciful names, and the cross-multiplication property has been concealed in an arbitrary rule.

The Rule of Three appears in Arabic and Latin works without explanation until the Renaissance. It was used in commercial arithmetic and occasionally was called the Merchant's Key or the Golden Rule. A popular seventeenth-century English arithmetic states: "The Rule of Three is commonly called The Golden Rule; and indeed it might be so termed; for as gold transcends all other metals, so doth this rule all others in Arithmetick." The Rule of Three often appeared in verse as a memory aid.

Arithmetic texts of the sixteenth and seventeenth centuries had pages or chapters called Practice. At that time, the term "practice" was used to mean commercial arithmetic usually involving the Rule of Three and other short processes for solving applied problems. Sometimes such problems were called Italian Practice because the problems often related to the methods developed in Italian commercial arithmetic.

"Multiplication is vexation,
Division is as bad;
The Rule of Three doth puzzle me,
And practice drives me mad."

Mother Goose Rhyme

Rational expressions are fractions in which the numerator and denominator are polynomials. All of the rules you learned in arithmetic about operating with fractions are going to be applied. For example, to add or subtract rational expressions, you need a common denominator, just as with fractions in arithmetic. To find common denominators and to multiply, divide, and simplify rational expressions, you are going to apply all the factoring skills you learned in Chapter 6. Thus, all those skills will be reinforced and, even if you had some difficulty with factoring, you probably will be very comfortable with factoring after Chapter 7.

The use of rational expressions opens the way to a variety of algebraic expressions, equations, and applications not available otherwise. In Section 7.4, you will learn to solve inequalities involving fractions, graph the solutions, and represent the solutions in interval notation in a manner similar to that discussed in Section 3.5. In Section 7.5, the applied problems involve fractions using some familiar formulas from a different point of view. You should find them particularly interesting.

7.1 Multiplication and Division with Rational Expressions

Objectives

After completing this section, you will be able to:

1. *Define rational expressions.*
2. *Raise rational expressions to higher terms.*
3. *Reduce rational expressions to lowest terms.*
4. *Multiply rational expressions.*
5. *Divide rational expressions.*

Basic Properties

Expressions such as

$$x-5, \quad x^2+1, \quad x^5-3x^4+2, \quad \text{and} \quad 2y^3+4y^2-8y+6$$

are polynomials. Any fraction formed with a polynomial as numerator and a polynomial as denominator is called a **rational expression**. Thus,

$$\frac{2x}{x^2+1}, \quad \frac{x+3}{x-5}, \quad \text{and} \quad \frac{y^3-3y^2}{y^2-7y+12}$$

are all rational expressions.

Rational Expression

*A **rational expression** is an expression of the form* $\frac{P}{Q}$ *(or in function notation,* $\frac{P(x)}{Q(x)}$*) where P and Q are polynomials and* $Q \neq 0$.

In the definition, the denominator $Q \neq 0$ means that for any rational expression we assume that **the variable will not equal any value that will cause a denominator to be 0.** For example, in the rational expression

$$\frac{P}{Q} = \frac{x^2 + 2x + 1}{3x - 6} \qquad Q = 3x - 6, \text{ and we assume that } 3x - 6 \neq 0 \text{ (or } x \neq 2\text{)}.$$

All of the basic properties of real numbers hold for rational expressions because rational expressions represent real numbers. The rules for operating with rational expressions are essentially the same as those for operating with fractions in arithmetic. That is, addition, subtraction, multiplication, and division with rational expressions are operations involving factoring and common denominators just as with fractions in arithmetic. The following rules for operating with rational numbers apply to operating with real numbers and rational expressions.

Arithmetic Rules for Rational Numbers (or Fractions)

*A **rational number** is a number that can be written in the fraction form* $\frac{a}{b}$ *where a and b are integers and* $b \neq 0$. *No denominator can be 0.*

The Fundamental Principle: $\frac{a}{b} = \frac{a \cdot k}{b \cdot k}$ *where* $k \neq 0$

*The **reciprocal** of* $\frac{a}{b}$ *is* $\frac{b}{a}$ *and* $\frac{a}{b} \cdot \frac{b}{a} = 1$.

Multiplication: $\frac{a}{b} \cdot \frac{c}{d} = \frac{a \cdot c}{b \cdot d}$

Division: $\frac{a}{b} \div \frac{c}{d} = \frac{a}{b} \cdot \frac{d}{c}$

Addition: $\frac{a}{b} + \frac{c}{b} = \frac{a+c}{b}$

Subtraction: $\frac{a}{b} - \frac{c}{b} = \frac{a-c}{b}$

Each rule can be restated by replacing a and b with P and Q where P and Q represent polynomials. In particular, the Fundamental Principle can be restated as follows:

Fundamental Principle of Fractions

If $\frac{P}{Q}$ *is a rational expression and K is a polynomial and* $K \neq 0$*, then*

$$\frac{P}{Q} = \frac{P}{Q} \cdot \frac{K}{K} = \frac{P \cdot K}{Q \cdot K}.$$

The Fundamental Principle can be used to build a rational expression to **higher terms** (for addition or subtraction) and to **reduce** a rational expression to **lower terms** (for multiplication or division). Just as with rational numbers, a rational expression is said to be **reduced to lowest terms** if the numerator and denominator have no common factors other than 1 and –1.

Example 1: Fundamental Principle

Use the Fundamental Principle to build up each expression as indicated. State any restrictions on the variable using the fact that no denominator can be 0.

a. $\frac{7}{8} = \frac{?}{24}$

Solution: Because $24 = 8 \cdot 3$

$$\frac{7}{8} = \frac{7 \cdot 3}{8 \cdot 3} = \frac{21}{24}$$

Building up a fraction to higher terms by using the Fundamental Principle.

b. $\frac{5x}{x+3} = \frac{?}{x^2 - x - 12}$

Solution: Because $x^2 - x - 12 = (x+3)(x-4)$,

$$\frac{5x}{x+3} = \frac{5x(x-4)}{(x+3)(x-4)}$$

$$= \frac{5x^2 - 20x}{x^2 - x - 12}$$

Building up a rational expression by using the Fundamental Principle ($x \neq -3$ and $x \neq 4$ because either of these values for x would make the denominator 0).

Use the Fundamental Principle to reduce each expression to lowest terms. State any restrictions on the variable using the fact that no denominator can be 0.

c. $\frac{2x-10}{3x-15}$

Solution: $\frac{2x-10}{3x-15} = \frac{2\cancel{(x-5)}}{3\cancel{(x-5)}} = \frac{2}{3} \quad (x \neq 5)$

Note that $x - 5$ is a common **factor**. The key word here is **factor**. We reduce using **factors** only.

d. $\frac{x^2-16}{x^3-64}$

Solution: $\frac{x^2-16}{x^3-64}=\frac{(x+4)\cancel{(x-4)}}{\cancel{(x-4)}(x^2+4x+16)}$

$=\frac{(x+4)}{(x^2+4x+16)} \quad (x \neq 4)$

Reduce; the common **factor** is $x-4$. Note that x^3-64 is the difference of two cubes.

The following properties are concerned with the placement of negative signs. These properties are particularly useful in addition and subtraction with rational expressions. In other words, these properties state that a negative sign can be in front of an expression, or with the denominator, or with the numerator, and the expression will have the same value. Additionally, negative signs can be introduced into a rational expression as long as they are placed in two positions, as shown. This means that we can use the form that best suits our purposes when simplifying particular expressions.

Negative Signs

$$-\frac{P}{Q}=\frac{P}{-Q}=\frac{-P}{Q} \quad \textit{and} \quad \frac{P}{Q}=\frac{-P}{-Q}=-\frac{-P}{Q}=-\frac{P}{-Q}$$

Example 2: Negative Signs

a. $-\frac{35}{5}=\frac{35}{-5}=\frac{-35}{5}=-7$

b. $-\frac{y-3}{y^2}=\frac{y-3}{-y^2}=\frac{-1\cdot(y-3)}{y^2} \quad (y \neq 0)$

c. $\frac{24}{6}=\frac{-24}{-6}=-\frac{-24}{6}=-\frac{24}{-6}=4$

d. $-\frac{5-x}{x-5}=\frac{-1\cdot(5-x)}{x-5}=\frac{\cancel{(x-5)}}{\cancel{x-5}}=1 \quad (x \neq 5)$

Closely related to the placement of negative signs is the following statement about opposites. If any number or algebraic expression is divided by its opposite, the quotient is -1.

Opposites

Teaching Notes:
To help convince your students that −1 is always the result you might want to have them pick **several pairs** of unequal values for x and a and evaluate $\frac{a-x}{x-a}$.

In general,

$$\frac{-P}{P} = -1 \qquad \text{if } P \neq 0$$

In particular,

$$\frac{a-x}{x-a} = -1 \qquad \text{if } x \neq a$$

Note carefully that $a - x$ and $x - a$ are opposites. Thus,

$$a - x = -x + a = -1\cdot(x-a) \text{ which leads to } \frac{a-x}{x-a} = \frac{-1\cdot(\cancel{x-a})}{\cancel{x-a}} = -1.$$

Example 3: Opposites

a. $\frac{x^2-2x-3}{-x^2+2x+3} = -1$ $\qquad x^2-2x-3 = -(-x^2+2x+3)$

$x \neq 3, -1$

b. $\frac{17-x}{x-17} = \frac{-(-17+x)}{x-17} = \frac{-1\cdot(\cancel{x-17})}{+1\cdot(\cancel{x-17})} = -1 \qquad x \neq 17$

Common Factors

Reduce only ***common factors****. Do not reduce terms unless they are* ***factors*** *common to both the numerator and denominator. Remember that* ***factors*** *imply multiplication.*

$\frac{x+\cancel{2}}{\cancel{2}}$ ← ***WRONG*** *2 is not a common factor.*

$\frac{\cancel{x^2}^{\,x} - \cancel{9}^{\,3}}{\cancel{x} - \cancel{3}}$ ← ***WRONG*** *3 and x are not common factors.*

$\frac{x^2-9}{x-3} = \frac{(x+3)(\cancel{x-3})}{(\cancel{x-3})}$ ← ***RIGHT*** *$x-3$ is a common factor.*

$\frac{2x+8}{2} = \frac{\cancel{2}(x+4)}{\cancel{2}}$ ← ***RIGHT*** *2 is a common factor.*

Multiplication of Rational Expressions

As stated earlier, multiplication of rational expressions is accomplished in the same manner as multiplication of rational numbers. For example, the products $\frac{2}{3}\cdot\frac{5}{9}$ and $\frac{15}{7}\cdot\frac{49}{65}$ can be found by multiplying the numerators and denominators, factoring, and reducing.

$$\frac{2}{3}\cdot\frac{5}{9}=\frac{2\cdot 5}{3\cdot 9}=\frac{10}{27} \quad \text{and} \quad \frac{15}{7}\cdot\frac{49}{65}=\frac{3\cdot\cancel{5}\cdot\cancel{7}\cdot 7}{\cancel{7}\cdot\cancel{5}\cdot 13}=\frac{21}{13}$$

The same techniques are used to multiply rational expressions.

$$\frac{2x}{x-6}\cdot\frac{x+5}{x-4}=\frac{2x(x+5)}{(x-6)(x-4)}=\frac{2x^2+10x}{x^2-10x+24}$$

$$\frac{y^2-4}{y^3}\cdot\frac{y^2-3y}{y^2-y-6}=\frac{\cancel{(y+2)}(y-2)\cdot\cancel{y}\cancel{(y-3)}}{\cancel{y^3}_{\,y^2}\cancel{(y-3)}\cancel{(y+2)}}=\frac{y-2}{y^2}$$

Multiplication of Rational Expressions

$$\frac{P}{Q}\cdot\frac{R}{S}=\frac{P\cdot R}{Q\cdot S} \qquad \textit{where } Q,\ S\neq 0$$

Example 4: Multiplication of Rational Expressions

Find the following products and reduce to lowest terms by factoring whenever possible. Assume that no denominator has a value of 0.

a. $\frac{5x^2y}{9xy^3}\cdot\frac{6x^3y^2}{15xy^4}=\frac{\cancel{5}\cdot 2\cdot\cancel{3}x^5y^3}{3\cdot 3\cdot\cancel{5}\cdot\cancel{3}x^2y^7}=\frac{2x^{5-2}y^{3-7}}{9}=\frac{2x^3y^{-4}}{9}=\frac{2x^3}{9y^4}$

b. $\frac{x}{x-2}\cdot\frac{x^2-4}{x^2}=\frac{\cancel{x}(x+2)\cancel{(x-2)}}{\cancel{(x-2)}\cancel{x^2}_{\,x}}=\frac{x+2}{x}$

c. $\frac{3x-3}{x^2+x}\cdot\frac{x^2+2x+1}{3x^2-6x+3}=\frac{\cancel{3}\cancel{(x-1)}\cancel{(x+1)^2}^{\,(x+1)}}{x\cancel{(x+1)}\cdot\cancel{3}\cancel{(x-1)^2}_{\,(x-1)}}=\frac{x+1}{x(x-1)}$

Division of Rational Expressions

The rational expression $\frac{S}{R}$ is the **reciprocal** of $\frac{R}{S}$. Just as in division with rational numbers, division with rational expressions is accomplished by performing multiplication by the reciprocal of the divisor.

Division of Rational Expressions

$$\frac{P}{Q} \div \frac{R}{S} = \frac{P}{Q} \cdot \frac{S}{R} \qquad \text{where } Q,\ R,\ S \neq 0$$

Example 5: Division of Rational Expressions

Find the following quotients and reduce to lowest terms by factoring whenever possible. Assume that no denominator has a value of 0. Write the final answer using only positive exponents.

a. $\frac{12x^2y}{10xy^2} \div \frac{3x^4y}{xy^3}$

Solution: $\frac{12x^2y}{10xy^2} \div \frac{3x^4y}{xy^3} = \frac{12x^2y}{10xy^2} \cdot \frac{xy^3}{3x^4y}$

$$= \frac{\cancel{2} \cdot 2 \cdot \cancel{3} x^3 y^4}{\cancel{2} \cdot 5 \cdot \cancel{3} x^5 y^3} = \frac{2x^{3-5}y^{4-3}}{5} = \frac{2x^{-2}y}{5} = \frac{2y}{5x^2}$$

b. $\frac{x^3 - y^3}{x^3} \div \frac{y-x}{xy}$

Solution: $\frac{x^3 - y^3}{x^3} \div \frac{y-x}{xy} = \frac{x^3 - y^3}{x^3} \cdot \frac{xy}{y-x}$

$$= \frac{\overset{-1}{\cancel{(x-y)}}(x^2 + xy + y^2)\cancel{x}y}{\underset{x^2}{\cancel{x^3}}\cancel{(y-x)}}$$

Note that $\frac{x-y}{y-x} = -1$

$$= \frac{-y(x^2 + xy + y^2)}{x^2}$$

Teaching Notes:
Some students may note that to leave both the numerator and denominator in factored form provides assurance that the rational expression is truly in reduced form.

c. $\dfrac{x^2-8x+15}{2x^2+11x+5} \div \dfrac{2x^2-5x-3}{4x^2-1}$

Solution: $\dfrac{x^2-8x+15}{2x^2+11x+5} \div \dfrac{2x^2-5x-3}{4x^2-1} = \dfrac{x^2-8x+15}{2x^2+11x+5} \cdot \dfrac{4x^2-1}{2x^2-5x-3}$

$$= \frac{\cancel{(x-3)}(x-5)(2x-1)\cancel{(2x+1)}}{(2x+1)(x+5)\cancel{(x-3)}\cancel{(2x+1)}}$$

$$= \frac{(x-5)(2x-1)}{(2x+1)(x+5)} = \frac{2x^2-11x+5}{(2x+1)(x+5)}$$

NOTES

As illustrated in the answer in Example 5c, generally the denominator will be left in factored form, and the numerator will be multiplied out. This form makes the results easier to add or subtract, as we will see in the next section. However, be aware that leaving the denominator in factored form is just an option, and multiplying out the denominator is not an error. Thus in Example 5c we can write the answer either as

$$\frac{2x^2-11x+5}{(2x+1)(x+5)} \text{ or as } \frac{2x^2-11x+5}{2x^2+11x+5}.$$

Practice Problems

Reduce to lowest terms. State any restrictions on the variables.

1. $\dfrac{5x+20}{7x+28}$ **2.** $\dfrac{4-2x}{2x-4}$ **3.** $\dfrac{x^2+x-2}{x^2+3x+2}$

Perform the following operations and simplify the results. Assume that no denominator is 0.

4. $\dfrac{x-7}{x^3} \cdot \dfrac{x^2}{49-x^2}$ **5.** $\dfrac{y^2-y-6}{y^2-5y+6} \cdot \dfrac{y^2-4}{y^2+4y+4}$

6. $\dfrac{x^3+3x}{2x+1} \div \dfrac{x^2+3}{x+1}$ **7.** $\dfrac{x^2+2x-3}{x^2-3x-10} \cdot \dfrac{2x^2-9x-5}{x^2-2x+1} \div \dfrac{4x+2}{x^2-x}$

Answers to Practice Problems: 1. $\dfrac{5}{7}$, $x \neq -4$ **2.** -1, $x \neq 2$ **3.** $\dfrac{x-1}{x+1}$, $x \neq -1, -2$ **4.** $\dfrac{-1}{x(x+7)}$ **5.** 1 **6.** $\dfrac{x^2+x}{2x+1}$ **7.** $\dfrac{x^2+3x}{2(x+2)}$

7.1 Exercises

1. $-12x^4y$ **2.** $18xy^3$
3. $4x^2+8x$
4. $5x^2+25x$
5. x^2+x-2
6. $5(x+2)$
7. $-8(x+2)$
8. $6(8-y)$
9. $7(x^2-4x+16)$
10. $12(x^2-3x+9)$
11. $\frac{3x}{4y}; x\neq 0; y\neq 0$
12. $\frac{2y^3}{3x}; x\neq 0, y\neq 0$
13. $\frac{2x^3}{3y^3}; x\neq 0; y\neq 0$
14. $\frac{3y^2}{4x^3}; x\neq 0, y\neq 0$
15. $\frac{1}{(x-3)}; x\neq 0, 3$
16. $\frac{3}{(x+5)}; x\neq 0, -5$
17. $7; x\neq 2$
18. $-1; x\neq 2$
19. $-\frac{3}{4}; x\neq 3$
20. $\frac{2x}{y}; x\neq -\frac{2}{3}, y\neq 0$
21. $-\frac{3}{4}; x\neq 3$
22. $\frac{1}{4x}; x\neq 0, -3y$
23. $\frac{x}{(x-1)}; x\neq -6, 1$
24. $\frac{(x-3)}{(x+1)}; x\neq 2, -1$
25. $\frac{x-y}{3x}; x\neq 0, -y$

In Exercises 1 – 10, build each rational expression to higher terms as indicated. Assume that no denominator equals 0.

1. $\frac{3x^2}{-8y^2}=\frac{?}{32x^2y^3}$ **2.** $\frac{3y}{16x^2}=\frac{?}{96x^3y^2}$ **3.** $\frac{4x}{x+4}=\frac{?}{(x+2)(x+4)}$

4. $\frac{5x}{x-3}=\frac{?}{(x-3)(x+5)}$ **5.** $\frac{x-1}{x+1}=\frac{?}{x^2+3x+2}$ **6.** $\frac{5}{9x^2-3x}=\frac{?}{9x^3+15x^2-6x}$

7. $\frac{8}{x-4}=\frac{?}{8+2x-x^2}$ **8.** $\frac{6}{3-y}=\frac{?}{y^2-11y+24}$ **9.** $\frac{7}{x+4}=\frac{?}{x^3+64}$

10. $\frac{12}{x+3}=\frac{?}{x^3+27}$

Reduce each rational expression in Exercises 11 – 30. State any restrictions on the variable using the fact that no denominator can equal 0.

11. $\frac{9x^2y^3}{12xy^4}$ **12.** $\frac{18xy^4}{27x^2y}$ **13.** $\frac{20x^5}{30x^2y^3}$ **14.** $\frac{15y^4}{20x^3y^2}$

15. $\frac{x}{x^2-3x}$ **16.** $\frac{3x}{x^2+5x}$ **17.** $\frac{7x-14}{x-2}$ **18.** $\frac{4-2x}{2x-4}$

19. $\frac{9-3x}{4x-12}$ **20.** $\frac{6x^2+4x}{3xy+2y}$ **21.** $\frac{3x-9}{12-4x}$ **22.** $\frac{x+3y}{4x^2+12xy}$

23. $\frac{x^2+6x}{x^2+5x-6}$ **24.** $\frac{x^2-5x+6}{x^2-x-2}$ **25.** $\frac{x^2-y^2}{3x^2+3xy}$ **26.** $\frac{x^2-4}{x^3-8}$

27. $\frac{x^3+64}{2x^2+x-28}$ **28.** $\frac{3x^2+14x-24}{18-9x-2x^2}$ **29.** $\frac{x^3-2x^2+5x-10}{x^3-8}$ **30.** $\frac{xy-3y+2x-6}{y^2-4}$

Perform the indicated operations in Exercises 31 – 84. Assume that no denominator equals 0.

31. $\frac{ax^2}{b}\cdot\frac{b^2}{x^2y}$ **32.** $\frac{18x^3}{5y^2}\cdot\frac{30y^3}{9x^4}$ **33.** $\frac{24x^3}{25y^2}\cdot\frac{10y^5}{18x}$ **34.** $\frac{16x^8}{3y^{11}}\cdot\frac{-21y^9}{10x^7}$

35. $\frac{x^2-9}{x^2+2x}\cdot\frac{x+2}{x-3}$ **36.** $\frac{16x^2-9}{3x^2-15x}\cdot\frac{6}{4x+3}$ **37.** $\frac{x^2+2x-3}{x^2+3x}\cdot\frac{x}{x+1}$

38. $\frac{4x+16}{x^2-16}\cdot\frac{x-4}{x}$ **39.** $\frac{x^2+6x-16}{x^2-64}\cdot\frac{1}{2-x}$ **40.** $\frac{4-x^2}{x^2-4x+4}\cdot\frac{3}{x+2}$

41. $\frac{x^2-5x+6}{x^2-4x}\cdot\frac{x-4}{x-3}$ **42.** $\frac{2x^2+x-3}{x^2+4x}\cdot\frac{2x+8}{x-1}$ **43.** $\frac{2x^2+10x}{3x^2+5x+2}\cdot\frac{6x+4}{x^2}$

44. $\frac{x+3}{x^2-16}\cdot\frac{x^2-3x-4}{x^2-1}$ **45.** $\frac{x}{x^2+7x+12}\cdot\frac{x^2-2x-24}{x^2-7x+6}$ **46.** $\frac{x^2-2x-3}{x+5}\cdot\frac{x^2-5x-14}{x^2-x-6}$

47. $\dfrac{8-2x-x^2}{x^2-2x}\cdot\dfrac{x-4}{x^2-3x-4}$

48. $\dfrac{3x^2+21x}{x^2-49}\cdot\dfrac{x^2-5x+4}{x^2+3x-4}$

49. $\dfrac{(x-2y)^2}{x^2-5xy+6y^2}\cdot\dfrac{x+2y}{x^2-4xy+4y^2}$

50. $\dfrac{4x^2+6x}{x^2+3x-10}\cdot\dfrac{x^2+4x-12}{x^2+5x-6}$

51. $\dfrac{2x^2+5x+2}{3x^2+8x+4}\cdot\dfrac{3x^2-x-2}{4x^3-x}$

52. $\dfrac{x^2+5x}{4x^2+12x+9}\cdot\dfrac{6x^2+7x-3}{x^2+10x+25}$

53. $\dfrac{x^2+x+1}{x^2-1}\cdot\dfrac{x^2-2x+1}{x^3-1}$

54. $\dfrac{x-2}{x^2-2x+4}\cdot\dfrac{x^3+8}{x^2-4x+4}$

55. $\dfrac{2x^2-7x+3}{x^2-9}\cdot\dfrac{3x^2+8x-3}{6x^2+x-1}$

56. $\dfrac{12x^2y}{9xy^9}\div\dfrac{4x^4y}{x^2y^3}$

57. $\dfrac{35xy^3}{24x^3y}\div\dfrac{15x^4y^3}{84xy^4}$

58. $\dfrac{45xy^4}{21x^2y^2}\div\dfrac{40x^4}{112xy^5}$

59. $\dfrac{x-3}{15x}\div\dfrac{4x-12}{5}$

60. $\dfrac{x-1}{6x+6}\div\dfrac{2x-2}{x^2+x}$

61. $\dfrac{7x-14}{x^2}\div\dfrac{x^2-4}{x^3}$

62. $\dfrac{6x^2-54}{x^4}\div\dfrac{x-3}{x^2}$

63. $\dfrac{x^2-25}{6x+30}\div\dfrac{x-5}{x}$

64. $\dfrac{2x-1}{x^2+2x}\div\dfrac{10x^2-5x}{6x^2+12x}$

65. $\dfrac{x+3}{x^2+3x-4}\div\dfrac{x+2}{x^2+x-2}$

66. $\dfrac{6x^2-7x-3}{x^2-1}\div\dfrac{2x-3}{x-1}$

67. $\dfrac{x^2-9}{2x^2+7x+3}\div\dfrac{x^2-3x}{2x^2+11x+5}$

68. $\dfrac{x^2-8x+15}{x^2-9x+14}\div\dfrac{x^2+4x-21}{x-1}$

69. $\dfrac{2x+1}{4x-x^2}\div\dfrac{4x^2-1}{x^2-16}$

70. $\dfrac{x^2-6x+9}{x^2-4x+3}\div\dfrac{2x^2-7x+3}{x^2-3x+2}$

71. $\dfrac{x^2-4x+4}{x^2+5x+6}\div\dfrac{x^2+2x-8}{x^2+7x+12}$

72. $\dfrac{x^2-x-6}{x^2+6x+8}\div\dfrac{x^2-4x+3}{x^2+5x+4}$

73. $\dfrac{x^2-x-12}{6x^2+x-9}\div\dfrac{x^2-6x+8}{3x^2-x-6}$

74. $\dfrac{6x^2+5x+1}{4x^3-3x^2}\div\dfrac{3x^2-2x-1}{3x^2-2x+1}$

75. $\dfrac{8x^2+2x-15}{3x^2+13x+4}\div\dfrac{2x^2+5x+3}{6x^2-x-1}$

76. $\dfrac{3x^2+13x+14}{4x^3-3x^2}\div\dfrac{6x^2-x-35}{4x^2+5x-6}$

77. $\dfrac{3x^2+2x}{9x^2-4}\div\dfrac{27x^3-8}{9x^2-6x+4}$

78. $\dfrac{x^3+2x^2}{x^3+64}\div\dfrac{4x^2}{x^2-4x+16}$

79. $\dfrac{6-11x-10x^2}{2x^2+x-3}\div\dfrac{5x^3-2x^2}{3x^2-5x+2}$

80. $\dfrac{x-6}{x^2-7x+6}\cdot\dfrac{x^2-3x}{x+3}\cdot\dfrac{x^2-9}{x^2-4x+3}$

81. $\dfrac{3x^2+11x+10}{2x^2+x-6}\cdot\dfrac{x^2+2x-3}{2x-1}\cdot\dfrac{2x-3}{3x^2+2x-5}$

82. $\dfrac{x^3+3x^2}{x^2+7x+12}\cdot\dfrac{2x^2+7x-4}{2x^2-x}\div\dfrac{2x^2-x-1}{x^2+4x-5}$

26. $\dfrac{(x+2)}{(x^2+2x+4)};x\neq 2$

27. $\dfrac{x^2-4x+16}{(2x-7)};x\neq -4,\dfrac{7}{2}$

28. $-\dfrac{(3x-4)}{(2x-3)};x\neq -6,\dfrac{3}{2}$

29. $\dfrac{x^2+5}{x^2+2x+4};x\neq 2$

30. $\dfrac{(x-3)}{(y-2)};y\neq -2,2$

31. $\dfrac{ab}{y}$ **32.** $\dfrac{12y}{x}$

33. $\dfrac{8x^2y^3}{15}$ **34.** $-\dfrac{56x}{5y^2}$

35. $\dfrac{x+3}{x}$ **36.** $\dfrac{8x-6}{x(x-5)}$

37. $\dfrac{x-1}{(x+1)}$ **38.** $\dfrac{4}{x}$

39. $-\dfrac{1}{x-8}$

40. $-\dfrac{3}{(x-2)}$

41. $\dfrac{x-2}{x}$

42. $\dfrac{2(2x+3)}{x}$

43. $\dfrac{4x+20}{x(x+1)}$

44. $\dfrac{(x+3)}{(x+4)(x-1)}$

45. $\dfrac{x}{(x+3)(x-1)}$

46. $\dfrac{x^2-6x-7}{(x+5)}$

47. $-\dfrac{x+4}{x(x+1)}$

48. $\dfrac{3x^2-12x}{(x-7)(x+4)}$

49. $\dfrac{x+2y}{(x-3y)(x-2y)}$ **50.** $\dfrac{4x^2+6x}{(x+5)(x-1)}$ **51.** $\dfrac{x-1}{x(2x-1)}$ **52.** $\dfrac{3x^2-x}{(2x+3)(x+5)}$ **53.** $\dfrac{1}{x+1}$

54. $\frac{x+2}{x-2}$ 55. $\frac{2x-1}{2x+1}$

56. $\frac{1}{3xy^6}$ 57. $\frac{49y^3}{6x^5}$

58. $\frac{6y^7}{x^4}$ 59. $\frac{1}{12x}$

60. $\frac{x}{12}$ 61. $\frac{7x}{x+2}$

62. $\frac{6x+18}{x^2}$ 63. $\frac{x}{6}$

64. $\frac{6}{5x}$ 65. $\frac{x+3}{x+4}$

66. $\frac{3x+1}{x+1}$ 67. $\frac{x+5}{x}$

68. $\frac{x^2-6x+5}{(x-7)(x-2)(x+7)}$

69. $-\frac{x+4}{x(2x-1)}$

70. $\frac{(x-2)}{(2x-1)}$ 71. $\frac{x-2}{x+2}$

72. $\frac{x+1}{x-1}$

73. $\frac{3x^3+8x^2-9x-18}{(6x^2+x-9)(x-2)}$

74. $\frac{6x^3-x^2+1}{x^2(4x-3)(x-1)}$

75. $\frac{8x^2-14x+5}{(x+4)(x+1)}$

83. $\frac{x^2+2x-3}{x^2+10x+21} \div \frac{x^2-7x-8}{x^2+6x+5} \cdot \frac{x^2-x-56}{x^2-3x-40}$ 84. $\frac{2x^2-5x+2}{4xy-2y+6x-3} \div \frac{xy-2y+3x-6}{2y^2+9y+9}$

Writing and Thinking About Mathematics

85. **a.** Define rational expression.
b. Give an example of a rational expression that is undefined for $x = -2$ and $x = 3$ and has a value of 0 for $x = 1$. Explain how you determined this expression.
c. Give an example of a rational expression that is undefined for $x = -5$ and never has a value of 0. Explain how you determined this expression.

86. Write the opposite of each of the following expressions.
a. $3 - x$ **b.** $2x - 7$ **c.** $x + 5$ **d.** $-3x - 2$

87. Given the rational function $f(x) = \frac{x-4}{x^2-100}$:
a. For what values, if any, will $f(x) = 0$?
b. For what values, if any, is $f(x)$ undefined?

88. The following "proof" that 2 = 0 contains an error. Discuss the error in your own words.

$$a = b$$
$$2a = 2b$$
$$2a - 2b = 0$$
$$2(a-b) = 0$$
$$\frac{2(\cancel{a-b})}{\cancel{a-b}} = \frac{0}{a-b}$$
$$2 = 0$$

85. **a.** Answers will vary. **b.** $\frac{(x-1)}{(x+2)(x-3)}$ **c.** $\frac{1}{(x+5)}$

86. **a.** $x-3$ **b.** $7-2x$ **c.** $-x-5$ **d.** $3x+2$

87. **a.** $x = 4$ **b.** $x = 10$, $x = -10$

88. Since $a = b$, $a - b = 0$. So the equation cannot be divided by $a - b$, since division by 0 is undefined.

Hawkes Learning Systems: Introductory & Intermediate Algebra

Defining Rational Expressions
Multiplication and Division of Rational Expressions

76. $\frac{x^2+4x+4}{x^2(2x-5)}$ 77. $\frac{9x^3-6x^2+4x}{(9x^2-12x+4)(9x^2+6x+4)}$ 78. $\frac{x+2}{4(x+4)}$ 79. $-\frac{3x-2}{x^2}$

80. $\frac{x^2-3x}{(x-1)(x-1)}$ 81. $\frac{x+3}{2x-1}$ 82. $\frac{x^2+5x}{(2x+1)}$ 83. $\frac{x-1}{x-8}$ 84. 1

7.2 Addition and Subtraction with Rational Expressions

Objectives

After completing this section, you will be able to:

1. *Find the least common multiple of a set of algebraic expressions.*
2. *Add rational expressions.*
3. *Subtract rational expressions.*

Adding and Subtracting Rational Expressions with Common Denominators

Teaching Notes: In introducing this section you will probably want to focus on the distinction between a *factor* and a *term*.

To add or subtract rational expressions with a common denominator, proceed just as you would with numerical fractions. That is, add (or subtract) the numerators and keep the common denominator. For example,

$$\frac{x}{x+2}+\frac{3}{x+2}=\frac{x+3}{x+2} \quad \text{and} \quad \frac{x}{x+2}-\frac{3}{x+2}=\frac{x-3}{x+2} \qquad (x \neq -2)$$

Just as with numerical fractions, the sum (or difference) should be reduced if possible. For example, in the following sum the result can be reduced.

$$\frac{x}{x^2-1}+\frac{1}{x^2-1}=\frac{x+1}{x^2-1} \qquad \text{Factor and reduce the sum.}$$

$$=\frac{\overset{1}{\cancel{(x+1)}}}{\cancel{(x+1)}(x-1)}=\frac{1}{x-1} \qquad x^2-1\neq 0\ (\text{or } x\neq \pm 1)$$

A difference such as $\frac{x^2}{x^2+4x+4}-\frac{2x+8}{x^2+4x+4}$ is found by subtracting the numerators and using the common denominator.

$$\frac{x^2}{x^2+4x+4}-\frac{2x+8}{x^2+4x+4}=\frac{x^2-(2x+8)}{x^2+4x+4}$$

$$=\frac{x^2-2x-8}{x^2+4x+4}$$

Again, the result can be reduced.

$$\frac{x^2-2x-8}{x^2+4x+4}=\frac{(x-4)\cancel{(x+2)}}{(x+2)\cancel{(x+2)}}$$

$$=\frac{x-4}{x+2} \quad (x\neq -2)$$

Addition and Subtraction of Rational Expressions

$$\frac{P}{Q}+\frac{R}{Q}=\frac{P+R}{Q} \quad \text{and} \quad \frac{P}{Q}-\frac{R}{Q}=\frac{P-R}{Q} \quad \text{where } Q \neq 0$$

Note that the expression $P-R$ indicates the difference of two polynomials and this will affect all of the signs in R. **A good idea is to put P in parentheses and R in parentheses so that all changes in sign will be done correctly.**

Example 1: Addition and Subtraction of Rational Expressions

Find the indicated sum or difference. Reduce if possible. Assume that no denominator is 0.

a. $$\frac{2x+1}{3x-3}+\frac{x+2}{3x-3}=\frac{(2x+1)+(x+2)}{3x-3}$$
$$=\frac{3x+3}{3x-3}=\frac{\cancel{3}(x+1)}{\cancel{3}(x-1)}=\frac{x+1}{x-1}$$

b. $$\frac{2x-5y}{x+y}-\frac{3x-7y}{x+y}=\frac{(2x-5y)-(3x-7y)}{x+y}$$
$$=\frac{2x-5y-3x+7y}{x+y}=\frac{-x+2y}{x+y}$$

NOTES

Common Error

Many students make a **mistake** in subtracting fractions by not subtracting the entire numerator. They make a mistake similar to the following.

WRONG → $$\frac{7}{x+6}-\frac{x+3}{x+6}=\frac{7-x+3}{x+6}=\frac{10-x}{x+6}$$ (Error: $7-x+3$)

By using parentheses, you can avoid such mistakes.

RIGHT → $$\frac{7}{x+6}-\frac{x+3}{x+6}=\frac{7-(x+3)}{x+6}=\frac{7-x-3}{x+6}=\frac{4-x}{x+6}$$

Finding the Least Common Multiple (LCM)

The rational expressions added and subtracted in Examples 1a and 1b had common denominators. To add or subtract expressions with different denominators, each expression must be built to higher terms with a common denominator. This common denominator is the **least common denominator** (**LCD**) and represents the **least common multiple** (**LCM**) of the denominators. The technique for finding the LCD is developed in the following discussion.

To Find the LCM for a Set of Polynomials

1. *Completely factor each polynomial (including the prime factors for numerical factors).*
2. *Form the product of all distinct factors that appear, using each factor the most number of times it appears in any one factorization.*

NOTES

Recall that the prime numbers are {2, 3, 5, 7, 11, 13, 17, 19, 23, 29, 31, 37, 41, 43, 47, 53,...}. Also, 0 and 1 are **not** prime numbers.

Consider the three terms in the set $\left\{18x^3, 24xy, 63\right\}$. To find the **LCM**:

1. Find the complete factorization of each term, including prime factors.

$$18x^3 = 2\cdot3\cdot3\cdot x^3 \longleftarrow \text{one 2, two 3's, } x^3$$
$$24xy = 2\cdot2\cdot2\cdot3\cdot x\cdot y \longleftarrow \text{three 2's, one 3, } x, y$$
$$63 = 3\cdot3\cdot7 \longleftarrow \text{two 3's, one 7}$$

2. Form a product using each prime factor and each variable the most number of times it appears in **any one** of the factorizations.

$$\text{LCM} = 2\cdot2\cdot2\cdot3\cdot3\cdot7\cdot x^3\cdot y \longleftarrow \text{three 2's, two 3's, one 7, } x^3, y$$
$$= 504x^3y$$

This product, $504x^3y$, is the least common multiple. This is the smallest number with the smallest positive exponents on the variables that is divisible by all three terms.

Now, use the same technique to find the LCM for the polynomial expressions

$$x^2+6x+9, \quad x^2-9, \quad \text{and} \quad 2x+6.$$

1. Factor each expression completely.

$$x^2+6x+9=(x+3)^2$$
$$x^2-9=(x+3)(x-3)$$
$$2x+6=2(x+3)$$

2. To determine the LCM, form the product of 2, $(x+3)^2$, and $(x-3)$. That is, use each factor the most number of times it appears in **any one** factorization.

$$\text{LCM} = 2(x+3)^2(x-3)$$

Adding and Subtracting Rational Expressions with Different Denominators

Use the following procedure when adding or subtracting rational expressions with different denominators.

Procedure for Adding or Subtracting Rational Expressions with Different Denominators

1. *Find the LCD (the LCM of the denominators).*
2. *Rewrite each fraction in an equivalent form with the LCD.*
3. *Add (or subtract) the numerators and keep the common denominator.*
4. *Reduce if possible.*

The procedure is outlined in detail in finding the following sum:

$$\frac{1}{x^2+6x+9}+\frac{1}{x^2-9}+\frac{1}{2x+6}$$

First, find the LCD which is the LCM as just discussed above:

$$\text{LCD} = \text{LCM} = 2(x+3)^2(x-3)$$

Next, by using the Fundamental Principle of Fractions, each rational expression is multiplied by 1 in a form that will give an equivalent expression with the desired denominator. Thus,

$$\frac{1}{x^2+6x+9}=\frac{1}{(x+3)^2}\cdot\frac{2(x-3)}{2(x-3)}=\frac{1\cdot 2(x-3)}{(x+3)^2\cdot 2(x-3)}$$

$$\frac{1}{x^2-9}=\frac{1}{(x+3)(x-3)}\cdot\frac{2(x+3)}{2(x+3)}=\frac{1\cdot 2(x+3)}{(x+3)(x-3)\cdot 2(x+3)}$$

$$\frac{1}{2x+6}=\frac{1}{2(x+3)}\cdot\frac{(x+3)(x-3)}{(x+3)(x-3)}=\frac{1\cdot(x+3)(x-3)}{2(x+3)\cdot(x+3)(x-3)}$$

Each rational expression now has the same denominator $2(x+3)^2(x-3)$, and the resulting fractions can be added.

$$\frac{1}{x^2+6x+9} + \frac{1}{x^2-9} + \frac{1}{2x+6}$$

$$= \frac{1}{(x+3)^2} + \frac{1}{(x+3)(x-3)} + \frac{1}{2(x+3)}$$

$$= \frac{1\cdot 2(x-3)}{(x+3)^2\cdot 2(x-3)} + \frac{1\cdot 2(x+3)}{(x+3)(x-3)\cdot 2(x+3)} + \frac{1\cdot (x+3)(x-3)}{2(x+3)\cdot (x+3)(x-3)}$$

$$= \frac{(2x-6)+(2x+6)+(x^2-9)}{2(x+3)^2(x-3)}$$

$$= \frac{x^2+4x-9}{2(x+3)^2(x-3)}$$

NOTES The denominator is left in factored form as a convenience for possibly reducing or adding to some other expression later. Denominators are left in factored form in the answers in the back of the text.

Example 2: Adding and Subtracting Rational Expressions

Perform the indicated operation. Reduce if possible. Assume that no denominator is 0.

a. $\frac{x}{x-3} + \frac{6}{x+4}$

Solution: Here, neither denominator can be factored so the LCD is the product of these factors. That is, LCD = $(x-3)(x+4)$.

$$\frac{x}{x-3} + \frac{6}{x+4} = \frac{x(x+4)}{(x-3)(x+4)} + \frac{6(x-3)}{(x+4)(x-3)}$$

$$= \frac{(x^2+4x)+(6x-18)}{(x-3)(x+4)}$$

$$= \frac{x^2+4x+6x-18}{(x-3)(x+4)}$$

$$= \frac{x^2+10x-18}{(x-3)(x+4)}$$

NOTES Again, the denominator is left in factored form as a convenience for possibly reducing or adding to some other expression later. You may choose to multiply these factors. Either form is correct.

continued on next page ...

b. $\dfrac{x}{x-5}-\dfrac{3}{5-x}$

Solution: Because each denominator is the opposite of the other, the numerator and denominator of one fraction can both be multiplied by –1. Then both denominators will be the same.
Note: $(5-x)(-1)=-5+x=x-5$.

$$\frac{x}{x-5}-\frac{3}{5-x}\cdot\frac{(-1)}{(-1)}=\frac{x}{x-5}-\frac{-3}{x-5}=\frac{x-(-3)}{x-5}=\frac{x+3}{x-5}$$

c. $\dfrac{x+5}{x-5}-\dfrac{100}{x^2-25}$

Solution:
$$\left.\begin{aligned}x-5&=x-5\\x^2-25&=(x+5)(x-5)\end{aligned}\right\rangle\ \text{LCD}=(x+5)(x-5)$$

$$\begin{aligned}\frac{x+5}{x-5}-\frac{100}{x^2-25}&=\frac{(x+5)(x+5)}{(x-5)(x+5)}-\frac{100}{(x+5)(x-5)}\\&=\frac{(x^2+10x+25)-100}{(x+5)(x-5)}\\&=\frac{x^2+10x+25-100}{(x+5)(x-5)}\\&=\frac{x^2+10x-75}{(x+5)(x-5)}=\frac{(x+15)\cancel{(x-5)}}{\cancel{(x-5)}(x+5)}=\frac{x+15}{x+5}\end{aligned}$$

d. $\dfrac{x+y}{(x-y)^2}+\dfrac{x}{2x^2-2y^2}$

Solution:
$$\left.\begin{aligned}(x-y)^2&=(x-y)^2\\2x^2-2y^2&=2(x+y)(x-y)\end{aligned}\right\rangle\ \text{LCD}=2(x-y)^2(x+y)$$

$$\begin{aligned}&\frac{x+y}{(x-y)^2}+\frac{x}{2x^2-2y^2}\\&=\frac{(x+y)\cdot2(x+y)}{(x-y)^2\cdot2(x+y)}+\frac{x(x-y)}{2(x-y)(x+y)(x-y)}\\&=\frac{2x^2+4xy+2y^2+x^2-xy}{2(x-y)^2(x+y)}\\&=\frac{3x^2+3xy+2y^2}{2(x-y)^2(x+y)}\end{aligned}$$

e. $\frac{3x-12}{x^2+x-20}-\frac{x^2+5x}{x^2+9x+20}$

Hint: In this problem, both expressions can be reduced before looking for the LCD.

Solution: $\frac{3x-12}{x^2+x-20}-\frac{x^2+5x}{x^2+9x+20}$

$$=\frac{3\cancel{(x-4)}}{(x+5)\cancel{(x-4)}}-\frac{x\cancel{(x+5)}}{\cancel{(x+5)}(x+4)}$$

$$=\frac{3}{x+5}-\frac{x}{x+4}$$

Now subtract these two expressions with LCD = $(x+5)(x+4)$.

$$\frac{3}{x+5}-\frac{x}{x+4}=\frac{3(x+4)}{(x+5)(x+4)}-\frac{x(x+5)}{(x+4)(x+5)}$$

$$=\frac{(3x+12)-(x^2+5x)}{(x+5)(x+4)}$$

$$=\frac{3x+12-x^2-5x}{(x+5)(x+4)}$$

$$=\frac{-x^2-2x+12}{(x+5)(x+4)}$$

f. $\frac{4}{x^2-1}+\frac{4}{x+1}+\frac{2}{1-x}$

Solution: $x^2-1=(x+1)(x-1)$

$x+1=x+1$

$1-x=-1(x-1)$

Note that $1-x=-1(x-1)$.

Rewrite the problem by changing $1-x$ to $-1(x-1)$. In this situation, the addition problem becomes a subtraction problem, and -1 is not considered to be part of the LCD.

$$\frac{4}{x^2-1}+\frac{4}{x+1}+\frac{2}{-1(x-1)}$$

$$=\frac{4}{(x+1)(x-1)}+\frac{4(x-1)}{(x+1)(x-1)}-\frac{2(x+1)}{(x-1)(x+1)}$$

LCD = $(x+1)(x-1)$

$$=\frac{4+4(x-1)-2(x+1)}{(x+1)(x-1)}$$

$$=\frac{4+4x-4-2x-2}{(x+1)(x-1)}$$

$$=\frac{2x-2}{(x+1)(x-1)}=\frac{2\cancel{(x-1)}}{(x+1)\cancel{(x-1)}}=\frac{2}{x+1}$$

continued on next page ...

g. $\dfrac{x+1}{xy-3y+4x-12}+\dfrac{x-3}{xy+6y+4x+24}$

Solution:

$$\left.\begin{aligned} xy-3y+4x-12 &= y(x-3)+4(x-3) \\ &= (x-3)(y+4) \\ xy+6y+4x+24 &= y(x+6)+4(x+6) \\ &= (x+6)(y+4) \end{aligned}\right\rangle \text{LCD} = (y+4)(x-3)(x+6)$$

$$\begin{aligned} \frac{x+1}{xy-3y+4x-12}+\frac{x-3}{xy+6y+4x+24} &= \frac{(x+1)(x+6)}{(y+4)(x-3)(x+6)}+\frac{(x-3)(x-3)}{(y+4)(x-3)(x+6)} \\ &= \frac{x^2+7x+6+x^2-6x+9}{(y+4)(x-3)(x+6)} \\ &= \frac{2x^2+x+15}{(y+4)(x-3)(x+6)} \end{aligned}$$

Practice Problems

1. 3 **2.** 7 **3.** 2 **4.** 2 **5.** 1 **6.** $x+3$ **7.** $\dfrac{2}{x-1}$

8. $\dfrac{2x^2+3x+4}{x^2-4}$

9. $\dfrac{14}{7-x}$ **10.** 6 **11.** 4

12. $\dfrac{23}{x-10}$

13. $\dfrac{x^2-x+1}{x^2+x-12}$

14. $\dfrac{5}{x-3}$ **15.** $\dfrac{x-2}{x+2}$

Perform the indicated operations and reduce if possible. Assume that no denominator is 0.

1. $\dfrac{5}{x-1}-\dfrac{4+x}{x-1}$

2. $\dfrac{5}{x+1}+\dfrac{10x}{x^2+4x+3}$

3. $\dfrac{1}{x^2+x}+\dfrac{4}{x^2}-\dfrac{2}{x^2-x}$

4. $\dfrac{x}{2x-1}-\dfrac{2}{1-2x}$

5. $\dfrac{x}{x^2-1}+\dfrac{1}{x-1}$

6. $\dfrac{x+3}{x^2+x-6}+\dfrac{x-2}{x^2+4x-12}$

7. $\dfrac{1}{y+2}-\dfrac{1}{y^3+8}$

8. $\dfrac{1}{1-y}+\dfrac{2}{y^2-1}$

16. $\dfrac{-2x+3}{(2x+1)(x-1)}$ **17.** $\dfrac{4x+5}{2(7x-2)}$ **18.** $\dfrac{x+9}{2(2x+5)}$ **19.** $\dfrac{6x+15}{(x+3)(x-3)}$ **20.** $\dfrac{x^2+3x+1}{(x-5)(x+2)}$

Answers to Practice Problems: **1.** -1 **2.** $\dfrac{15}{x+3}$ **3.** $\dfrac{3x^2-3x-4}{x^2(x+1)(x-1)}$ **4.** $\dfrac{x+2}{2x-1}$ **5.** $\dfrac{2x+1}{(x+1)(x-1)}$ **6.** $\dfrac{2x+4}{(x-2)(x+6)}$ **7.** $\dfrac{y^2-2y+3}{y^3+8}$ **8.** $\dfrac{-1}{y+1}$

7.2 Exercises

Perform the indicated operations and reduce if possible. Assume that no denominator is 0.

1. $\dfrac{3x}{x+4}+\dfrac{12}{x+4}$
2. $\dfrac{7x}{x+5}+\dfrac{35}{x+5}$
3. $\dfrac{x-1}{x+6}+\dfrac{x+13}{x+6}$
4. $\dfrac{3x-1}{2x-6}+\dfrac{x-11}{2x-6}$
5. $\dfrac{3x+1}{5x+2}+\dfrac{2x+1}{5x+2}$
6. $\dfrac{x^2+3}{x+1}+\dfrac{4x}{x+1}$
7. $\dfrac{x-5}{x^2-2x+1}+\dfrac{x+3}{x^2-2x+1}$
8. $\dfrac{2x^2+5}{x^2-4}+\dfrac{3x-1}{x^2-4}$
9. $\dfrac{13}{7-x}-\dfrac{1}{x-7}$
10. $\dfrac{6x}{x-6}+\dfrac{36}{6-x}$
11. $\dfrac{3x}{x-4}+\dfrac{16-x}{4-x}$
12. $\dfrac{20}{x-10}-\dfrac{3}{10-x}$
13. $\dfrac{x^2+2}{x^2+x-12}+\dfrac{x+1}{12-x-x^2}$
14. $\dfrac{10}{x^2-x-6}-\dfrac{5x}{6+x-x^2}$
15. $\dfrac{x^2+2}{x^2-4}-\dfrac{4x-2}{x^2-4}$
16. $\dfrac{2x+5}{2x^2-x-1}-\dfrac{4x+2}{2x^2-x-1}$
17. $\dfrac{x+3}{7x-2}+\dfrac{2x-1}{14x-4}$
18. $\dfrac{3x+1}{4x+10}+\dfrac{4-x}{2x+5}$
19. $\dfrac{5}{x-3}+\dfrac{x}{x^2-9}$
20. $\dfrac{x+1}{x^2-3x-10}+\dfrac{x}{x-5}$
21. $\dfrac{x}{x-1}-\dfrac{4}{x+2}$
22. $\dfrac{x-1}{3x-1}-\dfrac{8+4x}{x+2}$
23. $\dfrac{x+2}{x+3}-\dfrac{4}{3-x}$
24. $\dfrac{x-1}{4-x}+\dfrac{3x}{x+5}$
25. $\dfrac{x+2}{3x+9}+\dfrac{2x-1}{2x-6}$
26. $\dfrac{x}{4x-8}-\dfrac{3x+2}{3x+6}$
27. $\dfrac{3x}{6+x}-\dfrac{2x}{x^2-36}$
28. $\dfrac{3x-4}{x^2-x-20}-\dfrac{2}{5-x}$
29. $\dfrac{4x+1}{7-x}+\dfrac{x-1}{x^2-8x+7}$
30. $\dfrac{4}{x+5}-\dfrac{2x+3}{x^2+4x-5}$
31. $\dfrac{4x}{x^2+3x-28}+\dfrac{3}{x^2+6x-7}$
32. $\dfrac{3x}{x^2+2x+1}-\dfrac{x}{x^2+4x+4}$
33. $\dfrac{x+1}{x^2+4x+4}-\dfrac{x-3}{x^2-4}$
34. $\dfrac{x-4}{x^2-5x+6}+\dfrac{2x}{x^2-2x-3}$
35. $\dfrac{3x}{9-x^2}+\dfrac{5}{x^2-7x+12}$
36. $\dfrac{4x}{3x^2+4x+1}-\dfrac{x+4}{x^2+7x+6}$
37. $\dfrac{x-6}{7x^2-3x-4}+\dfrac{7-x}{7x^2+18x+8}$
38. $\dfrac{x+5}{9x^2-26x-3}-\dfrac{8x}{9x^2+11x+3}$
39. $\dfrac{x-3}{4x^2-5x-6}-\dfrac{4x+10}{2x^2+x-10}$
40. $\dfrac{2x+1}{8x^2-37x-15}+\dfrac{2-x}{8x^2+11x+3}$
41. $\dfrac{3x}{4-x}+\dfrac{7x}{x+4}-\dfrac{x-3}{x^2-16}$
42. $\dfrac{x}{x+3}+\dfrac{x+1}{3-x}+\dfrac{x^2+4}{x^2-9}$
43. $2-\dfrac{4x+1}{x-4}+\dfrac{x-3}{x^2-6x+8}$
44. $-4+\dfrac{1-2x}{x+6}+\dfrac{x^2+1}{x^2+4x-12}$

21. $\dfrac{x^2-2x+4}{(x+2)(x-1)}$
22. $\dfrac{-11x+3}{3x-1}$
23. $\dfrac{x^2+3x+6}{(x+3)(x-3)}$
24. $\dfrac{-2x^2+16x-5}{(4-x)(x+5)}$
25. $\dfrac{8x^2+13x-21}{6(x+3)(x-3)}$
26. $\dfrac{-9x^2+22x+16}{(4x-8)(3x+6)}$
27. $\dfrac{3x^2-20x}{(x+6)(x-6)}$
28. $\dfrac{5x+4}{(x-5)(x+4)}$
29. $\dfrac{-4x}{x-7}$
30. $\dfrac{2x-7}{(x+5)(x-1)}$
31. $\dfrac{4x^2-x-12}{(x+7)(x-4)(x-1)}$
32. $\dfrac{x(2x^2+10x+11)}{(x+1)(x+1)(x+2)(x+2)}$
33. $\dfrac{4}{(x+2)(x+2)(x-2)}$
34. $\dfrac{3x^2-7x-4}{(x-3)(x-2)(x+1)}$
35. $\dfrac{-3x^2+17x+15}{(x-3)(x-4)(x+3)}$
36. $\dfrac{x^2+11x-4}{(3x+1)(x+6)(x+1)}$
37. $\dfrac{4x-19}{(7x+4)(x-1)(x+2)}$

38. $\frac{-63x^3+264x^2+82x+15}{(9x+1)(x-3)(9x^2+11x+3)}$

39. $\frac{-7x-9}{(4x+3)(x-2)}$

40. $\frac{x^2+10x-9}{(8x+3)(x-5)(x+1)}$

41. $\frac{4x^2-41x+3}{(x+4)(x-4)}$

42. $\frac{x^2-7x+1}{(x-3)(x+3)}$

43. $\frac{-2x^2-4x+15}{(x-4)(x-2)}$

44. $\frac{-5x^2-11x+47}{(x+6)(x-2)}$

45. $\frac{x^2-4x-6}{(x+2)(x-2)(x-1)}$

46. $\frac{3x^2+26x-3}{(x+7)(x-3)(x+1)}$

47. $\frac{6x+2}{(x-1)(x+3)}$

48. $\frac{2x-51}{(x+5)(x-2)(x-5)}$

49. $\frac{5x+12}{(x+3)(x+1)}$

45. $\frac{2}{x^2-4}-\frac{3}{x^2-3x+2}+\frac{x-1}{x^2+x-2}$

46. $\frac{x}{x^2+4x-21}+\frac{1-x}{x^2+8x+7}+\frac{3x}{x^2-2x-3}$

47. $\frac{3(x+3)}{x^2-5x+4}+\frac{49}{12+x-x^2}+\frac{3x+21}{x^2+2x-3}$

48. $\frac{4}{x^2+3x-10}+\frac{3}{x^2-25}-\frac{5}{x^2-7x+10}$

49. $\frac{5x+22}{x^2+8x+15}+\frac{4}{x^2+4x+3}+\frac{6}{x^2+6x+5}$

50. $\frac{x+1}{2x^2-x-1}+\frac{2x}{2x^2+5x+2}-\frac{2x}{3x^2+4x-4}$

51. $\frac{x-6}{3x^2+10x+3}-\frac{2x}{5x^2-3x-2}+\frac{2x}{3x^2-2x-1}$

52. $\frac{x}{xy+x-2y-2}+\frac{x+2}{xy+x+y+1}$

53. $\frac{4x}{xy-3x+y-3}+\frac{x+2}{xy+2y-3x-6}$

54. $\frac{3y}{xy+2x+3y+6}+\frac{x}{x^2-2x-15}$

55. $\frac{2}{xy-4x-2y+8}+\frac{5y}{y^2-3y-4}$

56. $\frac{x+6}{x^2+x+1}-\frac{3x^2+x-4}{x^3-1}$

57. $\frac{2x-5}{8x^2-4x+2}+\frac{x^2-2x+5}{8x^3+1}$

58. $\frac{x+1}{x^3-3x^2+x-3}+\frac{x^2-5x-8}{x^4-8x^2-9}$

59. $\frac{x+4}{x^3-5x^2+6x-30}-\frac{x-7}{x^3-2x^2+6x-12}$

60. $\frac{x+2}{9x^2-6x+4}+\frac{10x-5x^2}{27x^3+8}-\frac{2}{3x+2}$

Writing and Thinking About Mathematics

61. Discuss the steps in the process you go through when adding two rational expressions. That is, discuss how you find the least common denominator when adding two fractions (rational expressions) and how you use this LCD in finding equivalent fractions that you can add.

50. $\frac{5x^3-x^2+6x-4}{(2x+1)(x-1)(x+2)(3x-2)}$

51. $\frac{9x^3-19x^2+22x+12}{(3x+1)(x+3)(5x+2)(x-1)}$

52. $\frac{2x^2+x-4}{(x-2)(y+1)(x+1)}$

Hawkes Learning Systems: Introductory & Intermediate Algebra

Addition and Subtraction of Rational Expressions

53. $\frac{5x+1}{(y-3)(x+1)}$

54. $\frac{2x+4xy-15y}{(x+3)(y+2)(x-5)}$

55. $\frac{5xy-8y+2}{(y-4)(y+1)(x-2)}$

56. $\frac{-2x+2}{(x^2+x+1)}$

57. $\frac{6x^2-12x+5}{2(2x+1)(4x^2-2x+1)}$

58. $\frac{2x^2-x-5}{(x-3)(x+3)(x^2+1)}$

59. $\frac{14x-43}{(x^2+6)(x-5)(x-2)}$

60. $\frac{-20x^2+30x-4}{(3x+2)(9x^2-6x+4)}$

61. Answers will vary.

7.3 Complex Fractions

Objectives

Teaching Notes: Students may not be prepared to write out as much detail in reducing complex fractions to lowest terms as they will need to. For many students, this content gets little or no attention in the beginning algebra course. You might want to let your students have some practice with "pure numerical" complex fractions.

After completing this section, you will be able to:

Simplify complex fractions and complex algebraic functions.

A **complex fraction** is a fraction in which the numerator or denominator is itself a fraction or the sum or difference of fractions. Examples of complex fractions are

$$\frac{\frac{6x}{5y^2}}{\frac{8x^2}{10y^3}}, \quad \frac{x+y}{x^{-1}+y^{-1}}, \quad \text{and} \quad \frac{\frac{1}{x+3}-\frac{1}{x}}{1+\frac{3}{x}}.$$

In the first example, the numerator and denominator are both single fractions; no sum or difference is indicated. To simplify this expression, we simply divide and reduce as with rational expressions.

Example 1: Complex Fraction

Simplify $\dfrac{\frac{6x}{5y^2}}{\frac{8x^2}{10y^3}}$

Solution: $\dfrac{\frac{6x}{5y^2}}{\frac{8x^2}{10y^3}} = \dfrac{6x}{5y^2} \div \dfrac{8x^2}{10y^3} = \dfrac{6x}{5y^2} \cdot \dfrac{10y^3}{8x^2} = \dfrac{\cancel{2} \cdot 3 \cdot \cancel{2} \cdot \cancel{5} \cancel{x} \cancel{y^3}^{\,y}}{\cancel{5} \cdot \cancel{2} \cdot \cancel{2} \cdot 2 \cancel{y^2} \cancel{x^2}_{\,x}} = \dfrac{3y}{2x}$

There are two methods for simplifying a complex fraction when a sum or difference of fractions is indicated. The method you choose depends on which method is easier to apply or better fits the particular problem.

To Simplify a Complex Fraction (First Method)

1. *Simplify the numerator and denominator separately so that the numerator and denominator are simple fractions.*
2. *Divide by multiplying by the reciprocal of the denominator.*

Example 2: First Method

a. $\dfrac{x+y}{x^{-1}+y^{-1}}$

Solution: $\dfrac{x+y}{x^{-1}+y^{-1}}=\dfrac{x+y}{\frac{1}{x}+\frac{1}{y}}$

Recall that $x^{-1}=\frac{1}{x}$ and $y^{-1}=\frac{1}{y}$.

$$=\frac{\frac{x+y}{1}}{\frac{1}{x}\cdot\frac{y}{y}+\frac{1}{y}\cdot\frac{x}{x}}$$

Add the two fractions in the denominator.

$$=\frac{\frac{x+y}{1}}{\frac{y}{xy}+\frac{x}{xy}}=\frac{\frac{x+y}{1}}{\frac{y+x}{xy}}$$

$$=\frac{\cancel{x+y}}{1}\cdot\frac{xy}{\cancel{y+x}}=\frac{xy}{1}=xy$$

Multiply by the reciprocal of the denominator.

b. $\dfrac{\frac{1}{x+3}-\frac{1}{x}}{1+\frac{3}{x}}$

Solution: $\dfrac{\frac{1}{x+3}-\frac{1}{x}}{1+\frac{3}{x}}=\dfrac{\frac{1\cdot x}{(x+3)\cdot x}-\frac{1(x+3)}{x(x+3)}}{\frac{x}{x}+\frac{3}{x}}$

Combine the fractions in the numerator and in the denominator separately.

Note that $1=\frac{x}{x}$.

$$=\frac{\frac{x-(x+3)}{x(x+3)}}{\frac{x+3}{x}}=\frac{\frac{x-x-3}{x(x+3)}}{\frac{x+3}{x}}$$

$$=\frac{-3}{\cancel{x}(x+3)}\cdot\frac{\cancel{x}}{x+3}=\frac{-3}{(x+3)^2}$$

Multiply by the reciprocal of the denominator.

To Simplify a Complex Fraction (Second Method)

1. *Find the LCM of all the denominators in the original numerator and denominator.*
2. *Multiply both the numerator and denominator by this LCM.*

Example 3: Second Method

a. $\frac{x+y}{x^{-1}+y^{-1}}$

Solution: $\frac{x+y}{x^{-1}+y^{-1}} = \frac{\frac{x+y}{1}}{\frac{1}{x}+\frac{1}{y}}$

$$= \frac{\left(\frac{x+y}{1}\right)xy}{\left(\frac{1}{x}+\frac{1}{y}\right)xy} = \frac{(x+y)xy}{\frac{1}{x}\cdot xy+\frac{1}{y}\cdot xy}$$

The LCM for $\{x, y, 1\}$ is xy.

$$= \frac{\cancel{(x+y)}xy}{\cancel{y+x}} = xy$$

Note that this multiplication can be done because the net effect is that the fraction is multiplied by 1.

b. $\frac{\frac{1}{x+3}-\frac{1}{x}}{1+\frac{3}{x}}$

Solution: $\frac{\frac{1}{x+3}-\frac{1}{x}}{1+\frac{3}{x}} = \frac{\left(\frac{1}{x+3}-\frac{1}{x}\right)\cdot x(x+3)}{\left(1+\frac{3}{x}\right)\cdot x(x+3)}$

The LCM for $\{x, x+3\}$ is $x(x+3)$.

$$= \frac{\frac{1}{\cancel{x+3}}\cdot x\cancel{(x+3)}-\frac{1}{\cancel{x}}\cdot \cancel{x}(x+3)}{1\cdot x(x+3)+\frac{3}{\cancel{x}}\cdot \cancel{x}(x+3)}$$

$$= \frac{x-(x+3)}{x(x+3)+3(x+3)} = \frac{x-x-3}{(x+3)(x+3)}$$

$$= \frac{-3}{(x+3)^2}$$

Simplifying Complex Algebraic Expressions

A complex algebraic expression (as stated earlier) is an expression that involves rational expressions and more than one operation. In simplifying such expressions, the rules for order of operations apply. The objective is to simplify the expression so that it is written in the form of **a single reduced rational expression**.

Example 4: Simplifying Complex Algebraic Expressions

Simplify the following expression.

$$\frac{4-x}{x+3}+\frac{x}{x+3}\div\frac{x}{x-3}$$

Solution: In a complex algebraic expression such as

$$\frac{4-x}{x+3}+\frac{x}{x+3}\div\frac{x}{x-3}$$

the Rules for Order of Operations indicate that the division is to be done first.

$$\frac{4-x}{x+3}+\frac{x}{x+3}\div\frac{x}{x-3}=\frac{4-x}{x+3}+\frac{\cancel{x}}{x+3}\cdot\frac{x-3}{\cancel{x}}=\frac{4-x}{x+3}+\frac{x-3}{x+3}$$

$$=\frac{4-x+x-3}{x+3}=\frac{1}{x+3}$$

7.3 Exercises

1. $\frac{4}{5xy}$ 2. $12xy$
3. $\frac{8}{7x^2y}$ 4. $\frac{6x}{y^4}$
5. $\frac{2x(x+3)}{2x-1}$
6. $\frac{x(x-2)}{2(x+3)}$
7. $\frac{7}{2(x+2)}$ 8. $\frac{2x-1}{2+3x}$
9. $\frac{x}{x-1}$ 10. $\frac{y}{2-y}$
11. $\frac{4x}{3(x+6)}$ 12. 3

Simplify each complex fraction in Exercises 1 – 36.

1. $\dfrac{\frac{2x}{3y^2}}{\frac{5x^2}{6y}}$
2. $\dfrac{\frac{6x^2}{5y}}{\frac{x}{10y^2}}$
3. $\dfrac{\frac{12x^3}{7y^2}}{\frac{3x^5}{2y}}$
4. $\dfrac{\frac{9x^2}{7y^3}}{\frac{3xy}{14}}$
5. $\dfrac{\frac{x+3}{2x}}{\frac{2x-1}{4x^2}}$
6. $\dfrac{\frac{x-2}{6x}}{\frac{x+3}{3x^2}}$
7. $\dfrac{\frac{3}{x}+\frac{1}{2x}}{1+\frac{2}{x}}$
8. $\dfrac{\frac{2x-1}{x}}{\frac{2}{x}+3}$
9. $\dfrac{1+\frac{1}{x}}{1-\frac{1}{x^2}}$
10. $\dfrac{\frac{2}{y}+1}{\frac{4}{y^2}-1}$
11. $\dfrac{\frac{1}{x}+\frac{1}{3x}}{\frac{x+6}{x^2}}$
12. $\dfrac{\frac{3}{x}-\frac{6}{x^2}}{\frac{x-2}{x^2}}$

13. $\frac{7x}{x+2}$ **14.** 3 **15.** $\frac{x}{6}$

16. $\frac{11}{2(1+4x)}$

17. $\frac{24y+9x}{18y-20x}$

18. $\frac{x}{x-1}$ **19.** $\frac{xy}{x+y}$

20. $\frac{1}{xy}$ **21.** $\frac{y+x}{y-x}$

22. $\frac{xy}{y-x}$ **23.** $\frac{2(x+1)}{x+2}$

24. $\frac{x-2}{x(x-3)}$

25. $\frac{x^2-3x}{x-4}$ **26.** $\frac{x+1}{x+3}$

27. $\frac{x+1}{2x-3}$

28. $\frac{-1}{x(x+h)}$

29. $-\frac{2x+h}{x^2(x+h)^2}$

30. $\frac{-1}{x(x+h)}$

31. $-(x-2y)(x-y)$

32. $\frac{2x}{x^2+1}$

33. $\frac{2-x}{x+2}$

34. $\frac{(x-3)(x^2-2x+4)}{(x-4)(x-2)(x+1)}$

35. $\frac{-x^2y^2}{x+y}$ **36.** $\frac{x-y}{x^2y^2}$

13. $\dfrac{\frac{7}{x}-\frac{14}{x^2}}{\frac{1}{x}-\frac{4}{x^3}}$ **14.** $\dfrac{\frac{3}{x}-\frac{6}{x^2}}{\frac{1}{x}-\frac{2}{x^2}}$ **15.** $\dfrac{\frac{x}{y}-\frac{1}{3}}{\frac{6}{y}-\frac{2}{x}}$ **16.** $\dfrac{\frac{3}{x}+\frac{5}{2x}}{\frac{1}{x}+4}$

17. $\dfrac{\frac{2}{x}+\frac{3}{4y}}{\frac{3}{2x}-\frac{5}{3y}}$ **18.** $\dfrac{1+x^{-1}}{1-x^{-2}}$ **19.** $\dfrac{1}{x^{-1}+y^{-1}}$ **20.** $\dfrac{x^{-1}+y^{-1}}{x+y}$

21. $\dfrac{x^{-1}+y^{-1}}{x^{-1}-y^{-1}}$ **22.** $\dfrac{x^{-1}+y^{-1}}{x^{-2}-y^{-2}}$ **23.** $\dfrac{2-\frac{4}{x}}{\frac{x^2-4}{x^2+x}}$ **24.** $\dfrac{\frac{1}{x}}{1-\frac{1}{x-2}}$

25. $\dfrac{x+\frac{3}{x-4}}{1-\frac{1}{x}}$ **26.** $\dfrac{1-\frac{4}{x+3}}{1-\frac{2}{x+1}}$ **27.** $\dfrac{1+\frac{4}{2x-3}}{1+\frac{x}{x+1}}$ **28.** $\dfrac{\frac{1}{x+h}-\frac{1}{x}}{h}$

29. $\dfrac{\frac{1}{(x+h)^2}-\frac{1}{x^2}}{h}$ **30.** $\dfrac{\left(2+\frac{1}{x+h}\right)-\left(2+\frac{1}{x}\right)}{h}$ **31.** $\dfrac{x^2-4y^2}{1-\frac{2x+y}{x-y}}$

32. $\dfrac{\frac{x+1}{x-1}-\frac{x-1}{x+1}}{\frac{x+1}{x-1}+\frac{x-1}{x+1}}$ **33.** $\dfrac{\frac{1}{x^2-1}-\frac{1}{x+1}}{\frac{1}{x-1}+\frac{1}{x^2-1}}$ **34.** $\dfrac{\frac{x}{x-4}-\frac{1}{x-1}}{\frac{x}{x-1}+\frac{2}{x-3}}$

35. $\dfrac{x-y}{x^{-2}-y^{-2}}$ **36.** $\dfrac{y^{-2}-x^{-2}}{x+y}$

Write each of the expressions as a single fraction reduced to lowest terms in Exercises 37 – 44.

37. $\frac{1}{x+1}-\frac{3}{2x}\cdot\frac{4x}{x+1}$ **38.** $\frac{4}{x}-\frac{2}{x^2-2x}\cdot\frac{x-2}{5}$ **39.** $\left(\frac{8}{x}-\frac{3}{4x}\right)\div\frac{4x+5}{x}$

40. $\left(\frac{2}{x}+\frac{5}{x-3}\right)\cdot\frac{2x-6}{x}$ **41.** $\frac{x}{x-1}-\frac{3}{x-1}\cdot\frac{x+2}{x}$ **42.** $\frac{x+3}{x+2}+\frac{x}{x+2}\div\frac{x^2}{x-3}$

43. $\frac{x-1}{x+4}+\frac{x-6}{x^2+3x-4}\div\frac{x-4}{x-1}$ **44.** $\frac{x}{x+3}-\frac{3}{x-5}\cdot\frac{x^2-3x-10}{x-2}$

37. $\frac{-5}{x+1}$ **39.** $\frac{29}{4(4x+5)}$ **41.** $\frac{x^2-3x-6}{x(x-1)}$ **43.** $\frac{x^2-4x-2}{(x-4)(x+4)}$

38. $\frac{18}{5x}$ **40.** $\frac{14x-12}{x^2}$ **42.** $\frac{x^2+4x-3}{x(x+2)}$ **44.** $\frac{-2x^2-17x-18}{(x+3)(x-2)}$

45. $\dfrac{R_1R_2}{R_2+R_1}$

46. $\dfrac{R_1R_2R_3}{R_2R_3+R_1R_3+R_1R_2}$

47. a. $\dfrac{8}{5}$ b. 1

c. $\dfrac{x^4+x^3+3x^2+2x+1}{x^3+x^2+2x+1}$

Writing and Thinking About Mathematics

45. In electronics, when resistors R_1 and R_2 are in series, one after the other, then the total resistance is the sum

$$R_{\text{total}} = R_1 + R_2.$$

When resistors are studied in parallel, the total resistance, R_{total}, of two resistors with resistance R_1 and R_2 can be found by using the formula $\dfrac{1}{R_{\text{total}}} = \dfrac{1}{R_1} + \dfrac{1}{R_2}$ which means that

$$R_{\text{total}} = \frac{1}{\frac{1}{R_1} + \frac{1}{R_2}}.$$

Simplify the complex fraction in the right hand side of the last formula in the form of a single fraction.

46. By using the concepts discussed in Exercise 45, write a single fraction for R_{total} if three resistors, R_1, R_2, and R_3 are in parallel.

47. Some complex fractions involve the sum (or difference) of complex fractions. Beginning with the "farthest" denominator, simplify each of the following expressions.

a. $1+\dfrac{1}{1+\dfrac{1}{1+\dfrac{1}{1+1}}}$ **b.** $2-\dfrac{1}{2-\dfrac{1}{2-\dfrac{1}{2-1}}}$ **c.** $x+\dfrac{1}{x+\dfrac{1}{x+\dfrac{1}{x+1}}}$

Hawkes Learning Systems: Introductory & Intermediate Algebra

Complex Fractions

7.4 Equations and Inequalities with Rational Expressions

Objectives

After completing this section, you will be able to:

1. *Solve equations containing rational expressions.*
2. *Solve inequalities containing rational expressions.*
3. *Graph the solutions for inequalities containing rational expressions.*

Solving Equations with Rational Expressions

To solve an equation that has fractions, such as $\frac{x}{5}-\frac{x}{2}=-6$, we can multiply both sides of the equation by the LCM of the denominators so that the new coefficients will be integers. The new equation will have integer coefficients and constants and be easier to work with. Thus,

$$\frac{x}{5}-\frac{x}{2}=-6$$

$$10\cdot\left(\frac{x}{5}-\frac{x}{2}\right)=10\cdot(-6) \qquad \text{10 is the LCM of the denominators.}$$

$$2x-5x=-60 \qquad \text{Use the distributive property.}$$

$$-3x=-60 \qquad \text{Simplify.}$$

$$x=20 \qquad \text{Divide both sides by } -3.$$

A **rational equation** is an equation that contains at least one rational expression. To solve rational equations we use the following procedure.

To Solve a Rational Equation

1. *Find the LCM of all the denominators of all the rational expressions in the equation.*
2. *Multiply both sides of the equation by this LCM. Use the distributive property if necessary.*
3. *Simplify both sides of the resulting equation.*
4. *Solve this equation.*
5. *Check each solution in the **original equation**.*
   (Remember that no denominator can be 0.)

Teaching Notes: This section may introduce some students to the fact that multiplying both sides of an equation with a variable term may not strictly result in an equivalent equation – an unsettling realization to these students. You might point out to your students that multiplying by variable terms may introduce extraneous solutions with the increased degree of the resulting polynomial. For example,
$x = 3$
$x \cdot x = 3 \cdot x$
$x^2 = 3x$
Now, $x^2 = 3x$ has two solutions.

Step 5 is critical when solving rational equations. **If both sides of an equation are multiplied by the same nonzero number or expression, the solution set of the new equation will contain the solutions to the original equation**. However, if the multiplication is by a variable expression, then the new equation may actually have more solutions than the original equation. These extra solutions are called **extraneous solutions** or **extraneous roots**. Therefore, it is absolutely necessary to check all solutions in the original equation to identify the actual solutions and any extraneous solutions.

Example 1: Solving Equations with Rational Expressions

Find the solution set for each of the following equations. Multiply both sides of each equation by the LCM of the denominators.

a. $\dfrac{x-5}{2x} = \dfrac{6}{3x}$ LCM $= 6x$.

Solution:
$$6x \cdot \left(\frac{x-5}{2x}\right) = 6x \cdot \left(\frac{6}{3x}\right) \qquad (x \neq 0)$$
$$3(x-5) = 2(6)$$
$$3x - 15 = 12$$
$$3x = 27$$
$$x = 9$$

Check:
$$\frac{9-5}{2 \cdot 9} \stackrel{?}{=} \frac{6}{3 \cdot 9}$$
$$\frac{4}{18} \stackrel{?}{=} \frac{6}{27}$$
$$\frac{2}{9} = \frac{2}{9}$$
The solution is $x = 9$.

b. $\dfrac{3}{x-6} = \dfrac{5}{x}$ LCM $= x(x-6)$

Solution:
$$x\cancel{(x-6)} \cdot \frac{3}{\cancel{x-6}} = \cancel{x}(x-6) \cdot \frac{5}{\cancel{x}} \qquad (x \neq 0, 6)$$
$$3x = 5x - 30$$
$$30 = 2x$$
$$15 = x$$

Check:
$$\frac{3}{15-6} \stackrel{?}{=} \frac{5}{15}$$
$$\frac{1}{3} = \frac{1}{3}$$
The solution is $x = 15$.

c. $\dfrac{2}{x^2-9}=\dfrac{1}{x^2}+\dfrac{1}{x^2-3x}$

Solution: First find the LCM of the denominators and then multiply both sides of the equation by the LCM.

$$\left.\begin{aligned} x^2-9 &= (x+3)(x-3) \\ x^2 &= x^2 \\ x^2-3x &= x(x-3) \end{aligned}\right\rangle \qquad \text{LCM}=x^2(x+3)(x-3)$$

$$x^2\cancel{(x+3)}\cancel{(x-3)}\cdot\frac{2}{\cancel{(x+3)}\cancel{(x-3)}}=\cancel{x^2}(x+3)(x-3)\cdot\frac{1}{\cancel{x^2}}+\overset{x}{\cancel{x^2}}(x+3)\cancel{(x-3)}\cdot\frac{1}{\cancel{x}\cancel{(x-3)}}$$

$$2x^2=(x+3)(x-3)+x(x+3)\ (x\neq 0,-3,3)$$

$$2x^2=x^2-9+x^2+3x$$

$$2x^2=2x^2+3x-9$$

$$9=3x$$

$$\cancel{3=x}$$ 3 is not allowed since no denominator can be 0.

There is no solution. The solution set is the empty set, $\varnothing$. The original equation is a contradiction. (Multiplying by the factor $x-3$ was, in effect, multiplying by 0.)

d. $\dfrac{1}{x-7}=\dfrac{2}{x^2-12x+35}+\dfrac{x}{x^2-5x}$

Solution: First find the LCM of the denominators and then multiply both sides of the equation by the LCM.

$$\left.\begin{aligned} x-7 &= x-7 \\ x^2-12x+35 &= (x-5)(x-7) \\ x^2-5x &= x(x-5) \end{aligned}\right\rangle \qquad \text{LCM}=x(x-5)(x-7)$$

$$x(x-5)\cancel{(x-7)}\cdot\frac{1}{\cancel{x-7}}=x\cancel{(x-5)}\cancel{(x-7)}\cdot\frac{2}{\cancel{(x-5)}\cancel{(x-7)}}+\cancel{x}\cancel{(x-5)}(x-7)\cdot\frac{x}{\cancel{x}\cancel{(x-5)}}$$

$$x(x-5)=2x+x(x-7) \qquad (x\neq 0,5,7)$$

$$x^2-5x=2x+x^2-7x$$

$$x^2-5x=x^2-5x$$

$$0=0$$

continued on next page ...

The equation $0 = 0$ is true for all real numbers. Therefore, x can be any real number with the exception of 0, 5, or 7. We indicated $x \neq 0, 5, 7$. These values are not included since each of them would give a 0 denominator. The implication is that all other values are allowed. The original equation is an **identity**.

e. The formula $C = \frac{5}{9}(F - 32)$ is solved for C and represents the relationship between temperature measured in degrees Celsius and degrees Fahrenheit. Solve this formula for F.

Solution:

$$C = \frac{5}{9}(F - 32) \quad \text{Write the formula.}$$

$$\frac{9}{5} \cdot C = \frac{9}{5} \cdot \frac{5}{9}(F - 32) \quad \text{Multiply both sides of the equation by } \frac{9}{5}.$$

$$\frac{9}{5}C = F - 32 \quad \text{Simplify.}$$

$$\frac{9}{5}C + 32 = F \quad \text{Add 32 to both sides.}$$

Thus, the formula solved for F is: $F = \frac{9}{5}C + 32$.

Solving Inequalities with Rational Expressions

In Chapter 3, we solved linear (or first-degree) inequalities and introduced intervals and interval notation. For example,

if $x - 4 \leq 0$

then $x \leq 4$ or x is in $(-\infty, 4]$.

Graphically,

Notice that the solution consists of the point where $x - 4 = 0$ ($x = 4$) and all the points to the left of this point. For linear inequalities, the solution always contains the points to one side or the other of the point where the inequality has value 0. We use this idea in the following discussion.

A rational inequality may involve the product or quotient of several first-degree expressions. For example, the inequality

$$\frac{x+3}{x-2} > 0$$

involves the two first-degree expressions $x + 3$ and $x - 2$.

The following procedure for solving such an inequality is based on the fact that an expression of the form $x - a$ changes sign when x has values on either side of a. That is, if $x < a$, then $x - a$ is negative; if $x > a$, then $x - a$ is positive.

The steps are as follows:

a. Find the points where each linear factor has value 0.

$$x+3=0 \qquad x-2=0$$

$$x=-3 \qquad x=2$$

b. Mark each of these points on a number line. (Consider these points as endpoints of intervals.)

Three intervals, $(-\infty,-3),(-3,2),\text{and}(2,\infty)$ are formed.

c. Choose any number from each interval as a **test value** to determine the sign of the expression for all values in that interval. Remember, we are not interested in the value of the expression, only whether it is positive (>0) or negative (<0).

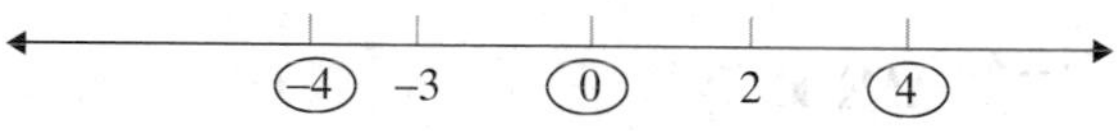

We have chosen the convenient test values –4, 0, and 4. Substituting these values into the original inequality gives the following results:

Results	Explanation
$\frac{x+3}{x-2}=\frac{-4+3}{-4-2}=\frac{1}{6}>0$	This means that $\frac{x+3}{x-2}$ is positive $(+)$ for all x in $(-\infty,-3)$.
$\frac{x+3}{x-2}=\frac{0+3}{0-2}=-\frac{3}{2}<0$	This means that $\frac{x+3}{x-2}$ is negative $(-)$ for all x in $(-3, 2)$.
$\frac{x+3}{x-2}=\frac{4+3}{4-2}=\frac{7}{2}>0$	This means that $\frac{x+3}{x-2}$ is positive $(+)$ for all x in $(2,\infty)$.
+++++ 0 (at –3) ----- 0 (at 2) +++++	

d. The solution to the inequality consists of all the intervals that indicate the desired sign: + (for > 0) or − (for < 0). The solution for $\frac{x+3}{x-2} > 0$ is

$$\textbf{all } x \textbf{ in } (-\infty, -3) \textbf{ or } (2, \infty).$$

In algebraic notation: $x < -3$ **or** $x > 2$; In set notation $x \in (-\infty, -3) \cup (2, \infty)$.
Graphically,

Example 2: Solving and Graphing Inequalities

Solve and graph the solution for each of the following inequalities.

a. $\frac{x+3}{x-2} < 0$

Solution: From the previous discussion, we know that $\frac{x+3}{x-2}$ is negative whenever x is in $(-3, 2)$. Similarly, $\frac{x+3}{x-2} < 0$ if $-3 < x < 2$.

Graphically,

b. $\frac{x+5}{x-4} \geq -1$

Solution:

$$\frac{x+5}{x-4} + 1 \geq 0 \qquad \text{One side must be 0.}$$

$$\frac{x+5}{x-4} + \frac{x-4}{x-4} \geq 0$$

$$\frac{2x+1}{x-4} \geq 0 \qquad \text{Simplify to get one fraction.}$$

Set each linear expression equal to 0 to find the interval endpoints.

$$2x+1=0 \qquad x-4=0$$

$$x=-\frac{1}{2} \qquad x=4$$

Test a value from each of the intervals:
$\left(-\infty, -\frac{1}{2}\right)$, $\left(-\frac{1}{2}, 4\right)$, and $(4, \infty)$

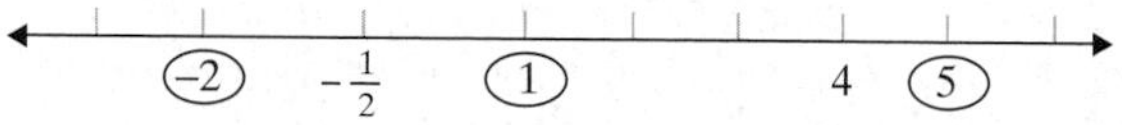

Using the values circled above, we obtain the following results:

Results	Explanation
$\frac{2(-2)+1}{-2-4}=\frac{-3}{-6}>0$	This means that $\frac{2x+1}{x-4}>0$ for all x in $\left(-\infty,-\frac{1}{2}\right)$.
$\frac{2(1)+1}{1-4}=\frac{3}{-3}<0$	This means that $\frac{2x+1}{x-4}<0$ for all x in $\left(-\frac{1}{2},4\right)$.
$\frac{2(5)+1}{5-4}=\frac{11}{1}>0$	This means that $\frac{2x+1}{x-4}>0$ for all x in $(4,\infty)$.

$\frac{2x+1}{x-4}=0$ if the numerator $2x+1=0$ or $x=-\frac{1}{2}$. Since $x-4\neq 0$, 4 is not included in the solution. Thus, the solution is all x in the interval $\left(-\infty,-\frac{1}{2}\right]$ or $(4,\infty)$. In algebraic notation, $x\leq -\frac{1}{2}$ or $x>4$.

Graphically,

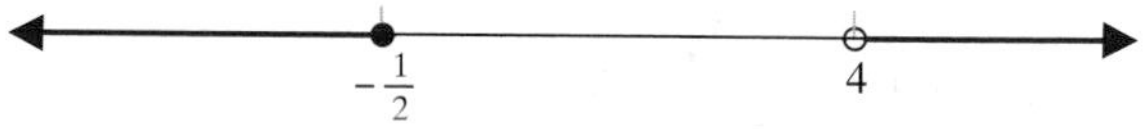

Teaching Notes:
You should inform your students (if they have not already asked) that not all rational expressions have linear factors as illustrated in this section. However, the method discussed is valid in general and the material on quadratic equations coming up in Chapter 11 will help cover some of the situations where factoring is not easily done.

We can summarize the technique for solving rational inequalities as follows:

Procedure for Solving Inequalities with Rational Expressions

1. *Simplify the inequality so that one side is 0 and on the other side both the numerator and denominator are in factored form.*
2. *Find the points where each linear factor is 0.*
3. *Mark each of these points on a number line.*
4. *Choose a number from each indicated interval as a test value.*
5. *The intervals where the test values satisfy the conditions of the inequality are the solution intervals.*
6. *Mark a solid circle for endpoints that are included and an open circle for endpoints that are not included.*

Notice that in the first step, we do **not** multiply by the denominator $x - 4$. The reason is that the variable expression is positive for some values of x and negative for other values of x. Therefore, if we did multiply by $x - 4$, we would not be able to determine whether the inequality should stay as $\geq$ or be reversed to $\leq$.

7.4 Exercises

Solve each equation in Exercises 1 – 34.

1. $\frac{4x}{7} = \frac{x+5}{3}$
2. $\frac{3x+1}{-4} = \frac{2x+1}{-3}$
3. $\frac{5x+2}{11x} = \frac{x-6}{4x}$
4. $\frac{x+3}{5x} = \frac{x-1}{6x}$
5. $\frac{5x}{4} - \frac{1}{2} = -\frac{3}{16}$
6. $\frac{x}{6} - \frac{1}{42} = \frac{1}{7}$
7. $\frac{4x}{3} - \frac{3}{4} = \frac{5x}{6}$
8. $\frac{x-2}{3} - \frac{x-3}{5} = \frac{13}{15}$
9. $\frac{2+x}{4} - \frac{5x-2}{12} = \frac{8-2x}{5}$
10. $\frac{8x+10}{5} = 2x+3-\frac{6x+1}{4}$
11. $\frac{2}{3x} = \frac{1}{4} - \frac{1}{6x}$
12. $\frac{x-4}{x} + \frac{3}{x} = 0$
13. $\frac{3}{8x} - \frac{7}{10} = \frac{1}{5x}$
14. $\frac{1}{x} - \frac{8}{21} = \frac{3}{7x}$
15. $\frac{3}{4x} - \frac{1}{2} = \frac{7}{8x} + \frac{1}{6}$
16. $\frac{7}{x-3} = \frac{6}{x-4}$
17. $\frac{2}{3x+2} = \frac{4}{5x+1}$
18. $\frac{-3}{2x+1} = \frac{4}{3x+1}$
19. $\frac{9}{5x-3} = \frac{5}{3x+7}$
20. $\frac{5x+2}{x-6} = \frac{11}{4}$
21. $\frac{x+9}{3x+2} = \frac{5}{8}$
22. $\frac{8}{2x+3} = \frac{9}{4x-5}$
23. $\frac{x}{x-4} - \frac{4}{2x-1} = 1$
24. $\frac{x}{x+3} + \frac{1}{x+2} = 1$
25. $\frac{x+2}{x+1} + \frac{x+2}{x+4} = 2$
26. $\frac{x-2}{x-3} + \frac{x-3}{x-2} = \frac{2x^2}{x^2-5x+6}$
27. $\frac{2}{4x-1} + \frac{1}{x+1} = \frac{3}{x+1}$
28. $\frac{3x-2}{15} - \frac{16-3x}{x+6} = \frac{x+3}{5}$
29. $\frac{x}{x-4} - \frac{12x}{x^2+x-20} = \frac{x-1}{x+5}$
30. $\frac{x-2}{x+4} - \frac{3}{2x+1} = \frac{x-7}{x+4}$
31. $\frac{3x+5}{3x+2} - \frac{4-2x}{3x^2+8x+4} = \frac{x+4}{x+2}$
32. $\frac{3}{3x-1} + \frac{1}{x+1} = \frac{4}{2x-1}$
33. $\frac{5}{2x+1} - \frac{1}{2x-1} = \frac{2}{x-2}$
34. $\frac{2}{x+1} + \frac{4}{2x-3} = \frac{4}{x-5}$

1. $x=7$ **2.** $x=1$
3. $x=-\frac{74}{9}$ **4.** $x=-23$
5. $x=\frac{1}{4}$ **6.** $x=1$
7. $x=\frac{3}{2}$ **8.** $x=7$
9. $x=4$
10. $x=\frac{15}{22}$ **11.** $x=\frac{10}{3}$
12. $x=1$
13. $x=\frac{1}{4}$ **14.** $x=\frac{3}{2}$
15. $x=-\frac{3}{16}$ **16.** $x=10$
17. $x=-3$
18. $x=-\frac{7}{17}$
19. $x=-39$
20. $x=-\frac{74}{9}$
21. $x=\frac{62}{7}$
22. $x=\frac{67}{14}$ **23.** $x=-3$
24. $x=-\frac{3}{2}$ **25.** $x=2$
26. $x=\frac{13}{10}$ **27.** $x=\frac{2}{3}$
28. $x=9$ **29.** $x=-2$
30. $x=1$
31. No solution

32. $x=\frac{1}{5}$ **33.** $x=\frac{7}{11}$

34. $x=\frac{11}{19}$

35. $x\le -4$ or $x>0$

36. $x\le 0$ or $x>4$

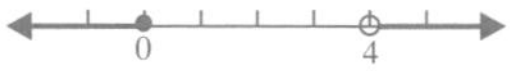

37. $x<-6$

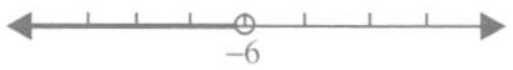

38. $x<-1$ or $x>3$

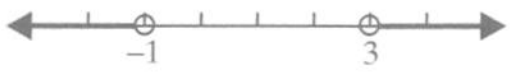

39. $x<-9$ or $x>-3$

40. $-\frac{3}{2}<x<4$

41. $2<x<\frac{5}{2}$

42. $x<-2$ or $x\ge\frac{4}{3}$

43. $x>7$

44. $\frac{1}{2}<x<2$

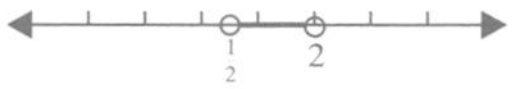

45. $\frac{7}{5}\le x<4$

46. $x<\frac{1}{4}$ or $x>\frac{5}{9}$

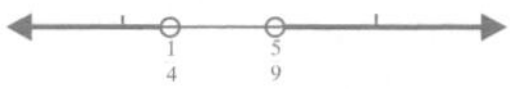

47. $-9<x<-\frac{4}{3}$

Solve and graph the solution set of each of the inequalities in Exercises 35 – 50.

35. $\frac{x+4}{2x}\ge 0$ **36.** $\frac{x}{x-4}\ge 0$ **37.** $\frac{x+6}{x^2}<0$

38. $\frac{3-x}{x+1}<0$ **39.** $\frac{x+3}{x+9}>0$ **40.** $\frac{2x+3}{x-4}<0$

41. $\frac{3x-6}{2x-5}<0$ **42.** $\frac{4-3x}{2x+4}\le 0$ **43.** $\frac{x+5}{x-7}\ge 1$

44. $\frac{x+4}{2x-1}>2$ **45.** $\frac{2x+5}{x-4}\le -3$ **46.** $\frac{3x+2}{4x-1}<3$

47. $\frac{5-2x}{3x+4}<-1$ **48.** $\frac{8-x}{x+5}<-4$ **49.** $\frac{x(x+4)}{x-3}\le 0$

50. $\frac{(x+3)(x-2)}{x+1}>0$

Solve each of the formulas in Exercises 51 – 60 for the indicated variables.

51. $S=\frac{a}{1-r}$; solve for r (formula used in mathematics)

52. $z=\frac{x-\overline{x}}{s}$; solve for x (formula used in statistics)

53. $z=\frac{x-\overline{x}}{s}$; solve for s (formula used in statistics)

54. $a_n=a_1+(n-1)d$; solve for d (formula used in mathematics)

55. $\frac{1}{R_{\text{total}}}=\frac{1}{R_1}+\frac{1}{R_2}$; solve for R_{total} (formula used in electronics)

56. $m=\frac{y-y_1}{x-x_1}$; solve for y (formula used for slope of a line)

57. $A=P+Pr$; solve for P (formula used for compound interest)

58. $v_{\text{ave}}=\frac{d_2-d_1}{t_2-t_1}$; solve for d_2 (formula for average velocity)

59. $y=\frac{ax+b}{cx+d}$; solve for x (formula used in mathematics)

60. $\frac{1}{x}=\frac{1}{t_1}+\frac{1}{t_2}$; solve for x (formula used in mathematics)

48. $-\frac{28}{3} < x < -5$

49. $x \le -4$ or $0 \le x < 3$

50. $-3 < x < -1$ or $x > 2$

51. $r = \frac{S-a}{S}$

52. $x = zs + \bar{x}$

53. $s = \frac{x-\bar{x}}{z}$

54. $d = \frac{a_n - a_1}{n-1}$

Writing and Thinking About Mathematics

In simplifying rational expressions, the result is a rational expression. However, in solving equations with rational expressions, the goal is to find a value (or values) for the variable that will make the equation a true statement. Many students confuse these two ideas. To avoid confusing the techniques for adding and subtracting rational expressions with the techniques for solving equations, simplify the expression in part (a) and solve the equation in part (b). Explain, in your own words, the differences in your procedures.

61. a. $\frac{10}{x} + \frac{31}{x-1} + \frac{4x}{x-1}$ **b.** $\frac{10}{x} + \frac{31}{x-1} = \frac{4x}{x-1}$

62. a. $\frac{-4}{x^2-16} + \frac{x}{2x+8} - \frac{1}{4}$ **b.** $\frac{-4}{x^2-16} + \frac{x}{2x+8} = \frac{1}{4}$

Hawkes Learning Systems: Introductory & Intermediate Algebra

Solving Equations Involving Rational Expressions
Solving Inequalities with Rational Expressions

55. $R_{\text{total}} = \frac{R_1 R_2}{R_1 + R_2}$

56. $y = m(x - x_1) + y_1$

57. $P = \frac{A}{1+r}$

58. $d_2 = v_{\text{ave}}(t_2 - t_1) + d_1$

59. $x = \frac{b-yd}{yc-a}$

60. $x = \frac{t_1 t_2}{t_1 + t_2}$

61. a. $\frac{4x^2 + 41x - 10}{x(x-1)}$

b. $x = \frac{1}{4}$, 10 and $x \ne 0, 1$

62. a. $\frac{x^2 - 8x}{4x^2 - 64}$

b. $x = 0, 8$ and $x \ne 4, -4$

7.5 Applications

Objectives

After completing this section, you will be able to:

Solve the following types of applied problems by using equations containing rational expressions:

a. fractions,

b. similar triangles,

c. jobs, and

d. distance-rate-time.

The following Strategy for Solving Word Problems is valid for all word problems that involve algebraic equations (or inequalities).

Strategy for Solving Word Problems

1. *Read the problem carefully. Read it several times if necessary.*
2. *Decide what is asked for and assign a variable to the unknown quantity.*
3. *Draw a diagram or set up a chart whenever possible.*
4. *Form an equation (or inequality) that relates the information provided.*
5. *Solve the equation (or inequality).*
6. *Check your solution with the wording of the problem to be sure it makes sense.*

We now introduce word problems involving rational expressions with problems relating the numerator and denominator of a fraction. Let one variable represent either the numerator or denominator, then write the equation to be solved using the information given.

Example 1: Fractions

a. The denominator of a fraction is 8 more than the numerator. If both the numerator and denominator are increased by 3, the resulting fraction is equal to $\frac{1}{2}$. Find the original fraction.

Solution: Reread the problem to be sure that you understand it. Assign variables to the unknown quantities.

continued on next page ...

Let $\quad n = \text{original numerator}$

$$n+8 = \text{original denominator}$$

$$\frac{n}{n+8} = \text{original fraction}$$

$$\frac{n+3}{(n+8)+3} = \frac{1}{2}$$

The numerator and the denominator are each increased by 3, making a new fraction that is equal to $\frac{1}{2}$.

$$\frac{n+3}{n+11} = \frac{1}{2}$$

$$2(n+11)\cdot\left(\frac{n+3}{n+11}\right) = 2(n+11)\cdot\frac{1}{2}$$

$$2n+6 = n+11$$

$$n = 5 \quad \leftarrow \text{Original numerator}$$

$$n+8 = 13 \quad \leftarrow \text{Original denominator}$$

Check: $\quad \frac{5+3}{13+3} \stackrel{?}{=} \frac{8}{16} \stackrel{?}{=} \frac{1}{2}$

The original fraction is $\frac{5}{13}$.

b. In the figure shown, $\Delta ABC \sim \Delta PQR$. Find the lengths of the sides AB and QR.

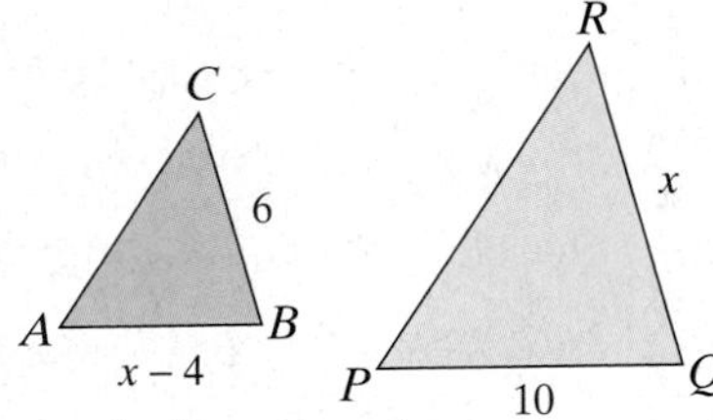

Solution: Remember that, in similar triangles, corresponding sides are proportional, so set up a proportion involving the corresponding sides and solve for x.

$$\frac{x-4}{10} = \frac{6}{x}$$

$$10x \cdot \frac{x-4}{10} = \frac{6}{x} \cdot 10x$$

$$x(x-4) = 6 \cdot 10$$

$$x^2 - 4x = 60$$

$$x^2 - 4x - 60 = 0$$

$$(x-10)(x+6) = 0$$

$$x - 10 = 0 \quad \text{or} \quad x + 6 = 0$$

$$x = 10 \qquad\qquad x = -6$$

Because the length of a triangle cannot be negative, the only acceptable solution is $x = 10$. Substituting 10 for x gives $AB = 10 - 4 = 6$ and $QR = 10$.

Problems involving jobs (sometimes called **work problems**) usually translate into equations involving rational expressions. The basic idea is to **represent what part of a job is done in one unit of time**. For example, if a manuscript was typed in 35 hours, what part was typed in 1 hour? Assuming an even typing speed, $\frac{1}{35}$ of the manuscript was typed in 1 hour. Similarly, if a boy can paint a fence in 2 days, then he can do $\frac{1}{2}$ the job in 1 day.

Example 2: Work Problems

a. A carpenter can build a certain type of patio cover in 6 hours. His partner takes 8 hours to build the same cover. How long would it take them working together to build this type of patio cover?

Solution: Let x = number of hours to build the cover working together

Person(s)	Time of Work (in Hours)	Part of Job Done in 1 Hour
Carpenter	6	$\frac{1}{6}$
Partner	8	$\frac{1}{8}$
Together	x	$\frac{1}{x}$

$$\underbrace{\text{Part done in 1 hr by carpenter}} + \underbrace{\text{Part done in 1 hr by partner}} = \underbrace{\text{Part done in 1 hr together}}$$

$$\frac{1}{6} + \frac{1}{8} = \frac{1}{x}$$

$$\frac{1}{6}(24x) + \frac{1}{8}(24x) = \frac{1}{x}(24x)$$ Multiply each term on both sides by $24x$, the LCM of the denominators.

$$4x + 3x = 24$$

$$7x = 24$$

$$x = \frac{24}{7}$$

Together, they can build the patio cover in $\frac{24}{7}$ hours, or $3\frac{3}{7}$ hours.

(Note that this answer is reasonable because the time is less than either person would take working alone.)

continued on next page ...

b. A man can wax his car three times as fast as his daughter can. Together they can do the job in 4 hours. How long does it take each of them working alone?

Solution: Let t = number of hours for man alone
$3t$ = number of hours for daughter alone

Person(s)	Time of Work (in Hours)	Part of Job Done in 1 Hour
Man	t	$\frac{1}{t}$
Daughter	$3t$	$\frac{1}{3t}$
Together	4	$\frac{1}{4}$

$$\underbrace{\text{part done by man alone in 1 hour}} + \underbrace{\text{part done by daughter alone in 1 hour}} = \underbrace{\text{part done working together in 1 hour}}$$

$$\frac{1}{t} \quad + \quad \frac{1}{3t} \quad = \quad \frac{1}{4}$$

$$\frac{1}{t}(12t)+\frac{1}{3t}(12t)=\frac{1}{4}(12t)$$

Multiply each term on both sides by $12t$, the LCM of the denominators.

$$12+4=3t$$
$$16=3t$$
$$\frac{16}{3}=t$$

Check: Man's part in 1 hr $\stackrel{?}{=} \frac{1}{t} = \frac{1}{\frac{16}{3}} = \frac{3}{16}$

Daughter's part in 1 hr $\stackrel{?}{=} \frac{1}{3t} = \frac{1}{3\cdot\frac{16}{3}} = \frac{1}{16}$

Man's part in 4 hr $\stackrel{?}{=} \frac{3}{16}\cdot 4 = \frac{3}{4}$

Daughter's part in 4 hr $\stackrel{?}{=} \frac{1}{16}\cdot 4 = \frac{1}{4}$

$\frac{3}{4}+\frac{1}{4}=1$ car waxed in 4 hours.

Working alone, the man takes $\frac{16}{3}$ or $5\frac{1}{3}$ hours, and his daughter takes 16 hours.

c. An inlet pipe on a swimming pool can be used to fill the pool in 36 hours. The drain pipe can be used to empty the pool in 40 hours. If the pool is $\frac{2}{3}$ filled and then the drain pipe is accidentally opened, how long from that time will it take to fill the pool?

Solution: Let t = hours to fill pool with both pipes open

Pipe(s)	Time of work (in Hours)	Part of Job Done in 1 Hour
Inlet	36	$\frac{1}{36}$
Outlet	40	$\frac{1}{40}$
Together	t	$\frac{1}{t}$

$$\underbrace{\begin{matrix}\text{part filled}\\ \text{by inlet pipe}\\ \text{in 1 hour}\end{matrix}}_{\frac{1}{36}} - \underbrace{\begin{matrix}\text{part emptied}\\ \text{by drain pipe}\\ \text{in 1 hour}\end{matrix}}_{\frac{1}{40}} = \underbrace{\begin{matrix}\text{part filled in 1}\\ \text{hour when both}\\ \text{pipes are open}\end{matrix}}_{\frac{1}{t}}$$

$$\frac{1}{36}(360t) - \frac{1}{40}(360t) = \frac{1}{t}(360t)$$

$$10t - 9t = 360$$

$$t = 360$$

However, 360 hours is the time it would take if the pool was empty at the beginning. Only $\frac{1}{3}$ this time will be used since the pool is $\frac{2}{3}$ filled.

$$\frac{1}{3} \cdot 360\text{ hr} = 120\text{ hr}$$

Check: Part filled in 120 hr $\stackrel{?}{=} \frac{1}{36}(120) = \frac{10}{3} = 3\frac{1}{3}$

Part drained in 120 hr $\stackrel{?}{=} \frac{1}{40}(120) = 3$

continued on next page ...

$3\frac{1}{3}$	Part filled
-3	Part drained
$\frac{1}{3}$	Part filled in 120 hr

It will take 120 hours to fill the remaining third of the pool.

Problems involving distance, rate, and time were discussed in Section 3.3. You may recall that the basic formula is $r \cdot t = d$. However, this relationship can also be stated in the forms $t = \frac{d}{r}$ and $r = \frac{d}{t}$.

If distance and rate are known or can be represented, then $t = \frac{d}{r}$ is the way to represent time. Similarly, if the distance and time are known or can be represented, then $r = \frac{d}{t}$ is the way to represent rate.

Example 3: Rate Problems

a. A man can row his boat 5 miles per hour on a lake. On a river, it takes him the same time to row 5 miles downstream as it does to row 3 miles upstream. What is the speed of the river current in miles per hour?

Solution:

Rate and distance are represented in the following table.

	Rate	Distance
Downstream	$5 + c$	5
Upstream	$5 - c$	3

c represents speed of the current.

Now, represent the time going downstream and coming back upstream in terms of the rate and distance. If the rate is in miles per hour, then the distance is in miles and the time is in hours.

	Rate	$t = \frac{d}{r}$	Distance
Downstream	$5 + c$	$\frac{5}{5+c}$	5
Upstream	$5 - c$	$\frac{3}{5-c}$	3

$$\frac{5}{5+c}=\frac{3}{5-c} \qquad \text{The times are equal.}$$

$$(5+c)(5-c)\cdot\frac{5}{5+c}=(5+c)(5-c)\cdot\frac{3}{5-c}$$

$$25-5c=15+3c$$

$$10=8c$$

$$c=\frac{5}{4} \text{ miles per hour}$$

Check: Time downstream $=\dfrac{5}{5+\frac{5}{4}}=\dfrac{5}{\frac{20}{4}+\frac{5}{4}}=\dfrac{5}{\frac{25}{4}}=5\cdot\dfrac{4}{25}=\dfrac{4}{5}$ hr.

Time upstream $=\dfrac{3}{5-\frac{5}{4}}=\dfrac{3}{\frac{20}{4}-\frac{5}{4}}=\dfrac{3}{\frac{15}{4}}=3\cdot\dfrac{4}{15}=\dfrac{4}{5}$ hr.

The times are equal. The rate of the river current is $\frac{5}{4}$ mph $\left(\text{or } 1\frac{1}{4} \text{ mph}\right)$.

b. If a passenger train travels three times as fast as a freight train, and the freight train takes 4 hours longer to travel 210 miles, what is the speed of each train?

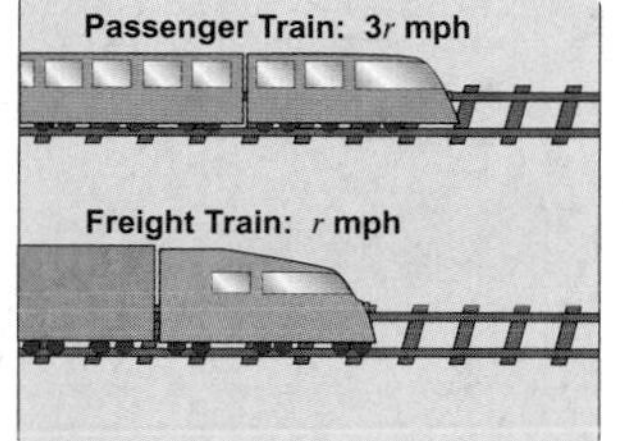

Solution: Let r = rate of freight train in miles per hour
$3r$ = rate of passenger train in miles per hour

	Rate	$t=\frac{d}{r}$	Distance
Freight	r	$\frac{210}{r}$	210
Passenger	$3r$	$\frac{210}{3r}$	210

(**Note:** If the rate is faster, then the time is shorter. Thus, the fraction $\frac{210}{3r}$ is smaller than the fraction $\frac{210}{r}$.)

continued on next page ...

$$\frac{210}{r} - \frac{210}{3r} = 4$$

$$\frac{210}{r} - \frac{70}{r} = 4$$

$$\frac{210}{r} \cdot r - \frac{70}{r} \cdot r = 4 \cdot r$$

$$210 - 70 = 4r$$

$$140 = 4r \qquad \text{The difference between their times is 4 hours.}$$

$$35 = r$$

$$105 = 3r$$

Check: Time for freight train $= \frac{210}{35} = 6\,\text{hr}$

Time for passenger train $= \frac{210}{105} = 2\,\text{hr}$

$6 - 2 = 4$ hours difference in time

The freight train travels at 35 mph, and the passenger train travels at 105 mph.

7.5 Exercises

1. 10
2. 9
3. 14
4. $\frac{6}{13}$
5. $\frac{12}{7}$
6. 36, 27

Solve the following word problems in Exercises 1 – 30.

1. If 4 is subtracted from a certain number and the difference is divided by 2, the result is 1 more than $\frac{1}{5}$ of the original number. Find the original number.

2. What number must be added to both numerator and denominator of $\frac{16}{21}$ to make the resulting fraction equal to $\frac{5}{6}$?

3. Find the number that can be subtracted from both numerator and denominator of the fraction $\frac{69}{102}$ so that the result is $\frac{5}{8}$.

4. The denominator of a fraction exceeds the numerator by 7. If the numerator is increased by 3 and the denominator is increased by 5, the resulting fraction is equal to $\frac{1}{2}$. Find the original fraction.

5. The numerator of a fraction exceeds the denominator by 5. If the numerator is decreased by 4 and the denominator is increased by 3, the resulting fraction is equal to $\frac{4}{5}$. Find the original fraction.

6. One number is $\frac{3}{4}$ of another number. Their sum is 63. Find the numbers.

7. 15, 9

8. 7, 12

9. 2, 5

10. 45

11. \$1500

12. 225 miles

13. 1875 miles

14. $2\frac{2}{5}$ hours

15. 52 mph, 48 mph

16. $1\frac{4}{5}$ hours

17. 45 minutes

18. 6 hours

19. 120 mph, 300 mph

20. 63 mph

7. The sum of two numbers is 24. If $\frac{2}{5}$ the larger number is equal to $\frac{2}{3}$ the smaller number, find the numbers.

8. One number exceeds another by 5. The sum of their reciprocals is equal to 19 divided by the product of the two numbers. Find the numbers.

9. One number is 3 less than another. The sum of their reciprocals is equal to 7 divided by the product of the two numbers. Find the numbers.

10. A manufacturer sold a group of shirts for \$1026. One-fifth of the shirts were priced at \$18 each, and the remainder at \$24 each. How many shirts were sold?

11. Luis spent $\frac{1}{5}$ of his monthly salary for rent and $\frac{1}{6}$ of his monthly salary for his car payment. If \$950 was left, what was his monthly salary?

12. It takes Rosa, traveling at 50 mph, 45 minutes longer to go a certain distance than it takes Maria traveling at 60 mph. Find the distance traveled.

13. It takes a plane flying at 450 mph 25 minutes longer to travel a certain distance than it takes a second plane to fly the same distance at 500 mph. Find the distance.

14. Toni needs 4 hours to complete the yard work. Her husband, Sonny, needs 6 hours to do the work. How long will the job take if they work together?

15. Beth can travel 208 miles in the same length of time it takes Anna to travel 192 miles. If Beth's speed is 4 mph greater than Anna's, find both rates.

16. Ben's secretary can address the weekly newsletters in $4\frac{1}{2}$ hours. Charlie's secretary needs only 3 hours. How long will it take if they both work on the job?

17. Working together, Greg and Cindy can clean the snow from the driveway in 20 minutes. It would have taken Cindy, working alone, 36 minutes. How long would it have taken Greg alone?

18. A carpenter and his partner can put up a patio cover in $3\frac{3}{7}$ hours. If the partner needs 8 hours to complete the patio alone, how long would it take the carpenter working alone?

19. A commercial airliner can travel 750 miles in the same length of time that it takes a private plane to travel 300 miles. The speed of the airliner is 60 mph more than twice the speed of the private plane. Find the speed of each aircraft.

20. Gabriela travels 350 miles at a certain speed. If the average speed had been 9 mph less, she could have traveled only 300 miles in the same length of time. What was the average rate of speed?

21. 250 mph, 500 mph

22. 6 hours

23. 50 mph

24. small: 9 hrs, large: 6 hrs

25. John: 11 hours Ralph: 22 hours, Denny: 33 hours

26. 4.5 days, 9 days

27. 1 min. and 20 sec.

28. boat: 14 mph, current: 2 mph

29. 1 mph

30. 45 minutes

21. A jet flies twice as fast as a propeller plane. On a 1500 mile trip, the propeller plane took 3 hours longer than the jet. Find the speed of each plane.

22. A family travels 18 miles downriver and returns. It takes 8 hours to make the round trip. Their rate in still water is twice the rate of the current. How long will the return trip take?

23. An airplane can fly 650 mph in calm air. If it can travel 2800 miles with the wind in the same time it can travel 2400 miles against the wind, find the wind speed.

24. Using a small inlet pipe, it takes 3 hours longer to fill a pool than if a larger pipe is used. If both are used, it takes $3\frac{3}{5}$ hours to fill the pool. Using each pipe alone, how long would it take to fill the pool?

25. John, Ralph, and Denny, working together, can clean the store in 6 hours. Working alone, Ralph takes twice as long to clean the store as does John. Denny needs three times as long as does John. How long would it take each man working alone?

26. A contractor hires two bulldozers to clear the trees from a 20-acre tract of land. One works twice as fast as the other. It takes them 3 days to clear the tract working together. How long would it take each of them alone?

27. The hot water tap can fill a given sink in 4 minutes. If the cold water tap is turned on as well, the sink fills in 1 minute. How long would it take for the cold water to fill the sink alone?

28. Francois went 36 miles downstream and returned. The round trip took $5\frac{1}{4}$ hours. Find the speed of the boat in still water and the speed of the current if the speed of the current is $\frac{1}{7}$ the speed of the boat.

29. Town A is 12 miles upstream from Town B (on the same side of the river). A motorboat that can travel 8 mph in still water leaves A and travels downstream toward B. At the same time, another boat that can travel 10 mph leaves B and travels upstream toward A. Each boat completes the trip at the same time. Find the rate of the current.

30. Paul and Nicole are picking raspberries for homemade jam. Nicole can fill a bucket in a half hour, but Paul is eating the raspberries that Nicole has picked at a rate of one bucket per 1.5 hours. How long does it take Nicole to fill her bucket?

31. $AC = 2,\ ST = 12$

32. $DF = 10,\ GH = 4$

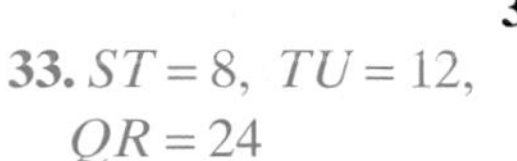

33. $ST = 8,\ TU = 12,\ QR = 24$

34. $LK = 9,\ JB = 7$

Exercises 31 – 34 each illustrate a pair of similar triangles. Find the lengths of the sides with unknown variables in each of these exercises.

31.

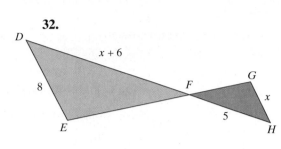

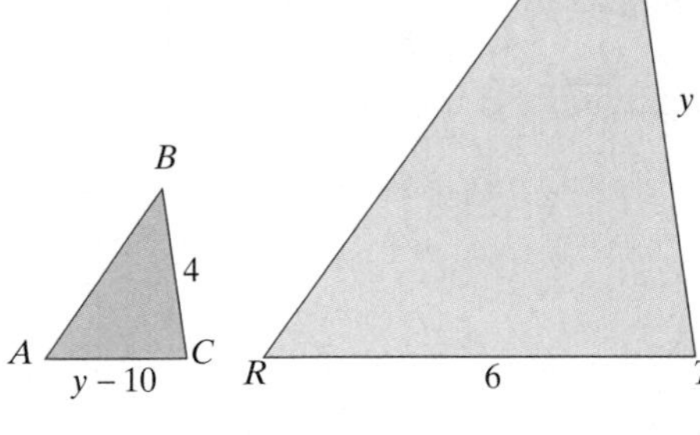

32.

D
x + 6
G
8
F
x
5
E
H

33.

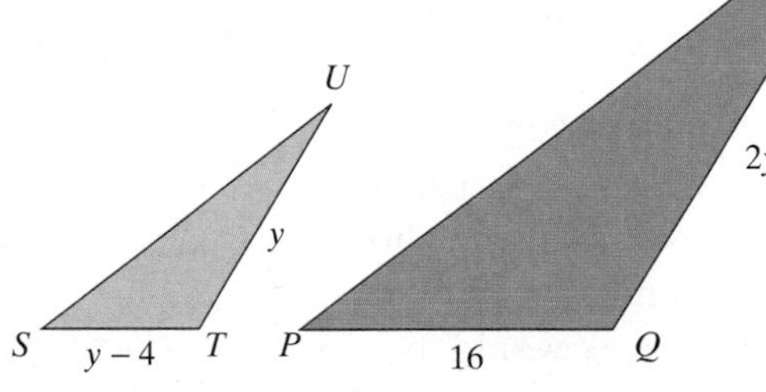

34.

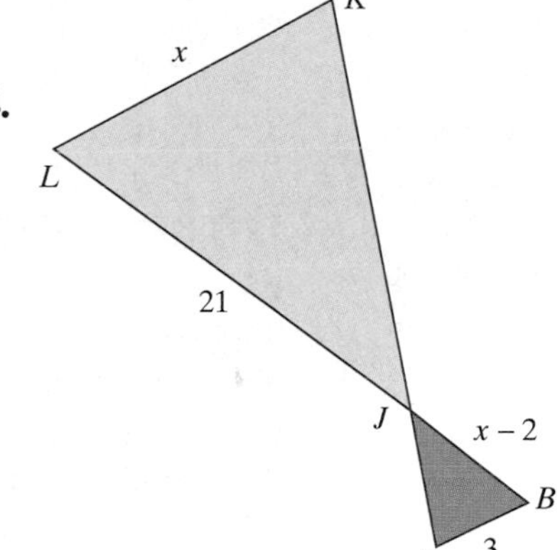

35. a. 5 and 7
b. 2 and 4

Writing and Thinking About Mathematics

35. If n is any integer, then $2n$ is an even integer and $2n + 1$ is an odd integer. Use these ideas to solve the following problems.

a. Find two consecutive odd integers such that the sum of their reciprocals is $\frac{12}{35}$.

b. Find two consecutive even integers such that the sum of the first and the reciprocal of the second is $\frac{9}{4}$.

Hawkes Learning Systems: Introductory & Intermediate Algebra

Applications Involving Rational Expressions

7.6 Additional Applications: Variation

Objectives

Teaching Notes:
You might want to make the case for the study of variation (and proportion) as being extremely important to the sciences and technology. Many major relationships in chemistry, physics, and astronomy began by recognizing a variation relationship.

After completing this section, you will be able to:

Solve applied problems using the principles of variation:

a. direct variation,

b. inverse variation, and

c. combined variation.

Direct Variation

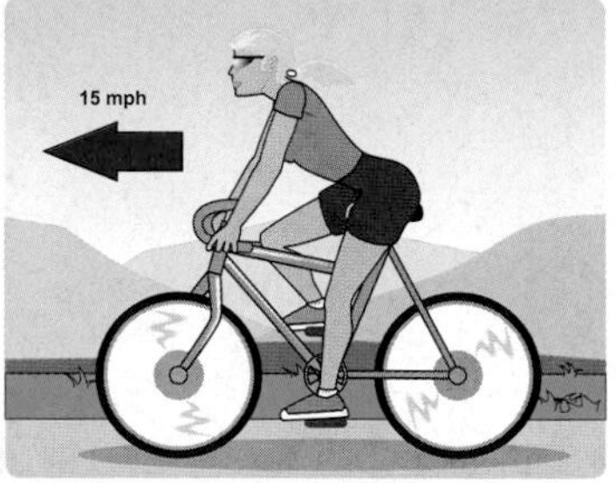

Suppose that you ride your bicycle at a steady rate of 15 miles per hour (not quite as fast as Lance Armstrong, but you are enjoying yourself). If you ride for 1 hour, the distance you travel would be 15 miles. If you ride for two hours, the distance you travel would be 30 miles. This relationship can be written in the form of the formula $d = 15t$ (or $\frac{d}{t} = 15$) where d is the distance traveled and t is the time in hours. We say that distance and time **vary directly** (or are in **direct variation** or are **directly proportional**). The term proportional implies that the ratio is constant. In this example, 15 is the constant and is called the **constant of variation**. When two variables vary directly, an increase in the value of one variable indicates an increase in the other, and the ratio of the two quantities is constant.

Direct Variation

A variable quantity y ***varies directly*** *as (or is* ***directly proportional to****) a variable x if there is a constant k such that*

$$\frac{y}{x} = k \text{ or } y = kx.$$

The constant k is called the ***constant of variation****.*

Example 1: Direct Variation

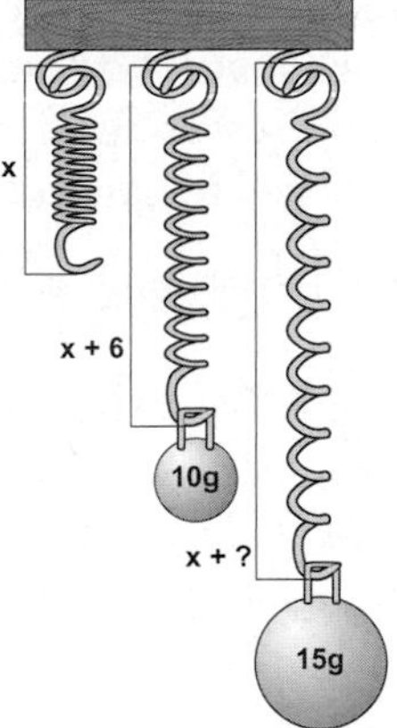

A spring will stretch a greater distance as more weight is placed on the end of the spring. The distance, *s*, the spring stretches varies directly as the weight, *w*, placed at the end of the spring. This is a property of springs studied in physics and is known as Hooke's Law. If a weight of 10 grams stretches a certain spring 6 centimeters, how far will the spring stretch with a weight of 15 g? (**Note:** We assume that the weight is not so great as to break the spring.)

Solution: Because the two variables are directly proportional, the relationship can be indicated with the formula

$s = k \cdot w$ where $s =$ distance spring stretches, $w =$ weight in cm,

and $k =$ constant of proportionality.

First substitute the given information to find the value for k. (The value of k will depend on the particular spring. Springs made of different material or which are wound more tightly, will have different values for k.)

$s = k \cdot w$

$6 = k \cdot 10$ Substitute the known values into the formula.

$\frac{3}{5} = k$ Find the value of k to substitute into the formula.

So, $s = \frac{3}{5}w$ The constant of proportionality is $\frac{3}{5}$ (or 0.6).

If $w = 15$, we have

$$s = \frac{3}{5} \cdot 15 = 9 \text{ cm}.$$

The spring will stretch 9 cm if a weight of 15 g is placed at its end.

Listed here are several formulas involving direct variation.

$d = \frac{3}{5}w$ Hooke's Law for a spring where $k = \frac{3}{5}$.

$C = 2\pi r$ The circumference of a circle varies directly as the radius.

$A = \pi r^2$ The area of a circle is directly proportional to the **square** of its radius.

$P = 625d$ Water pressure is proportional to the depth of the water.

Inverse Variation

When two variables vary in such a way that their product is constant, we say that the two variables **vary inversely** (or are **inversely proportional**). For example, if a gas is placed in a container (as in an automobile engine) and pressure is increased on the gas, then the product of the pressure and the volume of gas will remain constant. That is, pressure and volume are related by the formula $V \cdot P = k \left(\text{or } V = \frac{k}{P} \right)$.

Note that if a product of two variables is to remain constant, then an increase in the value of one variable must be accompanied by a decrease in the other. Or, in the case of a fraction with a constant numerator, if the denominator increases in value, then the fraction decreases in value. For the gas in an engine, an increase in pressure indicates a decrease in the volume of gas.

Inverse Variation

A variable quantity y ***varies inversely*** *as (or is* ***inversely proportional to****) a variable x if there is a constant k such that*

$$x \cdot y = k \text{ or } y = \frac{k}{x}.$$

The constant k is called the ***constant of variation.***

Example 2: Inverse Variation

The gravitational force, F, between an object and the earth is inversely proportional to the square of the distance, d, from the object to the center of the earth. Hence, we have the formula

$$F \cdot d^2 = k \text{ or } F = \frac{k}{d^2},$$

where F = force, d = distance, and k = constant of proportionality

(As the distance of an object from the earth becomes larger, the gravitational force exerted by the earth on the object becomes smaller.)

If an astronaut weighs 200 pounds on the surface of the earth, what will he weigh 100 miles above the earth? Assume that the radius of the earth is 4000 miles.

Solution: We know $F = 200 = 2 \times 10^2$ pounds
when $d = 4000 = 4 \times 10^3$ miles.

$$2\times10^2 = \frac{k}{\left(4\times10^3\right)^2}$$ Substitute and solve for k.

$$k = 2\times10^2 \times 16\times10^6 = 32\times10^8 = 3.2\times10^9$$

So, $$F = \frac{3.2\times10^9}{d^2}.$$

Let $d = 4100 = 4.1 \times 10^3$ miles. Then,

$$F = \frac{3.2\times10^9}{16.81\times10^6} \approx 0.190\times10^3 = 190 \text{ pounds.}$$

That is, 100 miles above the earth the astronaut will weigh 190 pounds.

Combined Variation

If a variable varies either directly or inversely with more than one other variable, the variation is said to be a **combined variation**. If the combined variation is all direct variation (the variables are multiplied), then it is called **joint variation**. For example, the volume of a cylinder varies jointly as its height and the square of its radius.

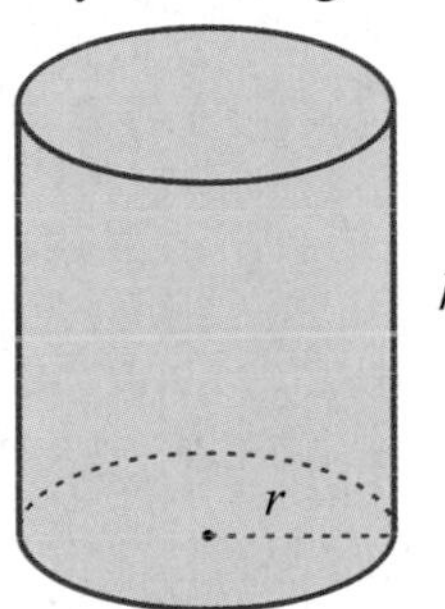

$V = kr^2h$ where r = radius, h = height, and k = constant of proportionality

For example, what is the value of k, the constant of proportionality, if a cylinder has the approximate measurements $V = 198$ cubic feet, $r = 3$ feet, and $h = 7$ feet?

Solution: $V = k \cdot r^2 \cdot h$ — V **varies jointly** as r^2 and h.

$198 = k \cdot 3^2 \cdot 7$ — Substitute the known values.

$\frac{198}{9 \cdot 7} = k$

$k = \frac{22}{7} \approx 3.14$ — We know from experience that $k = \pi$. Since the measurements are only approximate, the estimate for k is only approximate.

The formula is $V = \pi r^2 h$.

Example 3: Variations

a. If y is **directly proportional** to x^2, and $y = 9$ and $x = 2$, what is y when $x = 4$?

Solution:

$y = k \cdot x^2$ First substitute known values and solve for k.

$9 = k \cdot 2^2$ Use this value for k in the formula.

$\frac{9}{4} = k$

So, $y = \frac{9}{4}x^2.$

If $x = 4$, then

$$y = \frac{9}{4} \cdot 4^2$$
$$= 36.$$

b. The distance an object falls **varies directly** as the square of the time it falls (until it hits the ground and assuming little or no air resistance). If an object fell 64 feet in two seconds, how far would it have fallen by the end of 3 seconds?

Solution:

$d = k \cdot t^2$ where d = distance, t = time (in seconds), and k = constant of proportionality

$64 = k \cdot 2^2$ Substitute the known values and solve for k.

$16 = k$ Use this value for k in the formula.

So, $d = 16t^2$

$d = 16 \cdot 3^2 = 144$ feet

The object would have fallen 144 feet in 3 seconds.

c. The volume of a gas in a container **varies inversely** as the pressure on the gas. If a gas has a volume of 200 cubic inches under pressure of 5 pounds per square inch, what will be its volume if the pressure is increased to 8 pounds per square inch?

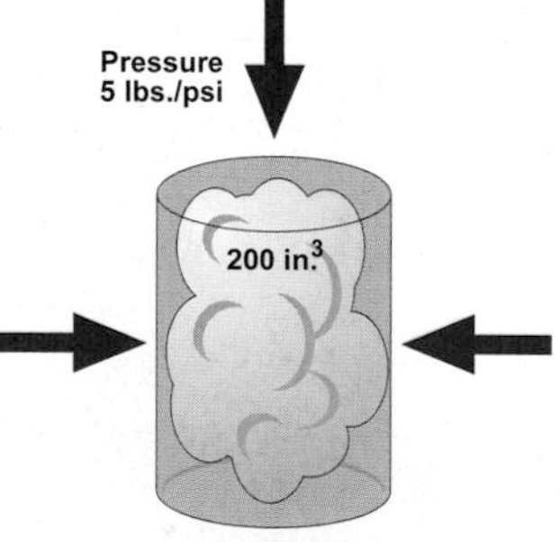

Solution:

$V = \frac{k}{P}$ where V = volume, P = pressure, and k = constant of proportionality

$200 = \frac{k}{5}$ Substitute the known values and solve for k.

$$k = 1000$$

So, $V = \frac{1000}{P}$ Substitute 1000 for k.

$$V = \frac{1000}{8} = 125 \text{ cu. in.}$$

The volume will be 125 cubic inches.

d. The illumination (in foot-candles, fc) of a light source **varies directly** as the intensity (in candlepower, cp) of the source and **inversely** as the square of the distance from the source. If a certain light source with intensity of 300 cp provides an illumination of 10 fc at a distance of 20 feet, what is the illumination at a distance of 40 feet? (**Note:** This is an illustration of combined variation.)

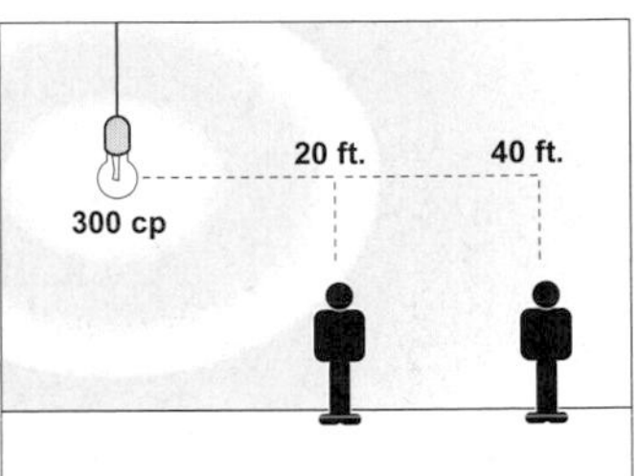

Solution: $I = \frac{k \cdot i}{d^2}$ Where I = illumination, i = intensity, and d = distance.

$$10 = \frac{k \cdot 300}{(20)^2}$$

$$k = \frac{400 \cdot 10}{300}$$

$$k = \frac{40}{3}$$ Value to be used in the formula.

So, $I = \frac{\frac{40}{3} \cdot i}{d^2}$

$$I = \frac{\frac{40}{3} \cdot 300}{(40)^2} = \frac{40 \cdot 100}{40 \cdot 40} = \frac{5}{2} = 2.5 \text{ fc}$$

The illumination at 40 feet is 2.5 fc.

7.6 Exercises

For Exercises 1 – 34, write an equation or formula that represents the general relationship indicated. Then use the given information to find the unknown value.

1. If y varies directly as x, and $y = 3$ when $x = 9$, find y if $x = 7$.
2. If y is directly proportional to x^2, and $y = 9$ when $x = 2$, what is y when $x = 4$?
3. If y varies inversely as x, and $y = 5$ when $x = 8$, find y if $x = 20$.
4. If y varies inversely as x^2, and $y = -8$ when $x = 2$, find y if $x = 3$.

1. $y = kx, \frac{7}{3}$

2. $y = kx^2, 36$

3. $y = \frac{k}{x}, 2$

4. $y = \frac{k}{x^2}, -\frac{32}{9}$

5. $y = \frac{k}{x}$, 10

6. $y = \frac{k}{x^3}$, 135

7. $y = k\sqrt{x}$, 36

8. $y = kx^2$, 180

9. $y = kx^3$, 24

10. $z = kxy$, 120

11. $z = kxy$, $-\frac{27}{5}$

12. $z = kx^2y$, $\frac{56}{3}$

13. $z = \frac{kx}{y^2}$, 40

14. $z = \frac{kx^3}{y^2}$, 81

15. $z = \frac{k\sqrt{x}}{y}$, 54

16. $z = kx^2y^3$, 384

17. $z = \frac{kx^2}{\sqrt{y}}$, 32

18. $s = \frac{k(r+t)}{w}$, $\frac{48}{5}$

19. $L = \frac{kmn}{P}$, 27

20. $F = \frac{k}{d^2}$, 232.3 pounds

21. $d = kt^2$, 256 feet

22. $V = \frac{k}{P}$, 200 cm^3

23. $s = kw$, 6 in.

24. $C = kd$, 4.71 feet

5. If y is inversely proportional to x, and $y = 5$ when $x = 4$, what is y when $x = 2$?

6. If y is inversely proportional to x^3, and $y = 40$ when $x = \frac{1}{2}$, what is y when $x = \frac{1}{3}$?

7. If y is proportional to the square root of x, and $y = 6$ when $x = \frac{1}{4}$, what is y when $x = 9$?

8. If y is proportional to the square of x, and $y = 80$ when $x = 4$, what is y when $x = 6$?

9. If y varies directly as x^3, and $y = 81$ when $x = 3$, find y if $x = 2$.

10. z varies jointly as x and y, and $z = 60$ when $x = 2$ and $y = 3$. Find z if $x = 3$ and $y = 4$.

11. z varies jointly as x and y, and $z = -6$ when $x = 5$ and $y = 8$. Find z if $x = 12$ and $y = 3$.

12. z varies jointly as x^2 and y, and $z = 20$ when $x = 2$ and $y = 3$. Find z if $x = 4$ and $y = \frac{7}{10}$.

13. z varies directly as x and inversely as y^2. If $z = 5$ when $x = 1$ and $y = 2$, find z if $x = 2$ and $y = 1$.

14. z varies directly as x^3 and inversely as y^2. If $z = 24$ when $x = 2$ and $y = 2$, find z if $x = 3$ and $y = 2$.

15. z varies directly as $\sqrt{x}$ and inversely as y. If $z = 24$ when $x = 4$ and $y = 3$, find z if $x = 9$ and $y = 2$.

16. z is jointly proportional to x^2 and y^3. If $z = 192$ when $x = 4$ and $y = 2$, find z when $x = 2$ and $y = 4$.

17. z varies directly as x^2 and inversely as $\sqrt{y}$. If $z = 108$ when $x = 6$ and $y = 4$, find z if $x = 4$ and $y = 9$.

18. s varies directly as the sum of r and t and inversely as w. If $s = 24$ when $r = 7$ and $t = 8$ and $w = 9$, find s if $r = 9$ and $t = 3$ and $w = 18$.

19. L varies jointly as m and n and inversely as p. If $L = 6$ when $m = 7$ and $n = 8$ and $p = 12$, find L if $m = 15$ and $n = 14$ and $p = 10$.

20. If an astronaut weighs 250 pounds on the surface of the earth, what will the astronaut weigh 150 miles above the earth? (Assume that the radius of the earth is 4000 miles, and round off to the nearest tenth.)

21. The distance a free falling object falls is directly proportional to the square of the time it falls (before it hits the ground). If an object fell 144 feet in 3 seconds, how far will it have fallen by the end of 4 seconds?

22. The volume of gas in a container is 300 cm^3 when the pressure on the gas is 20 g per cm^2. What will be the volume of the gas if the pressure is increased to 30 g per cm^2?

23. A hanging spring will stretch 5 in. if a weight of 10 lbs is placed at its end. How far will the spring stretch if the weight is increased to 12 lbs?

24. The circumference of a circle varies directly as the diameter. A circular pizza pie with a diameter of 1 foot has a circumference of 3.14 feet. What will be the circumference of a pizza pie with a diameter of 1.5 feet?

25. $A = kr^2$, 254.34 feet

26. $P = kg$, \$18.75

27. $C = kd$, 4.71 feet

28. $C = kd$, 4.71 feet

29. $E = \frac{kml}{A}$, 0.0073 cm

30. $E = \frac{kml}{A}$, 0.008 cm

31. $L = \frac{kwd^2}{l}$, 1890 lbs.

32. $L = \frac{kwd^2}{l}$, 16,000 lbs.

33. $F = \frac{km_1m_2}{d^2}$, 9×10^{-11} N

34. $F = \frac{km_1m_2}{d^2}$, $3.6 \cdot 10^{-10}$ N

25. The area of a circle varies directly as the square of its radius. A circular pizza pie with a diameter of 12 inches has an area of 113.04 in.^{2} What will be the area of a pizza pie with a diameter of 18 inches?

26. The total price, P, of gasoline purchased varies directly with the number of gallons purchased. If 10 gallons are purchased for \$12.50, what will be the price of 15 gallons?

27. Several triangles are to have the same area. In this set of triangles the height and base are inversely proportional. In one such triangle the height is 5 m and the base is 12 m. Find the height of the triangle in this set with a base of 10 m.

28. W varies jointly as x and y and inversely as z. If $W = 10$ when $x = 6$ and $y = 5$ and $z = 2$, find W if $x = 12$ and $y = 6$ and $z = 3$.

29. The elongation, E, in a wire when a mass, m, is hung at its free end varies jointly as the mass and the length, l, of the wire and inversely as the cross-sectional area, A, of the wire. The elongation is 0.0055 cm when a mass of 120 g is attached to a wire 330 cm long with a cross-sectional area of 0.4 sq cm. Find the elongation if a mass of 160 g is attached to the same wire.

30. When a mass of 240 oz. is suspended by a wire 49 inches long whose cross-sectional area is 0.035 sq. in., the elongation of the wire is 0.016 in. Find the elongation if the same mass is suspended by a 28-in. wire of the same material with a cross-sectional area of 0.04 sq. in. (See Exercise 29.)

31. The safe load, L, of a wooden beam supported at both ends varies jointly as the width, w, and the square of the depth, d, and inversely as the length, l. A wooden beam 2 in. wide, 8 in. deep, and 14 ft. long holds up 2400 lb. What load would a beam 3 in. × 6 in. × 15 ft. of the same material support?

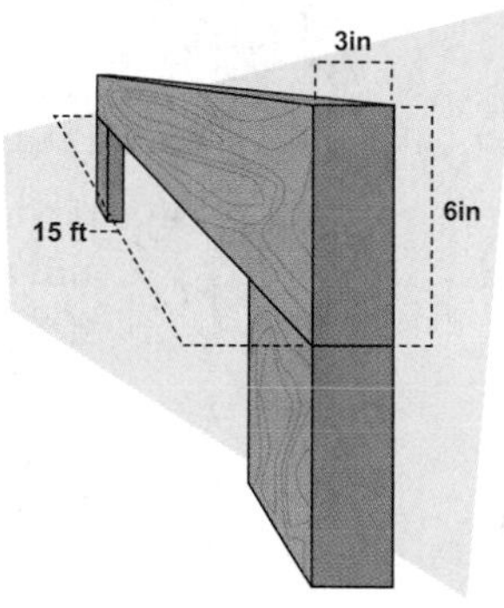

32. A 4 in. × 6 in. beam 12 ft. long supports a load of 4800 lb. What is the safe load of a beam of the same material that is 6 in. × 10 in. × 15 ft. long? (See Exercise 31.)

33. The gravitational force of attraction, F, between two bodies varies directly as the product of their masses, m_1 and m_2, and inversely as the square of the distance, d, between them. The gravitational force between a 5-kg mass and a 2-kg mass 1 m apart is 1.5×10^{-10} N. Find the force between a 24-kg mass and a 9-kg mass that are 6 m apart. (N represents a unit of force called a **newton**.)

34. In Exercise 33, what happens to the force if the distance between the bodies is cut in half?

35. 15,000 lbs.

36. 139.219 ft.^{2}

37. 25,200 lbs.

38. 192 lb per ft.^{2}

39. 1.8 cu. ft.

40. 700 lb per ft^2

Lifting Force

The lifting force (or lift), P, in pounds exerted by the atmosphere on the wings of an airplane is related to the area, A, of the wings in square feet and the speed (or velocity), v, of the plane in miles per hour by the formula

$\boldsymbol{P = kAv^2}$ where k is the constant of variation.

35. If the lift is 9600 lbs. for a wing area of 120 ft.^{2} and a speed of 80 mph, find the lift of the same airplane at a speed of 100 mph.

36. If the lift is 12,000 lbs. for a wing area of 110 ft.^{2} and a speed of 90 mph, what would be the necessary wing area to attain the same lift at 80 mph? (Round to the nearest thousandth.)

37. The lift for a wing of area 280 ft.^{2} is 34,300 lbs. when the plane is going 210 mph. What is the lift if the speed is decreased to 180 mph?

Pressure

Boyle's Law states that if the temperature of a gas sample remains the same, the pressure of the gas is related to the volume by the formula

$$P = \frac{k}{V} \quad \text{where } k \text{ is the constant of proportionality.}$$

38. The temperature of a gas remains the same. The pressure of the gas is 16 lbs. per ft.^{2} when the volume is 300 cu. ft. What will be the pressure if the gas is compressed into 25 cu. ft.?

39. A pressure of 1600 lbs. per ft.^{2} is exerted by 2 cu. ft. of air in a cylinder. If a piston is pushed into the cylinder until the pressure is 1800 lbs. per ft.^{2}, what will be the volume of the air? (Round to the nearest tenth.)

40. If 1000 cu. ft. of gas exerting a pressure of 140 lb. per ft.^{2} must be placed in a container which has a capacity of 200 cu. ft., what is the new pressure exerted by the gas?

41. 15 ohms

42. $5\frac{1}{3}$ ohms

43. 5.2 ohms

44. 384 rpm

45. 35 teeth

46. 36 teeth

Electricity

The resistance, R (in ohms), in a wire is given by the formula

$$R = \frac{kL}{d^2}$$

where k is the constant of variation, L is the length of the wire and d is the diameter.

41. The resistance of a wire 500 ft. long with a diameter of 0.01 in. is 20 ohms. What is the resistance of a wire 1500 ft. long with a diameter of 0.02 in.?

42. The resistance of a wire 100 ft. long and 0.01 in. in diameter is 8 ohms. What is the resistance of a piece of the same type of wire with a length of 150 ft. and a diameter of 0.015 in.?

43. The resistance is 2.6 ohms when the diameter of a wire is 0.02 in. and the wire is 10 ft. long. Find the resistance of the same type of wire with a diameter of 0.01 in. and a length of 5 ft.

Gears

If a gear with T_1 teeth is meshed with a gear having T_2 teeth, the numbers of revolutions of the two gears are related by the proportion

$$\frac{T_1}{T_2} = \frac{R_2}{R_1} \text{ or } T_1R_1 = T_2R_2$$

where R_1 and R_2 are the numbers of revolutions.

44. A gear having 96 teeth is meshed with a gear having 27 teeth. Find the speed (in revolutions per minute) of the small gear if the large one is traveling at 108 revolutions per minute.

45. What size gear (how many teeth) is needed to turn a pulley on a feed grinder at 24 revolutions per minute if it is driven by a 15 tooth gear turning at 56 revolutions per minute?

46. What size gear (how many teeth) is needed to turn the reel on a harvester at 20 revolutions per minute if it is driven by a 12-tooth gear turning at 60 revolutions per minute?

47. 900 lbs.

48. 4 ft.

49. $10\frac{2}{3}$ feet from the 120 lb. weight

Lever

If a lever is balanced with weight on opposite sides of the balance points, then the following proportion exists:

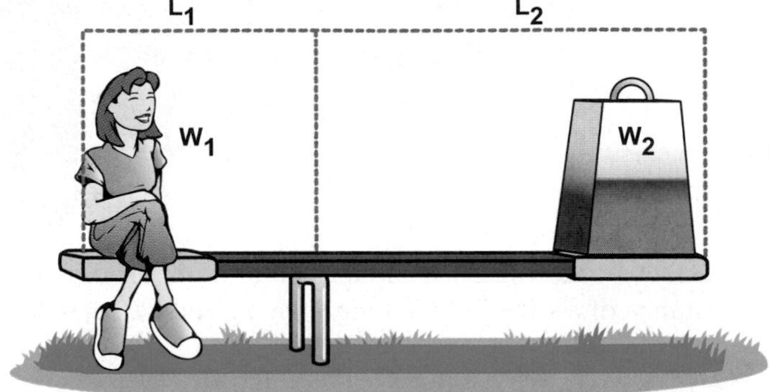

$$\frac{W_1}{W_2} = \frac{L_2}{L_1} \text{ or } W_1L_1 = W_2L_2$$

47. How much weight can be raised at one end of a bar 8 ft. long by the downward force of 60 lbs. when the balance point is $\frac{1}{2}$ ft. from the unknown weight?

48. A force of 40 lbs. at one end of a bar 5 ft. long is to balance 160 lbs. at the other end of the bar. Ignoring the weight of the bar, how far from the 40 lb. weight should a hole be drilled in the bar for a bolt to serve as a balance point?

49. Where should the balance point of a bar 12 ft. long be located if a 120 lb. force is to raise a load weighing 960 lbs.?

50. Answers will vary.

Writing and Thinking About Mathematics

50. Explain in your own words, the meaning of the terms
- **a.** direct variation,
- **b.** inverse variation,
- **c.** joint variation, and
- **d.** combined variation.

Discuss an example of each type of variation that you have observed in your daily life.

Hawkes Learning Systems: Introductory & Intermediate Algebra

Applications: Variation

Chapter 7 Index of Key Ideas and Terms

Section 7.1 Multiplication and Division with Rational Expressions

Rational Expression page 477

A **rational expression** is an expression of the form

$\frac{P}{Q}$ where P and Q are polynomials and $Q \neq 0$.

Arithmetic Rules for Rational Numbers (or Fractions) page 477

Fundamental Principle of Fractions page 478

If $\frac{P}{Q}$ is a rational expression and K is a polynomial

and $K \neq 0$, then $\frac{P}{Q} = \frac{P}{Q} \cdot \frac{K}{K} = \frac{P \cdot K}{Q \cdot K}$.

Negative Signs in Rational Expressions page 479

$$-\frac{P}{Q} = \frac{P}{-Q} = \frac{-P}{Q} \text{ and } \frac{P}{Q} = \frac{-P}{-Q} = -\frac{-P}{Q} = -\frac{P}{-Q}$$

Opposites in Rational Expressions page 480

In general, $\frac{-P}{P} = -1$ if $P \neq 0$.

In particular, $\frac{a-x}{x-a} = -1$ if $x \neq a$.

Multiplication of Rational Expressions page 481

$\frac{P}{Q} \cdot \frac{R}{S} = \frac{P \cdot R}{Q \cdot S}$ where $Q, S \neq 0$.

Division of Rational Expressions page 482

$\frac{P}{Q} \div \frac{R}{S} = \frac{P}{Q} \cdot \frac{S}{R}$ where $Q, R, S \neq 0$.

Section 7.2 Addition and Subtraction with Rational Expressions

Addition and Subtraction of Rational Expressions page 488

$$\frac{P}{Q}+\frac{R}{Q}=\frac{P+R}{Q} \text{ and } \frac{P}{Q}-\frac{R}{Q}=\frac{P-R}{Q} \text{ where } Q \neq 0.$$

Finding the LCM for a Set of Polynomials page 489

1. Completely factor each polynomial (including prime factors for numerical factors).
2. Form the product of all distinct factors that appear, using each factor the most number of times it appears in any one factorization.

Procedure for Adding or Subtracting Rational Expressions with Different Denominators page 490

1. Find the LCD (the LCM of the denominators).
2. Rewrite each fraction in an equivalent form with the LCD.
3. Add (or subtract) the numerators and keep the common denominator.
4. Reduce if possible.

Section 7.3 Complex Fractions

Complex Fraction page 497

A **complex fraction** is a fraction in which the numerator or denominator is a fraction or the sum or difference of fractions.

Simplifying Complex Fractions pages 497 - 499

Method 1 page 497

1. Simplify the numerator and denominator separately so that the numerator and denominator are simple fractions.
2. Divide by multiplying by the reciprocal of the denominator.

Method 2 page 499

1. Find the LCM of all the denominators in the original numerator and denominator.
2. Multiply both the numerator and denominator by this LCM.

Simplifying Complex Algebraic Expressions page 500

Section 7.4 Equations and Inequalities with Rational Expressions

To Solve a Rational Equation page 503

1. Find the LCM of all the denominators of all the rational expressions in the equation.
2. Multiply both sides of the equation by this LCM. Use the distributive property if necessary.
3. Simplify both sides of the resulting equation.
4. Solve this equation.
5. Check each solution in the **original equation**. (Remember that no denominator can be 0.)

Solving Inequalities with Rational Expressions page 509

1. Simplify the inequality so that one side is 0 and on the other side both the numerator and denominator are in factored form.
2. Find the points where each linear factor is 0.
3. Mark each of these points on a number line.
4. Choose a number from each indicated interval as a test value.
5. The intervals where the test values satisfy the conditions of the inequality are the solution intervals.
6. Mark a solid circle for endpoints that are included and an open circle for endpoints that are not included.

Section 7.5 Applications

Strategy for Solving Word Problems page 513

1. Read the problem carefully. Read it several times if necessary.
2. Decide what is asked for and assign a variable to the unknown quantity.
3. Draw a diagram or set up a chart whenever possible.
4. Form an equation (or inequality) that relates the information provided.
5. Solve the equation (or inequality).
6. Check your solution with the wording of the problem to be sure it makes sense.

Section 7.6 Additional Applications: Variation

Direct Variation page 524

A variable quantity y **varies directly** as (or is **directly proportional to**) a variable x if there is a constant k such that

$$\frac{y}{x} = k \text{ or } y = kx$$

The constant k is called the **constant of variation.**

Inverse Variation page 526

A variable quantity y **varies inversely** as (or is **inversely proportional to**) a variable x if there is a constant k such that

$$x \cdot y = k \text{ or } y = \frac{k}{x}$$

The constant k is called the **constant of variation.**

Combined Variation page 527

If a variable varies either directly or inversely with more than one other variable, the variation is said to be **combined variation.**

Joint Variation page 527

If a combined variation is all direct variation (the variables are multiplied), then it is called **joint variation.**

Chapter 7 Review

For a review of the topics and problems from Chapter 7, look at the following lessons from *Hawkes Learning Systems: Introductory & Intermediate Algebra.*

Defining Rational Expressions
Multiplication and Division of Rational Expressions
Addition and Subtraction of Rational Expressions
Complex Fractions
Solving Equations Involving Rational Expressions
Solving Inequalities with Rational Expressions
Applications Involving Rational Expressions
Applications: Variation

Chapter 7 Test

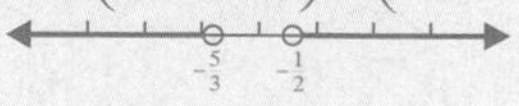

Assume that none of the denominators in the rational expressions on this test has a value of 0.

Write each expression in Exercises 1 and 2 in lowest terms.

1. $\dfrac{x^2+3x}{x^2+7x+12}$ **2.** $\dfrac{x^3-64}{16-x^2}$

3. Determine the missing numerator that will make the following rational expressions equivalent.

a. $\dfrac{x-5}{3-x}=\dfrac{?}{x-3}$ **b.** $\dfrac{x-1}{3x+1}=\dfrac{?}{(3x+1)(x+2)}$

Perform the indicated operations in Exercises 4 – 9. Reduce all answers to lowest terms.

4. $\dfrac{x+3}{x^2+3x-4}\cdot\dfrac{x^2+x-2}{x+2}$ **5.** $\dfrac{6x^2-x-2}{12x^2+5x-2}\div\dfrac{4x^2-1}{8x^2-6x+1}$

6. $\dfrac{x}{x^2+3x-10}+\dfrac{3x}{4-x^2}$ **7.** $\dfrac{x-4}{3x^2+5x+2}-\dfrac{x-1}{x^2-3x-4}$

8. $\dfrac{x^2-16}{x^2-4x}\cdot\dfrac{x^2}{x+4}\div\dfrac{x-1}{2x^2-2x}$ **9.** $\dfrac{x}{x+3}-\dfrac{x+1}{x-3}+\dfrac{x^2+4}{x^2-9}$

10. Simplify each of the following complex fractions.

a. $\dfrac{\frac{4}{3x}+\frac{1}{6x}}{\frac{1}{x^2}-\frac{1}{2x}}$ **b.** $\dfrac{1-4x^{-2}}{1+2x^{-1}}$

11. Simplify the expression in part (a) and solve the equation in part (b).

a. $\dfrac{3}{x}-\dfrac{2}{x+1}+\dfrac{5}{2x}$ **b.** $\dfrac{3}{x}-\dfrac{2}{x+1}=\dfrac{5}{2x}$

Solve each of the equations in Exercises 12 and 13.

12. $\dfrac{4}{7}-\dfrac{1}{2x}=1+\dfrac{1}{x}$ **13.** $\dfrac{4}{x+4}+\dfrac{3}{x-1}=\dfrac{1}{x^2+3x-4}$

Solve each of the inequalities in Exercises 14 and 15. Graph each solution set on a number line and write the answer in interval notation.

14. $\dfrac{2x+5}{x-3}\geq 0$ **15.** $\dfrac{x-3}{2x+1}<2$

1. $\dfrac{x}{x+4}$

2. $-\dfrac{x^2+4x+16}{x+4}$

3. a. $5-x$ **b.** x^2+x-2

4. $\dfrac{x+3}{x+4}$ **5.** $\dfrac{3x-2}{3x+2}$

6. $\dfrac{-2x^2-13x}{(x+5)(x+2)(x-2)}$

7. $\dfrac{-2x^2-7x+18}{(3x+2)(x-4)(x+1)}$

8. $2x^2$

9. $\dfrac{x^2-7x+1}{(x+3)(x-3)}$

10. a. $\dfrac{3x}{2-x}$ **b.** $\dfrac{x-2}{x}$

11. a. $\dfrac{7x+11}{2x(x+1)}$ **b.** $x=\dfrac{1}{3}$

12. $x=-\dfrac{7}{2}$

13. $x=-1$

14. $\left(-\infty,-\dfrac{5}{2}\right]\cup(3,\infty)$

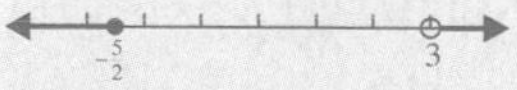

15. $\left(-\infty,-\dfrac{5}{3}\right)\cup\left(-\dfrac{1}{2},\infty\right)$

16. $\frac{2}{7}$

17. 4 hr.

18. 42 mph-Carlos; 57 mph-Mario

19. 11 mph

20. $\frac{400}{9}$

21. $\frac{15}{2}$ cm

22. 25,000 lbs.

23. 3 ohms

24. 6 hrs.

16. The denominator of fraction is three more than twice the numerator. If eight is added to both the numerator and the denominator, the resulting fraction is equal to $\frac{2}{3}$. Find the original fraction.

17. Sonya can clean the apartment in 6 hours. It takes Lucy 12 hours to clean it. If they work together, how long will it take them?

18. Mario can travel 228 miles in the same time that Carlos travels 168 miles. If Mario's speed is 15 mph faster than Carlos', find their rates.

19. Bob travels 4 miles upstream. In the same time, he could have traveled 7 miles downstream. If the speed of the current is 3 mph, find the speed of the boat in still water.

20. z varies directly as x^2 and inversely as $\sqrt{y}$. If $z = 24$ when $x = 3$ and $y = 4$, find z if $x = 5$ and $y = 9$.

21. Hooke's Law states that the distance a spring will stretch vertically is directly proportional to the weight placed at its end. If a particular spring will stretch 5 cm when a weight of 4 g is placed at its end, how far will the spring stretch if a weight of 6 g is placed at its end?

22. The lifting force on the wings of an airplane is given by the formula $P = kAv^2$. If the lift on the wing of a certain plane is 16,000 lbs. for a wing area of 180 ft.^{2} at a speed of 120 mph, find the lift of the same plane at a speed of 150 mph.

23. The resistance, R (in ohms), in a wire is given by the formula $R = \frac{kL}{d^2}$. The resistance of a wire with a diameter of 0.01 in. and 250 ft. long is 10 ohms. What is the resistance of a piece of the same type of wire with a length of 300 ft. and a diameter of 0.02 in.?

24. A swimming pool has two inlet pipes. One inlet pipe alone can fill the pool in 3 hours and the two pipes open together can fill the pool in 2 hours. How long would it take for the second inlet pipe alone to fill the pool?

Cumulative Review: Chapters 1 – 7

1. a. $\frac{19}{12}$ b. $-\frac{3}{14}$ c. $\frac{1}{4}$ d. $\frac{1}{12}$ 2. 5

3. $3x^2 - 5x - 28$

4. $4x^2 - 20x + 25$

5. $6x - 23$ 6. $10x + 4$ 7. $t = \frac{A-P}{Pr}$ 8. $r = \frac{C}{2\pi}$

9. $(2x + 5)(2x + 3)$

10. $2(x + 5)(x - 2)$

11. $(3x+2)(5x+4)$

12. $(x+4)(2x^2+3)$

13. $2(2x-3)(4x^2+6x+9)$

1. Perform the indicated operations and reduce.

a. $\frac{5}{6}+\frac{3}{4}$ **b.** $\frac{1}{2}-\frac{5}{7}$ **c.** $\frac{1}{2}\cdot\frac{1}{2}$ **d.** $\frac{7}{32}\div\frac{21}{8}$

2. Find 12.5% of 40.

14. $x^{-3}(2x+1)(x-3)$
15. a. $D = \{-1, 0, 5, 6\}$
b. $R = \{0, 5, 2\}$
c. It is not a function because the x-coordinate -1 has more than one corresponding y-coordinate.
16. a. 8 b. -1 c. $-\frac{59}{27}$
17. $y = \frac{3}{4}x + \frac{11}{2}$;

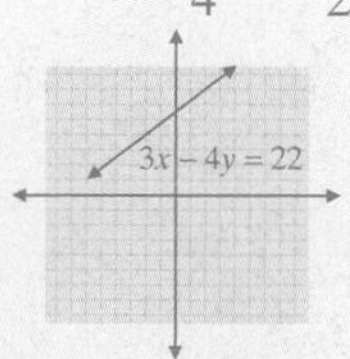

18. $y + 2x = -8$

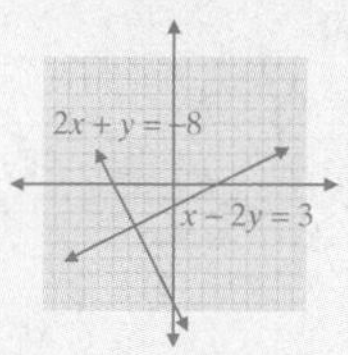

19. a. $(-2, 6]$

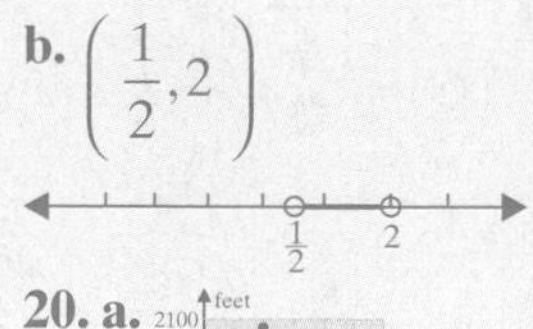

b. $\left(\frac{1}{2}, 2\right)$

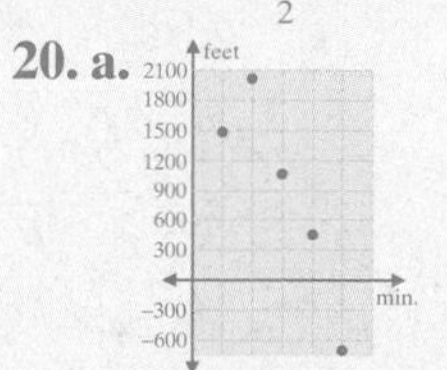

20. a.

b. 528, −968, −616, −1144
c. Juan is accelerating at a rate of 528 feet per minute; decelerating 968 feet per minute; decelerating 616 feet per minute; decelerating 1144 feet per minute.

In Exercises 3 – 6, perform the indicated operations and simplify the expressions.

3. $(3x+7)(x-4)$

4. $(2x-5)^2$

5. $-4(x+2)+5(2x-3)$

6. $x(x+7)-(x+1)(x-4)$

7. In the formula $A = P + Prt$, solve for t.

8. In the formula $C = 2\pi r$, solve for r.

Factor each expression in Exercises 9 and 10.

9. $4x^2+16x+15$

10. $2x^2+6x-20$

Factor each expression in Exercises 11 – 14 completely.

11. $15x^2+22x+8$

12. $2x^3+8x^2+3x+12$

13. $16x^3-54$

14. $2x^{-1}-5x^{-2}-3x^{-3}$

15. Given the relation $r = \{(-1,5), (0,2), (-1,0), (5,0), (6,0)\}$:

a. What is the domain of the relation?
b. What is the range of the relation?
c. Is the relation a function? Explain.

16. For the function $f(x) = x^3 - 2x^2 - 1$, find
a. $f(3)$ **b.** $f(0)$ **c.** $f\left(-\frac{2}{3}\right)$

17. Find the equation of the line that has slope $\frac{3}{4}$ and passes through the point $(-2, 4)$. Graph the line.

18. Find the equation of the line perpendicular to the line $x - 2y = 3$ and passing through the point $(-4, 0)$. Graph both lines.

19. Solve each of the following inequalities. Graph each solution set on a number line and write the answer in interval notation.

a. $\frac{3x-10}{x+2} \le 1$ **b.** $\frac{x+1}{2x-1} > 1$

20. The following table shows the number of feet that Juan drove his car from one minute to another. (Sometimes he backed up.)

a. Plot the points indicated in the table.
b. Calculate the slope of the line segments from point to point.
c. Interpret each slope.

Minutes	Feet
1	1496
2	2024
3	1056
4	440
5	−704

21. Function
22. Function
23. a. Not a function
b. Function;
$D = \{-7, -3, 0, 5, 6\}$;
$R = \{-2, 1, 4\}$
c. Function;
D: $(-\infty, \infty)$; R: $[0, \infty)$
24. $(-1.68, .326)$ and $(1.484, 1.594)$

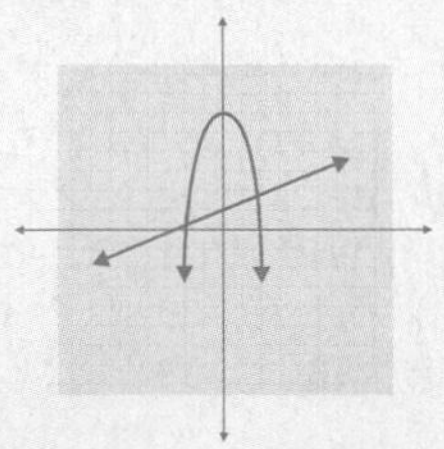

25. 6.8 **26.** 1.05×10^6
27. $8x^2$
28. $5x^3y^2$
29. $\frac{y^6}{9x^4}$ **30.** $\frac{x^6}{y^6}$
31. $\frac{3}{2x^4y}$
32. a. $x^2 + 9x + 20 = 0$
b. $x^3 - 6x^2 + 11x - 6 = 0$
33. $\frac{3}{5}x + 2y + \frac{y^2}{x}$
34. $1 - 3y + \frac{8}{7}y^2$
35. $x - 15 - \frac{1}{x+1}$
36. $2x^2 - x + 3 - \frac{2}{x+3}$
37. $4x^2 + 3x + 8 + \frac{15}{x-2}$
38. $2x^2 + x - 3 + \frac{9}{x+2}$
39. $x - 4 + \frac{3(2x-5)}{x^2-6}$

In Exercises 21 and 22, use the vertical line test to determine whether each of the graphs does or does not represent a function.

21.

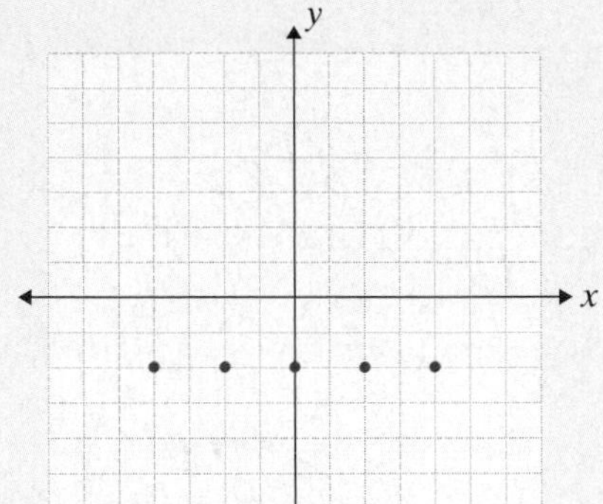

22.

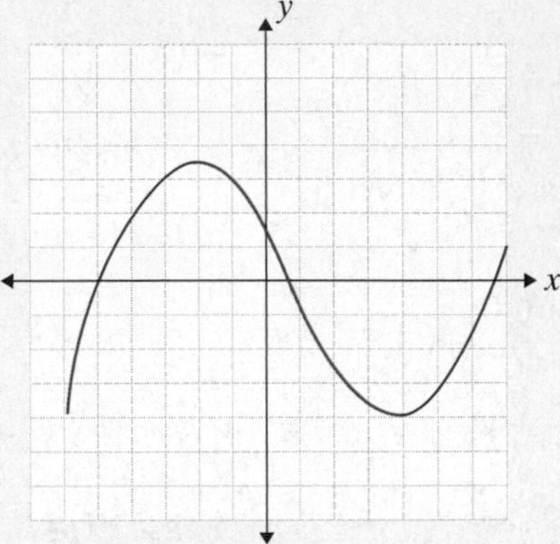

23. Three graphs are shown below. Use the vertical line test to determine whether or not each graph represents a function. If the graph represents a function, state its domain and range.

a.

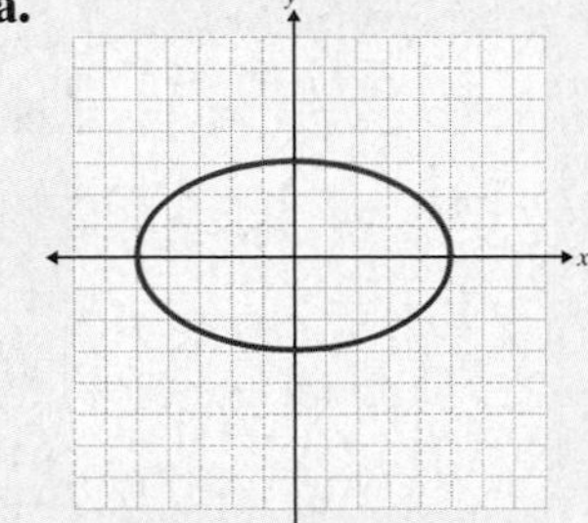

b.

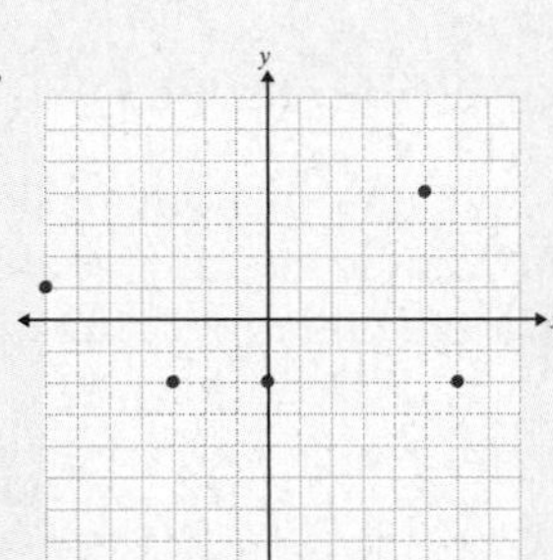

c.

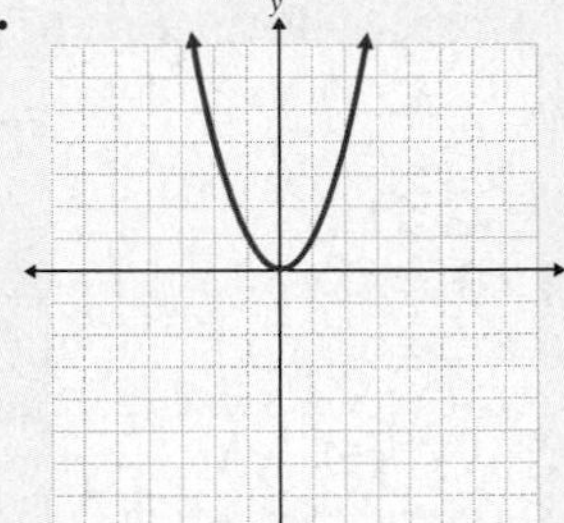

24. Use a graphing calculator to graph the two functions $f(x) = \frac{2}{5}x + 1$ and $g(x) = -2x^2 + 6$. Estimate the points of intersection.

Write Exercises 25 and 26 in scientific notation.

25. $(17000)(0.0004)$ **26.** $\frac{6300}{0.006}$

Simplify each expression in Exercises 27 – 31.

27. $\frac{32x^3y^2}{4xy^2}$ **28.** $\frac{15xy^{-1}}{3x^{-2}y^{-3}}$ **29.** $\left(3x^2y^{-3}\right)^{-2}$

30. $\left(\frac{x^4y}{x^2y^3}\right)^3$ **31.** $\left(\frac{5x^2y^{-1}}{2xy}\right)\left(\frac{3x^{-3}y}{5x^2y^0}\right)$

32. a. Find an equation that has $x = -5$ and $x = -4$ as roots.
b. Find an equation that has $x = 1$, $x = 2$, and $x = 3$ as roots.

40. $\frac{4x^2}{x-3}, x \neq 3, -\frac{1}{2}$

41. Does not reduce, $x \neq 2$

42. $x-3, x \neq -3$

43. $\frac{1}{x+1}; x \neq 0, -1$

44. $-\frac{1}{3}; x \neq 4$

45. $\frac{2}{3}; x \neq -3$

46. $\frac{x}{x+4}; x \neq -4, -3$

47. $\frac{x+5}{2(x-3)}; x \neq 3$

48. $x-y$

49. $\frac{4(4x-7)}{x(x-4)}$

50. $\frac{4}{(y+2)(y+3)}$

51. $\frac{x}{3x+3}$

52. $\frac{x+4}{3x}$

53. $\frac{3x(x+2)^2}{x+3}$

54. $\frac{2x+1}{x-1}$

55. $\frac{2x^2+14x-8}{(x+3)(x-1)(x-2)}$

56. $\frac{2(x+2)}{(x-1)(x+4)}$

57. $\frac{x-4}{(x+2)(x-2)}$

58. $\frac{2x-1}{(x+1)(x-1)}$

59. $\frac{x^2+15x-26}{(x-4)(x+2)(2x+3)}$

In Exercises 33 and 34, express each quotient as a sum of fractions and simplify if possible.

33. $\frac{3x^3y+10x^2y^2+5xy^3}{5x^2y}$ **34.** $\frac{7x^2y^2-21x^2y^3+8x^2y^4}{7x^2y^2}$

Divide in Exercises 35 – 39 by using the long division algorithm. Write the answer in the form $Q(x)+\frac{R(x)}{D(x)}$.

35. $(x^2-14x-16)\div(x+1)$ **36.** $\frac{2x^3+5x^2+7}{x+3}$ **37.** $\frac{4x^3-5x^2+2x-1}{x-2}$.

38. $\frac{2x^3+5x^2-x+3}{x+2}$ **39.** $\frac{x^3-4x^2+9}{x^2-6}$

In Exercises 40 – 47, reduce the rational expressions to lowest terms and indicate any restrictions on the variable.

40. $\frac{8x^3+4x^2}{2x^2-5x-3}$ **41.** $\frac{4x^2+5x+9}{x-2}$ **42.** $\frac{x^2-9}{x+3}$

43. $\frac{x}{x^2+x}$ **44.** $\frac{4-x}{3x-12}$ **45.** $\frac{2x+6}{3x+9}$

46. $\frac{x^2+3x}{x^2+7x+12}$ **47.** $\frac{x^2+2x-15}{2x^2-12x+18}$

Perform the indicated operations and simplify in Exercises 48 – 59. Assume that no denominator is 0.

48. $\frac{x^2}{x+y}-\frac{y^2}{x+y}$ **49.** $\frac{7}{x}+\frac{9}{x-4}$ **50.** $\frac{4}{y+2}-\frac{4}{y+3}$

51. $\frac{4x}{3x+3}-\frac{x}{x+1}$ **52.** $\frac{4x}{x-4}\div\frac{12x^2}{x^2-16}$ **53.** $\frac{x^2+3x+2}{x+3}\div\frac{x+1}{3x^2+6x}$

54. $\frac{2x+1}{x^2+5x-6}\cdot\frac{x^2+6x}{x}$ **55.** $\frac{8}{x^2+x-6}+\frac{2x}{x^2-3x+2}$

56. $\frac{x}{x^2+3x-4}+\frac{x+1}{x^2-1}$ **57.** $\frac{x+1}{x^2+4x+4}\div\frac{x^2-x-2}{x^2-2x-8}$

58. $\frac{x+1}{x^2+3x-4}\cdot\frac{2x^2+7x-4}{x^2-1}\div\frac{x+1}{x-1}$ **59.** $\frac{x}{x^2-2x-8}+\frac{x-3}{2x^2-5x-12}-\frac{2x-5}{2x^2+7x+6}$

Simplify the complex algebraic fractions in Exercises 60 – 62.

60. $\frac{\frac{3}{x}+\frac{1}{6x}}{\frac{7}{3x}}$ **61.** $\frac{\frac{1}{x}-\frac{1}{x^2}}{\frac{1}{x}+\frac{1}{x^2}}$ **62.** $\frac{\frac{4}{3x}+\frac{1}{6x}}{\frac{1}{x^2}-\frac{1}{2x}}$

60. $\frac{19}{14}$ **61.** $\frac{x-1}{x+1}$

62. $\frac{3x}{2-x}$ **63.** $-\frac{3}{5}$

64. $x=6,-5$

65. $x=2,5$

66. $x=2,-10$

67. $x=-45$

68. $x=-\frac{9}{8}$

69. $x=\frac{29}{11}$

70. $x=-\frac{4}{3}, x=4$

71. $\left(-\infty,-\frac{1}{4}\right]$

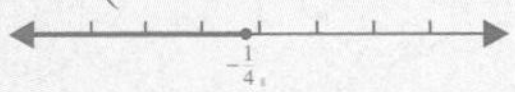

72. $\left(-\infty,-\frac{50}{3}\right)$

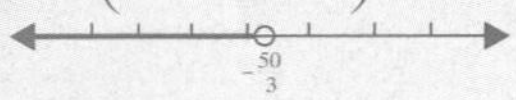

73. $(-\infty,12]$

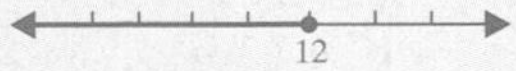

74. a. $\frac{9-2x}{x(x+3)}$

b. $x=\frac{9}{2}$

75. a. $\frac{6x^2-16x-20}{(x-4)(x+2)}$

b. $x=7$

76. The father takes $2\frac{2}{3}$ hrs. and the daughter takes 8 hrs.

77. 2.5 fc

78. $7\frac{9}{13}$ in.

79. 256 feet

80. 24 feet

81. 16, 18

Solve each of the equations in Exercises 63 – 67.

63. $4(x+2)-7=-2(3x+1)-3$

64. $(x+4)(x-5)=10$

65. $0=x^2-7x+10$

66. $x^2+8x-4=16$

67. $\frac{7}{2x-1}=\frac{3}{x+6}$

Solve each of the equations in Exercises 68 – 70.

68. $4(3x-1)=2(2x-5)-3$

69. $\frac{4x-1}{3}+\frac{x-5}{2}=2$

70. $\left|\frac{3x}{4}-1\right|=2$

Solve the inequalities in Exercises 71 – 73 and graph the solutions. Write the solutions in interval notation.

71. $x+4-3x\geq 2x+5$

72. $\frac{x}{5}-3.4>\frac{x}{2}+1.6$

73. $\frac{3x}{2}-3\leq 3+x$

74. a. Simplify the following expression: $\frac{3}{x}-\frac{5}{x+3}$

b. Solve the following equation: $\frac{3}{x}=\frac{5}{x+3}$

75. a. Simplify the following expression: $\frac{3x}{x+2}+\frac{2}{x-4}+3$

b. Solve the following equation: $\frac{3x}{x+2}+\frac{2}{x-4}=3$

76. A man can wax his car three times as fast as his daughter can. Together they can complete the job in 2 hours. How long would it take each of them working alone?

77. The illumination (in foot-candles, fc) of a light source varies inversely as the square of the distance from the source. If a certain light source provides an illumination of 10 fc at a distance of 20 ft., what is the illumination at a distance of 40 ft.?

78. If a hanging spring stretches 5 cm when a weight of 13 g is placed at its end, how far will the spring stretch if a weight of 20 g is placed at its end?

79. The distance an object falls (in feet) varies directly as the square of the time (in seconds) that it falls. If an object falls 16 feet in 1 sec., how far will it fall in 4 sec.?

80. A right triangle is formed when a ladder is leaned against a building. If the ladder is 26 ft. long and its bottom is placed 10 feet from the base of the building, how far up the building does the ladder reach?

81. The sum of the squares of two consecutive positive even integers is 580. What are the integers?

Review of Chapters 1 – 7

CHAPTER 8

Did You Know?

Throughout history, teachers of mathematics have tried to develop calculation methods that were easy to use or to memorize. One of the more interesting of these techniques is the Rule of Double False Position. As a student of algebra, it may seem strange to you that such a complicated method would be developed to solve a simple first-degree equation of the form $ax + b = 0$. But remember that you have modern symbolism at your disposal. To use the rule, we will make two guesses as to the solution of the equation. We shall designate the guesses as g_1 and g_2. Now we will let $e_1 = ag_1 + b$ and $e_2 = ag_2 + b$, where e_1 and e_2 represent the amount of error in our guesses. Then the solution is

$$x = \frac{e_1 g_2 - e_2 g_1}{e_1 - e_2}.$$

We now illustrate the Rule of Double False Position with an example: $3x + 6 = 0$. Suppose we guess that the solution is either 1 or 2. That is, let $g_1 = 1$ and $g_2 = 2$. Then $e_1 = 3(1) + 6 = 9$ and $e_2 = 3(2) + 6 = 12$, and the solution to the equation would be

$$\frac{9(2) - 12(1)}{9 - 12} = \frac{18 - 12}{-3} = \frac{6}{-3} = -2.$$

This unnecessarily complicated method of solving first-degree equations was taught until the nineteenth century. One of the most popular English texts of the sixteenth century, *The Grounde of Artes* by Robert Recorde (1510?–58), gives the Rule of Double False Position in verse:

Recorde

Gesse at this woorke as happe doth leade.
By chance to truthe you may procede.
And firste woorke by the question,
Although no truthe therein be don.
Suche falsehode is so good a grounde,
That truth by it will soone be founde.
From many bate to many mo,
From to fewe take to fewe also.
With to much ioyne to fewe againe,
To to fewe adde to manye plaine.
To crossewaies multiplye contrary kinde,
All truthe by falsehode for to fynde.

Students memorized the poem as a method of remembering the rule, which they generally did not understand. Can you figure out why making two false guesses can lead to the correct answer? After reviewing Chapters 1 – 7, you may be able to verify the Rule of Double False Position.

"But it should always be required that a mathematical subject not be considered exhausted until it has become intuitively evident".

Felix Klein (1849 – 1925)

Chapter 8 provides a review of the topics generally covered in a one semester beginning algebra course. Each section corresponds to a chapter in the text; and for more detailed discussions of particular topics, students should refer to the corresponding section number or page number listed in the left margin.

8.1 Chapter 1 Review: Real Numbers

Section 1.1 The Real Number Line and Absolute Value

Graph, p. 3
Coordinate, p. 3
Point, p. 3

Number lines are used to graph sets of numbers. The **graph** of a number is the point that corresponds to the number and the number is called the **coordinate** of the point. The graph of the set $A = \left\{-2.5, -1, 0, \frac{2}{3}, \sqrt{3}, \pi\right\}$ is shown in Figure 8.1.

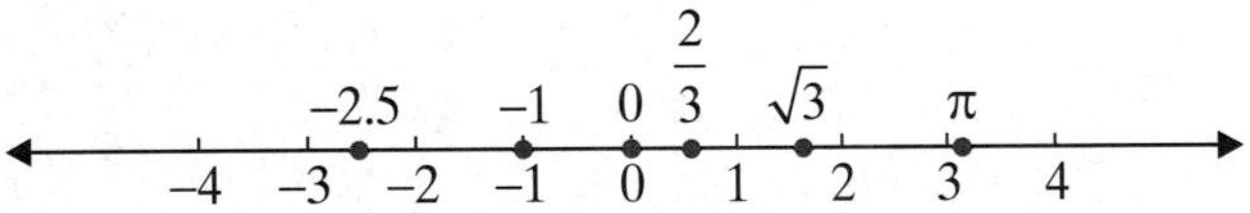

Figure 8.1

Various types of numbers are given names as shown below.

Types of Numbers

Relationships of real numbers, p. 6

Natural Numbers: $N = \{1, 2, 3, 4, 5, \ldots\}$

Whole Numbers: $W = \{0, 1, 2, 3, 4, 5, \ldots\}$

Integers: $Z = \{\ldots, -3, -2, -1, 0, 1, 2, 3, \ldots\}$

Rational Numbers: $Q = \left\{\begin{array}{l} \textit{numbers that can be written in the form } \frac{a}{b} \\ \textit{where } a \textit{ and } b \textit{ are integers and } b \neq 0 \end{array}\right\}$

In decimal form, rational numbers are either terminating decimals or infinite repeating decimals.

Irrational Numbers: {*Numbers that can be written as infinite nonrepeating decimals*}

Real Numbers: $R = \{\textit{All rational and irrational numbers}\}$

Real numbers, p. 5

All rational and irrational numbers are classified as **real numbers**, and number lines are called **real number lines**.

On a horizontal number line, smaller numbers are always to the left of larger numbers. The following table of symbols indicates ways in which numbers are compared.

Table of Symbols

Inequality symbols, p. 7

$=$	*is equal to*	$\neq$	*is not equal to*
$<$	*is less than*	$>$	*is greater than*
$\leq$	*is less than or equal to*	$\geq$	*is greater than or equal to*

Example 1: Inequalities

a. Determine whether each of the following statements is true or false.

$4 < -10$	False, since 4 is greater than –10 (or –10 is less than 4).
$6 \geq 6$	True, since 6 is equal to 6.
$-3.5 > -8.1$	True, since –3.5 is greater than –8.1 (or –8.1 is less than –3.5).

b. Graph the set of real numbers $\left\{-\frac{1}{2}, 0, 2, 3.1, 5\right\}$.

Solution:

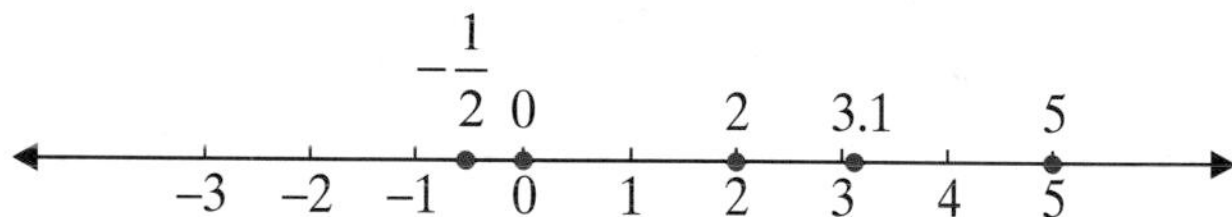

c. Graph the set of integers x, where $x < 4$.

Solution:

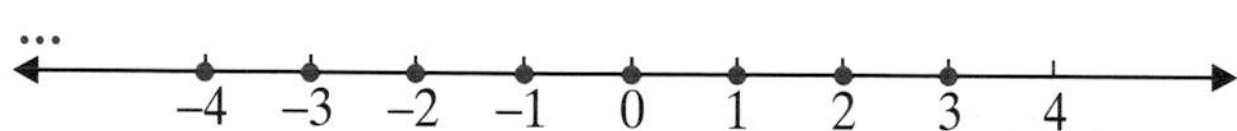

Absolute values, p. 9

The distance a number is from 0 on a number line is called its **absolute value** and is symbolized with two vertical lines as in $|x|$.

Example 2: Absolute Value

a. If $|x| = 6$, what are the possible values for x?

Solution: $x = 6$ or $x = -6$ since $|6| = 6$ and $|-6| = 6$.

continued on next page ...

b. If $|y| = -3.2$, what are the possible values for y?

Solution: There are no values of y for which $|y| = -3.2$. The absolute value is never negative.

c. If $|x| \le 5$, what are the possible integer values for x? Graph these integers on a real number line.

Solution: The integers must be within or equal to 5 units from 0. They are

$$\{-5, -4, -3, -2, -1, 0, 1, 2, 3, 4, 5\}$$

Sections 1.2 – 1.4

Operations with Integers

The formal statements of the rules for adding integers involve the use of absolute value. (**Note:** From past experience, you may have less formal ways of finding the sum of integers.)

Rules for Addition with Integers

Addition with Integers, Section 1.2

1. *To add two integers* ***with like signs****, add their absolute values and use the common sign.*

2. *To add two integers* ***with unlike signs****, subtract their absolute values (the smaller from the larger) and use the sign of the number with the larger absolute value.*

Example 3: Addition

Find each of the following sums.

Vertical addition, p. 17

a. $11 + (-7)$ **b.** $(-9) + (-12)$

c. $-23 + 8 + 2$ **d.** $68 + 72$

Solution: **a.** $11 + (-7) = 4$ **b.** $(-9) + (-12) = -21$

c. $-23 + 8 + 2 = -13$ **d.** $68 + 72 = 140$

Opposites, p. 3

To **subtract** numbers, the **opposite** of the number being subtracted is **added**. For example,

$$14 - (-2) = 14 + (+2) = 16 \text{ and } -10 - (+3) = -10 + (-3) = -13.$$

Additive inverse, p. 20

Note: The opposite of a number is also called its **additive inverse**. The sum of any number and its opposite is 0: symbolically $a + (-a) = 0$.

Subtraction

Subtraction with Integers, Section 1.3

For any integers a and b,

$$a - b = a + (-b)$$

In words: to ***subtract*** *any number, add its opposite.*

Example 4: Subtraction

Find the following differences.

a. $(-6) - (-5)$ **b.** $15 - (-2)$

c. $-12 - (10)$ **d.** $-13 - (-20)$

Solution: **a.** $(-6) - (-5) = -6 + (+5) = -1$ **b.** $15 - (-2) = 15 + (+2) = +17$

c. $-12 - (10) = -12 + (-10) = -22$ **d.** $-13 - (-20) = -13 + (+20) = +7$

(Remember that + signs are optional for positive answers, but negative signs must be used if a number is negative.)

If there are no parentheses in expressions involving signed numbers, the entire expression can be treated as addition of signed numbers.

Example 5: Signed Numbers

Find the value of each of the following expressions:

a. $8 - 5 - 3$ **b.** $-6 - 4 - (-3)$ **c.** $14 - 10 - 5 + 6$

Solution: **a.** $8 - 5 - 3 = 0$

b. $-6 - 4 - (-3) = -6 - 4 + (+3) = -10 + 3 = -7$

c. $14 - 10 - 5 + 6 = +5$

Example 11: Operations with Fractions

Simplify each of the following expressions by performing the indicated operations. Follow the Rules for Order of Operations and reduce all answers to lowest terms.

a. $\frac{3}{5}+\frac{4}{5}$ **b.** $\frac{7}{15}-\frac{2}{15}$ **c.** $\frac{3}{8}\cdot\frac{2}{5}+\frac{1}{15}$ **d.** $\frac{a}{2}+\frac{1}{3}+\frac{5}{6}$

Solution: **a.** $\frac{3}{5}+\frac{4}{5}=\frac{7}{5}$

b. $\frac{7}{15}-\frac{2}{15}=\frac{5}{15}=\frac{\cancel{5}\cdot 1}{3\cdot\cancel{5}}=\frac{1}{3}$

c. $\frac{3}{8}\cdot\frac{2}{5}+\frac{1}{15}=\frac{3}{\cancel{8}_4}\cdot\frac{\cancel{2}}{5}+\frac{1}{15}=\frac{3}{20}+\frac{1}{15}=\frac{3}{20}\cdot\frac{3}{3}+\frac{1}{15}\cdot\frac{4}{4}=\frac{9}{60}+\frac{4}{60}=\frac{13}{60}$

d. $\frac{a}{2}+\frac{1}{3}+\frac{5}{6}=\frac{3}{3}\cdot\frac{a}{2}+\frac{2}{2}\cdot\frac{1}{3}+\frac{5}{6}=\frac{3a+2+5}{6}=\frac{3a+7}{6}$

Section 1.8 Operations with Decimal Numbers

Place value, p. 61

Generally, decimal numbers are written with a decimal point and a place value system that indicate whole numbers to the left of the decimal point and fractions less than 1 to the right of the decimal point. Operating with decimal numbers involves the correct placement of the decimal point in the answer (sum, difference, product, or quotient).

Rules for Operating with Decimal Numbers

1. *To* ***add*** *(or* ***subtract****) two decimal numbers, align the decimal points in a vertical format and perform the addition (or subtraction). Place the decimal point in the answer in line with the other decimal points.*

2. *To* ***multiply*** *two decimal numbers, count the total number of places to the right of the decimal points in both numbers. Place the decimal point in the product so that the number of places to the right of the decimal point is this total.*

3. *To divide decimal numbers, move the decimal point in the divisor to the right to get a whole number. Move the decimal point in the dividend the same number of places. Divide as with whole numbers.*

Example 12: Decimal Numbers

Perform the indicated operations with decimal numbers.

a. $37.53 + 6.78 + 42.9$ **b.** $92.127 - 36.5$

c. $(5.35)(0.1)$ **d.** $7.44 \div 2.4$

Solution: **a.**

$$\begin{array}{r} 37.53 \\ 6.78 \\ +\ 42.90 \\ \hline 87.21 \end{array}$$

b.

$$\begin{array}{r} 92.127 \\ -\ 36.500 \\ \hline 55.627 \end{array}$$

c. $(5.35)(0.1) = 0.535$ **d.** $7.44 \div 2.4 = 3.1$

Example 13: Calculator

Use a calculator to change each of the following fractions into decimal form by dividing the numerator by the denominator. Write the quotients accurate to four decimal places.

Calculator commands, p. 68

a. $\frac{5}{8} = 0.625$ **b.** $\frac{6}{11} \approx 0.5455$ (rounded off to 4 places)

Section 1.9 Properties of Real Numbers

The various properties of real numbers under the operations of addition and multiplication are used throughout algebra and mathematics. They are summarized in the following table.

Properties of Real Numbers under the Operations of Addition and Multiplication *for any real numbers a, b, and c*		
Addition	**Property Name**	**Multiplication**
$a + b = b + a$	***Commutative***	$a \cdot b = b \cdot a$
$a + (b + c) = (a + b) + c$	***Associative***	$a \cdot (b \cdot c) = (a \cdot b) \cdot c$
$a + 0 = a$	***Identity***	$a \cdot 1 = a$
$a + (-a) = 0$	***Inverse***	$a \cdot \frac{1}{a} = 1 \ (a \neq 0)$
Zero Factor Property: $a \cdot 0 = 0 \cdot a = 0$		
Distributive Property: $a(b + c) = a \cdot b + a \cdot c$		

Example 14: State the Property

State the name of the real number property illustrated.

	Solution:
a. $15 + 4 = 4 + 15$	Commutative property of addition
b. $3(5 \cdot 8) = (3 \cdot 5) \cdot 8$	Associative property of multiplication
c. $5(x + 14) = 5x + 70$	Distributive property
d. $0(-5.73) = 0$	Zero factor property

8.1 Exercises

In Exercises 1 and 2, graph each set of numbers on a real number line.

1. $A = \left\{-5, -3.2, -0.3, \frac{1}{4}, 2\frac{1}{2}\right\}$ **2.** $B = \left\{-4.2, -3, -1.5, 0, 3\frac{1}{4}, 6\right\}$

3. Graph the set of integers that satisfy the following conditions.

a. $|x| = 7$ **b.** $|x| < 7$ **c.** $|x| > 7$

Determine whether each statement in Exercises 4 – 8 is true or false. If a statement is false, rewrite it in a form that is a true statement. (There may be more than one way to correct a statement.)

4. $|3.5| \le 3.5$ **5.** $|6| < -6$ **6.** $\frac{3}{5} > \frac{3}{4}$

7. $|-1.8| < -2$ **8.** $-|-5| \ge -5$

In Exercises 9 – 25, perform the indicated operations and simplify each answer.

9. $(-13) + 3$ **10.** $(-9) + (-6) + 5$ **11.** $-13 - 1 - 13$

12. $-15 - (-4)$ **13.** $36 - (-2)$ **14.** $-15.7 + (-13.3)$

15. $-\frac{1}{2} + \left(-\frac{1}{2}\right)$ **16.** $\frac{3}{16} - \left(-\frac{3}{8}\right)$ **17.** $-6.3 - 7.1 - 3.9$

18. $\frac{-54}{6}$ **19.** $\frac{78}{-39}$ **20.** $43(-3)$

21. $-6.1(-2)$ **22.** $3^2 \div (-9) + (4 - 2^2) - 10^2$ **23.** $5(13 - 15)^3 \cdot 8 \div 2$

24. $8 - 6[(-22) \div 11 \cdot 2 - (-5)]$ **25.** $10 \cdot 3 \div (2^2 - 5) + 14 - (-3)$

1. -5, -3.2, -0.3, $\frac{1}{4}$, $2\frac{1}{2}$

2. -4.2, -3, -1.5, 0, $3\frac{1}{4}$, 6

3.
a. -7, 7
b. -7, 0, 7
c.

4. True
5. False, $|6| > -6$
6. False, $\frac{3}{5} < \frac{3}{4}$
7. False, $|-1.8| > -2$
8. True
9. -10
10. -10
11. -27
12. -11
13. 38 **14.** -29
15. -1 **16.** $\frac{9}{16}$
17. -17.3
18. -9 **19.** -2
20. -129 **21.** 12.2 **22.** -101 **23.** -160 **24.** 2 **25.** -13

In Exercises 26 – 28, multiply or divide as indicated and reduce each answer to lowest terms.

26. $\frac{8x}{-40y} \cdot \frac{15x}{3y} \cdot \frac{2y^2}{x}$ **27.** $\frac{46}{7a} \div \frac{92a}{77}$ **28.** $\frac{25m}{21n^2} \div \frac{40mn}{9m^2} \cdot \frac{14n}{12}$

29. Find the prime factorization of each of the following numbers.
a. 180 **b.** 270 **c.** 10,000

30. Find the LCM (least common multiple) of each of the following sets of numbers.
a. 16, 20, 45 **b.** 25, 30, 36 **c.** 20, 70, 90, 200

31. A study shows that $\frac{3}{5}$ of the students at a certain school are women. If the school has an enrollment of 1200 students, how many are women? How many are men?

32. An airplane is carrying 120 passengers. This is $\frac{3}{4}$ of the capacity of the airplane.
a. Is the capacity of the airplane more or less than 120?
b. If you were to multiply 120 by $\frac{3}{4}$, would the product be more or less than 120?
c. What is the capacity of the airplane?

33. If the product of $\frac{5}{8}$ with another number is $\frac{9}{10}$, what is the other number?

In Exercises 34 – 45, perform the indicated operations. Reduce each answer to lowest terms.

34. $\frac{2}{15} - \frac{7}{15}$ **35.** $\frac{5}{11} + \frac{2}{11} + \frac{3}{11}$ **36.** $\frac{5}{12} + \frac{7}{15}$ **37.** $\frac{3}{14} - \frac{2}{21}$

38. $\frac{y}{3} + \frac{3}{4}$ **39.** $\frac{8}{9} - \frac{2}{x}$ **40.** $\frac{a}{2} - \frac{1}{3} - \frac{1}{6}$ **41.** $\frac{1}{8} - \frac{10}{x}$

42. $\frac{3}{8} \cdot \frac{2}{5} + \frac{4}{15}$ **43.** $\frac{5}{4a} \div \frac{5}{16a} - \frac{2b}{3} \cdot \frac{3}{5b}$ **44.** $\left(\frac{1}{6a} - \frac{2}{3a}\right) \div \left(\frac{5}{5b} + \frac{3}{3b}\right)$

45. $\left(\frac{1}{2x} - \frac{1}{5x}\right) \div \left(\frac{3}{7y} + \frac{1}{3y}\right)$

46. Write each of the following decimal numbers in words.
a. 65.3 **b.** 4.6175 **c.** 4500.0045

47. Round off each decimal number to the nearest hundredth.
a. 17.8546 **b.** 920.0067 **c.** 0.5789

48. Change each of the following fractions or mixed numbers to decimal form.
a. $\frac{8}{100}$ **b.** $3\frac{57}{1000}$ **c.** $\frac{3}{7}$

Answers:

26. $-2x$ **27.** $\frac{11}{2a^2}$ **28.** $\frac{5m^2}{16n^2}$
29. a. $2^2 \cdot 3^2 \cdot 5$ **b.** $2 \cdot 3^3 \cdot 5$ **c.** $2^4 \cdot 5^4$
30. a. 720 **b.** 900 **c.** 12600
31. 720 women, 480 men
32. a. more than 120 **b.** less than 120 **c.** 160
33. $\frac{36}{25}$ **34.** $\frac{-1}{3}$
35. $\frac{10}{11}$ **36.** $\frac{53}{60}$
37. $\frac{5}{42}$ **38.** $\frac{4y+9}{12}$
39. $\frac{8x-18}{9x}$
40. $\frac{a-1}{2}$
41. $\frac{x-80}{8x}$
42. $\frac{5}{12}$ **43.** $\frac{18}{5}$
44. $\frac{-b}{4a}$ **45.** $\frac{63y}{160x}$
46. a. Sixty-five and three tenths **b.** Four and six thousand one hundred seventy-five ten thousandths **c.** Four thousand five hundred and forty-five ten thousandths
47. a. 17.85 **b.** 920.01 **c.** 0.58
48. a. 0.08 **b.** 3.057 **c.** 0.42857

49. Jorge wanted to lose weight and went on a 6 week diet plan. He lost 5 lbs. during the first week, gained 1 lb. the second week, lost 4 lbs. the third week, lost 2 lbs. the fourth week, gained 3 lbs. the fifth week, and lost 5 lbs. the sixth week. If he weighed 195 lbs. when he started his diet, what did he weigh at the end of the six week period?

50. Ms. Mariani knew that the balance in her checking account was $2300. She made deposits of $560, $3000, and $575. She wrote checks for $650, $800, and $225. What was her new balance?

In Exercises 51 – 60, perform the indicated operations.

51. $244.5 + 63.22 + 8.34$
52. $29.42 + 17.2 + 73.6$
53. $67.8 - 33.9$
54. $189.73 - 97.86$
55. $-17.2 + 14.5$
56. $-83.57 - 14.2$
57. $(16.1)(5.3)$
58. $(-3.1)(-15.4)$
59. $5.1\overline{)82.11}$
60. $0.023\overline{)0.5152}$

In Exercises 61 – 68, use a calculator to find the values accurate to four decimal places.

61. $(3.5)^6$
62. $(-2.5)^5$
63. $78.956 \div 3.27$
64. $0.93 \div 15.61$
65. $\dfrac{93.56+7.4}{6.52}$
66. $\dfrac{7.15-14.68}{3.4+5.7}$
67. $\dfrac{-5-18}{-7.9-1.3}$
68. $(1.3)^5 + 17.5(16.81)$

In Exercises 69 – 76, name the property of real numbers illustrated.

69. $5 \cdot y = y \cdot 5$
70. $9 + (6 + x) = (9 + 6) + x$
71. $35 + 0 = 35$
72. $35 \cdot 1 = 35$
73. $6(x + y) = 6x + 6y$
74. $5 \cdot \dfrac{1}{5} = 1$
75. $3(xy) = 3x(y)$
76. $-14 + (+14) = 0$

77. Given the set of numbers $A = \left\{-2.48, -\dfrac{1}{3}, 0, 2.3, \pi, 4, \sqrt{7}, \dfrac{13}{5}\right\}$ tell which of the numbers are (a) integers, (b) rational numbers, (c) irrational numbers, and (d) real numbers.

78. Explain, in your own words, why division by 0 is not possible under any circumstances.

79. Use your calculator to change the decimal 0.74 to fraction form, reduced.

80. Use your calculator to change the sum 5.6 + 3.8 to fraction form, reduced.

49. 183 lbs. **50.** $4760
51. 316.06 **52.** 120.22
53. 33.9 **54.** 91.87
55. – 2.7 **56.** –97.77
57. 85.33 **58.** 47.74
59. 16.1 **60.** 22.4
61. 1838.2656
62. – 97.6563
63. 24.1456
64. 0.0596
65. 15.4847
66. – 0.8275
67. 2.5 **68.** 297.8879
69. Commutative of Multiplication
70. Associative of Addition
71. Identity of Addition
72. Identity of Multiplication
73. Distributive Property
74. Inverse Property of Multiplication
75. Associative Property of Multiplication
76. Inverse Property of Addition
77. a. $0, 4$
b. $-2.48, \dfrac{13}{5}, \dfrac{-1}{3}, 2.3, 0, 4$
c. $\pi, \sqrt{7}$
d. $-2.48, -\dfrac{1}{3}, 0, 2.3, \pi, 4, \sqrt{7}, \dfrac{13}{5}$
78. Answers will vary.
79. $\dfrac{37}{50}$ **80.** $\dfrac{47}{5}$

8.2 Chapter 2 Review: Algebraic Expressions, Linear Equations, and Applications

Section 2.1 Simplifying and Evaluating Algebraic Expressions

Any constant or variable or the indicated product and/or quotient of constants and powers of variables is called a **term**. Examples of terms are

$$-17, \quad 4x, \quad -7.3, \quad 2.5xy, \quad 13x^3y^2, \quad \text{and} \quad \frac{y^2}{3x}.$$

Coefficient, p. 92

In the term $6x^3$, the constant 6 is called the **numerical coefficient** of x^3 (or simply the **coefficient** of x^3).

If no number is written next to a variable, then the coefficient of the variable is understood to be 1. For example,

$$x = 1 \cdot x, \quad a^4 = 1 \cdot a^4, \text{ and } xy^2 = 1 \cdot xy^2.$$

A negative sign written next to a variable indicates a coefficient of –1. For example,

$$-x = -1 \cdot x, \quad -a^4 = -1 \cdot a^4, \text{ and } -xy^2 = -1 \cdot xy^2.$$

Like terms, p. 92
Algebraic expressions, p. 93

Like terms (or similar terms) are terms that are constants or terms that contain the same variables (if any) raised to the same powers. **Algebraic expressions** that indicate operations with terms such as

$$3x + 4y, \quad 15n - 2n + m^2, \text{ and } \frac{2x + 8x}{5}$$

are not terms. However, we would like to simplify such expressions if they contain like terms. That is, we want to **combine like terms**. By applying the distributive property, we can combine like terms by adding or subtracting the coefficients of like terms. For example,

$$6x + 8x = (6 + 8)x = 14x$$

and $$5x^2 - 12x^2 = (5 - 12)x^2 = -7x^2.$$

Example 1: Combine Like Terms

Simplify each expression by combining like terms whenever possible.

a. $3ab + 6ab - a - 5a$

b. $3x + 5 - 7x + x$

Solution: **a.** $3ab + 6ab - a - 5a = 9ab - 6a$

b. $3x + 5 - 7x + x = 5 - 3x$ (or $-3x + 5$)

Evaluating algebraic expressions, p. 95

In most cases, if an expression is to be evaluated, like terms should be combined first and then the resulting expression evaluated by following the Rules for Order of Operations. **Remember to use parentheses around negative numbers when substituting.**

Example 2: Evalute Algebraic Expressions

Simplify each expression by combining like terms. Then evaluate the resulting expression for $x = -2$, $y = 3$, and $a = 5$.

a. $3x^2 + 2x - 5 + x^2 - x + 10$ **b.** $4y^3 - 7a + 3y^3 - a^2 + 3a$

Solution: **a.** Simplify first: $3x^2 + 2x - 5 + x^2 - x + 10 = 4x^2 + x + 5$

Now evaluate: $4x^2 + x + 5 = 4(-2)^2 + (-2) + 5 = 16 - 2 + 5 = 19$

b. Simplify first: $4y^3 - 7a + 3y^3 - a^2 + 3a = 7y^3 - a^2 - 4a$

Now evaluate: $7y^3 - a^2 - 4a = 7(3)^3 - 5^2 - 4 \cdot 5 = 189 - 25 - 20 = 144$

Section 2.2 Translating English Phrases and Algebraic Expressions

Key words, p. 101

Key words, such as those in the following table, indicate operations with numbers and variables. Their meanings are used when translating English phrases into algebraic expressions, and vice versa.

Addition	Subtraction	Multiplication	Division
add	subtract (from)	multiply	divide
sum	difference	product	quotient
plus	minus	times	
more than	less than	twice	
increased by	decreased by	of (with fractions and percent)	
	less		

Example 3: Algebraic Expression to Phrase

Write an English phrase that indicates the meaning of each algebraic expression.

a. $15 - x$ **b.** $2(n - 4)$ **c.** $6a + 3a$

Solution: **a.** $15 - x$: 15 decreased by a number
b. $2(n - 4)$: twice the difference between a number and 4
c. $6a + 3a$: 6 times a number plus 3 times the same number

Example 4: Phrase to Algebraic Expression

Change each phrase into an equivalent algebraic expression.

a. the **quotient** of a number and 7
b. 8 **more than** 3 **times** a number
c. 5 **times** the **sum** of a number and 3

Solution: **a.** $\frac{n}{7}$ **b.** $3x + 8$ **c.** $5(y + 3)$

Sections 2.3 and 2.4 Solving Linear Equations

An **equation** is a statement that two algebraic expressions are equal. That is, both expressions represent the same number. If an equation contains a variable, the value (or set of values) that gives a true statement when substituted for the variable is called the **solution** (or **solution set**) to the equation. The process of finding the solution is called **solving the equation**. A **linear equation** (or **first-degree equation**) is an equation of the form $\boldsymbol{ax + b = c}$ where a, b, and c are constants and $a \neq 0$. To solve linear equations, we apply the Addition Property and the Multiplication Property.

Linear equation, p. 106

The Addition and Multiplication Properties

Addition property, p. 107

1. ***The Addition Property:***
   If A, B, and C are algebraic expressions,
   then the equations $A = B$ and $A + C = B + C$ are equivalent.

Multiplication property, p. 109

2. ***The Multiplication Property:***
   If A and B are algebraic expressions and C is any nonzero constant,
   then the equations $A = B$ and $A \cdot C = B \cdot C$ are equivalent.

The objective in solving a linear equation is to get the variable by itself on one side of the equation. The following procedure should be used to solve a linear equation.

Procedure for Solving an Equation That Simplifies to the Form ax = c

General procedure for solving linear equations, p. 119

1. *Combine any like terms on each side of the equation.*
2. *Use the Multiplication Property of Equality and multiply both sides of the equation by the reciprocal of the coefficient of the variable. (**Note:** This is the same as dividing both sides of the equation by the coefficient.) Thus, the coefficient of the variable will become +1.*
3. *Check your answer by substituting it into the original equation.*

Example 5: Solving Equations

Solve each of the following linear equations.

a. $5x + 14 = 29$ **b.** $11.6x - 17.5 = -12.86$

c. $\frac{3n}{14}+\frac{1}{2}=\frac{n}{14}+\frac{11}{14}$ **d.** $2(y-6) = 4(y+2) - 36$

Solution: a.

$$5x+14=29$$
$$5x+14-14=29-14$$
$$5x=15$$
$$\frac{\cancel{5}x}{\cancel{5}}=\frac{15}{5}$$
$$x=3$$

b.

$$11.6x-17.5=-12.86$$
$$11.6x-17.5+17.5=-12.86+17.5$$
$$11.6x=4.64$$
$$\frac{\cancel{11.6}x}{\cancel{11.6}}=\frac{4.64}{11.6}$$
$$x=0.4$$

c.

$$14\cdot\left(\frac{3n}{14}+\frac{1}{2}\right)=14\cdot\left(\frac{n}{14}+\frac{11}{14}\right)$$
$$14\cdot\frac{3n}{14}+14\cdot\frac{1}{2}=14\cdot\frac{n}{14}+14\cdot\frac{11}{14}$$
$$3n+7=n+11$$
$$3n+7-7=n+11-7$$
$$3n=n+4$$
$$3n-n=n+4-n$$
$$2n=4$$
$$\frac{\cancel{2}n}{\cancel{2}}=\frac{4}{2}$$
$$n=2$$

d.

$$2(y-6)=4(y+2)-36$$
$$2y-12=4y+8-36$$
$$2y-12=4y-28$$
$$2y-12+12=4y-28+12$$
$$2y=4y-16$$
$$2y-4y=4y-16-4y$$
$$-2y=-16$$
$$\frac{\cancel{-2}y}{\cancel{-2}}=\frac{-16}{-2}$$
$$y=8$$

Example 6: Absolute Value Equations

Absolute value equations, p. 122

Solve each of the following absolute value equations.

a. $|3x-7|=8$ **b.** $\left|\frac{2}{3}n+14\right|=0$

Solution: a. $|3x-7|=8$

Absolute value, p. 9

Using the definition of absolute value, we solve two linear equations:

$$3x-7=8 \qquad 3x-7=-8$$
$$3x-7+7=8+7 \qquad 3x-7+7=-8+7$$
$$3x=15 \qquad 3x=-1$$
$$\frac{\cancel{3}x}{\cancel{3}}=\frac{15}{3} \qquad \frac{\cancel{3}x}{\cancel{3}}=\frac{-1}{3}$$
$$x=5 \qquad x=-\frac{1}{3}$$

The two solutions are $x = 5$ and $x = -\frac{1}{3}$.

b. $\left|\frac{2}{3}n + 14\right| = 0$

Using the definition of absolute value, we solve one linear equation:

$$\frac{2}{3}n + 14 = 0$$

$$\frac{2}{3}n + 14 - 14 = 0 - 14$$

$$\frac{2}{3}n = -14$$

$$\frac{3}{2} \cdot \frac{2}{3}n = -14 \cdot \frac{3}{2}$$

$$n = -21$$

The solution is $n = -21$.

Sections 2.5 and 2.6

Applications

Pòlya's four-step process emphasizes the importance of organization in the problem solving procedure.

Pòlya's Four-Step Process for Solving Problems

1. *Understand the problem. (Read the problem carefully and be sure that you understand all the terms used.)*
2. *Devise a plan. (Set up an equation or a table or chart relating the information.)*
3. *Carry out the plan. (Perform any operations indicated in Step 2.)*
4. *Look back over the results. (Ask yourself if the answer seems reasonable and if you could solve similar problems in the future.)*

The types of problems discussed in Chapter 2 relate to geometry, number problems, consecutive integers, and percent. Four examples are given here for review.

Example 7: Geometry

The three sides of a triangle are x, $2x$, and $x + 5$ as shown in the figure.
If the perimeter of the triangle is 45 inches, what are the lengths of the sides?

Solution:

$$x+2x+x+5=45$$
$$4x+5=45$$
$$4x=40$$
$$x=10$$

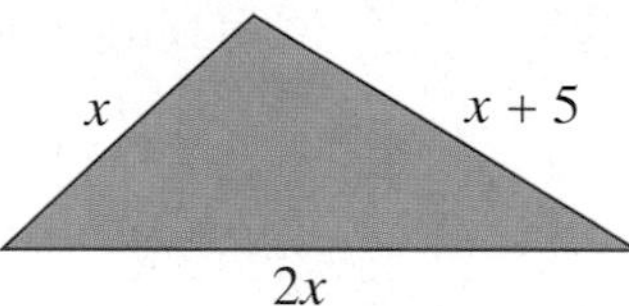

The lengths of the sides are 10 in., 20 in., and 15 in.

Example 8: Number Problems

Number problems, p. 129

Three times the sum of a number and 7 is equal to twice the number decreased by 42. What is the number?

Solution: Let n = the number.
Then,

$$3(n+7)=2n-42$$
$$3n+21=2n-42$$
$$3n=2n-63$$
$$n=-63$$

The number is -63.

Example 9: Consecutive Integers

Consecutive integers, p. 131

Find three consecutive even integers such that twice the sum of the first and second is 60 more than the third.

Solution: Let n = first integer
$n+2$ = second integer
$n+4$ = third integer
Then,

$$2(n+n+2)=n+4+60$$
$$2(2n+2)=n+64$$
$$4n+4=n+64$$
$$3n=60$$
$$n=20$$

The three consecutive even integers are: 20, 22, and 24.

Example 10: Percent

Percent, Section 2.6

A DVD player was marked on sale for $135.00. This was a discount of 20% off the original price. What was the original price?

Solution: Let $x =$ the original price.
Then, because the discount was 20%, the new price is 80% of the original price.

This gives,

$$0.80x = 135.00$$

$$x = 168.75$$

The original price was $168.75.

8.2 Exercises

1. $12x - 15y$
2. $-2a + 12b$
3. $4x - 4$
4. 19 **5.** $x^2 + 3x$
6. $17y^3 + 2y^2 + y - 6$
7. $2y - 20$ **8.** 7
9. $6a^3 + 10ab - 8a$
10. $13a^2b - 2ab - 4a$
11. 10 **12.** 92
13. 99.9 **14.** 0.1
15. $\frac{53}{8}$ **16.** $-\frac{86}{15}$
17. $-\frac{25}{12}$ **18.** 25.6
19. $\frac{145}{12}$ **20.** 1805
21. $5 + 2x$ **22.** $-3 + 7x$
23. $-6 + x$ **24.** $3x - 4$
25. $2(x + 5)$ **26.** $8x + 10$

In Exercises 1 – 10, simplify each expression by combining like terms.

1. $7x + 5x - 14y - y$ **2.** $-3a + 14b + a - 2b$ **3.** $-10 + 5x - x + 6$

4. $15 + 3y + y + 4 - 4y$ **5.** $6x^2 - 4x^2 + 3x - x^2$ **6.** $17y^3 + 2y^2 + y - 6$

7. $-4(y + 5) + 6y$ **8.** $\frac{4(5x - x)}{8} - 2x + 7$

9. $6ab + 7a^3 - a^3 + 4ab - 9a + a$ **10.** $3a^2b - 5ab + 9a^2b + a^2b + 3ab - 4a$

In Exercises 11 – 20, simplify each expression and then evaluate the result for x = –2 and y = 5.

11. $5x^2 - 8x^2 + 15 - 2x + 3$ **12.** $-y^3 + 5y^2 + 13y - 3y + 42$

13. $1.4y^2 + 2.5y^2 - 3.2 + 5.6$ **14.** $1.6x^2 + 2.3x - 7.5 + 1.8x^2 - 1.4$

15. $\frac{3}{4}x^2 - \frac{1}{2}x - x + \frac{5}{8}$ **16.** $\frac{7}{8}x^3 + \frac{1}{6}x - \frac{1}{2}x + \frac{3}{5}$

17. $\frac{5x}{6} + \frac{2x}{3} - \frac{x}{12} + \frac{3}{4}$ **18.** $0.18y^2 + y^2 - 1.3y + 2y - 7.4$

19. $\frac{3y}{9} + \frac{2y}{3} - \frac{y}{12} + \frac{3y}{2}$ **20.** $16y^3 - 8y + 42y^2 - 10y^3 + 9y$

In Exercises 21 – 30, translate each English phrase into an equivalent algebraic expression.

21. 5 more than twice a number **22.** 3 less than 7 times a number

23. 6 less than a number **24.** 3 times a number decreased by 4

25. twice the sum of a number and 5 **26.** the product of a number and 8 increased by 10

27. $3x - 20$ 28. $20 - 3x$
29. $\frac{x}{5}+x$ 30. $\frac{x}{7}+15$
31. 7 more than 8 times a number
32. 1.5 times a number decreased by 8
33. 5 times the sum of a number and 10
34. 4 times the difference of a number and 13
35. Sum of 6 times a number and 7 times the same number
36. Product of $\frac{1}{2}$ and a number increased by $\frac{3}{4}$
37. 5 less than the quotient of a number and 14
38. 3 less than the quotient of 16 and a number
39. 11 more than the product of –9 and a number
40. –2 times the sum of a number and 20
41. 0.21 (rounded to 2 decimals)
42. –2 43. 7
44. $\frac{17}{3}$ 45. $-\frac{3}{2}$
46. –14 47. –19
48. 1 49. –4

27. 20 less than three times a number

28. 20 less three times a number

29. the quotient of a number and 5 plus the number

30. 15 more than the quotient of a number and 7

In Exercises 31 – 40, translate each algebraic expression into an English phrase.

31. $7 + 8x$ **32.** $1.5x - 8$ **33.** $5(n + 10)$ **34.** $4(n - 13)$

35. $6a + 7a$ **36.** $\frac{1}{2}x+\frac{3}{4}$ **37.** $\frac{y}{14}-5$ **38.** $3-\frac{16}{n}$

39. $-9x + 11$ **40.** $-2(x + 20)$

Solve each of the equations in Exercises 41 – 60.

41. $3.1x = 0.64$ **42.** $4x + 3 = -5$ **43.** $17 = 5x - 18$

44. $-3 = 3y - 20$ **45.** $14 = -2y + 11$ **46.** $-16 = 4 - (6 - x)$

47. $5 - (x + 3) = 21$ **48.** $2x = -5x + 7$

49. $2y - 6 = 5y + 2 - y$ **50.** $n + 8 - 2n + 10 = 7n + 14$

51. $\frac{2}{3}y+\frac{1}{2}y=\frac{3}{4}y$ **52.** $1+\frac{1}{5}y=\frac{2}{15}y+\frac{3}{2}$

53. $4.6 + 0.6x = 1.4 - 0.2x$ **54.** $-2(5x + 1.2) - 0.5 = -6(x - 2)$

55. $2 + 3(4 - a) = 1.5 - 2(a + 3)$ **56.** $3(a + 0.4) - 0.2a = 1.8a + 2.7$

57. $|5a + 4| = 11$ **58.** $|-4x + 3| = 17$

59. $\left|\frac{1}{5}n-15\right|=0$ **60.** $|3y - 6| = -10$

Find the missing number or percent in Exercises 61 – 70.

61. 15% of 80 is _____. **62.** 125% of 90 is _____.

63. What is 72% of 100? **64.** What percent of 180 is 22.5?

65. 86% of what number is 129? **66.** 92% of what number is 69.92?

67. What percent of 64 is 40? **68.** Find 24% of 344.

69. Find 150% of 32. **70.** What percent of 50 is 75?

71. If the difference between $\frac{3}{4}$ and $\frac{7}{8}$ is divided by the sum of $\frac{2}{3}$ and $\frac{7}{15}$, what is the quotient?

72. From the sum of –10 and –13 subtract the product of –3 and –15.

73. You need to borrow \$5000 (fast) and your aunt is the only person who has the money and loves you dearly. But, she wants you to pay 6.5% interest and pay her back in 9 months. How much will you pay to her at the end of the 9 months?

74. Computers are on sale at a 35% discount. Sales tax is at 8.25%. What total amount will you pay for a computer that is originally priced at \$750?

75. Which is the better investment: (a) a profit of $700 on an investment of $10,000 or (b) a profit of $1200 on an investment of $15,000? Explain in terms of percent.

76. Find three consecutive odd integers such that the sum of the first and twice the second is equal to 15 more than twice the third.

77. Find four consecutive integers such that the sum of the first, second, and fourth is 64 more than the third.

78. If twice a certain number is decreased by 25, the result is 18 less than three times the number. What is the number?

79. The triangle shown here indicates that the sides can be represented by x, $x + 3$, and $2x - 1$. What is the length of each side if the perimeter is 34 feet?

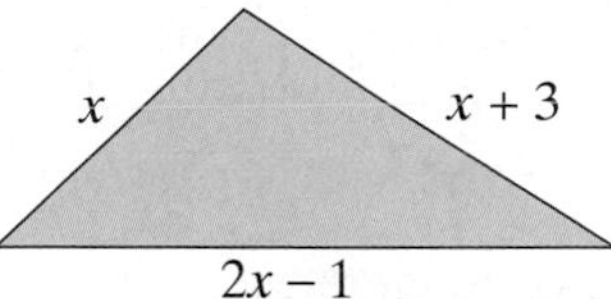

80. The perimeter of a rectangle is 210 meters. If the length is 15 more than twice the width, what are the dimensions of the rectangle?

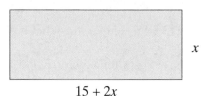

50. $\frac{1}{2}$ **51.** 0
52. $\frac{15}{2}$ **53.** –4
54. –3.725 **55.** 18.5
56. 1.5
57. $a = -3, \frac{7}{5}$
58. $x = 5, x = -\frac{7}{2}$
59. $n = 75$
60. No solution
61. 12
62. 112.5 **63.** 72
64. 12.5% **65.** 150
66. 76 **67.** 62.5%
68. 82.56 **69.** 48
70. 150% **71.** $\frac{-15}{136}$
72. –68
73. $5243.75
74. $527.72
75. a. 7%, **b.** 8%, b is better. 8% is better than 7%, so a profit of $1200 on an investment of $15,000 is better
76. 19, 21, 23
77. 31, 32, 33, 34
78. –7
79. 8, 11, 15
80. width = 30m, length = 75m

8.3 Chapter 3 Review: Formulas, Applications, and Linear Inequalities

Section 3.1 Working with Formulas

Formulas, p. 160

Formulas are general rules or principles stated mathematically. In working with formulas, we need to be aware that the variables generally stand for a specific type of quantity and capital letters and lowercase letters can have different meanings. For example, in the formula $d = rt$ we know that d is distance, r is rate, and t is time. In the formula $F = ma$, F represents force, m stands for mass, and a indicates acceleration.

Sometimes we simply want to evaluate a formula for a particular variable given data for other variables. In other cases, where a formula is solved for one of the variables, we want to solve the formula for one of the other variables. In these cases, we solve by treating the variables as constants in solving a linear equation.

Example 1: Evaluating Formulas

Evaluating formulas, p. 161

$C = \frac{5}{9}(F - 32)$ is the formula for changing from Fahrenheit to Celsius temperature.

Find the Celsius temperature that is the same as 50° F.

Solution: Substituting 50 for F and evaluating for C gives:

$$C = \frac{5}{9}(50 - 32) = \frac{5}{9}(18) = 10$$

Thus, 50° F = 10° C.

Example 2: Solving for Different Variables

Solving for different variables, p. 163

Solve each formula for the indicated variable.

a. $P = 2l + 2w$: solve for w.
b. $2x + 5y = 10$: solve for y.

Solution: **a.**

$$P = 2l + 2w$$
$$P - 2l = 2w$$
$$\frac{P - 2l}{2} = w$$

b.

$$2x + 5y = 10$$
$$5y = 10 - 2x$$
$$y = \frac{10 - 2x}{5}$$

Section 3.2 Formulas in Geometry

The formulas in geometry discussed in this text deal with perimeter, area, and volume. For formulas, such as the circumference and area of a circle, that involve π we use $\pi \approx 3.14$. Some of the common geometric formulas are listed here for easy reference. Other formulas may be found in Section 3.2.

Perimeter, p. 171

Perimeter or Circumference

$P = 4s$	Perimeter of a square
$P = 2l + 2w$	Perimeter of a rectangle
$C = 2\pi r$ or $C = \pi d$	Circumference of a circle
$P = a + b + c$	Perimeter of a triangle

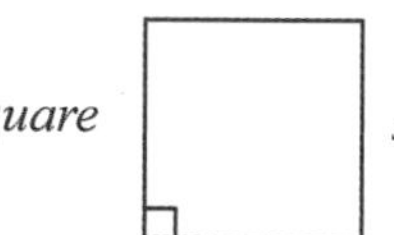

Rectangle
w
l

Area, p. 174

Area

$A = s^2$	Area of a square
$A = lw$	Area of a rectangle
$A = \pi r^2$	Area of a circle
$A = \frac{1}{2}bh$	Area of a triangle

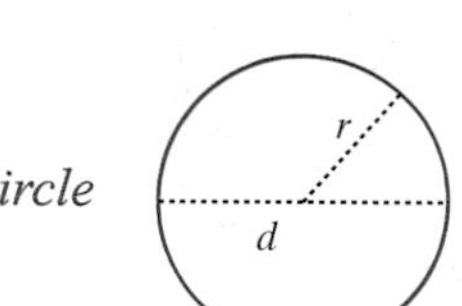

Volume, p. 176

Volume

$V = lwh$	Volume of a rectangular solid
$V = \pi r^2 h$	Volume of a cylinder
$V = \frac{4}{3}\pi r^3$	Volume of a sphere
$V = \frac{1}{3}\pi r^2 h$	Volume of a right circular cone

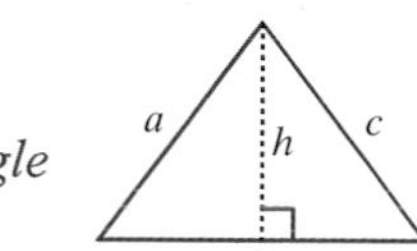

Example 3: Using Formulas

a. Find the circumference of a circle with diameter 10 cm.

b. Find the area of a triangle with base 20 in. and height 15 in.

c. Find the volume of a cylinder with radius 4 m and height 10 m.

Solution: **a.** Use the formula $C = \pi d$: $C \approx 3.14(10) = 31.4$ cm

b. Use the formula $A = \frac{1}{2}bh$: $A = \frac{1}{2} \cdot 20 \cdot 15 = 150 \text{ in.}^2$

c. Use the formula $V = \pi r^2 h$: $V \approx 3.14 \cdot 4^2 \cdot 10 = 502.4 \text{ m}^3$

Section 3.3 Applications

Algebra is an important tool used to solve many types of word problems. If you learn to think "algebraically," you will be able to analyze and solve problems that would otherwise seem very difficult or even impossible to you. Regardless of the type of word problem, four steps that are always useful are the following:

1. Understand the problem. (Analyze the wording and the vocabulary.)
2. Devise a plan. (Possibly set up an equation or a table or a diagram.)
3. Carry out the plan. (Solve the equation.)
4. Look back over the results. (See if the results are reasonable and consider the possibility of another approach.)

Example 4: Percent

a. Ms. Jones invested \$20,000, part at 5% and part at 6.5%. In one year, the interest from the 6.5% investment exceeds the interest from the 5% investment by \$380. How much did she invest at each rate?

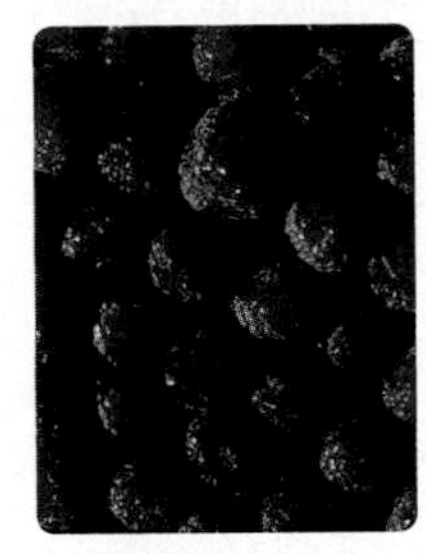

b. A farmer grows raspberries. They cost him \$0.90 a basket to produce. He is able to sell only 87% of those he produces. If he sells his raspberries at \$2.50 a basket, how many must he produce to make a profit of \$8287.50? How many did he actually sell?

Solution: **a.** Let x = amount invested at 6.5% = 0.065
then $20{,}000 - x$ = amount invested at 5% = 0.05.
This gives the following equation:

$$\begin{aligned} 0.065x - 0.05(20{,}000 - x) &= 380 \\ 0.065x - 1000 + 0.05x &= 380 \\ 0.115x &= 1380 \\ x &= 12000 \\ 20{,}000 - x &= 8000 \end{aligned}$$

She invested \$12,000 at 6.5% and \$8000 at 5%.

b. Use the formula: Revenue – Cost = Profit.
Let x = amount produced,
then $0.87x$ = amount actually sold.

$$\begin{aligned} 2.50(0.87x) - 0.90x &= 8287.50 \\ 2.175x - 0.90x &= 8287.50 \\ 1.275x &= 8287.50 \\ x &= 6500 \\ 0.87x = 0.87(6500) &= 5655 \end{aligned}$$

He produced 6500 baskets of raspberries and sold 5655 baskets.

Section 3.4 Ratios and Proportions

Ratio, p. 196

A **ratio** is a comparison of two quantities by division. The **ratio of *a* to *b*** can be written in fraction form $\frac{a}{b}$. The numerator and denominator should be labeled. For example, the ratio of 5 hits to 20 times at bat for a baseball player can be written as

$$\frac{5 \text{ hits}}{20 \text{ at bats}} \text{ and reduced to } \frac{1 \text{ hit}}{4 \text{ at bats}}.$$

Proportion, p. 198

A **proportion** is a statement that two ratios are equal. In symbols $\frac{a}{b} = \frac{c}{d}$ is a proportion.

True proportion, p. 199

In this form, a and d are called the **extremes** and b and c are called the **means**. **In a true proportion, the product of the extremes is equal to the product of the means**. This property allows us to test whether a proportion is true or false and allows us to solve for an unknown term.

Example 5: Proportions

a. Determine whether the proportion $\frac{2.5}{6} = \frac{3.5}{8}$ is true or false.

Solving for the unknown term, p. 200

b. Find the value of A in the proportion $\frac{3.5}{A} = \frac{2.5}{6}$.

Solution: **a.** The product of the extremes is $8 \cdot 2.5 = 20.0$.
The product of the means is $6 \cdot 3.5 = 21.0$.
Since these two products are not equal, the proportion is **false**.

b.
$$\frac{3.5}{A} = \frac{2.5}{6}$$
$$6 \cdot 3.5 = 2.5A$$
$$\frac{21.0}{2.5} = A$$
$$8.4 = A$$

Section 3.5 Linear Inequalities

Linear inequality, p. 213

Just as an equation of the form $ax + b = c$ is called a **linear equation**, an inequality of the form $ax + b < c$ is called a **linear inequality**. (**Note:** A linear inequality may have the symbol $\leq$, $>$, or $\geq$ instead of $<$.) To solve a linear inequality, perform the following procedures.

To Solve a Linear Inequality

1. *Simplify each side of the inequality by removing any grouping symbols and combining like terms.*
2. *Add the opposites of constants and/or variable expressions to both sides so that variables are on one side and constants are on the other.*
3. *Divide both sides by the coefficient of the variable and*
   a. leave the direction of the inequality unchanged if the coefficient is positive; or
   b. reverse the direction of the inequality if the coefficient is negative.

Also, if an inequality has three parts, it is essentially two inequalities in one expression. Isolate the variable by performing the same operations to each part of the inequality.

Example 6: Solving Linear Inequalities

Types of intervals, p. 212

Solve each of the following linear inequalities. Graph the solution set on a number line and write the answer in interval notation.

a. $4x+5\le 5x-1$

Linear inequalities with three parts, p. 217

b. $-15 < 2x + 3 < 23$

Solution: **a.**

$$4x+5\le 5x-1$$
$$4x\le 5x-6$$
$$-x\le -6$$
$$x\ge 6$$

In interval notation: $[6,\infty)$

Graphically:

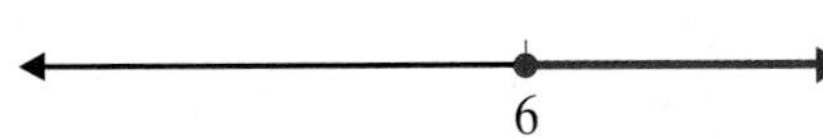

1. Square, $P = 4a$

Rectangle, $P = 2(l + w)$,

Circle, $P = 2\pi r$

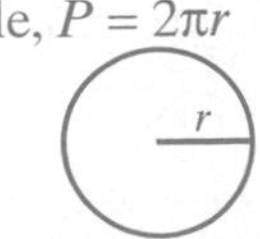

b.

$$-15<2x+3<23$$
$$-15-3<2x+3-3<23-3$$
$$-18<2x<20$$
$$-9<x<10$$

In interval notation: $(-9, 10)$

Graphically:

Example 7: Average

Applications of linear inequalities, p. 218

What is the smallest number that can be added to 80, 90, 88, and 95 so that the average will be 90 or more?

Average, p. 187

Solution: Let n = the unknown number

Then,

$$\frac{80+90+88+95+n}{5} \geq 90$$

$$353+n \geq 450$$

$$n \geq 97$$

The number must be 97 (or more).

8.3 Exercises

1. Write three formulas that you know about perimeter of geometric figures and sketch a diagram of each figure.

2. Write three formulas that you know about area of geometric figures and sketch a diagram of each figure.

3. Write three formulas that you know about volume of geometric figures and sketch a diagram of each figure.

4. Write the formula for finding simple interest and explain the meaning of each variable in the formula.

In Exercises 5 – 10, evaluate the formula by using the given information. (Use $\pi \approx 3.14$.)

5. $C = \frac{5}{9}(F-32)$: Find C for $F = 158$.

6. $F = \frac{4}{3}Av^2$: Find F for $A = 240$ and $v = 90$.

7. $A = \pi r^2$: Find A for $r = 3.5$.

8. $V = \pi r^2 h$: Find V for $r = 13$ and $h = 5.6$.

9. $A = P + Prt$: Find A for $P = 5000$, $r = 0.08$, and $t = 0.5$.

10. $S = \frac{a}{1-r}$: Find S for $a = 12$ and $r = \frac{1}{3}$.

Answers:

2. Rectangle, $A = l \cdot w$,

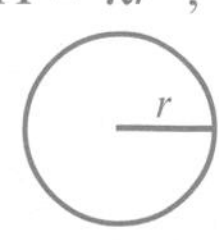

Circle, $A = \pi r^2$,

r

Parallelogram, $A = b \cdot h$,

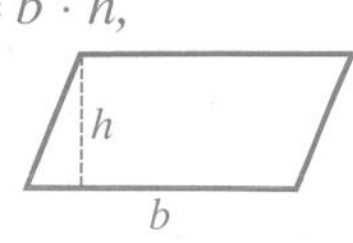

3. Rectangular solid $V = lwh$,

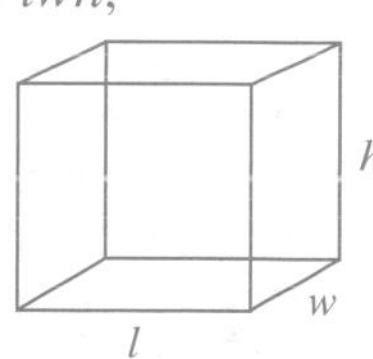

Right circular cylinder $V = \pi r^2 h$,

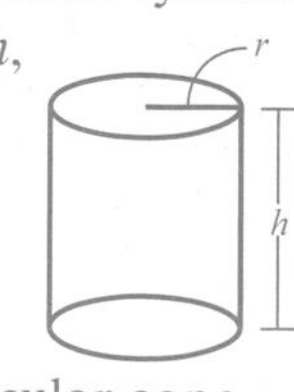

Right circular cone $V = \frac{1}{3}\pi r^2 h$,

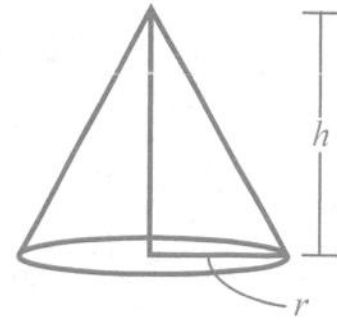

4. $SI = Prt$, where P = Principal, r = rate of interest, and t = time

5. 70
6. 2592000
7. 38.465
8. 2971.696
9. 5200
10. 18
11. $r=\frac{d}{t}$
12. $c=P-a-b$
13. $\beta=180-\alpha-\gamma$
14. $w=\frac{V}{lh}$
15. $\pi=\frac{3V}{4r^3}$
16. $h=\frac{2A}{b+c}$
17. $y=3x-14$
18. $y=\frac{20-x}{3}$
19. $n=\frac{L-a}{d}+1$
20. $m=\frac{y-b}{x}$
21. $a=\frac{84}{13}$
22. $y=60$
23. $w=2$
24. $A=7$
25. $x=25$
26. $x=-5$
27. $x=50$
28. $y=67$
29. $y=\frac{75}{14}$
30. $x=3$
31. $x=9,\ y=6$
32. $x=60^\circ,\ y=80^\circ$
33. $x=17.5,\ y=15$
34. $x=60^\circ,\ y=30^\circ,\ a=12$
35. 1962.5
36. 6.63 in. (or) $\frac{126}{19}$ in.
37. 2200 ft.^{2}
38. 55 m

In Exercises 11 – 20, solve for the indicated variable.

11. $d=rt$; solve for r.
12. $P=a+b+c$; solve for c.
13. $\alpha+\beta+\gamma=180$; solve for β.
14. $V=lwh$; solve for w.
15. $V=\frac{4}{3}\pi r^3$; solve for π.
16. $A=\frac{1}{2}h(b+c)$; solve for h.
17. $3x-y=14$; solve for y.
18. $x+3y=20$; solve for y.
19. $L=a+(n-1)d$; solve for n.
20. $y=mx+b$; solve for m.

Solve the proportions in Exercises 21 – 30.

21. $\frac{6}{a}=\frac{39}{42}$
22. $\frac{3}{10}=\frac{y}{200}$
23. $\frac{3}{4}=\frac{1\frac{1}{2}}{w}$
24. $\frac{A}{20}=\frac{35}{100}$
25. $\frac{24}{x+3}=\frac{6}{7}$
26. $\frac{x+13}{38}=\frac{4}{19}$
27. $\frac{2.5}{3.6}=\frac{x}{72}$
28. $\frac{2.7}{35}=\frac{5.4}{y+3}$
29. $\frac{2\frac{1}{3}}{5}=\frac{2\frac{1}{2}}{y}$
30. $\frac{5\frac{1}{3}}{16}=\frac{x}{9}$

In similar triangles, corresponding sides are proportional. The pairs of triangles in Exercises 31 – 34 are similar triangles. Find the values of x and y.

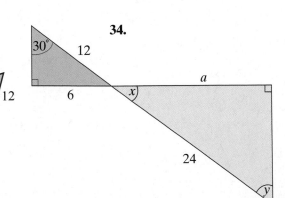

31. **32.**

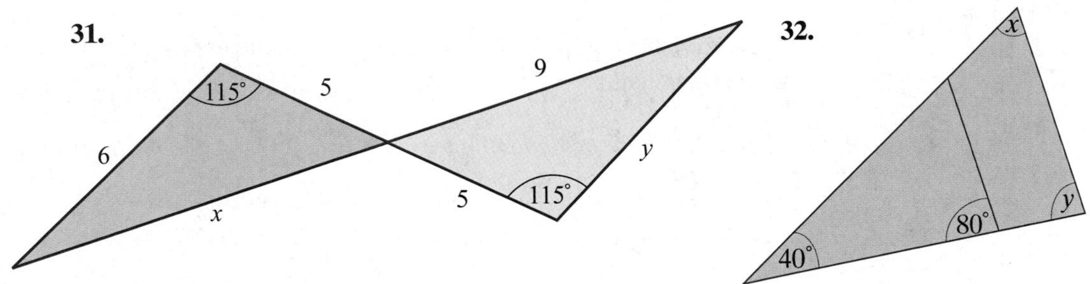

33. **34.**

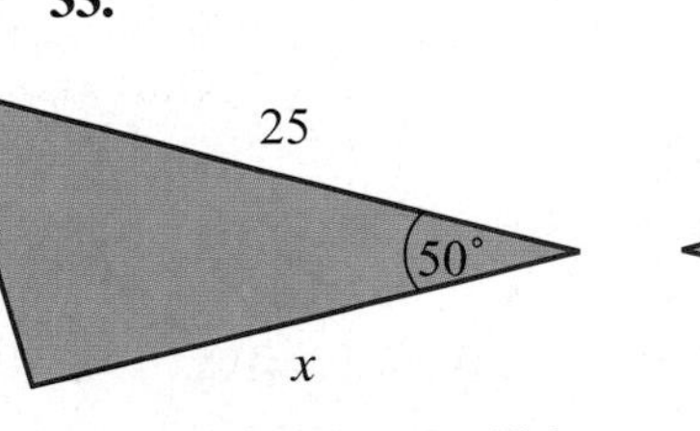

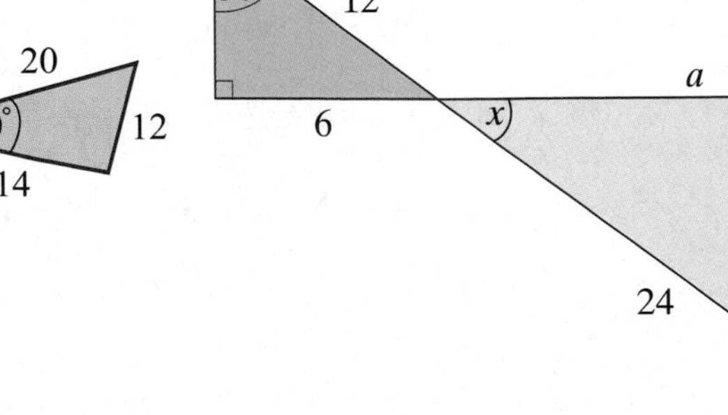

39. $P = 56$ cm, $A = 192$ cm
40. $P = 73.12$ cm, $A = 356.48$ cm
41. 3.75 in.
42. 75 feet
43. \$30,000 at 3.5% and \$20,000 at 10%
44. \$2522
45. $48.9\overline{3}$
46. 103
47. a. 64 feet per second, the object is going up since the velocity is positive.
b. −160 feet per second, the object is falling down since the velocity is negative.
48. \$15,200
49. $(-\infty, 2]$
50. $[3, \infty)$

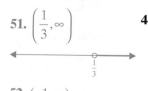

51. $\left(\frac{1}{3}, \infty\right)$
52. $(-1, \infty)$
53. $(-\infty, 0]$

35. The radius of a circle is 25 cm. Find the area. (Use $\pi \approx 3.14$.)

36. The area of a trapezoid is 63 square inches. One base is 9 in. long and the other is 10 in. long. Find the height of the trapezoid.

37. The perimeter of a rectangle is 188 feet. If the length is 50 feet, find the area of the rectangle.

38. An isosceles triangle has a base of 15 m. If the perimeter is 125 m, what is the length of each of the two equal sides?

In Exercises 39 and 40, find the perimeter and area for each figure. (Use $\pi \approx 3.14$.)

39. $P =$ ________ $A =$ ________

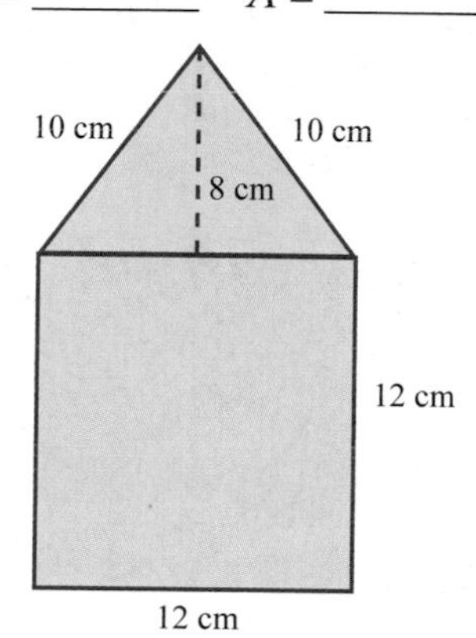

40. $P =$ ________ $A =$ ________

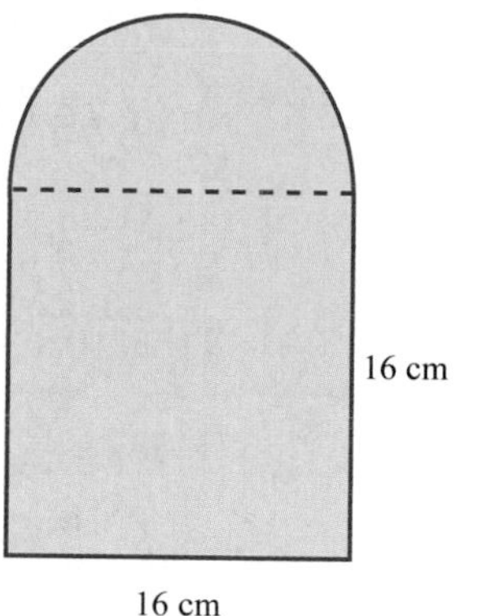

41. An architect plans to make a drawing of a building that uses a scale of 1.5 inches to 20 feet. If two points on the building are known to be 50 feet apart, how many inches apart on the drawing should these two points be?

42. To estimate the height of an office building, an engineer (6 feet tall) notices that his shadow is 2 feet long. At that same time, he measures the length of the shadow of the building to be 25 feet long. What is his estimate of the height of the building?

43. Maria has had \$50,000 invested for one year, some with a CD at 3.5% and the rest in the stock market which yielded 10% for the year. If her interest income last year was \$3050, how much did she have in the CD and how much in the stock?

44. A jeweler paid \$1455 for a watch. He wants to price the watch for sale so that he can give the customer a 25% discount on the selling price and still make a profit of 30% on his cost. What selling price should he mark for the watch?

45. The following speeds (in miles per hour) were clocked for 15 cars at a particularly dangerous intersection. Find the average speed of these cars.

45 55 42 50 62 58 40 35
50 56 48 60 36 52 45

46. Suppose that you have scores of 86, 93, 80, and 88 on four exams in your algebra class. What score will you need on the fifth exam to have an average of 90?

47. If an object is shot upward with an initial velocity v_0 feet per second, the velocity, v, in feet per second is given by the formula $v = v_0 - 32t$. If the initial velocity is 160 feet per second, find (a) the velocity at the end of 3 seconds and (b) the velocity at the end of 10 seconds. Explain the meanings of your answers.

48. The value, V, of an item after t years of "linear" depreciation is given by the formula $V = C - Crt$ where C is the original cost and r is the rate of depreciation expressed as a decimal. If you buy a car for \$38,000 and depreciate it linearly at a rate of 12%, what will be its value after 5 years?

Solve the linear inequalities in Exercises 49 – 66. Write each answer in interval notation and graph the solution set on a number line.

49. $3x + 5 - 3 \le 8$

50. $5x - 11 \ge 4$

51. $-2(x+3) < x - 7$

52. $-5(x-5) < 30$

53. $14y + 3.5 \le 2y + 3.5$

54. $6y + \frac{1}{2} > -2y + \frac{1}{2}$

55. $0 \le 3x + 1 \le 10$

56. $-5 < 5 - 5y < 13$

57. $0 \le \frac{1}{2}x + 6 < 2\frac{1}{2}$

58. $-4 < 10x + 20 \le 15$

59. $-8 < 2(3-x) < 16$

60. $-3 \le 3(y+4) \le 15$

61. $0 \le 5x + 3 \le 4.5$

62. $-2.3 < a + 0.5 \le 2.4$

63. $0 > -5y + 6 > -9$

64. $-12 \le -4x - 8 \le 0$

65. $-10 \le 5(4-x) < 20$

66. $-2 < 2y - 6 \le 4$

67. Find the volume of a circular cylinder with radius 10 in. and height 15 in.

68. Find the volume of a cone that has radius 10 in. and height 15 in.

69. If the volume of rectangular solid is 175 cm^3 and its base is a square 5 cm on each side, what is the height of the solid?

70. Find the volume of a sphere that has a radius of 3 feet.

71. What is the area between two circles with the same center if the radius of one circle is 5 m and the radius of the other is 3 m?

72. The circumference of a circle is 94.2 in. Find the area and the radius of the circle. (Use 3.14 as an approximation of π.)

73. The perimeter of a rectangle is 60 cm. If the length of the rectangle is 3 times the width, find the area. What are the length and width of the rectangle?

54. $(0, \infty)$

55. $\left[-\frac{1}{3}, 3\right]$

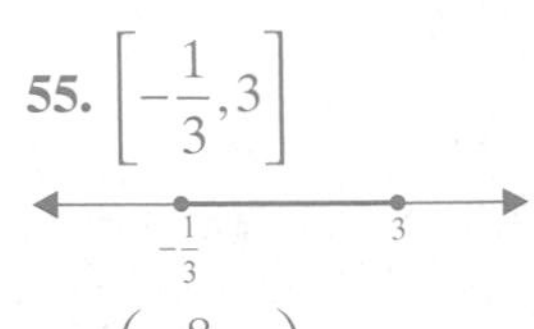

56. $\left(-\frac{8}{5}, 2\right)$

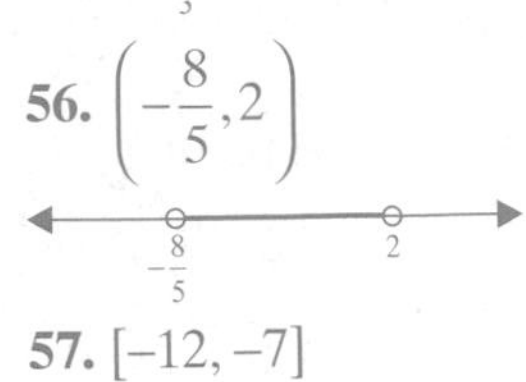

57. $[-12, -7]$

58. $\left(-\frac{12}{5}, -\frac{1}{2}\right]$

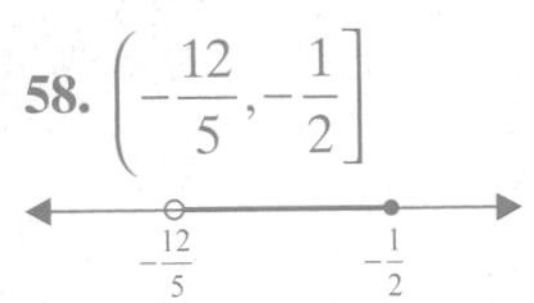

59. $(-5, 7)$

60. $[-5, 1]$

61. $\left[-\frac{3}{5}, \frac{3}{10}\right]$

62. $(-2.8, 1.9]$

63. $\left(\frac{6}{5}, 3\right)$

64. $[-2, 1]$

65. $(0, 6]$

66. $(2, 5]$

67. 4710 in.^{3}

68. 1570 in.^{3}

69. 7 cm

70. 113.04 ft.^{3}

71. 50.24 m^2

72. Area = 706.5 in.^{2}, radius = 15 in. **73.** Area = 168.75 cm^2, width = 7.5 cm, length = 22.5 cm

8.4 Chapter 4 Review: Straight Lines and Functions

Sections 4.1 and 4.2 The Cartesian Coordinate System and Graphing Linear Equations in Two Variables

Quadrants, p. 236

Ordered pairs, p. 234

Dependent variable, p. 235

Independent variable, p. 235

In the Cartesian coordinate system, the plane is separated into four **quadrants** by two perpendicular lines called **axes**. The **origin**, designated by the ordered pair (0, 0), is the point of intersection of the two lines. The horizontal line is called the ***x*-axis**, the vertical line is called the ***y*-axis**, and **ordered pairs** of real numbers are represented in the form (x, y). In an ordered pair such as (5, –2), 5 is called the **first coordinate** and –2 is called the **second coordinate**. Also, for an ordered pair (x, y) in general, y is called the **dependent variable** and x is called the **independent variable**. Figure 8.2 shows the quadrants, axes, and several ordered pairs of real numbers.

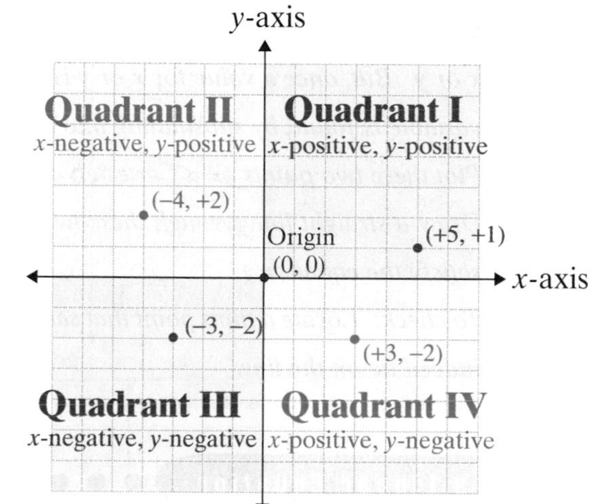

Figure 8.2

Example 1: Graphs of Ordered Pairs

The graph shows several points. Give the coordinates of each point and tell in which quadrant the point lies.

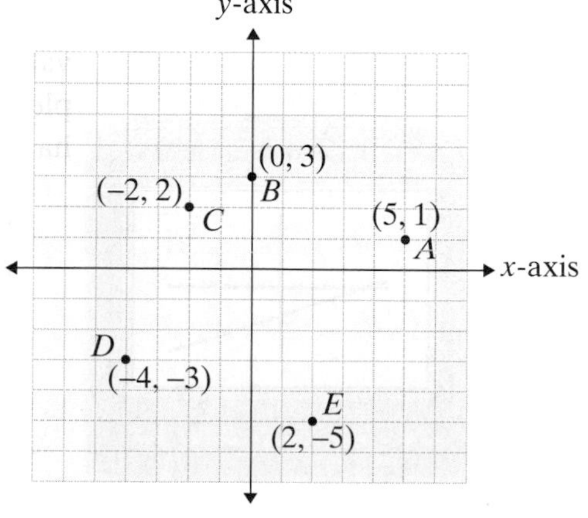

Solution:

The coordinates and quadrants are:
***A*:** (5, 1), quadrant I
***B*:** (0, 3), no quadrant
***C*:** (–2, 2), quadrant II
***D*:** (–4, –3), quadrant III
***E*:** (2, –5), quadrant IV

Vertical Line Test

Vertical line test, p. 290 *If **any** vertical line intersects the graph of a relation at more than one point, then the relation graphed is **not** a function.*

Example 8: The Vertical Line Test

Use the vertical line test to determine whether or not each graph represents a function.

a.

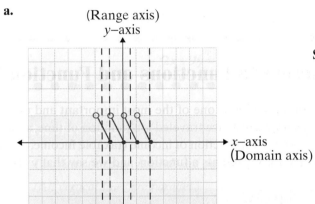

(Range axis)
y–axis
x–axis
(Domain axis)

Solution: The relation is a function. No vertical line will intersect the graph at more than one point. Several vertical lines are drawn to illustrate this. For this function, we see from the graph that $D = (-2, 2]$ and $R = [0, 2)$.

b.

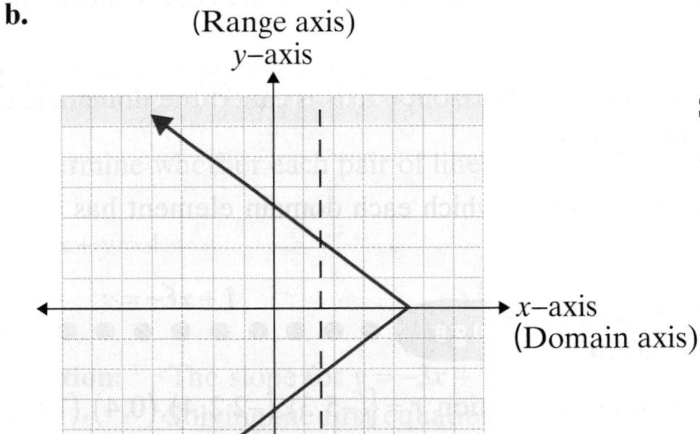

Solution: The relation is **not** a function. At least one vertical line (drawn) intersects the graph at more than one point.

Function notation, p. 295

Suppose that $y = 4x + 3$ is a given linear equation. Since the equation is solved for y it represents a linear function, and we can replace y with the function notation $f(x)$ (read "f of x") as follows:

$$y = 4x + 3 \qquad \text{can be written as} \qquad f(x) = 4x + 3.$$

In function notation, $f(2)$ indicates that x is to be replaced by 2. Thus, in this example, $f(2) = 4 \cdot 2 + 3 = 11$. The ordered pair $(2, 11)$ is the same as $(2, f(2))$.

Example 9: Function Notation

a. For the function $f(x) = -3x + 1$, find $f(-2)$ and $f(0)$.

b. For the function $g(x) = x^2 - 5x + 6$, find $g(2)$ and $g(5)$.

Solution:

a. $f(-2) = -3(-2) + 1 = 6 + 1 = 7$ and $f(0) = -3(0) + 1 = 1$

b. $g(2) = 2^2 - 5 \cdot 2 + 6 = 4 - 10 + 6 = 0$ and $g(5) = 5^2 - 5 \cdot 5 + 6 = 25 - 25 + 6 = 6$

Section 4.6 Graphing Linear Inequalities in Two Variables

Boundary line, p. 308

A straight line separates a plane into two **half-planes**. The points on one side of the line are in one half-plane, and the points on the other side of the line are in the other half-plane. The line itself is called the **boundary line**. The half-plane is **closed** if the line is included, otherwise it is **open**.

Example 10: Graphing a Linear Inequality

Two methods for graphing linear inequalities, p. 308

a. Graph the inequality $y - 2x \geq 1$.

b. Use a TI-83 Plus graphing calculator to graph the inequality $y < -\frac{2}{3}x + 1$.

Solution: **a.** Solving for y gives $y \geq 2x + 1$. So, we want the points above the line with the boundary line included.

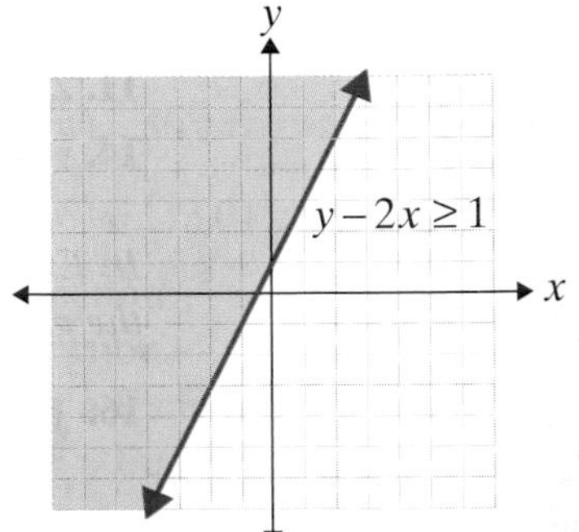

continued on next page ...

Example 2: Factoring Out a GCF

Factor each polynomial by finding the GCF.

a. $-4x^4+2x^3-8x^2$ **b.** $12a^4b^5-7a^2b^4+6a^6b^3$

Solution: **a.** $-4x^4+2x^3-8x^2=-2x^2\left(2x^2-x+4\right)$

When the leading coefficient is negative, factoring out a negative term is preferred.

b. $12a^4b^5-7a^2b^4+6a^6b^3=a^2b^3\left(12a^2b^2-7b+6a^4\right)$

We can treat an expression such as $y(x+5)-4(x+5)$ as a binomial with two "terms" separated by the – sign. In this case, we consider the binomial $(x+5)$ to be the common factor. Factoring out $(x+5)$ gives the following result:

$$y(x+5)-4(x+5)=(x+5)(y-4)$$

Factoring by grouping, p. 409

If an expression has four terms, we can sometimes find a common binomial factor by **grouping** the terms two at a time and looking for a common monomial factor in each "group." This method of factoring is called **factoring by grouping**.

Example 3: Factoring by Grouping

Factor each polynomial by grouping.

a. $xy+4x+6y+24$ **b.** $ab-5a-3b+15$

Solution: **a.** $xy+4x+6y+24=x(y+4)+6(y+4)=(y+4)(x+6)$

b. $ab-5a-3b+15=a(b-5)-3(b-5)=(b-5)(a-3)$

Sections 6.2 and 6.3

Special Factoring Techniques I & II

FOIL method, p. 418

The **FOIL method** (or **Trial and Error method**) of factoring trinomials is actually applying the FOIL method of multiplication of two binomials in reverse. Two basic forms are considered:

1. The leading coefficient is 1: x^2+bx+c
2. The leading coefficient is not 1: ax^2+bx+c

Example 4: Factoring by Using the FOIL Method

Factor each expression by using the FOIL method (trial and error method).

a. $x^2+8x+12$ **b.** $2x^2-7x-15$ **c.** $6y^2+26y+8$

Solution: **a.** The leading coefficient is 1. We know $6\cdot 2=12$ and $6+2=8$. Thus, we have

$$\mathbf{F}=x^2 \qquad \mathbf{L}=12$$

$$x^2+8x+12=(x+6)(x+2)$$

$$\mathbf{I}=6x \qquad \mathbf{O}=2x$$

b. Trial and error gives the following:

$$2x^2-7x-15=(2x+3)(x-5)$$

c. Remember to always look for a common monomial factor first.

$$6y^2+26y+8=2(3y^2+13y+4)=2(3y+1)(y+4)$$

The following formulas for special products were discussed in Chapter 5 and in Section 8.5. We rewrite them here in reverse form for factoring instead of multiplying.

Special Formulas for Factoring

Difference of squares, p. 414

I. *Difference of squares:* $X^2-A^2=(X+A)(X-A)$

Perfect square trinomial, p. 414

II. *Perfect square trinomial:* $X^2+2AX+A^2=(X+A)^2$

III. *Perfect square trinomial:* $X^2-2AX+A^2=(X-A)^2$

Difference of two cubes, p. 425

IV. *Difference of two cubes:* $X^3-A^3=(X-A)(X^2+AX+A^2)$

Sum of two cubes, p. 425

V. *Sum of two cubes:* $X^3+A^3=(X+A)(X^2-AX+A^2)$

Example 5: Special Factoring Techniques

Use your knowledge of the special factoring formulas and factoring monomials to completely factor each of the following expressions.

a. $12x^2 - 75$ **b.** $8y^3 + 125$

c. $xy^4 - 12xy^2 + 36x$ **d.** $9x^{-2} + 36x^{-4}$

Solution: **a.** $12x^2 - 75 = 3(4x^2 - 25) = 3(2x+5)(2x-5)$ Factoring out 3 and the difference of squares.

b. $8y^3 + 125 = (2y+5)(4y^2 - 10y + 25)$ Factoring the sum of two cubes.

c. $xy^4 - 12xy^2 + 36x = x(y^4 - 12y^2 + 36) = x(y^2 - 6)^2$ Factoring out x and a perfect square trinomial.

d. $9x^{-2} + 36x^{-4} = 9x^{-4}(x^2 + 4)$ Factoring out $9x^{-4}$.

(**Note:** –4 is the smaller exponent and $x^2 + 4$ is not factorable in the real number system.)

Section 6.4 Solving Quadratic Equations by Factoring

Equations that can be solved by factoring depend on the **zero factor property** of the number 0.

Zero Factor Property

Zero Factor Property, p. 433

If the product of two factors is 0, then one or both of the factors must be 0.

That is, if a and b are real numbers,

if $a \cdot b = 0$***, then*** $a = 0$ ***or*** $b = 0$.

Example 6: Solving an Equation by Using the Zero Factor Property

Use the zero factor property to solve the equation $(x - 6)(2x + 3) = 0$.

Solution: By the zero factor property, either $x - 6 = 0$ or $2x + 3 = 0$. Solving these two linear equations gives the two solutions to the original equation.

$$x - 6 = 0 \qquad\qquad 2x + 3 = 0$$

$$x = 6 \qquad\qquad 2x = -3$$

$$x = -\frac{3}{2}$$

These solutions can be checked by substituting them, one at a time, into the original equation.

We say that the solutions are $x = 6$ and $x = -\frac{3}{2}$ or that the solution set is $\left\{6, -\frac{3}{2}\right\}$.

Quadratic equation, p. 434

A second-degree polynomial, such as $2x^2 - 9x - 18$, is called a quadratic expression and the corresponding equation $2x^2 - 9x - 18 = 0$ is called a **quadratic equation**.

Quadratic Equation

An equation that can be written in the form

$$\boldsymbol{ax^2 + bx + c = 0} \qquad \textit{where a, b, and c are real numbers and } a \neq 0$$

is called a ***quadratic equation.***

Many polynomial equations, including quadratic equations, can be solved by factoring and using the zero factor property. The following list of steps outlines the procedure.

To Solve an Equation by Factoring

1. *Add or subtract terms so that one side of the equation is 0.*
2. *Factor the polynomial expression.*
3. *Set each factor equal to 0 and solve for the variable.*

Example 7: Solving Quadratic Equations

Solve the following quadratic equations by factoring.

a. $y^2 + 12y = -32$ **b.** $(x-3)(3x+2) = -8$

Solution: **a.** $y^2 + 12y = -32$

$y^2 + 12y + 32 = 0$	One side must be 0.
$(y+8)(y+4) = 0$	Factor.
$y + 8 = 0$ or $y + 4 = 0$	Set each factor equal to 0 and solve.
$y = -8$ $\quad$ $y = -4$	

continued on next page ...

b. $(x-3)(3x+2)=-8$

$3x^2-7x-6=-8$ Multiply using the FOIL method.

$3x^2-7x+2=0$ One side must be 0.

$(3x-1)(x-2)=0$ Factor.

$3x-1=0$ or $x-2=0$ Set each factor equal to 0 and solve.

$x=\frac{1}{3}$ $x=2$

Sections 6.5 and 6.6

Applications of Quadratic Equations

Several types of problems lead to quadratic equations. Among these are problems involving right triangles (the Pythagorean Theorem), other geometric figures, number problems, consecutive integers, and projectiles.

Example 8: Applications Using Quadratic Equations

a. One number is 5 more than twice another and the sum of their squares is 325. If both numbers are positive, what are the numbers?

Solution: Let x = smaller number. Then $2x + 5$ = larger number.

$x^2+(2x+5)^2=325$ The sum of the squares is 325.

$$x^2+4x^2+20x+25=325$$

$$5x^2+20x-300=0$$

$$5(x^2+4x-60)=0$$

$$5(x+10)(x-6)=0$$

$$x=-10 \text{ or } x=6$$

Since the numbers are positive, –10 is not a solution.
The solution is $x=6$

and $2x+5=17$.

Thus, the numbers are 6 and 17.

b. In a rectangular garden, the length is 34 yards greater than the width. If the diagonal of the garden is 2 yards longer than the length, what are the dimensions of the garden?

Solution: Let w = width. Then $w + 34$ = length and $w + 36$ = length of diagonal.

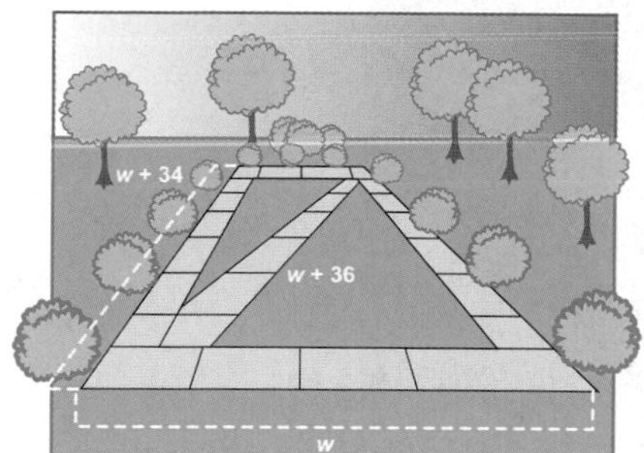

Using the Pythagorean Theorem gives:

$$w^2+(w+34)^2=(w+36)^2$$

$$w^2+w^2+68w+1156=w^2+72w+1296$$

$$w^2-4w-140=0$$

$$(w-14)(w+10)=0$$

$$w=14 \quad \text{or} \quad w=-10$$

Pythagorean theorem, p. 446

Because distance is positive, –10 is not a possible solution.

The solution is $w=14$

and $w+34=48$.

Thus, the garden is 14 yards wide and 48 yards long.

Section 6.7 Using a Graphing Calculator to Solve Equations and Absolute Values

The points where the graph of a function $y = f(x)$ crosses the x-axis are the values of x where y is equal to 0. These values of x are called the **zeros** of the function and are the **solutions to the equation** $f(x) = 0$. With the calculator as a tool, we can solve (or estimate the solutions of) equations of the form $f(x) = 0$ by graphing the corresponding function and finding the zeros of the function. The settings may need some WINDOW adjustment so that the display shows all of the points where the graph crosses the x-axis.

There are two basic strategies to solving equations by using the graphing calculator:

1. graph one function and find the zeros of the function, or
2. graph two functions and find the points of intersection of the functions.

Examples 9 and 10 illustrate these strategies and the steps to use with the graphing calculator.

Example 9: Solving an Equation by Using a Graphing Calculator

Use a graphing calculator to solve the equation $2x^2=-9x+5$.

Solution:

Strategy: Manipulate the equation so that one side is 0. Graph the indicated function on the nonzero side. The zeros of this function are the solutions of the original equation.

$$2x^2=-9x+5$$

$$2x^2+9x-5=0$$

continued on next page ...

Enter the function as follows:

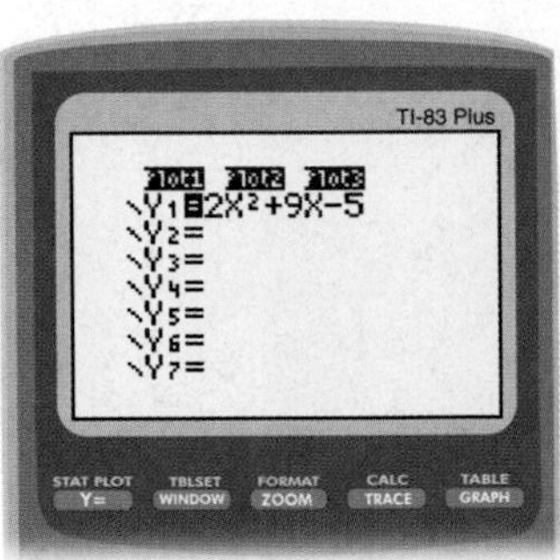

With the standard window the graph will appear as follows:

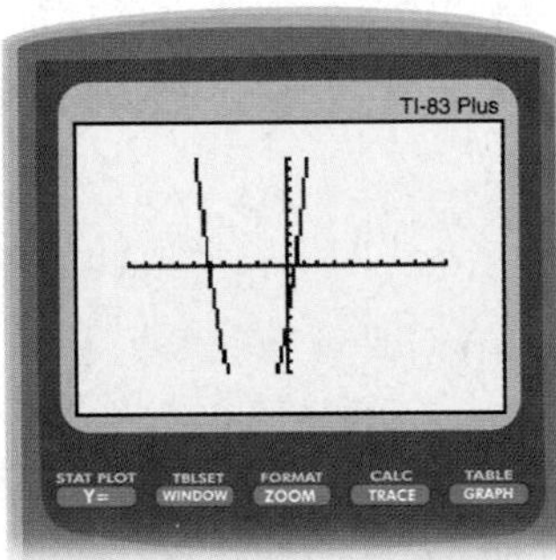

With the 2nd > CALC > **2:zero** sequence of commands you will find the following zeros (and therefore solutions to the equation):

$$x = -5 \text{ and } x = 0.5$$

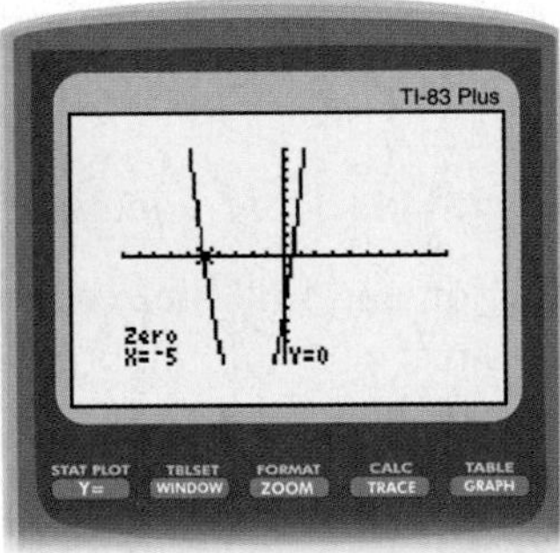

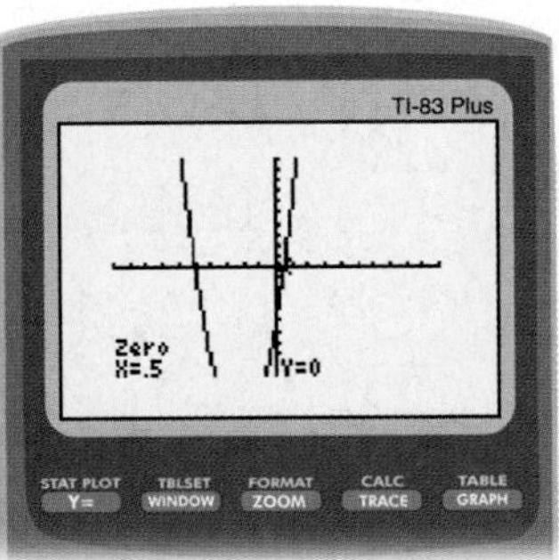

Example 10: Solving an Absolute Value Equation

Use a graphing calculator to solve the equation $|2x+1| = 7$.

Solution:

Strategy: Graph the function indicated on each side of the equation. This includes the constant function. Find the points of intersection of these two graphs. The x-values of these points are the solutions of the original equation. Remember that the absolute value command can be found in the MATH > **NUM** menu.

1. 15 **2.** 6
3. $6x^2y$ **4.** $12a^2b^2$
5. $10a^2bc$ **6.** $6xyz^2$
7. $-2x^2(x-1)$
8. $-5a(a^2+25)$
9. $6x^2y(5+8y^2+9y^3)$
10. $12a^2b^2(2a+3b-4ab)$
11. $2(x^2+9x+10)$
12. $3(3x-8)(x+3)$
13. $-(x+5)(x+5)$
14. $5(a^2+25)$
15. $(y+5)(y+1)$
16. $(2x+1)(2x+3)$
17. $(3x-2)(y+7)$
18. $(2x-3)(x+y)$
19. $(x^k+4)(x^{2k}-4x^k+16)$
20. $(a-5b^2)(a^2+5ab^2+25b^4)$
21. $-(3x-2)(2x-3)$
22. $-2x(x^2-2x+2)$
23. $(6x^k+5)(6x^k-5)$
24. $(y-3+7x)(y-3-7x)$

Enter the functions as follows:

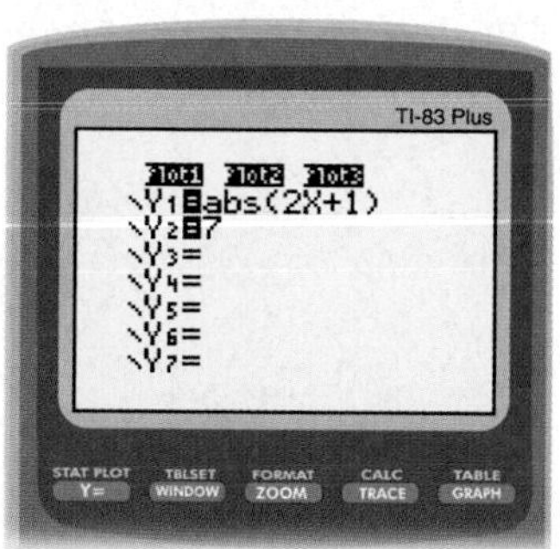

With the standard window the graphs will appear as follows:

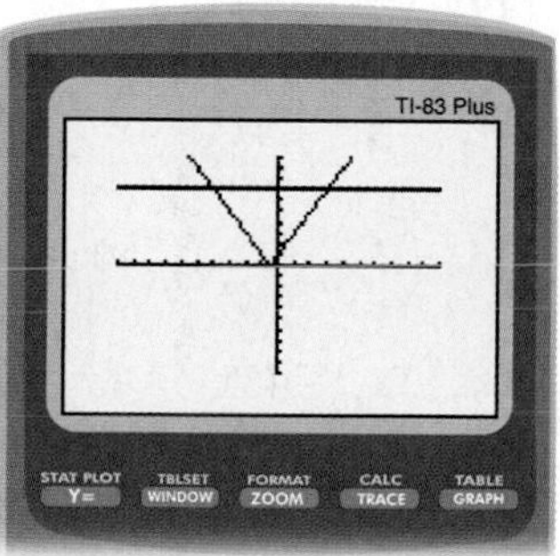

With the **2nd** > CALC > `5:intersect` sequence of commands you will find the following x-values of the points of intersection (and therefore solutions to the equation):

$$x=-4 \text{ and } x=3$$

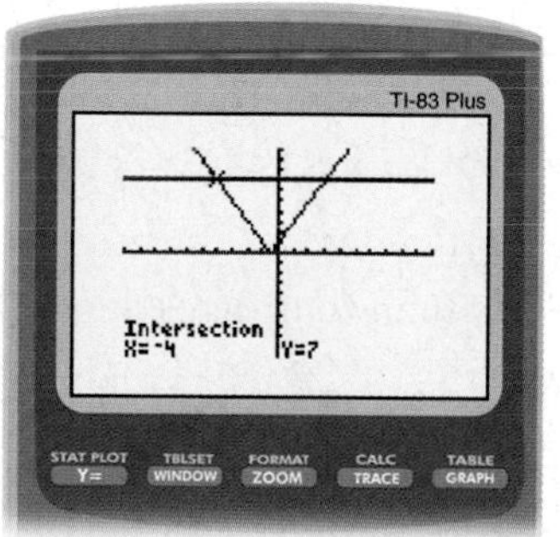

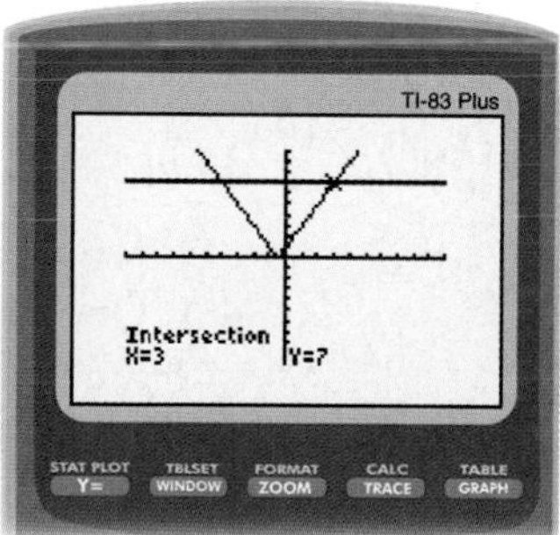

[**Note:** In general, the **TRACE** command gives only approximate values and the **zero** and **intersect** commands give exact values.]

8.6 Exercises

25. $(x-2+9y^2)(x-2-9y^2)$

Find the GCF for each of the sets of terms in Exercises 1 – 6.

1. 15, 45, 75

2. 30, 60, 96

3. $30x^2y,\ 48x^2y^3,\ 54x^2y^4$

4. $24a^3b^2,\ 36a^2b^3,\ 48a^3b^3$

5. $20a^5bc,\ 30a^3b^2c,\ 40a^2b^2c$

6. $36xyz^2,\ 42x^2yz^3,\ 60xy^2z^2$

In Exercises 7 – 30, factor completely, if possible.

7. $-2x^3+2x^2$ **8.** $-5a^3-125a$ **9.** $30x^2y+48x^2y^3+54x^2y^4$

10. $24a^3b^2+36a^2b^3-48a^3b^3$ **11.** $2x^2+18x+20$ **12.** $9x^2+3x-72$

13. $-x^2-10x-25$ **14.** $5a^2+125$ **15.** $(y+2)^2+2(y+2)-3$

16. $(2x-1)^2+6(2x-1)+8$ **17.** $3xy-2y+21x-14$ **18.** $2x^2+2xy-3x-3y$

19. $x^{3k}+64$ **20.** a^3-125b^6 **21.** $-6x^2+13x-6$

22. $-2x^3+4x^2-4x$ **23.** $36x^{2k}-25$ **24.** $(y^2-6y+9)-49x^2$

25. $(x^2-4x+4)-81y^4$ **26.** $36x^2+25$ **27.** $121a^{-2}-144a^{-4}$

28. $3x^{-3}-48x^{-5}$ **29.** y^3-12y^2+36y **30.** $8x^3-2x^2-15x$

Solve the equations in Exercises 31 – 45.

31. $-5x^2=30x$ **32.** $\frac{2x^2}{3}-10=\frac{4x}{3}$ **33.** $(4x+1)(2x-3)=0$

34. $(y-7)(5y-1)=0$ **35.** $\frac{6x^2}{5}=2-\frac{2x}{5}$ **36.** $1.5y^2=4.5y$

37. $8x^2+10x=3$ **38.** $3x^3+27x^2+42x=0$ **39.** $2x^3-14x^2+20x=0$

40. $a^2+2a=-1$ **41.** $(x-5)(x+1)=-5$ **42.** $\frac{1}{6}x^2-\frac{5}{3}x+\frac{25}{6}=0$

43. $(3x+1)(x+1)=18x-7$ **44.** $35x^2=x+6$ **45.** $144x^3=10x^2+50x$

46. One number is 15 less than four times another number. Their product is 250. Find the positive numbers.

47. The product of two consecutive positive integers is 132. Find the two integers.

48. The product of two consecutive odd positive integers is 575. What are the integers?

49. The difference between two positive numbers is 6. If twice the smaller is added to the square of the larger, the result is 348. What are the numbers?

50. In a right triangle the hypotenuse is known to be 50 cm and one of the legs is 10 cm longer than the other. What are the lengths of the legs?

26. $36x^2+25$
27. $a^{-4}(11a+12)(11a-12)$
28. $3x^{-5}(x-4)(x+4)$
29. $y(y-6)^2$
30. $x(4x+5)(2x-3)$
31. $x=0, x=-6$
32. $x=-3, x=5$
33. $x=\frac{3}{2}, x=-\frac{1}{4}$
34. $y=7, y=\frac{1}{5}$
35. $x=\frac{-1+\sqrt{61}}{6}$, $x=\frac{-1-\sqrt{61}}{6}$
36. $y=0, y=3$
37. $x=\frac{1}{4}, x=-\frac{3}{2}$
38. $x=0, x=-2, x=-7$
39. $x=0, x=2, x=5$
40. $a=-1$
41. $x=0, x=4$
42. $x=5$
43. $x=4, x=\frac{2}{3}$
44. $x=\frac{3}{7}, x=-\frac{2}{5}$
45. $x=\frac{5}{8}, x=\frac{-5}{9}, x=0$
46. 10, 25 **47.** 11, 12
48. 23, 25 **49.** 12, 18
50. 30 cm, 40 cm
51. a. 5 sec. **b.** 3.5 sec. **c.** 2 sec.
52. 16 feet
53. 20 yards, 25 yards

54. 60 feet

55. length = 12 in.
width = 10 in.

56. $x = 6, x = -6$

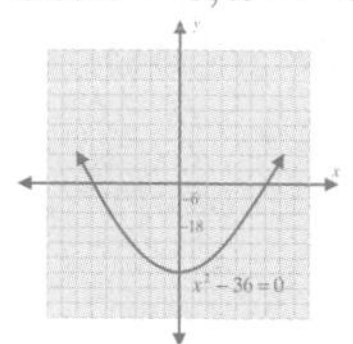

57. $x = -2.24, x = 2.24$

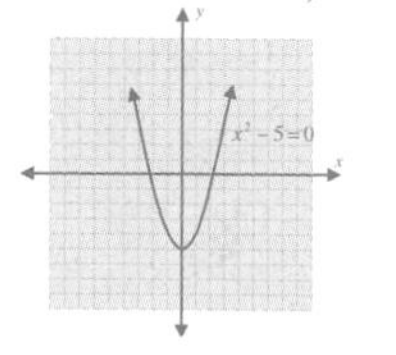

58. No values of x satisfy, as this function does not cross or touch the x-axis.

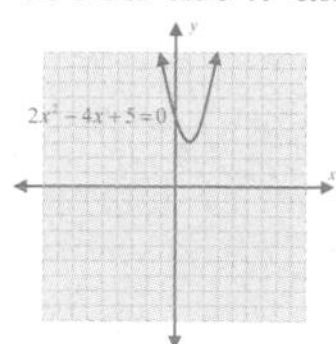

59. No values of x satisfy, as this function does not cross or touch the x-axis.

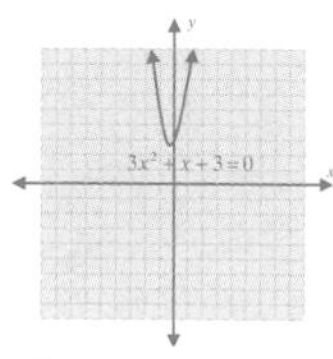

60. $x = 0, x = -4.47, x = 4.47$

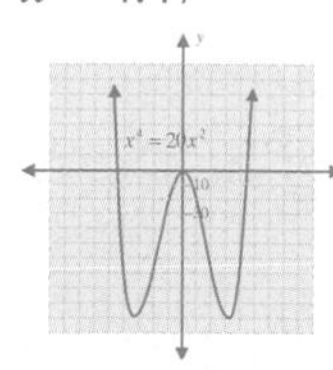

51. A ball is dropped from the edge of a cliff that is known to be 400 feet above the beach. The formula for finding the height of the ball above the beach in t seconds after it is dropped is $h = 400 - 16t^2$.

- **a.** In how many seconds will the ball hit the beach?
- **b.** How many seconds did it take for the ball to reach a height of 204 feet above the beach?
- **c.** How many seconds did it take for the ball to fall 64 feet?

52. The base of a triangle is 10 feet greater than the height. If the area of the triangle is 48 ft.^{2}, find the length of the base.

53. A rectangle has an area of 500 square yards and a perimeter of 90 yards. What are the dimensions of the rectangle?

54. A telephone pole is to have a guy wire attached at the top and anchored to a point on the ground at a distance of 19 feet less than half the height of the pole. If the wire is to be 1 foot longer than the height of the pole, what is the height of the pole?

55. A box is formed by cutting 2 inch squares from each corner of a rectangular piece of metal and folding up the edges. If the volume of the box is 96 cubic inches, what were the dimensions of the original rectangle if the length is 2 inches more than the width?

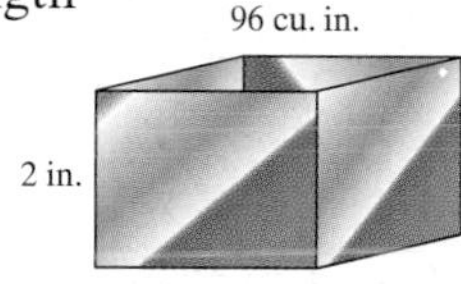

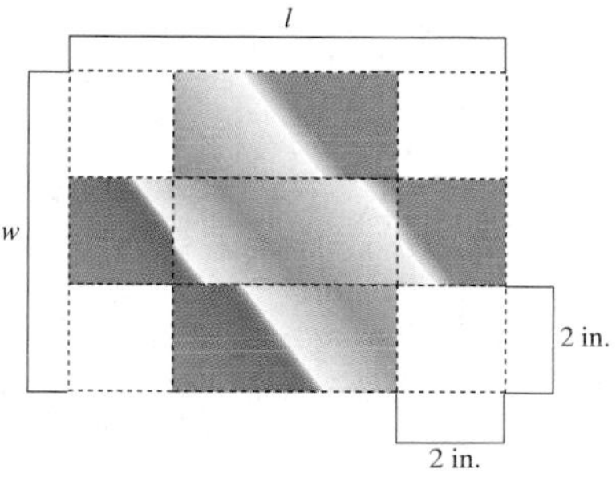

Use a graphing calculator to solve (or estimate the solutions of) the equations in Exercises 56 – 65. Sketch the graphs of the functions you used and find the solutions accurate to 2 decimal places.

56. $x^2 - 36 = 0$

57. $x^2 - 5 = 0$

58. $2x^2 - 4x + 5 = 0$

59. $3x^2 + x + 3 = 0$

60. $x^4 = 20x^2$

61. $8x - 5 = x^3$

62. $x(x-4)(x+2) = 0$

63. $|3 - 2x| = 11$

64. $|4x + 3| = 17$

65. $|x - 3| = |4 - x|$

61. $x = .66, x = 2.44, x = -3.1$

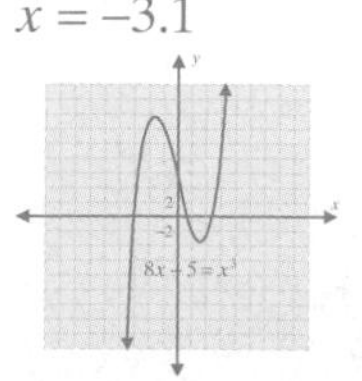

62. $x = 0, x = -2, x = 4$

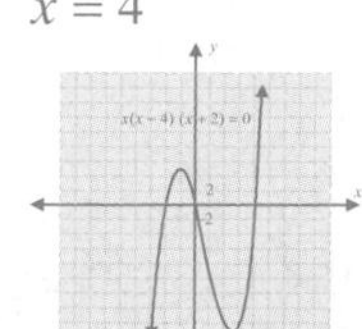

63. $x = -4, x = 7$

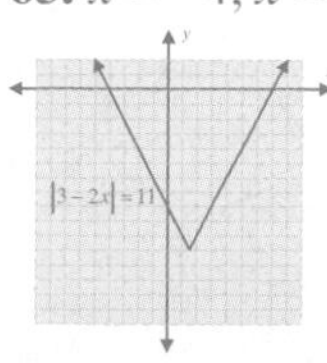

64. $x = 3.5, x = -5$

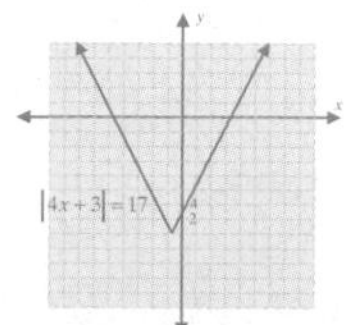

65. $x = 3.5$

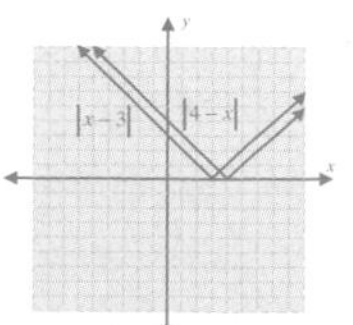

8.7 Chapter 7 Review: Rational Expressions

Section 7.1 Multiplication and Division with Rational Expressions

Rational expression, p. 477

Any fraction formed with a polynomial as numerator and a polynomial as denominator is called a **rational expression**.

Rational Expression

*A **rational expression** is an expression of the form* $\frac{P}{Q}$ *(or in function notation,* $\frac{P(x)}{Q(x)}$ *) where P and Q are polynomials and* $Q \neq 0$.

In the definition, the denominator $Q \neq 0$ means that the variable will not equal any value that will cause a denominator to be 0. For example, in the rational expression

$$\frac{P}{Q} = \frac{x^2 + 5x + 6}{2x - 10} \qquad Q = 2x - 10, \text{ and we assume that } 2x - 10 \neq 0 \ (\text{or } x \neq 5).$$

The operations of multiplication, division, addition, and subtraction with rational expressions are essentially the same as those for operating with fractions in arithmetic. That is, these operations involve factoring, reducing, and finding common denominators.

Basic Properties of Rational Expressions

Fundamental principle, p. 478

1. *The Fundamental Principle*

If $\frac{P}{Q}$ *is a rational expression and K is a polynomial and* $K \neq 0$, *then*

$$\frac{P}{Q} = \frac{P}{Q} \cdot \frac{K}{K} = \frac{P \cdot K}{Q \cdot K}.$$

Negative signs, p. 479

2. *Negative Signs*

$$-\frac{P}{Q} = \frac{P}{-Q} = \frac{-P}{Q} \quad \textit{and} \quad \frac{P}{Q} = \frac{-P}{-Q} = -\frac{-P}{Q} = -\frac{P}{-Q}$$

Opposites, p. 480

3. *Opposites*

In general, $\frac{-P}{P} = -1$ *if* $P \neq 0$. *In particular,* $\frac{a-x}{x-a} = -1$ *if* $x \neq a$.

Multiplication, p. 481

4. *Multiplication*

$$\frac{P}{Q} \cdot \frac{R}{S} = \frac{P \cdot R}{Q \cdot S} \qquad \text{where } Q, S \neq 0.$$

Division, p. 482

5. *Division*

$$\frac{P}{Q} \div \frac{R}{S} = \frac{P}{Q} \cdot \frac{S}{R} \qquad \text{where } Q, R, S \neq 0.$$

Example 1: Multiplication and Division with Rational Expressions

Find the indicated product or quotient and reduce to lowest terms. State any restrictions on the variables.

a. $\frac{x-5}{x^2}\cdot\frac{x^3}{25-x^2}$ **b.** $\frac{x^2-2x+1}{x+6}\cdot\frac{x^2+x-30}{x^2-1}$

c. $\frac{y+3}{y^2+3y-4}\div\frac{2y+6}{y^2-16}$ **d.** $\frac{x^3-27}{2x+1}\div\frac{x^2+3x+9}{2x^2+7x+3}$

Solution: **a.**
$$\frac{x-5}{x^2}\cdot\frac{x^3}{25-x^2}=\frac{\overset{-1}{\cancel{x-5}}}{\cancel{x^2}}\cdot\frac{\overset{x}{\cancel{x^3}}}{(5+x)\cancel{(5-x)}}=-\frac{x}{x+5} \qquad x\neq 0,5,-5$$

b.
$$\frac{x^2-2x+1}{x+6}\cdot\frac{x^2+x-30}{x^2-1}=\frac{\overset{x-1}{\cancel{(x-1)^2}}}{\cancel{x+6}}\cdot\frac{\cancel{(x+6)}(x-5)}{(x+1)\cancel{(x-1)}}$$
$$=\frac{(x-1)(x-5)}{x+1}=\frac{x^2-6x+5}{x+1} \qquad x\neq -6,1,-1$$

c.
$$\frac{y+3}{y^2+3y-4}\div\frac{2y+6}{y^2-16}=\frac{\cancel{y+3}}{\cancel{(y+4)}(y-1)}\cdot\frac{\cancel{(y+4)}(y-4)}{2\cancel{(y+3)}}$$
$$=\frac{y-4}{2(y-1)} \qquad y\neq 4,-4,1,-3$$

d.
$$\frac{x^3-27}{2x+1}\div\frac{x^2+3x+9}{2x^2+7x+3}=\frac{(x-3)\cancel{(x^2+3x+9)}}{\cancel{2x+1}}\cdot\frac{\cancel{(2x+1)}(x+3)}{\cancel{x^2+3x+9}}$$
$$=(x-3)(x+3)=x^2-9 \qquad x\neq -\frac{1}{2},-3$$

Section 7.2 Addition and Subtraction with Rational Expressions

To add or subtract rational expressions proceed just as you would with numerical fractions.

Addition and Subtraction of Rational Expressions

Addition and subtraction of rational expressions, p. 488

$$\frac{P}{Q}+\frac{R}{Q}=\frac{P+R}{Q} \quad \text{and} \quad \frac{P}{Q}-\frac{R}{Q}=\frac{P-R}{Q} \quad \text{where } Q\neq 0$$

9.2 Systems of Equations: Solutions by Substitution

Objectives

After completing this section, you will be able to:

Solve systems of linear equations by substitution.

The Substitution Method

As we discussed in Section 9.1, solving systems of linear equations by graphing is somewhat limited in accuracy. The graphs must be drawn very carefully and even then the points of intersection (if there are any) can be difficult to estimate accurately.

In this section, we will develop an algebraic method called the **method of substitution**. The objective in the substitution method is to eliminate one of the variables so that a new equation is formed with just one variable. If this new equation has one solution, then the system is **consistent**. If this new equation is never true, then the system is **inconsistent**. If this new equation is always true, then the system is **dependent**.

To Solve a System of Linear Equations by Substitution

1. *Solve one of the equations for one of the variables.*
2. *Substitute the resulting expression into the other equation.*
3. *Solve this new equation, if possible, and then substitute back into one of the original equations to find the value of the other variable. (This is known as* ***back substitution****.)*

To illustrate the substitution method, consider the following system:

$$\begin{cases} y = -2x + 5 \\ x + 2y = 1 \end{cases}$$

How would you substitute? Since the first equation is already solved for y, a reasonable substitution would be to put $-2x + 5$ for y in the second equation. Try this and see what happens.

First equation already solved for y: $\quad y = -2x + 5$

Substitute into second equation: $\quad x + 2y = 1$

$$x + 2(-2x + 5) = 1$$

We now have one equation in only one variable, namely x. The problem has been reduced from one of solving two equations in two variables to solving one equation in one variable. Solve this equation for x. Then find the corresponding y-value by substituting this x-value into **either of the two original equations**.

$$\begin{aligned} x+2(-2x+5)&=1 \\ x-4x+10&=1 \\ -3x&=-9 \\ x&=3 \end{aligned}$$

Substituting $x = 3$ into $y = -2x + 5$ gives

$$\begin{aligned} y&=-2\cdot3+5 \\ &=-6+5 \\ &=-1. \end{aligned}$$

Thus, the system is consistent and the solution to the system is $x = 3$ and $y = -1$, or the point $(3, -1)$.

Substitution is not the only algebraic technique for solving a system of linear equations. It does work in all cases but generally is used only when one of the equations is easily solved for one variable.

In the following examples, note how the results in Example 1c indicate that the system is inconsistent and the results in Example 1d indicate that the system is dependent.

Example 1: Solve by Substitution

Solve the following systems of linear equations by using the technique of substitution.

a. $\begin{cases} y=\dfrac{5}{6}x+1 \\ 2x+6y=7 \end{cases}$

Solution: $y=\dfrac{5}{6}x+1$ is already solved for y. Substituting into the other equation, we find

$$\begin{aligned} 2x+6y&=7 \\ 2x+6\left(\frac{5}{6}x+1\right)&=7 \\ 2x+5x+6&=7 \\ 7x&=1 \\ x&=\frac{1}{7} \end{aligned}$$

$$y=\frac{5}{6}\cdot\frac{1}{7}+1=\frac{5}{42}+\frac{42}{42}=\frac{47}{42}$$

continued on next page ...

The system is **consistent** and the solution is $\left(\frac{1}{7}, \frac{47}{42}\right)$, or $x = \frac{1}{7}$ and $y = \frac{47}{42}$.

b. $\begin{cases} x = -5 \\ y = 2x + 9 \end{cases}$

Solution: In the first equation, $x = -5$ is already solved for x. The equation represents a vertical line. Substituting $x = -5$ into the second equation we have,

$$y = 2x + 9$$
$$y = 2(-5) + 9 = -10 + 9 = -1.$$

The system is **consistent** and the solution is $x = -5$ and $y = -1$, or $(-5, -1)$.

c. $\begin{cases} 3x + y = 1 \\ 6x + 2y = 3 \end{cases}$

Solution: Solving for y in the first equation gives

$$3x + y = 1$$
$$y = -3x + 1.$$

Substituting for y in the second equation yields

$$6x + 2y = 3$$
$$6x + 2(-3x + 1) = 3$$
$$6x - 6x + 2 = 3$$
$$2 = 3.$$

This last equation, $2 = 3$, is never true. This tells us that the system is **inconsistent**. Graphically, the lines are **parallel** and there is no point of intersection. The system has no solution.

d. $\begin{cases} x - 2y = 1 \\ 3x - 6y = 3 \end{cases}$

Solution: Solving for x in the first equation gives

$$x - 2y = 1$$
$$x = 2y + 1.$$

Substituting for x in the second equation yields

$$3x - 6y = 3$$
$$3(2y + 1) - 6y = 3$$
$$6y + 3 - 6y = 3$$
$$3 = 3.$$

This last equation, $3 = 3$, is always true. This tells us that the system is **dependent**. Graphically, the two lines are the same and there are an infinite number of points of intersection. Therefore, the solution consists of all the points that satisfy either of the original equations, $x - 2y = 1$ or $3x - 6y = 3$.

e. $\begin{cases} x+y=5 \\ 0.2x+0.3y=0.9 \end{cases}$

Solution: Solving the first equation for y, we have

$$\begin{aligned} x+y&=5 \\ y&=5-x. \end{aligned}$$

Substituting for y in the second equation yields

$$\begin{aligned} 0.2x+0.3y&=0.9 \\ 0.2x+0.3(5-x)&=0.9 \\ 0.2x+1.5-0.3x&=0.9 \\ -0.1x&=-0.6 \\ \frac{-0.1x}{-0.1}&=\frac{-0.6}{-0.1} \\ x&=6 \\ y&=5-x=5-6=-1. \end{aligned}$$

Note that you could multiply each term by 10 to eliminate the decimal.

The system is **consistent** and the solution is $x = 6$ and $y = -1$, or $(6, -1)$.

Practice Problems

Solve the following systems using the technique of substitution.

1. $\begin{cases} x+y=3 \\ y=2x \end{cases}$ *2.* $\begin{cases} y=3x-1 \\ 2x+y=4 \end{cases}$ *3.* $\begin{cases} x+2y=-1 \\ x-4y=-4 \end{cases}$

9.2 Exercises

Solve the systems of linear equations in Exercises 1 – 38 by using the technique of substitution. If the system is inconsistent or dependent, say so in your answer.

1. $\begin{cases} x+y=6 \\ y=2x \end{cases}$ **2.** $\begin{cases} 5x+2y=21 \\ x=y \end{cases}$ **3.** $\begin{cases} x-7=3y \\ y=2x-4 \end{cases}$

1. (2, 4)
2. (3, 3)
3. (1, –2)

Answer to Practice Problems: **1.** $x=1, y=2$ **2.** $x=1, y=2$ **3.** $x=-2,\ y=\frac{1}{2}$

4. $(-1, 1)$
5. $(-6, -2)$
6. Inconsistent
7. $(4, 1)$
8. $(0, 3)$
9. Inconsistent
10. $(4, -5)$
11. $(3, 2)$
12. Inconsistent
13. Dependent
14. $(3, -2)$
15. $(2, 2)$
16. $(-8, 16)$
17. $\left(\frac{1}{2}, -4\right)$
18. $\left(2, \frac{1}{3}\right)$ **19.** $\left(2, \frac{5}{2}\right)$
20. $\left(\frac{7}{2}, -\frac{1}{2}\right)$
21. $\left(-2, \frac{1}{2}\right)$
22. $\left(\frac{17}{16}, \frac{-19}{16}\right)$
23. $(2, -10)$
24. $(-2, 1)$
25. $\left(\frac{5}{2}, \frac{1}{8}\right)$
26. $\left(\frac{-4}{5}, \frac{-7}{5}\right)$
27. $(4, -1)$
28. $\left(\frac{11}{7}, \frac{8}{7}\right)$
29. $\left(\frac{13}{5}, \frac{-39}{5}\right)$
30. $\left(2, \frac{5}{3}\right)$
31. $(10, 4)$
32. $(3, 3)$
33. $(9, -3)$
34. $(10, 20)$
35. $(6, -4)$
36. Inconsistent
37. $(-7.5, 0)$
38. Dependent

4. $\begin{cases} y = 3x + 4 \\ 2y = 3x + 5 \end{cases}$

5. $\begin{cases} x = 3y \\ 3y - 2x = 6 \end{cases}$

6. $\begin{cases} 4x = y \\ 4x - y = 7 \end{cases}$

7. $\begin{cases} x - 5y + 1 = 0 \\ x = 7 - 3y \end{cases}$

8. $\begin{cases} 2x + 5y = 15 \\ x = y - 3 \end{cases}$

9. $\begin{cases} 7x + y = 9 \\ y = 4 - 7x \end{cases}$

10. $\begin{cases} 3y + 5x = 5 \\ y = 3 - 2x \end{cases}$

11. $\begin{cases} 3x - y = 7 \\ x + y = 5 \end{cases}$

12. $\begin{cases} 4x - 2y = 5 \\ y = 2x + 3 \end{cases}$

13. $\begin{cases} 6x + 3y = 9 \\ y = 3 - 2x \end{cases}$

14. $\begin{cases} x - y = 5 \\ 2x + 3y = 0 \end{cases}$

15. $\begin{cases} 4x = 8 \\ 3x + y = 8 \end{cases}$

16. $\begin{cases} x + y = 8 \\ 3x + 2y = 8 \end{cases}$

17. $\begin{cases} y = 2x - 5 \\ 2x + y = -3 \end{cases}$

18. $\begin{cases} 2x + 3y = 5 \\ x - 6y = 0 \end{cases}$

19. $\begin{cases} 2y = 5 \\ 3x - 4y = -4 \end{cases}$

20. $\begin{cases} x + 5y = 1 \\ x - 3y = 5 \end{cases}$

21. $\begin{cases} 3x + 8y = -2 \\ x + 2y = -1 \end{cases}$

22. $\begin{cases} 4x - 4y = 9 \\ 3x + y = 2 \end{cases}$

23. $\begin{cases} 5x + 2y = -10 \\ 7x = 4 - y \end{cases}$

24. $\begin{cases} x - 2y = -4 \\ 3x + y = -5 \end{cases}$

25. $\begin{cases} x + 4y = 3 \\ 3x - 4y = 7 \end{cases}$

26. $\begin{cases} 3x - y = -1 \\ 7x - 4y = 0 \end{cases}$

27. $\begin{cases} x + 5y = -1 \\ 2x + 7y = 1 \end{cases}$

28. $\begin{cases} x + 3y = 5 \\ 3x + 2y = 7 \end{cases}$

29. $\begin{cases} 3x - 4y - 39 = 0 \\ 2x - y - 13 = 0 \end{cases}$

30. $\begin{cases} \frac{x}{3} + \frac{y}{5} = 1 \\ x + 6y = 12 \end{cases}$

31. $\begin{cases} \frac{x}{5} + \frac{y}{4} - 3 = 0 \\ \frac{x}{10} - \frac{y}{2} + 1 = 0 \end{cases}$

32. $\begin{cases} 6x - y = 15 \\ 0.2x + 0.5y = 2.1 \end{cases}$

33. $\begin{cases} x + 2y = 3 \\ 0.4x + y = 0.6 \end{cases}$

34. $\begin{cases} 0.2x - 0.1y = 0 \\ y = x + 10 \end{cases}$

35. $\begin{cases} 0.1x - 0.2y = 1.4 \\ 3x + y = 14 \end{cases}$

36. $\begin{cases} 3x - 2y = 5 \\ y = \frac{3}{2}x + 2 \end{cases}$

37. $\begin{cases} x = 2y - 7.5 \\ 2x + 4y = -15 \end{cases}$

38. $\begin{cases} \frac{1}{2}x + \frac{1}{3}y = 4 \\ 3x + 2y = 24 \end{cases}$

39. (20, 5)
40. (15, 10)
41. (2, 8)
42. $c = 120$, $t = 50$

For Exercises 39 – 42, a word problem is stated with equations given that represent a mathematical model for the problem. Solve the system by using the method of substitution. (Note: These exercises were also given in Section 9.1. Check to see that you arrived at the same answers by both methods.)

39. The sum of two numbers is 25 and their difference is 15. What are the two numbers?
Let x = one number and y = the other number.

The corresponding modeling system is $\begin{cases} x+y=25 \\ x-y=15 \end{cases}$.

40. The perimeter of a rectangle is 50 meters and the length is 5 meters longer than the width. Find the dimensions of the rectangle.
Let x = the length and y = the width.

The corresponding modeling system is $\begin{cases} 2x+2y=50 \\ x-y=5 \end{cases}$.

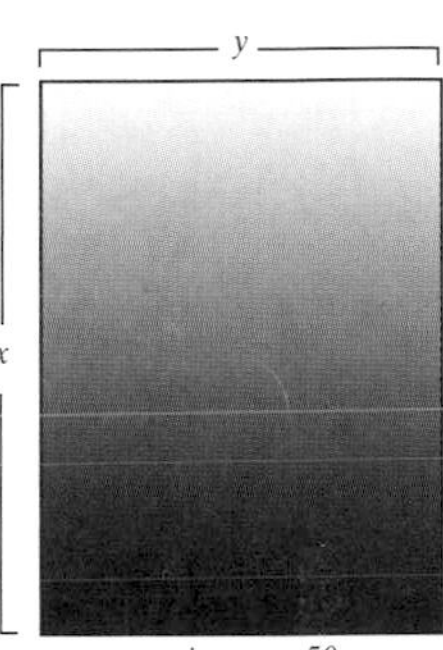

perimeter = 50m

41. Ten gallons of a salt solution consists of 30% salt. It is the result of mixing a 50% solution with a 25% solution. How many gallons of each of the mixing solutions were used?
Let x = the number of gallons of the 50% solution
and y = the number of gallons of the 25% solution.

The corresponding modeling system is $\begin{cases} x+y=10 \\ 0.50x+0.25y=0.30(10) \end{cases}$.

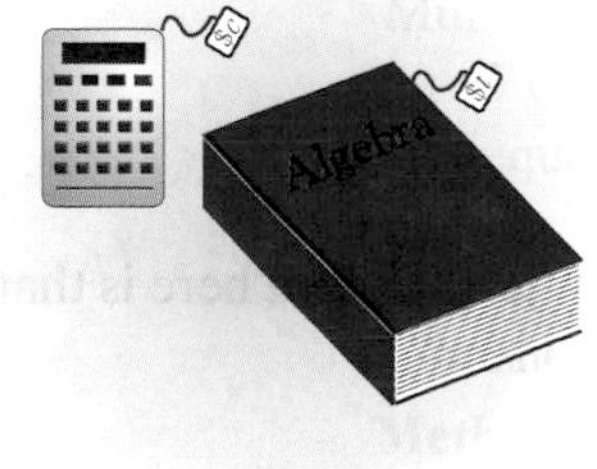

42. A student bought a calculator and a textbook for a course in algebra. He told his friend that the total cost was \$170 (without tax) and that the calculator cost \$20 more than twice the cost of the textbook. What was the cost of each item?
Let c = the cost of the calculator and t = the cost of the textbook.

The corresponding modeling system is $\begin{cases} c+t=170 \\ c=2t+20 \end{cases}$.

Hawkes Learning Systems: Introductory & Intermediate Algebra

Solving Systems of Linear Equations By Substitution

Example 5: Amounts and Costs

Three hot dogs and two orders of French fries cost \$10.80. Four hot dogs and four orders of fries cost \$16.60. What is the cost of a hot dog? What is the cost of an order of fries?

Solution: Let x = cost of one hot dog
and y = cost of one order of fries.
Then the system of linear equations is

$$\begin{cases} 3x+2y=10.80 & \text{Three hot dogs and two orders of French fries cost \$10.80.} \\ 4x+4y=16.60 & \text{Four hot dogs and four orders of fries cost \$16.60.} \end{cases}$$

Both equations are in standard form. Solve using the addition method.

$$\begin{cases} [-2]\ 3x+2y=10.80 \rightarrow & -6x-4y=-21.60 \\ \quad 4x+4y=16.60 \rightarrow & \underline{4x+4y=\quad 16.60} \end{cases}$$

$$\begin{aligned} -2x &= -5.00 \\ x &= 2.50 \end{aligned}$$

Substitute $x = 2.50$ into one of the original equations.

$$\begin{aligned} 3(2.50)+2y &= 10.80 \\ 7.50+2y &= 10.80 \\ 2y &= 3.30 \\ y &= 1.65 \end{aligned}$$

One hot dog costs \$2.50 and one order of fries costs \$1.65.

9.4 Exercises

1. $\begin{cases} x+y=56 \\ x-y=10 \end{cases}$; $x=33, y=23$

2. $\begin{cases} x+y=40 \\ 4x+2y=108 \end{cases}$; $x=14, y=26$

Solve each problem by setting up a system of two equations in two unknowns and solving the system.

1. The sum of two numbers is 56. Their difference is 10. Find the numbers.

2. The sum of two numbers is 40. The sum of twice the larger and 4 times the smaller is 108. Find the numbers.

3. The sum of two numbers is 36. Three times the smaller plus twice the larger is 87. Find the two numbers.

3. $\begin{cases} x+y=36 \\ 2x+3y=87 \end{cases}$;

$x=21, y=15$

4. $\begin{cases} x-y=17 \\ 4y=7+x \end{cases}$;

$x=25, y=8$

5. $\begin{cases} \frac{1}{3}(x+y)=4 \\ \frac{1}{2}(x-y)=4 \end{cases}$;

Rate of boat $(x)=10$ mph, Rate of current $(y)=2$ mph

6. $\begin{cases} 6(x+y)=1188 \\ 6(x-y)=\frac{2}{3}\cdot 1188 \end{cases}$;

Speed of the plane (x) $=165$ mph

Speed of the wind (y) $=33$ mph

7. $\begin{cases} x+y=3\frac{1}{2} \\ 52x+56y=190 \end{cases}$;

He traveled $1\frac{1}{2}$ hrs. at the first rate (x) and 2 hrs. at the second rate (y).

8. $\begin{cases} x+y=5\frac{3}{4} \\ 60x+55y=335 \end{cases}$;

First rate $(x)=3\frac{3}{4}$ hrs.,

Second rate $(y)=2$ hrs.

9. $\begin{cases} x-y=10 \\ \frac{200}{x}+\frac{200}{y}=9 \end{cases}$;

where x is his going rate and y is his return rate. His rate of speed to the city was 50 mph.

4. The difference between two numbers is 17. Four times the smaller is equal to 7 more than the larger. What are the numbers?

5. Ken makes a 4-mile motorboat trip downstream in 20 minutes $\left(\frac{1}{3}\text{hr.}\right)$. The return trip takes 30 minutes $\left(\frac{1}{2}\text{hr.}\right)$. Find the rate of the boat in still water and the rate of the current.

6. Mr. McKelvey finds that flying with the wind he can travel 1188 miles in 6 hours. However, when flying against the wind, he travels only $\frac{2}{3}$ of the distance in the same amount of time. Find the speed of the plane in still air and the wind speed.

7. Randy made a business trip of 190 miles. He averaged 52 mph for the first part of the trip and 56 mph for the second part. If the total trip took $3\frac{1}{2}$ hours, how long did he travel at each rate?

8. Marian drove to a resort 335 miles from her home. She averaged 60 mph for the first part of her trip and 55 mph for the second part. If her total driving time was $5\frac{3}{4}$ hours, how long did she travel at each rate?

9. Mr. Green traveled to a city 200 miles from his home to attend a meeting. Due to car trouble, his average speed returning was 10 mph less than his speed going. If the outbound trip was 4 hours and the return trip was 5 hours, at what rate of speed did he travel to the city?

10. Two trains leave Kansas City at the same time. One train travels east and the other travels west. The speed of the westbound train is 5 mph greater than the speed of the eastbound train. After 6 hours, they are 510 miles a part. Find the rate of each train. Assume the trains travel in a straight line in opposite directions.

11. Steve travels 4 times as fast as Fred. Traveling in opposite directions, they are 105 miles apart after 3 hours. Find their rates of travel.

12. Sue travels 5 mph less than twice as fast as June. Starting at the same point and traveling in the same direction, they are 80 miles apart after 4 hours. Find their speeds.

13. Mary and Linda live 324 miles apart. They start at the same time and travel toward each other. Mary's speed is 8 mph greater than Linda's. If they meet in 3 hours, find their speeds.

14. Two planes leave from points 1860 miles apart at the same time and travel toward each other (at slightly different altitudes, of course). If their rates are 220 mph and 400 mph, how soon will they meet?

10. $\begin{cases} x = y + 5 \\ 6(x + y) = 510 \end{cases}$;
Westbound train (x) = 45 mph, Eastbound train (y) = 40 mph

11. $\begin{cases} x = 4y \\ 3(x + y) = 105 \end{cases}$;
Steve (x) traveled at 28 mph and Fred (y) traveled at 7 mph

12. $\begin{cases} 4(S - J) = 80 \\ S = 2J - 5 \end{cases}$;
Sue's speed = 45 mph, June's speed = 25 mph

13. $\begin{cases} x = y + 8 \\ 3(x + y) = 324 \end{cases}$;
Mary's speed (x) was 58 mph and Linda's speed (y) was 50 mph.

14. $\begin{cases} 220x + 400y = 1860 \\ x = y \end{cases}$;
They will meet in 3 hrs.

15. $\begin{cases} x + y = \dfrac{8}{5} \\ 10x = 6y \end{cases}$;
He jogged about 12 miles.

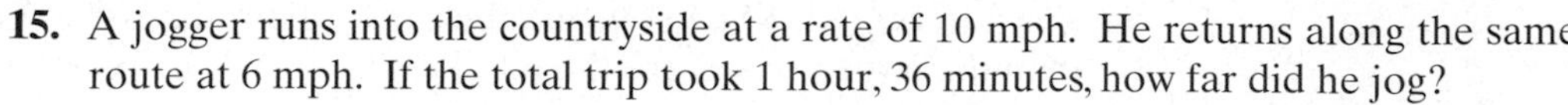

15. A jogger runs into the countryside at a rate of 10 mph. He returns along the same route at 6 mph. If the total trip took 1 hour, 36 minutes, how far did he jog?

16. A cyclist traveled to her destination at an average rate of 15 mph. By traveling 3 mph faster, she took 30 minutes less to return. What distance did she travel each way?

17. An airliner's average speed is $3\frac{1}{2}$ times the average speed of a private plane. Two hours after traveling in the same direction they are 580 miles apart. What is the average speed of each plane? (**Hint:** Since they are traveling in the same direction, the distance between them will be the difference of their distances.)

18. Sonja has some nickels and dimes. If she has 30 coins worth a total of \$2.00, how many of each type of coin does she have?

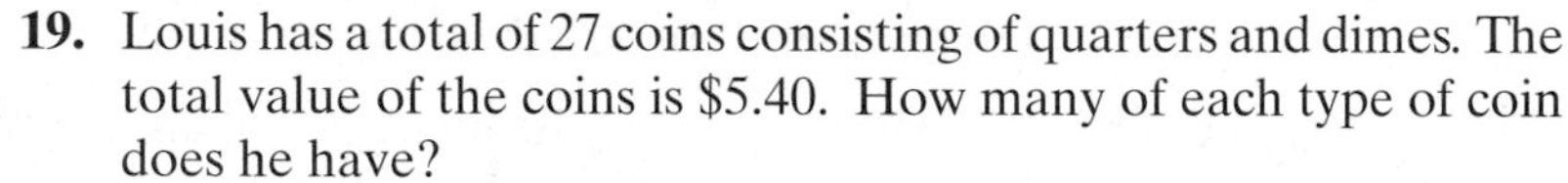

19. Louis has a total of 27 coins consisting of quarters and dimes. The total value of the coins is \$5.40. How many of each type of coin does he have?

20. Jill is 8 years older than her brother Curt. Four years from now, Jill will be twice as old as Curt. How old is each at the present time?

21. Two years ago, Anna was half as old as Beth. Eight years from now she will be two-thirds as old as Beth. How old are they now?

22. The length of a rectangle is 10 meters more than one-half the width. If the perimeter is 44 meters, what are the length and width?

23. The length of a rectangle is 1 meter less than twice the width. If each side is increased by 4 meters, the perimeter will be 116 meters. Find the length and the width of the original rectangle.

24. The line $y = mx + b$ passes through the two points $(1, 3)$ and $(5, 1)$. Find the equation of the line. (**Hint:** Substitute the values for x and y in the equation and solve the resulting system of equations for m and b.)

25. The line $y = mx + b$ passes through the two points $(-2, -1)$ and $(6, -7)$. Find the equation of the line. (**Hint:** Substitute the values for x and y in the equation and solve the resulting system of equations for m and b.)

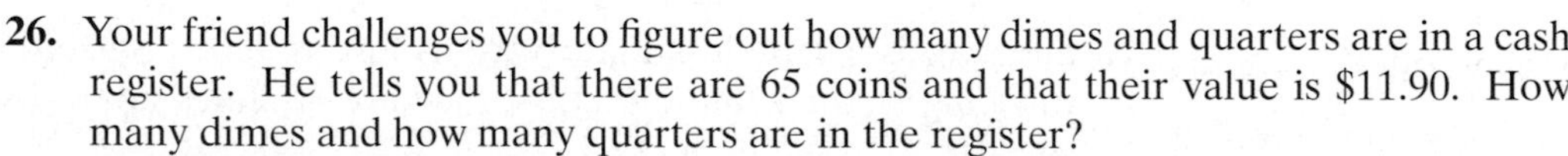

26. Your friend challenges you to figure out how many dimes and quarters are in a cash register. He tells you that there are 65 coins and that their value is \$11.90. How many dimes and how many quarters are in the register?

27. A bag contains pennies and nickels only. If there are 182 coins in all and their value is \$3.90, how many pennies and how many nickels are in the bag?

28. A Christmas charity party sold tickets for \$45.00 for adults and \$25.00 for children. The total number of tickets sold was 320 and the total for the ticket sales was \$13,000. How many adult and how many children's tickets were sold?

16. $\begin{cases} 15x = 18y \\ x - \frac{1}{2} = y \end{cases}$; where x is the time spent traveling to her destination and y is the time returning. She traveled 45 miles each way.

17. $\begin{cases} x = \frac{7}{2}y \\ 2(x - y) = 580 \end{cases}$; speed of airliner $(x) =$ 406 mph and speed of private plane $(y) = 116$ mph.

18. $\begin{cases} n + d = 30 \\ 0.05n + 0.1d = 2 \end{cases}$; $n = 20, d = 10$

19. $\begin{cases} q + d = 27 \\ 0.25q + 0.1d = 5.4 \end{cases}$; 18 quarters and 9 dimes.

20. $\begin{cases} J - C = 8 \\ J + 4 = 2(C + 4) \end{cases}$; Jill is 12 years old and Curt is 4 years old.

21. $\begin{cases} A - 2 = \frac{1}{2}(B - 2) \\ A + 8 = \frac{2}{3}(B + 8) \end{cases}$; Anna is 12 years old and Beth is 22 years old.

22. $\begin{cases} l = 10 + \frac{w}{2} \\ 2l + 2w = 44 \end{cases}$; $l = 14$ m and $w = 8$ m

29. Tickets for the local high school basketball game were priced at \$3.50 for adults and \$2.50 for students. If the income for one game was \$9550 and the attendance was 3500, how many adults and how many students attended that game?

30. The width of a rectangle is $\frac{3}{4}$ of its length. If the perimeter of the rectangle is 140 feet, what are the dimensions of the rectangle?

31. A farmer has 260 meters of fencing to build a rectangular corral. He wants the length to be 3 times as long as the width. What dimensions should he make his corral?

32. Joan went to a book sale on campus and bought paperback books for \$0.25 each and hardback books for \$1.75 each. If she bought a total of 15 books for \$11.25, how many of each type of book did she buy?

33. Admission to the baseball game is \$2.00 for general admission and \$3.50 for reserved seats. The receipts were \$36,250 for 12,500 paid admissions. How many of each ticket, general and reserved, were sold?

34. A men's clothing store sells two styles of sports jackets, one selling for \$95 and one selling for \$120. Last month, the store sold 40 jackets, with receipts totaling \$4250. How many of each style did the store sell?

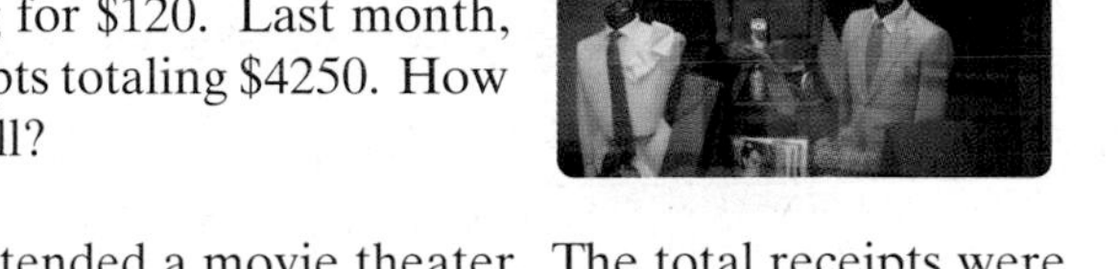

35. Seventy children and 160 adults attended a movie theater. The total receipts were \$620. One adult ticket and 2 children's tickets cost \$7. Find the price of each type of ticket.

36. Morton took some old newspapers and aluminum cans to the recycling center. Their total weight was 180 pounds. He received 1.5¢ per pound for the newspapers and 30¢ per pound for the cans. The total received was \$14.10. How many pounds of each did Morton have?

37. Frank bought 2 shirts and 1 pair of slacks for a total of \$55. If he had bought 1 shirt and 2 pairs of slacks, he would have paid \$68. What was the price of each shirt and each pair of slacks?

38. Four hamburgers and three orders of French fries cost \$5.15. Three hamburgers and five orders of fries cost \$5.10. What would one hamburger and one order of fries cost?

39. A small manufacturer produces two kinds of radios, model X and model Y. Model X takes 4 hours to produce and costs \$8 each to make. Model Y takes 3 hours to produce and costs \$7 each to make. If the manufacturer decides to allot a total of 58 hours and \$126 each week, how many of each model will be produced?

40. A furniture shop refinishes chairs. Employees use two methods to refinish a chair. Method I takes 1 hour and the material costs \$3. Method II takes $1\frac{1}{2}$ hours and the material costs \$1.50. Last week, they took 36 hours and spent \$60 refinishing chairs. How many did they refinish with each method?

23. $\begin{cases} l = 2w - 1 \\ 2(l+4)+2(w+4) = 116 \end{cases}$;
Length is 33 m and width is 17 m.

24. $\begin{cases} 3 = m(1)+b \\ 1 = m(5)+b; \end{cases}$;
$y = -\frac{1}{2}x + \frac{7}{2}$.

Writing and Thinking About Mathematics

41. A two digit number can be written as *ab*, where *a* and *b* are the digits. We do **not** mean that the digits are multiplied, but the value of the number is $10a + b$. For example, the two digit number 34 has a value of $10 \cdot 3 + 4$. Set up and solve a system of equations for the following problem:

The sum of the digits of a two digit number is 13. If the digits are reversed, then the value of the number is increased by 45. What is the number?

Hawkes Learning Systems: Introductory & Intermediate Algebra

Applications: Distance-Rate-Time, Number Problems, Amounts and Costs

25. $\begin{cases} -1 = m(-2)+b \\ -7 = m(6)+b \end{cases}$;
$y = -\frac{3}{4}x - \frac{5}{2}$.

26. $\begin{cases} q+d = 65 \\ 0.25q+0.1d = 11.9 \end{cases}$;
36 quarters and 29 dimes.

27. $\begin{cases} p+n = 182 \\ 0.01p+0.05n = 3.90 \end{cases}$;
$n = 52$ and $p = 130$.

28. $\begin{cases} a+c = 320 \\ 45a+25c = 13000 \end{cases}$;
250 adult tickets and 70 children's tickets were sold.

29. $\begin{cases} a+s = 3500 \\ 3.5a+2.5s = 9550 \end{cases}$;
800 adults and 2700 students attended.

30. $\begin{cases} w = \frac{3}{4}l \\ 2w+2l = 140 \end{cases}$;
Length is 40 ft. and width is 30 ft.

31. $\begin{cases} l = 3w \\ 2(l+w) = 260 \end{cases}$;
Length is 97.5 m and width is 32.5 m.

32. $\begin{cases} p + h = 15 \\ 0.25p + 1.75h = 11.25 \end{cases}$;
She bought 10 paperbacks and 5 hardbacks.

33. $\begin{cases} g+r = 12{,}500 \\ 2g+3.5r = 36{,}250 \end{cases}$;
5000 general admission and 7500 reserved tickets were sold.

34. $\begin{cases} 95x + 120y = 4250 \\ x + y = 40 \end{cases}$;
The store sold 22 of the \$95 jackets (x) and 18 of the \$120 jackets (y).

35. $\begin{cases} 70c + 160a = 620 \\ 2c + a = 7 \end{cases}$;
1 adult ticket is \$3 and 1 child's ticket is \$2.

36. $\begin{cases} .015n + .30c = 14.1 \\ n + c = 180 \end{cases}$;
Morton has 140 pounds newspapers and 40 pounds cans.

37. $\begin{cases} 2x + y = 55 \\ x + 2y = 68 \end{cases}$;
$x = \$14$ for shirts
$y = \$27$ for pair of slacks.

38. $\begin{cases} 4h + 3f = 5.15 \\ 3h + 5f = 5.10 \end{cases}$;
One hamburger costs \$0.95 and one order of fries costs \$ 0.45.

39. $\begin{cases} 4x + 3y = 58 \\ 8x + 7y = 126 \end{cases}$;
They produced 7 of Model *X* and 10 of Model *Y*.

40. $\begin{cases} x + \frac{3}{2}y = 36 \\ 3x + \frac{3}{2}y = 60 \end{cases}$;
They refinished 12 using Method I (x) and 16 using Method II (y).

41. $\begin{cases} a+b = 13 \\ 10b+a = 10a+b+45 \end{cases}$;
$a = 4; b = 9$;
The number is 49.

9.5 Applications: Interest and Mixture

Objectives

After completing this section, you will be able to:

1. *Solve applied problems related to interest by using systems of linear equations.*
2. *Solve applied problems related to mixture by using systems of linear equations.*

As we have seen throughout Chapter 9, systems of equations occur in many practical applications. In this section we will study two more types of applications: interest (involving money invested) and mixture. These applications can be "wordy" and you will need to study the following examples carefully to understand how to set up the system of equations.

Interest

People in business and banking know several formulas for calculating interest. The formula used depends on the method of payment (monthly or yearly) and the type of interest (simple or compound). Also, penalties for late payments and even penalties for early payments might be involved. In any case, the standard notation is to let P represent the principal (amount of money invested or borrowed), r represent the rate of interest (an annual rate), t represent the time, and I represent the interest.

In this section, we will use only the basic formula for simple interest $\boldsymbol{I = Prt}$ with interest calculated on an annual basis. **So, in the special case with $t = 1$, the formula becomes $I = Pr$.**

Example 1: Interest

a. James has two investment accounts, one pays 6% interest and the other pays 10% interest. He has \$1000 more in the 10% account than he has in the 6% account and the income from the 10% account exceeds the interest from the 6% account by \$260 each year. How much does he have in each account?

Solution: Let x = amount invested at 6%
and y = amount invested at 10%.

Then the system of equations is

$$\begin{cases} y - x = 1000 & y \text{ is larger than } x \text{ by \$1000.} \\ 0.10y - 0.06x = 260 & \text{Income from the 10\% account exceeds (is larger than) income from the 6\% account by \$260.} \end{cases}$$

continued on next page ...

We can solve the first equation for y: $y = x + 1000$.
Substituting for y in the second equation gives

$$
\begin{aligned}
0.10(x+1000)-0.06x &= 260 \\
10(x+1000) - 6x &= 26{,}000 \quad \text{Multiply by 100} \\
10x+10{,}000 - 6x &= 26{,}000 \\
4x &= 16{,}000 \\
x &= 4000
\end{aligned}
$$

Substitute $x = 4000$ into one of the original equations.

$$y - 4000 = 1000 \qquad \text{or} \qquad y = 4000 + 1000 = 5000$$

James has \$4000 invested at 6% and \$5000 invested at 10%.

b. Lila has \$7000 to invest. She decides to separate her funds into two investments. One yields an interest of 7%, and the other, 12%. If she wants an annual income from the investments to be \$690, how should she split the money? (**Note:** The higher interest account is considered more risky. Otherwise, she would put all her money in that account.)

Solution: Let x = amount to be invested at 7% and y = amount to be invested at 12%.

Then the system of equations is

$$\begin{cases} x + y = 7000 & \text{The sum of both amounts is \$7000.} \\ 0.07x + 0.12y = 690 & \text{The total interest is \$690.} \end{cases}$$

Both equations are in standard form. Multiplying the first equation by −7 and the second by 100 gives

$$\begin{cases} [-7] \quad x + y = 7000 \rightarrow -7x - 7y = -49{,}000 \\ [100] \quad 0.07x + 0.12y = 690 \rightarrow \underline{7x + 12y = 69{,}000} \end{cases}$$

$$
\begin{aligned}
5y &= 20{,}000 \\
y &= 4000
\end{aligned}
$$

Substitute $y = 4000$ into one of the original equations.

$$
\begin{aligned}
x + 4000 &= 7000 \\
x &= 3000
\end{aligned}
$$

She should invest \$3000 at 7% and \$4000 at 12%.

Mixture

Problems involving mixtures occur in physics and chemistry and in such places as candy stores or coffee shops. Two or more items of a different percentage of concentration of a chemical such as salt, chlorine, or antifreeze are to be mixed; or two or more types of coffee are to be mixed to form a final mixture that satisfies certain conditions of percentage of concentration.

The basic plan is to write an equation that deals with only one part of the mixture (such as the salt in the mixture). The following examples explain how this can be accomplished.

Example 2: Mixture

a. How many ounces each of a 10% salt solution and a 15% salt solution must be used to produce 50 ounces of a 12% salt solution?

Solution: Let x = amount of 10% solution
and y = amount of 15% solution.

	amount of solution ·	percent of salt =	amount of salt
10% solution	x	0.10	$0.10x$
15% solution	y	0.15	$0.15y$
12% solution	50	0.12	0.12(50)

continued on next page ...

Then the system of linear equations is

$$\begin{cases} x + y = 50 & \text{The sum of the two amounts must be 50 ounces.} \\ 0.10x + 0.15y = 0.12(50) & \text{The sum of the amount of salt from each solution equals the total amount of salt in the final solution.} \end{cases}$$

Multiplying the first equation by –10 and the second by 100 gives,

$$\begin{cases} [-10] \quad x + y = 50 \rightarrow -10x - 10y = -500 \\ [100] \quad 0.10x + 0.15y = 0.12(50) \rightarrow \underline{10x + 15y = 600} \end{cases}$$

$$5y = 100$$

$$y = 20$$

Substitute $y = 20$ into one of the original equations.

$$x + 20 = 50$$

$$x = 30.$$

Use 30 ounces of the 10% solution and 20 ounces of the 15% solution.

Teaching Notes:
Because "mixture" and "interest" appear to represent different contexts, students may not recognize that the structures of the equations are the same, or that the interest context is a mixture (amount of the interest is the part of the mixture we are interested in). You might emphasize the fact that both types of problems involve percent.

b. How many gallons of a 20% acid solution should be mixed with a 30% acid solution to produce 100 gallons of a 23% solution?

Solution: Let x = amount of 20% solution and y = amount of 30% solution.

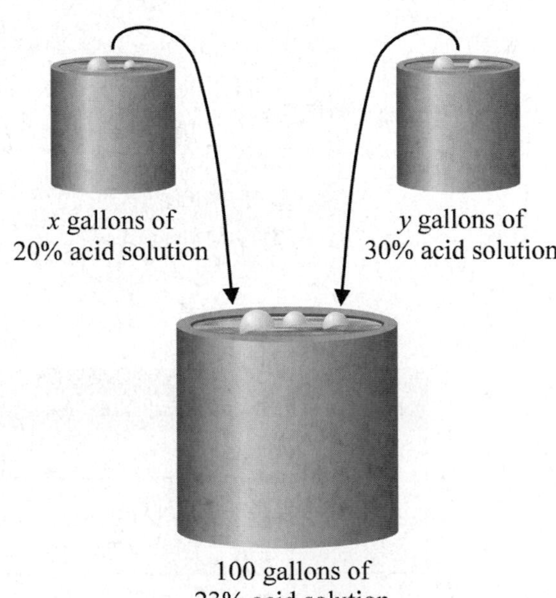

	amount of solution	· percent of acid	= amount of acid
20% solution	x	0.20	$0.20x$
30% solution	y	0.30	$0.30y$
23% solution	100	0.23	0.23(100)

Then the system of linear equations is

$$\begin{cases} x+y=100 & \text{The sum of the two amounts must be 100 gallons.} \\ 0.20x+0.30y=0.23(100) & \text{The sum of the amount of acid from each solution equals the total amount of acid in the final solution.} \end{cases}$$

Multiplying the first equation by –20 and the second by 100 gives,

$$\begin{cases} [-20] & x+\quad y=100 & \to -20x-20y=-2000 \\ [100] & 0.20x+0.30y=0.23(100) & \to \underline{20x+30y=2300} \end{cases}$$

$$10y=300$$

$$y=30$$

Substitute $y=30$ into one of the original equations.

$$x+30=100$$

$$x=70$$

Seventy gallons of the 20% solution should be added to 30 gallons of the 30% solution. This will produce 100 gallons of a 23% solution.

9.5 Exercises

1. \$5500 at 6%, \$3500 at 10%

Solve each problem by setting up a system of two equations in two unknowns and solving the system.

1. Carmen invested \$9000, part in a 6% passbook account and the rest in a 10% certificate account. If her annual interest was \$680, how much did she invest at each rate?

2. Mrs. Brown has \$12,000 invested. Part is invested at 6% and the remainder at 8%. If the interest from the 6% investment exceeds the interest from the 8% investment by \$230, how much is invested at each rate?

2. \$8500 at 6%, \$3500 at 8%
3. \$7400 at 5.5%, \$2600 at 6%
4. \$3200 at 3%, \$6300 at 6%

3. Ten thousand dollars is invested, part at 5.5% and part at 6%. The interest from the 5.5% investment exceeds the interest from the 6% investment by \$251. How much is invested at each rate?

4. On two investments totaling \$9500, Bill lost 3% on one and earned 6% on the other. If his net annual receipts were \$282, how much was each investment?

5. Marsha has money in two savings accounts. One rate is 8% and the other is 10%. If she has $200 more in the 10% account, how much is invested at 8% if the total interest is $101?

6. Money is invested at two rates. One rate is 9% and the other is 13%. If there is $700 more invested at 9%, find the amount invested at each rate if the annual interest is $239.

7. Frank has half of his investments in stock paying an 11% dividend and the other half in a debentured stock paying 13% interest. If his total annual interest is $840, how much does he have invested?

8. Betty invested some of her money at 12% interest. She invested $300 more than twice that amount at 10%. How much is invested at each rate if her income is $318 annually?

9. GFA invested some money in a development yielding 24% and $9000 less in a development yielding 18%. If the first investment produces $2820 more per year than the second, how much is invested in each development?

10. Judy invests a certain amount of money at 7% annual interest and three times that amount at 8%. If her annual income is $232.50, how much does she have invested at each rate?

11. Norman has a certain amount of money invested at 5% annual interest and $500 more than twice that amount invested in bonds yielding 7%. His total income from interest is $187. How much does he have invested at each rate?

12. A total of $6000 is invested, part at 8% and the remainder at 12%. How much is invested at each rate if the annual interest is $620?

13. Mr. Brown has $12,000 invested. Part is invested at 9% and the remainder at 11%. If the interest from the 9% investment exceeds the interest from the 11% investment by $380, how much is invested at each rate?

14. Eight thousand dollars is invested, part at 15% and the remainder at 12%. If the annual income from the 15% investment exceeds the income from the 12% investment by $66, how much is invested at each rate?

15. A metallurgist has one alloy containing 20% copper and another containing 70% copper. How many pounds of each alloy must he use to make 50 pounds of a third alloy containing 50% copper?

5. $450 at 8%, $650 at 10%

6. $1500 at 9%, $800 at 13%

7. $3500 in each or $7000 total

8. $900 at 12%, $2100 at 10%

9. $20,000 at 24%, $11,000 at 18%

10. $750 at 7%, $2250 at 8%

11. $800 at 5%, $2100 at 7%

12. $2500 at 8%, $3500 at 12%

13. $8500 at 9%, $3500 at 11%

14. $3800 at 15%, $4200 at 12%

15. 20 pounds of 20%, 30 pounds of 70%

16. 8 tons of 80%, 16 tons of 50%

16. A manufacturer has received an order for 24 tons of a 60% copper alloy. His stock contains only alloys of 80% copper and 50% copper. How much of each will he need to fill the order?

17. 20 ounces of 30%, 30 ounces of 20%

17. A tobacco shop wants 50 ounces of tobacco that is 24% rare Turkish blend. How much each of a 30% Turkish blend and a 20% Turkish blend will be needed?

18. 40 litres of 40%, 20 litres of 55%

18. How many liters each of a 40% acid solution and a 55% acid solution must be used to produce 60 liters of a 45% acid solution?

19. 450 pounds of 35%, 1350 pounds of 15%

19. A dairy man wants to mix a 35% protein supplement and a standard 15% protein ration to make 1800 pounds of a high-grade 20% protein ration. How many pounds of each should he use?

20. 440 pounds of 70%, 200 pounds of 54%

20. To meet the government's specifications, an alloy must be 65% aluminum. How many pounds each of a 70% aluminum alloy and a 54% aluminum alloy will be needed to produce 640 pounds of the 65% aluminum alloy?

21. 20 pounds of 40%, 30 pounds of 15%

21. A meat market has ground beef that is 40% fat and extra lean ground beef that is only 15% fat. How many pounds of each will be needed to obtain 50 pounds of lean ground beef that is 25% fat?

22. Regular costs \$1.10 per gallon, Premium costs \$1.25 per gallon

22. George decides to mix grades of gasoline in his truck. He puts in 8 gallons of regular and 12 gallons of premium for a total cost of \$23.80. If premium gasoline costs \$0.15 more per gallon than regular, what was the price of each grade of gasoline?

23. 15 pounds of cashews, 5 pounds of peanuts

23. Cashews cost \$0.80 a pound and peanuts cost \$0.68 a pound. A caterer needs a 20 pound mixture of cashews and peanuts for an upcoming event. She wants to pay \$0.77 per pound. How many pounds of each type of nut should she buy?

24. 2 pounds at \$3.10, 3 pounds at \$5.60

24. Jennifer wants to create a mixture of two different flavored coffee beans. She wishes to create a mixture of 5 pounds and spend only \$23.00. The two flavors she wants cost \$3.10 and \$5.60. How many pounds of each flavor should she buy?

Hawkes Learning Systems: Introductory & Intermediate Algebra

Applications: Interest and Mixture

Chapter 9 Index of Key Ideas and Terms

Section 9.1 Systems of Equations: Solutions by Graphing

System of Linear Equations page 626

A pair of linear equations considered together is called a **system of linear equations** (or a **set of simultaneous equations**).

Solutions by Graphing page 627

The two lines related to a system of equations can be graphed.

- **a.** If the two lines intersect in a single point, the point is the solution to the system. The system is said to be **consistent**. page 628
- **b.** If the two lines are parallel (do not intersect), the system is said to be **inconsistent** and there are no solutions. pages 628 - 629
- **c.** If the two lines coincide (are the same line), the system is said to be **dependent** and there are an infinite number of solutions. page 629

Section 9.2 Systems of Equations: Solutions by Substitution

Solving a System by Substitution page 638

1. Solve one of the equations for one of the variables.
2. Substitute the resulting expression into the other equation.
3. Solve this new equation, if possible, and then substitute back into one of the original equations to find the value of the other variable. (This is known as **back substitution**.)

Section 9.3 Systems of Equations: Solutions by Addition

Solutions by Addition pages 644 - 647

1. Write the equations in standard form.
2. Arrange the equations vertically so that like terms are aligned.
3. Multiply one or both of the equations by a constant so that the coefficients of one of the variables are opposites.
4. Add the like terms.
5. Solve the resulting equation.
6. Substitute the results back into one of the original equations to find the value of the other variable.

continued on next page ...

Section 9.3 Systems of Equations: Solutions by Addition (continued)

Guidelines for Deciding which Method to Use in Solving a System of Linear Equations page 650

1. The graphing method is helpful in "seeing" the geometric relationship between the lines and finding approximate solutions. A calculator can be very helpful here.
2. Both the substitution method and the addition method give exact solutions.
3. The substitution method may be reasonable and efficient if one of the coefficients of one of the variables is 1.
4. In general, the method of addition will prove to be most efficient.

Section 9.4 Applications: Distance-Rate-Time, Number Problems, Amounts and Costs

Applications

Distance-Rate-Time, Number Problems, Amounts and Costs pages 654 - 658

Section 9.5 Applications: Interest and Mixture

Applications

Interest and Mixture pages 663 - 667

Chapter 9 Review

For a review of the topics and problems from Chapter 9, look at the following lessons from *Hawkes Learning Systems: Introductory & Intermediate Algebra.*

Solving Systems of Linear Equations by Graphing
Solving Systems of Linear Equations by Substitution
Solving Systems of Linear Equations by Addition
Applications: Distance-Rate-Time, Number Problems, Amounts and Costs
Applications: Interest and Mixture

Chapter 9 Test

In Exercises 1 and 2, determine which of the points satisfy the given system of linear equations.

1. $\begin{cases} 3x-7y=5 \\ 5x-2y=-11 \end{cases}$

a. $(1,8)$
b. $(4,1)$
c. $(-3,-2)$
d. $(13,10)$

2. $\begin{cases} x-2y=7 \\ 2x-3y=5 \end{cases}$

a. $(0,3)$
b. $(7,0)$
c. $(1,-1)$
d. $(-11,-9)$

For Exercises 3 – 13 solve as directed. In each exercise state whether the system is (a) consistent, (b) inconsistent, or (c) dependent.

Solve the systems in Exercises 3 – 6 by graphing.

3. $\begin{cases} y=2-5x \\ x-y=6 \end{cases}$

4. $\begin{cases} x-y=3 \\ 2x+3y=11 \end{cases}$

5. $\begin{cases} 2x+y=16 \\ 0.1x+0.01y=0.64 \end{cases}$

6. $\begin{cases} 25x+18y=70 \\ 3.5x-2.7y=4 \end{cases}$

Solve the systems in Exercises 7 and 8 by using substitution.

7. $\begin{cases} 5x-2y=0 \\ y=3x+4 \end{cases}$

8. $\begin{cases} x=\frac{1}{3}y-4 \\ 2x+\frac{3}{2}y=5 \end{cases}$

Solve the systems in Exercises 9 and 10 by using the method of additon.

9. $\begin{cases} -2x+3y=6 \\ 4x+y=1 \end{cases}$

10. $\begin{cases} 2x-4y=9 \\ 3x+6y=8 \end{cases}$

Solve the systems in Exercises 11 – 13 by using any method.

11. $\begin{cases} x+y=2 \\ y=-2x-1 \end{cases}$

12. $\begin{cases} 6x+2y-8=0 \\ y=-3x \end{cases}$

13. $\begin{cases} 7x+5y=-9 \\ 6x+2y=6 \end{cases}$

14. Determine the values of a and b such that the straight line $ax+by=11$ passes through the two points $(1,-3)$ and $(2,5)$.

1. c
2. d
3. Consistent, $\left(\frac{4}{3}, \frac{-14}{3}\right)$

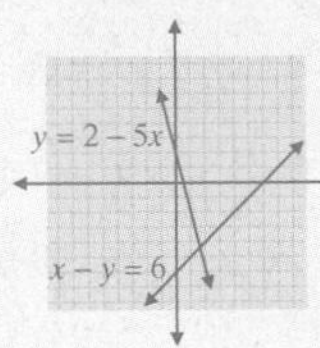

4. Consistent, $(4,1)$

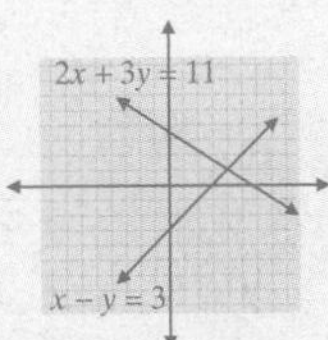

5. $x=6, y=4$

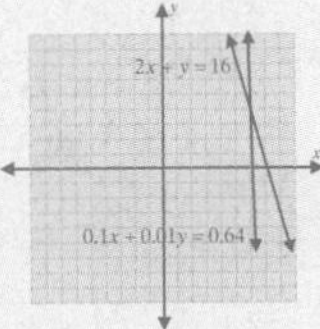

6. $x=2,\ y=1.\overline{11}$

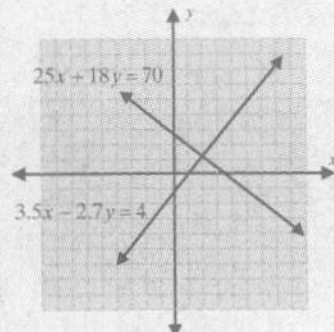

7. Consistent
$x=-8, y=-20$
8. Consistent
$x=-2, y=6$
9. Consistent
$x=-\frac{3}{14}, y=\frac{13}{7}$
10. Consistent
$x=\frac{43}{12}, y=-\frac{11}{24}$
11. Consistent
$x=-3, y=5$
12. Inconsistent
13. Consistent
$x=3, y=-6$
14. $a=8, b=-1$

15. Speed of the boat = 14 mph; Speed of the current = 2 mph

16. Pen price = \$0.79; Pencil price = \$0.08

17. \$1600 at 8% and \$960 at 6%

18. 1600 lbs. of 83% and 400 lbs. of 68%

19. 13 in. by 17 in.

20. nickels = 45, quarters = 60

21. 80, 15

22. Scott is 9 years old and Clayton is 3 years old.

23. 4 gallons of 65%, 10 gallons of 30%

In Exercises 15 – 23, solve each problem by setting up a system of two equations in two unknowns and solving the system.

15. Pete's boat can travel 48 miles upstream in 4 hours. The return trip takes 3 hours. Find the speed of the boat in still water and the speed of the current.

16. Eight pencils and two pens cost \$2.22. Three pens and four pencils cost \$2.69. What is the price of each pen and each pencil?

17. Gary has two investments yielding a total annual interest of \$185.60. The amount invested at 8% is \$320 less than twice the amount invested at 6%. How much is invested at each rate?

18. A metallurgist needs 2000 pounds of an alloy that is 80% copper. In stock, he has only alloys of 83% copper and 68% copper. How many pounds of each must be used?

19. The perimeter of a rectangle is 60 inches and the length is 4 inches longer than the width. Find the dimensions of the rectangle.

20. Sonia has a bag of coins with only nickels and quarters. She wants you to figure out how many of each type of coin she has and tells you that she has 105 coins and that the value of the coins is \$17.25. Tell her that you know algebra and determine how many nickels and how many quarters she has.

21. The difference between two numbers is 65. Six times the smaller number equals 10 more than the larger number. What are the two numbers?

22. Scott is 6 years older than his brother Clayton. Three years from now, Scott will be twice as old as Clayton. How old is each at the present time?

23. Fourteen gallons of a salt solution consists of 40% salt. It is the result of mixing a 65% solution with a 30% solution. How many gallons of each of the mixing solutions were used?

Cumulative Review: Chapters 1 – 9

1. -165
2. 28
3. 55
4. 8750
5. $x = \frac{1}{5}$
6. $x = \frac{11}{4}$
7. $y = -7$
8. $a = \frac{25}{11}$
9. $x \geq 13$

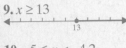

13

10. $-5 \leq x < -4.2$

−5 −4.2

11. $t = \frac{v-k}{g}$
12. $y = \frac{3x-6}{4}$
13. $m = -\frac{1}{2},\ b = \frac{11}{4}$

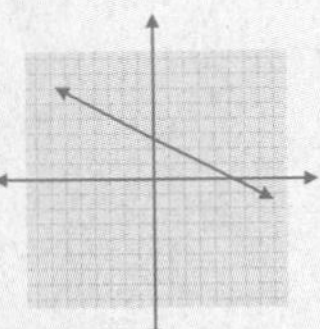

14. $x - 3y = -5$

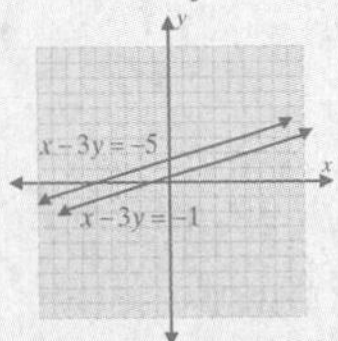

15. $4x + 5y - 2 = 0$

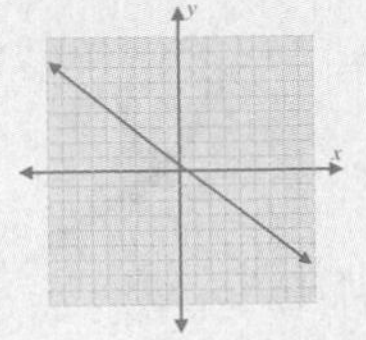

Use the rules for order of operations to evaluate the expressions in Exercises 1 and 2.

1. $-20 + 15 \div (-5) \cdot 2^3 - 11^2$

2. $24 \div 4 \cdot 6 - 36 \cdot 2 \div 3^2$

3. What is 110% of 50?

4. Find $\frac{7}{8}$ of 10,000.

Solve each of the equations in Exercises 5 – 8.

5. $2(x - 7) + 14 = -3x + 1$

6. $\frac{5x}{6} - \frac{2}{3} = \frac{x}{2} + \frac{1}{4}$

7. $2.3y - 1.6 = 3(1.2y + 2.5)$

8. $\frac{a}{5} + \frac{a}{7} = \frac{a}{2} - \frac{5}{14}$

In Exercises 9 and 10, solve the inequality and graph the solution set on a real number line.

9. $3x - 14 \geq 25$

10. $0 \leq 2x + 10 < 1.6$

In Exercises 11 and 12, solve for the indicated variable.

11. $v = k + gt$; solve for t.

12. $3x - 4y = 6$; solve for y.

13. For the equation $2x + 4y = 11$, determine the slope, m, and the y-intercept, b. Then use a graphing calculator to graph the line.

14. Write an equation for the line parallel to the line $x - 3y = -1$ and passing through the point $(1, 2)$. Use a graphing calculator to graph both lines.

15. Find an equation in standard form for the line that has slope $m = -\frac{4}{5}$ and contains the point $(-2, 2)$. Then graph the line.

16. Find an equation in standard form for the line that passes through the point $(3, -4)$ and is parallel to the line $2x - y = 5$. Graph both lines.

17. Find an equation in standard form for the line that passes through the point $(7, 4)$ and is perpendicular to the line $6x + y = 13$. Graph both lines.

18. Find an equation in slope-intercept form for the line which contains the point $(-9, 3)$ and is perpendicular to the line that passes through the points $(4, 8)$ and $(12, 4)$. Graph both lines.

19. Determine the values of a and b such that the straight line $ax + by = 14$ passes through the two points $(-1, 19)$ and $(2, 4)$.

20. Determine which of the points lie on both of the lines in the given system of equations.

$$\begin{cases} x - 2y = 6 \\ y = \frac{1}{2}x - 3 \end{cases}$$

a. $(0, 6)$ **b.** $(6, 0)$ **c.** $(2, -2)$ **d.** $\left(3, -\frac{3}{2}\right)$

21. Use a graphing calculator to estimate the solution to the following system of linear equations: $\begin{cases} y = 5x - 3 \\ 2x + y = 10 \end{cases}$

In Exercises 22 – 27, use a graphing calculator to estimate the solutions to each system of equations. You may need to estimate the answers.

22. $\begin{cases} x + y = 5.2 \\ 2x - y = -1.5 \end{cases}$ **23.** $\begin{cases} -x + 2y = 6 \\ 3x + 2y = -4 \end{cases}$

24. $\begin{cases} 2x + 3y = 18 \\ 3x - 2y = -12 \end{cases}$ **25.** $\begin{cases} y = \frac{1}{3}x + 5 \\ y = -2x + 1 \end{cases}$

26. $\begin{cases} 2y = 9x - 16 \\ 4y = 12 - 2x \end{cases}$ **27.** $\begin{cases} 8x = y - 1 \\ 14 = 16x + 3y \end{cases}$

For Exercises 28 – 46 solve as directed. In each exercise state whether the system is (a) consistent, (b) inconsistent, or (c) dependent.

Solve the systems in Exercises 28 – 31 by graphing.

28. $\begin{cases} 2x + 3y = 4 \\ 3x - y = 6 \end{cases}$ **29.** $\begin{cases} 5x + 2y = 3 \\ y = 4 \end{cases}$ **30.** $\begin{cases} 3x + y = 3 \\ x - 3y = 4 \end{cases}$ **31.** $\begin{cases} y = 4x - 6 \\ 8x - 2y = -4 \end{cases}$

Solve the systems in Exercises 32 – 35 by using the method of substitution.

32. $\begin{cases} x + y = -4 \\ 2x + 7y = 2 \end{cases}$ **33.** $\begin{cases} x = 2y \\ y = \frac{1}{2}x + 9 \end{cases}$ **34.** $\begin{cases} 4x + 3y = 8 \\ x + \frac{3}{4}y = 2 \end{cases}$ **35.** $\begin{cases} 2x + y = 0 \\ 7x + 6y = -10 \end{cases}$

Solve the systems in Exercises 36 – 39 using the method of addition.

36. $\begin{cases} 2x + y = 7 \\ 2x - y = 1 \end{cases}$ **37.** $\begin{cases} 3x - 2y = 9 \\ x - 2y = 11 \end{cases}$ **38.** $\begin{cases} 2x + 4y = 9 \\ 3x + 6y = 8 \end{cases}$ **39.** $\begin{cases} x + 5y = 10 \\ y = 2 - \frac{1}{5}x \end{cases}$

16. $2x - y - 10 = 0$

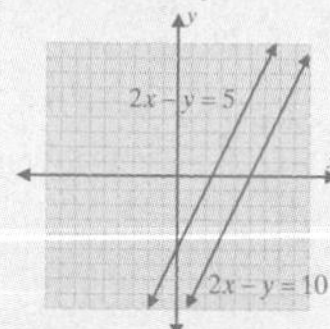

17. $x - 6y = -17$

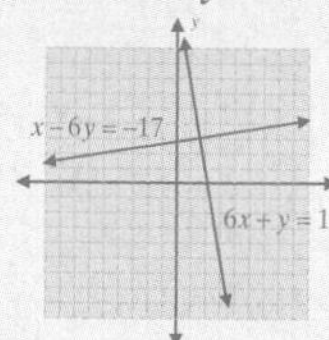

18. $y = 2x + 21$; original line is $y = -\frac{1}{2}x + 10$

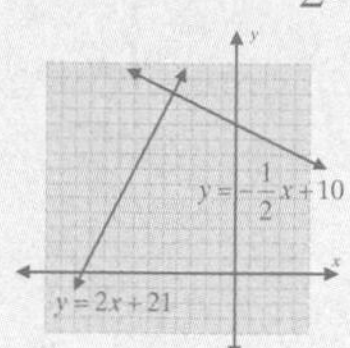

19. $a = 5, b = 1$

20. b, c, d

21. $x = \frac{13}{7}, y = \frac{44}{7}$

22. $x = \frac{37}{30}, y = \frac{119}{30}$

23. $x = -\frac{5}{2}, y = \frac{7}{4}$

24. $x = 0, y = 6$

25. $x = -\frac{12}{7}, y = \frac{31}{7}$

26. $x = \frac{11}{5}, y = \frac{19}{10}$

27. $x = \frac{11}{40}, y = \frac{16}{5}$

28. Consistent, $(2, 0)$

29. Consistent, $(-1, 4)$

30. Consistent, $\left(\frac{13}{10}, -\frac{9}{10}\right)$

Solve the systems in Exercises 40 – 46 by using any method.

40. $\begin{cases} 3y - x = 7 \\ x - 2y = -2 \end{cases}$ **41.** $\begin{cases} x - \frac{2}{5}y = \frac{4}{5} \\ \frac{3}{4}x + \frac{3}{4}y = \frac{5}{4} \end{cases}$ **42.** $\begin{cases} x + 3y = 7 \\ 5x - 2y = 1 \end{cases}$ **43.** $\begin{cases} 3x - 5y = 17 \\ x + 2y = 4 \end{cases}$

44. $\begin{cases} 2x - y = 10 \\ x + 2y = 5 \end{cases}$ **45.** $\begin{cases} y = 3x + 2 \\ -6x + 2y = 4 \end{cases}$ **46.** $\begin{cases} y = \frac{1}{2}x + 3 \\ 6x - 2y = 3 \end{cases}$

Solve each system of equations in Exercises 47 and 48 by using an algebraic method.

47. $\begin{cases} 3x - 4y = -25 \\ 2x + y = 9 \end{cases}$ **48.** $\begin{cases} x + 8y = -22 \\ 3x - y = -9 \end{cases}$

In Exercises 49 and 50, write an equation for the line determined by the two given points. Use the formula $y = mx + b$ to set up a system of equations with m and b as the unknowns.

49. $(3, -1), (2, 6)$ **50.** $(-2, 5), (4, -3)$

51. a. Two angles are **complementary** if the sum of their measures is 90°. Find two complementary angles such that one is 10° more than three times the other.

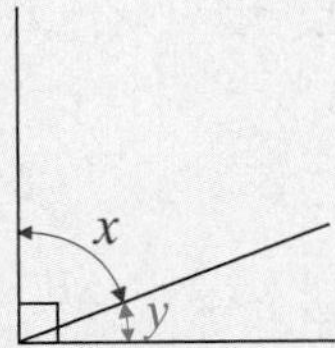

b. Two angles are **supplementary** if the sum of their measures is 180°. Find two supplementary angles such that one is 30° less than one-half of the other.

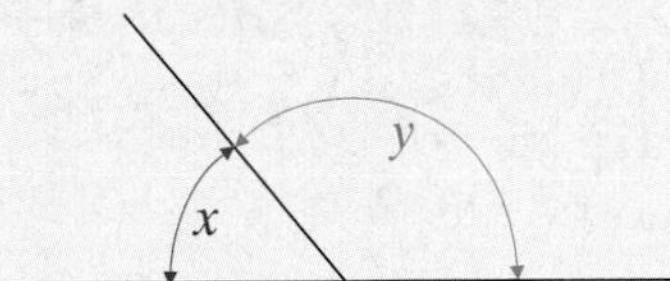

52. Determine the values of *a* and *b* such that the straight line $ax + by = 9$ passes through the two points $(-1, -3)$ and $(1, -1.5)$.

53. Sylvester has two investments that total \$100,000. One investment earns interest at 6% and the other investment earns interest at 8%. If the total amount of interest from the two investments is \$6700 in one year, how much money does he have invested at each rate?

31. Inconsistent
32. Consistent, $(-6, 2)$
33. Inconsistent
34. Dependent
35. Consistent, $(2, -4)$
36. Consistent, $(2, 3)$
37. Consistent, $(-1, -6)$
38. Inconsistent
39. Dependent
40. Consistent, $(8, 5)$
41. Consistent, $\left(\frac{22}{21}, \frac{13}{21}\right)$
42. Consistent, $(1, 2)$
43. Consistent, $\left(\frac{54}{11}, \frac{-5}{11}\right)$
44. Consistent, $(5, 0)$
45. Dependent
46. Consistent, $\left(\frac{9}{5}, \frac{39}{10}\right)$
47. $x = 1, y = 7$
48. $x = -\frac{94}{25}, y = -\frac{57}{25}$
49. $y = -7x + 20$
50. $y = -\frac{4}{3}x + \frac{7}{3}$
51. a. The measures of the angles are 20° and 70°.
b. The measures of the angles are 40° and 140°.
52. $a = 3, b = -4$
53. \$35,000 at 8% & \$65,000 at 6%

54. 6 ounces of 10% mixture & 2 ounces of 30% mixture

54. A chemist needs 8 ounces of a mixture that is 15% iodine. He has a 10% mixture and a 30% mixture in stock. How many ounces of each mixture must he use to get the mixture he wants?

55. Ten 25¢ stamps and five 34¢ stamps.

55. Bernice's secretary bought 15 stamps, some 25¢ and some 34¢. If he spent $4.20, how many of each kind did he buy?

56. speed of the boat =10 mph and speed of the current = 2 mph.

56. A boat can travel 24 miles downstream in 2 hours. The return trip takes 3 hours. Find the speed of the boat in still water and the speed of the current.

57. length = 14 yards width = 11 yards.

57. The perimeter of a rectangle is 50 yards and the width is only 3 yards less than the length. What are the dimensions of the rectangle?

58. A = 4; B = 18

58. A company manufactures two kinds of dresses, Model A and Model B. Each Model A dress takes 4 hours to produce and each costs $18. Each Model B takes 2 hours to produce and each costs $7. If during a week there were 52 hours of production and costs of $198, how many of each model were produced?

59. Eastbound train is traveling at 40 mph, Westbound train is traveling at 45 mph.

59. Two trains leave Kansas City at the same time. One train travels east and the other travels west. The speed of the westbound train is 5 mph greater than the speed of the eastbound train. After 6 hours, they are 510 miles apart. Find the rate of each train. (Assume that the trains travel in a straight line in opposite directions.)

60. train's speed = 60 mph; speed of airplane = 230 mph

60. An airplane can travel 1035 miles in the same time that a train travels 270 miles. The speed of the plane is 50 mph more than three times the speed of the train. Find the speed of each.

61. 60, 24

61. Two numbers are in a ratio of 2 to 5. If their sum is 84, find the numbers.

62. 3 mph

62. A family travels 18 miles downriver and returns. It takes 8 hours to make the round trip. Their rate in still water is twice the rate of the current. Find the rate of the current.

63. $x = 3, y = 11$

64. $\frac{2}{3}$ hr or 40 minutes

65. 120 sq yds

63. The sum of two numbers is 14. Twice the larger number added to three times the smaller is equal to thirty-one. Find both numbers.

64. Emily started walking to a town 10 miles away at a rate of 3 mph. After walking part of the way, she got a ride on a bus. The bus traveled at an average rate of 48 mph, and Emily reached the town 50 minutes after she started. How long did she walk?

65. LeAnn is a carpet layer. She has agreed to carpet a house for $18 per square yard. The carpet will cost her $11 per square yard, and she knows that there is approximately 12% waste in cutting and matching. If she plans to make a profit of $580.80, how many yards of carpet will she buy?

66. $N = 950t - 850$

67. { (−5, −2), (−3, −2), (−2, 4), (1, −1), (2, 4) }; $D = \{-5, -3, -2, 1, 2\}$; $R = \{-2, -1, 4\}$; It is a function.

68. Not a function

69. Function

70. **a.** −22 **b.** 23 **c.** 3.5

71. **a.** −10 **b.** −1 **c.** 80

66. In a biology experiment, Owen observed that the bacteria count in a culture was approximately 100 at the end of 1 hour. At the end of 3 hours, the bacteria count was about 2000. Write a linear equation describing the bacteria count, N, in terms of the time, t.

67. List the set of ordered pairs corresponding to the points on the graph. Give the domain and range, and indicate if the relation is or is not a function.

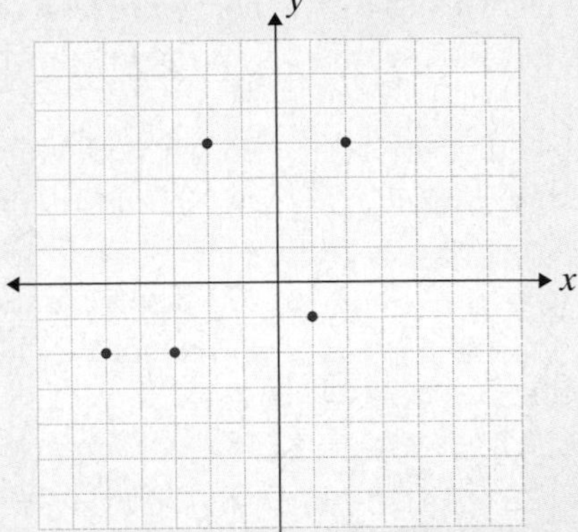

In Exercises 68 and 69, use the vertical line test to determine whether each of the graphs does or does not represent a function.

68.

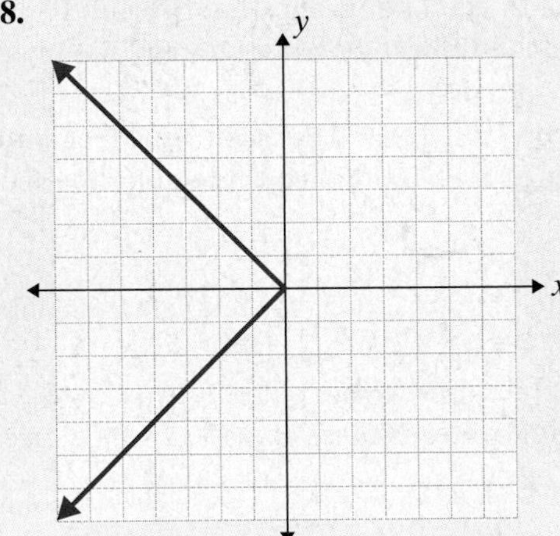

69.

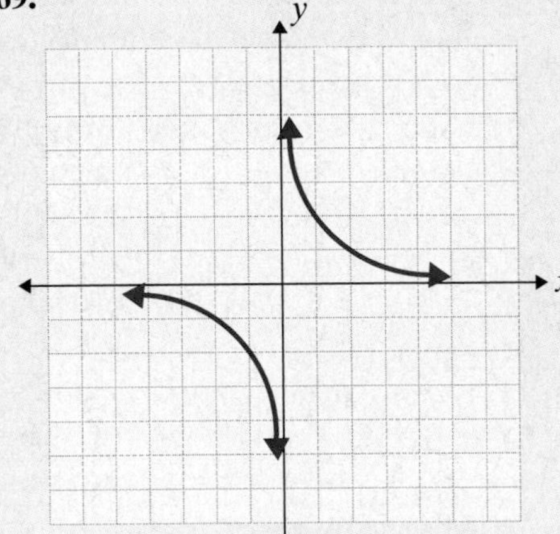

70. Given that $F(x) = 3x - 7$, find each of the following:

a. $F(-5)$ **b.** $F(10)$ **c.** $F(3.5)$

71. Given that $f(x) = 2x^2 + 3x - 10$, find each of the following:

a. $f(0)$ **b.** $f(-3)$ **c.** $f(6)$

Roots, Radicals, and Complex Numbers

CHAPTER 10

Did You Know?

An important method of reasoning related to mathematical proofs is proof by contradiction. See if you can follow the reasoning in the following proof that $\sqrt{2}$ is an irrational number.

We need the following statement (which can be proven algebraically):

> The square of an integer is even if and only if the integer is even.

Proof: $\sqrt{2}$ is either an irrational number or a rational number.

Suppose that $\sqrt{2}$ is a rational number and

	$\frac{a}{b}=\sqrt{2}$	Where a and b are integers and $\frac{a}{b}$ is reduced.
	$\frac{a^2}{b^2}=2$	Square both sides.
	$a^2=2b^2$	This means a^2 is an even integer.
So	$a=2n$	Since a^2 is even, a must be even.
	$a^2=4n^2$	Square both sides.
	$a^2=4n^2=2b^2$	Substitution.
	$2n^2=b^2$	This means b^2 is an even integer.

Therefore, b is an even integer.

But if a and b are both even, 2 is a common factor.

This contradicts the statement that $\frac{a}{b}$ is reduced.

Thus, our original supposition that $\sqrt{2}$ is rational is false, and $\sqrt{2}$ is an irrational number.

"The number of grains of sand on the beach at Coney Island is much less than a googol—10,000,000,000,000,000,000,000,000,000,000,000,000,000,000,000,000,000,000,000,000,000,000,000,000,000,000,000,000,000,000,000,000,000,000,000."

Edward Kasner

In this chapter, we will discuss expressions with exponents and their close relationship with radical expressions such as square roots ($\sqrt{x}$) and cube roots ($\sqrt[3]{x}$). This relationship allows translation from one type of expression to the other with relative ease and a choice for the form of an answer that best suits the purposes of the problem. For example, in higher level mathematics courses, particularly in calculus, an expression with a square root such as $\sqrt{x^2+1}$ may be changed into an equivalent expression with a fractional exponent such as $\left(x^2+1\right)^{\frac{1}{2}}$. Operations learned in calculus can be performed on expressions in the fractional exponent form. Then, if desired, answers can be changed back into a form with radical notation.

An important concern is under what conditions an expression with fractional exponents (or radicals) will be defined to be a real number. This concern leads to the definition of a new category of numbers called complex numbers. These numbers include the real numbers and a type of number called imaginary numbers. The term "imaginary" is somewhat unfortunate and misleading because these numbers have many practical applications and are no more imaginary than any other type of number. In fact, at one time negative numbers were thought of as "imaginary" since they represent an impossible quantity. However, complex numbers are particularly useful in electrical engineering and hydrodynamics. As we will see in Chapter 11, complex numbers evolve quite naturally as solutions to second-degree equations.

10.1 Roots and Radicals

Objectives

After completing this section, you will be able to:

1. *Evaluate square roots.*
2. *Simplify expressions with square roots.*
3. *Evaluate cube roots.*
4. *Simplify expressions with cube roots.*
5. *Rationalize denominators of fractional expressions.*

You are probably familiar with the concept of **square roots** and the **square root symbol** (or **radical sign**) ($\sqrt{}$) from your work in beginning algebra and our discussions of real numbers in Chapter 1. For example, the **radical notation** for the square root of 2 is $\sqrt{2}$, and $\sqrt{3}$ represents the square root of 3. In this section, we discuss the meanings of square roots and cube roots and methods for simplifying expressions with square roots and cube roots. In Section 10.2, we will expand these ideas to n^{th} roots in general and the meanings of fractional exponents (rational exponents).

Square Roots: $\sqrt{a}$

A number is squared when it is multiplied by itself. The exponent 2 is used to indicate squares. For example,

$$6^2 = 6 \cdot 6 = 36 \text{ and } (-15)^2 = (-15) \cdot (-15) = 225.$$

If an integer is squared, the result is called a **perfect square**. For understanding and easy reference, Table 10.1 shows the squares of the integers from 1 to 20 and thus the perfect squares from 1 to 400.

Squares of Integers from 1 to 20 (Perfect Squares)										
Integers (n)	1	2	3	4	5	6	7	8	9	10
Perfect Squares (n^2)	1	4	9	16	25	36	49	64	81	100
Integers (n)	11	12	13	14	15	16	17	18	19	20
Perfect Squares (n^2)	121	144	169	196	225	256	289	324	361	400

Table 10.1

Now, we want to reverse the process of squaring. That is, given a number, we want to find a number that when squared will result in the given number. This is called **finding a square root** of the given number. In general,

if $b^2 = a$, then b is a square root of a.

For example,

- because $5^2 = 25$, then 5 is a **square root** of 25 and we write $\sqrt{25} = 5$.
- because $9^2 = 81$, then 9 is a **square root** of 81 and we write $\sqrt{81} = 9$.

Terminology

*The symbol $\sqrt{\ }$ is called a **radical sign**.*

*The number under the radical sign is called the **radicand**.*

*The complete expression, such as $\sqrt{64}$, is called a **radical** or **radical expression**.*

Every positive real number has two square roots, one positive and one negative. The positive square root is called the **principal square root**. For example,

- because $(8)^2 = 64$, then $\sqrt{64} = 8$ ⟵ the **principal square root**
- because $(-8)^2 = 64$, then $-\sqrt{64} = -8$ ⟵ the **negative square root**

The number 0 has only one square root, namely 0.

Square Root

If a is a nonnegative real number and b is a real number such that

*$b^2 = a$, then b is called a **square root** of a.*

If b is nonnegative, then we write

$\sqrt{a} = b$ ⟵ *b is called the **principal square root***

and $-\sqrt{a} = -b$ ⟵ *$-b$ is called the **negative square root**.*

Example 1: Evaluating Square Roots

(**Note:** As these examples illustrate, the radicand may be an integer, a fraction, or a decimal number.)

Evaluate each of the following radical expressions.

a. $\sqrt{121}$ **b.** $\sqrt{\frac{16}{25}}$ **c.** $-\sqrt{0.0036}$ **d.** $\sqrt{16+9}$ **e.** $\sqrt{16}+\sqrt{9}$

Solutions:

a. $\sqrt{121} = 11$ The principal square root of 121 is 11 because $11^2 = 121$.

b. $\sqrt{\frac{16}{25}} = \frac{4}{5}$ The principal square root of $\frac{16}{25}$ is $\frac{4}{5}$ because $\left(\frac{4}{5}\right)^2 = \frac{16}{25}$.

c. $-\sqrt{0.0036} = -0.06$ The negative square root of 0.0036 is -0.06 because $(-0.06)^2 = (-0.06)(-0.06) = 0.0036$.

d. $\sqrt{16+9} = \sqrt{25} = 5$ Note that the sum under the radical sign is found first, then the square root is found. In this situation, the radical sign is similar to parentheses.

e. $\sqrt{16}+\sqrt{9} = 4 + 3 = 7$ Note that as illustrated in Examples 1d and 1e, in general, $\sqrt{a+b} \neq \sqrt{a}+\sqrt{b}$.

Example 2: Using a Calculator to Find Square Roots

Teaching Notes:
You might help your students become aware of the subtleties of their calculators by squaring as follows:

14.14213562^2
199.999999

(which is not 200). However, if they proceed to find the square root and then square, the calculator will give 200:

$\sqrt{200}$
14.14213562

Ans^2
200

Many real numbers have square roots that are not terminating decimals or fractions. The square roots of these numbers are infinite nonrepeating decimals (irrational numbers). Your calculator will find these values accurate to as many as nine decimal places.

Follow the steps outlined here with your calculator and the $\sqrt{}$ key to find the following square roots accurate to four decimal places.

a. $\sqrt{200}$ **b.** $-\sqrt{35}$

Solutions:

a. **Step 1:** Press 2nd and then the $\sqrt{}$ key.

Step 2: Enter 200) and then press .

The display on the screen should appear as follows:

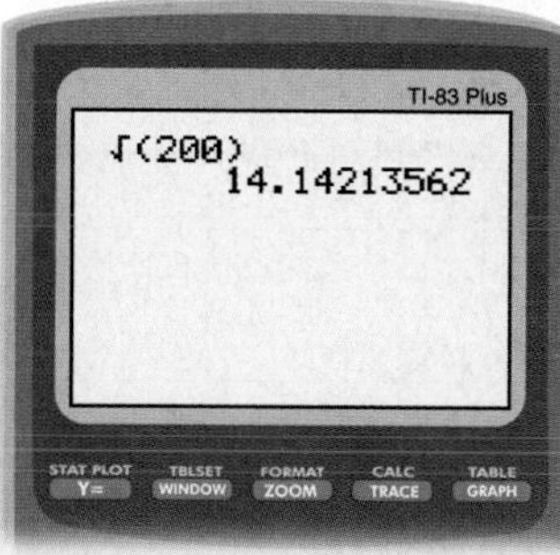

Thus, $\sqrt{200} \approx 14.1421$ accurate to four decimal places.

b. **Step 1:** Press the (–) key (located next to ENTER).

Step 2: Press 2nd and then the $\sqrt{}$ key.

Step 3: Enter 35) and then press ENTER.

The display on the screen should appear as follows:

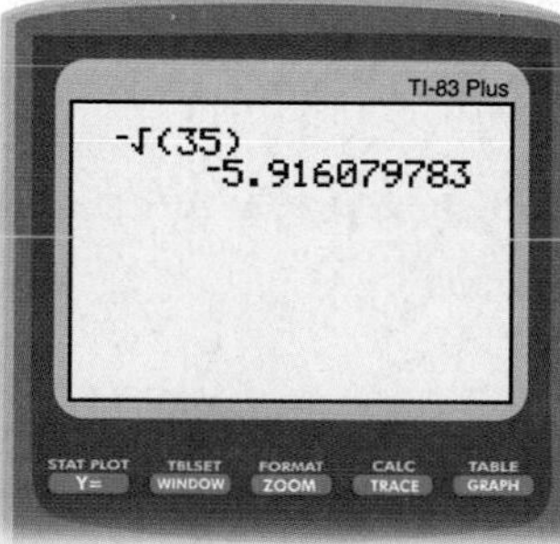

Thus, $-\sqrt{35} \approx -5.9161$ accurate to four decimal places.

Teaching Notes: Encourage your students to ponder their reaction to having stumbled upon the realization that the square root of a negative number may occur in computation, but has no place on the real number line.

Next, consider the square root of a negative number. For example, what is the value of $\sqrt{-49}$? There is no real number whose square is –49. The square of every real number is nonnegative. Thus, $\sqrt{-49}$ is a not a real number. In general,

$\sqrt{x}$ is not a real number if x is negative ($x < 0$).

We will discuss these nonreal numbers (part of the complex number system) later in this chapter.

Simplifying Expressions with Square Roots

Various roots can be related to solutions of equations, and we want such numbers to be in a **simplified form** for easier calculations and algebraic manipulations. We need the two properties of radicals stated here for square roots. (Similar properties are true for other roots.)

Properties of Square Roots

*If a and b are **positive** real numbers, then*

1. $\sqrt{ab} = \sqrt{a}\sqrt{b}$ **2.** $\sqrt{\dfrac{a}{b}} = \dfrac{\sqrt{a}}{\sqrt{b}}$

As an example, we know that 144 is a perfect square and $\sqrt{144} = 12$. However, in a situation where you may have forgotten this, you can proceed as follows using Property 1 of Square Roots:

$$\sqrt{144} = \sqrt{36} \cdot \sqrt{4} = 6 \cdot 2 = 12$$

Similarly, using Property 2, we can write

$$\sqrt{\frac{49}{36}} = \frac{\sqrt{49}}{\sqrt{36}} = \frac{7}{6}.$$

Simplest Form

*A square root is considered to be in **simplest form** when the radicand has no perfect square as a factor.*

The number 648 is not a perfect square, and to simplify $\sqrt{648}$ we can use Property 1 of Square Roots and any of the following three approaches.

Approach 1: Factor 648 as $36 \cdot 18$ because 36 is a perfect square. Then,

$$\sqrt{648} = \sqrt{36 \cdot 18} = \sqrt{36} \cdot \sqrt{18} = 6\sqrt{18}.$$

However, $6\sqrt{18}$ is **not in simplest form** because 18 has a perfect square factor, 9. Thus to complete the process, we have

$$\sqrt{648} = 6\sqrt{18} = 6\sqrt{9 \cdot 2} = 6\sqrt{9} \cdot \sqrt{2} = 6 \cdot 3\sqrt{2} = 18\sqrt{2}.$$

Approach 2: Note that 324 is a perfect square factor of 648 and $648 = 324 \cdot 2$.

$$\sqrt{648} = \sqrt{324 \cdot 2} = \sqrt{324} \cdot \sqrt{2} = 18\sqrt{2}$$

Approach 3: Use prime factors.

$$\begin{aligned}\sqrt{648} &= \sqrt{12 \cdot 54} \\ &= \sqrt{2 \cdot 2 \cdot 3 \cdot 2 \cdot 3 \cdot 3 \cdot 3} \\ &= \sqrt{2 \cdot 2 \cdot 3 \cdot 3 \cdot 3 \cdot 3} \cdot \sqrt{2} \\ &= 2 \cdot 3 \cdot 3 \cdot \sqrt{2} \\ &= 18\sqrt{2}\end{aligned}$$

Of these three approaches, the second appears to be the easiest because it has the fewest steps. However, "seeing" the largest perfect square factor may be difficult. If you do not immediately see a perfect square factor, then proceed to find other factors or prime factors as illustrated.

Example 3: Simplifying Expressions with Square Roots

Simplify the following square roots.

a. $\sqrt{48}$ **b.** $\sqrt{\dfrac{75}{16}}$ **c.** $\dfrac{2-\sqrt{20}}{2}$

Solutions:

a. $\sqrt{48} = \sqrt{16 \cdot 3} = \sqrt{16} \cdot \sqrt{3} = 4\sqrt{3}$ Find the largest perfect square factor.

b. $\sqrt{\dfrac{75}{16}} = \dfrac{\sqrt{75}}{\sqrt{16}} = \dfrac{\sqrt{25 \cdot 3}}{\sqrt{16}} = \dfrac{5\sqrt{3}}{4}$

c. $\dfrac{2-\sqrt{20}}{2} = \dfrac{2-\sqrt{4 \cdot 5}}{2} = \dfrac{2-2\sqrt{5}}{2} = \dfrac{2}{2} - \dfrac{2\sqrt{5}}{2} = 1-\sqrt{5}$

[or factoring and reducing: $\dfrac{2-2\sqrt{5}}{2} = \dfrac{\cancel{2}\left(1-\sqrt{5}\right)}{\cancel{2}} = 1-\sqrt{5}$]

Simplifying Square Roots with Variables

Now, we consider simplifying square root expressions that contain variables. To simplify the expression $\sqrt{x^2}$ we must consider the fact that we do not know whether x represents a positive number ($x > 0$) or a negative number ($x < 0$) along with the two squares

$$x \cdot x = x^2 \text{ and } (-x)(-x) = x^2.$$

For example,

$$\text{If } x = 5, \text{ then } \sqrt{x^2} = \sqrt{5^2} = \sqrt{25} = 5 = x.$$

$$\text{But, if } x = -5, \text{ then } \sqrt{x^2} = \sqrt{(-5)^2} = \sqrt{25} = 5 \neq x.$$

$$\text{In fact, if } x = -5, \text{ then } \sqrt{x^2} = \sqrt{(-5)^2} = \sqrt{25} = 5 = |-5| = |x|.$$

Thus, simplifying radical expressions with variables involves more detailed analysis than simplifying radical expressions with only constants. The following definition indicates the correct way to simplify $\sqrt{x^2}$.

Square Root of x^2

If x is a real number, then $\sqrt{x^2} = |x|$.

Note: *If* $x \geq 0$ *is given, then we can write* $\sqrt{x^2} = x$.

Example 4: Simplifying Square Root Expressions with Variables

Simplify each of the following radical expressions.

a. $\sqrt{16x^2}$ **b.** $\sqrt{72a^2}$ **c.** $\sqrt{12x^2y^2}$

Solutions:

a. $\sqrt{16x^2} = \sqrt{16}\sqrt{x^2} = 4|x|$

b. $\sqrt{72a^2} = \sqrt{72}\sqrt{a^2} = \sqrt{36}\sqrt{2}\sqrt{a^2} = 6\sqrt{2}|a|$

c. $\sqrt{12x^2y^2} = \sqrt{12}\sqrt{x^2}\sqrt{y^2} = \sqrt{4}\sqrt{3}\sqrt{x^2}\sqrt{y^2} = 2\sqrt{3}|x||y|$ (or $2\sqrt{3}|xy|$)

When expressions with the same base are multiplied, the exponents are added. Thus, if an expression is multiplied by itself, the exponents (if there are any) will be doubled and, therefore, even. This means that, to find the square root of an expression with even exponents, the exponents can be divided by 2.

For example,

$$x^2 \cdot x^2 = x^4 \qquad a^3 \cdot a^3 = a^6 \qquad y^5 \cdot y^5 = y^{10}$$

$$\downarrow \qquad\qquad \downarrow \qquad\qquad \downarrow$$

and

$$\sqrt{x^4} = x^2 \qquad \sqrt{a^6} = \left|a^3\right| \qquad \sqrt{y^{10}} = \left|y^5\right|.$$

To find the square root of an expression with odd exponents, factor the expression into two terms, one with exponent 1 and the other with an even exponent. For example,

$$x^3 = x^2 \cdot x \quad \text{and} \quad y^9 = y^8 \cdot y$$

which means that

$$\sqrt{x^3} = \sqrt{x^2 \cdot x} = \sqrt{x^2} \cdot \sqrt{x} = |x| \cdot \sqrt{x} \quad \text{and} \quad \sqrt{y^9} = \sqrt{y^8 \cdot y} = y^4\sqrt{y}.$$

Square Roots of Expressions with Even and Odd Exponents

For any real number x and positive integer m,

$$\sqrt{x^{2m}} = \left|x^m\right| \quad \text{and} \quad \sqrt{x^{2m+1}} = \left|x^m\right|\sqrt{x}\,.$$

Note that the absolute value sign is necessary only if m is odd. If m is even, then x^m is positive and the absolute value sign is not necessary. Thus, the absolute value sign is necessary only if m is odd.

Also, note that for $\sqrt{x^{2m+1}}$ to be defined as real, x cannot be negative.

Note: *If m is any integer, then 2m is even and 2m + 1 is odd.*

Example 5: Simplifying Radical Expressions with Variables

Simplify each of the following radical expressions.

a. $\sqrt{81x^4}$ **b.** $\sqrt{64x^5y}$, assume that $x,y \ge 0$ **c.** $\sqrt{18a^4b^6}$ **d.** $\sqrt{\dfrac{9a^{13}}{b^4}}$

Solutions:

a. $\sqrt{81x^4} = 9x^2$ The exponent 4 is divided by 2 and x can be positive or negative.

b. $\sqrt{64x^5y} = \sqrt{64 \cdot x^4 \cdot x \cdot y} = 8x^2\sqrt{xy}$ Assuming that $x,\ y \ge 0$

c. $\sqrt{18a^4b^6} = \sqrt{9 \cdot 2 \cdot a^4 \cdot b^6} = 3\sqrt{2}a^2\left|b^3\right|$ Each exponent is divided by 2.

d. $\sqrt{\dfrac{9a^{13}}{b^4}} = \dfrac{\sqrt{9 \cdot a^{12} \cdot a}}{\sqrt{b^4}} = \dfrac{3a^6\sqrt{a}}{b^2}$ We have assumed that $b \ne 0$ and $a \ge 0$. Thus, both $\sqrt{a^{13}}$ and $\sqrt{a}$ are real.

Cube Roots: $\sqrt[3]{a}$

For understanding and easy reference, the cubes of the integers from 1 to 10 are shown in Table 10.2. These perfect cubes occur frequently in the exercises and should be memorized.

Cubes of Integers from 1 to 10 (Perfect Cubes)										
Integers (n)	1	2	3	4	5	6	7	8	9	10
Perfect Cubes (n^3)	1	8	27	64	125	216	343	512	729	1000

Table 10.2

Cube Root

If a and b are real numbers such that

$b^3 = a$, *then b is called the* ***cube root*** *of a.*

We write $\sqrt[3]{a} = b$ ⟵ *the* ***cube root.***

NOTES In the cube root expression $\sqrt[3]{a}$ the number 3 is called the **index**. In a square root expression such as $\sqrt{a}$ the index is understood to be 2 and is not written. That is, $\sqrt{a}$ and $\sqrt[2]{a}$ have the same meaning. We will discuss indices in more detail in Section 10.2.

Example 6: Evaluating Cube Roots

Teaching Notes: You might explain that entering the right parenthesis in the expression $\sqrt[3]{(500)}$ is optional. However, there are times when not entering the parenthesis (i.e. closing the expression properly) will lead to incorrect answers. Therefore, entering the right parenthesis is a sort of safety measure.

Find the value of each of the following cube roots.

a. $\sqrt[3]{216}$ **b.** $\sqrt[3]{-8}$ **c.** $\sqrt[3]{500}$

Solutions:

a. $\sqrt[3]{216} = 6$ As we can see in Table 10.2, $6^3 = 216$ and 216 is a perfect cube.

b. $\sqrt[3]{-8} = -2$ Note that the cube root of a negative number is negative. In this example $\sqrt[3]{-8} = -2$ because $(-2)^3 = -8$.

c. $\sqrt[3]{500}$ 500 is not a perfect cube.

A TI-83 Plus calculator can be used to estimate the value by using the following steps.

Step 1: Press MATH and press . (This will select $\sqrt[3]{(}$.)

Step 2: Enter 500) and press .

The display should appear as follows:

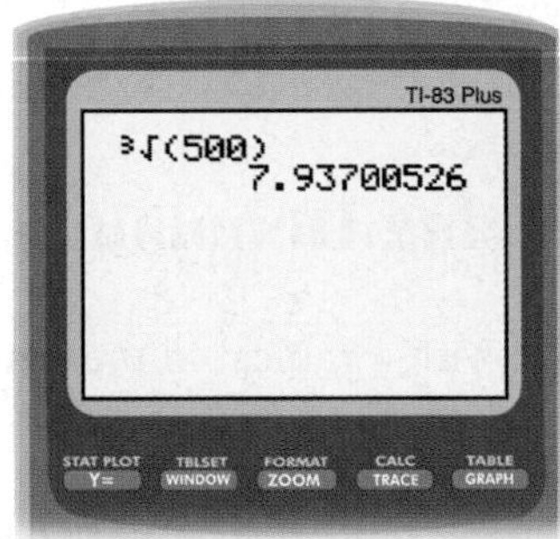

Thus, the cube root of 500 is approximately 7.93700526 (or 7.9370 to four decimal places).

Simplifying Cube Roots

When simplifying expressions with cube roots, we need to be aware of perfect cube numbers and variables with exponents that are multiples of 3. (Multiples of 3 are 3, 6, 9, 12, 15, and so on.) Thus, exponents are divided by 3 in simplifying cube root expressions.

Simplest Form

A cube root is considered to be in ***simplest form*** *when the radicand has no perfect cube as a factor.*

Example 7: Simplifying Expressions with Cube Roots

Simplify each of the following cube root expressions by finding the largest perfect cube factor.

a. $\sqrt[3]{54x^6}$ **b.** $\sqrt[3]{-40x^4y^{13}}$ **c.** $\sqrt[3]{250a^8b^{11}}$

Solutions:

a. $\sqrt[3]{54x^6} = \sqrt[3]{27 \cdot 2 \cdot x^6}$
$= 3x^2\sqrt[3]{2}$

Note that 27 is a perfect cube and the exponent 6 is divided by 3 because we are finding a cube root.

b. $\sqrt[3]{-40x^4y^{13}} = \sqrt[3]{(-8) \cdot 5 \cdot x^3 \cdot x \cdot y^{12} \cdot y}$
$= -2xy^4\sqrt[3]{5xy}$

Note that –8 is a perfect cube because $(-2)^3 = -8$, and each variable expression is separated so that one exponent is a multiple of 3. In simplifying, these exponents are divided by 3.

continued on next page ...

c. $\sqrt[3]{250a^8b^{11}} = \sqrt[3]{125 \cdot 2 \cdot a^6 \cdot a^2 \cdot b^9 \cdot b^2}$
$= 5a^2b^3\sqrt[3]{2a^2b^2}$

Note that 125 is a perfect cube and each variable expression is separated so that one exponent is a multiple of 3.

Rationalizing Denominators in Radical Expressions

An expression with a radical in the denominator may not be in the simplest form for further algebraic manipulation or operations. If this is the case, then we may want to rationalize the denominator. That is, we want to find an equivalent fraction in which the denominator does not have a radical. The numerator may still have a radical in it, but a rational denominator definitely makes arithmetic with radicals much easier. (**Note:** In this section we will deal with radicals that are square roots or cube roots. Other roots will be discussed in Section 10.2.)

To Rationalize the Denominator of a Radical Expression

1. *If the denominator contains a square root, multiply both the numerator and denominator by a square root. Choose this square root so that the denominator will be a perfect square.*

2. *If the denominator contains a cube root, multiply both the numerator and denominator by a cube root. Choose this cube root so that the denominator will be a perfect cube.*

Example 8: Rationalizing Denominators

Simplify each of the following radical expressions so that the denominator is a rational expression. Assume that each variable is positive.

a. $\sqrt{\dfrac{5}{4x}}$ **b.** $\dfrac{7}{\sqrt[3]{32y}}$ **c.** $\sqrt{\dfrac{18a^2b}{30ac}}$

Solutions:

a. Multiply the numerator and denominator by $\sqrt{x}$ because $4x \cdot x = 4x^2$ and $4x^2$ is a perfect square expression.

$$\sqrt{\frac{5}{4x}} = \frac{\sqrt{5}}{\sqrt{4x}} = \frac{\sqrt{5} \cdot \sqrt{x}}{\sqrt{4x} \cdot \sqrt{x}} = \frac{\sqrt{5x}}{\sqrt{4x^2}} = \frac{\sqrt{5x}}{2x}$$

Or, multiply by $\dfrac{x}{x}$ under the radical first as follows:

$$\sqrt{\frac{5}{4x}} = \sqrt{\frac{5}{4x} \cdot \frac{x}{x}} = \frac{\sqrt{5x}}{\sqrt{4x^2}} = \frac{\sqrt{5x}}{2x}$$

b. Multiply the numerator and denominator by $\sqrt[3]{2y^2}$ because $32y \cdot 2y^2 = 64y^3$ and $64y^3$ is a perfect cube expression since $(4y)^3 = 64y^3$.

$$\frac{7}{\sqrt[3]{32y}} = \frac{7 \cdot \sqrt[3]{2y^2}}{\sqrt[3]{32y} \cdot \sqrt[3]{2y^2}} = \frac{7\sqrt[3]{2y^2}}{\sqrt[3]{64y^3}} = \frac{7\sqrt[3]{2y^2}}{4y}$$

c. In this case we can simplify under the radical first and then rationalize the denominator.

$$\sqrt{\frac{18a^2b}{30ac}} = \sqrt{\frac{3ab}{5c}} = \frac{\sqrt{3ab} \cdot \sqrt{5c}}{\sqrt{5c} \cdot \sqrt{5c}} = \frac{\sqrt{15abc}}{\sqrt{25c^2}} = \frac{\sqrt{15abc}}{5c}$$

Practice Problems

Use a calculator to find the value of each number accurate to four decimal places.

1. $\sqrt{40}$ **2.** $\sqrt[3]{40}$

Simplify the following radical expressions. Assume that x may be positive or negative, but that a and b must be positive.

3. $\sqrt{80x^3}$ **4.** $\sqrt[3]{-27x^9}$ **5.** $\sqrt{18x^2}$ **6.** $\sqrt{128a^2b^5}$

Simplify each expression so that the denominator is a rational expression.

7. $\sqrt{\frac{3}{8a^2}}$ **8.** $\frac{\sqrt[3]{4ab}}{\sqrt[3]{2a^2b^4}}$

10.1 Exercises

1. 0.02 2. 0.05
3. 32 4. 26.8328
5. 67.0820 6. 7
7. 0.02 8. 36.8403
9. 2.3208 10. 4.6416

In Exercises 1 – 10, use a calculator to find the value of each radical accurate to four decimal places.

1. $\sqrt{0.0004}$ **2.** $\sqrt{0.0025}$ **3.** $\sqrt{1024}$ **4.** $\sqrt{720}$ **5.** $\sqrt{4500}$

6. $\sqrt[3]{343}$ **7.** $\sqrt[3]{0.000008}$ **8.** $\sqrt[3]{50,000}$ **9.** $\sqrt[3]{12.5}$ **10.** $\sqrt[3]{100}$

Answers to Practice Problems: 1. 6.3246 **2.** 3.4200 **3.** $4|x|\sqrt{5x}$ **4.** $-3x^3$ **5.** $3|x|\sqrt{2}$ **6.** $8ab^2\sqrt{2b}$ **7.** $\frac{\sqrt{6}}{4a}$ **8.** $\frac{\sqrt[3]{2a^2}}{ab}$

11. $2\sqrt{3}$ **12.** $3\sqrt{2}$
13. $7\sqrt{2}$ **14.** $6\sqrt{6}$
15. $-9\sqrt{2}$ **16.** $-3\sqrt{3}$
17. $2\sqrt[3]{2}$ **18.** $2\sqrt[3]{5}$
19. $3\sqrt[3]{4}$ **20.** $-3\sqrt[3]{2}$
21. Nonreal
22. Nonreal
23. $2yx^5\sqrt{6x}$
24. $2x^7y\sqrt{5xy}$
25. $ac\sqrt[3]{a^2b^2}$
26. $-y^2\sqrt[3]{x}$
27. Nonreal
28. Nonreal
29. $2x^3y^4$ **30.** $8ab^9$
31. $9ab^2\sqrt[3]{ab^2}$
32. $5\sqrt[3]{x^2y^2}$
33. $5xy^3\sqrt{5x}$
34. $2x^2y^2\sqrt{2x}$
35. $2bc\sqrt{3ac}$
36. $3abc^2\sqrt{5b}$
37. $-5x^2y^3z^4\sqrt{3}$
38. $-10xyz\sqrt{2}$
39. $2xy^2z^3\sqrt[3]{3x^2y}$
40. $5x^2y^3z^5\sqrt[3]{2}$
41. $5|y|$ **42.** $-9|x|$
43. $-8|a^3|$
44. $3|x||y|\sqrt{2}$
45. $4x^2y^4\sqrt{2}$
46. $-2xy^2\sqrt[3]{3}$
47. $3b^3\sqrt[3]{4a}$
48. $-4a^4$
49. $3xy^2\sqrt[3]{3x^2y}$

In Exercises 11 – 40, simplify each radical expression. Assume that the variables are positive. Write "nonreal" if the expression does not represent a real number.

11. $\sqrt{12}$ **12.** $\sqrt{18}$ **13.** $\sqrt{98}$ **14.** $\sqrt{216}$ **15.** $-\sqrt{162}$

16. $-\sqrt{27}$ **17.** $\sqrt[3]{16}$ **18.** $\sqrt[3]{40}$ **19.** $\sqrt[3]{108}$ **20.** $\sqrt[3]{-54}$

21. $\sqrt{-25}$ **22.** $\sqrt{-100}$ **23.** $\sqrt{24x^{11}y^2}$ **24.** $\sqrt{20x^{15}y^3}$

25. $\sqrt[3]{a^5b^2c^3}$ **26.** $\sqrt[3]{-xy^6}$ **27.** $\sqrt{-4x^5}$ **28.** $\sqrt{-9a^2}$

29. $\sqrt[3]{8x^9y^{12}}$ **30.** $\sqrt[3]{512a^3b^{27}}$ **31.** $\sqrt[3]{729a^4b^8}$ **32.** $\sqrt[3]{125x^2y^2}$

33. $\sqrt{125x^3y^6}$ **34.** $\sqrt{8x^5y^4}$ **35.** $\sqrt{12ab^2c^3}$ **36.** $\sqrt{45a^2b^3c^4}$

37. $-\sqrt{75x^4y^6z^8}$ **38.** $-\sqrt{200x^2y^2z^2}$ **39.** $\sqrt[3]{24x^5y^7z^9}$ **40.** $\sqrt[3]{250x^6y^9z^{15}}$

In Exercises 41 – 50, simplify each radical expression. Assume that the variables may be positive or negative.

41. $\sqrt{25y^2}$ **42.** $-\sqrt{81x^2}$ **43.** $-\sqrt{64a^6}$ **44.** $\sqrt{18x^2y^2}$ **45.** $\sqrt{32x^4y^8}$

46. $\sqrt[3]{-24x^3y^6}$ **47.** $\sqrt[3]{108ab^9}$ **48.** $\sqrt[3]{-64a^{12}}$ **49.** $\sqrt[3]{81x^5y^7}$ **50.** $\sqrt[3]{54a^4b^2}$

In Exercises 51 – 62, simplify each expression so that the denominator is a rational expression. Assume that the variables are positive.

51. $\sqrt{\frac{1}{64x}}$ **52.** $\sqrt{\frac{81}{x}}$ **53.** $-\sqrt{\frac{2}{3y}}$ **54.** $-\sqrt{\frac{25}{x^3}}$ **55.** $\frac{\sqrt{5y^2}}{\sqrt{8x}}$

56. $\frac{\sqrt{4x}}{\sqrt{3y^2}}$ **57.** $\frac{\sqrt{16y^2}}{\sqrt{2y^3}}$ **58.** $\frac{\sqrt{24a^3b}}{\sqrt{6ab^2}}$ **59.** $\sqrt[3]{\frac{2y^3}{27x^2}}$ **60.** $\sqrt[3]{\frac{7x}{2y^4}}$

61. $\sqrt[3]{\frac{6a^2}{25b}}$ **62.** $\frac{\sqrt[3]{x^5}}{\sqrt[3]{9xy}}$

50. $3a\sqrt[3]{2ab^2}$

51. $\frac{\sqrt{x}}{8x}$ 52. $\frac{9\sqrt{x}}{x}$

53. $\frac{-\sqrt{6y}}{3y}$ 54. $\frac{-5\sqrt{x}}{x^2}$

55. $\frac{y\sqrt{10x}}{4x}$ 56. $\frac{2\sqrt{3x}}{3y}$

57. $\frac{2\sqrt{2y}}{y}$ 58. $\frac{2a\sqrt{b}}{b}$

59. $\frac{y\sqrt[3]{2x}}{3x}$

60. $\frac{\sqrt[3]{28xy^2}}{2y^2}$

61. $\frac{\sqrt[3]{30a^2b^2}}{5b}$

62. $\frac{x\sqrt[3]{3xy^2}}{3y}$

63. $\sqrt{x^2}$ represents the positive square root.

64. When $a < 0$

65. A cube root has no restrictions.

66. See page 690.

Writing and Thinking About Mathematics

63. Explain, in your own words, why we cannot just say that $\sqrt{x^2} = x$. That is, why do we write $\sqrt{x^2} = |x|$?

64. Under what conditions is the expression $\sqrt{a}$ not a real number?

65. Explain why the expression $\sqrt[3]{y}$ is a real number regardless of whether $y > 0$ or $y < 0$ or $y = 0$.

66. Explain the technique for rationalizing the denominator when the denominator is a radical with index greater than 2.

Hawkes Learning Systems: Introductory & Intermediate Algebra

Evaluating Radicals
Simplifying Radicals
Division of Radicals

10.2 Rational Exponents

Objectives

After completing this section, you will be able to:

1. *Understand the meaning of* n^{th} *root.*
2. *Simplify expressions using the properties of rational exponents.*
3. *Evaluate expressions of the form* $a^{\frac{m}{n}}$ *with a calculator.*

n^{th} Roots: $\sqrt[n]{a} = a^{\frac{1}{n}}$

In Section 10.1 we restricted our discussions to radicals involving square roots and cube roots. In this section we will expand on those ideas by discussing radicals indicating n^{th} roots in general and how to relate radical expressions to expressions with rational (fractional) exponents. For example, the fifth root of x can be written in radical form as $\sqrt[5]{x}$ and with a fractional exponent as $x^{\frac{1}{5}}$.

To understand roots in general, consider the following analysis (assuming that $b > 0$):

For square roots,	if $b^2 = a$, then $b = \sqrt{a}$ (or $b = a^{\frac{1}{2}}$).
For cube roots,	if $b^3 = a$, then $b = \sqrt[3]{a}$ (or $b = a^{\frac{1}{3}}$).
For fourth roots,	if $b^4 = a$, then $b = \sqrt[4]{a}$ (or $b = a^{\frac{1}{4}}$).
For n^{th} roots,	if $b^n = a$, then $b = \sqrt[n]{a}$ (or $b = a^{\frac{1}{n}}$).

(**Note:** In this discussion, we assume that $n \neq 0$.)

For example:

Because $2^4 = 16$ we can say that $\sqrt[4]{16} = 2$ (or $2 = 16^{\frac{1}{4}}$).

Because $3^5 = 243$ we can say that $\sqrt[5]{243} = 3$ (or $3 = 243^{\frac{1}{5}}$).

The following notation is used for all radical expressions.

Radical Notation

If n is a positive integer and $b^n = a$, *then* $\mathbf{b = \sqrt[n]{a} = a^{\frac{1}{n}}}$ *(assuming* $\sqrt[n]{a}$ *is a real number).*

The expression $\sqrt[n]{a}$ *is called a* ***radical.***

The symbol $\sqrt[n]{}$ *is called a* ***radical sign.***

n is called the ***index.***

a is called the ***radicand.***

*(**Note:** If no index is given, it is understood to be 2. For example,* $\sqrt{3} = \sqrt[2]{3} = 3^{\frac{1}{2}}$*.)*

Special Note About the Index *n*:

For the expression $\sqrt[n]{a}$ *(or* $a^{\frac{1}{n}}$*) to be a real number:*

1. *n can be any index when a is nonnegative.*
2. *n must be odd when a is negative.*

(If a is negative and n is even, then $\sqrt[n]{a}$ *is nonreal.)*

Example 1: Principal n^{th} Root

a. $49^{\frac{1}{2}} = \sqrt{49} = 7$ because $7^2 = 49$.

b. $81^{\frac{1}{4}} = \sqrt[4]{81} = 3$ because $3^4 = 81$.

c. $(-8)^{\frac{1}{3}} = \sqrt[3]{-8} = -2$ because $(-2)^3 = -8$.

d. $(0.00001)^{\frac{1}{5}} = \sqrt[5]{0.00001} = 0.1$ because $(0.1)^5 = 0.00001$.

e. $(-16)^{\frac{1}{2}} = \sqrt{-16}$ is not a real number. (Any even root of a negative number is non-real.)

Example 2: Odd Powers

Use a TI-83 Plus calculator to find the value of each of the following roots accurate to four decimal places.

a. $\sqrt[5]{200}$ **b.** $\sqrt[6]{1.25}$

continued on next page ...

Solutions:

a. To find $\sqrt[5]{200}$ proceed as follows:

Step 1: Enter 5. (Note: This 5 is the index.)

Step 2: Press MATH.

Step 3: Choose 5: $\sqrt[x]{\ }$.

Step 4: Enter 200 and press .

The display will appear as follows:

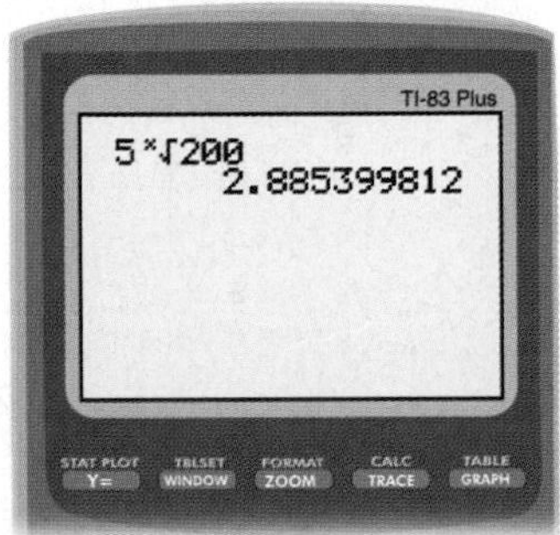

Thus, $\sqrt[5]{200} \approx 2.8854$ accurate to four decimal places.

(**Note:** Another approach is to use the fractional exponent as follows: 200 ^ (1/5). Be sure to put parentheses around the exponent. The exponent form will give the same result as the radical form.)

b. To find $\sqrt[6]{1.25}$ proceed as follows:

Step 1: Enter 6. (**Note:** This 6 is the index.)

Step 2: Press MATH.

Step 3: Choose 5: $\sqrt[x]{\ }$.

Step 4: Enter 1.25 and press ENTER.

The display will appear as follows:

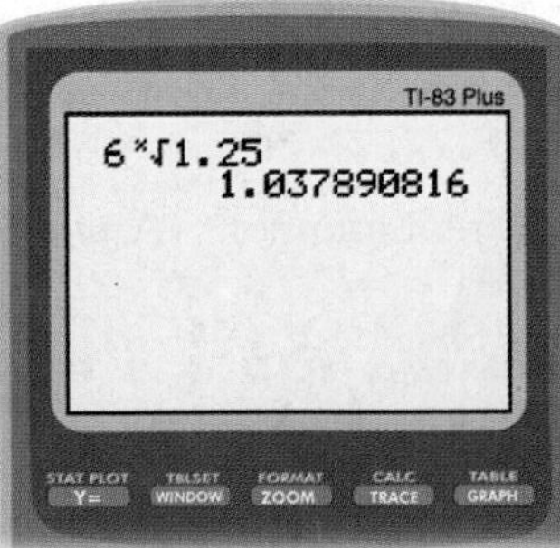

Thus, $\sqrt[6]{1.25} \approx 1.0379$ accurate to four decimal places.

(**Note:** Again, the exponent form 1.25 ^ (1/6) will give the same result. Be sure to put the exponent in parentheses.)

Rational Exponents of the Form $\frac{m}{n}$: $\sqrt[n]{a^m} = a^{\frac{m}{n}}$

In Chapter 5 we discussed the properties of exponents using only integer exponents. These same properties of exponents apply to rational exponents (fractional exponents) as well and are repeated here for emphasis and easy reference.

Summary of Properties of Exponents

For nonzero real numbers a and b and rational numbers m and n,

1. ***The Exponent 1:*** $a = a^1$ (*a is any real number.*)
2. ***The Exponent 0:*** $a^0 = 1 \quad (a \neq 0)$
3. ***Product Rule:*** $a^m \cdot a^n = a^{m+n}$
4. ***Quotient Rule:*** $\frac{a^m}{a^n} = a^{m-n}$
5. ***Power Rule:*** $\left(a^m\right)^n = a^{mn}$
6. ***Negative Exponents:*** $a^{-n} = \frac{1}{a^n}, \quad \frac{1}{a^{-n}} = a^n$
7. ***Power Rule for Products:*** $(ab)^n = a^n b^n$
8. ***Power Rule for Fractions:*** $\left(\frac{a}{b}\right)^n = \frac{a^n}{b^n}$

Teaching Notes: Highlight the fact that m and n are not restricted to being positive integers – that they may be negative and that they may be rational numbers. Another very handy property of exponents (due to the commutative property of multiplication) is $(a^m)^n = (a^n)^m$. For example, $\left(81^3\right)^{\frac{1}{4}}$ is much easier to simplify if written as $\left(81^{\frac{1}{4}}\right)^3$.

Now consider the problem of evaluating the expression $8^{\frac{2}{3}}$ where the exponent, $\frac{2}{3}$, is of the form $\frac{m}{n}$. By using the Power Rule for exponents, we can write

$$8^{\frac{2}{3}} = \left(8^{\frac{1}{3}}\right)^2 = (2)^2 = 4$$

or, $$8^{\frac{2}{3}} = (8^2)^{\frac{1}{3}} = (64)^{\frac{1}{3}} = 4.$$

The result is the same with either approach. That is, we can take the cube root first and then square the answer. Or, we can square first and then take the cube root. In general, for an exponent of the form $\frac{m}{n}$, taking the nth root first and then raising this root to the power m is easier because the numbers are smaller.

For example,

$$81^{\frac{3}{4}} = \left(81^{\frac{1}{4}}\right)^3 = (3)^3 = 27$$

is easier to calculate and work with than

$$81^{\frac{3}{4}} = (81^3)^{\frac{1}{4}} = (531,441)^{\frac{1}{4}} = 27.$$

The fourth root of 81 is more commonly known than the fourth root of 531,441.

The General Form $a^{\frac{m}{n}}$

If n is a positive integer and m is any integer and $a^{\frac{1}{n}}$ *is a real number, then*

$$a^{\frac{m}{n}} = \left(a^{\frac{1}{n}}\right)^m = (a^m)^{\frac{1}{n}}.$$

In radical notation:

$$a^{\frac{m}{n}} = \left(\sqrt[n]{a}\right)^m = \sqrt[n]{a^m}$$

Example 3: Conversion

Assume that each variable represents a positive real number.
Each expression is changed to an equivalent expression in radical notation.

a. $x^{\frac{2}{3}} = \sqrt[3]{x^2}$ Note that the index, 3, is the denominator in the rational exponent.

b. $3x^{\frac{4}{5}} = 3\sqrt[5]{x^4}$ Note that the coefficient, 3, is not affected by the exponent.

c. $-a^{\frac{3}{2}} = -\sqrt{a^3}$ Note that -1 is the understood coefficient.

d. $\sqrt[6]{a^5} = a^{\frac{5}{6}}$ Note that the index is the denominator of the rational exponent.

e. $5\sqrt{x} = 5x^{\frac{1}{2}}$ Note that, in a square root, the index is understood to be 2.

f. $-\sqrt[3]{4} = -4^{\frac{1}{3}}$ Note that the coefficient, -1, is not affected by the exponent. Also, we could write $-4^{\frac{1}{3}} = -1 \cdot 4^{\frac{1}{3}}$.

Simplifying Expressions with Rational Exponents

Expressions with rational exponents such as

$$x^{\frac{2}{3}} \cdot x^{\frac{1}{6}}, \frac{x^{\frac{3}{4}}}{x^{\frac{1}{3}}}, \text{ and } \left(2a^{\frac{1}{4}}\right)^3$$

can be simplified by using the properties of exponents.

NOTES Unless otherwise stated, we will assume, for the remainder of this chapter, that all variables represent non-negative real numbers.

Example 4: Simplifying Expressions with Rational Exponents

Each expression is simplified by using one or more of the rules of exponents.

a. $x^{\frac{2}{3}} \cdot x^{\frac{1}{6}} = x^{\frac{2}{3}+\frac{1}{6}}$ Find a common denominator and add the exponents.

$$= x^{\frac{4}{6}+\frac{1}{6}} = x^{\frac{5}{6}}$$

b. $\frac{x^{\frac{3}{4}}}{x^{\frac{1}{3}}} = x^{\frac{3}{4}-\frac{1}{3}}$ Subtract the exponents.

$$= x^{\frac{9}{12}-\frac{4}{12}} = x^{\frac{5}{12}}$$

c. $\left(2a^{\frac{1}{4}}\right)^3 = 2^3 \cdot a^{\frac{1}{4}\cdot 3} = 8a^{\frac{3}{4}}$

d. $\left(27y^{-\frac{9}{10}}\right)^{-\frac{1}{3}} = 27^{-\frac{1}{3}} \cdot y^{-\frac{9}{10}\left(-\frac{1}{3}\right)}$

$$= \frac{y^{\frac{3}{10}}}{27^{\frac{1}{3}}} = \frac{y^{\frac{3}{10}}}{3}$$

Multiply the exponents of y and reduce the fraction to $\frac{3}{10}$.

e. $(-36)^{-\frac{1}{2}} = \frac{1}{(-36)^{\frac{1}{2}}}$ This is not a real number because $(-36)^{\frac{1}{2}} = \sqrt{-36}$ is not real.

f. $9^{\frac{2}{4}} = 9^{\frac{1}{2}} = 3$ The exponent can be reduced as long as the expression is real.

g. $\left(\frac{49x^6y^{-2}}{z^{-4}}\right)^{\frac{1}{2}} = \frac{49^{\frac{1}{2}}x^3y^{-1}}{z^{-2}}$ **Study this example carefully.**

$$= \frac{7x^3z^2}{y}$$

Example 5 shows how to use fractional exponents to simplify rather complicated looking radical expressions. The results may seem surprising at first.

Example 5: Simplifying Radical Notation by Changing to Exponential Notation

Each radical expression is changed to an equivalent expression with exponential notation and simplified in this form. Then, the result is returned to radical notation.

a. $\sqrt[4]{\sqrt[3]{x}} = \left(\sqrt[3]{x}\right)^{\frac{1}{4}} = \left(x^{\frac{1}{3}}\right)^{\frac{1}{4}} = x^{\frac{1}{12}} = \sqrt[12]{x}$ Note that $\frac{1}{3} \cdot \frac{1}{4} = \frac{1}{12}$.

b. $\sqrt[3]{a}\sqrt{a} = a^{\frac{1}{3}} \cdot a^{\frac{1}{2}} = a^{\frac{1}{3}+\frac{1}{2}} = a^{\frac{2}{6}+\frac{3}{6}} = a^{\frac{5}{6}} = \sqrt[6]{a^5}$

c. $\dfrac{\sqrt{x^3}\sqrt[3]{x^2}}{\sqrt[5]{x^2}} = \dfrac{x^{\frac{3}{2}} \cdot x^{\frac{2}{3}}}{x^{\frac{2}{5}}} = \dfrac{x^{\frac{3}{2}+\frac{2}{3}}}{x^{\frac{2}{5}}} = \dfrac{x^{\frac{13}{6}}}{x^{\frac{2}{5}}}$

$= x^{\frac{13}{6}-\frac{2}{5}} = x^{\frac{65}{30}-\frac{12}{30}} = x^{\frac{53}{30}}$

$= x^{\frac{30}{30}} \cdot x^{\frac{23}{30}} = x\sqrt[30]{x^{23}}$

Evaluating Roots with a TI-83 Plus Calculator (The ^ key)

The up arrow key (or caret key) ^ on the TI-83 Plus calculator (and most graphing calculators) is used to indicate exponents. By using this key, roots of real numbers can be calculated with up to nine digit accuracy. To set the number of decimal places you wish in any calculations, press the MODE key and highlight the digit opposite the word **FLOAT** that indicates the desired accuracy. If no digit is highlighted, then the accuracy will be to nine decimal places (in some cases ten decimal places).

To Find the Value of $a^{\frac{m}{n}}$ with a Calculator

Step 1: *Enter the value of the base, a.*

Step 2: *Press the up arrow key* ^.

Step 3: *Enter the fractional exponent enclosed in parentheses.*
(This exponent may be positive or negative.)

Step 4: *Press* ENTER.

Example 6: Using the TI-83 Plus

Find the value of each numerical expression accurate to 4 decimal places. Use the MODE key and at **FLOAT** highlight **4**.

a. $125^{\frac{4}{3}}$

Solution:

Step 1: Enter the base, 125.

Step 2: Press the up arrow key ^.

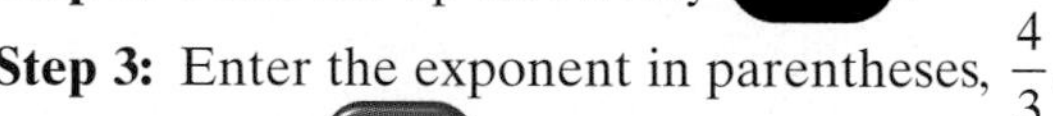

Step 3: Enter the exponent in parentheses, $\frac{4}{3}$.

Step 4: Press ENTER.

The display should read as follows:

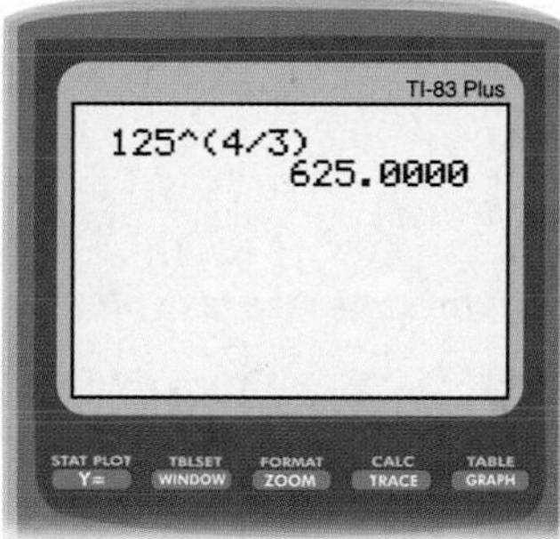

(**Note:** This particular result can be found by using our knowledge of roots as follows: $125^{\frac{4}{3}} = \left(125^{\frac{1}{3}}\right)^4 = 5^4 = 625$.)

b. $36^{\frac{3}{5}}$

Solution:

Step 1: Enter the base, 36.

Step 2: Press the up arrow key ^.

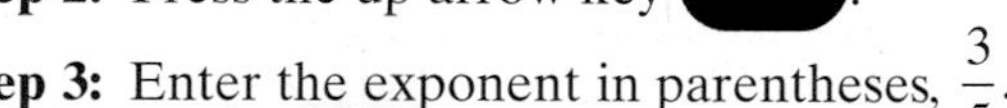

Step 3: Enter the exponent in parentheses, $\frac{3}{5}$.

Step 4: Press ENTER.

The display should read as follows:

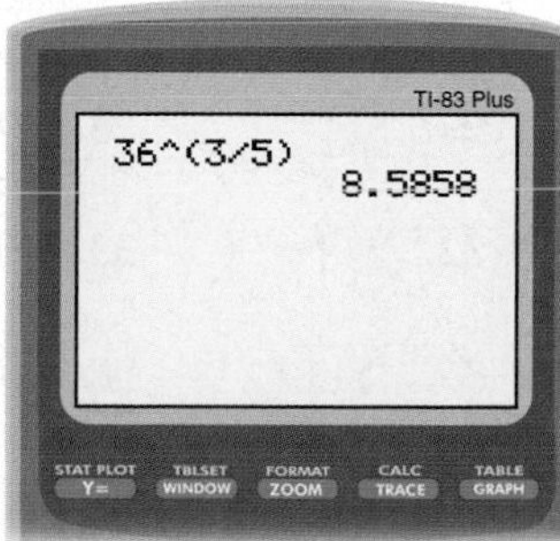

Practice Problems

Simplify each of the following expressions. Leave the answers with rational exponents.

1. $64^{\frac{2}{3}}$ 2. $x^{\frac{3}{4}} \cdot x^{\frac{1}{5}} \cdot x^{\frac{1}{2}}$ 3. $\dfrac{x^{\frac{1}{6}} \cdot y^{\frac{1}{2}}}{x^{\frac{1}{3}} \cdot y^{\frac{1}{4}}}$ 4. $\left(\dfrac{16x^{-\frac{1}{6}}}{x^{-\frac{2}{3}}}\right)^{\frac{1}{4}}$ 5. $-81^{\frac{1}{4}}$

Simplify each radical expression and leave the answers in simplest radical form.

6. $\sqrt[4]{x} \cdot \sqrt{x}$ 7. $\sqrt[5]{\sqrt[3]{a^2}}$ 8. $\dfrac{\sqrt{36x}}{\sqrt[3]{8x^2}}$

Use a TI-83 Plus calculator to find the following values accurate to 4 decimal places.

9. $128^{\frac{1}{5}}$ 10. $100^{-\frac{1}{4}}$

10.2 Exercises

1. 3 **2.** 11 **3.** $\frac{1}{10}$ **4.** $\frac{1}{5}$ **5.** −512 **6.** Nonreal **7.** Nonreal **8.** $-\frac{2}{5}$ **9.** $\frac{3}{7}$ **10.** $\frac{5}{4}$ **11.** −4 **12.** 16 **13.** −5 **14.** $-\frac{1}{6}$ **15.** $\frac{1}{4}$ **16.** $\frac{5}{2}$ **17.** $-\frac{27}{8}$ **18.** $\frac{1}{4}$ **19.** $\frac{9}{16}$ **20.** $\frac{3}{8}$ **21.** $\frac{2}{5}$ **22.** $-\frac{1}{1000}$ **23.** $\frac{1}{81}$ **24.** 64 **25.** $-\frac{1}{16,807}$ **26.** 8.5499 **27.** 2187 **28.** 10,000,000 **29.** 4.6416 **30.** 99.6055 **31.** 0.0131 **32.** 0.0922 **33.** 158.7401 **34.** 0.2236

Simplify each numerical expression in Exercises 1 – 25.

1. $9^{\frac{1}{2}}$ **2.** $121^{\frac{1}{2}}$ **3.** $100^{-\frac{1}{2}}$ **4.** $25^{-\frac{1}{2}}$ **5.** $-64^{\frac{3}{2}}$

6. $(-64)^{\frac{3}{2}}$ **7.** $\left(-\frac{4}{25}\right)^{\frac{1}{2}}$ **8.** $-\left(\frac{4}{25}\right)^{\frac{1}{2}}$ **9.** $\left(\frac{9}{49}\right)^{\frac{1}{2}}$ **10.** $\left(\frac{225}{144}\right)^{\frac{1}{2}}$

11. $(-64)^{\frac{1}{3}}$ **12.** $64^{\frac{2}{3}}$ **13.** $(-125)^{\frac{1}{3}}$ **14.** $(-216)^{-\frac{1}{3}}$ **15.** $8^{-\frac{2}{3}}$

16. $\left(\frac{8}{125}\right)^{-\frac{1}{3}}$ **17.** $-\left(\frac{16}{81}\right)^{-\frac{3}{4}}$ **18.** $\left(-\frac{1}{32}\right)^{\frac{2}{5}}$ **19.** $\left(\frac{27}{64}\right)^{\frac{2}{3}}$ **20.** $3 \cdot 16^{-\frac{3}{4}}$

21. $2 \cdot 25^{-\frac{1}{2}}$ **22.** $-100^{-\frac{3}{2}}$ **23.** $\left[(-27)^{\frac{2}{3}}\right]^{-2}$ **24.** $\left[\left(\frac{1}{32}\right)^{\frac{2}{5}}\right]^{-3}$ **25.** $-49^{-\frac{5}{2}}$

In Exercises 26 – 40, use a calculator to find the value of each numerical expression accurate to 4 decimal places.

26. $25^{\frac{2}{3}}$ **27.** $81^{\frac{7}{4}}$ **28.** $100^{\frac{7}{2}}$ **29.** $100^{\frac{1}{3}}$ **30.** $250^{\frac{5}{6}}$

31. $18^{-\frac{3}{2}}$ **32.** $24^{-\frac{3}{4}}$ **33.** $2000^{\frac{2}{3}}$ **34.** $\sqrt[4]{0.0025}$ **35.** $\sqrt[4]{3600}$

36. $\sqrt[5]{35.4}$ **37.** $\sqrt[10]{1.8}$ **38.** $\sqrt[6]{4500}$ **39.** $\sqrt[9]{72}$ **40.** $\sqrt[5]{0.00032}$

Answers to Practice Problems: 1. 16 **2.** $x^{\frac{29}{20}}$ **3.** $\dfrac{y^{\frac{1}{4}}}{x^{\frac{1}{6}}}$ **4.** $2x^{\frac{1}{8}}$ **5.** −3 **6.** $\sqrt[4]{x^3}$ **7.** $\sqrt[15]{a^2}$ **8.** $\dfrac{3\sqrt[6]{x^5}}{x}$ **9.** 2.6390 **10.** 0.3162

35. 7.7460 **36.** 2.0408
37. 1.0605 **38.** 4.0632
39. 1.6083 **40.** 0.2
41. $8x$ **42.** $81x^2$
43. $3a^2$ **44.** $\frac{1}{2a^{\frac{3}{4}}}$
45. $8x^{\frac{5}{2}}$
46. $3x^{\frac{11}{3}}$ **47.** $5a^{\frac{13}{6}}$
48. $a^{\frac{1}{15}}$ **49.** $a^{\frac{8}{5}}$
50. $x^{\frac{7}{12}}$ **51.** $x^{\frac{1}{2}}$
52. $a^{\frac{5}{9}}$ **53.** $a^{\frac{7}{6}}$
54. $\frac{1}{a^{\frac{9}{8}}}$ **55.** $a^{\frac{1}{4}}$ **56.** 1
57. $\frac{y^{\frac{1}{2}}}{x^{\frac{4}{3}}}$ **58.** $\frac{a^2}{b^{\frac{6}{5}}}$
59. $\frac{1}{a^{\frac{3}{4}}b^{\frac{1}{2}}}$ **60.** $8x^{\frac{3}{2}}y$
61. $\frac{x^{\frac{3}{2}}}{16y^{\frac{2}{5}}}$ **62.** a^5
63. $\frac{x^2y^4}{z^4}$ **64.** $\frac{y^{\frac{3}{2}}z^2}{x}$
65. $\frac{c^3}{3ab^2}$ **66.** $\frac{a^2}{3b^{\frac{1}{2}}}$
67. $\frac{8b^{\frac{9}{4}}}{a^3c^3}$ **68.** $-3a^{\frac{2}{3}}bc$
69. $x^{\frac{1}{4}}y^{\frac{5}{4}}$ **70.** $\frac{y^{\frac{2}{3}}}{x^{\frac{5}{6}}}$
71. $\frac{y^{\frac{2}{3}}}{50x^{\frac{4}{3}}}$ **72.** $\frac{5a^{\frac{7}{5}}}{8b^{\frac{23}{10}}}$
73. $\frac{6a}{b^2}$ **74.** $\frac{x^{\frac{7}{4}}}{y^{\frac{3}{4}}}$ **75.** $\frac{b^{\frac{5}{12}}}{a^{\frac{11}{12}}}$

Simplify each algebraic expression in Exercises 41 – 77. Leave the answers in rational exponent form.

41. $\left(2x^{\frac{1}{3}}\right)^3$ **42.** $\left(3x^{\frac{1}{2}}\right)^4$ **43.** $(9a^4)^{\frac{1}{2}}$ **44.** $(16a^3)^{-\frac{1}{4}}$ **45.** $8x^2\cdot x^{\frac{1}{2}}$

46. $3x^3\cdot x^{\frac{2}{3}}$ **47.** $5a^2\cdot a^{-\frac{1}{3}}\cdot a^{\frac{1}{2}}$ **48.** $a^{\frac{2}{3}}\cdot a^{-\frac{3}{5}}\cdot a^0$ **49.** $\frac{a^2}{a^{\frac{2}{5}}}$ **50.** $\frac{x^{\frac{3}{4}}}{x^{\frac{1}{6}}}$

51. $\frac{x^{\frac{2}{5}}}{x^{-\frac{1}{10}}}$ **52.** $\frac{a^{\frac{2}{3}}}{a^{\frac{1}{9}}}$ **53.** $\frac{a^{\frac{1}{2}}}{a^{-\frac{2}{3}}}$ **54.** $\frac{a^{\frac{3}{4}}\cdot a^{\frac{1}{8}}}{a^2}$ **55.** $\frac{a^{\frac{1}{2}}\cdot a^{-\frac{3}{4}}}{a^{-\frac{1}{2}}}$

56. $\frac{x^{\frac{2}{3}}\cdot x^{\frac{4}{3}}}{x^2}$ **57.** $\frac{x^{\frac{2}{3}}y}{x^2y^{\frac{1}{2}}}$ **58.** $\frac{a^{\frac{3}{2}}b^{\frac{4}{5}}}{a^{-\frac{1}{2}}b^2}$ **59.** $\frac{a^{\frac{3}{4}}b^{-\frac{1}{3}}}{a^{\frac{3}{2}}b^{\frac{1}{6}}}$ **60.** $\left(2x^{\frac{1}{2}}y^{\frac{1}{3}}\right)^3$

61. $\left(4x^{-\frac{3}{4}}y^{\frac{1}{5}}\right)^{-2}$ **62.** $\left(a^{\frac{1}{2}}a^{\frac{1}{3}}\right)^6$ **63.** $(-x^3y^6z^{-6})^{\frac{2}{3}}$ **64.** $\left(\frac{x^2y^{-3}}{z^4}\right)^{-\frac{1}{2}}$

65. $\left(\frac{27a^3b^6}{c^9}\right)^{-\frac{1}{3}}$ **66.** $(81a^{-8}b^2)^{-\frac{1}{4}}$ **67.** $\left(\frac{16a^{-4}b^3}{c^4}\right)^{\frac{3}{4}}$ **68.** $\left(\frac{-27a^2b^3}{c^{-3}}\right)^{\frac{1}{3}}$

69. $\frac{\left(x^{\frac{1}{4}}y^{\frac{1}{2}}\right)^3}{x^{\frac{1}{2}}y^{\frac{1}{4}}}$ **70.** $\frac{\left(x^{\frac{1}{2}}y\right)^{-\frac{1}{3}}}{x^{\frac{2}{3}}y^{-1}}$ **71.** $\frac{(8x^2y)^{-\frac{1}{3}}}{\left(5x^{\frac{1}{3}}y^{-\frac{1}{2}}\right)^2}$ **72.** $\frac{(25a^4b^{-1})^{\frac{1}{2}}}{\left(2a^{\frac{1}{5}}b^{\frac{3}{5}}\right)^3}$

73. $\left(\frac{5a^{-3}}{21b^2}\right)^{-1}\cdot\left(\frac{49a^4}{100b^{-8}}\right)^{-\frac{1}{2}}$ **74.** $\left(\frac{x^2y^{\frac{1}{3}}}{x^{\frac{1}{2}}y^{\frac{3}{2}}}\right)^{\frac{1}{2}}\cdot\left(\frac{x^{-\frac{1}{2}}y^{\frac{2}{3}}}{x^{-1}y^{\frac{3}{4}}}\right)^2$

75. $\left(\frac{a^{-3}b^{\frac{1}{3}}}{a^{\frac{1}{2}}b}\right)^{\frac{1}{2}}\cdot\left(\frac{ab^{\frac{1}{2}}}{a^{-\frac{2}{3}}b^{-1}}\right)^{\frac{1}{2}}$ **76.** $\left(\frac{a^3b^{-2}}{ab^4}\right)^{\frac{1}{6}}\cdot\left(\frac{a^{\frac{1}{5}}b^{\frac{1}{3}}}{a^{-\frac{1}{2}}}\right)^3$ **77.** $\frac{\left(27xy^{\frac{1}{2}}\right)^{\frac{1}{3}}}{\left(25x^{-\frac{1}{2}}y\right)^{\frac{1}{2}}}\cdot\frac{\left(x^{\frac{1}{2}}y\right)^{\frac{1}{6}}}{\left(16x^{\frac{1}{3}}y\right)^{\frac{1}{2}}}$

In Exercises 78 – 91, simplify each expression by first changing to an equivalent expression with rational exponents. Rewrite the answer in simplified radical form.

78. $\sqrt[3]{a}\cdot\sqrt{a}$ **79.** $\sqrt[3]{x^2}\cdot\sqrt[5]{x^3}$ **80.** $\frac{\sqrt[4]{y^3}}{\sqrt[6]{y}}$ **81.** $\frac{\sqrt[4]{x}}{\sqrt[3]{x^4}}$ **82.** $\frac{\sqrt[3]{x^2}\sqrt[5]{x^4}}{\sqrt{x^3}}$

76. $a^{\frac{73}{30}}$ **77.** $\dfrac{3x^{\frac{1}{2}}}{20y^{\frac{2}{3}}}$

78. $\sqrt[6]{a^5}$ **79.** $x\sqrt[15]{x^4}$

80. $\sqrt[12]{x^7}$ **81.** $\dfrac{1}{x\sqrt[12]{x}}$

82. $\dfrac{1}{\sqrt[30]{x}}$ **83.** $\sqrt[12]{a^5}$

84. $\sqrt[6]{y}$ **85.** $\sqrt[10]{x}$

86. $\sqrt[9]{x}$ **87.** $\sqrt[4]{a}$

83. $\dfrac{a\sqrt[4]{a}}{\sqrt[3]{a}\sqrt{a}}$ **84.** $\sqrt{\sqrt[3]{y}}$ **85.** $\sqrt[5]{\sqrt{x}}$ **86.** $\sqrt[3]{\sqrt[3]{x}}$ **87.** $\sqrt{\sqrt{a}}$

88. $\sqrt[15]{(7a)^5}$ **89.** $\left(\sqrt[4]{a^3b^6c}\right)^{12}$ **90.** $\sqrt[5]{\sqrt[4]{\sqrt[3]{x}}}$ **91.** $\left(\sqrt[3]{a^4bc^2}\right)^{15}$

Writing and Thinking About Mathematics

92. Is $\sqrt[5]{a}\cdot\sqrt{a}$ the same as $\sqrt[5]{a^2}$? Explain why or why not.

Hawkes Learning Systems: Introductory & Intermediate Algebra

Rational Exponents

88. $\sqrt[3]{7a}$ **89.** $a^9b^{18}c^3$

90. $\sqrt[60]{x}$ **91.** $a^{20}b^5c^{10}$

92. No: $\sqrt[5]{a}\cdot\sqrt{a}=a^{\frac{7}{10}}$

$\sqrt[5]{a^2}=a^{\frac{2}{5}}$

$a^{\frac{7}{10}}\neq a^{\frac{2}{5}}$

10.3 Arithmetic with Radicals

Objectives

After completing this section, you will be able to:

1. *Perform arithmetic operations with radical expressions.*
2. *Rationalize the denominators of radicals.*

Teaching Notes:
That we cannot combine two exact real numbers into a simple, singular numerical form may be difficult for some students to follow. You may want to emphasize the fact that, for example, $2\sqrt{3}+2\sqrt{7}$ is a single real number.

Addition and Subtraction with Radical Expressions

To find the sum $2x^2 + 3x^2 - 8x^2$, you can use the distributive property and write

$$2x^2 + 3x^2 - 8x^2 = (2+3-8)x^2$$
$$= -3x^2.$$

Recall that the terms $2x^2$, $3x^2$, and $-8x^2$ are called **like terms** because each term contains the same variable expression, x^2. Similarly,

$$2\sqrt{5}+3\sqrt{5}-8\sqrt{5} = (2+3-8)\sqrt{5}$$
$$= -3\sqrt{5}$$

and $2\sqrt{5}$, $3\sqrt{5}$, and $-8\sqrt{5}$ are called **like radicals** because each term contains the same radical expression, $\sqrt{5}$. The terms $2\sqrt{3}$ and $2\sqrt{7}$ are not like radicals because the radical expressions are not the same, and neither expression can be simplified. Therefore, a sum such as

$$2\sqrt{3}+2\sqrt{7}$$

cannot be simplified. That is, the terms cannot be combined.

In some cases, radicals that are not like radicals can be simplified, and the results may lead to like radicals. For example, $4\sqrt{12}$, $\sqrt{75}$, and $-\sqrt{108}$ are not like radicals. However, simplification of each radical allows the sum of these radicals to be found as follows:

$$4\sqrt{12}+\sqrt{75}-\sqrt{108} = 4\sqrt{4\cdot 3}+\sqrt{25\cdot 3}-\sqrt{36\cdot 3}$$
$$= 4\cdot 2\sqrt{3}+5\sqrt{3}-6\sqrt{3}$$
$$= (8+5-6)\sqrt{3}$$
$$= 7\sqrt{3}$$

Example 1: Radicals with Positive Variables

Perform the indicated operation and simplify, if possible. Assume that all variables are positive.

a. $\sqrt{32x}+\sqrt{18x}$

Solution: $\sqrt{32x}+\sqrt{18x}=\sqrt{16\cdot 2x}+\sqrt{9\cdot 2x}$

$$=4\sqrt{2x}+3\sqrt{2x}$$

$$=7\sqrt{2x}$$

b. $\sqrt{12}+\sqrt{18}+\sqrt{27}$

Solution: $\sqrt{12}+\sqrt{18}+\sqrt{27}=\sqrt{4\cdot 3}+\sqrt{9\cdot 2}+\sqrt{9\cdot 3}$

$$=2\sqrt{3}+3\sqrt{2}+3\sqrt{3}$$

$$=5\sqrt{3}+3\sqrt{2}$$

Note that $\sqrt{3}$ and $\sqrt{2}$ are **not** like radicals. Therefore, the last expression cannot be simplified.

c. $\sqrt[3]{5x}-\sqrt[3]{40x}$

Solution: $\sqrt[3]{5x}-\sqrt[3]{40x}=\sqrt[3]{5x}-\sqrt[3]{8\cdot 5x}$

$$=\sqrt[3]{5x}-2\sqrt[3]{5x}$$

$$=(1-2)\sqrt[3]{5x}$$

$$=-\sqrt[3]{5x}$$

d. $x\sqrt{4y^3}-5\sqrt{x^2y^3}$

Solution: $x\sqrt{4y^3}-5\sqrt{x^2y^3}=x\sqrt{4y^2}\sqrt{y}-5\sqrt{x^2y^2}\sqrt{y}$

$$=2xy\sqrt{y}-5xy\sqrt{y}$$

$$=-3xy\sqrt{y}$$

Multiplication with Radical Expressions

To find a product such as $(\sqrt{3}+5)(\sqrt{3}-7)$ treat the two expressions as two binomials and multiply just as with polynomials. For example, using the FOIL method, we get

$$(\sqrt{3}+5)(\sqrt{3}-7)=(\sqrt{3})^2+5\sqrt{3}-7\sqrt{3}+5(-7)$$

$$=(\sqrt{3})^2+5\sqrt{3}-7\sqrt{3}-35$$

$$=3-2\sqrt{3}-35$$

$$=-32-2\sqrt{3}$$

(**Note:** Remember if a and b are positive, then $\sqrt{a}\cdot\sqrt{b}=\sqrt{ab}$. Refer to page 684 for more detail.)

Example 2: Multiplication

Multiply and simplify the following expressions.

a. $(3\sqrt{7}-2)(\sqrt{7}+3)$

Solution: $(3\sqrt{7}-2)(\sqrt{7}+3)=3(\sqrt{7})^2-2\sqrt{7}+9\sqrt{7}-2(3)$

$$=21-2\sqrt{7}+9\sqrt{7}-6$$
$$=21-6-2\sqrt{7}+9\sqrt{7}$$
$$=15+7\sqrt{7}$$

b. $(\sqrt{6}+\sqrt{2})^2$

Solution: $(\sqrt{6}+\sqrt{2})^2=(\sqrt{6})^2+2\sqrt{6}\sqrt{2}+(\sqrt{2})^2$ $\qquad (a+b)^2=a^2+2ab+b^2$

$$=6+2\sqrt{12}+2$$
$$=8+2\sqrt{4}\sqrt{3}$$
$$=8+4\sqrt{3}$$

c. $(\sqrt{2x}+5)(\sqrt{2x}-5)$

Solution: $(\sqrt{2x}+5)(\sqrt{2x}-5)=(\sqrt{2x})^2-(5)^2$ $\qquad (a+b)(a-b)=a^2-b^2$

$$=2x-25$$

Rationalizing Denominators of Rational Expressions

We discussed **rationalizing the denominator** in Section 10.1. In that discussion, all denominators were single terms. If the denominator has a radical expression with **a sum or difference involving square roots** such as

$$\frac{2}{4-\sqrt{2}} \text{ or } \frac{12}{3+\sqrt{5}}$$

then the method of rationalizing the denominator must be changed. In this situation, think of the denominator in the form of $a-b$ or $a+b$. Thus,

$$\text{if } 4-\sqrt{2}=a-b \text{ then } 4+\sqrt{2}=a+b,$$

$$\text{and if } 3+\sqrt{5}=a+b, \text{ then } 3-\sqrt{5}=a-b.$$

The two expressions $(a-b)$ and $(a+b)$ are called **conjugates** of each other, and the product $(a-b)(a+b)$ results in the difference of two squares.

$$(a-b)(a+b)=a^2-b^2$$

Thus, even if either a or b (or both) is a square root, the difference of the squares will not contain a square root. With this in mind, we can proceed as follows to rationalize a denominator with square roots.

Rationalizing Denominators with Sums or Differences of Square Roots

1. *If the denominator is of the form a – b, multiply both the numerator and denominator by its conjugate a + b.*

2. *If the denominator is of the form a + b, multiply both the numerator and denominator by its conjugate a – b.*

In either case, the new denominator becomes

$a^2 - b^2$, the difference of two squares

and this denominator is a rational number or a rational expression.

Example 3: Rationalization

Simplify the following expressions by rationalizing the denominators.

a. $\frac{2}{4-\sqrt{2}}$

Solution: Multiply the numerator and denominator by $4+\sqrt{2}$.

$$\frac{2}{4-\sqrt{2}} = \frac{2(4+\sqrt{2})}{(4-\sqrt{2})(4+\sqrt{2})}$$

If $a-b=4-\sqrt{2}$, then $a+b=4+\sqrt{2}$.

$$= \frac{2(4+\sqrt{2})}{4^2-(\sqrt{2})^2}$$

The denominator is the difference of two squares.

$$= \frac{2(4+\sqrt{2})}{16-2}$$

The denominator is a rational number.

$$= \frac{2(4+\sqrt{2})}{14} = \frac{4+\sqrt{2}}{7}$$

Note that the numerator is now irrational. However, this is generally preferred to having an irrational denominator.

b. $\frac{31}{6+\sqrt{5}}$

Solution: Multiply the numerator and denominator by $6-\sqrt{5}$.

$$\frac{31}{6+\sqrt{5}} = \frac{31(6-\sqrt{5})}{(6+\sqrt{5})(6-\sqrt{5})}$$

$$= \frac{31(6-\sqrt{5})}{36-5}$$

$$= \frac{\cancel{31}(6-\sqrt{5})}{\cancel{31}} = 6-\sqrt{5}$$

c. $\dfrac{1}{\sqrt{7}-\sqrt{2}}$

Solution: Multiply the numerator and denominator by $\sqrt{7}+\sqrt{2}$.

$$\begin{aligned}\frac{1}{\sqrt{7}-\sqrt{2}} &= \frac{1\left(\sqrt{7}+\sqrt{2}\right)}{\left(\sqrt{7}-\sqrt{2}\right)\left(\sqrt{7}+\sqrt{2}\right)}\\ &= \frac{\left(\sqrt{7}+\sqrt{2}\right)}{7-2}\\ &= \frac{\sqrt{7}+\sqrt{2}}{5}\end{aligned}$$

d. $\dfrac{6}{1-\sqrt{x}}$

Solution: $$\begin{aligned}\frac{6}{1-\sqrt{x}} &= \frac{6\left(1+\sqrt{x}\right)}{\left(1-\sqrt{x}\right)\left(1+\sqrt{x}\right)}\\ &= \frac{6\left(1+\sqrt{x}\right)}{1-x}\end{aligned}$$

e. $\dfrac{x-y}{\sqrt{x}-\sqrt{y}}$

Solution: $$\begin{aligned}\frac{x-y}{\sqrt{x}-\sqrt{y}} &= \frac{(x-y)\left(\sqrt{x}+\sqrt{y}\right)}{\left(\sqrt{x}-\sqrt{y}\right)\left(\sqrt{x}+\sqrt{y}\right)}\\ &= \frac{\cancel{(x-y)}\left(\sqrt{x}+\sqrt{y}\right)}{\cancel{(x-y)}}\\ &= \sqrt{x}+\sqrt{y}\end{aligned}$$

Evaluating Radical Expressions with a Calculator

In Sections 10.1 and 10.2, we showed how to use a TI-83 Plus calculator to evaluate expressions in radical form and in exponential form. These same basic techniques are used to evaluate numerical expressions that contain sums, differences, products, and quotients of radicals. Be careful to use parentheses to ensure that the rules for order of operations are maintained. In particular, sums and differences in numerators and denominators of fractions must be enclosed in parentheses.

Example 4: Using a Calculator

Use a TI-83 Plus calculator to evaluate each expression accurate to 4 decimal places.

a. $3+2\sqrt{5}$

b. $(\sqrt{2}+5)(\sqrt{2}-5)$

c. $\dfrac{3}{\sqrt{6}-\sqrt{2}}$

Solutions:

The displays should appear as follows:

a. $3+2\sqrt{5}$

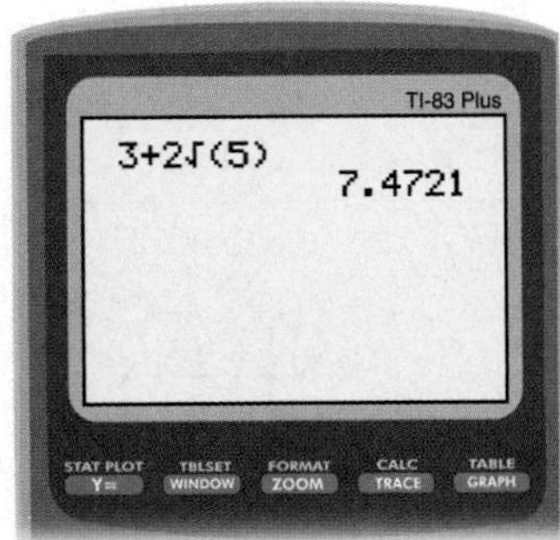

b. $(\sqrt{2}+5)(\sqrt{2}-5)$

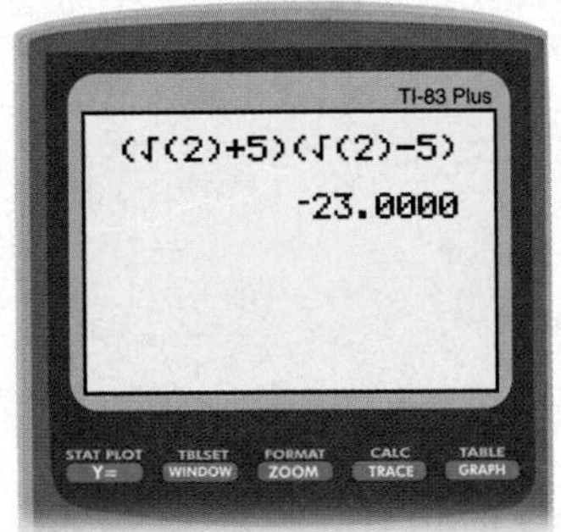

Note that the right parenthesis on 2 must be included. Otherwise, the calculator will interpret the expression as $\sqrt{(2+5)}$ $\left(\text{or } \sqrt{7}\right)$, which is not intended.

c. $\dfrac{3}{\sqrt{6}-\sqrt{2}}$

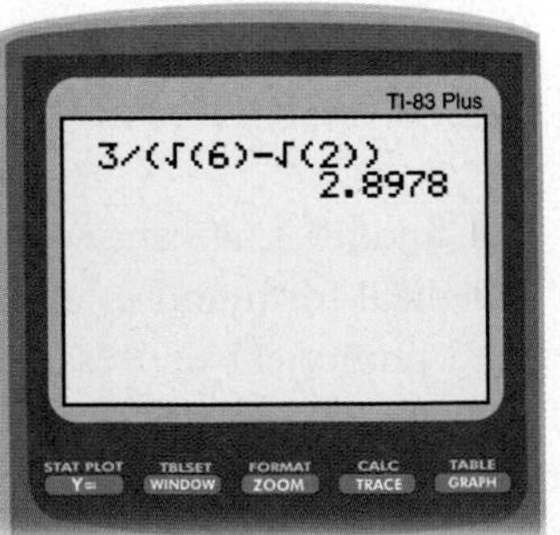

Count the parentheses in pairs.

Practice Problems

Simplify the following expressions. Assume that all variables are positive.

1. $2\sqrt{10}-6\sqrt{10}$
2. $\sqrt{5}+\sqrt{45}-\sqrt{15}$
3. $\sqrt{8x}-3\sqrt{2x}+\sqrt{18x}$
4. $\sqrt[3]{x^5}+x\sqrt[3]{27x^2}$
5. $\sqrt[3]{x^3y^6z}+4xy^2\sqrt[3]{z}$
6. $(\sqrt{6}-2\sqrt{5})(\sqrt{6}+\sqrt{5})$
7. $\dfrac{4}{\sqrt{2}+\sqrt{6}}$
8. $\dfrac{x-5}{\sqrt{x}-\sqrt{5}}$
9. $(\sqrt{3}+\sqrt{2})^2$
10. $(3+\sqrt{2})^2$
11. $(\sqrt{3}+\sqrt{8})^2$
12. $\dfrac{\sqrt{5}-3\sqrt{2}}{\sqrt{6}+\sqrt{10}}$

1. $-6\sqrt{2}$ **2.** $7\sqrt{11}$
3. $5\sqrt{x}$ **4.** $9\sqrt[3]{xy}$
5. $-3\sqrt[3]{7x^2}$
6. $-4\sqrt[3]{4x}$ **7.** $10\sqrt{3}$
8. $-7\sqrt{3}$ **9.** $-7\sqrt{2}$
10. $6\sqrt{2}-\sqrt{3}$
11. $20\sqrt{3}$
12. $3\sqrt{7}+8\sqrt{10}$
13. $3\sqrt{6}+7\sqrt{3}$
14. $8\sqrt{3x}$
15. $20x\sqrt{5x}$
16. $-7y\sqrt{2x}$
17. $14\sqrt{5}-3\sqrt{7}$
18. $15\sqrt{3}+4\sqrt{6}$
19. $11\sqrt{2x}+14\sqrt{3x}$
20. $-7\sqrt[3]{3x^2}-10\sqrt[3]{6x^2}$
21. $4\sqrt[3]{5}-13\sqrt[3]{2}$
22. $2x\sqrt{y}-y\sqrt{xy}$
23. $x^2\sqrt{2x}$
24. $-2xy\sqrt{y}$
25. $4x^2y\sqrt{x}$
26. $13+2\sqrt{2}$
27. $3x-9\sqrt{3x}+8$
28. $2-2\sqrt{7}$
29. $24+10\sqrt{2x}+2x$
30. $44-39\sqrt{3}$
31. 2 **32.** $11+4\sqrt{10}$
33. $13+4\sqrt{10}$
34. $38+12\sqrt{10}$ **35.** -3
36. $\sqrt{10}+\sqrt{15}-\sqrt{6}-3$

10.3 Exercises

Perform the indicated operations and simplify for Exercises 1 – 45. Assume that all variables are positive.

1. $\sqrt{2}-7\sqrt{2}$
2. $6\sqrt{11}+4\sqrt{11}-3\sqrt{11}$
3. $2\sqrt{x}+4\sqrt{x}-\sqrt{x}$
4. $8\sqrt[3]{xy}-3\sqrt[3]{xy}+4\sqrt[3]{xy}$
5. $9\sqrt[3]{7x^2}-4\sqrt[3]{7x^2}-8\sqrt[3]{7x^2}$
6. $12\sqrt[3]{4x}-10\sqrt[3]{4x}-6\sqrt[3]{4x}$
7. $2\sqrt{3}+4\sqrt{12}$
8. $2\sqrt{48}-3\sqrt{75}$
9. $2\sqrt{18}+\sqrt{8}-3\sqrt{50}$
10. $2\sqrt{12}+\sqrt{72}-\sqrt{75}$
11. $5\sqrt{48}+2\sqrt{45}-3\sqrt{20}$
12. $3\sqrt{28}-\sqrt{63}+8\sqrt{10}$
13. $2\sqrt{96}+\sqrt{147}-\sqrt{150}$
14. $7\sqrt{12x}-4\sqrt{27x}+\sqrt{108x}$
15. $6\sqrt{45x^3}+\sqrt{80x^3}-\sqrt{20x^3}$
16. $2\sqrt{18xy^2}+\sqrt{8xy^2}-3y\sqrt{50x}$
17. $\sqrt{125}-\sqrt{63}+3\sqrt{45}$
18. $5\sqrt{48}+2\sqrt{24}-\sqrt{75}$
19. $\sqrt{32x}+7\sqrt{12x}+\sqrt{98x}$
20. $\sqrt[3]{81x^2}-5\sqrt[3]{48x^2}-5\sqrt[3]{24x^2}$
21. $\sqrt[3]{16}-5\sqrt[3]{54}+2\sqrt[3]{40}$
22. $x\sqrt{y}+\sqrt{x^2y}-\sqrt{xy^3}$
23. $x\sqrt{2x^3}-3\sqrt{8x^5}+x\sqrt{72x^3}$
24. $x\sqrt{y^3}-2\sqrt{x^2y^3}-y\sqrt{x^2y}$

Answers to Practice Problems: 1. $-4\sqrt{10}$ **2.** $4\sqrt{5}-\sqrt{15}$ **3.** $2\sqrt{2x}$ **4.** $4x\sqrt[3]{x^2}$ **5.** $5xy^2\sqrt[3]{z}$ **6.** $-4-\sqrt{30}$ **7.** $\sqrt{6}-\sqrt{2}$ **8.** $\sqrt{x}+\sqrt{5}$ **9.** $5+2\sqrt{6}$ **10.** $11+6\sqrt{2}$ **11.** $11+4\sqrt{6}$ **12.** $\dfrac{5\sqrt{2}+6\sqrt{3}-6\sqrt{5}-\sqrt{30}}{4}$

37. $6+\sqrt{30}-2\sqrt{3}-\sqrt{10}$

38. 58 **39.** $5-\sqrt{33}$

40. $x-2\sqrt{6x}-18$

41. $49x-2$

42. $x+10\sqrt{xy}+25y$

43. $9x+6\sqrt{xy}+y$

44. $4x-9\sqrt{xy}-9y$

45. $12x+7\sqrt{2xy}+2y$

46. $\sqrt{2}-1$

47. $-\dfrac{3\sqrt{3}+15}{22}$

48. $-\dfrac{\sqrt{15}+4\sqrt{3}}{11}$

49. $\dfrac{3\sqrt{6}-\sqrt{42}}{2}$

50. $2\sqrt{3}-2\sqrt{2}$

51. $4\sqrt{5}+4\sqrt{3}$

52. $\dfrac{\sqrt{35}+\sqrt{15}}{4}$

53. $\dfrac{-5\sqrt{2}-4\sqrt{5}}{3}$

54. $\dfrac{\sqrt{6}}{3}$ **55.** $\dfrac{59+21\sqrt{5}}{44}$

56. $-\dfrac{5+\sqrt{21}}{2}$

57. $8+3\sqrt{6}$

58. $\dfrac{2x+y\sqrt{x}-y^2}{x-y^2}$

59. $\dfrac{x^2+4x\sqrt{y}+4y}{x^2-4y}$

60. $\dfrac{y\sqrt{x}-y\sqrt{2y}}{x-2y}$

61. $\dfrac{1}{\sqrt{7}+2}$

25. $x\sqrt{9x^3y^2}-5x^2\sqrt{xy^2}+6y\sqrt{x^5}$ **26.** $(3+\sqrt{2})(5-\sqrt{2})$

27. $(\sqrt{3x}-8)(\sqrt{3x}-1)$ **28.** $(2\sqrt{7}+4)(\sqrt{7}-3)$

29. $(6+\sqrt{2x})(4+\sqrt{2x})$ **30.** $(5\sqrt{3}-2)(2\sqrt{3}-7)$

31. $(\sqrt{6}+2)(\sqrt{6}-2)$ **32.** $(3\sqrt{2}+\sqrt{5})(\sqrt{2}+\sqrt{5})$

33. $(\sqrt{5}+2\sqrt{2})^2$ **34.** $(2\sqrt{5}+3\sqrt{2})^2$

35. $(3\sqrt{5}+4\sqrt{3})(3\sqrt{5}-4\sqrt{3})$ **36.** $(\sqrt{2}+\sqrt{3})(\sqrt{5}-\sqrt{3})$

37. $(\sqrt{6}+\sqrt{5})(\sqrt{6}-\sqrt{2})$ **38.** $(3\sqrt{7}+\sqrt{5})(3\sqrt{7}-\sqrt{5})$

39. $(\sqrt{11}+\sqrt{3})(\sqrt{11}-2\sqrt{3})$ **40.** $(\sqrt{x}+\sqrt{6})(\sqrt{x}-3\sqrt{6})$

41. $(7\sqrt{x}+\sqrt{2})(7\sqrt{x}-\sqrt{2})$ **42.** $(\sqrt{x}+5\sqrt{y})^2$

43. $(3\sqrt{x}+\sqrt{y})^2$ **44.** $(4\sqrt{x}+3\sqrt{y})(\sqrt{x}-3\sqrt{y})$

45. $(2\sqrt{2x}+\sqrt{y})(3\sqrt{2x}+2\sqrt{y})$

In Exercises 46 – 60, rationalize the denominator and simplify if possible.

46. $\dfrac{1}{\sqrt{2}+1}$ **47.** $\dfrac{3}{\sqrt{3}-5}$ **48.** $\dfrac{\sqrt{3}}{\sqrt{5}-4}$ **49.** $\dfrac{\sqrt{6}}{\sqrt{7}+3}$

50. $\dfrac{2}{\sqrt{2}+\sqrt{3}}$ **51.** $\dfrac{8}{\sqrt{5}-\sqrt{3}}$ **52.** $\dfrac{\sqrt{5}}{\sqrt{7}-\sqrt{3}}$ **53.** $\dfrac{\sqrt{10}}{\sqrt{5}-2\sqrt{2}}$

54. $\dfrac{2-\sqrt{6}}{\sqrt{6}-3}$ **55.** $\dfrac{7+2\sqrt{5}}{7-\sqrt{5}}$ **56.** $\dfrac{\sqrt{3}+\sqrt{7}}{\sqrt{3}-\sqrt{7}}$ **57.** $\dfrac{2\sqrt{3}+\sqrt{2}}{\sqrt{3}-\sqrt{2}}$

58. $\dfrac{2\sqrt{x}-y}{\sqrt{x}-y}$ **59.** $\dfrac{x+2\sqrt{y}}{x-2\sqrt{y}}$ **60.** $\dfrac{y}{\sqrt{x}+\sqrt{2y}}$

In Exercises 61 – 70, rationalize the numerator (by using the same technique used to rationalize the denominator in Exercises 46 – 60) and simplify if possible.

61. $\dfrac{\sqrt{7}-2}{3}$ **62.** $\dfrac{\sqrt{5}+1}{2}$ **63.** $\dfrac{\sqrt{15}+\sqrt{3}}{6}$ **64.** $\dfrac{\sqrt{10}-\sqrt{2}}{-8}$

65. $\dfrac{\sqrt{x}+\sqrt{5}}{x-5}$ **66.** $\dfrac{\sqrt{y}+\sqrt{2}}{y-2}$ **67.** $\dfrac{\sqrt{2y}-\sqrt{x}}{x}$ **68.** $\dfrac{3\sqrt{x}-y}{3x}$

69. $\dfrac{\sqrt{2+h}-\sqrt{2}}{h}$ **70.** $\dfrac{\sqrt{5+h}-\sqrt{5}}{h}$

In Exercises 71 – 80, use a calculator to find the value of each expression accurate to 5 decimal places.

71. $13-\sqrt{75}$ **72.** $5-\sqrt{67}$ **73.** $\sqrt{900}+\sqrt{2.56}$ **74.** $\sqrt{1600}-\sqrt{1.69}$

75. $(\sqrt{7}+8)(\sqrt{7}-8)$ **76.** $(\sqrt{3}+\sqrt{2})(\sqrt{5}-\sqrt{10})$ **77.** $(6\sqrt{8}+5\sqrt{7})(3\sqrt{39}-2\sqrt{27})$

78. $\dfrac{19}{35-\sqrt{60}}$ **79.** $\dfrac{\sqrt{5}}{1+2\sqrt{5}}$ **80.** $\dfrac{\sqrt{10}-\sqrt{2}}{\sqrt{10}+\sqrt{2}}$

Writing and Thinking About Mathematics

81. One of the most studied and interesting visual and numerical concepts in algebra is the **Golden Ratio**. Ancient Greeks thought (and many people still do) that a rectangle was most aesthetically pleasing to the eye if the ratio of its length to its width is the Golden Ratio (about 1.618). In fact, the Parthenon, built by Greeks in the fifth century B.C. utilizes the Golden Ratio. A rectangle is "golden" if its length, l, and width, w, satisfy the equation $\dfrac{l}{w}=\dfrac{w}{l-w}$.

a. By letting $w=1$ unit in the equation above, we get the equation $\dfrac{l}{1}=\dfrac{1}{l-1}$. Solve this equation for l and find the algebraic expression for the golden ratio (a positive number).

b. Suppose that an architect is constructing a building with a rectangular front that is to be 60 feet high. About how long should the front be if he wants the appearance of a golden rectangle?

c. Look at the two rectangles shown here. Which seems most pleasing to your eye? Measure the length and width of each rectangle and see if you chose the golden rectangle.

Answers (margin):

62. $\dfrac{2}{\sqrt{5}-1}$

63. $\dfrac{2}{\sqrt{15}-\sqrt{3}}$

64. $\dfrac{-1}{\sqrt{10}+\sqrt{2}}$

65. $\dfrac{1}{\sqrt{x}-\sqrt{5}}$

66. $\dfrac{1}{\sqrt{y}-\sqrt{2}}$

67. $\dfrac{2y-x}{x\sqrt{2y}+x\sqrt{x}}$

68. $\dfrac{9x-y^2}{9x\sqrt{x}+3xy}$

69. $\dfrac{1}{\sqrt{2+h}+\sqrt{2}}$

70. $\dfrac{1}{\sqrt{5+h}+\sqrt{5}}$

71. 4.33975
72. –3.18535
73. 31.60000
74. 38.70000
75. –57.00000
76. –2.91410
77. 251.94353
78. 0.69714
79. 0.40863
80. 0.38197
81. a. $l^2-l-1=0$
b. 97.08 feet
c. The yellow rectangle is "golden".

Hawkes Learning Systems: Introductory & Intermediate Algebra

Addition and Subtraction of Radicals
Multiplication of Radicals

10.4 Functions with Radicals

Objectives

After completing this section, you will be able to:

1. *Recognize radical functions.*
2. *Evaluate radical functions.*
3. *Find the domain and range of radical functions.*
4. *Graph radical functions.*

The concept of functions is among the most important and useful ideas in all of mathematics. We introduced functions in Chapter 4 and used the function notation, $P(x)$ (read "P of x"), with polynomials in Chapters 5 and 6. In this section we will expand the function concept to include **radical functions** (functions with radicals). More functions will be discussed in Chapters 12 and 13. For review and easy reference, we restate the definitions of relations and functions and the vertical line test as stated in Chapter 4.

Relation

*A **relation** is a set of ordered pairs of real numbers.*

*The **domain, D**, of a relation is the set of all first coordinates in the relation.*

*The **range, R**, of a relation is the set of all second coordinates in the relation.*

Function

*A **function** is a relation in which each domain element has exactly one corresponding range element.*

Functions have the following two characteristics:

1. *A function is a relation in which each first coordinate appears only once.*
2. *A function is a relation in which no two ordered pairs have the same first coordinate.*

Vertical Line Test

*If **any** vertical line intersects the graph of a relation at more than one point, then the relation graphed is **not** a function.*

Evaluating Radical Functions

We have used the ordered pair notation (x, y) to represent points on the graphs of relations and functions. For example,

$y = 2x - 5$ represents a linear function and its graph is a straight line.

We define radical functions (functions with radical expressions) as follows.

Radical Function

*A **radical function** is a function of the form $y = \sqrt[n]{g(x)}$ in which the radicand contains a variable expression.*

*The **domain** of such a function depends on the index, n:*

1. *If n is an even number, the domain is the set of all x such that $g(x) \geq 0$.*
2. *If n is an odd number, the domain is the set of all real numbers, $(-\infty, \infty)$.*

Examples of radical functions are

$$y = 3\sqrt{x}, \quad f(x) = \sqrt{2x+3}, \quad y = \sqrt[3]{x-7}.$$

The functions

$$y = \sqrt{2}x \text{ and } f(x) = x^2 + \sqrt{3}$$

are **not** radical functions because the radicand does not contain the variable.

The function notation, $f(x)$ (read "f of x"), is very useful when evaluating a function for a particular value of x. For example,

$f(9)$ means to substitute 9 for x in the function.

Thus,

if $f(x) = \sqrt{x-5}$,

then $f(9) = \sqrt{9-5} = \sqrt{4} = 2$.

We can use a calculator to find decimal approximations. Such approximations are particularly helpful when estimating the locations of points on a graph. For example,

if $f(x) = \sqrt{x-5}$, then $f(8) = \sqrt{3} \approx 1.7321$ and $f(25) = \sqrt{20} = 2\sqrt{5} \approx 4.4721$.

Example 1: Evaluating a Radical Function

Complete each table by finding each value of the function for the related value of x.

a. $f(x)=3\sqrt{x}$

x	$f(x)$
0	?
4	?
6	?

b. $y=\sqrt[3]{x-7}$

x	y
7	?
6	?
–1	?

Solutions:

a.

x	$f(x)$
0	$3\sqrt{0}=0$
4	$3\sqrt{4}=3\cdot 2=6$
6	$3\sqrt{6}\approx 7.3485$ ← with a calculator

b.

x	y
7	$\sqrt[3]{7-7}=\sqrt[3]{0}=0$
6	$\sqrt[3]{6-7}=\sqrt[3]{-1}=-1$
–1	$\sqrt[3]{-1-7}=\sqrt[3]{-8}=-2$

Example 2: Domain of a Radical Function

Determine the domain of each radical function:

a. $y=\sqrt{2x+3}$ **b.** $f(x)=\sqrt[3]{x-5}$

Solutions:

a. $y=\sqrt{2x+3}$

Because the index is 2 (understood), an even number, the radicand must be nonnegative. Thus, we have

$$2x+3\ge 0$$

$$2x\ge -3$$

$$x\ge -\frac{3}{2}$$

The domain is the interval $\left[-\frac{3}{2},\infty\right)$.

b. $f(x)=\sqrt[3]{x-5}$

Because the index is 3, an odd number, the radicand may be any real number. The domain is $(-\infty,\infty)$.

Graphing Radical Functions

To graph a radical function, we need to be aware of its domain and to plot at least a few points to see the nature of the resulting curve. Example 3 shows how to proceed, at least in the beginning, to graph the radical function $y=\sqrt{x+5}$.

Example 3: Graphing a Radical Function

Teaching Notes: Reinforce the need for interpreting $y=\sqrt{x+5}$ as using the principal square root to build the table of values for graphing purposes.

Graph the function $y=\sqrt{x+5}$.

Solution:

For the domain we have

$$x+5\geq 0$$
$$x\geq -5.$$

To see the nature of the graph we select a few values for x in the domain and find the corresponding values of y:

x	y
–5	0
–4	1
–3	$\sqrt{2}\approx 1.41$
0	$\sqrt{5}\approx 2.24$
4	3

Now we plot these points on a graph.

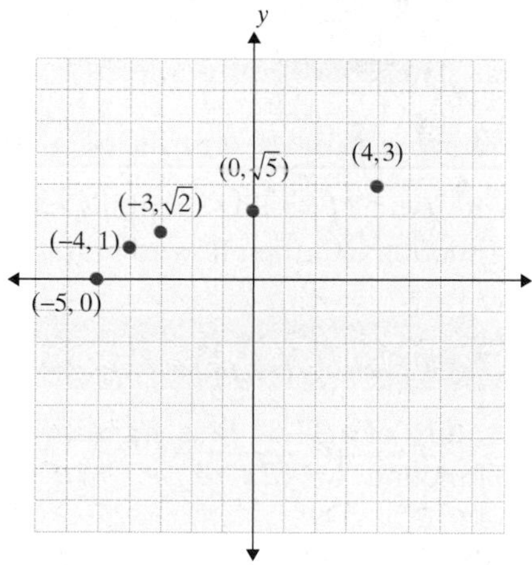

Next we complete the graph by drawing a smooth curve that passes through the selected points. This is the graph of the function.

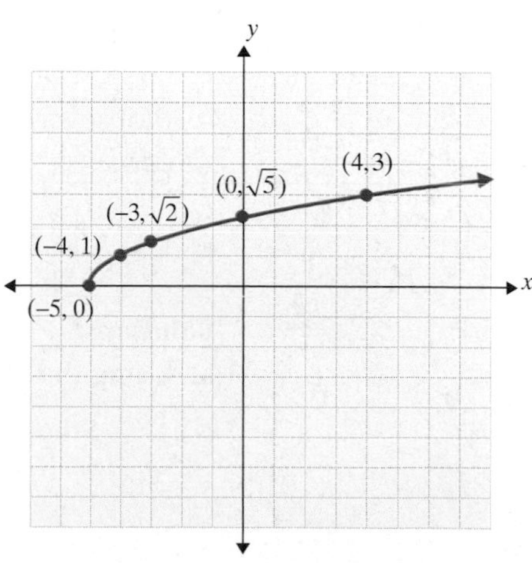

continued on next page ...

To use a TI-83 Plus graphing calculator to graph this function,

Step 1: Press Y= and enter the function as follows:

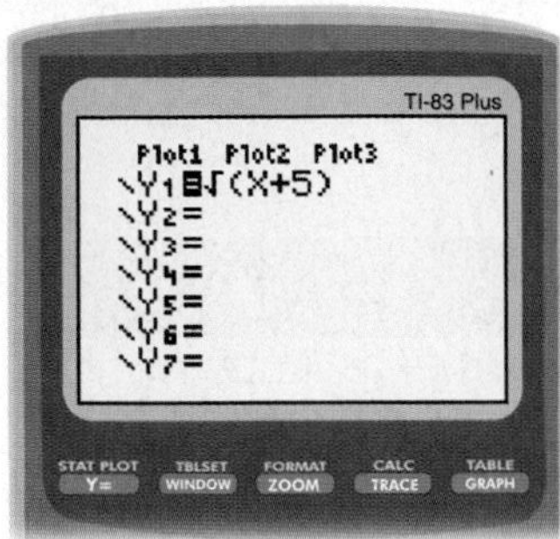

Step 2: Press GRAPH. (You may need to adjust the WINDOW.)

The result will be the graph as shown here:

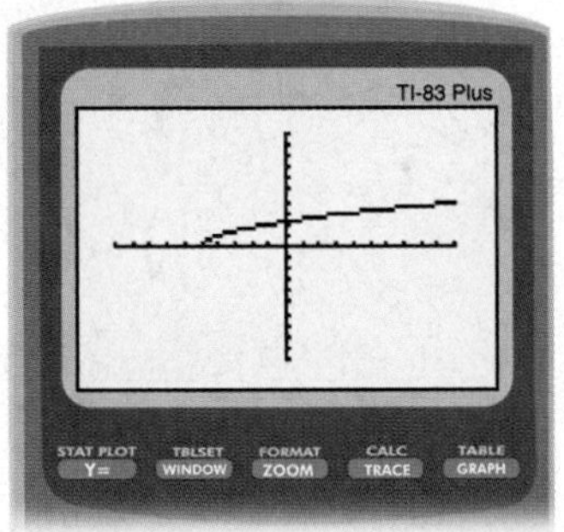

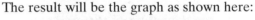

Example 4 shows how to use a TI-83 Plus graphing calculator to find a lot of points on the graph of a radical function and then how to graph the function.

Example 4: Using a TI-83 Plus to Graph a Radical Function

a. Use the TABLE feature of a TI-83 Plus graphing calculator to locate many points on the graph of the function $y = \sqrt[3]{2x-3}$.

b. Plot several points on a graph (approximately) and then connect them with a smooth curve.

c. Use a TI-83 Plus graphing calculator to graph the function.

Solutions:

a. Using the TABLE feature of a TI-83 Plus:

Step 1: Press Y= and enter the function as follows:

1. Press MATH.
2. Choose **4 :** $\sqrt[3]{\ }$.
3. Enter $(2x - 3)$ and press ENTER.

Step 2: Press TBLSET (which is 2nd WINDOW) and set the display as shown here:

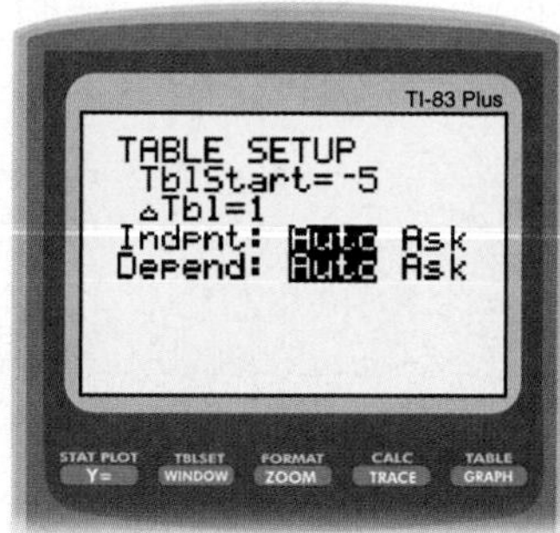

Step 3: Press TABLE (which is 2nd GRAPH) and the display will appear as follows:

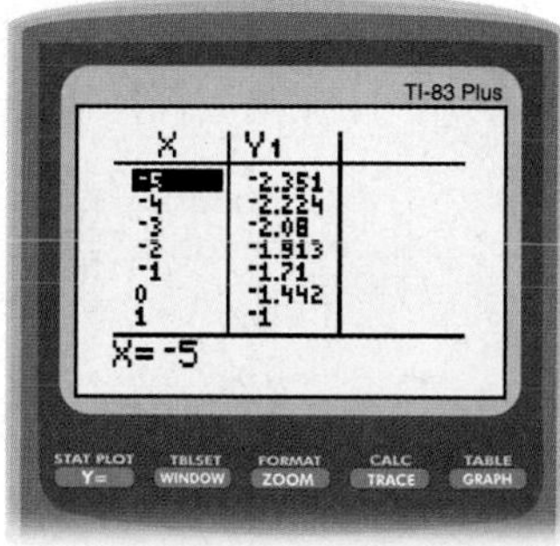

b. You may scroll up and down the display to find as many points as you like. A few are shown here to see the nature of the graph.

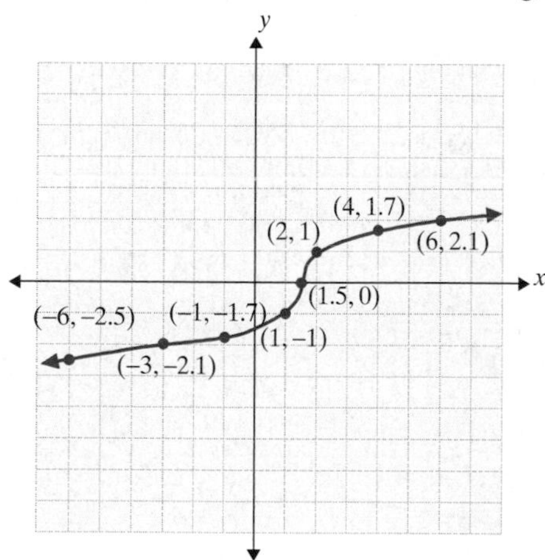

c. Press GRAPH and the display will appear with the curve as follows:

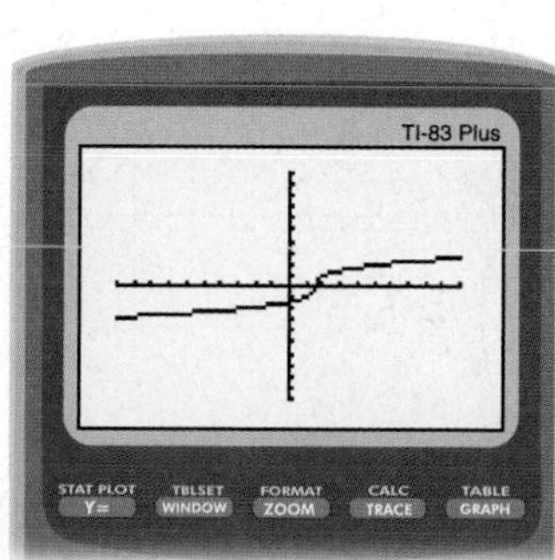

10.4 Exercises

1. a. $\sqrt{5}$ and 2.2361
b. $\sqrt{9}$ and 3.0000
c. $\sqrt{50}$ and 7.0711
d. $\sqrt{4}$ and 2.0000

2. a. $\sqrt{5}$ and 2.2361
b. $\sqrt{11}$ and 3.3166
c. $\sqrt{25}$ and 5.0000
d. $\sqrt{12.2}$ and 3.4928

3. a. $\sqrt[3]{27}$ and 3.0000
b. $\sqrt[3]{-1}$ and −1.0000
c. $\sqrt[3]{-8}$ and −2.0000
d. $\sqrt[3]{24}$ and 2.8845

4. a. $\sqrt[3]{0}$ and 0
b. $\sqrt[3]{8}$ and 2.0000
c. $\sqrt[3]{0.001}$ and 0.1000
d. $\sqrt[3]{6.5}$ and 1.8663

5. $[-8, \infty)$

6. $\left[\frac{1}{2}, \infty\right)$

7. $\left(-\infty, \frac{1}{2}\right]$

8. $\left(-\infty, \frac{1}{3}\right]$

9. $(-\infty, \infty)$

10. $(-\infty, \infty)$

11. $[0, \infty)$

12. $(-\infty, 7]$

13. $(-\infty, \infty)$

14. $(-\infty, \infty)$

15. E
16. D
17. B
18. F
19. A
20. C

In Exercises 1 – 4, write answers in both radical notation and decimal notation (accurate to 4 decimal places).

1. Given $f(x) = \sqrt{2x+1}$, find
a. $f(2)$ **b.** $f(4)$ **c.** $f(24.5)$ **d.** $f(1.5)$

2. Given $f(x) = \sqrt{5-3x}$, find
a. $f(0)$ **b.** $f(-2)$ **c.** $f\left(-\frac{20}{3}\right)$ **d.** $f(-2.4)$

3. Given $g(x) = \sqrt[3]{x+6}$, find
a. $g(21)$ **b.** $g(-7)$ **c.** $g(-14)$ **d.** $g(18)$

4. Given $h(x) = \sqrt[3]{4-x}$, find
a. $h(4)$ **b.** $h(-4)$ **c.** $h(3.999)$ **d.** $h(-2.5)$

In Exercises 5 – 14, use interval notation to indicate the domain of each radical function.

5. $y = \sqrt{x+8}$ **6.** $y = \sqrt{2x-1}$ **7.** $y = \sqrt{2.5-5x}$ **8.** $y = \sqrt{1-3x}$

9. $f(x) = \sqrt[3]{x+4}$ **10.** $f(x) = \sqrt[3]{6x}$ **11.** $g(x) = \sqrt[4]{x}$ **12.** $g(x) = \sqrt[4]{7-x}$

13. $y = \sqrt[5]{4x-1}$ **14.** $y = \sqrt[5]{8+x}$

Match the function given in Exercises 15 – 20 with the graph of that function (A) – (F).

15. $y = \sqrt{x-2}$ **16.** $y = \sqrt{2-x}$ **17.** $y = -\sqrt{x-3}$

18. $y = -\sqrt{3-x}$ **19.** $y = \sqrt{x+4}$ **20.** $y = \sqrt{x-4}$

A.

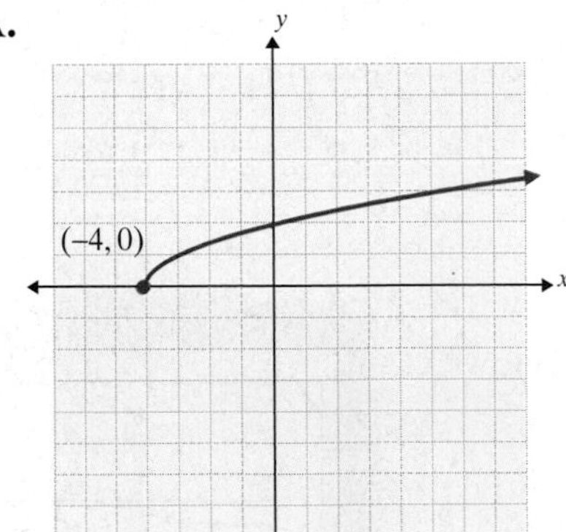

B.

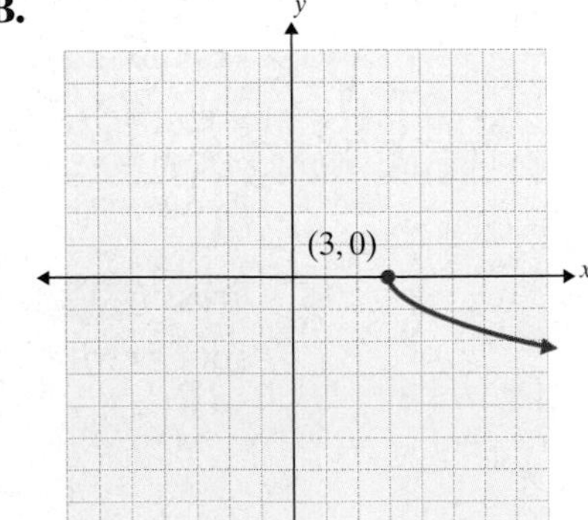

C.

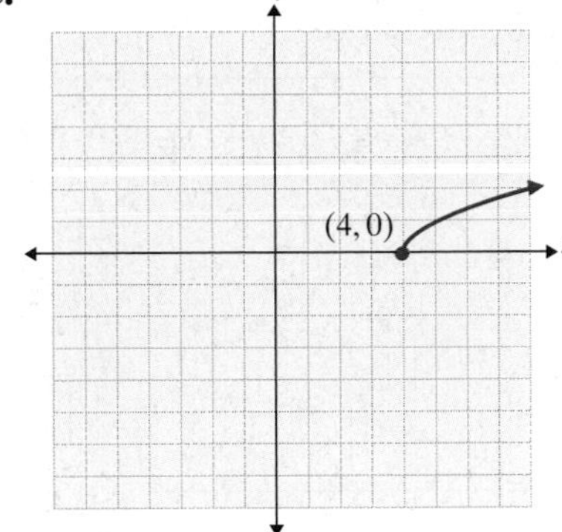

D.

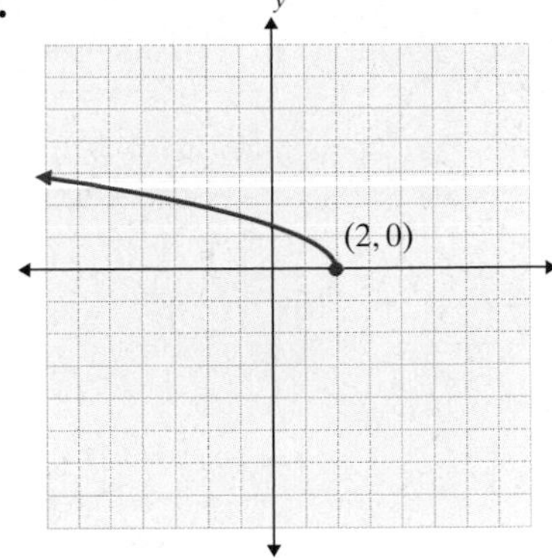

E.

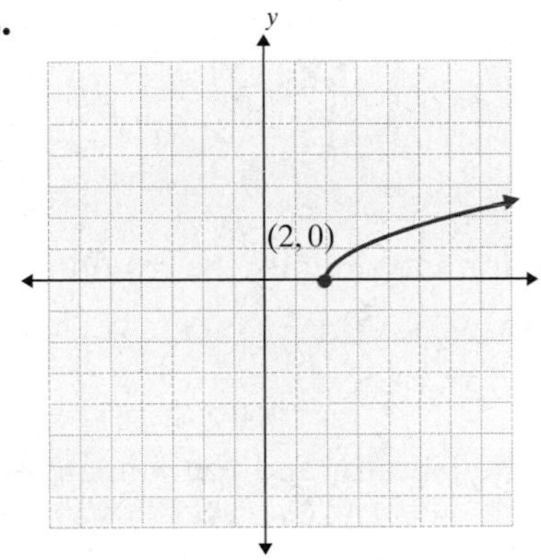

F.

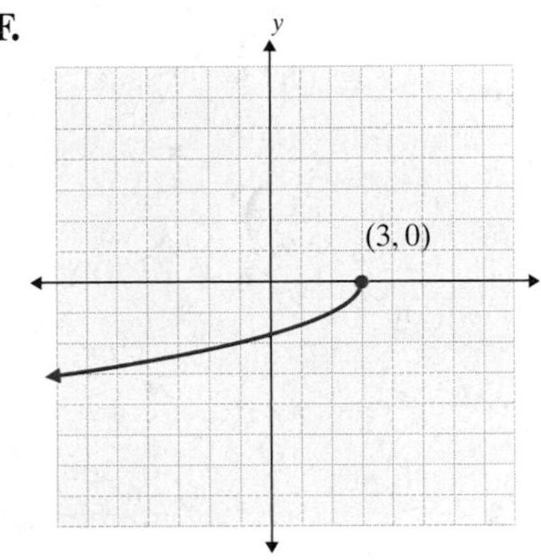

In Exercises 21 – 25, find and label at least 5 points on the graph of the function and then sketch the graph of the function.

21. $f(x)=\sqrt[3]{x+2}$ **22.** $g(x)=\sqrt[3]{x-6}$ **23.** $y=\sqrt[4]{x}$

24. $y=\sqrt[4]{2x+8}$ **25.** $y=\sqrt[5]{x-1}$

Use a TI-83 Plus graphing calculator to graph each of the functions in Exercises 26 – 35.

26. $y=3\sqrt{x+2}$ **27.** $y=2\sqrt{3-x}$ **28.** $f(x)=-\sqrt{x+2.5}$

29. $f(x)=-\sqrt{3-x}$ **30.** $y=-\sqrt[3]{x+2}$ **31.** $y=\sqrt[3]{3x+4}$

32. $g(x)=-\sqrt{2x}$ **33.** $g(x)=-\sqrt[4]{x+5}$ **34.** $y=\sqrt[4]{2x+6}$

35. $y=\sqrt[5]{x+7}$

21.

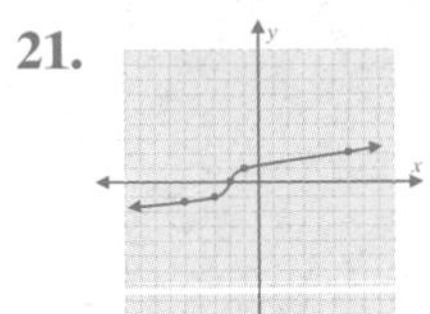

22.

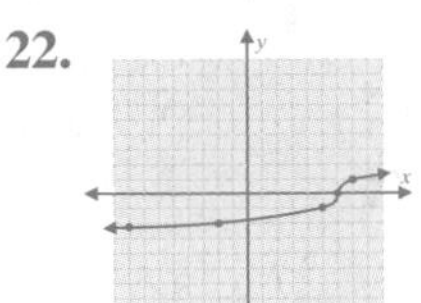

23.

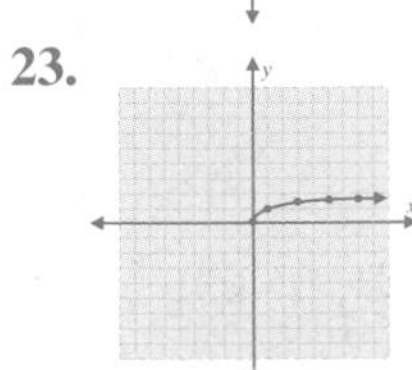

24.

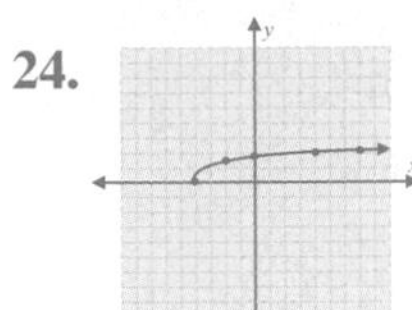

25.

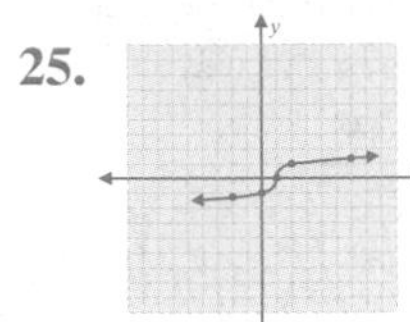

26.

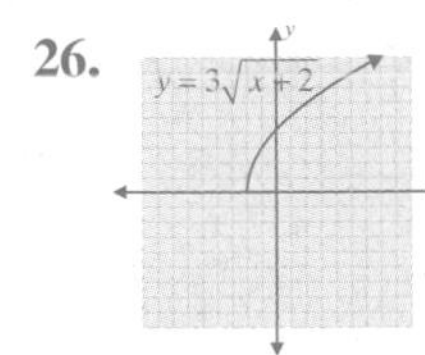

27.

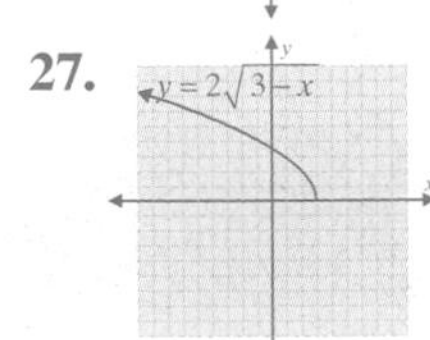

28.

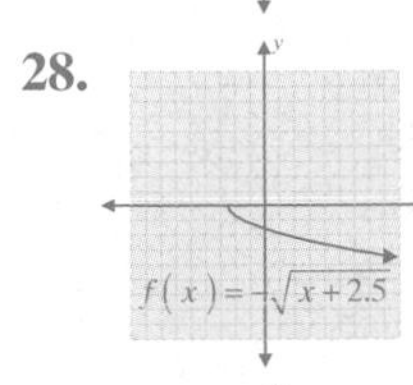

29.

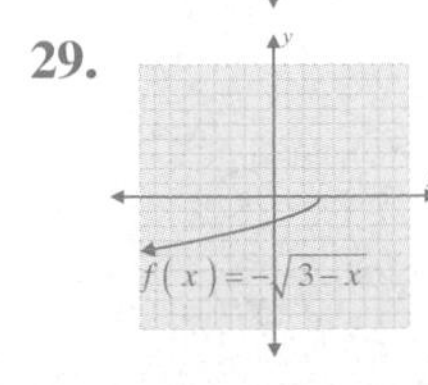

30.

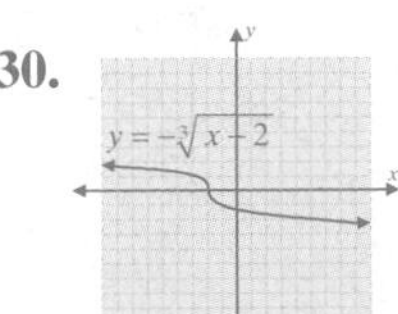

31.

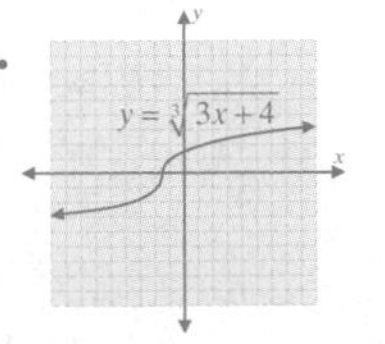

32.

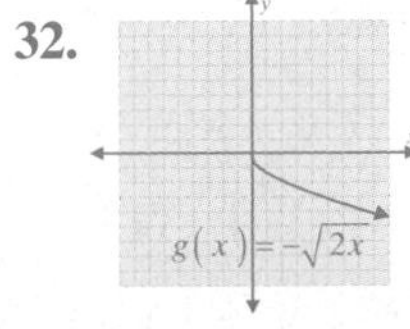

33.

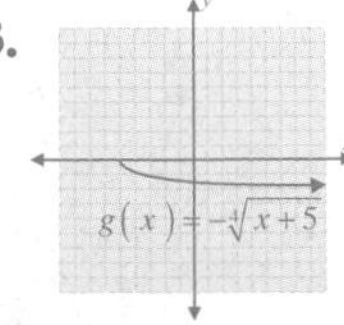

34.

35.

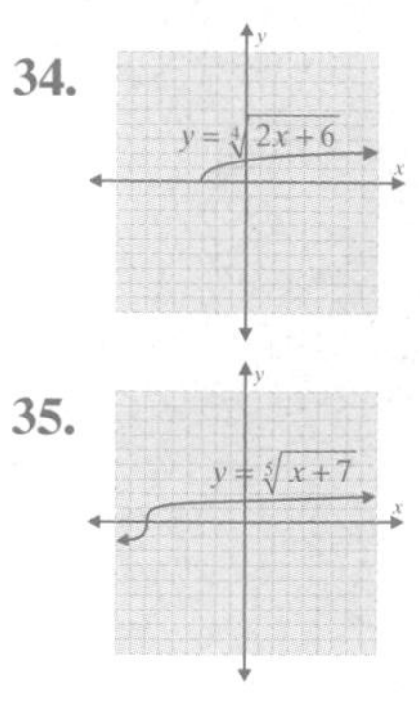

36. a. $\dfrac{1}{\sqrt{3+h}+\sqrt{3}}$

b. Slope of the line connecting $(3+h, f(3+h))$ and $(3, f(3))$

c. Line becomes more tangential.

d. $\dfrac{1}{2\sqrt{3}}$; represents the slope of the line tangent to $f(x)$ at $x = 3$.

Writing and Thinking About Mathematics

36. The graph of the radical function $f(x)=\sqrt{x}$ is shown with two values of x on the x-axis, 3 and $3 + h$.

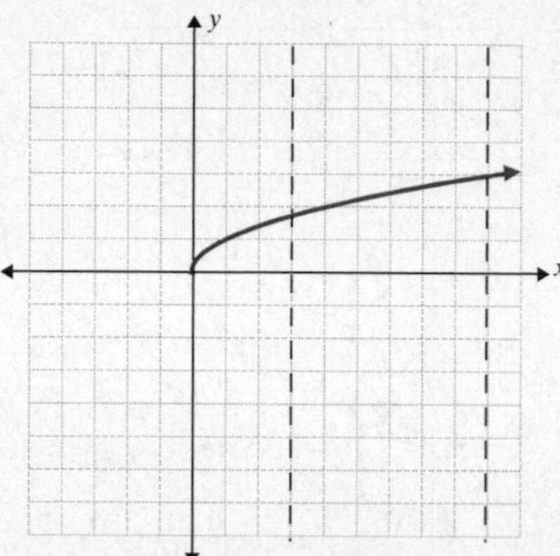

a. Rationalize the numerator and simplify the resulting expression

$$\frac{f(3+h)-f(3)}{h}=\frac{\sqrt{3+h}-\sqrt{3}}{h}$$

by multiplying both the numerator and denominator by the conjugate of the numerator.

b. What do you think that this expression represents graphically? (**Hint:** Two points determine a line.)

c. Using your result from part **b.**, what do you see happening on the graph if the value of h shrinks slowly to 0?

d. Using your analysis from part **c.**, what happens to the value of your simplified expression in part **a.** and what do you think this value represents?

Hawkes Learning Systems: Introductory & Intermediate Algebra

Functions with Radicals

10.5 Introduction to Complex Numbers

Objectives

After completing this section, you will be able to:

1. *Identify the real parts and the imaginary parts of complex numbers.*
2. *Simplify square roots of negative numbers.*
3. *Add and subtract complex numbers.*
4. *Solve linear equations with complex numbers by setting the real and imaginary parts equal.*

One of the properties of real numbers is that the square of any real number is nonnegative. That is, for any real number x, $x^2 \geq 0$. The square roots of negative numbers, such as $\sqrt{-4}$ and $\sqrt{-5}$, are not real numbers. However, they can be defined by expanding the real number system into the system of **complex numbers**.

Complex numbers include all the real numbers and the even roots of negative numbers. In Chapter 11, we will see how these numbers occur as solutions to quadratic equations. At first such numbers seem to be somewhat impractical because they are difficult to picture in any type of geometric setting and they are not solutions to the types of word problems that are familiar. However, complex numbers do occur quite naturally in trigonometry and higher level mathematics and have practical applications in such fields as electrical engineering.

The first step in the development of complex numbers is to define $\sqrt{-1}$.

$\sqrt{-1}$

$$i = \sqrt{-1} \text{ and } i^2 = \left(\sqrt{-1}\right)^2 = -1$$

With this definition, the following definition of the square root of any negative number can be made.

$\sqrt{-a}$

$$\sqrt{-a} = \sqrt{a} \cdot \sqrt{-1} = \sqrt{a}\, i$$

Example 1: $\sqrt{-a}$

a. $\sqrt{-25} = \sqrt{-1}\sqrt{25} = i \cdot 5 = 5i$

Note: $(5i)^2 = 5^2 i^2 = 25(-1) = -25$

b. $\sqrt{-36} = \sqrt{-1}\sqrt{36} = i \cdot 6 = 6i$

c. $\sqrt{-24} = \sqrt{-1}\sqrt{4 \cdot 6} = i \cdot 2 \cdot \sqrt{6} = 2\sqrt{6}\,i$ (or $2i\sqrt{6}$)

d. $\sqrt{-45} = \sqrt{-1}\sqrt{9 \cdot 5} = i \cdot 3 \cdot \sqrt{5} = 3\sqrt{5}\,i$ (or $3i\sqrt{5}$)

Note: We can write $2\sqrt{6}\,i$ and $3\sqrt{5}\,i$ as long as we take care not to include the i under the radical sign.

Complex Numbers

A ***complex number*** *is a number of the form* $\boldsymbol{a + bi}$*, where a and b are real numbers.*

a is called the ***real part*** *and b is called the* ***imaginary part****.*

If b = 0, then a + bi = a + 0i = a is a ***real number****.*

If a = 0, then a + bi = 0 + bi = bi is called a ***pure imaginary number*** *(or an* ***imaginary number****).*

Complex Number: $\boldsymbol{a + bi}$

real part → a; *imaginary part* → b

NOTES The term "imaginary" is somewhat unfortunate. Complex numbers and imaginary numbers are no more "imaginary" than any other type of number. In fact, all the types of numbers that we have studied (whole numbers, integers, rational numbers, irrational numbers, and real numbers) are products of human imagination.

Example 2: Real and Imaginary Parts

Identify the real and imaginary parts of each complex number.

a. $4 - 2i$ — 4 is the real part; –2 is the imaginary part.

b. $\sqrt{5} + 3\sqrt{2}\,i$ — $\sqrt{5}$ is the real part; $3\sqrt{2}$ is the imaginary part.

c. $7 = 7 + 0i$ — 7 is the real part; 0 is the imaginary part. (Remember, if $b = 0$, the complex number is a real number.)

d. $-\sqrt{3}\,i = 0 - \sqrt{3}\,i$ — 0 is the real part; $-\sqrt{3}$ is the imaginary part. (If $a = 0$ and $b \neq 0$, then the complex number is a pure imaginary number.)

In general, if a is a real number, then we can write $a = a + 0i$. This means that a is a complex number. Thus, **every real number is a complex number**. Figure 10.1 illustrates the relationships among the various types of numbers we study.

Teaching Notes:
There are alternative models that illustrate the subset relationship with intersecting and non-intersecting sets of numbers. Students may not pick up on the subset relationships shown in Figure 10.1, and you may know of another diagram that clarifies the subset relationships. Seeing more than one approach helps many students.

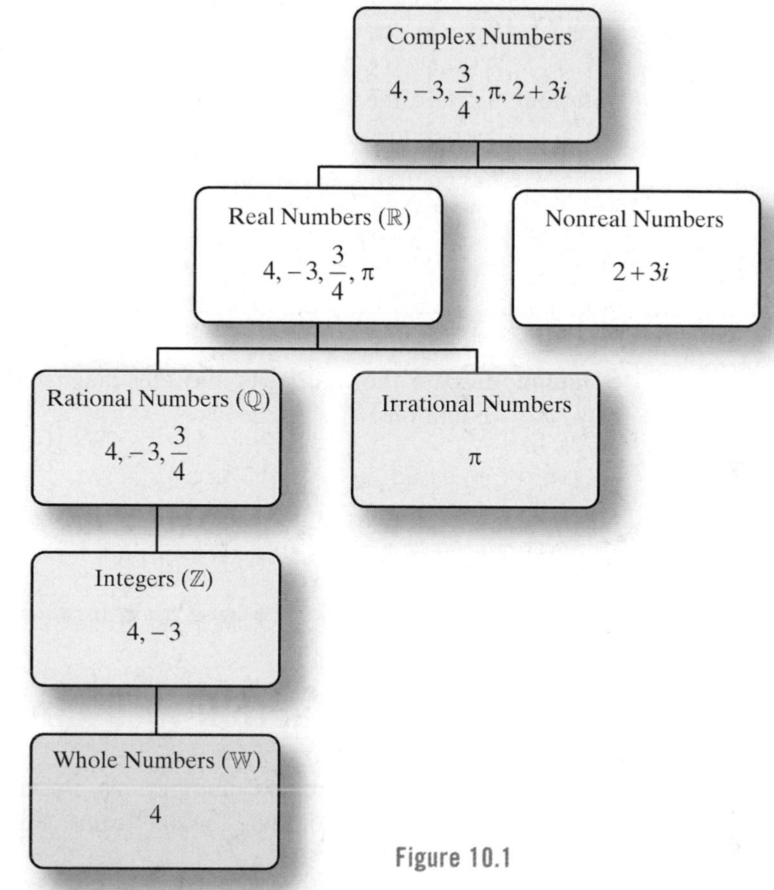

Figure 10.1

If two complex numbers are equal, then the real parts are equal and the imaginary parts are equal. For example, if

$$x + yi = 7 + 2i$$

then

$$x = 7 \quad \text{and} \quad y = 2.$$

This relationship can be used to solve equations involving complex numbers.

Equality of Complex Numbers

For complex numbers $a + bi$ and $c + di$,

$$\boldsymbol{a + bi = c + di} \quad \textit{if and only if} \quad \boldsymbol{a = c \text{ and } b = d}.$$

Example 3: Solving Equations

Solve each equation for the unknown numbers.

a. $(x + 3) + 2yi = 7 - 6i$

Solution: Equate the real parts and the imaginary parts and solve the resulting equations.

$$\begin{aligned} x + 3 &= 7 \\ x &= 4 \end{aligned} \quad \text{and} \quad \begin{aligned} 2y &= -6 \\ y &= -3 \end{aligned}$$

b. $2y + 3 - 8i = 9 + 4xi$

Solution: Equate the real parts and the imaginary parts and solve the resulting equations.

$$\begin{aligned} 2y + 3 &= 9 \\ 2y &= 6 \\ y &= 3 \end{aligned} \quad \text{and} \quad \begin{aligned} -8 &= 4x \\ -2 &= x \end{aligned}$$

Addition and Subtraction with Complex Numbers

Adding and subtracting complex numbers is similar to adding and subtracting polynomials. That is, simply combine the like terms. For example,

$$\begin{aligned} (2 + 3i) + (9 - 8i) &= 2 + 9 + 3i - 8i \\ &= (2 + 9) + (3 - 8)i \\ &= 11 - 5i \end{aligned}$$

Similarly,

$$\begin{aligned} (5 - 2i) - (6 + 7i) &= 5 - 2i - 6 - 7i \\ &= (5 - 6) + (-2 - 7)i \\ &= -1 - 9i \end{aligned}$$

Addition and Subtraction with Complex Numbers

For complex numbers $a + bi$ and $c + di$

$$(\mathbf{a} + \mathbf{b}\boldsymbol{i}) + (\mathbf{c} + \mathbf{d}\boldsymbol{i}) = (\mathbf{a} + \mathbf{c}) + (\mathbf{b} + \mathbf{d})\boldsymbol{i}$$

and

$$(\mathbf{a} + \mathbf{b}\boldsymbol{i}) - (\mathbf{c} + \mathbf{d}\boldsymbol{i}) = (\mathbf{a} - \mathbf{c}) + (\mathbf{b} - \mathbf{d})\boldsymbol{i}$$

Example 4: Addition and Subtraction with Complex Numbers

Find each sum or difference as indicated.

a. $(6 - 2i) + (1 - 2i)$

Solution: $(6 - 2i) + (1 - 2i) = (6 + 1) + (-2 - 2)i$
$= 7 - 4i$

b. $\left(-8-\sqrt{2}\,i\right)-\left(-8+\sqrt{2}\,i\right)$

Solution: $\left(-8-\sqrt{2}\,i\right)-\left(-8+\sqrt{2}\,i\right)=-8-\sqrt{2}\,i+8-\sqrt{2}\,i$

$=-8+8-\sqrt{2}\,i-\sqrt{2}\,i$

$=0+\left(-\sqrt{2}-\sqrt{2}\right)i$

$=-2\sqrt{2}\,i$

NOTES

When an expression with a radical is the coefficient of *i*, be sure that the radical sign does not cover the *i*. For example, in a case such as $-2\sqrt{2}i$ the square root symbol might extend too far. To avoid this error and confusion, you may choose to write the expression in the form $-2i\sqrt{2}$.

c. $\left(\sqrt{3}-2i\right)+\left(1+\sqrt{5}\,i\right)$

Solution: $\left(\sqrt{3}-2i\right)+\left(1+\sqrt{5}\,i\right)=\sqrt{3}+1-2i+\sqrt{5}\,i$

$=\left(\sqrt{3}+1\right)+\left(\sqrt{5}-2\right)i$

Note: Here, the coefficients do not simplify, and the real part is $\sqrt{3}+1$ and the imaginary part is $\sqrt{5}-2$.

Practice Problems

1. *Find the imaginary part and the real part of* $2-\sqrt{39}i$.

Add or subtract as indicated. Simplify your answers.

2. $(-7+\sqrt{3i})+(5-2i)$
3. $(4 + i) - (5 + 2i)$

Solve for x and y.

4. $x+yi=\sqrt{2}-7i$
5. $3y + (x - 7)i = -9 + 2i$

Answers to Practice Problems: **1.** Imaginary part is $-\sqrt{39}$ and real part is 2 **2.** $-2+\left(-2+\sqrt{3}\right)i$ **3.** $-1-i$
4. $x=\sqrt{2}$ and $y=-7$ **5.** $x=9$ and $y=-3$

10.5 Exercises

Find the imaginary part and the real part of each of the complex numbers in Exercises 1 – 10.

1. $4 - 3i$
2. $6+\sqrt{3}i$
3. $-11+\sqrt{2}i$
4. $\frac{3}{4}+i$
5. $\frac{2}{3}+\sqrt{17}i$
6. $\frac{4}{7}i$
7. $\frac{4+7i}{5}$
8. $\frac{2-i}{4}$
9. $\frac{3}{8}$
10. $-\sqrt{5}+\frac{\sqrt{2}}{2}i$

Simplify the radicals Exercises 11 – 24.

11. $\sqrt{-49}$
12. $\sqrt{-121}$
13. $-\sqrt{-64}$
14. $\sqrt{-169}$
15. $3\sqrt{147}$
16. $\sqrt{128}$
17. $2\sqrt{-150}$
18. $4\sqrt{-99}$
19. $-2\sqrt{-108}$
20. $2\sqrt{175}$
21. $\sqrt{242}$
22. $\sqrt{-192}$
23. $\sqrt{-1000}$
24. $\sqrt{-243}$

Find each sum or difference as indicated in Exercises 25 – 46.

25. $(2 + 3i) + (4 - i)$
26. $(7 - i) + (3 + 6i)$
27. $(4 + 5i) - (3 - 2i)$
28. $(-3 + 2i) - (6 + 2i)$
29. $-3i + (2 - 3i)$
30. $(7 + 5i) + (6 - 2i)$
31. $(8 + 9i) - (8 - 5i)$
32. $(-6 + i) - (2 + 3i)$
33. $(\sqrt{5}-2i)+(3-4i)$
34. $(4+3i)-(\sqrt{2}+3i)$
35. $(7+\sqrt{6}i)+(-2+i)$
36. $(\sqrt{11}+2i)+(5-7i)$
37. $(\sqrt{3}+\sqrt{2}i)-(5+\sqrt{2}i)$
38. $(\sqrt{5}+\sqrt{3}i)+(1-i)$
39. $(5+\sqrt{-25})-(7+\sqrt{-100})$
40. $(1+\sqrt{-36})-(-4-\sqrt{-49})$
41. $(13-3\sqrt{-16})+(-2-4\sqrt{-1})$
42. $(7+\sqrt{-9})-(3-2\sqrt{-25})$
43. $(4 + i) + (-3 - 2i) - (-1 - i)$
44. $(-2 - 3i) + (6 + i) - (2 + 5i)$
45. $(7 + 3i) + (2 - 4i) - (6 - 5i)$
46. $(-5 + 7i) + (4 - 2i) - (3 - 5i)$

Solve the equations in Exercises 47 – 60 for x and y.

47. $x + 3i = 6 - yi$
48. $2x - 8yi = -2 + 4yi$
49. $\sqrt{5}-2i = y+xi$
50. $\frac{2}{3}-2yi = 2x+\frac{4}{5}$
51. $\sqrt{2}+i-3 = x+yi$
52. $\sqrt{5}i-3+4i = x+yi$
53. $2x + 3 + 6i = 7 - (y + 2)i$
54. $x + yi + 8 = 2i + 4 - 3yi$
55. $x + 2i = 5 - yi - 3 - 4i$
56. $3x + 2 - 7i = i - 2yi + 5$
57. $2 + 3i + x = 5 - 7i + yi$
58. $11i - 2x + 4 = 10 - 3i + 2yi$
59. $2x - 2yi + 6 = 6i - x + 2$
60. $x + 4 - 3x + i = 8 + yi$

1. Real part is 4, imaginary part is –3 **2.** Real part is 6, imaginary part is $\sqrt{3}$ **3.** Real part is –11, imaginary part is $\sqrt{2}$ **4.** Real part is $\frac{3}{4}$, imaginary part is 1 **5.** Real part is $\frac{2}{3}$, imaginary part is $\sqrt{17}$ **6.** Real part is 0, imaginary part is $\frac{4}{7}$ **7.** Real part is $\frac{4}{5}$, imaginary part is $\frac{7}{5}$ **8.** Real part is $\frac{1}{2}$, imaginary part is $-\frac{1}{4}$ **9.** Real part is $\frac{3}{8}$, imaginary part is 0. **10.** Real part is $-\sqrt{5}$, imaginary part is $\frac{\sqrt{2}}{2}$ **11.** $7i$ **12.** $11i$ **13.** $-8i$ **14.** $13i$ **15.** $21\sqrt{3}$ **16.** $8\sqrt{2}$ **17.** $10i\sqrt{6}$ **18.** $12i\sqrt{11}$ **19.** $-12i\sqrt{3}$ **20.** $10\sqrt{7}$ **21.** $11\sqrt{2}$ **22.** $8i\sqrt{3}$ **23.** $10i\sqrt{10}$ **24.** $9i\sqrt{3}$ **25.** $6 + 2i$ **26.** $10 + 5i$ **27.** $1 + 7i$ **28.** -9 **29.** $2 - 6i$ **30.** $13 + 3i$ **31.** $14i$ **32.** $-8 - 2i$ **33.** $(3+\sqrt{5})-6i$ **34.** $4-\sqrt{2}$ **35.** $5+(\sqrt{6}+1)i$ **36.** $(\sqrt{11}+5)-5i$ **37.** $\sqrt{3}-5$ **38.** $(\sqrt{5}+1)+(\sqrt{3}-1)i$

39. $-2-5i$ **40.** $5+13i$
41. $11-16i$ **42.** $4+13i$
43. 2 **44.** $2-7i$
45. $3+4i$ **46.** $-4+10i$
47. $x=6, y=-3$
48. $x=-1, y=0$
49. $x=-2, y=\sqrt{5}$
50. $x=-\frac{1}{15}, y=0$
51. $x=\sqrt{2}-3, y=1$
52. $x=-3, y=4+\sqrt{5}$
53. $x=2, y=-8$
54. $x=-4, y=\frac{1}{2}$
55. $x=2, y=-6$

Writing and Thinking About Mathematics

61. Answer the following questions and give a brief explanation of your answer.
  a. Is every real number a complex number?
  b. Is every complex number a real number?

62. If you can, list 5 numbers that do and 5 numbers that do not fit each of the following categories.
  a. rational number
  b. integer
  c. real number
  d. pure imaginary number
  e. complex number
  f. irrational number

Hawkes Learning Systems: Introductory & Intermediate Algebra

Complex Numbers

56. $x=1, y=4$
57. $x=3, y=10$
58. $x=-3, y=7$
59. $x=-\frac{4}{3}, y=-3$
60. $x=-2, y=1$
61. a. Yes
  b. No
62. Answers will vary.

10.6 Multiplication and Division with Complex Numbers

Objectives

After completing this section, you will be able to:

1. *Multiply complex numbers.*
2. *Divide complex numbers.*
3. *Simplify powers of i.*

Multiplication with Complex Numbers

The product of two complex numbers can be found using the same procedure as in multiplying two binomials. This is similar to multiplying binomial expressions with the sums and differences of radicals as we did in Section 10.3. **Remember that $i^2 = -1$.** For example,

$$\begin{aligned}(3+5i)(2+i) &= (3+5i)2 + (3+5i)i \\ &= 6 + 10i + 3i + 5i^2 \\ &= 6 + 13i - 5 && 5i^2 = 5(-1) = -5 \\ &= 1 + 13i\end{aligned}$$

Multiplication with Complex Numbers

For complex numbers $a + bi$ and $c + di$,

$$(a+bi)(c+di) = (ac-bd) + (bc+ad)i$$

This definition is an application of the FOIL method of multiplying two binomials. **Memorizing the definition is not recommended.** An easier approach is simply to use the FOIL method for each product and do as many steps as you can mentally. **Remember that $i^2 = -1$.**

Example 1: Multiplication with Complex Numbers

Find the following products.

a. $(6+3i)(2-7i)$

Solution: $(6+3i)(2-7i) = 12 + 6i - 42i - 21i^2$

$= 12 - 36i + 21$

$= 33 - 36i$

b. $(\sqrt{2}-i)(\sqrt{2}-i)$

Solution: $(\sqrt{2}-i)(\sqrt{2}-i)=(\sqrt{2})^2-\sqrt{2}\cdot i-\sqrt{2}\cdot i+i^2$

$=2-2\sqrt{2}i-1$

$=1-2\sqrt{2}i$

c. $(-1+i)(2-i)$

Solution: $(-1+i)(2-i)=-2+2i+i-i^2$

$=-2+3i+1$

$=-1+3i$

Common Error

Remember that $\sqrt{a}\cdot\sqrt{b}=\sqrt{ab}$ *only if a and b are nonnegative real numbers. Applying this rule to negative real numbers can lead to an error. The error can be avoided by first changing the radicals to imaginary form.*

WRONG	RIGHT
$\sqrt{-6}\cdot\sqrt{-2}=\sqrt{12}$	$\sqrt{-6}\cdot\sqrt{-2}=\sqrt{6}\,i\cdot\sqrt{2}\,i$
$=\sqrt{4}\cdot\sqrt{3}$	$=\sqrt{12}\,i^2$
$=2\sqrt{3}$	$=2\sqrt{3}(-1)$
	$=-2\sqrt{3}$

Division with Complex Numbers

The two complex numbers $a + bi$ and $a - bi$ are called **complex conjugates** or simply **conjugates** of each other. As the following steps show, **the product of two complex conjugates will always be a non-negative real number**.

$$\begin{aligned}(a+bi)(a-bi) &= (a+bi)a+(a+bi)(-bi)\\ &= a^2+abi-abi-b^2i^2\\ &= a^2-b^2i^2\\ &= a^2+b^2\end{aligned}$$

The resulting product, $a^2 + b^2$, is a real number, and it is non-negative since it is the sum of the squares of real numbers.

The form $a + bi$ is called the **standard form** of a complex number. The standard form allows easy identification of the real and imaginary parts. Thus,

$$\frac{1+3i}{5}=\frac{1}{5}+\frac{3}{5}i \text{ in standard form.}$$

The real part is $\frac{1}{5}$ and the imaginary part is $\frac{3}{5}$.

A fraction formed with complex numbers, such as $\frac{1+i}{2-3i}$, indicates division of the numerator by the denominator. However, we do not divide these numbers in the usual sense. The objective is to find an equivalent expression that is in standard form, $a + bi$.

To write the fraction $\frac{1+i}{2-3i}$ in standard form, multiply both the numerator and denominator by $2 + 3i$ and simplify. This will give a positive real number in the denominator.

$$\frac{1+i}{2-3i} = \frac{(1+i)(2+3i)}{(2-3i)(2+3i)} \qquad 2+3i \text{ is the conjugate of the denominator.}$$

$$= \frac{2+2i+3i+3i^2}{2^2-6i+6i-3^2i^2}$$

$$= \frac{2+5i-3}{2^2+3^2} = \frac{-1+5i}{13}$$

$$= -\frac{1}{13} + \frac{5}{13}i$$

To Write a Fraction with Complex Numbers in Standard Form

1. *Multiply both the numerator and denominator by the complex conjugate of the denominator.*
2. *Simplify the resulting products in both the numerator and denominator.*
3. *Write the simplified result in standard form.*

Remember the following special product. We restate it here to emphasize its importance.

$$(a+bi)(a-bi) = a^2 + b^2$$

Example 2: Division with Complex Numbers

Write the following fractions in standard form.

a. $\frac{4}{-1-5i}$

Solution:
$$\frac{4}{-1-5i} = \frac{4(-1+5i)}{(-1-5i)(-1+5i)}$$

$$= \frac{-4+20i}{(-1)^2+(5)^2}$$

$$= \frac{-4+20i}{26}$$

$$= -\frac{4}{26} + \frac{20}{26}i$$

$$= -\frac{2}{13} + \frac{10}{13}i$$

b. $\dfrac{\sqrt{3}+i}{\sqrt{3}-i}$

Solution: $\dfrac{\sqrt{3}+i}{\sqrt{3}-i}=\dfrac{(\sqrt{3}+i)(\sqrt{3}+i)}{(\sqrt{3}-i)(\sqrt{3}+i)}$

$$=\frac{3+2\sqrt{3}i+i^2}{(\sqrt{3})^2-i^2}=\frac{2+2\sqrt{3}i}{3+1}$$

$$=\frac{2+2\sqrt{3}i}{4}=\frac{2}{4}+\frac{2\sqrt{3}i}{4}$$

$$=\frac{1}{2}+\frac{\sqrt{3}i}{2}$$

c. $\dfrac{6+i}{i}$

Solution: $\dfrac{6+i}{i}=\dfrac{(6+i)(-i)}{i(-i)}$

Since $i = 0 + i$ and $-i = 0 - i$, the number $-i$ is the conjugate of i.

$$=\frac{-6i-i^2}{-i^2}=\frac{-6i+1}{1}$$

$$=1-6i$$

d. $\dfrac{\sqrt{2}+i}{-\sqrt{2}+i}$

Solution: $\dfrac{\sqrt{2}+i}{-\sqrt{2}+i}=\dfrac{(\sqrt{2}+i)(-\sqrt{2}-i)}{(-\sqrt{2}+i)(-\sqrt{2}-i)}$

$$=\frac{-(\sqrt{2})^2-\sqrt{2}i-\sqrt{2}i-i^2}{(-\sqrt{2})^2-(i)^2}$$

$$=\frac{-2-2\sqrt{2}i+1}{2+1}=\frac{-1-2\sqrt{2}i}{3}$$

$$=-\frac{1}{3}-\frac{2\sqrt{2}}{3}i$$

Powers of *i*

The powers of i form an interesting pattern. Regardless of the particular integer exponent, there are only four possible values for any power of i:

$$i,\quad -1,\quad -i,\quad \text{and}\quad 1$$

The fact that these are the only four possibilities for powers of i becomes apparent from studying the following powers.

$$\begin{aligned}
i^1 &= i \\
i^2 &= -1 \\
i^3 &= i^2 \cdot i = -1 \cdot i = -i \\
i^4 &= i^2 \cdot i^2 = (-1)(-1) = 1 \\
i^5 &= i^4 \cdot i = +1 \cdot i = i \\
i^6 &= i^4 \cdot i^2 = (1)(-1) = -1 \\
i^7 &= i^4 \cdot i^3 = (1)(-i) = -i \\
i^8 &= i^4 \cdot i^4 = (1)(1) = 1
\end{aligned}$$

Higher powers of i can be simplified by using the fact that when i is raised to a power that is a multiple of 4, the result is 1. Thus, if n is a positive integer, then

$$\begin{aligned}
i^{4n} &= (i^4)^n = 1^n = 1 \\
i^{4n+1} &= i^{4n} \cdot i = 1 \cdot i = i \\
i^{4n+2} &= i^{4n} \cdot i^2 = 1 \cdot (-1) = -1 \\
i^{4n+3} &= i^{4n} \cdot i^3 = 1 \cdot (-i) = -i
\end{aligned}$$

Example 3: Powers of i

a. $i^{45} = i^{44} \cdot i = (i^4)^{11} \cdot i = 1^{11} \cdot i = i$

b. $i^{59} = i^{56} \cdot i^3 = (i^4)^{14} \cdot i^2 \cdot i = 1^{14} \cdot (-1)i = -i$

c. $i^{-6} = \dfrac{1}{i^6} = \dfrac{1}{i^4 \cdot i^2} = \dfrac{1}{1(-1)} = \dfrac{1}{-1} = -1$

Practice Problems

Write each of the following numbers in standard form.

1. $-2i(3 - i)$ **2.** $(2 + 4i)(1 + i)$ **3.** i^{13}

4. i^{-2} **5.** $\dfrac{2}{1+5i}$ **6.** $\dfrac{7+i}{2-i}$

1. $16 + 24i$
2. $-21 + 12i$
3. $-7\sqrt{2} + 7i$
4. $3 + 2\sqrt{3}i$
5. $3 + 12i$
6. $-28 - 24i$
7. $1 - i\sqrt{3}$
8. $-4 + 2i\sqrt{5}$
9. $5\sqrt{2} + 10i$
10. $3 + 2i\sqrt{3}$
11. $2 + 8i$

10.6 Exercises

In Exercises 1 – 70, write the numbers in standard form. Assume k is a positive integer.

1. $8(2 + 3i)$ **2.** $-3(7 - 4i)$ **3.** $-7(\sqrt{2} - i)$ **4.** $\sqrt{3}(\sqrt{3} + 2i)$

5. $3i(4 - i)$ **6.** $-4i(6 - 7i)$ **7.** $-i(\sqrt{3} + i)$ **8.** $2i(\sqrt{5} + 2i)$

12. $5+44i$ 13. $-7-11i$
14. $20+0i$ 15. $13+0i$
16. $34+13i$ 17. $-3-7i$
18. $-24+70i$
19. $5+12i$
20. $41+0i$ 21. $5-i\sqrt{3}$
22. $13+i\sqrt{5}$
23. $21+0i$
24. $4+4i\sqrt{7}$
25. $23-10i\sqrt{2}$
26. $61+0i$
27. $(2+\sqrt{10})+(2\sqrt{2}-\sqrt{5})i$
28. $(8\sqrt{3}-3)+(4+6\sqrt{3})i$
29. $(9-\sqrt{30})+(3\sqrt{5}+3\sqrt{6})i$
30. $(6-\sqrt{6})-(3\sqrt{3}+2\sqrt{2})i$
31. $0+3i$ 32. $0-7i$
33. $0-\frac{5}{4}i$ 34. $0+\frac{3}{2}i$
35. $-\frac{1}{4}+\frac{1}{2}i$
36. $-\frac{4}{3}-i$
37. $-\frac{4}{5}+\frac{8}{5}i$
38. $\frac{35}{29}+\frac{14}{29}i$
39. $\frac{24}{25}+\frac{18}{25}i$
40. $-\frac{48}{37}+\frac{8}{37}i$
41. $-\frac{1}{13}+\frac{5}{13}i$
42. $-\frac{6}{5}-\frac{2}{5}i$
43. $-\frac{1}{29}-\frac{12}{29}i$

9. $5i(2-\sqrt{2}i)$ **10.** $\sqrt{3}i(2-\sqrt{3}i)$ **11.** $(5+3i)(1+i)$ **12.** $(2+7i)(6+i)$

13. $(-3+5i)(-1+2i)$ **14.** $(6+2i)(3-i)$ **15.** $(2-3i)(2+3i)$

16. $(4+3i)(7-2i)$ **17.** $(-2+5i)(i-1)$ **18.** $(5+7i)^2$

19. $(3+2i)^2$ **20.** $(4+5i)(4-5i)$ **21.** $(\sqrt{3}+i)(\sqrt{3}-2i)$

22. $(2\sqrt{5}+3i)(\sqrt{5}-i)$ **23.** $(4+\sqrt{5}i)(4-\sqrt{5}i)$ **24.** $(\sqrt{7}+3i)(\sqrt{7}+i)$

25. $(5-\sqrt{2}i)(5-\sqrt{2}i)$ **26.** $(7+2\sqrt{3}i)(7-2\sqrt{3}i)$ **27.** $(\sqrt{5}+2i)(\sqrt{2}-i)$

28. $(2\sqrt{3}+i)(4+3i)$ **29.** $(3+\sqrt{5}i)(3+\sqrt{6}i)$ **30.** $(2-\sqrt{3}i)(3-\sqrt{2}i)$

31. $\frac{-3}{i}$ **32.** $\frac{7}{i}$ **33.** $\frac{5}{4i}$ **34.** $\frac{-3}{2i}$ **35.** $\frac{2+i}{-4i}$

36. $\frac{3-4i}{3i}$ **37.** $\frac{-4}{1+2i}$ **38.** $\frac{7}{5-2i}$ **39.** $\frac{6}{4-3i}$ **40.** $\frac{-8}{6+i}$

41. $\frac{2i}{5-i}$ **42.** $\frac{-4i}{1+3i}$ **43.** $\frac{2-i}{2+5i}$ **44.** $\frac{6+i}{3-4i}$ **45.** $\frac{2-3i}{-1+5i}$

46. $\frac{-3+i}{7-2i}$ **47.** $\frac{1+4i}{\sqrt{3}+i}$ **48.** $\frac{9-2i}{\sqrt{5}+i}$ **49.** $\frac{\sqrt{3}+2i}{\sqrt{3}-2i}$ **50.** $\frac{\sqrt{6}-3i}{\sqrt{6}+3i}$

51. i^{13} **52.** i^{20} **53.** i^{30} **54.** i^{15} **55.** i^{-3}

56. $(i)^{-5}$ **57.** $(i)^{4k}$ **58.** i^{4k+2} **59.** i^{4k+3} **60.** i^{4k+1}

61. $(x+3i)(x-3i)$ **62.** $(y+5i)(y-5i)$ **63.** $(x+\sqrt{2}i)(x-\sqrt{2}i)$

64. $(2x+\sqrt{7}i)(2x-\sqrt{7}i)$ **65.** $(\sqrt{5}y+2i)(\sqrt{5}y-2i)$ **66.** $(y-\sqrt{3}i)(y+\sqrt{3}i)$

67. $[(x+2)+6i][(x+2)-6i]$ **68.** $[(x+1)-\sqrt{8}i][(x+1)+\sqrt{8}i]$

69. $[(y-3)+2i][(y-3)-2i]$ **70.** $[(x-1)+5i][(x-1)-5i]$

Writing and Thinking About Mathematics

71. Explain why the product of every nonzero complex number and its conjugate is a positive real number.

72. Explain why $\sqrt{-4}\cdot\sqrt{-4}\neq 4$. What is the correct value of $\sqrt{-4}\cdot\sqrt{-4}$?

73. What condition is necessary for the conjugate of a complex number, $a+bi$, to be equal to the reciprocal of this number?

Answers to Practice Problems: **1.** $-2-6i$ **2.** $-2+6i$ **3.** i **4.** -1 **5.** $\frac{1}{13}-\frac{5}{13}i$ **6.** $\frac{13}{5}+\frac{9}{5}i$

Hawkes Learning Systems: Introductory & Intermediate Algebra

Multiplication and Division of Complex Numbers

44. $\frac{14}{25}+\frac{27}{25}i$

45. $-\frac{17}{26}-\frac{7}{26}i$

46. $-\frac{23}{53}+\frac{1}{53}i$

47. $\frac{4+\sqrt{3}}{4}+\left(\frac{4\sqrt{3}-1}{4}\right)i$

48. $\left(-\frac{1}{3}+\frac{3\sqrt{5}}{2}\right)-\left(\frac{3}{2}+\frac{\sqrt{5}}{3}\right)i$

49. $-\frac{1}{7}+\frac{4\sqrt{3}}{7}i$

50. $-\frac{1}{5}-\frac{2\sqrt{6}}{5}i$

51. $0+i$ **52.** $1+0i$
53. $-1+0i$ **54.** $0-i$
55. $0+i$ **56.** $0-i$
57. $1+0i$ **58.** $-1+0i$
59. $0-i$ **60.** $0+i$
61. x^2+9 **62.** y^2+25
63. x^2+2 **64.** $4x^2+7$
65. $5y^2+4$ **66.** y^2+3
67. $x^2+4x+40$
68. x^2+2x+9
69. $y^2-6y+13$
70. $x^2-2x+26$
71. Answers will vary
72. Answers will vary, –4
73. $a^2+b^2=1$

Chapter 10 Index of Key Ideas and Terms

Section 10.1 Roots and Radicals

Perfect Squares (Table of) page 681

Square Roots page 681

If $b^2 = a$, then b is a **square root** of a.

Terminology page 681

The symbol $\sqrt{}$ is called a **radical sign**.
The number under the radical sign is called the **radicand**.
The complete expression, such as $\sqrt{64}$, is called
a **radical** or **radical expression**.

Square Root page 682

If a is a nonnegative real number and b is a real number such that $b^2 = a$, then b is called a **square root** of a.
If b is nonnegative, then we write

$$\sqrt{a} = b \leftarrow b \text{ is called the } \textbf{principal square root}$$

and $-\sqrt{a} = -b \leftarrow -b$ is called the **negative square root**.

Simplest Form of Square Roots page 684

A square root is considered to be in simplest form
when the radicand has no perfect square as a factor.

Square Roots of Expressions with Even and Odd Exponents page 687

For any real number x and positive integer m,

$$\sqrt{x^{2m}} = |x^m| \quad \text{and} \quad \sqrt{x^{2m+1}} = |x^m|\sqrt{x}.$$

Note that the absolute value sign is necessary only if m is odd.
Also, note that for $\sqrt{x^{2m+1}}$ to be defined as real, x cannot be negative.

Comment: If m is any integer, then $2m$ is even and $2m + 1$ is odd.

Perfect Cubes (Table of) page 688

Cube Root page 688

If a and b are real numbers such that $b^3 = a$, then b is called the **cube root** of a. We write $\sqrt[3]{a} = b \leftarrow$ the **cube root**.

Simplest Form of Cube Roots page 689

A cube root is considered to be in simplest form
when the radicand has no perfect cube as a factor.

continued on next page ...

Section 10.1 Roots and Radicals (continued)

To Rationalize the Denominator of a Radical Expression (with Square Roots or Cube Roots) page 690

1. If the denominator contains a square root, multiply both the numerator and denominator by a square root. Choose this square root so that the denominator will be a perfect square.
2. If the denominator contains a cube root, multiply both the numerator and denominator by a cube root. Choose this cube root so that the denominator will be a perfect cube.

Section 10.2 Rational Exponents

Radical Notation page 695

If n is a positive integer and $b^n = a$, then
$b = \sqrt[n]{a} = a^{\frac{1}{n}}$ (assuming $\sqrt[n]{a}$ is a real number).

The expression $\sqrt[n]{a}$ is called a **radical.**

The symbol $\sqrt[n]{\ }$ is called a **radical sign.**
n is called the **index.**
a is called the **radicand.**
(**Note:** If no index is given, it is understood to be 2.

For example, $\sqrt{3} = \sqrt[2]{3} = 3^{\frac{1}{2}}$.)

Special Note About the Index n page 695

For the expression $\sqrt[n]{a}$ (or $a^{\frac{1}{n}}$) to be a real number:
1. n can be any index when a is nonnegative.
2. n must be odd when a is negative.
(If a is negative and n is even, then $\sqrt[n]{a}$ is nonreal.)

Properties of Exponents page 697

For nonzero real numbers a and b and rational numbers m and n,

The Exponent 1:	$a = a^1$ (a is any real number.)
The Exponent 0:	$a^0 = 1$ $(a \neq 0)$
Product Rule:	$a^m \cdot a^n = a^{m+n}$
Quotient Rule:	$\dfrac{a^m}{a^n} = a^{m-n}$
Power Rule:	$(a^m)^n = a^{mn}$
Negative Exponents:	$a^{-n} = \dfrac{1}{a^n}$, $\dfrac{1}{a^{-n}} = a^n$

continued on next page ...

Section 10.2 Rational Exponents (continued)

Power Rule for Products: $(ab)^n = a^n b^n$

Power Rule for Fractions: $\left(\frac{a}{b}\right)^n = \frac{a^n}{b^n}$

The General Form $a^{\frac{m}{n}}$ page 698

If n is a positive integer and m is any integer and $a^{\frac{1}{n}}$ is a real number, then

$$a^{\frac{m}{n}} = \left(a^{\frac{1}{n}}\right)^m = \left(a^m\right)^{\frac{1}{n}}.$$

In radical notation:

$$a^{\frac{m}{n}} = \left(\sqrt[n]{a}\right)^m = \sqrt[n]{a^m}$$

To Find the Value of $a^{\frac{m}{n}}$ with a Calculator page 700

Step 1: Enter the value of the base, a.

Step 2: Press the up arrow key ^ .

Step 3: Enter the fractional exponent enclosed in parentheses. (This exponent may be positive or negative.)

Step 4: Press ENTER.

Section 10.3 Arithmetic with Radicals

Addition and Subtraction with Radical Expressions page 705

Like radicals

Multiplication with Radical Expressions page 706

Rationalizing Denominators of Rational Expressions page 708

If the denominator is a sum or difference with square roots:

1. If the denominator is of the form $a - b$, multiply both the numerator and denominator by $a + b$.
2. If the denominator is of the form $a + b$, multiply both the numerator and denominator by $a - b$.

Evaluating Radical Expressions with a Calculator page 709

Section 10.4 Functions with Radicals

Radical Function page 715

A **radical function** is a function of the form $y = \sqrt[n]{g(x)}$ in which the radicand contains the variable.

The **domain** of such a function depends on the index, n:

1. If n is an even number, the domain is the set of all x such that $g(x) \geq 0$.
2. If n is an odd number, the domain is the set of all real numbers $(-\infty, \infty)$.

Section 10.5 Introduction to Complex Numbers

Complex Numbers page 723

$i = \sqrt{-1}$ and $i^2 = -1$

If a is a positive real number, then $\sqrt{-a} = \sqrt{a} \cdot \sqrt{-1} = \sqrt{a}\, i$.

Standard Form: $a + bi$ page 724

A **complex number** is a number of the form $a + bi$, where a and b are real numbers.

a is called the **real part** and b is called the **imaginary part**.

If $b = 0$, then $a + bi = a + 0i = a$ is a real number.

If $a = 0$, then $a + bi = 0 + bi = bi$ is called a **pure imaginary number** (or an **imaginary number**).

Equality of Complex Numbers page 725

For complex numbers $a + bi$ and $c + di$,

$a + bi = c + di$ if and only if $a = c$ and $b = d$.

Addition and Subtraction with Complex Numbers page 726

For complex numbers $a + bi$ and $c + di$

$$(a+bi)+(c+di)=(a+c)+(b+d)i$$

and $$(a+bi)-(c+di)=(a-c)+(b-d)i.$$

Section 10.6 Multiplication and Division with Complex Numbers

Multiplication with Complex Numbers page 730

For complex numbers $a+bi$ and $c+di$

$$(a+bi)(c+di)=(ac-bd)+(bc+ad)i.$$

Complex Conjugates page 731

$a+bi$ and $a-bi$ are complex conjugates

The product of complex conjugates will always be a nonnegative real number: $(a+bi)(a-bi)=a^2+b^2$.

To Simplify a Fraction with Complex Numbers page 732

1. Multiply both the numerator and denominator by the complex conjugate of the denominator.
2. Simplify the resulting products in both the numerator and denominator.
3. Write the simplified result in standard form.

Powers of *i*: i^n pages 733 - 734

$$i^{4n}=(i^4)^n=1^n=1$$
$$i^{4n+1}=i^{4n}\cdot i=1\cdot i=i$$
$$i^{4n+2}=i^{4n}\cdot i^2=1\cdot(-1)=-1$$
$$i^{4n+3}=i^{4n}\cdot i^3=1\cdot(-i)=-i$$

Chapter 10 Review

For a review of the topics and problems from Chapter 10, look at the following lessons from *Hawkes Learning Systems: Introductory & Intermediate Algebra.*

Evaluating Radicals
Simplifying Radicals
Division of Radicals
Rational Exponents
Addition and Subtraction of Radicals
Multiplication of Radicals
Functions with Radicals
Complex Numbers
Multiplication and Division of Complex Numbers

Chapter 10 Test

Simplify each expression in Exercises 1 – 5. Assume that all variables are positive.

1. $(-8)^{\frac{2}{3}}$ **2.** $(9)^{\frac{-3}{2}}$ **3.** $4x^{\frac{1}{2}} \cdot x^{\frac{2}{3}}$

4. $\left(49x^{\frac{1}{2}}y^{\frac{-2}{3}}\right)^{\frac{1}{2}}$ **5.** $\left(\frac{16x^{-4}y}{y^{-1}}\right)^{\frac{3}{4}}$

6. Write $(2x)^{\frac{2}{3}}$ in radical notation. **7.** Write $\sqrt[6]{8x^2y^4}$ in exponential notation.

Simplify the expressions in Exercises 8 – 11.

8. $\sqrt[3]{x^2} \cdot \sqrt[4]{x}$ **9.** $\sqrt{112}$ **10.** $\sqrt[3]{48x^2y^5}$ **11.** $\sqrt{\frac{5y^2}{8x^3}}$

In Exercises 12 – 17, perform the indicated operations and simplify. Assume that all variables are positive.

12. $2\sqrt{75} + 3\sqrt{27} - \sqrt{12}$ **13.** $\sqrt{16x^3} + \sqrt{9x^3} - \sqrt{36x}$

14. $\sqrt[3]{24} + 2\sqrt[3]{81}$ **15.** $\left(\sqrt{3} - \sqrt{2}\right)^2$

16. $5x\sqrt{y^3} - 2\sqrt{x^2y^3} - 4y\sqrt{x^2y}$ **17.** $\left(6 + \sqrt{3x}\right)\left(5 - 2\sqrt{3x}\right)$

18. Rationalize the denominator and simplify the quotient.

a. $\frac{2}{\sqrt{3} - \sqrt{5}}$ **b.** $\frac{1-x}{1-\sqrt{x^3}}$

19. Find the domain of each of the following radical functions. Write the answer in interval notation.

a. $y = \sqrt{3x+4}$ **b.** $f(x) = \sqrt[3]{2x+5}$

20. Use a TI-83 Plus graphing calculator to graph each of the following radical functions. Sketch the graph on your test paper and label 3 points.

a. $f(x) = -\sqrt{x+3}$ **b.** $y = \sqrt[3]{x-4}$

In Exercises 21 – 24, perform the indicated operations and write the results in standard form.

21. $(5+8i)+(11-4i)$ **22.** $\left(2+3\sqrt{-4}\right)-\left(7-2\sqrt{-25}\right)$

23. $(4+3i)(2-5i)$ **24.** $\frac{2+i}{3+2i}$

25. Solve for x and y: $(2x+3i)-(6+2yi) = 5-3i$

26. Write i^{23} in the standard form, $a + bi$.

27. Find the product and simplify: $(x+2i)(x-2i)$

1. 4 **2.** $\frac{1}{27}$ **3.** $4x^{\frac{7}{6}}$

4. $\frac{7x^{\frac{1}{4}}}{y^{\frac{1}{3}}}$ **5.** $\frac{8y^{\frac{3}{2}}}{x^3}$

6. $\sqrt[3]{4x^2}$ **7.** $2^{\frac{1}{2}}x^{\frac{1}{3}}y^{\frac{2}{3}}$

8. $\sqrt[12]{x^{11}}$ **9.** $4\sqrt{7}$

10. $2y\sqrt[3]{6x^2y^2}$

11. $\frac{y}{4x^2}\sqrt{10x}$

12. $17\sqrt{3}$

13. $(7x-6)\sqrt{x}$

14. $8\sqrt[3]{3}$

15. $5-2\sqrt{6}$

16. $-xy\sqrt{y}$

17. $30-7\sqrt{3x}-6x$

18. a. $-\left(\sqrt{3}+\sqrt{5}\right)$

b. $\frac{1+\sqrt{x^3}}{1+x+x^2}$

19. a. $\left[-\frac{4}{3}, \infty\right)$

b. $(-\infty, \infty)$

20. a.

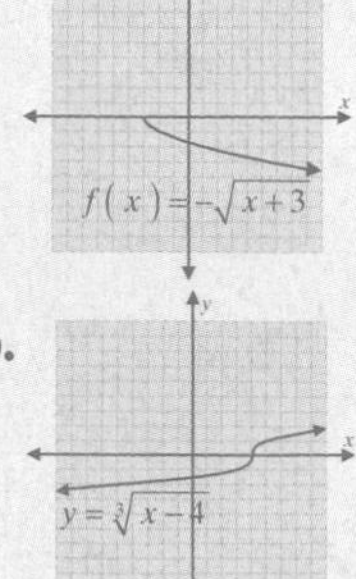

b.

21. $16+4i$ **22.** $-5+16i$

23. $23-14i$

24. $\frac{8}{13}-\frac{1}{13}i$

25. $x=\frac{11}{2}, y=3$

26. $0-i$

27. x^2+4

28. $x^2+6x+12$

29. Answers will vary

30. 0.1250

28. Find the product and simplify: $\left(x+3-\sqrt{3}i\right)\left(x+3+\sqrt{3}i\right)$

29. Explain, in your own words, why the product of a nonzero complex number and its conjugate will always result in a positive real number.

In Exercises 30 – 32, use a calculator to find the value of each expression accurate to 4 decimal places.

30. $32^{\frac{-3}{5}}$ **31.** $\left(\sqrt{2}+6\right)\left(\sqrt{2}-1\right)$ **32.** $\frac{\sqrt{3}-\sqrt{5}}{\sqrt{7}-\sqrt{10}}$

Cumulative Review: Chapters 1 – 10

31. 3.0711

32. 0.9758

1. $2x^3-4x^2+8x-4$

2. $6x^2+19x-7$

3. $-5x^2+18x+8$

4. $n=\frac{s-a}{d}+1$ or $\frac{s-a+d}{d}$

5. $p=\frac{A}{1+rt}$

6. $(4x+3)(3x-4)$

7. $(7+2x)(4-x)$

8. $5(x-4)(x^2+4x+16)$

9. $(x+4)(x+1)(x-1)$

10. $\frac{4x^2-12x-1}{(x+3)(x-2)(x-5)}$

11. $\frac{4x^2+6x+17}{(2x+1)(2x-1)(x+3)}$

12. $\frac{x^2-1}{2x-4}$ 13. $\frac{x+6}{x-1}$

14. $x=-\frac{5}{7}$ 15. $x=\frac{83}{4}$

Perform the indicated operations in Exercises 1 – 3.

1. $\left(x^2+7x-5\right)-\left(-2x^3+5x^2-x-1\right)$ **2.** $(2x+7)(3x-1)$

3. $(5x+2)(4-x)$

Solve for the indicated variable in Exercises 4 and 5.

4. Solve $s=a+(n-1)d$ for n **5.** Solve $A=p+prt$ for p

In Exercises 6 – 9, factor completely.

6. $12x^2-7x-12$ **7.** $28+x-2x^2$ **8.** $5x^3-320$

9. x^3+4x^2-x-4

Perform the indicated operations in Exercises 10 and 11 and reduce if possible.

10. $\frac{x+1}{x^2+x-6}+\frac{3x-2}{x^2-2x-15}$ **11.** $\frac{2x+5}{4x^2-1}-\frac{2-x}{2x^2+7x+3}$

Simplify the complex fractions in Exercises 12 and 13.

12. $\frac{1-\frac{1}{x^2}}{\frac{2}{x}-\frac{4}{x^2}}$ **13.** $\frac{x+2-\frac{12}{x+3}}{x-5+\frac{16}{x+3}}$

Solve each equation in Exercises 14 and 15.

14. $\frac{3}{x}+\frac{2}{x+5}=\frac{8}{3x}$ **15.** $\frac{9}{x+7}+\frac{3x}{x^2+4x-21}=\frac{8}{x-3}$

Solve the inequalities in Exercises 16 and 17 and graph the solutions on a real number line. Write the solutions in interval notation.

16. $\frac{5x-3}{2x+4}\le 0$ **17.** $\frac{x-5}{3x-1}\ge 1$

18. V (volume) varies inversely as P (pressure) when a gas is enclosed in a container. In a particular situation, the volume of gas is 25 in.^3 when a force of 10 pounds is exerted. What would be the volume of gas if a force of 15 pounds were to be used?

19. The resistance, R (in ohms), in a wire is directly proportional to the length, L, and inversely proportional to the square of the diameter of the wire. The resistance of a wire 500 ft. long with a diameter of 0.01 in. is 20 ohms. What is the resistance of a wire of the same type that is 200 ft. long?

In Exercises 20 – 22, solve each quadratic equation by factoring.

20. $x^2 - 13x - 48 = 0$ **21.** $x = 2x^2 - 6$ **22.** $0 = 15x^2 - 11x + 2$

23. Determine the values of a and b such that the straight line $ax + by = -5$ passes through the two points $(-3, -2.5)$ and $(1, 2.5)$.

24. Solve the following system of linear equations by graphing both equations and locating the point of intersection.

$$\begin{cases} 3x - 2y = 7 \\ x + 3y = -5 \end{cases}$$

25. Solve the following system of linear equations by using the substitution method.

$$\begin{cases} 3x - y = -4 \\ 2x + 2y = 0 \end{cases}$$

26. Solve the following system of linear equations by using the addition method.

$$\begin{cases} -2x + 5y = 10 \\ 6x - 2y = 30 \end{cases}$$

27. **a.** Find an equation that has $x = 4$ and $x = 7$ as roots.
b. Find an equation that has $x = -2$, $x = 12$, and $x = 1$ as roots.

28. Use a graphing calculator to find the solution set to the following system of equations: $\begin{cases} x + y = 4 \\ 3x - y = 2 \end{cases}$

---

16. $\left(-2, \frac{3}{5}\right]$

17. $\left[-2, \frac{1}{3}\right)$

18. $\frac{50}{3} \text{ in.}^3$ or $16.6\overline{6} \text{ in.}^3$

19. 8 ohms

20. $(x-16)(x+3) = 0$
$x = -3$ and 16

21. $(2x+3)(x-2) = 0$
$x = -\frac{3}{2}$ and 2

22. $(5x-2)(3x-1) = 0$
$x = \frac{1}{3}$ and $\frac{2}{5}$

23. $a = 5$, $b = -4$

24. $(1, -2)$

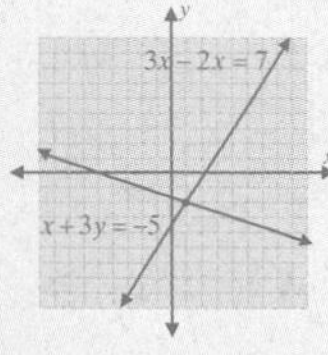

25. $x = -1$, $y = 1$

26. $x = \frac{85}{13}$, $y = \frac{60}{13}$

27. a. $x^2 - 11x + 28 = 0$

b. $x^3 - 11x^2 - 14x + 24 = 0$

28. $(1.5, 2.5)$

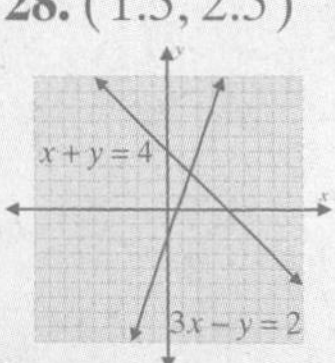

29. $x-9+\frac{21x-24}{x^2+2x-1}$

30. x^2-4x+3

31. a.

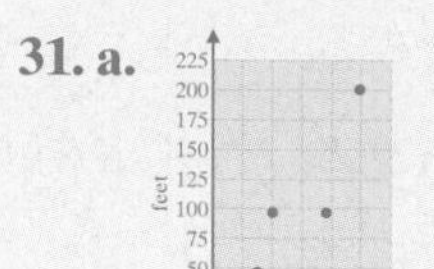

b. $m_1=8, m_2=24,$ $m_3=0, m_4=\frac{104}{5}$

c. Rate increased 8 ft./sec. (fps), then increased 24 fps then stayed constant, then increased $\frac{104}{5}$ fps.

32. a. Function, $D=(-\infty,\infty)$, $R=(-\infty,0]$

b. Not a function

c. Function, $D=(-\infty,\infty)$, $R=[-1,1]$

33. 15,17

34. $x=1$

35. $x=\pm\sqrt{12}\approx\pm3.46$, $\pm\sqrt{6}\approx\pm2.45$

In Exercises 29 and 30, divide by using long division. Write the answer in the form $Q(x)+\frac{R(x)}{D(x)}$.

29. $\dfrac{x^3-7x^2+2x-15}{x^2+2x-1}$

30. $\dfrac{x^3-8x^2+19x-12}{x-4}$

31. The following table shows the number of feet that Linda skied downhill from one second to another. (She never skied uphill, but she did stop occasionally.)

Distance Traveled (in feet)	Time Elapsed (in seconds)
24	3
48	6
96	8
96	15
200	20

a. Plot the points indicated in the table.

b. Calculate the slope of the line segments from point to point.

c. Interpret each slope.

32. The graphs of three curves are shown. Use the vertical line test to determine whether or not each graph represents a function. If the graph represents a function, state its domain and range.

a.

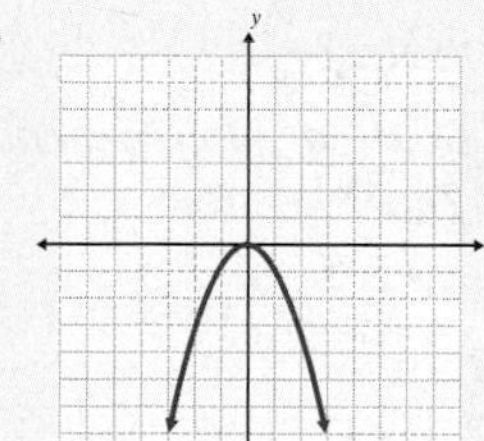

b.

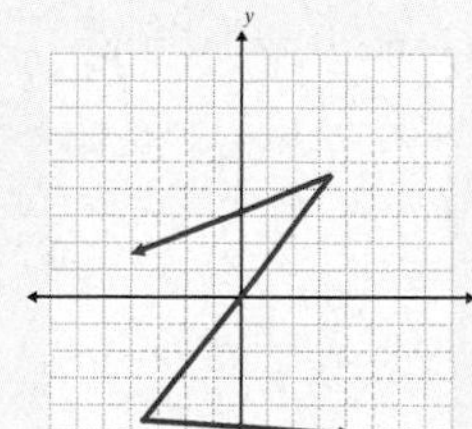

c.

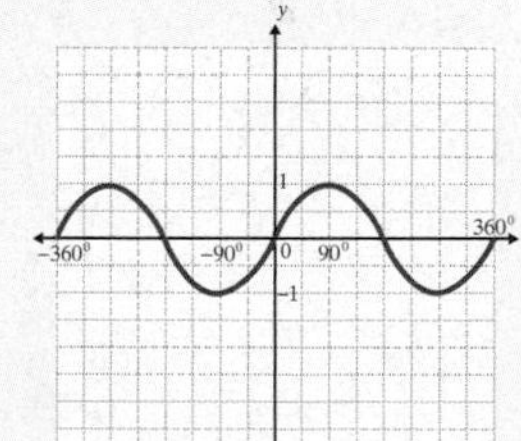

33. The sum of the squares of two consecutive positive odd integers is 514. What are the integers?

Use a graphing calculator to solve (or estimate the solutions of) the equations in Exercises 34 and 35. Sketch the graphs that you use and find any approximations accurate to two decimal places.

34. $x^3=3x^2-3x+1$

35. $|x^2-9|=3$

36. $\frac{1}{16}$

37. $x^{\frac{7}{3}}$

38. $12\sqrt{2}$

39. $2x^2y^3\sqrt[3]{2y}$

40. $14-2i$

41. $-\frac{8}{17}-\frac{19}{17}i$

42. 10 mph

43. 7.5 hours

44. $35,000 in 6% and $15,000 in 10%

45. 60 pounds of first type ($1.25 candy) and 40 pounds of second type ($2.50 candy)

46. $x \approx -1.372$
$x \approx 4.372$

47. $x = 1$
$x \approx -2.303$
$x \approx 1.303$

48. $x = -4$
$x = 1$
$x = 5$

49. 279,936

50. 0.04000

51. 1.30268

52. 6.45522

Simplify the expressions in Exercises 36 – 39. Assume that all variables are positive.

36. $8^{\frac{-4}{3}}$

37. $\left(x^{\frac{1}{2}} \cdot x^{\frac{2}{3}}\right)^2$

38. $\sqrt{288}$

39. $\sqrt[3]{16x^6y^{10}}$

Perform the indicated operations in Exercises 40 and 41 and simplify. Write your answers in standard form.

40. $(2+2i)(3-4i)$

41. $\frac{4-3i}{1+4i}$

42. Susan traveled 25 miles downstream. In the same length of time, she could have traveled 15 miles upstream. If the speed of the current is 2.5 mph, find the speed of Susan's boat in still water.

43. Robin can prepare a monthly sales report in 5 hours. If Mac helps her, together they can prepare the report in 3 hours. How long would it take Mac if he worked alone?

44. Harold has $50,000 that he wants to invest in two accounts. One pays 6% interest, and the other (at a higher risk) pays 10% interest. If he wants a $3600 annual return on these two investments, how much should he put into each account?

45. A grocer plans to make up a special mix of two popular kinds of candy for Halloween. He wants to mix a total of 100 pounds to sell for $1.75 per pound. Individually, the two types sell for $1.25 and $2.50 per pound. How many pounds of each of the two kinds should he put in the mix?

In Exercises 46 – 48, use a graphing calculator to graph each function and use the TRACE and ZOOM features to estimate the x-intercepts.

46. $y = x^2 - 3x - 6$

47. $y = x^3 - 4x + 3$

48. $y = -x^3 + 2x^2 + 19x - 20$

In Exercises 49 – 52, use a calculator to find the value of each expression accurate to 5 decimal places.

49. $36^{\frac{7}{2}}$

50. $125^{\frac{-2}{3}}$

51. $\sqrt{5}\left(\sqrt{21}-4\right)$

52. $\frac{13+\sqrt{12}}{5-\sqrt{6}}$

Quadratic Equations

CHAPTER 11

Did You Know?

The quadratic formula is a general method for solving second-degree equations of the form $ax^2 + bx + c = 0$, where a, b, and c can be any real numbers. The quadratic formula is a very old formula; it was known to Babylonian mathematicians around 2000 B.C. However, Babylonian and, later, Greek mathematicians always discarded negative solutions of quadratic equations because they felt that these solutions had no physical meaning. Greek mathematicians always tried to interpret their algebraic problems from a geometrical viewpoint and hence the development of the geometric method of "completing the square".

Consider the following equation: $x^2 + 6x = 7$. A geometric figure is constructed having areas x^2, $3x$ and $3x$.

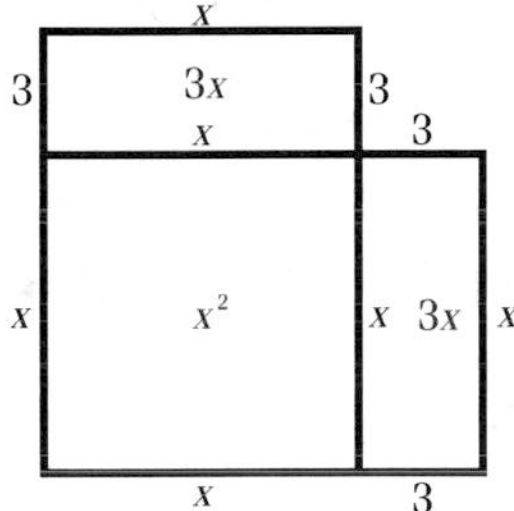

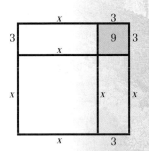

Note that to make the figure a square, one must add a 3 by 3 section (area = 9). Thus, 9 must be added to both sides of the equation to restore equality. Therefore,

$$\begin{aligned} x^2 + 6x &= 7 && \text{Original equation} \\ x^2 + 6x + 9 &= 7 + 9 && \text{Adding 9 to complete the square} \\ x^2 + 6x + 9 &= 16. \end{aligned}$$

So, the square with side $x + 3$ now has an area of 16 square units. Therefore, the sides must be of length 4 units, which means $x + 3 = 4$. Hence, $x = 1$.

Note that the solution set of the original equation is actually $\{1, -7\}$, since $(-7)^2 + 6(-7) = 49 - 42 = 7$. Thus, the Greek mathematicians "lost" the negative solution because of their strictly geometric interpretation of quadratic equations. There were, therefore, many quadratic equations that the Greek mathematicians could not solve because both solutions were negative numbers or nonreal (complex) numbers. Negative solutions to equations were almost completely ignored until the early 1500's, during the Renaissance.

"What is the difference between method and device? A method is a device which you used twice."

George Pòlya in *How to Solve It*

Quadratic equations appear in one form or another in almost every course in mathematics and in many courses in related fields such as business, biology, engineering, and computer science. In this chapter, we will discuss three techniques for solving quadratic equations: factoring, completing the square, and the quadratic formula. Factoring has already been discussed in Section 6.4 and, when possible, is generally considered the method of first choice because it is easier to apply, and factoring quadratic expressions is useful in other mathematical situations. However, the quadratic formula is very important and should be memorized as it works in all cases. A part of the formula, called the discriminant, gives ready information about the nature of the solutions. Additionally, the formula is easy to use in computer programs.

The wide variety of applications presented in Section 11.3 illustrates the practicality of knowing how to recognize and solve quadratic equations.

11.1 Quadratic Equations: Completing the Square

Objectives

After completing this section, you will be able to:

1. *Solve quadratic equations by factoring.*
2. *Solve quadratic equations by using the definition of square root.*
3. *Solve quadratic equations by completing the square.*

Review of Solving Equations by Factoring

Not every polynomial can be factored so that the factors have integer coefficients, and not every polynomial equation can be solved by factoring. However, when the solutions of a polynomial equation can be found by factoring, the method depends on the **zero factor property**, which is restated here for easy reference.

Zero Factor Property

If the product of two factors is 0, then one or both of the factors must be zero. Symbolically, for factors a and b,

$$\text{if } a \cdot b = 0, \text{ then } a = 0 \text{ or } b = 0.$$

Also as discussed in Section 6.4, polynomial equations of second-degree are called **quadratic equations** and, since these are the equations of interest in this chapter, the definition is restated here.

Quadratic Equation

An equation that can be written in the form

$$ax^2 + bx + c = 0 \qquad \text{where } a, b, \text{ and } c \text{ are real numbers and } a \neq 0$$

is called a ***quadratic equation****.*

The procedure for solving quadratic equations by factoring involves making sure that one side of the equation is 0 and then applying the zero factor property. The following list of steps outlines the procedure.

To Solve an Equation By Factoring

1. *Add or subtract terms so that* ***one side of the equation is 0****.*
2. *Factor the polynomial expression.*
3. *Set each factor equal to 0 and solve each of the resulting equations.*

*(**Note:** If two of the factors are the same, then the solution is said to be a* ***double root*** *or a* ***root of multiplicity two****.)*

NOTES

We will see throughout this chapter that quadratic equations may have nonreal solutions.

Also, you will need to remember that the sum of two squares can be factored as complex conjugates. For example,

$$x^2 + 9 = (x + 3i)(x - 3i).$$

Example 1: Factorization

Solve the following equations by factoring.

a. $x^2 - 15x = -50$

Solution: $x^2 - 15x = -50$

$x^2 - 15x + 50 = 0$	Add 50 to both sides. **One side must be 0.**
$(x-5)(x-10) = 0$	Factor the left-hand side.
$x - 5 = 0$ or $x - 10 = 0$	Set each factor equal to 0.
$x = 5 \qquad x = 10$	Solve each linear equation.

Check: $5^2 - 15 \cdot 5 \stackrel{?}{=} -50 \qquad 10^2 - 15 \cdot 10 \stackrel{?}{=} -50$

$25 - 75 \stackrel{?}{=} -50 \qquad 100 - 150 \stackrel{?}{=} -50$

$-50 = -50 \qquad -50 = -50$

continued on next page ...

b. $x^2 - 8x = -16$

Solution: $x^2 - 8x = -16$

$x^2 - 8x + 16 = 0$ Add 16 to both sides.

$(x-4)^2 = 0$ The trinomial is a perfect square.

$x - 4 = 0$ The two factors are the same.

$x = 4$ The solution is a **double root**.

Check: $4^2 - 8 \cdot 4 \stackrel{?}{=} -16$

$16 - 32 \stackrel{?}{=} -16$

$-16 = -16$

c. $x^2 + 4 = 0$

Solution: $x^2 + 4 = 0$

$(x+2i)(x-2i) = 0$

$x + 2i = 0$ or $x - 2i = 0$

$x = -2i$ $\quad$ $x = 2i$

Note that $x^2 + 4$ is the sum of two squares and can be factored into the product of conjugates of complex numbers.

Check: $(-2i)^2 + 4 \stackrel{?}{=} 0$ $\quad$ $(2i)^2 + 4 \stackrel{?}{=} 0$

$4i^2 + 4 \stackrel{?}{=} 0$ $\quad$ $4i^2 + 4 \stackrel{?}{=} 0$

$-4 + 4 \stackrel{?}{=} 0$ $\quad$ $-4 + 4 \stackrel{?}{=} 0$

$0 = 0$ $\quad$ $0 = 0$

Teaching Notes:
You might want to explain how to use fractional exponents in solving equations. This may help your students understand solving higher degree equations later. For example,

$x^2 = 13$

$\left(x^2\right)^{\frac{1}{2}} = 13^{\frac{1}{2}}$

$|x| = \sqrt{13}$

$x = \pm\sqrt{13}$

Using the Definition of Square Root and the Square Root Property

Consider the equation

$$x^2 = 13.$$

The definition of square root, $\left[\sqrt{x^2} = |x|\right]$, leads to two solutions, as follows:

Taking the square root of both sides of the equation gives:

$$|x| = \sqrt{13}.$$

So,

$$x = \sqrt{13} \quad \text{or} \quad x = -\sqrt{13}.$$

For both solutions, we write

$$x = \pm\sqrt{13}.$$

Similarly, for the equation

$$(x-3)^2 = 5$$

the definition of square root gives

$$x-3 = \pm\sqrt{5}$$

which leads to the two equations and the two solutions, as follows:

$$x-3 = \sqrt{5} \qquad \text{or} \qquad x-3 = -\sqrt{5}$$

$$x = 3+\sqrt{5} \qquad\qquad x = 3-\sqrt{5}$$

We can write the two solutions in the form

$$x = 3 \pm \sqrt{5}.$$

These examples illustrate how the definition of square root can be used to solve some quadratic equations. In particular, if one side of the equation is a squared expression and the other side is a constant, we can simply take the square roots of each side. However, we must keep in mind that the square roots of negative numbers are nonreal complex numbers and will involve the number i. We have the following **Square Root Property**.

Square Root Property

If $x^2 = c$, then $x = \pm\sqrt{c}$.

If $(x-a)^2 = c$, then $x-a = \pm\sqrt{c}$ (or $x = a \pm \sqrt{c}$).

Example 2: Solving Quadratic Equations by Using the Square Root Property

Solve the following quadratic equations.

a. $x^2 = -25$

Solution: $x^2 = -25$

$$x = \pm\sqrt{-25}$$

$$x = \pm 5i$$

b. $(y+4)^2 = 8$

Solution: $(y+4)^2 = 8$

$$y+4 = \pm\sqrt{8}$$

$$y = -4 \pm 2\sqrt{2}$$

Completing the Square

Recall that a perfect square trinomial (Section 5.6) is the result of squaring a binomial. Our objective here is to find the third term of a perfect square trinomial when the first two terms are given. This is called **completing the square**. We will find this procedure useful in solving quadratic equations and in developing the quadratic formula.

Study the following examples to help in your understanding.

Perfect Square Trinomials		Equal Factors		Square of a Binomial
$x^2 - 8x + 16$	=	$(x-4)(x-4)$	=	$(x-4)^2$
$x^2 + 20x + 100$	=	$(x+10)(x+10)$	=	$(x+10)^2$
$x^2 - 9x + \frac{81}{4}$	=	$\left(x-\frac{9}{2}\right)\left(x-\frac{9}{2}\right)$	=	$\left(x-\frac{9}{2}\right)^2$
$x^2 - 2hx + h^2$	=	$(x-h)(x-h)$	=	$(x-h)^2$
$x^2 + 2hx + h^2$	=	$(x+h)(x+h)$	=	$(x+h)^2$.

Teaching Notes: Some of the *power* of algebra can be illustrated by pointing out how the general representation of the perfect square trinomial $x^2 + 2hx + h^2$ leads to the procedure for completing the square. This is another example of the advantage in learning to "read" the algebraic form of expressions and equations.

The last two examples are in the form of formulas. We see two things in each case:

1. The leading coefficient (the coefficient of x^2) is 1.
2. The constant term is the square of $\frac{1}{2}$ of the coefficient of x.

   For example, $\frac{1}{2}(2h) = h$ and the square of this result is the constant h^2.

What constant should be added to $x^2 - 16x$ to get a perfect square trinomial? By following the ideas just discussed, we find that $\frac{1}{2}(-16) = -8$ and $(-8)^2 = 64$. Therefore, to complete the square, we add 64. Thus,

$$x^2 - 16x + 64 = (x-8)^2.$$

By adding 64, we have completed the square for $x^2 - 16x$.

Example 3: Completing the Square

Add the constant that will complete the square for each expression, and write the new expression as the square of a binomial.

a. $x^2 + 10x$

Solution: $x^2 + 10x + \underline{\quad\quad} = (\quad\quad)^2$

$$\frac{1}{2}(10) = 5 \text{ and } (5)^2 = 25$$

So, add 25: $x^2 + 10x + 25 = (x+5)^2$

b. $x^2 - 7x$

Solution: $x^2 - 7x + \_\_\_\_ = (\quad)^2$

$$\frac{1}{2}(-7) = -\frac{7}{2} \text{ and } \left(-\frac{7}{2}\right)^2 = \frac{49}{4}$$

$$\text{So, add } \frac{49}{4}\text{: } x^2 - 7x + \frac{49}{4} = \left(x - \frac{7}{2}\right)^2$$

Solving Quadratic Equations by Completing the Square

Now we want to use the process of completing the square to help in solving quadratic equations. This technique involves the following steps.

To Solve a Quadratic Equation by Completing the Square

1. *If necessary, divide or multiply both sides of the equation so that the leading coefficient (the coefficient of x^2) is 1.*
2. *If necessary, isolate the constant term on one side of the equation.*
3. *Find the constant that completes the square of the polynomial and add this constant to both sides.*
4. *Use the Square Root Property to find the solutions of the equation.*

Example 4: Solving Quadratic Equations by Completing the Square

Solve the following quadratic equations by completing the square.

a. $x^2 - 8x = 25$

Solution:

$$x^2 - 8x = 25$$

The coefficient of x^2 is already 1.

$$x^2 - 8x + 16 = 25 + 16$$

$\frac{1}{2}(-8) = -4$ and $(-4)^2 = 16$.
Therefore, add 16 to both sides.

$$(x-4)^2 = 41$$

$$x - 4 = \sqrt{41} \quad \text{or} \quad x - 4 = -\sqrt{41}$$

Use the Square Root Property.

$$x = 4 + \sqrt{41} \qquad x = 4 - \sqrt{41}$$

There are two real solutions: $4 + \sqrt{41}$ and $4 - \sqrt{41}$. We write $x = 4 \pm \sqrt{41}$.

continued on next page ...

b. $3x^2 + 6x - 15 = 0$

Solution: $3x^2 + 6x - 15 = 0$

$$\frac{3x^2}{3} + \frac{6x}{3} - \frac{15}{3} = \frac{0}{3}$$

Divide each term by 3. **The leading coefficient must be 1.**

$$x^2 + 2x - 5 = 0$$

$$x^2 + 2x = 5$$

$$x^2 + 2x + 1 = 5 + 1$$

Isolate the constant term and complete the square: $\frac{1}{2}(2) = 1$ and $1^2 = 1$. Therefore, add 1 to both sides.

$$(x+1)^2 = 6$$

$$x + 1 = \pm\sqrt{6}$$

$$x = -1 \pm \sqrt{6}$$

c. $2x^2 + 2x - 7 = 0$

Solution: $2x^2 + 2x - 7 = 0$

$$x^2 + x - \frac{7}{2} = 0$$

Divide each term by 2 so that the leading coefficient will be 1.

$$x^2 + x = \frac{7}{2}$$

Isolate the constant term and complete the square: $\frac{1}{2}(1) = \frac{1}{2}$ and $\left(\frac{1}{2}\right)^2 = \frac{1}{4}$.

$$x^2 + x + \frac{1}{4} = \frac{7}{2} + \frac{1}{4}$$

$$\left(x + \frac{1}{2}\right)^2 = \frac{15}{4}$$

$$x + \frac{1}{2} = \pm\sqrt{\frac{15}{4}}$$

$$x = -\frac{1}{2} \pm \frac{\sqrt{15}}{2}$$

$$x = \frac{-1 \pm \sqrt{15}}{2}$$

d. $x^2 - 2x + 13 = 0$

Solution: $x^2 - 2x + 13 = 0$

$$x^2 - 2x = -13$$

$$x^2 - 2x + 1 = -13 + 1$$

$$(x-1)^2 = -12$$

$$x - 1 = \pm\sqrt{-12} = \pm i\sqrt{12} = \pm 2i\sqrt{3}$$

$$x = 1 \pm 2i\sqrt{3}$$

The solutions are nonreal complex numbers.

Writing Equations with Known Roots

In Section 6.4, we found equations with known roots by setting the product of factors equal to 0 and simplifying. The same method is applied here with roots that are nonreal and roots that involve radicals.

Example 5: Equations with Known Roots

Find polynomial equations that have the given roots.

a. $y = 3 + 2i$ and $y = 3 - 2i$

Solution: $y = 3 + 2i$ $\quad$ $y = 3 - 2i$

$y - 3 - 2i = 0$ $\quad$ $y - 3 + 2i = 0$ $\quad$ Get 0 on one side of each equation.

Set the product of the two factors equal to 0 and simplify.

$$[y-3-2i][y-3+2i]=0$$

$$[(y-3)-2i][(y-3)+2i]=0$$

Regroup the terms to represent the product of complex conjugates. This makes the multiplication easier.

$$(y-3)^2-4i^2=0$$

$$y^2-6y+9+4=0 \qquad i^2=-1$$

$$y^2-6y+13=0$$

This equation has two solutions: $y = 3 + 2i$ and $y = 3 - 2i$

b. $x=5-\sqrt{2}$ and $x=5+\sqrt{2}$

Solution: $x=5-\sqrt{2}$ $\quad$ $x=5+\sqrt{2}$

$x-5+\sqrt{2}=0$ $\quad$ $x-5-\sqrt{2}=0$ $\quad$ Get 0 on one side of each equation.

Set the product of the two factors equal to 0 and simplify.

$$\left[x-5+\sqrt{2}\right]\left[x-5-\sqrt{2}\right]=0$$

$$\left[(x-5)+\sqrt{2}\right]\left[(x-5)-\sqrt{2}\right]=0$$

Regroup the terms to make the multiplication easier.

$$(x-5)^2-\left(\sqrt{2}\right)^2=0$$

$$x^2-10x+25-2=0$$

$$x^2-10x+23=0$$

This equation has two solutions: $x=5-\sqrt{2}$ and $x=5+\sqrt{2}$.

c. $x=3+i\sqrt{5}$ and $x=3-i\sqrt{5}$

Solution: $x=3+i\sqrt{5}$ $\quad$ $x=3-i\sqrt{5}$

$x-3-i\sqrt{5}=0$ $\quad$ $x-3+i\sqrt{5}=0$ $\quad$ Get 0 on one side of each equation.

continued on next page ...

1. $x^2-12x+\underline{36}=(x-6)^2$

2. $y^2+14y+\underline{49}=(y+7)^2$

3. $x^2+6x+\underline{9}=(x+3)^2$

4. $x^2+8x+\underline{16}=(x+4)^2$

5. $x^2-5x+\underline{\frac{25}{4}}=\left(x-\frac{5}{2}\right)^2$

6. $x^2+7x+\underline{\frac{49}{4}}=\left(x+\frac{7}{2}\right)^2$

7. $y^2+y+\underline{\frac{1}{4}}=\left(y+\frac{1}{2}\right)^2$

8. $x^2+\frac{1}{2}x+\underline{\frac{1}{16}}=\left(x+\frac{1}{4}\right)^2$

9. $x^2+\frac{1}{3}x+\underline{\frac{1}{36}}=\left(x+\frac{1}{6}\right)^2$

Set the product of the two factors equal to 0 and simplify.

$$\left[x-3-i\sqrt{5}\right]\left[x-3+i\sqrt{5}\right]=0$$

$$\left[(x-3)-i\sqrt{5}\right]\left[(x-3)+i\sqrt{5}\right]=0$$ Regroup the terms to make the multiplication easier.

$$(x-3)^2-\left(i\sqrt{5}\right)^2=0$$

$$x^2-6x+9-\left(i\sqrt{5}\right)^2=0$$

$$x^2-6x+9-i^2\left(\sqrt{5}\right)^2=0$$

$$x^2-6x+9-(-1)(5)=0$$ Remember, $i^2=-1$.

$$x^2-6x+14=0$$ This equation has two solutions: $x=3+i\sqrt{5}$ and $x=3-i\sqrt{5}$.

Practice Problems

10. $y^2+\frac{3}{4}y+\underline{\frac{9}{64}}=\left(y+\frac{3}{8}\right)^2$

11. $x=\pm12$ **12.** $x=\pm13$

13. $x=\pm5i$

14. $x=\pm2i\sqrt{6}$

15. $x=\pm3i\sqrt{2}$

Solve each of the following quadratic equations by completing the square.

1. $2x^2+5x-3=0$
2. $x^2+2x+2=0$
3. $x^2-24x+72=0$
4. $x^2-3x+1=0$
5. $3x^2-6x+15=0$
6. *Find a quadratic equation that has the roots* $x=2\pm3i$.

11.1 Exercises

16. $x=5,\ x=-1$

17. $x=9,\ x=-1$

18. $x=\pm\sqrt{5}$

19. $x=\pm2\sqrt{3}$

20. $x=-3\pm\sqrt{3}$

21. $x=1\pm\sqrt{5}$

22. $x=3\pm2i$

23. $x=-8\pm3i$

24. $x=-2\pm i\sqrt{7}$

25. $x=5\pm i\sqrt{10}$

26. $x=1,\ x=-5$

27. $x=1,\ x=-7$

28. $y=-1\pm\sqrt{6}$

29. $x=4\pm\sqrt{13}$

Add the correct constant to complete the square in Exercises 1 – 10, then factor the trinomial as indicated.

1. x^2-12x+ ____ $=(\quad)^2$ **2.** y^2+14y+ ____ $=(\quad)^2$ **3.** x^2+6x+ ____ $=(\quad)^2$

4. x^2+8x+ ____ $=(\quad)^2$ **5.** x^2-5x+ ____ $=(\quad)^2$ **6.** x^2+7x+ ____ $=(\quad)^2$

7. y^2+y+ ____ $=(\quad)^2$ **8.** $x^2+\frac{1}{2}x+$ ____ $=(\quad)^2$ **9.** $x^2+\frac{1}{3}x+$ ____ $=(\quad)^2$

10. $y^2+\frac{3}{4}y+$ ____ $=(\quad)^2$

Solve the equations in Exercises 11 – 25.

11. $x^2-144=0$ **12.** $x^2-169=0$ **13.** $x^2+25=0$

14. $x^2+24=0$ **15.** $x^2+18=0$ **16.** $(x-2)^2=9$

17. $(x-4)^2=25$ **18.** $x^2=5$ **19.** $x^2=12$

Answers to Practice Problems: **1.** $x=-3,\ x=\frac{1}{2}$ **2.** $x=-1\pm i$ **3.** $x=12\pm6\sqrt{2}$ **4.** $x=\frac{3\pm\sqrt{5}}{2}$ **5.** $x=1\pm2i$ **6.** $x^2-4x+13=0$

30. $x=5\pm\sqrt{22}$

31. $z=-2\pm\sqrt{6}$

32. $x = 3 \pm i$
33. $x = -1 \pm i\sqrt{5}$
34. $x = 1,\ x = 11$
35. $y = 5 \pm \sqrt{21}$
36. $z = \frac{-3 \pm \sqrt{29}}{2}$
37. $x = \frac{5 \pm \sqrt{5}}{2}$
38. $x = \frac{-5 \pm \sqrt{17}}{2}$
39. $x = \frac{-1 \pm i\sqrt{7}}{2}$
40. $x = 1 \pm 2i$
41. $x = -2 \pm \sqrt{7}$
42. $y = \frac{-3 \pm i\sqrt{3}}{2}$
43. $x = -1 \pm i\sqrt{3}$
44. $x = 2,\ x = -3$
45. $y = 1,\ y = -\frac{4}{3}$
46. $x = \frac{5 \pm \sqrt{10}}{3}$
47. $x = \frac{-7 \pm \sqrt{17}}{8}$
48. $y = \frac{-5 \pm \sqrt{61}}{6}$
49. $x = \frac{1 \pm i\sqrt{11}}{4}$
50. $x = \frac{-1 \pm i\sqrt{11}}{6}$
51. $y = \frac{-3 \pm i\sqrt{11}}{2}$
52. $x = -4,\ x = -\frac{1}{2}$
53. $x = 2 \pm \sqrt{2}$

20. $2(x+3)^2 = 6$ **21.** $3(x-1)^2 = 15$ **22.** $(x-3)^2 = -4$
23. $(x+8)^2 = -9$ **24.** $(x+2)^2 = -7$ **25.** $(x-5)^2 = -10$

Solve the quadratic equations in Exercises 26 – 55 by completing the square.

26. $x^2 + 4x - 5 = 0$ **27.** $x^2 + 6x - 7 = 0$ **28.** $y^2 + 2y = 5$
29. $x^2 + 3 = 8x$ **30.** $x^2 - 10x + 3 = 0$ **31.** $z^2 + 4z = 2$
32. $x^2 - 6x + 10 = 0$ **33.** $x^2 + 2x + 6 = 0$ **34.** $x^2 + 11 = 12x$
35. $y^2 - 10y + 4 = 0$ **36.** $z^2 + 3z - 5 = 0$ **37.** $x^2 - 5x + 5 = 0$
38. $x^2 + 5x + 2 = 0$ **39.** $x^2 + x + 2 = 0$ **40.** $x^2 - 2x + 5 = 0$
41. $x^2 = 3 - 4x$ **42.** $3y^2 + 9y + 9 = 0$ **43.** $4x^2 + 8x + 16 = 0$
44. $x^2 = 6 - x$ **45.** $3y^2 = 4 - y$ **46.** $3x^2 - 10x + 5 = 0$
47. $7x + 2 = -4x^2$ **48.** $3y^2 + 5y - 3 = 0$ **49.** $4x^2 - 2x + 3 = 0$
50. $2x + 2 = -6x^2$ **51.** $5y^2 + 15y + 25 = 0$ **52.** $2x^2 + 9x + 4 = 0$
53. $2x^2 - 8x + 4 = 0$ **54.** $3 = 3x - 6x^2$ **55.** $4x^2 + 20x + 32 = 0$

For Exercises 56 – 70, write a quadratic equation with integer coefficients that has the given roots.

56. $x = \sqrt{7}, x = -\sqrt{7}$ **57.** $x = \sqrt{5}, x = -\sqrt{5}$
58. $x = 1 + \sqrt{3}, x = 1 - \sqrt{3}$ **59.** $z = 2 + \sqrt{2}, z = 2 - \sqrt{2}$
60. $y = -2 + \sqrt{5}, y = -2 - \sqrt{5}$ **61.** $x = 1 + 2\sqrt{3}, x = 1 - 2\sqrt{3}$
62. $x = 4i, x = -4i$ **63.** $x = 7i, x = -7i$
64. $y = i\sqrt{6}, y = -i\sqrt{6}$ **65.** $y = i\sqrt{5}, y = -i\sqrt{5}$
66. $x = 2 + i, x = 2 - i$ **67.** $x = -3 + 2i, x = -3 - 2i$
68. $x = 1 + i\sqrt{2}, x = 1 - i\sqrt{2}$ **69.** $x = 2 + i\sqrt{3}, x = 2 - i\sqrt{3}$
70. $x = -5 + 2i\sqrt{6}, x = -5 - 2i\sqrt{6}$

Writing and Thinking About Mathematics

71. Explain, in your own words, the steps involved in the process of solving a quadratic equation by completing the square.

54. $x = \frac{1 \pm i\sqrt{7}}{4}$ 55. $x = \frac{-5 \pm i\sqrt{7}}{2}$ 56. $x^2 - 7 = 0$ 57. $x^2 - 5 = 0$ 58. $x^2 - 2x - 2 = 0$

Hawkes Learning Systems: Introductory & Intermediate Algebra

SQE: The Square Root Method
SQE: Completing the Square

59. $z^2 - 4z + 2 = 0$ 60. $y^2 + 4y - 1 = 0$ 61. $x^2 - 2x - 11 = 0$ 62. $x^2 + 16 = 0$ 63. $x^2 + 49 = 0$ 64. $y^2 + 6 = 0$ 65. $y^2 + 5 = 0$ 66. $x^2 - 4x + 5 = 0$ 67. $x^2 + 6x + 13 = 0$ 68. $x^2 - 2x + 3 = 0$ 69. $x^2 - 4x + 7 = 0$ 70. $x^2 + 10x + 49 = 0$ 71. Answers will vary

11.2 Quadratic Equations: The Quadratic Formula

Objectives

After completing this section, you will be able to:

1. *Determine the nature of the solutions (one real, two real, or two nonreal) for quadratic equations by using the discriminant.*
2. *Solve quadratic equations by using the quadratic formula.*

The **quadratic formula** gives the roots of any quadratic equation in terms of the coefficients a, b, and c of the **general quadratic equation**

$$ax^2 + bx + c = 0.$$

Therefore, if you have memorized the quadratic formula, you can solve any quadratic equation by simply substituting the coefficients into the formula. To develop the quadratic formula, we solve the general quadratic equation by completing the square as follows:

$ax^2 + bx + c = 0$	The general quadratic equation.
$x^2 + \frac{b}{a}x + \frac{c}{a} = \frac{0}{a}$	Divide each term of the equation by a. Since $a \neq 0$ this is permissible.
$x^2 + \frac{b}{a}x = -\frac{c}{a}$	Add $-\frac{c}{a}$ to both sides of the equation.
$x^2 + \frac{b}{a}x + \frac{b^2}{4a^2} = \frac{b^2}{4a^2} - \frac{c}{a}$	$\frac{1}{2}\left(\frac{b}{a}\right) = \frac{b}{2a}$ and $\left(\frac{b}{2a}\right)^2 = \frac{b^2}{4a^2}$. Add $\frac{b^2}{4a^2}$ to both sides of the equation.
$\left(x + \frac{b}{2a}\right)^2 = \frac{b^2}{4a^2} - \frac{4ac}{4a^2}$	Factor the left side. $4a^2$ is the common denominator on the right side.
$\left(x + \frac{b}{2a}\right)^2 = \frac{b^2 - 4ac}{4a^2}$	Simplify.
$x + \frac{b}{2a} = \pm\sqrt{\frac{b^2 - 4ac}{4a^2}}$	Use the Square Root Property.
$x + \frac{b}{2a} = \pm\frac{\sqrt{b^2 - 4ac}}{2a}$	Simplify.
$x = -\frac{b}{2a} \pm \frac{\sqrt{b^2 - 4ac}}{2a}$	Solve for x.
$x = \frac{-b \pm \sqrt{b^2 - 4ac}}{2a}$	**This equation is called the Quadratic Formula.**

NOTES

Note about the coefficient a

For convenience and without loss of generality, in the development of the quadratic formula (and in the examples and exercises) the leading coefficient, a, is positive. If a is a negative number, we can multiply both sides of the equation by -1. This will make the leading coefficient positive without changing any solutions of the original equation.

The Quadratic Formula

Teaching Notes:
Early on, remind students that the "±" symbol is an economial way of representing two solutions simultaneously.

For the general quadratic equation

$$ax^2 + bx + c = 0 \qquad \text{where } a \neq 0$$

the solutions are

$$\boldsymbol{x = \frac{-b \pm \sqrt{b^2 - 4ac}}{2a}}.$$

The quadratic formula should be memorized.

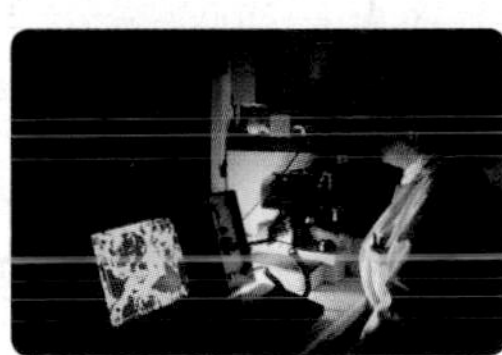

Applications of quadratic equations are found in such fields as economics, business, computer science, and chemistry, and in almost all branches of mathematics. Most instructors assume that their students know the quadratic formula and how to apply it. You should recognize that the importance of the quadratic formula lies in the fact that it allows you to solve **any** quadratic equation.

Example 1: Quadratic Formula

Solve the following quadratic equations by using the quadratic formula.

a. $x^2 - 5x + 3 = 0$

Solution: Substitute $a = 1$, $b = -5$, and $c = 3$ into the formula:

$$x = \frac{-b \pm \sqrt{b^2 - 4ac}}{2a} = \frac{-(-5) \pm \sqrt{(-5)^2 - 4 \cdot 1 \cdot 3}}{2 \cdot 1}$$

$$= \frac{5 \pm \sqrt{25 - 12}}{2} = \frac{5 \pm \sqrt{13}}{2}$$

Thus, the solutions are $\frac{5+\sqrt{13}}{2}$ and $\frac{5-\sqrt{13}}{2}$.

continued on next page ...

b. $7x^2 - 2x + 1 = 0$

Solution: Substitute $a = 7, b = -2$, and $c = 1$ into the formula:

$$x = \frac{-b \pm \sqrt{b^2 - 4ac}}{2a} = \frac{-(-2) \pm \sqrt{(-2)^2 - 4 \cdot 7 \cdot 1}}{2 \cdot 7}$$

$$= \frac{2 \pm \sqrt{4 - 28}}{14}$$

$$= \frac{2 \pm \sqrt{-24}}{14}$$

$$= \frac{2 \pm 2i\sqrt{6}}{14}$$

$$= \frac{\cancel{2}\left(1 \pm i\sqrt{6}\right)}{\cancel{2} \cdot 7}$$ Factor and reduce.

$$= \frac{1 \pm i\sqrt{6}}{7}$$ The solutions are nonreal complex numbers.

$$= \frac{1 + i\sqrt{6}}{7}, \quad \frac{1 - i\sqrt{6}}{7}$$

c. $\frac{3}{4}x^2 - \frac{1}{2}x = \frac{1}{3}$

Solution: Multiply each term by the LCM, 12, so that the coefficients will be integers. The quadratic formula is easier to use with integer coefficients.

$$12 \cdot \frac{3}{4}x^2 - 12 \cdot \frac{1}{2}x = 12 \cdot \frac{1}{3}$$

$$9x^2 - 6x = 4$$

$$9x^2 - 6x - 4 = 0$$ **To apply the formula, one side must be 0.**

$$x = \frac{-(-6) \pm \sqrt{(-6)^2 - 4(9)(-4)}}{2 \cdot 9}$$

$$= \frac{6 \pm \sqrt{36 + 144}}{18}$$

$$= \frac{6 \pm \sqrt{180}}{18} = \frac{6 \pm 6\sqrt{5}}{18}$$

$$= \frac{\cancel{6}\left(1 \pm \sqrt{5}\right)}{\cancel{6} \cdot 3} = \frac{1 \pm \sqrt{5}}{3}$$ Factor out 6 and reduce.

$$= \frac{1 + \sqrt{5}}{3}, \quad \frac{1 - \sqrt{5}}{3}$$

COMMON ERROR

NOTES

Many students make a mistake when simplifying fractions by dividing the denominator into only one of the terms in the numerator.

WRONG $\frac{4+\cancel{2}\sqrt{3}}{\cancel{2}}=4+\sqrt{3}$

The correct method is to divide both terms by the denominator or to factor out a common factor in the numerator and then reduce.

RIGHT $\frac{4+2\sqrt{3}}{2}=\frac{4}{2}+\frac{2\sqrt{3}}{2}=2+\sqrt{3}$

RIGHT $\frac{4+2\sqrt{3}}{2}=\frac{\cancel{2}\left(2+\sqrt{3}\right)}{\cancel{2}}=2+\sqrt{3}$

In Example 2, the equation is third-degree (a cubic equation) and one of the factors is quadratic. The quadratic formula can be applied to this factor.

Example 2: Cubic Equation

Solve the following cubic equation using the quadratic formula.

$$2x^3-10x^2+6x=0$$

Solution: $2x^3-10x^2+6x=0$

$2x\left(x^2-5x+3\right)=0$		Factor out $2x$.
$2x=0$ or	$x^2-5x+3=0$	Set each factor equal to 0.
$x=0$	$x=\frac{5\pm\sqrt{13}}{2}$	Solve each equation. (The quadratic equation was solved in Example 1a by using the quadratic formula.)

The Discriminant

The expression $\boldsymbol{b^2-4ac}$, the part of the quadratic formula that lies under the radical sign, is called the **discriminant**. The discriminant identifies the kind of numbers that are solutions to a quadratic equation. Assuming a, b, and c are all real numbers, there are three possibilities: the discriminant is either positive, negative, or zero.

In Example 1a $\left(x^2-5x+3=0\right)$, the discriminant was positive, $b^2-4ac=(-5)^2-4(1)(3)=13$, and there were **two** real solutions: $x=\frac{5\pm\sqrt{13}}{2}$. In Example 1b, the discriminant was negative, $b^2-4ac=(-2)^2-4(7)(1)=-24$, and there were **two** nonreal solutions: $x=\frac{1\pm i\sqrt{6}}{7}$.

The discriminant gives the following information:

Discriminant	Nature of Solutions
$b^2 - 4ac > 0$	Two real solutions
$b^2 - 4ac = 0$	One real solution, $x = \frac{-b \pm 0}{2a} = -\frac{b}{2a}$
$b^2 - 4ac < 0$	Two nonreal solutions

In the case where $b^2 - 4ac = 0$, we say $x = -\frac{b}{2a}$ is a **double root**. Additionally, if the discriminant is a perfect square, the equation is factorable.

Example 3: Finding the Discriminant

Find the discriminant and determine the nature of the solutions to each of the following quadratic equations.

a. $3x^2 + 11x - 7 = 0$

Solution: $b^2 - 4ac = 11^2 - 4(3)(-7)$

$= 121 + 84$

$= 205 > 0$

There are two real solutions.

b. $x^2 + 6x + 9 = 0$

Solution: $b^2 - 4ac = 6^2 - 4(1)(9)$

$= 36 - 36$

$= 0$

There is one real solution.

c. $x^2 + 1 = 0$

Solution: Here $b = 0$. We could write $x^2 + 0x + 1 = 0$.

$b^2 - 4ac = 0^2 - 4(1)(1)$

$= 0 - 4$

$= -4$

There are two nonreal solutions.

Example 4: Using the Discriminant

a. Determine the values for k so that $x^2 + 8x - k = 0$ will have one real solution.

Hint: Set the discriminant equal to 0 and solve the equation for k.

Solution: $b^2 - 4ac = 8^2 - 4(1)(-k) = 0$

$64 + 4k = 0$

$4k = -64$

$k = -16$

Check: $x^2+8x-(-16)\stackrel{?}{=}0$

$$x^2+8x+16\stackrel{?}{=}0$$

$$(x+4)^2\stackrel{?}{=}0$$

$$x=-4$$

There is only one real solution. Thus, –4 is a double root.

b. Determine the values for k so that $kx^2-8x+4=0$ will have two nonreal solutions.

Solution: $b^2-4ac=64-4(k)(4)$

$$64-4(k)(4)<0$$

$$64-16k<0$$

$$-16k<-64$$

$$k>4$$

Thus, if k is any real number greater than 4, the discriminant will be negative and the equation will have two nonreal soutions.

Practice Problems

Solve each of the following quadratic equations by using the quadratic formula.

1. $x^2+2x-4=0$ **2.** $2x^2-3x+4=0$ **3.** $5x^2-x-4=0$

4. $\frac{1}{4}x^2-\frac{1}{2}x=-\frac{1}{4}$ **5.** $3x^2+5=0$

11.2 Exercises

1. 68, two real solutions
2. 5, two real solutions
3. 0, one real solution
4. –11, two nonreal solutions
5. –44, two nonreal solutions
6. –23, two nonreal solutions
7. 4, two real solutions

Find the discriminant and determine the nature of the solutions to each quadratic equation in Exercises 1 – 12.

1. $x^2+6x-8=0$ **2.** $x^2+3x+1=0$ **3.** $x^2-8x+16=0$

4. $x^2+3x+5=0$ **5.** $4x^2+2x+3=0$ **6.** $3x^2-x+2=0$

7. $5x^2+8x+3=0$ **8.** $4x^2+12x+9=0$ **9.** $100x^2-49=0$

10. $9x^2+121=0$ **11.** $3x^2+x+1=0$ **12.** $5x^2-3x-2=0$

Answers to Practice Problems: **1.** $x=-1\pm\sqrt{5}$ **2.** $x=\frac{3\pm i\sqrt{23}}{4}$ **3.** $x=1,\ x=\frac{-4}{5}$ **4.** $x=1$ **5.** $x=\frac{\pm i\sqrt{15}}{3}$

8. 0, one real solution
9. 19,600, two real solutions
10. –4356, two nonreal solutions
11. –11, two nonreal solutions
12. 49, two real solutions
13. $k<16$ **14.** $k<\frac{25}{4}$
15. $k=\frac{81}{4}$ **16.** $k=\frac{49}{4}$
17. $k>3$ **18.** $k<-2$
19. $k>-\frac{1}{36}$ **20.** $k<3$
21. $k=\frac{49}{48}$ **22.** $k=\frac{1}{8}$
23. $k>\frac{4}{3}$ **24.** $k>\frac{9}{8}$
25. $x=\frac{-3\pm\sqrt{29}}{2}$
26. $x=\frac{7\pm\sqrt{61}}{2}$
27. $x=\frac{5\pm\sqrt{17}}{2}$
28. $x=-1,\ x=-3$
29. $x=\frac{-7\pm\sqrt{33}}{4}$
30. $x=\frac{-1\pm\sqrt{7}}{3}$
31. $x=1,\ x=-\frac{1}{6}$
32. $x=\frac{-1\pm\sqrt{65}}{8}$
33. $x=\pm\sqrt{\frac{4}{3}}$
34. $x=\frac{-3\pm\sqrt{2}}{7}$
35. $x=\frac{9\pm\sqrt{65}}{2},\ x=0$
36. $x=\frac{11\pm\sqrt{133}}{2},\ x=0$
37. $x=\frac{-3\pm\sqrt{5}}{2},\ x=0$

In Exercises 13 – 24, find the indicated values for k.

13. Determine the values for k so that $x^2 - 8x + k = 0$ will have two real solutions.

14. Determine the values for k so that $x^2 + 5x + k = 0$ will have two real solutions.

15. Determine the values for k so that $x^2 + 9x + k = 0$ will have one real solution.

16. Determine the values for k so that $x^2 - 7x + k = 0$ will have one real solution.

17. Determine the values for k so that $kx^2 - 6x + 3 = 0$ will have two nonreal solutions.

18. Determine the values for k so that $kx^2 + 4x - 2 = 0$ will have two nonreal solutions.

19. Determine the values for k so that $kx^2 + x - 9 = 0$ will have two real solutions.

20. Determine the values for k so that $kx^2 + 6x + 3 = 0$ will have two real solutions.

21. Determine the values for k so that $kx^2 + 7x + 12 = 0$ will have one real solution.

22. Determine the values for k so that $kx^2 - 2x + 8 = 0$ will have one real solution.

23. Determine the values for k so that $3x^2 + 4x + k = 0$ will have two nonreal solutions.

24. Determine the values for k so that $2x^2 + 3x + k = 0$ will have two nonreal solutions.

Solve the equations in Exercises 25 – 52. You may use any of the techniques discussed for solving quadratic equations: factoring, completing the square, or the quadratic formula.

25. $x^2 + 3x - 5 = 0$ **26.** $x^2 = 7x + 3$ **27.** $x^2 - 5x + 2 = 0$

28. $x^2 + 4x + 3 = 0$ **29.** $2x^2 + 7x + 2 = 0$ **30.** $3x^2 + 2x - 2 = 0$

31. $6x^2 = 5x + 1$ **32.** $4x^2 + x - 4 = 0$ **33.** $3x^2 - 4 = 0$

34. $7x^2 + 6x + 1 = 0$ **35.** $x^3 - 9x^2 + 4x = 0$ **36.** $x^3 - 8x^2 = 3x^2 + 3x$

37. $x^3 + 3x^2 + x = 0$ **38.** $4x^3 + 10x^2 - 3x = 0$ **39.** $x^2 - 3x - 4 = 0$

40. $9x^2 - 6x + 1 = 0$ **41.** $2x^2 + 8x + 9 = 0$ **42.** $3x^2 + 7x - 4 = 0$

43. $x^2 - 7 = 0$ **44.** $3x^2 - 6x + 4 = 0$ **45.** $x^2 + 4x = x - 2x^2$

46. $3x^2 + 4x = 0$ **47.** $5x^2 - 7x + 5 = 0$ **48.** $4x^2 - 5x + 3 = 0$

49. $6x^2 + 2x - 20 = 0$ **50.** $10x^2 + 35x + 30 = 0$ **51.** $4x^2 + 9 = 0$

52. $3x^2 - 8x + 6 = 0$

38. $x=\frac{-5\pm\sqrt{37}}{4},\ x=0$ **39.** $x=-1,\ x=4$ **40.** $x=\frac{1}{3}$ **41.** $x=\frac{-4\pm i\sqrt{2}}{2}$ **42.** $x=\frac{-7\pm\sqrt{97}}{6}$

43. $x=\pm\sqrt{7}$ **44.** $x=\frac{3\pm i\sqrt{3}}{3}$ **45.** $x=0,\ x=-1$ **46.** $x=0,\ x=-\frac{4}{3}$ **47.** $x=\frac{7\pm i\sqrt{51}}{10}$

In Exercises 53 – 60, first multiply each side of the equation by the LCM of the denominator to get integer coefficients and then solve the resulting equation.

53. $3x^2 - 4x + \frac{1}{3} = 0$

54. $\frac{3}{4}x^2 - 2x + \frac{1}{8} = 0$

55. $\frac{3}{7}x^2 - \frac{1}{2}x + 1 = 0$

56. $2x^2 + 3x + \frac{5}{4} = 0$

57. $\frac{1}{2}x^2 - x + \frac{1}{4} = 0$

58. $\frac{2}{3}x^2 - \frac{1}{3}x + \frac{1}{2} = 0$

59. $\frac{1}{4}x^2 + \frac{7}{8}x + \frac{1}{2} = 0$

60. $\frac{5}{12}x^2 - \frac{1}{2}x - \frac{1}{4} = 0$

In Exercises 61 – 68, solve the quadratic equations by using the quadratic formula and your calculator. Write the solutions accurate to 4 decimal places.

61. $0.02x^2 - 1.26x + 3.14 = 0$

62. $0.5x^2 + 0.07x - 5.6 = 0$

63. $\sqrt{2}x^2 - \sqrt{3}x - \sqrt{5} = 0$

64. $x^2 - 2\sqrt{10}x + 10 = 0$

65. $0.3x^2 + \sqrt{2}x + 0.72 = 0$

66. $\sqrt[3]{4}x^2 - \sqrt[4]{2}x - \sqrt{11} = 0$

67. $x^2 + 2\sqrt{15} + 15 = 0$

68. $0.05x^2 - \sqrt{30} = 0$

Writing and Thinking About Mathematics

69. Find an equation of the form $Ax^4 + Bx^2 + C = 0$ that has the four roots ± 2 and ± 3. Explain how you arrived at this equation.

70. The surface area of a circular cylinder can be found with the following formula:

$S = 2\pi r^2 + 2\pi rh$ where r is the radius of the cylinder and h is the height. Estimate the radius of a circular cylinder of height 30 cm and surface area 300 cm^2. Explain how you used your knowledge of quadratic equations.

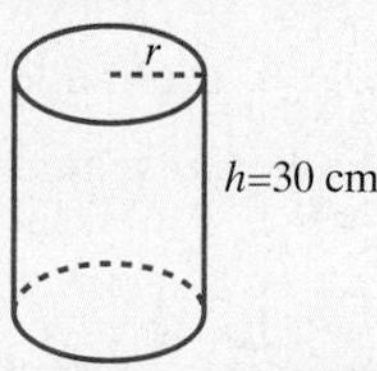

Answers:

48. $x = \frac{5 \pm i\sqrt{23}}{8}$

49. $x = -2,\ x = \frac{5}{3}$

50. $x = -\frac{3}{2},\ x = -2$

51. $x = \pm\frac{3}{2}i$

52. $x = \frac{4 \pm i\sqrt{2}}{3}$

53. $x = \frac{2 \pm \sqrt{3}}{3}$

54. $x = \frac{8 \pm \sqrt{58}}{6}$

55. $x = \frac{7 \pm i\sqrt{287}}{12}$

56. $x = \frac{-3 \pm i}{4}$

57. $x = \frac{2 \pm \sqrt{2}}{2}$

58. $x = \frac{1 \pm i\sqrt{11}}{4}$

59. $x = \frac{-7 \pm \sqrt{17}}{4}$

60. $x = \frac{3 \pm 2\sqrt{6}}{5}$

61. $x = 60.4007, 2.5993$

62. $x = 3.2774, -3.4174$

63. $x = 2.0110, -0.7862$

64. $x = 3.1623$

65. $x = -0.5806, -4.1334$

66. $x = -1.1186, 1.8678$

67. $x = -4.7693i, +4.7693i$

68. $x = -10.4664, 10.4664$

69. $x^4 - 13x^2 + 36$, answers will vary

70. $r \approx 1.515$ cm.

Hawkes Learning Systems: Introductory & Intermediate Algebra

SQE: The Quadratic Formula

11.3 Applications

Objectives

After completing this section, you will be able to:

Solve applied problems by using quadratic equations.

The following Strategy for Solving Word Problems, given in Section 2.3 and again in Section 3.3, is a valid approach to solving word problems at all levels.

Strategy for Solving Word Problems

1. Understand the problem.

- *a. Read the problem carefully. (Read it several times if necessary.)*
- *b. If it helps, restate the problem in your own words.*

2. Devise a plan.

- *a. Decide what is asked for; assign a variable to the unknown quantity. Label this variable so you know exactly what it represents.*
- *b. Draw a diagram or set up a chart whenever possible.*
- *c. Write an equation that relates the information provided.*

3. Carry out the plan.

- *a. Study your picture or diagram for insight into the solution.*
- *b. Solve the equation.*

4. Look back over the results.

- *a. Does your solution make sense in terms of the wording of the problem?*
- *b. Check your solution in the equation.*

The problems in this section can be solved by setting up quadratic equations and then solving these equations by factoring, completing the square, or using the quadratic formula.

The Pythagorean Theorem

The **Pythagorean Theorem** is one of the most interesting and useful ideas in mathematics. We discussed the Pythagorean Theorem in Section 6.5 and do so again here because problems with right triangles often generate quadratic equations.

In a **right triangle**, one of the angles is a right angle (measures 90º), and the side opposite this angle (the longest side) is called the **hypotenuse**. The other two sides are called **legs**.

The Pythagorean Theorem

In a right triangle, the square of the hypotenuse is equal to the sum of the squares of the legs.

$$c^2 = a^2 + b^2$$

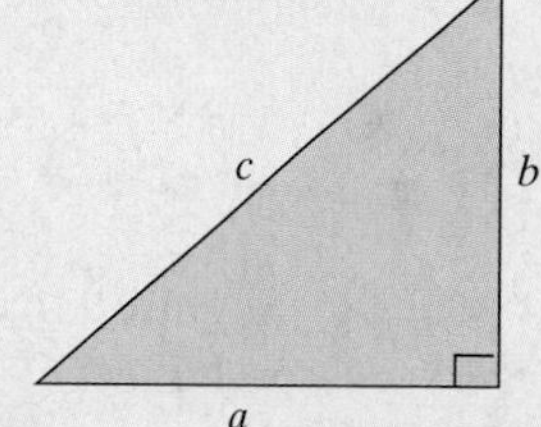

Example 1: The Pythagorean Theorem

Teaching Notes:
Some students may tend to believe that all negative solutions are irrelevant or not applicable. Emphasis on the context and interpretation of any application is highly recommended. Students need to understand that algebraic solutions may or may not be related to the problem that generated the original equation.

The length of a rectangular field is 6 meters more than its width. If the diagonal foot path is 30 meters, what are the dimensions of the field?

Solution: Let w = width
$w + 6$ = length

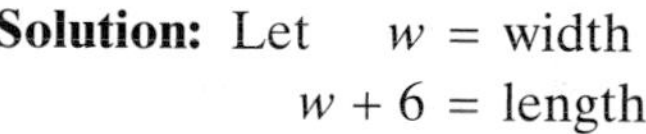

$$(w+6)^2 + w^2 = 30^2 \quad \text{Use the Pythagorean Theorem.}$$

$$w^2 + 12w + 36 + w^2 = 900$$

$$2w^2 + 12w - 864 = 0$$

$$w^2 + 6w - 432 = 0$$

$$(w+24)(w-18) = 0$$

~~$w = -24$~~ or $w = 18$ A negative number does not fit the conditions of the problem.

w = 18 meters
$w + 6$ = 24 meters

The length is 24 meters and the width is 18 meters.

Projectiles

The formula $h = -16t^2 + v_0 t + h_0$ is used in physics and relates to the height of a projectile such as a thrown ball, a bullet, or a rocket.

h = height of object, in feet
t = time object is in the air, in seconds
v_0 = beginning velocity, in feet per second
h_0 = beginning height ($h_0 = 0$ if the object is initially at ground level.)

Example 2: Projectiles

A bullet is fired straight up from ground level with a muzzle velocity of 320 ft. per sec.

a. When will the bullet hit the ground?
b. When will the bullet be 1200 ft. above the ground?

Solution: In this problem, $v_0 = 320$ ft. per sec.
and $h_0 = 0$

a. The bullet hits the ground when $h = 0$.

$$h = -16t^2 + v_0 t + h_0$$

$$0 = -16t^2 + 320t + 0$$

$$0 = t^2 - 20t \qquad \text{Divide both sides by } -16.$$

$$0 = t(t-20) \qquad \text{Factor.}$$

$$t = 0 \quad \text{or} \quad t = 20$$

The bullet hits the ground in 20 seconds. The solution $t = 0$ confirms the fact that the bullet was fired from the ground.

b. Let $h = 1200$

$$1200 = -16t^2 + 320t$$

$$0 = -16t^2 + 320t - 1200$$

$$0 = t^2 - 20t + 75$$

$$0 = (t-5)(t-15)$$

$$t = 5 \quad \text{or} \quad t = 15$$

Both solutions are meaningful. The bullet is at 1200 ft. twice; once in 5 sec. going up and once in 15 sec. coming down.

Example 3: Geometry

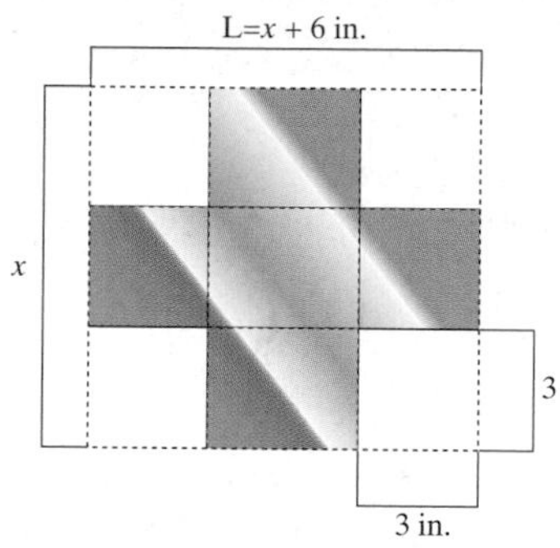

A rectangular sheet of copper is 6 in. longer than it is wide. An open box was made by cutting 3 in. squares at each corner and folding up the sides. If the box has a volume of 336 cu in., what were the dimensions of the sheet of copper?

Solution: Let x = width of copper sheet
then $x + 6$ = length of copper sheet

The volume of the box is equal to 3 times its width times its length.

$[V = lwh]$

$$3x(x-6) = 336$$

$$3x^2 - 18x = 336$$

$$3x^2 - 18x - 336 = 0$$

$$x^2 - 6x - 112 = 0$$

$$(x+8)(x-14) = 0$$

$$\cancel{x=-8} \text{ or } x = 14$$

$$x + 6 = 20$$

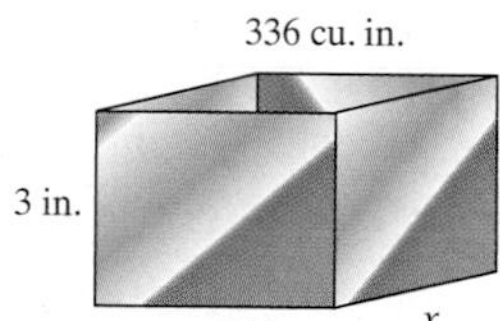

The length of the sheet is 20 in., and the width is 14 in.

Cost Per Person

In the following example, note that the **cost per person** is found by dividing the total cost by the number of people going to the tournament. The cost per person changes because the total cost remains fixed but the number of people changes.

Example 4: Cost Per Person

The members of a bowling club were going to fly commercially to a tournament at a total cost of $2420, which was to be divided equally among the members. At the last minute, two of the members decided to fly their own private planes. The cost to the remaining members increased $11 each. How many members flew commercially?

Solution: Let x = number of club members
then $x - 2$ = number of club members that flew commercially

$$\underbrace{\text{final cost per member}} - \underbrace{\text{initial cost per member}} = \underbrace{\text{difference in cost per member}}$$

$$\frac{2420}{x-2} - \frac{2420}{x} = 11$$

continued on next page ...

$$\cancel{x(x-2)}\frac{2420}{\cancel{x-2}}-\cancel{x}(x-2)\frac{2420}{\cancel{x}}=x(x-2)\cdot 11$$

$$2420x-2420(x-2)=11x(x-2)$$

$$2420x-2420x+4840=11x^2-22x$$

$$0=11x^2-22x-4840$$

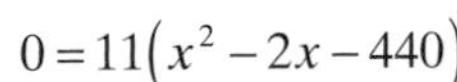

$$0=11\left(x^2-2x-440\right)$$

$$0=11(x-22)(x+20)$$

$$x=22 \text{ or } \cancel{x=-20}$$

$$x-2=20$$

(−20 does not fit the conditions. That is, the number of people in a club is a positive number.)

Check: Final cost per member $=\dfrac{2420}{20}=\$121$

Initial cost per member $=\dfrac{2420}{22}=\$110$

$\$121-\$110=\$11$ Difference in cost per member.

Twenty members flew commercially.

11.3 Exercises

1. $6, 13$
2. $81, -9$ or $64, 8$
3. $4, 14$
4. $7, 17$ or $-\frac{17}{2}, -14$
5. $-6, -11$
6. $1+\sqrt{3}$
7. $-5-4\sqrt{3}$
8. $\frac{3-3\sqrt{2}}{2}$
9. $\frac{1+\sqrt{33}}{4}$
10. $7, 15, 46$
11. $7\sqrt{2}$ cm, $7\sqrt{2}$ cm

1. A positive integer is one more than twice another. Their product is 78. Find the two integers.

2. One number is equal to the square of another. Find the numbers if their sum is 72.

3. Find two positive numbers whose difference is 10 and whose product is 56.

4. One number is three more than twice a second number. Their product is 119. Find the numbers.

5. The sum of two numbers is −17. Their product is 66. Find the numbers.

6. Find a positive real number such that its square is equal to twice the number increased by 2.

7. Find a negative real number such that the square of the sum of the number and 5 is equal to 48.

8. The square of a negative real number is decreased by 2.25 and the result is equal to 3 times the number. What is the number?

9. Twice the square of a positive real number is equal to 4 more than the number. What is the number?

10. The sum of three positive integers is 68. The second is one more than twice the first. The third is three less than the square of the first. Find the integers.

11. A right triangle has two equal sides. The hypotenuse is 14 centimeters. Find the length of the sides.

12. $3\sqrt{5}$ m, $6\sqrt{5}$ m

13. Mel: 30 mph, John: 40 mph

14. 27 meters

15. 4 feet×10 feet

16. 3 inches

17. 5 m × 8 m

18. 32 cm × 12 cm × 4 cm

19. 70

20. 40 seats

21. 3 m × 9 m

22. 17 inches, 6 inches

23. 7 cm×9 cm

24. 5 amperes

25. 2 amperes or 8 amperes

12. The length of one leg of a right triangle is twice the length of the second leg. The hypotenuse is 15 meters. Find the lengths of the two legs.

13. Mel and John leave Desert Point at the same time. Mel drives north and John drives east. Mel's average speed is 10 mph slower than John's. At the end of one hour they are 50 miles apart. Find the average speed of each driver.

14. A flag pole was bent over at a point $\frac{4}{9}$ of the distance from its base to the top. The top of the pole reached a point on the ground 9 meters from the base of the pole. What was the original height of the pole?

15. The length of a rectangle is 2 feet less than three times the width. If the area of the rectangle is 40 square feet, find the dimensions.

16. A picture 9 in. wide and 12 in. long is surrounded by a frame of uniform width. The area of the frame only is 162 sq. in. Find the width of the frame.

17. A rectangle is 3 meters longer than it is wide. If the width is doubled and the length decreased by 4 meters, the area is unchanged. Find the original dimensions.

18. A rectangular piece of cardboard twice as long as it is wide has a small square 4 cm by 4 cm cut from each corner. The edges are then folded up to form an open box with a volume of 1536 cu cm. What are the dimensions of the box? (See Example 3.)

19. An orchard has 2030 trees. The number of trees in each row exceeds twice the number of rows by 12. How many trees are in each row?

20. A rectangular auditorium seats 960 people. The number of seats in each row exceeds the number of rows by 16. Find the number of seats in each row.

21. The perimeter of a rectangle is 24 meters and its area is 27 square meters. Find the dimensions of the rectangle.

22. The area of a rectangle is 102 square inches and the perimeter of the rectangle is 46 inches. Find the length and width.

23. The length of a rectangle is 2 cm greater than its width. If the length and the width are each increased by 3 cm, the area is increased by 57 sq cm. Find the dimensions of the original rectangle.

24. A 40-volt generator with a resistance of 4 ohms delivers power externally of $40I - 4I^2$ watts, where I is the current measured in amperes. Find the current needed for the generator to deliver 100 watts of power.

25. Find the current needed for the 40-volt generator in Exercise 24 to deliver 64 watts of power.

26. 12 signs

27. a. $307.20
b. 45 cents or 35 cents

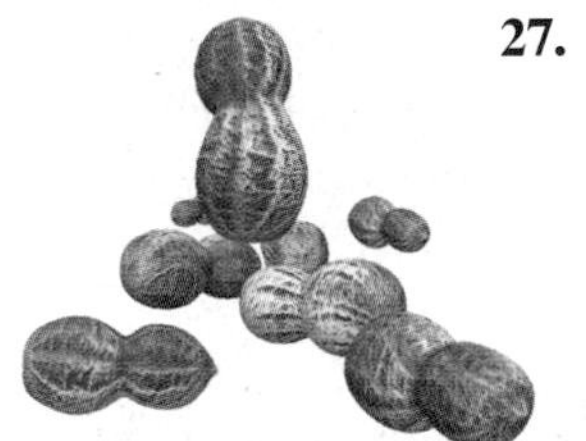

28. $12

29. $13 and $22

30. 12 people

31. 64 mph

32. 6 mph

33. Sam: 7.5 hours, Bob: 12.5 hours

34. 3 hours and 6 hours

35. 150 mph

36. 10 lb. of Grade A, 20 lb. of Grade B

26. Vince operates a small sign-making business. He finds that if he charges x dollars for each sign, he sells $40 - x$ signs per week. What is the least number of signs he can sell to have an income of $336 in one week?

27. Sam operates a peanut stand. He estimates that he can sell 600 bags of peanuts per day if he charges 50¢ for each bag. He determines that he can sell 20 more bags for each 1¢ reduction in price.
a. What would his revenue be if he charged 48¢ per bag?
b. What should he charge in order to have receipts of $315?

28. It costs Ms. Snow $3 to build a picture frame. She estimates that if she charges x dollars each, she can sell $60 - x$ frames per week. What is the lowest price necessary to make a profit of $432 each week?

29. J.B. bought some shirts and pants. He bought two more pairs of pants than shirts. He spent $154 on pants and $65 for shirts. Find the price of each type of clothing if the price of a pair of pants exceeds the price of a shirt by $9.

30. The Piton Rock Climbing Club planned a climbing expedition. The total cost was $900, which was to be divided equally among the members going. While practicing, three members fell and were hurt so they were unable to go. If the cost per person increased by $15, how many people went on the expedition?

31. Mark traveled 240 miles to a convention. Later, his wife Ann drove up to meet him. Ann's average speed exceeded Mark's by 4 mph, and the trip took her 15 minutes less time. Find Ann's speed.

32. In two hours, a motorboat can travel 8 miles down a river and return 4 miles back. If the river flows at a rate of 2 miles per hour, how fast can the boat travel in still water?

33. It takes Bob 5 hours longer to assemble a machine than it does Sam. If Bob only works for as long as it takes Sam to assemble a machine (and therefore Bob has not finished his job), Sam can complete the rest of Bob's assembly in 3 hours. How long would it take each man working alone to assemble the machine? (**Hint:** Represent the total job by the number 1.)

34. Two employees together can prepare a large order in 2 hours. Working alone, one employee takes three hours longer than the other. How long does it take each person working alone?

35. Fern can fly her plane 240 miles against the wind in the same time it takes her to fly 360 miles with the wind. The speed of the plane in still air is 30 mph more than four times the speed of the wind. Find the speed of the plane in still air.

36. A grocer mixes $9.00 worth of Grade A coffee with $12.00 worth of Grade B coffee to obtain 30 pounds of a blend. If Grade A costs 30¢ a pound more than Grade B, how many pounds of each were used?

37. 190 reserved, 150 general

38. 12 hours

39. a. 6.75 sec.
b. 3 sec., 3.75 sec.

40. a. 10 sec.
b. 5 sec.

41. a. 3.5 sec.
b. 144 ft.

42. a. 7 sec.
b. 2 sec., 12 sec.

43. 14.1 cm

44. 10.9087 inches

37. The Andersonville Little Theater Group sold 340 tickets to their spring production. Receipts from the sale of reserved tickets were \$855. Receipts from general admission tickets were \$375. How many of each type ticket were sold if the cost of a reserved ticket is \$2 more than a general admission ticket?

38. It takes a young man 2 hours longer to build a wall than it does his father. After the son has worked for 1 hour, his father joins him and they finish the job together in 6 more hours. How long would it take the father working alone?

In Exercises 39 – 42, use the formula $h = -16t^2 + v_0t + h_0$.

39. A ball is thrown vertically from ground level with an initial speed of 108 ft. per sec.
a. When will the ball hit the ground?
b. When will the ball be 180 ft. above the ground?

40. A ball is thrown vertically from the ground with an initial speed of 160 ft. per sec.
a. When will the ball strike the ground?
b. When will the ball be 400 ft. above the ground?

41. A stone is dropped from a platform 196 ft. high.
a. When will it hit the ground?
b. How far will it fall during the third second of time? **Hint:** Since the stone is dropped, $v_0 = 0$.

42. An arrow is shot vertically upward from a platform 40 ft. high at a rate of 224 ft. per sec.
a. When will the arrow be 824 ft. above the ground?
b. When will it be 424 ft. above the ground?

43. If a triangle is inscribed in a circle so that one side of the triangle is a diameter of the circle, the triangle will be a right triangle (every time). If an isosceles triangle (two sides equal) is inscribed in this manner in a circle with diameter 20 cm, find the length of the two equal sides to the nearest tenth of a centimeter.

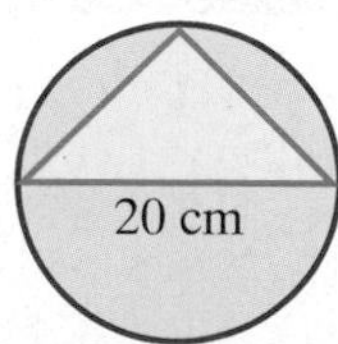

44. If a triangle is inscribed in a semicircle such that one side of the triangle is the diameter of a circle with radius 6 in., and one side of the triangle is 5 in., what is the length of the third side? (See Exercise 43.)

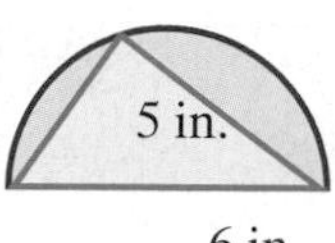

In Exercises 45 – 50, use your calculator to find the answers accurate to two decimal places.

45. a. 94.20 ft., 706.50 ft.^{2}
b. 84.84 ft., 449.86 ft.^{2}

46. a. 127.28 ft.
b. 127.28 ft.

47. a. No
b. Home plate
c. No

48. 855.86 ft.

49. 8.49 cm

50. 2.24 miles

51.
$$ax^2+bx+c=0$$
$$ax^2+bx=-c$$
$$x^2+\frac{b}{a}x=-\frac{c}{a}$$
$$\left(x^2+\frac{b}{a}x+\frac{b^2}{4a^2}\right)=-\frac{c}{a}+\frac{b^2}{4a^2}$$
$$\left(x+\frac{b}{2a}\right)^2=-\frac{c}{a}+\frac{b^2}{4a^2}$$
$$x+\frac{b}{2a}=\pm\sqrt{-\frac{c}{a}+\frac{b^2}{4a^2}}$$
$$x=-\frac{b}{2a}\pm\frac{\sqrt{-4ac+b^2}}{2a}$$
$$x=\frac{-b\pm\sqrt{b^2-4ac}}{2a}$$

45. A square is said to be inscribed in a circle if each corner of the square lies on the circle. (Use $\pi \approx 3.14$.)

a. Find the circumference and area of a circle with diameter 30 feet.

b. Find the perimeter and area of a square inscribed in the circle.

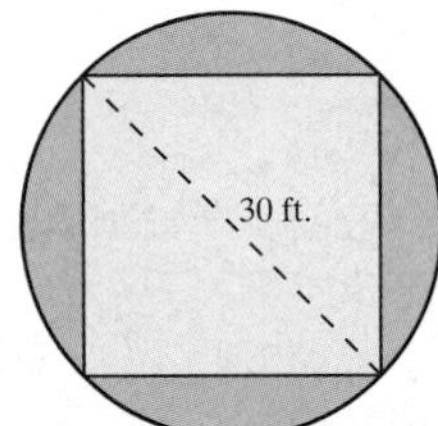

46. The shape of a baseball infield is a square with sides 90 feet long.

a. Find the distance (to the nearest hundredth of a foot) from home plate to second base.

b. Find the distance (to the nearest hundredth of a foot) from first base to third base.

47. The distance from home plate to the pitcher's mound for the field in question 46 is 60.5 feet.

a. Is the pitcher's mound exactly half way between home plate and second base?

b. If not, which base is it closer to, home plate or second base?

c. Do the two diagonals of the square intersect at the pitcher's mound?

48. The GE Building in New York is 850 feet tall (70 stories). At a certain time of day, the building casts a shadow 100 feet long. Find the distance from the top of the building to the tip of the shadow (to the nearest hundredth of a foot).

49. To create a square inside a square, a quilting pattern requires four triangular pieces like the one shaded in the figure shown here. If the square in the center measures 12 centimeters on a side, and the two legs of each traingle are of equal length, how long are the legs of each triangle, to the nearest hundredth of a centimeter?

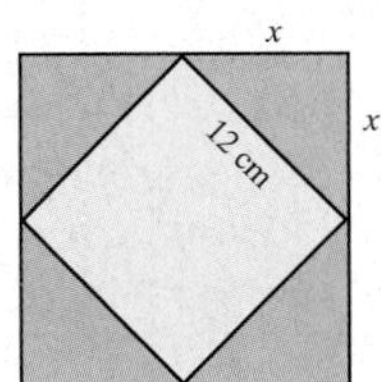

50. If an airplane passes directly over your head at an altitude of 1 mile, how far (to the nearest hundredth of a mile) is the airplane from your position after it has flown 2 miles farther at the same altitude?

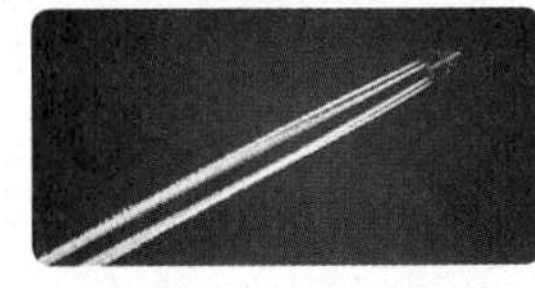

Writing and Thinking About Mathematics

51. Develop the Quadratic Formula by using the technique of completing the square with the general quadratic equation $ax^2 + bx + c = 0$.

Hawkes Learning Systems: Introductory & Intermediate Algebra

Applications: Quadratic

11.4 Equations with Radicals

Objectives

After completing this section, you will be able to:

Solve equations that contain one or more radical expressions.

Teaching Notes:
Up to this point in algebra, students have solved equations by creating a series of equivalent equations. You might want to include a discussion about the procedures of this section that introduce additional variable factors into the series of equations, thereby creating equations that may not be entirely equivalent.

Each of the following equations involves at least one radical expression:

$$x+3=\sqrt{x+5} \qquad \sqrt{x}-\sqrt{2x-14}=1 \qquad \sqrt[3]{x+1}=5$$

If the radicals are square roots, we solve by squaring both sides of the equations. If the radical is some other root and this root can be isolated on one side of the equation, we solve by raising both sides of the equation to the integer power corresponding to the root. For example, with a cube root both sides are raised to the third power.

Squaring both sides of an equation may introduce new solutions. For example, the first-degree equation $x=-3$ has only one solution, namely, -3. However, squaring both sides gives the quadratic equation

$$x^2=(-3)^2 \qquad \text{or} \qquad x^2=9.$$

The quadratic equation $x^2=9$ has two solutions, 3 and -3. Thus, a new solution that is not a solution to the original equation has been introduced. Such a solution is called an **extraneous solution**.

When both sides of an equation are raised to a power, an extraneous solution may be introduced. Be sure to check all solutions in the original equation.

The following examples illustrate a variety of situations involving radicals. The steps used are related to the following general method.

Method for Solving Equations with Radicals

Step 1: *Isolate one of the radicals on one side of the equation. (An equation may have more than one radical.)*

Step 2: *Raise both sides of the equation to the power corresponding to the index of the radical.*

Step 3: *If the equation still contains a radical, repeat Steps 1 and 2.*

Step 4: *Solve the equation after all the radicals have been eliminated.*

Step 5: *Be sure to check all possible solutions in the original equation and eliminate any extraneous solutions.*

19. $x = 4$
20. $x = 8$
21. $x = 0$
22. $x = 9$
23. $x = 5$
24. $x = -3$
25. No solution
26. $x = 5$
27. $x = 4$

34. $\sqrt{2x-5} - 2 = \sqrt{x-2}$ **35.** $\sqrt{2x-3} + \sqrt{x+3} = 6$ **36.** $\sqrt{2x+3} - \sqrt{x+5} = 1$

37. $\sqrt[3]{4+3x} = -2$ **38.** $\sqrt[3]{2+9x} = 9$ **39.** $\sqrt[3]{5x+4} = 4$

40. $\sqrt[3]{7x+1} = -5$ **41.** $\sqrt{2x+1} = -4$ **42.** $\sqrt{3x-5} = -2$

43. $\sqrt[4]{2x+1} = 3$ **44.** $\sqrt[4]{x-6} = 2$

Writing and Thinking About Mathematics

45. Explain, in your own words, why in general $(a+b)^2 \neq a^2 + b^2$.

Hawkes Learning Systems: Introductory & Intermediate Algebra

Solving Radical Equations

28. No solution **29.** $x = -1,\ x = 3$ **30.** $x = 4$ **31.** $x = 5$ **32.** $x = 4$

33. $x = 2$ **34.** $x = 27$ **35.** $x = 6$ **36.** $x = 11$ **37.** $x = -4$

38. $x = \frac{727}{9}$ **39.** $x = 12$ **40.** $x = -18$ **41.** No solution **42.** No solution

43. $x = 40$ **44.** $x = 22$ **45.** $(a+b)^2 = (a+b)(a+b)$
$= a^2 + 2ab + b^2$
$\neq a^2 + b^2$

11.5 Equations in Quadratic Form

Objectives

After completing this section, you will be able to:

1. *Solve the equations that can be written in quadratic form by appropriate substitutions.*
2. *Solve equations that contain rational expressions.*

Teaching Notes:
This technique might be introduced to students as a next step in the effort to solve a special family of higher order equations. Point out that solving equations of higher order requires various techniques and many have been around for hundreds of years.

The general quadratic equation is $ax^2 + bx + c = 0$, where $a \neq 0$.

The equations

$$x^4 - 7x^2 + 12 = 0 \quad \text{and} \quad x^{\frac{2}{3}} - 4x^{\frac{1}{3}} - 21 = 0$$

are **not** quadratic equations, but they are in **quadratic form** because the degree of the middle term is one-half the degree of the first term. Specifically,

$$\frac{1}{2}(4) = 2 \quad \text{and} \quad \frac{1}{2}\left(\frac{2}{3}\right) = \frac{1}{3}$$

first term exponent, middle term exponent; first term exponent, middle term exponent

Equations in quadratic form can be solved by using the quadratic formula or by factoring just as if they were quadratic equations. In each case, a substitution can be made to clarify the problem. Try to follow these suggestions:

Solving Equations in Quadratic Form by Substitution

1. *Look at the middle term.*
2. *Substitute a first-degree variable, say, u, for the variable expression in the middle term.*
3. *Substitute the square of this variable, u^2, for the variable expression in the first term.*
4. *Solve the resulting quadratic equation for u.*
5. *Substitute the results "back" for u in the beginning substitution and solve for the original variable.*

The following examples illustrate how such a substitution may help. Study these examples carefully and note the variety of algebraic manipulations used.

Example 1: Substitution

Solve the following equations. Equations a - d are in quadratic form, and a substitution will help.

Teaching Notes: You may want to discuss solving Examples 1a - 1d by factoring directly without substitution. This will probably depend on the ability level and mathematical sophistication of your students.

a. $x^4 - 7x^2 + 12 = 0$

Solution: $x^4 - 7x^2 + 12 = 0$

$u^2 - 7u + 12 = 0$ Substitute $u = x^2$ and $u^2 = x^4$.

$(u-3)(u-4) = 0$ Solve for u by factoring.

$u = 3$ or $u = 4$

$x^2 = 3$ or $x^2 = 4$ Now substitute back: x^2 for u.

$x = \pm\sqrt{3}$ $x = \pm 2$ Solve quadratic equations for x.

There are four solutions: $\sqrt{3}, -\sqrt{3}$, 2, and -2.

b. $x^{\frac{2}{3}} - 4x^{\frac{1}{3}} - 21 = 0$

Solution: $x^{\frac{2}{3}} - 4x^{\frac{1}{3}} - 21 = 0$

$u^2 - 4u - 21 = 0$ Let $u = x^{\frac{1}{3}}$ and $u^2 = x^{\frac{2}{3}}$.

$(u-7)(u+3) = 0$ Solve for u by factoring.

$u = 7$ or $u = -3$

$x^{\frac{1}{3}} = 7$ or $x^{\frac{1}{3}} = -3$ Substitute back: $x^{\frac{1}{3}}$ for u.

$\left(x^{\frac{1}{3}}\right)^3 = 7^3$ or $\left(x^{\frac{1}{3}}\right)^3 = (-3)^3$ Cube both sides.

$x = 343$ $x = -27$

There are two solutions: 343 and –27.

c. $x^{-4} - 7x^{-2} + 10 = 0$

Solution: $x^{-4} - 7x^{-2} + 10 = 0$

$u^2 - 7u + 10 = 0$ Let $u = x^{-2}$ and $u^2 = x^{-4}$.

$(u-2)(u-5) = 0$ Solve for u by factoring.

$u = 2$ or $u = 5$

$x^{-2} = 2$ $x^{-2} = 5$ Substitute back: x^{-2} for u.

$\frac{1}{x^2} = 2$ $\frac{1}{x^2} = 5$ Remember $x^{-2} = \frac{1}{x^2}$.

$x^2 = \frac{1}{2}$ $x^2 = \frac{1}{5}$ Reciprocals.

$x = \pm\sqrt{\frac{1}{2}}$ $x = \pm\sqrt{\frac{1}{5}}$

$x = \pm\frac{1}{\sqrt{2}}$ $x = \pm\frac{1}{\sqrt{5}}$

Rationalizing denominators, we have $x = \pm\frac{\sqrt{2}}{2}$ $\quad x = \pm\frac{\sqrt{5}}{5}$

There are four solutions: $\frac{\sqrt{2}}{2}, \frac{-\sqrt{2}}{2}, \frac{\sqrt{5}}{5}, \frac{-\sqrt{5}}{5}$.

d. $(x+2)^2 - (x+2) - 12 = 0$

Solution: $(x+2)^2 - (x+2) - 12 = 0$

$u^2 - u - 12 = 0$ Let $u = x + 2$.

$(u-4)(u+3) = 0$ Solve for u by factoring.

$u = 4$ or $u = -3$

$x + 2 = 4$ $\quad x + 2 = -3$ Substitute back: $x + 2$ for u.

$x = 2$ $\quad x = -5$

There are two solutions: 2 and –5.

e. $x^5 - 16x = 0$

Solution: $x^5 - 16x = 0$

$x(x^4 - 16) = 0$ Factor out the common term x.

$x(x^2+4)(x^2-4) = 0$ Factor the difference of two squares.

$x = 0$ or $x^2 = -4$ or $x^2 = 4$

$x = \pm 2i$ $\quad x = \pm 2$

There are five solutions: $0, 2i, -2i, 2, -2$.

f. $\frac{2}{3x-1} + \frac{1}{x+1} = \frac{x}{x+1}$

Solution: This equation is not in quadratic form. However, multiplying both sides of the equation by the LCM of the denominators gives a quadratic equation.

$$(x+1)\cancel{(3x-1)}\cdot\frac{2}{\cancel{3x-1}} + \cancel{(x+1)}(3x-1)\cdot\frac{1}{\cancel{x+1}} = \cancel{(x+1)}(3x-1)\cdot\frac{x}{\cancel{x+1}}$$

$$2(x+1) + 3x - 1 = (3x-1)x$$

$$2x + 2 + 3x - 1 = 3x^2 - x$$

$$0 = 3x^2 - 6x - 1$$

$$x = \frac{6 \pm \sqrt{36 - 4\cdot 3(-1)}}{6}$$

$$= \frac{6 \pm \sqrt{48}}{6}$$

$$= \frac{6 \pm 4\sqrt{3}}{6}$$

$$= \frac{3 \pm 2\sqrt{3}}{3}$$

1. $x = \pm 2,\ x = \pm 3$
2. $x = \pm 2,\ x = \pm 5$
3. $x = \pm 2, x = \pm\sqrt{5}$
4. $y = \pm 3,\ y = \pm\sqrt{2}$
5. $y = \pm\sqrt{7}, y = \pm 2i$
6. $y = \pm\sqrt{3},\ y = \pm 2i$

Practice Problems

Solve the following equations.

1. $x - x^{\frac{1}{2}} - 2 = 0$

   (Let $u = x^{\frac{1}{2}}$ and $u^2 = x$.)

2. $x^4 + 16x^2 = -48$

   (Let $u = x^2$ and $u^2 = x^4$.)

3. $\dfrac{3(x-2)}{x-1} = \dfrac{2(x+1)}{x-2} + 2$

11.5 Exercises

7. $y = \pm\sqrt{5}, y = \pm i\sqrt{5}$
8. $x = \frac{1}{5}, x = \frac{1}{7}$
9. $z = \frac{1}{6}, z = -\frac{1}{4}$
10. $x = \pm\frac{5}{2}i, x = 0$
11. $x = 4, x = \frac{25}{4}$
12. $x = \frac{1}{4}, x = 1$
13. $x = 1, x = 4$
14. $y = 1, y = 9$
15. $x = \frac{1}{8}, x = -8$
16. $x = -\frac{1}{8}, x = \frac{8}{27}$
17. $x = \frac{1}{25}$
18. $x = \frac{2}{3}, x = \frac{1}{2}$
19. $x = -\frac{1}{3}, x = \frac{3}{8}$
20. $y = 9, y = \frac{1}{4}$
21. $x = 0, x = -27, x = -8$
22. $x = 0, x = 64, x = 49$
23. $x = 1, x = 2$
24. $x = 3, x = -2$
25. $x = -3, x = -\frac{7}{2}$
26. $x = 0, x = \frac{6}{5}$

Solve the following equations.

1. $x^4 - 13x^2 + 36 = 0$
2. $x^4 - 29x^2 + 100 = 0$
3. $x^4 - 9x^2 + 20 = 0$
4. $y^4 - 11y^2 + 18 = 0$
5. $y^4 - 3y^2 - 28 = 0$
6. $y^4 + y^2 - 12 = 0$
7. $y^4 - 25 = 0$
8. $x^{-2} - 12x^{-1} + 35 = 0$
9. $z^{-2} - 2z^{-1} - 24 = 0$
10. $16x^3 + 100x = 0$
11. $2x - 9x^{\frac{1}{2}} + 10 = 0$
12. $2x - 3x^{\frac{1}{2}} + 1 = 0$
13. $x^3 - 9x^{\frac{3}{2}} + 8 = 0$
14. $y^3 - 28y^{\frac{3}{2}} + 27 = 0$
15. $2x^{\frac{2}{3}} + 3x^{\frac{1}{3}} - 2 = 0$
16. $2x^{\frac{-2}{3}} + x^{\frac{-1}{3}} - 6 = 0$
17. $x^{-1} + 5x^{\frac{-1}{2}} - 50 = 0$
18. $2x^{-2} - 7x^{-1} + 6 = 0$
19. $3x^{-2} + x^{-1} - 24 = 0$
20. $3y^{-1} - 7y^{\frac{-1}{2}} + 2 = 0$
21. $3x^{\frac{5}{3}} + 15x^{\frac{4}{3}} + 18x = 0$
22. $2x^2 - 30x^{\frac{3}{2}} + 112x = 0$
23. $(3x - 5)^2 + (3x - 5) - 2 = 0$
24. $(x - 1)^2 + (x - 1) - 6 = 0$
25. $(2x + 3)^2 + 7(2x + 3) + 12 = 0$
26. $(5x - 4)^2 + 2(5x - 4) - 8 = 0$
27. $(x - 3)^2 - 2(x - 3) - 15 = 0$
28. $(x + 4)^2 - 2(x + 4) = 3$
29. $(2x + 1)^2 + (2x + 1) = 0$
30. $(x + 7)^2 + 5(x + 7) = 50$
31. $x^4 - 2x^2 + 2 = 0$
32. $x^4 - 4x^2 + 5 = 0$
33. $x^4 - 2x^2 + 10 = 0$
34. $x^4 + 16 = 0$
35. $x^4 - 4x^2 + 7 = 0$
36. $x^4 - 6x^2 + 11 = 0$
37. $x^{-4} - 6x^{-2} + 5 = 0$
38. $3x^{-4} - 5x^{-2} + 2 = 0$
39. $3x^{-4} + 25x^{-2} - 18 = 0$
40. $2x^{-4} + 3x^{-2} - 20 = 0$
41. $\dfrac{2}{4x-1} + \dfrac{1}{x+1} = \dfrac{-x}{x+1}$
42. $\dfrac{3x-2}{15} - \dfrac{16-3x}{x+6} = \dfrac{x+3}{5}$
43. $\dfrac{2x}{x-4} - \dfrac{12x}{x^2+x-20} = \dfrac{x-1}{x+5}$
44. $\dfrac{x+1}{x+3} + \dfrac{2x-1}{x-2} = \dfrac{12x-2}{x^2+x-6}$
45. $\dfrac{x+5}{3x+2} - \dfrac{4-2x}{3x^2+8x+4} = \dfrac{x+4}{x+2}$
46. $\dfrac{x+5}{3x+4} + \dfrac{16x^2+5x+6}{3x^2-2x-8} = \dfrac{4x}{x-2}$
47. $\dfrac{4x+1}{x-6} - \dfrac{3x^2-8x+20}{2x^2-13x+6} = \dfrac{3x+7}{2x-1}$

Answers to Practice Problems: **1.** $x = 4$ **2.** $x = \pm 2i, x = \pm 2i\sqrt{3}$ **3.** $x = -3 \pm \sqrt{19}$

48. $\frac{3x+2}{x+3}+\frac{22x-31}{x^2-x-12}=\frac{3(x+4)}{x+3}$

49. $\frac{5(x-10)}{x-7}=\frac{2(x+1)}{x-4}+3$

50. $2+\frac{2-x}{x+2}=\frac{x-3}{x+5}$

Writing and Thinking About Mathematics

51. Consider the following equation: $x-x^{\frac{1}{2}}-6=0$
In your own words, explain why, even though it is in quadratic form, this equation has only one solution.

Hawkes Learning Systems: Introductory & Intermediate Algebra

Equations in Quadratic Form

---

27. $x=0, x=8$
28. $x=-1, x=-5$
29. $x=-1, x=-\frac{1}{2}$
30. $x=-17, x=-2$
31. $x=\pm\sqrt{1+i}, x=\pm\sqrt{1-i}$ **32.** $x=\pm\sqrt{2+i}, x=\pm\sqrt{2-i}$ **33.** $x=\pm\sqrt{1+3i}, x=\pm\sqrt{1-3i}$ **34.** $x=\pm 2\sqrt{i}, x=\pm 2i\sqrt{i}$
35. $x=\pm\sqrt{2\pm i\sqrt{3}}$
36. $x=\pm\sqrt{3\pm i\sqrt{2}}$
37. $x=\pm\frac{1}{5}\sqrt{5}, x=\pm 1$
38. $x=\pm\frac{1}{2}\sqrt{6}, x=\pm 1$
39. $x=\pm\frac{1}{2}\sqrt{6}, x=\pm\frac{1}{3}i$ **40.** $x=\pm\frac{1}{2}i, x=\pm\frac{1}{5}\sqrt{10}$ **41.** $x=-\frac{1}{4}$ **42.** $x=9$ **43.** $x=-4, x=1$ **44.** $x=-\frac{1}{3}, x=3$
45. $x=-\frac{1}{2}$ **46.** $x=-\frac{2}{5}$ **47.** $x=-7, x=-\frac{3}{2}$ **48.** $x=-\frac{3}{4}$ **49.** $x=\frac{26}{5}$ **50.** $x=-3$ **51.** Because of the square root, the solution must be positive.

Chapter 11 Index of Key Ideas and Terms

Section 11.1 Quadratic Equations: Completing the Square

Zero Factor Property page 748

If the product of two factors is 0, then one or both of the factors must be 0. Symbolically, for factors a and b,

if $a \cdot b = 0$, then $a = 0$ or $b = 0$.

Quadratic Equation page 749

An equation that can be written in the form

$ax^2 + bx + c = 0$ where a, b, and c are real numbers and $a \neq 0$

is called a **quadratic equation.**

To Solve an Equation by Factoring page 749

1. Add or subtract terms so that **one side of the equation is 0**.
2. Factor the polynomial expression.
3. Set each factor equal to 0 and solve each of the resulting equations. (**Note:** If two of the factors are the same, then the solution is said to be a **double root** or a **root of multiplicity two.**)

Square Root Property page 751

If $x^2 = c$, then $x = \pm\sqrt{c}$.

If $(x-a)^2 = c$, then $x - a = \pm\sqrt{c}$ (or $x = a \pm \sqrt{c}$).

To Solve a Quadratic Equation by Completing the Square page 753

1. If necessary, divide or multiply both sides of the equation so that the leading coefficient (the coefficient of x^2) is 1.
2. If necessary, isolate the constant term on one side of the equation.
3. Find the constant that completes the square of the polynomial and add this constant to both sides.
4. Use the Square Root Property to find the solutions of the equation.

Writing Equations with Known Roots pages 755 - 756

Section 11.2 Quadratic Equations: The Quadratic Formula

Quadratic Formula page 759

For the general quadratic equation $ax^2 + bx + c = 0$ where $a \neq 0$

the solutions are $\boldsymbol{x = \frac{-b \pm \sqrt{b^2 - 4ac}}{2a}}$.

Discriminant pages 761 - 763

The expression $\boldsymbol{b^2 - 4ac}$, that part of the quadratic formula that lies under the radical sign, is called the **discriminant**.

If $b^2 - 4ac > 0$, there are two real solutions.
If $b^2 - 4ac = 0$, there is one real solution.
If $b^2 - 4ac < 0$, there are two nonreal solutions.

Section 11.3 Applications

Applications

Pythagorean Theorem pages 766 - 767
Projectiles page 768
Geometry page 769
Cost Per Person pages 769 - 770

The Pythagorean Theorem page 767

In a right triangle, the square of the hypotenuse is equal to the sum of the squares of the legs.

Section 11.4 Equations with Radicals

Equations with Radicals page 775

Method for Solving Equations with Radicals

Step 1: Isolate one of the radicals on one side of the equation. (An equation may have more than one radical.)

Step 2: Raise both sides of the equation to the power corresponding to the index of the radical.

Step 3: If the equation still contains a radical, repeat Steps 1 and 2.

Step 4: Solve the equation after all the radicals have been eliminated.

Step 5: Be sure to check all possible solutions in the original equation and eliminate any extraneous solutions.

Section 11.5 Equations in Quadratic Form

Equations in Quadratic Form page 781

Solving Equations in Quadratic Form by Substitution

1. Look at the middle term.
2. Substitute a first-degree variable, say, u, for the variable expression in the middle term.
3. Substitute the square of this variable, u^2, for the variable expression in the first term.
4. Solve the resulting quadratic equation for u.
5. Substitute the results "back" for u in the beginning substitution and solve for the original variable.

Chapter 11 Review

For a review of the topics and problems from Chapter 11, look at the following lessons from *Hawkes Learning Systems: Introductory & Intermediate Algebra.*

SQE: The Square Root Method
SQE: Completing the Square
SQE: The Quadratic Formula
Applications: Quadratic
Solving Radical Equations
Equations in Quadratic Form

Chapter 11 Test

1. a. ± 4
b. $\pm 4i$

2. $x = -\frac{1}{2}, x = 0$

3. a. $x^2 - 30x + \underline{225} = (x-15)^2$

b. $x^2 + 5x + \frac{25}{4} = \left(x + \frac{5}{2}\right)^2$

4. a. $x^2 + 8 = 0$
b. $x^2 - 2x - 4 = 0$

5. $x = -2 \pm \sqrt{3}$

6. 73, two real solutions

7. $k = \pm 2\sqrt{6}$

8. $x = \frac{-1 \pm i\sqrt{7}}{4}$

9. $x = \frac{3 \pm \sqrt{41}}{4}$

10. $x = \frac{2 \pm i\sqrt{2}}{2}$

11. $x = 1$

12. $x = 0,\ x = 36$

13. $x = \pm 1,\ x = \pm 3$

14. $x = \frac{3}{2},\ x = -1$

15. $x = 0,\ x = 1$

16. Triangle is not a right triangle because $6^2 + 8^2 \neq 11^2$.

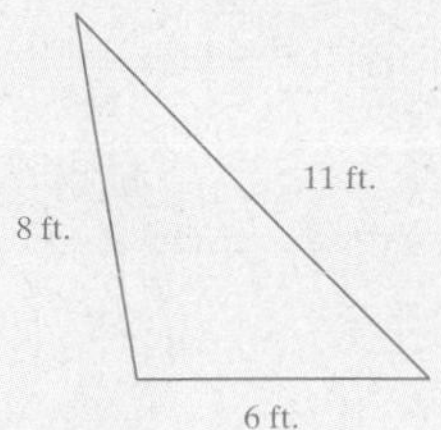

Solve the equations in Exercises 1 and 2 by factoring.

1. a. $x^2 - 16 = 0$
b. $x^2 + 16 = 0$

2. $4x^3 = -4x^2 - x$

3. Add the constant that will complete the square in each expression and write the new expression as the square of a binomial.

a. $x^2 - 30x + \_\_\_\_ = (\quad)^2$ **b.** $x^2 + 5x + \_\_\_\_ = (\quad)^2$

4. Write a quadratic equation in the form $ax^2 + bx + c = 0$ that

a. has the two numbers $\pm 2i\sqrt{2}$ as roots.

b. has the two numbers $1 \pm \sqrt{5}$ as roots.

5. Solve the following equation by completing the square. Show all the steps.

$$x^2 + 4x + 1 = 0$$

6. What is the discriminant of the quadratic equation $4x^2 + 5x - 3 = 0$? Without finding the roots, tell how many roots the equation has and what type of number they are.

7. By using the discriminant, determine the values for k so that the equation $2x^2 - kx + 3 = 0$ will have exactly one real root.

Solve the equations in Exercises 8 – 15 by using any method.

8. $2x^2 + x + 1 = 0$ **9.** $2x^2 - 3x - 4 = 0$ **10.** $2x^2 + 3 = 4x$

11. $\sqrt{x+8} - 2 = x$ **12.** $\sqrt{2x+9} = \sqrt{x} + 3$ **13.** $x^4 = 10x^2 - 9$

14. $3x^{-2} + x^{-1} - 2 = 0$ **15.** $\frac{2x}{x-3} - \frac{2}{x-2} = 1$

16. Determine whether a triangle with sides of 6 ft., 8 ft., and 11 ft. is a right triangle. Explain your answer in detail. Sketch a graph of the triangle and label the sides.

17. A person standing at the edge of a cliff 112 ft. above the beach throws a ball into the air with a velocity of 96 ft. per sec. (Use the formula $h = -16t^2 + v_0t + h_0$.)

a. When will the ball hit the beach?

b. When will the ball be 64 ft. above the beach?

18. The length of a rectangle is 4 inches longer than the width. If the diagonal is 20 inches long, what are the dimensions of the rectangle?

19. Sandy made a business trip to a city 200 miles away and then returned home. Her average speed on the return trip was 10 mph less than her average speed going. If her total travel time was 9 hours, what was her average rate in each direction?

17. a. $t = 7$ seconds
b. $t = 6.464$ seconds
18. 12 inches, 16 inches
19. towards home 50 mph, away 40 mph
20. 0.0714, -175.0714
21. Height = 20 ft.
Base = 10 ft.
22. Length = 8 ft.
Brace = 10 ft.
23. 4.5 seconds
24. 21 members

20. Use your calculator and your knowledge of the quadratic formula to estimate the solutions to the following quadratic equation: $0.02x^2 + 3.5x - 0.25 = 0$

21. The sail of a sailboat is triangular shaped. Its base is 10 feet less than its height and it has an area of 100 ft.^{2}. What are the base and height of the sailboat?

22. A diagonal brace supports the length of a rectangular dining room table. The width of the table is 6 feet and the length of the diagonal brace is 2 feet longer than the length of the table. Find the length of the dining room table and the length of the brace.

23. A ball is thrown upward with a velocity of 40 feet per second from the top of a 144-foot building. How long after being thrown will the ball hit the ground?

24. The members of a white water rafting club decided to go on a rafting trip that costs a total of $2352, which was to be divided equally among the members. At the last minute 3 members dropped out of the trip. The cost to the remaining members increased by $14 each. How many members are to go on the trip?

Cumulative Review: Chapters 1 – 11

1. $x = 6$ 2. $x = \frac{7}{8}$

3. $\left[-\frac{19}{2}, \infty\right)$

$-\frac{19}{2}$

4. $\left(-\infty, \frac{7}{5}\right]$

$\frac{7}{5}$

5. $y = -\frac{3}{2}x + 3$

6. $y = -\frac{3}{2}x + 10$

7. $\frac{1}{x^5}$ 8. x^3y^6

9. $\frac{9y^{\frac{1}{2}}}{x^2}$ 10. $\frac{27x^2}{8y}$

11. $-3x^2y^2\sqrt[3]{y^2}$

12. $2x^2y^3\sqrt[4]{2xy^3}$

13. $\frac{9}{2}\sqrt{2}$ 14. $\frac{4}{3}\sqrt{3}$

Solve the equations in Exercises 1 and 2.

1. $7(2x - 5) = 5(x + 3) + 4$

2. $(2x + 1)(x - 4) = (2x - 3)(x + 6)$

Solve the inequalities in Exercises 3 and 4 and graph the solutions. Write the solutions in interval notation.

3. $4(x+3)-1 \geq 2(x-4)$

4. $\frac{7}{2}x + 3 \leq x + \frac{13}{2}$

Solve for the indicated variable in Exercises 5 and 6.

5. $3x + 2y = 6$ for y

6. $\frac{3}{4}x + \frac{1}{2}y = 5$ for y

Simplify Exercises 7 – 14. Assume that all variables are positive.

7. $(4x^{-3})(2x)^{-2}$

8. $\frac{x^{-2}y^4}{x^{-5}y^{-2}}$

9. $\left(27x^{-3}y^{\frac{3}{4}}\right)^{\frac{2}{3}}$

10. $\left(\frac{9x^{\frac{4}{3}}}{4y^{\frac{2}{3}}}\right)^{\frac{3}{2}}$

11. $\sqrt[3]{-27x^6y^8}$

12. $\sqrt[4]{32x^9y^{15}}$

13. $\frac{\sqrt{72}}{3} + 5\sqrt{\frac{1}{2}}$

14. $\frac{1}{2}\sqrt{\frac{4}{3}} + 3\sqrt{\frac{1}{3}}$

15. $4x^2+17x-15=0$

16. $x^2-2x-19=0$

17. $x=-1,\ y=-2$

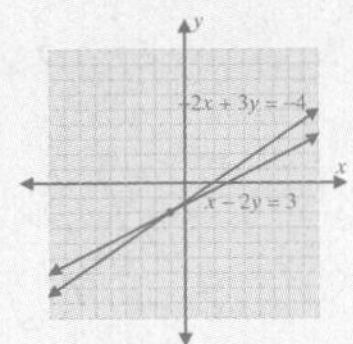

18. $x=5,\ y=4$

19. Inconsistent

20. $(-1.3,\ 3.1)$

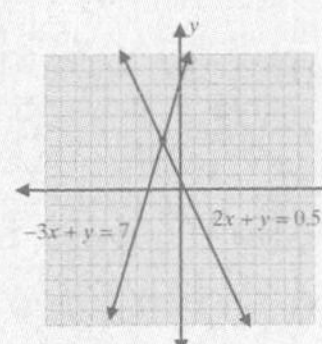

21. $x=-\frac{3}{2},\ x=\frac{2}{5}$

22. $x=\frac{-7\pm\sqrt{33}}{8}$

23. $x=-1$

24. $x=-2,\ x=\frac{4}{3}$

25. $x^2-12x+26-\frac{32}{x+2}$

26. $x^3-8x^2+4x-\frac{2}{x-2}$

27. **a.** 5.9136 **b.** 14.8306 **c.** −5.8284

28. $5\left(\sqrt{3}-\sqrt{2}\right)$

29. $(x+9)\left(\sqrt{x}-3\right)$

30. $\sqrt{5}\left(\sqrt{6}-1\right)$

In Exercises 15 and 16, find an equation with integer coefficients that has the indicated roots.

15. $x=\frac{3}{4}, x=-5$

16. $x=1-2\sqrt{5}, x=1+2\sqrt{5}$

17. Solve the following system graphically. $\begin{cases}-2x+3y=-4\\ x-2y=3\end{cases}$

18. Solve the following system of linear equations by using the substitution method.

$$\begin{cases}3x-2y=7\\ -2x+y=-6\end{cases}$$

19. Solve the following system of linear equations by using the addition method.

$$\begin{cases}-6x+2y=5\\ 3x-y=1\end{cases}$$

20. Use a graphing calculator to find the solution set to the following system of linear equations.

$$\begin{cases}2x+y=0.5\\ -3x+y=7\end{cases}$$

Solve the equations in Exercises 21 – 24.

21. $10x^2+11x-6=0$

22. $4x^2+7x+1=0$

23. $\sqrt{x+5}-2=x+1$

24. $8x^{-2}-2x^{-1}-3=0$

In Exercises 25 and 26, divide by using long division. Write the answer in the form $Q(x)+\frac{R(x)}{D(x)}$.

25. $\frac{x^3-10x^2+2x+20}{x+2}$

26. $\frac{x^4-10x^3+20x^2-8x-2}{x-2}$

27. Use a calculator to estimate the value of each number accurate to 4 decimal places.

a. $\sqrt{6}+2\sqrt{3}$ **b.** $\sqrt{2}\left(1+3\sqrt{10}\right)$ **c.** $\frac{\sqrt{2}+2}{\sqrt{2}-2}$

Rationalize the denominator and simplify each expression in Exercises 28 – 30.

28. $\frac{5}{\sqrt{2}+\sqrt{3}}$

29. $\frac{x^2-81}{3+\sqrt{x}}$

30. $\frac{\sqrt{125}}{1+\sqrt{6}}$

In Exercises 31 – 33, perform the indicated operations and write the results in the standard form a + bi.

31. $(2 + 5i) + (2 - 3i)$ **32.** $(2+5i)(2-3i)$ **33.** $\dfrac{2+5i}{2-3i}$

Use a graphing calculator and trace and zoom features to graph the functions in Exercises 34 – 37 and estimate the x-intercepts.

34. $f(x) = 2x^2 - 5$ **35.** $y = -x^3 + 2x^2 - 1$

36. $g(x) = x^4 - x^2 + 8$ **37.** $h(x) = x^3 + 3x + 2$

Use a graphing calculator to graph each of the functions in Exercises 38 – 40. State the domain and range of each function.

38. $f(x) = \sqrt{x-3}$ **39.** $g(x) = \sqrt{1-x}$ **40.** $y = -\sqrt{x+1}$

41. A car rental agency rents 200 cars per day at a rate of $30 per day for each car. For each $1 increase in the daily rate, the owners have found that they will rent 5 fewer cars per day. What daily rate would give total receipts of $6125? (**Hint:** Let x = the number of $1 increases.)

42. Find the dimensions of a rectangle that has an area of 520 m^2 and a perimeter of 92 m.

43. Find three consecutive even integers such that the square of the first added to the product of the second and third gives a result of 368.

44. The base of a triangle is 3 cm more than twice its altitude. If the area of the triangle is 76 cm^2, find the lengths of the base and altitude of the triangle.

45. What is the vertical line test for functions? Why does it work?

46. A rectangular auditorium seats 300 people. The number of seats in each row exceeds the number of rows by 20. Find the number of seats in each row.

47. A rock is thrown upward with a velocity of 20 feet per second from the top of a 24 feet high cliff, and it does not hit the cliff of the way back down. When will the rock be 7 feet from ground level? (Round your answer to the nearest tenth.)

31. $4+2i$

32. $19+4i$

33. $\dfrac{-11+16i}{13}$

34. x-int: $(-1.58, 0)$ and $(1.58, 0)$

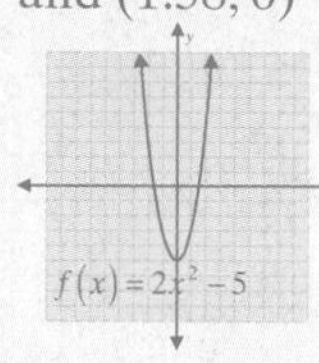

35. x-int: $(-0.62, 0)$, $(1, 0)$, and $(1.62, 0)$

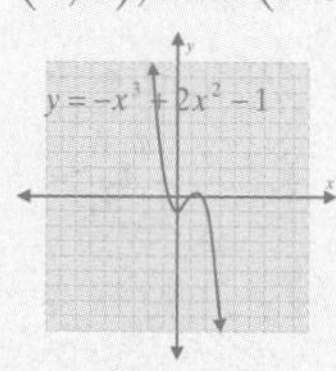

36. x-int: none

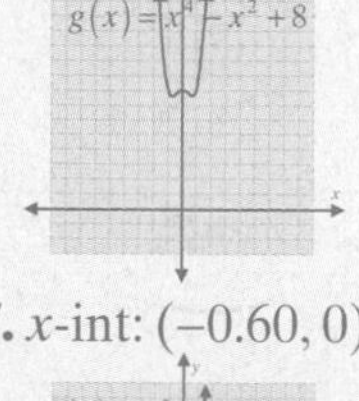

37. x-int: $(-0.60, 0)$

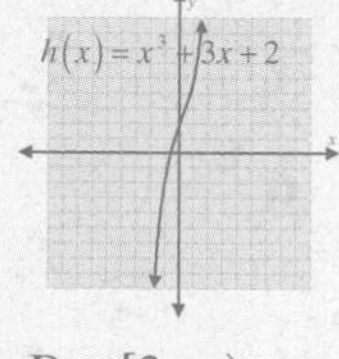

38. $D = [3, \infty)$
$R = [0, \infty)$

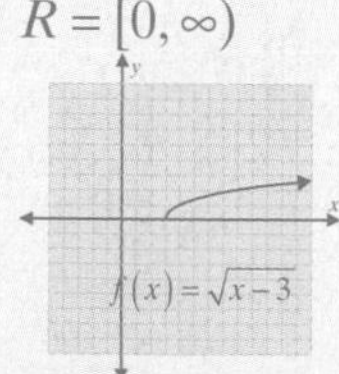

39. $D = (-\infty, 1]$
$R = [0, \infty)$

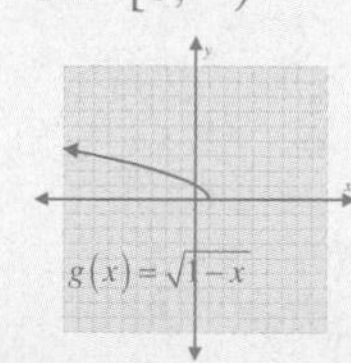

40. $D = [-1, \infty)$
$R = (-\infty, 0]$

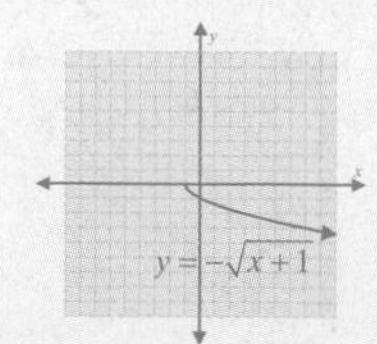

41. $35 **42.** 20, 26 **43.** 12, 14, 16

44. length of the base = 19 cm, altitude = 8 cm

45. If any vertical line intersects the graph of a relation at more than one point, then the relation graphed is not a function.

46. 30 seats **47.** 1.8 seconds

Quadratic Functions and Conic Sections

CHAPTER 12

Did You Know?

Euler

The mathematician who invented the notation for functions, $f(x)$, which you will study in this chapter, was Leonhard Euler (1707–1783) of Switzerland. Euler was one of the most prolific mathematical researchers of all time, and he lived during a period in which mathematics was making great progress. He studied mathematics, theology, medicine, astronomy, physics, and oriental languages before he began a career as a court philosopher-mathematician. His professional life was spent at St. Petersburg Academy by invitation of Catherine I of Russia, at the Berlin Academy under Frederick the Great of Prussia, and again at the St. Petersburg Academy under Catherine the Great. The collected works of Euler fill 80 volumes, and for almost 50 years after Euler's death, the publications of the St. Petersburg Academy continued to include articles by him.

Euler was blind the last 17 years of his life, but he continued his mathematical research by writing on a large slate and dictating to a secretary. He was responsible for the conventionalization of many mathematical symbols such as $f(x)$ for function notation, i for $\sqrt{-1}$, e for the base of the natural logarithms, π for the ratio of circumference to diameter of a circle, and Σ for the summation symbol.

From the age of 20 to his death, Euler was busy adding to knowledge in every branch of mathematics. He wrote with modern symbolism, and his work in calculus was particularly outstanding. Euler had a rich family life, having had 13 children, and he not only contributed to mathematics but reformed the Russian system of weights and measures, supervised the government pension system in Prussia and the government geographic office in Russia, designed canals, and worked in many areas of physics, including acoustics and optics. It was said of Euler, by the French academician François Arago, that he could calculate without apparent effort "just as men breathe and eagles sustain themselves in the air."

An interesting story is told about Euler's meeting with the French philosopher Diderot at the Russian court. Diderot had angered the czarina by his antireligious views, and Euler was called to the court to debate Diderot. Diderot was told that the great mathematician Euler had an algebraic proof that God existed. Euler walked in towards Diderot and said, "Monsieur, $\dfrac{a+b^n}{n}=x$, therefore God exists, respond." Diderot, who had no understanding of algebra, was unable to respond.

"I have not hesitated in 1900, at the Congress of Mathematicians in Paris, to call the nineteenth century the century of the theory of functions."

Vito Volterra (1860 – 1940)

The speed at which you drive your car is a **function** of how far you depress the accelerator; your energy level is a **function** of the amount and type of food you eat; your grade in this class is a **function** of the quality time you spend studying. Obviously, the concept of a **function** is present in many aspects of our daily lives. In this text we quantify these ideas by dealing only with functions involving pairs of real numbers. As we have seen in Chapters 4, 6, and 10, this restriction allows us to analyze functions in terms of their graphs in the Cartesian coordinate system. In this chapter, we will be particularly interested in a special category of function called quadratic functions whose graphs are parabolas (curves that, among other things, can be used to describe the paths of projectiles).

Throughout the remainder of the text, we continue to build a list of general properties of functions that form the basis for all our work with functions and lead to the topics of exponential and logarithmic functions in Chapter 13. The function notation, $f(x)$, is particularly useful in understanding, evaluating, and manipulating functions. Mathematical models for applications depend on the concept of functions and use this function notation. For example, we might represent the depreciated value in dollars of a piece of equipment as $f(t) = 10{,}000 + 4000 \cdot 2^{-0.3t}$ where t is time in years. Or, the profit in dollars from selling x items might be a more familiar polynomial function such as $P(x) = 500 + 3x - 2x^2$. In any case we will find that the concept of functions and related topics, such as graphs and domain and range, provides the basic tools for understanding many applied problems.

12.1 Quadratic Functions: Parabolas

Objectives

After completing this section, you will be able to:

1. *Graph a parabola by finding the zeros (if any) and completing the square, if necessary, to determine the vertex, range, and line of symmetry.*
2. *Solve applied problems by using quadratic functions.*

Quadratic Functions: $y = ax^2 + bx + c,\ a \neq 0$

We have studied various types of functions: linear functions, polynomial functions, and functions with radicals. In each case we have been interested in the corresponding graphs, the concepts of domain and range and the points where the graph crosses the x-axis (**zeros of the function**). Also, recall that the vertical line test can be used to tell whether or not a graph represents a function.

Vertical Line Test

If ***any*** *vertical line intersects a graph in more than one point, then the relation graphed is* ***not*** *a function.*

In this section, we expand our interest in functions to include a detailed analysis of **quadratic functions**, functions that are represented by quadratic expressions. For example, consider the function

$$y = x^2 - 4x + 3.$$

What is the graph of this function? Since the equation is not linear, the graph will not be a straight line. The nature of the graph can be investigated by plotting several points (See Figure 12.1).

x	$x^2 - 4x + 3 = y$
-1	$(-1)^2 - 4(-1) + 3 = 8$
0	$0^2 - 4(0) + 3 = 3$
$\frac{1}{2}$	$\left(\frac{1}{2}\right)^2 - 4\left(\frac{1}{2}\right) + 3 = \frac{5}{4}$
1	$1^2 - 4(1) + 3 = 0$
2	$2^2 - 4(2) + 3 = -1$
3	$3^2 - 4(3) + 3 = 0$
$\frac{7}{2}$	$\left(\frac{7}{2}\right)^2 - 4\left(\frac{7}{2}\right) + 3 = \frac{5}{4}$
4	$4^2 - 4(4) + 3 = 3$
5	$5^2 - 4(5) + 3 = 8$

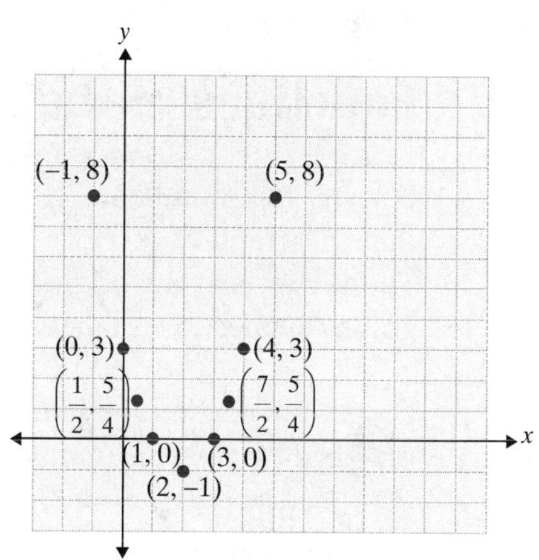

Figure 12.1

The complete graph of $y = x^2 - 4x + 3$ is shown in Figure 12.2. The curve is called a **parabola**. The point $(2, -1)$ is the "turning point" of the parabola and is called the **vertex** of the parabola. The line $x = 2$ is the **line of symmetry** or **axis of symmetry** for the parabola. That is, the curve is a "mirror image" of itself with respect to the line $x = 2$.

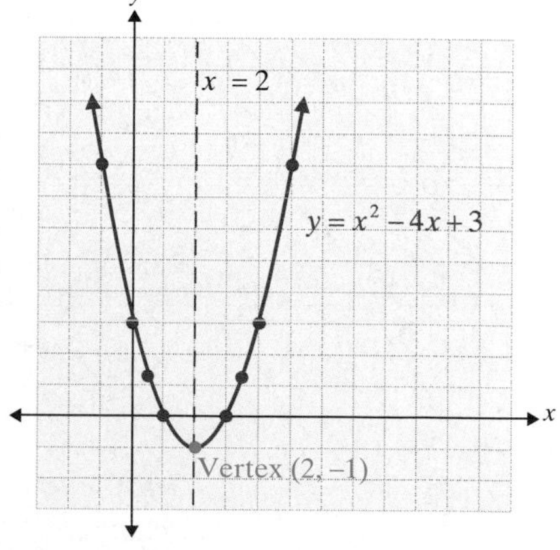

Figure 12.2

Vertex is $(2, -1)$.
$y = x^2 - 4x + 3$ is a parabola.
$x = 2$ is the line of symmetry.

Quadratic Function

Any function that can be written in the form

$$y = ax^2 + bx + c$$

where a, b, c are real constants and $a \neq 0$ *is a* ***quadratic function****.*

The graph of every quadratic function is a parabola. The position of the parabola, its shape, and whether it "opens up" or "opens down" can be determined by investigating the function itself. For convenience, we will refer to parabolas that open up or down as **vertical parabolas**. Parabolas that open left or right will be called **horizontal parabolas**. As we will see in Section 12.4, **horizontal parabolas do not represent functions**.

We will discuss quadratic functions in each of the following five forms where a, b, c, h, and k are constants:

$$y = ax^2 \qquad y = ax^2 + k \qquad y = a(x-h)^2 \qquad y = a(x-h)^2 + k \qquad y = ax^2 + bx + c$$

Functions of the form $y = ax^2$

For any real number x, $x^2 \geq 0$. So, $ax^2 \geq 0$ if $a > 0$ and $ax^2 \leq 0$ if $a < 0$. This means that the graph of $y = ax^2$ is "above" the x-axis if $a > 0$ and "below" the x-axis if $a < 0$. The **vertex** is at the origin (0, 0) in either of these cases and is the one point where each graph touches (or is tangent to) the x-axis.

For all quadratic functions, the **domain** is the set of all real numbers. That is, x can be replaced by any real number and there will be one corresponding y-value. The **range** of the function depends on the value of a. If $a > 0$, then $y \geq 0$. If $a < 0$, then $y \leq 0$.

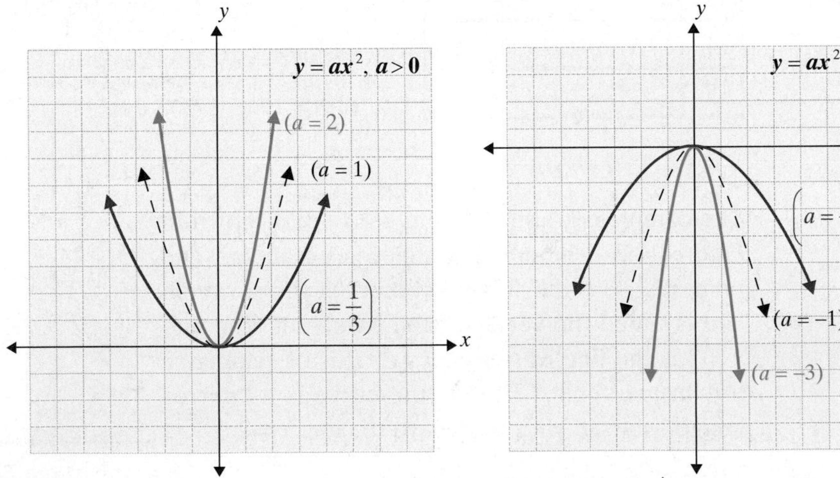

Domain: $\{x \mid x \text{ is any real number}\}$;
Range: $\{y \mid y \geq 0\}$
In interval notation:
Domain: $(-\infty, \infty)$; Range: $[0, \infty)$

Domain: $\{x \mid x \text{ is any real number}\}$;
Range: $\{y \mid y \leq 0\}$
In interval notation:
Domain: $(-\infty, \infty)$; Range: $(-\infty, 0]$

Figure 12.3

Figure 12.3 illustrates several properties of quadratic functions of the form $y = ax^2$. If $a > 0$, the parabola "opens upward." If $a < 0$, the parabola "opens downward." The bigger $|a|$ is, the narrower the opening; the smaller $|a|$ is, the wider the opening. The line $x = 0$ (the *y*-axis) is the line of symmetry.

Functions of the form $y = ax^2 + k$

Adding k to ax^2 simply changes each *y*-value of $y = ax^2$ by k units (increase if k is positive, decrease if k is negative). That is, the graph of $y = ax^2 + k$ can be found by "sliding or shifting" the graph of $y = ax^2$ up k units if $k > 0$ or down $|k|$ units if $k < 0$ (Figure 12.4).

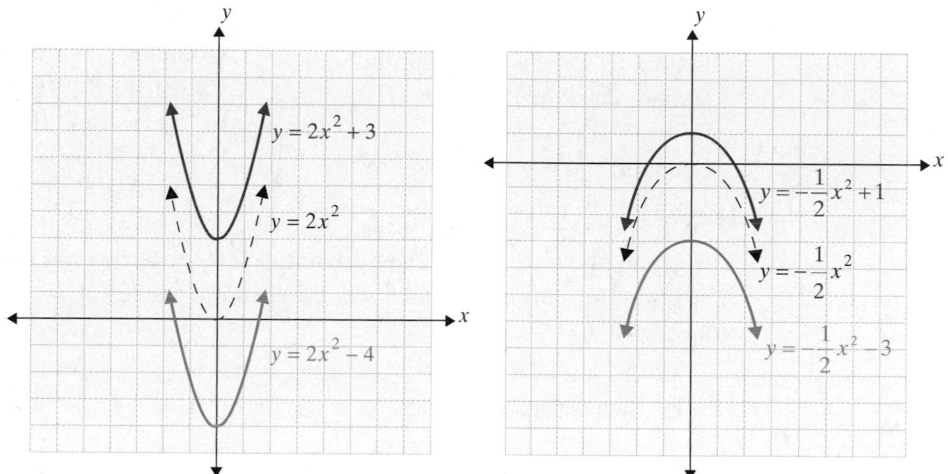

Domain: $\{x \mid x \text{ is any real number}\}$;
Range: $\{y \mid y \geq k\}$
In interval notation:
Domain: $(-\infty, \infty)$; Range: $[k, \infty)$

Domain: $\{x \mid x \text{ is any real number}\}$;
Range: $\{y \mid y \leq k\}$
In interval notation:
Domain: $(-\infty, \infty)$; Range: $(-\infty, k]$

Figure 12.4

The vertex of $y = ax^2 + k$ is at the point $(0, k)$. The graph of $y = ax^2 + k$ is a **vertical shift** (or **vertical translation**) of the graph of $y = ax^2$. The line $x = 0$ (the *y*-axis) is the line of symmetry just as with equations of the form $y = ax^2$.

Functions of the form $y = a(x - h)^2$

We know that $(x-h)^2 \geq 0$. So, if $a > 0$, then $y = a(x-h)^2 \geq 0$. If $a < 0$, then $y = a(x-h)^2 \leq 0$. Also notice that $ax^2 = 0$ when $x = 0$, and $a(x-h)^2 = 0$ when $x = h$. Thus the vertex is at $(h, 0)$, and the parabola "opens upward" if $a > 0$ and "opens downward" if $a < 0$. (Figure 12.5).

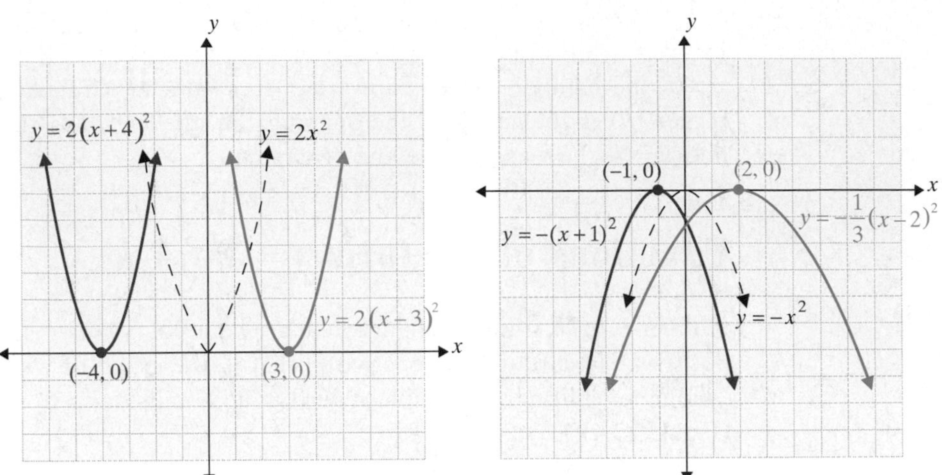

Domain: $\{x \mid x \text{ is any real number}\}$;
Range: $\{y \mid y \geq 0\}$
In interval notation:
Domain: $(-\infty, \infty)$; Range: $[0, \infty)$

Domain: $\{x \mid x \text{ is any real number}\}$;
Range: $\{y \mid y \leq 0\}$
In interval notation:
Domain: $(-\infty, \infty)$; Range: $(-\infty, 0]$

Figure 12.5

The graph of $y = a(x-h)^2$ is a **horizontal shift** (or **horizontal translation**) of the graph of $y = ax^2$. The shift is to the right if $h > 0$ and to the left if $h < 0$. As a special comment, note that if $h = -3$, then

$$y = a(x-h)^2$$

$$\text{gives } y = a(x-(-3))^2$$

$$\text{or } y = a(x+3)^2.$$

Thus, if h is negative, the expression $(x-h)^2$ appears with a plus sign. If h is positive, the expression $(x-h)^2$ appears with a minus sign. In either case, the line $x = h$ is the **line of symmetry**.

Example 1: Quadratic Functions

Graph the following quadratic functions. Set up a table of values for x and y as an aid and choose values of x on each side of the line of symmetry. Find the line of symmetry and the vertex and state the domain and range of each function.

a. $y = 2x^2 - 3$

Solution: Line of symmetry is $x = 0$. [The parabola opens upward since a is positive.]

Vertex is at $(0, -3)$. Domain: $\{x \mid x \text{ is any real number}\}$ or $(-\infty, \infty)$

Range: $\{y \mid y \geq -3\}$ or $[-3, \infty)$

x	y
0	−3
$\frac{1}{2}$	$-\frac{5}{2}$
$-\frac{1}{2}$	$-\frac{5}{2}$
1	−1
−1	−1
2	5
−2	5

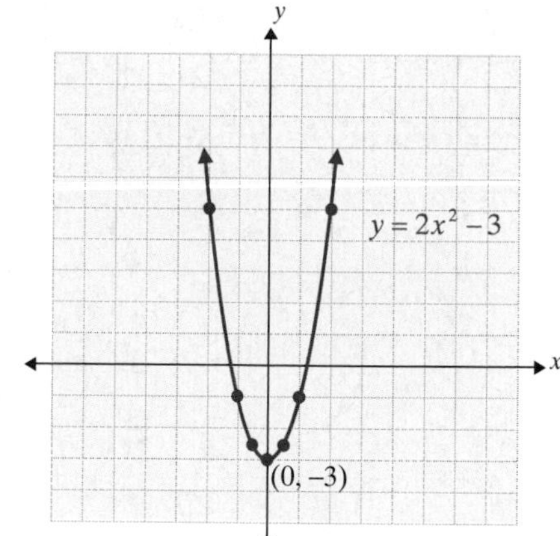

b. $y = -\left(x - \frac{5}{2}\right)^2$

Solution: Line of symmetry is $x = \frac{5}{2}$. [The parabola opens down since a is negative.]
Vertex is at $\left(\frac{5}{2}, 0\right)$. Domain: $\{x \mid x \text{ is any real number}\}$ or $(-\infty, \infty)$
Range: $\{y \mid y \le 0\}$ or $(-\infty, 0]$

x	y
$\frac{5}{2}$	0
2	$-\frac{1}{4}$
3	$-\frac{1}{4}$
1	$-\frac{9}{4}$
4	$-\frac{9}{4}$
0	$-\frac{25}{4}$
5	$-\frac{25}{4}$

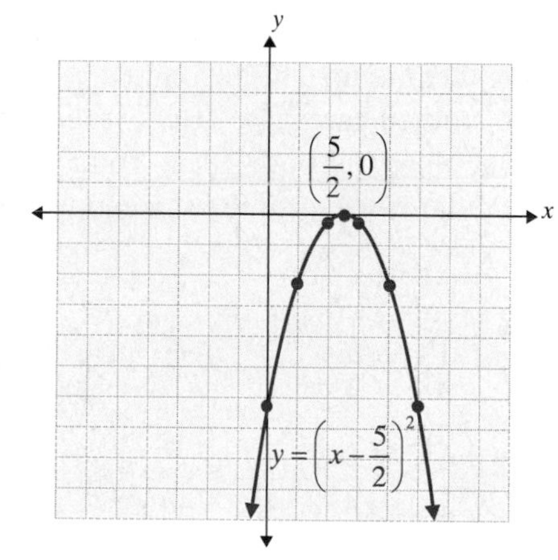

Functions of the form $y = a(x - h)^2 + k$ and $y = ax^2 + bx + c$

The graphs of equations of the form

$$y = a(x - h)^2 + k$$

combine both the vertical shift of k units and the horizontal shift of h units. The vertex is at (h, k). For example, the graph of the function $y = -2(x - 3)^2 + 5$ is a shift of the graph of $y = -2x^2$ up 5 units and to the right 3 units and has its vertex at (3, 5).

The graph of $y=\left(x+\frac{1}{2}\right)^2-2$ is the same as the graph of $y=x^2$ but is shifted left $\frac{1}{2}$ unit and down 2 units. The vertex is at $\left(-\frac{1}{2},-2\right)$ (Figure 12.6).

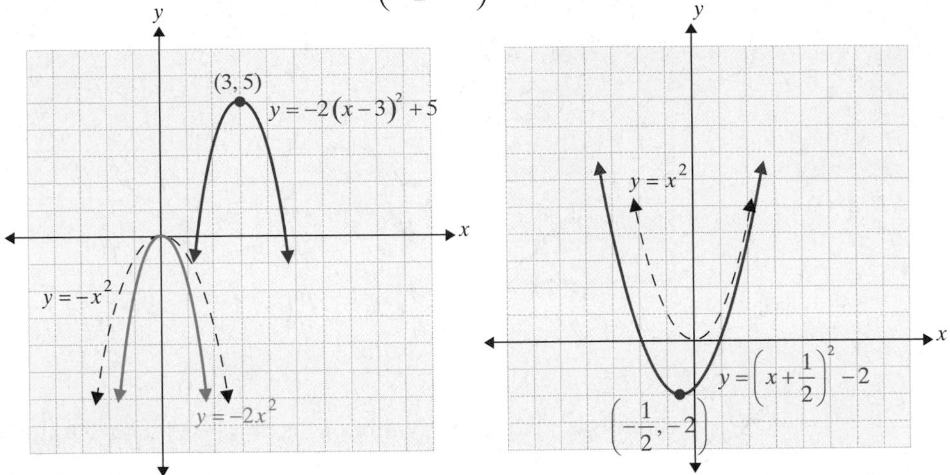

Figure 12.6

Teaching Notes: This is another good place in this development to make the case for learning to "read" information from an equation so that key characteristics can be known about these functions before (or without) graphing.

The general form of a quadratic function is $y=ax^2+bx+c$. However, this form does not give as much information about the graph of the corresponding parabola as the form $y=a(x-h)^2+k$. Therefore, to easily find the vertex, line of symmetry, and range, and to graph the parabola, we want to change the general form $y=ax^2+bx+c$ into the form $y=a(x-h)^2+k$. This can be accomplished by completing the square using the following technique. This technique is also useful in other courses in mathematics and should be studied carefully. (Be aware that you are not solving an equation. You do not "do something" to both sides. You are **changing the form** of a function.) **Note:** A graphing calculator will give the same graph regardless of the form of the function.

$y=ax^2+bx+c$	Write the function.
$=a\left(x^2+\frac{b}{a}x\right)+c$	Factor a from just the first two terms.
$=a\left(x^2+\frac{b}{a}x+\frac{b^2}{4a^2}-\frac{b^2}{4a^2}\right)+c$	Complete the square of $x^2+\frac{b}{a}x$. $\frac{1}{2}\cdot\frac{b}{a}=\frac{b}{2a}$ and $\left(\frac{b}{2a}\right)^2=\frac{b^2}{4a^2}$. Add and subtract $\frac{b^2}{4a^2}$ inside the parentheses.
$=a\left(x^2+\frac{b}{a}x+\frac{b^2}{4a^2}\right)-\frac{b^2}{4a}+c$	Multiply $a\left(\frac{-b^2}{4a^2}\right)$ and write this term outside the parentheses.
$=a\left(x+\frac{b}{2a}\right)^2+\frac{4ac-b^2}{4a}$	Write the square of the binomial and simplify the fraction to get the form $y=a(x-h)^2+k$.

In terms of the coefficients a, b, and c,

$$x = -\frac{b}{2a} \text{ is the line of symmetry}$$

and

$$(h,k) = \left(-\frac{b}{2a}, \frac{4ac - b^2}{4a}\right) \text{ is the vertex.}$$

NOTES

Rather than memorize the formula for the coordinates of the vertex, you should just remember that the x-coordinate of the vertex is $x = -\frac{b}{2a}$. Substituting this value for x in the function will give the y-value for the vertex.

Zeros of a Quadratic Function

The points where a parabola crosses the x-axis, if any, are the x-intercepts. This is where $y = 0$. These points are called the **zeros of the function**. We find these points by substituting 0 for y and solving the resulting quadratic equation:

$y = ax^2 + bx + c$ **quadratic function**

$0 = ax^2 + bx + c$ **quadratic equation**

If the solutions are nonreal complex numbers, then the graph does not cross the x-axis. It is either entirely above the x-axis or entirely below the x-axis.

The following examples illustrate how to apply all our knowledge about quadratic functions.

Example 2: Zeros of a Function

a. $y = x^2 - 6x + 1$

Solution: Find the zeros of the function, the line of symmetry, the vertex, the domain, the range, and graph the parabola.

$x^2 - 6x + 1 = 0$

$$\begin{aligned} x &= \frac{6 \pm \sqrt{36-4}}{2} \\ &= \frac{6 \pm \sqrt{32}}{2} \\ &= \frac{6 \pm 4\sqrt{2}}{2} \\ &= 3 \pm 2\sqrt{2} \end{aligned}$$

continued on next page ...

Change the form of the function for easier graphing.

$$y = x^2 - 6x + 1$$
$$= (x^2 - 6x + 9 - 9) + 1$$
$$= (x^2 - 6x + 9) - 9 + 1$$
$$= (x - 3)^2 - 8$$

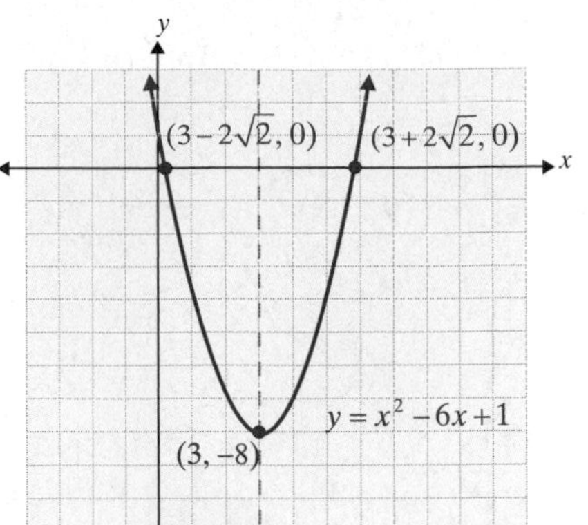

Line of symmetry is $x = 3$. Vertex: (3, –8).
Domain: $\{x \mid x \text{ is any real number}\}$ or $(-\infty, \infty)$;
Range: $\{y \mid y \geq -8\}$ or $[-8, \infty)$.

b. $y = -x^2 - 4x + 2$

Solution: Find the zeros of the function, the line of symmetry, the vertex, the domain, the range, and graph the parabola.

$$-x^2 - 4x + 2 = 0$$

$$x = \frac{4 \pm \sqrt{16+8}}{-2}$$
$$= \frac{4 \pm \sqrt{24}}{-2}$$
$$= \frac{4 \pm 2\sqrt{6}}{-2}$$
$$= -2 \pm \sqrt{6}$$

The zeros are $-2 \pm \sqrt{6}$.

Change the form of the function for easier graphing.
$y = -x^2 - 4x + 2$

$= -(x^2 + 4x) + 2$	Factor -1 from the first two terms only.
$= -(x^2 + 4x + 4 - 4) + 2$	Add $0 = 4 - 4$ inside the parentheses.
$= -(x^2 + 4x + 4) + 4 + 2$	Multiply $-1(-4)$ and put this outside the parentheses.
$= -(x + 2)^2 + 6$	

Line of symmetry is $x = -2$. Vertex: (–2, 6).
Domain: $\{x \mid x \text{ is any real number}\}$ or $(-\infty, \infty)$;
Range: $\{y \mid y \leq 6\}$ or $(-\infty, 6]$.

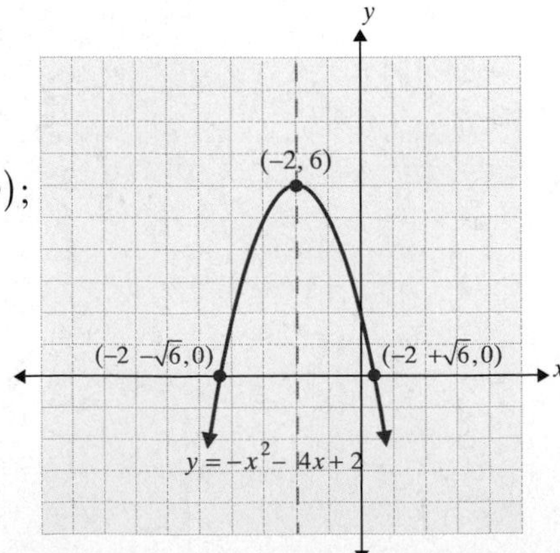

c. $y = 2x^2 - 6x + 5$

Teaching Notes:
You might want to show your students the following approach to changing to the form $y = a(x-h)^2 + k$ where completing the square is done with a leading coefficient of 1.

For example,

$$y = 2x^2 - 6x + 5$$
$$\frac{y}{2} = x^2 - 3x + \frac{5}{2}$$
$$\frac{y}{2} - \frac{5}{2} = x^2 - 3x$$
$$\frac{y}{2} - \frac{5}{2} + \frac{9}{4} = x^2 - 3x + \frac{9}{4}$$
$$\frac{y}{2} - \frac{1}{4} = \left(x - \frac{3}{2}\right)^2$$
$$\frac{y}{2} = \left(x - \frac{3}{2}\right)^2 + \frac{1}{4}$$
$$y = 2\left(x - \frac{3}{2}\right)^2 + \frac{1}{2}$$

Solution: Find the zeros of the function, the line of symmetry, the vertex, the domain, the range, and graph the parabola.

$$2x^2 - 6x + 5 = 0$$

$$x = \frac{6 \pm \sqrt{36-40}}{4} = \frac{6 \pm \sqrt{-4}}{4} \qquad \text{Quadratic formula}$$

There are no real zeros since the discriminant is negative. The graph will not cross the x-axis. Now, change the form of the function for easier graphing.

$y = 2x^2 - 6x + 5$

$= 2(x^2 - 3x) + 5$ Factor 2 from the first two terms only.

$= 2\left(x^2 - 3x + \frac{9}{4} - \frac{9}{4}\right) + 5$ Add $0 = \frac{9}{4} - \frac{9}{4}$ inside the parentheses.

$= 2\left(x^2 - 3x + \frac{9}{4}\right) + 2\left(-\frac{9}{4}\right) + 5$ Multiply $2\left(-\frac{9}{4}\right)$ and put this outside the parentheses.

$= 2\left(x - \frac{3}{2}\right)^2 + \frac{1}{2}$ Simplify: $2\left(-\frac{9}{4}\right) + 5 = -\frac{9}{2} + \frac{10}{2} = \frac{1}{2}$

Line of symmetry is $x = \frac{3}{2}$. Vertex: $\left(\frac{3}{2}, \frac{1}{2}\right)$

Domain: $\{x \mid x \text{ is any real number}\}$ or $(-\infty, \infty)$;

Range: $\left\{y \mid y \geq \frac{1}{2}\right\}$ or $\left[\frac{1}{2}, \infty\right)$.

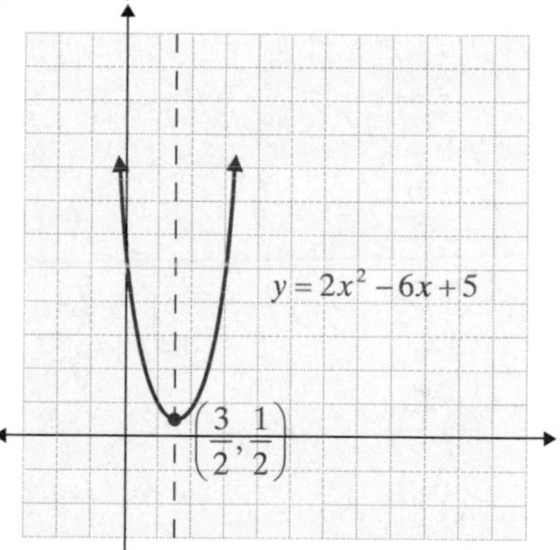

Maximum and Minimum Values

The vertex of a vertical parabola is either the lowest point or the highest point on the parabola.

Minimum and Maximum Values

For a parabola with an equation in the form $y = a(x-h)^2 + k$,

1. *If $a > 0$, then (h, k) is the lowest point and $y = k$ is called the* ***minimum value*** *of the function.*
2. *If $a < 0$, then (h, k) is the highest point and $y = k$ is called the* ***maximum value*** *of the function.*

If the function is in the general quadratic form $y = ax^2 + bx + c$, then the maximum or minimum value can be found by letting $x = -\frac{b}{2a}$ and solving for y.

The concepts of maximum and minimum values of a function help not only in graphing but also in solving many types of applications. Applications involving quadratic functions are discussed here. Other types of applications are discussed in Chapter 14 and in more advanced courses in mathematics.

Example 3: Minimum and Maximum Values

a. A sandwich company sells hot dogs at the local baseball stadium for \$3.00 each and sells 2000 hot dogs per game. The company estimates that each time the price is raised by 25¢, they will sell 100 fewer hot dogs.

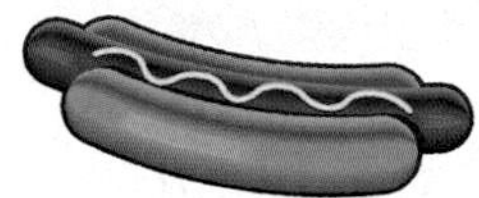

i. What price should they charge to maximize their revenue (income) per game?
ii. What will be the maximum revenue?

Solution: Let x = number of 25¢ increases in price.

Then $3.00 + 0.25x$ = price per hot dog
and $2000 - 100x$ = number of hot dogs sold.
Revenue = (price per unit)·(number of units sold)

So,
$$R = (3.00 + 0.25x)(2000 - 100x)$$
$$= 6000 + 500x - 300x - 25x^2$$
$$= 6000 + 200x - 25x^2$$

The revenue is represented by a quadratic function and the maximum revenue occurs at the point where

$$x = -\frac{b}{2a} = -\frac{200}{-50} = 4.$$

For $x = 4$,

price per hot dog $= 3.00 + 0.25(4) = \$4.00$

and Revenue $= R = (4)(2000 - 400) = \$6400.$

Thus, the company will make its maximum revenue of \$6400 by charging \$4 per hot dog.

1. $x = 0, (0, -4),$
$\{x | x \text{ is real}\},$
$\{y | y \geq -4\}$
2. $x = 0, (0, 7),$
$\{x | x \text{ is real}\},$
$\{y | y \geq 7\}$
3. $x = 0, (0, 9),$
$\{x | x \text{ is real}\},$
$\{y | y \geq 9\}$
4. $x = 0, (0, -1),$
$\{x | x \text{ is real}\},$
$\{y | y \geq -1\}$
5. $x = 0, (0, 1),$
$\{x | x \text{ is real}\},$
$\{y | y \leq 1\}$

b. A rancher is going to build three sides of a rectangular corral next to a river. He has 240 feet of fencing and wants to enclose the maximum area possible inside the corral. What are the dimensions of the corral with the maximum area and what is this area?

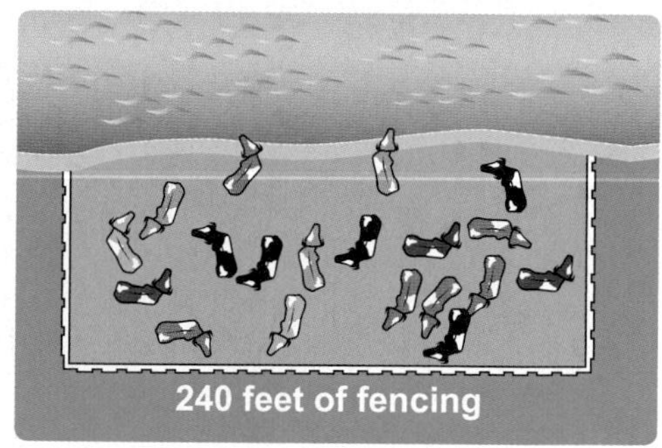

Solution: Let $x =$ length of one of the two equal sides of the rectangle.
Then $240 - 2x =$ length of third side of the rectangle.
Since the area is the length times the width, the area is represented by the quadratic function $A = x(240 - 2x) = 240x - 2x^2$, the maximum area occurs at the point where $x = -\frac{b}{2a} = -\frac{240}{-4} = 60.$

Two sides of the rectangle are 60 feet and the third side is $240 - 2x = 120$ feet. The maximum area possible is $60(120) = 7200 \text{ ft}^2$.

Practice Problems

6. $x = 0, (0, -6),$
$\{x | x \text{ is real}\},$
$\{y | y \leq -6\}$

1. *Write the function $y = 2x^2 - 4x + 3$ in the form $y = a(x-h)^2 + k$.*
2. *Find the zeros of the function $y = x^2 - 7x + 10$.*
3. *Find the vertex and the range of the function $y = -x^2 + 4x - 5$.*

12.1 Exercises

7. $x = 0, \left(0, \frac{1}{5}\right),$
$\{x | x \text{ is real}\},$
$\left\{y \middle| y \leq \frac{1}{5}\right\}$
8. $x = 0, \left(0, \frac{7}{8}\right),$
$\{x | x \text{ is real}\},$
$\left\{y \middle| y \geq \frac{7}{8}\right\}$
9. $x = -1, (-1, 0),$
$\{x | x \text{ is real}\},$
$\{y | y \geq 0\}$

For each of the quadratic functions in Exercises 1 – 20, determine the line of symmetry, the vertex, and the domain and range.

1. $y = 3x^2 - 4$ **2.** $y = \frac{2}{3}x^2 + 7$ **3.** $y = 7x^2 + 9$

4. $y = 5x^2 - 1$ **5.** $y = -4x^2 + 1$ **6.** $y = -2x^2 - 6$

7. $y = -\frac{3}{4}x^2 + \frac{1}{5}$ **8.** $y = \frac{5}{3}x^2 + \frac{7}{8}$ **9.** $y = (x+1)^2$

10. $y = (x-1)^2$ **11.** $y = -\frac{2}{3}(x-4)^2$ **12.** $y = -5(x+2)^2$

13. $y = 2(x+3)^2 - 2$ **14.** $y = 4(x-5)^2 + 1$ **15.** $y = \frac{3}{4}(x+2)^2 - 6$

Answers to Practice Problems: 1. $y = 2(x-1)^2 + 1$ **2.** $x = 5, x = 2$ **3.** Vertex: $(2, -1)$, Range: $y \leq -1$

16. $y=-2(x+1)^2-4$ **17.** $y=-\frac{1}{2}\left(x-\frac{3}{2}\right)^2+\frac{7}{2}$ **18.** $y=-\frac{5}{3}\left(x-\frac{9}{2}\right)^2+\frac{3}{4}$

19. $y=\frac{1}{4}\left(x-\frac{4}{5}\right)^2-\frac{11}{5}$ **20.** $y=\frac{10}{3}\left(x+\frac{7}{8}\right)^2-\frac{9}{16}$

21. Graph the function $y=x^2$. Then, without additional computation, graph the following translations.

a. $y=x^2-2$ **b.** $y=(x-3)^2$ **c.** $y=-(x-1)^2$ **d.** $y=5-(x+1)^2$

22. Graph the function $y=2x^2$. Then, without additional computation, graph the following translations.

a. $y=2x^2-3$ **b.** $y=2(x-4)^2$ **c.** $y=-2(x+1)^2$ **d.** $y=-2(x+2)^2-4$

23. Graph the function $y=\frac{1}{2}x^2$. Then, without additional computation, graph the following translations.

a. $y=\frac{1}{2}x^2+3$ **b.** $y=\frac{1}{2}(x+2)^2$ **c.** $y=-\frac{1}{2}x^2$ **d.** $y=\frac{1}{2}(x-1)^2-4$

24. Graph the function $y=\frac{1}{4}x^2$. Then, without additional computation, graph the following translations.

a. $y=-\frac{1}{4}x^2$ **b.** $y=\frac{1}{4}x^2-5$ **c.** $y=\frac{1}{4}(x+4)^2$ **d.** $y=2-\frac{1}{4}(x+2)^2$

Rewrite each of the quadratic functions in Exercises 25 – 40 in the form $y=a(x-h)^2+k$*. Find the vertex, range, and zeros of each function. Graph the function.*

25. $y=2x^2-4x+2$ **26.** $y=-3x^2+12x-12$ **27.** $y=x^2-2x-3$

28. $y=x^2-4x+5$ **29.** $y=x^2+6x+5$ **30.** $y=x^2-8x+12$

31. $y=2x^2-8x+5$ **32.** $y=2x^2-6x+5$ **33.** $y=-3x^2-12x-9$

34. $y=3x^2-6x-1$ **35.** $y=5x^2-10x+8$ **36.** $y=-4x^2+16x-11$

37. $y=-x^2-5x-2$ **38.** $y=x^2+3x-1$ **39.** $y=2x^2+7x+5$

40. $y=2x^2+x-3$

10. $x=1, (1,0)$, $\{x \mid x \text{ is real}\}$, $\{y \mid y \geq 0\}$

11. $x=4, (4,0)$, $\{x \mid x \text{ is real}\}$, $\{y \mid y \leq 0\}$

12. $x=-2, (-2,0)$, $\{x \mid x \text{ is real}\}$, $\{y \mid y \leq 0\}$

13. $x=-3, (-3,-2)$, $\{x \mid x \text{ is real}\}$, $\{y \mid y \geq -2\}$

14. $x=5, (5,1)$, $\{x \mid x \text{ is real}\}$, $\{y \mid y \geq 1\}$

15. $x=-2, (-2,-6)$, $\{x \mid x \text{ is real}\}$, $\{y \mid y \geq -6\}$

16. $x=-1, (-1,-4)$, $\{x \mid x \text{ is real}\}$, $\{y \mid y \leq -4\}$

17. $x=\frac{3}{2}, \left(\frac{3}{2},\frac{7}{2}\right)$, $\{x \mid x \text{ is real}\}$, $\left\{y \mid y \leq \frac{7}{2}\right\}$

18. $x=\frac{9}{2}, \left(\frac{9}{2},\frac{3}{4}\right)$, $\{x \mid x \text{ is real}\}$, $\left\{y \mid y \leq \frac{3}{4}\right\}$

19. $x=\frac{4}{5}, \left(\frac{4}{5},-\frac{11}{5}\right)$, $\{x \mid x \text{ is real}\}$, $\left\{y \mid y \geq -\frac{11}{5}\right\}$

20. $x=-\frac{7}{8}, \left(-\frac{7}{8},-\frac{9}{16}\right)$, $\{x \mid x \text{ is real}\}$, $\left\{y \mid y \geq -\frac{9}{16}\right\}$

21. a. **b.**

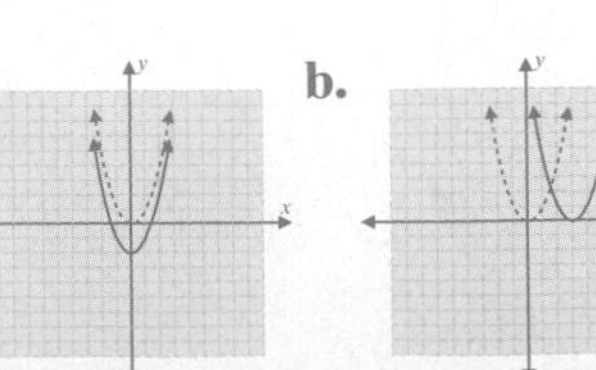

c.

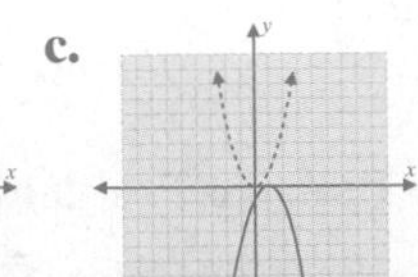

d.

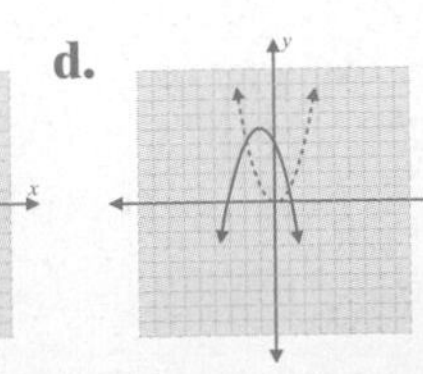

22. a.
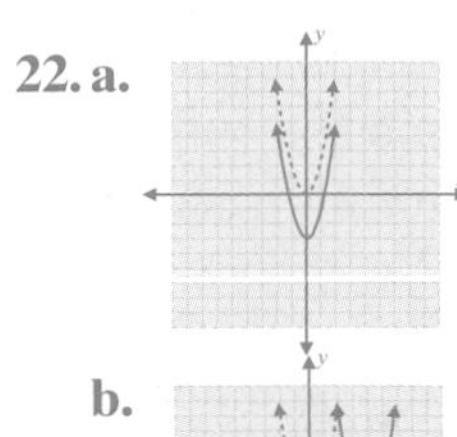

b.
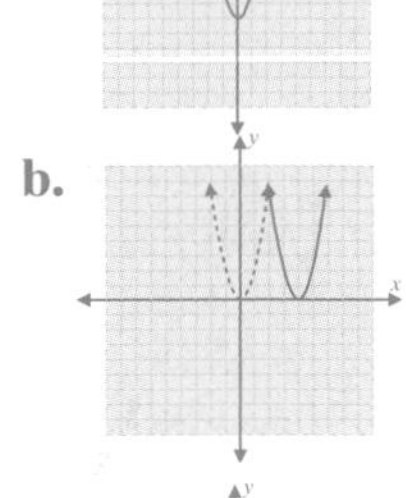

c.
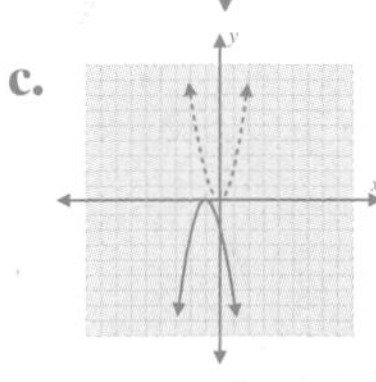

d.
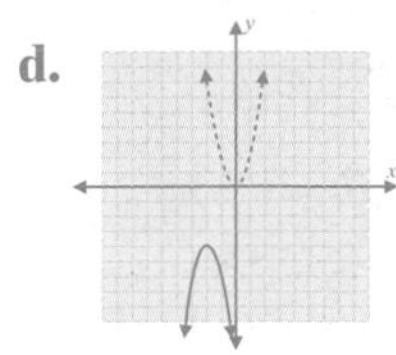

23. a.
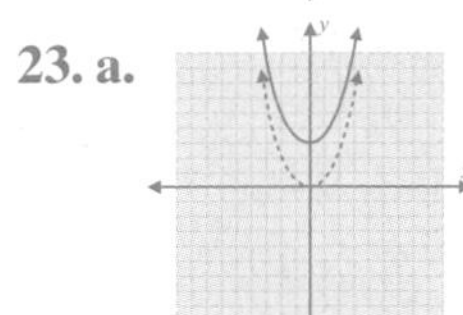

b.
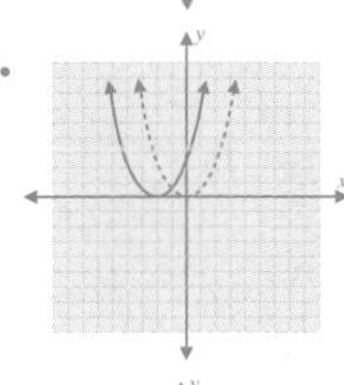

c.
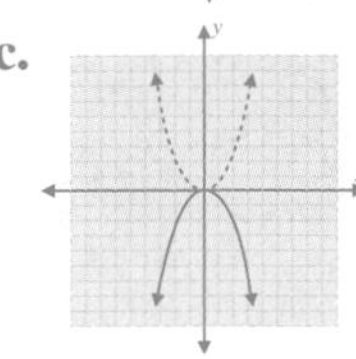

d.
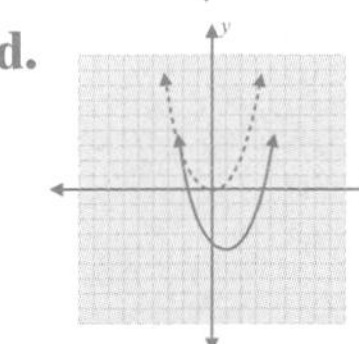

24. a.
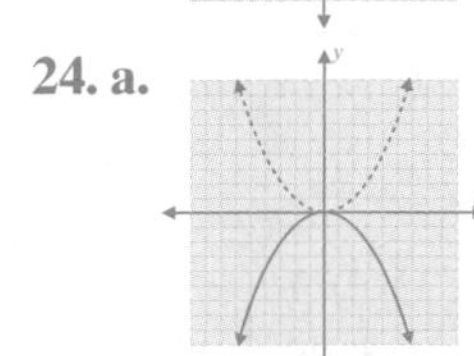

In Exercises 41–44, graph the two given functions and answer the following questions:

a. *Are the graphs the same?*

b. *Do the functions have the same zeros?*

c. *Briefly discuss your interpretation of the results in parts (a) and (b).*

41. $\begin{cases} y = x^2 - 3x - 10 \\ y = -x^2 + 3x + 10 \end{cases}$

42. $\begin{cases} y = x^2 - 5x + 6 \\ y = -x^2 + 5x - 6 \end{cases}$

43. $\begin{cases} y = 2x^2 - 5x - 3 \\ y = -2x^2 + 5x + 3 \end{cases}$

44. $\begin{cases} y = -4x^2 - 15x + 4 \\ y = 4x^2 + 15x - 4 \end{cases}$

In Exercises 45 – 48, use the function $h = -16t^2 + v_0 t + h_0$ where h is the height of the object after time t, v_0 is the initial velocity, and h_0 is the initial height.

45. A ball is thrown vertically upward from the ground with an initial velocity of 112 ft. per sec.

a. When will the ball reach its maximum height?

b. What will be the maximum height?

46. A ball is thrown vertically upward from the ground with an initial velocity of 104 ft. per sec.

a. When will the ball reach its maximum height?

b. What will be the maximum height?

47. A stone is projected vertically upward from a platform that is 32 ft. high at a rate of 128 ft. per sec.

a. When will the stone reach its maximum height?

b. What will be the maximum height?

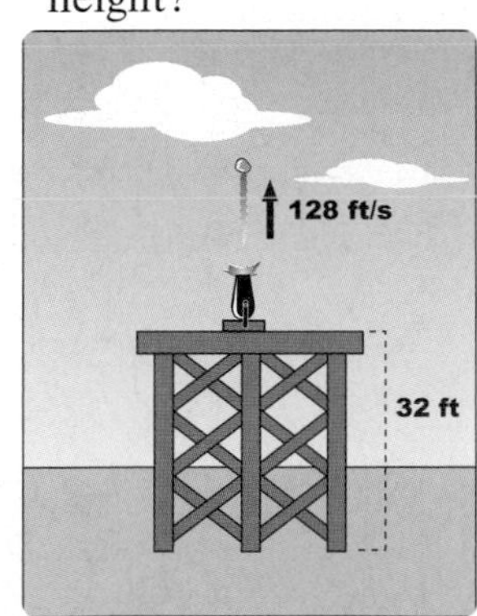

48. A stone is projected vertically upward from a platform that is 20 ft. high at a rate of 160 ft. per sec.

a. When will the stone reach its maximum height?

b. What will be the maximum height?

49. A store owner estimates that by charging x dollars each for a certain lamp, he can sell $40 - x$ lamps each week. What price will give him maximum receipts?

50. A retailer sells radios. He estimates that by selling them for x dollars each, he will be able to sell $100 - x$ radios each month.

a. What price will yield maximum revenue?

b. What will be the maximum revenue?

51. Ms. Richey can sell 72 picture frames each month if she charges \$24 each. She estimates that for each \$1 increase in price, she will sell 2 fewer frames.

a. Find the price that will yield maximum revenue.

b. What will be the maximum revenue?

24. b.

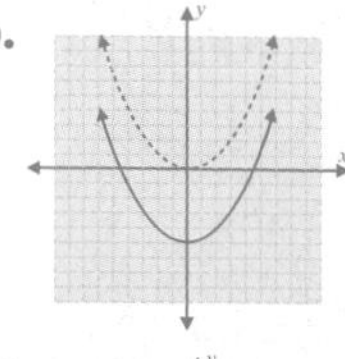

c.

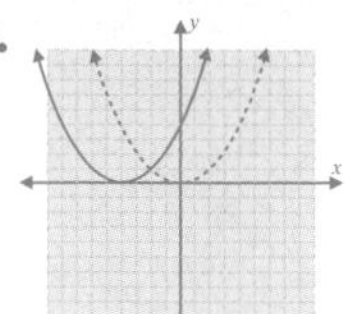

d.

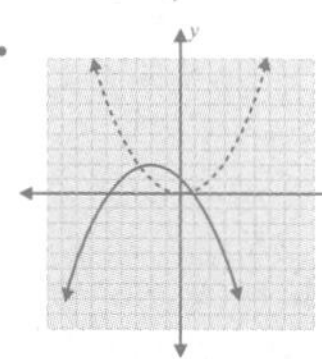

25. $y = 2(x-1)^2$
Vertex $= (1, 0)$
$R_f = \{y \mid y \geq 0\}$
Zeros: $x = 1$

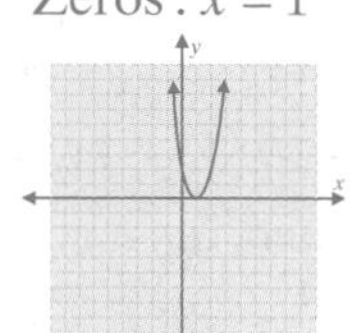

52. A contractor is to build a brick wall 6 feet high to enclose a rectangular garden. The wall will be on three sides of the rectangle because the fourth side is a building. The owner wants to enclose the maximum area but wants to pay for only 150 feet of wall. What dimensions should the contractor make the garden?

In Exercises 53 – 58, use a graphing calculator to graph each function and use the ***ZOOM*** *and* ***TRACE*** *features of the calculator to estimate the zeros of the function (if any) and the coordinates of the vertex.*

53. $y = x^2 - 2x - 2$ **54.** $y = 3x^2 + x - 1$ **55.** $y = -2x^2 + 2x + 5$

56. $y = -x^2 - 2x + 7$ **57.** $y = x^2 + 3x + 3$ **58.** $y = -4x^2 - x - 6$

Writing and Thinking About Mathematics

59. Discuss the following features of the general quadratic function $y = ax^2 + bx + c$.
a. What type of curve is its graph?
b. What is the value of x at the vertex of the parabola?
c. What is the equation of the line of symmetry?
d. Does the graph always cross the x-axis? Explain.

60. Discuss the discriminant of the general quadratic equation $y = a(x-h)^2 + k$ and how the value of the discriminant is related to the graph of the corresponding quadratic function $y = ax^2 + bx + c$.

61. Discuss the domain and range of a quadratic function in the form $y = a(x-h)^2 + k$.

Hawkes Learning Systems: Introductory & Intermediate Algebra

Graphing Parabolas

26. $y = -3(x-2)^2$
Vertex $= (2, 0)$
$R_f = \{y \mid y \leq 0\}$
Zeros: $x = 2$

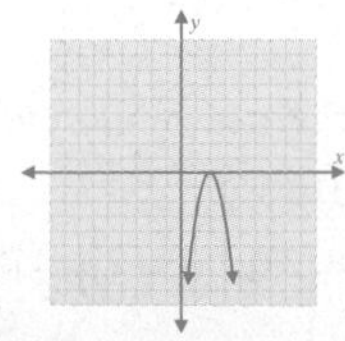

27. $y = (x-1)^2 - 4$
Vertex $= (1, -4)$
$R_f = \{y \mid y \geq -4\}$
Zeros: $x = 3, x = -1$

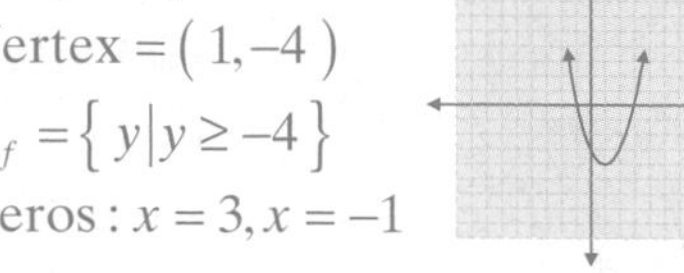

28. $y = (x-2)^2 + 1$
Vertex $= (2, 1)$
$R_f = \{y \mid y \geq 1\}$
Zeros: none

29. $y = (x+3)^2 - 4$
Vertex $= (-3, -4)$
$R_f = \{y \mid y \geq -4\}$
Zeros: $x = -5, x = -1$

30. $y = (x-4)^2 - 4$
Vertex $= (4, -4)$
$R_f = \{y \mid y \geq -4\}$
Zeros: $x = 2, x = 6$

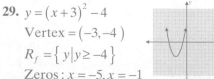

31. $y = 2(x-2)^2 - 3$
Vertex $= (2, -3)$
$R_f = \{y \mid y \geq -3\}$
Zeros: $x = \dfrac{4 \pm \sqrt{6}}{2}$

32. $y = 2\left(x - \dfrac{3}{2}\right)^2 + \dfrac{1}{2}$
Vertex $= \left(\dfrac{3}{2}, \dfrac{1}{2}\right)$
$R_f = \left\{y \mid y \geq \dfrac{1}{2}\right\}$
Zeros: none

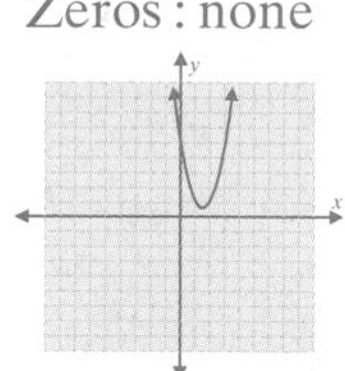

33. $y = -3(x+2)^2 + 3$
Vertex $= (-2, 3)$
$R_f = \{y \mid y \leq 3\}$
Zeros: $x = -1, x = -3$

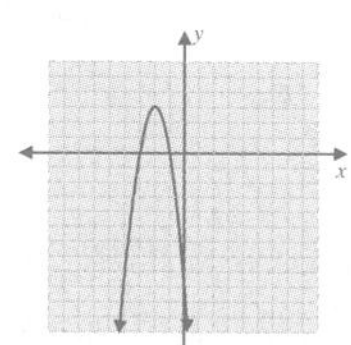

34. $y = 3(x-1)^2 - 4$
Vertex $= (1, -4)$
$R_f = \{y \mid y \geq -4\}$
Zeros: $x = \dfrac{3 \pm 2\sqrt{3}}{3}$

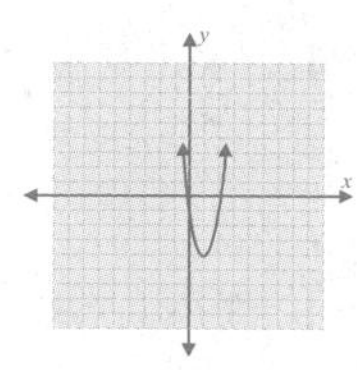

35. $y = 5(x-1)^2 + 3$
Vertex $= (1, 3)$
$R_f = \{y \mid y \geq 3\}$
Zeros: none

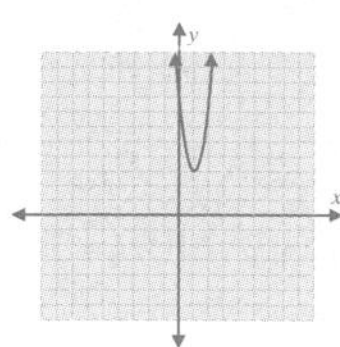

36. $y = -4(x-2)^2 + 5$
Vertex $= (2, 5)$
$R_f = \{y \mid y \leq 5\}$
Zeros: $x = \dfrac{4 \pm \sqrt{5}}{2}$

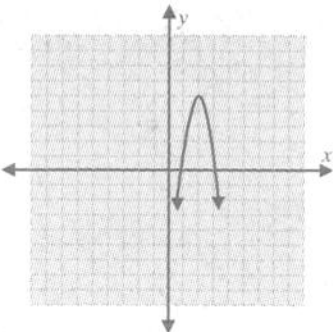

37. $y = -\left(x + \dfrac{5}{2}\right)^2 + \dfrac{17}{4}$
Vertex $= \left(-\dfrac{5}{2}, \dfrac{17}{4}\right)$
$R_f = \left\{y \mid y \leq \dfrac{17}{4}\right\}$
Zeros: $x = \dfrac{-5 \pm \sqrt{17}}{2}$

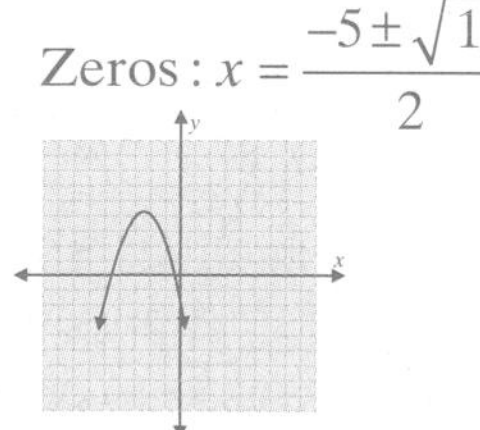

38. $y = \left(x + \dfrac{3}{2}\right)^2 - \dfrac{13}{4}$
Vertex $= \left(-\dfrac{3}{2}, -\dfrac{13}{4}\right)$
$R_f = \left\{y \mid y \geq -\dfrac{13}{4}\right\}$
Zeros: $x = \dfrac{-3 \pm \sqrt{13}}{2}$

39. $y = 2\left(x + \dfrac{7}{4}\right)^2 - \dfrac{9}{8}$
Vertex $= \left(-\dfrac{7}{4}, -\dfrac{9}{8}\right)$

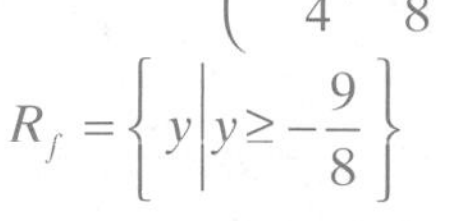

$R_f = \left\{y \mid y \geq -\dfrac{9}{8}\right\}$
Zeros: $x = -1, x = -\dfrac{5}{2}$

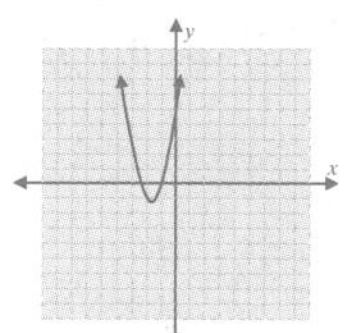

40. $y = 2\left(x + \dfrac{1}{4}\right)^2 - \dfrac{25}{8}$
Vertex $= \left(-\dfrac{1}{4}, -\dfrac{25}{8}\right)$
$R_f = \left\{y \mid y \geq -\dfrac{25}{8}\right\}$
Zeros: $x = 1, x = -\dfrac{3}{2}$

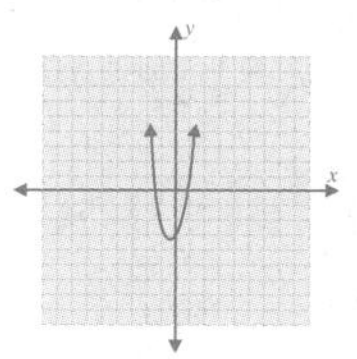

41. **a.** No **b.** Yes
c. Answers will vary.

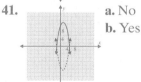

42. **a.** No **b.** Yes

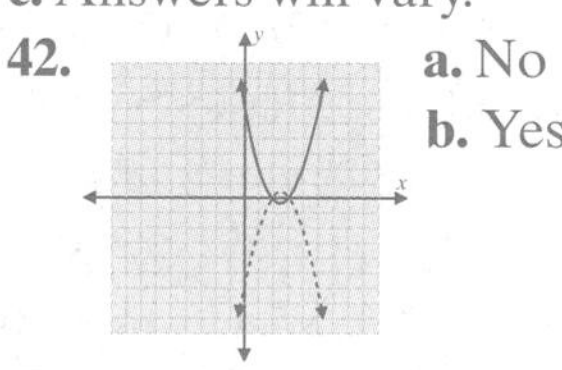

c. Answers will vary.

43. **a.** No **b.** Yes

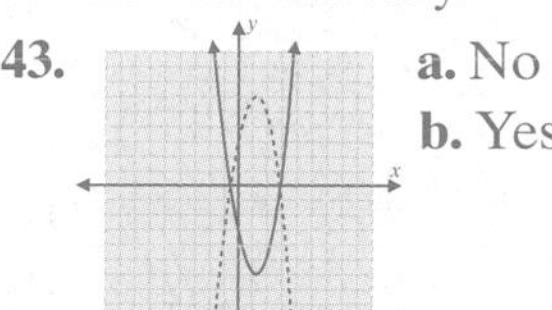

c. Answers will vary.

44. **a.** No **b.** Yes

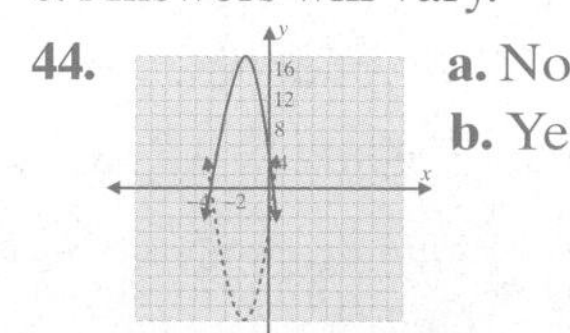

c. Answers will vary.

45. **a.** $3\frac{1}{2}$ sec. **b.** 196 ft.
46. **a.** $3\frac{1}{4}$ sec. **b.** 169 ft.
47. **a.** 4 sec. **b.** 288 ft.
48. **a.** 5 sec. **b.** 420 ft.
49. $20
50. **a.** $50 **b.** $2500
51. **a.** $30 **b.** $1800
52. 37.5 ft. × 75 ft.
53. zeros: $x \approx 2.732$, $x \approx -0.732$; vertex $(1, -3)$
54. zeros: $x \approx -0.768$, $x \approx 0.434$; vertex $\left(-\dfrac{1}{6}, -\dfrac{13}{12}\right)$
55. zeros: $x \approx 2.158$, $x \approx -1.158$; vertex $\left(\dfrac{1}{2}, \dfrac{11}{2}\right)$
56. zeros: $x \approx -3.828$, $x \approx 1.828$; vertex $(-1, 8)$
57. no real zeros; vertex $\left(-\dfrac{3}{2}, \dfrac{3}{4}\right)$
58. no real zeros; vertex $\left(-\dfrac{1}{8}, -\dfrac{95}{16}\right)$
59. **a.** a parabola
b. $x = -\dfrac{b}{2a}$
c. $x = h$
d. No. Answers will vary.
60. discriminant > 0 - graph crosses x-axis twice; discriminant < 0 - graph is entirely above or entirely below the x-axis; discriminant $= 0$ - graph is tangent to the x-axis.
61. Domain $= (-\infty, \infty)$. For $|a| > 0$, Range $= [k, \infty)$. For $|a| < 0$, Range $= (-\infty, k]$

12.2 Quadratic and Other Inequalities

Objectives

After completing this section, you will be able to:

1. *Solve quadratic and other inequalities.*
2. *Solve higher degree inequalities.*
3. *Graph the solutions for inequalities on real number lines.*

Teaching Notes:
Many common situations routinely and naturally use the inequality concept. You might want to remind students that common problems dealing with budgets, capacities, sales quotas, profit, fluid pressure, memberships, etc. are more likely expressed in terms of "at least", "at most", "not greater than", "not less than", "minimum", and "maximum", rather than discrete numbers.

In Section 3.5, we solved first-degree (or linear) inequalities and, in Section 7.4, we solved inequalities involving rational expressions. In this section we will solve polynomial inequalities of second-degree or higher with emphasis on second-degree or quadratic inequalities. For example,

$$x^2+3x+2\geq 0,\qquad x^2-2x>8,\qquad \text{and}\qquad x^3+4x^2-5x<0.$$

We will use two basic methods: solving algebraically by factoring or by using the quadratic formula, and solving by using a graphing calculator.

Polynomial Inequalities: Solved Algebraically

The technique used is based on the same concepts that were used in Section 7.4. That is, the sign of a factor in the form $(x-a)$ changes for x on either side of a. In other words,

if $x>a$, then $(x-a)$ is positive.

if $x<a$, then $(x-a)$ is negative.

The procedure for solving polynomial inequalities algebraically involves getting 0 on one side of the inequality and then factoring (if possible) the polynomial on the other side. The values that make each factor 0 are used to determine intervals over which the polynomial is positive or negative.

To Solve a Polynomial Inequality

1. *Arrange the terms so that one side of the inequality is 0.*
2. *Factor the algebraic expression, if possible, and find the points (numbers) where each factor is 0. (Use the quadratic formula, if necessary.)*
3. *Mark these points on a number line.*
4. *Test one point from each interval to determine the sign of the polynomial expression for all points in that interval.*
5. *The solution consists of those intervals where the test points satisfy the original inequality.*

The following examples illustrate the technique.

Example 1: Solving Polynomial Inequalities Algebraically

Solve the following inequalities by factoring and using a number line. Graph the solution set on a number line.

a. $x^2 - 2x > 8$

Solution: $x^2 - 2x > 8$

$x^2 - 2x - 8 > 0$ Add -8 to both sides so that one side is 0.

$(x+2)(x-4) > 0$ Factor.

Set each factor equal to 0 to find endpoints of intervals.

$x+2=0$ $\quad$ $x-4=0$

$x=-2$ $\quad$ $x=4$

Mark these points on a number line and test one point from each of the intervals formed.

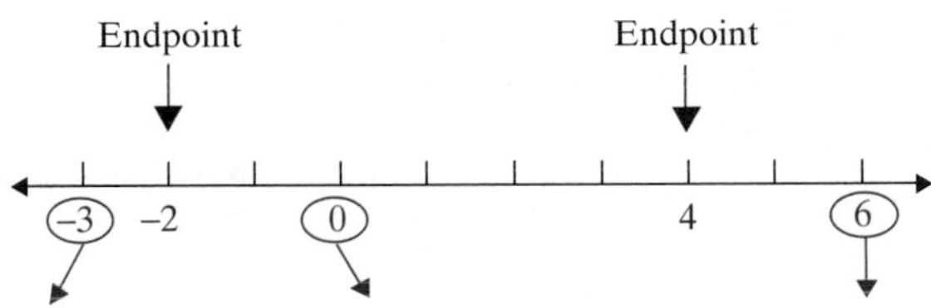

Test $x=-3$:

$(-3+2)(-3-4)$

$=(-1)(-7)=7>0$

This means that $(x+2)(x-4)>0$ if $x<-2$.

Test $x=0$:

$(0+2)(0-4)$

$=(2)(-4)=-8<0$

This means that $(x+2)(x-4)<0$ if $-2<x<4$.

Test $x=6$:

$(6+2)(6-4)$

$=(8)(2)=16>0$

This means that $(x+2)(x-4)>0$ if $x>4$.

−3 −2 0 4 6

The solution is:

(algebraic notation) $x<-2$ or $x>4$ $\quad$ or $\quad$ (interval notation) x is in $(-\infty,-2) \cup (4,\infty)$

b. $2x^2 + 15 \leq 13x$

Solution: $2x^2 + 15 \leq 13x$

$2x^2 - 13x + 15 \leq 0$ Add $-13x$ to both sides so that one side is 0.

$(2x-3)(x-5) \leq 0$ Factor.

continued on next page ...

Set each factor equal to 0 to locate interval endpoints.

$$2x-3=0 \qquad x-5=0$$

$$x=\frac{3}{2} \qquad x=5$$

Test one point from each interval formed.

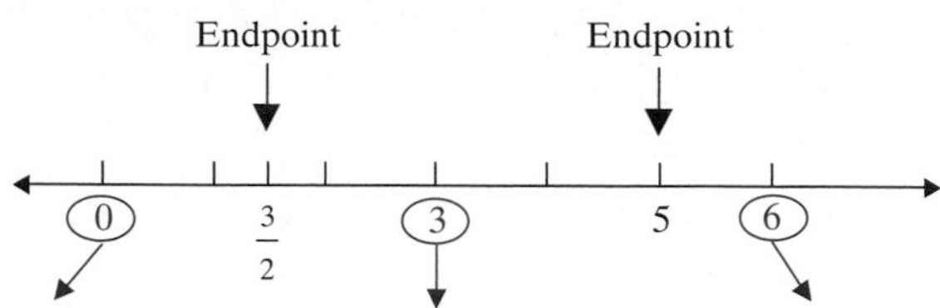

Test $x=0$:

$$(2\cdot 0-3)(0-5)$$
$$=(-3)(-5)=15>0$$

This means that $(2x-3)(x-5)>0$ if $x<\frac{3}{2}$.

Test $x=3$:

$$(2\cdot 3-3)(3-5)$$
$$=(3)(-2)=-6<0$$

This means that $(2x-3)(x-5)<0$ if $\frac{3}{2}<x<5$.

Test $x=6$:

$$(2\cdot 6-3)(6-5)$$
$$=(9)(1)=9>0$$

This means that $(2x-3)(x-5)>0$ if $x>5$.

The solution includes both endpoints since the inequality ($\le$) includes 0:

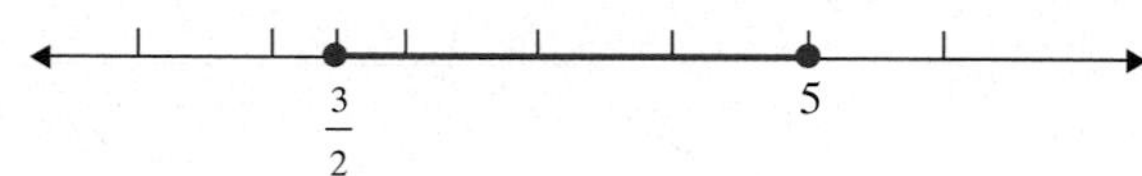

The solution is:

(algebraic notation)	or	(interval notation)
$\frac{3}{2}\le x\le 5$		x is in $\left[\frac{3}{2},5\right]$

c. $x^3+4x^2-5x<0$

Solution:

$$x^3+4x^2-5x<0$$
$$x\left(x^2+4x-5\right)<0$$
$$x(x+5)(x-1)<0$$

Set each factor equal to 0 to locate interval endpoints.

$$x=0 \qquad x+5=0 \qquad x-1=0$$
$$x=-5 \qquad x=1$$

Test one point from each interval formed.

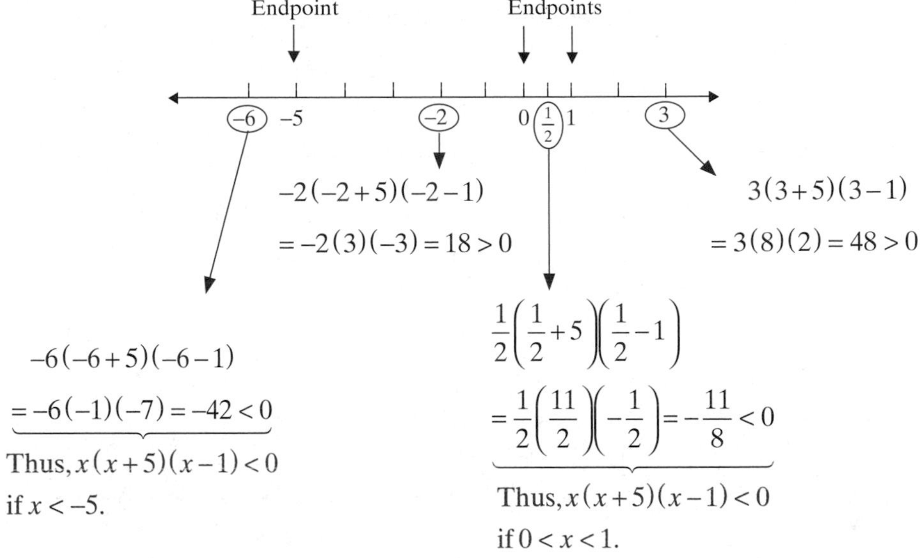

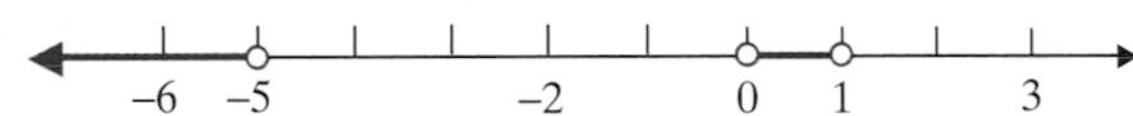

The solution is:

(algebraic notation)	or	(interval notation)
$x<-5$ or $0<x<1$		x is in $(-\infty,-5) \cup (0,1)$

d. $x^2-2x-1>0$

Solution: The quadratic expression x^2-2x-1 will not factor with integer coefficients. Use the quadratic formula and find the roots of the equation $x^2-2x-1=0$. Then use these roots as endpoints for the intervals. The test points can themselves be integers.

$$x^2-2x-1=0$$

$$x=\frac{2\pm\sqrt{(-2)^2-4(1)(-1)}}{2} \quad \text{Use the quadratic formula.}$$

$$x=\frac{2\pm\sqrt{4+4}}{2}$$

$$x=\frac{2\pm2\sqrt{2}}{2}=1\pm\sqrt{2}$$

The endpoints are $x=1-\sqrt{2}$ and $x=1+\sqrt{2}$.

continued on next page ...

Test one point from each interval in the expression $x^2 - 2x - 1$.
Note: With a calculator, you can determine that $1-\sqrt{2} \approx -0.414$ and $1+\sqrt{2} \approx 2.414$.

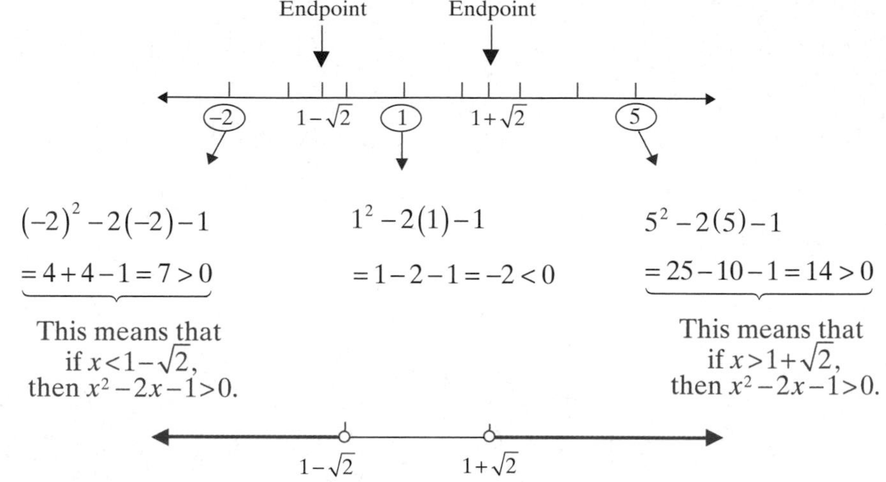

The solution is:

(algebraic notation) $x < 1-\sqrt{2}$ or $x > 1+\sqrt{2}$ or   (interval notation) x is in $\left(-\infty, 1-\sqrt{2}\right) \cup \left(1+\sqrt{2}, \infty\right)$

e. $x^2 - 2x + 13 > 0$

Solution: To find where $x^2 - 2x + 13 > 0$, use the quadratic formula:

$$x = \frac{2 \pm \sqrt{(-2)^2 - 4(1)(13)}}{2} = \frac{2 \pm \sqrt{-48}}{2}$$

$$= \frac{2 \pm 4i\sqrt{3}}{2} = 1 \pm 2i\sqrt{3}$$

Since these values are nonreal, the polynomial is either always positive or always negative for real values of x. Therefore, we only need to test one point. If that point satisfies the inequality, then the solution is all real numbers. Otherwise, there is no solution. In this example, we test $x = 0$ because the polynomial is easy to evaluate for $x = 0$.

$$0^2 - 2(0) + 13 = 13 > 0$$

Since the real number 0 satisfies the inequality, the solution is all real numbers. In interval notation, we write $(-\infty, \infty)$. Graphically,

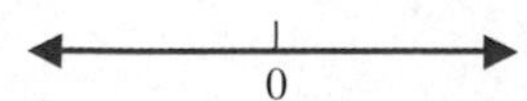

Quadratic Inequalities: Solved with a Graphing Calculator

In Section 3.5, we solved linear inequalities with a graphing calculator by shading on one side of a straight line. To solve quadratic (or other polynomial) inequalities we will simply graph the corresponding function and find intervals over which the function is positive or negative by noting where the graph is above the x-axis and where it is below the x-axis.

To Solve a Polynomial Inequality with a Graphing Calculator

1. *Arrange the terms so that one side of the inequality is 0.*
2. *Set the quadratic (or other polynomial) equal to y and graph the function.*
   [Be sure to set the ***WINDOW*** *so that all of the zeros are easily seen.]*
3. *Use the* ***TRACE*** *key or* ***CALC*** *key to approximate the zeros of the function.*
4. **a.** *If the inequality is of the form $y > 0$, then the solution consists of those intervals of x where the graph is above the x-axis.*
   b. *If the inequality is of the form $y < 0$, then the solution consists of those intervals of x where the graph is below the x-axis.*
5. *Endpoints of the intervals are included if the inequality is $y \geq 0$ or $y \leq 0$.*

Example 2: Solving a Polynomial Inequality with a Graphing Calculator

Solve the following inequalities by using a graphing calculator. Graph the solution set on a number line.

a. $x^2 - 2x > 8$

Solution: This inequality has been solved algebraically in Example 1a. We repeat the solution here to show how the two methods are related.

$x^2 - 2x > 8$

$x^2 - 2x - 8 > 0$ Add -8 to both sides so that one side is 0.

$y = x^2 - 2x - 8$ Set the quadratic expression equal to y.

Now graph the function (in this case a parabola).

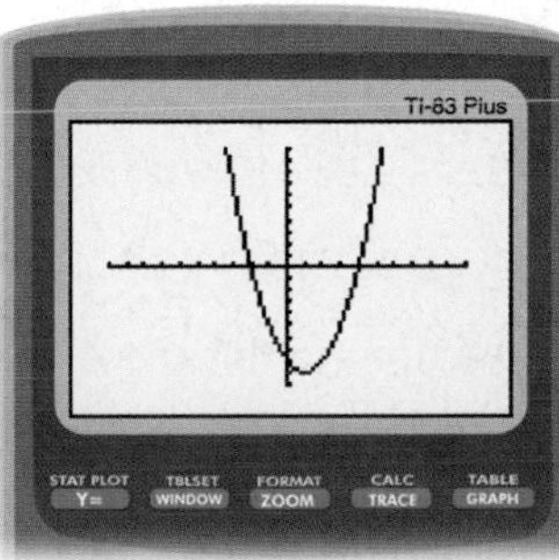

continued on next page ...

Pressing **2nd** CALC > `2:zero` and following the directions for `LeftBound?` and `RightBound?` and `Guess?` will give $x=-2$ and $x=4$ (the same two values for x that we found in Example 1a). Because we want to know where $y>0$, we look for the intervals on the x-axis where the parabola is above the x-axis.

The solution set is (as we expected) $(-\infty,-2)$ or $(4,\infty)$.

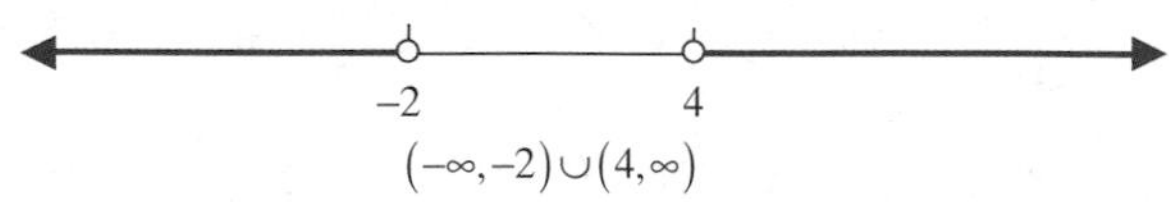

$(-\infty,-2)\cup(4,\infty)$

CAUTION: Choose the Left Bound and Right Bound so that only one zero is between them. Otherwise, you will get an ERROR message.

b. $2x^2+3x-10\le 0$

Solution: $2x^2+3x-10\le 0$

$y=2x^2+3x-10$ Set the expression equal to y.

Now graph the function (in this case a parabola).

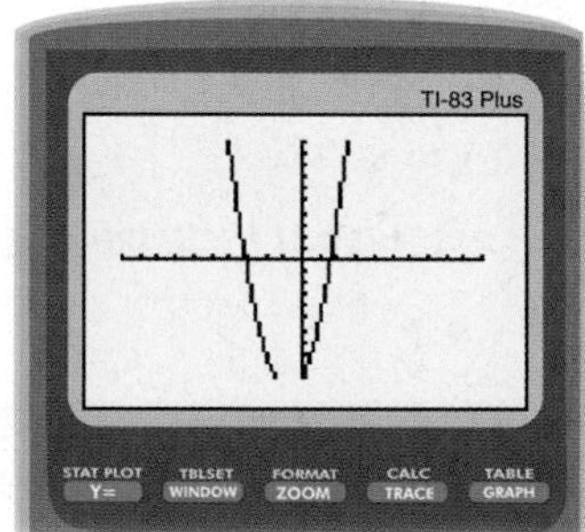

Pressing **2nd** CALC > `2:zero` and following the directions for `Left Bound?` and `RightBound?` and `Guess?` will give the approximate values $x=-3.1085$ and $x=1.6085$. Because we want to know where $y\le 0$, we look for the intervals on the x-axis where the parabola is below the x-axis.

The solution is the closed interval $[-3.1085, 1.6085]$ (with the understanding that we have only 4 decimal place accuracy).

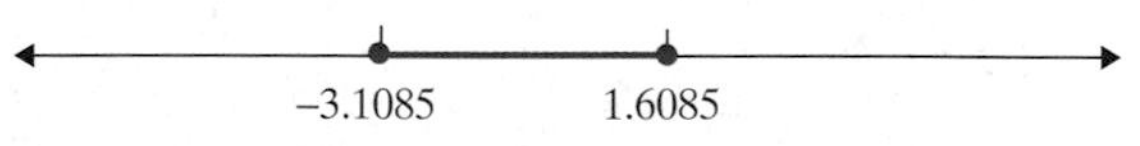

c. $x^3+2x^2-11x-12>0$

Solution: $x^3+2x^2-11x-12>0$

$y=x^3+2x^2-11x-12$ Set the expression equal to y.

Now graph the function (in this case not a parabola).

1. $(-2,6)$

2. $(-\infty,-4)\cup(2,\infty)$

3. $\left(-\infty,\frac{2}{3}\right)\cup(5,\infty)$

4. $\left[-1,-\frac{1}{4}\right]$

5. $(-\infty,-7]\cup\left[\frac{5}{2},\infty\right)$

6. $\left[\frac{3}{5},3\right]$

7. $\left[-2,-\frac{1}{3}\right]$

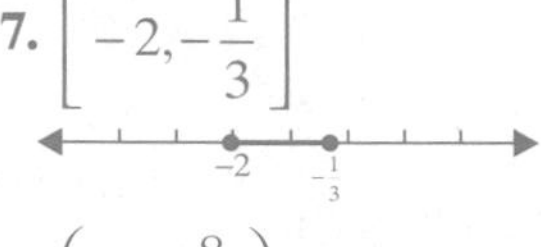

8. $\left(-\infty,\frac{8}{3}\right)\cup(4,\infty)$

9. $\left(-\infty,-\frac{4}{3}\right)\cup(0,5)$

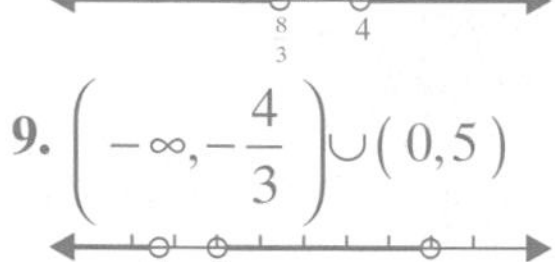

10. $(-\infty,-4)\cup\left(-\frac{5}{2},1\right)$

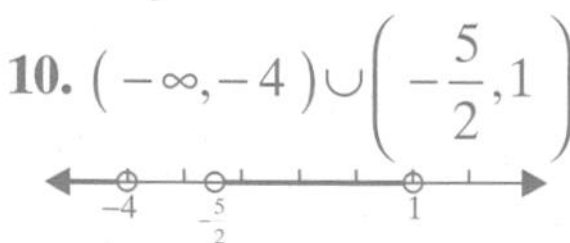

11. $x=-2$

12. $(-\infty,-2)\cup\left(\frac{6}{5},\infty\right)$

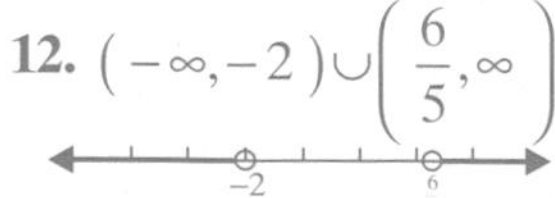

13. $\left(-\infty,-\frac{5}{2}\right)\cup(3,\infty)$

14. $\left(-\infty,-\frac{2}{3}\right)\cup\left(\frac{1}{2},\infty\right)$

15. $\left(-\frac{1}{4},\frac{3}{2}\right)$

16. $\left(-2,\frac{5}{2}\right)$

17. $\left(-\infty,\frac{1}{2}\right]\cup[2,\infty)$

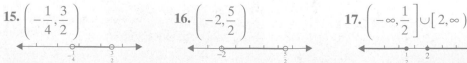

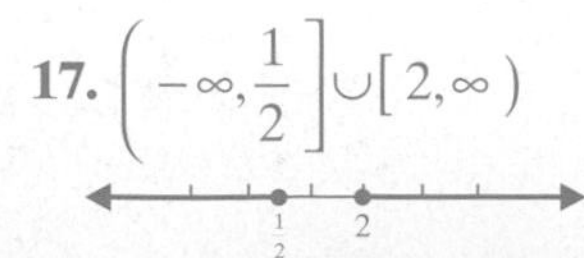

18. $\left(-\frac{3}{5},2\right)$

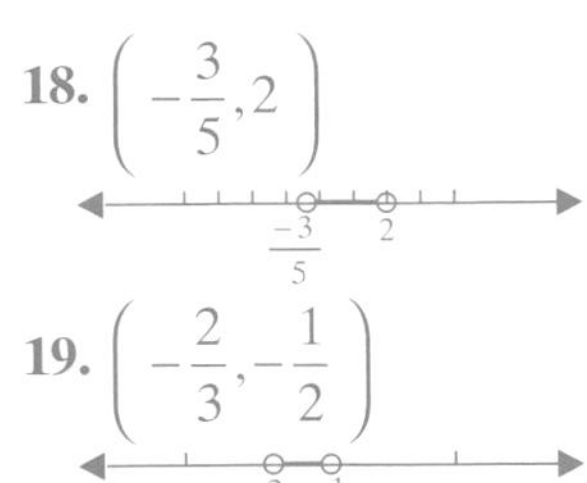

19. $\left(-\frac{2}{3},-\frac{1}{2}\right)$

20. $\left(-\infty,\frac{1}{3}\right]\cup\left[3,\infty\right)$

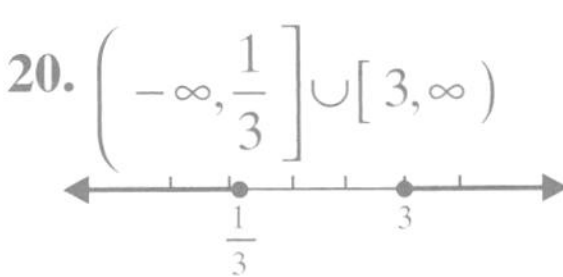

21. $\left(-\infty,\frac{5}{2}\right)\cup\left(\frac{5}{2},\infty\right)$

22. $\left[-\frac{2}{3},\frac{7}{5}\right]$

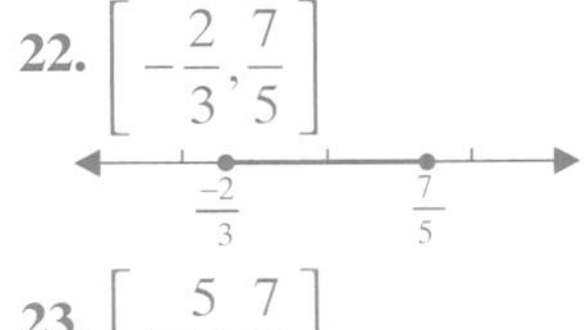

23. $\left[-\frac{5}{2},\frac{7}{4}\right]$

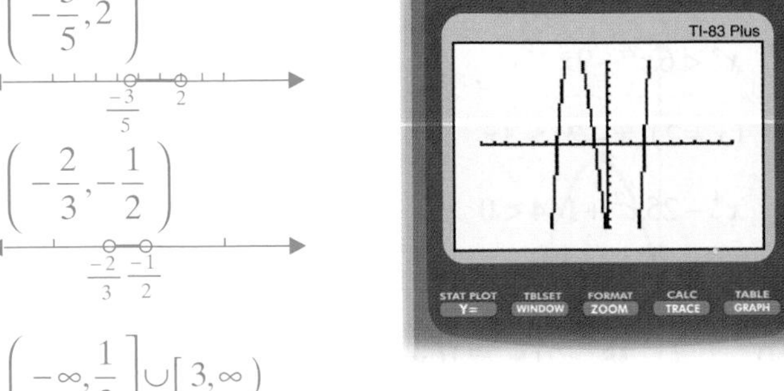

Pressing 2nd CALC > `2:zero` and following the directions for `LeftBound?` and `RightBound?` and `Guess?` will give the three values $x=-4$, $x=-1$, and $x=3$.

Because we want to know where $y>0$, we look for the intervals on the x-axis where the curve is above the x-axis.

The solution is the union of two intervals $(-4,-1)\cup(3,\infty)$.

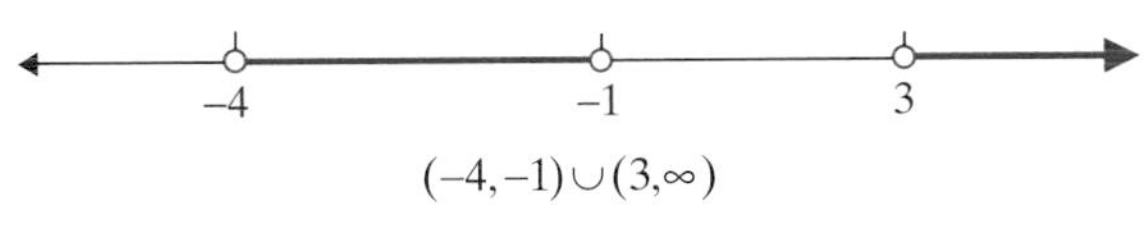

12.2 Exercises

24. $(-7,0)\cup(1,\infty)$

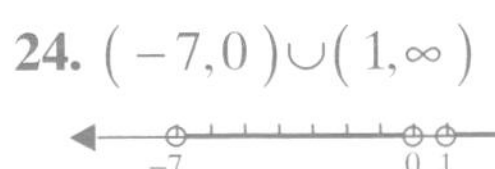

25. $(-1,0)\cup(3,\infty)$

26. $(-\infty,0)$

27. $(0,1)\cup(4,\infty)$

28. $[0,1]\cup[3,\infty)$

29. $(-\infty,-1)\cup(4,\infty)$

30. $(-3,2)$

31. $(-\infty,-2)\cup(-1,1)\cup(2,\infty)$

*In Exercises 1 – 50, solve the inequalities algebraically and graph each solution set on a number line. Write the answers in interval notation. (**Note:** You may need to use the quadratic formula to find endpoints of intervals.)*

1. $(x-6)(x+2)<0$

2. $(x+4)(x-2)>0$

3. $(3x-2)(x-5)>0$

4. $(4x+1)(x+1)\le 0$

5. $(x+7)(2x-5)\ge 0$

6. $(x-3)(5x-3)\le 0$

7. $(3x+1)(x+2)\le 0$

8. $(x-4)(3x-8)>0$

9. $x(3x+4)(x-5)<0$

10. $(x-1)(x+4)(2x+5)<0$

11. $x^2+4x+4\le 0$

12. $5x^2+4x-12>0$

13. $2x^2>x+15$

14. $6x^2+x>2$

15. $8x^2<10x+3$

16. $2x^2<x+10$

17. $2x^2-5x+2\ge 0$

18. $15y^2-21y-18<0$

19. $6y^2+7y<-2$

20. $3x^2+3\ge 10x$

21. $4z^2-20z+25>0$

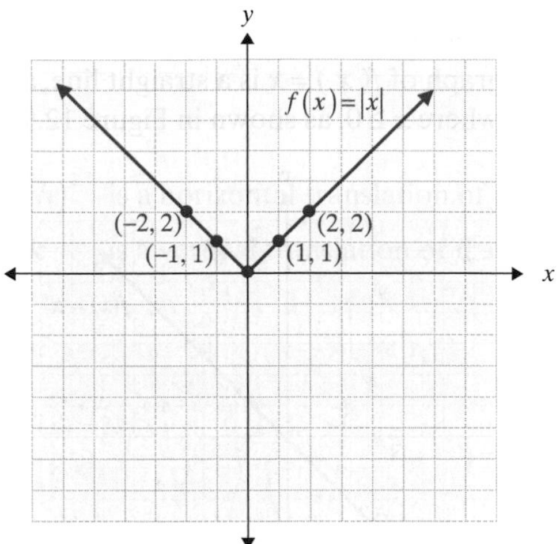

Figure 12.11

The following examples illustrate the graphs of horizontal and vertical translations of the graph of the function $f(x) = |x|$.

Example 3: Horizontal and Vertical Translations

Graph each of the following functions.

a. $y = |x - 3| + 2$

Solution: Here $(h, k) = (3, 2)$, so there is a horizontal translation of 3 units and a vertical translation of 2 units. In effect, $(3, 2)$ is the vertex of the graph just as $(0, 0)$ is the vertex for $y = |x|$. The points $(2, 3)$ and $(4, 3)$ are shown to make sure that the graph is in the right position. You should check that both points satisfy the function.

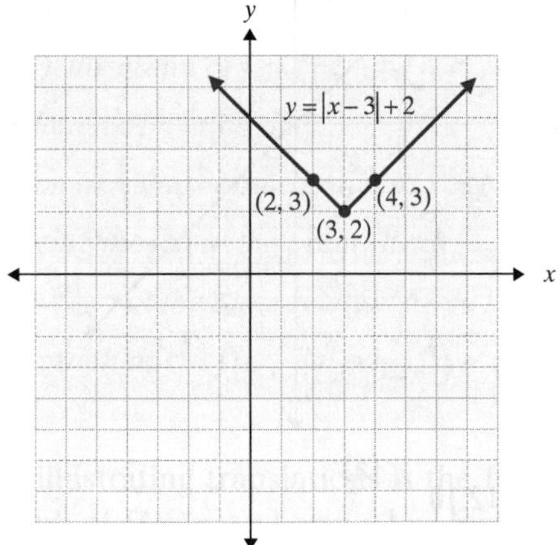

b. $y = |x + 4| - 1$

Solution: Here $(h, k) = (-4, -1)$, so there is a horizontal translation of -4 and a vertical translation of -1. The effect is that the vertex is now at the point $(-4, -1)$.

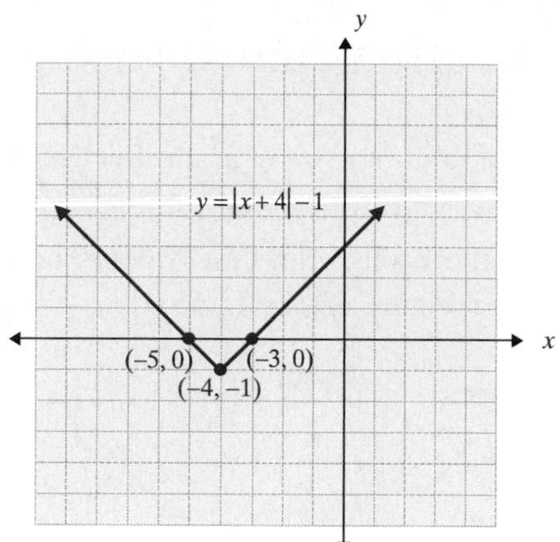

c. $y = |x + 2|$

Solution: Here $(h, k) = (-2, 0)$, so there is a horizontal translation of –2 units. There is no vertical translation.

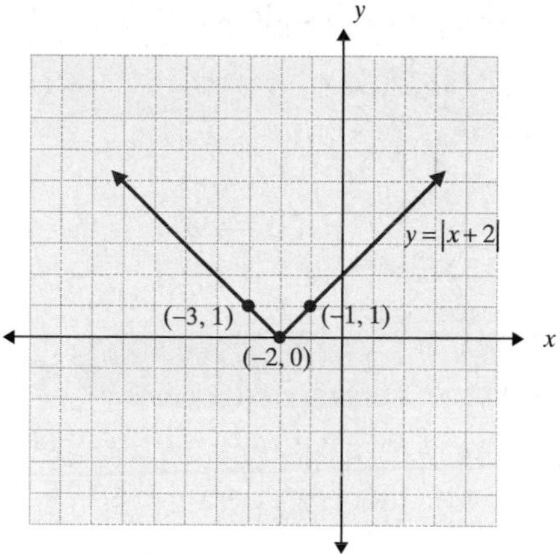

d. $y = |x + 4| - 7.$

Solution: Here $(h, k) = (-4, -7)$, so the graph of $y = |x|$ is translated –4 units horizontally and –7 units vertically.

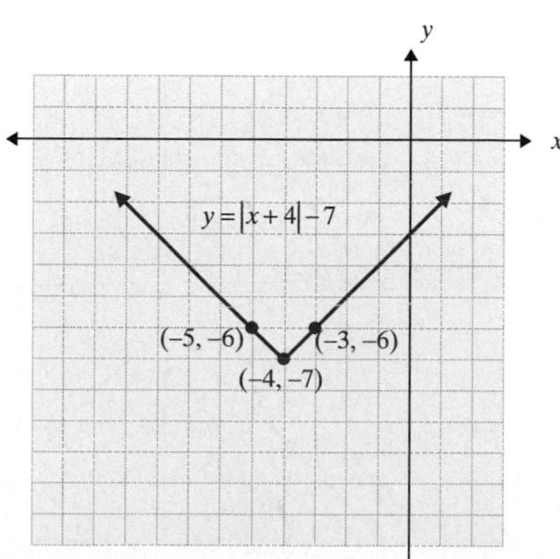

Reflections and Translations

The graph of $y = -f(x)$ is a **reflection across the x-axis** of the graph of $y = f(x)$. In the case of $y = |x|$ and the graph of $y = -|x|$, each graph is the mirror image of the other across the x-axis. The first "opens" upward and the second "opens" downward as illustrated in Figure 12.12.

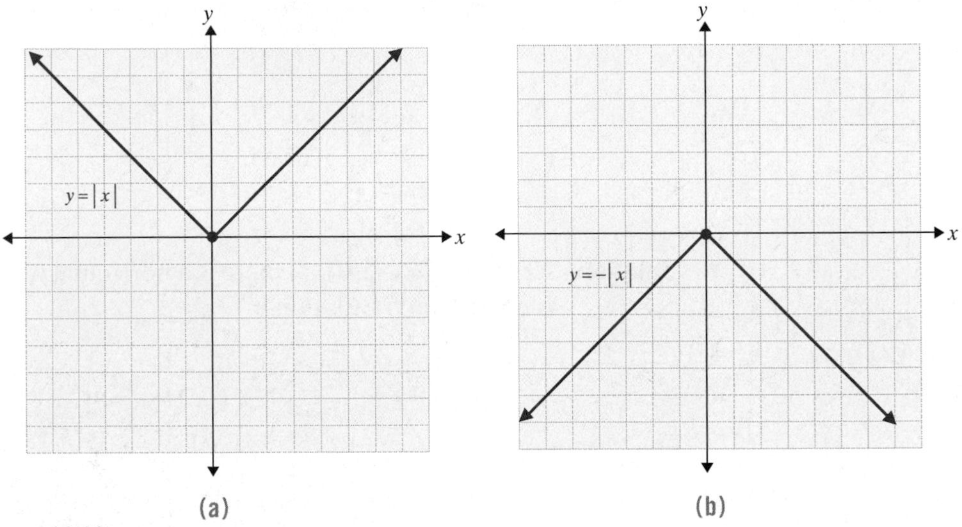

Figure 12.12

Example 4: Reflections and Translations

Graph the function $y = -|x+2| + 5$.

Solution: Here $(h,k) = (-2,5)$, and the graph is reflected across the x-axis and "opens" downward. We show step-by-step how to "arrive" at the graph. (You should do these steps mentally and graph only the last step.)

Step 1:
Graph the reflection $y = -|x|$.

Step 2:
Translate the graph horizontally 2 units left.

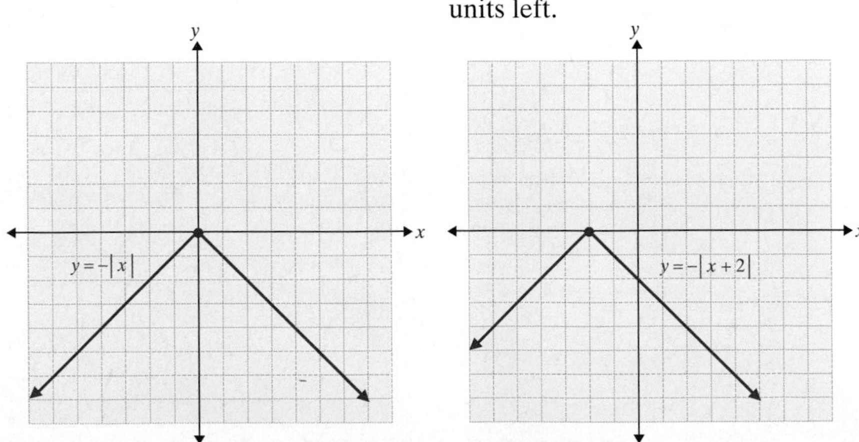

Step 3:
Translate the graph vertically 5 units.

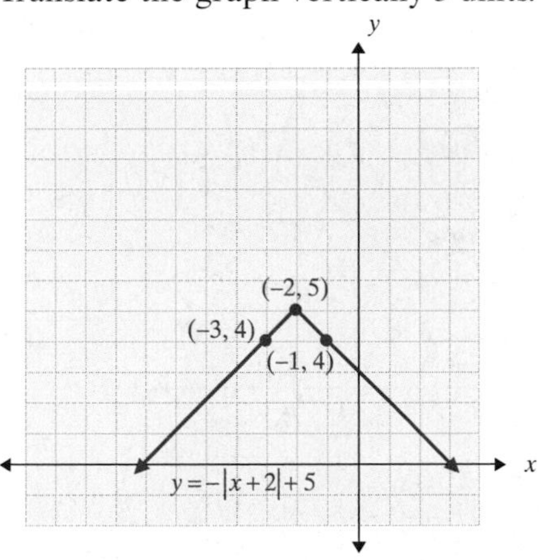

Example 5 illustrates how the concept of translation can be applied to the graph of any function if the graph of the function is known.

Example 5: Translations of Functions

a. Graph the function $y=\sqrt{x-2}+1$. The graph of $y=\sqrt{x}$ is given.

Solution: If $y=\sqrt{x}$ is written $y=f(x)$, then $y=\sqrt{x-2}+1$ is the same as $y=f(x-2)+1$. So, $(h, k) = (2, 1)$, and there is a horizontal translation of 2 units and a vertical translation of 1 unit.

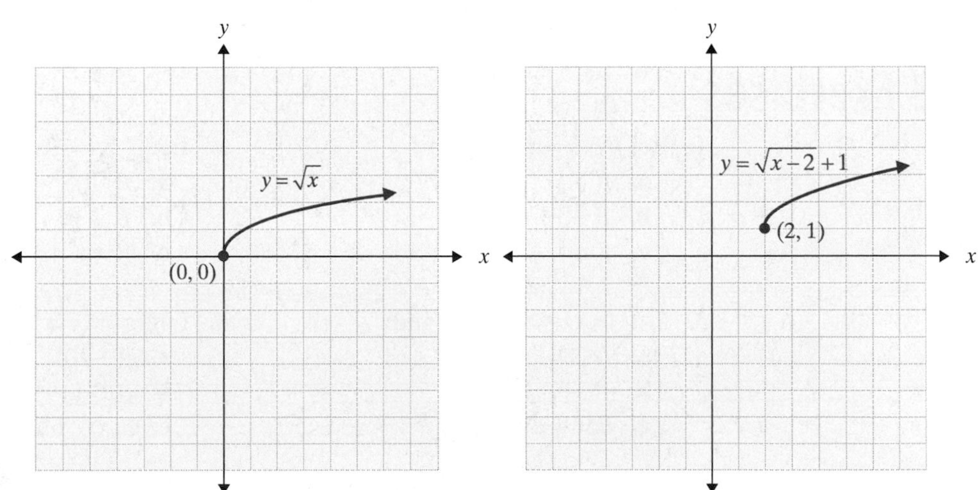

continued on next page ...

b. The graph of $y = f(x)$ is given. Draw the graph $y = f(x - 3) - 2$.
Solution: Here $(h, k) = (3, -2)$, so translate horizontally 3 units and vertically -2 units. (Add 3 to each x-value and -2 to each y-value.)

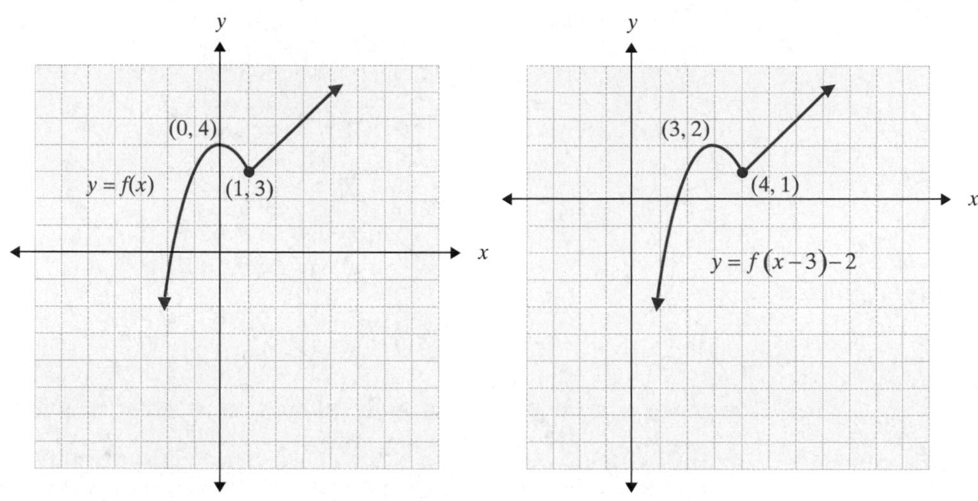

Practice Problems

1. *For $f(x) = x^2 - 5$, find:*
   a. $f(0)$
   b. $f(a)$
   c. $f(a + 2)$

2. *If $g(x) = 3x + 7$, find:*
   a. $g(0)$
   b. $g(x + h)$
   c. $\dfrac{g(x+h) - g(x)}{h}$

12.3 Exercises

1. a. 12 b. 4 c. $a + 8$
2. a. -3 b. 11 c. $2x + 5$
3. a. 4 b. 31 c. $x^2 - 4x - 1$
4. a. 17 b. 26 c. $x^2 - 6x + 10$

1. Let $f(x) = x + 7$. Find:
   a. $f(5)$
   b. $f(-3)$
   c. $f(a + 1)$

2. Let $f(x) = 2x - 3$. Find:
   a. $f(0)$
   b. $f(7)$
   c. $f(x + 4)$

3. Let $f(x) = x^2 - 5$. Find:
   a. $f(3)$
   b. $f(-6)$
   c. $f(x - 2)$

4. Let $g(x) = x^2 + 1$. Find:
   a. $g(-4)$
   b. $g(5)$
   c. $g(x - 3)$

Answers to Practice Problems: **1. a.** $f(0) = -5$ **b.** $f(a) = a^2 - 5$ **c.** $f(a+2) = (a+2)^2 - 5 = a^2 + 4a - 1$

2. a. $g(0) = 7$ **b.** $g(x + h) = 3(x + h) + 7$ **c.** $\dfrac{g(x+h) - g(x)}{h} = 3$

5. a. 1 b. $4a + 5$
c. $4x + 4h - 3$ d. 4
6. a. 7 b. $1 - 2a$
c. $5 - 2x - 2h$ d. -2
7. a. 0 b. $a^2 - 6a + 5$
c. $x^2 + 2xh + h^2 - 4$
d. $2x + h$
8. a. 0 b. $1 - a^2 + 2a$
c. $2 - x^2 - 2xh - h^2$
d. $-2x - h$
9. a. -3 b. $2a^2 - 8a + 5$
c. $2x^2 + 4xh + 2h^2 - 3$
d. $4x + 2h$
10. a. 44
b. $3a^2 + 11a + 10$
c. $3x^2 + 6xh + 3h^2 - x - h$
d. $6x + 3h - 1$

11. 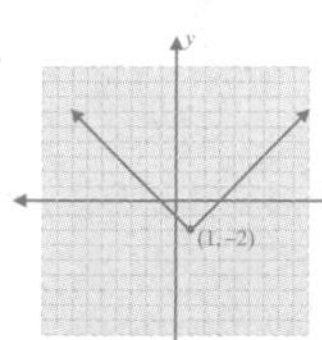

12. 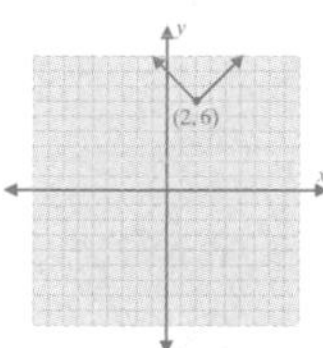

13. 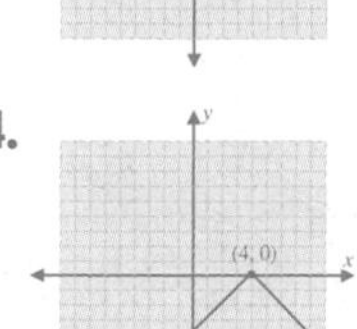

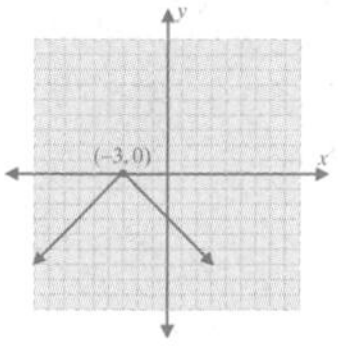

14. 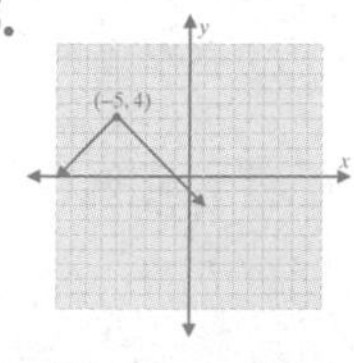

15. (−5, 4)

5. Let $g(x) = 4x - 3$. Find:
a. $g(1)$
b. $g(a + 2)$
c. $g(x + h)$
d. $\dfrac{g(x+h) - g(x)}{h}$

6. Let $f(x) = 5 - 2x$. Find:
a. $f(-1)$
b. $f(a + 2)$
c. $f(x + h)$
d. $\dfrac{f(x+h) - f(x)}{h}$

7. Let $f(x) = x^2 - 4$. Find:
a. $f(-2)$
b. $f(a - 3)$
c. $f(x + h)$
d. $\dfrac{f(x+h) - f(x)}{h}$

8. Let $g(x) = 2 - x^2$. Find:
a. $g(\sqrt{2})$
b. $g(a - 1)$
c. $g(x + h)$
d. $\dfrac{g(x+h) - g(x)}{h}$

9. Let $f(x) = 2x^2 - 3$. Find:
a. $f(0)$
b. $f(a - 2)$
c. $f(x + h)$
d. $\dfrac{f(x+h) - f(x)}{h}$

10. Let $f(x) = 3x^2 - x$. Find:
a. $f(4)$
b. $f(a + 2)$
c. $f(x + h)$
d. $\dfrac{f(x+h) - f(x)}{h}$

Using the graph of $y = |x|$, graph the functions in Exercises 11 – 20 without additional computation. (See Example 3.)

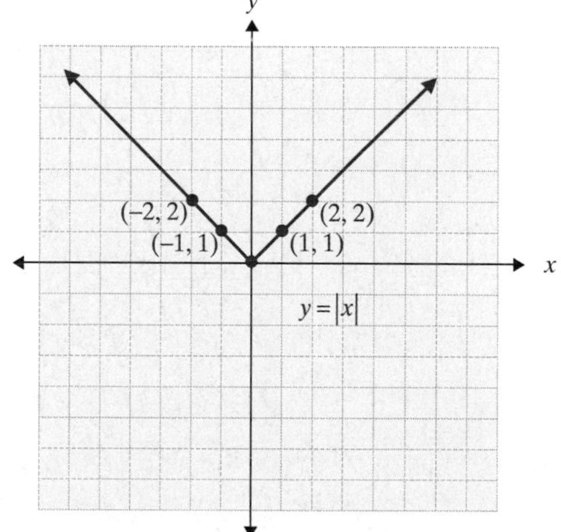

11. $y = |x - 1| - 2$ **12.** $y = |x - 2| + 6$ **13.** $y = -|x + 3|$

14. $y = -|x - 4|$ **15.** $y = -|x + 5| + 4$ **16.** $y = \left|x + \frac{3}{4}\right| - 3$

17. $y = |x - 3| + 5$ **18.** $y = |x + 2| - 3$ **19.** $y = \left|x + \frac{1}{2}\right| - \frac{3}{2}$

20. $y = \frac{5}{2}\left|x - \frac{2}{3}\right|$

16.

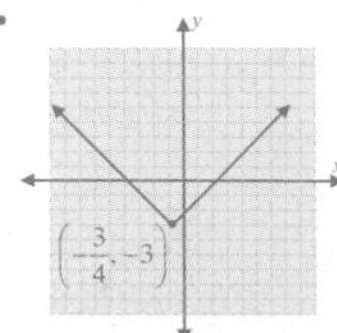

17.

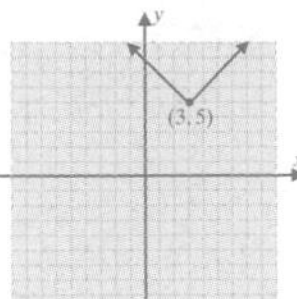

18.

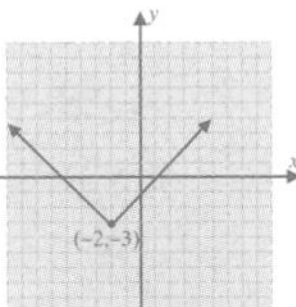

19.

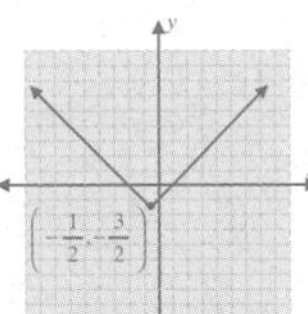

20.

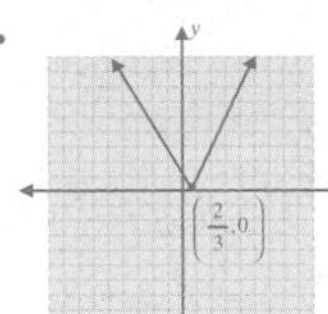

21.

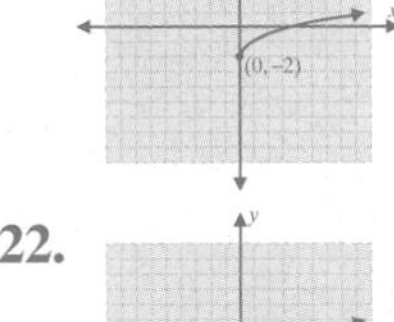

22.

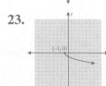

23.

24.

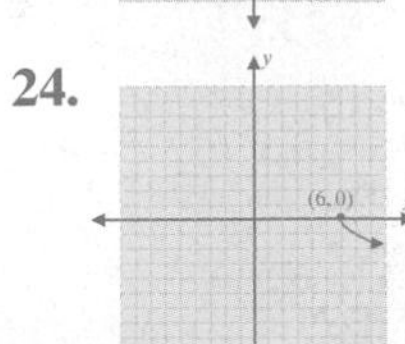

Using the graph of $y = \sqrt{x}$, *graph the functions in Exercises 21 – 30 without additional computation. (See Example 5.)*

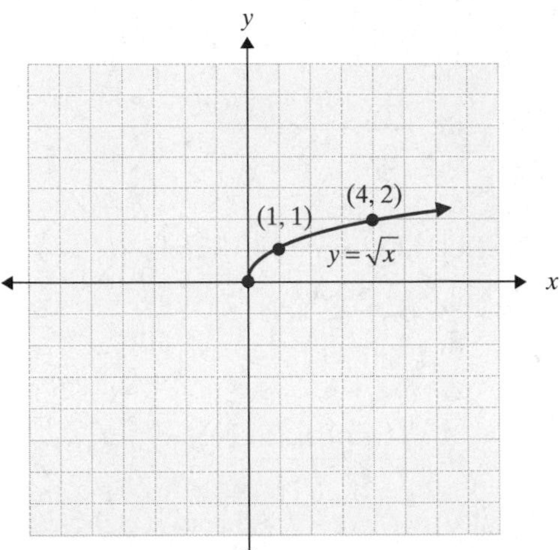

21. $y = \sqrt{x} - 2$ **22.** $y = \sqrt{x} + 1$ **23.** $y = -\sqrt{x+1}$ **24.** $y = -\sqrt{x-6}$

25. $y = \sqrt{x-4} - 3$ **26.** $y = \sqrt{x-2} - 4$ **27.** $y = \sqrt{x-3} + \frac{1}{2}$ **28.** $y = \sqrt{x + \frac{3}{2}} + 2$

29. $y = 5 + \sqrt{x+2}$ **30.** $y = \sqrt{x+4} - 3$

Using the graph of $y = \frac{1}{x}$, *graph the functions in Exercises 31 – 40 without additional computation.*

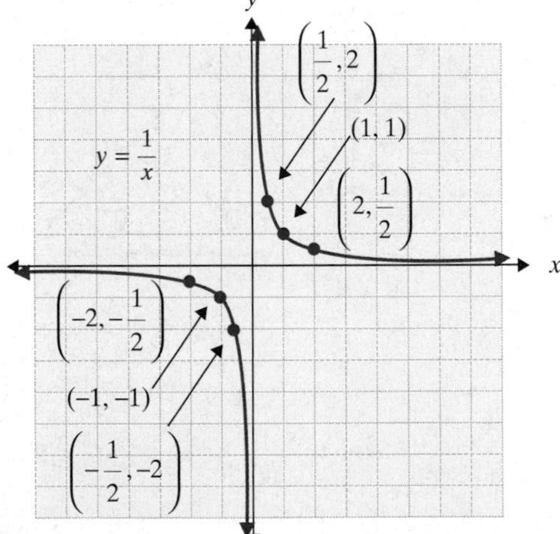

31. $y = \frac{1}{x} - 3$ **32.** $y = \frac{1}{x} + 5$ **33.** $y = \frac{1}{x-1}$ **34.** $y = \frac{1}{x+2}$

35. $y = \frac{1}{x-3} + 1$ **36.** $y = \frac{1}{x+5} - 2$ **37.** $y = \frac{1}{x+1} - 4$ **38.** $y = \frac{1}{x-2} + 3$

39. $y = \frac{1}{x+4} - 5$ **40.** $y = \frac{1}{x-5} + 2$

25.

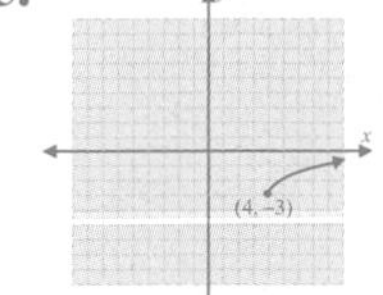

26.

27.

28.

29.

30.

31.

32.

33.

Using the graph $y = f(x)$, graph the functions in Exercises 41– 50 without additional computation.

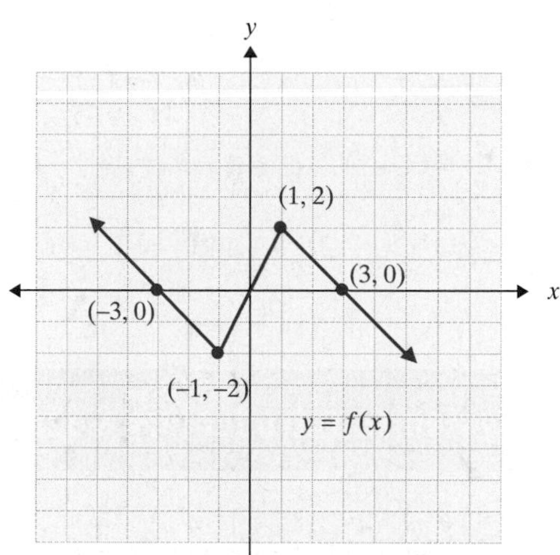

41. $y = f(x) - 1$ **42.** $y = f(x) + 2$ **43.** $y = f(x - 3)$ **44.** $y = f(x + 1)$

45. $y = -f(x)$ **46.** $y = -f(x - 4)$ **47.** $y = f(x + 5) + 3$

48. $y = f(x - 1) + 5$ **49.** $y = f(x + 2) - 4$ **50.** $y = f(x + 3) + 2$

In Exercises 51 – 54, match each equation with its corresponding graph.

51. $y = x^2 - 3$ **52.** $y = -x^2 + 5$ **53.** $y = (x - 1)^2 + 1$ **54.** $y = (x + 3)^2 - 2$

i.

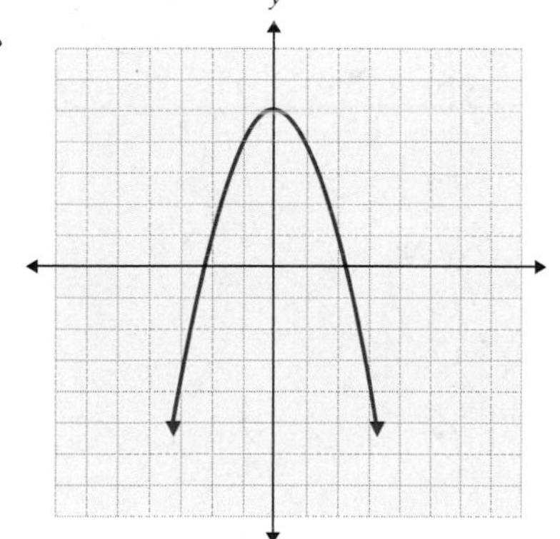

ii.

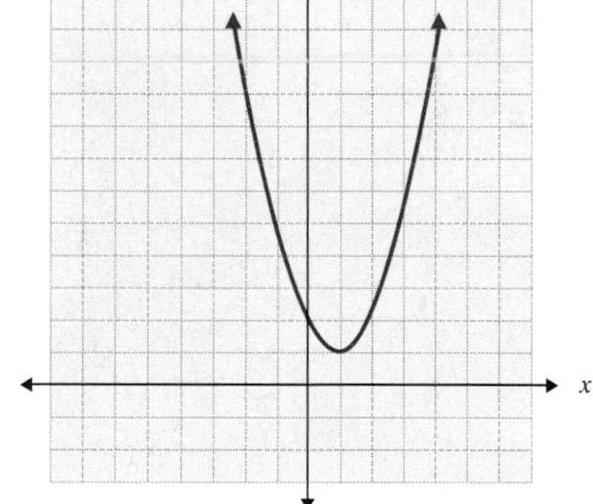

iii.

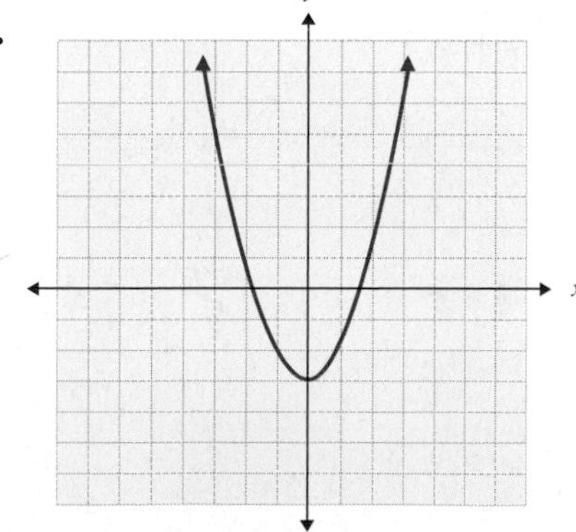

iv.

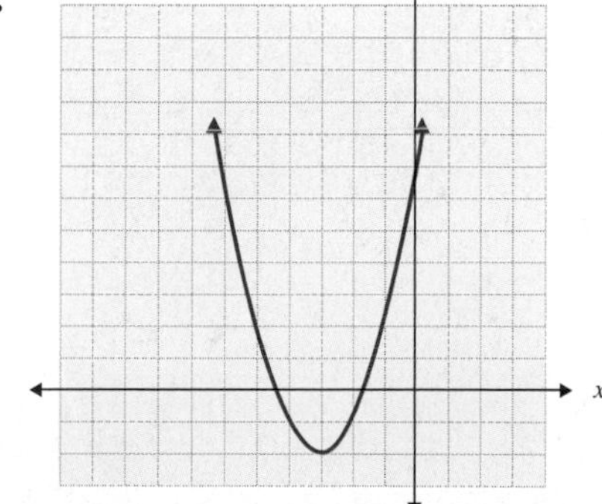

34.

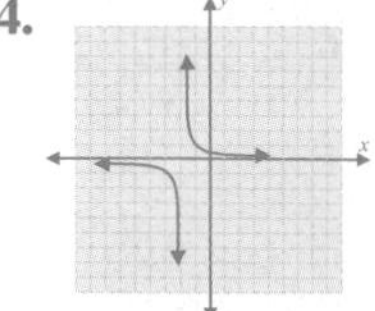

35.

In Exercises 55 – 60, use a graphing calculator to graph each pair of functions on the same set of axes.

55. $y = 2x^2$ and $y = -3x^2$

56. $y = x^2 + 5$ and $y = (x-1)^2$

57. $y = (x+1)^2 - 4$ and $y = x^2 - 4$

58. $y = 2(x+3)^2 - 4$ and $y = 2x^2 + 3$

59. $y = -3(x-2)^2 + 1$ and $y = -x^2 + 1$

60. $y = 4x^2 + 4x - 4$ and $y = x^2 + x - 1$

Hawkes Learning Systems: Introductory & Intermediate Algebra

Function Notation and Translations

36.

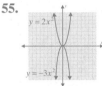

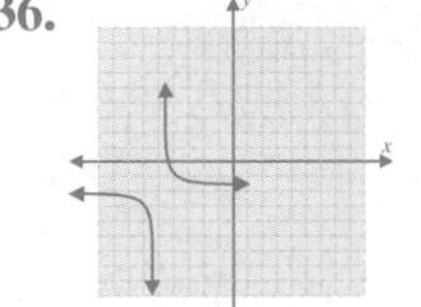

37.

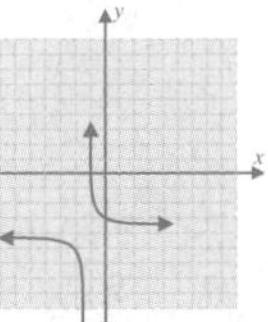

38.

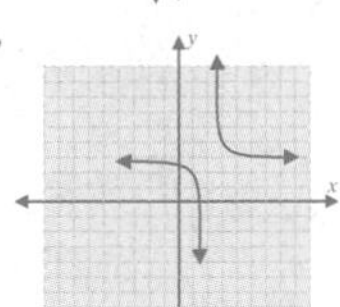

39.

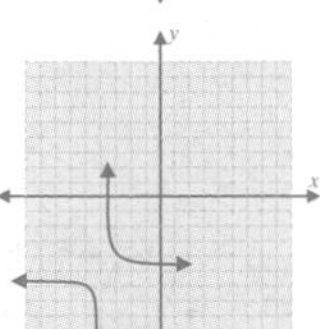

40.

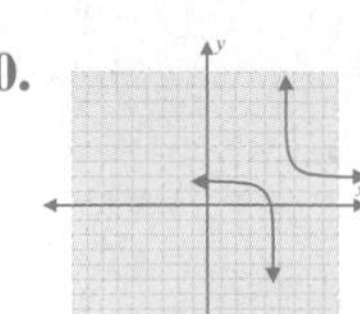

41.

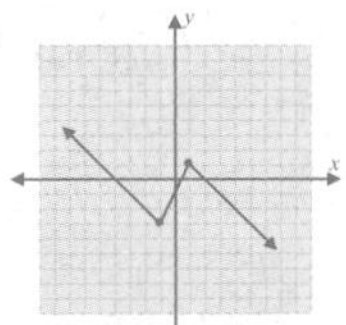

42.

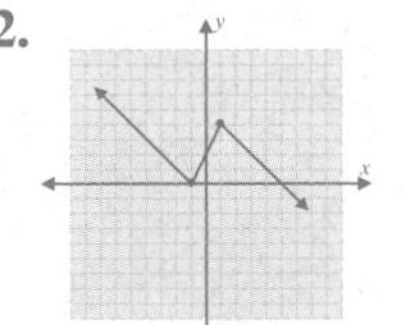

43.

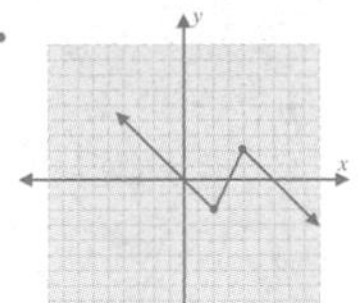

44.

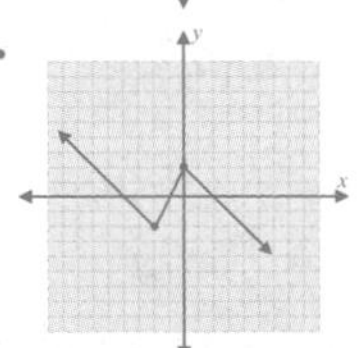

45.

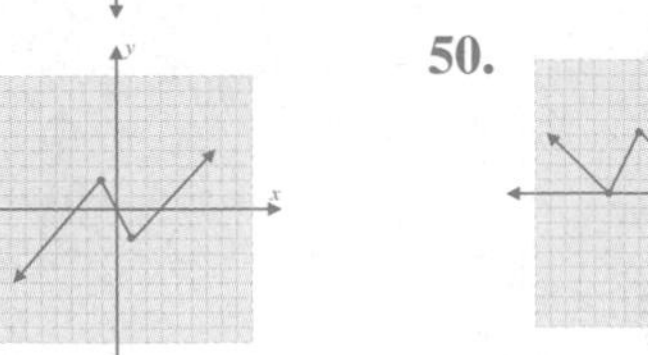

46.

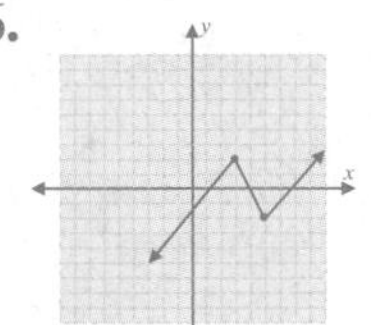

47.

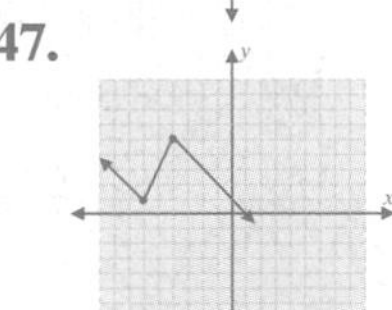

48.

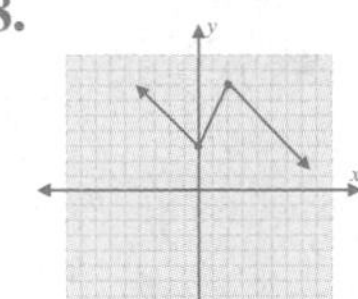

49.

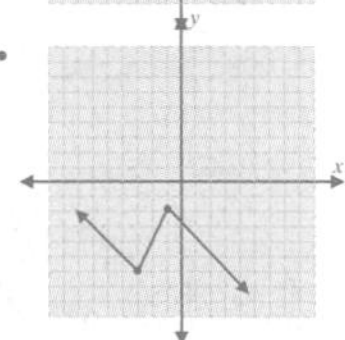

50.

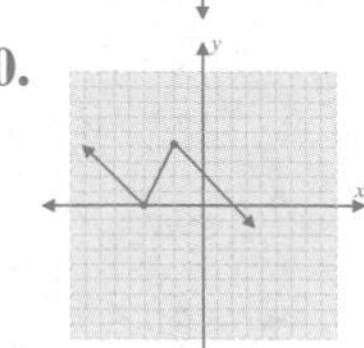

51. iii **52.** i **53.** ii **54.** iv

55.

56.

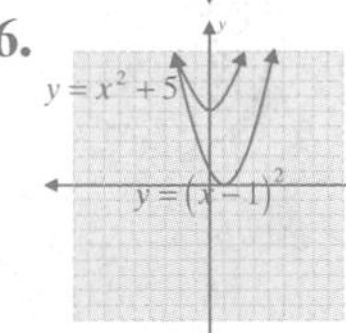

57.

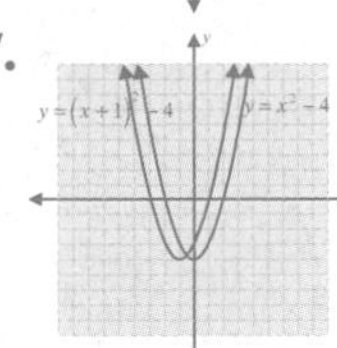

58.

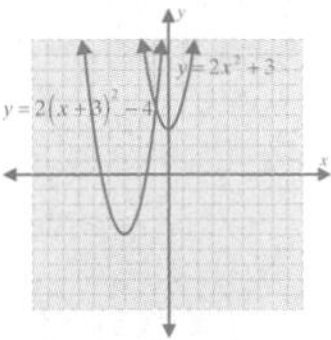

59.

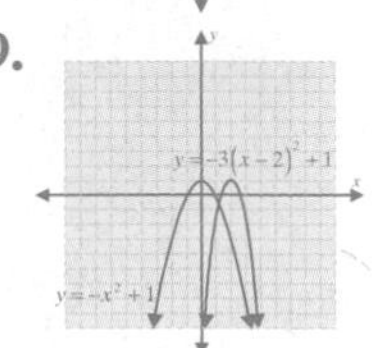

60.

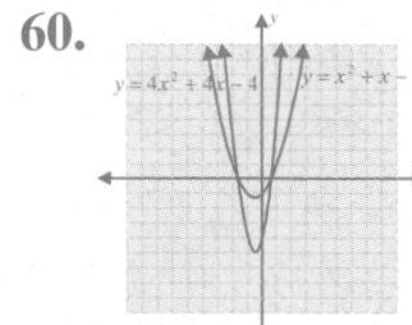

12.4 Parabolas as Conic Sections

Objectives

After completing this section, you will be able to:

1. *Graph parabolas with lines of symmetry parallel to the x-axis.*

2. *Find the vertices, y-intercepts, and lines of symmetry for parabolas that open left or right.*

Conic sections are curves in a plane that are found when the plane intersects a cone. Four such sections are the circle, ellipse, parabola, and hyperbola, as shown in Figure 12.13 respectively.

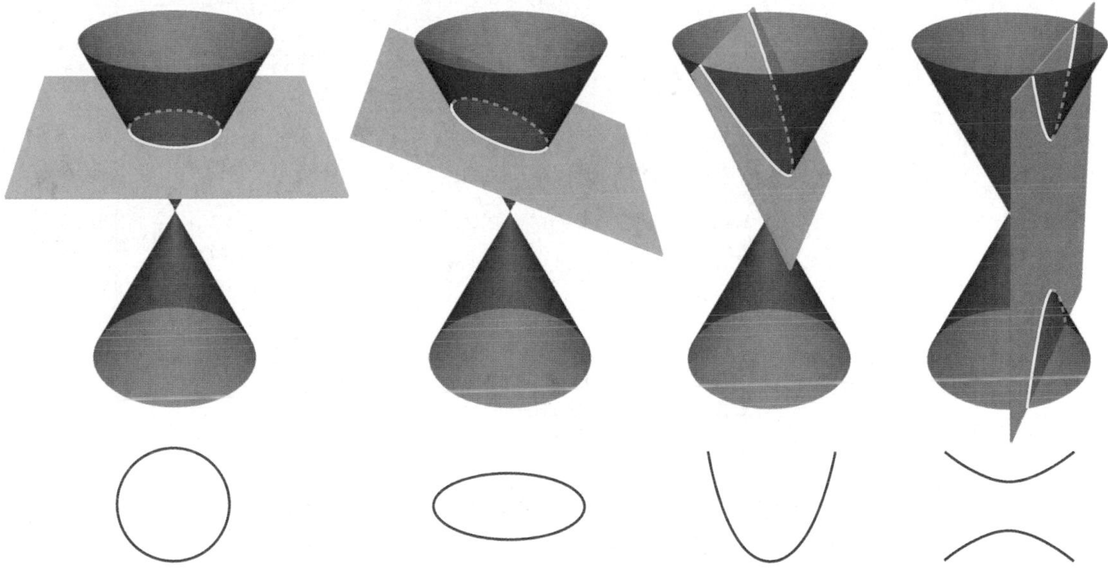

Figure 12.13

The corresponding equations for these conic sections are called quadratic equations because they are second-degree in x and/or y. After some practice you will be able to look at one of these equations and tell immediately what type of curve it represents and where the curve is located with respect to a Cartesian coordinate system. The technique is similar to that used in Chapter 4 in discussing straight lines. By looking at a linear equation, possibly with some algebraic manipulations, you can identify the slope of the line, its y-intercept, and where it is located. By looking at a quadratic equation, you will be able to identify the type of curve it represents and, additionally, you will be able to

a. identify the center and radius for a circle;
b. identify the center and intercepts for an ellipse;
c. identify the vertex and line of symmetry for a parabola; and
d. identify the vertices and asymptotes for a hyperbola.

Parabolas

As discussed in Section 12.1, the equations of **quadratic functions** are of the basic form $\boldsymbol{y = ax^2}$, and the corresponding graphs are parabolas. From the general view of **conic sections**, not all parabolas are functions. Parabolas that open upward or downward are functions, but those that open to the left or to the right are not functions.

The basic form for equations of parabolas that open left or right is $\boldsymbol{x = ay^2}$, and several graphs of equations of this type are shown in Figure 12.14. As the vertical line test will confirm, these graphs do not represent functions.

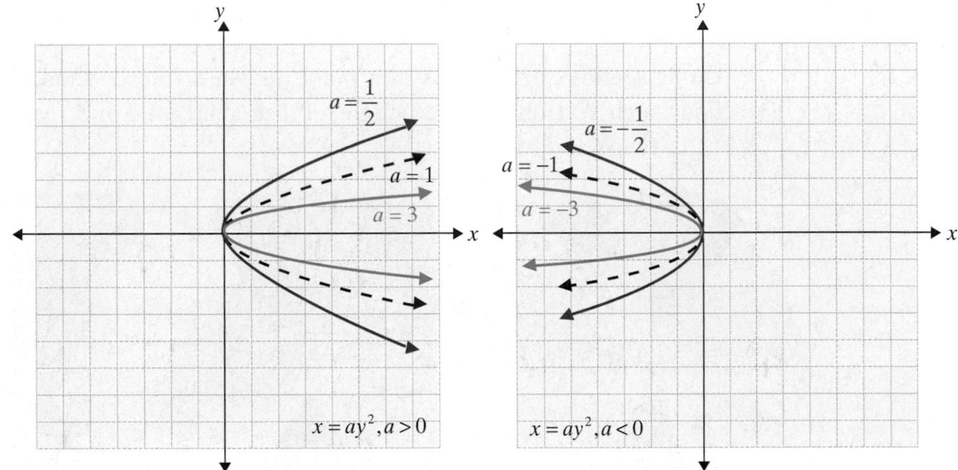

Figure 12.14

In general, the equations of **vertical parabolas** (parabolas that open upward or downward) are in the form

$$y = ax^2 + bx + c \quad \text{or} \quad y = a(x-h)^2 + k \quad (\text{where } a \neq 0)$$

and the parabolas open down (if $a < 0$) or up (if $a > 0$) with vertex at (h, k). The line $x = h$ is the line of symmetry.

By exchanging the roles of x and y, the equations of **horizontal parabolas** (parabolas that open to the left or right) can be written in the following form.

Equations of Horizontal Parabolas

Equations of horizontal parabolas (parabolas that open to the left or right) are of the form

$$\boldsymbol{x = ay^2 + by + c} \quad \text{or} \quad \boldsymbol{x = a(y-k)^2 + h} \qquad (\text{where } a \neq 0).$$

The parabola opens left if a is negative and right if a is positive.

The vertex is at ***(h, k)****.*

The line $\boldsymbol{y = k}$ *is the line of symmetry.*

In a manner similar to the discussion in Section 12.3, adding h to the right hand side and replacing y with $(y-k)$ in the equation $x=ay^2$ gives an equation whose graph is a horizontal translation of h units and a vertical translation of k units of the graph of $x=ay^2$.

For example, the graph of $x=2(y-3)^2-1$ is shown in Figure 12.15 with a table of y- and x-values. The vertex is at $(h,k)=(-1,3)$, and the line of symmetry is $y=3$. The y-values in the table are chosen on each side of the line of symmetry.

y	$2(y-3)^2-1=x$
3	$2(3-3)^2-1=-1$
4	$2(4-3)^2-1=1$
2	$2(2-3)^2-1=1$
5	$2(5-3)^2-1=7$
1	$2(1-3)^2-1=7$

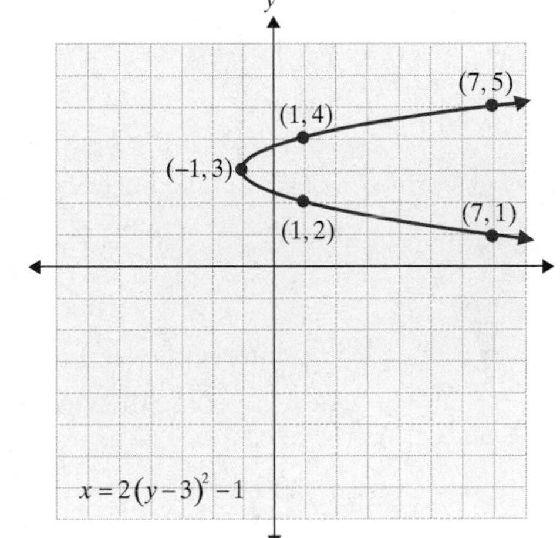

Figure 12.15

The graph of an equation of the form $x=ay^2+by+c$ can be found by completing the square (as in Section 12.3) and writing the equation in the form

$$x=a(y-k)^2+h.$$

Also, by setting $x=0$ and solving the following quadratic equation

$$0=ay^2+by+c$$

we can determine at what points, if any, the graph intersects the **y-axis**. These points are called **y-intercepts**.

Example 1: Horizontal Parabolas

a. For $x=y^2-6y+6$, find the vertex, the points where the graph intersects the y-axis, and the line of symmetry. Then sketch the graph.

Solution: To find the vertex, complete the square:

$$x=y^2-6y+6$$

$$x=(y^2-6y+9)-9+6$$

$$x=(y-3)^2-3$$

continued on next page ...

Teaching Notes:
You will probably want to remind your students that they can use their calculators to estimate the values of the intercepts when radicals are involved. For example, in 1a they should understand how to determine that $3+\sqrt{3}\approx 4.7$ and $3-\sqrt{3}\approx 1.3$.

To find the y-intercepts, let $x=0$:

$$y^2-6y+6=x$$

$$y^2-6y+6=0$$

$$y=\frac{6\pm\sqrt{(-6)^2-4\cdot1\cdot6}}{2}$$

$$y=\frac{6\pm\sqrt{12}}{2}$$

$$y=3\pm\sqrt{3}$$

Since $a=1$, the parabola has the same shape as $x=y^2$. The vertex is at $(-3,3)$.

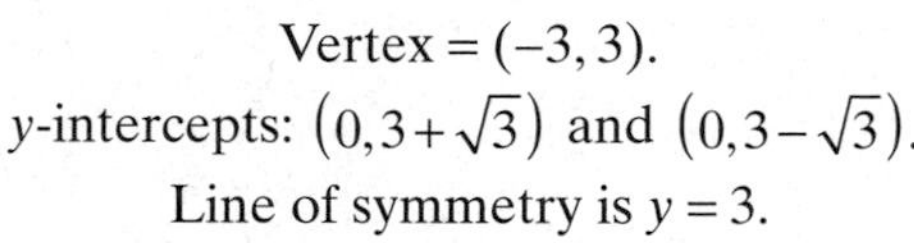

Vertex = $(-3, 3)$.
y-intercepts: $(0,3+\sqrt{3})$ and $(0,3-\sqrt{3})$.
Line of symmetry is $y=3$.

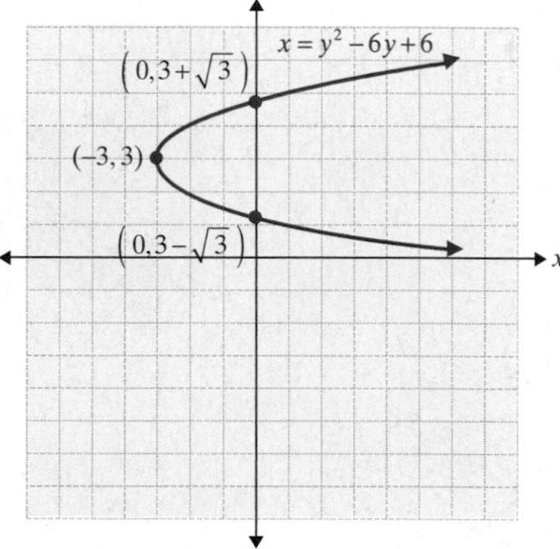

b. For $x=-2y^2-4y+6$, locate the vertex, the y-intercepts, and the line of symmetry. Then sketch the graph.

Solution: To find the vertex, complete the square:

$$x=-2y^2-4y+6$$

$$x=-2(y^2+2y)+6$$

$$x=-2(y^2+2y+1-1)+6$$

$$x=-2(y^2+2y+1)+2+6$$

$$x=-2(y^2+2y+1)+8$$

$$x=-2(y+1)^2+8$$

To find the y-intercepts, let $x=0$:

$$-2y^2-4y+6=0$$

$$-2(y^2+2y-3)=0$$

$$-2(y+3)(y-1)=0$$

$$y=-3 \quad\text{or}\quad y=1$$

Since $a = -2$, the graph opens to the left and is slightly narrower than $x = y^2$.
The vertex is at $(8,-1)$.

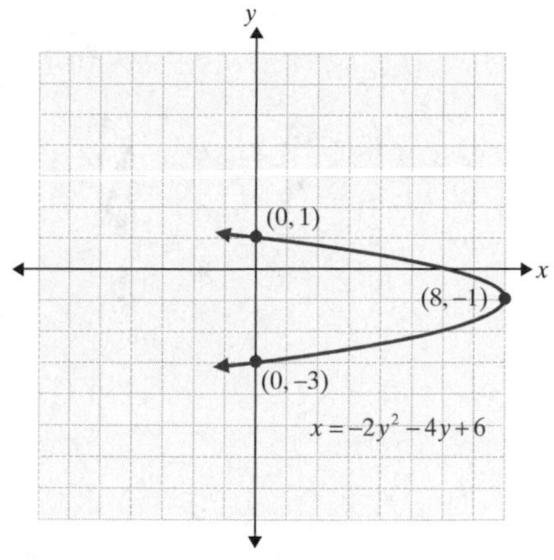

Vertex $= (8,-1)$.
The y-intercepts are $(0,-3)$ and $(0,1)$.
Line of symmetry is $y = -1$.

Using a Calculator to Graph Horizontal Parabolas

Horizontal parabolas are not functions and the graphing calculator is designed to graph only functions. Therefore to graph a horizontal parabola, solve the given equation for y. For example, from the previous discussion we know that the graph of the equation $x = y^2 - 2$ is a horizontal parabola opening right with vertex at $(-2,0)$.

To use a graphing calculator, we must first solve for y since equations must be entered with y (to the first power) on the left hand side. By using the definition of square root, we can find two functions that we will designate as y_1 and y_2 as follows:

$$x = y^2 - 2$$

$$y^2 = x + 2 \qquad \text{First solve for } y^2.$$

$$\left.\begin{aligned} y_1 &= \sqrt{x+2} \\ y_2 &= -\sqrt{x+2} \end{aligned}\right\} \quad \text{Solving for } y \text{ gives two equations that represent two functions.}$$

Graphing these equations individually gives the upper and lower halves of the parabola.

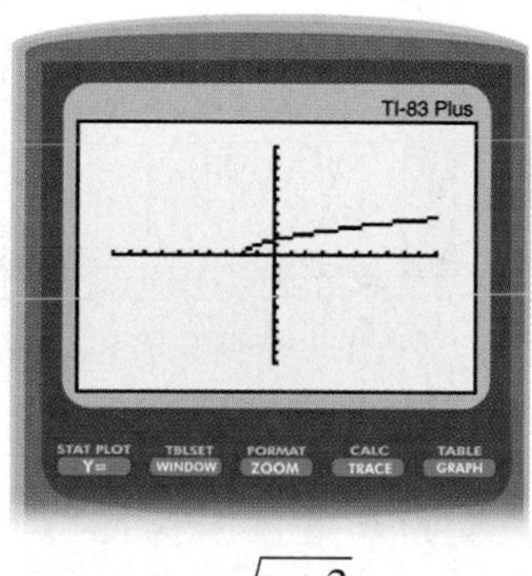

$y_1 = \sqrt{x+2}$
(upper half)

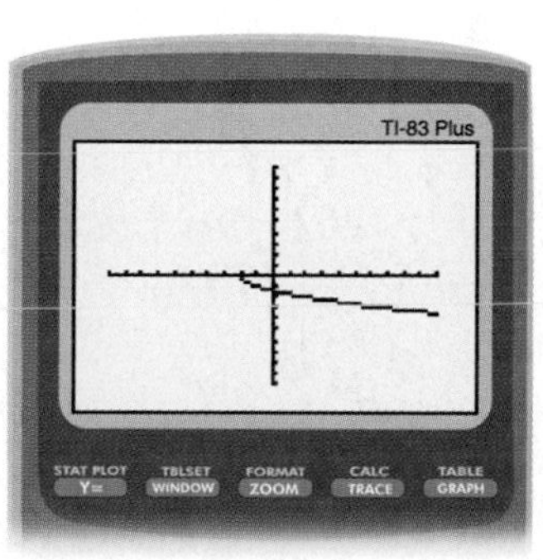

$y_2 = -\sqrt{x+2}$
(lower half)

Graphing both halves at the same time gives the entire parabola $x = y^2 - 2$.

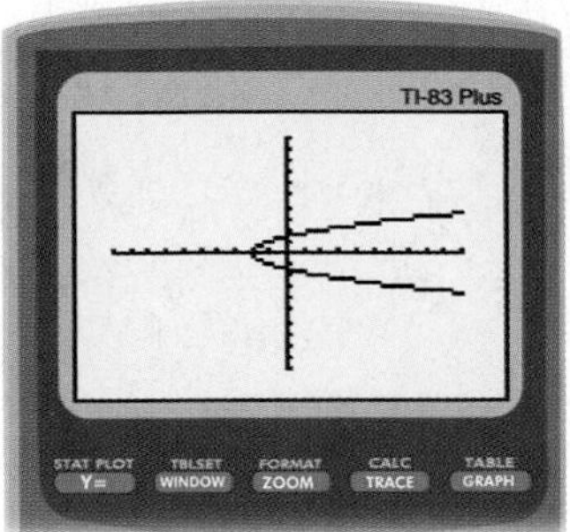

1. a. $(4, 0)$ **b.** none
c. $y = 0$

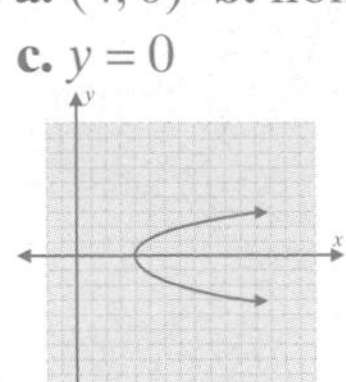

Example 2: Horizontal Parabola

Use a graphing calculator to graph the horizontal parabola $x = y^2 - 4y + 5$. Estimate the y-intercepts by using the TRACE and ZOOM features of the calculator.

Solution: To solve for y, complete the square and use the definition of square root as follows.

$$x = y^2 - 4y + 5$$

$$y^2 - 4y = x - 5$$

$$y^2 - 4y + 4 = x - 5 + 4$$

$$(y-2)^2 = x - 1$$

$$y - 2 = \pm\sqrt{x-1}$$

$$\left.\begin{aligned} y_1 &= \sqrt{x-1} + 2 \\ y_2 &= -\sqrt{x-1} + 2 \end{aligned}\right\} \quad \text{Graph both of these equations.}$$

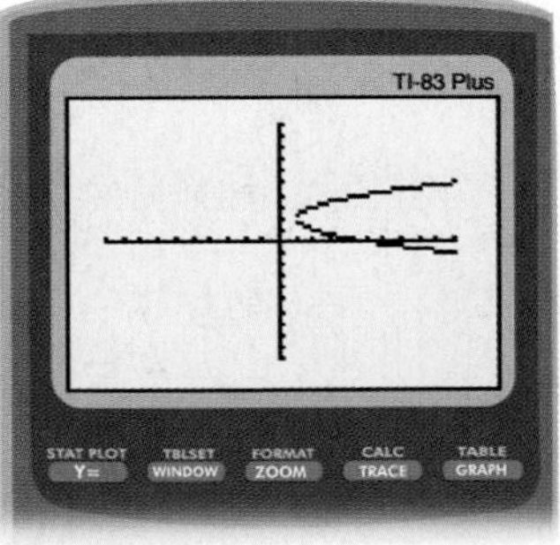

From this graph, we can determine that there are no y-intercepts.

2. a. $(-5, 0)$
b. $(0, \sqrt{5}), (0, -\sqrt{5})$
c. $y = 0$

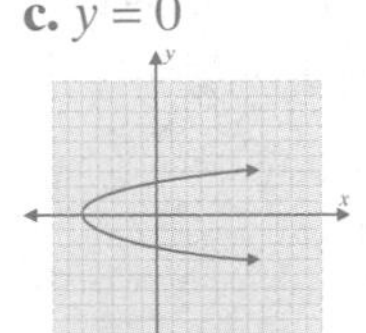

3. a. $(-3, 0)$
b. $(0, \sqrt{3}), (0, -\sqrt{3})$
c. $y = 0$

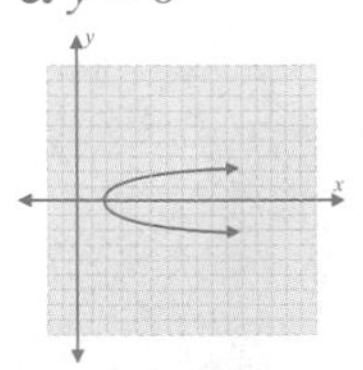

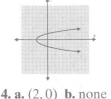

4. a. $(2, 0)$ **b.** none
c. $y = 0$

5. a. $(3, 0)$ **b.** none
c. $y = 0$

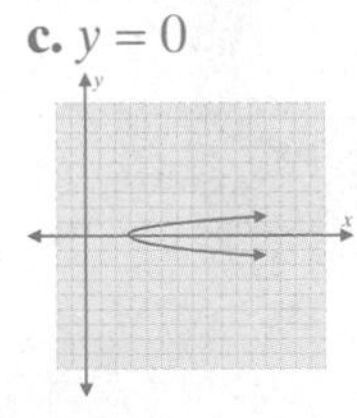

Practice Problems

1. *Write the equation* $x=-y^2-10y-24$ *in the form* $x=a(y-k)^2+h$.
2. *Find the vertex, y-intercepts, and line of symmetry for the curve* $x=y^2-4$.
3. *Find the y-intercepts for the curve* $x=y^2+2y+2$.

6. a. $(1,0)$ **b.** none **c.** $y=0$

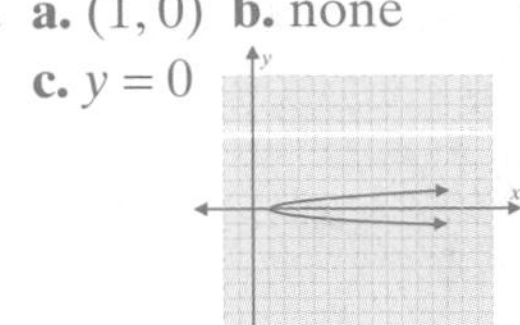

12.4 Exercises

7. a. $(0,3)$ **b.** $(0,3)$ **c.** $y=3$

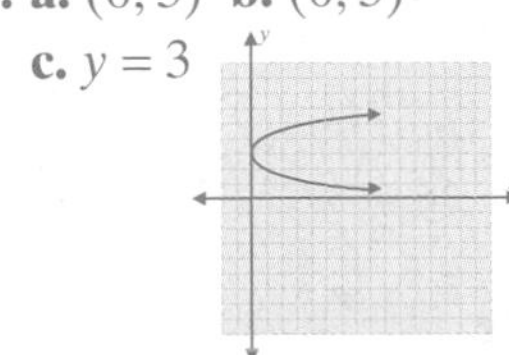

8. a. $(0,2)$ **b.** $(0,2)$ **c.** $y=2$

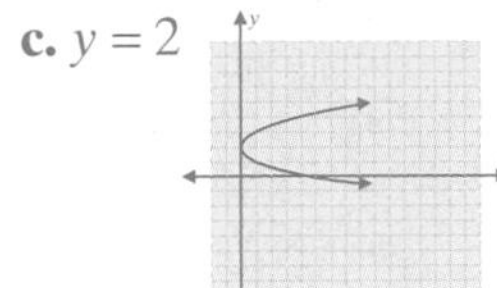

9. a. $(4,-2)$ **b.** none **c.** $y=-2$

10. a. $(-3,5)$ **b.** $\left(0,5+\sqrt{3}\right)$, $\left(0,5-\sqrt{3}\right)$ **c.** $y=5$

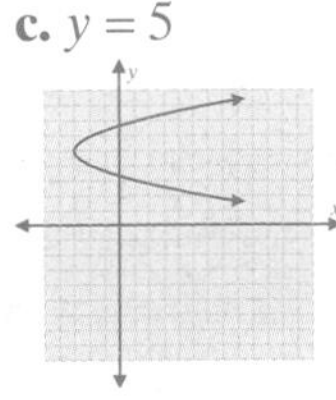

11. a. $(-1,1)$ **b.** $(0,0),(0,2)$ **c.** $y=1$

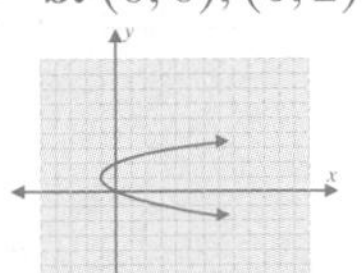

In each of the Exercises 1 – 30, find the vertex, y-intercepts, and the line of symmetry; then draw the graph.

1. $x=y^2+4$ **2.** $x=y^2-5$ **3.** $x+3=y^2$

4. $x-2=y^2$ **5.** $x=2y^2+3$ **6.** $x=3y^2+1$

7. $x=(y-3)^2$ **8.** $x=(y-2)^2$ **9.** $x-4=(y+2)^2$

10. $x+3=(y-5)^2$ **11.** $x+1=(y-1)^2$ **12.** $x-5=(y-3)^2$

13. $x=y^2+4y+4$ **14.** $x=-y^2+10y-25$ **15.** $x=y^2-8y+16$

16. $x=y^2+6y+1$ **17.** $y=-x^2-4x+5$ **18.** $y=x^2+5x+6$

19. $y=x^2+6x+5$ **20.** $y=x^2-2x-5$ **21.** $x=-y^2+4y-3$

22. $x=y^2+8y+10$ **23.** $y=2x^2+x-1$ **24.** $y=-2x^2+x+3$

25. $x=3y^2+6y-5$ **26.** $x=3y^2+5y+2$ **27.** $x=-2y^2+5y-2$

28. $x=4y^2-4y-15$ **29.** $y=4x^2-12x+9$ **30.** $y=-5x^2+10x+2$

Use a graphing calculator to graph each of the parabolas in Exercises 31 – 40. Use the TRACE and ZOOM features of the calculator to estimate the y-intercepts of the parabolas.

31. $x=2y^2-3$ **32.** $x=-3y^2+1$ **33.** $x=-y^2+2y$

34. $x=y^2-5y$ **35.** $x=2y^2+y+1$ **36.** $x=-y^2-4y+1$

37. $x=4y^2+8y-7$ **38.** $x=3y^2+3y+2$ **39.** $x=-2y^2+4y+3$

40. $x=-5y^2-10y-4$

Answers to Practice Problems: 1. $x=-(y+5)^2+1$ **2.** Vertex: $(-4,0)$, y-intercepts: $(0,2)$ and $(0,-2)$, Line of symmetry: $y=0$ **3.** There are no y-intercepts. The solutions to the equation $0=y^2+2y+2$ are nonreal.

12.a. $(5, 3)$ **b.** none
c. $y = 3$

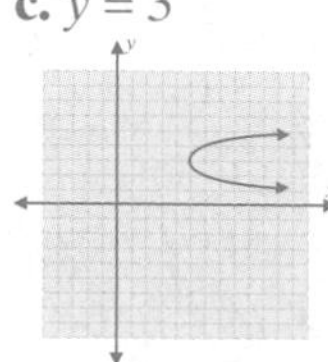

In Exercises 41 – 44, use your knowledge of parabolas and equations to match the equation with the graph.

41. $x = 2(y-3)^2 + 4$

42. $x = -(y+1)^2 + 5$

43. $x = y^2 - 6$

44. $x = -y^2 - 1$

13.a. $(0, -2)$ **b.** $(0, -2)$
c. $y = -2$

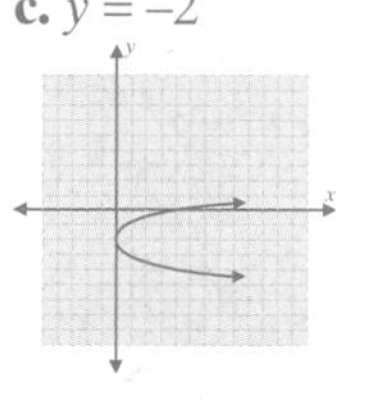

i.

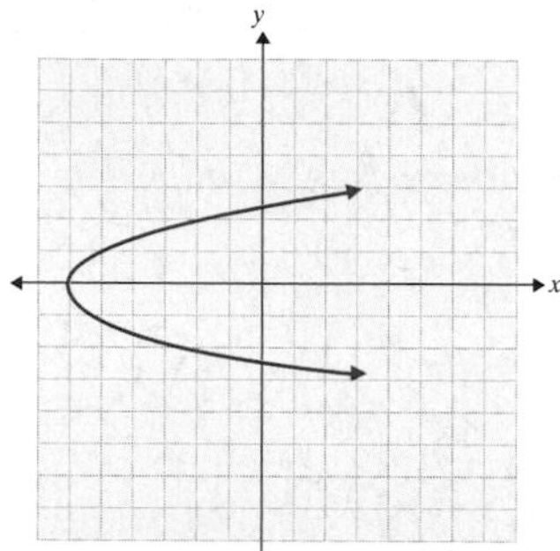

ii.

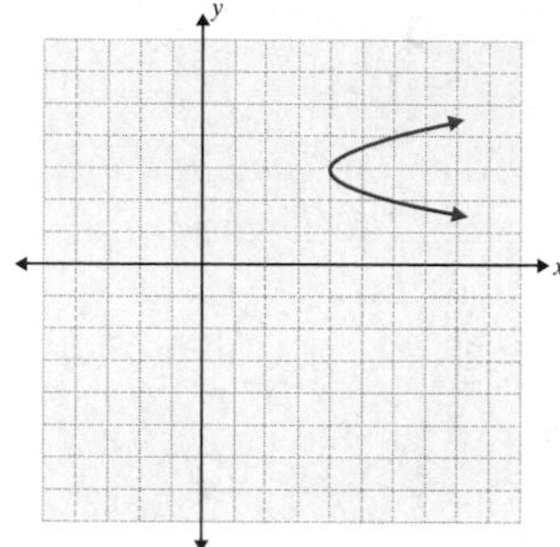

14.a. $(0, 5)$ **b.** $(0, 5)$
c. $y = 5$

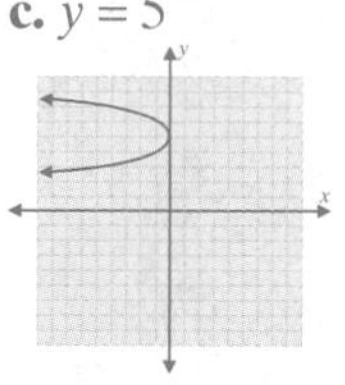

iii.

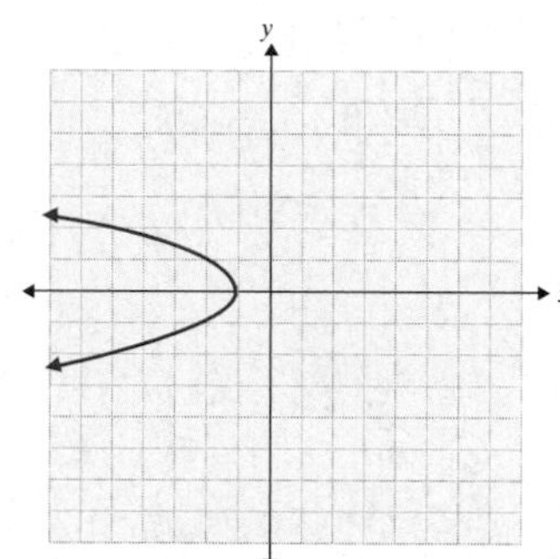

iv.

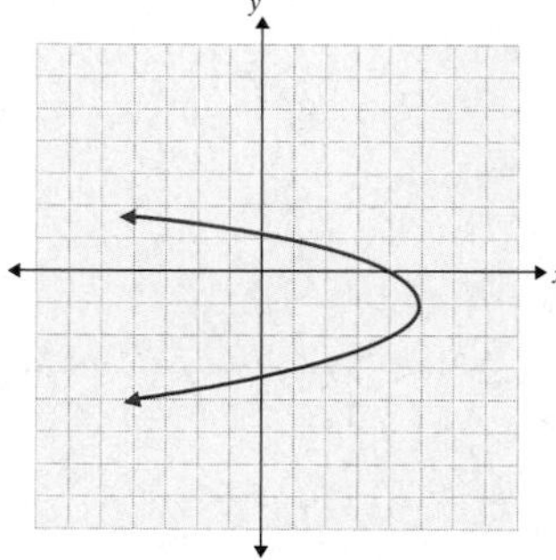

15.a. $(0, 4)$ **b.** $(0, 4)$
c. $y = 4$

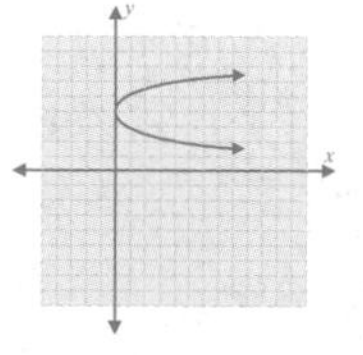

Writing and Thinking About Mathematics

45. For $x = ay^2 + by + c$ we know that the graph of the parabola opens to the right if $a > 0$ and to the left if $a < 0$. Discuss what values of a will cause the parabola to be "wider" and "narrower".

Hawkes Learning Systems: Introductory & Intermediate Algebra

Parabolas as Conic Sections

16.a. $(-8, -3)$
b. $\left(0, -3+2\sqrt{2}\right)$, $\left(0, -3-2\sqrt{2}\right)$
c. $y = -3$

17.a. $(-2, 9)$
b. $(0, 5)$
c. $x = -2$

18. a. $\left(-\frac{5}{2},-\frac{1}{4}\right)$
b. $(0,6)$
c. $x=-\frac{5}{2}$

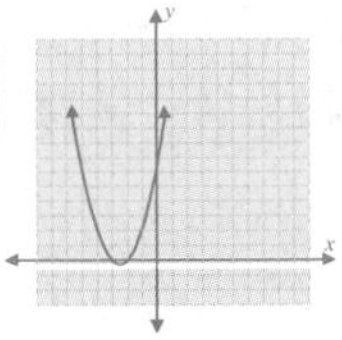

19. a. $(-3,-4)$
b. $(0,5)$
c. $x=-3$

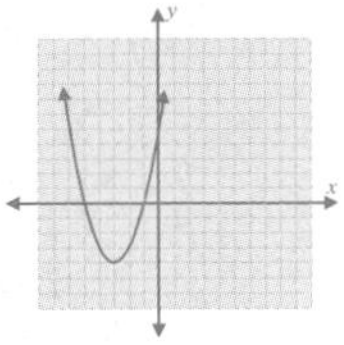

20. a. $(1,-6)$
b. $(0,-5)$
c. $x=1$

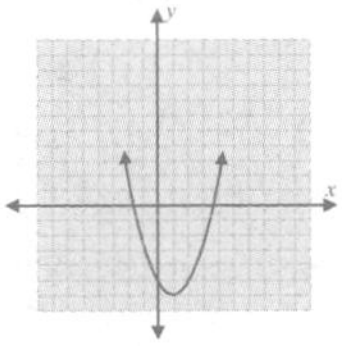

21. a. $(1,2)$
b. $(0,1)$, $(0,3)$
c. $y=2$

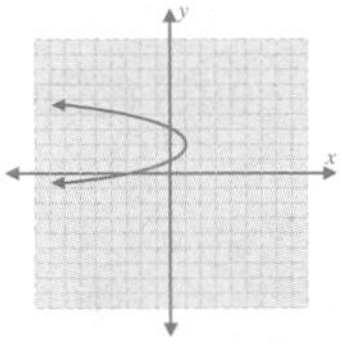

22. a. $(-6,-4)$
b. $\left(0,-4+\sqrt{6}\right)$, $\left(0,-4-\sqrt{6}\right)$
c. $y=-4$

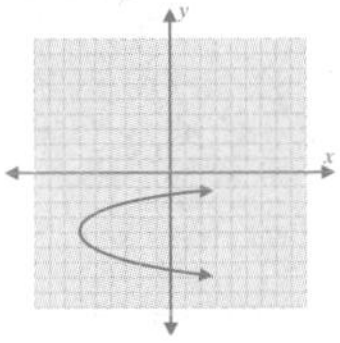

23. a. $\left(-\frac{1}{4},-\frac{9}{8}\right)$
b. $(0,-1)$
c. $x=-\frac{1}{4}$

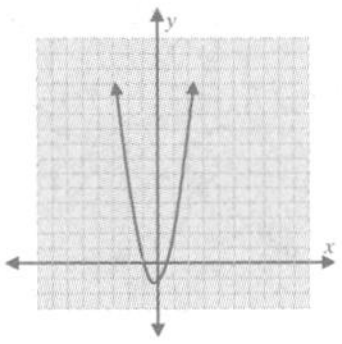

24. a. $\left(\frac{1}{4},\frac{25}{8}\right)$
b. $(0,3)$
c. $x=\frac{1}{4}$

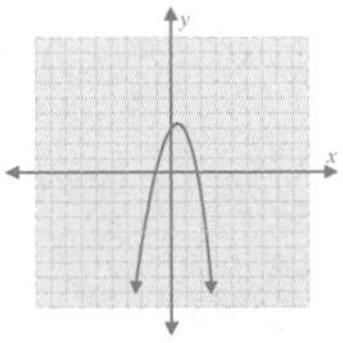

25. a. $(-8,-1)$
b. $\left(0,-1+\frac{2\sqrt{6}}{3}\right)$, $\left(0,-1-\frac{2\sqrt{6}}{3}\right)$
c. $y=-1$

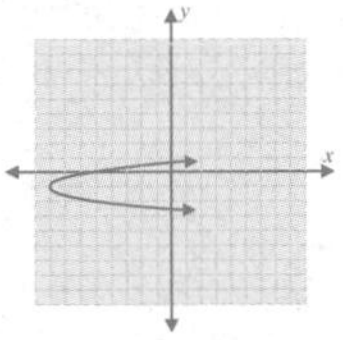

26. a. $\left(-\frac{1}{12},-\frac{5}{6}\right)$
b. $\left(0,-\frac{2}{3}\right)$, $(0,-1)$
c. $y=-\frac{5}{6}$

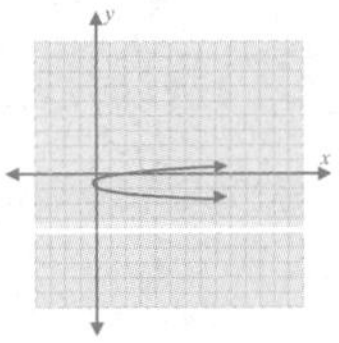

27. a. $\left(\frac{9}{8},\frac{5}{4}\right)$
b. $\left(0,\frac{1}{2}\right)$, $(0,2)$
c. $y=\frac{5}{4}$

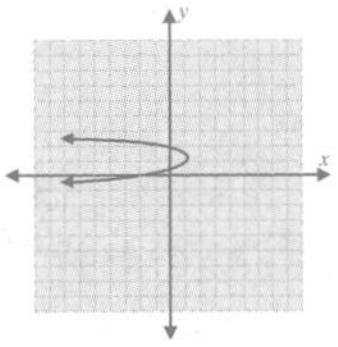

28. a. $\left(-16,\frac{1}{2}\right)$
b. $\left(0,\frac{5}{2}\right)$, $\left(0,-\frac{3}{2}\right)$
c. $y=\frac{1}{2}$

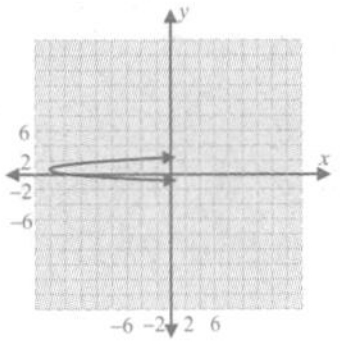

29. a. $\left(\frac{3}{2},0\right)$
b. $(0,9)$
c. $x=\frac{3}{2}$

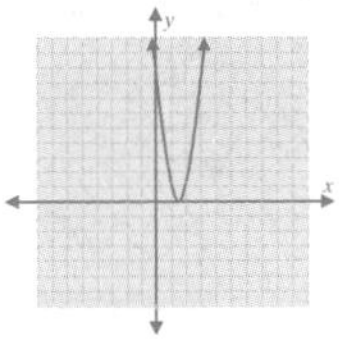

30. a. $(1,7)$
b. $(0,2)$
c. $x=1$

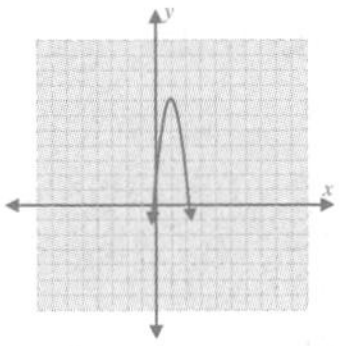

31. 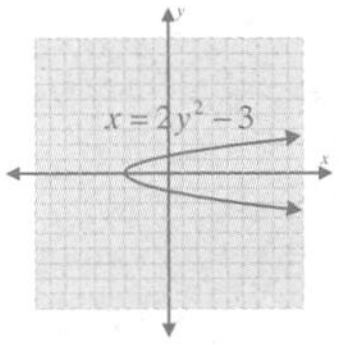 $x=2y^2-3$ — $(0,1.225)$, $(0,-1.225)$

32. $x=-3y^2+1$ — $(0,0.577)$, $(0,-0.577)$

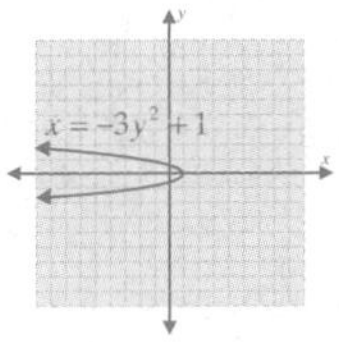

33. $x=-y^2+2y$ — $(0,2)$, $(0,0)$

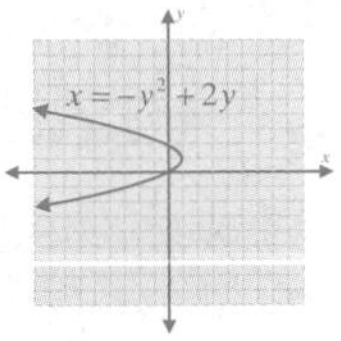

34. $x=y^2-5y$ — $(0,5)$, $(0,0)$

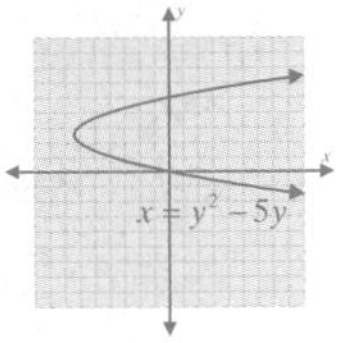

35. $x=2y^2+y+1$ — no y-intercept

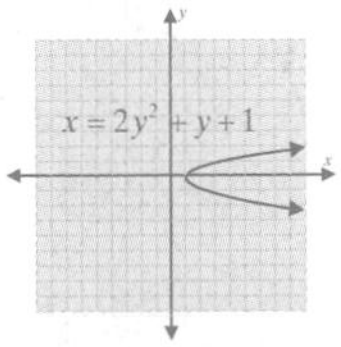

36. $x=-y^2-4y+1$ — $(0,0.236)$, $(0,-4.236)$

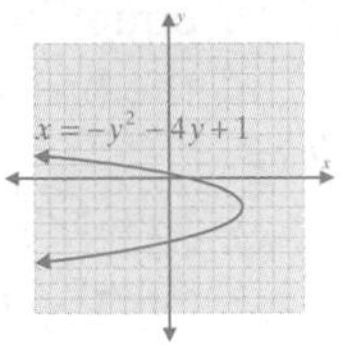

37. $x=4y^2+8y-7$ — $(0,0.658)$, $(0,-2.658)$

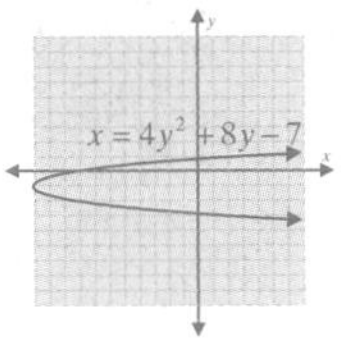

38. $x=3y^2+3y+2$ — no y-intercept

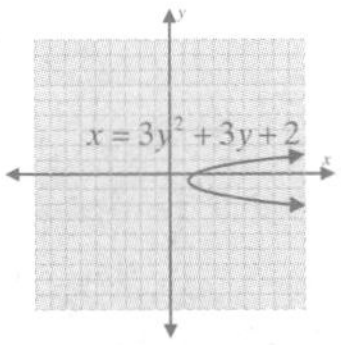

39. $x=-2y^2+4y+3$ — $(0,2.581)$, $(0,-0.581)$

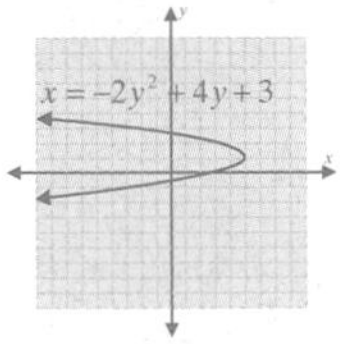

40. $x=-5y^2-10y-4$ — $(0,-0.553)$, $(0,-1.447)$

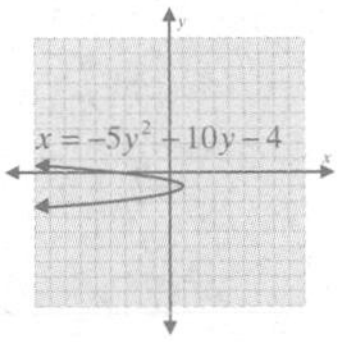

41. ii
42. iv
43. i
44. iii
45. The smaller $|a|$ is, the more open the graph will be.

12.5 Distance Formula, Midpoint Formula, and Circles

Objectives

After completing this section, you will be able to:

1. *Find the distance between any two points in a plane.*
2. *Write the equations of a circle given its center and radius.*
3. *Graph circles centered at the point (h, k).*

Distance Between Two Points

The formula for the distance between two points in a plane is needed to develop the equations of circles. The **Pythagorean Theorem**, previously discussed in Section 6.5, is the basis for the formula and is repeated here for easy reference.

In a right triangle, the square of the length of the hypotenuse is equal to the sum of the squares of the lengths of the two legs.

$$a^2 + b^2 = c^2$$

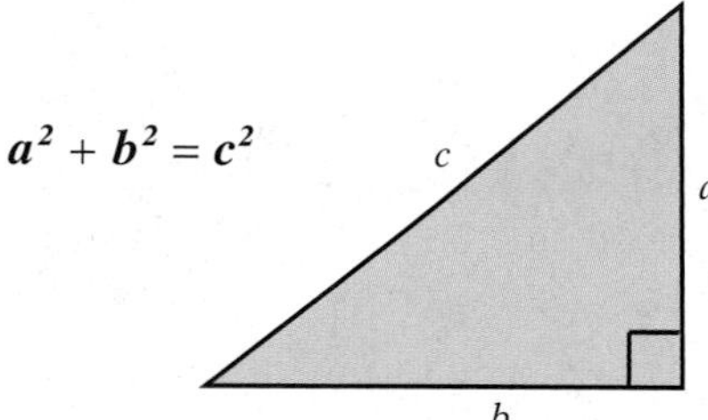

To find the distance between the two points $P(-1,2)$ and $Q(5,6)$, as shown in Figure 12.16(a), form a right triangle, as shown in Figure 12.16(b), and find the lengths of the sides a and b. Then, using a and b and the Pythagorean Theorem, we can find the length of the hypotenuse, which is the distance between the two points.

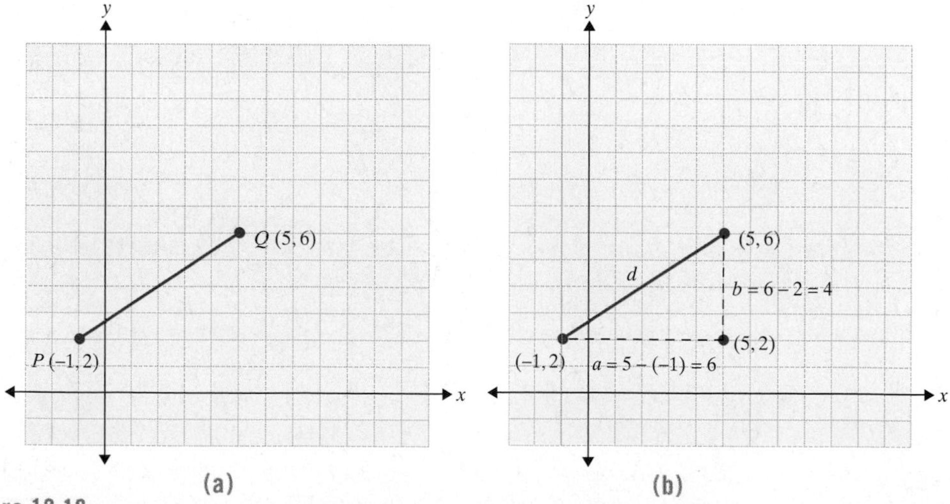

Figure 12.16

From Figure 12.16(b), $d^2 = a^2 + b^2$

$$= 6^2 + 4^2$$
$$= 36 + 16$$
$$= 52$$
$$d = \sqrt{52}$$
$$= 2\sqrt{13}$$

By going directly to the Pythagorean Theorem, the distance can be represented by the formula

$$d = \sqrt{a^2 + b^2}.$$

More generally, we can write the formula for d involving the coordinates of two points $P(x_1, y_1)$ and $Q(x_2, y_2)$ as illustrated in Figure 12.17. With $a = |x_2 - x_1|$ and $b = |y_2 - y_1|$, the distance formula is

$$d = \sqrt{(x_2 - x_1)^2 + (y_2 - y_1)^2}.$$

Teaching Notes:
You will probably want to reinforce the fact (with a few examples) that the order of the points (or the order of the subtraction) doesn't matter because of the squaring.

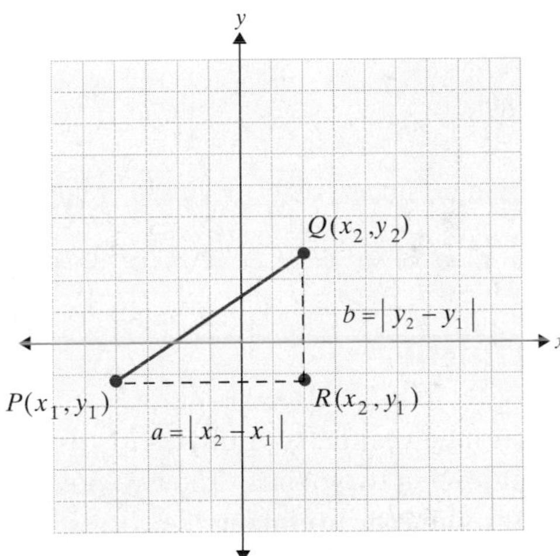

Figure 12.17

Note that in Figure 12.17, the calculations for a and b involve absolute value. These absolute values guarantee nonnegative values for a and b to represent the lengths of legs. In the distance formula, the absolute values are disregarded because $x_2 - x_1$ and $y_2 - y_1$ are squared. **In the actual calculations of *d*, be sure to add the squares before taking the square root.**

Example 1: The Distance Formula

Each of the following examples illustrates how to use the distance formula:

$$d = \sqrt{(x_2 - x_1)^2 + (y_2 - y_1)^2}.$$

a. Find the distance between the two points (3, 4) and (–2, 7).

Solution: $d = \sqrt{[3-(-2)]^2 + (4-7)^2}$

$$= \sqrt{5^2 + (-3)^2} = \sqrt{25+9} = \sqrt{34}$$

continued on next page ...

b. Determine whether or not the triangle determined by the three points, $A\ (-5,-1)$, $B\ (2,1)$, and $C\ (0,7)$ is a right triangle.

Solution: Find the lengths of the three line segments $\overline{AB}$, $\overline{AC}$, and $\overline{BC}$, and decide whether or not the Pythagorean Theorem is satisfied.

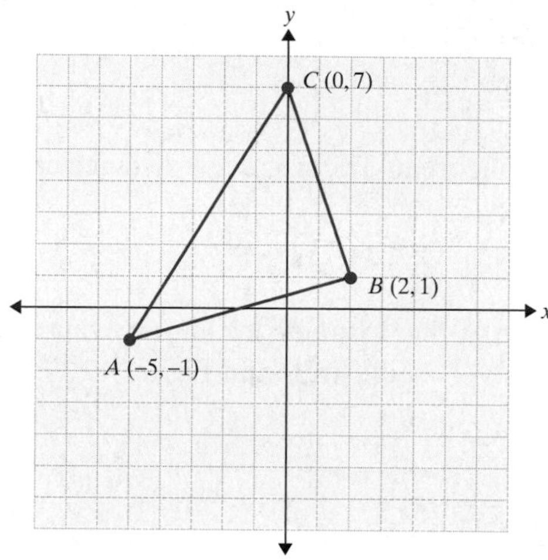

$$AB = \sqrt{(-5-2)^2 + (-1-1)^2} = \sqrt{(-7)^2 + (-2)^2}$$
$$= \sqrt{49+4} = \sqrt{53}$$
$$AC = \sqrt{(-5-0)^2 + (-1-7)^2} = \sqrt{(-5)^2 + (-8)^2}$$
$$= \sqrt{25+64} = \sqrt{89}$$
$$BC = \sqrt{(2-0)^2 + (1-7)^2} = \sqrt{(2)^2 + (-6)^2}$$
$$= \sqrt{4+36} = \sqrt{40}$$

The longest side is $AC = \sqrt{89}$.

The triangle is **not** a right triangle since $(\sqrt{89})^2 \neq (\sqrt{53})^2 + (\sqrt{40})^2$ or $89 \neq 53 + 40$.

The Midpoint Formula

Another useful formula that involves the coordinates of two points is that for finding the **midpoint** of the segment joining two points. The two points are called **endpoints** of the segment. The midpoint is found by **averaging** the corresponding coordinates of the endpoints.

Midpoint Formula

The formula for the midpoint between two points $P(x_1, y_1)$ and $Q(x_2, y_2)$ is

$$\left(\frac{x_1 + x_2}{2}, \frac{y_1 + y_2}{2}\right).$$

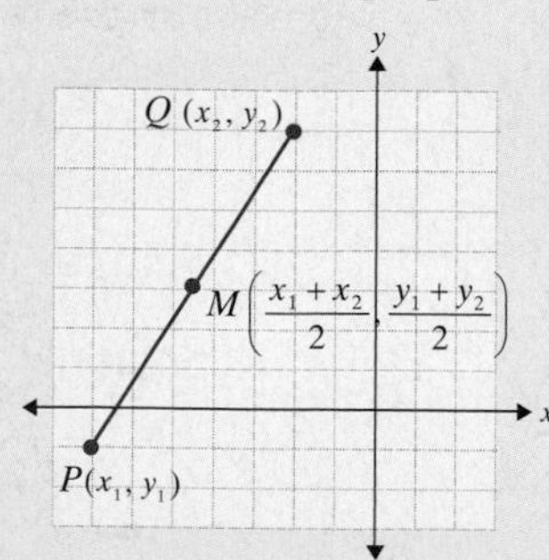

Example 2: Midpoint Formula

Find the coordinates of the midpoint of the line segment joining the two points $P(-4, 6)$ and $Q(1, 2)$.

Solution: The midpoint is $\left(\frac{-4+1}{2}, \frac{6+2}{2}\right) = \left(-\frac{3}{2}, 4\right)$.

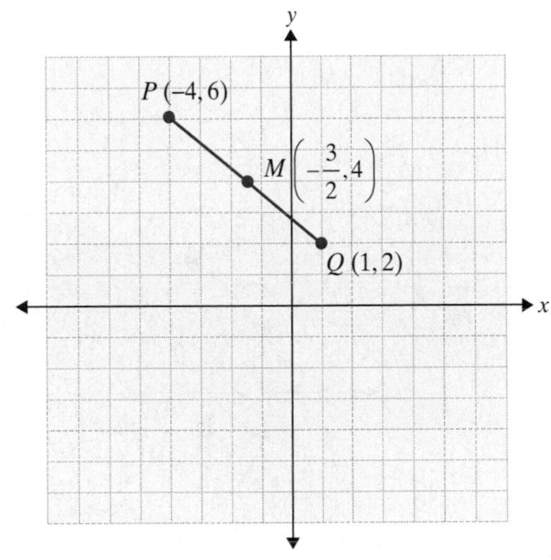

Equations of Circles

Circles and the terms related to circles (**center**, **radius**, and **diameter**) are defined as follows:

Circle, Center, Radius, and Diameter

*A **circle** is the set of all points in a plane that are a fixed distance from a fixed point.*

*The fixed point is called the **center** of the circle.*

*The distance from the center to any point on the circle is called the **radius** of the circle.*

*The distance from one point on the circle to another point on the circle measured through the center is called the **diameter** of the circle.*

Note: *The diameter is twice the length of the radius.*

radius
diameter
center

Teaching Notes: While discussing terms related to circles, you might want to point out the dual use of some terms. "The" radius most often refers to a length, *per se*, while "a" radius most often refers to a particular line segment. (Similarly with diameter.)

The distance formula is used to find the equation of a circle. For example, to find the equation of the circle with its center at the origin $(0, 0)$ and radius 5, for any point on the circle (x, y) the distance from (x, y) to $(0, 0)$ must be 5.

Therefore, using the distance formula,

$$\sqrt{(x_2 - x_1)^2 + (y_2 - y_1)^2} = d$$

$$\sqrt{(x-0)^2 + (y-0)^2} = 5$$

$$\sqrt{x^2 + y^2} = 5$$

$$x^2 + y^2 = 25 \qquad \text{Square both sides.}$$

Thus, as shown in Figure 12.18, all points on the circle satisfy the equation $x^2 + y^2 = 25$.

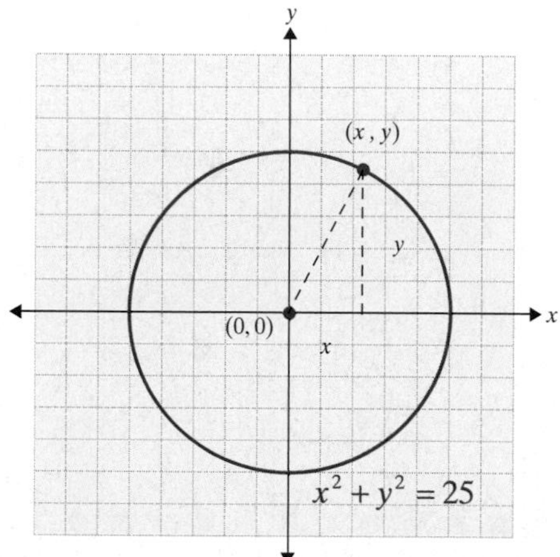

Figure 12.18

In general, any point (x, y) on a circle with center at (h, k) and radius $r > 0$ must satisfy the equation

$$\sqrt{(x-h)^2 + (y-k)^2} = r.$$

Squaring both sides of this equation gives the **standard form** for the equation of a circle:

$$(\boldsymbol{x} - \boldsymbol{h})^2 + (\boldsymbol{y} - \boldsymbol{k})^2 = \boldsymbol{r}^2.$$

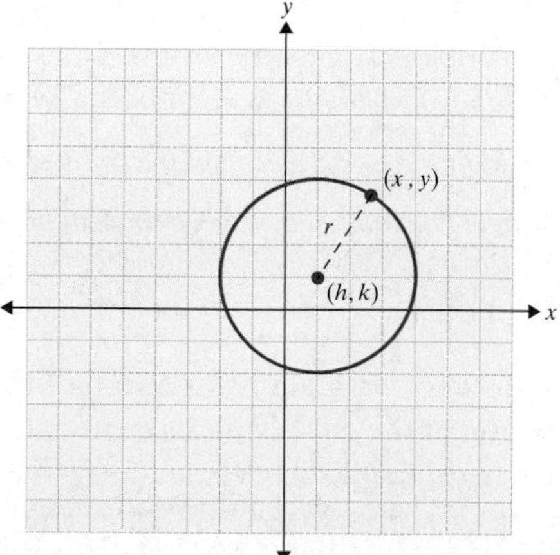

Figure 12.19

The equation for a circle of radius r with **center at the origin** is

$$x^2 + y^2 = r^2.$$

By thinking of this circle translated to have **center at (h, k)**, we can get the standard form by substituting $(x - h)$ for x and $(y - k)$ for y:

$$(x-h)^2 + (y-k)^2 = r^2.$$

Example 3: Equations of Circles

a. Find the equation of the circle with center at the origin and radius $\sqrt{3}$. Are the points $(\sqrt{2},1)$ and $(1,2)$ on the circle?

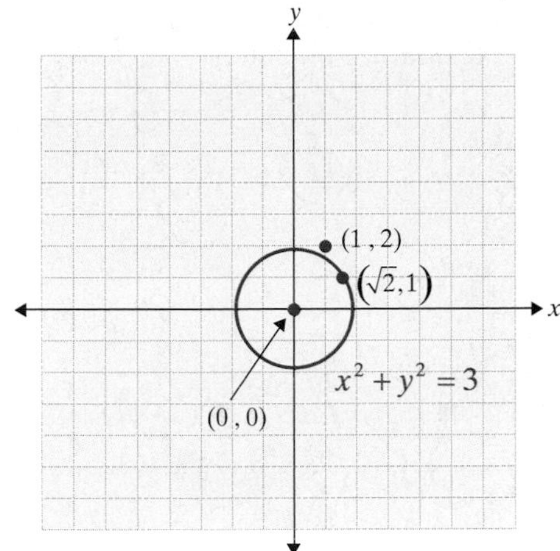

Solution: The equation is $x^2 + y^2 = 3$.

To determine whether or not the points $(\sqrt{2},1)$ and $(1,2)$ are on the circle, substitute each of these points into the equation.
Substituting $(\sqrt{2},1)$ gives $(\sqrt{2})^2 + (1)^2 = 2 + 1 = 3$.
Substituting $(1,2)$ gives $(1)^2 + (2)^2 = 1 + 4 = 5 \neq 3$.
Therefore, $(\sqrt{2},1)$ is on the circle, but $(1,2)$ is not on the circle.

b. Find the equation of the circle with center at $(5, 2)$ and radius 3. Is the point $(5, 5)$ on the circle?

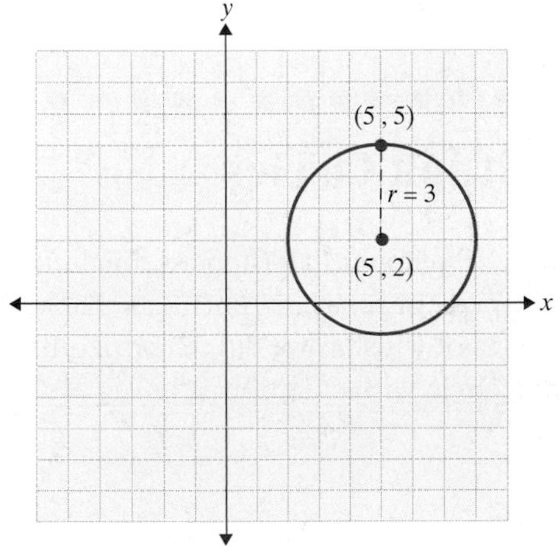

Solution: The equation is $(x-5)^2 + (y-2)^2 = 9$.

Substituting $(5, 5)$ gives $(5-5)^2 + (5-2)^2 = 0^2 + 3^2 = 9$.
Therefore, $(5, 5)$ is on the circle.

continued on next page ...

c. Show that $x^2 + y^2 - 8x + 2y = 0$ represents a circle. Find its center and radius. Then graph the circle.

Solution: Rearrange the terms and complete the square for $x^2 - 8x$ and $y^2 + 2y$.

$$x^2 + y^2 - 8x + 2y = 0$$

$$x^2 - 8x + y^2 + 2y = 0$$

$$x^2 - 8x + 16 + y^2 + 2y + 1 = 16 + 1 \qquad \text{Add 16 and 1 to both sides.}$$

Completes the square

$$(x-4)^2 + (y+1)^2 = 17 \qquad \text{Equation of a circle}$$

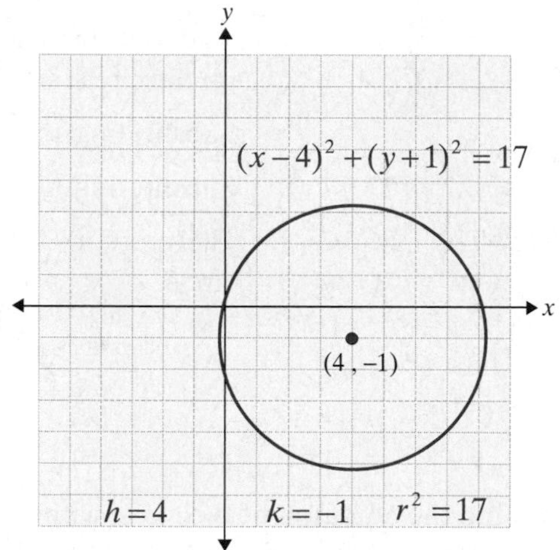

Center is at $(4, -1)$.

Radius $= \sqrt{17}$.

Using a Graphing Calculator to Graph Circles

The equation of a circle does not represent a function. The upper half (upper semicircle) and the lower half (lower semicircle) do, however, represent separate functions. Therefore, to graph a circle, solve the equation for two values of *y* just as we did for horizontal parabolas in Section 12.4. For example, consider the circle with equation $x^2 + y^2 = 4$.

$$x^2 + y^2 = 4$$

$$y^2 = 4 - x^2 \qquad \text{First solve for } y^2.$$

$$\left.\begin{aligned} y_1 &= \sqrt{4 - x^2} \\ y_2 &= -\sqrt{4 - x^2} \end{aligned}\right\} \qquad \text{Solving for } y \text{ gives two equations that represent two functions.}$$

Graphing both of these functions gives the figure pictured in Figure 12.20.

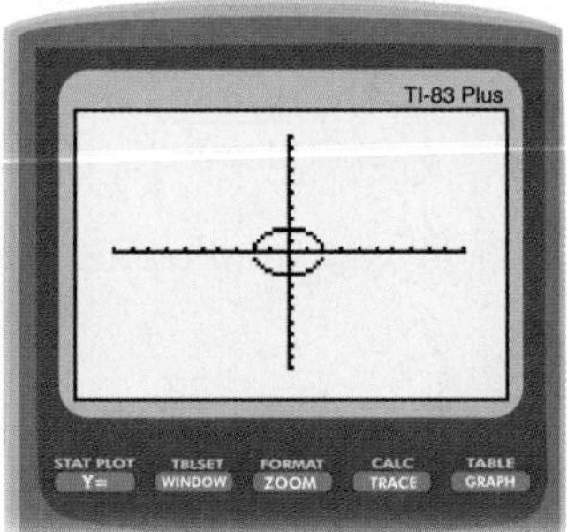

Figure 12.20

The screen on a TI calculator is rectangular and not a square. The ranges for the standard viewing window are from –10 for both *X*min and *Y*min to 10 for both *X*max and *Y*max. That is, the horizontal scale (x values) and the vertical scale (y values) are the same. Since the rectangular screen is in the approximate ratio of 2 to 3, the graph of a circle using the standard WINDOW will appear flattened as the circle did in Figure 12.20.

To get a more realistic picture of a circle press the WINDOW key and set the **`Xmin`** and **`Xmax`** values to –6 and 6, respectively. Set the **`Ymin`** and **`Ymax`** values to –4 and 4, respectively. Since the numbers 4 and 6 are in the ratio of 2 to 3, the screen is said to show a "square window," and the graphs of y_1 and y_2 will now give a more realistic picture of a circle as shown in Figure 12.21. Alternatively, pressing the ZOOM key and choosing option **`5:ZSquare`** automatically "squares" the window.

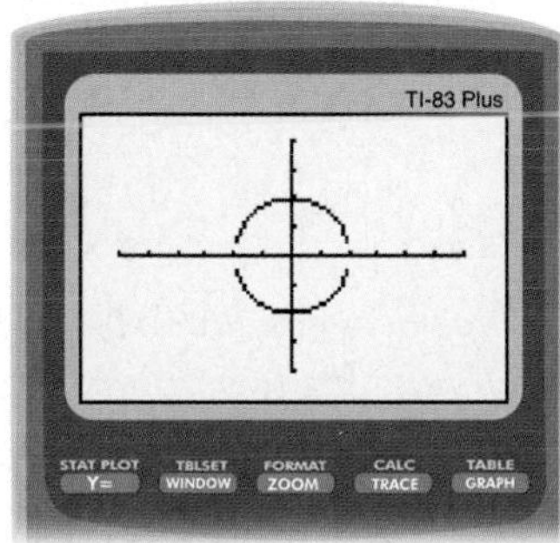

Figure 12.21

Example 4: Using a Graphing Calculator

Use a graphing calculator with a "square window" to graph the circle $x^2 + y^2 = 9$.

Solution: Set the WINDOW scales to –6 and 6 for **`Xmin`** and **`Xmax`** and –4 and 4 for **`Ymin`** and **`Ymax`**, respectively.

continued on next page ...

1. 5;(4, 5.5) **2.** 13;(3.5, 6)
3. 13;(3, 4.5) **4.** 5;(−4, 1.5)
5. $\sqrt{29}$;(2, 4.5)
6. $5\sqrt{2}$;(0.5, −1.5)
7. 3;(5.5, −3) **8.** 7;(1.5, 6)
9. $\sqrt{13}$;(6, −3.5)
10. $\sqrt{85}$;(7, −0.5)
11. 17;(−3, −4.5)
12. 13;(0.5, 2)
13. $x^2+y^2=16$
14. $x^2+y^2=36$
15. $x^2+y^2=3$
16. $x^2+y^2=7$
17. $x^2+y^2=11$
18. $x^2+y^2=13$
19. $x^2+y^2=\frac{4}{9}$
20. $x^2+y^2=\frac{49}{16}$
21. $x^2+(y-2)^2=4$
22. $x^2+(y-5)^2=25$
23. $(x-4)^2+y^2=1$
24. $(x+3)^2+y^2=16$
25. $(x+2)^2+y^2=8$
26. $(x-5)^2+y^2=2$
27. $(x-3)^2+(y-1)^2=36$
28. $(x+1)^2+(y-2)^2=25$
29. $(x-3)^2+(y-5)^2=12$
30. $(x-4)^2+(y+2)^2=14$
31. $(x-7)^2+(y-4)^2=10$
32. $(x+3)^2+(y-2)^2=7$

Solving for y^2 gives: $y^2=9-x^2$

Solving for y_1 and y_2 gives: $\left.\begin{aligned} y_1 &= \sqrt{9-x^2} \\ y_2 &= -\sqrt{9-x^2} \end{aligned}\right\}$

Graphing both y_1 and y_2 gives the following graph of the circle.

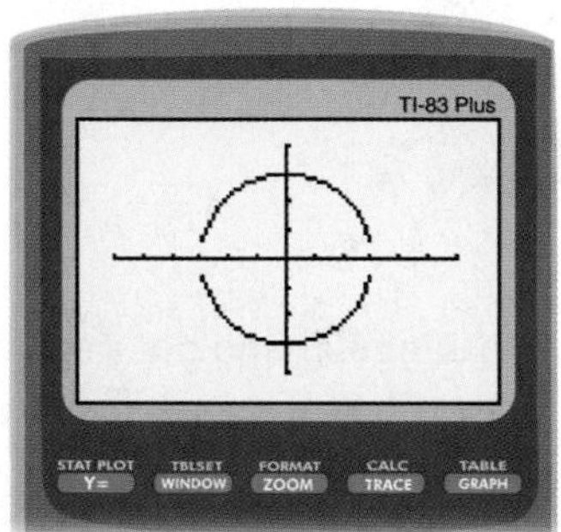

Practice Problems

1. *Find the equation of the circle with center at $(-2,3)$ and radius 6.*
2. *Write the equation in standard form and find the center and radius for the circle with equation $x^2+y^2+6y=7$.*
3. *Find the distance between the two points $(5,3)$ and $(-1,-3)$.*

12.5 Exercises

In Exercises 1 – 12, find the distance between the two given points and the coordinates of the midpoint of the line segment joining the two points.

1. $(2,4),(6,7)$ **2.** $(1,0),(6,12)$ **3.** $(-3,2),(9,7)$

4. $(-6,3),(-2,0)$ **5.** $(1,7),(3,2)$ **6.** $(-2,1),(3,-4)$

7. $(4,-3),(7,-3)$ **8.** $(-2,6),(5,6)$ **9.** $(5,-2),(7,-5)$

10. $(6,4),(8,-5)$ **11.** $(-7,3),(1,-12)$ **12.** $(3,8),(-2,-4)$

Find equations for each of the circles in Exercises 13 – 32.

13. Center $(0,0)$; $r=4$ **14.** Center $(0,0)$; $r=6$ **15.** Center $(0,0)$; $r=\sqrt{3}$

Answers to Practice Problems: **1.** $(x+2)^2+(y-3)^2=36$ **2.** $x^2+(y+3)^2=16$; center at $(0,-3)$ and radius 4 **3.** $\sqrt{72}=6\sqrt{2}$

33. $x^2 + y^2 = 9$
Center: $(0,0), r = 3$

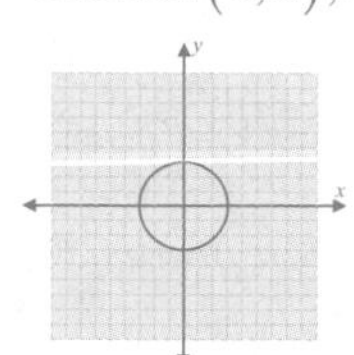

34. $x^2 + y^2 = 16$
Center: $(0,0), r = 4$

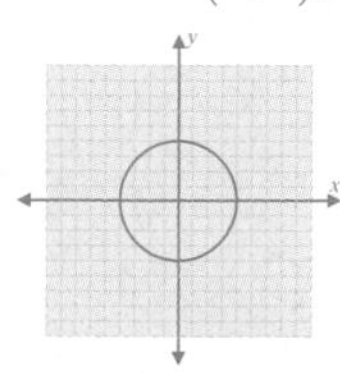

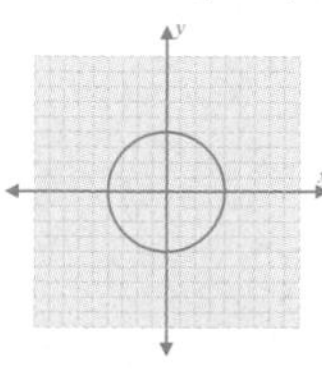

35. $x^2 + y^2 = 49$
Center: $(0,0), r = 7$

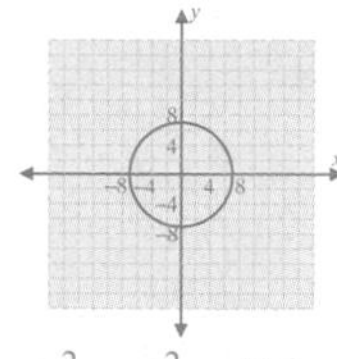

36. $x^2 + y^2 = 25$
Center: $(0,0), r = 5$

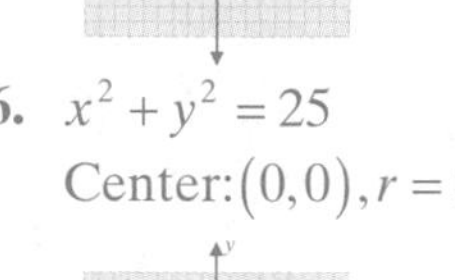

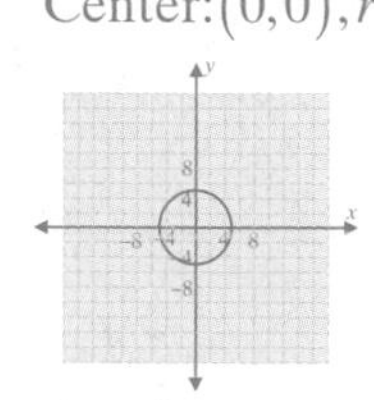

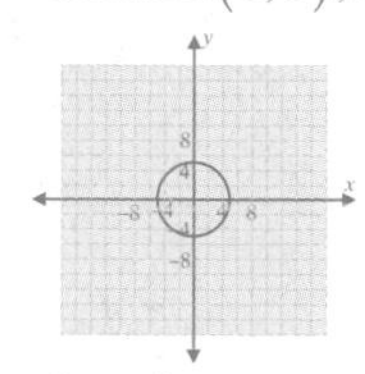

37. $x^2 + y^2 = 18$
Center: $(0,0), r = 3\sqrt{2}$

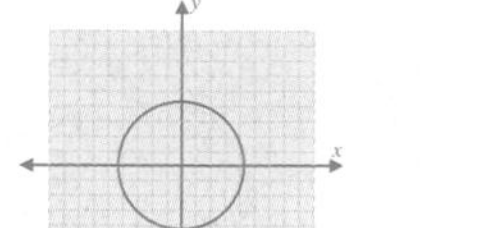

38. $x^2 + y^2 = 12$
Center: $(0,0), r = 2\sqrt{3}$

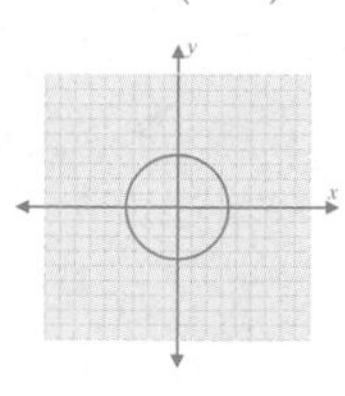

16. Center $(0,0)$; $r = \sqrt{7}$ **17.** Center $(0,0)$; $r = \sqrt{11}$ **18.** Center $(0,0)$; $r = \sqrt{13}$

19. Center $(0,0)$; $r = \frac{2}{3}$ **20.** Center $(0,0)$; $r = \frac{7}{4}$ **21.** Center $(0,2)$; $r = 2$

22. Center $(0,5)$; $r = 5$ **23.** Center $(4,0)$; $r = 1$ **24.** Center $(-3,0)$; $r = 4$

25. Center $(-2,0)$; $r = \sqrt{8}$ **26.** Center $(5,0)$; $r = \sqrt{2}$ **27.** Center $(3,1)$; $r = 6$

28. Center $(-1,2)$; $r = 5$ **29.** Center $(3,5)$; $r = \sqrt{12}$

30. Center $(4,-2)$; $r = \sqrt{14}$ **31.** Center $(7,4)$; $r = \sqrt{10}$

32. Center $(-3,2)$; $r = \sqrt{7}$

Write each of the equations in Exercises 33 – 48 in standard form. Find the center and radius of the circle and then sketch the graph.

33. $x^2 + y^2 = 9$ **34.** $x^2 + y^2 = 16$ **35.** $x^2 = 49 - y^2$

36. $y^2 = 25 - x^2$ **37.** $x^2 + y^2 = 18$ **38.** $x^2 + y^2 = 12$

39. $x^2 + y^2 + 2x = 8$ **40.** $x^2 + y^2 - 4x = 12$ **41.** $x^2 + y^2 - 4y = 0$

42. $x^2 + y^2 + 6x = 0$ **43.** $x^2 + y^2 + 2x + 4y = 11$

44. $x^2 + y^2 - 4x + 10y + 20 = 0$ **45.** $x^2 + y^2 + 4x + 4y - 8 = 0$

46. $x^2 + y^2 - 6x - 8y + 9 = 0$ **47.** $x^2 + y^2 - 4x - 6y + 5 = 0$

48. $x^2 + y^2 + 10x - 2y + 14 = 0$

In Exercises 49 and 50, use the Pythagorean Theorem to decide if the triangle determined by the given points is a right triangle.

49. $A(1,-2)$, $B(7,1)$, $C(5,5)$ **50.** $A(-5,-1)$, $B(2,1)$, $C(-1,6)$

In Exercises 51 and 52, show that the triangle determined by the given points is an isosceles triangle (has two equal sides).

51. $A(1,1)$, $B(5,9)$, $C(9,5)$ **52.** $A(1,-4)$, $B(3,2)$, $C(9,4)$

In Exercises 53 and 54, show that the triangle determined by the given points is an equilateral triangle (all sides equal).

53. $A(1,0)$, $B(3,\sqrt{12})$, $C(5,0)$ **54.** $A(0,5)$, $B(0,-3)$, $C(\sqrt{48},1)$

39. $(x+1)^2+y^2=9$
Center: $(-1,0), r=3$

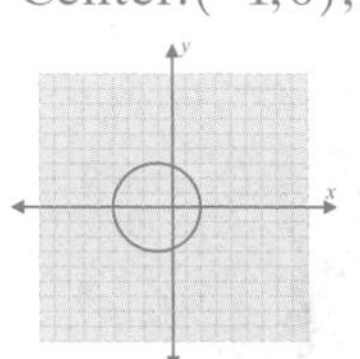

40. $(x-2)^2+y^2=16$
Center: $(2,0), r=4$

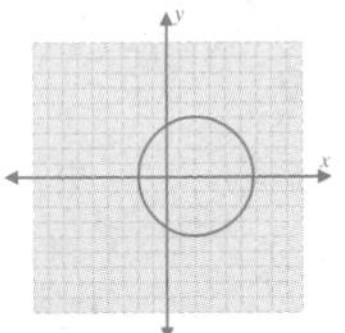

41. $x^2+(y-2)^2=4$
Center: $(0,2), r=2$

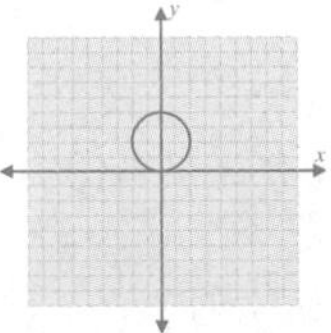

42. $(x+3)^2+y^2=9$
Center: $(-3,0), r=3$

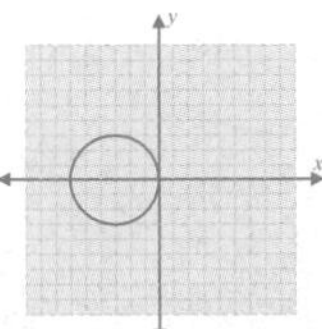

43. $(x+1)^2+(y+2)^2=16$
Center: $(-1,-2), r=4$

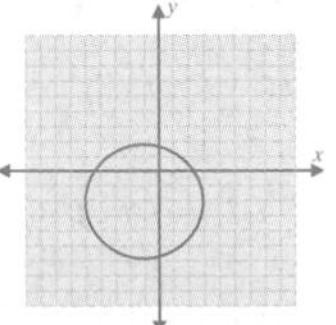

44. $(x-2)^2+(y+5)^2=9$
Center: $(2,-5), r=3$

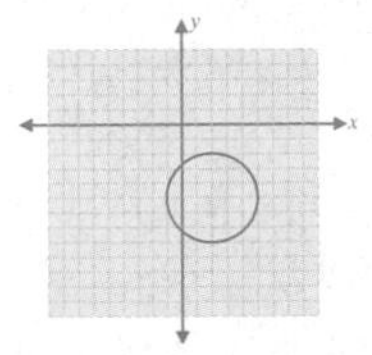

In Exercises 55 and 56, show that the diagonals (AC and BD) of the rectangle ABCD are equal.

55. $A(2,-2), B(2,3), C(8,3), D(8,-2)$

56. $A(-1,1), B(-1,4), C(4,4), D(4,1)$

In Exercises 57 – 60, find the perimeter of the triangle determined by the given points.

57. $A(-5,0), B(3,4), C(0,0)$

58. $A(-6,-1), B(-3,3), C(6,4)$

59. $A(-2,5), B(3,1), C(2,-2)$

60. $A(1,4), B(-3,3), C(-1,7)$

In Exercises 61 – 64, use a graphing calculator to graph the circles. Be sure to set a square window.

61. $x^2+y^2=16$

62. $x^2+y^2=25$

63. $(x+3)^2+y^2=49$

64. $(x-2)^2+(y-5)^2=100$

Writing and Thinking About Mathematics

65. For a given line and a point not on the line, a parabola is defined as the set of all points that are the same distance from the point and the line. The point is called the focus and the line is called the directrix. See the figure below.

a. Suppose that (x,y) is any point on a parabola and $(0,p)$ is the focus. Find the distance from (x,y) to the focus.

b. Suppose that (x,y) is any point on the same parabola in part **a.** and the line $y=-p$ is the directrix. Find the distance from (x,y) to the directrix.

c. Show that the equation of the parabola is $x^2=4py$.

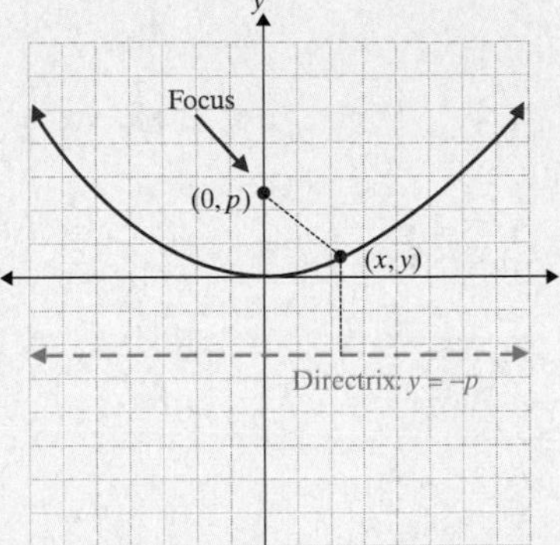

66. Using the equation developed in Exercise 65, find the equation of the parabola with focus at $(0,2)$ and line $y=-2$ as directrix. Draw the graph.

continued on next page ...

45. $(x+2)^2+(y+2)^2=16$
Center: $(-2,-2), r=4$

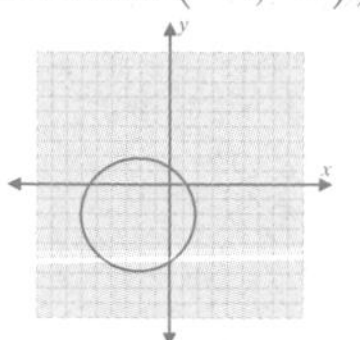

46. $(x-3)^2+(y-4)^2=16$
Center: $(3,4), r=4$

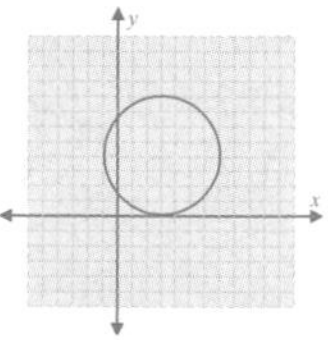

47. $(x-2)^2+(y-3)^2=8$
Center: $(2,3), r=2\sqrt{2}$

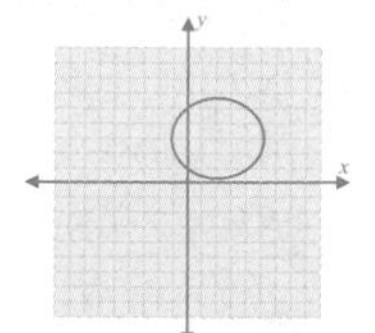

67. For a given line and a point not on the line, a parabola is defined as the set of all points that are the same distance from the point and the line. The point is called the focus and the line is called the directrix. See the figure below.

a. Suppose that (x,y) is any point on a parabola and $(p,0)$ is the focus. Find the distance from (x,y) to the focus.

b. Suppose that (x,y) is any point on the same parabola in part **a.** and the line $x=-p$ is the directrix. Find the distance from (x,y) to the directrix.

c. Show that the equation of the parabola is $y^2=4px$.

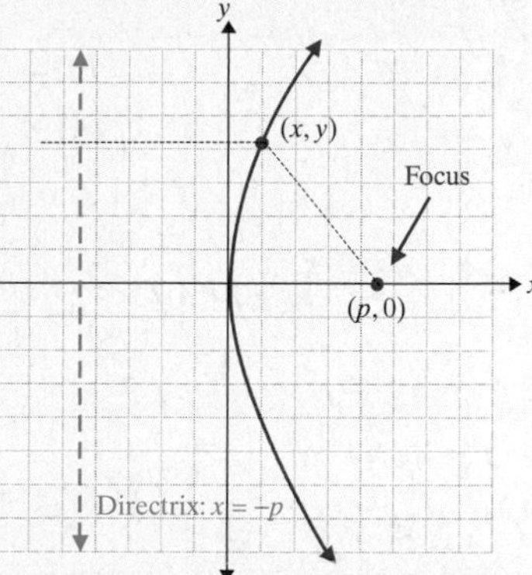

68. Using the equation developed in Exercise 67, find the equation of the parabola with focus at $(-3,0)$ and line $x=3$ as directrix. Draw the graph.

Hawkes Learning Systems: Introductory & Intermediate Algebra

The Distance Formula and Circles

48. $(x+5)^2+(y-1)^2=12$
Center: $(-5,1), r=2\sqrt{3}$

49. $|\overline{AB}|^2=45, |\overline{AC}|^2=65$
$|\overline{BC}|^2=20$
$|\overline{AB}|^2+|\overline{BC}|^2=|\overline{AC}|^2$
Right Triangle

50. $|\overline{AB}|^2=53$
$|\overline{AC}|^2=65$
$|\overline{BC}|^2=34$
not a right triangle

51. $|\overline{AB}|=|\overline{AC}|=4\sqrt{5}$

52. $|\overline{AB}|=|\overline{BC}|=2\sqrt{10}$

53. $|\overline{AB}|=|\overline{AC}|=|\overline{BC}|=4$

54. $|\overline{AB}|=|\overline{AC}|=|\overline{BC}|=8$

55. $|\overline{AC}|=|\overline{BD}|=\sqrt{61}$

56. $|\overline{AC}|=|\overline{BD}|=\sqrt{34}$

57. $10+4\sqrt{5}$

58. $18+\sqrt{82}$

59. $\sqrt{41}+\sqrt{65}+\sqrt{10}$

60. $\sqrt{17}+\sqrt{13}+2\sqrt{5}$

61.

62.

63.

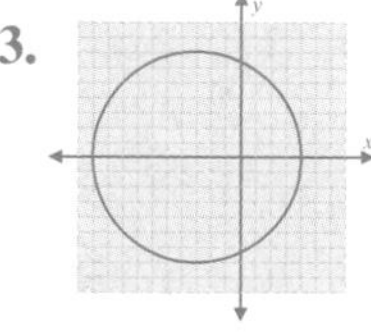

64.

65. **a.** $d=\sqrt{x^2+(y-p)^2}$ **b.** $d=y+p$
c. $y+p=\sqrt{x^2+(y-p)^2}$
$(y+p)^2=x^2+(y-p)^2$
$y^2+2py+p^2=x^2+y^2-2py+p^2$
$x^2=4py$

66. $y=\frac{1}{8}x^2$

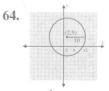

67. **a.** $d=\sqrt{(x-p)^2+y^2}$
b. $d=x+p$
c. $x+p=\sqrt{(x-p)^2+y^2}$
$(x+p)^2=(x-p)^2+y^2$
$x^2+2px+p^2=x^2-2px+p^2+y^2$
$y^2=4px$

68. $x=-\frac{1}{12}y^2$

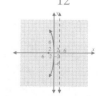

12.6 Ellipses and Hyperbolas

Objectives

After completing this section, you will be able to:

1. *Graph ellipses centered at the origin or at the point (h, k).*
2. *Graph hyperbolas centered at the origin or at the point (h, k).*
3. *Find the equations for the asymptotes of hyperbolas.*

Equations of Ellipses

An **ellipse** is the set of all points in a plane the sum of whose distances from two fixed points is constant. Each of the fixed points is called a **focus** (plural **foci**). The **center** of an ellipse is the point midway between the foci. Ellipses have many practical applications in the sciences, particularly in astronomy. For example, the planets in our solar system have elliptical orbits and the sun is a focus of each ellipse.

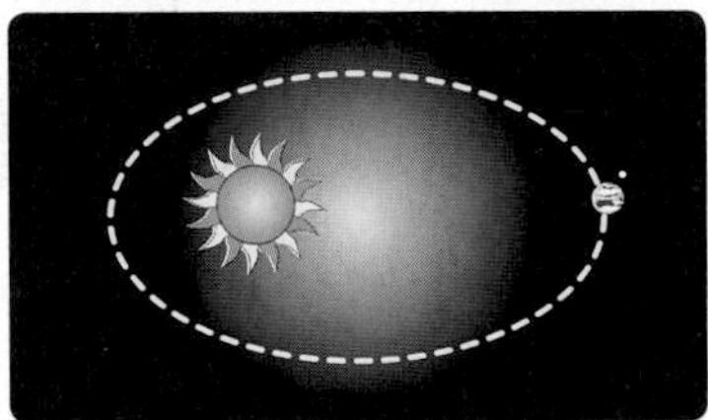

An ellipse with its center at the origin and foci at $(-c, 0)$ and $(c, 0)$ and x-intercepts at $(-a, 0)$ and $(a, 0)$ and y-intercepts at $(0, -b)$ and $(0, b)$ is shown in Figure 12.22 below.

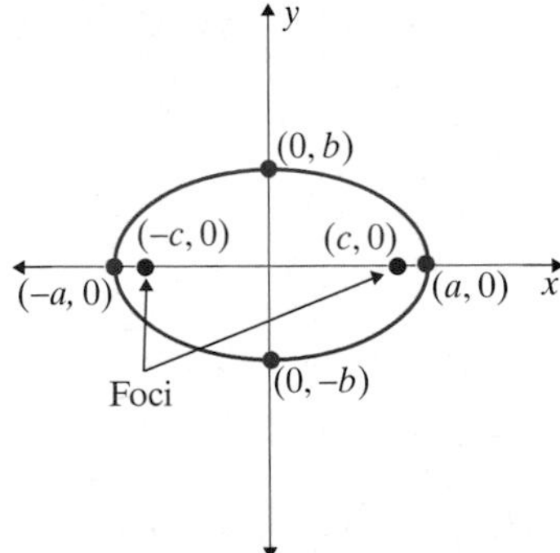

Figure 12.22

As an example, consider the equation

$$\frac{x^2}{25}+\frac{y^2}{9}=1.$$

Several points that satisfy this equation are given on the following page in tabular form and are graphed in Figure 12.23.

x	y
5	0
–5	0
0	3
0	–3
3	$\frac{12}{5}$
3	$-\frac{12}{5}$
–3	$\frac{12}{5}$
–3	$-\frac{12}{5}$

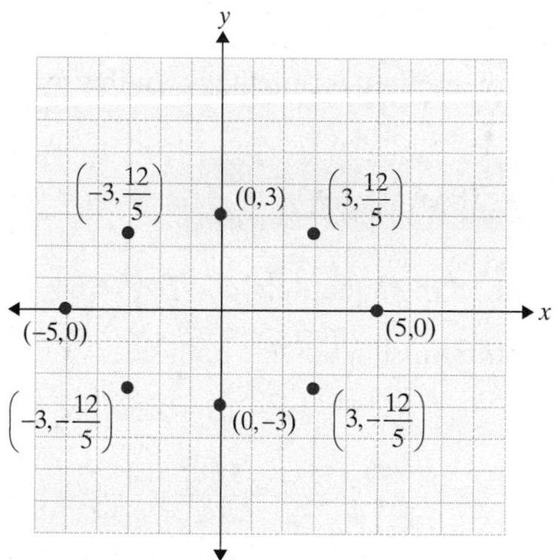

Figure 12.23

Joining the points in Figure 12.23 with a smooth curve, we get the graph of the **ellipse** shown in Figure 12.24. The points $(5,0)$ and $(-5,0)$ are the x-intercepts, and the points $(0,3)$ and $(0,-3)$ are the y-intercepts.

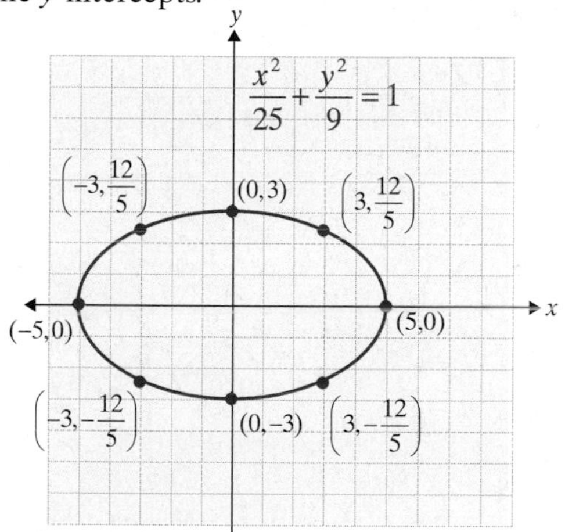

Figure 12.24

Equation of an Ellipse

The standard form for the equation of an ***ellipse*** *with its center at the origin is*

$$\frac{x^2}{a^2}+\frac{y^2}{b^2}=1 \qquad \text{where} \qquad a^2 \geq b^2.$$

The points $(a,0)$ and $(-a,0)$ are the x-intercepts.

The points $(0,b)$ and $(0,-b)$ are the y-intercepts.

The segment of length 2a joining the x-intercepts is called the ***major axis.***

The segment of length 2b joining the y-intercepts is called the ***minor axis.***

Note: Example 2 illustrates a second form, $\frac{x^2}{b^2}+\frac{y^2}{a^2}=1$, and corresponding adjustments in the related terminology. In this form, the major axis is along the y-axis.

Example 1: Equation of an Ellipse

Graph the equation $4x^2+16y^2=64$.

Solution: First divide both sides of the given equation by 64 to find the standard form.

$$4x^2+16y^2=64$$

$$\frac{4x^2}{64}+\frac{16y^2}{64}=\frac{64}{64}$$

$$\frac{x^2}{16}+\frac{y^2}{4}=1$$

The curve is an ellipse. The endpoints of the major axis are (–4, 0) and (4, 0). The endpoints of the minor axis are (0, –2) and (0, 2).

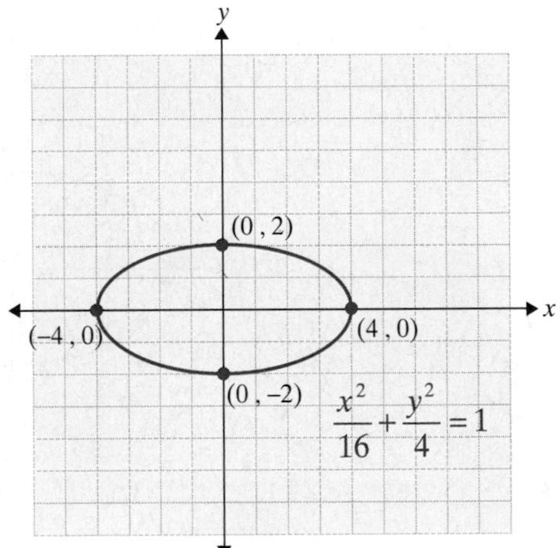

In the general discussion and in Example 1, the major axis is horizontal and the minor axis is vertical. For ellipses in standard form, the larger denominator is treated as a^2 and the smaller denominator as b^2. Thus, if the larger denominator is below y^2, then the major axis is vertical and the minor axis is horizontal. This situation is illustrated in Example 2.

Example 2: The Major and Minor Axes

Graph the equation $\frac{x^2}{1}+\frac{y^2}{9}=1$.

Solution: The equation is in standard form. However, since the larger denominator, 9, is below y^2, the major axis is vertical. The ellipse is elongated along the y-axis. The points $(0,-3)$ and $(0,3)$ are the endpoints of the major axis while $(-1,0)$ and $(1,0)$ are the endpoints of the minor axis.

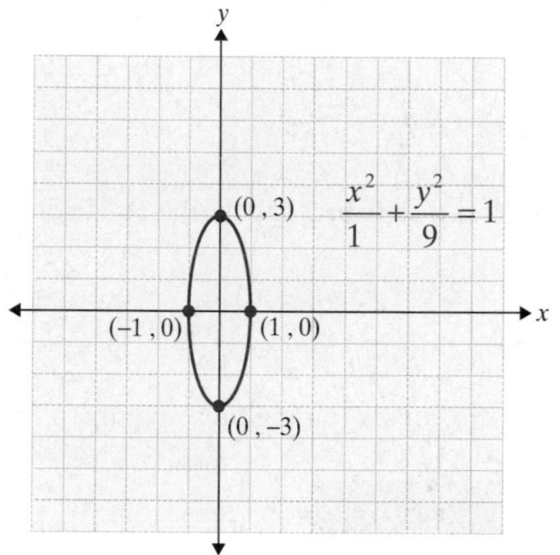

In the equation of an ellipse,

$$\frac{x^2}{a^2}+\frac{y^2}{b^2}=1 \qquad \left(\text{or } \frac{x^2}{b^2}+\frac{y^2}{a^2}=1\right),$$

the coefficients for x^2 and y^2 are both positive. If one of these coefficients is negative, then the equation represents a **hyperbola**.

Equations of Hyperbolas

A **hyperbola** is the set of all points in a plane such that the absolute value of the difference of the distances to two fixed points is constant. Each of the fixed points is called a **focus**. The graph of a hyperbola with its **center** (the point midway between the foci) at the origin and foci at $(-c, 0)$ and $(c, 0)$ and x-intercepts at $(-a, 0)$ and $(a, 0)$ is shown in Figure 12.25.

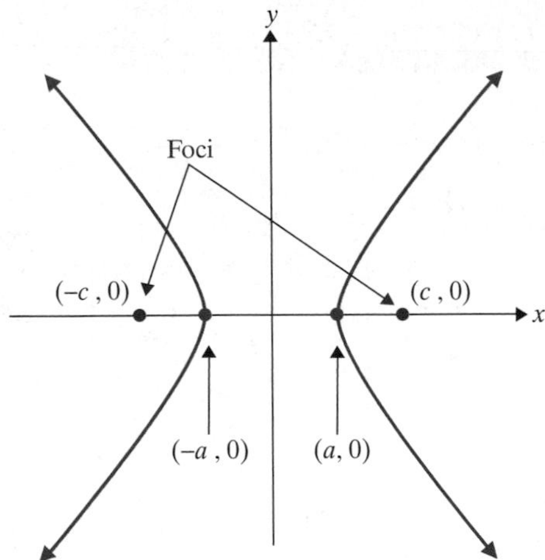

Figure 12.25

Several points that satisfy the equation

$$\frac{x^2}{25} - \frac{y^2}{9} = 1$$

and the curves joining these points (a hyperbola) are shown in Figure 12.26.

x	y
5	0
−5	0
7	$\frac{6\sqrt{6}}{5}$
7	$\frac{-6\sqrt{6}}{5}$
−7	$\frac{6\sqrt{6}}{5}$
−7	$\frac{-6\sqrt{6}}{5}$

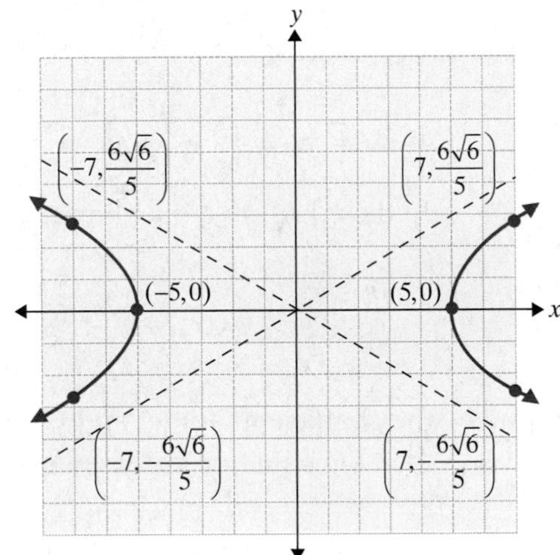

Figure 12.26

The two dotted lines shown in Figure 12.26 are called **asymptotes**. These lines are not part of the hyperbola, but they serve as guidelines for the graph because the curve gets closer and closer to these lines without ever touching them. The equations of these lines are

$$y = \frac{3}{5}x \quad \text{and} \quad y = -\frac{3}{5}x.$$

Standard Form for Equations of Hyperbolas

In general, there are ***two standard forms*** *for equations of hyperbolas with their* ***centers at the origin****:*

1. $\frac{x^2}{a^2}-\frac{y^2}{b^2}=1$

x-intercepts (vertices) at (a, 0) and (–a, 0); no y-intercepts;

Asymptotes: $y=\frac{b}{a}x$ *and* $y=-\frac{b}{a}x$*;*

The curve "opens" left and right.

2. $\frac{y^2}{a^2}-\frac{x^2}{b^2}=1$

y-intercepts (vertices) at (0, a) and (0, –a); no x-intercepts;

Asymptotes: $y=\frac{a}{b}x$ *and* $y=-\frac{a}{b}x$*;*

The curve "opens" up and down.

Geometrical Aid for Sketching Asymptotes

The asymptotes $y=\frac{a}{b}x$ and $y=-\frac{a}{b}x$ pass through the diagonals of the **fundamental rectangle** formed by joining the points $(a, 0)$, $(-a, 0)$, $(0, b)$, and $(0, -b)$. Fundamental rectangles are shown in Examples 3a and 3b.

Example 3: Asymptotes

a. Graph the curve $x^2-4y^2=4$.

Solution: Write the equation in standard form by dividing by 4 which yields

$$\frac{x^2}{4}-\frac{y^2}{1}=1.$$

Here, $a^2=4$ and $b^2=1$. So, using $a=2$ and $b=1$, the asymptotes are $y=\frac{1}{2}x$ and $y=-\frac{1}{2}x$. Vertices are $(2, 0)$ and $(-2, 0)$. The curve "opens" left and right. Notice that the asymptotes pass through the diagonals of the fundamental rectangle.

continued on next page ...

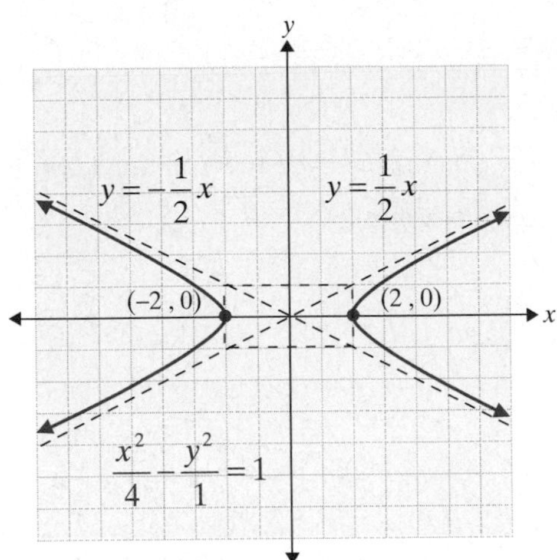

b. Graph the curve $\frac{y^2}{1}-\frac{x^2}{4}=1$.

Solution: First locate the asymptotes and the y-intercepts, then sketch the curve. Here, $a = 1$ and $b = 2$. The asymptotes are $y=\frac{1}{2}x$ and $y=-\frac{1}{2}x$. The y-intercepts are $(0,1)$ and $(0,-1)$. The curve "opens" up and down.

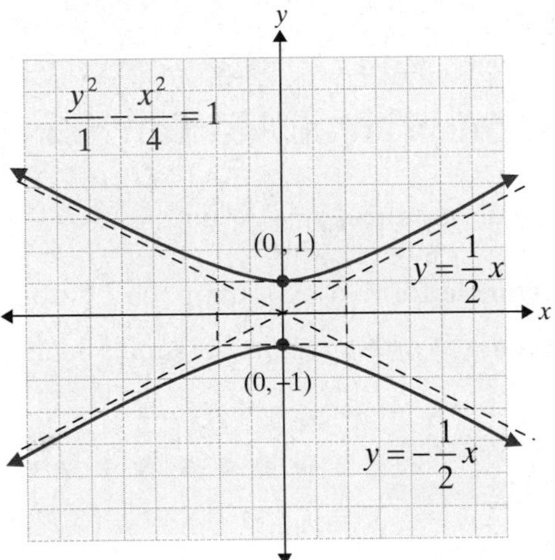

Ellipses and Hyperbolas with Centers at (h, k)

With our knowledge of translations discussed in Section 12.3, we know that replacing x with $x - h$ in an equation gives the graph a horizontal shift of h units, and replacing y with $y - k$ in an equation gives the graph a vertical shift of k units. We used these ideas in Section 12.5 when we discussed the equations and graphs of circles with centers at (h, k), points other than the origin.

For example,

$x^2 + y^2 = 16$ Equation of the circle with center at $(0, 0)$ and radius 4

$(x-1)^2 + (y-3)^2 = 16$ Equation of the circle with center at $(1, 3)$ and radius 4

The same procedure can be used to obtain the equations of ellipses and hyperbolas with centers at (h, k). That is, the equation of an ellipse and the equation of a hyperbola with center at (h, k) can be found by replacing x with $x - h$ and y with $y - k$ in the standard forms of the equations.

Ellipse with Center at (*h*, *k*)

The equation of an ellipse with its center at (h, k) is

$$\frac{(x-h)^2}{a^2} + \frac{(y-k)^2}{b^2} = 1 \quad or \quad \frac{(x-h)^2}{b^2} + \frac{(y-k)^2}{a^2} = 1 \quad \text{where } a^2 \geq b^2.$$

Note: *a and b are distances from (h, k) to the vertices. (See Figure 12.27.)*

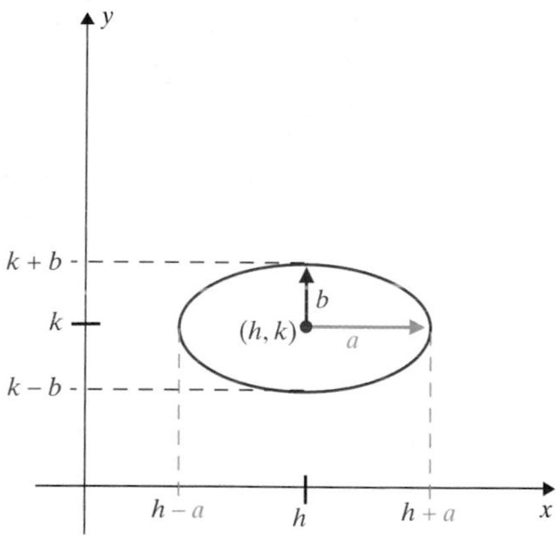

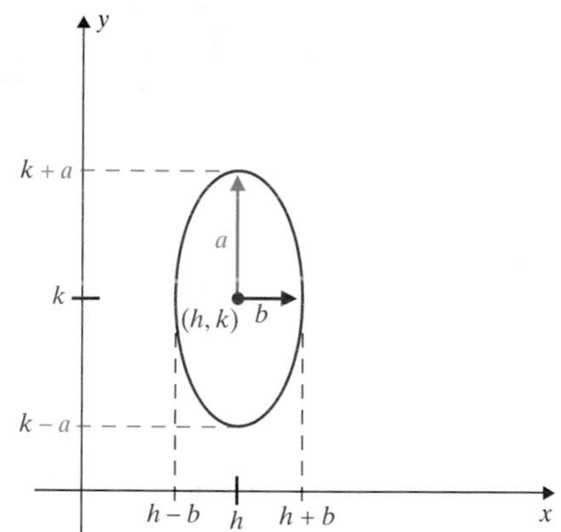

Figure 12.27

Example 4: Ellipse with Center at (*h*, *k*)

Graph the ellipse $\frac{(x+2)^2}{16} + \frac{(y-1)^2}{9} = 1$.

Solution: The graph of $\frac{x^2}{16} + \frac{y^2}{9} = 1$ is translated 2 units left and 1 unit up so that the center is at $(-2, 1)$ with $a = 4$ and $b = 3$. The graph is shown here with the center and vertices labeled.

continued on next page ...

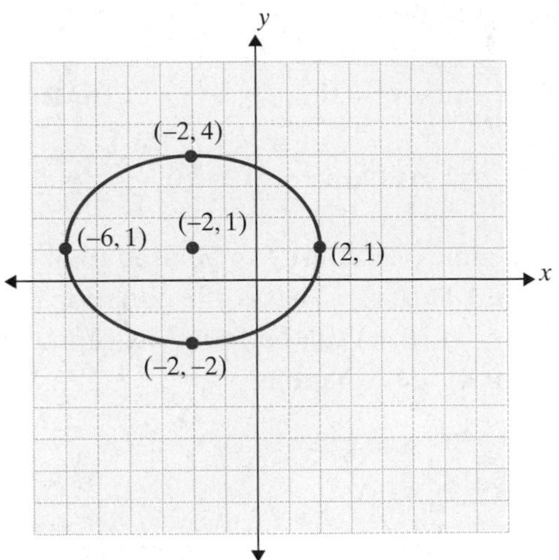

Hyperbola with Center at (h, k)

The equation of a hyperbola with its center at (h, k) is

$$\frac{(x-h)^2}{a^2}-\frac{(y-k)^2}{b^2}=1 \quad or \quad \frac{(y-k)^2}{a^2}-\frac{(x-h)^2}{b^2}=1 .$$

Note: *a and b are used as in the standard form but are measured from (h,k). (See Figure 12.28.)*

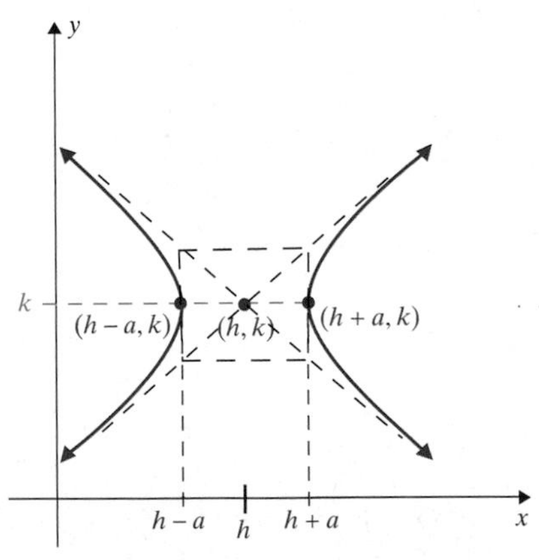

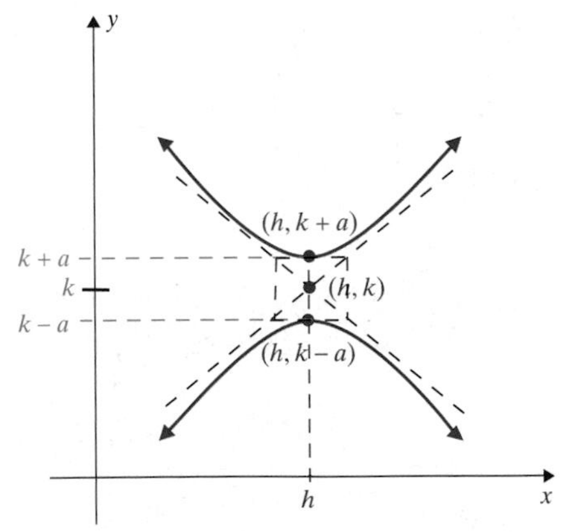

Figure 12.28

Answers to Practice Problems: **1.** $\frac{x^2}{9}+\frac{y^2}{2}=1$; Major axis: 6, Minor axis: $2\sqrt{2}$ **2.** $\frac{x^2}{9}-\frac{y^2}{1}=1$; asymptotes: $y=\frac{1}{3}x$, $y=-\frac{1}{3}x$ **3.**

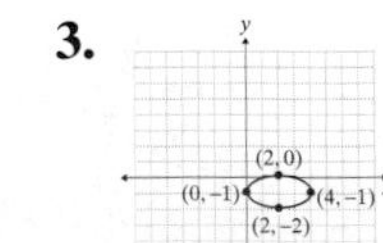

4.

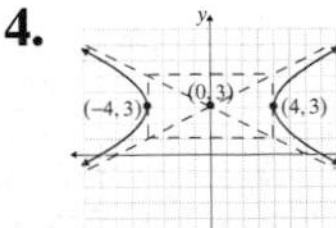

1. $\frac{x^2}{36}+\frac{y^2}{4}=1$

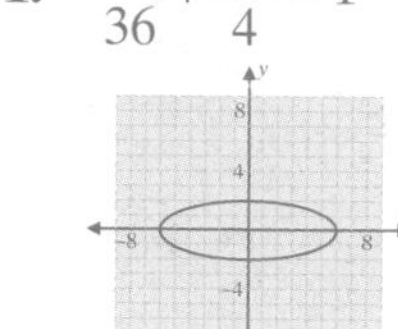

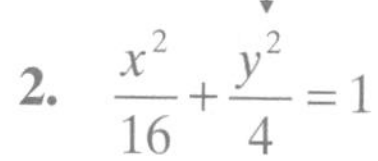

2. $\frac{x^2}{16}+\frac{y^2}{4}=1$

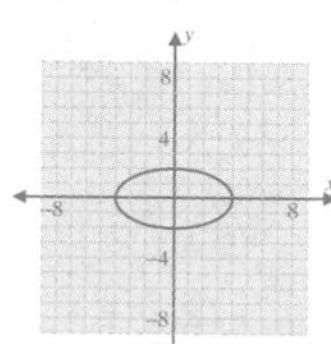

3. $\frac{x^2}{25}+\frac{y^2}{4}=1$

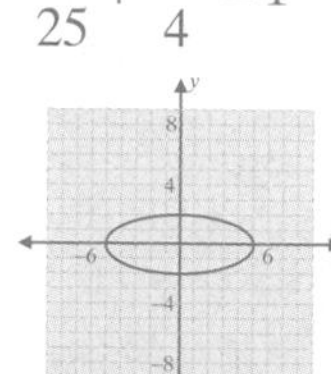

4. $\frac{x^2}{9}+\frac{y^2}{4}=1$

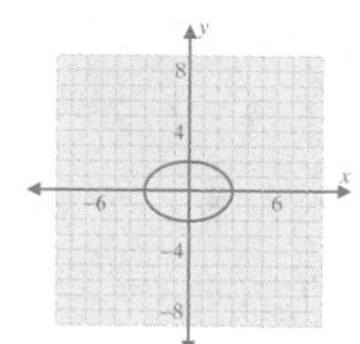

5. $\frac{x^2}{1}+\frac{y^2}{16}=1$

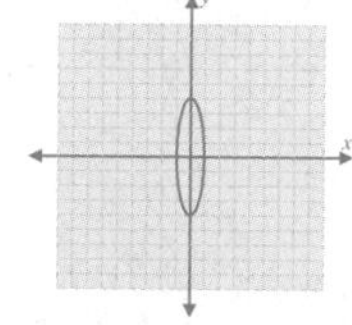

Example 5: Hyperbola with Center at (h, k)

Graph the hyperbola $\frac{(x-3)^2}{25}-\frac{(y+4)^2}{36}=1$.

Solution: The graph of $\frac{x^2}{25}-\frac{y^2}{36}=1$ is translated 3 units right and 4 units down so that the center is at $(3,-4)$ with $a=5$ and $b=6$. The graph is shown here with the asymptotes shown and center and vertices labeled.

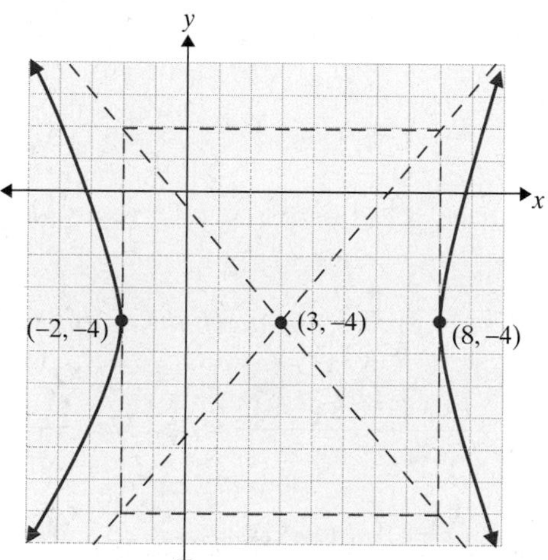

Practice Problems

1. Write the equation $2x^2+9y^2=18$ in standard form. State the length of the major axis and the length of the minor axis.

2. Write the equation $x^2-9y^2=9$ in standard form. Write the equations of the asymptotes.

3. Graph the ellipse $\frac{(x-2)^2}{4}+\frac{(y+1)^2}{1}=1$.

4. Graph the hyperbola $\frac{x^2}{16}-\frac{(y-3)^2}{4}=1$.

12.6 Exercises

Write each of the equations in Exercises 1 – 30 in standard form, then sketch the graph. For hyperbolas, graph the asymptotes as well.

1. $x^2+9y^2=36$	**2.** $x^2+4y^2=16$	**3.** $4x^2+25y^2=100$
4. $4x^2+9y^2=36$	**5.** $16x^2+y^2=16$	**6.** $25x^2+9y^2=36$
7. $x^2-y^2=1$	**8.** $x^2-y^2=4$	**9.** $9x^2-y^2=9$

6. $\dfrac{x^2}{\left(\frac{36}{25}\right)} + \dfrac{y^2}{4} = 1$

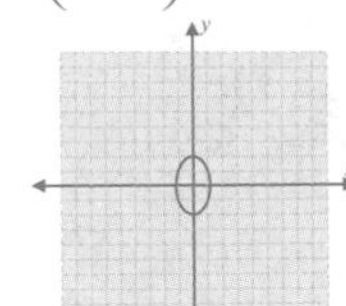

7. $\dfrac{x^2}{1} - \dfrac{y^2}{1} = 1$

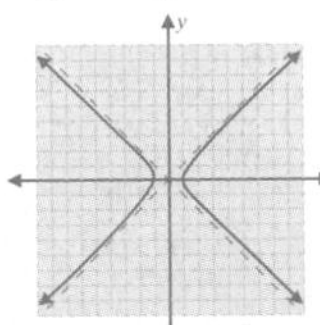

8. $\dfrac{x^2}{4} - \dfrac{y^2}{4} = 1$

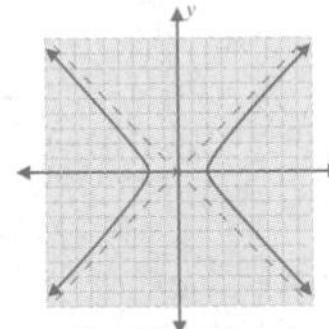

9. $\dfrac{x^2}{1} - \dfrac{y^2}{9} = 1$

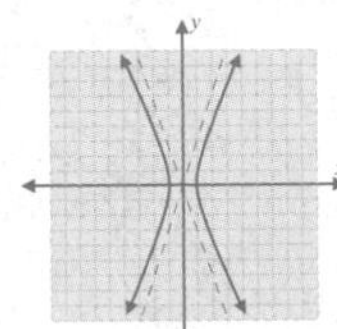

10. $\dfrac{x^2}{1} - \dfrac{y^2}{4} = 1$

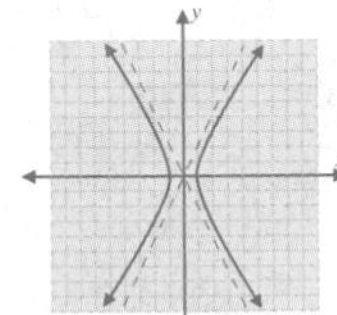

11. $\dfrac{x^2}{9} - \dfrac{y^2}{4} = 1$

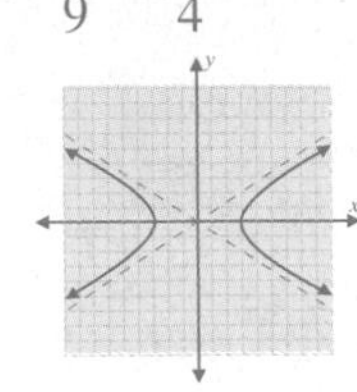

10. $4x^2 - y^2 = 4$ **11.** $4x^2 - 9y^2 = 36$ **12.** $9x^2 - 16y^2 = 144$

13. $2x^2 + y^2 = 8$ **14.** $3x^2 + y^2 = 12$ **15.** $x^2 + 5y^2 = 20$

16. $x^2 + 7y^2 = 28$ **17.** $y^2 - x^2 = 9$ **18.** $y^2 - x^2 = 16$

19. $y^2 - 2x^2 = 8$ **20.** $y^2 - 3x^2 = 12$ **21.** $y^2 - 2x^2 = 18$

22. $y^2 - 5x^2 = 20$ **23.** $3x^2 + 2y^2 = 18$ **24.** $4x^2 + 3y^2 = 12$

25. $4x^2 + 5y^2 = 20$ **26.** $3x^2 + 8y^2 = 48$ **27.** $3x^2 - 5y^2 = 75$

28. $4x^2 - 7y^2 = 28$ **29.** $3y^2 - 4x^2 = 36$ **30.** $9y^2 - 8x^2 = 72$

In Exercises 31 – 36, match the graph with the given equation.

31. $\dfrac{(x-1)^2}{4} + \dfrac{(y-3)^2}{25} = 1$ **32.** $\dfrac{(x+1)^2}{4} + \dfrac{(y+3)^2}{25} = 1$ **33.** $\dfrac{(x-1)^2}{25} + \dfrac{(y-3)^2}{4} = 1$

34. $\dfrac{(x+1)^2}{25} + \dfrac{(y+3)^2}{4} = 1$ **35.** $\dfrac{(x+1)^2}{25} - \dfrac{(y+3)^2}{4} = 1$ **36.** $\dfrac{(y+3)^2}{4} - \dfrac{(x+1)^2}{25} = 1$

A.

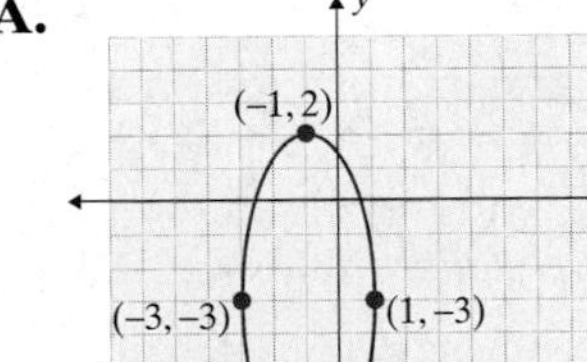

B.

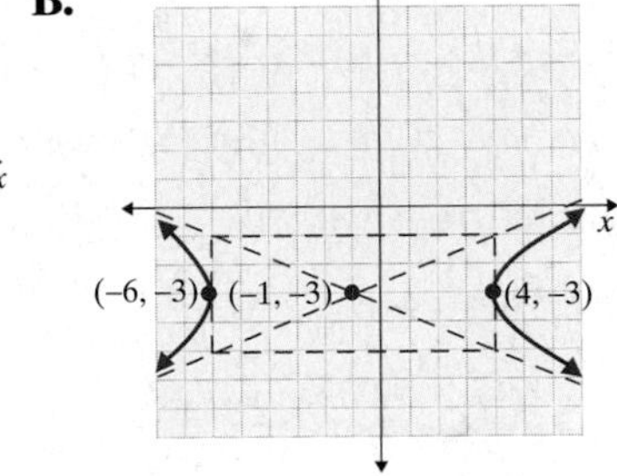

C.

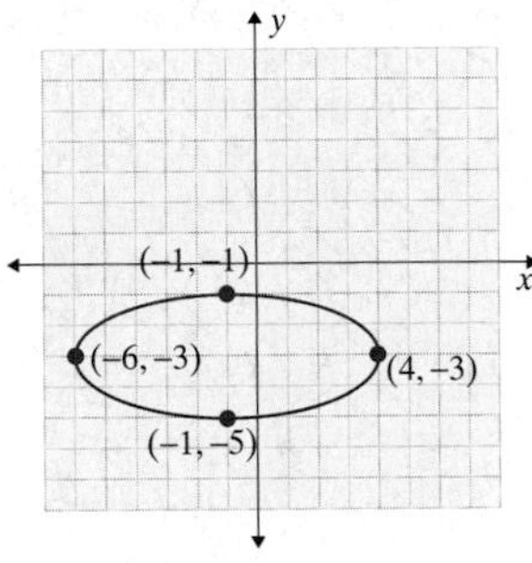

D.

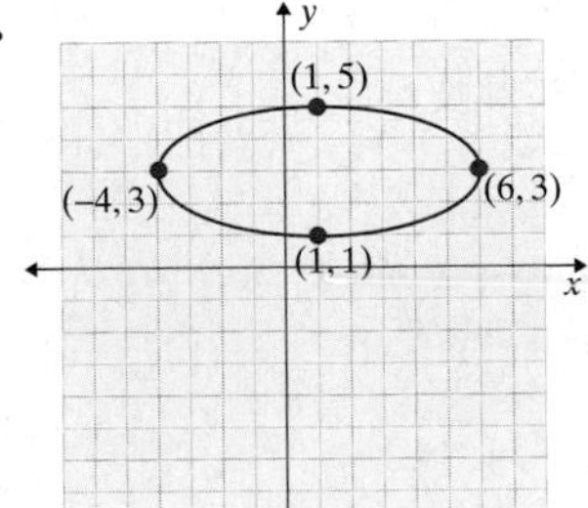

E.

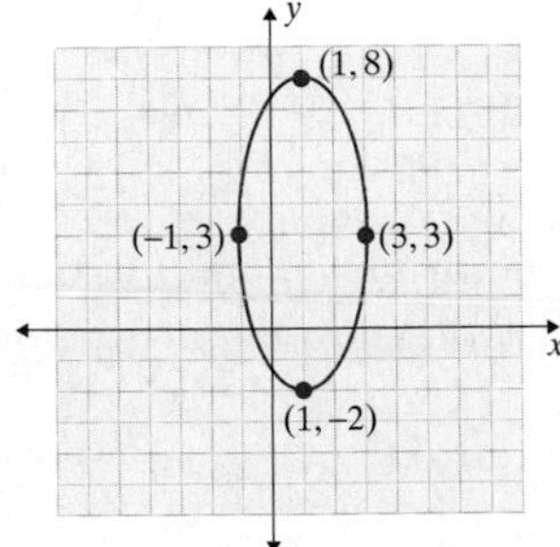

F.

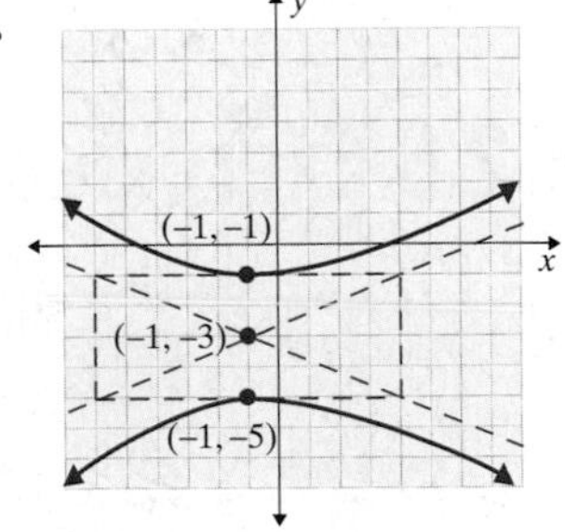

12. $\frac{x^2}{16}-\frac{y^2}{9}=1$

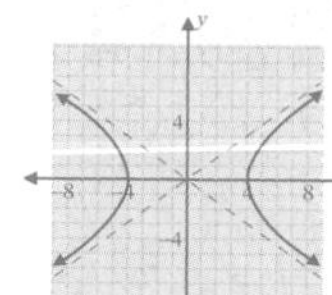

13. $\frac{x^2}{4}+\frac{y^2}{8}=1$

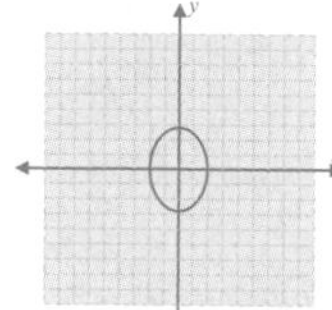

14. $\frac{x^2}{4}+\frac{y^2}{12}=1$

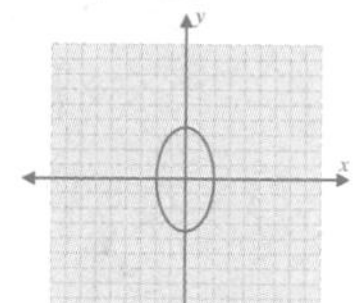

15. $\frac{x^2}{20}+\frac{y^2}{4}=1$

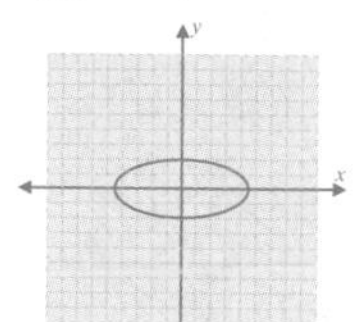

16. $\frac{x^2}{28}+\frac{y^2}{4}=1$

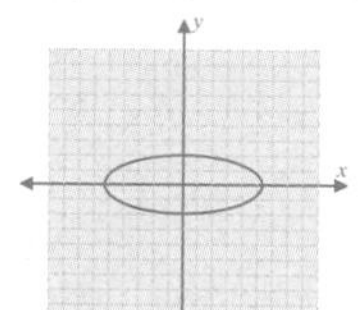

17. $\frac{y^2}{9}-\frac{x^2}{9}=1$

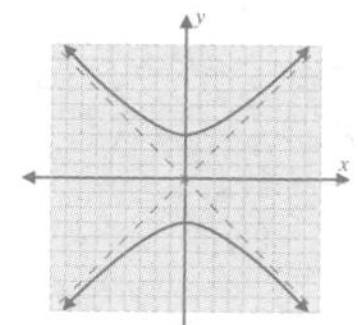

In Exercises 37 – 42, use your knowledge of translations to graph each of the following equations. These graphs are ellipses and hyperbolas with centers at points other than the origin.

37. $\frac{(x-2)^2}{25}+\frac{(y-1)^2}{9}=1$ **38.** $\frac{(x+1)^2}{16}+\frac{(y-4)^2}{1}=1$ **39.** $\frac{(x+5)^2}{1}-\frac{(y+2)^2}{16}=1$

40. $\frac{(x-4)^2}{9}-\frac{(y-3)^2}{36}=1$ **41.** $\frac{(x+1)^2}{49}+\frac{(y-6)^2}{100}=1$ **42.** $\frac{(y-2)^2}{9}-\frac{(x+2)^2}{4}=1$

Writing and Thinking About Mathematics

43. The definition of an ellipse is given in the text as follows:
An ellipse is the set of all points in a plane the sum of whose distances from two fixed points is constant.

a. Draw an ellipse by proceeding as follows:

Step 1: Place two thumb tacks in a piece of cardboard.

Step 2: Select a piece of string slightly longer than the distance between the two tacks.

Step 3: Tie the string to each thumb tack and stretch the string taut by using a pencil.

Step 4: Use the pencil to trace the path of an ellipse on the cardboard by keeping the string taut. (The length of the string represents the fixed distance from points on the ellipse to the two foci.)

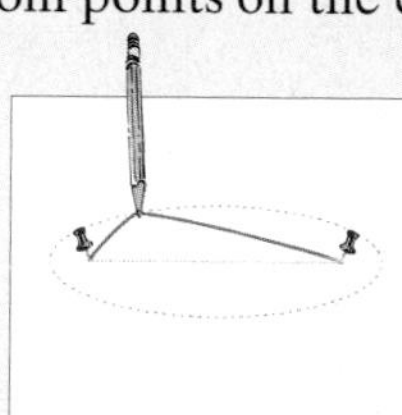

b. Show that the equation of an ellipse with foci at $(-c, 0)$ and $(c, 0)$, center at the origin, and $2a$ as the constant sum of the lengths to the foci can be written in the form

$$\frac{x^2}{a^2}+\frac{y^2}{a^2-c^2}=1.$$

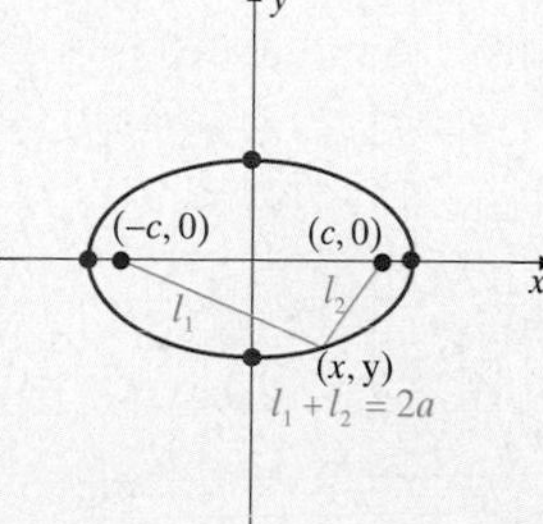

c. In the equation in part (b), substitute $b^2 = a^2 - c^2$ to get the standard form for the equation of an ellipse. Show that the points $(0, -b)$ and $(0, b)$ are the y-intercepts and a is the distance from each y-intercept to a focus.

Hawkes Learning Systems: Introductory & Intermediate Algebra

Ellipses and Hyperbolas

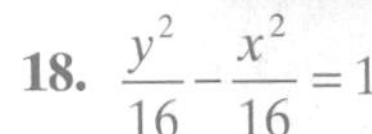

18. $\frac{y^2}{16} - \frac{x^2}{16} = 1$

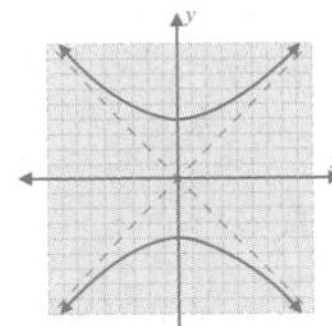

19. $\frac{y^2}{8} - \frac{x^2}{4} = 1$

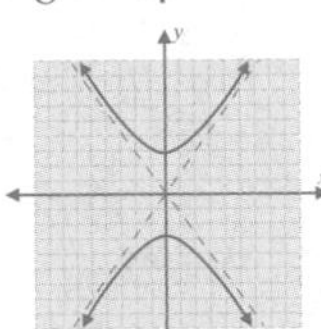

20. $\frac{y^2}{12} - \frac{x^2}{4} = 1$

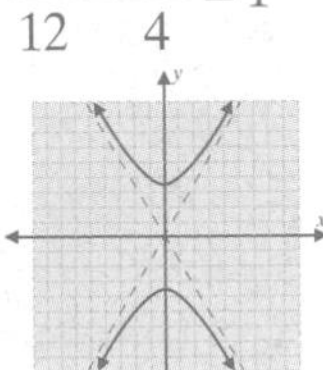

21. $\frac{y^2}{18} - \frac{x^2}{9} = 1$

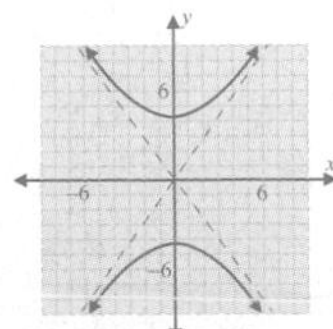

22. $\frac{y^2}{20} - \frac{x^2}{4} = 1$

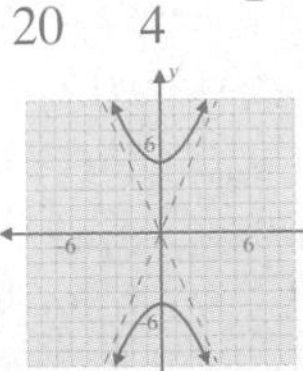

23. $\frac{x^2}{6} + \frac{y^2}{9} = 1$

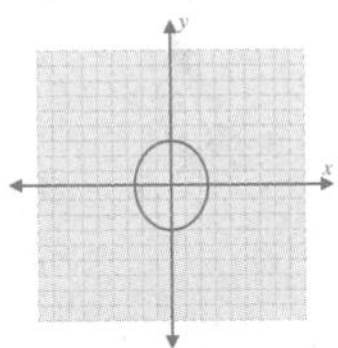

24. $\frac{x^2}{3} + \frac{y^2}{4} = 1$

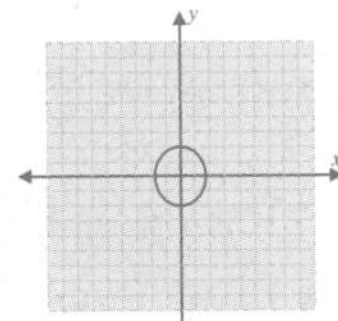

25. $\frac{x^2}{5} + \frac{y^2}{4} = 1$

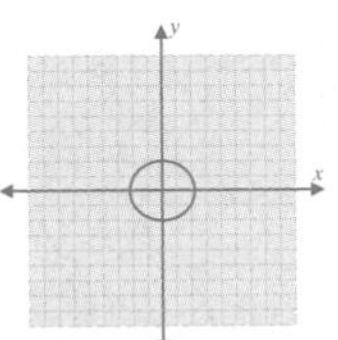

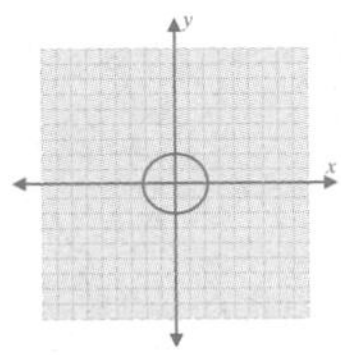

26. $\frac{x^2}{16} + \frac{y^2}{6} = 1$

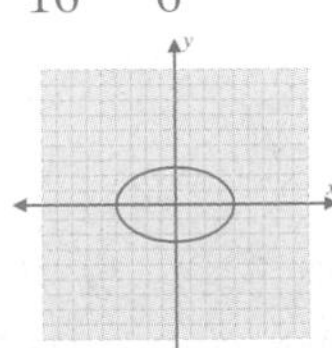

27. $\frac{x^2}{25} - \frac{y^2}{15} = 1$

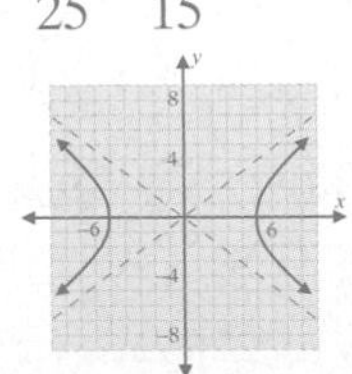

28. $\frac{x^2}{7} - \frac{y^2}{4} = 1$

29. $\frac{y^2}{12} - \frac{x^2}{9} = 1$

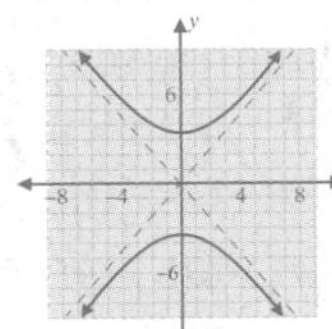

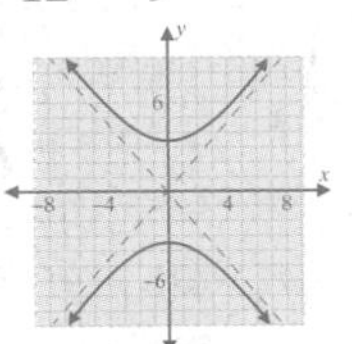

30. $\frac{y^2}{8} - \frac{x^2}{9} = 1$

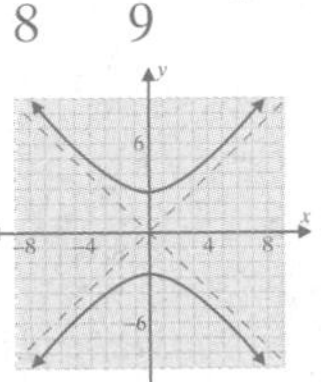

31. E
32. A
33. D
34. C
35. B
36. F

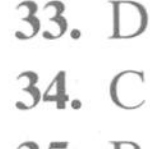

37.

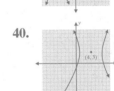

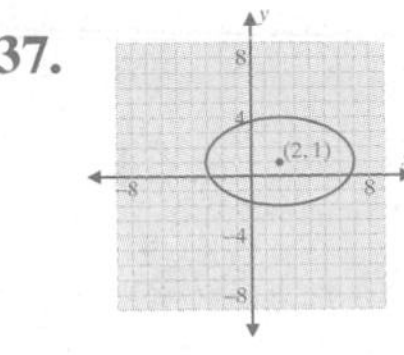

38.

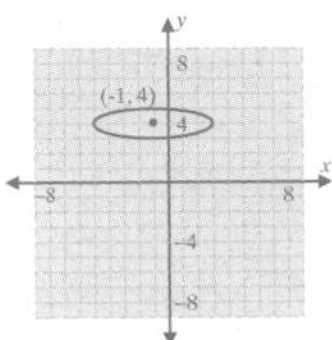

39.

40.

41.

42.

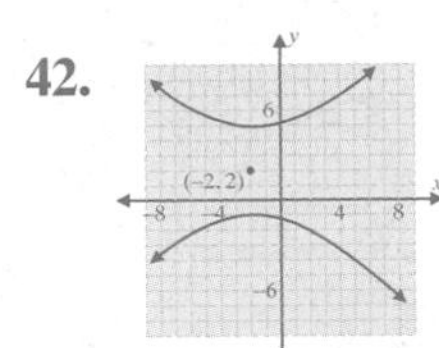

43. b. Set up the equation $\sqrt{(x+c)^2+(y-0)^2}$ $+\sqrt{(x-c)^2+(y-0)^2}=2a$ and square both sides twice and simplify.

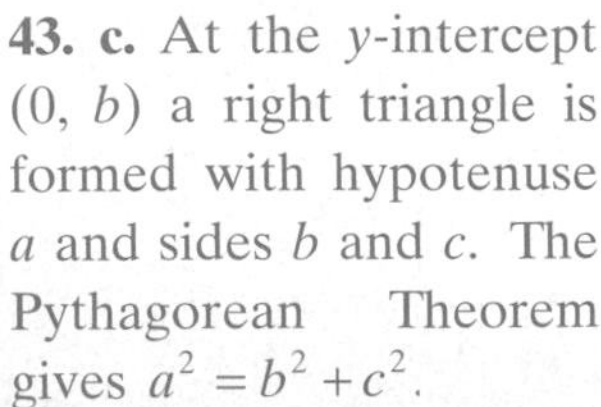

43. c. At the y-intercept $(0, b)$ a right triangle is formed with hypotenuse a and sides b and c. The Pythagorean Theorem gives $a^2 = b^2 + c^2$.

12.7 Nonlinear Systems of Equations

Objectives

After completing this section, you will be able to:

Solve systems of either two quadratic equations or one quadratic and one linear equation in two variables.

The equations for the conic sections that we have discussed all have at least one term that is second-degree. These equations are called **quadratic equations**. (Only the equations for parabolas of the form $y = ax^2 + bx + c$ are **quadratic functions**.) A summary of the equations with their related graphs is shown in Figure 12.29.

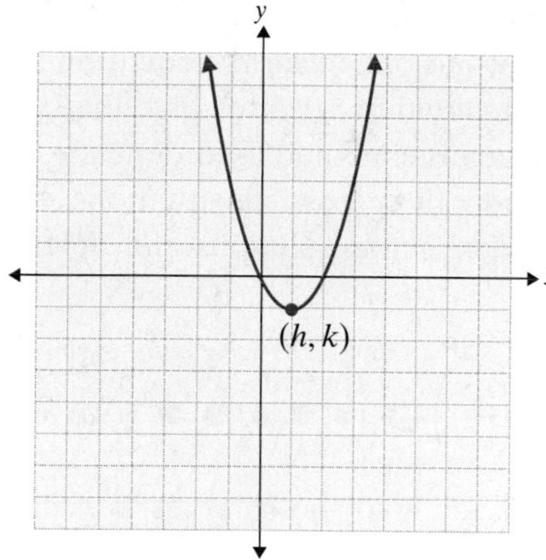

Parabola: $y = a(x-h)^2 + k$
$a > 0$, opens upward
$a < 0$, opens downward

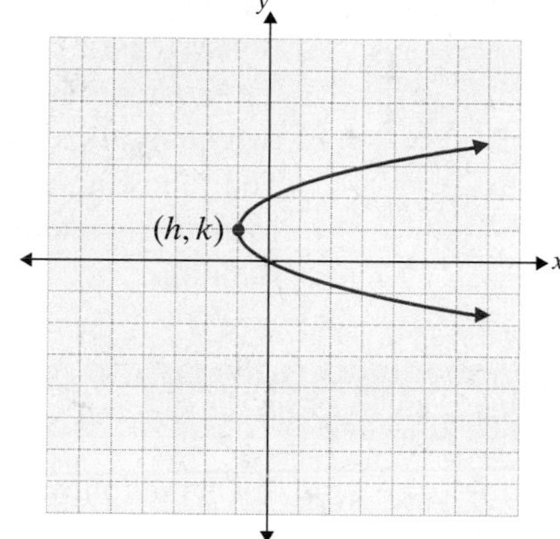

Parabola: $x = a(y-k)^2 + h$
$a > 0$, opens right
$a < 0$, opens left

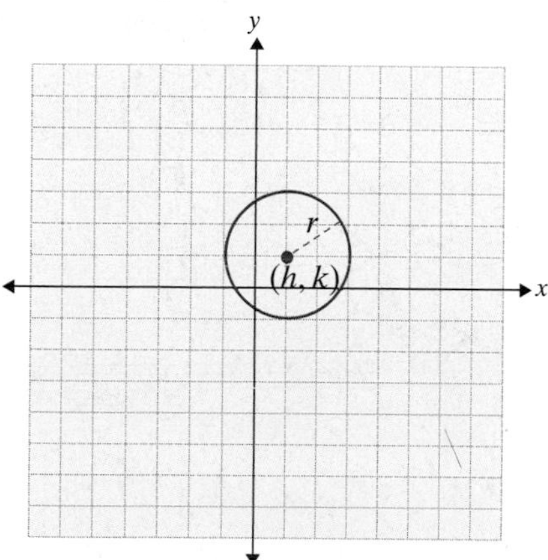

Circle: $(x-h)^2 + (y-k)^2 = r^2$

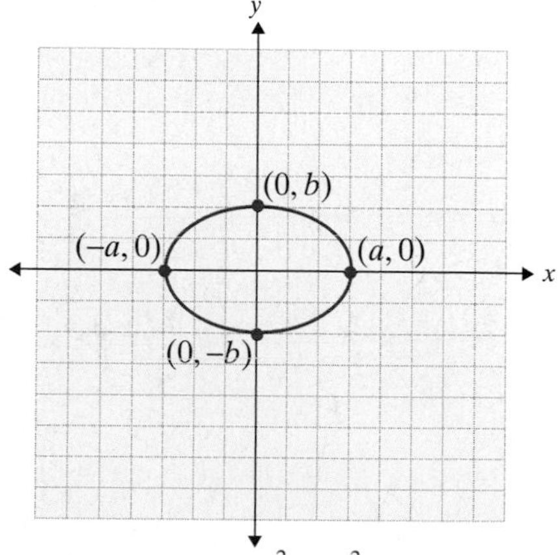

Ellipse: $\frac{x^2}{a^2} + \frac{y^2}{b^2} = 1$

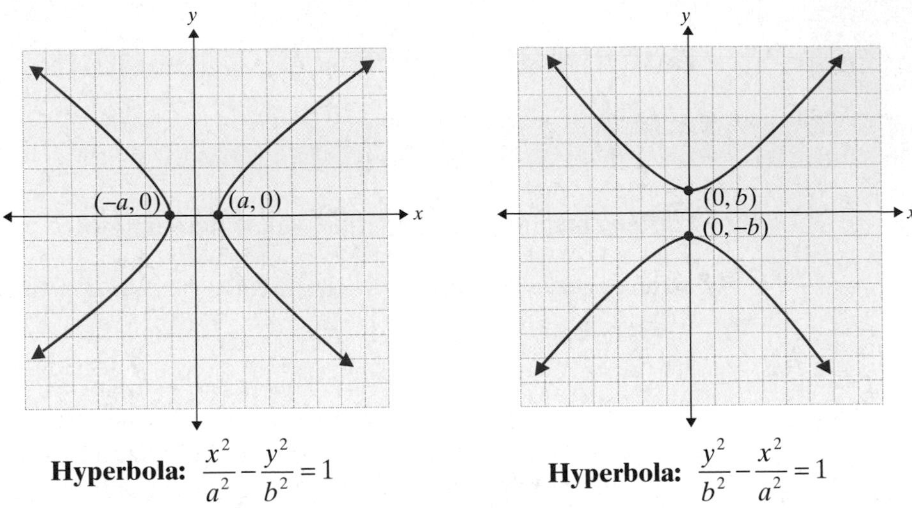

Hyperbola: $\dfrac{x^2}{a^2} - \dfrac{y^2}{b^2} = 1$ **Hyperbola:** $\dfrac{y^2}{b^2} - \dfrac{x^2}{a^2} = 1$

Figure 12.29

If a system of two equations has one quadratic equation and one linear equation, then the method of substitution should be used to solve the system. If the system involves two quadratic equations, then the method used depends on the form of the equations. The following examples show three possible situations. The graphs of the curves are particularly useful for approximating solutions and determining the exact number of solutions.

Example 1: Graphing Curves

Solve the following systems and graph both curves in each system.

a. $\begin{cases} x^2 + y^2 = 25 \\ x + y = 5 \end{cases}$

Solution: Solve $x + y = 5$ for y (or x). Then substitute into the other equation.

$$y = 5 - x$$

$$x^2 + (5 - x)^2 = 25$$

$$x^2 + 25 - 10x + x^2 = 25$$

$$2x^2 - 10x = 0$$

$$2x(x - 5) = 0 \qquad \text{Now solve for } x.$$

$$\begin{cases} x = 0 \\ y = 5 - 0 = 5 \end{cases} \quad \text{or} \quad \begin{cases} x = 5 \\ y = 5 - 5 = 0 \end{cases}$$

The solutions (points of intersection) are (0, 5) and (5, 0).

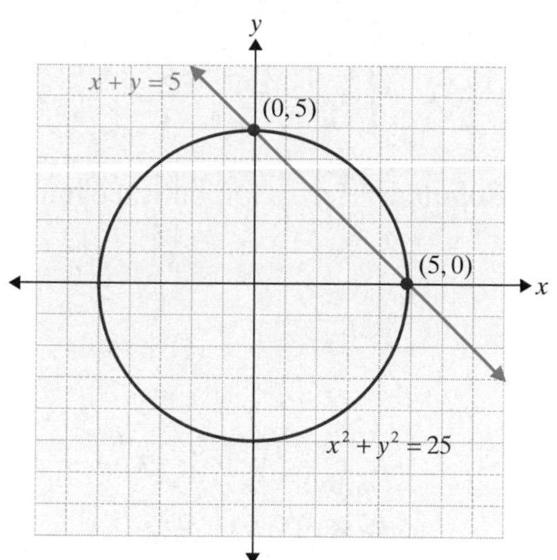

b. $\begin{cases} x + y = -7 \\ y = x^2 - 4x - 5 \end{cases}$

Solution: Solve the linear equation for y, then substitute. (In this case, the quadratic equation is already solved for y, and the substitution could be made the other way.)

$$y = -x - 7$$
$$-x - 7 = x^2 - 4x - 5$$
$$0 = x^2 - 3x + 2$$
$$0 = (x-2)(x-1)$$

$$\begin{cases} x = 2 \\ y = -2 - 7 = -9 \end{cases} \quad \text{or} \quad \begin{cases} x = 1 \\ y = -1 - 7 = -8 \end{cases}$$

The solutions are $(2, -9)$ and $(1, -8)$.

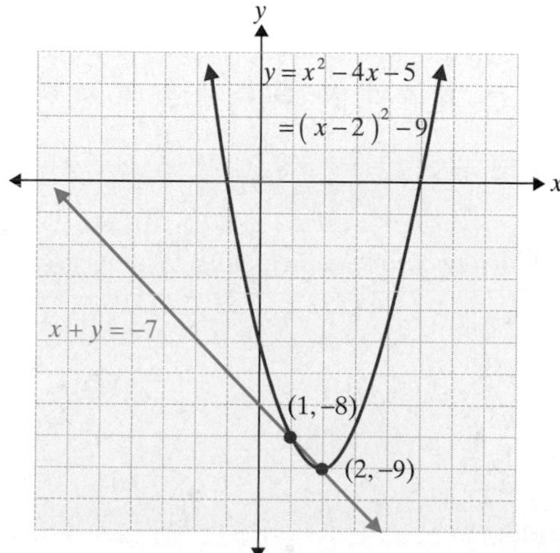

continued on next page ...

c. $\begin{cases} x^2 - y^2 = 4 \\ x^2 + y^2 = 36 \end{cases}$

Solution: Here addition will eliminate y^2.

$$\begin{array}{rcrcr} x^2 & - & y^2 & = & 4 \\ x^2 & + & y^2 & = & 36 \\ \hline 2x^2 & & & = & 40 \\ x^2 & & & = & 20 \end{array}$$

$x^2 = \pm\sqrt{20} = \pm 2\sqrt{5}$

if $x = 2\sqrt{5}$: $20 + y^2 = 36$
$y^2 = 16$
$y = \pm 4$

if $x = -2\sqrt{5}$: $20 + y^2 = 36$
$y^2 = 16$
$y = \pm 4$

There are four points of intersection:
$(2\sqrt{5},4), (2\sqrt{5},-4), (-2\sqrt{5},4), (-2\sqrt{5},-4)$

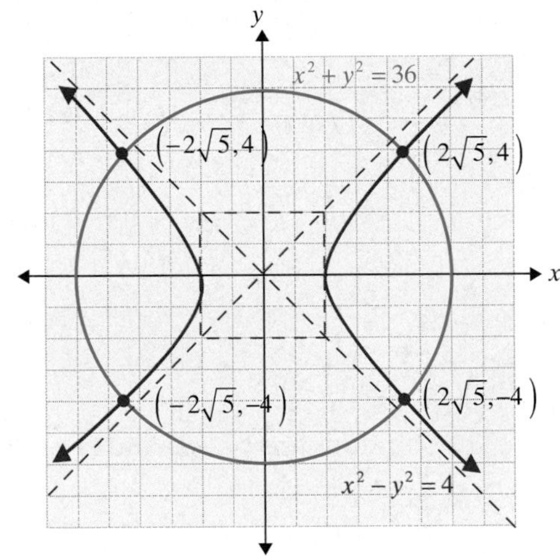

Practice Problems

Solve each of the following systems algebraically.

1. $\begin{cases} y = x^2 - 4 \\ x - y = 2 \end{cases}$

2. $\begin{cases} x^2 + y^2 = 72 \\ x = y^2 \end{cases}$

Answers to Practice Problems: 1. $(-1, -3)$ and $(2, 0)$ **2.** $(8, 2\sqrt{2})$ and $(8, -2\sqrt{2})$

12.7 Exercises

Solve each of the systems of equations in Exercises 1 – 16. Sketch the graphs.

1. $\begin{cases} y = x^2 + 1 \\ 2x + y = 4 \end{cases}$ **2.** $\begin{cases} y = 3 - x^2 \\ x + y = -3 \end{cases}$ **3.** $\begin{cases} y = 2 - x \\ y = (x-2)^2 \end{cases}$

4. $\begin{cases} x^2 + y^2 = 25 \\ y + x + 5 = 0 \end{cases}$ **5.** $\begin{cases} x^2 + y^2 = 20 \\ x - y = 2 \end{cases}$ **6.** $\begin{cases} x^2 - y^2 = 16 \\ 3x + 5y = 0 \end{cases}$

7. $\begin{cases} y = x - 2 \\ x^2 = y^2 + 16 \end{cases}$ **8.** $\begin{cases} x^2 + 3y^2 = 12 \\ x = 3y \end{cases}$ **9.** $\begin{cases} x^2 + y^2 = 9 \\ x^2 - y^2 = 9 \end{cases}$

10. $\begin{cases} x^2 + y^2 = 9 \\ x^2 - y + 3 = 0 \end{cases}$ **11.** $\begin{cases} 4x^2 + y^2 = 25 \\ 3x - y^2 + 3 = 0 \end{cases}$ **12.** $\begin{cases} x^2 - 4y^2 = 9 \\ x + 2y^2 = 3 \end{cases}$

13. $\begin{cases} x^2 + y^2 + 4x - 2y = 4 \\ x + y = 2 \end{cases}$ **14.** $\begin{cases} x^2 - y^2 = 9 \\ x^2 + y^2 - 2x - 3 = 0 \end{cases}$ **15.** $\begin{cases} x^2 - y^2 = 5 \\ x^2 + 4y^2 = 25 \end{cases}$

16. $\begin{cases} 2x^2 - 3y^2 = 6 \\ 2x^2 + y^2 = 22 \end{cases}$

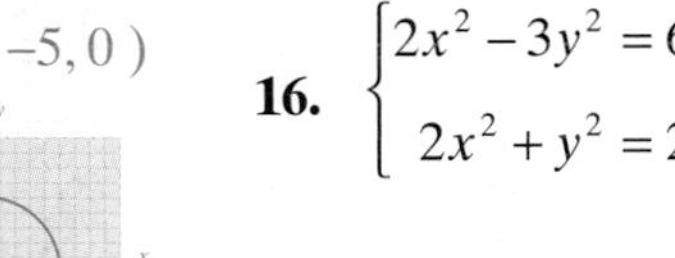

Solve each of the systems in Exercises 17 – 30.

17. $\begin{cases} x^2 - y^2 = 20 \\ x^2 - 9y = 0 \end{cases}$ **18.** $\begin{cases} x^2 + 5y^2 = 16 \\ x^2 + y^2 = 4x \end{cases}$ **19.** $\begin{cases} x^2 + y^2 = 10 \\ x^2 + y^2 - 4y + 2 = 0 \end{cases}$

20. $\begin{cases} x^2 + y^2 = 20 \\ 4x + 8 = y^2 \end{cases}$ **21.** $\begin{cases} 2x^2 - y^2 = 7 \\ 2x^2 + y^2 = 29 \end{cases}$ **22.** $\begin{cases} y = x^2 + 2x + 2 \\ 2x + y = 2 \end{cases}$

23. $\begin{cases} 4y + 10x^2 + 7x - 8 = 0 \\ 6x - 8y + 1 = 0 \end{cases}$ **24.** $\begin{cases} x^2 + y^2 - 4x + 6y + 3 = 0 \\ 2x - y - 2 = 0 \end{cases}$

25. $\begin{cases} x^2 + y^2 - 4y = 16 \\ x - y = 0 \end{cases}$ **26.** $\begin{cases} 4x^2 + y^2 = 11 \\ y = 4x^2 - 9 \end{cases}$ **27.** $\begin{cases} x^2 - y^2 - 2y = 22 \\ 2x + 5y + 5 = 0 \end{cases}$

28. $\begin{cases} x^2 + y^2 - 6y = 0 \\ 2x^2 - y^2 + 15 = 0 \end{cases}$ **29.** $\begin{cases} y = x^2 - 2x + 3 \\ y = -x^2 + 2x + 3 \end{cases}$ **30.** $\begin{cases} y^2 = x^2 - 5 \\ 4x^2 - y^2 = 32 \end{cases}$

1. $(-3, 10), (1, 2)$

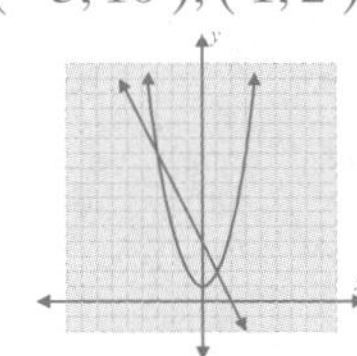

2. $(-2, -1), (3, -6)$

3. $(1, 1), (2, 0)$

4. $(0, -5), (-5, 0)$

5. $(-2, -4), (4, 2)$

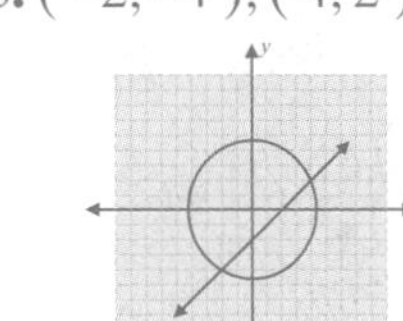

6. $(-5, 3), (5, -3)$

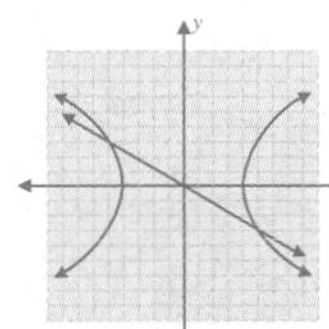

7. $(5, 3)$

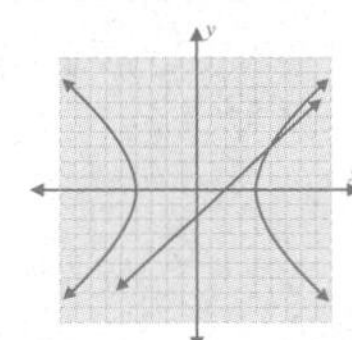

8. (3, 1), (−3, −1)

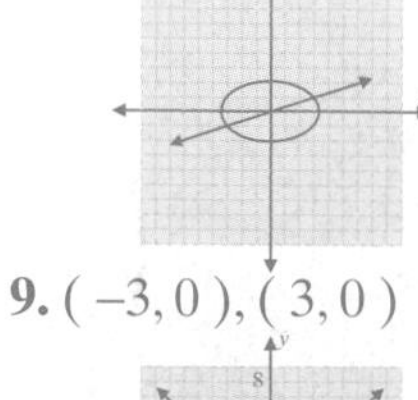

9. (−3, 0), (3, 0)

In Exercises 31 – 36, use a graphing calculator to graph and estimate the solutions to the systems of equations.

31. $\begin{cases} y = x^2 + 3 \\ x + y = 3 \end{cases}$

32. $\begin{cases} y = 1 - x^2 \\ x + y = -4 \end{cases}$

33. $\begin{cases} y = 3 - 2x \\ y = (x-1)^2 \end{cases}$

34. $\begin{cases} x^2 + y^2 = 36 \\ y = x + 5 \end{cases}$

35. $\begin{cases} x^2 + y^2 = 10 \\ x - y = 1 \end{cases}$

36. $\begin{cases} x^2 + y^2 = 4 \\ x^2 - y^2 = 3 \end{cases}$

Hawkes Learning Systems: Introductory & Intermediate Algebra

Nonlinear Systems of Equations

10. (0, 3)

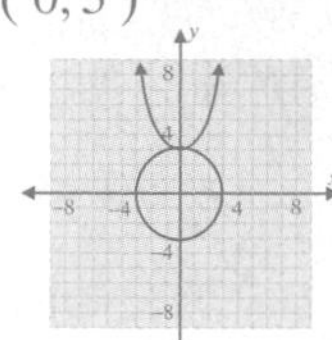

11. (2, 3), (2, −3)

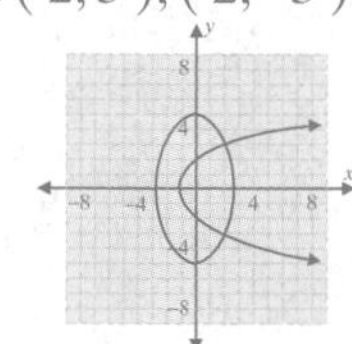

12. (−5, −2), (−5, 2), (3, 0)

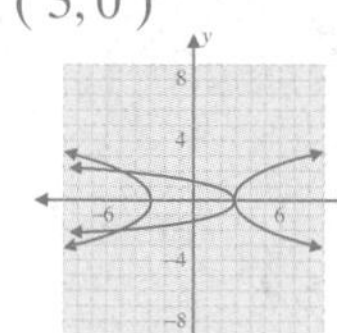

13. (−2, 4), (1, 1)

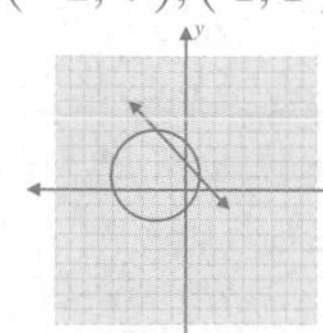

14. (3, 0)

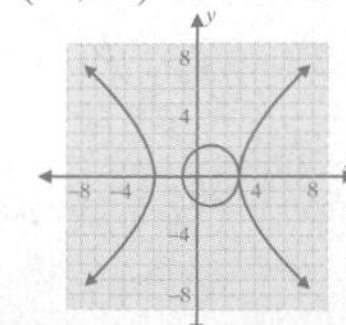

15. (−3, −2), (−3, 2), (3, −2), (3, 2)

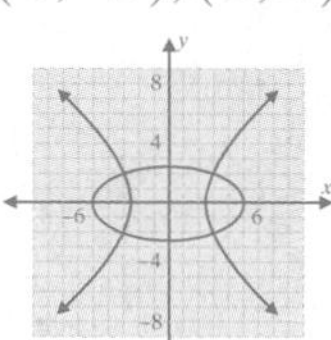

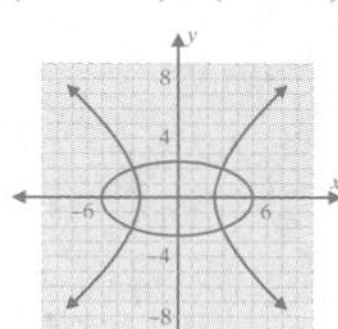

16. (−3, −2), (−3, 2), (3, −2), (3, 2)

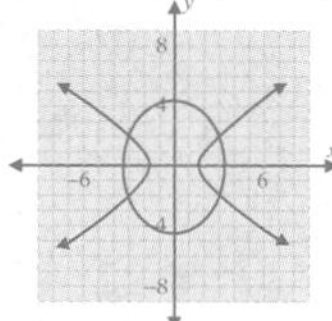

17. $\left(-3\sqrt{5}, 5\right), \left(-6, 4\right), \left(3\sqrt{5}, 5\right), \left(6, 4\right)$

18. $\left(1, -\sqrt{3}\right), \left(1, \sqrt{3}\right), \left(4, 0\right)$

19. (1, 3), (−1, 3)

20. (2, 4), (2, −4)

21. $\left(-3, \sqrt{11}\right), \left(-3, -\sqrt{11}\right), \left(3, \sqrt{11}\right), \left(3, -\sqrt{11}\right)$

22. (−4, 10), (0, 2)

23. $\left(-\frac{3}{2}, -1\right), \left(\frac{1}{2}, \frac{1}{2}\right)$

24. (1, 0), (−1, −4)

25. (4, 4), (−2, −2)

26. $\left(\frac{1}{2}\sqrt{7}, -2\right), \left(-\frac{1}{2}\sqrt{7}, -2\right), \left(\frac{1}{2}\sqrt{10}, 1\right), \left(-\frac{1}{2}\sqrt{10}, 1\right)$

27. (5, −3), (−5, 1)

28. $\left(\sqrt{5}, 5\right), \left(-\sqrt{5}, 5\right)$

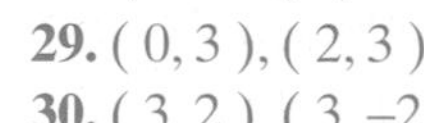

29. (0, 3), (2, 3)

30. (3, 2), (3, −2), (−3, 2), (−3, −2)

31. (0, 3), (−1, 4)

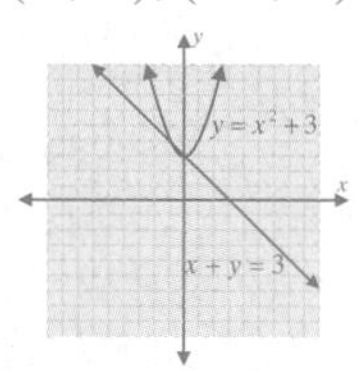

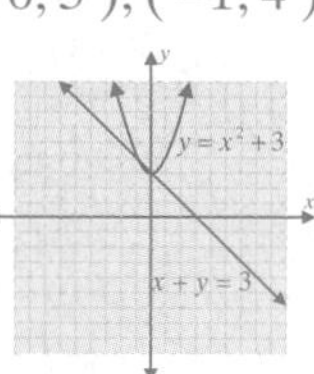

32. (−1.791, −2.209), (2.791, −6.791)

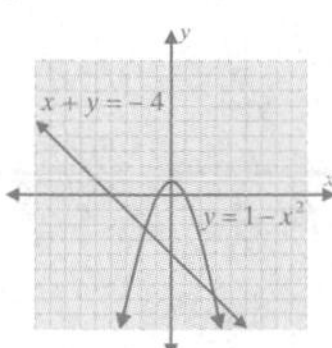

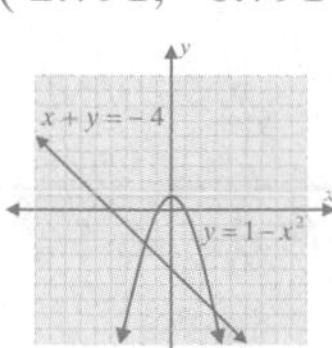

33. (−1.414, 5.828), (1.414, 0.172)

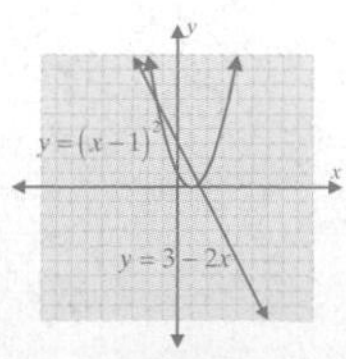

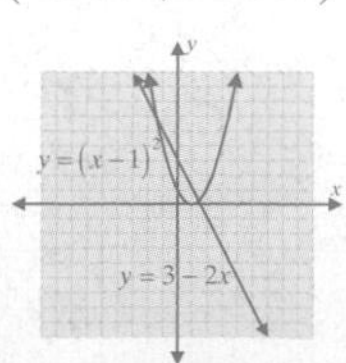

34. (0.928, 5.928), (−5.928, −0.928)

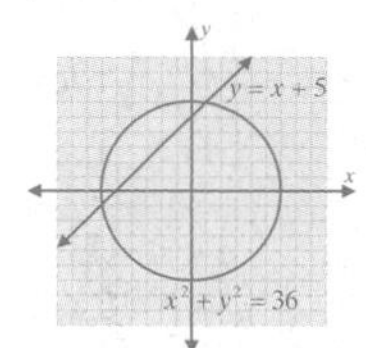

35. (2.679, 1.679), (−1.679, −2.679)

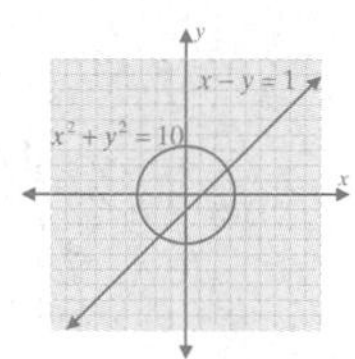

36. (1.871, 0.707), (1.871, −0.707), (−1.871, 0.707), (−1.871, −0.707)

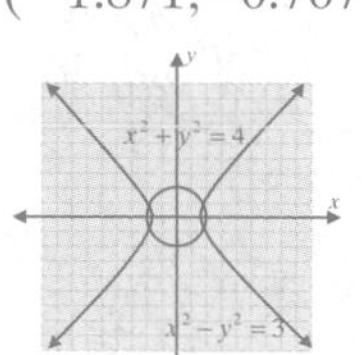

Chapter 12 Index of Key Ideas and Terms

Section 12.1 Quadratic Functions: Parabolas

Function page 794

Zeros of a Function page 794

The zeros of a function are the values of x when the graph crosses the x-axis.

Vertical Line Test page 795

If **any** vertical line intersects a graph in more than one point, then the relation graphed is **not** a function.

Parabola page 795

The graph of every quadratic function is a parabola.

Quadratic Function page 796

$y = ax^2 + bx + c$

Equations of Parabolas

$y = ax^2$ page 796

$y = ax^2 + k$ page 797

$y = a(x-h)^2$ pages 797 - 799

$y = a(x-h)^2 + k$ pages 799 - 801

$y = ax^2 + bx + c$ pages 799 - 801

Line of Symmetry page 801

The line $x = -\frac{b}{2a}$ is the line of symmetry of the graph of the quadratic function $y = ax^2 + bx + c$.

Vertex page 801

The vertex of the graph of $y = ax^2 + bx + c$ is at the point $\left(-\frac{b}{2a}, \frac{4ac - b^2}{4a}\right)$.

Zeros (of a Quadratic Function) page 801

Maximum and Minimum Values pages 803 - 805

For a parabola with equation of the form $y = a(x-h)^2 + k$, page 803

1. If $a > 0$, then (h, k) is the lowest point and $y = k$ is called the **minimum value** of the function.
2. If $a < 0$, then (h, k) is the highest point and $y = k$ is called the **maximum value** of the function.

Section 12.2 Quadratic and Other Inequalities

To Solve a Polynomial Inequality page 810

1. Arrange the terms so that one side of the inequality is 0.
2. Factor the algebraic expression, if possible, and find the points (numbers) where each factor is 0. (Use the quadratic formula, if necessary.)
3. Mark these points on a number line.
4. Test one point from each interval to determine the sign of the polynomial expression for all points in that interval.
5. The solution consists of those intervals where the test points satisfy the original inequality.

To Solve a Polynomial Inequality with a Graphing Calculator page 815

1. Arrange the terms so that one side of the inequality is 0.
2. Set the quadratic (or other polynomial) equal to y and graph the function. [Be sure to set the **WINDOW** so that all of the zeros are easily seen.]
3. Use the **TRACE** key or **CALC** key to approximate the zeros of the function.
4. **a.** If the inequality is of the form $y > 0$, then the solution consists of those intervals of x where the graph is above the x-axis.
   b. If the inequality is of the form $y < 0$, then the solution consists of those intervals of x where the graph is below the x-axis.
5. Endpoints of the intervals are included if the inequality is $y \geq 0$ or $y \leq 0$.

Section 12.3 *f(x)* Notation and Translations

***f*(*x*) Notation and Evaluating Functions** page 820

Difference Quotient page 822

The formula $\frac{f(x+h)-f(x)}{h}$ is called the difference quotient.

Horizontal and Vertical Translations page 824

Given the graph of $y = f(x)$, the graph of $y = f(x-h)+k$ is

1. a horizontal translation of h units, and
2. a vertical translation of k units of the graph of $y = f(x)$.

Reflections and Translations pages 828 - 830

Section 12.4 Parabolas as Conic Sections

Conic Sections page 835

Circles, ellipses, parabolas, and hyperbolas are conic sections.

Horizontal Parabolas page 836

Equations of horizontal parabolas (parabolas that open to the left or right) are of the form

$x = ay^2 + by + c$ or $x = a(y-k)^2 + h$ where $a \neq 0$.

The parabola opens left if $a < 0$ and right if $a > 0$.

The vertex is at (h, k).

The line $y = k$ is the line of symmetry.

Using a Graphing Calculator to Graph Horizontal Parabolas pages 839 - 840

Section 12.5 Distance Formula, Midpoint Formula, and Circles

Distance Between Two Points pages 844 - 846

The Pythagorean Theorem page 844

The Distance Formula: $d = \sqrt{(x_2 - x_1)^2 + (y_2 - y_1)^2}$ page 845

Midpoint Formula: $\left(\frac{x_1 + x_2}{2}, \frac{y_1 + y_2}{2} \right)$ page 846

Circle page 847

A **circle** is the set of all points in a plane that are a fixed distance from a fixed point.

The fixed point is called the **center** of the circle.

Radius page 847

The distance from the center to any point on a circle is called the **radius** of the circle.

Diameter page 847

The distance from one point on a circle to another point on the circle measured through the center is called the **diameter** of the circle.

Standard form for Equation of a Circle page 848

$(x-h)^2 + (y-k)^2 = r^2$

Center at (h, k) and radius r.

Using a Graphing Calculator to Graph Circles pages 850 - 852

Section 12.6 Ellipses and Hyperbolas

Ellipse page 857

The standard form for the equation of an ellipse with its center at the origin is

$$\frac{x^2}{a^2}+\frac{y^2}{b^2}=1 \text{ where } a^2 \geq b^2.$$

The points $(a, 0)$ and $(-a, 0)$ are the x-intercepts.
The points $(0, b)$ and $(0, -b)$ are the y-intercepts.
The line segment of length $2a$ joining the x-intercepts is called the **major axis**.
The line segment of length $2b$ joining the y-intercepts is called the **minor axis**.

Hyperbola page 861

In general, there are two standard forms for equations of hyperbolas with their centers at the origin.

1. $\frac{x^2}{a^2}-\frac{y^2}{b^2}=1$

x-intercepts (vertices) at $(a, 0)$ and $(-a, 0)$; no y-intercepts

Asymptotes: $y=\frac{b}{a}x$ and $y=-\frac{b}{a}x$

The curve "opens" left and right.

2. $\frac{y^2}{a^2}-\frac{x^2}{b^2}=1$

y-intercepts (vertices) at $(0, a)$ and $(0, -a)$; no x-intercepts

Asymptotes: $y=\frac{a}{b}x$ and $y=-\frac{a}{b}x$

The curve "opens" up and down.

Section 12.7 Nonlinear Systems of Equations

Nonlinear Systems of Equations pages 869 - 872

Chapter 12 Review

For a review of the topics and problems from Chapter 12, look at the following lessons from *Hawkes Learning Systems: Introductory & Intermediate Algebra.*

Graphing Parabolas
Solving Quadratic and Other Inequalities
Function Notation and Translations
Parabolas as Conic Sections
The Distance Formula and Circles
Ellipses and Hyperbolas
Nonlinear Systems of Equations

1. $y=(x-3)^2-1$
Vertex: $(3,-1)$
Axis: $x=3$
Domain: $(-\infty,\infty)$
Range: $y\geq -1$
Zeros: $x=2,4$

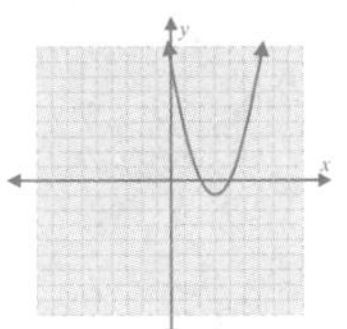

2. $y=-2\left(x-\frac{3}{2}\right)^2+\frac{15}{2}$
Vertex: $\left(\frac{3}{2},\frac{15}{2}\right)$
Axis: $x=\frac{3}{2}$
Domain: $(-\infty,\infty)$
Range: $y\leq\frac{15}{2}$
Zeros: $x=\frac{3\pm\sqrt{15}}{2}$

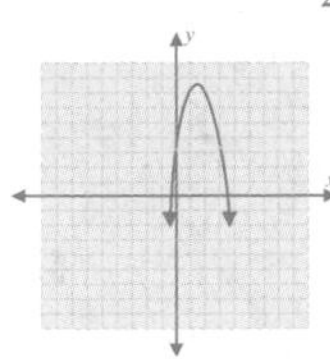

3. $y=2(x-3)^2-9$
Vertex: $(3,-9)$
Axis: $x=3$
Domain: $(-\infty,\infty)$
Range: $y\geq -9$
Zeros: $x=\frac{6\pm3\sqrt{2}}{2}$

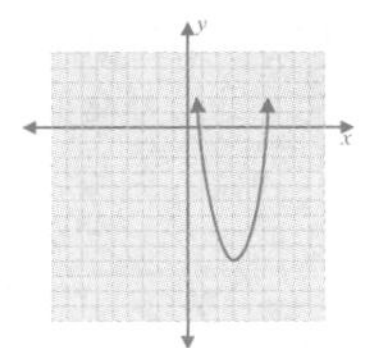

4. **a.** 13
b. $2x^2+6$
c. $2x^2+4x+7$
d. $4x+2h$

5. $(-\infty,-3]\cup[5,\infty)$

6. $\left(-4,-\frac{1}{2}\right)$

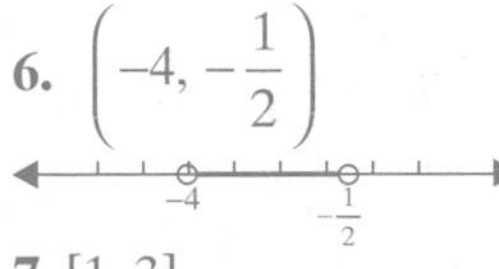

7. $[1,3]$

8. $(-0.6697, 1.4231)\cup(5.2466,\infty)$

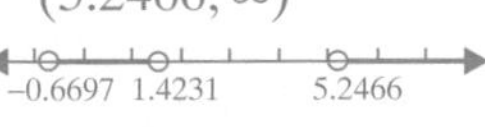

9.

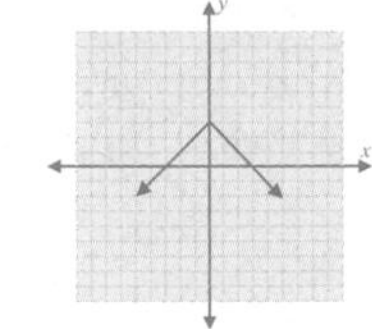

10.

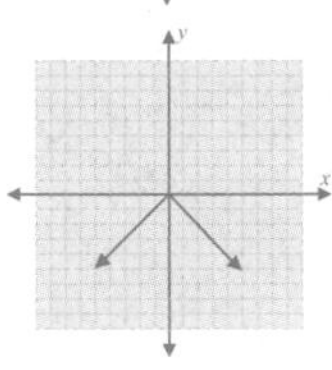

11.

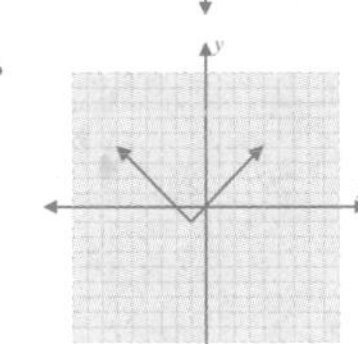

12. -25

13. $5\frac{1}{2}$ in. by $5\frac{1}{2}$ in.

14. vertex: $(-5,0)$
y-intercept: $(0,\sqrt{5}),(0,-\sqrt{5})$
line of symmetry: $y=0$

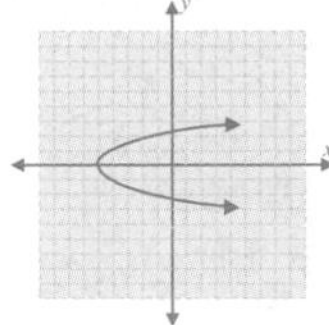

15. vertex: $\left(-\frac{25}{4},-\frac{3}{2}\right)$
y-intercept: $(0,-4),(0,1)$
line of symmetry: $y=-\frac{3}{2}$

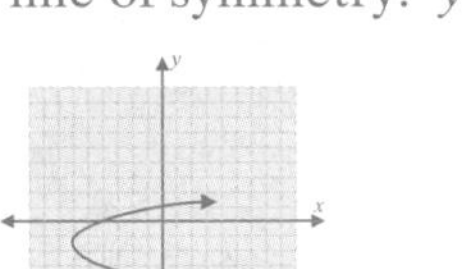

16. $3\sqrt{10}$

17. $|\overline{AB}|^2=52,$

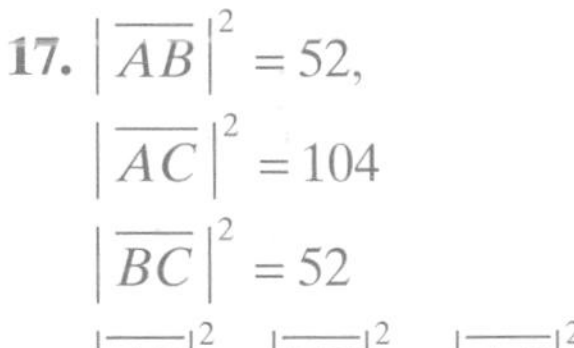

$|\overline{AC}|^2=104$
$|\overline{BC}|^2=52$
$|\overline{AB}|^2+|\overline{BC}|^2=|\overline{AC}|^2$

18. $(x+3)^2+(y+1)^2=25$

Chapter 12 Test

In Exercises 1 – 3, write the quadratic function in the form $y=a(x-h)^2+k$. Find the vertex, axis of symmetry, domain, range, and zeros. Graph the function.

1. $y=x^2-6x+8$ **2.** $y=-2x^2+6x+3$ **3.** $y=2x^2-12x+9$

4. For the function $f(x)=2x^2+5$, find:

a. $f(-2)$

b. $f(x)+1$

c. $f(x+1)$

d. $\dfrac{f(x+h)-f(x)}{h}$

In Exercises 5 and 6, solve the quadratic inequalities algebraically. Write the solution set in interval notation and graph the solution set on a number line.

5. $x^2-2x\geq 15$ **6.** $2x^2+9x+4<0$

In Exercises 7 and 8, solve each inequality by using a graphing calculator. Show a sketch of the corresponding graph, write the solution set in interval notation, and graph the solution set on a number line. (Estimate endpoints of intervals accurate to 4 decimal places.)

7. $x^2\leq 4x-3$ **8.** $x^3-6x^2+3x+5>0$

Use the graph of $y=f(x)$ shown below to graph the functions indicated in Exercises 9 – 11.

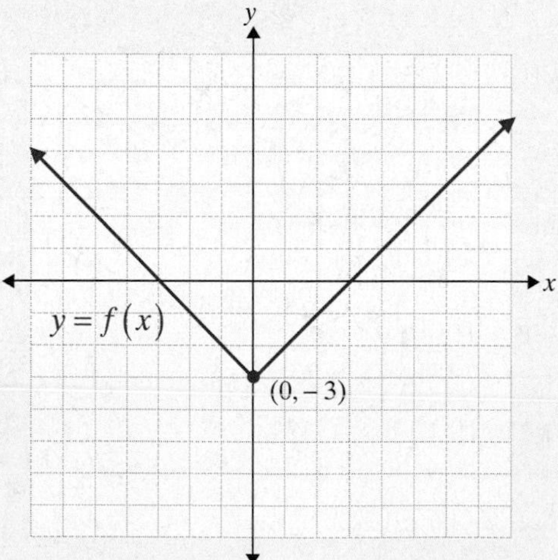

9. $y=-f(x)$ **10.** $y=-f(x)-3$ **11.** $y=f(x+1)+2$

19. $x^2+(y-1)^2=9$

Center: $(0,1), r=3$

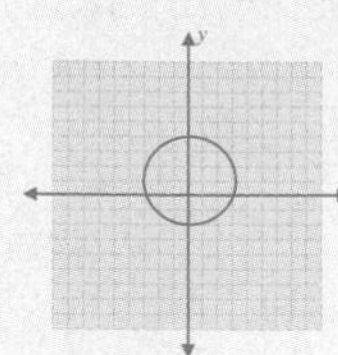

20. $\dfrac{x^2}{4}-\dfrac{y^2}{9}=1$

$y=\dfrac{3}{2}x, y=-\dfrac{3}{2}x$

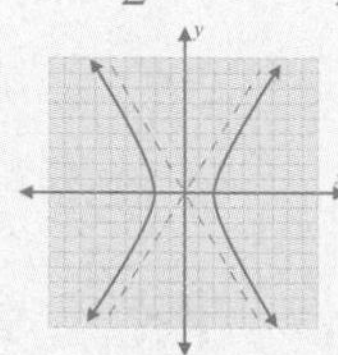

21. $\dfrac{x^2}{9}+\dfrac{y^2}{\frac{9}{4}}=1$

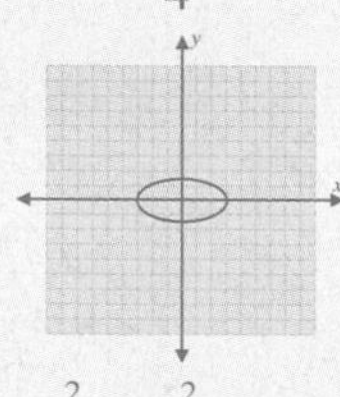

22. $\dfrac{y^2}{9}-\dfrac{x^2}{16}=1$

$y=\dfrac{3}{4}x, y=-\dfrac{3}{4}x$

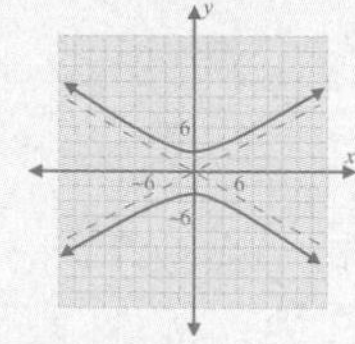

23. $\dfrac{x^2}{4}+\dfrac{y^2}{25}=1$

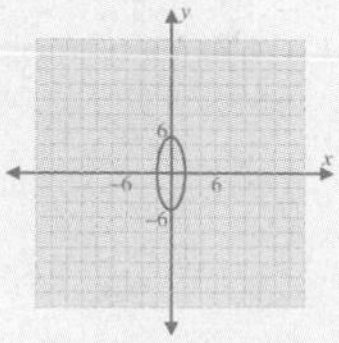

24.

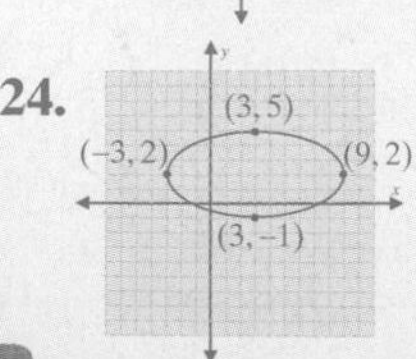

25.

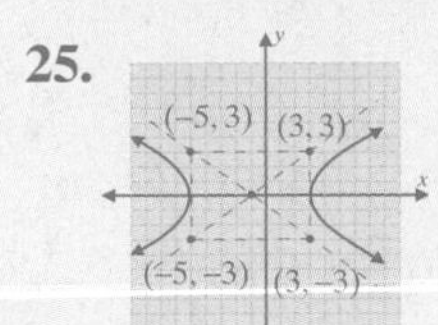

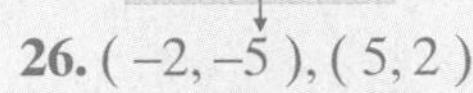

26. $(-2,-5),(5,2)$

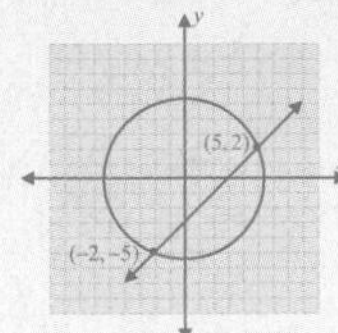

27. $(0,\sqrt{2}),(0,-\sqrt{2}),$
$(-2,0)$

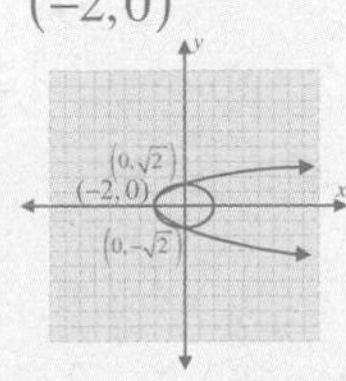

28. $(4,3),(4,-3)$
$(-4,3),(-4,-3)$

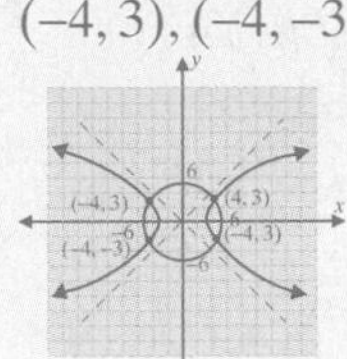

29. Maximum Area = 11,250 square feet
Dimensions = 75 x 150

30. Maximum Area = 6050 square feet
Dimensions = 55 x 110

31. Kim drives 60 mph and her sister drives 70 mph

12. One number exceeds another by 10. Find the minimum product of the two numbers.

13. The perimeter of a rectangle is 22 inches. Find the dimensions that will maximize the area.

In Exercises 14 and 15, find the vertex, the y-intercepts, and the line of symmetry. Graph the curve.

14. $x = y^2 - 5$

15. $x + 4 = y^2 + 3y$

16. Find the distance between the two points $(5,-2)$ and $(-4,1)$.

17. Show that the triangle determined by the points $A(-4,-3), B(0,3), C(6,-1)$ is a right triangle.

18. Find the equation for the circle with center at $(-3,-1)$ and radius 5.

19. Find the center and radius of the circle with equation $x^2 + y^2 - 2y - 8 = 0$, then sketch the graph.

In Exercises 20 – 23, write the equation in standard form, then sketch the graph. If the graph is a hyperbola, write the equations and graph the asymptotes.

20. $9x^2 - 4y^2 = 36$

21. $x^2 + 4y^2 = 9$

22. $16y^2 - 9x^2 = 144$

23. $25x^2 + 4y^2 = 100$

24. Graph the ellipse and label the vertices: $\frac{(x-3)^2}{36} + \frac{(y-2)^2}{9} = 1$

25. Graph the hyperbola and the asymptotes: $\frac{(x+1)^2}{16} - \frac{y^2}{9} = 1$

In Exercises 26 – 28, graph each pair of equations, and then solve the system.

26.

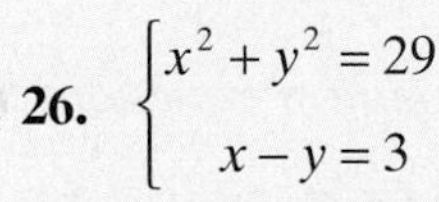

$$\begin{cases} x^2 + y^2 = 29 \\ x - y = 3 \end{cases}$$

27. $$\begin{cases} x^2 + 2y^2 = 4 \\ x = y^2 - 2 \end{cases}$$

28. $$\begin{cases} x^2 + y^2 = 25 \\ x^2 - y^2 = 7 \end{cases}$$

29. The back of Mica's property is a creek. Mica would like to build a fence along the other three sides to create a pasture for her two horses. If she has 300 feet of material, what is the maximum area that the pasture can be? What are the dimensions of the maximum area?

30. A parking lot is going to add three sides of a fence onto the side of an existing fence. The length of fence material is 220 feet, which is much less than the length of the existing fence. What are the maximum square feet that the parking lot can be? What are the dimensions of the maximum area?

31. Kim takes half an hour longer than her sister to drive 210 miles between two cities. If Kim drives 10 miles slower than her sister, at what speeds are the two going?

Cumulative Review: Chapters 1 – 12

1. 1 **2.** $\dfrac{9x^8}{4y^8}$

3. $5x^{\frac{3}{4}}$ **4.** $\dfrac{8y^{\frac{3}{5}}}{x}$

5. $\sqrt[3]{(7x^3y)^2} = x^2\sqrt[3]{49y^2}$

6. $(32x^6y)^{1/3} = 2x^2(4y)^{1/3}$

7. $2(x+3)(x^2-3x+9)$

8. $x^{-2}(2x-5)(x+7)$

9. $(x-4)(x^2+3)$

10. $13\sqrt{3}$

11. $2\sqrt{3x}+10\sqrt{6}$

12. $5x-2y=-16$

13. $3x-4y=16$

14. $\dfrac{2x^2+x+4}{(2x+3)(x-4)(3x-2)}$

15. -1

16. $x=\dfrac{-1\pm\sqrt{7}}{3}$

17. $x=-9$ **18.** $x=2$

19. $x=-2, y=-3$

20. $x=\dfrac{28}{13}, y=\dfrac{10}{13}$

21. $y=\dfrac{1}{2}(x+0)^2-3$

vertex: $(0,-3)$

axis : $x=0$,

$D_f=(-\infty,\infty)$

$R_f=\{y|y\geq -3\}$,

Zeros: $x=\pm\sqrt{6}$

Simplify each of the expressions in Exercises 1 – 4. Assume all variables are positive.

1. $\dfrac{x^{-3}\cdot x}{x^2\cdot x^{-4}}$ **2.** $\left(\dfrac{2x^{-1}y^2}{3x^3y^{-2}}\right)^{-2}$ **3.** $5x^{1/2}\cdot x^{1/4}$ **4.** $\left(4x^{-2/3}y^{2/5}\right)^{3/2}$

5. Write $(7x^3y)^{2/3}$ in radical notation. **6.** Write $\sqrt[3]{32x^6y}$ in exponential notation.

Completely factor each expression in Exercises 7 – 9.

7. $2x^3+54$ **8.** $2+9x^{-1}-35x^{-2}$ **9.** $x^3-4x^2+3x-12$

Perform the indicated operations and simplify in Exercises 10 and 11.

10. $2\sqrt{12}+5\sqrt{108}-7\sqrt{27}$ **11.** $3\sqrt{48x}-2\sqrt{75x}+5\sqrt{24}$

12. Find an equation for the line parallel to $5x-2y=8$ and passing through $(-2,3)$.

13. Find an equation for the line perpendicular to $4x+3y=8$ and passing through $(4,-1)$.

Perform the indicated operations in Exercises 14 and 15.

14. $\dfrac{x}{2x^2-5x-12}-\dfrac{x+1}{6x^2+5x-6}$ **15.** $\dfrac{x^2+2x-3}{x^2+x-2}\div\dfrac{9-x^2}{x^2-x-6}$

Solve each of the equations in Exercises 16 – 18.

16. $3x^2+2x-2=0$ **17.** $\dfrac{5}{x-3}-\dfrac{3}{x+2}=\dfrac{1}{x^2-x-6}$ **18.** $\sqrt{x+14}-2=x$

Use any algebraic method to solve the systems of linear equations in Exercises 19 and 20.

19. $\begin{cases}2x-3y=5\\-5x+y=7\end{cases}$ **20.** $\begin{cases}5x-y=10\\3x+2y=8\end{cases}$

In Exercises 21 and 22, write the quadratic function in the form $y=a(x-h)^2+k$. Find the vertex, axis of symmetry, domain, range, and zeros. Graph the function.

21. $y=\dfrac{1}{2}x^2-3$ **22.** $y=-x^2+4x-4$

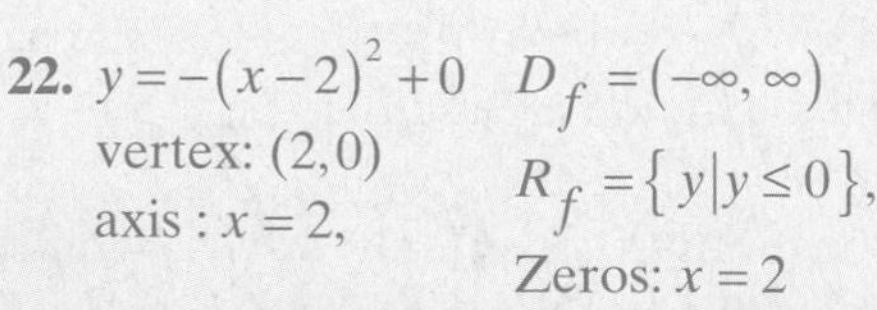

22. $y=-(x-2)^2+0$ $D_f=(-\infty,\infty)$

vertex: $(2,0)$ $R_f=\{y|y\leq 0\}$,

axis : $x=2$, Zeros: $x=2$

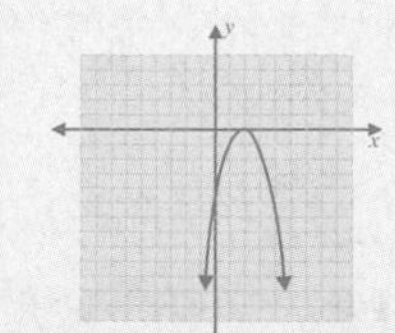

23. 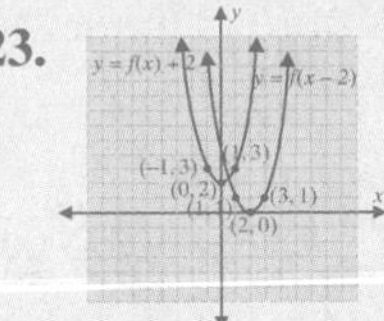

24. $x \geq 5, -1 \leq x < 0$

25. 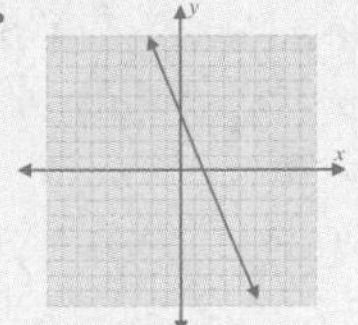

26. 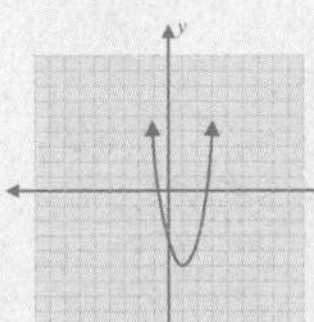

27. 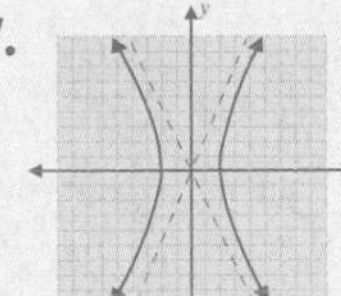

28. 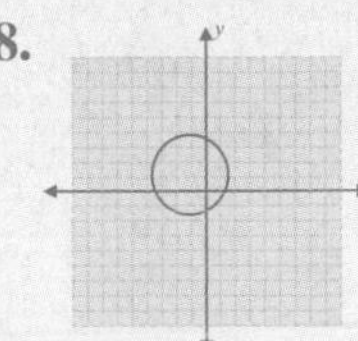

29. a. 5
b. $4x - 11$
c. $4x - 4$
d. 4

30. a. 13
b. $2x^2 + 3x - 3$
c. $2x^2 - 5x + 1$
d. $4x + 2h + 3$

31. \$3600 at 7%; \$5400 at 8%

32. 36

33. a. $t = 2\frac{1}{2}$ sec
b. 148 ft.

34. \$15

23. Given the graph of $y = f(x)$ shown here,

a. draw the graph of $y = f(x-2)$

b. draw the graph of $y = f(x)+2$

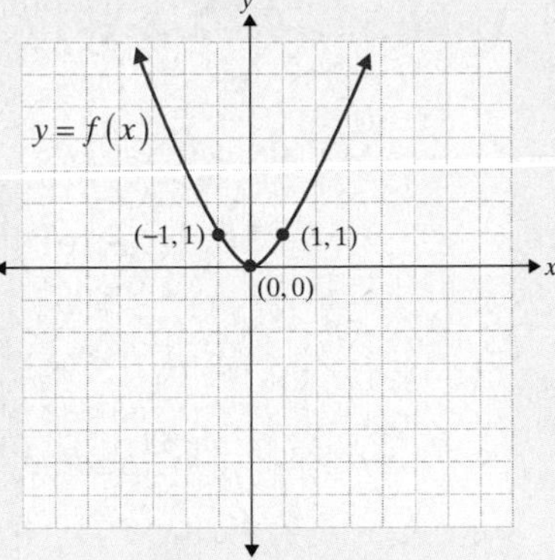

24. Solve the inequality $\frac{x^2 - 4x - 5}{x} \geq 0$.

Graph each of the equations in Exercises 25 – 28.

25. $5x + 2y = 8$

26. $y = 2x^2 - 4x - 3$

27. $4x^2 - y^2 = 16$

28. $x^2 + 2x + y^2 - 2y = 4$

29. For $f(x) = 4x - 7$, find:

a. $f(3)$

b. $f(x-1)$

c. $f(x)+3$

d. $\frac{f(x+h) - f(x)}{h}$

30. For $g(x) = 2x^2 + 3x - 1$, find:

a. $g(2)$

b. $g(x) - 2$

c. $g(x-2)$

d. $\frac{g(x+h) - g(x)}{h}$

31. Lisha has \$9000 to invest in two different accounts. One account pays interest at the rate of 7%; the other pays at the rate of 8%. If she wants her annual interest to total \$684, how much should she invest at each rate?

32. The average of a number and its square root is 21. Find the number.

33. The height of a ball projected vertically is given by the function $h = -16t^2 + 80t + 48$ where h is the height in feet and t is the time in seconds.

a. When will the ball reach maximum height?

b. What will be the maximum height?

34. A store owner estimates that by charging x dollars for a certain shirt, he can sell $60 - 2x$ shirts each week. What price will give him maximum receipts?

Use a graphing calculator to graph the functions in Exercises 35 – 38 and estimate the zeros of each function.

35. $y = 2x - 5$ **36.** $y = x^2 - 3$ **37.** $y = -x^2 + 5$

38. $y = (x+3)^2 - 6$

In Exercises 39 and 40, use a graphing calculator to graph each function. Estimate (accurate to 4 decimal places) the solutions to the equations and write the solutions to the inequalities in interval notation.

39. $f(x) = x^2 - 3x - 10$

a. $x^2 - 3x - 10 = 0$

b. $x^2 - 3x - 10 > 0$

c. $x^2 - 3x - 10 < 0$

40. $f(x) = -2x^2 + 3x + 5$

a. $-2x^2 + 3x + 5 = 0$

b. $-2x^2 + 3x + 5 > 0$

c. $-2x^2 + 3x + 5 < 0$

In Exercises 41 and 42, use a graphing calculator to graph both equations and estimate the solutions to the system.

41. $\begin{cases} x^2 + y^2 = 35 \\ x + y = 4 \end{cases}$ **42.** $\begin{cases} 3x^2 + 4y^2 = 12 \\ x = y^2 - 3 \end{cases}$

43. Divide the following by using long division. Write the answer in the form $Q(x) + \dfrac{R(x)}{D(x)}$.

$$\frac{2x^3 - 8x^2 + 10x - 50}{x - 5}$$

44. State the Pythagorean Theorem.

45. Given that $g(x) = \sqrt{x - 7}$, state the domain and range of g and sketch the graph of the function.

35.

$y = 2x - 5$; $(2.5, 0)$

36.

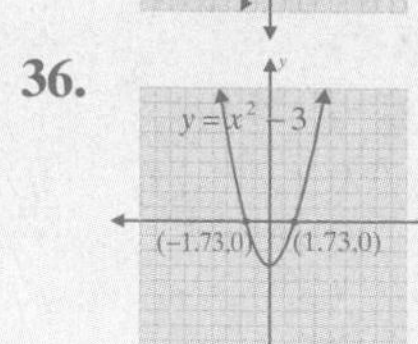

37.

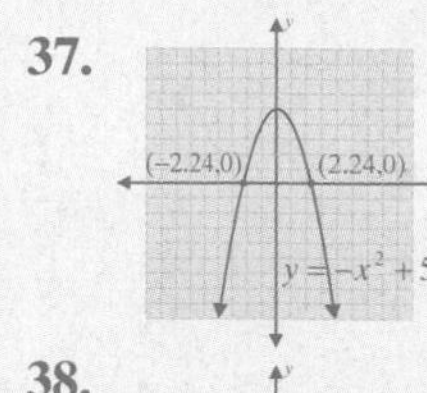

38.

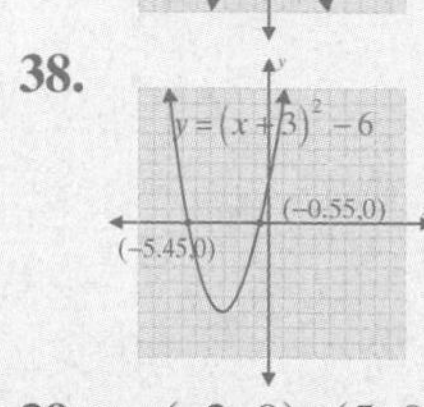

39. a. $(-2, 0), (5, 0)$
b. $(-\infty, -2) \cup (5, \infty)$
c. $(-2, 5)$

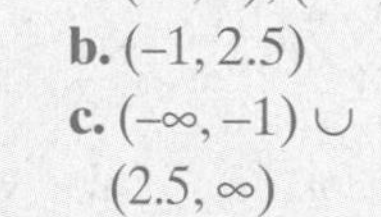

40. a. $(-1, 0), (2.5, 0)$
b. $(-1, 2.5)$
c. $(-\infty, -1) \cup (2.5, \infty)$

41.

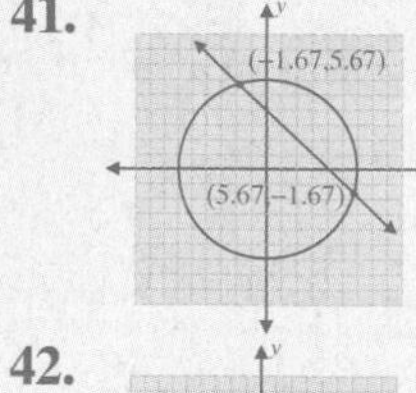

42.

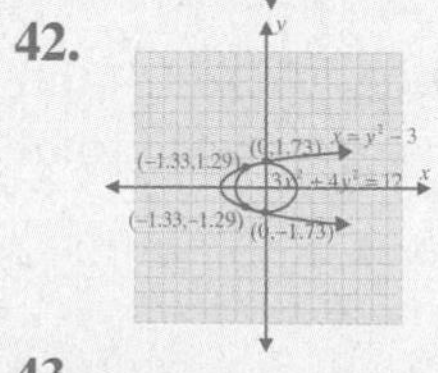

43. $2x^2 + 2x + 20 - \dfrac{50}{x - 5}$

44. $a^2 + b^2 = c^2$

45. Domain: $[7, \infty)$,
Range: $[0, \infty)$

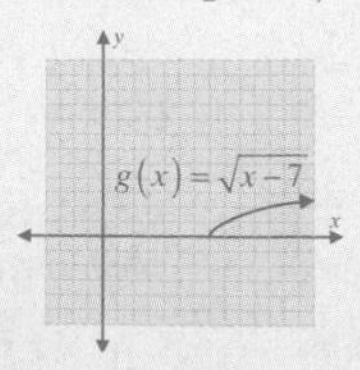

Exponential and Logarithmic Functions

CHAPTER 13

Did You Know?

In this chapter, you will study exponential functions and their related inverses, logarithmic functions. You will also see how the use of logarithms can simplify calculations involving multiplication and division. Although electronic calculators have made calculation with logarithms obsolete, it is still important to study the logarithmic functions because they have many applications other than computing.

Napier

The inventor of logarithms was John Napier (1550–1617). Napier was a Scottish nobleman, Laird of Merchiston Castle, a stronghold on the outskirts of the town of Edinburgh. An eccentric, Napier was intensely involved in the political and religious struggles of his day. He had interests in many areas, including mathematics. In 1614, Napier published his "Description of the Laws of Logarithms," and thus he is given credit for first publishing and popularizing the idea of logarithms. Napier used a base close to the number e for his system, and natural logarithms (base e) are often called **Napierian logarithms**. Napier soon saw that a base of 10 would be more appropriate for calculations since our decimal number system uses base 10. Napier began work on a base-10 system but was unable to complete it before his death. Henry Briggs (1561–1630) completed Napier's work, and base-10 logarithms are often called **Briggsian logarithms** in his honor.

Napier's interest in simplifying calculations was based on the need at that time to do many calculations by hand for astronomical and scientific research. He also invented the forerunner of the slide rule and predicted tanks, submarines, and other advanced war technology. Napier's remarkable ingenuity led the local people to consider him either crazy or a dealer in the black art of magic.

A particularly amusing story is told of Napier's method of identifying which of his servants was stealing from him. He told his servants that his black rooster would identify the thief. Each servant was sent alone into a darkened room to pet the rooster on the back. Napier had coated the back of the rooster with soot, and the guilty servant came out of the room with clean hands.

Napier was a staunch Presbyterian, and he felt that his claim to immortality would be an attack that he had written on the Catholic Church. The scientific community more correctly judged that logarithms would be his one great contribution.

The invention of logarithms: "by shortening the labors doubled the life of the astronomer."

Pierre de Laplace (1749 – 1827)

n	$\left(1+\frac{1}{n}\right)$	$\left(1+\frac{1}{n}\right)^n$
1	$\left(1+\frac{1}{1}\right)=2$	$2^1=2$
2	$\left(1+\frac{1}{2}\right)=1.5$	$(1.5)^2=2.25$
5	$\left(1+\frac{1}{5}\right)=1.2$	$(1.2)^5=2.48832$
10	$\left(1+\frac{1}{10}\right)=1.1$	$(1.1)^{10}=2.59374246$
100	$\left(1+\frac{1}{100}\right)=1.01$	$(1.01)^{100}=2.704813829$
1000	$\left(1+\frac{1}{1000}\right)=1.001$	$(1.001)^{1000}=2.716923932$
10,000	$\left(1+\frac{1}{10,000}\right)=1.0001$	$(1.0001)^{10,000}=2.718145927$
100,000	$\left(1+\frac{1}{100,000}\right)=1.00001$	$(1.00001)^{100,000}=2.718268237$
↓		↓
∞		$e=2.718281828459...$

Table 13.2

Teaching Notes: Have students notice how little change is occurring in this sequence of values (3rd column) and that this suggests that there might be a boundary ("limit") to the sequence – and there is – *e*! Stress that the concept of limit presented in the definition of *e* is precise, but beyond the scope of this text.

The Number *e*

The Number *e*

The number e is defined to be

$$e=\lim_{n\to\infty}\left(1+\frac{1}{n}\right)^n=2.718281828459\ldots$$

Now, to show how to find the formula for compounding interest continuously, we rewrite the formula for compound interest as follows:

$$A=P\left(1+\frac{r}{n}\right)^{nt}=P\left(1+\frac{r}{n}\right)^{\frac{n}{r}\cdot rt}=P\left[\left(1+\frac{1}{\frac{n}{r}}\right)^{\frac{n}{r}}\right]^{rt}$$

Now, substituting $m=\frac{n}{r}$, we can write

$$A=P\left[\left(1+\frac{1}{m}\right)^{m}\right]^{rt}.$$

Since r is a constant, the value of $m=\frac{n}{r}$ approaches ∞ as n approaches ∞. This means that the expression in brackets approaches e as $n\to\infty$, and the formula for **continuously compounded interest** is

$$A=Pe^{rt}.$$

Example 3: Continuously Compounded Interest

Find the value of \$1000 invested at 6% for 3 years if interest is compounded continuously. (Use e = 2.718281828, or on the TI-83 the number e is in yellow and can be found by pressing 2nd and then pressing the ÷ key.)

Solution:

$$\begin{aligned} A &= Pe^{rt} \\ &= 1000e^{0.06(3)} \\ &= 1000(2.718281828)^{0.18} \\ &= 1000(1.197217363) \\ &\approx \$1197.22 \end{aligned}$$

Comparing Examples 2c and 3, we see that there is only 2¢ difference in A when compounding interest daily or continuously over 3 years at 6%.

Practice Problems

1. *Sketch the graph of the exponential function $f(x)=2\cdot 3^{x}$ and label 3 points on the graph.*
2. *Sketch the graph of the exponential decay function $y=0.5\cdot 2^{-x}$ and label 3 points on the graph.*
3. *Find the value of \$5000 invested at 8% for 10 years if interest is (a) compounded monthly, (b) compounded continuously.*

Answers to Practice Problems: **1.** **2.**

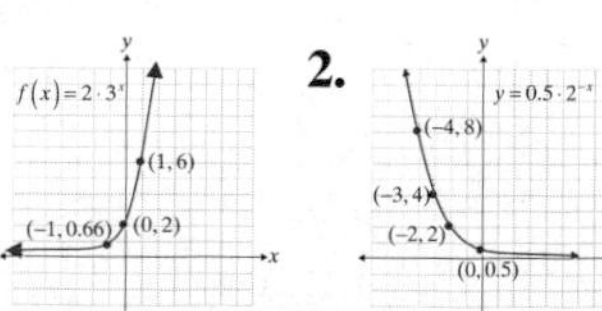

3. a. \$11,098.20 **b.** \$11,127.70

13.3 Exercises

1.

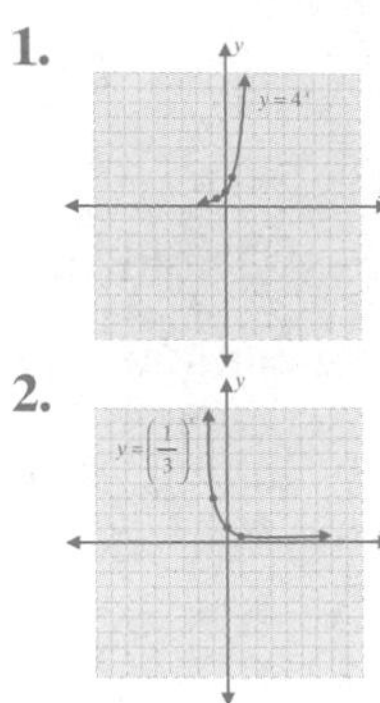

Sketch the graph of each of the exponential functions in Exercises 1 – 20, and label three points on each graph. (In some exercises you may need to use your knowledge of horizontal and vertical shifts.)

1. $y = 4^x$ **2.** $y = \left(\frac{1}{3}\right)^x$ **3.** $y = \left(\frac{1}{5}\right)^x$ **4.** $y = 5^x$

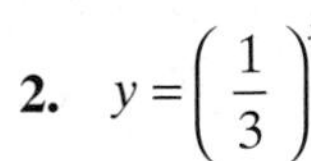

5. $y = 10^x$ **6.** $y = \left(\frac{2}{3}\right)^x$ **7.** $y = \left(\frac{5}{2}\right)^x$ **8.** $y = \left(\frac{1}{2}\right)^{-x}$

9. $y = 2^{x-1}$ **10.** $y = 3^{x+1}$ **11.** $f(x) = 2^x + 1$ **12.** $f(x) = 2^{x+1}$

13. $f(x) = 3^{2x}$ **14.** $f(x) = 2^{0.5x}$ **15.** $g(x) = 0.5 \cdot 3^x - 1$

16. $g(x) = 10^{-x} - 3$ **17.** $g(x) = -2^{-x}$ **18.** $g(x) = 10^{0.5x}$

19. $y = 3 \cdot \left(\frac{1}{2}\right)^{0.2x}$ **20.** $y = -4 \cdot \left(\frac{1}{3}\right)^{x-1}$

21. If $f(t) = 3 \cdot 4^t$ what is the value of $f(2)$?

22. Use your calculator to find the value (to the nearest hundredth) of $f(2)$ if $f(x) = 27.3 \cdot e^{-0.4x}$.

23. For $f(x) = 3 \cdot 10^{2x}$, find the value of $f(0.5)$.

24. Use your calculator to find the value of $f(9)$ if $f(t) = 2000 \cdot e^{0.08t}$. What does this value indicate to you about investing money?

25. Use your calculator to find the value of $f(22)$ if $f(t) = 2000 \cdot e^{0.05t}$. What does this value indicate to you about investing money?

26. A biologist knows that in the laboratory, bacteria in a culture grow according to the function $y = y_0 \cdot 5^{0.2t}$, where y_0 is the initial number of bacteria present and t is time measured in hours. How many bacteria will be present in a culture at the end of 5 hours if there were 5000 present initially?

27. In Exercise 26, how many bacteria were present initially if at the end of 15 hours, there were 2,500,000 bacteria present?

28. Four thousand dollars is deposited into a savings account at the rate of 8% per year. Find the total amount, A, on deposit at the end of 5 years if the interest is compounded:

a. annually
b. semiannually
c. quarterly
d. daily
e. continuously

2.
3.
4.
5.
6.
7.
8.
9.

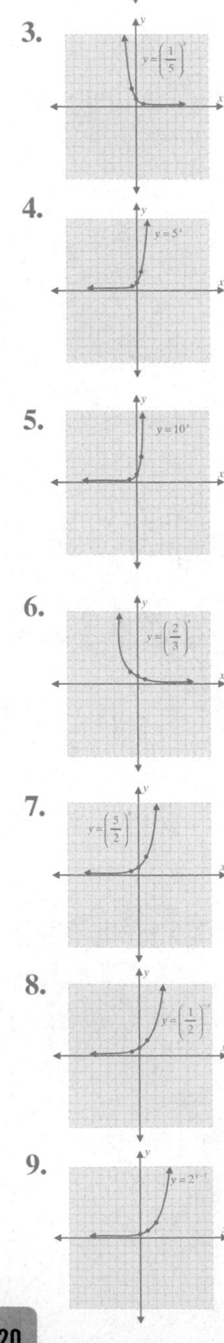

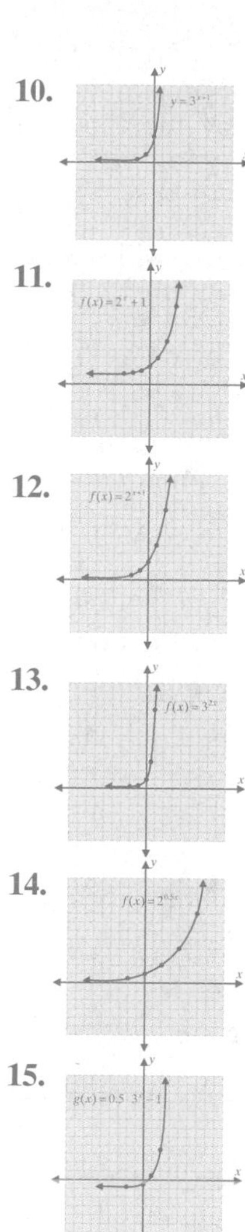

29. Find the amount, A, in a savings account if \$2000 is invested at 7% for 4 years and the interest is compounded:

a. annually
b. semiannually
c. quarterly
d. daily
e. continuously

30. Find the value of \$1800 invested at 6% for 3 years if the interest is compounded continuously.

31. Find the value of \$2500 invested at 5% for 5 years if the interest is compounded continuously.

32. The revenue function is given by $R(x) = x \cdot p(x)$ dollars, where x is the number of units sold and $p(x)$ is the unit price. If $p(x) = 25(2)^{\frac{-x}{5}}$, find the revenue if 15 units are sold.

33. In Exercise 32, if $p(x) = 40(3)^{\frac{-x}{6}}$, find the revenue if 12 units are sold.

34. A radio station knows that during an intense advertising campaign, the number of people, N, who will hear a commercial is given by $N = A(1 - 2^{-0.05t})$, where A is the number of people in the broadcasting area and t is the number of hours the commercial has been run. If there are 500,000 people in the area, how many will hear a commercial during the first 20 hours?

35. Statistics show that the fractional part of flashlight batteries, P, that are still good after t hours of use is given by $P = 4^{-0.02t}$. What fractional part of the batteries are still operating after 150 hours of use?

36. If a principal, P, is invested at a rate, r (expressed as a decimal), compounded continuously, the interest earned is given by $I = A - P$. How much interest will be earned in 20 years on an investment of \$10,000 invested at 10% and compounded continuously?

37. In Exercise 36, find the interest earned in 20 years on \$10,000 invested at 5% and compounded continuously. Explain why the interest earned at 5% is not just one-half of the interest earned at 10% in Exercise 36.

38. The value of a machine, V, at the end of t years is given by $V = C(1-r)^t$, where C is the original cost and r is the rate of depreciation. Find the value of a machine at the end of 4 years if the original cost was \$1200 and $r = 0.20$.

39. In Exercise 38, find the value of a machine at the end of 3 years if the original cost was \$2000 and $r = 0.15$.

40. Use a graphing calculator to graph each of the following functions. In each case the x-axis is a horizontal asymptote. Explain why the graphing calculator does not seem to indicate this fact.

a. $y = e^x$

b. $y = e^{-x}$

c. $y = e^{-x^2}$

17.
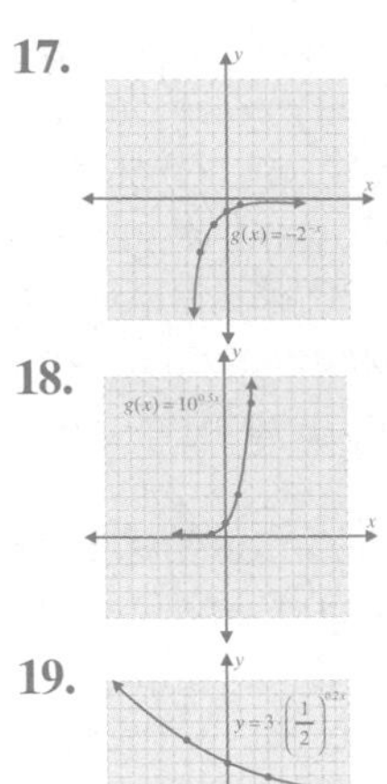

18.

19.

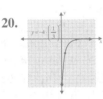

20.

21. 48 22. 12.27
23. 30 24. 4108.866
25. 6008.332
26. 25,000 bacteria
27. 20,000 bacteria
28. a. \$5877.31
b. \$5920.98
c. \$5943.79
d. \$5967.03
e. \$5967.30

41. The depreciation of a \$20,000 car that is t years old is given by the function $D(t) = 20000\left(\frac{2}{3}\right)^t$. Determine the value of the car after the owner has had it for 3 years. (Round your answer to the nearest cent.) For what value of t will the car be worth less than \$2000?

42. The demand for a product is given by the function $p = 4000\left(1 - \frac{6}{6 + e^{-.001x}}\right)$ where x is the number of units of the product sold and p is the price of the number of units sold. What is the price when x = 300 units? when x = 600 units? when x = 1000 units?

43. The population of bunnies in a certain area increased by the function $P(t) = 100e^{0.53t}$ where t was the number of months. (t = 0 corresponds to January of the previous year and the values of t ranged from 0 – 11). What was the population of bunnies at the end of March? How many bunnies existed at the end of the year?

Writing and Thinking About Mathematics

44. Discuss, in your own words, how the graph of each of the following functions is related to the graph of the exponential function $y = b^x$.

a. $y = a \cdot b^x$ **b.** $y = b^{x-h}$ **c.** $y = b^x + k$

45. Discuss, in your own words, the symmetrical relationship of the graphs of the two exponential functions $y = 10^x$ and $y = 10^{-x}$.

46. Discuss, in your own words, the symmetrical relationship of the graphs of the two exponential functions $y = 10^x$ and $y = -10^x$.

Hawkes Learning Systems: Introductory & Intermediate Algebra

Exponential Functions and the Number *e*

29. a. \$2621.59
b. \$2633.62
c. \$2639.86
d. \$2646.19
e. \$2646.26
30. \$2154.99
31. \$3210.06

32. \$46.88
33. \$53.33
34. 250,000 people
35. $\frac{1}{64}$
36. \$63,890.56
37. \$17,182.82, see page 918
38. \$491.52
39. \$1228.25

40. a.
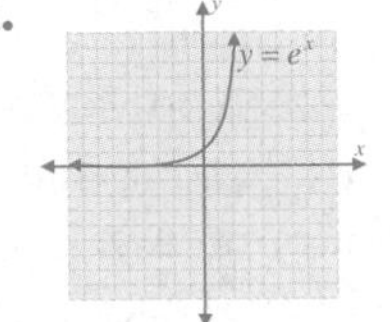
b.
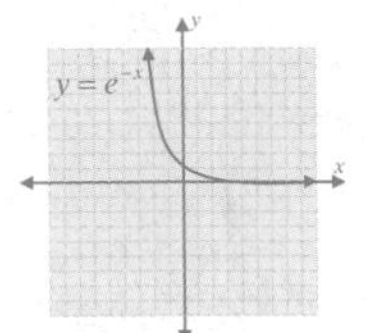
c.
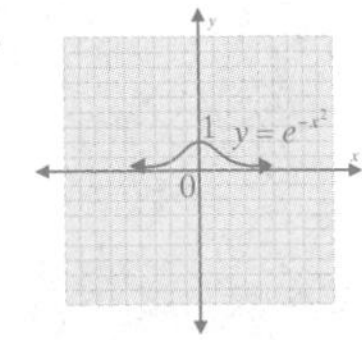

41. Value after 3 years = \$5925.93, t = 6 years
42. Price @ 300 units = \$439.60, Price @ 600 units = \$335.21, Price @ 1000 units = \$231.08,
43. Pop. up to March = 288 bunnies. Pop. at end of year = 34,035 bunnies.
44. – 46. Answers will vary.

13.4 Logarithmic Functions

Objectives

After completing this section, you will be able to:

1. *Write exponential expressions in logarithmic form.*
2. *Write logarithmic expressions in exponential form.*
3. *Graph exponential functions and logarithmic functions on the same set of axes.*
4. *Use a calculator to find the values of common logarithms and natural logarithms.*
5. *Use a calculator to find inverse logarithms for common logarithms and for natural logarithms.*

Exponential functions of the form $y = b^x$ are 1-1 functions and, therefore, have inverses. The inverse functions of exponential functions are called **logarithmic functions.** To find the inverse of a function, we interchange the x and y and solve for y. Thus, for the function

$$y = b^x$$

interchanging x and y gives the inverse function

$$x = b^y.$$

Figure 13.15 shows the graphs of these two functions with $b > 1$.

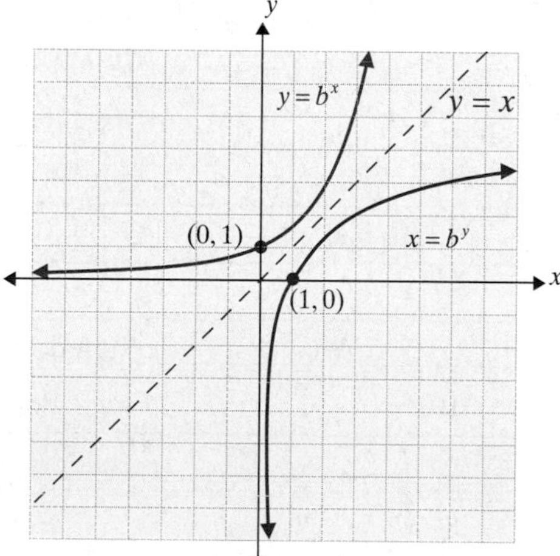

Figure 13.15

Now we need to solve the equation $x = b^y$ for y. But there is no algebraic technique for doing this. Mathematicians have simply created a name for this new function and called it a **logarithm**, abbreviated as **log**. This means that **the inverse of an exponential function is a logarithmic function.**

We have the following relationship:

$$\text{If} \quad f(x) = b^x$$

$$\text{then} \quad f^{-1}(x) = \log_b x.$$

The properties of inverse functions give the following two important results:

$$f\left(f^{-1}(x)\right) = b^{f^{-1}(x)} = b^{\log_b x} = x$$

and

$$f^{-1}\left(f(x)\right) = \log_b\left(f(x)\right) = \log_b\left(b^x\right) = x.$$

Logarithm

For $b > 0$ and $b \neq 1$,

$$x = b^y \textit{ if and only if } y = \log_b x.$$

Thus, a logarithm is the name of an exponent, and the equations

$$x = b^y \text{ and } y = \log_b x$$

are equivalent.

Example 1: Translations from Exponential Form to Logarithmic Form

	Exponential Form	Logarithmic Form		
a.	$2^3 = 8$	$\log_2 8 = 3$	←	3 is the logarithm.
b.	$2^4 = 16$	$\log_2 16 = 4$	←	4 is the logarithm.
c.	$10^3 = 1000$	$\log_{10} 1000 = 3$	←	3 is the logarithm.
d.	$10^4 = 10{,}000$	$\log_{10} 10{,}000 = 4$	←	4 is the logarithm.
e.	$2^0 = 1$	$\log_2 1 = 0$	←	0 is the logarithm.
f.	$3^0 = 1$	$\log_3 1 = 0$	←	0 is the logarithm.
g.	$10^1 = 10$	$\log_{10} 10 = 1$	←	1 is the logarithm.
h.	$5^1 = 5$	$\log_5 5 = 1$	←	1 is the logarithm.
i.	$2^{-2} = \frac{1}{4}$	$\log_2 \frac{1}{4} = -2$	←	–2 is the logarithm.
j.	$10^{-1} = \frac{1}{10}$	$\log_{10} \frac{1}{10} = -1$	←	–1 is the logarithm.

REMEMBER, a logarithm is an exponent. For example,

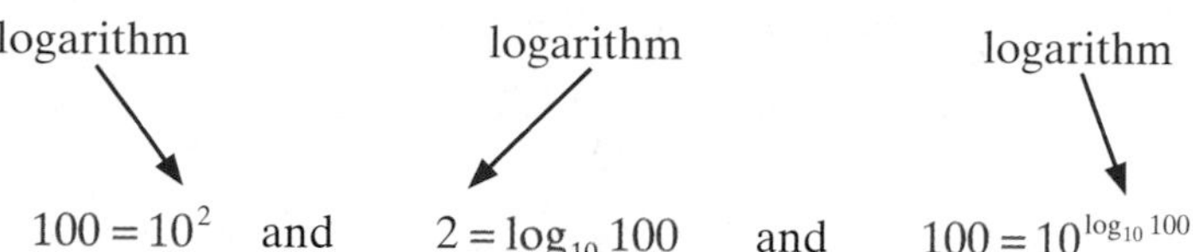

$100 = 10^2$ and $2 = \log_{10} 100$ and $100 = 10^{\log_{10} 100}$

are all equivalent. In words,

2 is the **exponent** of the base 10 to get $100\left(10^2 = 100\right)$; and

2 is the **logarithm** base 10 of 100 $\left(2 = \log_{10} 100\right)$.

Evaluating Logarithms

Logarithmic expressions can often be evaluated by changing them to the equivalent exponential form. Several cases are illustrated in Example 2. Note that parentheses are used as in $y = \log_b(x)$ if there is any doubt as to what we are finding the logarithm of or what a base is.

Example 2: Evaluating Logarithms

a. Evaluate $\log_2 32$.

Solution: Let $\log_2 32 = x$.

Then $2^x = 32$

$2^x = 2^5$

$x = 5$

Thus, $\log_2 32 = 5$.

b. Find the value of $\log_{10}(0.01)$.

Solution: Let $\log_{10}(0.01) = x$.

Then $10^x = 0.01$

$10^x = \dfrac{1}{100}$

$10^x = 10^{-2}$

$x = -2$

Thus, $\log_{10}(0.01) = -2$

continued on next page ...

c. Evaluate $\log_4 8$.

Solution: Let $\log_4 8 = x$.

Then $4^x = 8$

$$(2^2)^x = 2^3$$

$$2^{2x} = 2^3$$

$$2x = 3$$

$$x = \frac{3}{2}$$

Thus, $\log_4 8 = \frac{3}{2}$.

d. Find the value of x if $\log_{16} x = \frac{3}{4}$.

Solution: $\log_{16} x = \frac{3}{4}$ $\qquad$ $\frac{3}{4}$ is the **logarithm** of x.

Then $x = 16^{\frac{3}{4}}$ $\qquad$ $\frac{3}{4}$ is the **exponent** of the base, 16.

$$x = (2^4)^{\frac{3}{4}}$$

$$x = 2^3 = 8$$

Thus, $\log_{16} 8 = \frac{3}{4}$.

Three Basic Properties of Logarithms

We have previously discussed the following two properties of exponents:

$$b^0 = 1 \text{ and } b^1 = b.$$

Since exponents are logarithms, these same two properties can be stated in logarithmic form. Thus,

$$\log_b 1 = 0 \quad \text{because} \quad b^0 = 1$$

and $$\log_b b = 1 \quad \text{because} \quad b^1 = b.$$

Also, the definition of logarithm states that if $x = b^y$, then

$$y = \log_b x.$$

By substituting for y,

$$y = \log_b x$$

$$\downarrow$$

$$x = b^y$$

we get

$$x = b^{\log_b x}.$$

This equation indicates that $\log_b x$ is the exponent of b that will give x as a result. In summary, we have the following three basic properties of logarithms.

Properties of Logarithms

For $b > 0$, $b \neq 1$, and $x > 0$,

1. $\log_b 1 = 0$ — *The logarithm of 1 is always 0.*
2. $\log_b b = 1$ — *The logarithm of the base is always 1.*
3. $b^{\log_b x} = x$

REMEMBER: A logarithm is an exponent.

Example 3: Properties of Logarithms

a. $\log_3 1 = 0$ — Property 1: $\log_b 1 = 0$. The logarithm of 1 is 0.

b. $\log_8 8 = 1$ — Property 2: $\log_b b = 1$. The logarithm of the base is 1.

c. $10^{\log_{10} 20} = 20$ — Property 3: $b^{\log_b x} = x$.

d. $10^{\log_{10} 300} = 300$ — Property 3: $b^{\log_b x} = x$.

A calculator will show that $\log_{10} 20 = 1.301029996$ and $\log_{10} 300 = 2.477121255$.

Note: The equal sign is used even though the decimals are irrational numbers, infinite nonrepeating decimals. So, Examples 3c and 3d can be written as

$$10^{1.301029996} = 20 \quad \text{and} \quad 10^{2.477121255} = 300.$$

Graphs of Logarithmic Functions

Because logarithmic functions are the inverses of exponential functions, the graphs of logarithmic functions can be found by reflecting the corresponding exponential functions across the line $y = x$. This was illustrated in general for $b > 1$ in Figure 13.15.

Figure 13.16a shows how the graph of $y = \log_2 x$ is related to the graph of $y = 2^x$ and Figure 13.16b shows how the graph of $y = 10^x$ is related to the graph of $y = \log_{10} x$. Note that in the graphs of both logarithmic functions, the logarithms are negative for $0 < x < 1$.

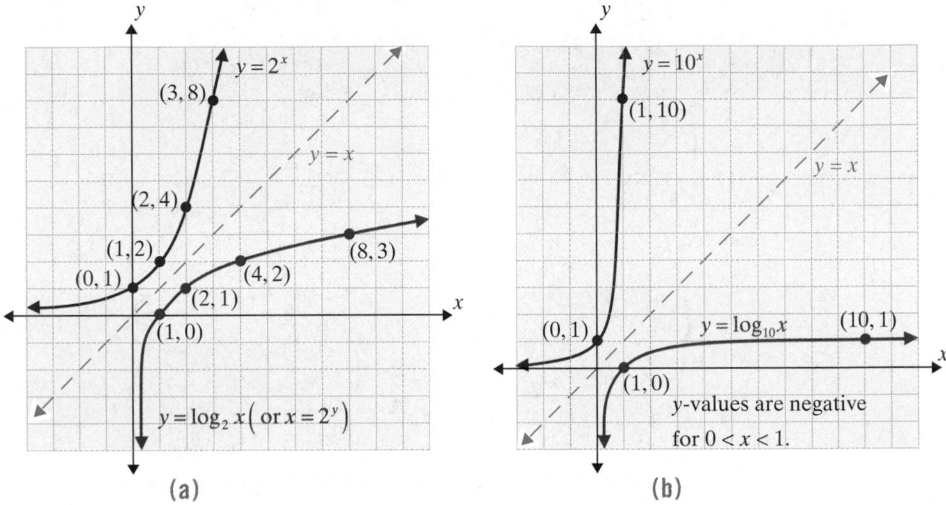

Figure 13.16

Notice that points on the graphs of inverse functions can be found by reversing the coordinates of ordered pairs.

Recall that the domain and range of a function and its inverse are interchanged. Thus, for exponential functions and logarithmic functions, we have the following:

For the exponential function $y = b^x$

the domain is all real x, and
the range is all $y > 0$. (The graph is above the x-axis.)
Horizontal asymptote

For the logarithmic function $y = \log_b x \; (\text{or } x = b^y)$,

the domain is all $x > 0$, and
the range is all real y. (The graph is to the right of the y-axis.)
Vertical asymptote

Common Logarithms (Base-10 Logarithms)

Base-10 logarithms are called **common logarithms**. The notation $\log x$ is used to indicate common logarithms. That is,

$$\log x = \log_{10} x.$$

Using a Calculator

Finding values of common logarithms on a TI-83 is a simple three-step process:

1. *Press the* LOG *key.* `log(` *will appear on the display.*
2. *Enter the number and a right hand parenthesis* `)`.
3. *Press* ENTER.

Example 4: Finding Logarithms Using a Calculator

Use a calculator to find the values of the following common logarithms. To check your understanding, write your estimate of each value on a piece of paper before you use the calculator.

a. log 200
b. log 50,000
c. log 0.0006

Solutions: a. $\log 200 = 2.301029996$ Note that this means $10^{2.301029996} = 200$.
b. $\log 50{,}000 = 4.698970004$
c. $\log 0.0006 = -3.22184875$ Note that a logarithm can be negative.

The domain of any logarithmic function is the set of positive real numbers. Negative numbers and 0 are not in the domain. **The logarithm of a negative number or zero is undefined**. For example, $\log_{10}(-2)$ is undefined. That is, $10^x = -2$ is impossible with real exponents. If you try to find the logarithm of a negative number with a calculator, then an error message will appear. For example, enter LOG (–2) on the TI-83 and you will get the error message shown here. (If you select `2: Goto`, the cursor will move to the location of the error.)

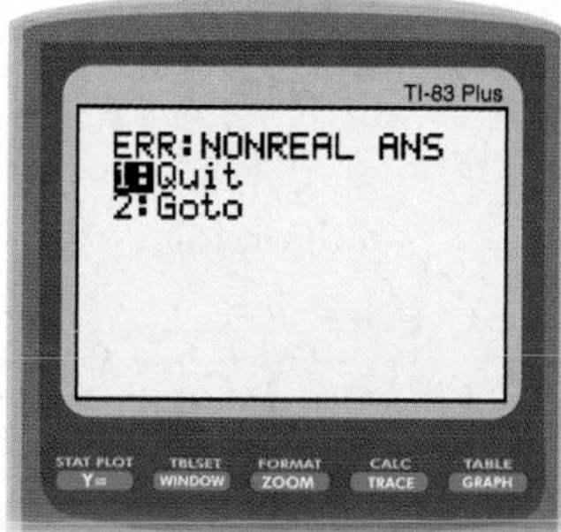

However, logarithms are exponents and they may be negative or 0. Thus,

$$10^{-2} = \frac{1}{100} \quad \text{and} \quad -2 = \log_{10}\frac{1}{100}$$

$$10^0 = 1 \quad \text{and} \quad 0 = \log_{10} 1.$$

If $\log x = N$, then we know that $x = 10^N$. The number x is called the **inverse log of** $\boldsymbol{N}$.

Using a TI-83 Plus Calculator to Find the Inverse Log of *N*

*Finding the **inverse log of N** is a three-step process:*

1. *Press* 2nd *and* LOG. *The expression* `10^(` *will appear on the display.*

2. *Enter the value of N and a right-hand parenthesis*) .
3. *Press* ENTER.

Example 5: Finding the Inverse Log of *N*

Use a TI-83 Plus calculator to find the value of x (which is the ***inverse log of N***). To check your understanding, write your estimate of x on paper before you use the calculator.

a. $\log x = 5$
b. $\log x = -2$
c. $\log x = 2.4142$
d. $\log x = 16.5$

Solutions: **a.** For $\log x = 5$, the calculator shows $x = 10^5 = 100{,}000$.

b. For $\log x = -2$, the calculator shows $x = 10^{-2} = 0.01$.

c. For $\log x = 2.4142$, the calculator shows $x = 10^{2.4142} = 259.5374301$.

d. For $\log x = 16.5$, the calculator shows $x = 10^{16.5} = 3.16227766E16$.

The letter E in the solution is the calculator version of scientific notation. Thus, $3.16227766E16 = 3.16227766 \cdot 10^{16}$.

Natural Logarithms (Base-*e* Logarithms)

Base-e logarithms are called natural logarithms. The notation $\ln x$ (read "natural log of x") is used to indicate that e is the base. That is,

$$\ln x = \log_e x$$

Figure 13.17 shows the graphs of the two functions $y = e^x$ and $y = \ln x$.

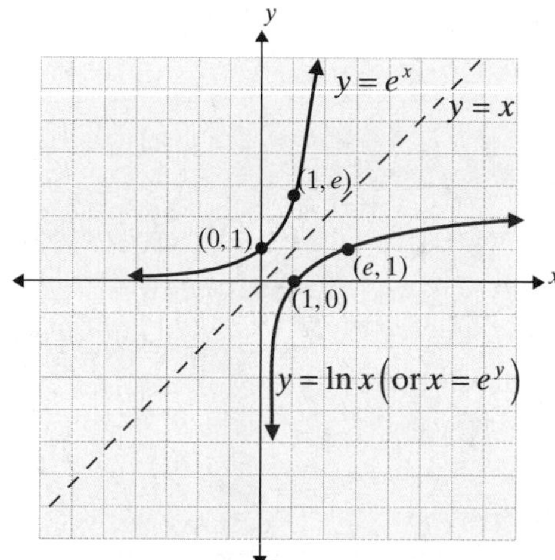

Figure 13.17

Notice again that points on inverse functions can be found by interchanging the coordinates of points.

Using a Calculator

Finding values of natural logarithms on a TI-83 is similar to finding common logarithms:

1. *Press the* LN *key.* `ln(` *will appear on the display.*
2. *Enter the number and a right hand parenthesis* `)` .
3. *Press* ENTER.

Example 6: Natural Logarithms

Use a calculator to find the following natural logarithms. To check your understanding, write your estimate of each value on a piece of paper before you use the calculator.

a. $\ln 1$
b. $\ln 3$
c. $\ln(-15)$
d. $\ln 0.02$

Solutions: **a.** $\ln 1 = 0$ This means that $e^0 = 1$.

b. $\ln 3 = 1.098612289$

c. $\ln(-15) = \text{error}$ There are no logarithms of negative numbers.

d. $\ln 0.02 = -3.912023005$ This means that $e^{-3.912023005} = 0.02$.

If $\ln x = N$, then we know that $x = e^N$. The number x is called **the inverse ln of N**. As the following procedure shows, finding the inverse ln is similar to finding the inverse log.

Using a TI-83 Plus Calculator to Find the Inverse Ln of *N*

*Finding the **inverse ln of N** is a three-step process:*

1. *Press* 2nd *and* LN. *The expression* `e^(` *will appear on the display.*

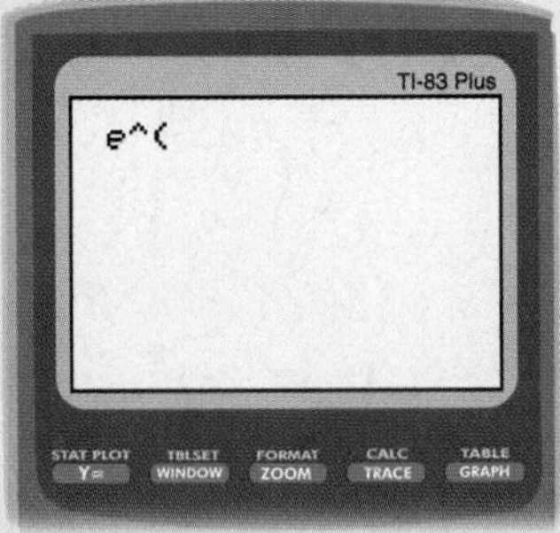

2. *Enter the value of N and a right-hand parenthesis* `)` .
3. *Press* ENTER.

Example 7: Finding the Inverse ln of *N*

Use a TI-83 Plus calculator to find the value of x in each of the following expressions. To check your understanding, write an estimate of x on your paper before you use the calculator.

a. $\ln x = 3$ **b.** $\ln x = -1$ **c.** $\ln x = -0.1$ **d.** $\ln x = 50$

Solutions: **a.** For $\ln x = 3$, the calculator shows $x = e^3 = 20.08553692$.

b. For $\ln x = -1$, the calculator shows $x = e^{-1} = 0.367879441$.

c. For $\ln x = -0.1$, the calculator shows $x = e^{-0.1} = 0.904837418$.

d. For $\ln x = 50$, the calculator shows $x = e^{50} = 5.184705529E21$.

Practice Problems

1. $\log_7 49 = 2$
2. $\log_3 27 = 3$
3. $\log_5 \frac{1}{25} = -2$
4. $\log_{10} 100 = \log 100 = 2$
5. $\log_2 \frac{1}{32} = -5$
6. $\log_\pi 1 = 0$
7. $\log_{2/3} \frac{4}{9} = 2$
8. $\log 23 = k$
9. $\ln 17 = x$
10. $\ln 11.6 = k$
11. $\log 10 = 1$
12. $\ln 1 = 0$
13. $3^2 = 9$
14. $5^3 = 125$
15. $9^{1/2} = 3$
16. $b^{2/3} = 4$
17. $7^{-1} = \frac{1}{7}$
18. $\left(\frac{1}{2}\right)^{-3} = 8$
19. $e^{1.74} = N$
20. $e^x = 42.3$
21. $b^4 = 18$
22. $b^{10} = 39$
23. $n^x = y^2$
24. $b^x = a$ **25.** $x = 16$
26. $x = 81$ **27.** $x = 2$
28. $x = -3$ **29.** $x = \frac{1}{6}$
30. $x = 2$
31. $x = 11$ **32.** $x = \frac{1}{27}$

1. *Express the given equation in logarithmic form.*

a. $4^2 = 16$ **b.** $10^3 = 1000$ **c.** $5^{-1} = \frac{1}{5}$

2. *Express the given equation in exponential form.*

a. $\log_2 x = -1$ **b.** $\log_5 x = 2$ **c.** $\ln x = 3$

3. *Find the value of x:*

a. $x = \ln(0.11)$ **b.** $x = \log(1.5)$ **c.** $x = \log_2 32$

4. *Find the value of x:*

a. $\log x = 2.46$ **b.** $\ln x = -2$ **c.** $\log_4 x = 2.5$

13.4 Exercises

In Exercises 1 – 12, express each equation in logarithmic form.

1. $7^2 = 49$ **2.** $3^3 = 27$ **3.** $5^{-2} = \frac{1}{25}$ **4.** $10^2 = 100$

5. $2^{-5} = \frac{1}{32}$ **6.** $1 = \pi^0$ **7.** $\left(\frac{2}{3}\right)^2 = \frac{4}{9}$ **8.** $10^k = 23$

9. $17 = e^x$ **10.** $e^k = 11.6$ **11.** $10^1 = 10$ **12.** $e^0 = 1$

In Exercises 13 – 24, express each equation in exponential form.

13. $\log_3 9 = 2$ **14.** $\log_5 125 = 3$ **15.** $\log_9 3 = \frac{1}{2}$ **16.** $\log_b 4 = \frac{2}{3}$

17. $\log_7 \frac{1}{7} = -1$ **18.** $\log_{\frac{1}{2}} 8 = -3$ **19.** $\ln N = 1.74$ **20.** $\ln 42.3 = x$

21. $\log_b 18 = 4$ **22.** $\log_b 39 = 10$ **23.** $\log_n y^2 = x$ **24.** $\log_b a = x$

Solve Exercises 25 – 40 by first changing each equation to exponential form.

25. $\log_4 x = 2$ **26.** $\log_3 x = 4$ **27.** $\log_{14} 196 = x$ **28.** $\log_5 \frac{1}{125} = x$

29. $\log_{36} x = -\frac{1}{2}$ **30.** $\log_x 32 = 5$ **31.** $\log_x 121 = 2$ **32.** $\log_{81} x = -\frac{3}{4}$

Answers to Practice Problems: 1. a. $\log_4 16 = 2$ **b.** $\log_{10} 1000 = 3$ **c.** $\log_5 \left(\frac{1}{5}\right) = -1$ **2. a.** $x = 2^{-1}$ **b.** $x = 5^2$ **c.** $x = e^3$ **3. a.** -2.207 **b.** 0.176 **c.** 5 **4. a.** 288.4 **b.** 0.135 **c.** 32

33. $x = 32$
34. $x = \frac{3}{2}$ 35. $x = -2$
36. $x = 3.7$
37. $x = 1.52$
38. $x = 2$ 39. $x = 3$
40. $x = 4$
41.
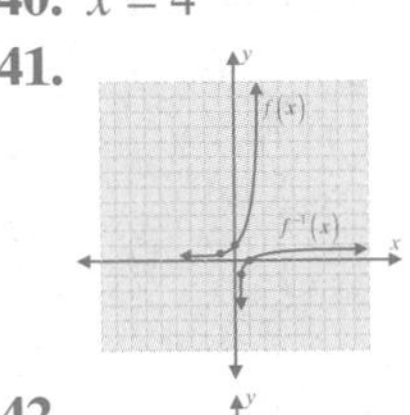
42.
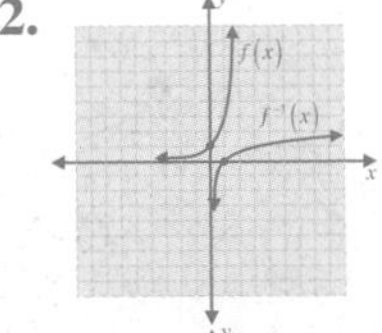
43.
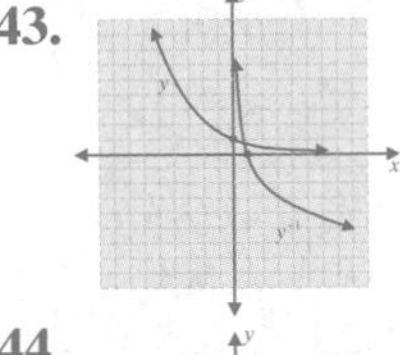
44.
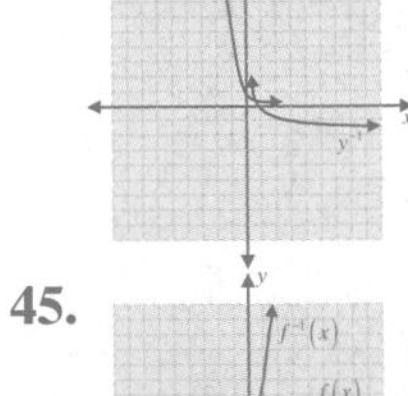
45.
46.
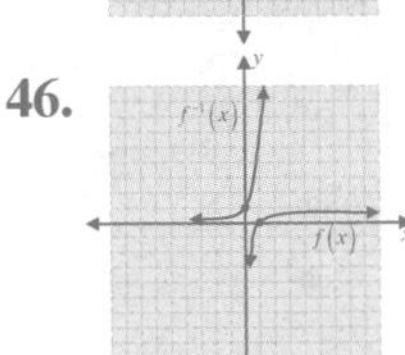
47.
48.
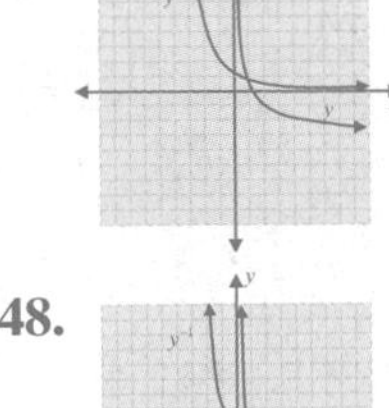

33. $\log_8 x = \frac{5}{3}$ **34.** $\log_{25} 125 = x$ **35.** $\log_3 \frac{1}{9} = x$ **36.** $\log_8 8^{3.7} = x$

37. $\log_{10} 10^{1.52} = x$ **38.** $\log_5 5^{\log_5 25} = x$ **39.** $\log_4 4^{\log_2 8} = x$ **40.** $\log_p p^{\log_3 81} = x$

In Exercises 41 – 50, graph each function and its inverse on the same set of axes.

41. $f(x) = 6^x$ **42.** $f(x) = 2^x$ **43.** $y = \left(\frac{2}{3}\right)^x$ **44.** $y = \left(\frac{1}{4}\right)^x$

45. $f(x) = \log_4 x$ **46.** $f(x) = \log_5 x$ **47.** $y = \log_{\frac{1}{2}} x$ **48.** $y = \log_{\frac{1}{3}} x$

49. $y = \log_8 x$ **50.** $y = \log_7 x$

Use a calculator to evaluate the logarithms in Exercises 51 – 62.

51. log 173 **52.** log 396 **53.** log 88.4 **54.** log 0.0061

55. log 0.0573 **56.** log (–8.47) **57.** ln 37.5 **58.** ln 96

59. ln (–14.9) **60.** ln 157.6 **61.** ln 0.00461 **62.** ln 0.0139

Use a calculator to find the value of x in each equation in Exercises 63 –74.

63. $\log x = 2.31$ **64.** $\log x = -3$ **65.** $\log x = -1.7$ **66.** $\log x = 4.1$
67. $2 \log x = -0.038$ **68.** $5 \log x = 9.4$ **69.** $\ln x = 5.17$ **70.** $\ln x = 4.9$
71. $\ln x = -8.3$ **72.** $\ln x = 6.74$ **73.** $0.2 \ln x = 0.0079$ **74.** $3 \ln x = -0.066$

In Exercises 75 – 78, use a graphing calculator to graph each function. State the domain and range of each function.

75. $f(x) = \log(x+1)$ **76.** $f(x) = 1 + \log x$ **77.** $f(x) = \log(-x)$

78. $f(x) = -\log x$

79. Consider the function $y = Ce^x$. Discuss the following:
   a. the domain of the function
   b. the range of the function
   c. any asymptotes of the graph of the function
   Give C two different values and sketch both of the graphs.

80. Consider the function $y = Ce^{-x}$. Discuss the following:
   a. the domain of the function
   b. the range of the function
   c. any asymptotes of the graph of the function
   Give C two different values and sketch both of the graphs.

49. **50.**

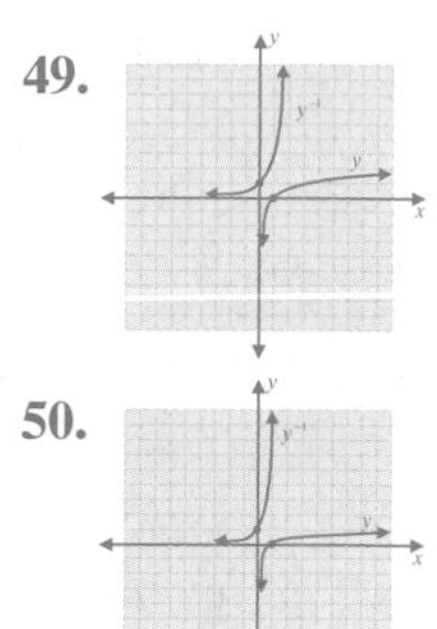

51. 2.23805
52. 2.59770
53. 1.94645

Writing and Thinking About Mathematics

81. Discuss, in your own words, how the graph of each of the following functions is related to the graph of the logarithmic function $y = \log_b x$.

a. $y = a \cdot \log_b x$ **b.** $y = \log_b(x - h)$ **c.** $y = \log_b(x) + k$

82. Discuss, in your own words, the symmetrical relationship of the graphs of the two logarithmic functions $y = 10^x$ and $y = \log x$.

83. Discuss, in your own words, the symmetrical relationship of the graphs of the two logarithmic functions $y = \log x$ and $y = -\log x$.

Hawkes Learning Systems: Introductory & Intermediate Algebra

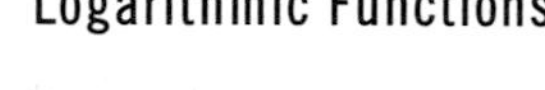

Logarithmic Functions

54. –2.21467
55. –1.24185
56. Error
57. 3.62434
58. 4.56435
59. Error
60. 5.06006
61. –5.37953
62. –4.27587
63. 204.17379
64. 0.0010000
65. 0.0199526
66. 12589.25412
67. 0.95719
68. 75.85776
69. 175.91484
70. 134.28978
71. 0.00024852
72. 845.56074
73. 1.04029
74. 0.97824

75. 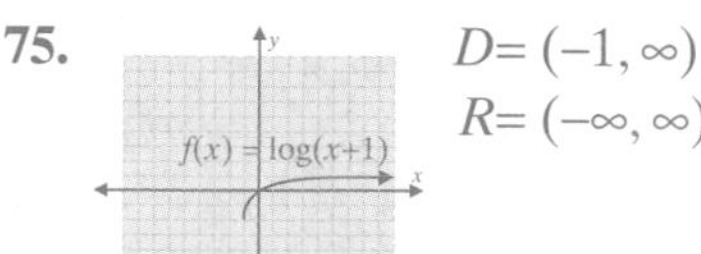

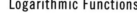

$D = (-1, \infty)$
$R = (-\infty, \infty)$

76. 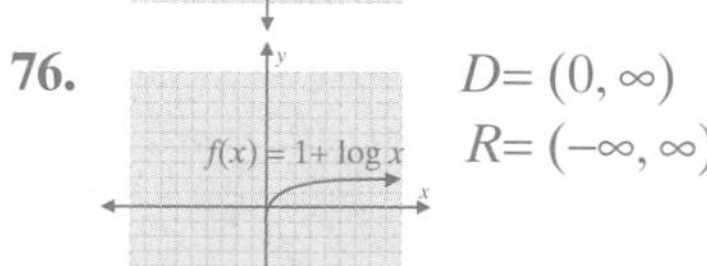

$D = (0, \infty)$
$R = (-\infty, \infty)$

77. 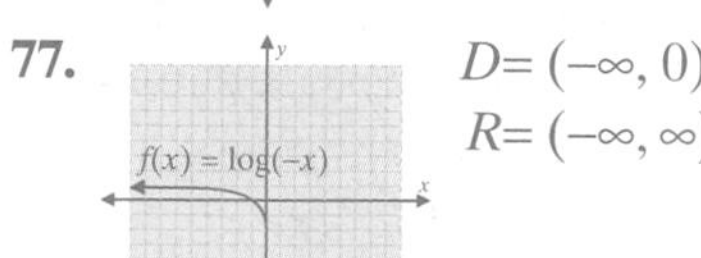

$D = (-\infty, 0)$
$R = (-\infty, \infty)$

78. 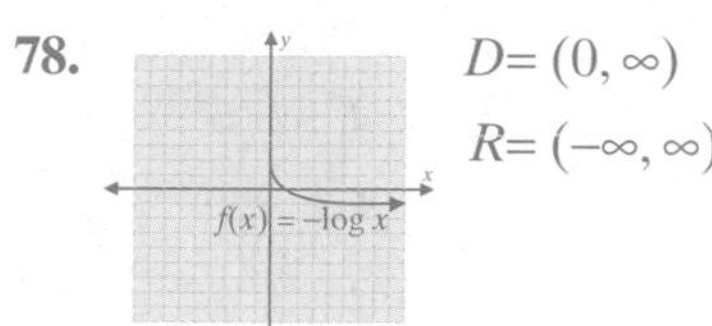

$D = (0, \infty)$
$R = (-\infty, \infty)$

79. 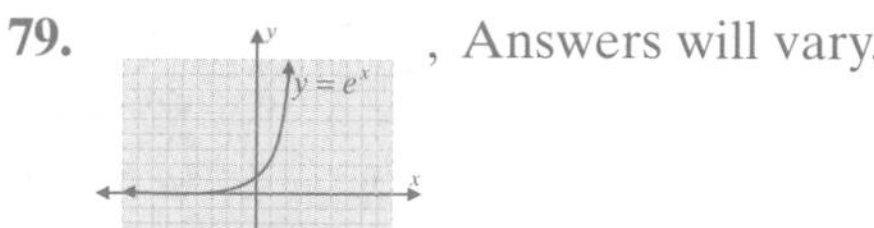

, Answers will vary.

80. 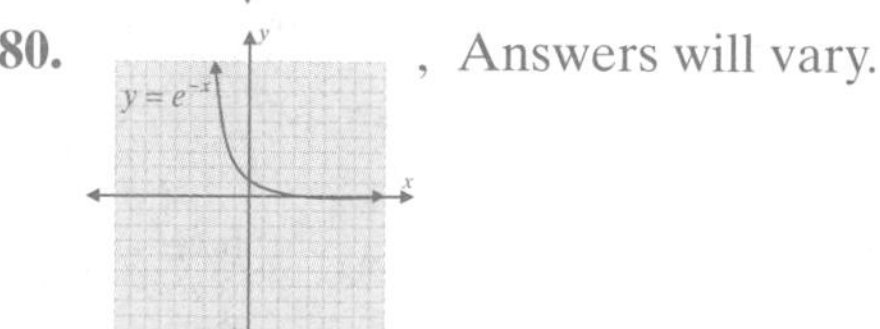

, Answers will vary.

81. Answers will vary.
82. Answers will vary.
83. Answers will vary.

13.5 Properties of Logarithms

Objectives

After completing this section, you will be able to:

1. *Evaluate logarithms using the properties of logarithms.*
2. *Write logarithmic expressions as the sums and/or differences of logarithms.*
3. *Write the sums and/or differences of logarithms as single logarithmic expressions.*

Teaching Notes:
It doesn't hurt to review the product rule for like bases before getting into this section, as the product rule is the same in this context. Similarly, the power rule and quotient rule are the same.

Although calculators are certainly effective in giving numerical evaluations, they generally do not simplify or solve equations involving exponential or logarithmic expressions. (You might be aware that there are calculators with computer algebra systems, CAS, that do simplify expressions and solve equations.) In this section, we will discuss several properties of logarithms that are helpful in solving equations and simplifying expressions involving logarithms and/or exponential functions.

The following three basic properties have been discussed:

1. $\log_b 1 = 0$
2. $\log_b b = 1$
3. $b^{\log_b x} = x$

With these three properties as a basis, we now develop three more properties (or rules) for logarithms: the Product Rule, the Quotient Rule, and the Power Rule.

The Product Rule

Because logarithms are exponents, their properties are similar to those of exponents. In fact, the properties of exponents are used to prove the rules for logarithms. Consider the following analysis.

We know that $\quad 6 = 2 \cdot 3$

Using Property 3, we have

$$10^{\log 6} = 6, \quad 10^{\log 2} = 2, \quad \text{and} \quad 10^{\log 3} = 3.$$

Thus, we can write

$$\begin{array}{ccccccc} 6 & = & 2 & \cdot & 3 & & \\ \downarrow & & \downarrow & & \downarrow & & \\ 10^{\log 6} & = & 10^{\log 2} & \cdot & 10^{\log 3} & = & 10^{\log 2 + \log 3} \end{array}$$

Equating the exponents on the left and right, the result is:

$$\begin{array}{lccccc} & \log 6 = \log(2\cdot 3) = & \log 2 & + & \log 3 \\ & \downarrow & \downarrow & & \downarrow \\ \text{Numerically,} & 0.7782 = & 0.3010 & + & 0.4771 \end{array}$$

Note: For convenience, we will round off logarithms to 4 decimal place accuracy in the remainder of this chapter.

This technique can be used to prove the following Product Rule for Logarithms.

Product Rule for Logarithms

For $b > 0$, $b \neq 1$, and $x, y > 0$,

$$\log_b xy = \log_b x + \log_b y$$

In words, the logarithm of a product is equal to the sum of the logarithms of the factors.

Proof of the Product Rule

$$\overbrace{b^{\log_b xy} = xy}^{\text{Property 3}} = \overbrace{x\cdot y = b^{\log_b x}\cdot b^{\log_b y}}^{\text{Property 3 again}} = \overbrace{b^{\log_b x + \log_b y}}^{\text{Add exponents}}$$

Thus,

$$b^{\log_b xy} = b^{\log_b x + \log_b y}$$

Equating the exponents gives the Product Rule for Logarithms:

$$\log_b xy = \log_b x + \log_b y$$

Example 1: Using the Product Rule

A calculator can be used to show the following results:

a. $\log(1000) = \log(10\cdot 100) = \log 10 + \log 100 = 1 + 2 = 3$

b. $\log(30) = \log(5\cdot 6) = \log 5 + \log 6 \approx 0.6990 + 0.7782 = 1.4772$

c. $\ln(16) = \ln(2\cdot 8) = \ln 2 + \ln 8 \approx 0.6931 + 2.0794 = 2.7725$

Note: Of course other numbers could be used in these products. The results will be the same.

The Quotient Rule

Now consider the problem of finding the logarithm of a quotient. For example to find $\log\frac{3}{2}$, we could proceed as follows:

$$\begin{aligned} 10^{\log\left(\frac{3}{2}\right)} &= \frac{3}{2} \\ &= \frac{10^{\log 3}}{10^{\log 2}} \\ &= 10^{\log 3 - \log 2} \end{aligned}$$

Equating exponents gives

$$\begin{aligned} \log\frac{3}{2} &= \log 3 - \log 2 \\ &\approx 0.4771 - 0.3010 \\ &= 0.1761 \end{aligned}$$

This result can be verified by using a calculator to find log (1.5).

These ideas lead to the following Quotient Rule for Logarithms. The proof is left as an exercise for the student.

Quotient Rule for Logarithms

For $b > 0$, $b \neq 1$, and $x,y > 0$,

$$\log_b \frac{x}{y} = \log_b x - \log_b y.$$

In words, the logarithm of a quotient is equal to the difference between the logarithm of the numerator and the logarithm of the denominator.

Example 2: Using the Quotient Rule

A calculator can be used to show the following results:

a. $\log\left(\frac{1}{2}\right) = \log 1 - \log 2 \approx 0 - 0.3010 = -0.3010$

b. $\log\left(\frac{4}{3}\right) = \log 4 - \log 3 \approx 0.6021 - 0.4771 = 0.1250$

c. $\ln\left(\frac{15}{2}\right) = \ln 15 - \ln 2 \approx 2.7081 - 0.6931 = 2.0150$

NOTES

Because we are rounding off individual logarithms to 4 places, there may be a slight difference in answers if a calculator is used to find the logarithm of the original number. For example, a calculator will show that

$\ln\left(\frac{15}{2}\right) = \ln(7.5) = 2.0149$ to 4 decimal places.

The Power Rule

The next property of logarithms involves a number raised to a power and the multiplication of exponents. For example,

$$\begin{aligned} 10^{\log(2^3)} &= 2^3 \\ &= \left[10^{\log 2}\right]^3 \\ &= 10^{3(\log 2)} \end{aligned}$$

Equating exponents gives

$$\log(2^3) = 3(\log 2).$$

These ideas lead to the following Power Rule for Logarithms.

Power Rule for Logarithms

For $b > 0$, $b \neq 1$, $x > 0$, and any real number r,

$$\log_b x^r = r \cdot \log_b x.$$

In words, the logarithm of a number raised to a power is equal to the product of the exponent and the logarithm of the number.

Proof of the Power Rule for Logarithms

$$\overbrace{b^{\log_b(x^r)} = x^r = \left[b^{\log_b x}\right]^r}^{\text{Property 3 twice}} = \overbrace{b^{r \cdot \log_b x}}^{\text{Multiply exponents}}$$

Equating the exponents gives the result called the Power Rule for Logarithms:

$$\log_b x^r = r \cdot \log_b x$$

Example 3: Using the Power Rule

A calculator can be used to show the following results:

a. $\log(\sqrt{2}) = \log(2)^{\frac{1}{2}} = \frac{1}{2}\log(2) \approx \frac{1}{2}(0.3010) = 0.1505$

continued on next page ...

b. $\log(25) = \log(5^2) = 2\log 5 \approx 2(0.6990) = 1.3979$

c. $\ln(8) = \ln(2^3) = 3 \cdot \ln 2 \approx 3(0.6931) = 2.0794$

Table 13.3 summarizes the properties of logarithms for base b. **To emphasize that b could be e, we list the properties separately for natural logarithms.**

Properties of Logarithms			
For $b > 0$, $b \neq 1$, $x,y > 0$ and any real number r:		For natural logarithms:	
1.	$\log_b 1 = 0$	**1.**	$\ln 1 = 0$
2.	$\log_b b = 1$	**2.**	$\ln e = 1$
3.	$x = b^{\log_b x}$	**3.**	$x = e^{\ln x}$
4.	Product Rule: $\log_b xy = \log_b x + \log_b y$	**4.**	Product Rule: $\ln xy = \ln x + \ln y$
5.	Quotient Rule: $\log_b \frac{x}{y} = \log_b x - \log_b y$	**5.**	Quotient Rule: $\ln \frac{x}{y} = \ln x - \ln y$
6.	$\log_b x^r = r \cdot \log_b x$	**6.**	$\ln x^r = r \cdot \ln x$

Table 13.3

Example 4: Properties of Logarithms

Use the properties of logarithms to write each expression as a sum and/or difference of logarithmic expressions.

a. $\log 2x^3$

Solution: $\log 2x^3 = \log 2 + \log x^3$ — Product Rule

$= \log 2 + 3\log x$ — Power Rule

b. $\log\left(\frac{ab^2}{c}\right)$

Solution: $\log\left(\frac{ab^2}{c}\right) = \log(ab^2) - \log c$ — Quotient Rule

$= \log a + \log b^2 - \log c$ — Product Rule

$= \log a + 2\log b - \log c$ — Power Rule

c. $\ln(xy)^{-3}$

Solution: $\ln(xy)^{-3} = -3\ln(xy)$ — Power Rule

$= -3(\ln x + \ln y)$ — Product Rule

d. $\ln\left(\sqrt{3x}\right)$

Solution: $\ln\left(\sqrt{3x}\right) = \ln(3x)^{\frac{1}{2}} = \frac{1}{2}\ln(3x)$ Power Rule

$= \frac{1}{2}(\ln 3 + \ln x)$ Product Rule

Example 5: Single Logarithm

Use the properties of logarithms to write each expression as a single logarithm.

a. $2\log_b x - 3\log_b y$

Solution: $2\log_b x - 3\log_b y = \log_b x^2 - \log_b y^3$ Power Rule

$= \log_b\left(\frac{x^2}{y^3}\right)$ Quotient Rule

b. $\frac{1}{2}\ln 4 + \ln 5 - \ln y$

Solution: $\frac{1}{2}\ln 4 + \ln 5 - \ln y = \ln 4^{\frac{1}{2}} + \ln 5 - \ln y$ Power Rule

$= \ln\left(4^{\frac{1}{2}} \cdot 5\right) - \ln y$ Product Rule

$= \ln\frac{10}{y}$ Quotient Rule $\left(4^{\frac{1}{2}} = 2\right)$

c. $\log(x+1) + \log(x-1)$

Solution: $\log(x+1) + \log(x-1) = \log(x+1)(x-1)$ Product Rule

$= \log\left(x^2 - 1\right)$

d. $\ln\sqrt{x} + \ln\sqrt[3]{x}$

Solution: $\ln\sqrt{x} + \ln\sqrt[3]{x} = \ln(x)^{\frac{1}{2}} + \ln(x)^{\frac{1}{3}} = \frac{1}{2}\ln x + \frac{1}{3}\ln x$ Power Rule

$= \left(\frac{1}{2} + \frac{1}{3}\right)\ln x$ Algebra

$= \frac{5}{6}\ln x$

1. 5 **2.** 2 **3.** −2
4. −3 **5.** $\frac{1}{2}$ **6.** $\frac{3}{2}$
7. 10 **8.** 17 **9.** $\sqrt{3}$
10. 5
11. log 5 + 4 log x
12. log 3 + 2 log x + log y
13. ln 2 − 3 ln x + ln y
14. ln x + 2 ln y − ln z

NOTES

A Note About Common Misunderstandings:

There is no logarithmic property for a sum or a difference. That is, there is no property that will relate to $\log_b(x+y)$ and no property that will relate to $\log_b(x-y)$.

Also,

$$\log_b \frac{x}{y} \neq \frac{\log_b x}{\log_b y}$$

$$\log_b x \cdot \log_b y \neq \log_b x + \log_b y.$$

Practice Problems

15. log 2 + log x − 3 log y
16. log x + log y − log 4 − log z
17. 2 ln x − ln y − ln z
18. ln x +2 ln y − 2 ln z

1. *Write $\log(x^3y)$ as a sum and/or difference of logarithmic expressions.*
2. *Write the expression $2\ln 5 + \ln x - \ln 3$ as a single logarithm.*
3. *Write $2\log x - \log(x+1)$ as a single logarithm.*

13.5 Exercises

19. − 2 log x − 2 log y
20. 8 log x + 4 log y
21. $\frac{1}{3}\log x + \frac{2}{3}\log y$
22. $\frac{1}{2}\log 2 + \frac{3}{2}\log x + \frac{1}{2}\log y$
23. $\frac{1}{2}\ln x + \frac{1}{2}\ln y - \frac{1}{2}\ln z$
24. $\frac{2}{3}\ln x - \frac{1}{3}\ln y$
25. $\log 21 + 2\log x + \frac{2}{3}\log y$
26. $\log 15 - \frac{1}{2}\log x + \frac{1}{3}\log y$

In Exercises 1 – 30, use your knowledge of logarithms and exponents to find the value of each expression.

1. $\log_2 32$	**2.** $\log_3 9$	**3.** $\log_4 \frac{1}{16}$	**4.** $\log_5 \frac{1}{125}$
5. $\log_3 \sqrt{3}$	**6.** $\log_2 \sqrt{8}$	**7.** $5^{\log_5 10}$	**8.** $3^{\log_3 17}$
9. $6^{\log_6 \sqrt{3}}$	**10.** $e^{\ln 5}$	**11.** $\log 5x^4$	**12.** $\log 3x^2y$
13. $\ln 2x^{-3}y$	**14.** $\ln xy^2z^{-1}$	**15.** $\log \frac{2x}{y^3}$	**16.** $\log \frac{xy}{4z}$
17. $\ln \frac{x^2}{yz}$	**18.** $\ln \frac{xy^2}{z^2}$	**19.** $\log(xy)^{-2}$	**20.** $\log(x^2y)^4$
21. $\log\sqrt[3]{xy^2}$	**22.** $\log\sqrt{2x^3y}$	**23.** $\ln\sqrt{\frac{xy}{z}}$	**24.** $\ln\sqrt[3]{\frac{x^2}{y}}$
25. $\log 21x^2y^{\frac{2}{3}}$	**26.** $\log 15x^{-\frac{1}{2}}y^{\frac{1}{3}}$	**27.** $\log\frac{x}{\sqrt{x^3y^5}}$	**28.** $\log\frac{1}{\sqrt{x^4y}}$
29. $\ln\left(\frac{x^3y^2}{z}\right)^{-3}$	**30.** $\ln\left(\frac{x^{-\frac{1}{2}}y}{z^2}\right)^{-2}$		

Answers to Practice Problems: 1. 3 log x + log y **2.** $\ln\frac{25x}{3}$ **3.** $\log\frac{x^2}{x+1}$

27. $-\frac{1}{2}\log x-\frac{5}{2}\log y$

28. $-2\log x-\frac{1}{2}\log y$

29. $-9\ln x-6\ln y+3\ln z$

30. $\ln x-2\ln y+4\ln z$

31. $\ln\frac{9x}{5}$ 32. $\ln\frac{15}{x}$

33. $\log\frac{7x^2}{8}$

34. $\log 24y$

35. $\log x^2y$

36. $\log xy^3$ 37. $\ln\frac{x^3}{y^2}$

38. $\ln\frac{y^3}{\sqrt{x}}$ 39. $\ln\sqrt{\frac{x}{y}}$

40. $\ln\sqrt[3]{\frac{x}{y^2}}$ 41. $\log\frac{xz}{y}$

42. $\log\frac{xy^2}{\sqrt{z}}$

43. $\log\frac{x}{y^2z^2}$

44. $\log\sqrt[3]{\frac{z^2}{x^2y}}$

Use the properties of logarithms to write each expression in Exercises 31 – 52 as a single logarithm.

31. $2\ln 3+\ln x-\ln 5$

32. $\frac{1}{2}\ln 25+\ln 3-\ln x$

33. $\log 7-\log 8+2\log x$

34. $\log 4+\log 6+\log y$

35. $2\log x+\log y$

36. $\log x+3\log y$

37. $3\ln x-2\ln y$

38. $3\ln y-\frac{1}{2}\ln x$

39. $\frac{1}{2}(\ln x-\ln y)$

40. $\frac{1}{3}(\ln x-2\ln y)$

41. $\log x-\log y+\log z$

42. $\log x+2\log y-\frac{1}{2}\log z$

43. $\log x-2\log y-2\log z$

44. $-\frac{2}{3}\log x-\frac{1}{3}\log y+\frac{2}{3}\log z$

45. $\log x+\log(2x+1)$

46. $\log(x+3)+\log(x-3)$

47. $\ln(x-1)+\ln(x+3)$

48. $\ln(3x+1)+2\ln x$

49. $\log(x^2-2x-3)-\log(x-3)$

50. $\log(x-4)-\log(x^2-2x-8)$

51. $\log(x+6)-\log(2x^2+9x-18)$

52. $\log(3x^2+5x-2)-\log(3x-1)$

Writing and Thinking About Mathematics

53. Prove the Quotient Rule for Logarithms: For $b>0, b\neq 1$, and $x,y>0$,

$$\log_b\frac{x}{y}=\log_b x-\log_b y.$$

54. Prove the following property of logarithms: For $b>0,\ b\neq 1,$ and $x>0,\ \log_b b^x=x.$

Hawkes Learning Systems: Introductory & Intermediate Algebra

Properties of Logarithmic Functions

45. $\log(2x^2+x)$

46. $\log(x^2-9)$

47. $\ln(x^2+2x-3)$

48. $\ln x^2(3x+1)$

49. $\log\frac{x^2-2x-3}{x-3}$

50. $\log\frac{x-4}{x^2-2x-8}$

51. $\log\frac{x+6}{2x^2+9x-18}$

52. $\log\frac{3x^2+5x-2}{3x-1}$

53. Answers will vary.

54. Answers will vary.

13.6 Logarithmic and Exponential Equations

Objectives

After completing this section, you will be able to:

1. *Solve exponential equations in which the bases are not necessarily the same.*
2. *Solve logarithmic equations without necessarily changing to exponential form.*
3. *Use the change-of-base formula and a calculator to evaluate logarithmic expressions.*

Solving Exponential Equations with the Same Base

Both exponential and logarithmic functions involve the meaning of expressions with exponents that are real numbers, including irrational numbers such as $\sqrt{2}$ and π. That is, we are familiar with expressions such as 2^3 and 3^{-2}, but you need to be aware that we have been using (in the definitions of exponential functions and logarithmic functions and in their graphs) the meaning and value of expressions with real exponents such as $2^{\sqrt{2}}$, 2^{π}, and $10^{1.4}$. To be complete in our understanding of exponents, we state, without proof, that all the properties of exponents that have been discussed are valid for real exponents with a positive base.

Properties of Real Exponents

If a and b are positive real numbers and x and y are any real numbers, then:

1. $b^0 = 1$
2. $b^{-x} = \dfrac{1}{b^x}$
3. $b^x \cdot b^y = b^{x+y}$
4. $\dfrac{b^x}{b^y} = b^{x-y}$
5. $\left(b^x\right)^y = b^{xy}$
6. $\left(ab\right)^x = a^x b^x$
7. $\left(\dfrac{a}{b}\right)^x = \dfrac{a^x}{b^x}$

In this section we first show how to solve exponential equations with expressions that have the same base, then exponential equations with different bases, and finally equations that involve logarithms. We need the following properties of equations containing exponents and logarithms to help in solving the related equations.

Properties of Equations with Exponents and Logarithms

For $b > 0,\ b \neq 1$,

1. *If $b^x = b^y$, then $x = y$.*
2. *If $x = y$ then $b^x = b^y$.*
3. *If $\log_b x = \log_b y$, then $x = y$ ($x > 0$ and $y > 0$).*
4. *If $x = y$, then $\log_b x = \log_b y$ ($x > 0$ and $y > 0$).*

Example 1 shows how Property 1 can be used to solve exponential equations with expressions that have the same base.

Example 1: Solving Equations for *x* when the Base is the Same

Solve each equation for x.

a. $3^{x-2} = 3^{\frac{1}{2}}$

Solution: $3^{x-2} = 3^{\frac{1}{2}}$ Both bases are 3.

$x - 2 = \frac{1}{2}$ Since the bases are the same, the exponents must be equal.

$x = \frac{5}{2}$ Solve for x.

b. $2^{x^2-7} = 2^{6x}$

Solution: $2^{x^2-7} = 2^{6x}$ Both bases are 2.

$x^2 - 7 = 6x$ The exponents are equal because the bases are the same.

$x^2 - 6x - 7 = 0$ Solve for x.

$(x-7)(x+1) = 0$

$x = 7$ or $x = -1$

c. $8^{4-2x} = 4^{x+2}$

Solution: $8^{4-2x} = 4^{x+2}$ Here, the bases are different.

$\left(2^3\right)^{4-2x} = \left(2^2\right)^{x+2}$ Rewrite both sides so that the bases are the same.

$2^{12-6x} = 2^{2x+4}$ Use the property $\left(b^x\right)^y = b^{xy}$.

continued on next page ...

$12-6x=2x+4$ The exponents must be equal.

$8=8x$ Solve for x.

$1=x$

Solving Exponential Equations with Different Bases

Example 2 illustrates how to solve exponential equations that have exponential expressions with different bases. The technique is to use Property 4 of equations and take the log (or ln) of the expressions on both sides of the equation or to use the definition of logarithm as an exponent. (As we have seen in Example 1, if the bases are the same there is no need to deal with logarithms.)

Example 2: Solving Exponential Equations with Different Bases

Solve each of the following exponential equations by taking the log (or ln) of both sides of the equation or by using the definition of logarithm as an exponent.

a. $10^{3x}=2.1$

Solution: Since the base of $3x$ is 10, we can solve by taking the log of both sides.

$$10^{3x}=2.1$$

$$\log 10^{3x}=\log 2.1 \qquad \text{Take the log of both sides.}$$

$$3x\log 10=\log 2.1 \qquad \text{Power Rule}$$

$$3x=\log 2.1 \qquad \log 10=1$$

$$x=\frac{\log 2.1}{3}$$

Using a calculator,

$$x=\frac{\log 2.1}{3}\approx\frac{0.3222}{3}\approx 0.1074$$

We could also have simply used the definition of a logarithm as an exponent and stated directly

$$10^{3x}=2.1$$

$$3x=\log 2.1 \qquad \text{By the definition of logarithm}$$

$$x=\frac{\log 2.1}{3}$$

b. $e^{0.2x} = 50$

Solution: Using the definition of a logarithm as an exponent, we have

$$e^{0.2x} = 50$$

$$0.2x = \ln 50 \qquad \text{By the definition of logarithm}$$

$$x = \frac{\ln 50}{0.2}$$

Using a calculator,

$$x = \frac{\ln 50}{0.2} \approx \frac{3.912}{0.2} = 19.56$$

c. $6^x = 18$

Solution: The base is 6, not 10 or e, but we can solve by taking the **log** of both sides or by taking the **ln** of both sides. The result is the same.

Taking the log of both sides:

$$6^x = 18$$

$$\log 6^x = \log 18$$

$$x \cdot \log 6 = \log 18$$

$$x = \frac{\log 18}{\log 6}$$

Using a calculator,

$$x = \frac{\log 18}{\log 6} \approx \frac{1.2553}{0.7782}$$

$$= 1.6131$$

Taking the ln of both sides:

$$6^x = 18$$

$$\ln 6^x = \ln 18$$

$$x \cdot \ln 6 = \ln 18$$

$$x = \frac{\ln 18}{\ln 6}$$

Using a calculator

$$x = \frac{\ln 18}{\ln 6} \approx \frac{2.8904}{1.7918}$$

$$= 1.6131$$

d. $5^{2x-1} = 10^x$

Solution:

$$5^{2x-1} = 10^x$$

$$\log 5^{2x-1} = \log 10^x \qquad \text{Take the log of both sides.}$$

$$(2x-1)\log 5 = x \log 10 \qquad \text{Power Rule}$$

$$2x \cdot \log 5 - 1 \cdot \log 5 = x \qquad \log 10 = 1$$

$$2x \log 5 - x = \log 5 \qquad \text{Arrange } x\text{-terms on one side.}$$

$$x(2\log 5 - 1) = \log 5 \qquad \text{Factor out the } x.$$

$$x = \frac{\log 5}{2\log 5 - 1}$$

As a decimal approximation,

$$x = \frac{\log 5}{2\log 5 - 1} \approx \frac{0.6990}{2(0.6990) - 1} \approx 1.7563$$

Solving Equations with Logarithms

All the various properties of logarithms can be used to solve equations that involve logarithms. Remember that logarithms are defined only for positive real numbers, so each answer should be checked in the original equation.

Example 3: Solving Equations with Logarithms

Use the properties of logarithms to solve the following equations.

a. $\log(5x) = 3$

Solution: $\log(5x) = 3$

$5x = 10^3$ Definition of logarithm

$5x = 1000$

$x = 200$

b. $\log(x-1) + \log(x-4) = 1$

Solution: $\log(x-1) + \log(x-4) = 1$

$\log((x-1)(x-4)) = 1$ Product Rule

$(x-1)(x-4) = 10^1$ Change to exponential form using base 10.

$x^2 - 5x + 4 = 10$

$x^2 - 5x - 6 = 0$

$(x-6)(x+1) = 0$ Solve by factoring.

$x = 6$ or ~~$x = -1$~~ Checking $x = -1$ yields $\log(-1-1) = \log(-2)$, which is undefined.

c. $\log x - \log(x-1) = \log 3$

Solution: $\log x - \log(x-1) = \log 3$

$\log\left(\frac{x}{x-1}\right) = \log 3$ Quotient Rule

$\frac{x}{x-1} = 3$ If $\log_b x = \log_b y$, then $x = y$.

$x = 3(x-1)$ Solve for x.

$x = 3x - 3$

$3 = 2x$

$\frac{3}{2} = x$

d. $\ln(x^2 - x - 6) - \ln(x+2) = 2$

Solution: $\ln(x^2 - x - 6) - \ln(x+2) = 2$

$\ln\left(\frac{x^2 - x - 6}{x+2}\right) = 2$ Quotient Rule

$\ln\left(\frac{(x+2)(x-3)}{(x+2)}\right) = 2$ Factor the numerator.

$\ln(x-3) = 2$ Simplify.

$x - 3 = e^2$ Change to exponential form with base e.

$x = 3 + e^2$

Or, using a calculator, $x = 3 + e^2 \approx 3 + 7.3891 = 10.3891$

Change-of-Base

Because a calculator can be used to evaluate common logarithms and natural logarithms, we have restricted most of the examples to base 10 or base e expressions. If an equation involves logarithms of other bases, the following discussion shows how to rewrite each logarithm using any base you choose.

Change-of-Base

$$\log_b x = \frac{\log_a x}{\log_a b}$$

The change-of-base formula can be derived by using properties of logarithms as follows:

$b^{\log_b x} = x$	Property 3 in Section 13.5
$\log_a\left(b^{\log_b x}\right) = \log_a x$	Take the log(base a) of both sides.
$\log_b x\left(\log_a b\right) = \log_a x$	By the Power Rule using $\log_b x$ as the exponent r.
$\log_b x = \dfrac{\log_a x}{\log_a b}$	Divide both sides by $\log_a b$ to arrive at the change-of-base formula.

Example 4: Change-of-Base

Use the change-of-base formula to evaluate the expressions in (a) and (b) and to solve the equation in (c).

a. $\log_2 3.42$

Solution: This expression can be evaluated by using either base 10 or base e since both are easily available on a calculator.

$$\log_2 3.42 = \frac{\ln 3.42}{\ln 2} \approx \frac{1.2296}{0.6931} = 1.7741 \qquad \text{Using rounded values.}$$

(The student can show that $\dfrac{\log 3.42}{\log 2}$ gives the same result.)

In exponential form: $2^{1.7741} \approx 3.42$

b. $\log_3 0.3333$

Solution: $\log_3 0.3333 = \dfrac{\log 0.3333}{\log 3} \approx \dfrac{-0.4772}{0.4771} = -1.0002$

c. $5^x = 16$

Solution: Because the base is 5, we can take $\log_5$ of both sides. (This method is not necessary, but it does show how the change-of-base formula can be used.)

$$5^x = 16$$

$$\log_5\left(5^x\right) = \log_5 16$$

$$x = \log_5 16$$

$$x = \frac{\ln 16}{\ln 5} \approx 1.7227$$

1. $x=11$ **2.** $x=5$
3. $x=9$ **4.** $x=3$
5. $x=\frac{1}{6}$ **6.** $x=\frac{7}{2}$
7. $x=\frac{7}{12}$ **8.** $x=\frac{6}{5}$
9. $x=\frac{7}{2}$ **10.** $x=\frac{8}{3}$

Practice Problems

Solve each of the following equations.

1. $4^x=64$

2. $10^x=64$

3. $2^{3x-1}=0.1$

4. $15\log x=45.15$

5. $\ln\left(x^2-x-6\right)-\ln\left(x-3\right)=1$

13.6 Exercises

11. $x=-2$ **12.** $x=-5$
13. $x=-3$ **14.** $x=-8$
15. $x=4, x=-1$
16. $x=\pm 3$
17. $x=-1, x=\frac{3}{2}$
18. $x=-1, x=-2$
19. $x=-2, x=3$
20. $x=-1, x=-4$
21. $x=2$ **22.** $x=2$
23. $x=0, x=-\frac{3}{2}$
24. $x=0, x=-\frac{5}{2}$
25. $x=3, x=-1$
26. $x=3,-1$
27. $x\approx 0.7154$
28. $x\approx 0.9934$
29. $x\approx 7.5098$
30. $x\approx -4.0572$
31. $x\approx -3.648$
32. $x\approx -0.6879$
33. $x\approx 24.7312$
34. $x\approx -7.7003$
35. $t\approx -653.0008$
36. $t\approx 1.3037$
37. $t\approx -1.5193$
38. $t\approx 0.9061$
39. $x\approx 3.322$

Use the properties of exponents and logarithms to solve each of the equations in Exercises 1 – 84.

1. $2^4\cdot 2^7=2^x$ **2.** $3^7\cdot 3^{-2}=3^x$ **3.** $\left(3^5\right)^2=3^{x+1}$

4. $\left(5^x\right)^2=5^6$ **5.** $\left(2^x\right)^3=\sqrt{2}$ **6.** $\dfrac{10^4\cdot 10^{\frac{1}{2}}}{10^x}=10$

7. $\left(10^2\right)^x=\dfrac{10\cdot 10^{\frac{2}{3}}}{10^{\frac{1}{2}}}$ **8.** $2^{5x}=4^3$ **9.** $\left(25\right)^x=5^3\cdot 5^4$

10. $7^{3x}=49^4$ **11.** $10^x\cdot 10^8=100^3$ **12.** $8^{x+3}=2^{x-1}$

13. $27^x=3\cdot 9^{x-2}$ **14.** $100^{2x+1}=1000^{x-2}$ **15.** $2^{3x+5}=2^{x^2+1}$

16. $10^{x^2+x}=10^{x+9}$ **17.** $10^{2x^2+3}=10^{x+6}$ **18.** $3^{x^2+5x}=3^{2x-2}$

19. $\left(3^{x+1}\right)^x=\left(3^{x+3}\right)^2$ **20.** $\left(10^x\right)^{x+3}=\left(10^{x+2}\right)^{-2}$ **21.** $3^x=9$

22. $2^{5x-8}=4$ **23.** $4^{x^2}=\left(\frac{1}{2}\right)^{3x}$ **24.** $25^{x^2+2x}=5^{-x}$

25. $5^{2x-x^2}=\frac{1}{125}$ **26.** $10^{x^2-2x}=1000$ **27.** $10^{3x}=140$

28. $10^{2x}=97$ **29.** $10^{0.32x}=253$ **30.** $10^{-0.48x}=88.6$

31. $4.10^{-0.94x}=126.2$ **32.** $3\cdot 10^{-2.1x}=83.5$ **33.** $e^{0.03x}=2.1$

34. $e^{-0.5x}=47$ **35.** $e^{-0.006t}=50.3$ **36.** $e^{4t}=184$

37. $3e^{-0.12t}=3.6$ **38.** $5e^{2.4t}=44$ **39.** $2^x=10$

40. $3^{x-2}=100$ **41.** $5^{2x}=\frac{1}{100}$ **42.** $7^{2x-3}=10$

43. $5^{1-x}=1$ **44.** $4^{2x+5}=0.01$ **45.** $4^{2-3x}=0.1$

Answers to Practice Problems: **1.** $x=3$ **2.** $x\approx 1.8062$ **3.** $x\approx -0.7740$ **4.** $x=10^{3.01}\approx 1023.2930$ **5.** No solution

40. $x \approx 6.1918$
41. $x \approx -1.4307$
42. $x \approx 2.0916$
43. $x = 1$
44. $x \approx -4.161$
45. $x \approx 1.2203$
46. $x \approx 1.2058$
47. $x \approx -1.6467$
48. $x = -\frac{2}{5}$
49. $x \approx 1.1292$
50. $x \approx 4.3219$
51. $x \approx 4.854$
52. $x \approx 0.9741$
53. $x \approx 1.252$
54. $x \approx 1.5472$
55. $x \approx 25.1189$
56. $x \approx 25118.8643$
57. $x \approx 31.6228$
58. $x \approx 31622776.6017$
59. $x = 0.0001$
60. $x \approx \pm 5.6234 \times 10^{-21}$
61. $x \approx 4.953$
62. $x \approx 0.2231$
63. $x \approx \pm 0.3329$
64. $x \approx \pm 1079754999.46$
65. $x = 3$ **66.** $x = 5$
67. $x = 100$ **68.** $x = 10$
69. $x = 6$ **70.** $x = 6$
71. $x = 20$
72. $x \approx 3.1893$
73. $x = \frac{25}{2}$
74. No solution
75. $x = 1001$
76. No solution
77. No solution
78. $x = 105$
79. $x \approx 1.1353$
80. $x = \sqrt{2}$
81. $x \approx 22.0855$

46. $14^{3x-1} = 10^3$ **47.** $12^{2x+7} = 10^4$ **48.** $12^{5x+2} = 1$

49. $7^x = 9$ **50.** $2^x = 20$ **51.** $3^{x-2} = 23$

52. $5^{2x} = 23$ **53.** $6^{2x-1} = 14.8$ **54.** $4^{7-3x} = 26.3$

55. $5\log x = 7$ **56.** $3\log x = 13.2$ **57.** $4\log x - 6 = 0$

58. $2\log x - 15 = 0$ **59.** $4\log x^{\frac{1}{2}} + 8 = 0$ **60.** $\frac{2}{3}\log x^{\frac{2}{3}} + 9 = 0$

61. $5\ln x - 8 = 0$ **62.** $2\ln x + 3 = 0$ **63.** $\ln x^2 + 2.2 = 0$

64. $\ln x^2 - 41.6 = 0$ **65.** $\log x + \log 2x = \log 18$

66. $\log(x+4) + \log(x-4) = \log 9$ **67.** $\log x^2 - \log x = 2$

68. $\log x + \log x^2 = 3$ **69.** $\ln(x-3) + \ln x = \ln 18$

70. $\ln(x^2 - 3x + 2) - \ln(x-1) = \ln 4$ **71.** $\log(x-15) + \log x = 2$

72. $\log(3x-5) + \log(x-1) = 1$ **73.** $\log(2x-17) = 2 - \log x$

74. $\log(x-3) - 1 = \log(x+1)$ **75.** $\log(x^2 + 2x - 3) = 3 + \log(x+3)$

76. $\log(x^2 - 9) - \log(x-3) = -2$ **77.** $\log(x^2 - x - 12) + 2 = \log(x-4)$

78. $\log(x^2 - 4x - 5) - \log(x+1) = 2$ **79.** $\ln(x^2 + 4x - 5) - \ln(x+5) = -2$

80. $\ln(x+1) + \ln(x-1) = 0$ **81.** $\ln(x^2 - 4) - \ln(x+2) = 3$

82. $\ln(x^2 + 2x - 3) = 1 + \ln(x-1)$ **83.** $\log\sqrt[3]{x^2 + 2x + 20} = \frac{2}{3}$

84. $\log\sqrt{x^2 - 24} = \frac{3}{2}$

Use the Change-of-Base formula to evaluate each of the expressions or solve the equations in Exercises 85 – 100.

85. $\log_3 12$ **86.** $\log_4 36$ **87.** $\log_5 1.68$ **88.** $\log_{11} 39.6$

89. $\log_8 0.271$ **90.** $\log_7 0.849$ **91.** $\log_{15} 739$ **92.** $\log_2 14.2$

93. $\log_{20} 0.0257$ **94.** $\log_9 2.384$ **95.** $2^x = 5$ **96.** $3^{2x} = 10$

97. $5^{x-1} = 30$ **98.** $9^{2x-1} = 100$ **99.** $4^{3-x} = 20$ **100.** $6^{3x-4} = 25$

82. No solution
83. $x = -10, x = 8$
84. $x = \pm 32$
85. 2.2619
86. 2.5850
87. 0.3223
88. 1.5342
89. −0.6279
90. −0.0841
91. 2.4391
92. 3.8278

Writing and Thinking About Mathematics

101. Solve the following equation for x two different ways: $a^{2x-1} = 1$.

102. Rewrite each of the following expressions as products:
a. 5^{x+2} **b.** 3^{x-2}

103. Explain, in your own words, why $7 \cdot 7^x \neq 49^x$. Show each of the expressions $7 \cdot 7^x$ and 49^x as a single exponential expression with base 7.

Hawkes Learning Systems: Introductory & Intermediate Algebra

Exponential and Logarithmic Equations

93. −1.2222
94. 0.3954
95. 2.3219
96. 1.048
97. 3.1133
98. 1.548
99. 0.839
100. 1.9322
101. Answers will vary.
102. **a.** $5^x \cdot 5^2$ **b.** $3^x \cdot 3^{-2}$
103. Answers will vary, 7^{1+x}, 7^{2x}

13.7 Applications

Objectives

After completing this section, you will be able to:

Solve applied problems by using logarithms and exponential equations.

In Section 13.3, we found that the number e appears in a surprisingly natural way in the formula for continuously compounding interest

$$A = Pe^{rt}$$

which was developed from the formula for compounding n times per year:

$$A = P\left(1+\frac{r}{n}\right)^{nt}$$

There are many formulas that involve exponential functions. A few are shown here and in the exercises.

$A = A_0e^{-0.04t}$ This is a law for decomposition of radium where t is in centuries.

$A = A_0e^{-0.1t}$ This is one law for skin healing where t is measured in days.

$A = A_o2^{-t/5600}$ This law is used for carbon-14 dating to determine the age of fossils where t is measured in years.

$T = Ae^{-kt} + C$ This is Newton's law of cooling where C is the constant temperature of the surrounding medium. The values of A and k depend on the particular object that is cooling.

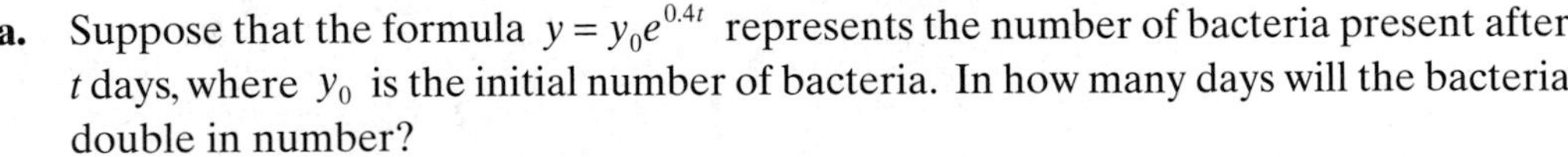

Example 1: Exponential Growth and Exponential Decay

a. Suppose that the formula $y = y_0e^{0.4t}$ represents the number of bacteria present after t days, where y_0 is the initial number of bacteria. In how many days will the bacteria double in number?

Solution: $y = y_0e^{0.4t}$

$2y_0 = y_0e^{0.4t}$ $2y_0$ is double the initial number present.

$2 = e^{0.4t}$ Divide both sides by y_0.

$\ln 2 = 0.4t$

$$t = \frac{\ln 2}{0.4} \approx \frac{0.6931}{0.4}$$

$$t = 1.73$$

The number of bacteria will double in approximately 1.73 days. Note that this number is completely independent of the number of bacteria initially present. That is, if $y_0 = 10$ or $y_0 = 1000$ the doubling time is the same, namely 1.73 days.

b. Suppose that the room temperature is 70°, and the temperature of a cup of tea is 150° when it is placed on the table. In 5 minutes, the tea cools to 120°. How long will it take for the tea to cool to 100° ?

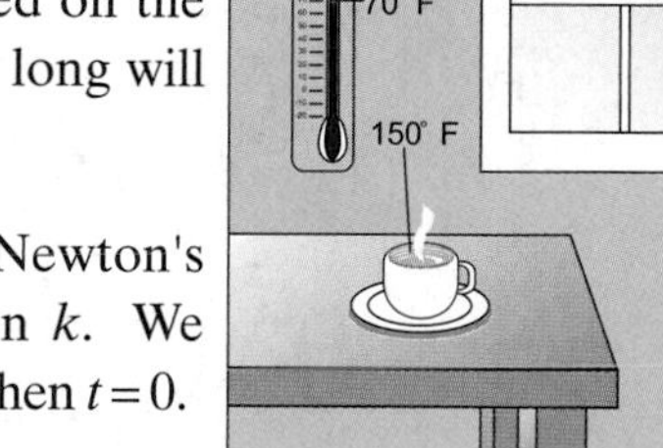

Solution: Using the formula $T = Ae^{-kt} + C$ (Newton's law of cooling), first find A and then k. We know that $C = 70°$ and that $T = 150°$ when $t = 0$. Find A by substituting these values:

$$150 = Ae^{-k(0)} + 70$$

$$150 = A \cdot 1 + 70 \qquad e^{-k(0)} = e^0 = 1$$

$$80 = A$$

Therefore, the formula can be written as $T = 80e^{-kt} + 70$.
Since $T = 120°$ when $t = 5$, substituting these values allows us to find k:

$$120 = 80e^{-k(5)} + 70$$

$$50 = 80e^{-5k}$$

$$\frac{50}{80} = e^{-5k}$$

$$\ln\frac{5}{8} = \ln e^{-5k} \qquad \text{Take the natural log of both sides.}$$

$$\ln 0.625 = -5k$$

$$k = \frac{\ln 0.625}{-5} \approx \frac{-0.4700}{-5} = 0.0940$$

The formula can now be written as $T = 80e^{-0.0940t} + 70$.
With all the constants in the formula known, we can find t when $T = 100°$.

$$100 = 80e^{-0.0940t} + 70$$

$$30 = 80e^{-0.0940t}$$

$$\frac{30}{80} = e^{-0.0940t}$$

$$\ln e^{-0.0940t} = \ln\frac{3}{8} \qquad \text{Take the natural log of both sides.}$$

$$-0.0940t = \ln 0.375$$

continued on next page ...

$$t = \frac{\ln 0.375}{-0.0940}$$
$$= \frac{-0.9808}{-0.0940} = 10.43 \text{ minutes}$$

The tea will cool to 100° in about 10.43 minutes.

c. If \$1000 is invested at a rate of 6% compounded continuously, in how many years will it grow to \$5000?

Solution:
$$A = Pe^{rt}$$
$$5000 = 1000e^{0.06t}$$
$$5 = e^{0.06t}$$
$$\ln 5 = 0.06t$$
$$t = \frac{\ln 5}{0.06} \approx \frac{1.6094}{0.06}$$
$$t \approx 26.82$$

\$1000 will grow to \$5000 in approximately 26.82 years.

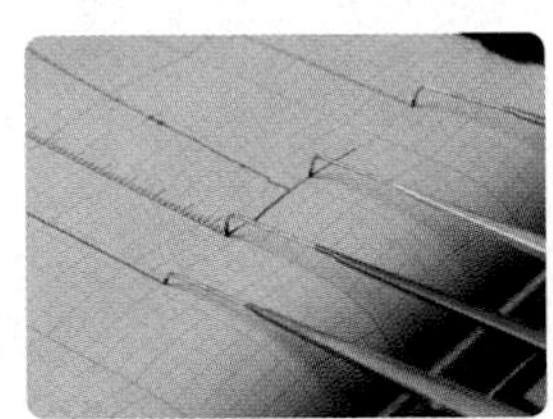

d. The magnitude of an earthquake is measured on the **Richter scale** as a logarithm of the intensity of the shock wave. For magnitude R and intensity I, the formula is $R = \log I$. The 1994 earthquake in Northridge, California measured 6.7 on the Richter scale. What was the intensity of this earthquake?

Solution: Substitute 6.7 for R in the formula and solve for I:
$$6.7 = \log I$$
$$I = 10^{6.7}$$

e. The Long Beach earthquake in 1933 measured 6.2 on the Richter scale. How much stronger was the Northridge earthquake than the 1933 Long Beach earthquake?

Solution: The comparative sizes of the quakes can be found by finding the ratio of the intensities. For the Long Beach quake, $6.2 = \log I$ and $I = 10^{6.2}$. Therefore, the ratio of the two intensities is

$$\frac{I \text{ for Northridge}}{I \text{ for Long Beach}} = \frac{10^{6.7}}{10^{6.2}} = 10^{0.5} \approx 3.2$$

Thus, the Northridge earthquake had an intensity about 3.2 times the Long Beach earthquake.

f. The **half-life** of a substance is the time needed for the substance to decay to one-half of its original amount. The half-life of radioactive radium is 1620 years. If 10 grams are present today, how many grams will remain in 500 years?

Solution: The model for radioactive decay is $y = y_0e^{-kt}$. Since the half-life is 1620 years, if we assume $y_0 = 10$ g, then y would be 5 g after 1620 years. We solve for k as follows:

$$5 = 10e^{-k(1620)} \qquad \text{Substitute } y = 5, y_0 = 10, \text{and } t = 1620.$$

$$\frac{5}{10} = e^{-1620k} \qquad \text{Solve for } k.$$

$$-1620k = \ln(0.5) \qquad \text{Take the natural log of both sides.}$$

$$k = \frac{\ln(0.5)}{-1620} \approx \frac{-0.6931}{-1620} = 0.0004279$$

The model is $y = 10e^{-0.0004279t}$

Substituting $t = 500$ gives

$$y = 10e^{(-0.0004279)(500)} = 10e^{-0.21395} = 10(0.8074) = 8.074$$

Thus, there will still be about 8.1 g of the radioactive radium remaining after 500 years.

13.7 Exercises

1. $4027.51
2. $4726.34
3. 13.9 years
4. 11.6 years
5. $f = \frac{3}{10}$
6. 23.1 hours

1. If $2000 is invested at the rate 7% compounded continuously, what will be the balance after 10 years?

2. Find the amount of money that will be accumulated in a savings account if $3200 is invested at 6.5% for 6 years and the interest is compounded continuously.

3. How long does it take $1000 to double if it is invested at 5% compounded continuously?

4. Four thousand dollars is invested at 6% compounded continuously. How long will it take for the balance to be $8000?

5. The reliability of a certain type of flashlight battery is given by $f = e^{-0.03x}$, where f is the fractional part of the batteries produced that last x hours. What fraction of the batteries produced are good after 40 hours of use?

6. From Exercise 5, how long will at least one-half of the batteries last?

7. 1.73 hours
8. 0.05 ml
9. 8166 bees
10. 7.4 hours
11. 12.28 lbs. per sq. in.
12. 11 days
13. 2350 years
14. 15.4 hours
15. 2.3 days

16. 39.7 minutes
17. 8.75 years
18. 7 years
19. 9 years
20. 12 years
21. a. 13.86 years
b. 6.93 years
22. a. 27.5 years
b. 13.7 years

7. The concentration of a drug in the blood stream is given by $C = C_0e^{-0.8t}$, where C_0 is the initial dosage and t is the time in hours elapsed after administering the dose. If 20 mg of a drug is given, how much time elapses until 5 mg of the drug remains?

8. Using the formula in Exercise 7, determine the amount of insulin present after 3 hours if 0.60 ml are given.

9. A swarm of bees grows according to the formula $P = P_0e^{0.35t}$, where P_0 is the number present initially and t is the time in days. How many bees will be present in 6 days if there were 1000 present initially?

10. If inversion of raw sugar is given by $A = A_0e^{-0.03t}$, where A_0 is the initial amount and t is the time in hours, how long will it take for 1000 lbs. of raw sugar to be reduced to 800 lbs.?

11. Atmospheric pressure P is related to the altitude h by the formula $P = P_0e^{-0.00004h}$, where P_0, the pressure at sea level, is approximately 15 lb. per sq. in. Determine the pressure at 5000 ft.

12. One law for skin healing is $A = A_0e^{-0.1t}$, where A is the number of sq. cm of unhealed area after t days and A_0 is the number of sq. cm of the original wound. Find the number of days needed to reduce the wound to one-third the original size.

13. A radioactive substance decays according to $A = A_0e^{-0.0002t}$, where A_0 is the initial amount and t is the time in years. If $A_0 = 640$ grams, find the time for A to decay to 400 grams.

14. A substance decays according to $A = A_0e^{-0.045t}$, where t is in hours and A_0 is the initial amount. Determine the half-life of the substance.

15. An employee is learning to assemble remote-control units. The number of units per day he can assemble after t days of training is given by $N = 80\left(1 - e^{-0.3t}\right)$. How many days of training will be needed before the employee is able to assemble 40 units per day?

16. The temperature of a carrot cake is 350° when it is removed from the oven. The temperature in the room is 72°. In 10 minutes, the cake cools to 280°. How long will it take for the cake to cool to 160°?

17. How long does it take $10,000 to double if it is invested at 8% compounded quarterly?

18. If $1000 is deposited at 6% compounded monthly, how long will it take before the balance is $1520?

19. The value of a machine, V, at the end of t years is given by $V = C(1-r)^t$, where C is the original cost of the machine and r is the rate of depreciation. A machine that originally cost $12,000 in now valued at $3800. How old is the machine if $r = 0.12$?

20. Using the formula in Exercise 19, determine the age of a machine valued at $5800 if its original value was $18,000 and $r = 0.09$.

21. If a principal P is doubled, then $A = 2P$. Use the formula for continuous compounding to find the time that a principal will double in value if the rate of interest is

a. 5% **b.** 10%

(Note that the time for doubling the principal is completely independent of the principal itself.)

23. 294.41 days
24. 5600 years
25. 100
26. 6.3
27. 8.64 million
28. 1083, 1272, 1562
29. 2083, 2437, 2744
30. **a.** 6.07
b. 3.2×10^{-5}

22. If a principal P is tripled, then $A=3P$. Use the formula for continuous compounding to find the time that a principal will triple in value if the interest rate is

a. 4% **b.** 8%

(Note that the time for tripling the principal is completely independent of the principal itself.)

23. Radioactive iodine has a half-life of 60 days. If an accident occurs at a nuclear plant and 30 grams of radioactive iodine are present, in how many days will 1 gram be present?

24. The formula $A=A_0 2^{-t/5600}$ is used for carbon-14 dating to determine the age of fossils where t is measured in years. Determine the half-life of carbon-14.

25. The 1906 earthquake in San Francisco measured 8.6 on the Richter scale. In 1971, an earthquake in the San Fernando Valley measured 6.6 on the Richter scale. How many times greater was the 1906 earthquake than the 1971 earthquake?

26. In 1985, an earthquake in Mexico measured 8.1 on the Richter scale. How many times greater was this earthquake than the one in Landers, California in 1992 that measured 7.3 on the Richter scale?

27. Population does not generally grow in a linear fashion. In fact, population of many species grows exponentially, at least for a limited time. Using the exponential model $y=y_0e^{kt}$ for population growth, estimate the population of a state in 2020 if the population was 5 million in 1990 and 6 million in 2000. (Assume that t is measured in years and $t=0$ corresponds to 1990.)

28. Suppose that a lake is stocked with 500 fish, and biologists predict that the population of these fish will be approximated by the function $P(t)=500\ln(2t+e)$ where t is measured in years. What will be the fish population in 3 years? in 5 years? in 10 years?

29. Sales representatives of a new type of computer predict that sales can be approximated by the function $S(t)=1000+500\ln(3t+e)$ where t is measured in years. What are the predicted sales in 2 years? in 5 years? in 10 years?

30. In chemistry, the pH of a solution is a measure of the acidity or alkalinity of a solution. Water has a pH of 7 and, in general, acids have a pH less than 7 and alkaline solutions have a pH greater than 7. The model for pH is $\text{pH}=-\log\left[H^+\right]$ where $\left[H^+\right]$ is the hydrogen ion concentration in moles per liter of a solution.

a. Find the pH of a solution with a hydrogen ion concentration of 8.6×10^{-7}.

b. Find the hydrogen ion concentration $\left[H^+\right]$ of a solution if the pH of the solution is 4.5. Write the answer in scientific notation.

Hawkes Learning Systems: Introductory & Intermediate Algebra

Applications: Exponential and Logarithmic Functions

Chapter 13 Index of Key Ideas and Terms

Section 13.1 Algebra of Functions

Algebraic Operations with Functions page 887

If $f(x)$ and $g(x)$ represent two functions and x is a value in the domain of both functions, then

1. Sum of Two Functions: $(f+g)(x)=f(x)+g(x)$

2. Difference of Two Functions: $(f-g)(x)=f(x)-g(x)$

3. Product of Two Functions: $(f\cdot g)(x)=f(x)\cdot g(x)$

4. Quotient of Two Functions: $\left(\frac{f}{g}\right)(x)=\frac{f(x)}{g(x)}$ where $g(x)\neq 0$

Graphing the Sum of Two Functions pages 889 - 891

Using a Graphing Calculator to Graph the Sum of Two Functions pages 891 - 892

Section 13.2 Composition of Functions and Inverse Functions

Composition of Two Functions page 896

Composite Functions page 897

For two functions f and g, the **composite function** $f \circ g$ is defined as follows: $(f\circ g)(x)=f(g(x))$.

Domain of $f \circ g$: The domain of $f \circ g$ consists of those values of x in the domain of g for which $g(x)$ is in the domain of f.

One-to-One Functions page 899

A function is a **one-to-one function** (or **1-1 function**) if for each value of y in the range there is only one corresponding value of x in the domain.

Horizontal Line Test page 899

A function is one-to-one if no horizontal line intersects the graph of the function in more than one point.

continued on next page ...

Section 13.2 Composition of Functions and Inverse Functions (continued)

Inverse Functions page 901

If f is a 1-1 function with ordered pairs of the form (x, y), then its **inverse function**, denoted as f^{-1}, is also a 1-1 function with ordered pairs of the form (y, x).

Inverse Functions page 903

If f and g are one-to-one functions and

$f(g(x)) = x$ for all x in D_g

$g(f(x)) = x$ for all x in D_f

then f and g are **inverse functions**.
That is, $g = f^{-1}$ and $f = g^{-1}$.

To Find the Inverse of a 1-1 Function page 905

1. Let $y = f(x)$. (In effect, substitute y for $f(x)$.)
2. Interchange x and y.
3. In the new equation, solve for y in terms of x.
4. Substitute $f^{-1}(x)$ for y. (This new function is the inverse of f.)

Section 13.3 Exponential Functions

Exponential Functions page 911

An exponential function is a function of the form $f(x) = b^x$ where $b > 0, b \neq 1$, and x is any real number.

General Concepts of Exponential Functions page 914

For $b > 1$:

1. $b^x > 0$.
2. b^x increases to the right and is called an **exponential growth function**.
3. $b^0 = 1$, so $(0, 1)$ is on the graph.
4. b^x approaches the x-axis for negative values of x. (The x-axis is a horizontal asymptote. See Figure 13.12.)

For $0 < b < 1$:

1. $b^x > 0$.
2. b^x decreases to the right and is called an **exponential decay function**.
3. $b^0 = 1$, so $(0, 1)$ is on the graph.
4. b^x approaches the x-axis for positive values of x. (The x-axis is a horizontal asymptote. See Figure 13.14.)

Compound Interest page 916

Formula for compound interest: $A = P\left(1 + \frac{r}{n}\right)^{nt}$

continued on next page ...

Section 13.3 Exponential Functions (continued)

Interest Compounded Continuously page 919

Formula for compounding continuously: $A = Pe^{rt}$

The Number *e* page 918

The number e is defined to be $e = \lim_{n\to\infty}\left(1+\frac{1}{n}\right)^n = 2.718281828459\ldots$

n	$\left(1+\frac{1}{n}\right)$	$\left(1+\frac{1}{n}\right)^n$
1	$\left(1+\frac{1}{1}\right)=2$	$(2)^1=2$
2	$\left(1+\frac{1}{2}\right)=1.5$	$(1.5)^2=2.25$
5	$\left(1+\frac{1}{5}\right)=1.2$	$(1.2)^5=2.48832$
10	$\left(1+\frac{1}{10}\right)=1.1$	$(1.1)^{10}=2.59374246$
100	$\left(1+\frac{1}{100}\right)=1.01$	$(1.01)^{100}=2.704813829$
1000	$\left(1+\frac{1}{1000}\right)=1.001$	$(1.001)^{1000}=2.716923932$
10,000	$\left(1+\frac{1}{10,000}\right)=1.0001$	$(1.0001)^{10,000}=2.718145927$
100,000	$\left(1+\frac{1}{100,000}\right)=1.00001$	$(1.00001)^{100,000}=2.718268237$
$\downarrow$		$\downarrow$
∞		$e=2.718281828459\ldots$

Section 13.4 Logarithmic Functions

Logarithm page 923

The inverse of an exponential function is a logarithmic function.

Logarithm page 924

For $b>0$, and $b\neq 1$, $x=b^y$ if and only if $y=\log_b x$.

continued on next page ...

Section 13.4 Logarithmic Functions (continued)

Three Basic Properties of Logarithms page 927

For $b > 0$, $b \neq 1$, and $x > 0$

1. $\log_b 1 = 0$
2. $\log_b b = 1$
3. $b^{\log_b x} = x$

Graphs of Logarithmic Functions pages 927 - 928

Common Logarithms (Base-10 Logarithms) page 928

Using a calculator to find common logarithms page 929

Inverse Logarithms page 929

Using a calculator to find inverse logarithms page 930

Natural Logarithms (Base-*e* Logarithms) page 930

Using a calculator to find natural logarithms page 931

Using a calculator to find inverse natural logarithms page 932

Section 13.5 Properties of Logarithms

Properties of Logarithms page 940

For $b > 0, b \neq 1, x,y > 0$ and any real number r:

1. $\log_b 1 = 0$
2. $\log_b b = 1$
3. $x = b^{\log_b x}$
4. Product Rule:
   $\log_b xy = \log_b x + \log_b y$
5. Quotient rule:
   $\log_b \dfrac{x}{y} = \log_b x - \log_b y$
6. $\log_b x^r = r \cdot \log_b x$

For natural logarithms:

1. $\ln 1 = 0$
2. $\ln e = 1$
3. $x = e^{\ln x}$
4. Product Rule:
   $\ln xy = \ln x + \ln y$
5. Quotient rule:
   $\ln \dfrac{x}{y} = \ln x - \ln y$
6. $\ln x^r = r \cdot \ln x$

Section 13.6 Logarithmic and Exponential Equations

For $b > 0, b \neq 1$,

1. If $b^x = b^y$, then $x = y$.
2. If $x = y$, then $b^x = b^y$.
3. If $\log_b x = \log_b y$, then $x = y$ ($x > 0$ and $y > 0$).
4. If $x = y$, then $\log_b x = \log_b y$ ($x > 0$ and $y > 0$).

Change-of-Base formula: $\log_b x = \dfrac{\log_a x}{\log_a b}$

Section 13.7 Applications

Exponential growth
Exponential decay
Newton's law of cooling
Interest
Earthquakes
Half-life

Chapter 13 Review

For a review of the topics and problems from Chapter 13, look at the following lessons from *Hawkes Learning Systems: Introductory & Intermediate Algebra.*

Algebra of Functions
Composition of Functions and Inverse Functions
Exponential Functions and the Number *e*
Logarithmic Functions
Properties of Logarithmic Functions
Exponential and Logarithmic Equations
Applications: Exponential and Logarithmic Functions

Chapter 13 Test

1. a. $\sqrt{x-3}+x^2+1$

b. $\sqrt{x-3}-x^2-1$

c. $\left(\sqrt{x-3}\right)\cdot\left(x^2+1\right)$

d. $\dfrac{\sqrt{x-3}}{x^2+1}$ **e.** a. $x \geq 3$

b. $x \geq 3$ c. $x \geq 3$

d. $x \geq 3$

2. a. $-4x^2+1$

b. $-8x^2+40x-47$

3. a. Not inverse to each other

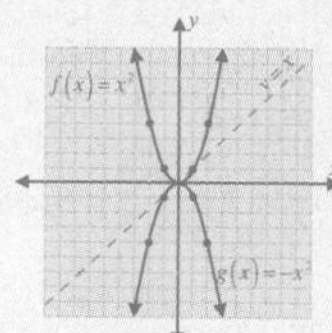

b. Inverse to each other

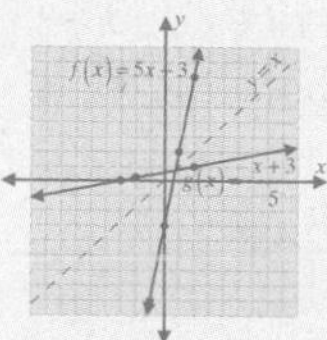

c. Inverse to each other

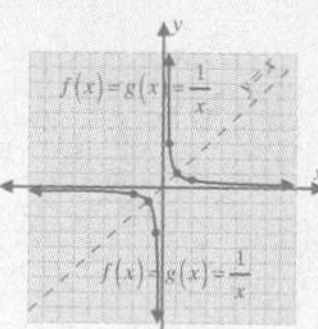

4. $f^{-1}(x)=\dfrac{1}{x}+2$

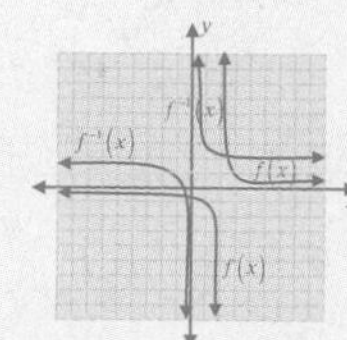

You may use a graphing calculator as an aid for any of the problems on this test. If you are to graph a function, sketch the graph on the test paper and label any points as requested.

1. Given the two functions,

$f(x)=\sqrt{x-3}$ and $g(x)=x^2+1$,

find

a. $(f+g)(x)$ **b.** $(f-g)(x)$

c. $(f\cdot g)(x)$ **d.** $\left(\dfrac{f}{g}\right)(x)$

e. State the domain of each function in parts (a) – (d).

2. If $f(x)=2x-5$ and $g(x)=3-2x^2$, find

a. $f[g(x)]$ **b.** $g[f(x)]$

3. Determine, algebraically, whether or not each pair of functions are inverses of each other. Graph each pair of functions on the same set of axes and show the line $y=x$ as a dotted line on each graph.

a. $f(x)=x^2$ and $g(x)=-x^2$

b. $f(x)=5x-3$ and $g(x)=\dfrac{x+3}{5}$

c. $f(x)=\dfrac{1}{x}$ and $g(x)=\dfrac{1}{x}$

4. Find $f^{-1}(x)$ if $f(x)=\dfrac{1}{x-2}$. Graph both functions.

5. Sketch the graph of $y=4^x$ and label three points on the graph.

6. Solve each of the following equations for x.

a. $7^3\cdot 7^x=7^{-1}$

b. $6^{x-1}=36^{x+1}$

7. A scientist knows that a certain strain of bacteria grows according to the function $y=y_0\cdot 3^{0.25t}$, where t is a measurement in hours. If she starts a culture with 5000 bacteria, how many will be present after 6 hours?

8. Write the following equations in logarithmic form:

a. $10^5=100{,}000$ **b.** $\left(\dfrac{1}{2}\right)^{-3}=8.$

9. Write the following equations in exponential form:

a. $\ln x=4$ **b.** $\log_3\dfrac{1}{9}=-2.$

10. Solve the following equations by first changing them to exponential form.

a. $\log_7 x=3.$ **b.** $\log_9 27=x.$

11. Find the inverse of the function

$y=\left(\dfrac{1}{2}\right)^x$. Graph both the function and its inverse on the same set of axes. Label three points on each graph and show the line $y=x$ as a dotted line on the graph.

12. Use a calculator to find the value of x.

a. $x=\log 579$ **b.** $5\ln x=9.35$

13. Write each expression as the sum and/or difference of logarithms.

a. $\ln\left(x^2-25\right)$ **b.** $\log\sqrt[3]{\dfrac{x^2}{y}}$

14. Write each expression in the form of a single logarithm.

a. $\ln(x+5)+\ln(x-4)$

b. $\log\sqrt{x}+\log x^2-\log 5x$

5.
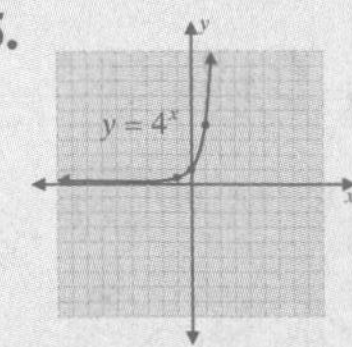

6. a. $x=-4$ **b.** $x=-3$
7. 25,981
8. a. $\log_{10} 100{,}000=5$
b. $\log_{\frac{1}{2}} 8=-3$
9. a. $e^4=x$ **b.** $\frac{1}{9}=3^{-2}$
10. a. $x=343$ **b.** $x=\frac{3}{2}$
11.
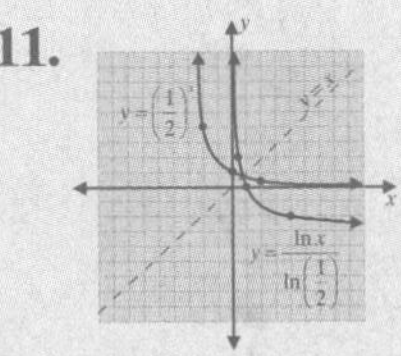

12. a. 2.763 **b.** 6.488
13. a. $\ln(x+5)+\ln(x-5)$
b. $\frac{2}{3}\log x-\frac{1}{3}\log y$
14. a. $\ln(x^2+x-20)$
b. $\log\left(\frac{x^2\sqrt{x}}{5}\right)$
15. $x=-2+\log 283\approx 0.4518$

Solve the equations in Exercises 15 – 19.

15. $10^{x+2}=283$ **16.** $2e^{0.24x}=26$ **17.** $4^x=12$

18. $\log(2x+3)-\log(x+1)=0$ **19.** $\ln(x^2+3x-4)-\ln(x+4)=3$

20. If \$1000 is invested at 7% compounded continuously, when will the amount be \$3800?

21. A substance decomposes according to $A=A_0e^{-0.0035t}$, where t is measured in years and A_0 is the initial amount.

a. How long will it take for 800 grams to decompose to 500 grams?
b. What is the half-life of this substance?

22. The temperature of a cheese cake is 450° when it is removed from the oven. The room temperature is 70°. In 20 minutes, the cheese cake cools to 260°. How long does it take for the cheese cake to cool to 100°?

23. A substance decays according to $A=10\left(\frac{1}{2}\right)^{\frac{t}{1620}}$, where t is the number of years and A is the initial amount. What is the half-life of the substance?

24. Inflation is the continual rise in the general price level. If the annual inflation rate averages 4.2% over the next 8 years, the cost, C, of goods and services during the range of those 8 years is given by the function $C(t)=P(1.042)^t$ where t is the time in years and P is the current price. If the price to fill up your car tank is \$25.60, estimate what the price will be in 8 years.

16. $x=\frac{\ln 13}{.24}\approx 10.69$ **17.** $x=\frac{\ln 12}{\ln 4}\approx 1.79$ **18.** No solution **19.** $x=1+e^3\approx 21.09$ **20.** 19.07 years
21. a. 134.3 years **b.** 198 years **22.** about 73.26 minutes **23.** half-life = 5730 years **24.** \$35.58

Cumulative Review: Chapters 1 – 13

1. 1 **2.** $-\frac{x+3}{x+1}$
3. $\frac{2}{x+4}$ **4.** $\frac{x-4}{x+1}$
5. $\frac{x^{\frac{1}{6}}}{y^{\frac{1}{3}}}$ **6.** $\frac{x^{\frac{1}{2}}}{2y^2}$
7. $x^2y^{\frac{3}{2}}$

Perform the indicated operations in Exercises 1 – 4 and simplify the results.

1. $\frac{2x^2+7x+3}{x^2-3x-18}\cdot\frac{x^2-x-30}{2x^2+11x+5}$ **2.** $\frac{9-x^2}{x^2+7x+6}\div\frac{x-3}{x+6}$

3. $\frac{1}{x-1}+\frac{x-6}{x^2+3x-4}$ **4.** $\frac{2x}{2x+3}-\frac{7x+12}{2x^2+5x+3}$

Simplify Exercises 5 and 6. Assume all variables are positive.

5. $\frac{x^{\frac{2}{3}}y^{\frac{1}{3}}}{x^{\frac{1}{2}}y^{\frac{2}{3}}}$ **6.** $(x^2y^{-1})^{\frac{1}{2}}(4xy^3)^{-\frac{1}{2}}$

8. $x^{\frac{2}{3}}y$

Change each expression in Exercises 7 and 8 to an equivalent exponential expression. Assume variables are positive.

9. Dependent, $x-5y=10$

7. $\sqrt{x^4y^3}$ **8.** $\sqrt[3]{x^2y^3}$

10. $x=2,\ y=5$

11. Vertex: $(3,-11)$
Range: $y\geq -11$
Zeros: $x=3\pm\sqrt{11}$

Use any algebraic method to solve the systems of linear equations in Exercises 9 and 10.

9. $\begin{cases} x-5y=10 \\ y=0.2x-2 \end{cases}$ **10.** $\begin{cases} 2x-y=-1 \\ x+2y=12 \end{cases}$

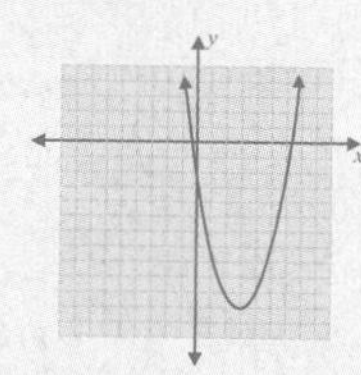

In Exercises 11 and 12, find the vertex, range, and zeros for the quadratic functions. Graph each function.

12. Vertex: $(-2,-5)$
Range: $(y\geq -5)$
Zeros: $x=-2\pm\frac{\sqrt{10}}{2}$

11. $y=x^2-6x-2$ **12.** $y=2x^2+8x+3$

Solve each of the equations in Exercises 13 – 18.

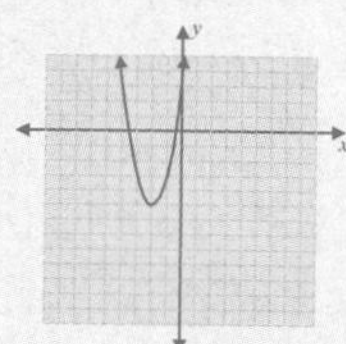

13. $x^2+5x-2=0$ **14.** $x^4-13x^2+36=0$

15. $x-2=\sqrt{x+10}$ **16.** $\frac{2}{x-2}+\frac{3}{x-1}=1$

13. $x=\frac{-5\pm\sqrt{33}}{2}$

17. $6^{3x+5}=55$ **18.** $\ln(x^2-7x+10)-\ln(x-2)=2.5$

14. $x=\pm2,\ x=\pm3$
15. $x=6$
16. $x=4\pm\sqrt{6}$
17. $x\approx-0.9212$

Find equations for each of the circles indicated in Exercises 19 and 20.

19. Center$(0,0)$; $r=2\sqrt{3}$ **20.** Center$(1,2)$; $r=3$

18. $x=5+e^{2.5}\approx17.18$

Solve each of the systems of equations in Exercises 21 – 23. Graph both curves and label the points of intersection.

19. $x^2+y^2=12$
20. $(x-1)^2+(y-2)^2=9$

21. $\begin{cases} x^2+y^2=10 \\ 2x+y=1 \end{cases}$ **22.** $\begin{cases} 4x^2+y^2=13 \\ x+y=2 \end{cases}$ **23.** $\begin{cases} 4x^2+y^2=16 \\ y=x^2-4 \end{cases}$

21.

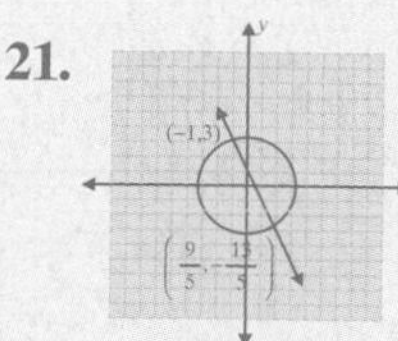

22.

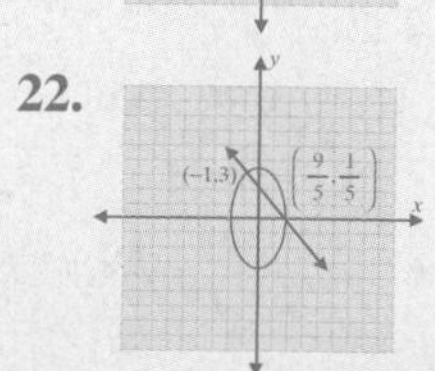

23. 

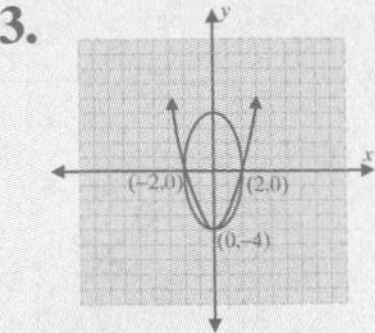

24. Is a function

25. Is a function

26. Is not a function

27. Is a function

28. $\{(-2, 13), (-1, 8), (0, 5), (1, 4), (2, 5)\}$

29. $\{(-2, -28), (-1, -6), (0, 0), (1, -4), (2, -12)\}$

Use the vertical line test to determine whether or not each of the graphs in Exercises 24 – 27 represents a function.

24. **25.**

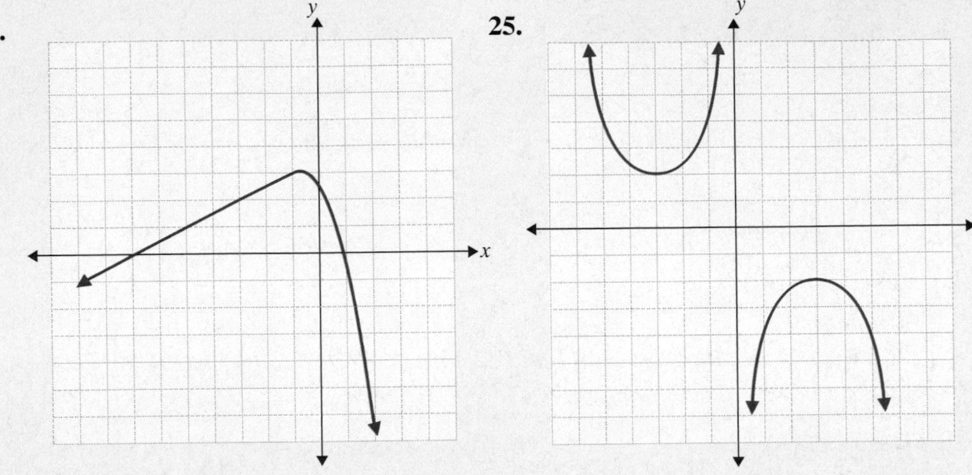

26.

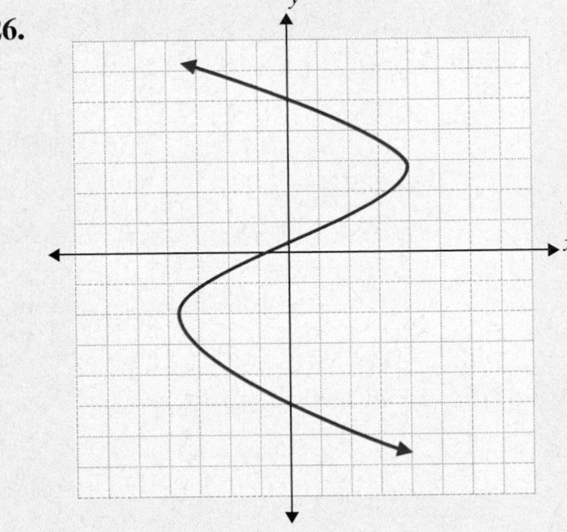

27.

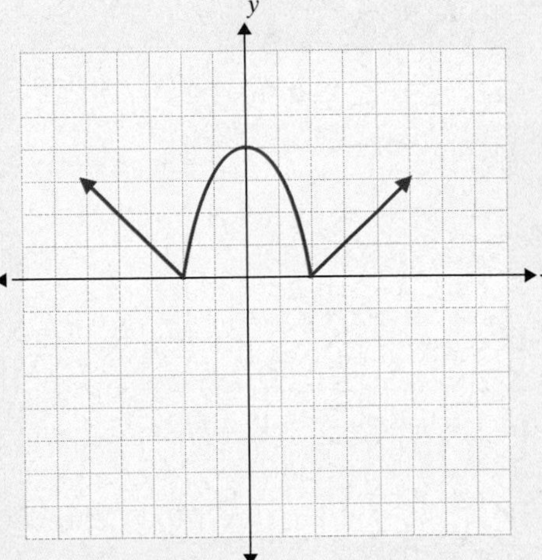

In Exercises 28 and 29, write the function as a set of ordered pairs for the given equation and domain.

28. $y = x^2 - 2x + 5$

$D = \{-2, -1, 0, 1, 2\}$

29. $y = x^3 - 5x^2$

$D = \{-2, -1, 0, 1, 2\}$

30.

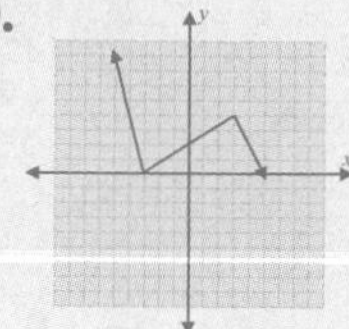

31.

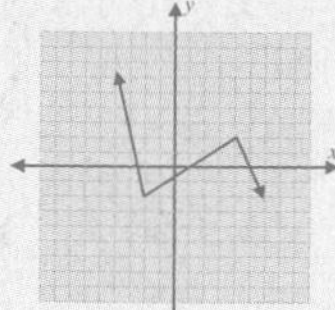

32.

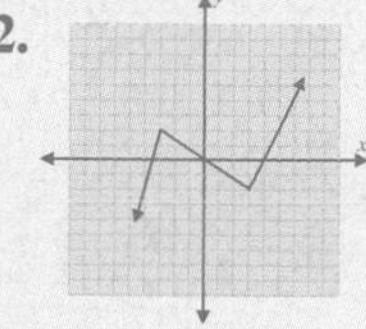

33. Is not a 1-1 function

34. Is not a 1-1 function

35. Is a 1-1 function

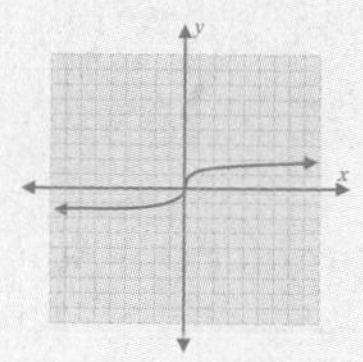

36. Is a 1-1 function

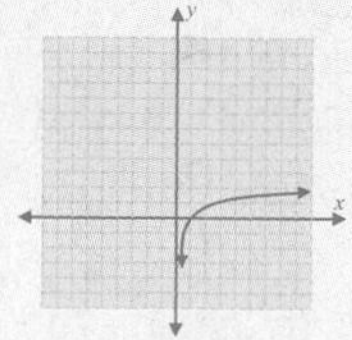

Using the following graph of $y = f(x)$, *graph the functions in Exercises 30 – 32.*

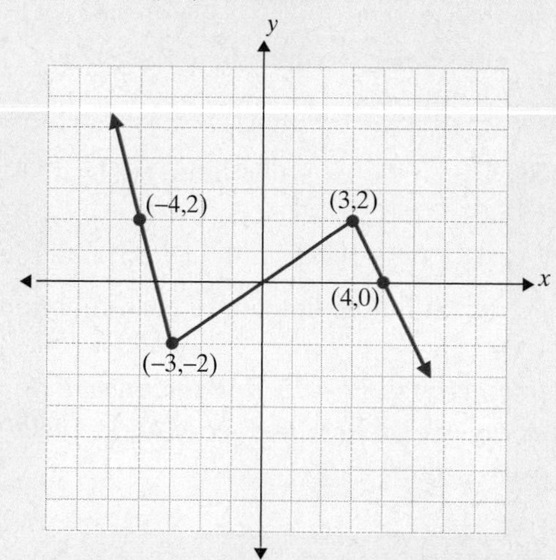

30. $y = f(x) + 2$ **31.** $y = f(x-1)$ **32.** $y = -f(x)$

Which of the functions in Exercises 33 – 36 are 1-1 functions? If a function is 1-1, draw the graph of its inverse.

33.

34.

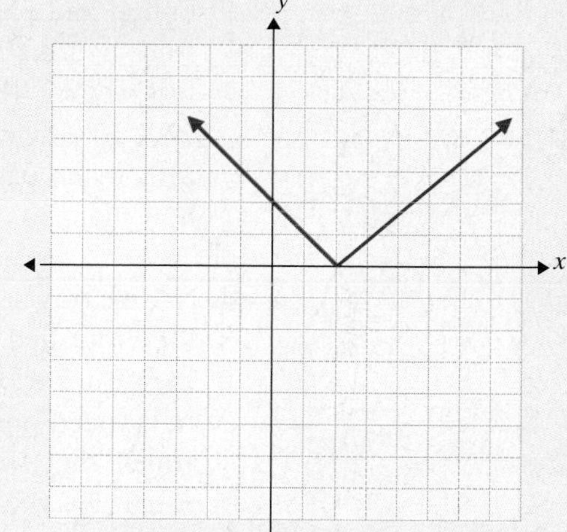

35.

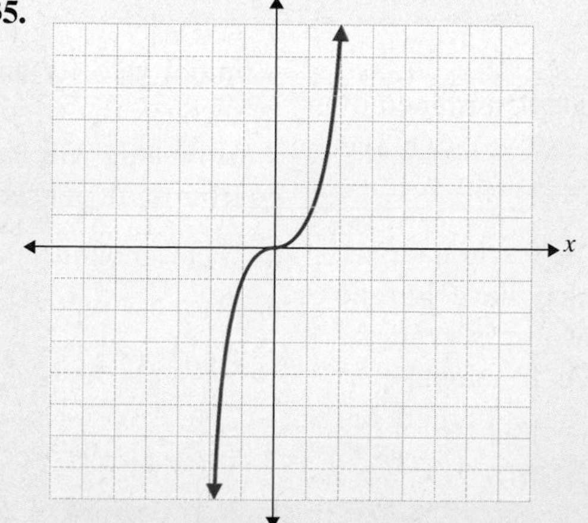

36.

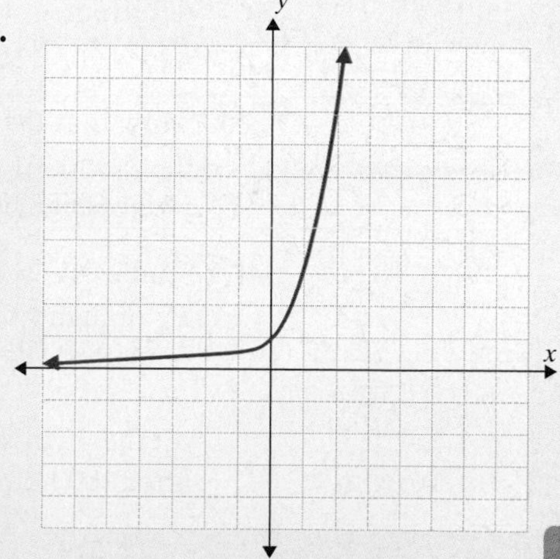

37. a. $x^2+x+\sqrt{x+2}$

b. $x^2+x-\sqrt{x+2}$

c. $\left(x^2+x\right)\sqrt{x+2}$

d. $\dfrac{x^2+x}{\sqrt{x+2}}$

e. a. $x \geq -2$

b. $x \geq -2$ c. $x \geq -2$

d. $x > -2$

38. a. x^2+2x+4

b. x^2-2x+2

c. $2x^3+x^2+6x+3$

d. $\dfrac{x^2+3}{2x+1}$

e. a. All real numbers

b. All real numbers

c. All real numbers

d. All real numbers except $x=-\dfrac{1}{2}$

39.

40.

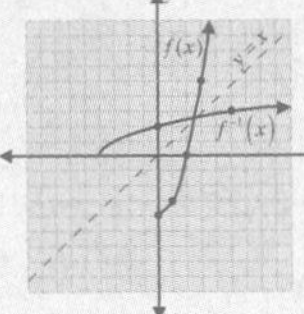

41.

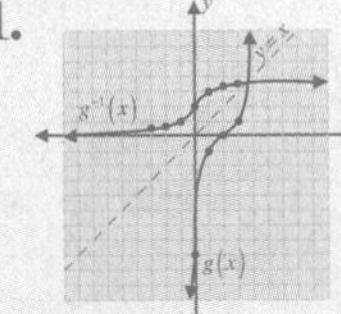

42.

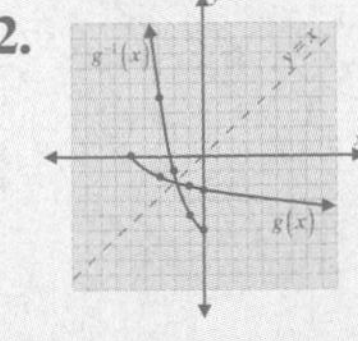

37. Given the two functions, $f(x)=x^2+x$ and $g(x)=\sqrt{x+2}$, find

a. $(f+g)(x)$ **b.** $(f-g)(x)$ **c.** $(f\cdot g)(x)$ **d.** $\left(\dfrac{f}{g}\right)(x)$

e. State the domain of each function in parts (a) – (d).

38. Given the two functions, $f(x)=x^2+3$ for $x \geq 0$ and $g(x)=2x+1$ for $x \geq 0$, find

a. $(f+g)(x)$ **b.** $(f-g)(x)$ **c.** $(f\cdot g)(x)$ **d.** $\left(\dfrac{f}{g}\right)(x)$

e. State the domain of each function in parts (a) – (d).

In Exercises 39 – 42, find the inverse of each function and graph both the function and its inverse on the same set of axes. Include the graph of the line y = x as a dotted line in each graph.

39. $f(x)=3x+2$

40. $f(x)=x^2-4$ for $x \geq 0$

41. $g(x)=(x-2)^3$

42. $g(x)=-\sqrt{x+5}$

43. A book is available in both clothbound and paperback. A bookstore sold a total of 43 books during the week. The total receipts were \$297.50. If clothbound books sell for \$12.50 and paperbacks sell for \$4.50, how many of each were sold?

44. Nadine traveled 8 miles upstream and then returned. Her average speed on the return trip was 6 mph faster than her speed upstream. If her total travel time was 2.8 hours, find her rate each way.

45. Find the perimeter of the triangle determined by the points $A(1, 3)$, $B(-2, 2)$, $C(4, 6)$.

46. Show that the triangle determined by the points $A(-3, 1)$, $B(0, -5)$, $C(1, 3)$ is a right triangle.

47. Studies show that the fractional part, P, of light bulbs that have burned out after t hours of use is given by $P=1-2^{-0.03t}$. What fractional part of the light bulbs have burned out after 100 hours of use?

48. If P dollars are invested at a rate, r (expressed as a decimal), and compounded k times a year, the amount, A, due at the end of t years is given by $A=P\left(1+\dfrac{r}{k}\right)^{kt}$ dollars. Find A if \$500 is invested at 10% compounded quarterly for 2 years.

49. Radium decomposes according to $A=A_0e^{-0.04t}$, where t is measured in centuries and A_0 is the initial amount. Determine the half-life for radium.

50. When friction is used to stop the motion of a wheel, the velocity may be given by $V=V_0e^{-0.35t}$, where is V_0 the initial velocity and t is the number of seconds the friction has been applied. How long will it take to slow a wheel from 75 ft. per sec. to 15 ft. per sec.?

43. 13 clothbound, 30 paperback

44. 10 mph, 4 mph **45.** 14.616

46. $\left|\overline{AB}\right|^2+\left|\overline{AC}\right|^2=\left|\overline{BC}\right|^2=65$ **47.** $\dfrac{7}{8}$ **48.** \$609.20

49. 17.33 centuries **50.** 4.6 sec.

CHAPTER 14

Systems of Linear Equations II

Did You Know?

As we will see in Chapter 14, a rectangular array of numbers is called a **matrix**. This term was first used in 1850 by James Joseph Sylvester. However, the Chinese text Nine Chapters on the Mathematical Art written during the Han Dynasty between 200 BC and 100 BC contains a problem and its solution clearly related to the matrix methods of solving a set of linear equations that we will be discussing in Chapter 14. The problem was the following:

Sylvester

There are three types of corn, of which three bundles of the first, two of the second, and one of the third make 39 measures. Two of the first, three of the second and one of the third make 34 measures. And one of the first, two of the second and three of the third make 26 measures. How many measures of corn are contained of one bundle of each type?

The author, remarkably, sets up a rectangular array of the numerical coefficients of the corresponding system of three linear equations on a "counting board" as follows:

1	2	3
2	3	2
3	1	1
26	34	39

Then the author, remember this is about 200 BC, has the reader multiply the middle column by 3 and subtract the right column *as many times as possible*. Then the right column is to be subtracted *as many times as possible* from 3 times the first column. This results in the following arrangement:

0	0	3
4	5	2
8	1	1
39	24	39

In the next step, the middle column is subtracted *as many times as possible* from 5 times the first column, giving the following result:

0	0	3
0	5	2
36	1	1
99	24	39

From this last arrangement the solution can be found as $\frac{99}{36}$ for the third type of corn, and as we will see in Chapter 14, **back substitution** will give the values for the other two types of corn.

This method for solving a system of linear equations is now known as **Gaussian elimination** [Karl Friedrich Gauss (1777 – 1855)]. Also, as you will see, we now write the coefficients in a horizontal arrangement instead of the vertical arrangement shown in the Chinese version.

ref: http://www.ualr.edu/~lasmoller.matrices.html
http://www-history.mcs.st-and.ac.uk/history/HistTopics/Matrices_and_determinants.html

"Through and through the world is infested with quantity: To talk sense is to talk quantities. It is no use saying the nation is large–How large? It is no use saying that radium is scarce–How scarce? You cannot evade quantity. You may fly to poetry and music, and quantity and number will face you in your rhythms and your octaves."

Alfred North Whitehead (1861 – 1947)

Many applications involve two (or more) quantities and, by using two (or more) variables, a set of two or more equations can be formed by using the given information. If the equations are linear, then the set of equations is called a **system of linear equations**. The purpose of this chapter is to develop more techniques for solving systems of linear equations.

As we have seen in Chapter 9, graphing systems of two equations in two variables is helpful in visualizing the relationships between the equations. However, this approach is somewhat limited in finding solutions since numbers might be quite large, or solutions might involve fractions that must be estimated on the graph. Therefore, algebraic techniques are necessary to accurately solve systems of linear equations. Graphing in three dimensions will be left to later courses.

Two ideas probably new to students at this level are matrices and determinants. Matrices and determinants provide powerful general approaches to solving large systems of equations with many variables. Discussions in this chapter will be restricted to two linear equations in two variables and three linear equations in three variables.

14.1 Review of Systems of Linear Equations: Two Variables

Objectives

After completing this section, you will be able to:

Solve systems of linear equations in two variables using three methods:

a. *graphing,*

b. *substitution, and*

c. *addition.*

Two (or more) linear equations considered at one time are said to form a **system of equations** or a **set of simultaneous equations**. For example, consider the following system of two equations

$$\begin{cases} 2x + y = 5 \\ x - y = 1 \end{cases}.$$

Each equation has an infinite number of solutions. That is, there is an infinite number of ordered pairs that satisfy each equation. But, the question of interest is, "Are there any ordered pairs that satisfy both equations at the same time?" In this example, the answer is yes. The ordered pair (2, 1) satisfies both equations:

$$2(2) + 1 = 5 \quad \text{substituting into the first equation}$$
$$2 - 1 = 1 \quad \text{substituting into the second equation}$$

There are several questions that need to be answered.

How do we find the solution, if there is one?
Will there always be a solution to a system of linear equations?
Can there be more than one solution?

Graphing linear equations provides initial insight to the answers to these questions. However, as the chapter progresses we will see that other algebraic techniques are even more informative.

Table 14.1 illustrates the three possibilities for a system of two linear equations in two variables. The system can be

1. **consistent** (has exactly one solution)
2. **inconsistent** (has no solution)
3. **dependent** (has an infinite number of solutions)

System	Graph	Intersection	Terms
$\begin{cases} 2x+y=5 \\ x-y=1 \end{cases}$	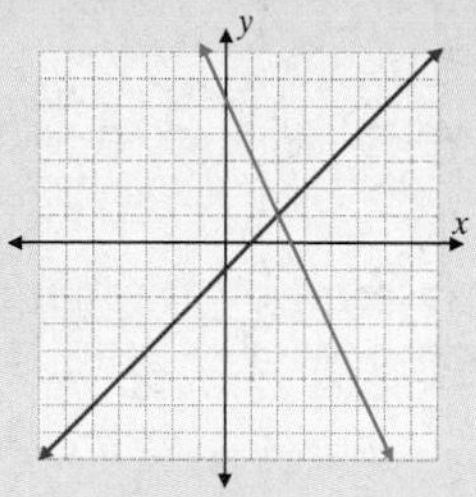	(2,1) or $\begin{cases} x=2 \\ y=1 \end{cases}$ (The lines intersect at one point.)	**Consistent**
$\begin{cases} 3x-2y=2 \\ 6x-4y=-4 \end{cases}$	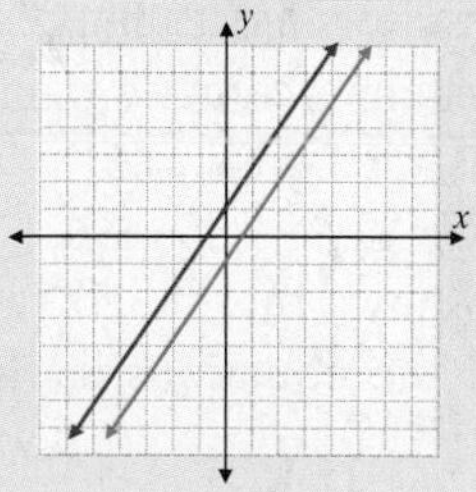	No Solution (The lines are parallel.)	**Inconsistent**
$\begin{cases} 2x-4y=6 \\ x-2y=3 \end{cases}$	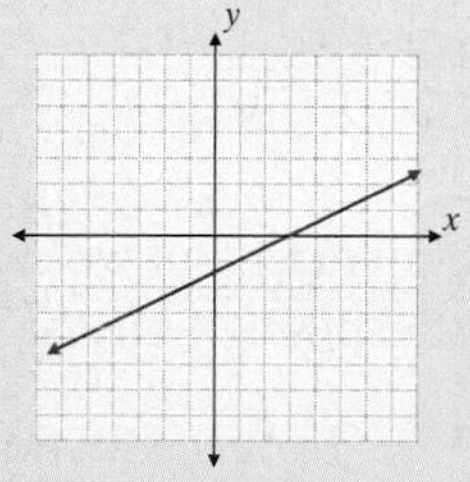	Any ordered pair that satisfies $x-2y=3$ (The lines are the same line. There are an infinite number of solutions. $\{(x,y)\mid x-2y=3\}$.)	**Dependent**

Table 14.1

The three basic methods for solving a system of two linear equations in two variables that will be discussed in this section are:

1. **graphing** (as illustrated in Table 14.1)
2. **substitution** (algebraically substituting from one equation into the other)
3. **addition** (combining like terms from both equations)

The Graphing Method

Teaching Notes:
It should be made clear to students that graphing is useful for estimating and studying possiblities of solutions to systems of equations, especially prior to determining exact solutions through algebraic techniques.

To solve a system of two linear equations in two variables by graphing,

1. graph both lines on the same set of axes, and
2. observe the point of intersection (if there is one).

a. If the slopes of the two lines are different, then the lines will intersect in one and only one point. The system is consistent and has a single point solution.
b. If the lines are distinct and have the same slope, then the lines will be parallel and the system is inconsistent. The system will have no solution.
c. If the lines are the same line, the system is dependent and all the points on both lines constitute the solution.

Solving by graphing can involve estimating the solutions whenever the intersection of the two lines is at a point not represented by a pair of integers. (There is nothing wrong with this technique. Just be aware that at times it can lack accuracy. See Example 1b.)

Example 1: Graphing

Solve each of the following systems by graphing.

a. $\begin{cases} x+y=6 \\ y=x+4 \end{cases}$

Solution: The two lines intersect at the point $(1, 5)$. The system is consistent and the solution is $x = 1$ and $y = 5$.

Check: Substitution shows that $(1, 5)$ satisfies both of the equations in the system.

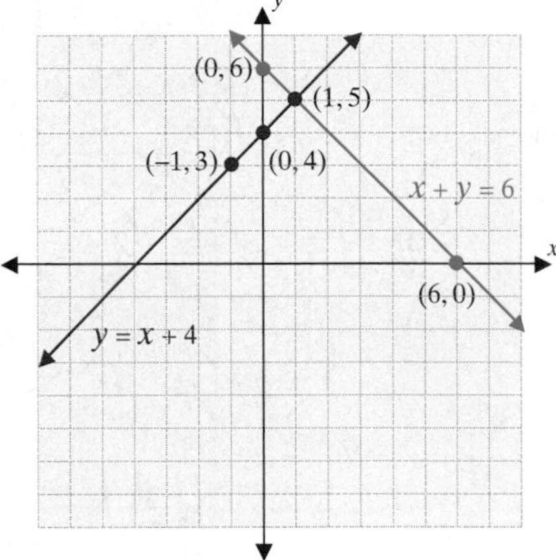

b. $\begin{cases} x-3y=4 \\ 2x+y=3 \end{cases}$

Solution: The two lines intersect, but we can only estimate the point of intersection at 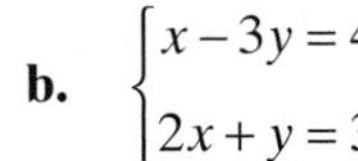 $\left(2, -\frac{1}{2}\right)$. In this situation be aware that, while graphing gives a good "estimate," finding exact solutions to the system is not likely.

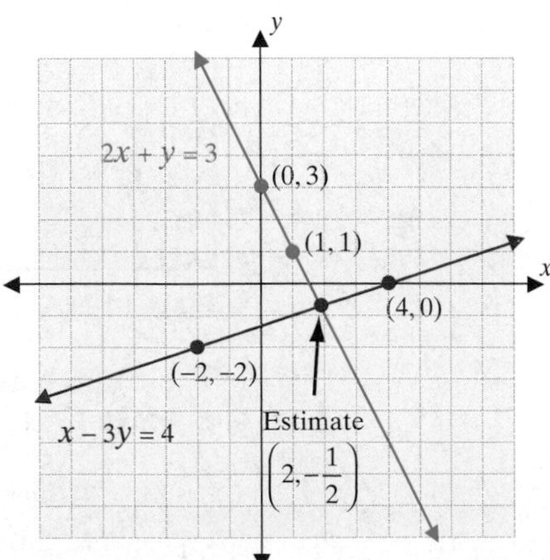

Check: Substituting $x=2$ and $y=-\frac{1}{2}$ gives:

$$2-3\left(-\frac{1}{2}\right)\stackrel{?}{=}4 \quad \text{and} \quad 2(2)+\left(-\frac{1}{2}\right)\stackrel{?}{=}3$$

$$\frac{7}{2}\neq 4 \qquad\qquad \frac{7}{2}\neq 3$$

Thus, checking shows that the estimated solution $\left(2,-\frac{1}{2}\right)$ does not satisfy either equation. The estimated point of intersection is just that – an estimate. The following discussion develops an algebraic technique that gives the exact solution as $\left(\frac{13}{7},-\frac{5}{7}\right)$.

The Substitution Method

The objective in the substitution method is to eliminate one of the variables so that a new equation is formed with just one variable. If this new equation has one solution, then the system is **consistent**. If this new equation is never true, then the system is **inconsistent**. If this new equation is always true, then the system is **dependent**.

To Solve a System of Linear Equations by Substitution

1. *Solve one of the equations for one of the variables.*
2. *Substitute the resulting expression into the other equation.*
3. *Solve this new equation, if possible, and then substitute back into one of the original equations to find the value of the other variable. (This is known as **back substitution**.)*

Example 2: Substitution

Solve the following systems of equations by substitution.

a. $\begin{cases} x-3y=4 \\ 2x+y=3 \end{cases}$ **b.** $\begin{cases} 6x+3y=14 \\ 2x+y=-3 \end{cases}$

continued on next page ...

a. $\begin{cases} x-3y=4 \\ 2x+y=3 \end{cases}$

Solution:

$$x-3y=4$$

$$x=4+3y \qquad \text{Solve the first equation for } x.$$

$$2(4+3y)+y=3 \qquad \text{Substitute } 4+3y \text{ for } x \text{ in the second equation.}$$

$$8+6y+y=3 \qquad \text{Solve the new equation for } y.$$

$$7y=-5$$

$$y=-\frac{5}{7}$$

To find x, we "**back substitute**" $-\frac{5}{7}$ for y in one of the original equations.

$$x-3\left(-\frac{5}{7}\right)=4$$

$$x=4-\frac{15}{7}=\frac{13}{7}$$

The system is consistent, and the solution is $x=\frac{13}{7}$ and $y=-\frac{5}{7}$, or $\left(\frac{13}{7},-\frac{5}{7}\right)$.

In this example, the second equation could have been solved for y and the substitution made into the first equation. The solution would have been the same.

Teaching Notes:
To convince the students that it doesn't matter which equation or which variable is used, you might want to have the students solve for y in the second equation in Example 2a and substitute back into the first equation. Emphasize that they should choose whichever equation and variable that are easiest to deal with.

NOTES

Either equation can be solved for either variable and the result substituted into the other equation. For simplicity, we generally solve for the variable that has a coefficient of 1 if there is such a variable.

b. $\begin{cases} 6x+3y=14 \\ 2x+y=-3 \end{cases}$

Solution:

$$2x+y=-3$$

$$y=-3-2x \qquad \text{Solve the second equation for } y.$$

$$6x+3(-3-2x)=14 \qquad \text{Substitute } -3-2x \text{ for } y \text{ in the first equation}$$

$$6x-9-6x=14 \qquad \text{Solve for } x.$$

$$-9=14$$

The variable x is eliminated and this last equation is never true.
Therefore, the system is **inconsistent**.
There is no solution to this system of equations.

The Addition Method

In arithmetic, we can show that if $a = b$ and $c = d$, then $a + c = b + d$. Similarly, in algebra, we can show that any solution to **both** linear equations

$$a_1x + b_1y = c_1$$

and

$$a_2x + b_2y = c_2$$

will also be a solution to the equation

$$k_1(a_1x + b_1y) + k_2(a_2x + b_2y) = k_1c_1 + k_2c_2$$

where k_1 and k_2 are not both 0.

Thus, to find a common solution to the original two equations, form a new equation by combining like terms of the two equations. As with the substitution method, the objective is to eliminate one of the variables so that the new equation has just one variable. If this new equation has one solution, then the system is **consistent**. If this new equation is never true, then the system is **inconsistent**. If this new equation is always true, then the system is **dependent**. The procedure is outlined as follows:

To Solve a System of Linear Equations by Addition

1. *Write the equations one under the other so that like terms are aligned.*
2. *Multiply all terms of one equation by a constant (and possibly all terms of the other equation by another constant) so that two like terms have opposite coefficients.*
3. *Add like terms and solve the resulting equation, if possible. Then, back substitute into one of the original equations to find the value of the other variable.*

NOTES

When solving a system of equations, two equations may be interchanged. That is, the order in which the equations are written has no bearing on the solutions. If for some reason you think that writing one equation before another will make the system easier to solve, you may do so without affecting the solutions.

Example 3: Addition

Solve the following systems of equations by addition.

a. $\begin{cases} 4x+3y=1 \\ 5x+y=-7 \end{cases}$

Solution: Multiply each term in the second equation by –3 so that the y-coefficients will be opposites. Add like terms to eliminate y. Solve for x.

[**Note:** We could have eliminated x by multiplying the first equation by –5 and the second equation by 4 then adding like terms. The result will be the same. Because the coefficient for y in the second equation is +1, it was less work to multiply this one equation by –3.]

$$\begin{cases} & 4x + 3y = 1 \\ [-3] & 5x + y = -7 \end{cases} \qquad \begin{array}{rcrcr} 4x & + & 3y & = & 1 \\ -15x & - & 3y & = & 21 \\ \hline -11x & & & = & 22 \\ x & & & = & -2 \end{array}$$

Now, back substitute $x=-2$ in one of the original equations and solve for y.

$$5(-2)+y=-7$$
$$-10+y=-7$$
$$y=3$$

The system is consistent, and the solution is $x=-2$ and $y=3$, or $(-2, 3)$.

[**Check:** As a thorough check, substitute the solution, in this case (–2, 3), into **both** of the original equations.]

b. $\begin{cases} 2x=-17-3y \\ 3x-\dfrac{51}{2}=4y \end{cases}$

Solution: Rewriting the equations in standard form yields $\begin{cases} 2x+3y=-17 \\ 3x-4y=\dfrac{51}{2} \end{cases}$.

Multiply each term in the first equation by –3 and each term in the second equation by 2 so that the x-coefficients will be opposites.

[**Note:** We could just as easily have eliminated y by multiplying the first equation by 4 and the second equation by 3 then adding like terms. The result will be the same.]

$$\begin{cases} [-3] & 2x + 3y = -17 \\ [2] & 3x - 4y = \dfrac{51}{2} \end{cases} \qquad \begin{array}{rcrcr} -6x & - & 9y & = & 51 \\ 6x & - & 8y & = & 51 \\ \hline & & -17y & = & 102 \\ & & y & = & -6 \end{array}$$

Now, back substitute $y = -6$ in one of the original equations and solve for x.

$$\begin{aligned} 2x + 3(-6) &= -17 \\ 2x &= -17 + 18 \\ 2x &= 1 \\ x &= \frac{1}{2} \end{aligned}$$

The system is consistent. The solution is $x = \frac{1}{2}$ and $y = -6$, or $\left(\frac{1}{2}, -6\right)$.

c. $\begin{cases} 3x - \frac{1}{2}y = 6 \\ 6x - y = 12 \end{cases}$

Solution: Multiply the first equation by –2.

$$\begin{cases} [-2] \quad 3x - \frac{1}{2}y = 6 \\ \qquad\; 6x - y = 12 \end{cases} \qquad \begin{array}{rcrcr} -6x & + & y & = & -12 \\ 6x & - & y & = & 12 \\ \hline & & 0 & = & 0 \end{array}$$

The last equation, $0 = 0$, is always true so the system is **dependent**. There are an infinite number of solutions. Any ordered pair that satisfies the equation $6x - y = 12$ (or the equation $3x - \frac{1}{2}y = 6$ since both equations are equivalent) is a solution to the system, written in set-builder notation as $\{(x, y) \mid 6x - y = 12\}$.

Solving a System of Linear Equations by Using a TI-83 Plus Calculator

Consider the system of two linear equations $\begin{cases} x + y = 4 \\ 3x - 2y = 7 \end{cases}$.

To solve this system by using a TI-83 Plus graphing calculator, we can proceed as follows:

Step 1: To be able to use the calculator's function mode, solve each equation for y:

$$\begin{cases} y = 4 - x \\ y = \frac{3}{2}x - \frac{7}{2} \end{cases}$$

Step 2: Press the Y= key and enter the two functions for Y1 and Y2 as shown here:

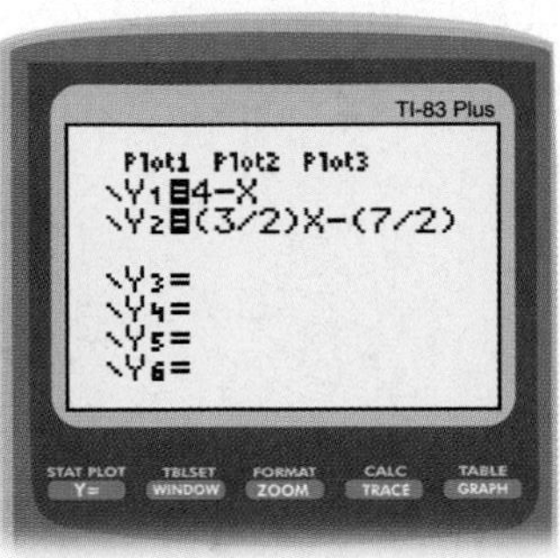

Step 3: Press GRAPH.
(You may need to check the WINDOW to be sure that both lines appear.)

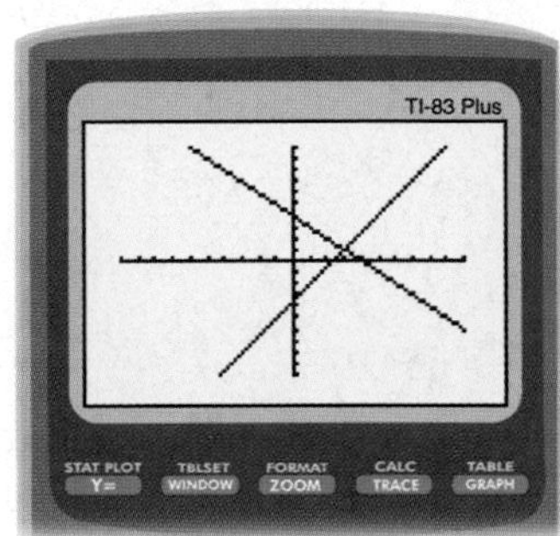

Teaching Notes: Some students may want to use the TRACE command. Certainly it is easier to use, but you will need to point out that, because of the pixels, tracing will seldom give exact answers.

Step 4: Press 2nd CALC
Choose `5: intersect`.
Move the cursor to one of the lines and press ENTER in response to the question `First curve?`.
Move the cursor to the second line and press ENTER in response to the question `Second curve?`.
Move the cursor near the point of intersection and press ENTER in response to the question `Guess?`.

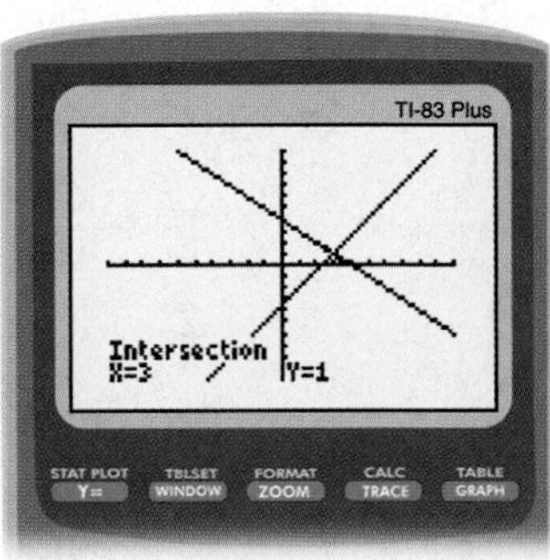

We see that the solution is $x = 3$ and $y = 1$.

[**Note:** In this case the solution shown is exact. In many cases the solution shown will be only an estimate. Thus, even with a calculator, the graphing method is limited.]

Practice Problems

Solve each of the following systems of linear equations algebraically.

1. $\begin{cases} y = 3x + 4 \\ 2x + y = -1 \end{cases}$ **2.** $\begin{cases} 2x - 3y = 0 \\ 6x + 3y = 4 \end{cases}$ **3.** $\begin{cases} 4x + y = 3 \\ 4x + y = 2 \end{cases}$

14.1 Exercises

1. (6, 0)

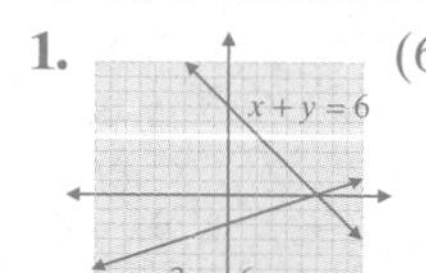
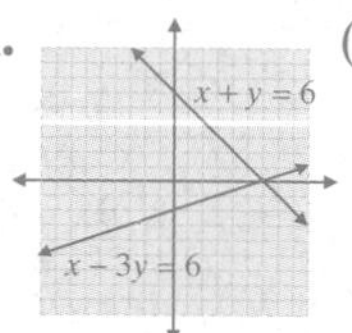

2. (3, −1)

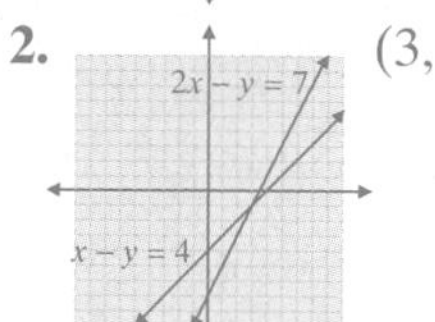

3. ∅

4. (2, 2)

5. (28, 22)

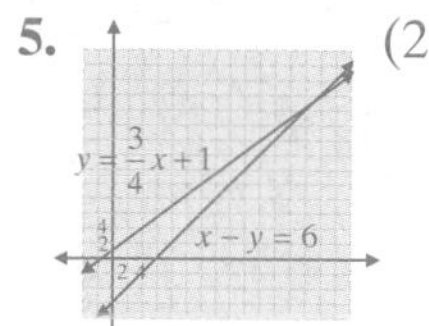

6. 3x + 2y = 5

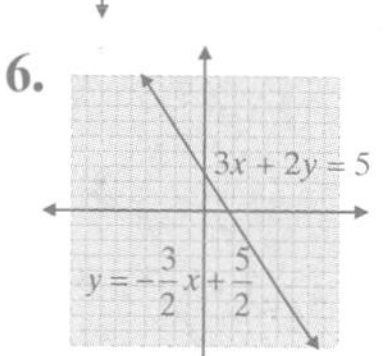

7. (−1, −1)

Solve each of the systems in Exercises 1 – 10 by graphing.

1. $\begin{cases} x+y=6 \\ x-3y=6 \end{cases}$ **2.** $\begin{cases} 2x-y=7 \\ x-y=4 \end{cases}$ **3.** $\begin{cases} y=-2x \\ 2x+y=3 \end{cases}$ **4.** $\begin{cases} x=y \\ 4x+3y=14 \end{cases}$

5. $\begin{cases} y=\frac{3}{4}x+1 \\ x-y=6 \end{cases}$ **6.** $\begin{cases} 3x+2y=5 \\ y=-\frac{3}{2}x+\frac{5}{2} \end{cases}$ **7.** $\begin{cases} 2x+y+3=0 \\ 3x+4y+7=0 \end{cases}$

8. $\begin{cases} 3x+4y=6 \\ x-y=\frac{1}{2} \end{cases}$ **9.** $\begin{cases} x-2y=10 \\ 2x-3y=15 \end{cases}$ **10.** $\begin{cases} 5x-3y=7 \\ 4x=-2y+3 \end{cases}$

Use the Substitution Method or the Addition Method to solve the systems of linear equations in Exercises 11 – 36. State whether each system is consistent, inconsistent, or dependent.

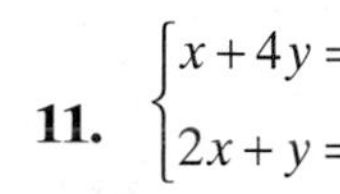

11. $\begin{cases} x+4y=6 \\ 2x+y=5 \end{cases}$ **12.** $\begin{cases} 2x+y=0 \\ x-2y=-10 \end{cases}$ **13.** $\begin{cases} 5x-y=-2 \\ x+2y=-7 \end{cases}$ **14.** $\begin{cases} 7x-y=18 \\ x+2y=9 \end{cases}$

15. $\begin{cases} x+2y=3 \\ 4x+8y=8 \end{cases}$ **16.** $\begin{cases} 2x+3y=3 \\ x+4y=4 \end{cases}$ **17.** $\begin{cases} 6x+2y=16 \\ 3x+y=8 \end{cases}$ **18.** $\begin{cases} 4x-y=18 \\ 3x+5y=2 \end{cases}$

19. $\begin{cases} y=3x+3 \\ y=-2x+8 \end{cases}$ **20.** $\begin{cases} x=5-4y \\ x=2y-7 \end{cases}$ **21.** $\begin{cases} 2x+y=4 \\ 4x+5y=11 \end{cases}$

22. $\begin{cases} x+6y=4 \\ 2x+3y=5 \end{cases}$ **23.** $\begin{cases} 3x+4y=6 \\ x-8y=9 \end{cases}$ **24.** $\begin{cases} 3x+5y=3 \\ 9x-y=-7 \end{cases}$

25. $\begin{cases} 2x=5y-1 \\ 4x-10y=0 \end{cases}$ **26.** $\begin{cases} 6x+2y=5 \\ 2x+y=1 \end{cases}$ **27.** $\begin{cases} 4x+12y=5 \\ 5x-6y=1 \end{cases}$

28. $\begin{cases} 2x-3y=18 \\ 5x+4y=-1 \end{cases}$ **29.** $\begin{cases} x+y=7 \\ 2x+3y=16 \end{cases}$ **30.** $\begin{cases} 5x-7y=8 \\ 3x+11y=-12 \end{cases}$

31. $\begin{cases} 6x-y=15 \\ 0.2x+0.5y=2.1 \end{cases}$ **32.** $\begin{cases} 3x+y=14 \\ 0.1x-0.2y=1.4 \end{cases}$ **33.** $\begin{cases} x+y=12 \\ 0.05x+0.25y=1.6 \end{cases}$

Answers to Practice Problems: **1.** $x=-1, y=1$ **2.** $x=\frac{1}{2}, y=\frac{1}{3}$ **3.** Inconsistent

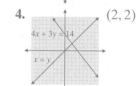

34. $\begin{cases} x+y=20 \\ 0.1x+2.5y=3.8 \end{cases}$

35. $\begin{cases} 0.6x+0.5y=5.9 \\ 0.8x+0.4y=6 \end{cases}$

36. $\begin{cases} 0.5x-0.3y=7 \\ 0.3x+0.4y=2 \end{cases}$

*In Exercises 37 – 44, use a graphing calculator and the ZOOM, TRACE, and CALC features to estimate the solutions to the systems of linear equations. (**HINT:** The CALC and 5: intersect commands may give the most accurate answers. Remember to solve each equation for y. Use both Y1 and Y2 in the* Y= *menu.)*

37. $\begin{cases} 2x+y=4 \\ x-y=6 \end{cases}$

38. $\begin{cases} 3x-y=-22 \\ 2x-y=-18 \end{cases}$

39. $\begin{cases} 4x-y=4 \\ y=-x \end{cases}$

40. $\begin{cases} x+y=-6 \\ 4x-y=6 \end{cases}$

41. $\begin{cases} x-3y=5 \\ -2x+y=1 \end{cases}$

42. $\begin{cases} x+\frac{1}{3}y=0 \\ 6x-y=4 \end{cases}$

43. $\begin{cases} 2x+3y=4 \\ x+2y=6 \end{cases}$

44. $\begin{cases} x+3y=5 \\ 2x+3y=8 \end{cases}$

8.

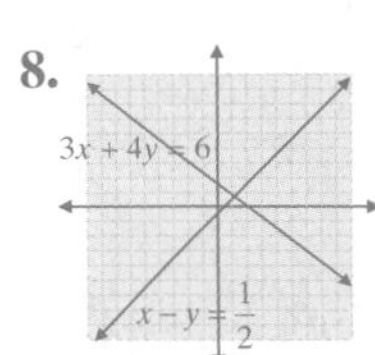

(1.1, 0.6)

9.

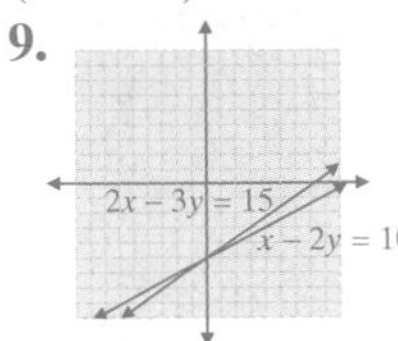

(0, –5)

10.

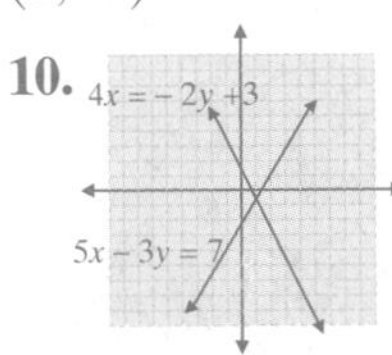

(1.04, –0.6)

11. $x=2,\ y=1$, consistent

12. $x=-2,\ y=4$, consistent

13. $x=-1,\ y=-3$, consistent

14. $x=3,\ y=3$, consistent

15. No solution, inconsistent

16. $x=0,\ y=1$, consistent

17. Infinite solutions, dependent

18. $x=4,\ y=-2$, consistent

19. $x=1,\ y=6$, consistent

20. $x=-3,\ y=2$, consistent

21. $x=\frac{3}{2},\ y=1$, consistent

22. $x=2,\ y=\frac{1}{3}$, consistent

23. $x=3,\ y=-\frac{3}{4}$, consistent

24. $x=-\frac{2}{3},\ y=1$, consistent

25. No solution, inconsistent

26. $x=\frac{3}{2},\ y=-2$, consistent

27. $x=\frac{1}{2}, y=\frac{1}{4}$, consistent

28. $x=3,\ y=-4$, consistent

29. $x=5,\ y=2$, consistent

30. $x=\frac{1}{19}, y=-\frac{21}{19}$, consistent

31. $x=3,\ y=3$, consistent

32. $x=6,\ y=-4$, consistent

33. $x=7,\ y=5$, consistent

34. $x=\frac{77}{4}, y=\frac{3}{4}$, consistent

35. $x=4,\ y=7$, consistent

36. $x=\frac{340}{29}, y=-\frac{110}{29}$, consistent

37. $x=\frac{10}{3},\ y=-\frac{8}{3}$

38. $x=-4,\ y=10$

39. $x=\frac{4}{5},\ y=-\frac{4}{5}$

40. $x=0,\ y=-6$

41. $x=-\frac{8}{5},\ y=-\frac{11}{5}$

42. $x=\frac{4}{9},\ y=-\frac{4}{3}$

43. $x=-10,\ y=8$

44. $x=3,\ y=\frac{2}{3}$

Hawkes Learning Systems: Introductory & Intermediate Algebra

Review: Solving Systems of Linear Equations – Two Variables

14.2 Applications

Objectives

After completing this section, you will be able to:

Solve applied problems by using systems of two linear equations in two variables.

In Chapters 2 and 3, a variety of word problems were solved by using one variable and one equation. However, many applications can be solved, and in fact are easier to solve, by using two variables and two equations. That is, the information given in a problem can be represented with a system of two linear equations. Then the system can be solved using the methods discussed in the previous section.

Example 1: Mixture

A manufacturer receives an order for 30 tons of a 40% copper alloy. He has only 20% alloy and 50% alloy in stock. How much of each will he need to fill the order?

Solution: Let x = amount of 20% alloy in tons
y = amount of 50% alloy in tons

Form two equations based on the information given.

Teaching Notes: Point out to your students that, in Examples 1 – 3, one of the equations involves the total amount of something (number of tons, amount of money, total hours worked). The other equation involves a percent or rate. This might help them in setting up the equations for similar problems.

Phrase from problem	Equation formed
The total order is 30 tons.	$x + y = 30$
The total amount of copper is 40% of 30.	$0.20x + 0.50y = 0.40(30)$

Now solve the system. We will use the addition method.

$$\begin{cases} [-2] & x + y = 30 \\ [10] & 0.20x + 0.50y = 0.40(30) \end{cases} \qquad \begin{aligned} -2x - 2y &= -60 \\ 2x + 5y &= 120 \\ \hline 3y &= 60 \\ y &= 20 \end{aligned}$$

Substituting $y = 20$ into the first equation yields

$$x + 20 = 30$$
$$x = 10.$$

The manufacturer will need 10 tons of the 20% alloy and 20 tons of the 50% alloy.

Example 2: Interest

A savings and loan company pays 7% interest on long-term savings, and a high-risk stock indicates that it should yield 12% interest. If a woman has $40,000 to invest and wants an annual income of $3550 from her investments, how much should she put in the savings and loan and how much in the stock?

Solution: Let x = amount invested at 7%
y = amount invested at 12%

Phrase from problem	Equation formed
The total invested is $40,000.	$x+y=40,000$
The total interest is $3,550.	$0.07x+0.12y=3,550$

$$\begin{cases} [-7] & x + y = 40,000 \\ [100] & 0.07x + 0.12y = 3,550 \end{cases}$$

$$\begin{array}{rcr} -7x - 7y &=& -280,000 \\ 7x + 12y &=& 355,000 \\ \hline 5y &=& 75,000 \\ y &=& 15,000 \end{array}$$

Back substituting gives:

$$x+15,000=40,000$$
$$x=25,000$$

She should put $25,000 in the long-term savings at 7% and $15,000 in stock at 12%.

Example 3: Work

Working his way through school, Richard works two part-time jobs for a total of 25 hours a week. Job A pays $4.50 per hour and job B pays $6.30 per hour. How many hours did he work at each job the week he made $137.70?

Solution: Let x = number of hours at job A at $4.50 per hour
y = number of hours at job B at $6.30 per hour

Phrase from problem	Equation formed
The total hours worked is 25.	$x+y=25$
The total earnings were $137.70.	$4.50x+6.30y=137.70$

$$\begin{cases} [-45] & x + y = 25 \\ [10] & 4.50x + 6.30y = 137.70 \end{cases}$$

$$\begin{array}{rcr} -45x - 45y &=& -1125 \\ 45x + 63y &=& 1377 \\ \hline 18y &=& 252 \\ y &=& 14 \end{array}$$

Back substitute to obtain:

$$x+14=25$$
$$x=11$$

Richard worked 11 hours at job A and 14 hours at job B.

Example 4: Algebra

Determine the value of a and b such that the straight line $ax + by = 11$ passes through the point $(3, -1)$ and has slope $-\frac{5}{4}$.

Teaching Notes: Without getting "too mathematical" or formal, it might be helpful to have students think of applied situations in which two or more conditions need to be considered when designing or doing something. (For example: the weight of a car for safety versus fuel economy).

Solution: Here, the unknown quantities are a and b, not x and y.

Since the point $(3, -1)$ is on the line, substitute $x = 3$ and $y = -1$ into the equation $ax + by = 11$.

$$3a - b = 11 \qquad \text{A linear equation in } a \text{ and } b$$

To find the slope in terms of a and b, write the equation $ax + by = 11$ in slope-intercept form.

$$ax + by = 11$$
$$by = -ax + 11$$
$$y = -\frac{a}{b}x + \frac{11}{b}$$

Thus, the slope is $-\frac{a}{b}$. Therefore,

$$-\frac{a}{b} = -\frac{5}{4} \qquad \text{Since the slope is given as } -\frac{5}{4}$$
$$\text{or} \qquad 4a = 5b$$
$$\text{or} \qquad 4a - 5b = 0 \qquad \text{Another linear equation in } a \text{ and } b$$

Now we have two linear equations in a and b so we can use the addition method to solve the system of equations.

$$\begin{cases} [-5] \quad 3a - b = 11 \\ \quad 4a - 5b = 0 \end{cases} \qquad \begin{array}{rcrcr} -15a & + & 5b & = & -55 \\ 4a & - & 5b & = & 0 \\ \hline -11a & & & = & -55 \\ a & & & = & 5 \end{array}$$

Back substituting $a = 5$ yields:

$$3(5) - b = 11$$
$$-b = -4$$
$$b = 4$$

Thus, $a = 5$ and $b = 4$, so the line $5x + 4y = 11$ passes through the point $(3, -1)$ and has slope $-\frac{5}{4}$.

NOTES

The equation $\frac{a}{b} = \frac{5}{4}$ does not necessarily mean that $a = 5$ and $b = 4$. It means that the **ratio** of a to b is 5 to 4. For example, if $a = 10$ and $b = 8$, then the ratio of a to b is still 5 to 4.

14.2 Exercises

1. 23, 79

2. 50, 37

3. 80°, 100°

4. 75°, 15°

5. 100 yards × 45 yards

6. Side of square is 15 ft., side of hexagon is 10 ft.

7. 40 liters of 12%, 50 liters of 30%

8. 20 lbs. of 40%, 30 lbs. of 15%

1. The sum of two integers is 102, and the larger number is 10 more than three times the smaller. Find the two integers.

2. The difference between two integers is 13, and their sum is 87. What are the two integers?

3. Two angles are supplementary if the sum of their measures is 180°. Find two supplementary angles such that the smaller is 30° more than one-half of the larger.

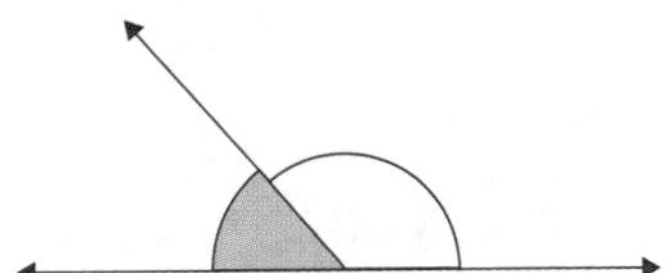

4. Two angles are complementary if the sum of their measures is 90°. Find two complementary angles such that one is 15° less than six times the other.

5. At present, the length of a rectangular soccer field is 55 yards longer than the width. The city council is thinking of rearranging the area containing the soccer field into two square playing fields. A math teacher on the council decided to test the council members' mathematical skills. (You know how math teachers are.) He told them that if the width of the current field were to be increased by 5 yards and the length cut in half, the field would be a square. What are the dimensions of the field currently?

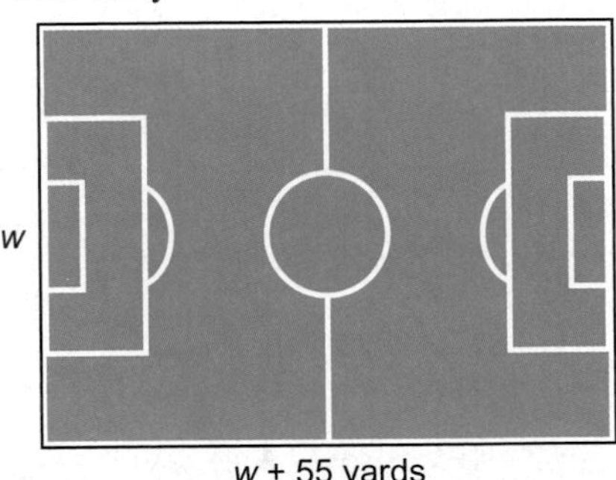

6. Consider a square and a regular hexagon (a six-sided figure with sides of equal length). One side of the square is 5 feet longer than a side of the hexagon, and the two figures have the same perimeter. What are the lengths of the sides of each figure?

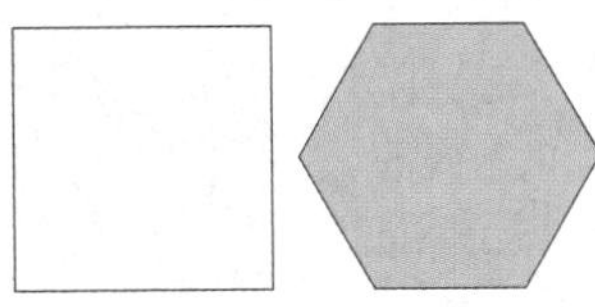

7. How many liters each of a 12% iodine solution and a 30% iodine solution must be used to produce a total mixture of 90 liters of a 22% iodine solution?

8. A meat market has ground beef that is 40% fat and extra lean ground beef that is only 15% fat. How many pounds of each (ground beef and extra lean) must be ground together to get a total of 50 pounds of "lean" ground beef that is 25% fat?

9. 240 gal. of 5%, 120 gal. of 2%

10. 27 oz. of 25%, 9 oz. of 45%

11. $87,000 in bonds, $37,000 in certificates

12. $20,000 at 6%, $28,000 at 10%

13. 325 at $3.50/share, 175 at $6.00/share

14. 4.5 hours at 100 units/hr., 7 hours at 75 units/hr.

15. 40 lbs. at $3.90/lb., 30 lbs. at $2.50/lb.

16. 30 of the 33 cent stamps, 50 of the 55 cent stamps

17. 16 lbs. at $.70/lb., 4 lbs. at $1.30/lb.

18. 6 dimes, 14 quarters.

9. A dairy needs 360 gallons of milk containing 4% butterfat. How many gallons each of milk containing 5% butterfat and milk containing 2% butterfat must be used to obtain the desired 360 gallons?

10. A druggist has two solutions of alcohol. One is 25% alcohol, the other is 45% alcohol. He wants to mix these two solutions to get 36 ounces that will be 30% alcohol. How many ounces of each of these two solutions should he mix together?

11. Roxanne inherited $124,000 from her Uncle Jake. She invested a portion in bonds and the remainder in a long-term certificate account. The amount invested in bonds was $24,000 less than 3 times the amount invested in certificates. How much was invested in bonds and how much in certificates?

12. Sang has invested $48,000, part at 6% and the rest in a higher risk investment at 10%. How much did she invest at each rate to receive $4000 in interest after one year?

13. An investor bought 500 shares of stock, some at $3.50 per share and some at $6.00 per share. If the total cost was $2187.50, how many shares of each stock did the investor buy?

14. A manufacturing plant is going to use two different stamping machines to complete an order of 975 units. One produces 100 units per hour, while the other produces 75 units per hour. How long must each machine operate to complete the order if the faster machine needs to be shut down for two and one-half hours for repairs?

15. A confectioner is going to mix candy worth $3.90 per pound with candy worth $2.50 per pound to obtain 70 pounds of candy worth $3.30 per pound. How many pounds of each kind should she use?

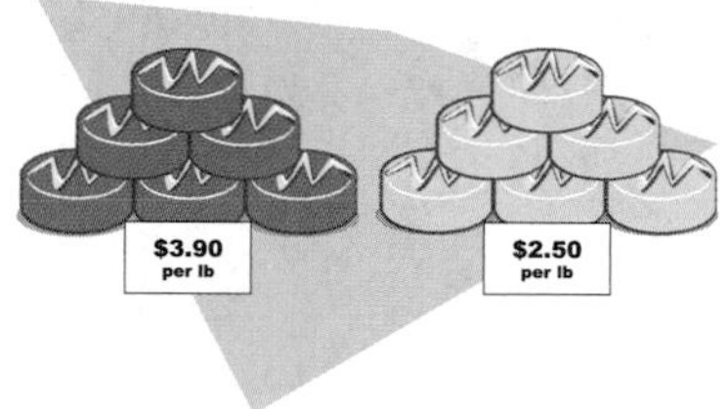

16. The postal service charges 33 cents for letters that weigh 1 ounce or less and 22 cents more for letters that weigh between 1 and 2 ounces. Jeffery, testing his father's math skills, gave his father $37.40 and asked him to purchase 80 stamps for his stamp collection, some 33-cent stamps and some 55-cent stamps. How many of each type of stamp did he buy?

17. A grocer wants to mix two kinds of nuts. One kind sells for 70 cents per pound, and the other sells for $1.30 per pound. He wants to mix a total of 20 pounds and sell it for 82 cents per pound. How many pounds of each kind should he use in the new mix?

18. Inez has 20 coins consisting of dimes and quarters. How many of each type does she have if all together she has $4.10?

19. \$5.50 for paperback, \$9.00 for hardback

20. 1670 votes for winner, 620 votes for loser

21. 150 legislators voted "for" the bill

22. Approximately 4.46 hours

23. 6:00 P.M.

24. Approximately 56.67 mph and 61.67 mph

25. Commercial jet: 300 mph, Private: 125 mph

26. $a = 3, b = 1$

27. $a = -2, b = 3$

28. $a = 2, b = -3$

19. The bookstore can buy a popular book with either paper back or hard back cover. A hard back book costs \$3.50 more than the paper back book. What is the cost of each if 90 paper back books cost the same as 55 hard back books?

20. In an election, the winner received 430 votes more than twice as many votes as the loser. If there was a total of 2290 votes cast, how many did each candidate receive?

21. A bill was defeated in the legislature by 50 votes. If one-fifth of those voting against the bill had voted for it, the bill would have passed by 30 votes. How many legislators voted for the bill?

22. Andrea made a trip of 440 kilometers. She averaged 54 kilometers per hour for the first part of the trip and 80 kilometers per hour for the second part. If the total trip took 6 hours, how long was she traveling at 80 kilometers per hour?

23. A boat left Dana Point Marina at 11:00 A.M. traveling at 10 knots (nautical miles per hour). Two hours later, a Coast Guard boat left the same marina traveling at 14 knots trying to catch the first boat. If both boats traveled the same course, at what time did the Coast Guard captain anticipate overtaking the other boat?

24. Two cars are to start at the same place in Knoxville and travel in opposite directions (assume in straight lines). The drivers know that one driver drives an average of 5 mph faster than the other. (They have been married for 20 years.) They have agreed to stop and call each other after driving for 3 hours. In the telephone conversation they realize that they are 355 miles apart. What was the average speed of each driver?

25. A private jet flies the same distance in 6 hours that a commercial jet flies in 2.5 hours. If the speed of the commercial jet was 75 mph less than three times the speed of the private jet, find the speed of each jet.

26. Determine a and b such that the line with equation $ax + by = 7$ passes through the two points $(2, 1)$ and $(-1, 10)$.

27. Determine a and b such that the line with equation $ax + by = 6$ passes through the points $(-6, -2)$ and $(3, 4)$.

28. Determine a and b such that the line with equation $ax + by = 4$ passes through the point $(5, 2)$ and has a slope of $\frac{2}{3}$.

29. $a = 10, b = -2$

30. 700 of Model A
1000 of Model B

31. \$11.20/hr. labor,
\$4.80/lb. materials

32. 78 chairs using Method I, 44 chairs using Method II

33. 9 lbs. of Ration I, 2 lbs. of Ration II

34. 55°, 55°, 70°

35. 65°, 65°, 50°

29. Determine a and b such that the line with equation $ax + by = -4$ contains the point $(-1, -3)$ and has a slope of 5.

30. A manufacturer produces two models of the same toy, Model A and Model B. Model A takes 4 hours to produce and costs \$8 each. Model B takes 3 hours to produce and costs \$7 each. If the manufacturer allots a total of 5800 hours and \$12,600 for production each week, how many of each model will be produced?

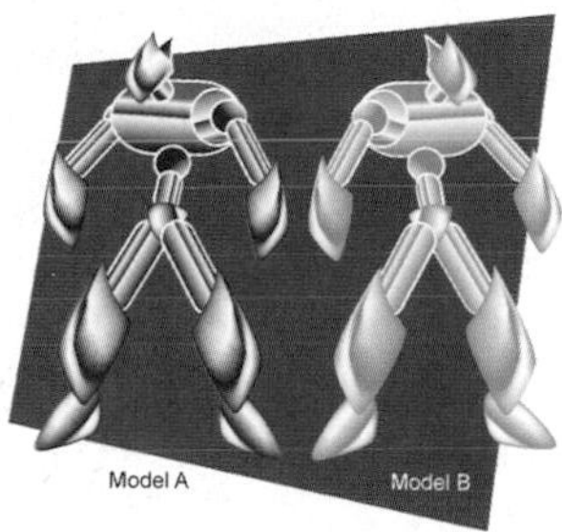

31. A company manufactures two products. One requires 2.5 hours of labor, 3 pounds of raw materials, and costs \$42.40 each to produce. The second product requires 4 hours of labor, 4 pounds of raw materials, and costs \$64 each to produce. Find the cost of labor per hour and the cost of raw materials per pound.

32. A furniture shop refinishes chairs. Employees use two methods to refinish a chair. Method I takes 1 hour, and the material costs \$6. Method II takes an hour and a half, and the material costs \$3. Next week, they plan to spend 144 hours in labor and \$600 in material for refinishing chairs. How many chairs should they plan to refinish with each method?

33. A large feed lot uses two feed supplements, Ration I and Ration II. Each pound of Ration I contains 4 units of protein and 2 units of carbohydrates. Each pound of Ration II contains 3 units of protein and 6 units of carbohydrates. If the dietary requirement calls for 42 units of protein and 30 units of carbohydrates, how many pounds of each ration should be used to satisfy the requirements?

34. The sum of the measures of the three angles of a triangle is 180°. In an isosceles triangle, two of the angles have the same measure. What are the measures of the angles of an isosceles triangle in which one angle measures 15° more than each of the other two equal angles?

35. The sum of the measures of the three angles of a triangle is 180°. In an isosceles triangle, two of the angles have the same measure. What are the measures of the angles of an isosceles triangle in which each of the two equal angles measures 15° more than the third angle?

Hawkes Learning Systems: Introductory & Intermediate Algebra

Applications: Systems of Equations

14.3 Systems of Linear Equations: Three Variables

Objectives

After completing this section, you will be able to :

1. *Solve systems of linear equations in three variables.*
2. *Solve applied problems by using systems of linear equations in three variables.*

The equation $2x + 3y - z = 16$ is called a **linear equation in three variables**. The general form is

$$Ax + By + Cz = D \text{ where } A, B, \text{ and } C \text{ are not all } 0.$$

The solutions to such equations are called **ordered triples** and are of the form (x_0, y_0, z_0) or $x = x_0$, $y = y_0$, and $z = z_0$. One ordered triple that satisfies the equation $2x + 3y - z = 16$ is $(1, 4, -2)$. To check this, substitute $x = 1$, $y = 4$, and $z = -2$ into the equation to see if the result is 16:

$$\begin{aligned} 2(1) + 3(4) - (-2) &= 2 + 12 + 2 \\ &= 16. \end{aligned}$$

There are an infinite number of ordered triples that satisfy any linear equation in three variables in which at least two of the coefficients are nonzero. Any two values may be substituted for two of the variables, and then the value for the third variable can be calculated. For example, by letting $x = -1$ and $y = 5$, we find:

$$\begin{aligned} 2(-1) + 3(5) - z &= 16 \\ -2 + 15 - z &= 16 \\ -z &= 3 \\ z &= -3. \end{aligned}$$

Hence, the ordered triple $(-1, 5, -3)$ satisfies the equation $2x + 3y - z = 16$.

Graphs can be drawn in three dimensions by using a coordinate system involving three mutually perpendicular number lines labeled as the x-axis, y-axis, and z-axis. Three planes are formed: the xy-plane, the xz-plane, and the yz-plane. The three axes separate space into eight regions called **octants**. You can "picture" the first octant as the region bounded by the floor of a room and two walls with the axes meeting in a corner. The floor is the xy-plane. The axes can be ordered in a "right-hand" or "left-hand" format. Figure 14.1 shows the point represented by the ordered triple $(2, 3, 1)$ in a right-hand system.

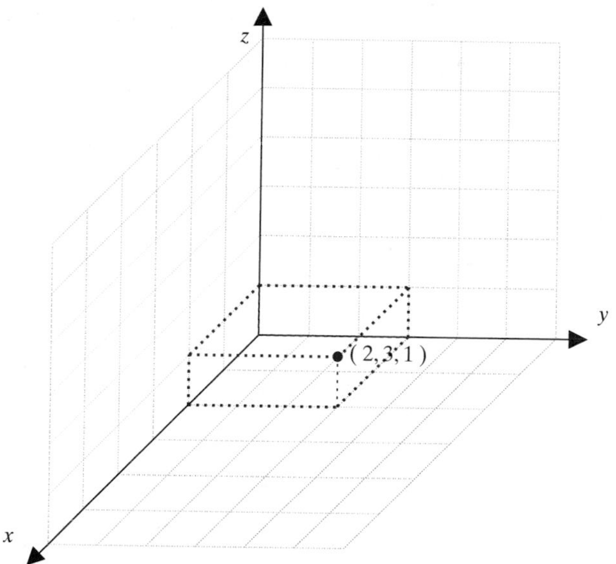

Figure 14.1

The graphs of linear equations in three variables are planes in three dimensions. A portion of the graph of $2x + 3y - z = 16$ appears in Figure 14.2.

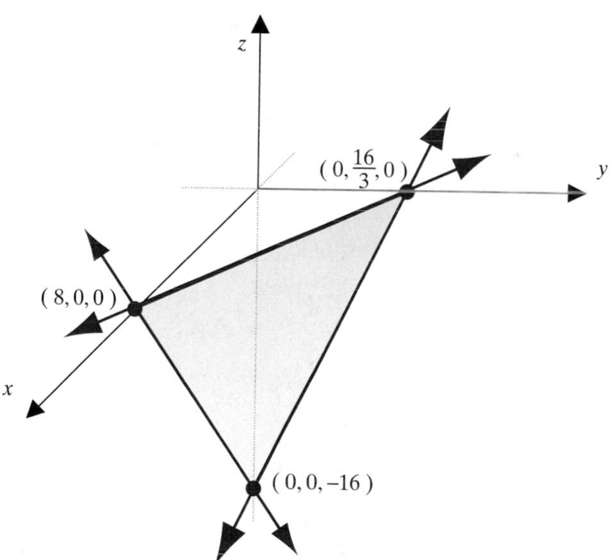

Figure 14.2

Two distinct planes will either be parallel or they will intersect. If they intersect, their intersection will be a straight line. If three distinct planes intersect, they will intersect in a straight line, or they will intersect in a single point represented by an ordered triple.

The graphs of systems of three linear equations in three variables can be both interesting and informative, but they can be difficult to sketch and the points of intersection difficult to estimate. Also, most graphing calculators are limited to graphs in two dimensions, so they are not useful in graphically analyzing systems of linear equations in three variables.

Therefore, in this text, only algebraic techniques for solving these systems will be discussed. Figure 14.3 illustrates four different possibilities for the relative positions of three planes.

Teaching Notes:
Have students notice commonly created instances of two planes coming together to form a line, (for example, a wall and floor) and three planes coming together to form a point (for example, a wall, floor, and adjacent wall.)

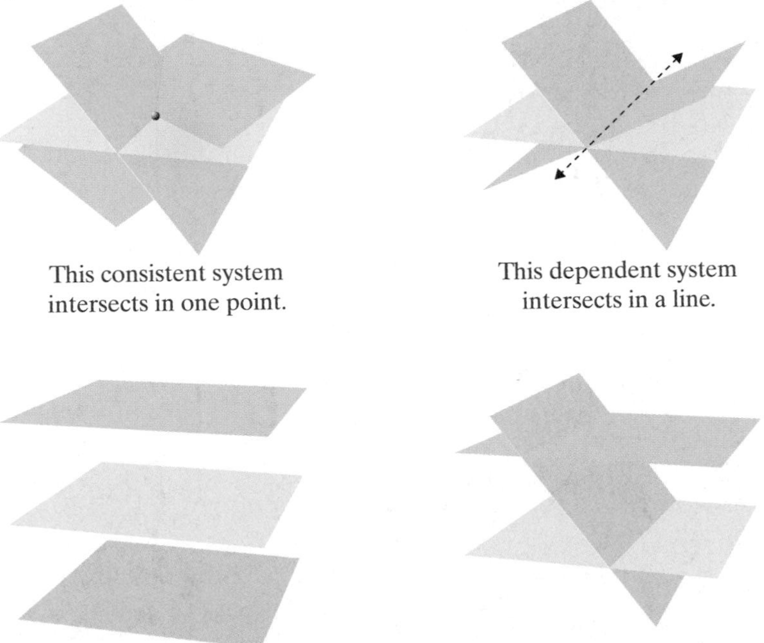

This consistent system intersects in one point.

This dependent system intersects in a line.

These two systems are both inconsistent. All three planes do not have a common intersection.

Figure 14.3

To Solve Three Linear Equations in Three Variables

Step 1: *Select two equations and eliminate one variable by using the addition method.*

Step 2: *Select a different pair of equations and eliminate the* ***same*** *variable.*

Step 3: *Steps 1 and 2 give* ***two*** *linear equations in* ***two*** *variables. Solve these equations by either addition or substitution as reviewed in Section 14.1.*

Step 4: *Back substitute the values found in Step 3 into any one of the original equations to find the value of the third variable.*

The solution possibilities for a system of three equations in three variables are as follows:

1. There will be exactly one ordered triple solution.
   (Graphically, the three planes intersect in one point.)
2. There will be an infinite number of solutions.
   (Graphically, the three planes intersect in a line or are the same plane.)
3. There will be no solutions.
   (Graphically, the three planes have no points in common.)

The technique is illustrated with the following system:

$$\begin{cases} 2x + 3y - z = 16 & \text{(I)} \\ x - y + 3z = -9 & \text{(II)} \\ 5x + 2y - z = 15 & \text{(III)} \end{cases}$$

Step 1: Using equations (I) and (II), eliminate y.
[**Note:** We could just as easily have chosen to eliminate x or z. To be sure that you understand the process you might want to solve the system by first eliminating x and then again by first eliminating z. In any case, the answer will be the same.]

$$\begin{cases} \ 2x + 3y - z = 16 \\ [3]\ x - y + 3z = -9 \end{cases} \qquad \begin{array}{rcrcrcr} 2x & + & 3y & - & z & = & 16 \\ 3x & - & 3y & + & 9z & = & -27 \\ \hline 5x & & & + & 8z & = & -11 \end{array}$$

Step 2: Using a different pair of equations, (II) and (III), eliminate the same variable y.

$$\begin{cases} [2]\ x - y + 3z = -9 \\ \ 5x + 2y - z = 15 \end{cases} \qquad \begin{array}{rcrcrcr} 2x & - & 2y & + & 6z & = & -18 \\ 5x & + & 2y & - & z & = & 15 \\ \hline 7x & & & + & 5z & = & -3 \end{array}$$

Step 3: Using the results of Steps 1 and 2, solve the two equations for x and z.

$$\begin{cases} [-7]\ 5x + 8z = -11 \\ [5]\ 7x + 5z = -3 \end{cases} \qquad \begin{array}{rcrcr} -35x & - & 56z & = & 77 \\ 35x & + & 25z & = & -15 \\ \hline & & -31z & = & 62 \\ & & z & = & -2 \end{array}$$

Back substitute $z = -2$ into the equation $5x + 8z = -11$ to find x.

$$\begin{aligned} 5x + 8(-2) &= -11 \\ 5x &= 5 \\ x &= 1 \end{aligned}$$

Step 4: Using $x = 1$ and $z = -2$, back substitute to find y.

$$\begin{aligned} 1 - y + 3(-2) &= -9 \quad \text{Using equation (II)} \\ -y &= -4 \\ y &= 4 \end{aligned}$$

The solution is $(1, 4, -2)$ or $x = 1$, $y = 4$, and $z = -2$. The solution can be checked by substituting the results into **all three** of the original equations.

$$\begin{cases} 2(1) + 3(4) - (-2) = 16 & \text{(I)} \\ 1 - (4) + 3(-2) = -9 & \text{(II)} \\ 5(1) + 2(4) - (-2) = 15 & \text{(III)} \end{cases}$$

Example 1: Three Variables (Consistent System)

Solve the following systems of linear equations:

$$\begin{cases} x-y+2z=-4 & \text{(I)} \\ 2x+3y+z=\dfrac{1}{2} & \text{(II)} \\ x+4y-2z=4 & \text{(III)} \end{cases}$$

Solution: Using equations (I) and (III), eliminate z.

$$\begin{array}{rcrcrcr} x & - & y & + & 2z & = & -4 \\ x & + & 4y & - & 2z & = & 4 \\ \hline 2x & + & 3y & & & = & 0 \end{array}$$

Using equations (I) and (II), eliminate z.

$$\begin{cases} & x - y + 2z = -4 \\ [-2] & 2x + 3y + z = \dfrac{1}{2} \end{cases} \qquad \begin{array}{rcrcrcr} x & - & y & + & 2z & = & -4 \\ -4x & - & 6y & - & 2z & = & -1 \\ \hline -3x & - & 7y & & & = & -5 \end{array}$$

Eliminate the variable x using the two equations in x and y.

$$\begin{cases} [3] & 2x + 3y = 0 \\ [2] & -3x - 7y = -5 \end{cases} \qquad \begin{array}{rcrcr} 6x & + & 9y & = & 0 \\ -6x & - & 14y & = & -10 \\ \hline & & -5y & = & -10 \\ & & y & = & 2 \end{array}$$

Back substituting to find x yields:

$$\begin{aligned} 2x + 3(2) &= 0 \\ 2x &= -6 \\ x &= -3 \end{aligned}$$

Finally, using $x=-3$ and $y=2$, back substitute into (I).

$$\begin{aligned} -3-2+2z &= -4 \\ 2z &= 1 \\ z &= \frac{1}{2} \end{aligned}$$

The solution is $\left(-3,2,\dfrac{1}{2}\right)$.

The solution can be checked by substituting $\left(-3,2,\dfrac{1}{2}\right)$ into **all three** of the original equations.

$$\begin{cases} -3-2+2\left(\frac{1}{2}\right)=-4 & \text{(I)} \\ 2(-3)+3(2)+\frac{1}{2}=\frac{1}{2} & \text{(II)} \\ -3+4(2)-2\left(\frac{1}{2}\right)=4 & \text{(III)} \end{cases}$$

Example 2: Three Variables (Inconsistent System)

Solve the following systems of linear equations:

$$\begin{cases} 3x-5y+z=6 & \text{(I)} \\ x-y+3z=-1 & \text{(II)} \\ 2x-2y+6z=5 & \text{(III)} \end{cases}$$

Solution: Using equations (I) and (II), eliminate z.

$$\begin{cases} [-3]\ 3x-5y+z=6 \\ x-y+3z=-1 \end{cases} \qquad \begin{array}{rcrcrcr} -9x & + & 15y & - & 3z & = & -18 \\ x & - & y & + & 3z & = & -1 \\ \hline -8x & + & 14y & & & = & -19 \end{array}$$

Using equations (II) and (III), eliminate z.

$$\begin{cases} [-2]\ x-y+3z=-1 \\ 2x-2y+6z=5 \end{cases} \qquad \begin{array}{rcrcrcr} -2x & + & 2y & - & 6z & = & 2 \\ 2x & - & 2y & + & 6z & = & 5 \\ \hline & & & & 0 & = & 7 \end{array}$$

This last equation is false, and the system **does not have a solution**.

Example 3: Three Variables (Application)

A cash register contains \$341 in \$20, \$5, and \$2 bills. There are twenty-eight bills in all and three more twos than fives. How many bills of each kind are there?

Solution: Let x = number of \$20 bills
y = number of \$5 bills
z = number of \$2 bills

$$\begin{cases} x+y+z=28 & \text{(I) There are twenty-eight bills.} \\ 20x+5y+2z=341 & \text{(II) The total value is \$341.} \\ z=y+3 & \text{(III) There are three more twos than fives.} \end{cases}$$

continued on next page ...

Using equations (I) and (II), eliminate x.

$$\begin{cases} [-20] \quad x + y + z = 28 \\ 20x + 5y + 2z = 341 \end{cases} \qquad \begin{array}{rcrcrcr} -20x & - & 20y & - & 20z & = & -560 \\ 20x & + & 5y & + & 2z & = & 341 \\ \hline & - & 15y & - & 18z & = & -219 \end{array}$$

We rewrite equation (III) in the form $y - z = -3$ and use this equation along with the results just found:

$$\begin{cases} [15] \quad y - z = -3 \\ -15y - 18z = -219 \end{cases} \qquad \begin{array}{rcrcr} 15y & - & 15z & = & -45 \\ -15y & - & 18z & = & -219 \\ \hline & & -33z & = & -264 \\ & & z & = & 8 \end{array}$$

Back substituting to solve for y gives:

$$y - 8 = -3$$
$$y = 5$$

Now we can use the values $z = 8$ and $y = 5$ to substitute into equation (I).

$$x + 5 + 8 = 28$$
$$x = 15$$

There are fifteen \$20 bills, five \$5 bills, and eight \$2 bills.

Practice Problems

Solve the following system of linear equations:

$$\begin{cases} 2x + y + z = 4 \\ x + 2y + z = 1 \\ 3x + y - z = -3 \end{cases}$$

14.3 Exercises

Solve each of the systems of equations in Exercises 1 – 20. State which systems have no solution or an infinite number of solutions.

1. $x = 1, y = 0, z = 1$
2. $x = 2, y = 7, z = 4$
3. $x = 1, y = 2, z = -1$

1. $\begin{cases} x + y - z = 0 \\ 3x + 2y + z = 4 \\ x - 3y + 4z = 5 \end{cases}$

2. $\begin{cases} x - y + 2z = 3 \\ -6x + y + 3z = 7 \\ x + 2y - 5z = -4 \end{cases}$

3. $\begin{cases} 2x - y - z = 1 \\ 2x - 3y - 4z = 0 \\ x + y - z = 4 \end{cases}$

Answers to Practice Problems: $x = 1, y = -2, z = 4$

4. $x = 1, y = 3, z = 3$
5. Infinite solutions
6. $x = 1, y = -3, z = 2$
7. $x = -2, y = 9, z = 1$
8. $x = 1, y = 2, z = -1$
9. $x = 4, y = 1, z = 1$
10. Infinite solutions
11. $x = -2, y = 3, z = 1$
12. $x = 1, y = 2, z = 1$
13. No solution
14. $x = -1, y = 1, z = 2$
15. Infinite solutions
16. No solution
17. $x = \frac{1}{2}, y = \frac{1}{3}, z = -1$
18. $x = 1, y = -1, z = 3$
19. $x = 2, y = 1, z = -3$
20. $x = \frac{5}{2}, y = \frac{2}{3}, z = -\frac{1}{2}$
21. 34, 6, 27
22. 56, 84, 49
23. 11 nickels, 7 dimes, 5 quarters
24. 18 ones, 16 fives, 12 tens

4. $\begin{cases} y + z = 6 \\ x + 5y - 4z = 4 \\ x - 3y + 5z = 7 \end{cases}$

5. $\begin{cases} x + y - 2z = 4 \\ 2x + y = 1 \\ 5x + 3y - 2z = 6 \end{cases}$

6. $\begin{cases} 2y + z = -4 \\ 3x + 4z = 11 \\ x + y = -2 \end{cases}$

7. $\begin{cases} x - y + 5z = -6 \\ x + 2z = 0 \\ 6x + y + 3z = 0 \end{cases}$

8. $\begin{cases} x - y + 2z = -3 \\ 2x + y - z = 5 \\ 3x - 2y + 2z = -3 \end{cases}$

9. $\begin{cases} y + z = 2 \\ x + z = 5 \\ x + y = 5 \end{cases}$

10. $\begin{cases} x - y - 2z = 3 \\ x + 2y + z = 1 \\ 3y + 3z = -2 \end{cases}$

11. $\begin{cases} 2x - y + 5z = -2 \\ x + 3y - z = 6 \\ 4x + y + 3z = -2 \end{cases}$

12. $\begin{cases} 2x - y + 5z = 5 \\ x - 2y + 3z = 0 \\ x + y + 4z = 7 \end{cases}$

13. $\begin{cases} 3x + y + 4z = -6 \\ 2x + 3y - z = 2 \\ 5x + 4y + 3z = 2 \end{cases}$

14. $\begin{cases} 2x + y - z = -3 \\ -x + 2y + z = 5 \\ 2x + 3y - 2z = -3 \end{cases}$

15. $\begin{cases} x - 2y + z = 4 \\ x - y - 4z = 1 \\ 2x - 4y + 2z = 8 \end{cases}$

16. $\begin{cases} 2x - 2y + 3z = 4 \\ x - 3y + 2z = 2 \\ x + y + z = 1 \end{cases}$

17. $\begin{cases} 2x - 3y + z = -1 \\ 6x - 9y - 4z = 4 \\ 4x + 6y - z = 5 \end{cases}$

18. $\begin{cases} x + y + z = 3 \\ 2x - y - 2z = -3 \\ 3x + 2y + z = 4 \end{cases}$

19. $\begin{cases} 2x + 3y + z = 4 \\ 3x - 5y + 2z = -5 \\ 4x - 6y + 3z = -7 \end{cases}$

20. $\begin{cases} x + 6y + z = 6 \\ 2x + 3y - 2z = 8 \\ 2x + 4z = 3 \end{cases}$

21. The sum of three integers is 67. The sum of the first and second integers exceeds the third by 13. The third integer is 7 less than the first. Find the three integers.

22. The sum of three integers is 189. The first integer is 28 less than the second. The second integer is 21 less than the sum of the first and third integers. Find the three integers.

23. Sally is trying to get her brother Robert to learn to think algebraically. She tells him that she has 23 coins in her purse, including nickels, dimes, and quarters. She has two more dimes than quarters, and the total value of the coins is $2.50. How many of each kind of coin does she have?

24. A wallet contains $218 in $10, $5, and $1 bills. There are forty-six bills in all and four more fives than tens. How many bills of each kind are there?

25. $a = 1, b = 3, c = -2$

26. $a = 2, b = 1, c = -2$

27. 19 cm, 24 cm, 30 cm

28. bananas: $.60/lb., apples: $1.60/lb., grapes: $3.60/lb.

29. home: $90,000, lot: $22,000, improve: $11,000

30. $2.80

31. savings: $30,000, bonds: $55,000, stocks: $15,000

32. Pepsico: $9,000, IBM: $5,000, Microsoft: $16,000

33. 100°, 30°, 50°

25. Find values for a, b, and c so that the points (−1, −4), (2, 8), and (−2, −4) lie on the graph of the function $y = ax^2 + bx + c$.

26. Find values for a, b, and c so that the points (1, 1), (−3, 13), and (0, −2) lie on the graph of the function $y = ax^2 + bx + c$.

27. The perimeter of a triangle is 73 cm. The longest side is 13 cm less than the sum of the other two sides. The shortest side is 11 cm less than the longest side. Find the lengths of the three sides.

28. At Tony's Fruit Stand, 4 pounds of bananas, 2 pounds of apples, and 3 pounds of grapes cost $16.40. Five pounds of bananas, 4 pounds of apples, and 2 pounds of grapes cost $16.60. Two pounds of bananas, 3 pounds of apples, and 1 pound of grapes cost $9.60. Find the price per pound of each kind of fruit.

29. The Marshalls are having a house built. The cost of building the house is $24,000 more than three times the cost of the lot. The cost of the landscaping, sidewalks, and upgrades is one-half the cost of the lot. If the total cost is $123,000, what is the cost of each part of the construction (the home, the lot, and the improvements)?

30. At the Happy Burger Drive-In, you can buy 2 hamburgers, 1 chocolate shake, and 2 orders of fries, or 3 hamburgers and 1 order of fries, for $9.50. One hamburger, 2 chocolate shakes, and 1 order of fries cost $7.30. How much does a hamburger cost?

31. Kirk inherited $100,000 dollars from his aunt and decided to invest in three different accounts: savings, bonds, and stocks. His bond account was $10,000 more than three times the stock account. At the end of the first year, the savings returned 5%, the bonds 8%, and the stocks 10% for total interest of $7400. How much did he invest in each account?

32. Melissa has saved a total of $30,000 and wants to invest in three different stocks: Pepsico, IBM, and Microsoft. She wants the Pepsico amount to be $1000 less than twice the IBM amount and the Microsoft amount to be $2000 more than the total in the other two stocks. How much should she invest in each stock?

33. The sum of the measures of the three angles of a triangle is 180°. In one particular triangle, the largest angle is 10° more than three times the smallest angle, and the third angle is one-half the largest angle. What are the measures of the three angles?

34. 300 main floor, 200 balcony, and 80 mezzanine

35. 3 liters of 10%, 4.5 liters of 30%, 1.5 liters of 40%

36. No

37. A = 2, B = –1, C = 3

38. A = 4, B = 2, C = –1

34. The local theater has three types of seats for Broadway plays: main floor, balcony, and mezzanine. Main floor tickets are \$60, balcony tickets are \$45, mezzanine tickets are \$30. On one particular night the sales totaled \$29,400. Main floor sales were 20 more than the total of balcony and mezzanine sales. Balcony sales were 40 more than two times mezzanine sales. How many of each type of ticket were sold?

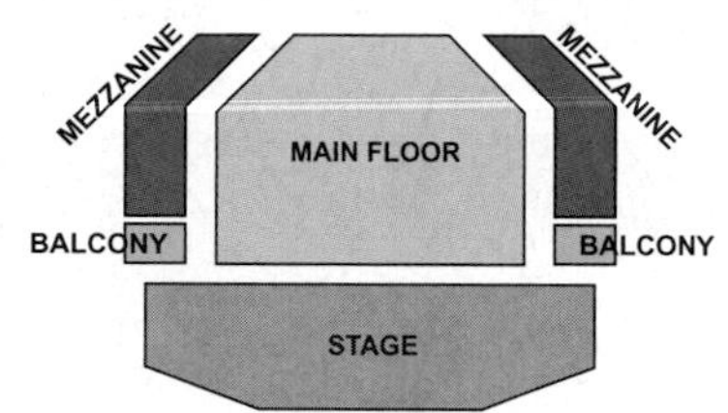

35. A chemist wants to mix 9 liters of a 25% acid solution. Because of limited amounts on hand, the mixture is to come from three different solutions, one with 10% acid, another with 30% acid, and a third with 40% acid. The amount of the 10% solution must be twice the amount of the 40% solution, and the amount of the 30% solution must equal the total amount of the other two solutions. How much of each solution must be used?

Writing and Thinking About Mathematics

36. Is it possible for three linear equations in three unknowns to have exactly two solutions? Explain your reasoning in some detail.

37. In geometry, we know that three non-collinear points determine a plane. (That is, if three points are not on a line, then there is a unique plane that contains all three points.) Find the values of A, B, and C (and therefore the equation of the plane) given $Ax + By + Cz = 3$ and the three points on the plane $(0, 3, 2)$, $(0, 0, 1)$ and $(-3, 0, 3)$. Sketch the plane in three dimensions as best you can by locating the three given points.

38. As stated in Exercise 37, three non-collinear points determine a plane. Find the values of A, B, and C (and therefore the equation of the plane) given $Ax + By + Cz = 10$ and the three points on the plane $(2, 0, -2)$, $(3, -1, 0)$ and $(-1, 5, -4)$. Sketch the plane in three dimensions as best you can by locating the three given points.

Hawkes Learning Systems: Introductory & Intermediate Algebra

Solving Systems of Linear Equations with Three Variables

14.4 Matrices and Gaussian Elimination

Objectives

After completing this section, you will be able to:

1. *Transform a matrix into triangular form by using elementary row operations.*
2. *Solve systems of linear equations by using the Gaussian elimination method.*

Matrices

A rectangular array of numbers is called a **matrix** (plural **matrices**). Matrices are usually named with capital letters, and each number in the matrix is called an **entry**. Entries written horizontally are said to form a **row**, and entries written vertically are said to form a **column**. The matrix A shown below has two rows and three columns and is a **2 × 3 matrix** (read "two by three matrix"). We say that the **dimension** of the matrix is two by three (or 2×3). Similarly, if a matrix has three rows and three columns then its dimension is 3×3.

$$A = \begin{bmatrix} 3 & 4 & 0 \\ 7 & -2 & 5 \end{bmatrix} \qquad \begin{bmatrix} 3 & 4 & 0 \\ 7 & -2 & 5 \end{bmatrix} \begin{matrix} \leftarrow \text{Row 1} \\ \leftarrow \text{Row 2} \end{matrix} \qquad \begin{bmatrix} 3 & 4 & 0 \\ 7 & -2 & 5 \end{bmatrix}$$

Column 1, Column 2, Column 3 (arrows pointing to the columns of the third matrix)

Three more examples are:

$$B = \begin{bmatrix} 5 & -1 \\ 2 & 3 \end{bmatrix} \qquad C = \begin{bmatrix} 5 & -1 & 0 & 7 \\ 2 & 3 & 2 & 8 \\ 1 & -3 & 0 & 6 \end{bmatrix} \qquad D = \begin{bmatrix} 0 & 4 \\ 1 & 6 \\ -1 & 3 \end{bmatrix}$$

2×2 matrix 3×4 matrix 3×2 matrix

A matrix with the same number of rows as columns is called a **square matrix**. Matrix B (shown above) is a square 2×2 matrix.

Elementary Row Operations

Matrices have many uses and are generated from various types of problems because they allow data to be presented in a systematic and orderly manner. (Business majors may want to look up a topic called Markov chains.)

Teaching Notes:
To generate interest in matrices, you might want to informally discuss and give simple examples of adding, subtracting, and multiplying with matrices and writing entries in the general form a_{ij}. Also, let students know that applications abound in physics and calculus with row or column matrices used to represent vectors. Most students find abstract material more worth their time and effort if they can see uses in future courses. Some of your students may even have seen or used Markov chains in a business or economics course and will want to discuss these applications.

Also, matrices can sometimes be added, subtracted, and multiplied. Some square matrices have inverses, much the same as multiplicative inverses for real numbers. Matrix methods of solving systems of linear equations can be done manually or with graphing calculators and computers. These topics are presented in courses such as finite mathematics and linear algebra.

In this text we will see that matrices can be used to solve systems of linear equations in which the equations are written in standard form. The two matrices derived from such a system are the **coefficient matrix** (made up of the coefficients of the variables) and the **augmented matrix** (including the coefficients and the constant terms). For example:

System	**Coefficient Matrix**	**Augmented Matrix**
$\begin{cases} x-y+z=-6 \\ 2x+3y=17 \\ x+2y+2z=7 \end{cases}$	$\begin{bmatrix} 1 & -1 & 1 \\ 2 & 3 & 0 \\ 1 & 2 & 2 \end{bmatrix}$	$\left[\begin{array}{ccc\|c} 1 & -1 & 1 & -6 \\ 2 & 3 & 0 & 17 \\ 1 & 2 & 2 & 7 \end{array}\right]$

Note that 0 is the entry in the second row, third column of both matrices. This 0 corresponds to the missing z-variable in the second equation. The second equation could have been written $2x + 3y + 0z = 17$.

In Sections 9.3 and 14.3, systems of linear equations were solved by the addition method and back substitution. In solving these systems, we can make any of the following three manipulations **without changing the solution set of the system.**

The system $\begin{cases} x-y+z=-6 \\ 2x+3y+0z=17 \\ x+2y+2z=7 \end{cases}$ is used here to illustrate some possibilities.

1. Any two equations may be interchanged.

$$\begin{cases} 2x+3y+0z=17 \\ x-y+z=-6 \\ x+2y+2z=7 \end{cases}$$

Here we have interchanged the first two equations.

2. All terms of any equation may be multiplied by a constant.

$$\begin{cases} -2x+2y-2z=12 \\ 2x+3y+0z=17 \\ x+2y+2z=7 \end{cases}$$

Here we have multiplied each term of the first equation by –2.

3. All terms of any equation may be multiplied by a constant and these new terms may be added to like terms of another equation. (The original equation remains unchanged.)

$$\begin{cases} x - y + z = -6 \\ 2x + 3y + 0z = 17 \\ 0x + 3y + z = 13 \end{cases}$$

Here we have multiplied the first equation by –1 (mentally) and added the results to the third equation.

When dealing with matrices, the three corresponding operations are called **elementary row operations**. These operations are listed below and illustrated in Example 1. Follow the steps outlined in Example 1 carefully, and note how the row operations are indicated, such as $\frac{1}{2}$R3 to indicate that all numbers in row 3 are multiplied by $\frac{1}{2}$. (Reasons for using these row operations are discussed under **Gaussian Elimination** on page 1005.)

Elementary Row Operations

1. *Interchange two rows.*

2. *Multiply a row by a nonzero constant.*

3. *Add a multiple of a row to another row.*

If any elementary row operation is applied to a matrix, the new matrix is said to be ***row-equivalent*** *to the original matrix.*

Example 1: Coefficient and Augmented Matrices

a. For the system $\begin{cases} y + z = 6 \\ x + 5y - 4z = 4 \\ 2x - 6y + 10z = 14 \end{cases}$

write the corresponding coefficient matrix and the corresponding augmented matrix.

Solution: Coefficient Matrix

$$\begin{bmatrix} 0 & 1 & 1 \\ 1 & 5 & -4 \\ 2 & -6 & 10 \end{bmatrix}$$

Augmented Matrix

$$\left[\begin{array}{ccc|c} 0 & 1 & 1 & 6 \\ 1 & 5 & -4 & 4 \\ 2 & -6 & 10 & 14 \end{array}\right]$$

b. In the augmented matrix, interchange rows 1 and 2 and multiply row 3 by $\frac{1}{2}$.

Solution:
$$\begin{matrix} R2 \rightarrow \\ R1 \rightarrow \\ \frac{1}{2}R3 \rightarrow \end{matrix} \left[\begin{array}{ccc|c} 1 & 5 & -4 & 4 \\ 0 & 1 & 1 & 6 \\ 1 & -3 & 5 & 7 \end{array}\right]$$

c. For the system $\begin{cases} x - y = 5 \\ 3x + 4y = 29 \end{cases}$

write the corresponding coefficient matrix and the corresponding augmented matrix.

Solution: Coefficient Matrix

$$\begin{bmatrix} 1 & -1 \\ 3 & 4 \end{bmatrix}$$

Augmented Matrix

$$\left[\begin{array}{cc|c} 1 & -1 & 5 \\ 3 & 4 & 29 \end{array}\right]$$

d. In the augmented matrix in Example 1c, add –3 times row 1 to row 2.

Solution: $R2 - 3 \cdot R1 \rightarrow \left[\begin{array}{cc|c} 1 & -1 & 5 \\ 0 & 7 & 14 \end{array}\right]$ Mentally $\left[\begin{array}{cc|c} 1 & -1 & 5 \\ 3-3(1) & 4-3(-1) & 29-3(5) \end{array}\right]$

Note that row 1 is left unchanged.

General Notation for a Matrix and a System of Equations

Notation with a small number to the right and below a variable is called **subscript** notation. For example, a_1 is read "*a* sub one" and b_3 is read "*b* sub three."

a_1 ← subscript; ↖ variable b_3 ← subscript; ↖ variable

With matrices we use capital letters to name the matrix and **double subscript** notation with corresponding lower case letters to indicate both the row and column location of an entry. For example, in a matrix *A* the entries will be designated as shown below.

a_{11} is read "*a* sub one one" and indicates the entry in the first row and first column;
a_{12} is read "*a* sub one two" and indicates the entry in the first row and second column;
a_{13} is read "*a* sub one three" and indicates the entry in the first row and third column;
a_{21} is read "*a* sub two one" and indicates the entry in the second row and first column;
and so on.

NOTES

We will see in dealing with polynomials later that a_{11} can be read simply as "a sub eleven". However, with matrices, we need to indicate the row and column corresponding to the entry. If there are more than nine rows or columns, then commas are used to separate the numbers as $a_{10,10}$. You will see the commas in use on your calculator.

With double subscript notation we can write the general form of a 2 × 3 matrix A and a 3 × 3 matrix B as follows:

$$A = \begin{bmatrix} a_{11} & a_{12} & a_{13} \\ a_{21} & a_{22} & a_{23} \end{bmatrix} \qquad B = \begin{bmatrix} b_{11} & b_{12} & b_{13} \\ b_{21} & b_{22} & b_{23} \\ b_{31} & b_{32} & b_{33} \end{bmatrix}$$

We will use this notation when discussing the use of calculators to define matrices and to operate with matrices. The general form of a system of three linear equations might use this notation in the following way:

$$\begin{cases} a_{11}x + a_{12}y + a_{13}z = k_1 \\ a_{21}x + a_{22}y + a_{23}z = k_2 \\ a_{31}x + a_{32}y + a_{33}z = k_3 \end{cases}$$

Teaching Notes:
You might want to illustrate a matrix in lower triangular form to satisfy students' curiosity.

A matrix is in **upper triangular form** (or just **triangular form** for our purposes) if its entries in the lower left triangular region are all 0's. The entries with the same numbers in the double subscript such as b_{11}, b_{22}, and b_{33} are said to form the **main diagonal** of a matrix. Thus, if all the entries below the main diagonal of a matrix are all 0's, the matrix is in triangular form as shown below:

$$B = \begin{bmatrix} b_{11} & b_{12} & b_{13} \\ 0 & b_{22} & b_{23} \\ 0 & 0 & b_{33} \end{bmatrix}$$

The upper triangular form of a matrix is also called the **row echelon form** (or **ref form**), and we will see that a graphing calculator can be used to change a matrix into the ref form.

Gaussian Elimination

Another method of solving a system of linear equations is the **Gaussian elimination** method (named after the famous German mathematician Karl Friedrich Gauss, 1777 – 1855). This method makes use of augmented matrices and elementary row operations. The objective is to transform an augmented matrix into triangular form (or ref form) and then use back substitution to find the values of the variables. The method is outlined as follows:

Strategy for Gaussian Elimination

1. *Write the augmented matrix for the system.*
2. *Use elementary row operations to transform the matrix into triangular form.*
3. *Solve the corresponding system of equations by using back substitution.*

The following examples illustrate the method. Study the steps and the corresponding comments carefully.

Example 2: Gaussian Elimination

a. Solve the following system of linear equations by using the Gaussian elimination method with back substitution.

$$\begin{cases} 2x+4y=-6 \\ 5x-\ \ y=7 \end{cases}$$

Solution:

Step 1: Write the augmented matrix.
[The following steps show how to use elementary row operations to get the matrix in triangular form (or ref form) with 0 in the lower left corner.]

$$\left[\begin{array}{cc|c} 2 & 4 & -6 \\ 5 & -1 & 7 \end{array}\right]$$

Step 2: Multiply row 1 by $\frac{1}{2}$ so that the entry in the upper left corner will be 1. This will help to get 0 below the 1 in the next step.

$$\left[\begin{array}{cc|c} 2 & 4 & -6 \\ 5 & -1 & 7 \end{array}\right] \qquad \frac{1}{2}\text{R1} \rightarrow \left[\begin{array}{cc|c} 1 & 2 & -3 \\ 5 & -1 & 7 \end{array}\right]$$

Step 3: To get 0 in the lower left corner, add –5 times row 1 to row 2.

$$\left[\begin{array}{cc|c} 1 & 2 & -3 \\ 5 & -1 & 7 \end{array}\right] \qquad \begin{array}{r} \\ \text{R2}-5\cdot\text{R1} \rightarrow \end{array} \left[\begin{array}{cc|c} 1 & 2 & -3 \\ 0 & -11 & 22 \end{array}\right]$$

continued on next page ...

Step 4: The triangular matrix in Step 3 represents the following system of linear equations.

$$\begin{cases} x+2y=-3 \\ 0x-11y=22 \end{cases}$$

Solving the last equation for y gives:

$$-11y=22$$
$$y=-2$$

Back substitute to find the value for x.

$$x+2(-2)=-3$$
$$x-4=-3$$
$$x=1$$

Thus, the solution is $x=1$ and $y=-2$. Or we can write $(1,-2)$.

b. Solve the following system of linear equations by using the Gaussian elimination method with back substitution.

$$\begin{cases} 2x-3y-z=-4 \\ -x+2y+z=6 \\ x-y+2z=14 \end{cases}$$

Solution:

Step 1: Write the augmented matrix.

$$\left[\begin{array}{ccc|c} 2 & -3 & -1 & -4 \\ -1 & 2 & 1 & 6 \\ 1 & -1 & 2 & 14 \end{array}\right]$$

Step 2: Exchange row 1 and row 3 so that the entry in the upper left corner will be 1.

$$\left[\begin{array}{ccc|c} 1 & -1 & 2 & 14 \\ -1 & 2 & 1 & 6 \\ 2 & -3 & -1 & -4 \end{array}\right]$$

Step 3: To get the 0's under the 1 in Column 1, add row 1 to row 2 and add -2 times row 1 to row 3.

$$\begin{array}{r} \\ R2+R1 \rightarrow \\ R3-2\cdot R1 \rightarrow \end{array} \left[\begin{array}{ccc|c} 1 & -1 & 2 & 14 \\ 0 & 1 & 3 & 20 \\ 0 & -1 & -5 & -32 \end{array}\right]$$

Step 4: Add row 2 to row 3 to arrive at the triangular form.

$$\begin{array}{r} \\ \\ R3 + R2 \rightarrow \end{array} \left[\begin{array}{ccc|c} 1 & -1 & 2 & 14 \\ 0 & 1 & 3 & 20 \\ 0 & 0 & -2 & -12 \end{array}\right]$$

Step 5: The triangle matrix in Step 4 represents the following system of linear equations:

$$\begin{cases} x - y + 2z = 14 \\ \quad y + 3z = 20 \\ \quad\quad -2z = -12 \end{cases}$$

Solving the last equation for z gives:

$$-2z = -12$$
$$z = 6$$

Back substitution into the equation $y + 3z = 20$ gives:

$$y + 3(6) = 20$$
$$y = 2$$

Back substitution into the equation $x - y + 2z = 14$ gives:

$$x - 2 + 2(6) = 14$$
$$x = 4$$

Thus, the solution is $x = 4$, $y = 2$, and $z = 6$. Or, we can write the solution in the form of an ordered triple as $(4, 2, 6)$.

Teaching Notes: You may or may not want to allow your students to explore the reduced row echelon form, `rref(` in the MATRIX MATH menu. This will lead directly to answers but may take away from their understanding of elementary row operations. You can explain how programmers must understand how to use elementary row operations to arrive at the ref form and the rref form used in calculators and in computers.

If the final matrix, in triangular form, has a row with all entries 0, then the system has an infinite number of solutions.

For example, solving the system $\begin{cases} x + 3y = 8 \\ 2x + 6y = 16 \end{cases}$ will result in the matrix $\left[\begin{array}{cc|c} 1 & 3 & 8 \\ 0 & 0 & 0 \end{array}\right]$.

The last line indicates that $0x + 0y = 0$ which is always true. Therefore, the solution to the system is the set of all solutions of the equation $x + 3y = 8$. The system is **dependent**.

If the triangular form of the augmented matrix shows the coefficient entries in one or more rows to be all 0's and the constant not 0, then the system has no solution.

For example, the last row of the augmented matrix $\left[\begin{array}{ccc|c} 1 & 2 & 2 & 7 \\ 0 & 1 & 3 & 6 \\ 0 & 0 & 0 & 15 \end{array}\right]$ indicates that $0 = 15$.

Since this is not true, the system has no solution. That is, the system is **inconsistent**.

Using the TI-83 Plus Calculator to Solve a System of Linear Equations

The TI-83 Plus calculator can be used to define and operate with matrices. The Gaussian elimination technique can be employed by the calculator by entering the coefficients and constants of a system of linear equations as an augmented matrix and then having the calculator reduce the matrix to ref form.

Pressing the matrix key (spelled MATRX on the keyboard) (found by pressing 2nd x^{-1}) will give the menu shown here:

Pressing the right arrow and moving to **MATH** will give the following choices:

Pressing the right arrow again and moving to **EDIT** will give the following choices:

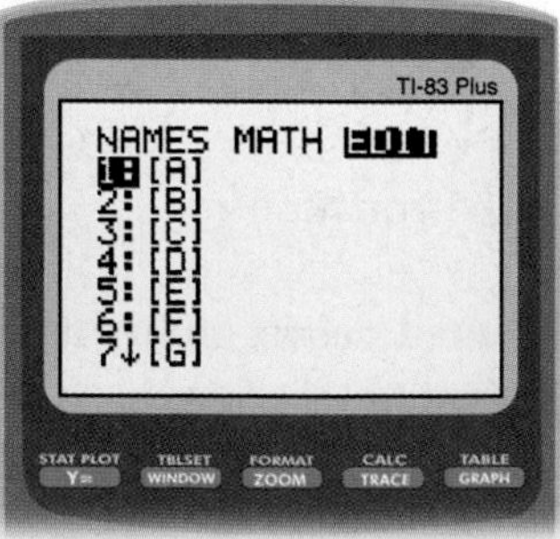

Example 3 shows how to use the calculator to solve a system of three linear equations in three variables. Study each step carefully.

Example 3: System of Equations

Use a TI-83 Plus calculator to solve the following system of linear equations.

$$\begin{cases} x+2y+z=1 \\ -x+y+z=-6 \\ 4x-y+3z=-1 \end{cases}$$

Solution:

Step 1: Press the MATRX key and move to the **EDIT** menu.

Press ENTER. The following display will appear.

Step 2: The augmented matrix is a 3 × 4 matrix. So, in the top line enter 3, press ENTER, enter 4, press ENTER and the display will appear as follows:

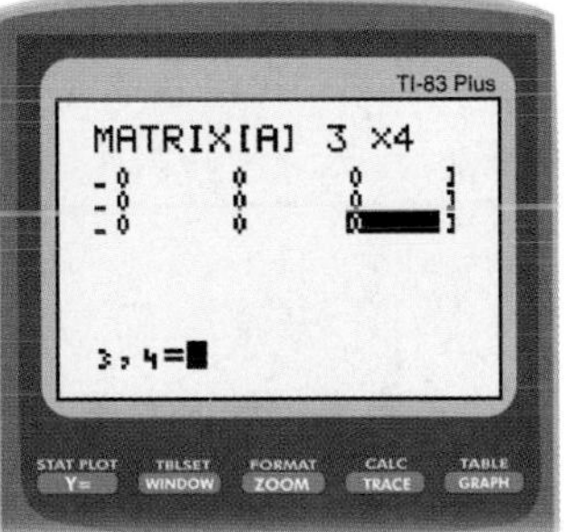

(**Note:** If other numbers are already present on the display, just type over them. The calculator will adjust automatically.)

Step 3: Move the cursor to the upper left entry position and enter the coefficients and constants in the matrix. As you enter each number press ENTER and the cursor will automatically move to the next position in the matrix. Note that the double subscripts appear at the bottom of the display as each number is entered.

continued on next page ...

The final display for matrix $[A]$ should appear as follows:

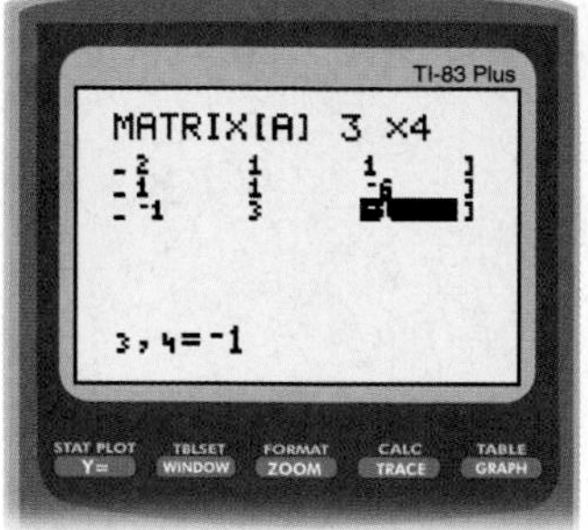

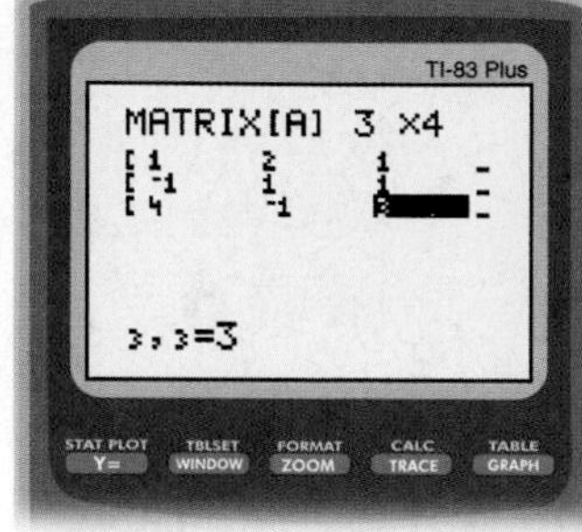

Note: The display only shows three columns at a time.

Step 4: Press 2nd QUIT; press MATRX again; go to **MATH**; move the cursor down to **A: ref(**; press ENTER. The display will appear as follows:

Step 5: Press MATRX again; press ENTER; enter a right parenthesis) ; press the MATH key on the keyboard; choose **1:>Frac** by pressing ENTER. The display will appear as follows:

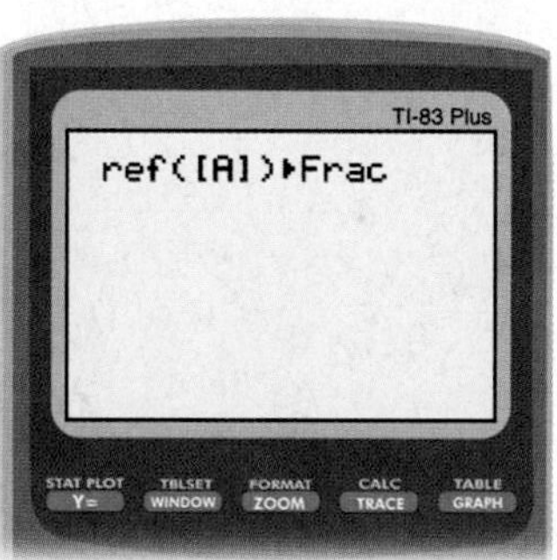

Step 6: Press ENTER and the ref form of the matrix will appear as follows:

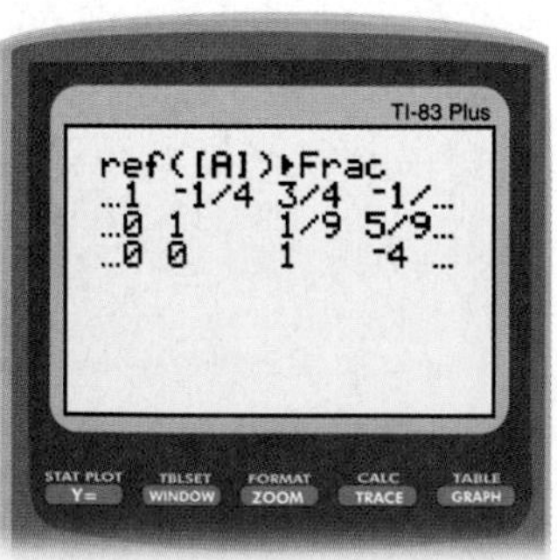

With back substitution we get the following solution: $x = 3, y = 1, z = -4$.

Practice Problems

Solve the following system of linear equations by using the Gaussian elimination method with back substitution.

$$\begin{cases} x-2y+3z=4 \\ 2x+y=0 \\ 3x+y-z=-4 \end{cases}$$

14.4 Exercises

1. $\begin{bmatrix} 2 & 2 \\ 5 & -1 \end{bmatrix}, \left[\begin{array}{cc|c} 2 & 2 & 13 \\ 5 & -1 & 10 \end{array}\right]$

2. $\begin{bmatrix} 1 & 4 \\ 2 & -3 \end{bmatrix}, \left[\begin{array}{cc|c} 1 & 4 & -1 \\ 2 & -3 & 7 \end{array}\right]$

3. $\begin{bmatrix} 7 & -2 & 7 \\ -5 & 3 & 0 \\ 0 & 4 & 11 \end{bmatrix},$ $\left[\begin{array}{ccc|c} 7 & -2 & 7 & 2 \\ -5 & 3 & 0 & 2 \\ 0 & 4 & 11 & 8 \end{array}\right]$

4. $\begin{bmatrix} -8 & 2 & -1 \\ 2 & 0 & 3 \\ -4 & -2 & 5 \end{bmatrix},$ $\left[\begin{array}{ccc|c} -8 & 2 & -1 & 6 \\ 2 & 0 & 3 & -3 \\ -4 & -2 & 5 & 13 \end{array}\right]$

5. $\begin{bmatrix} 3 & 1 & -1 & 2 \\ 1 & -1 & 2 & -1 \\ 0 & 2 & 5 & 1 \\ 1 & 3 & 0 & 3 \end{bmatrix},$ $\left[\begin{array}{cccc|c} 3 & 1 & -1 & 2 & 6 \\ 1 & -1 & 2 & -1 & -8 \\ 0 & 2 & 5 & 1 & 2 \\ 1 & 3 & 0 & 3 & 14 \end{array}\right]$

In Exercises 1 – 6, form the coefficient matrix and the augmented matrix for the given systems of linear equations.

1. $\begin{cases} 2x+2y=13 \\ 5x-y=10 \end{cases}$ **2.** $\begin{cases} x+4y=-1 \\ 2x-3y=7 \end{cases}$ **3.** $\begin{cases} 7x-2y+7z=2 \\ -5x+3y=2 \\ 4y+11z=8 \end{cases}$

4. $\begin{cases} -8x+2y-z=6 \\ 2x+3z=-3 \\ -4x-2y+5z=13 \end{cases}$ **5.** $\begin{cases} 3x+y-z+2w=6 \\ x-y+2z-w=-8 \\ 2y+5z+w=2 \\ x+3y+3w=14 \end{cases}$ **6.** $\begin{cases} 4x+y+3z-2w=13 \\ x-2y+z-4w=-3 \\ x+y+4z+2w=12 \\ -2x+3y-z-3w=5 \end{cases}$

In Exercises 7 – 10, write the system of linear equations represented by each of the augmented matrices. Use x, y, and z as the variables.

7. $\left[\begin{array}{cc|c} -3 & 5 & 1 \\ -1 & 3 & 2 \end{array}\right]$ **8.** $\left[\begin{array}{cc|c} 3 & -1 & 5 \\ -2 & 10 & 9 \end{array}\right]$ **9.** $\left[\begin{array}{ccc|c} 1 & 3 & 4 & 1 \\ 2 & -3 & -2 & 0 \\ 1 & 1 & 0 & -4 \end{array}\right]$ **10.** $\left[\begin{array}{ccc|c} 2 & -9 & 14 & 0 \\ -3 & 0 & -8 & 5 \\ 2 & -6 & 1 & 3 \end{array}\right]$

In Exercises 11 – 26, use the Gaussian elimination method with back substitution to solve the given system of linear equations.

11. $\begin{cases} x+2y=3 \\ 2x-y=-4 \end{cases}$ **12.** $\begin{cases} 4x+3y=5 \\ -x-2y=0 \end{cases}$ **13.** $\begin{cases} -8x+2y=6 \\ x-2y=1 \end{cases}$

14. $\begin{cases} 2x+y=-2 \\ 4x+3y=-2 \end{cases}$ **15.** $\begin{cases} x-3y+2z=11 \\ -2x+4y+z=-3 \\ x-2y+3z=12 \end{cases}$ **16.** $\begin{cases} x+2y-z=6 \\ 3x-y+2z=9 \\ x+y+z=6 \end{cases}$

Answers to Practice Problems: $x=-1, y=2, z=3$

6. $\begin{bmatrix} 4 & 1 & 3 & -2 \\ 1 & -2 & 1 & -4 \\ 1 & 1 & 4 & 2 \\ -2 & 3 & -1 & -3 \end{bmatrix}$, $\left[\begin{array}{cccc|c} 4 & 1 & 3 & -2 & 13 \\ 1 & -2 & 1 & -4 & -3 \\ 1 & 1 & 4 & 2 & 12 \\ -2 & 3 & -1 & -3 & 5 \end{array}\right]$

7. $\begin{cases} -3x+5y=1 \\ -x+3y=2 \end{cases}$

8. $\begin{cases} 3x-y=5 \\ -2x+10y=9 \end{cases}$

9. $\begin{cases} x+3y+4z=1 \\ 2x-3y-2z=0 \\ x+y=-4 \end{cases}$

10. $\begin{cases} 2x-9y+14z=0 \\ -3x-8z=5 \\ 2x-6y+z=3 \end{cases}$

11. $x=-1, y=2$
12. $x=2, y=-1$
13. $x=-1, y=-1$
14. $x=-2, y=2$
15. $x=-1, y=-2, z=3$
16. $x=3, y=2, z=1$
17. $x=1, y=0, z=1$
18. $x=2, y=-1, z=1$
19. $x=2, y=1, z=-1$
20. $x=-2, y=9, z=1$
21. $x=-2, y=-1, z=5$
22. $x=-2, y=0, z=1$
23. $x=1, y=2, z=1$
24. $x=-2, y=3, z=1$
25. $x=1, y=-3, z=2$
26. $x=4, y=1, z=1$
27. 52, 40, 77
28. bacon: \$3.09/lb., eggs: \$4.03/doz., bread: \$1.40/loaf

17. $\begin{cases} x+2y+3z=4 \\ x-y-z=0 \\ 4x-3y+z=5 \end{cases}$

18. $\begin{cases} x+y-2z=-1 \\ 3x+4y-2z=0 \\ x-y+z=4 \end{cases}$

19. $\begin{cases} x-y-2z=3 \\ x+2y-z=5 \\ 2x-3y-2z=3 \end{cases}$

20. $\begin{cases} x-y+5z=-6 \\ x+2z=0 \\ 6x+y+3z=0 \end{cases}$

21. $\begin{cases} x-3y-z=-4 \\ 3x-2y+z=1 \\ -2x+y+2z=13 \end{cases}$

22. $\begin{cases} 2x-y-5z=-9 \\ x-3y+2z=0 \\ 3x+2y+10z=4 \end{cases}$

23. $\begin{cases} x-2y+3z=0 \\ x+y+4z=7 \\ 2x-y+5z=5 \end{cases}$

24. $\begin{cases} 2x-y+5z=-2 \\ 4x+y+3z=-2 \\ x+3y-z=6 \end{cases}$

25. $\begin{cases} 3x+4z=11 \\ x+y=-2 \\ 2y+z=-4 \end{cases}$

26. $\begin{cases} y+z=2 \\ x+y=5 \\ x+z=5 \end{cases}$

For Exercises 27 – 30, set up a system of linear equations that represents the information and solve the system using Gaussian elimination.

27. The sum of three integers is 169. The first integer is twelve more than the second integer. The third integer is fifteen less than the sum of the first and second integers. What are the integers?

28. Julie bought a pound of bacon, a dozen eggs, and a loaf of bread. The total cost was \$8.52. The eggs cost \$0.94 more than the bacon. The combined cost of the bread and eggs was \$2.34 more than the cost of the bacon. Find the cost of each item.

29. A pizzeria sells three sizes of pizzas: small, medium, and large. The pizzas sell for \$6.00, \$8.00, and \$9.50, respectively. One evening they sold 68 pizzas for a total of \$528.00. If they sold twice as many medium sized pizzas as large-sized pizzas, how many of each size did they sell?

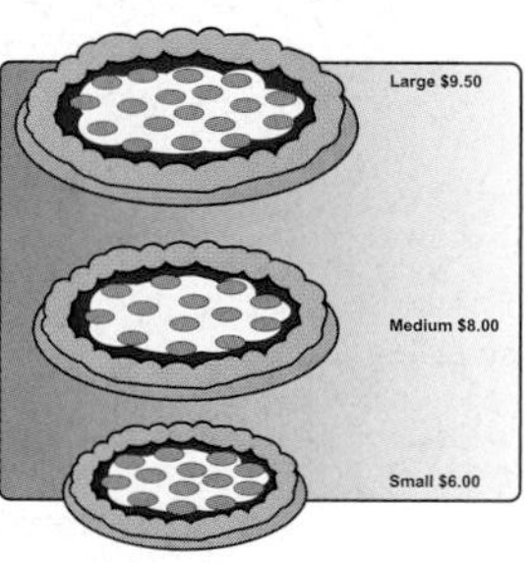

30. An investment firm is responsible for investing \$250,000 from an estate according to three conditions in the will of the deceased. The money is to be invested in three accounts paying 6%, 8%, and 11% interest. The amount invested in the 6% account is to be \$5000 more than the total invested in the other two accounts, and the total annual interest for the first year is to be \$19,250. How much is the firm supposed to invest in each account?

29. 20 small, 32 medium, 16 large
30. \$127,500 at 6%, \$62,500 at 8%, \$60,000 at 11%
31. $x = 0, y = -4$
32. $x = -4, y = 3$
33. $x = 2, y = 1, z = 7$
34. $x = -\frac{67}{45}, y = \frac{82}{45}, z = \frac{7}{90}$
35. $x = \frac{13}{12}, y = \frac{5}{4}, z = \frac{8}{3}$
36. $x = 3, y = 2, z = 0$
37. $x = -\frac{1}{3}, y = 1, z = \frac{16}{3}$
38. $x = \frac{1}{12}, y = \frac{8}{3}, z = -\frac{5}{3}, w = -\frac{13}{12}$
39. Answers will vary.

Use your graphing calculator to solve the systems of linear equations in Exercises 31 – 38.

31. $\begin{cases} x + y = -4 \\ 2x + 3y = -12 \end{cases}$

32. $\begin{cases} 2x + 3y = 1 \\ x - 5y = -19 \end{cases}$

33. $\begin{cases} x + y + z = 10 \\ 2x - y + z = 10 \\ -x + 2y + 2z = 14 \end{cases}$

34. $\begin{cases} 2x + y + 2z = -1 \\ x - y + 4z = -3 \\ 3x - y + \frac{1}{2}z = -\frac{25}{4} \end{cases}$

35. $\begin{cases} x - 3y + z = 0 \\ 2x + 2y - z = 2 \\ x + y + z = 5 \end{cases}$

36. $\begin{cases} x + 5y = 13 \\ 2x + z = 6 \\ 4y - z = 8 \end{cases}$

37. $\begin{cases} x - 2y - 2z = -13 \\ 2x + y - z = -5 \\ x + y + z = 6 \end{cases}$

38. $\begin{cases} x + y + z + w = 0 \\ x - y - z + w = -2 \\ 3x + 3y - z - w = 11 \\ y - 2z = 6 \end{cases}$

Writing and Thinking About Mathematics

39. Suppose that Gaussian elimination with a system of three linear equations in three unknowns results in the following triangular matrix. Discuss how you can use back substitution to find that the system has an infinite number of solutions. That is, the system is dependent. (**HINT:** Solve the second equation for z.)

$$\begin{bmatrix} 1 & 2 & -1 & 4 \\ 0 & 3 & 1 & 2 \\ 0 & 0 & 0 & 0 \end{bmatrix}$$

Hawkes Learning Systems: Introductory & Intermediate Algebra

Matrices and Gaussian Elimination

14.5 Determinants

Objectives

After completing this section, you will be able to:

1. *Evaluate 2 × 2 and 3 × 3 determinants.*
2. *Solve equations involving determinants.*

As was discussed in Section 14.4, a rectangular array of numbers is called a matrix, and matrices arise in connection with solving systems of linear equations such as:

$$\begin{cases} a_{11}x + a_{12}y = k_1 \\ a_{21}x + a_{22}y = k_2 \end{cases}$$

The **matrix of the coefficients** (or the **coefficient matrix**) is

$$A = \begin{bmatrix} a_{11} & a_{12} \\ a_{21} & a_{22} \end{bmatrix}.$$

If a matrix is **square** (the number of rows is equal to the number of columns), then there is a number associated with the matrix called its **determinant**. In this section, we will show how to evaluate determinants, and, in the next section, we will show how determinants can be used to solve systems of linear equations by using a method called **Cramer's Rule**.

Determinant

A ***determinant*** *is a real number associated with a square array of real numbers and is indicated by enclosing the array between two vertical bars. For a matrix A, the corresponding determinant is designated as det(A) and is read "determinant of A."*

Examples of determinants are:

(a) For the matrix, $A = \begin{bmatrix} 3 & 4 \\ 7 & -2 \end{bmatrix}$, $\det(A) = \begin{vmatrix} 3 & 4 \\ 7 & -2 \end{vmatrix}$

(b) For the matrix, $B = \begin{bmatrix} 1 & 6 & -3 \\ 4 & 5 & 5 \\ -1 & -1 & -1 \end{bmatrix}$, $\det(B) = \begin{vmatrix} 1 & 6 & -3 \\ 4 & 5 & 5 \\ -1 & -1 & -1 \end{vmatrix}$

Example (a) is a 2 × 2 determinant and has two rows and two columns.
Example (b) is a 3 × 3 determinant and has three rows and three columns.

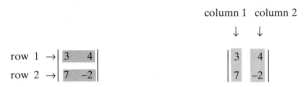

A 4 × 4 determinant has four rows and four columns. A determinant may be of any size $n \times n$ where n is a positive integer and $n \geq 2$. In this text, the discussion will be restricted to 2 × 2 and 3 × 3 determinants, and the entries will be real numbers. [Huge matrices and determinants (1000 × 1000 or larger) are common in industry, and their values are calculated by computers. Even then, someone must understand the algebraic techniques to be able to write the necessary programs.]

Every determinant with real entries has a real value. The method for finding the value of 3 × 3 determinants involves finding the value of 2 × 2 determinants. Determinants of larger matrices can be evaluated by using techniques similar to those shown here. Their applications occur in higher mathematics such as linear algebra and differential equations.

Value of a 2×2 Determinant

For the square matrix, $A = \begin{bmatrix} a_{11} & a_{12} \\ a_{21} & a_{22} \end{bmatrix}$, $\quad \boldsymbol{det(A)} = \begin{vmatrix} a_{11} & a_{12} \\ a_{21} & a_{22} \end{vmatrix} = a_{11}a_{22} - a_{21}a_{12}.$

As the definition indicates and the following examples illustrate, the value of a 2 × 2 determinant is the **product of the numbers in the diagonal containing the term in the first row, first column, minus the product of the numbers in the other diagonal**.

Example 1: 2×2 Determinant

Evaluate the following 2 × 2 determinants.

a. $\begin{vmatrix} 3 & 4 \\ 7 & -2 \end{vmatrix} = 3(-2) - 7(4) = -6 - 28 = -34$

b. $\begin{vmatrix} -5 & -\frac{1}{2} \\ 6 & 3 \end{vmatrix} = -5(3) - 6\left(-\frac{1}{2}\right) = -15 + 3 = -12$

c. $\begin{vmatrix} 1 & 7 \\ 2 & 14 \end{vmatrix} = 1(14) - 2(7) = 14 - 14 = 0$

One method of evaluating 3 × 3 determinants is called **expanding by minors**. In this method, **one row is chosen** and each entry in that row has a minor. Each minor is found by mentally crossing out both the row and column (shown here in the shaded regions) that contain that entry. The minors of the entries in the first row are illustrated here:

$$\begin{vmatrix} a_{11} & a_{12} & a_{13} \\ a_{21} & a_{22} & a_{23} \\ a_{31} & a_{32} & a_{33} \end{vmatrix} \longrightarrow \begin{vmatrix} a_{22} & a_{23} \\ a_{32} & a_{33} \end{vmatrix} \longleftarrow \text{minor of } a_{11}$$

$$\begin{vmatrix} a_{11} & a_{12} & a_{13} \\ a_{21} & a_{22} & a_{23} \\ a_{31} & a_{32} & a_{33} \end{vmatrix} \longrightarrow \begin{vmatrix} a_{21} & a_{23} \\ a_{31} & a_{33} \end{vmatrix} \longleftarrow \text{minor of } a_{12}$$

$$\begin{vmatrix} a_{11} & a_{12} & a_{13} \\ a_{21} & a_{22} & a_{23} \\ a_{31} & a_{32} & a_{33} \end{vmatrix} \longrightarrow \begin{vmatrix} a_{21} & a_{22} \\ a_{31} & a_{32} \end{vmatrix} \longleftarrow \text{minor of } a_{13}$$

To find the value of a determinant (of any dimension other than 2 × 2), first choose a row (or column) and find the product of each entry in that row (or column) with its corresponding minor. Then the value is determined by adding these products with appropriate adjustments of alternating signs of the minors. We say that the determinant has been expanded by that row (or column). The following illustrates how to find the value of a 3 × 3 determinant by expanding by the first row.

Value of a 3×3 Determinant

Teaching Notes:
A good in-class activity is to assign different groups of students a row to use in the determinant (i.e., using different minors) and demonstrate that the value of this determinant is the same, no matter which row is used.

For the square matrix, $A = \begin{bmatrix} a_{11} & a_{12} & a_{13} \\ a_{21} & a_{22} & a_{23} \\ a_{31} & a_{32} & a_{33} \end{bmatrix}$, $\mathbf{det(A)} = \begin{vmatrix} \mathbf{a_{11}} & \mathbf{a_{12}} & \mathbf{a_{13}} \\ a_{21} & a_{22} & a_{23} \\ a_{31} & a_{32} & a_{33} \end{vmatrix}$

$$= \mathbf{a_{11}}(\text{minor of } a_{11}) - \mathbf{a_{12}}(\text{minor of } a_{12}) + \mathbf{a_{13}}(\text{minor of } a_{13})$$

$$= \mathbf{a_{11}}\begin{vmatrix} a_{22} & a_{23} \\ a_{32} & a_{33} \end{vmatrix} - \mathbf{a_{12}}\begin{vmatrix} a_{21} & a_{23} \\ a_{31} & a_{33} \end{vmatrix} + \mathbf{a_{13}}\begin{vmatrix} a_{21} & a_{22} \\ a_{31} & a_{32} \end{vmatrix}$$

NOTES

CAUTION: The negative sign in the middle term of the expansion (representing −1 times a_{12}) is a critical part of the method and is a source of error for many students. **Be careful**.

Each minor is multiplied by its corresponding entry and +1 or −1 according to the pattern illustrated in Figure 14.4. [**Note:** The signs alternate and this pattern can be extended to apply to any $n \times n$ determinant.]

$$\begin{vmatrix} + & - & + \\ - & + & - \\ + & - & + \end{vmatrix}$$

Figure 14.4

For example, the value of a 3 × 3 determinant can be found by expanding by the minors of the second row as follows:

$$\det(A) = -a_{21}\,(\text{minor of } a_{21}) + a_{22}\,(\text{minor of } a_{22}) - a_{23}\,(\text{minor of } a_{23})$$

Note the use of the alternating + and – signs from the pattern in Figure 14.4. You may want to try this for practice with some of the exercises.

NOTES

There are methods other than expanding by minors for evaluating 3 × 3 determinants. Your instructor may wish to show you some of these, but they will not be discussed in the text. The advantage of learning to expand by minors is that this method can be used for evaluating higher order determinants.

Example 2: 3×3 Determinant

Evaluate the following 3 × 3 determinants.

a. $\begin{vmatrix} 5 & 1 & -4 \\ 2 & 6 & 3 \\ 2 & 2 & 1 \end{vmatrix}$ Using Row 1, mentally delete the shaded region.

Solution: $= 5\begin{vmatrix} 5 & 1 & -4 \\ 2 & 6 & 3 \\ 2 & 2 & 1 \end{vmatrix} - 1\begin{vmatrix} 5 & 1 & -4 \\ 2 & 6 & 3 \\ 2 & 2 & 1 \end{vmatrix} - 4\begin{vmatrix} 5 & 1 & -4 \\ 2 & 6 & 3 \\ 2 & 2 & 1 \end{vmatrix}$

$$\begin{aligned}
&= 5\begin{vmatrix} 6 & 3 \\ 2 & 1 \end{vmatrix} - 1\begin{vmatrix} 2 & 3 \\ 2 & 1 \end{vmatrix} - 4\begin{vmatrix} 2 & 6 \\ 2 & 2 \end{vmatrix} \\
&= 5(6 \cdot 1 - 2 \cdot 3) - 1(2 \cdot 1 - 2 \cdot 3) - 4(2 \cdot 2 - 2 \cdot 6) \\
&= 5(6 - 6) - 1(2 - 6) - 4(4 - 12) \\
&= 5(0) - 1(-4) - 4(-8) \\
&= 0 + 4 + 32 \\
&= 36
\end{aligned}$$

continued on next page ...

b. $\begin{vmatrix} 6 & -2 & 4 \\ 1 & 7 & 0 \\ -3 & 2 & -1 \end{vmatrix}$

Solution: $= 6\begin{vmatrix} 7 & 0 \\ 2 & -1 \end{vmatrix} + 2\begin{vmatrix} 1 & 0 \\ -3 & -1 \end{vmatrix} + 4\begin{vmatrix} 1 & 7 \\ -3 & 2 \end{vmatrix}$

$= 6(-7-0) + 2(-1-0) + 4(2+21)$ After some practice, many of these steps can be done mentally.

$= -42 - 2 + 92$

$= 48$

Example 3: Equations with Determinants

Solve the following equation for x: $\begin{vmatrix} 2 & 3 & 0 \\ 6 & x & 5 \\ 1 & -2 & 9 \end{vmatrix} = 53.$

Solution: First, evaluate the determinant.

$$\begin{vmatrix} 2 & 3 & 0 \\ 6 & x & 5 \\ 1 & -2 & 9 \end{vmatrix} = 2\begin{vmatrix} x & 5 \\ -2 & 9 \end{vmatrix} - 3\begin{vmatrix} 6 & 5 \\ 1 & 9 \end{vmatrix} + 0\begin{vmatrix} 6 & x \\ 1 & -2 \end{vmatrix}$$

$$= 2(9x+10) - 3(54-5) + 0$$

$$= 18x + 20 - 162 + 15$$

$$= 18x - 127$$

Now solve the equation.

$$18x - 127 = 53$$

$$18x = 180$$

$$x = 10$$

The technique of expanding by minors may be used (with appropriate adjustments) to evaluate any $n \times n$ determinant. For example, in a 4×4 determinant, the minors of the entries in a particular row will be 3×3 determinants. Also, there are techniques for simplifying determinants and there are rules for arithmetic with determinants. The general rules governing these operations are discussed in courses in precalculus mathematics, finite mathematics, and linear algebra.

Using the TI-83 Plus Calculator to Evaluate a Determinant

A TI-83 Plus calculator (and other graphing calculators) can be used to find the value of the determinant of a square matrix. The determinant command, **1: det(** is found as the first entry in the MATRIX MATH menu. Example 4 shows, in a step-by-step format, how to find the determinant of a given 3×3 matrix.

Example 4: Evaluating Determinants with a Calculator

Use a TI-83 Plus calculator to find the value of det(A) for the matrix

$$A = \begin{bmatrix} 2 & 5 & 7 \\ 3 & 1 & 0 \\ 4 & 0 & 3 \end{bmatrix}$$

Step 1: Press MATRX, go to the `EDIT` menu and enter the numbers in the matrix A. The display should appear as follows:

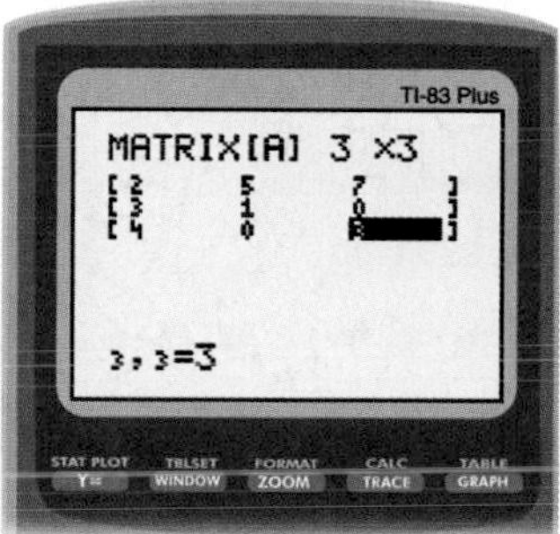

Step 2: Press QUIT then MATRX again and go to the `MATH` menu. On the `MATH` menu choose `1: det(` and press ENTER. The display should appear as follows:

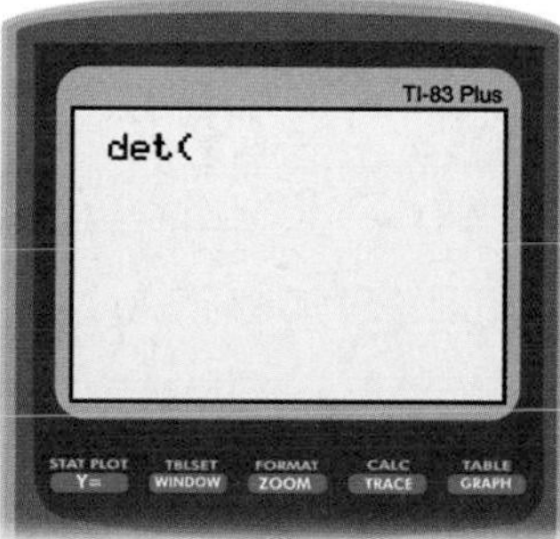

continued on next page ...

Step 3: Press MATRX again and on the **NAMES** menu choose `1:[A] 3×3`. Press ENTER and the display should appear as follows:

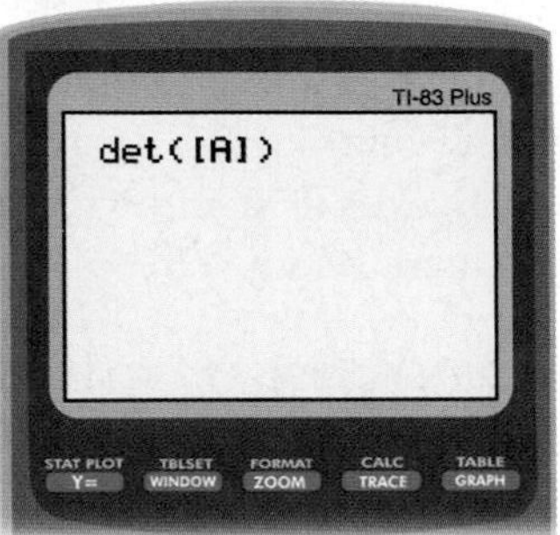

Step 4: Press ENTER and the display should appear as follows with the answer:

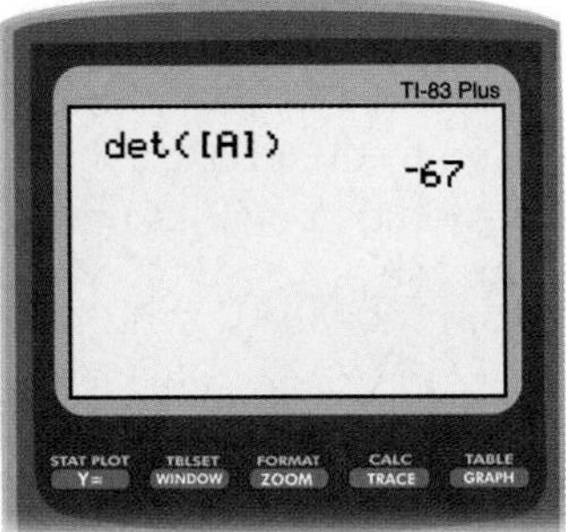

Practice Problems

Evaluate each of the following determinants.

1. $\begin{vmatrix} -3 & 2 \\ 4 & 7 \end{vmatrix}$ **2.** $\begin{vmatrix} 6 & 3 \\ 4 & 2 \end{vmatrix}$ **3.** $\begin{vmatrix} 1 & 4 & 0 \\ 2 & -1 & 5 \\ 0 & 7 & -1 \end{vmatrix}$

Use a TI-83 Plus calculator to find the value of the determinant.

4. $\begin{vmatrix} 5 & -1 & 3 \\ 0 & 4 & 2 \\ -3 & 1 & 3 \end{vmatrix}$

14.5 Exercises

1. −22
2. 11
3. −212
4. −76

In Exercises 1 – 4, the matrix A is given. Find det(A).

1. $A = \begin{bmatrix} 2 & 7 \\ 4 & 3 \end{bmatrix}$ **2.** $A = \begin{bmatrix} 7 & 3 \\ 8 & 5 \end{bmatrix}$ **3.** $A = \begin{bmatrix} -5 & 2 & 1 \\ 4 & 8 & 0 \\ -2 & 3 & 5 \end{bmatrix}$ **4.** $A = \begin{bmatrix} -6 & 5 & -3 \\ 4 & 0 & -1 \\ -2 & 7 & -2 \end{bmatrix}$

Answers to Practice Problems: 1. −29 **2.** 0 **3.** −26 **4.** 92

5. 11
6. –48
7. 3
8. –17
9. 47
10. 14
11. 36
12. 39
13. –3
14. –23
15. –4
16. 24
17. –25
18. –36
19. 20
20. –45
21. $x = 7$
22. $x = 9$
23. $x = -7$
24. $x = 2$
25. $x = -3$
26. $2x + 4y = 14$
27. $2x - 7y = -11$
28. $10x + 4y = 24$
29. $A = 1$

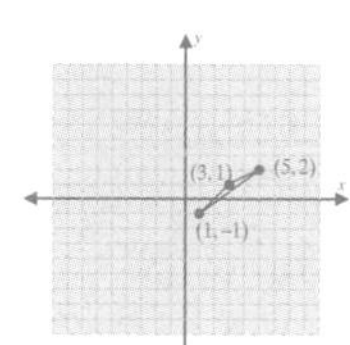

30. $A = \frac{7}{2}$

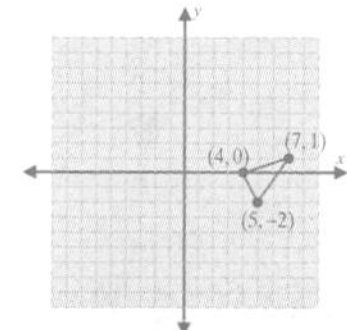

31. $A = \frac{31}{2}$

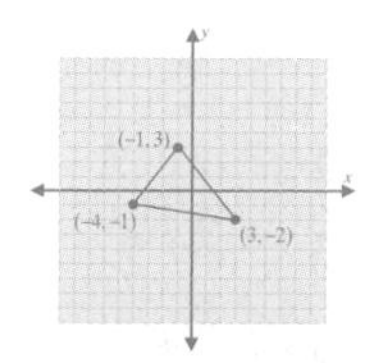

Evaluate the determinants in Exercises 5 – 20.

5. $\begin{vmatrix} 1 & 3 \\ -2 & 5 \end{vmatrix}$ **6.** $\begin{vmatrix} 7 & 2 \\ 3 & -6 \end{vmatrix}$ **7.** $\begin{vmatrix} 6 & 3 \\ -11 & -5 \end{vmatrix}$ **8.** $\begin{vmatrix} 2 & 3 \\ 3 & -4 \end{vmatrix}$

9. $\begin{vmatrix} 9 & 4 \\ 4 & 7 \end{vmatrix}$ **10.** $\begin{vmatrix} 3 & -4 \\ 8 & -6 \end{vmatrix}$ **11.** $\begin{vmatrix} 0 & -1 & 2 \\ 3 & 5 & -7 \\ -3 & 4 & 1 \end{vmatrix}$ **12.** $\begin{vmatrix} 1 & 0 & -1 \\ -2 & 3 & 5 \\ 6 & -3 & 4 \end{vmatrix}$

13. $\begin{vmatrix} 1 & -1 & 2 \\ -2 & 5 & -7 \\ 6 & 4 & 1 \end{vmatrix}$ **14.** $\begin{vmatrix} 2 & -1 & -3 \\ 5 & 9 & 4 \\ 7 & 6 & -2 \end{vmatrix}$ **15.** $\begin{vmatrix} 2 & 1 & 3 \\ 3 & 4 & 5 \\ 1 & 7 & 2 \end{vmatrix}$

16. $\begin{vmatrix} -3 & 2 & 1 \\ 1 & -4 & -1 \\ 2 & 5 & 3 \end{vmatrix}$ **17.** $\begin{vmatrix} 2 & 1 & -1 \\ 4 & 3 & 2 \\ 1 & 5 & 5 \end{vmatrix}$ **18.** $\begin{vmatrix} 6 & 7 & 1 \\ 0 & 3 & 3 \\ 4 & 1 & -5 \end{vmatrix}$

19. $\begin{vmatrix} 3 & -1 & -1 \\ 2 & 4 & 1 \\ -1 & 1 & 2 \end{vmatrix}$ **20.** $\begin{vmatrix} 2 & 3 & 2 \\ 1 & -1 & 5 \\ 0 & 5 & 1 \end{vmatrix}$

Use the method for evaluating determinants to solve the equations for x in Exercises 21 – 25.

21. $\begin{vmatrix} 1 & 3 & 4 \\ 2 & x & 3 \\ 1 & 3 & 5 \end{vmatrix} = 1$ **22.** $\begin{vmatrix} -2 & -1 & 1 \\ x & 1 & -1 \\ 4 & 3 & -2 \end{vmatrix} = 7$ **23.** $\begin{vmatrix} 1 & x & x \\ 2 & -2 & 1 \\ -1 & 3 & 2 \end{vmatrix} = 0$

24. $\begin{vmatrix} x & x & 1 \\ 1 & 5 & 0 \\ 0 & 1 & -2 \end{vmatrix} = -15$ **25.** $\begin{vmatrix} 3 & 1 & -2 \\ 1 & x & 4 \\ 2 & x & 0 \end{vmatrix} = 38$

The equation $\begin{vmatrix} x & y & 1 \\ x_1 & y_1 & 1 \\ x_2 & y_2 & 1 \end{vmatrix} = 0$ *is an equation of the line passing through the two points* $P_1(x_1, y_1)$ *and* $P_2(x_2, y_2)$*. Find an equation for the line determined by the pairs of points given in Exercises 26 – 28.*

26. $(3, 2), (-1, 4)$ **27.** $(-2, 1), (5, 3)$ **28.** $(4, -4), (0, 6)$

The area of the triangle having the vertices $P_1(x_1, y_1), P_2(x_2, y_2)$ *and* $P_3(x_3, y_3)$ *is given by the absolute value of the expression* $\frac{1}{2}\begin{vmatrix} x_1 & y_1 & 1 \\ x_2 & y_2 & 1 \\ x_3 & y_3 & 1 \end{vmatrix}$*. In Exercises 29 – 31, draw the triangle with the given points as vertices and then find the area of the triangle.*

29. $(3, 1), (5, 2), (1, -1)$ **30.** $(4, 0), (7, 1), (5, -2)$ **31.** $(-1, 3), (-4, -1), (3, -2)$

32. The points fall on a horizontal or vertical line (either the x values are the same, or the y values are the same).

32. Explain, in your own words, the position of the three points $P_1(x_1, y_1)$, $P_2(x_2, y_2)$, and $P_3(x_3, y_3)$ if the expression $\frac{1}{2}\begin{vmatrix} x_1 & y_1 & 1 \\ x_2 & y_2 & 1 \\ x_3 & y_3 & 1 \end{vmatrix}$ has a value of 0. **Hint:** Refer to the discussion before Exercises 29 – 31.

In Exercises 33 – 35, use a graphing calculator to find the value of the determinant.

33. –25

34. –46.804

35. –33.28

33. $\begin{vmatrix} 3 & -4 & 6 \\ 2 & 4 & -1 \\ 7 & 9 & -1 \end{vmatrix}$

34. $\begin{vmatrix} 2.1 & 3.5 & -3.4 \\ 2.6 & 5.0 & 1.2 \\ -1.0 & 3.4 & 9.3 \end{vmatrix}$

35. $\begin{vmatrix} 1.6 & \frac{1}{2} & -5.9 \\ 0.7 & \frac{3}{4} & 1.7 \\ 5.0 & 8.2 & -4.1 \end{vmatrix}$

36. 0

37. 0

38. a. it will be the opposite
b. it will be the opposite

Writing and Thinking About Mathematics

36. Suppose that in a 2 × 2 determinant two rows are identical. What will be the value of this determinant? Give two specific examples and a general example to back up your conclusion.

37. Suppose that in a 3 × 3 determinant one row is all 0's. What will be the value of this determinant? Give two specific examples and a general example to back up your conclusion.

38. (a) Suppose that in a 2 × 2 determinant two rows (or columns) are switched. How will the value of this new determinant relate to the value of the original determinant?

(b) Suppose that in a 3 × 3 determinant two rows (or columns) are switched. How will the value of this new determinant relate to the value of the original determinant?
Give two specific examples and a general example to back up your conclusion in each case.

Hawkes Learning Systems: Introductory & Intermediate Algebra

Determinants

14.6 Determinants and Systems of Linear Equations: Cramer's Rule

Objectives

After completing this section, you will be able to:

Solve systems of linear equations by using Cramer's Rule.

Cramer's Rule is a method that uses determinants for solving systems of linear equations. To explain the method and how these determinants are generated, we first illustrate the solution to a system of linear equations by the addition method and do not simplify the indicated products and sums of the coefficients. We will see that these products and sums can be represented as determinants.

Consider the following system of linear equations **with the equations in standard form:**

$$\begin{cases} 2x+3y=-5 \\ 4x+y=5 \end{cases}$$

Eliminating y gives:

$$\begin{cases} [1] & 2x & + & 3y & = & -5 \\ [-3] & 4x & + & y & = & 5 \end{cases}$$

$$\begin{aligned} 1(2x)+1(3y)&=1(-5) \\ -3(4x)-3(1y)&=-3(5) \\ \hline \left[1(2)-3(4)\right]x&=1(-5)-3(5) \\ x&=\frac{1(-5)-3(5)}{1(2)-3(4)} \end{aligned}$$

Eliminating x gives:

$$\begin{cases} [-4] & 2x+3y=-5 \\ [2] & 4x+y=5 \end{cases}$$

$$\begin{aligned} -4(2x)-4(3y)&=-4(-5) \\ 2(4x)+2(1y)&=2(5) \\ \hline \left[2(1)-4(3)\right]y&=2(5)-4(-5) \\ y&=\frac{2(5)-4(-5)}{2(1)-4(3)} \end{aligned}$$

Notice that the denominators for both x and y are the same number. This number is the value of the determinant of the coefficients. (Remember that the equations are in standard form.)

$$\text{Determinant of coefficients } = D = \begin{vmatrix} 2 & 3 \\ 4 & 1 \end{vmatrix} = 2 \cdot 1 - 4 \cdot 3 = -10$$

In determinant form, the numerators are

$$D_x = \begin{vmatrix} -5 & 3 \\ 5 & 1 \end{vmatrix} = 1(-5) - 3(5) = -20$$

and

$$D_y = \begin{vmatrix} 2 & -5 \\ 4 & 5 \end{vmatrix} = 2(5) - 4(-5) = 30.$$

Therefore, the values for x and y can be written in fraction form using determinants as follows:

$$x = \frac{D_x}{D} = \frac{-20}{-10} = 2 \qquad \text{and} \qquad y = \frac{D_y}{D} = \frac{30}{-10} = -3$$

The determinant D_x is formed as follows:

1. Form D, the determinant of the coefficients.
2. Replace the coefficients of x with the corresponding constants on the right hand side of the equations.

The determinant D_y is formed as follows:

1. Form D, the determinant of the coefficients.
2. Replace the coefficients of y with the corresponding constants on the right hand side of the equations.

Cramer's Rule is stated here only for 2 × 2 systems (systems of two linear equations in two variables) and 3 × 3 systems (systems of three linear equations in three variables). However, Cramer's Rule applies to all $n \times n$ systems of linear equations.

Cramer's Rule for 2 × 2 Matrices

For the system $\begin{cases} a_{11}x + a_{12}y = \mathbf{k_1} \\ a_{21}x + a_{22}y = \mathbf{k_2} \end{cases}$,

where

$$D = \begin{vmatrix} a_{11} & a_{12} \\ a_{21} & a_{22} \end{vmatrix} \qquad D_x = \begin{vmatrix} \mathbf{k_1} & a_{12} \\ \mathbf{k_2} & a_{22} \end{vmatrix} \qquad \text{and} \qquad D_y = \begin{vmatrix} a_{11} & \mathbf{k_1} \\ a_{21} & \mathbf{k_2} \end{vmatrix},$$

if $D \neq 0$, *then*

$$x = \frac{D_x}{D} \quad \text{and} \quad y = \frac{D_y}{D}$$

is the unique solution to the system.

Cramer's Rule for 3 × 3 Matrices

For the system $\begin{cases} a_{11}x + a_{12}y + a_{13}z = \mathbf{k_1} \\ a_{21}x + a_{22}y + a_{23}z = \mathbf{k_2}, \\ a_{31}x + a_{32}y + a_{33}z = \mathbf{k_3} \end{cases}$

where

$$D = \begin{vmatrix} a_{11} & a_{12} & a_{13} \\ a_{21} & a_{22} & a_{23} \\ a_{31} & a_{32} & a_{33} \end{vmatrix}$$

$$D_x = \begin{vmatrix} \mathbf{k_1} & a_{12} & a_{13} \\ \mathbf{k_2} & a_{22} & a_{23} \\ \mathbf{k_3} & a_{32} & a_{33} \end{vmatrix} \quad D_y = \begin{vmatrix} a_{11} & \mathbf{k_1} & a_{13} \\ a_{21} & \mathbf{k_2} & a_{23} \\ a_{31} & \mathbf{k_3} & a_{33} \end{vmatrix} \quad \textit{and} \quad D_z = \begin{vmatrix} a_{11} & a_{12} & \mathbf{k_1} \\ a_{21} & a_{22} & \mathbf{k_2} \\ a_{31} & a_{32} & \mathbf{k_3} \end{vmatrix}$$

if $D \neq 0$*, then*

$$x = \frac{D_x}{D}, \quad y = \frac{D_y}{D}, \quad \text{and} \quad z = \frac{D_z}{D},$$

is the unique solution to the system.

Possibilities if $D = 0$

Teaching Notes:
The two cases for $D = 0$ are, of course, entirely consistent with the rationale for the two cases in which division by zero is not allowed. This may help students distinguish the two different outcomes for the case of $D = 0$.

If $D = 0$:

For the 2 × 2 Case	***For the 3 × 3 Case***
1. *If either* $D_x \neq 0$ *or* $D_y \neq 0$*, the system is* ***inconsistent and there is no solution.***	**1.** *If* $D_x \neq 0$ *or* $D_y \neq 0$ *or* $D_z \neq 0$*, the system is* ***inconsistent and has no solution.***
2. *If both* $D_x = 0$ *and* $D_y = 0$*, the system is* ***dependent and has an infinite number of solutions.***	**2.** *If* $D_x = 0$ *and* $D_y = 0$ *and* $D_z = 0$*, the system is* ***dependent and has an infinite number of solutions.***

Example 1: Cramer's Rule

Using Cramer's Rule, solve the following systems of linear equations. The solutions are not checked here, but they can be checked by substituting the solutions into all of the equations in the system.

a. $\begin{cases} 2x+y=3 \\ 3x-2y=5 \end{cases}$

Solution: $D=\begin{vmatrix} 2 & 1 \\ 3 & -2 \end{vmatrix}=-7,\qquad D_x=\begin{vmatrix} 3 & 1 \\ 5 & -2 \end{vmatrix}=-11,\qquad D_y=\begin{vmatrix} 2 & 3 \\ 3 & 5 \end{vmatrix}=1$

$$x=\frac{D_x}{D}=\frac{-11}{-7}=\frac{11}{7}\qquad y=\frac{D_y}{D}=\frac{1}{-7}=-\frac{1}{7}$$

b. $\begin{cases} 2x+2y=8 \\ -x+3y=-8 \end{cases}$

Solution: $D=\begin{vmatrix} 2 & 2 \\ -1 & 3 \end{vmatrix}=8,\qquad D_x=\begin{vmatrix} 8 & 2 \\ -8 & 3 \end{vmatrix}=40,\qquad D_y=\begin{vmatrix} 2 & 8 \\ -1 & -8 \end{vmatrix}=-8$

$$x=\frac{D_x}{D}=\frac{40}{8}=5\qquad y=\frac{D_y}{D}=\frac{-8}{8}=-1$$

c. $\begin{cases} x+2y+3z=3 \\ 4x+5y+6z=1 \\ 7x+8y+9z=0 \end{cases}$

Solution: $D=\begin{vmatrix} 1 & 2 & 3 \\ 4 & 5 & 6 \\ 7 & 8 & 9 \end{vmatrix}=1\begin{vmatrix} 5 & 6 \\ 8 & 9 \end{vmatrix}-2\begin{vmatrix} 4 & 6 \\ 7 & 9 \end{vmatrix}+3\begin{vmatrix} 4 & 5 \\ 7 & 8 \end{vmatrix}$

$$=1(-3)-2(-6)+3(-3)=0$$

$$D_x=\begin{vmatrix} 3 & 2 & 3 \\ 1 & 5 & 6 \\ 0 & 8 & 9 \end{vmatrix}=3\begin{vmatrix} 5 & 6 \\ 8 & 9 \end{vmatrix}-2\begin{vmatrix} 1 & 6 \\ 0 & 9 \end{vmatrix}+3\begin{vmatrix} 1 & 5 \\ 0 & 8 \end{vmatrix}$$

$$=3(-3)-2(9)+3(8)=-3$$

The system has no solution because $\boldsymbol{D=0}$ **and** $\boldsymbol{D_x\neq 0}$.

d. $\begin{cases} x+y+3z=7 \\ 2x-y-3z=-4 \\ 5x-2y=-5 \end{cases}$

Solution: $D=\begin{vmatrix} 1 & 1 & 3 \\ 2 & -1 & -3 \\ 5 & -2 & 0 \end{vmatrix}=1\begin{vmatrix} -1 & -3 \\ -2 & 0 \end{vmatrix}-1\begin{vmatrix} 2 & -3 \\ 5 & 0 \end{vmatrix}+3\begin{vmatrix} 2 & -1 \\ 5 & -2 \end{vmatrix}=-18$

$$D_x=\begin{vmatrix} 7 & 1 & 3 \\ -4 & -1 & -3 \\ -5 & -2 & 0 \end{vmatrix}=7\begin{vmatrix} -1 & -3 \\ -2 & 0 \end{vmatrix}-1\begin{vmatrix} -4 & -3 \\ -5 & 0 \end{vmatrix}+3\begin{vmatrix} -4 & -1 \\ -5 & -2 \end{vmatrix}=-18$$

$$D_y=\begin{vmatrix} 1 & 7 & 3 \\ 2 & -4 & -3 \\ 5 & -5 & 0 \end{vmatrix}=1\begin{vmatrix} -4 & -3 \\ -5 & 0 \end{vmatrix}-7\begin{vmatrix} 2 & -3 \\ 5 & 0 \end{vmatrix}+3\begin{vmatrix} 2 & -4 \\ 5 & -5 \end{vmatrix}=-90$$

$$D_z=\begin{vmatrix} 1 & 1 & 7 \\ 2 & -1 & -4 \\ 5 & -2 & -5 \end{vmatrix}=1\begin{vmatrix} -1 & -4 \\ -2 & -5 \end{vmatrix}-1\begin{vmatrix} 2 & -4 \\ 5 & -5 \end{vmatrix}+7\begin{vmatrix} 2 & -1 \\ 5 & -2 \end{vmatrix}=-6$$

$$x=\frac{-18}{-18}=1 \qquad y=\frac{-90}{-18}=5 \qquad z=\frac{-6}{-18}=\frac{1}{3}$$

NOTES

The determinants shown in Examples 1c and 1d are expanded by the first row. However, you should remember that any row or column can be used in the expansion as long as the corresponding adjustments in the + and – signs are used with the minors. This may be particularly useful when a row or column has one or more 0's because multiplication by 0 will always give 0 and this will reduce the time needed for the expansion.

Practice Problems

1. *Solve the following system using Cramer's Rule.*

$\begin{cases} 2x-y=11 \\ x+y=-2 \end{cases}$

2. *Find D_x for the following system.*

$\begin{cases} x+2y+z=0 \\ 2x+y-2z=5 \\ 3x-y+z=-3 \end{cases}$

Answers to Practice Problems: **1.** $x=3,\ y=-5$ **2.** $D_x=0$

14.6 Exercises

1. $x = 4, y = 3$
2. $x = -6, y = 7$
3. $x = \frac{2}{3}, y = -\frac{1}{4}$
4. $x = 4, y = 3$
5. No solution
6. Infinite solutions
7. $x = -\frac{1}{4}, y = \frac{3}{2}$
8. $x = -\frac{1}{4}, y = -\frac{1}{3}$
9. $x = \frac{31}{17}, y = \frac{2}{17}$
10. $x = \frac{29}{19}, y = -\frac{6}{19}$
11. $x = \frac{39}{44}, y = \frac{41}{44}$
12. $x = \frac{42}{43}, y = \frac{9}{43}$
13. $x = \frac{18}{7}, y = -\frac{3}{7}$
14. $x = \frac{137}{131}, y = \frac{85}{131}$
15. $x = -\frac{7}{61}, y = \frac{266}{183}$
16. $x = \frac{93}{71}, y = \frac{18}{71}$

For Exercises 1 – 30, use Cramer's Rule to solve the following systems of linear equations.

1. $\begin{cases} 2x - 5y = -7 \\ 3x - 2y = 6 \end{cases}$ **2.** $\begin{cases} 3x + 5y = 17 \\ x + 3y = 15 \end{cases}$ **3.** $\begin{cases} 6x - 4y = 5 \\ 3x + 8y = 0 \end{cases}$ **4.** $\begin{cases} 3x + 4y = 24 \\ 2x + y = 11 \end{cases}$

5. $\begin{cases} 3x + y = 1 \\ -9x - 3y = 2 \end{cases}$ **6.** $\begin{cases} 4x + 8y = 12 \\ 3x + 6y = 9 \end{cases}$ **7.** $\begin{cases} 12x + 4y = 3 \\ -10x + 3y = 7 \end{cases}$ **8.** $\begin{cases} 4x - 9y = 2 \\ 8x - 15y = 3 \end{cases}$

9. $\begin{cases} 2x + 3y = 4 \\ 3x - 4y = 5 \end{cases}$ **10.** $\begin{cases} 5x + 2y = 7 \\ 2x - 3y = 4 \end{cases}$ **11.** $\begin{cases} 7x + 3y = 9 \\ 4x + 8y = 11 \end{cases}$ **12.** $\begin{cases} 5x - 9y = 3 \\ 11x + 6y = 12 \end{cases}$

13. $\begin{cases} 6x - 13y = 21 \\ 5x - 12y = 18 \end{cases}$ **14.** $\begin{cases} 10x + 7y = 15 \\ 13x - 4y = 11 \end{cases}$ **15.** $\begin{cases} 8x - 9y = -14 \\ 15x + 6y = 7 \end{cases}$ **16.** $\begin{cases} 17x - 5y = 21 \\ 4x + 3y = 6 \end{cases}$

17. $\begin{cases} 0.8x + 0.3y = 4 \\ 0.9x - 1.2y = 5 \end{cases}$ **18.** $\begin{cases} 0.4x + 0.7y = 3 \\ 0.5x + y = 6 \end{cases}$ **19.** $\begin{cases} 1.6x - 4.5y = 1.5 \\ 0.4x + 1.2y = 3.1 \end{cases}$

20. $\begin{cases} 2.3x + 1.8y = 4.6 \\ 0.8x - 1.4y = 3.2 \end{cases}$ **21.** $\begin{cases} x - 2y - z = -7 \\ 2x + y + z = 0 \\ 3x - 5y + 8z = 13 \end{cases}$ **22.** $\begin{cases} 2x + 3y + z = 0 \\ 5x + y - 2z = 9 \\ 10x - 5y + 3z = 4 \end{cases}$

23. $\begin{cases} 5x - 4y + z = 17 \\ x + y + z = 4 \\ -10x + 8y - 2z = 11 \end{cases}$ **24.** $\begin{cases} 9x + 10y = 2 \\ 2x + 6z = 4 \\ -3y + 3z = 1 \end{cases}$ **25.** $\begin{cases} 2x - 3y - z = -4 \\ -x + 2y + z = 6 \\ x - y + 2z = 14 \end{cases}$

26. $\begin{cases} 2x - 3y - z = 4 \\ x - 2y - z = 1 \\ x - y + 2z = 9 \end{cases}$ **27.** $\begin{cases} 3x + 2y + z = 5 \\ 2x + y - 2z = 4 \\ 5x + 3y - z = 9 \end{cases}$ **28.** $\begin{cases} 8x + 3y + 2z = 15 \\ 3x + 5y + z = -4 \\ 2x + 3y = -7 \end{cases}$

29. $\begin{cases} 2x - y + 3z = 1 \\ 5x + 2y - z = 2 \\ x - 2y + 5z = 2 \end{cases}$ **30.** $\begin{cases} 2x + 3y + 2z = -5 \\ 2x - 2y + z = -1 \\ 5x + y + z = 1 \end{cases}$

For Exercises 31 – 34, set up a system of linear equations that represents the information, then solve the system by using Cramer's Rule.

31. The three sides of a triangle are related as follows: the perimeter is 43 feet, the second side is 5 feet more than twice the first side, and the third side is 3 feet less than the sum of the other two sides. Find the lengths of the three sides of the triangle.

32. Joel loves candy bars and ice cream, and they have fat and calories as follows: each candy bar contains 5 g of fat and 280 calories; each serving of ice cream contains 10 g of fat and 150 calories. How many candy bars and how many servings of ice cream did he eat the week that he consumed 85 g of fat and 2300 calories from these two foods?

33. A financial advisor has $6 million to invest for her clients. She chooses, for one month, to invest in mutual funds and technology stocks. If the mutual funds earned 2% and the stocks earned 4% for a total of $170,000 in earnings for the month, how much money did she invest in each type of investment?

34. A farmer plants corn, wheat, and soybeans and rotates the planting each year on his 500-acre farm. In one particular year, the profits were: $120 per acre for corn, $100 per acre for wheat, and $80 per acre for soybeans. He planted twice as many acres with corn as with soybeans. How many acres did he plant with each crop the year he made a total profit of $51,800?

17. $x = \frac{210}{41}, y = -\frac{40}{123}$

18. $x = -24, y = 18$

19. $x = \frac{525}{124}, y = \frac{109}{93}$

20. $x = \frac{610}{233}, y = -\frac{184}{233}$

21. $x = -2, y = 1, z = 3$

22. $x = 1, y = 0, z = -2$

23. No solution

24. $x = -\frac{4}{17}, y = \frac{7}{17}, z = \frac{38}{51}$

25. $x = 4, y = 2, z = 6$

26. $x = 2, y = -1, z = 3$

27. Infinite solutions

28. $x = 1, y = -3, z = 8$

29. $x = -\frac{2}{3}, y = \frac{11}{3}, z = 2$

30. $x = \frac{23}{27}, y = -\frac{5}{27}, z = -\frac{83}{27}$

31. 6 feet, 17 feet, 20 feet

32. 5 candy bars and 6 servings of ice cream

33. $3,500,000 in mutual funds, $2,500,000 in stocks

34. 180 acres corn, 230 acres wheat, 90 acres soybean

Hawkes Learning Systems: Introductory & Intermediate Algebra

Determinants and Systems of Linear Equations: Cramer's Rule

14.7 Graphing Systems of Linear Inequalities

Objectives

After completing this section, you will be able to:

Solve systems of linear inequalities graphically.

In some branches of mathematics, in particular a topic called (interestingly enough) game theory, the solution to a very sophisticated problem can involve the set of points that satisfy a system of several **linear inequalities**. In business these ideas relate to problems such as minimizing the cost of shipping goods from several warehouses to distribution outlets. In this section we will consider graphing the solution sets to only two inequalities. We will leave the problem solving techniques to another course.

First, we review the ideas related to systems of equations discussed in Section 14.1. Systems of two linear equations were solved by using three methods: graphing, substitution, and addition. We found that such systems can be:

a. consistent (one point satisfies both equations, and the lines intersect in one point),

b. inconsistent (no point satisfies both equations, and the lines are parallel), or

c. dependent (an infinite number of points satisfy both equations, and the lines are the same).

Table 14.2 shows an example of each case.

System	Graph	Intersection	Terms
$\begin{cases} y = 3x - 1 \\ y = -x + 3 \end{cases}$	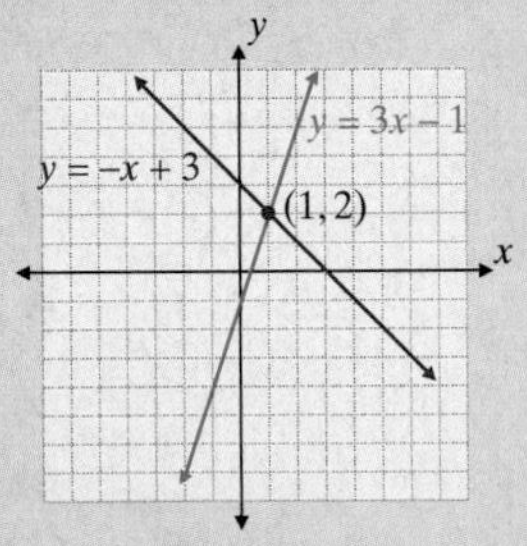	*One point: (1, 2)*	***Consistent***
$\begin{cases} 2x + y = 4 \\ y = -2x + 1 \end{cases}$	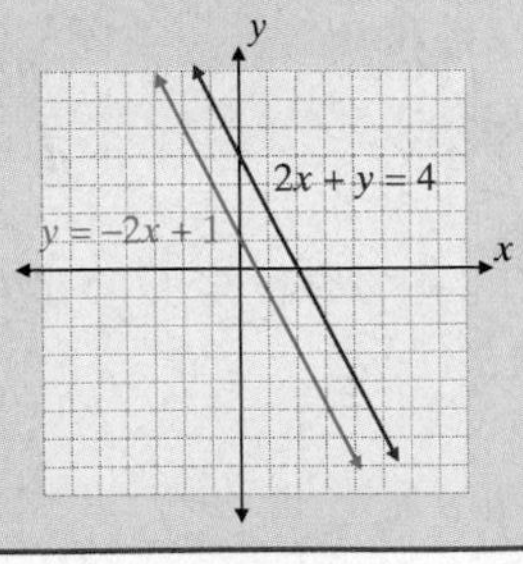	*No points;* *lines are parallel*	***Inconsistent***

System	Graph	Intersection	Terms
$\begin{cases} y=-x+5 \\ 2y+2x=10 \end{cases}$	y = −x + 5 2y + 2x = 10 x y	*Infinite number of points; lines are the same.*	***Dependent***

Table 14.2

In this section, we will develop techniques for graphing (and therefore solving) **systems of two linear inequalities**. The solution set (if there are any solutions) to a system of two linear inequalities consists of the points in the intersection of two half-planes and portions of boundary lines indicated by the inequalities. We know that a straight line separates a plane into two **half-planes**. The line itself is called the **boundary line**, and the boundary line may be included (the half-plane is **closed**) or the boundary line may not be included (the half-plane is **open**). The following procedure may be used to solve a system of linear inequalities.

To Solve a System of Two Linear Inequalities

1. *Graph the boundary lines for both half-planes.*
2. *Shade the region that is common to both of these half-planes.*
   (This region is called the ***intersection*** *of the two half-planes.)*
3. *To check, pick one test-point in the intersection and verify that it satisfies both inequalities.*

(***Note:*** *If there is no intersection, then the system is inconsistent and has no solution.)*

Example 1: Graphing Systems of Linear Inequalities

a. Graph the points that satisfy the system of inequalities $\begin{cases} x \le 2 \\ y \ge -x+1 \end{cases}$.

Solution: **Step 1:** For $x \le 2$, the points are to the left of and on the line $x = 2$.

Step 2: For $y \ge -x+1$, the points are above and on the line $y = -x + 1$.

continued on next page ...

Step 3: Shade only the region with points that satisfy both inequalities. In this case, we test the point $(0,3)$:

$0 \le 2$ A true statement

$3 \ge -0+1$ A true statement

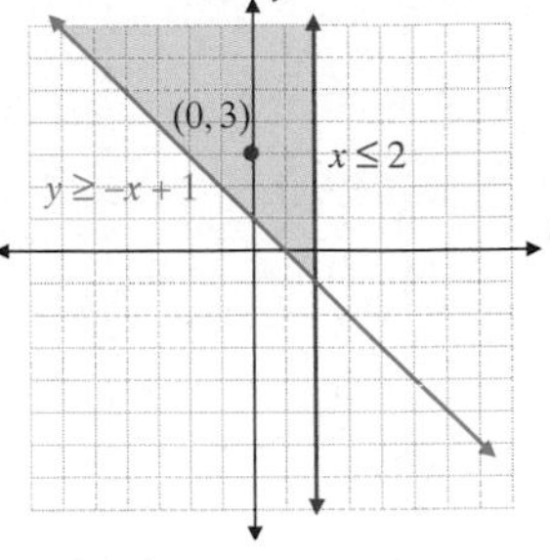

b. Solve the system of linear inequalities graphically: $\begin{cases} 2x+y \le 6 \\ x+y<4 \end{cases}$.

Solution: **Step 1:** Solve each inequality for y: $\begin{cases} y \le -2x+6 \\ y<-x+4 \end{cases}$.

Step 2: For $y \le -2x+6$, the points are below and on the line $y = -2x + 6$.

Step 3: For $y < -x + 4$, the points are below but not on the line $y = -x + 4$.

Step 4: Shade only the region with points that satisfy both inequalities. Note that the line $y = -x + 4$ is dashed. In this case, we test the point $(0,0)$.

$2 \cdot 0 + 0 \le 6$ A true statement

$0 + 0 < 4$ A true statement

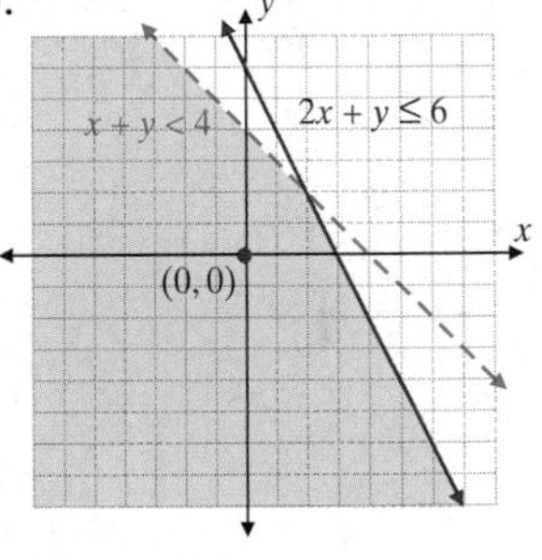

c. Solve the system of linear inequalities graphically: $\begin{cases} y \ge x \\ y \le x+2 \end{cases}$.

Solution: For $y \ge x$, the points are above and on the line $y = x$.
For $y \le x + 2$, the points are below and on the line $y = x + 2$.
The solution set consists of the boundary lines and the region between them.

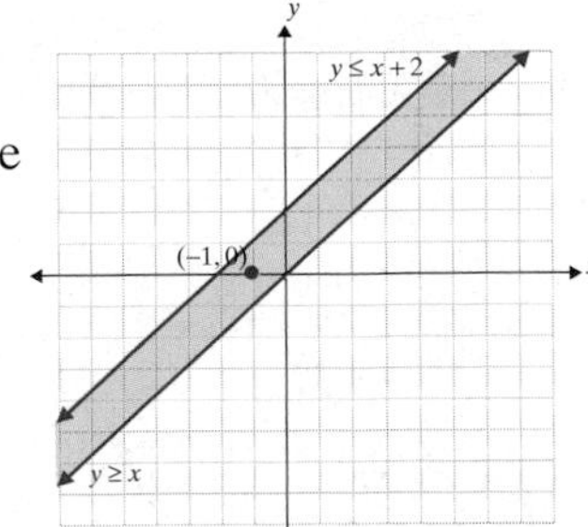

Note: When the boundary lines are parallel there are two possibilities:

1. The common region will be in the form of a strip between two lines (as in this example), or
2. There will be no common region and the solution set will be the empty set, ∅.

Using a TI-83 Plus Graphing Calculator to Graph Systems of Linear Inequalities

To graph a system of linear inequalities with a TI-83 Plus graphing calculator, first solve each inequality for y and then enter both of the corresponding functions after pressing the Y= key. By setting the graphing symbol to the left of Y1 and Y2 to the desired form and then pressing GRAPH, the desired region will be graphed as a cross-hatched area on the display (assuming that the window is set correctly). The following example shows how this can be done.

Example 2: Graphing Systems of Linear Inequalities

Use a TI-83 Plus graphing calculator to graph the following system of linear inequalities: $\begin{cases} 2x+y<4 \\ 2x-y\le 0 \end{cases}$.

Solution: **Step 1:** Solve each inequality for y: $\begin{cases} y<-2x+4 \\ y\ge 2x \end{cases}$.

(**Note:** Solving $2x-y\le 0$ for y can be written as $2x\le y$ and then as $y\ge 2x$.)

Step 2: Press the Y= key and enter both functions and the corresponding symbols as they appear here:

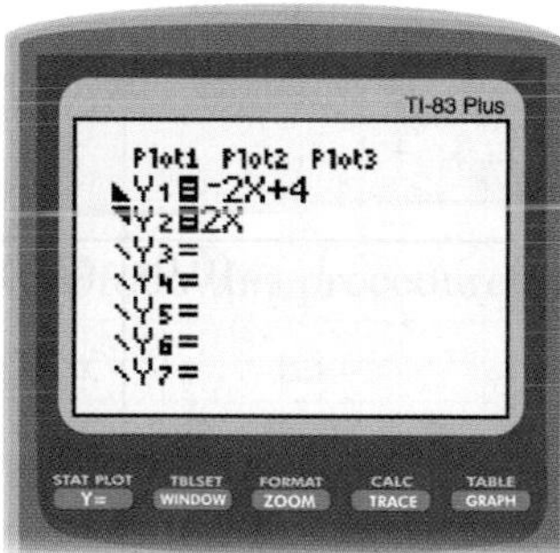

Step 3: Press GRAPH. The display should appear as follows. The solution is the cross-hatched region.

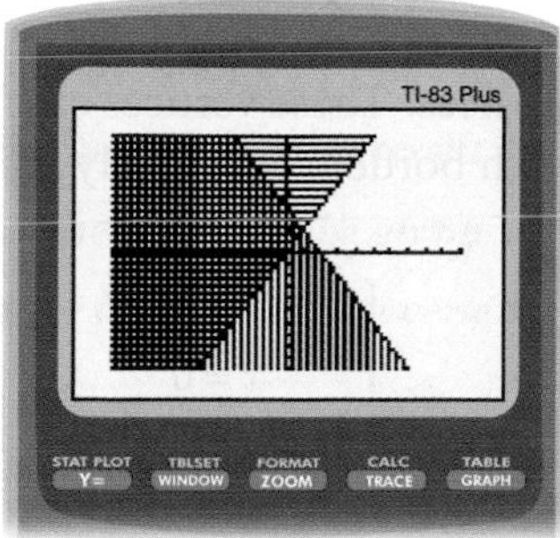

Use a graphing calculator to solve the systems of linear inequalities in Exercises 25 – 35.

25. $\begin{cases} y \ge 0 \\ 3x - 5y \le 10 \end{cases}$ **26.** $\begin{cases} 3x + 2y \le 15 \\ 2x + 5y \ge 10 \end{cases}$ **27.** $\begin{cases} 4x - 3y \ge 6 \\ 3x - y \le 3 \end{cases}$ **28.** $\begin{cases} y \le 0 \\ 3x + y \le 11 \end{cases}$

29. $\begin{cases} 3x - 4y \ge -6 \\ 3x + 2y \le 12 \end{cases}$ **30.** $\begin{cases} 3y \le 2x \\ x + 2y \le 11 \end{cases}$ **31.** $\begin{cases} x + y \le 8 \\ 3x - 2y \ge -6 \end{cases}$ **32.** $\begin{cases} x + y \le 7 \\ 2x - y \le 8 \end{cases}$

33. $\begin{cases} y \le x \\ y < 2x + 1 \end{cases}$ **34.** $\begin{cases} x - y \ge -2 \\ 4x - y < 16 \end{cases}$ **35.** $\begin{cases} y \ge x \\ y \le x + 7 \end{cases}$

In Exercises 36 – 41, solve the systems of inequalities graphically.

36. $\begin{cases} x \ge 0 \\ 3x - 5y \le 10 \\ 4x + 3y \le 23 \end{cases}$ **37.** $\begin{cases} y \ge 0 \\ 4x - 3y \ge -6 \\ 3x - y \le 3 \end{cases}$ **38.** $\begin{cases} x \ge 0 \\ y \ge 0 \\ 3x - 4y \ge -6 \\ 3x + 2y \le 12 \end{cases}$

39. $\begin{cases} x \ge 0 \\ y \ge 0 \\ x + y \le 8 \\ 3x - 2y \ge -6 \end{cases}$ **40.** $\begin{cases} x \ge 0 \\ y \ge 0 \\ y \le 2x + 1 \\ x + y \le 7 \\ 2x - y \le 8 \end{cases}$ **41.** $\begin{cases} x \ge 0 \\ y \ge 0 \\ y \le x \\ 2x + 5y \le 21 \\ 2x + y \le 17 \end{cases}$

Find the maximum value of the function F, subject to the conditions indicated by the system of inequalities in Exercises 42– 47.

42. $F = 5x - 2y$ $\begin{cases} x \ge 0 \\ y \ge 0 \\ 2x - y \ge -3 \\ 3x + y \le 24 \end{cases}$ **43.** $F = -3x - y$ $\begin{cases} x \ge 0 \\ y \ge 0 \\ x - y \ge -2 \\ 4x - y \le 16 \end{cases}$ **44.** $F = 4x + 3y$ $\begin{cases} x \ge 0 \\ y \ge 0 \\ x + 3y \le 18 \\ 4x + y \le 28 \end{cases}$

45. $F = 5x - 2y$ $\begin{cases} x \ge 0 \\ y \ge 0 \\ x \le 6 \\ x - 2y \ge -6 \\ 2x + 3y \le 18 \end{cases}$ **46.** $F = -3x - y$ $\begin{cases} x \ge 0 \\ y \ge 0 \\ y \le x + 2 \\ x + 2y \le 17 \\ 4x - y \le 32 \end{cases}$ **47.** $F = 4x + 3y$ $\begin{cases} x \ge 0 \\ y \ge 0 \\ 3y \le x + 12 \\ x + 2y \le 13 \\ 3x + y \le 24 \end{cases}$

26.
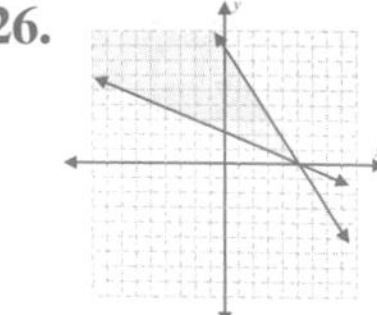

27.
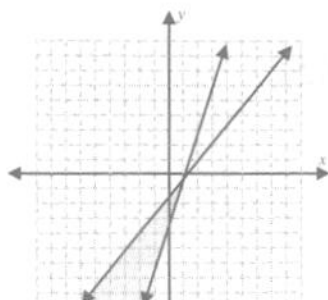

28.
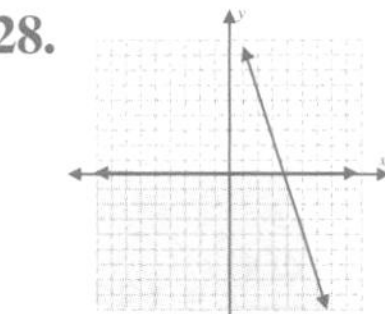

29.
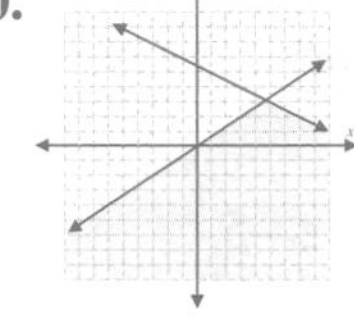

30.
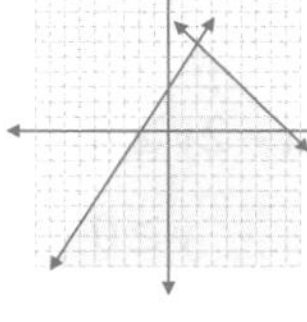

31.
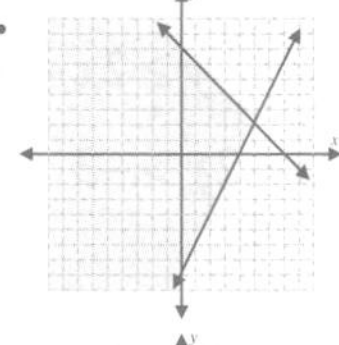

32.
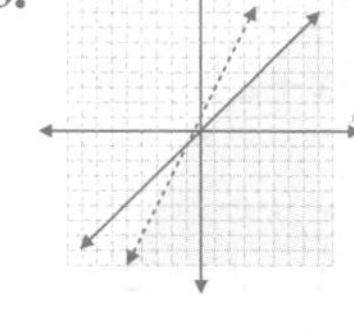

33.
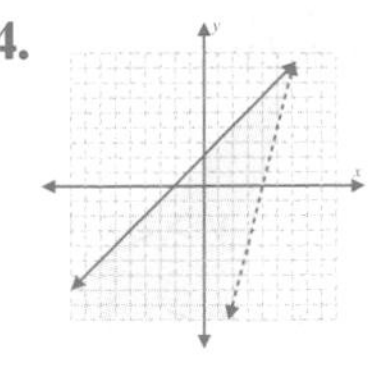

34.

35.
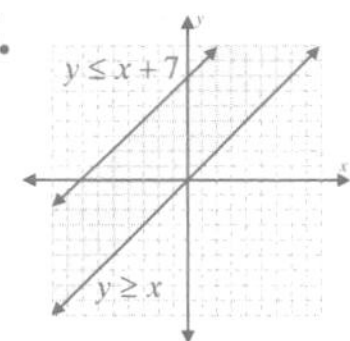

36.
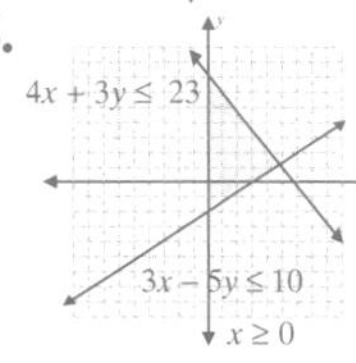

37.
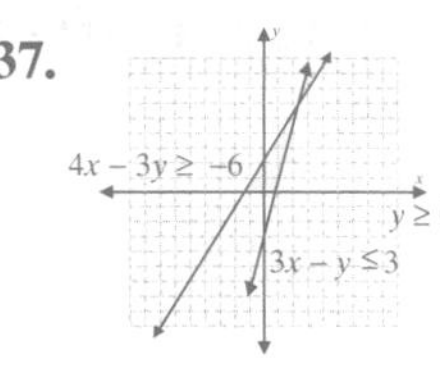

38.
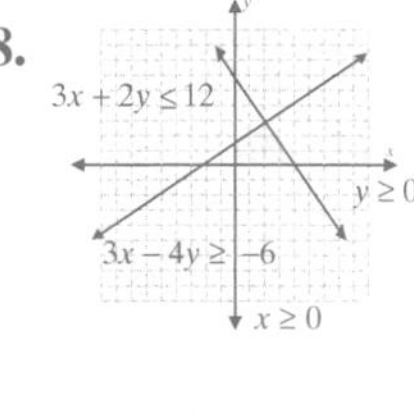

39.
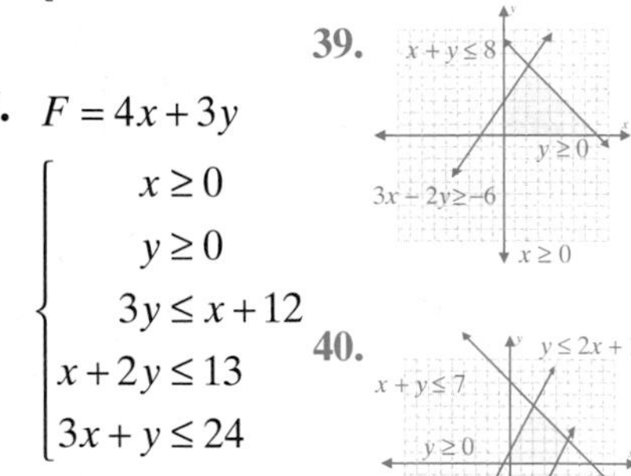

40.

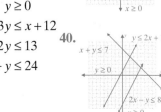

41. 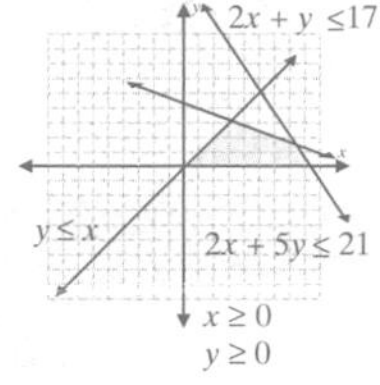

42. $F=40$

43. $F=-2$

44. $F=36$

45. $F=26$

46. $F=0$

47. $F=37$

48. \$15.20

49. \$225

50. \$3840

51. \$546

52. a. 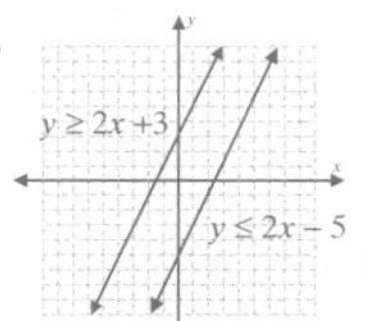

b. 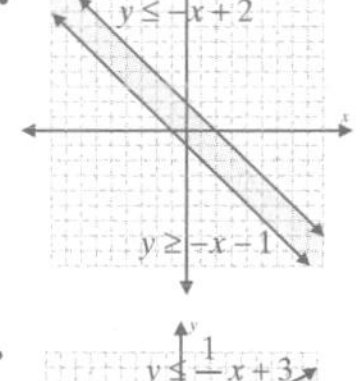

c. 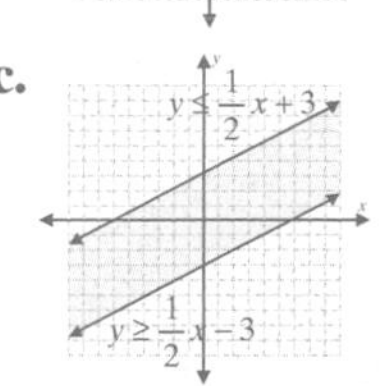

48. A small retail butcher shop sells two grades of hamburger. Grade A costs \$2.70 per pound and Grade B costs \$2.10 per pound. The shop's daily order can be no more than 60 pounds and can cost no more than \$45.60. If the profit is \$0.45 per pound on Grade B and \$0.90 per pound on Grade A, what is the maximum daily profit on hamburger?

49. A boy builds two kinds of model airplanes. Model X takes him 4 hours to produce, costs \$8.00, and sells for \$15.00. Model Y takes 3 hours to produce, costs \$7.00, and sells for \$12.00. If he allots a total of 58 hours and \$126.00 per month to produce model airplanes, find his maximum monthly income from the model planes.

50. Larry sells sound systems. The Boom system takes 2 hours to install and 0.5 hours to adjust. The Voom system takes 1.5 hours to install and 0.75 hours to adjust. Larry has decided to spend no more than 30 hours per week for installation and no more than 12 hours per week for adjustments. He sells at least 8 systems per week. Each of the Boom and Voom systems sell for \$160 and \$240, respectively. Find his maximum weekly income.

51. An antique shop specializes in refinishing antique chairs. They use two types of finishes. Type 1 finishes take 2 hours and material costs \$6.00. Type 2 finishes take 3 hours and material costs \$3.00. The shop charges \$30 per chair for Type 1 finishes and \$36 per hour for Type 2 finishes. If they allow no more than 40 hours per week for labor and no more than \$84 for material, what is their maximum income?

Writing and Thinking About Mathematics

52. Example 1c discusses a system of two linear inequalities in which the boundary lines are parallel. Describe, in your own words, how you might test whether or not you have graphed the correct solution set. Solve the following systems graphically and indicate how your method of testing works in each case.

a. $\begin{cases} y\le 2x-5 \\ y\ge 2x+3 \end{cases}$ **b.** $\begin{cases} y\le -x+2 \\ y\ge -x-1 \end{cases}$ **c.** $\begin{cases} y\le \frac{1}{2}x+3 \\ y\ge \frac{1}{2}x-3 \end{cases}$

Hawkes Learning Systems: Introductory & Intermediate Algebra

Systems of Linear Inequalities

Chapter 14 Index of Key Ideas and Terms

Section 14.1 Review of Systems of Linear Equations: Two Variables

Systems of Linear Equations: Two Variables page 972

Two (or more) linear equations considered at one time are said to form a **system of linear equations** or a set of **simultaneous equations**.

Simultaneous Equations page 972

Consistent page 973

If a system of linear equations has a unique solution, it is said to be **consistent**.

Inconsistent page 973

If a system has no solutions, it is said to be **inconsistent**.

Dependent page 973

If a system has an infinite number of solutions, it is said to be **dependent**.

Solving a System by Graphing page 974

To solve a system of two linear equations by graphing, graph both equations on the same set of axes and observe the point of intersection of the two lines (if there is one).

Solving a System by Substitution page 975

1. Solve one of the equations for one of the variables.
2. Substitute the resulting expression into the other equation.
3. Solve this new equation, if possible, and then substitute back into one of the original equations to find the value of the other variable. (This is known as **back substitution**.)

Solving a System by Addition page 977

1. Write the equations one under the other so that like terms are aligned.
2. Multiply all terms of one equation by a constant (and possibly the terms of the other equation by another constant). so that two like terms have opposite coefficients.
3. Add like terms and solve the resulting equation, if possible. Then, back substitute into one of the original equations to find the value of the other variable.

Section 14.2 Applications

Applications

Mixture page 983
Interest page 984
Work page 984
Algebra page 985

Section 14.3 Systems of Linear Equations: Three Variables

Ordered Triples page 990

An **ordered triple** is an ordering of three real numbers in the form (x, y, z).

Graphs in Three Dimensions page 990

Three mutually perpendicular axes labeled as the x-axis, y-axis, and z-axis are used to separate space into eight regions called **octants**.

Solving a System of Three Linear Equations page 992

Step 1: Select two equations and eliminate one variable by using the addition method.
Step 2: Select a different pair of equations and eliminate the **same** variable.
Step 3: Steps 1 and 2 give **two** linear equations in **two** variables. Solve these equations by either addition or substitution as discussed in Section 14.1.
Step 4: Back substitute the values found in Step 3 into any one of the original equations to find the value of the third variable.

Section 14.4 Matrices and Gaussian Elimination

Matrices page 1000

Entries, Rows, Columns, Dimensions, Square

Coefficient Matrix page 1001

A matrix formed from the coefficients of the variables in a system of linear equations is called a **coefficient matrix**.

Augmented Matrix page 1001

A matrix that includes the coefficients and the constant terms is called an **augmented matrix**.

continued on next page ...

Section 14.4 Matrices and Gaussian Elimination (continued)

Elementary Row Operations page 1002

There are three elementary row operations with matrices:

1. Interchange two rows.
2. Multiply a row by a nonzero constant.
3. Add a multiple of a row to another row.

If any elementary row operation is applied to a matrix, the new matrix is said to be **row-equivalent** to the original matrix.

Upper Triangular Form (or ref Form) page 1004

If all the entries below the main diagonal of a matrix are 0's, the matrix is said to be in **upper triangular form**.

Gaussian Elimination page 1005

To solve a system of linear equations by using Gaussian elimination:

1. Write the augmented matrix for the system.
2. Use elementary row operations to transform the matrix into triangular form.
3. Solve the corresponding system of equations by using back substitution.

Using a TI-83 Plus Calculator to Solve a System of Linear Equations pages 1008 - 1010

Section 14.5 Determinants

Determinants page 1014

A **determinant** is a real number associated with a square array of real numbers and is indicated by enclosing the array between two vertical bars. For a matrix A, the corresponding determinant is designated as $\det(A)$.

Value of a 2 × 2 Determinant page 1015

For the square matrix $A = \begin{bmatrix} a_{11} & a_{12} \\ a_{21} & a_{22} \end{bmatrix}$,

$$\det(A) = \begin{vmatrix} a_{11} & a_{12} \\ a_{21} & a_{22} \end{vmatrix} = a_{11}a_{22} - a_{21}a_{12}.$$

continued on next page ...

Section 14.5 Determinants (continued)

Value of a 3 × 3 Determinant page 1016

For the square matrix $A = \begin{bmatrix} a_{11} & a_{12} & a_{13} \\ a_{21} & a_{22} & a_{23} \\ a_{31} & a_{32} & a_{33} \end{bmatrix}$,

$$\det(A) = \begin{vmatrix} a_{11} & a_{12} & a_{13} \\ a_{21} & a_{22} & a_{23} \\ a_{31} & a_{32} & a_{33} \end{vmatrix}$$

$$= a_{11}(\text{minor of } a_{11}) - a_{12}(\text{minor of } a_{12}) + a_{13}(\text{minor of } a_{13})$$

$$= a_{11}\begin{vmatrix} a_{22} & a_{23} \\ a_{32} & a_{33} \end{vmatrix} - a_{12}\begin{vmatrix} a_{21} & a_{23} \\ a_{31} & a_{33} \end{vmatrix} + a_{13}\begin{vmatrix} a_{21} & a_{22} \\ a_{31} & a_{32} \end{vmatrix}$$

Sign Table for Minors of a 3 × 3 Determinant page 1017

$$\begin{vmatrix} + & - & + \\ - & + & - \\ + & - & + \end{vmatrix}$$

Using a TI-83 Plus Calculator to Evaluate a Determinant pages 1019 - 1020

Section 14.6 Determinants and Systems of Linear Equations: Cramer's Rule

Cramer's Rule page 1024

For 2 × 2 Matrices

For the system $\begin{cases} a_{11}x + a_{12}y = \boldsymbol{k_1} \\ a_{21}x + a_{22}y = \boldsymbol{k_2} \end{cases}$, where

$$D = \begin{vmatrix} a_{11} & a_{12} \\ a_{21} & a_{22} \end{vmatrix} \quad D_x = \begin{vmatrix} \boldsymbol{k_1} & a_{12} \\ \boldsymbol{k_2} & a_{22} \end{vmatrix} \quad \text{and} \quad D_y = \begin{vmatrix} a_{11} & \boldsymbol{k_1} \\ a_{21} & \boldsymbol{k_2} \end{vmatrix},$$

if $D \neq 0$, then $x = \dfrac{D_x}{D}$ and $y = \dfrac{D_y}{D}$

is the unique solution to the system.

continued on next page ...

Section 14.6 Determinants and Systems of Linear Equations: Cramer's Rule (continued)

For 3 × 3 Matrices page 1025

For the system $\begin{cases} a_{11}x + a_{12}y + a_{13}z = \mathbf{k_1} \\ a_{21}x + a_{22}y + a_{23}z = \mathbf{k_2} \\ a_{31}x + a_{32}y + a_{33}z = \mathbf{k_3} \end{cases}$, where $D = \begin{vmatrix} a_{11} & a_{12} & a_{13} \\ a_{21} & a_{22} & a_{23} \\ a_{31} & a_{32} & a_{33} \end{vmatrix}$

$$D_x = \begin{vmatrix} \mathbf{k_1} & a_{12} & a_{13} \\ \mathbf{k_2} & a_{22} & a_{23} \\ \mathbf{k_3} & a_{32} & a_{33} \end{vmatrix} \quad D_y = \begin{vmatrix} a_{11} & \mathbf{k_1} & a_{13} \\ a_{21} & \mathbf{k_2} & a_{23} \\ a_{31} & \mathbf{k_3} & a_{33} \end{vmatrix} \quad \text{and} \quad D_z = \begin{vmatrix} a_{11} & a_{12} & \mathbf{k_1} \\ a_{21} & a_{22} & \mathbf{k_2} \\ a_{31} & a_{32} & \mathbf{k_3} \end{vmatrix}$$

if $D \neq 0$, then $x = \dfrac{D_x}{D}$, $y = \dfrac{D_y}{D}$, and $z = \dfrac{D_z}{D}$,
is the unique solution to the system.

Possibilities if $D = 0$ page 1025

For the 2 × 2 Case

1. If either $D_x \neq 0$ or $D_y \neq 0$, the system is **inconsistent and there is no solution.**
2. If both $D_x = 0$ and $D_y = 0$, the system is **dependent and has an infinite number of solutions.**

For the 3 × 3 Case

1. If $D_x \neq 0$ or $D_y \neq 0$ or $D_z \neq 0$, the system is **inconsistent and has no solution.**
2. If $D_x = 0$ and $D_y = 0$ and $D_z = 0$, the system is **dependent and has an infinite number of solutions.**

Section 14.7 Graphing Systems of Linear Inequalities

Solving a System of Linear Inequalities page 1031

To solve a system of two linear inequalities:

1. Graph both half-planes.
2. Shade the region that is common to both of these half-planes. (This region is called the **intersection** of the two half-planes.)
3. To check, pick one test-point in the intersection and verify that it satisfies both inequalities.
   (**Note:** If there is no intersection, then the system is inconsistent and has no solution.)

continued on next page ...

Section 14.7 Graphing Systems of Linear Inequalities (continued)

To solve a linear programming problem:

1. Graph the set of inequalities that represent the conditions.
2. Find each vertex of the region in the graph.
3. Evaluate the objective function at each vertex and choose the maximum (or minimum) value.

Chapter 14 Review

For a review of the topics and problems from Chapter 14, look at the following lessons from *Hawkes Learning Systems: Introductory & Intermediate Algebra.*

Review: Solving Systems of Linear Equations – Two Variables
Applications: Systems of Equations
Solving Systems of Linear Equations with Three Variables
Matrices and Gaussian Elimination
Determinants
Determinants and Systems of Linear Equations: Cramer's Rule
Systems of Linear Inequalities

Chapter 14 Test

1. $x = 3, y = -1$

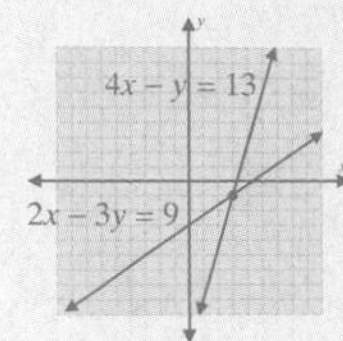

2. $x = 7, y = 2$
3. Inconsistent, No solution
4. $x = \frac{1}{2}, y = \frac{1}{4}$
5. $x = 11, y = -7$
6. $a = -3, b = 4$
7. 8 ft. × 23 ft.
8. $x = -1, y = 1, z = -2$
9. Inconsistent, No Solution
10. a. $\begin{bmatrix} 1 & 2 & -3 \\ 1 & -1 & -1 \\ 1 & 3 & 2 \end{bmatrix}$ 3×3

    $\left[\begin{array}{ccc|c} 1 & 2 & -3 & -11 \\ 1 & -1 & -1 & 2 \\ 1 & 3 & 2 & -4 \end{array}\right]$ 3×4

    b. $x = 1, y = -3, z = 2$
11. 60 33-cent stamps, 20 55-cent stamps, 10 78-cent stamps.
12. 42
13. 5

1. Solve the system by graphing: $\begin{cases} 4x - y = 13 \\ 2x - 3y = 9 \end{cases}$.

Solve each system of linear equations in Exercises 2 – 5 by the substitution method or the addition method. State which systems are dependent or inconsistent.

2. $\begin{cases} x + y = 9 \\ x - y = 5 \end{cases}$ **3.** $\begin{cases} 6x + 3y = 5 \\ 4x + 2y = -3 \end{cases}$ **4.** $\begin{cases} 7x - 6y = 2 \\ 5x + 2y = 3 \end{cases}$ **5.** $\begin{cases} 2x + 3y = 1 \\ 5x + 7y = 6 \end{cases}$

6. Determine a and b such that the line $ax + by = 17$ passes through the two points $(-3, 2)$ and $(1, 5)$.

7. The length of a rectangle is 7 ft. more than twice its width. The perimeter is 62 ft. Find the dimensions of the rectangle.

8. Solve the following system of equations algebraically: $\begin{cases} x - 2y - 3z = 3 \\ x + y - z = 2 \\ 2x - 3y - 5z = 5 \end{cases}$.

9. Solve the following system of equations algebraically: $\begin{cases} x + 2y - 2z = 0 \\ x - y + z = 2 \\ -x + 4y - 4z = -8 \end{cases}$.

10. For the following system of equations:

a. write the coefficient matrix and the augmented matrix and state the dimension of each, and

b. solve the system by using the Gaussian elimination method.

$$\begin{cases} x + 2y - 3z = -11 \\ x - y - z = 2 \\ x + 3y + 2z = -4 \end{cases}$$

11. Kimberly bought 90 stamps in denominations of 33¢, 55¢, and 78¢. To test her daughter, who is taking an algebra class, she said that she bought three times as many 33¢ stamps as 55¢ stamps and that the total cost of the stamps was \$38.60. How many stamps of each denomination did she buy?

Evaluate the determinants in Exercises 12 and 13.

12. $\begin{vmatrix} 6 & -3 \\ 4 & 5 \end{vmatrix}$ **13.** $\begin{vmatrix} 1 & 3 & 2 \\ 2 & 5 & 1 \\ 0 & 2 & 1 \end{vmatrix}$

14. $x = 6$

Solve for x in Exercises 14 and 15.

15. $x = \frac{17}{3}$

14. $\begin{vmatrix} 3 & x \\ 5 & 7 \end{vmatrix} = -9$

15. $\begin{vmatrix} -2 & 3 & 1 \\ 1 & 3 & x \\ 2 & 3 & x \end{vmatrix} = 14$

16. $x = -\frac{78}{5}, y = \frac{38}{5}$

Solve Exercises 16 and 17 by using Cramer's Rule.

17. $x = -\frac{5}{26}, y = \frac{33}{26}, z = \frac{9}{26}$

16. $\begin{cases} 3x + 8y = 14 \\ 2x + 7y = 22 \end{cases}$

17. $\begin{cases} x + 2y - z = 2 \\ x - 4y - 5z = -7 \\ x + 3y + 4z = 5 \end{cases}$

18. $50°, 60°, 70°$

18. The sum of the measures of the three angles of a triangle is 180°. If the largest angle is 40° less than the sum of the other two, and the middle angle is 40° less than twice the smallest angle, find the measures of the angles.

19.

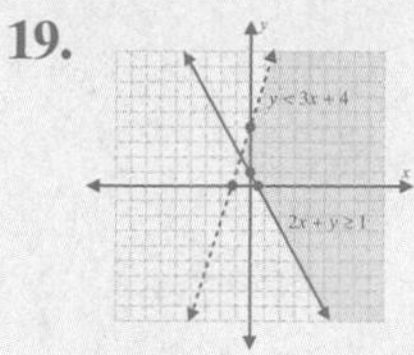

19. Solve the following system of linear inequalities graphically:

$$\begin{cases} y < 3x + 4 \\ 2x + y \geq 1 \end{cases}$$

Use a TI-83 Plus calculator to answer Exercises 20 and 21.

20. $x = -\frac{53}{11}, y = \frac{29}{11}, z = -\frac{2}{11}$

20. Solve the system of equations using the row echelon form command, ref:

$$\begin{cases} x - 3y - 4z = -12 \\ 3x + 4y + \frac{1}{2}z = -4 \\ -x - y + z = 2 \end{cases}$$

21. $D_x = -53$

21. Find the value of the determinant D_x for the system of equations in Exercise 20.

22. $F = \frac{34}{3}$

22. Maximize the function $F = x + 4y - 2$ subject to the following conditions:

$$\begin{cases} x \geq 0 \\ y \geq 0 \\ 2x + y \geq 8 \\ x + y \leq 6 \\ x + 2y \leq 8 \end{cases}$$

Cumulative Review: Chapters 1 – 14

1. **a.** $-3, 0, 1$
 b. $-3, -\frac{1}{2}, 0, \frac{5}{8}, 1$
 c. $-\sqrt{13}, \sqrt{2}, \pi$
 d. $-\sqrt{13}, -3, -\frac{1}{2}, 0, \frac{5}{8}, 1, \sqrt{2}, \pi$
2. 0
3. 320
4. $x = 2$
5. $x = -6$
6. $x = -\frac{29}{2}$ or $x = \frac{27}{2}$
7. $x = -\frac{2}{3}$
8. $m = \frac{2gK}{v^2}$
9. 4
 $(-\infty, 4)$
10. −144
 $[-144, \infty)$
11. $\frac{1}{x^8}$ **12.** $\frac{x^5}{y}$
13. $\frac{6}{x^2}$
14. 2.7×10^{-3}
15. **a.** $m = -\frac{1}{2}$
 b. $x + 2y = -1$
 c.

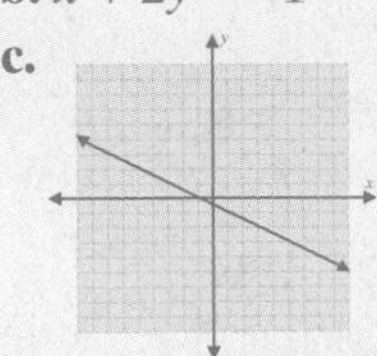

1. Given the set of numbers $\left\{-\sqrt{13}, -3, -\frac{1}{2}, 0, \frac{5}{8}, 1, \sqrt{2}, \pi\right\}$, list those in the set that are

a. integers
b. rational numbers
c. irrational numbers
d. real numbers

In Exercises 2 and 3, simplify each expression using the rules for order of operations.

2. $16 + (3^2 - 7^2) \div 5 \cdot 2$ **3.** $15 - 3(4^2 - 10^3 \div 2^3 + 6 \cdot 3) + 2^5$

Solve each equation in Exercises 4 – 7.

4. $4(x + 2) - (8 - 2x) = 12$ **5.** $\frac{3x - 2}{8} = \frac{x}{4} - 1$

6. $|2x + 1| = 28$ **7.** $|3x + 2| + 4 = 4$

8. Solve the formula $K = \frac{mv^2}{2g}$ for m.

Solve the inequalities in Exercises 9 and 10. Write the solutions in interval notation then graph each solution on a real number line.

9. $5x - 13 > 7(x - 3)$ **10.** $\frac{x}{3} - 21 \le \frac{x}{2} + 3$

In Exercises 11 – 13, simplify the expressions by using the rules for exponents.

11. $(x^2y^3)^{-1}(x^{-2}y)^3$ **12.** $\frac{x^2y^{-3}}{x^{-3}y^{-2}}$ **13.** $\left(\frac{5x^2y^2}{3xy^3}\right)^{-1}\left(\frac{10x^{-1}}{y}\right)$

14. Write each number in scientific notation and simplify:

$$\frac{810{,}000 \times 0.00014}{42{,}000}$$

15. Given the two points $(-5, 2)$ and $(5, -3)$:

a. Find the slope of the line that passes through the two points.
b. Find the equation of the line.
c. Graph the line.

16. Find the equation of the line that is parallel to the line $2x - 5y = 1$ and passes through the origin.

16. $y = \frac{2}{5}x$

17. x-intercepts: $(-3.45, 0)$ and $(1.45, 0)$ maximum: $(-1, 6)$

18. $(-1.65, -0.29)$ and $(3.65, 10.29)$

19. $x = 1, y = 4$

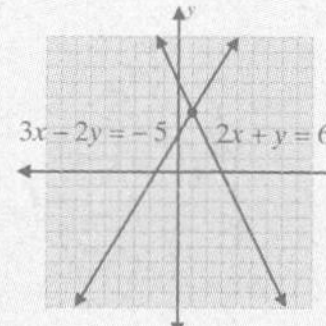

20. $x = 1, y = 3$

21. $x = 4, y = 3, z = 2$

22. $x = 2, y = \frac{2}{5}, z = -\frac{1}{2}$

23. 3 batches of Choc-O-Nut and 2 batches of Chocolate Krunch

24. \$4200 at 7%, \$2800 at 8%

25. $a = 2, b = 3, c = 4$

17. Graph the function $y = 5 - 2x - x^2$ on a graphing calculator. With the trace and zoom features, estimate the x-intercepts and the maximum value for y on the curve.

18. Using a graphing calculator, graph the two functions $y = x^2 - 3$ and $y = 2x + 3$ and estimate their points of intersection.

19. Solve the system of linear equations by graphing.

$$\begin{cases} 2x + y = 6 \\ 3x - 2y = -5 \end{cases}$$

20. Solve the system of linear equations by the addition method.

$$\begin{cases} x + 3y = 10 \\ 5x - y = 2 \end{cases}$$

21. Solve the system of linear equations by using the Gaussian elimination method.

$$\begin{cases} x - 3y + 2z = -1 \\ -2x + y + 3z = 1 \\ x - y + 4z = 9 \end{cases}$$

22. Solve the system of linear equations by using Cramer's Rule.

$$\begin{cases} 3x - 5y + 2z = 3 \\ 2x + 2z = 3 \\ -x + 5y - 4z = 2 \end{cases}$$

23. Karl makes two kinds of cookies. Choc-O-Nut requires 4 oz. of peanuts for each 10 oz. of chocolate chips. Chocolate Krunch requires 12 oz. of peanuts per 8 oz. of chocolate chips. How many batches of each can he make if he has 46 oz. of chocolate chips and 36 oz. of peanuts?

24. Alicia has \$7000 invested, some at 7% and the remainder at 8%. After one year, the interest from the 7% investment exceeds the interest from the 8% investment by \$70. How much is invested at each rate?

25. The points (0, 4), (–2, 6), and (1, 9) lie on the curve described by the function $y = ax^2 + bx + c$. Find the values of a, b, and c.

Use a TI-83 Plus calculator to answer Exercises 26 – 32.

26. Solve the system $\begin{cases} 6x + y = 0 \\ -3x + 2y = -15 \end{cases}$ by graphing.

26. $(1,-6)$

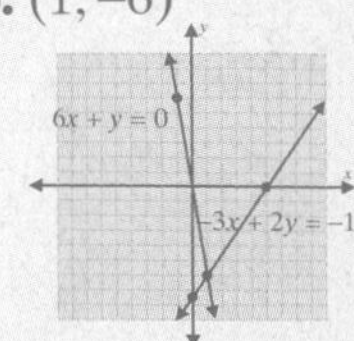

27. $x=\dfrac{8}{5},\ y=\dfrac{27}{10}$

28. $x=-3,\ y=0,\ z=5$

29. $x=1,\ y=1$

30. $x=1,\ y=2,\ z=3$

31.

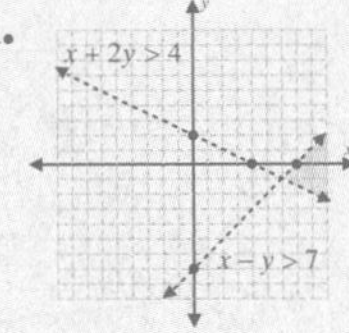

32.

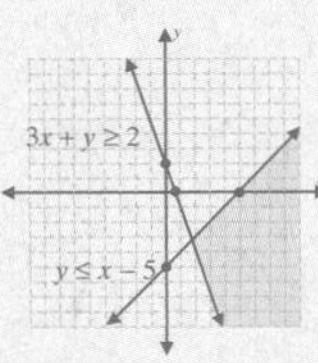

27. Solve the system of equations by using the row echelon form command, ref:

$$\begin{cases} x+2y=7 \\ -x+3y=\dfrac{13}{2} \end{cases}$$

28. Solve the system of equations by using the row echelon form command, ref:

$$\begin{cases} x+y-z=-8 \\ 2x-3y=-6 \\ x+y+z=2 \end{cases}$$

29. Solve the system of equations $\begin{cases} x-y=0 \\ 2x+y=3 \end{cases}$ by using Cramer's rule.

30. Solve the system of equations $\begin{cases} x+y-z=0 \\ 2x+3z=11 \\ 3y-z=3 \end{cases}$ by using Cramer's rule.

31. Solve the following system of linear inequalities graphically:

$$\begin{cases} x+2y>4 \\ x-y>7 \end{cases}$$

32. Solve the following system of linear inequalities graphically:

$$\begin{cases} y\le x-5 \\ 3x+y\ge 2 \end{cases}$$

Sequences, Series, and the Binomial Theorem

CHAPTER 15

Did You Know?

One of the outstanding mathematicians of the Middle Ages was Leonardo Fibonacci (Leonardo, son of Bonaccio), also known as Leonardo of Pisa (c. 1175–1250). Fibonacci's name is attached to an interesting sequence of numbers $1, 1, 2, 3, 5, 8, 13, \ldots. x, y, x + y, \ldots$, the so-called Fibonacci sequence, where each term after the first two is obtained by adding the preceding two terms together. The sequence of numbers arises from a problem found in Fibonacci's writings. How many pairs of rabbits can be produced from a single pair in a year if every month each pair begets a new pair that from the second month on becomes productive? The answer to this odd problem is the sum of the first 12 terms of the Fibonacci sequence. Can you verify this?

The Fibonacci sequence itself has been found to have many beautiful and interesting properties. For example, of mathematical interest is the fact that any two successive terms in the sequence are relatively prime; that is, their greatest common divisor is one. In the world of nature, the terms of the Fibonacci sequence also appear. Spirals formed by natural objects such as centers of daisies, pine cone scales, pineapple scales, and leaves generally have two sets of spirals, one clockwise, one counterclockwise. Each set is made up of a specific number of spirals in each direction, the number of spirals being adjacent terms in the Fibonacci sequence. For example, in pine cone scales, 5 spiral one way and 8 spiral the other; on pineapples, 8 one way and 13 the other.

Fibonacci

Leonardo Fibonacci was best known for the texts he wrote in which he introduced the Hindu-Arabic numeral system. He participated in the mathematical tournaments held at the court of the emperor Frederick I, and he used some of the challenge problems in the book he wrote. Fibonacci traveled widely and became acquainted with the different arithmetic systems in use around the Mediterranean. His most famous text, *Liber Abaci* (1202), combined arithmetic and elementary algebra with an emphasis on commercial applied problems. His attempt to reform and improve the study of mathematics in Europe was not too successful; he seemed to be ahead of his time. But his books did much to introduce Hindu-Arabic notation into Europe, and his sequence has provided interesting problems throughout the history of mathematics. In the United States, a Fibonacci Society exists to study the properties of this mysterious and intriguing sequence of numbers.

"There is no branch of mathematics, however abstract, which may not some day be applied to the phenomena of the real world."

Nicolas Ivanovich Lobachevsky (1793 – 1856)

Chapter 15 provides an introduction to a powerful notation using the Greek letter Σ, capital sigma. With this Σ-notation and a few basic properties, we will develop some algebraic formulas related to sums of numbers and, in some cases, even infinite sums. The concept of having the sum of an infinite amount of numbers equal to some finite number introduces the idea of limits. Consider adding fractions in the following manner:

$$\frac{1}{2}=\frac{1}{2};\ \frac{1}{2}+\frac{1}{4}=\frac{3}{4};\ \frac{1}{2}+\frac{1}{4}+\frac{1}{8}=\frac{7}{8};\ \frac{1}{2}+\frac{1}{4}+\frac{1}{8}+\frac{1}{16}=\frac{15}{16};\ \frac{1}{2}+\frac{1}{4}+\frac{1}{8}+\frac{1}{16}+\frac{1}{32}=\frac{31}{32}$$

Continuing to add fractions in this manner, in which the denominators are successive powers of two, will give sums that get closer and closer to 1. We say that the sums "approach" 1, and that 1 is the **limit** of the sum. These fascinating ideas are discussed in Section 15.4 and are fundamental in the development of calculus and higher level mathematics.

Other topics in this chapter – permutations, combinations, and the Binomial Theorem – find applications in courses in probability and statistics as well as in more advanced courses in mathematics and computer science.

15.1 Sequences

Objectives

After completing this section, you will be able to:

1. *Write several terms of a sequence given the formula for its general term.*
2. *Find the formula for the general term of a sequence given several terms.*
3. *Determine whether a sequence is increasing, decreasing, or neither.*

In mathematics, a **sequence** is a list of numbers that occur in a certain order. Each number in the sequence is called a **term** of the sequence, and a sequence may have a finite number of terms or an infinite number of terms. For example,

2, 4, 6, 8, 10, 12, 14, 16, 18	is a **finite sequence** consisting of positive even integers less than 20.
3, 6, 9, 12, 15, 18, . . .	is an **infinite sequence** consisting of the multiples of 3.

The infinite sequence of the multiples of 3 can be described in the following way:

For any positive integer n, the corresponding number in the list is $3n$. Thus, we know that

$3\cdot 6=18$ and 18 is the 6th number in the sequence,
$3\cdot 7=21$ and 21 is the 7th number in the sequence,
$3\cdot 8=24$ and 24 is the 8th number in the sequence, and so on.

Infinite Sequence

*An **infinite sequence** (or a **sequence**) is a function that has the positive integers as its domain.*

Note: A finite sequence will be so indicated. The term sequence, used alone, indicates an infinite sequence.

Consider the function $f(n)=\frac{1}{2^n}$ where n is any positive integer.

For this function,

$$f(1)=\frac{1}{2^1}=\frac{1}{2}$$

$$f(2)=\frac{1}{2^2}=\frac{1}{4}$$

$$f(3)=\frac{1}{2^3}=\frac{1}{8}$$

$$f(4)=\frac{1}{2^4}=\frac{1}{16}$$

$$\vdots$$

$$f(n)=\frac{1}{2^n}$$

$$\vdots$$

Or, using ordered pair notation,

$$f=\left\{\left(1,\frac{1}{2}\right),\left(2,\frac{1}{4}\right),\left(3,\frac{1}{8}\right),\left(4,\frac{1}{16}\right),\ldots,\left(n,\frac{1}{2^n}\right),\ldots\right\}.$$

The **terms** of the sequence are the numbers

$$\frac{1}{2},\frac{1}{4},\frac{1}{8},\frac{1}{16},\ldots,\frac{1}{2^n},\ldots$$

Because the order of terms corresponds to the positive integers, it is customary to indicate a sequence by writing only its terms. In general discussions and formulas, a sequence may be indicated with subscript notation as

$$a_1,\ a_2,\ a_3,\ a_4,\ \ldots,\ a_n,\ \ldots$$

The general term a_n is called the **n^{th} term** of the sequence. The entire sequence can be denoted by writing the nth term in braces as in $\{a_n\}$. Thus,

$$\{a_n\} \quad \text{and} \quad a_1, a_2, a_3, a_4, \ldots, a_n, \ldots$$

are both representations of the sequence with

a_1 as the first term,
a_2 as the second term,
a_3 as the third term,
⋮
a_n as the nth term,
⋮

Example 1: Sequences

a. Write the first three terms of the sequence $\left\{\frac{n}{n+1}\right\}$.

Solution:

$$a_1 = \frac{1}{1+1} = \frac{1}{2}$$

$$a_2 = \frac{2}{2+1} = \frac{2}{3}$$

$$a_3 = \frac{3}{3+1} = \frac{3}{4}$$

b. If $\{b_n\} = \{2n-1\}$, find b_1, b_2, b_3, and b_{50}.

Solution:

$$b_1 = 2 \cdot 1 - 1 = 1$$

$$b_2 = 2 \cdot 2 - 1 = 3$$

$$b_3 = 2 \cdot 3 - 1 = 5$$

$$b_{50} = 2 \cdot 50 - 1 = 99$$

c. Determine a_n if the first five terms of $\{a_n\}$ are $0, 3, 8, 15, 24$.

Solution: In this case, study the numbers carefully and make an intelligent guess. That is, keep trying a variety of formulas until you find one that "fits" the given terms. (Questions of this type relating to number patterns are common on intelligence tests.)

The formula is $a_n = n^2 - 1$.

Checking:

$$a_1 = 1^2 - 1 = 0$$
$$a_2 = 2^2 - 1 = 3$$
$$a_3 = 3^2 - 1 = 8$$
$$a_4 = 4^2 - 1 = 15$$
$$a_5 = 5^2 - 1 = 24$$

Although the formula for a_n may not be obvious, with practice it becomes easier to find.

Teaching Notes:
There are a variety of examples you can use to have students try to discover a numerical pattern, and thus, translate the pattern into an algebraic formula based on "n". It is important that students build a confidence level in being able to express (discover) the pattern, and then to write a formula that defines it mathematically.

An **alternating sequence** is one in which the terms alternate in sign. That is, if one term is positive, then the next term is negative. Alternating sequences generally involve the expression $(-1)^n$. Example 1d illustrates such a sequence.

d. Write the first five terms of the sequence in which $a_n = \dfrac{(-1)^n}{n}$.

Solution: $a_1 = \dfrac{(-1)^1}{1} = -1$

$$a_2 = \frac{(-1)^2}{2} = \frac{1}{2}$$
$$a_3 = \frac{(-1)^3}{3} = -\frac{1}{3}$$
$$a_4 = \frac{(-1)^4}{4} = \frac{1}{4}$$
$$a_5 = \frac{(-1)^5}{5} = -\frac{1}{5}$$

● ●

As Example 1d illustrates, while the domain of a sequence consists of the positive integers, some (or all) of the terms of a sequence may be negative.

In subscript notation, the term a_{n+1} is the term following a_n, and this term is found by substituting $n + 1$ for n in the formula for the general term. For example,

$$\text{if } a_n = \frac{1}{3n}, \text{ then } a_{n+1} = \frac{1}{3(n+1)} = \frac{1}{3n+3}.$$

Similarly,

$$\text{if } b_n = n^2, \text{ then } b_{n+1} = (n+1)^2.$$

Increasing and Decreasing Sequences

If the terms of a sequence grow increasingly smaller, then the sequence is said to be **decreasing**. If the terms grow successively larger, then the sequence is said to be **increasing**. The following definitions state these ideas algebraically.
Note: A sequence may be neither decreasing nor increasing.

Decreasing Sequence

A sequence $\{a_n\}$ *is*

decreasing *if* $a_n > a_{n+1}$ *for all n.*

(Successive terms become smaller.)

Increasing Sequence

A sequence $\{a_n\}$ *is*

increasing *if* $a_n < a_{n+1}$ *for all n.*

(Successive terms become larger.)

Example 2: Increasing and Decreasing Sequences

Determine whether each of the following sequences is increasing, decreasing, or neither.

a. $\{a_n\}=\left\{\frac{1}{2^n}\right\}$

Solution: Write the formula for terms a_n and a_{n+1} and compare them algebraically.

$$a_n = \frac{1}{2^n} \quad \text{and} \quad a_{n+1} = \frac{1}{2^{n+1}}$$

Note that $2^{n+1} > 2^n$ for all positive integer values of n.

Therefore, $\frac{1}{2^n} > \frac{1}{2^{n+1}}$. So we have $a_n > a_{n+1}$, and $\{a_n\}$ is **decreasing**.

b. $\{b_n\}=\{n+3\}$

Solution: $b_n = n+3$ and $b_{n+1} = (n+1)+3 = n+4$

Since $n + 3 < n + 4$, we have $b_n < b_{n+1}$, and $\{b_n\}$ is **increasing**.

c. $\{c_n\} = \{(-1)^n\}$

Solution: The first five terms of the sequence are $-1, 1, -1, 1, -1$. The sequence is alternating and cannot be increasing or decreasing. In general,

$$c_n = (-1)^n \qquad \text{and} \qquad c_{n+1} = (-1)^{n+1}.$$

The value of c_n depends on whether n is even or odd.

If n is even, then $n + 1$ is odd:

$c_n = (-1)^n = 1$ and $c_{n+1} = (-1)^{n+1} = -1$, indicating $c_n > c_{n+1}$.

If n is odd, then $n + 1$ is even:

$c_n = (-1)^n = -1$ and $c_{n+1} = (-1)^{n+1} = 1$, indicating $c_n < c_{n+1}$.

Therefore, the sequence is **neither increasing nor decreasing**.

Practice Problems

Write the first three terms of each sequence.

1. $\{n^2\}$ *2.* $\{2n+1\}$ *3.* $\left\{\frac{1}{n+1}\right\}$

4. Find a formula for the general term of sequence $-1, 1, 3, 5, 7, \ldots$

15.1 Exercises

Write the first four terms of each of the sequences in Exercises 1 – 15.

1. $\{2n-1\}$ **2.** $\{4n+1\}$ **3.** $\left\{1+\frac{1}{n}\right\}$

4. $\left\{\frac{n+3}{n+1}\right\}$ **5.** $\{n^2+n\}$ **6.** $\{n-n^2\}$

7. $\{2^n\}$ **8.** $\left\{\left(\frac{1}{2}\right)^n\right\}$ **9.** $\{(-1)^n(n^2+1)\}$

10. $\left\{(-1)^n\left(\frac{n}{n+1}\right)\right\}$ **11.** $\left\{(-1)^n\left(\frac{1}{2n+3}\right)\right\}$ **12.** $\{(-1)^{n-1}(3^n)\}$

13. $\{2^n-n^2\}$ **14.** $\left\{\frac{n(n-1)}{2}\right\}$ **15.** $\left\{\frac{1+(-1)^n}{2}\right\}$

Find a formula for the general term of each sequence in Exercises 16 – 25.

16. $2, 5, 8, 11, 14, \ldots$ **17.** $5, 9, 13, 17, 21, \ldots$ **18.** $6, 12, 18, 24, 30, \ldots$

1. $1, 3, 5, 7$
2. $5, 9, 13, 17$
3. $2, \frac{3}{2}, \frac{4}{3}, \frac{5}{4}$
4. $2, \frac{5}{3}, \frac{3}{2}, \frac{7}{5}$
5. $2, 6, 12, 20$
6. $0, -2, -6, -12$
7. $2, 4, 8, 16$
8. $\frac{1}{2}, \frac{1}{4}, \frac{1}{8}, \frac{1}{16}$
9. $-2, 5, -10, 17$
10. $-\frac{1}{2}, \frac{2}{3}, -\frac{3}{4}, \frac{4}{5}$
11. $-\frac{1}{5}, \frac{1}{7}, -\frac{1}{9}, \frac{1}{11}$
12. $3, -9, 27, -81$
13. $1, 0, -1, 0$
14. $0, 1, 3, 6$
15. $0, 1, 0, 1$
16. $\{3n-1\}$
17. $\{4n+1\}$

Answers to Practice Problems: **1.** $1, 4, 9$ **2.** $3, 5, 7$ **3.** $\frac{1}{2}, \frac{1}{3}, \frac{1}{4}$ **4.** $a_n = 2n-3$

19. $1, -3, 5, -7, 9, \ldots$ **20.** $-3, 7, -11, 15, -19, \ldots$ **21.** $1, 4, 9, 16, 25, \ldots$

22. $5, 10, 20, 40, 80, \ldots$ **23.** $\frac{1}{3}, \frac{1}{4}, \frac{1}{5}, \frac{1}{6}, \frac{1}{7}, \ldots$ **24.** $\frac{1}{2}, \frac{1}{4}, \frac{1}{8}, \frac{1}{16}, \frac{1}{32}, \ldots$

25. $2, 5, 10, 17, 26, \ldots$

For each of the sequences in Exercises 26 – 31, determine whether it is increasing or decreasing. Justify your answer by comparing a_n with a_{n+1}.

26. $\{n+4\}$ **27.** $\{1-2n\}$ **28.** $\left\{\frac{1}{n+3}\right\}$

29. $\left\{\frac{1}{3^n}\right\}$ **30.** $\left\{\frac{2n+1}{n}\right\}$ **31.** $\left\{\frac{n}{n+1}\right\}$

Write the finite sequence described by Exercises 32 – 35, then answer the question.

32. A certain automobile costs \$29,000 new and depreciates at a rate of $\frac{3}{10}$ of its current value each year. What will be its value after 3 years?

33. A ball is dropped from a height of 250 centimeters. Each time it bounces, it rises to $\frac{2}{5}$ of its previous height. How high will it rise after the fourth bounce?

34. A culture of bacteria triples every day. If there were 100 bacteria in the original culture, how many would be present after 4 days?

35. A university is experiencing a declining enrollment of 3% per year. If the present enrollment is 20,000, what is the projected enrollment after 5 years?

Writing and Thinking About Mathematics

36. Use the following formula to generate the first 5 terms of the sequence: $a_{n+2} = a_n + a_{n+1}$ where $a_1 = 1$ and $a_2 = 1$. (This sequence is the famous Fibonacci sequence.) (**Note:** Formulas of this type that use previous answers are said to be **recursive**.)

Answers:

18. $\{6n\}$
19. $\{(-1)^{n+1}(2n-1)\}$
20. $\{(-1)^n(4n-1)\}$
21. $\{n^2\}$
22. $\{5(2^{n-1})\}$
23. $\left\{\frac{1}{n+2}\right\}$
24. $\left\{\frac{1}{2^n}\right\}$
25. $\{n^2+1\}$
26. Increasing
27. Decreasing
28. Decreasing
29. Decreasing
30. Decreasing
31. Increasing
32. $\left\{29{,}000 \cdot \left(\frac{7}{10}\right)^n\right\}$; \$9947 when $n=3$
33. $\left\{250\left(\frac{2}{5}\right)^n\right\}$; 6.4 cm when $n=4$
34. $\{100(3^n)\}$; 8100 bacteria when $n=4$
35. $\left\{20{,}000 \cdot \left(\frac{97}{100}\right)^n\right\}$; 17,175 students when $n=5$
36. 1, 1, 2, 3, 5

Hawkes Learning Systems: Introductory & Intermediate Algebra

Sequences

15.2 Sigma Notation

Objectives

Teaching Notes:
You might want to make this Σ-notation palatable in terms of the need for the economy of expressing a longer number of addends and the convenience when a general form of the term is known. Have students write out several terms of sequences for which the general formula for a term is known – build the felt need for the convenience of the Σ-notation.

After completing this section, you will be able to:

1. *Write sums using Σ-notation.*
2. *Find the values of sums written in Σ-notation.*

Finding the sum of a finite number of terms of a sequence is the same as finding the sum of a finite sequence. Such a sum is called a **partial sum** and can be indicated by using **sigma notation** with the Greek letter capital sigma, Σ. (As we will see in Section 15.4, this notation can be used to indicate the sum of an entire sequence by using the symbol for infinity, ∞.)

Partial Sums Using Sigma Notation

The ***n*****th partial sum,** S_n*, of the first n terms of a sequence* $\{a_n\}$ *is*

$$S_n = \sum_{k=1}^{n} a_k = a_1 + a_2 + a_3 + \ldots + a_n$$

k is called the ***index of summation****, and k takes the integer values 1, 2, 3, . . . , n.*

n is the ***upper limit of summation****, and 1 is the* ***lower limit of summation.***

To understand the concept of partial sums, consider the sequence $\left\{\frac{1}{n}\right\}$ and the following partial sums:

$$S_1 = a_1 = \frac{1}{1}$$

$$S_2 = a_1 + a_2 = \frac{1}{1} + \frac{1}{2}$$

$$S_3 = a_1 + a_2 + a_3 = \frac{1}{1} + \frac{1}{2} + \frac{1}{3}$$

$$\vdots$$

$$S_n = a_1 + a_2 + a_3 + \ldots + a_n = \frac{1}{1} + \frac{1}{2} + \frac{1}{3} + \ldots + \frac{1}{n}$$

Note: In some cases, the lower limit of summation in sigma notation may be an integer other than 1. Also, letters other than *k* may be used as the index of summation. The lower case letters *i*, *j*, *k*, *l*, *m*, and *n* are commonly used.

For example, the sum of the second through sixth terms of the sequence $\{n^2\}$ can be written in sigma notation as

$$\sum_{i=2}^{6} i^2 = 2^2 + 3^2 + 4^2 + 5^2 + 6^2.$$

If the number of terms is large, then three dots are used to indicate missing terms after a pattern has been established with the first three or four terms. For example,

$$S_{100} = \sum_{k=1}^{100} (k-1) = 0 + 1 + 2 + 3 + \ldots + 99.$$

Example 1: Sigma Notation

Write the indicated sums of the terms and find the value of each sum.

a. $\sum_{k=1}^{4} k^3$

Solution: $\sum_{k=1}^{4} k^3 = 1^3 + 2^3 + 3^3 + 4^3 = 1 + 8 + 27 + 64 = 100$

b. $\sum_{k=5}^{9} (-1)^k k$

Solution: $\sum_{k=5}^{9} (-1)^k k = (-1)^5 5 + (-1)^6 6 + (-1)^7 7 + (-1)^8 8 + (-1)^9 9$

$$= -5 + 6 - 7 + 8 - 9 = -7$$

Properties of Σ-Notation

The following properties of Σ-notation are useful in developing systematic methods for finding sums of certain types of finite and infinite sequences.

Properties of Σ-Notation

For sequences $\{a_n\}$ and $\{b_n\}$ and any real number c:

I. $\sum_{k=1}^{n} a_k = \sum_{k=1}^{i} a_k + \sum_{k=i+1}^{n} a_k$ *for any i,* $1 \le i \le n-1$

II. $\sum_{k=1}^{n} (a_k + b_k) = \sum_{k=1}^{n} a_k + \sum_{k=1}^{n} b_k$

III. $\sum_{k=1}^{n} c a_k = c \sum_{k=1}^{n} a_k$

IV. $\sum_{k=1}^{n} c = nc$

These properties follow directly from the associative, commutative, and distributive properties for sums of real numbers.

I. $$\sum_{k=1}^{n} a_k = a_1 + a_2 + \ldots + a_i + a_{i+1} + a_{i+2} + \ldots + a_n$$
$$= (a_1 + a_2 + \ldots + a_i) + (a_{i+1} + a_{i+2} + \ldots + a_n)$$
$$= \sum_{k=1}^{i} a_k + \sum_{k=i+1}^{n} a_k$$

II. $$\sum_{k=1}^{n} (a_k + b_k) = (a_1 + b_1) + (a_2 + b_2) + \ldots + (a_n + b_n)$$
$$= (a_1 + a_2 + \ldots + a_n) + (b_1 + b_2 + \ldots + b_n)$$
$$= \sum_{k=1}^{n} a_k + \sum_{k=1}^{n} b_k$$

III. $$\sum_{k=1}^{n} ca_k = ca_1 + ca_2 + \ldots + ca_n$$
$$= c(a_1 + a_2 + \ldots + a_n)$$
$$= c\sum_{k=1}^{n} a_k$$

IV. $$\sum_{k=1}^{n} c = \underbrace{c + c + c + \ldots + c}_{c \text{ appears } n \text{ times}} = nc$$

Example 2: Properties of Σ-Notation

a. If $\sum_{k=1}^{7} a_k = 40$ and $\sum_{k=1}^{30} a_k = 75$, find $\sum_{k=8}^{30} a_k$.

Solution: Since $\sum_{k=1}^{7} a_k + \sum_{k=8}^{30} a_k = \sum_{k=1}^{30} a_k$, then
$$40 + \sum_{k=8}^{30} a_k = 75$$
$$\sum_{k=8}^{30} a_k = 35$$

b. If $\sum_{k=1}^{50} 3a_k = 600$, find $\sum_{k=1}^{50} a_k$.

Solution: Since $\sum_{k=1}^{50} 3a_k = 3\sum_{k=1}^{50} a_k$, then
$$3\sum_{k=1}^{50} a_k = 600$$
$$\sum_{k=1}^{50} a_k = 200$$

Practice Problems

1. *Write the indicated sum of the terms and find the value of the sum:* $\sum_{k=1}^{4}(k^2-1)$
2. *Write the sum 10 + 12 + 14 + 16 + 18 in Σ-notation.*
3. $\sum_{k=1}^{5}a_k = 20$ *and* $\sum_{k=6}^{10}a_k = 30$. *Find* $\sum_{k=1}^{10}2a_k$.

15.2 Exercises

For each of the sequences given in Exercises 1 – 10, write out the partial sums S_1, S_2, S_3, *and* S_4 *and evaluate each partial sum.*

1. $\{3k-1\}$ **2.** $\{2k+5\}$ **3.** $\left\{\frac{k}{k+1}\right\}$ **4.** $\left\{\frac{k+1}{k}\right\}$

5. $\{(-1)^{k-1}k^2\}$ **6.** $\{(-1)^k k^3\}$ **7.** $\left\{\frac{1}{2^k}\right\}$ **8.** $\left\{\left(\frac{2}{3}\right)^k\right\}$

9. $\left\{\left(-\frac{2}{3}\right)^k\right\}$ **10.** $\{k^2-k\}$

Write the sums in Exercises 11 – 26 in expanded form and evaluate.

11. $\sum_{k=1}^{5}2k$ **12.** $\sum_{k=1}^{11}k(k-1)$ **13.** $\sum_{k=2}^{6}(k+3)$ **14.** $\sum_{k=9}^{11}(2k+1)$

15. $\sum_{k=2}^{4}\frac{1}{k}$ **16.** $\sum_{k=1}^{3}\frac{1}{2k}$ **17.** $\sum_{k=1}^{3}2^k$ **18.** $\sum_{k=10}^{15}(-1)^k$

19. $\sum_{k=4}^{8}k^2$ **20.** $\sum_{k=1}^{4}k^3$ **21.** $\sum_{k=3}^{6}(9-2k)$ **22.** $\sum_{k=2}^{7}(4k-1)$

23. $\sum_{k=2}^{5}(-1)^k(k^2+k)$ **24.** $\sum_{k=1}^{6}(-1)^k(k^2-2)$

25. $\sum_{k=1}^{5}\frac{k}{k+1}$ **26.** $\sum_{k=3}^{5}(-1)^k\left(\frac{k+1}{k^2}\right)$

Write the sums in Exercises 27 – 35 in sigma notation.

27. 1 + 3 + 5 + 7 + 9 **28.** 16 + 25 + 36 + 49

29. −1 + 1 + (−1) + 1 + (−1) **30.** 4 + 7 + 10 + 13 + 16

31. $\frac{1}{8}-\frac{1}{27}+\frac{1}{64}-\frac{1}{125}+\frac{1}{216}$ **32.** $\frac{1}{8}+\frac{1}{16}+\frac{1}{32}+\frac{1}{64}+\frac{1}{128}$

Answers to Practice Problems: 1. 0 + 3 + 8 + 15 = 26 **2.** $\sum_{k=5}^{9}2k$ or $\sum_{k=1}^{5}(2k+8)$ **3.** 100

Answers (margin):

1. 2, 7, 15, 26
2. 7, 16, 27, 40
3. $\frac{1}{2},\frac{7}{6},\frac{23}{12},\frac{163}{60}$
4. $2,\frac{7}{2},\frac{29}{6},\frac{73}{12}$
5. 1, −3, 6, −10
6. −1, 7, −20, 44
7. $\frac{1}{2},\frac{3}{4},\frac{7}{8},\frac{15}{16}$
8. $\frac{2}{3},\frac{10}{9},\frac{38}{27},\frac{130}{81}$
9. $-\frac{2}{3},-\frac{2}{9},-\frac{14}{27},-\frac{26}{81}$
10. 0, 2, 8, 20
11. 2 + 4 + 6 + 8 + 10 = 30
12. 0 + 2 + 6 + 12 + 20 + 30 + 42 + 56 + 72 + 90 + 110 = 440
13. 5 + 6 + 7 + 8 + 9 = 35
14. 19 + 21 + 23 = 63
15. $\frac{1}{2}+\frac{1}{3}+\frac{1}{4}=\frac{13}{12}$
16. $\frac{1}{2}+\frac{1}{4}+\frac{1}{6}=\frac{11}{12}$
17. 2 + 4 + 8 = 14
18. 1 − 1 + 1 − 1 + 1 − 1 = 0
19. 16 + 25 + 36 + 49 + 64 = 190
20. 1 + 8 + 27 + 64 = 100
21. 3 + 1 + (−1) + (−3) = 0
22. 7 + 11 + 15 + 19 + 23 + 27 = 102

23. $6+(-12)+20+(-30)=-16$

24. $1+2+(-7)+14+(-23)+34=21$

25. $\frac{1}{2}+\frac{2}{3}+\frac{3}{4}+\frac{4}{5}+\frac{5}{6}=\frac{71}{20}$

26. $-\frac{4}{9}+\frac{5}{16}-\frac{6}{25}=-\frac{1339}{3600}$

27. $\sum_{k=1}^{5}(2k-1)$

28. $\sum_{k=4}^{7}k^2$

29. $\sum_{k=1}^{5}(-1)^k$

30. $\sum_{k=1}^{5}(3k+1)$

31. $\sum_{k=2}^{6}(-1)^k\left(\frac{1}{k^3}\right)$

32. $\sum_{k=3}^{7}\frac{1}{2^k}$

33. $\sum_{k=4}^{15}\frac{k}{k+1}$

34. $\sum_{k=3}^{7}(k^2-1)$

35. $\sum_{k=5}^{12}\frac{k+1}{k^2}$

36. 39 **37.** 7

38. 57 **39.** 48

40. −1 **41.** 10

42. 98 **43.** 295

44. 55 **45.** −152

46. $\sum_{k=11}^{15}-2k+\sum_{k=1}^{5}3k$

33. $\frac{4}{5}+\frac{5}{6}+\frac{6}{7}+\ldots+\frac{15}{16}$

34. $8+15+24+35+48$

35. $\frac{6}{25}+\frac{7}{36}+\frac{8}{49}+\frac{9}{64}+\ldots+\frac{13}{144}$

Find the indicated sums in Exercises 36 – 45.

36. $\sum_{k=1}^{14}a_k=18$ and $\sum_{k=1}^{14}b_k=21$. Find $\sum_{k=1}^{14}(a_k+b_k)$.

37. $\sum_{k=1}^{19}a_k=23$ and $\sum_{k=1}^{19}b_k=16$. Find $\sum_{k=1}^{19}(a_k-b_k)$.

38. $\sum_{k=1}^{15}a_k=19$. Find $\sum_{k=1}^{15}3a_k$.

39. $\sum_{k=1}^{25}a_k=63$ and $\sum_{k=1}^{11}a_k=15$. Find $\sum_{k=12}^{25}a_k$.

40. $\sum_{k=1}^{18}a_k=41$ and $\sum_{k=1}^{18}b_k=62$. Find $\sum_{k=1}^{18}(3a_k-2b_k)$.

41. $\sum_{k=1}^{21}a_k=-68$ and $\sum_{k=1}^{21}b_k=39$. Find $\sum_{k=1}^{21}(a_k+2b_k)$.

42. $\sum_{k=1}^{16}a_k=56$ and $\sum_{k=17}^{40}a_k=42$. Find $\sum_{k=1}^{40}a_k$.

43. $\sum_{k=13}^{29}a_k=84$ and $\sum_{k=1}^{29}a_k=143$. Find $\sum_{k=1}^{12}5a_k$.

44. $\sum_{k=1}^{20}b_k=34$ and $\sum_{k=1}^{20}(2a_k+b_k)=144$. Find $\sum_{k=1}^{20}a_k$.

45. $\sum_{k=1}^{27}a_k=46$ and $\sum_{k=1}^{10}a_k=122$. Find $\sum_{k=11}^{27}2a_k$.

Writing and Thinking About Mathematics

46. Use the sum of two Σ-notations to represent the following sum:
$-22+3-24+6-26+9-28+12-30+15$.

Hawkes Learning Systems: Introductory & Intermediate Algebra

Sigma Notation

15.3 Arithmetic Sequences

Objectives

After completing this section, you will be able to:

1. *Determine whether or not a sequence is arithmetic.*
2. *Find the general term for an arithmetic sequence.*
3. *Find the sum of the first n terms of an arithmetic sequence.*

There are many types of sequences studied in higher levels of mathematics. In the next two sections, we will discuss two types of sequences: **arithmetic sequences** and **geometric sequences**. In this discussion, sigma notation is used, and formulas for finding sums are developed. For arithmetic sequences, we can find sums of only a finite number of terms. For geometric sequences, we can find sums of a finite number of terms and, in some special cases, we define the sum of an infinite number of terms.

Arithmetic Sequences

The sequences

$$3, 5, 7, 9, 11, 13, \ldots$$

$$4, 5, 6, 7, 8, 9, \ldots$$

$$-2, -5, -8, -11, -14, -17, \ldots$$

all have a common characteristic. This characteristic is that **any two consecutive terms in each sequence have the same difference**.

$$\underbrace{3, 5}_{2}, \underbrace{5, 7}_{2}, \underbrace{7, 9}_{2}, \underbrace{9, 11}_{2}, \underbrace{11, 13}_{2}, \ldots \qquad 5-3=2,\ 7-5=2,\ 9-7=2, \text{ and so on}$$

$$\underbrace{4, 5}_{1}, \underbrace{5, 6}_{1}, \underbrace{6, 7}_{1}, \underbrace{7, 8}_{1}, \underbrace{8, 9}_{1}, \ldots \qquad 5-4=1,\ 6-5=1,\ 7-6=1, \text{ and so on}$$

$$\underbrace{-2, -5}_{-3}, \underbrace{-5, -8}_{-3}, \underbrace{-8, -11}_{-3}, \underbrace{-11, -14}_{-3}, \underbrace{-14, -17}_{-3}, \ldots \qquad -5-(-2)=-3,\ -8-(-5)=-3, \text{ and so on}$$

Such sequences are called **arithmetic sequences** or **arithmetic progressions**.

Arithmetic sequences are closely related to linear functions. To see this relationship we can plot the points of an arithmetic sequence and note that the rise from one point to the next is the difference, d, which is the slope of a line passing through all of the points. See Figure 15.1 as an illustration with a positive value for d.

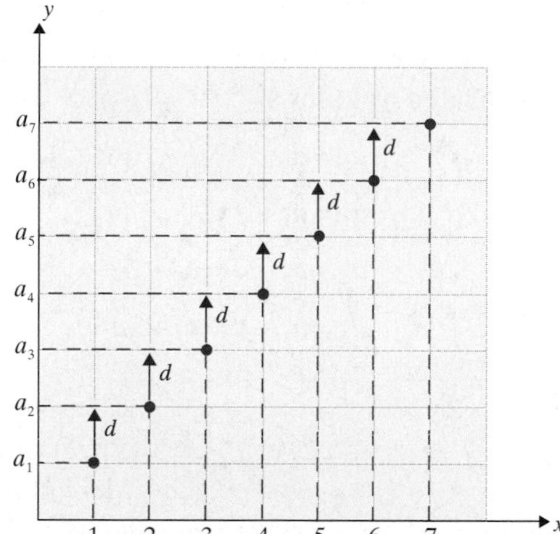

Figure 15.1

Arithmetic Sequence

*A sequence $\{a_n\}$ is called an **arithmetic sequence** (or **arithmetic progression**) if, for any natural number k,*

$$a_{k+1} - a_k = d \qquad \text{where } d \text{ is a constant.}$$

*d is called the **common difference.***

Example 1: Arithmetic Sequence

a. Show that the sequence $\{ 2n - 3 \}$ is arithmetic by finding d.

Solution: $a_k = 2k - 3$ and $a_{k+1} = 2(k+1) - 3 = 2k - 1$

$$a_{k+1} - a_k = (2k-1) - (2k-3) = 2k - 1 - 2k + 3 = 2$$

So, $d = 2$, and the sequence $\{ 2n - 3 \}$ is arithmetic.

b. Show that the sequence $\{ n^2 \}$ is not arithmetic.

Solution: Since $a_3 = 3^2$ and $a_2 = 2^2$ and $a_1 = 1^2$,

$$a_3 - a_2 = 9 - 4 = 5 \text{ and } a_2 - a_1 = 4 - 1 = 3.$$

Therefore, there is no common difference between consecutive terms, and $\{ n^2 \}$ is **not arithmetic.**

Note that in Example 1b, we only needed to show that there was **at least one** case in which the difference between two sets of consecutive terms was not the same. However, in Example 1a, showing truth in one case, or even 100 cases, is not enough. We used the general case to show truth in **every** case.

If the first term is a_1 and the common difference is d, then the arithmetic sequence can be indicated as follows:

$a_1 = a_1$	first term
$a_2 = a_1 + d$	second term
$a_3 = a_2 + d = a_1 + 2d$	third term
$a_4 = a_3 + d = a_1 + 3d$	fourth term
$\vdots$	$\vdots$
$a_n = a_{n-1} + d = a_1 + (n-1)d$	n^{th} term

The nth term of an Arithmetic Sequence

Teaching Notes:
Working through several examples of arithmetic sequences should remove any "mystery" about the addition of the $(n-1)d$ term.

If $\{a_n\}$ is an arithmetic sequence, then the nth term has the form

$$\boldsymbol{a_n = a_1 + (n-1)d}$$

where d is the common difference between the terms.

Example 2: The nth Term

Having students make absolute sense of formulas, whenever possible, should help them look beyond trying to memorize such equations.

a. If in an arithmetic sequence, $a_1 = 5$ and $d = 3$, find a_{16}.

Solution: To find the 16th term, let $n = 16$ in the formula $a_n = a_1 + (n-1)d$:

$$a_{16} = a_1 + 15d = 5 + 15 \cdot 3 = 50$$

b. Find the 20th term of the arithmetic sequence whose first three terms are –2, 8, and 18.

Solution: In this case, $a_1 = -2$ and $a_2 = 8$.

Since the sequence is arithmetic, $d = a_2 - a_1 = 8 - (-2) = 10$.

To find the 20th term, let $n = 20$ in the formula $a_n = a_1 + (n-1)d$:

$$a_{20} = -2 + (20-1)10 = -2 + 190 = 188$$

c. Find a_1 and d for the arithmetic sequence in which $a_3 = 6$ and $a_{21} = -48$.

Solution: Using the formula $a_n = a_1 + (n-1)d$ and solving simultaneous equations, we have

$$-48 = a_1 + 20d \qquad \begin{array}{rcrcr} -48 & = & a_1 & + & 20d \\ -6 & = & -a_1 & - & 2d \\ \hline -54 & = & & & 18d \\ -3 & = & & & d \end{array}$$

$$6 = a_1 + 2d$$

Then

$$\begin{array}{rcl} 6 & = & a_1 + 2(-3) \\ 12 & = & a_1 \end{array}$$

So, $a_1 = 12$ and $d = -3$.

Partial Sums of Arithmetic Sequences

Consider the problem of finding the sum $S = \sum_{k=1}^{6}(4k-1)$. We can, of course, write all the terms and then add them:

$$S = \sum_{k=1}^{6}(4k-1) = 3+7+11+15+19+23 = 78$$

However, to understand how the general formula is developed, we first write the sum and then write the sum again with the terms in reverse order. Adding vertically gives the same sum six times:

$$\begin{array}{rcrcrcrcrcrcr} S & = & 3 & + & 7 & + & 11 & + & 15 & + & 19 & + & 23 \\ S & = & 23 & + & 19 & + & 15 & + & 11 & + & 7 & + & 3 \\ \hline 2S & = & 26 & + & 26 & + & 26 & + & 26 & + & 26 & + & 26 \\ 2S & = & 6 & \cdot & 26 \\ S & = & 78 \end{array}$$

Using this same procedure with general terms in the subscript notation, we can develop the formula for the sum of any finite arithmetic sequence. Suppose that the n terms are

$$a_1,\ a_2 = a_1 + d,\ a_3 = a_1 + 2d,\ \ldots,\ a_{n-1} = a_n - d,\ a_n.$$

Thus, writing the terms in both ascending order and descending order and adding vertically, we have

$$\begin{array}{rcccccccccccc} S = & a_1 & + & (a_1+d) & + & (a_1+2d) & + \ldots + & (a_n-2d) & + & (a_n-d) & + & a_n \\ S = & a_n & + & (a_n-d) & + & (a_n-2d) & + \ldots + & (a_1+2d) & + & (a_1+d) & + & a_1 \\ \hline 2S = & (a_1+a_n) & + & (a_1+a_n) & + & (a_1+a_n) & + \ldots + & (a_1+a_n) & + & (a_1+a_n) & + & (a_1+a_n) \end{array}$$

(a_1+a_n) appears n times

$$2S = n(a_1+a_n)$$

$$S = \frac{n}{2}(a_1+a_n)$$

Partial Sums of Arithmetic Sequences

The nth ***partial sum***, S_n, *of the first n terms of an arithmetic sequence* $\{a_n\}$ *is*

$$S_n = \sum_{k=1}^{n} a_k = \frac{n}{2}(a_1+a_n).$$

A special case of an arithmetic sequence is $\{n\}$ and the corresponding sum of the first n terms:

$$\sum_{k=1}^{n} k = 1+2+3+\ldots+n$$

In this case, n = the number of terms, $a_1 = 1$, and $a_n = n$, so

$$\sum_{k=1}^{n} k = \frac{n}{2}(1+n).$$

Gauss

German mathematician Karl Friedrich Gauss (1777 – 1855) understood and applied this sum at the age of 7 in order to solve an arithmetic problem given to him and his classmates as "busy" work. Gauss probably observed the following pattern when told to find the sum of the whole numbers from 1 to 100:

$$1 + 2 + 3 + \ldots + 98 + 99 + 100$$

101 (3 + 98)

101 (2 + 99)

101 (1 + 100)

He saw that 101 was a sum 50 times. Thus, to find the sum he simply multiplied $101 \cdot 50 = 5050$. Not bad for a 7-year old.

Example 3: The Sum of a Finite Arithmetic Sequence

First show that the corresponding sequence is an arithmetic sequence by finding $a_{k+1} - a_k = d$. Then find the indicated sum using the formula.

a. $\sum_{k=1}^{75} k = 1 + 2 + 3 + \ldots + 75$

Solution: This is an example of the special case just discussed where the sequence is { k }, and the upper limit of summation is $n = 75$.

$$\sum_{k=1}^{75} k = \frac{75}{2}(1+75)$$
$$= \frac{75}{2}(76)$$
$$= 2850$$

b. $\sum_{k=1}^{50} 3k = 3 + 6 + 9 + \ldots + 150$

Solution: $a_k = 3k$ and $a_{k+1} = 3(k+1) = 3k + 3$

$$a_{k+1} - a_k = (3k+3) - 3k = 3 = d$$

So, { $3k$ } is an arithmetic sequence.

$$\sum_{k=1}^{50} 3k = \frac{50}{2}(3+150)$$ Here $n = 50$, $a_1 = 3$, and $a_{50} = 150$.
$$= 25(153)$$
$$= 3825$$

Also, Property III of Section 15.2 can be used to find the sum.

$$\sum_{k=1}^{50} 3k = 3\sum_{k=1}^{50} k$$ Property III of Section 15.2
$$= 3 \cdot \frac{50}{2}(1+50)$$ Here $n = 50$, $a_1 = 1$, and $a_{50} = 50$.
$$= 3 \cdot 25 \cdot 51$$
$$= 3825$$

c. $\sum_{k=1}^{70} (-2k+5) = 3 + 1 + (-1) + (-3) + \ldots + (-135)$

Solution: $a_k = -2k + 5$ and $a_{k+1} = -2(k+1) + 5 = -2k + 3,$

$$a_{k+1} - a_k = (-2k+3) - (-2k+5) = 3 - 5 = -2 = d$$

So, { $-2k + 5$ } is an arithmetic sequence.

$$\sum_{k=1}^{70} (-2k+5) = \frac{70}{2}[3 + (-135)]$$ Here $n = 70$, $a_1 = 3$, $a_{70} = -135$.
$$= 35(-132)$$
$$= -4620$$

continued on next page ...

Also, Properties II, III, and IV of Section 15.2 and the sum of a finite arithmetic sequence from this section can be used to find the sum.

$$\sum_{k=1}^{70}(-2k+5)=\sum_{k=1}^{70}-2k+\sum_{k=1}^{70}5 \qquad \text{Property II}$$

$$=-2\sum_{k=1}^{70}k+\sum_{k=1}^{70}5 \qquad \text{Property III}$$

$$=-2\cdot\frac{70}{2}(1+70)+70\cdot 5 \qquad \text{Property IV and the sum of a Finite Arithmetic Sequence}$$

$$=-2\cdot 35\cdot 71+350$$

$$=-4970+350$$

$$=-4620$$

Example 4: Arithmetic Sequence Application

Suppose that you are offered two jobs by the same company. The first job has a starting salary of \$25,000, with a "guaranteed" raise of \$2000 per year. The second job starts at \$30,000 with a "guaranteed" raise of \$1200 per year.

a. What would be your salary on the 10th year of each of these jobs?
b. If you were to stay 10 years with the company, which job would pay the most in total salary?

Solution:

Since the salary would increase the same amount each year, the yearly salaries form arithmetic sequences and we can use the corresponding formulas for a_{10} and S_{10}.

a. First job: $a_{10}=a_1+(10-1)d=25{,}000+9(2000)=\$43{,}000$
Second job: $a_{10}=a_1+(10-1)d=30{,}000+9(1200)=\$40{,}800$

You would be making a higher salary on the first job at the end of 10 years.

b. First job: $=5(25{,}000+43{,}000)=\$340{,}000$
Second job: $=5(30{,}000+40{,}800)=\$354{,}000$

At least for 10 years, the second job would pay more in total salary.

Practice Problems

1. *Show that the sequence $\{3n+5\}$ is arithmetic by finding d.*
2. *Find the 40th term of the arithmetic sequence with 1, 6, and 11 as its first three terms.*
3. *Find* $\sum_{k=1}^{50}(3k+5)$.

15.3 Exercises

1. Arithmetic sequence $d=3,\{3n-1\}$
2. Arithmetic sequence $d=4,\{4n-7\}$
3. Arithmetic sequence $d=-2,\{9-2n\}$
4. Arithmetic sequence $d=1,\{n+4\}$
5. Not an arithmetic sequence
6. Not an arithmetic sequence
7. Arithmetic sequence $d=-4,\{10-4n\}$
8. Arithmetic sequence $d=-5,\{9-5n\}$
9. Arithmetic sequence $d=\frac{1}{2},\left\{\frac{n-1}{2}\right\}$
10. Arithmetic sequence $d=\frac{1}{3},\left\{\frac{n+5}{3}\right\}$
11. $1,3,5,7,9$; arithmetic sequence
12. $3,2,1,0,-1$; arithmetic sequence
13. $-1,4,-7,10,-13$; not an arithmetic sequence
14. $\frac{3}{2},3,\frac{9}{2},6,\frac{15}{2}$; arithmetic sequence
15. $-1,-7,-13,-19,-25$; arithmetic sequence
16. $\frac{1}{2},\frac{1}{3},\frac{1}{4},\frac{1}{5},\frac{1}{6}$; not an arithmetic sequence
17. $\frac{20}{3},\frac{19}{3},6,\frac{17}{3},\frac{16}{3}$; arithmetic sequence

Determine which of the sequences in Exercises 1 – 10 are arithmetic. Find the common difference and the nth term for each arithmetic sequence.

1. $2, 5, 8, 11, \ldots$ **2.** $-3, 1, 5, 9, \ldots$ **3.** $7, 5, 3, 1, \ldots$ **4.** $5, 6, 7, 8, \ldots$

5. $1, 2, 3, 5, 8, \ldots$ **6.** $2, 4, 8, 16, \ldots$ **7.** $6, 2, -2, -6, \ldots$ **8.** $4, -1, -6, -11, \ldots$

9. $0, \frac{1}{2}, 1, \frac{3}{2}, \ldots$ **10.** $2, \frac{7}{3}, \frac{8}{3}, 3, \ldots$

In Exercise 11 – 20, write the first five terms of the sequence and determine which of the sequences are arithmetic.

11. $\{2n-1\}$ **12.** $\{4-n\}$ **13.** $\{(-1)^n(3n-2)\}$ **14.** $\left\{n+\frac{n}{2}\right\}$

15. $\{5-6n\}$ **16.** $\left\{\frac{1}{n+1}\right\}$ **17.** $\left\{7-\frac{n}{3}\right\}$ **18.** $\{(-1)^{n+1}(2n+1)\}$

19. $\left\{\frac{1}{2n}\right\}$ **20.** $\left\{\frac{2}{3}n-\frac{7}{3}\right\}$

Find the general term, a_n, for each of the arithmetic sequences in Exercises 21 – 30.

21. $a_1=1,\ d=\frac{2}{3}$ **22.** $a_1=9,\ d=-\frac{1}{3}$ **23.** $a_1=7,\ d=-2$ **24.** $a_1=-3,\ d=\frac{4}{5}$

25. $a_1=10,\ a_3=13$ **26.** $a_1=6,\ a_5=4$ **27.** $a_{10}=13,\ a_{12}=3$ **28.** $a_5=7,\ a_9=19$

29. $a_{13}=60,\ a_{23}=75$ **30.** $a_{11}=54,\ a_{29}=180$

In Exercises 31 – 38, $\{a_n\}$ is an arithmetic sequence.

31. $a_1=8,\ a_{11}=168$. Find a_{15}. **32.** $a_1=17,\ a_9=-55$. Find a_{20}.

33. $a_6=8,\ a_4=2$. Find a_{18}. **34.** $a_{16}=12,\ a_7=30$. Find a_9.

35. $a_{13}=34,\ d=2,\ a_n=22$. Find n. **36.** $a_4=20,\ d=3,\ a_n=44$. Find n.

37. $a_{10}=41,\ d=4,\ a_n=77$. Find n. **38.** $a_3=15,\ d=-\frac{3}{2},\ a_n=6$. Find n.

Find the indicated sums in Exercises 39 – 54 by using the formula for arithmetic sequences.

39. $-2+0+2+4+\ldots+24$ **40.** $3+6+9+\ldots+33$

41. $1+6+11+16+\ldots+46$ **42.** $5+9+13+17+\ldots+49$

Answers to Practice Problems: 1. $d=3$ **2.** $a_{40}=196$ **3.** 4075

43. $\sum_{k=1}^{9}(3k-1)$

44. $\sum_{k=1}^{12}(4-5k)$

45. $\sum_{k=1}^{11}(4k-3)$

46. $\sum_{k=1}^{10}(2k+7)$

47. $\sum_{k=1}^{13}\left(\frac{2k}{3}-1\right)$

48. $\sum_{k=1}^{28}(8k-5)$

49. $\sum_{k=7}^{15}\left(k+\frac{k}{3}\right)$

50. $\sum_{k=8}^{21}\left(9-\frac{k}{3}\right)$

51. If $\sum_{k=1}^{33}a_k=-12$, find $\sum_{k=1}^{33}(5a_k+7)$.

52. If $\sum_{k=1}^{15}a_k=60$, find $\sum_{k=1}^{15}(-2a_k-5)$.

53. If $\sum_{k=1}^{100}(-3a_k+4)=700$, find $\sum_{k=1}^{100}a_k$.

54. If $\sum_{k=1}^{50}(2b_k-5)=32$, find $\sum_{k=1}^{50}b_k$.

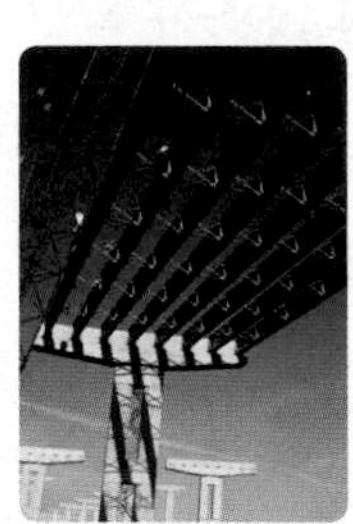

55. On a certain project, a construction company was penalized for taking more than the contractual time to finish the project. The company forfeited \$75 the first day, \$90 the second day, \$105 the third day, and so on. How many additional days were needed if the total penalty was \$1215?

56. The rungs of a ladder decrease uniformly in length from 84 cm to 46 cm. What is the total length of the wood in the rungs if there are 25 of them?

57. It is estimated that a certain piece of property, now valued at \$48,000, will appreciate as follows: \$1400 the first year, \$1450 the second year, \$1500 the third year, and so on. On this basis, what will be the value of the property after 10 years?

58. How many blocks are there in a pile if there are 19 in the first layer, 17 in the second layer, 15 in the third layer, and so on, with only 1 block on the top layer?

Writing and Thinking About Mathematics

59. Explain why an alternating sequence (one in which the terms alternate being positive and negative) cannot be an arithmetic sequence.

Hawkes Learning Systems: Introductory & Intermediate Algebra

Arithmetic Sequences

Answers (margin):

18. $3,-5,7,-9,11$; not an arithmetic sequence

19. $\frac{1}{2},\frac{1}{4},\frac{1}{6},\frac{1}{8},\frac{1}{10}$; not an arithmetic sequence

20. $-\frac{5}{3},-1,-\frac{1}{3},\frac{1}{3},1$; arithmetic sequence

21. $\left\{\frac{2n+1}{3}\right\}$

22. $\left\{\frac{28-n}{3}\right\}$

23. $\{9-2n\}$

24. $\left\{\frac{4n-19}{5}\right\}$

25. $\left\{\frac{17+3n}{2}\right\}$

26. $\left\{\frac{13-n}{2}\right\}$

27. $\{63-5n\}$

28. $\{3n-8\}$

29. $\left\{\frac{81+3n}{2}\right\}$

30. $\{7n-23\}$

31. 232

32. −154

33. 44

34. 26

35. 7

36. 12

37. 19

38. 9

39. 154

40. 198

41. 235

42. 324

43. 126

44. −342

45. 231

46. 180

47. $\frac{143}{3}$

48. 3108

49. 132

50. $\frac{175}{3}$

51. 171

52. −195

53. −100

54. 141

55. 9 days

56. 1625 cm

57. \$64,250

58. 100 blocks

59. Answers will vary.

15.4 Geometric Sequences and Series

Objectives

After completing this section, you will be able to:

1. *Determine whether or not a sequence is geometric.*
2. *Find the general term for a geometric sequence.*
3. *Find the specified terms of geometric sequences.*
4. *Find the sum of the first n terms of a geometric sequence.*
5. *Find the sum of an infinite geometric series.*

Geometric Sequences

Arithmetic sequences are characterized by having the property that any two consecutive terms have the same difference. **Geometric sequences** are characterized by having the property that **any two consecutive terms are in the same ratio**. That is, if consecutive terms are divided, the ratio will be the same regardless of which two consecutive terms are divided. Consider the three sequences

$$\frac{1}{2}, \frac{1}{4}, \frac{1}{8}, \frac{1}{16}, \frac{1}{32}, \ldots$$

$$3, 9, 27, 81, 243, \ldots$$

$$-9, 3, -1, \frac{1}{3}, -\frac{1}{9}, \ldots$$

As the following patterns show, each of these sequences has a **common ratio** when consecutive terms are divided.

$$\underbrace{\frac{1}{2}, \frac{1}{4}}_{\frac{1}{2}}, \underbrace{\frac{1}{4}, \frac{1}{8}}_{\frac{1}{2}}, \underbrace{\frac{1}{8}, \frac{1}{16}}_{\frac{1}{2}}, \underbrace{\frac{1}{16}, \frac{1}{32}}_{\frac{1}{2}}, \ldots \qquad \frac{\frac{1}{4}}{\frac{1}{2}} = \frac{1}{2}, \frac{\frac{1}{8}}{\frac{1}{4}} = \frac{1}{2}, \frac{\frac{1}{16}}{\frac{1}{8}} = \frac{1}{2}, \text{ and so on}$$

$$\underbrace{3, 9}_{3}, \underbrace{9, 27}_{3}, \underbrace{27, 81}_{3}, \underbrace{81, 243}_{3}, \ldots \qquad \frac{9}{3} = 3, \frac{27}{9} = 3, \frac{81}{27} = 3, \text{ and so on}$$

$$-9, 3, -1, \frac{1}{3}, -\frac{1}{9}, \ldots$$

with ratio $-\frac{1}{3}$ between each pair of consecutive terms:

$$\frac{3}{-9} = -\frac{1}{3}, \frac{-1}{3} = -\frac{1}{3}, \frac{\frac{1}{3}}{-1} = -\frac{1}{3}, \text{ and so on}$$

Therefore, these sequences are geometric sequences or geometric progressions.

Geometric Sequence

A sequence $\{a_n\}$ is called a **geometric sequence** *(or* **geometric progression***) if for any positive integer k,*

$$\frac{a_{k+1}}{a_k} = r \qquad \text{where } r \text{ is constant, and } r \neq 0.$$

r is called a ***common ratio.***

Teaching Notes: You might want to deal with the problem in a more open-ended way after showing these examples. What if we are given a sequence and want to determine if it is geometric? Students might be guided through an exploration strategy that includes trying to show there is a common ratio r, but if there is difficulty in resolving the algebra, perhaps they need to change the approach and suspect that no such r-value exists for the given sequence. Then, have them choose several terms and test them.

Example 1: Geometric Sequence

a. Show that the sequence $\left\{\frac{1}{2^n}\right\}$ is geometric by finding r.

Solution: $a_k = \frac{1}{2^k}$ and $a_{k+1} = \frac{1}{2^{k+1}}$

$$\frac{a_{k+1}}{a_k} = \frac{\frac{1}{2^{k+1}}}{\frac{1}{2^k}} = \frac{1}{2^k \cdot 2} \cdot \frac{2^k}{1} = \frac{1}{2} = r$$

Note that $2^{k+1} = 2^k \cdot 2^1$.

b. Show that the sequence $\{n^2\}$ is not geometric.

Solution: We want to show that different pairs of consecutive terms have different ratios. For this sequence, $a_3 = 3^2, a_2 = 2^2$, and $a_1 = 1^2$.

So,

$$\frac{a_3}{a_2} = \frac{3^2}{2^2} = \frac{9}{4} \text{ and } \frac{a_2}{a_1} = \frac{2^2}{1^2} = 4.$$

Since $\frac{9}{4} \neq 4$, there is no common ratio between consecutive terms, and $\{n^2\}$ is **not geometric.**

If the first term is a_1 and the common ratio is r, then the geometric sequence can be indicated as follows:

$a_1 = a_1$	$\rightarrow$	a_1	first term
$\dfrac{a_2}{a_1} = r$	$\rightarrow$	$a_2 = a_1 r$	second term
$\dfrac{a_3}{a_2} = r$	$\rightarrow$	$a_3 = a_2 r = (a_1 r) r = a_1 r^2$	third term
$\dfrac{a_4}{a_3} = r$	$\rightarrow$	$a_4 = a_3 r = (a_1 r^2) r = a_1 r^3$	fourth term
$\vdots$	$\vdots$		$\vdots$
$\dfrac{a_n}{a_{n-1}} = r$	$\rightarrow$	$a_n = a_{n-1} \cdot r = (a_1 r^{n-2}) r = a_1 r^{n-1}$	n^{th} term
$\vdots$	$\vdots$		$\vdots$

The nth Term of a Geometric Sequence

If $\{a_n\}$ is a geometric sequence, then the nth term has the form

$$a_n = a_1 r^{n-1}$$

where r is the common ratio.

Example 2: The nth Term of a Geometric Sequence

a. If in a geometric sequence, $a_1 = 4$ and $r = -\frac{1}{2}$, find a_8.

Solution: $a_8 = a_1 r^7 = 4\left(-\frac{1}{2}\right)^7 = 2^2\left(-\frac{1}{2^7}\right) = -\frac{1}{2^5} = -\frac{1}{32}$

b. Find the seventh term of the following geometric sequence: $3, \frac{3}{2}, \frac{3}{4}, \ldots$

Solution: Find r using the formula $r = \frac{a_{k+1}}{a_k}$ with $a_1 = 3$ and $a_2 = \frac{3}{2}$.

$$r = \frac{a_2}{a_1} = \frac{\frac{3}{2}}{3} = \frac{3}{2} \cdot \frac{1}{3} = \frac{1}{2}$$

Now, the seventh term is $a_7 = a_1 r^{7-1} = 3\left(\frac{1}{2}\right)^6 = \frac{3}{64}$.

continued on next page ...

c. Find a_1 and r for the geometric sequence in which $a_5 = 2$ and $a_7 = 4$.

Solution: Using the formula $a_n = a_1 r^{n-1}$, we get

$$2 = a_1 r^4 \text{ and } 4 = a_1 r^6.$$

Now, dividing gives

$$\frac{\cancel{a_1} \cancel{r^6}^{\,r^2}}{\cancel{a_1}\cancel{r^4}} = \frac{4}{2}$$

$$r^2 = 2$$

$$r = \pm\sqrt{2}$$

Using these values for r and the fact that $a_5 = 2$, we can find a_1.

For $r = \sqrt{2}$:

$$2 = a_1\left(\sqrt{2}\right)^4$$

$$2 = a_1 \cdot 4$$

$$\frac{1}{2} = a_1$$

For $r = -\sqrt{2}$:

$$2 = a_1\left(-\sqrt{2}\right)^4$$

$$2 = a_1 \cdot 4$$

$$\frac{1}{2} = a_1$$

There are two geometric sequences with $a_5 = 2$ and $a_7 = 4$. In both cases, $a_1 = \frac{1}{2}$. The two possibilities are

$$a_1 = \frac{1}{2} \text{ and } r = \sqrt{2}$$

or $$a_1 = \frac{1}{2} \text{ and } r = -\sqrt{2}.$$

The Sum of the Terms of a Geometric Sequence

The following discussion illustrates the method for finding the formula for the sum of the first n terms of a geometric sequence.

To find $S = \sum_{k=1}^{6} \frac{1}{3^k}$, we can write all the terms and then add them.

$$S = \sum_{k=1}^{6} \frac{1}{3^k} = \frac{1}{3} + \frac{1}{3^2} + \frac{1}{3^3} + \frac{1}{3^4} + \frac{1}{3^5} + \frac{1}{3^6}$$

$$= \frac{3^5 + 3^4 + 3^3 + 3^2 + 3 + 1}{3^6} = \frac{364}{729}$$

Or, we can first indicate the sum by writing the terms in order. Then indicate $\frac{1}{3}$ of the sum and multiply each term by $\frac{1}{3}$ (the common ratio). Writing these two sums in a vertical format and subtracting gives the following results.

$$\begin{array}{rcccccccccccccc} S & = & \frac{1}{3} & + & \frac{1}{3^2} & + & \frac{1}{3^3} & + & \frac{1}{3^4} & + & \frac{1}{3^5} & + & \frac{1}{3^6} & & \\ \frac{1}{3}S & = & & & \frac{1}{3^2} & + & \frac{1}{3^3} & + & \frac{1}{3^4} & + & \frac{1}{3^5} & + & \frac{1}{3^6} & + & \frac{1}{3^7} \\ \hline S-\frac{1}{3}S & = & \frac{1}{3} & - & 0 & - & 0 & - & 0 & - & 0 & - & 0 & - & \frac{1}{3^7} \end{array}$$

$$\left(1-\frac{1}{3}\right)S = \frac{1}{3}-\frac{1}{3^7} \qquad \text{Factor out the } S.$$

$$S = \frac{\frac{1}{3}-\frac{1}{3^7}}{1-\frac{1}{3}} = \frac{\frac{1}{3}-\left(\frac{1}{3}\right)^7}{1-\frac{1}{3}}$$

Teaching Notes:
At some point, it might be instructional to point out, if you haven't already, that the infinitely repeating decimals (e.g., $.\overline{3}$) are expandable as an infinite sum of terms having a common ratio. Their familiarity, at least with the numbers, might help them understand a finite sum from an infinite number of terms.

This procedure is certainly not necessary when only a few terms are to be added. However, it does illustrate the general method for finding a formula for the sum of the first n terms of any geometric sequence.

Suppose that the n terms are

$$a_1,\ a_2 = a_1r,\ a_3 = a_1r^2,\ldots,\ a_{n-1} = a_1r^{n-2},\ a_n = a_1r^{n-1}.$$

Thus, the sum can be written

$$S = a_1 + a_1r + a_1r^2 + \ldots + a_1r^{n-2} + a_1r^{n-1}$$

$$rS = a_1r + a_1r^2 + a_1r^3 + \ldots + a_1r^{n-1} + a_1r^n \qquad \text{Multiply each term by } r.$$

$$S - rS = a_1 - a_1r^n \qquad \text{Subtract.}$$

$$(1-r)S = a_1(1-r^n) \qquad \text{Factor.}$$

$$S = \frac{a_1(1-r^n)}{1-r} \qquad \text{Simplify.}$$

Partial Sums of Geometric Sequences

The nth ***partial sum***, S_n, *of the first n terms of a geometric sequence* $\{a_n\}$ *is*

$$\boldsymbol{S_n = \sum_{k=1}^{n} a_k = \frac{a_1(1-r^n)}{1-r}} \quad \text{where } r \neq 1.$$

Example 3: Partial Sums of Geometric Sequences

First show that the corresponding sequence is a geometric sequence by finding $\frac{a_{k+1}}{a_k} = r$.

Then find the indicated sum by using the formula $\sum_{k=1}^{n} a_k = \frac{a_1(1-r^n)}{1-r}$.

a. $\sum_{k=1}^{10} \frac{1}{2^k}$

Solution: Represent both a_k and a_{k+1} and find the ratio of these two terms.

$$a_k = \frac{1}{2^k} \text{ and } a_{k+1} = \frac{1}{2^{k+1}}$$

$$\frac{a_{k+1}}{a_k} = \frac{\frac{1}{2^{k+1}}}{\frac{1}{2^k}} = \frac{1}{2^{k+1}} \cdot \frac{2^k}{1} = \frac{1}{2 \cdot \cancel{2^k}} \cdot \frac{\cancel{2^k}}{1} = \frac{1}{2} = r$$

So, $\left\{\frac{1}{2^n}\right\}$ is a geometric sequence and

$$\sum_{k=1}^{10} \frac{1}{2^k} = \frac{\frac{1}{2}\left(1-\left(\frac{1}{2}\right)^{10}\right)}{1-\frac{1}{2}} = \frac{\frac{1}{2}\left(1-\frac{1}{1024}\right)}{\frac{1}{2}} = \frac{1023}{1024}.$$

b. $\sum_{k=1}^{5} (-1)^k \cdot 3^{\frac{k}{2}}$

Solution: Represent both a_k and a_{k+1} and find the ratio of these two terms.

$$\frac{a_{k+1}}{a_k} = \frac{(-1)^{k+1} \cdot 3^{(k+1)/2}}{(-1)^k \cdot 3^{k/2}} = \frac{\cancel{(-1)^k}(-1) \cdot \cancel{3^{k/2}} \cdot 3^{1/2}}{\cancel{(-1)^k} \cdot \cancel{3^{k/2}}}$$

$$= (-1) \cdot 3^{1/2} = -\sqrt{3} = r$$

So, $\{(-1^k) \cdot 3^{k/2}\}$ is a geometric sequence.

$$\sum_{k=1}^{5} (-1)^k \cdot 3^{k/2} = \frac{(-1) \cdot 3^{1/2} \cdot \left(1-\left(-\sqrt{3}\right)^5\right)}{1-\left(-\sqrt{3}\right)} = \frac{-\sqrt{3}\left(1+9\sqrt{3}\right)}{1+\sqrt{3}}$$

c. The parents of a small child decide to deposit \$1000 annually at the first of each year for 20 years for their child's education. If interest is compounded annually at 8%, what will be the value of the deposits after 20 years? (This type of investment is called an **annuity**.)

Solution: The formula for interest compounded annually is $A = P(1+r)^t$ where A is the amount in the account, r is the annual interest rate (in decimal form), and t is the time (in years).

The first deposit of \$1000 will earn interest for 20 years:

$$A_{20} = 1000(1+0.08)^{20} = 1000(1.08)^{20}$$

The second deposit will earn interest for 19 years:

$$A_{19} = 1000(1.08)^{19}$$

$$\vdots$$

The last deposit will earn interest for only one year:

$$A_1 = 1000(1.08)^1$$

The accumulated value of all deposits (plus interest) is the sum of the 20 terms of a geometric sequence:

Value at the end of twenty years:

$$= A_1 + A_2 + \cdots + A_{20} = \sum_{k=1}^{20} 1000(1.08)^k$$

$$= 1000(1.08)^1 + 1000(1.08)^2 + \cdots + 1000(1.08)^{20}$$

$$= \frac{1000(1.08)\left[1-(1.08)^{20}\right]}{1-1.08} \qquad \text{where } a_1 = 1000(1.08) \text{ and } r = 1.08$$

$$= \frac{1080[1-4.660957]}{-0.08}$$

$$= 49{,}423 \qquad \text{Rounded to the nearest dollar}$$

Thus, the accumulated value of the annuity is \$49,423.

Geometric Series

The indicated sum of all the terms (an infinite number of terms) of a sequence is called a **series**. A thorough study of series is a part of calculus. In this text, we will be concerned only with special cases of geometric series. In the following definition, the symbol ∞ (read "infinity") is used to indicate that the number of terms is unbounded. The symbol ∞ does not represent a number.

Infinite Series

The indicated sum of all terms of a sequence is called an ***infinite series*** *(or a* ***series****). For a sequence* $\{a_n\}$*, the corresponding series can be written as follows:*

$$\sum_{k=1}^{\infty}(a_k) = a_1 + a_2 + a_3 + \ldots + a_n + \ldots$$

For geometric sequences in the case where $|r| < 1$, it can be shown, in higher level mathematics, that r^n approaches 0 as n approaches infinity. This does not mean that r^n is ever equal to 0, only that it gets closer and closer to 0 as n becomes larger and larger. In symbols, we write

$$r^n \to 0 \quad \text{as} \quad n \to \infty.$$

Thus, we have the following result if $|r| < 1$:

$$S_n = \frac{a_1(1-r^n)}{1-r} \to \frac{a_1(1-0)}{1-r} = \frac{a_1}{1-r} \qquad \text{as } n \to \infty.$$

Theorem

If $\{a_n\}$ *is a geometric sequence and* $|r| < 1$*, then sum of the infinite series is*

$$S = \sum_{k=1}^{\infty}(a_k) = a_1 + a_1r + a_1r^2 + \ldots = \frac{a_1}{1-r}.$$

Example 4: Infinite Geometric Series

Find the sum of each of the following geometric series.

a. $$\sum_{k=1}^{\infty}\left(\frac{2}{3}\right)^{k-1} = \left(\frac{2}{3}\right)^0 + \left(\frac{2}{3}\right)^1 + \left(\frac{2}{3}\right)^2 + \left(\frac{2}{3}\right)^3 + \ldots$$
$$= 1 + \frac{2}{3} + \frac{4}{9} + \frac{8}{27} + \ldots$$

Solution: Here, $a_1 = 1$ and $r = \frac{2}{3}$. Substitution in the formula yields

$$S = \frac{1}{1-\frac{2}{3}} = \frac{1}{\frac{1}{3}} = 3.$$

b. $0.3333\ldots = 0.\overline{3}$ Recall that the bar over the 3 indicates a repeating pattern of digits in the decimal.

Solution: $0.33333\ldots = 0.3 + 0.03 + 0.003 + 0.0003 + 0.00003 + \ldots$

This format shows that the decimal number can be interpreted as a geometric series with $a_1 = 0.3 = \frac{3}{10}$ and $r = 0.1 = \frac{1}{10}$.

Applying the formula gives

$$S = \frac{\frac{3}{10}}{1-\frac{1}{10}} = \frac{\frac{3}{10}}{\frac{9}{10}} = \frac{3}{10} \cdot \frac{10}{9} = \frac{1}{3}.$$

In this way, an infinite repeating decimal can be converted to fraction form:

$$0.33333\ldots = \frac{1}{3}$$

c. $0.99999\ldots = 0.\overline{9}$

Solution: As shown in Example 4b, we can interpret the decimal number

$$0.99999\ldots = 0.9 + 0.09 + 0.009 + 0.0009 + \ldots$$

as a geometric series with $a_1 = 0.9 = \frac{9}{10}$ and $r = 0.1 = \frac{1}{10}$.

Applying the formula gives

$$S = \frac{\frac{9}{10}}{1-\frac{1}{10}} = \frac{\frac{9}{10}}{\frac{9}{10}} = \frac{9}{10} \cdot \frac{10}{9} = 1.$$

This very interesting result shows that the infinite decimal notation 0.99999… is just another way of writing 1.

In fact, we can prove the results of the following pattern:

$$0.11111\ldots = \frac{1}{9}$$

$$0.22222\ldots = \frac{2}{9}$$

$$\vdots$$

$$0.99999\ldots = \frac{9}{9} = 1$$

continued on next page ...

1. Not a geometric sequence
2. Geometric sequence $r = 2, \left\{ \frac{1}{12}\left(2^{n-1}\right) \right\}$
3. Geometric sequence $r = -\frac{1}{2}, \left\{ 3\left(-\frac{1}{2}\right)^{n-1} \right\}$
4. Not a geometric sequence
5. Geometric sequence $r = \frac{3}{8}, \left\{ \frac{32}{27}\left(\frac{3}{8}\right)^{n-1} \right\}$
6. Geometric sequence $r = \frac{2}{3}, \left\{ 18\left(\frac{2}{3}\right)^{n-1} \right\}$
7. Not a geometric sequence
8. Geometric sequence $r = -\frac{2}{3}, \left\{ \left(-\frac{2}{3}\right)^{n-1} \right\}$
9. Geometric sequence $r = -\frac{1}{4}, \left\{ 48\left(-\frac{1}{4}\right)^{n-1} \right\}$
10. Not a geometric sequence
11. $9, -27, 81, -243$; geometric sequence
12. $\frac{6}{5}, \frac{12}{25}, \frac{24}{125}, \frac{48}{625}$; geometric sequence
13. $\frac{2}{3}, \frac{4}{3}, 2, \frac{8}{3}$; not a geometric sequence
14. $\frac{2}{7}, -\frac{4}{49}, \frac{8}{343}, -\frac{16}{2401}$; geometric sequence

d. $5 - 1 + \frac{1}{5} - \frac{1}{25} + \frac{1}{125} - \frac{1}{625} + \ldots$

Solution: Here, $a_1 = 5$ and $r = -\frac{1}{5}$. A geometric series that alternates in sign will always have a negative value for r. Substitution in the formula gives

$$S = \frac{5}{1 - \left(-\frac{1}{5}\right)} = \frac{5}{1 + \frac{1}{5}} = \frac{5}{\frac{6}{5}} = \frac{5}{1} \cdot \frac{5}{6} = \frac{25}{6}.$$

Practice Problems

1. *Show that the sequence* $\left\{ \frac{(-1)^n}{3^n} \right\}$ *is geometric by finding r.*
2. *If in a geometric series,* $a_1 = 0.1$ *and* $r = 2$*, find* a_6.
3. *Find the sum* $\sum_{k=1}^{5} \frac{1}{2^k}$.
4. *Represent the decimal* $0.\overline{4}$ *as a series using Σ-notation.*
5. *Find the sum of the series in Problem 4.*

15.4 Exercises

Which of the sequences in Exercises 1 – 10 are geometric? Find the common ratio for each of the geometric sequences and write a formula for the nth term.

1. $2, 4, 6, 8, \ldots$
2. $\frac{1}{12}, \frac{1}{6}, \frac{1}{3}, \frac{2}{3}, \ldots$
3. $3, -\frac{3}{2}, \frac{3}{4}, -\frac{3}{8}, \ldots$
4. $5, 9, 13, 17, \ldots$
5. $\frac{32}{27}, \frac{4}{9}, \frac{1}{6}, \frac{1}{16}, \ldots$
6. $18, 12, 8, \frac{16}{3}, \ldots$
7. $\frac{14}{3}, \frac{2}{3}, \frac{2}{15}, \frac{2}{45}, \ldots$
8. $1, -\frac{2}{3}, \frac{4}{9}, -\frac{8}{27}, \ldots$
9. $48, -12, 3, -\frac{3}{4}, \ldots$
10. $4, -8, 12, -16, \ldots$

In Exercises 11 – 20 write the first four terms of the sequence and determine which of the sequences are geometric.

11. $\left\{(-3)^{n+1}\right\}$
12. $\left\{ 3\left(\frac{2}{5}\right)^n \right\}$
13. $\left\{ \frac{2}{3}n \right\}$
14. $\left\{ (-1)^{n+1}\left(\frac{2}{7}\right)^n \right\}$

Answers to Practice Problems: **1.** $r = -\frac{1}{3}$ **2.** $a_6 = 3.2$ **3.** $\frac{31}{32}$ **4.** $\sum_{k=1}^{\infty} \frac{4}{10^k}$ **5.** $\frac{4}{9}$

15. $-\frac{8}{5}, \frac{32}{25}, -\frac{128}{125}, \frac{512}{625}$; geometric sequence

16. $\frac{3}{2}, \frac{5}{4}, \frac{9}{8}, \frac{17}{16}$; not a geometric sequence

17. $3\sqrt{2}, 6, 6\sqrt{2}, 12$; geometric sequence

18. $2, \frac{5}{2}, \frac{10}{3}, \frac{17}{4}$; not a geometric sequence

19. $0.3, -0.09, 0.027, -0.0081$; geometric sequence

20. $6, \frac{3}{5}, \frac{3}{50}, \frac{3}{500}$; geometric sequence

21. $\left\{3(2)^{n-1}\right\}$

22. $\left\{-2\left(\frac{1}{5}\right)^{n-1}\right\}$

23. $\left\{\frac{1}{3}\left(-\frac{1}{2}\right)^{n-1}\right\}$

24. $\left\{5\left(\sqrt{2}\right)^{n-1}\right\}$

25. $\left\{\left(\sqrt{2}\right)^{n-1}\right\}$

26. $\left\{\frac{19}{27}(3)^{n-1}\right\}$

27. $\left\{\frac{1}{3}(3)^{n-1}\right\}$

28. $\left\{10\left(\frac{1}{2}\right)^{n-1}\right\}$

29. $\left\{-5\left(-\frac{3}{4}\right)^{n-1}\right\}$

30. $\left\{2(3)^{n-1}\right\}$

31. $\frac{1}{4}$ 32. $\frac{5}{16}$

15. $\left\{2\left(-\frac{4}{5}\right)^n\right\}$ **16.** $\left\{1+\frac{1}{2^n}\right\}$ **17.** $\left\{3(2)^{n/2}\right\}$ **18.** $\left\{\frac{n^2+1}{n}\right\}$

19. $\left\{(-1)^{n-1}(0.3)^n\right\}$ **20.** $\left\{6(10)^{1-n}\right\}$

Find the general term a_n for each of the geometric sequences in Exercises 21 – 30.

21. $a_1 = 3,\ r = 2$ **22.** $a_1 = -2,\ r = \frac{1}{5}$ **23.** $a_1 = \frac{1}{3},\ r = -\frac{1}{2}$

24. $a_1 = 5,\ r = \sqrt{2}$ **25.** $a_3 = 2,\ a_5 = 4,\ r > 0$ **26.** $a_4 = 19,\ a_5 = 57$

27. $a_2 = 1,\ a_4 = 9$ **28.** $a_2 = 5,\ a_5 = \frac{5}{8}$ **29.** $a_3 = -\frac{45}{16},\ r = -\frac{3}{4}$

30. $a_4 = 54,\ r = 3$

In Exercises 31 – 38, $\{a_n\}$ is a geometric sequence.

31. $a_1 = -32,\ a_6 = 1$. Find a_8. **32.** $a_1 = 20,\ a_6 = \frac{5}{8}$. Find a_7.

33. $a_1 = 18,\ a_7 = \frac{128}{81}$. Find a_5. **34.** $a_1 = -3,\ a_5 = -48$. Find a_7.

35. $a_3 = \frac{1}{2},\ a_7 = \frac{1}{32}$. Find a_4. **36.** $a_5 = 48,\ a_8 = -384$. Find a_9.

37. $a_1 = -2,\ r = \frac{2}{3},\ a_n = -\frac{16}{27}$. Find n. **38.** $a_1 = \frac{1}{9},\ r = \frac{3}{2},\ a_n = \frac{27}{32}$. Find n.

In Exercises 39 – 56, find the indicated sums.

39. $3 + 9 + 27 + \ldots + 243$ **40.** $-2 + 4 - 8 + 16$ **41.** $8 + 4 + 2 + \ldots + \frac{1}{64}$

42. $3 + 12 + 48 + \ldots + 3072$ **43.** $\sum_{k=1}^{3} -3\left(\frac{3}{4}\right)^k$ **44.** $\sum_{k=1}^{6} \left(\frac{-5}{3}\right)\left(\frac{1}{2}\right)^k$

45. $\sum_{k=1}^{5} \left(\frac{2}{3}\right)^k$ **46.** $\sum_{k=1}^{6} \left(\frac{1}{3}\right)^k$ **47.** $\sum_{k=4}^{7} 5\left(\frac{1}{2}\right)^k$

48. $\sum_{k=3}^{6} -7\left(\frac{3}{2}\right)^k$ **49.** $\sum_{k=1}^{\infty} \left(\frac{3}{4}\right)^{k-1}$ **50.** $\sum_{k=1}^{\infty} \left(\frac{5}{8}\right)^{k-1}$

51. $\sum_{k=1}^{\infty} \left(-\frac{1}{2}\right)^k$ **52.** $\sum_{k=1}^{\infty} \left(-\frac{2}{5}\right)^k$ **53.** $0.\overline{2}$

54. $0.\overline{6}$ **55.** $0.\overline{36}$ **56.** $0.\overline{81}$

57. Sue deposits \$800 annually at the first of each year for 10 years. If the interest is compounded annually at 9%, what will be the value of the deposit at the end of 10 years?

58. If \$1200 is deposited annually at the first of each year for 8 years, what will be the value of the deposit if the interest is compounded annually at 10%?

59. An automobile that costs \$8500 new depreciates at a rate of 20% of its value each year. What is its value after 3 years?

60. The radiator of a car contains 20 liters of water. Five liters are drained off and replaced by antifreeze. Then 5 liters of the mixture are drained off and replaced by antifreeze, and so on. This process is continued until six drain-offs and replacements have been made. How much antifreeze is in the final mixture?

61. A substance decays at a rate of $\frac{2}{5}$ of its weight per day. How much of the substance will be present after 4 days if initially there are 500 grams?

62. A ball rebounds to a height that is $\frac{3}{4}$ of its original height. How high will it rise after the fourth bounce if it is dropped from a height of 24 meters?

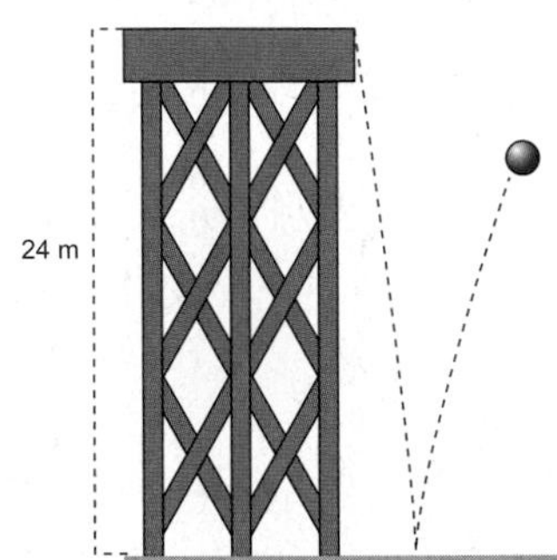

Writing and Thinking About Mathematics

63. Graph the first 8 partial sums of each geometric series as points to show how the sum of the series approaches a certain value. Show this value as a horizontal line in the graph.

a. $\sum_{k=1}^{\infty}\left(\frac{1}{2}\right)^{k-1}$ **b.** $\sum_{k=1}^{\infty}\frac{(-1)^{k+1}}{3^k}$

64. Consider the infinite series $4\cdot\sum_{k=1}^{\infty}\frac{(-1)^{k-1}}{2k-1}$. Write out several (at least 10 to 15) of the partial sums and their values until you can tell what number the partial sums "seem" to be approaching. What is this number?

65. Explain why there is no formula for finding the sum of an infinite geometric series when $|r|>1$.

Hawkes Learning Systems: Introductory & Intermediate Algebra

Geometric Sequences and Series

33. $\frac{32}{9}$ **34.** -192
35. $\frac{1}{4}$ or $-\frac{1}{4}$ **36.** 768
37. $n=4$ **38.** $n=6$
39. 363 **40.** 10
41. $\frac{1023}{64}$ **42.** 4095
43. $-\frac{333}{64}$ **44.** $-\frac{105}{64}$
45. $\frac{422}{243}$ **46.** $\frac{364}{729}$
47. $\frac{75}{128}$ **48.** $-\frac{12285}{64}$
49. 4 **50.** $\frac{8}{3}$
51. $-\frac{1}{3}$ **52.** $-\frac{2}{7}$
53. $\frac{2}{9}$ **54.** $\frac{2}{3}$
55. $\frac{4}{11}$ **56.** $\frac{9}{11}$
57. \$13,248.23
58. \$15,095.37
59. \$4352
60. 16.44 liters
61. 12.8 grams
62. 7.59 meters
63. a.

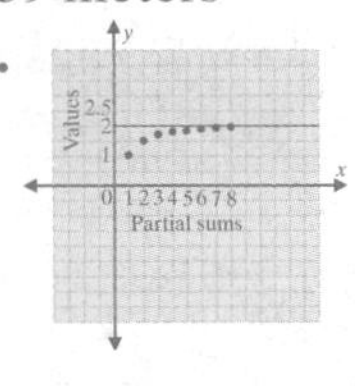

63. b.

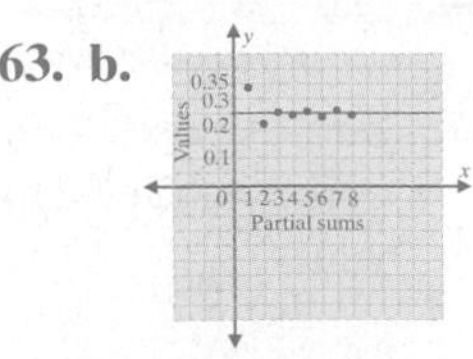

64. π
65. In the formula for S_n, $r^n\to\infty$ (or $r^n\to-\infty$) if $|r|>1$

15.5 The Binomial Theorem

Objectives

After completing this section, you will be able to:

1. *Calculate factorials.*
2. *Expand binomials using the Binomial Theorem.*
3. *Find specified terms in binomial expressions.*

Factorials

The objective in this section is to develop a formula stated as the **Binomial Theorem** (and sometimes called **binomial expansion**) that will allow you to write products such as

$$(a+b)^3,\quad (x+y)^7,\quad \text{and}\quad (x+5)^8$$

without having to multiply the binomial factors. For example, instead of multiplying three factors as follows,

$$\begin{aligned}(a+b)^3 &= (a+b)(a+b)(a+b)\\ &= (a^2+2ab+b^2)(a+b)\\ &= a^3+2a^2b+ab^2+a^2b+2ab^2+b^3\\ &= a^3+3a^2b+3ab^2+b^3\end{aligned}$$

knowledge of the Binomial Theorem will allow you to go directly to the final polynomial.

Before discussing the theorem itself, we need to understand the concept of **factorial**. For example, 6! (read "six factorial") represents the product of the positive integers from 6 to 1. Thus,

$$6! = 6\cdot5\cdot4\cdot3\cdot2\cdot1 = 720.$$

Also,

$$10! = 10\cdot9\cdot8\cdot7\cdot6\cdot5\cdot4\cdot3\cdot2\cdot1 = 3{,}628{,}800.$$

n Factorial ($n!$)

For any positive integer n,

$$n! = n(n-1)(n-2)\ldots 3\cdot 2\cdot 1$$

n! is read as "n factorial."

To evaluate an expression such as

$$\frac{7!}{6!}$$

do **not** evaluate each factorial. Instead, write the factorials as products and reduce the fraction.

$$\frac{7!}{6!} = \frac{7\cdot \cancel{6}\cdot \cancel{5}\cdot \cancel{4}\cdot \cancel{3}\cdot \cancel{2}\cdot \cancel{1}}{\cancel{6}\cdot \cancel{5}\cdot \cancel{4}\cdot \cancel{3}\cdot \cancel{2}\cdot \cancel{1}} = 7$$

Note that $n! = (n)(n-1)(n-2)\ldots(3)(2)(1)$

and $(n-1)! = (n-1)(n-2)(n-3)\ldots(3)(2)(1).$

So $n! = n(n-1)!$

In particular, $\dfrac{7!}{6!} = \dfrac{7\cdot(6!)}{6!} = 7.$

Also, for work with formulas involving factorials, zero factorial is **defined** to be 1.

0 Factorial

$$0! = 1$$

Using a Calculator

Factorials can be calculated with the TI-83 calculator by pressing the MATH *key and going to the menu under* **PRB**. *The fourth item in the list is the factorial symbol,* **!**. *For example, 6! can be calculated as follows:*

1. *Enter 6.*
2. *Press the* MATH *key.*
3. *Go to the* **PRB** *heading and press 4. (***6 !** *will appear on the display.)*
4. *Press* ENTER *and* **720** *will appear on the display.*

Example 1: Factorials

Simplify the following expressions.

a. $\dfrac{11!}{8!}$

Solution: $\dfrac{11!}{8!}=\dfrac{11\cdot 10\cdot 9\cdot(\cancel{8}\cdot\cancel{7}\cdot\cancel{6}\cdot\cancel{5}\cdot\cancel{4}\cdot\cancel{3}\cdot\cancel{2}\cdot\cancel{1})}{(\cancel{8}\cdot\cancel{7}\cdot\cancel{6}\cdot\cancel{5}\cdot\cancel{4}\cdot\cancel{3}\cdot\cancel{2}\cdot\cancel{1})}=990$

or $\dfrac{11!}{8!}=\dfrac{11\cdot 10\cdot 9\cdot 8!}{8!}=990$

b. $\dfrac{n!}{(n-2)!}$

Solution: $\dfrac{n!}{(n-2)!}=\dfrac{n(n-1)\cancel{(n-2)!}}{\cancel{(n-2)!}}=n(n-1)$

c. $\dfrac{30!}{28!2!}$

Solution: $\dfrac{30!}{28!2!}=\dfrac{\overset{15}{\cancel{30}}\cdot 29\cdot\cancel{28!}}{\cancel{28!}\cdot\cancel{2}\cdot 1}=15\cdot 29=435$

The expression in Example 1c can be written in the following notation.

$$\binom{30}{2}=\frac{30!}{28!2!}\quad\text{and}\quad\binom{30}{28}=\frac{30!}{2!28!}$$

Binomial Coefficient $\binom{n}{r}$

For non-negative integers n and r, with $0\le r\le n$, we define

$$\binom{n}{r}=\frac{n!}{(n-r)!\,r!}.$$

Because this quantity appears repeatedly in the Binomial Theorem, $\binom{n}{r}$ is often called a ***Binomial Coefficient****. It is read "the combination of n things taken r at a time."*

To get a formula for $\binom{n}{n-r}$, we apply the formula for $\binom{n}{r}$ and replace r with $n-r$. Thus,

$$\binom{n}{n-r} = \frac{n!}{(n-(n-r))!(n-r)!} = \frac{n!}{(n-n+r)!(n-r)!}$$

$$= \frac{n!}{r!(n-r)!} = \binom{n}{r}$$

Thus,

$$\binom{n}{n-r} = \binom{n}{r}.$$

Example 2: $\binom{n}{r}$

Evaluate the following.

a. $\binom{8}{2}$ and $\binom{8}{6}$

Solution: $\binom{8}{2} = \frac{8!}{6!2!} = \frac{\overset{4}{\cancel{8}} \cdot 7 \cdot \cancel{6!}}{\cancel{6!} \cdot \cancel{2}} = 28$ $\qquad$ $\binom{8}{6} = \frac{8!}{2!6!} = \frac{\overset{4}{\cancel{8}} \cdot 7 \cdot \cancel{6!}}{\cancel{2} \cdot \cancel{6!}} = 28$

b. $\binom{17}{0}$

Solution: $\binom{17}{0} = \frac{17!}{17!0!} = \frac{1}{1} = 1$

Using a Calculator

Other notations for $\binom{n}{r}$ are, ${}_nC_r$, C_r^n and $C(n, r)$. In particular, the notation ${}_nC_r$ is found in the TI-83 calculator. The "C" in these notations stands for "combination" for reasons which will be discussed in Section 15.7. Expressions of the form $\binom{n}{r}$ can be calculated with the TI-83 by pressing the MATH *key and going to the menu under* `PRB`. *The third item in the list is the symbol* ${}_{\mathtt{n}}\mathtt{C}_{\mathtt{r}}$ *read "the combination of n things taken r at a time." For example, $\binom{8}{2}$ can be calculated as follows:*

1. *Enter 8.*
2. *Press the* MATH *key.*
3. *Go to the* `PRB` *heading and press 3. (* $8\,{}_{\mathtt{n}}\mathtt{C}_{\mathtt{r}}$ *will appear on the display.)*
4. *Enter 2.*
5. *Press* ENTER *and* `28` *will appear on the display.*

The Binomial Theorem

The expansions of the binomial $a + b$ from $(a+b)^0$ to $(a+b)^5$ are shown here.

$$
\begin{aligned}
(a+b)^0 &= 1 \\
(a+b)^1 &= a+b \\
(a+b)^2 &= a^2+2ab+b^2 \\
(a+b)^3 &= a^3+3a^2b+3ab^2+b^3 \\
(a+b)^4 &= a^4+4a^3b+6a^2b^2+4ab^3+b^4 \\
(a+b)^5 &= a^5+5a^4b+10a^3b^2+10a^2b^3+5ab^4+b^5
\end{aligned}
$$

Teaching Notes:
There are several numerical patterns that the instructor could encourage students to discover before moving to the binomial theorem, especially as set up for this example. With the difficult pattern already shown in the example, most students can examine the pattern and write the theorem, if a little guidance is provided. The easier pattern is with the exponents. The other pattern is with the denominator factorials.

Three patterns are evident.

1. In each case, the **powers of a decrease by 1** in each term, and the **powers of b increase by 1** in each term.
2. In each term, the sum of the exponents is equal to the exponent on $(a + b)$.
3. A pattern, called Pascal's Triangle, is formed from the coefficients.

Pascal's Triangle

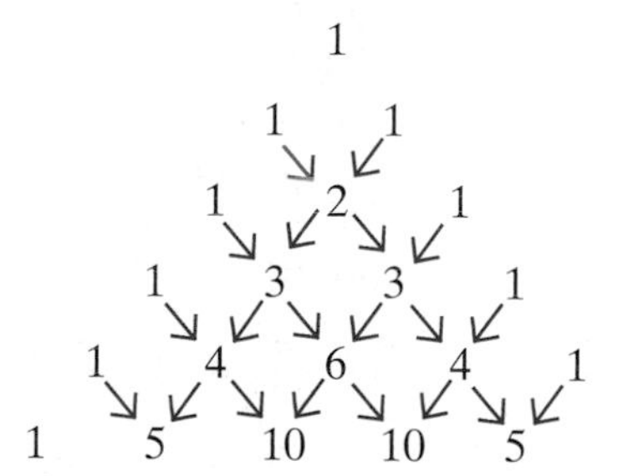

Pascal

In each case, the first and last coefficients are 1, and the other coefficients are the sum of the two numbers above to the left and above to the right of that coefficient. Thus, for $(a+b)^6$, we can construct another row of the triangle as follows:

1 5 10 10 5 1

1 6 15 20 15 6 1

and

$$(a+b)^6 = a^6+6a^5b+15a^4b^2+20a^3b^3+15a^2b^4+6ab^5+b^6.$$

Note that the coefficients can be written in factorial notation as follows:

$$\binom{6}{0}=\frac{6!}{0!6!}=1 \qquad \binom{6}{1}=\frac{6!}{1!5!}=6 \qquad \binom{6}{2}=\frac{6!}{2!4!}=15 \qquad \binom{6}{3}=\frac{6!}{3!3!}=20$$

$$\binom{6}{4}=\frac{6!}{4!2!}=15 \qquad \binom{6}{5}=\frac{6!}{5!1!}=6 \qquad \binom{6}{6}=\frac{6!}{6!0!}=1$$

So, the expansion can be written in the following form:

$$(a+b)^6=\binom{6}{0}a^6+\binom{6}{1}a^5b+\binom{6}{2}a^4b^2+\binom{6}{3}a^3b^3+\binom{6}{4}a^2b^4+\binom{6}{5}ab^5+\binom{6}{6}b^6.$$

This last form is the form used in the statement of the Binomial Theorem, stated here without proof.

The Binomial Theorem

$$(a+b)^n=\binom{n}{0}a^n+\binom{n}{1}a^{n-1}b+\binom{n}{2}a^{n-2}b^2+\ldots+\binom{n}{k}a^{n-k}b^k+\ldots+\binom{n}{n}b^n$$

In Σ–notation,

$$(\boldsymbol{a}+\boldsymbol{b})^n=\sum_{k=0}^{n}\binom{n}{k}\boldsymbol{a}^{n-k}\boldsymbol{b}^k$$

NOTES

1. There are $n+1$ terms in $(\boldsymbol{a}+\boldsymbol{b})^n$.

2. In each term of $(\boldsymbol{a}+\boldsymbol{b})^n$, the sum of the exponents of $\boldsymbol{a}$ and $\boldsymbol{b}$ is $\boldsymbol{n}$.

Example 3: The Binomial Theorem

a. Expand $(x+3)^5$ by using the Binomial Theorem.

Solution: $(x+3)^5=\displaystyle\sum_{k=0}^{5}\binom{5}{k}x^{5-k}3^k$

$$=\binom{5}{0}x^5+\binom{5}{1}x^4\cdot 3+\binom{5}{2}x^3\cdot 3^2+\binom{5}{3}x^2\cdot 3^3+\binom{5}{4}x\cdot 3^4+\binom{5}{5}3^5$$

$$=1\cdot x^5+5\cdot x^4\cdot 3+10\cdot x^3\cdot 9+10\cdot x^2\cdot 27+5\cdot x\cdot 81+1\cdot 243$$

$$=x^5+15x^4+90x^3+270x^2+405x+243$$

b. Expand $\left(y^2-1\right)^6$ by using the Binomial Theorem.

Solution: $$\left(y^2-1\right)^6=\sum_{k=0}^{6}\binom{6}{k}\left(y^2\right)^{6-k}(-1)^k$$

$$=\binom{6}{0}\left(y^2\right)^6+\binom{6}{1}\left(y^2\right)^5(-1)^1+\binom{6}{2}\left(y^2\right)^4(-1)^2$$

$$+\binom{6}{3}\left(y^2\right)^3(-1)^3+\binom{6}{4}\left(y^2\right)^2(-1)^4$$

$$+\binom{6}{5}\left(y^2\right)^1(-1)^5+\binom{6}{6}(-1)^6$$

$$=1\cdot y^{12}+6\cdot y^{10}(-1)+15\cdot y^8(+1)+20\cdot y^6(-1)$$

$$+15\cdot y^4(+1)+6\cdot y^2(-1)+1(+1)$$

$$=y^{12}-6y^{10}+15y^8-20y^6+15y^4-6y^2+1$$

c. Find the sixth term of the expansion of $\left(2x-\frac{1}{3}\right)^{10}$.

Solution: Since $\left(2x-\frac{1}{3}\right)^{10}=\sum_{k=0}^{10}\binom{10}{k}(2x)^{10-k}\left(-\frac{1}{3}\right)^k$, and the sum begins with $k=0$, the sixth term will occur when $k=5$.

$$\binom{10}{5}(2x)^{10-5}\left(-\frac{1}{3}\right)^5=\frac{10!}{5!5!}(2x)^5\left(-\frac{1}{3}\right)^5$$

$$=\overset{28}{\cancel{252}}\cdot 32x^5\left(-\frac{1}{\underset{27}{\cancel{243}}}\right)=\frac{-896x^5}{27}$$

The sixth term is $\frac{-896x^5}{27}$.

d. Find the fourth term of the expansion of $\left(x+\frac{1}{2}y\right)^8$.

Solution: $$\left(x+\frac{1}{2}y\right)^8=\sum_{k=0}^{8}\binom{8}{k}x^{8-k}\left(\frac{1}{2}y\right)^k$$

The fourth term occurs when $k=3$.

$$\binom{8}{3}x^{8-3}\left(\frac{1}{2}y\right)^3=\frac{8!}{3!5!}\cdot x^5\cdot\frac{1}{8}y^3=7x^5y^3$$

The fourth term is $7x^5y^3$.

continued on next page ...

e. Using the binomial expansion, approximate $(0.99)^4$ to the nearest thousandth.

Solution: $(0.99)^4 = (1-0.01)^4 = \sum_{k=0}^{4}\binom{4}{k}(1)^{4-k}(-0.01)^k$

$$= \binom{4}{0}\cdot 1^4 + \binom{4}{1}\cdot 1^3\cdot(-0.01) + \binom{4}{2}\cdot 1^2\cdot(-0.01)^2 + \binom{4}{3}\cdot 1\cdot(-0.01)^3 + \binom{4}{4}\cdot(-0.01)^4$$

$$= 1 + 4(-0.01) + 6(-0.01)^2 + 4(-0.01)^3 + 1(-0.01)^4$$

$$= 1 - 0.04 + 0.0006 - 0.000004 + (\text{small term})$$

$$\approx 0.9606$$

$$= 0.961 \quad (\text{to the nearest thousandth})$$

Practice Problems

1. *Simplify* $\frac{10!}{7!}$.
2. *Evaluate* $\binom{20}{2}$.
3. *Expand* $(x+2)^4$ *by using the Binomial Theorem.*
4. *Find the third term of the expansion of* $(2x-1)^7$.

15.5 Exercises

Simplify the expressions in Exercises 1 – 16.

1. $\frac{8!}{6!}$
2. $\frac{11!}{7!}$
3. $\frac{3!8!}{10!}$
4. $\frac{5!7!}{8!}$
5. $\frac{5!4!}{6!}$
6. $\frac{7!4!}{10!}$
7. $\frac{n!}{n}$
8. $\frac{n!}{(n-3)!}$
9. $\frac{(k+3)!}{k!}$
10. $\frac{n(n+1)!}{(n+2)!}$
11. $\binom{6}{3}$
12. $\binom{5}{4}$
13. $\binom{7}{3}$
14. $\binom{8}{5}$
15. $\binom{10}{0}$
16. $\binom{6}{2}$

Answers to Practice Problems: **1.** 720 **2.** 190 **3.** $x^4+8x^3+24x^2+32x+16$ **4.** $672x^5$

1. 56 **2.** 7920
3. $\frac{1}{15}$ **4.** 15
5. 4 **6.** $\frac{1}{30}$
7. $(n-1)!$
8. $n(n-1)(n-2)$
9. $(k+3)(k+2)(k+1)$
10. $\frac{n}{n+2}$ **11.** 20
12. 5 **13.** 35
14. 56 **15.** 1
16. 15
17. $x^7+7x^6y+21x^5y^2+35x^4y^3$
18. $x^{11}+11x^{10}y+55x^9y^2+165x^8y^3$
19. $x^9+9x^8+36x^7+84x^6$
20. $x^{12}+12x^{11}+66x^{10}+220x^9$
21. $x^5+15x^4+90x^3+270x^2$
22. $x^6-12x^5+60x^4-160x^3$
23. $x^6+12x^5y+60x^4y^2+160x^3y^3$
24. $x^5+15x^4y+90x^3y^2+270x^2y^3$
25. $2187x^7-5103x^6y+5103x^5y^2-2835x^4y^3$
26. $1024x^{10}-5120x^9y+11520x^8y^2-15360x^7y^3$
27. $x^{18}-36x^{16}y+576x^{14}y^2-5376x^{12}y^3$

Write the first four terms of the expansions in Exercises 17 – 28.

17. $(x+y)^7$ **18.** $(x+y)^{11}$ **19.** $(x+1)^9$ **20.** $(x+1)^{12}$

21. $(x+3)^5$ **22.** $(x-2)^6$ **23.** $(x+2y)^6$ **24.** $(x+3y)^5$

25. $(3x-y)^7$ **26.** $(2x-y)^{10}$ **27.** $(x^2-4y)^9$ **28.** $(x^2-2y)^7$

Using the Binomial Theorem, expand the expressions in Exercises 29 – 40.

29. $(x+y)^6$ **30.** $(x+y)^8$ **31.** $(x-1)^7$ **32.** $(x-1)^9$

33. $(3x+y)^5$ **34.** $(2x+y)^6$ **35.** $(x+2y)^4$ **36.** $(x+3y)^5$

37. $(3x-2y)^4$ **38.** $(5x+2y)^3$ **39.** $(3x^2-y)^5$ **40.** $(x^2+2y)^4$

Find the specified term in each of the expressions in Exercises 41 – 46.

41. $(x-2y)^{10}$, fifth term **42.** $(x+3y)^{12}$, third term

43. $(2x+3)^{11}$, fourth term **44.** $\left(x-\frac{y}{2}\right)^9$, seventh term

45. $(5x^2-y^2)^{12}$, tenth term **46.** $(2x^2+y^2)^{15}$, eleventh term

Approximate the value of each expression in Exercises 47 – 53 correct to the nearest thousandth.

47. $(1.01)^6$ **48.** $(0.96)^8$ **49.** $(0.97)^7$ **50.** $(1.02)^{10}$

51. $(2.3)^5$ **52.** $(2.8)^6$ **53.** $(0.98)^8$

Writing and Thinking About Mathematics

54. Factor the polynomial: $x^4 + 8x^3 + 24x^2 + 32x + 16$.

Hawkes Learning Systems: Introductory & Intermediate Algebra

The Binomial Theorem

28. $x^{14}-14x^{12}y+84x^{10}y^2-280x^8y^3$
29. $x^6+6x^5y+15x^4y^2+20x^3y^3+15x^2y^4+6xy^5+y^6$
30. $x^8+8x^7y+28x^6y^2+56x^5y^3+70x^4y^4+56x^3y^5+28x^2y^6+8xy^7+y^8$
31. $x^7-7x^6+21x^5-35x^4+35x^3-21x^2+7x-1$
32. $x^9-9x^8+36x^7-84x^6+126x^5-126x^4+84x^3-36x^2+9x-1$
33. $243x^5+405x^4y+270x^3y^2+90x^2y^3+15xy^4+y^5$
34. $64x^6+192x^5y+240x^4y^2+160x^3y^3+60x^2y^4+12xy^5+y^6$
35. $x^4+8x^3y+24x^2y^2+32xy^3+16y^4$
36. $x^5+15x^4y+90x^3y^2+270x^2y^3+405xy^4+243y^5$
37. $81x^4-216x^3y+216x^2y^2-96xy^3+16y^4$
38. $125x^3+150x^2y+60xy^2+8y^3$
39. $243x^{10}-405x^8y+270x^6y^2-90x^4y^3+15x^2y^4-y^5$
40. $x^8+8x^6y+24x^4y^2+32x^2y^3+16y^4$ **41.** $3360x^6y^4$ **42.** $594x^{10}y^2$
43. $1,140,480x^8$ **44.** $\frac{21}{16}x^3y^6$
45. $-27,500x^6y^{18}$
46. $96,096x^{10}y^{20}$ **47.** 1.062 **48.** 0.721
49. 0.808 **50.** 1.219 **51.** 64.363
52. 481.890 **53.** 0.851 **54.** $(x+2)$

15.6 Permutations

Objectives

After completing this section, you will be able to:

1. *Evaluate expressions representing permutations.*
2. *Solve applied problems involving permutations.*
3. *Solve applied problems involving the Fundamental Principle of Counting.*

The Fundamental Principle of Counting

Teaching Notes: Other commonly referenced examples include maximum numbers of automobile registration tags, social security numbers, zip codes, telephone numbers, PINs for bank accounts, and credit card numbers.

Many problems in statistics require a systematic approach to counting the number of ways several decisions can be made in succession (or several events can occur in succession). For example, if a seven-person board of education must elect a president and a vice-president from its own membership, how many ways can this be done? The reasoning is as follows:

The presidency can be filled by any one of 7.
After the president is elected, the vice-presidency can be filled by any one of 6.
The number of ways these two decisions can be made in succession is $7 \cdot 6 = 42$.

The procedure for counting the number of independent successive decisions (or events) is based on the **Fundamental Principle of Counting**.

The Fundamental Principle of Counting

If an event E_1 can occur in m_1 ways, an event E_2 can occur in m_2 ways, . . . , and an event E_k can occur in m_k ways, then the total number of ways that all events may occur is the product $m_1 \cdot m_2 \cdot \ldots \cdot m_k$.

Example 1: Fundamental Principle of Counting

A home contractor offers two basic house plans, each with two possible arrangements for the garage, four color combinations, and three types of landscaping. How many "different" homes can the contractor build?

Solution: The Fundamental Principle of Counting is used with each of the options considered as a decision or event.

$$\underset{\text{plans}}{\underset{\uparrow}{\mathbf{2}}} \cdot \underset{\text{garages}}{\underset{\uparrow}{\mathbf{2}}} \cdot \underset{\text{colors}}{\underset{\uparrow}{\mathbf{4}}} \cdot \underset{\text{landscaping}}{\underset{\uparrow}{\mathbf{3}}} = \mathbf{48}$$

He can build 48 "different" homes.

Permutations

A concept closely related to the Fundamental Principle of Counting is that of counting the number of ways that elements can be arranged (ordered). Each ordering of a set of elements is called a **permutation**.

Permutation

*A **permutation** is an arrangement (or ordering) of the elements of a set.*

In how many ways can the four letters a, b, c, and d be arranged? That is, how many permutations are there for the four letters a, b, c, and d? All the permutations are listed here.

abcd	bacd	cabd	dabc
abdc	badc	cadb	dacb
acbd	bcad	cbad	dbac
acdb	bcda	cbda	dbca
adbc	bdac	cdab	dcab
adcb	bdca	cdba	dcba

The number of permutations is the product

$$4 \cdot 3 \cdot 2 \cdot 1 = 4! = 24.$$

Five letters, such as a, b, c, d, and e, can be arranged in $5 \cdot 4 \cdot 3 \cdot 2 \cdot 1 = 5! = 120$ ways. Finding the number of ways that n elements can be ordered is an application of the Fundamental Principle of Counting and can be indicated with the factorial notation.

Number of Permutations of n Elements

There are $\boldsymbol{n \cdot (n-1) \cdot \ldots \cdot 2 \cdot 1 = n!}$ ***permutations of n elements.***

(That is, n elements can be arranged in n! ways.)

Not all permutation problems use all the elements of a set. For example, suppose that a 5-digit number (no digits are repeated) is to be formed by choosing five digits from the set $\{1,2,3,4,5,6,7,8\}$. How many such 5-digit numbers can be formed? The Fundamental Principle of Counting can be used in the following way.

Leave 5 spaces for the 5 digits.	_ _ _ _ _
8 digits are "eligible" for the first spot.	8 _ _ _ _
Once a digit is chosen for the first spot, there are 7 digits "eligible" for the second spot.	$8 \cdot 7$ _ _ _
Continuing with the same reasoning gives	$8 \cdot 7 \cdot 6 \cdot 5 \cdot 4$

Thus, $8 \cdot 7 \cdot 6 \cdot 5 \cdot 4 = 6720$ different 5-digit numbers can be formed using 8 digits. We say that there are 6720 permutations of 8 digits taken 5 at a time.

$${}_8P_5 = 8 \cdot 7 \cdot 6 \cdot 5 \cdot 4 = 6720$$

The notation ${}_8P_5$ is read "the number of permutations of 8 elements taken 5 at a time."

Using the factorial notation,

$${}_8P_5 = 8 \cdot 7 \cdot 6 \cdot 5 \cdot 4 = 8 \cdot 7 \cdot 6 \cdot 5 \cdot 4\left(\frac{3 \cdot 2 \cdot 1}{3 \cdot 2 \cdot 1}\right)$$

$$= \frac{8 \cdot 7 \cdot 6 \cdot 5 \cdot 4 \cdot 3 \cdot 2 \cdot 1}{3 \cdot 2 \cdot 1} = \frac{8!}{3!} = 6720$$

Number of Permutations of *n* Elements Taken *r* at a Time

The symbol ${}_nP_r$ denotes the number of permutations of n elements taken r at a time.

$${}_nP_r = n(n-1)(n-2)\ldots(n-r+1) = \frac{n!}{(n-r)!}$$

NOTES

Other notations for permutations are

$$P_r^n \text{ and } P(n,r).$$

Example 2: Number of Permutations of *n* Elements Taken *r* at a Time

a. A sailor has 7 different flags with which he can signal. How many signals can he send using only 3 flags?

Solution: ${}_7P_3 = \dfrac{7!}{(7-3)!} = \dfrac{7!}{4!} = 7 \cdot 6 \cdot 5 = 210$

He can send 210 different signals using 3 flags.

b. If the digits 1, 2, 3, 4, 5, and 6 are used to form three-digit numbers, how many numbers can be formed:
i. if digits may not be repeated, and
ii. if digits may be repeated?

Solution: i. $_6P_3 = \dfrac{6!}{(6-3)!} = \dfrac{6!}{3!} = 6 \cdot 5 \cdot 4 = 120$

ii. Since any of the digits can be used in more than one position, this part of the problem does not involve permutations. Using the Fundamental Principle of Counting, $6 \cdot 6 \cdot 6 = 216$.

There are 120 three-digit numbers if the digits may not be repeated and 216 three-digit numbers if digits may be repeated.

How many permutations are there of the letters in the word BEEHIVE? The three E's are not different from each other (that is, they are not distinct), and changing only these letters around would account for 3! permutations. Thus, if N represents the number of distinct permutations, $N \cdot 3! = 7!$ because there are seven letters involved.

This gives

$$N = \frac{7!}{3!} = 7 \cdot 6 \cdot 5 \cdot 4 = 840.$$

The following theorem is stated without proof.

Theorem for Distinct Permutations

If in a collection of n elements, m_1 are of one kind, m_2 are of another kind, ..., m_r are of still another kind, and

$$n = m_1 + m_2 + \ldots + m_r$$

then the total number of distinct permutations of the n elements is

$$N = \frac{n!}{m_1! \cdot m_2! \cdot \ldots \cdot m_r!}$$

Example 3: Permutations

Find the number of distinct permutations in the word MISSISSIPPI.

Solution: There are four **S**'s, four **I**'s, two **P**'s, and one **M**.

$$\frac{11!}{4!4!2!1!} = \frac{11 \cdot 10 \cdot 9 \cdot 8 \cdot 7 \cdot 6 \cdot 5 \cdot 4 \cdot 3 \cdot 2 \cdot 1}{4 \cdot 3 \cdot 2 \cdot 1 \cdot 4 \cdot 3 \cdot 2 \cdot 1 \cdot 2 \cdot 1 \cdot 1} = 34,650$$

Using a Calculator

Permutations can be calculated with the TI-83 calculator by pressing the MATH *key and going to the menu under* **PRB**. *The second item in the list is the symbol* $_{n}P_{r}$. *For example,* $_{7}P_{3}$ *can be calculated as follows:*

1. *Enter 7.*
2. *Press the* MATH *key.*
3. *Go to the* **PRB** *heading and press 2. (* $7\,_{n}P_{r}$ *will appear on the display.)*
4. *Enter 3.*
5. *Press* ENTER *and* **210** *will appear on the display.*

Practice Problems

1. *Evaluate* $_{6}P_{6}$.
2. *Evaluate* $_{6}P_{2}$.
3. *How many even numbers with four digits can be formed using the digits 1, 2, 3, 5, 7 if no repetitions are allowed?*
4. *Find the number of distinct permutations of the letters in the word SCIENTIFIC.*

15.6 Exercises

1. 840 **2.** 336
3. 24 **4.** 30
5. 120 **6.** 7

Evaluate the permutations in Exercises 1 – 10.

1. $_{7}P_{4}$ **2.** $_{8}P_{3}$ **3.** $_{4}P_{4}$ **4.** $_{6}P_{2}$ **5.** $_{5}P_{4}$

6. $_{7}P_{1}$ **7.** $_{9}P_{6}$ **8.** $_{11}P_{9}$ **9.** $_{10}P_{8}$ **10.** $_{9}P_{4}$

11. A football team of eleven players is electing a captain and a most valuable player. In how many ways can this be done if the awards must be given to two different players?

12. In how many ways can eleven girls be chosen for nine positions in a chorus?

13. There are eight men available for three outfield positions on the baseball team. If each man can play any position, in how many ways can the outfield positions be filled?

14. A president, a vice-president, a secretary, and a treasurer are to be selected for an organization of 20 members. In how many ways can these four people be selected?

7. 60,480 **8.** 19,958,400
9. 1,814,400 **10.** 3,024
11. 110 **12.** 19,958,400
13. 336 **14.** 116,280

Answers to Practice Problems: **1.** 720 **2.** 30 **3.** 24 **4.** 302,400

15. 72

16. 45

17. 360

18. 180

19. 64

20. 340

21. 576

22. a. 5040 b. 288

23. 288

24. 72

25. 69,300

26. 120

15. A luxury automobile is available in 6 body colors, 3 different vinyl tops, and 4 choices of interior. How many different cars must the dealer stock if he wishes to have one of each model?

16. In a subdivision, there are five basic floor plans. Each floor plan has three different exterior designs and three plans for landscaping the yard. How many different houses could be built?

17. How many four-digit numbers can be formed from the digits 1, 2, 5, 6, 7, and 9 if no repetitions are allowed?

18. How many **odd** numbers with four digits can be formed from the digits 1, 2, 3, 4, 6, and 7 with no repetitions allowed?

19. How many numbers of not more than four digits can be formed from the digits 2, 5, 7, and 9 if no repetitions are allowed?

20. How many numbers of not more than four digits can be formed from the digits 2, 5, 7, and 9 if repetitions are allowed?

21. A builder recently hired four superintendents and four foremen. He assigned a superintendent and a foreman to each of his four projects. In how many ways can this be done?

22. In how many ways can three math books and four English books be arranged on a shelf if:

a. they may be placed in any position?

b. the math books are together and the English books are together?

23. In how many ways can four algebra texts and three geometry texts be arranged on a shelf, keeping the subjects together?

24. In Exercise 23, if two of the algebra books are identical and two of the geometry books are identical, in how many ways can the books be arranged, keeping those on each subject together?

25. There are eleven flags that are displayed together, one above another. How many signals are possible if four of the flags are blue, two are red, three are yellow, and two are white?

Writing and Thinking About Mathematics

26. How many different ways can 6 people be seated at a round table?

Hawkes Learning Systems: Introductory & Intermediate Algebra

Permutations

15.7 Combinations

Objectives

After completing this section, you will be able to:

1. *Evaluate expressions representing combinations.*
2. *Solve applied problems involving combinations.*

Permutations indicate the order or arrangement of elements. Arrangements or groupings of elements in which order is not a concern are called **combinations**. For example, if you have ten books and are trying to arrange seven of these books on a shelf, alphabetized from left to right, then the number of arrangements is a permutation problem. However, if you are trying to decide which seven of the ten books to take on vacation, then the number of choices is a combination problem. Other examples of problems involving combinations are how many ways a committee can be formed or how many ways a hand of cards can be formed. In these cases, order is not involved.

Combination

Teaching Notes: Instructors should not forget that the word "combination" brings some ambiguity with it because the word is used generally in different ways that may be contrary to the mathematical concept. For example, with "combination locks", the "combination" is an ordered set of numbers. Some students will need more time than others in restricting the mathematical meaning of "combination".

*A **combination** is a collection of some (or all) of the elements of a set without regard to the order of the elements.*

If n distinct elements are given and a combination of r elements is to be selected, then the total number of combinations of n elements taken r at a time is symbolized ${}_nC_r$. Since each combination of r elements has $r!$ permutations, the product $r! \cdot {}_nC_r$ represents the number of permutations of n elements taken r at a time. Thus,

$$r! \cdot {}_nC_r = {}_nP_r$$

$$\text{or} \qquad {}_nC_r = \frac{{}_nP_r}{r!} = \frac{n!}{r!(n-r)!}.$$

Note that the formula ${}_nC_r$ for combinations is exactly the same as $\binom{n}{r}$ for the coefficients in the Binomial Theorem.

The relationship between combinations and permutations can be illustrated by considering the four letters a, b, c, and d and listing both ${}_4C_3$ and ${}_4P_3$.

$$\text{Combinations } \left({}_4C_3 \right) \quad \{\text{abc} \quad \text{abd} \quad \text{acd} \quad \text{bcd}\}$$

$$\text{Permutations } \left({}_4P_3 \right) \quad \begin{Bmatrix} \text{abc} & \text{abd} & \text{acd} & \text{bcd} \\ \text{acb} & \text{adb} & \text{adc} & \text{bdc} \\ \text{bac} & \text{bad} & \text{cad} & \text{cbd} \\ \text{bca} & \text{bda} & \text{cda} & \text{cdb} \\ \text{cab} & \text{dab} & \text{dac} & \text{dbc} \\ \text{cba} & \text{dba} & \text{dca} & \text{dcb} \end{Bmatrix}$$

Notice that each combination of 3 elements has $3! = 6$ corresponding permutations. Thus,

$$3! \cdot {}_4C_3 = 6 \cdot 4 = 24 = {}_4P_3.$$

Example 1: Combinations

a. In how many ways can a hand of 5 cards be dealt from a deck of 52 cards?

Solution: $${}_{52}C_5 = \frac{52!}{5!47!} = \frac{52 \cdot 51 \cdot \overset{5}{\cancel{50}} \cdot 49 \cdot \overset{4}{\cancel{48}}}{\cancel{5} \cdot \cancel{4} \cdot \cancel{3} \cdot \cancel{2} \cdot \cancel{1}} = 2,598,960$$

There are 2,598,960 possible hands of 5 cards.

b. A committee of 6 people is to be chosen from a group of 40 members. How many "different" committees are there?

Solution: $${}_{40}C_6 = \frac{40!}{6!34!} = \frac{\overset{2}{\cancel{40}} \cdot 39 \cdot 38 \cdot 37 \cdot \overset{1}{\cancel{36}} \cdot 35}{\cancel{6} \cdot \cancel{5} \cdot \cancel{4} \cdot \cancel{3} \cdot \cancel{2} \cdot \cancel{1}} = 3,838,380$$

There are 3,838,380 "different" committees.

These ideas can also be used in conjunction with the Fundamental Principle of Counting, as the following example illustrates.

Example 2: Combinations

A Senate committee of 6 members must be chosen with 3 Democrats and 3 Republicans. The eligible members are 10 Democrats and 8 Republicans. How many possible ways can such a committee be formed?

Solution: $${}_{10}C_3 = \frac{10!}{3!7!} = \frac{10 \cdot 9 \cdot 8}{3 \cdot 2 \cdot 1} = 120 \text{ groups of Democrats}$$

$${}_8C_3 = \frac{8!}{3!5!} = \frac{8 \cdot 7 \cdot 6}{3 \cdot 2 \cdot 1} = 56 \text{ groups of Republicans}$$

continued on next page ...

Using the Fundamental Principle of Counting,

$$_{10}C_3 \cdot {}_8C_3 = 120 \cdot 56 = 6720.$$

There are 6720 possible ways that this committee can be formed.

Practice Problems

1. *In how many ways can a jury of 6 men and 6 women be selected from a group of 10 men and 14 women?*
2. *A student senate committee of 5 people is to be chosen from a list of 20 students. In how many ways can this be done?*

15.7 Exercises

1. 35 **2.** 70
3. 1 **4.** 56
5. 84 **6.** 10
7. 6 **8.** 1
9. 120 **10.** 126
11. 816
12. 286
13. 792
14. 1365
15. 190
16. 21, 35, figures may vary
17. 30

18. 2520
19. 108

Evaluate the combinations in Exercises 1 – 10.

1. $_7C_3$ **2.** $_8C_4$ **3.** $_4C_4$ **4.** $_8C_5$ **5.** $_9C_6$

6. $_5C_3$ **7.** $_6C_1$ **8.** $_5C_5$ **9.** $_{10}C_7$ **10.** $_9C_4$

11. A committee of three is selected from the eighteen members of an organization. In how many ways may the committee be chosen?

12. You are permitted to answer any ten questions out of thirteen. In how many different ways can you make your ten selections?

13. If each girl can play any of the positions on a basketball team, how many different starting lineups of five can be formed from a team of twelve girls?

14. A ski club has fifteen members who desire to be on the four-man ski team. How many different ski teams can be formed from the members of the club?

15. Twenty people all shake hands with one another. How many different handshakes occur?

16. How many straight lines are determined by seven distinct points if no three points are collinear? How many triangles are formed? Sketch a picture of this geometric situation.

17. Five women and four men are candidates for the debate team. If the team consists of one woman and two men, in how many ways may the team be chosen?

18. In how many ways can a committee of two Republicans and three Democrats be selected from a group of seven Republicans and ten Democrats?

19. In a dozen eggs, three are spoiled. In how many ways can you select four eggs and get two spoiled ones?

Answers to Practice Problems: **1.** 630,630 **2.** 15,504

20. 26, 460
21. 14,700

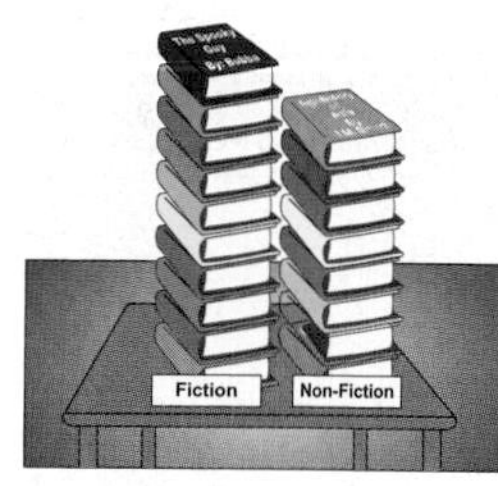

22. 1400
23. 66,528
24. 840
25. 600
26. 158, 184, 000
27.
a.

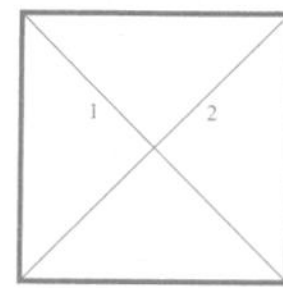

b.

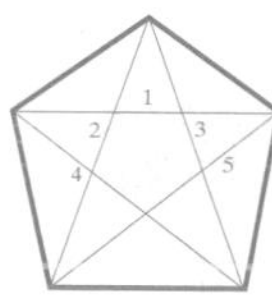

c.

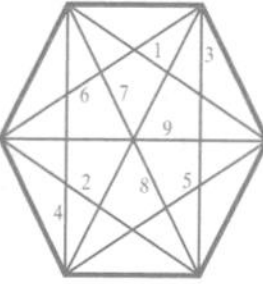

28. 18,009,460
29. 45
30. 117,600, permutation

20. A department store wishes to fill ten positions with four men and six women. In how many ways can these positions be filled if nine men and ten women have applied for the jobs?

21. A reading list consists of ten books of fiction and eight of nonfiction. In how many ways can a student select four fiction and four nonfiction books?

22. On a geometry test, there are eight theorems and six constructions to choose from. If you are required to do four theorems and three constructions, in how many ways could you work the test?

23. Sandy is packing for a trip. She plans to take five blouses and three skirts. In how many different ways can she make her selections if she has twelve blouses and nine skirts to choose from?

24. The high school Student Government Committee is composed of three boys and two girls. If there are eight boys and six girls eligible, how many different committees could be formed?

25. On a shelf, there are five English books, six algebra books, and three geometry books. Two English books, three algebra books, and two geometry books are selected. How many different selections are there?

26. In California, license plates are formed by a single digit followed by three letters followed by three digits. If the first digit cannot be 0, and all letters and digits can be repeated, how many license plates can be formed? (This, of course, does not allow for "vanity" plates.)

27. Find the number of diagonals of each of the following polygons. (A diagonal is a line segment connecting any two nonadjacent corners.) Sketch each figure and its diagonals.
a. square **b.** pentagon
c. hexagon

28. A lotto drawing consists of a random selection of 6 numbered balls from a set numbered from 1 to 51. In how many ways can six balls be selected?

29. A box contains balls numbered from 0 to 9. If two balls are selected and the balls are placed side by side to form a two digit number, how many of the numbers so formed will be odd?

30. A combination lock has numbers from 1 to 50. If no number may be selected twice, how many combinations to open the lock are possible that use three numbers? (**Hint:** Is this a permutation problem or a combination problem?)

Hawkes Learning Systems: Introductory & Intermediate Algebra

Combinations

Chapter 15 Index of Key Ideas and Terms

Section 15.1 Sequences

Sequences: $\{a_n\}$ page 1053

An infinite sequence (or a sequence) is a function that has the positive integers as its domain.

Terms of a sequence page 1052

Alternating Sequence page 1055

An alternating sequence is one in which the terms alternate in sign. Alternating sequences generally involve the expression $(-1)^n$.

Decreasing Sequence page 1056

A sequence $\{a_n\}$ is decreasing if $a_n > a_{n+1}$ for all n.

Increasing Sequence page 1056

A sequence $\{a_n\}$ is increasing if $a_n < a_{n+1}$ for all n.

Section 15.2 Sigma Notation

Partial Sum page 1059

Sigma Notation: $\sum$ page 1059

Partial Sums Using Sigma Notation page 1059

The nth **partial sum**, S_n, of the first n terms of a sequence $\{a_n\}$ is

$$S_n = \sum_{k=1}^{n} a_k = a_1 + a_2 + a_3 + \ldots + a_n.$$

k is called the **index of summation**, and k takes the integer values $1, 2, 3, \ldots, n$.

n is the **upper limit of summation**, and 1 is the **lower limit of summation.**

continued on next page ...

Section 15.2 Sigma Notation (continued)

Properties of Sigma Notation page 1060

For sequences $\{a_n\}$ and $\{b_n\}$ and any real number c:

I. $\sum_{k=1}^{n} a_k = \sum_{k=1}^{i} a_k + \sum_{k=i+1}^{n} a_k$ for any i, $1 \le i \le n-1$

II. $\sum_{k=1}^{n} (a_k + b_k) = \sum_{k=1}^{n} a_k + \sum_{k=1}^{n} b_k$

III. $\sum_{k=1}^{n} ca_k = c\sum_{k=1}^{n} a_k$

IV. $\sum_{k=1}^{n} c = nc$

Section 15.3 Arithmetic Sequences

Arithmetic Sequences page 1065

A sequence $\{a_n\}$ is called an arithmetic sequence (or arithmetic progression) if, for any positive integer k, $a_{k+1} - a_k = d$ where d is a constant. d is called the common difference.

n^{th} Term of an Arithmetic Sequence page 1066

If $\{a_n\}$ is an arithmetic sequence, then the nth term has the form

$a_n = a_1 + (n-1)d$ where d is the common difference between the terms.

Partial Sums of Arithmetic Sequences page 1068

If $\{a_n\}$ is an arithmetic sequence, then the sum of the first n terms

is $S_n = \sum_{k=1}^{n} a_k = \frac{n}{2}(a_1 + a_n)$.

Section 15.4 Geometric Sequences and Series

Geometric Sequences page 1074

A sequence $\{a_n\}$ is called a geometric sequence, (or geometric progression) if for any positive integer k, $\frac{a_{k+1}}{a_k} = r$ where r is constant and $r \ne 0$. r is called the common ratio.

continued on next page ...

Section 15.4 Geometric Sequences and Series (continued)

n^{th} Term of a Geometric Sequence page 1075

If $\{a_n\}$ is a geometric sequence, then the nth term has the form $a_n = a_1 r^{n-1}$ where r is the common ratio.

Partial Sums of Geometric Sequences page 1077

If $\{a_n\}$ is a geometric sequence, then the sum of the first n terms can be written as $S_n = \sum_{k=1}^{n} a_k = \dfrac{a_1(1-r^n)}{1-r}$ where $r \neq 1$.

Infinite Series (or Series) page 1080

The indicated sum of all the terms of a sequence is called an **infinite series** (or a **series**). For a sequence $\{a_n\}$, the corresponding series can be written as follows:

$$\sum_{k=1}^{\infty}(a_k) = a_1 + a_2 + a_3 + \ldots + a_n + \ldots$$

Sum of a Geometric Series page 1080

If $\{a_n\}$ is a geometric sequence and $|r| < 1$, then the sum of the infinite geometric series is

$$S = \sum_{k=1}^{\infty}(a_k) = a_1 + a_1 r + a_1 r^2 + \ldots = \frac{a_1}{1-r}.$$

Section 15.5 The Binomial Theorem

Factorial ($n!$) page 1086

For any positive integer n, $n! = n(n-1)(n-2)\ldots 3 \cdot 2 \cdot 1$.
$n!$ is read as "n factorial."

$0! = 1$ page 1086

Using a Calculator to Calculate Factorials page 1086

Binomial Coefficient $\binom{n}{r}$ page 1087

For non-negative integers n and r, with $0 \le r \le n$, we define

$$\binom{n}{r} = \frac{n!}{(n-r)!r!}.$$

Because this quantity appears repeatedly in the Binomial Theorem, $\binom{n}{r}$ is often called a Binomial Coefficient. It is read "the combination of n things taken r at a time."

continued on next page ...

Section 15.5 The Binomial Theorem (continued)

$$(a+b)^n = \binom{n}{0}a^n + \binom{n}{1}a^{n-1}b + \binom{n}{2}a^{n-2}b^2 + \ldots + \binom{n}{k}a^{n-k}b^k + \ldots + \binom{n}{n}b^n$$

In Σ-notation, $(a+b)^n = \sum_{k=0}^{n}\binom{n}{k}a^{n-k}b^k$.

Section 15.6 Permutations

If an event E_1 can occur in m_1 ways, an event E_2 can occur in m_2 ways, . . . , and an event E_k can occur in m_k ways, then the total number of ways that all events may occur is the product $m_1 \cdot m_2 \cdot \ldots \cdot m_k$.

A permutation is an arrangement (or ordering) of the elements of a set.

There are $\boldsymbol{n \cdot (n-1) \cdot \ldots \cdot 2 \cdot 1 = n!}$ **permutations of *n* elements.** (That is, n elements can be arranged in $n!$ ways.)

The symbol ${}_nP_r$ denotes the number of permutations of n elements taken r at a time.

$${}_nP_r = n(n-1)(n-2)\ldots(n-r+1) = \frac{n!}{(n-r)!}$$

If in a collection of n elements, m_1 are of one kind, m_2 are of another kind, ... m_r are of still another kind, and

$$n = m_1 + m_2 + \ldots + m_r$$

then the total number of distinct permutations of the n elements is

$$N = \frac{n!}{m_1! \cdot m_2! \cdot \ldots \cdot m_r!}.$$

Section 15.7 Combinations

Combinations page 1100

A combination is a collection of some (or all) of the elements of a set without regard to the order of the elements.

$$r! \cdot {}_nC_r = {}_nP_r$$

or $${}_nC_r = \frac{{}_nP_r}{r!} = \frac{n!}{r!(n-r)!}$$

Chapter 15 Review

For a review of the topics and problems from Chapter 15, look at the following lessons from *Hawkes Learning Systems: Introductory & Intermediate Algebra.*

Sequences
Sigma Notation
Arithmetic Sequences
Geometric Sequences and Series
The Binomial Theorem
Permutations
Combinations

Chapter 15 Test

1. $\frac{1}{4}, \frac{1}{7}, \frac{1}{10}, \frac{1}{13}, \ldots$; neither

2. $\left\{\frac{n}{2n+1}\right\}$

3. 26

4. $a_n = 8 - 2n$

5. 243

6. $\frac{1}{8}$

7. $a_n = \left(\sqrt{3}\right)^{n-2}$

8. $\frac{21}{16}$

9. 68

10. $-\frac{40}{729}$

11. $\frac{15}{100} + \frac{15}{10{,}000} + \ldots = \frac{5}{33}$

12. 74

13. \$13,050.16

14. 462

15. $32x^5 - 80x^4y + 80x^3y^2 - 40x^2y^3 + 10xy^4 - y^5$

16. $5670x^4y^4$

17. 151,200

18. 792

19. 720

20. 168

21. Answers will vary.

1. Write the first four terms of the sequence $\left\{\frac{1}{3n+1}\right\}$ and determine whether the sequence is arithmetic, geometric, or neither.

2. Find a general term for the sequence $\frac{1}{3}, \frac{2}{5}, \frac{3}{7}, \frac{4}{9}, \ldots$

In Exercises 3 – 5, $\{a_k\}$ is an arithmetic sequence. Find the indicated quantity.

3. $a_1 = 5,\ d = 3$. Find a_8.

4. $a_2 = 4,\ a_7 = -6$. Find a_n.

5. $a_4 = 22,\ a_7 = 37$. Find $\sum_{k=1}^{9} a_k$.

In Exercises 6 – 8, $\{a_k\}$ is a geometric sequence. Find the indicated quantity.

6. $a_1 = 8,\ r = \frac{1}{2}$. Find a_7.

7. $a_4 = 3,\ a_6 = 9$. Find a_n.

8. $a_2 = \frac{1}{3},\ a_5 = \frac{1}{24}$. Find $\sum_{k=1}^{6} a_k$.

Find each of the sums in Exercises 9 and 10.

9. $\sum_{k=1}^{8} (3k - 5)$

10. $\sum_{k=3}^{6} 2\left(-\frac{1}{3}\right)^k$

11. Write the decimal number $0.\overline{15}$ in the form of an infinite series and find its sum in the form of a proper fraction.

12. If $\sum_{k=1}^{50} a_k = 88$ and $\sum_{k=1}^{19} a_k = 14$, find $\sum_{k=20}^{50} a_k$.

13. A customer intends to buy a new car for \$25,000 and anticipates that it will depreciate at a rate of 15% of its value each year. What will be the value of the car in 4 years when he wants to trade it in for another new car?

14. Evaluate $\binom{11}{5}$.

15. Use the Binomial Theorem to expand $(2x - y)^5$.

16. Write the fifth term of the expansion of $(x + 3y)^8$.

17. Find the value of ${}_{10}P_6$.

18. Find the value of ${}_{12}C_7$.

19. There are ten players available for the three different outfield positions on a baseball team. Assuming each player can play any of the positions, in how many ways can the outfield be filled?

20. In a box there are eleven apples, three of which are spoiled. In how many ways could you select three good apples and one bad apple?

21. Explain, in your own words, why defining 0! to be 1 makes sense in mathematics.

Cumulative Review: Chapters 1 – 15

1. $11x+3$

2. $-x^2+3x+5$

3. $(4x+3)(16x^2-12x+9)$

4. $(3x-5)(2x+9)$

5. $(2x+1)(5x+7)(x-2)$

6. $\frac{11x+12}{15}$

7. $\frac{-6x-4}{(x+4)(x-4)(x-1)}$

8. $\frac{1}{x^2}$ **9.** $\frac{8^{1/2}y^{1/3}}{x^2}$

10. $\frac{2x^{\frac{4}{3}}}{3y}$ **11.** $3\sqrt{3x}$

12. $\frac{-13+11i}{29}$

13. $x>3$

3

14. $-\frac{5}{2}<x<\frac{9}{4}$

$-\frac{5}{2}$ $\frac{9}{4}$

15. $x=\frac{26}{7}$

16. $x=\frac{-2\pm i\sqrt{2}}{2}$

17. $x=4$

18. $x=\frac{39}{17}$

19. $x=5$

20. $x=e^{2.4}\approx 11.02$

21. No solution

Simplify each expression in Exercises 1 and 2.

1. $10x-\left[2x+(13-4x)-(11-3x)\right]+(2x+5)$

2. $(x^2+5x-2)-(3x^2-5x-3)+(x^2-7x+4)$

Factor completely in Exercises 3 – 5.

3. $64x^3+27$

4. $6x^2+17x-45$

5. $5x^2(2x+1)-3x(2x+1)-14(2x+1)$

Perform the indicated operations and simplify in Exercises 6 – 8.

6. $\frac{x+3}{3}+\frac{2x-1}{5}$

7. $\frac{x}{x^2-16}-\frac{x+1}{x^2-5x+4}$

8. $\frac{x^2-9}{x^4+6x^3}\div\frac{x^3-2x^2-3x}{x^2+7x+6}\cdot\frac{x^2}{x+3}$

Simplify each expression in Exercises 9 – 12. Assume that all variables are positive.

9. $\left(\frac{8x^{-1}y^{\frac{1}{3}}}{x^3y^{-\frac{1}{3}}}\right)^{\frac{1}{2}}$

10. $\sqrt[3]{\frac{8x^4}{27y^3}}$

11. $\sqrt{12x}-\sqrt{75x}+2\sqrt{27x}$

12. $\frac{1+3i}{2-5i}$

Solve the inequalities in Exercises 13 and 14. Graph the solution sets on real number lines.

13. $6(2x-3)+(x-5)>4(x+1)$

14. $8x^2+2x-45<0$

Solve the equations in Exercises 15 – 21.

15. $4(x-7)+2(3x+2)=3x+2$

16. $2x^2+4x+3=0$

17. $x-\sqrt{x}-2=0$

18. $\frac{1}{2x}+\frac{5}{x+3}=\frac{8}{3x}$

19. $2\sqrt{6-x}=x-3$

20. $5\ln x=12$

21. $|7-3x|-2=-4$

22. Solve the formula for n: $P=\frac{A}{1+ni}$

22. $n = \frac{A - P}{Pi}$

23. $x = 2, y = -5$

24. $x = 0, y = 2, z = -1$

25. a. $y = \frac{x + 7}{2}$

b. $2x^2 - 5$ c. $2x + 1$

26.

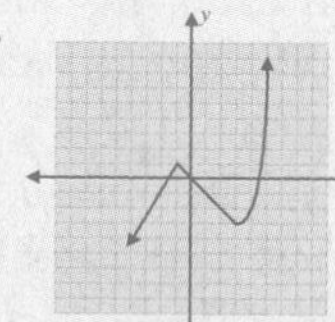

27.

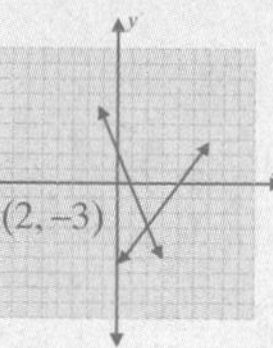

(2, –3)

28.

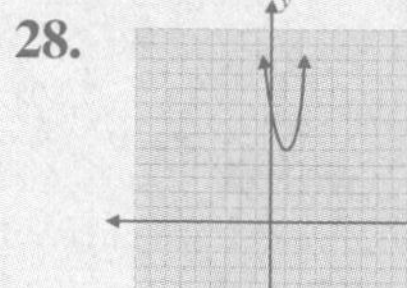

29.

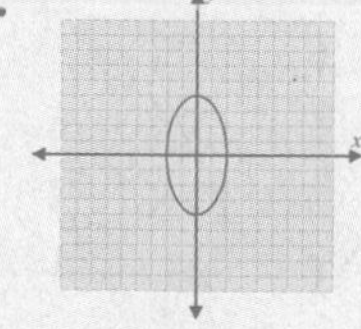

30.

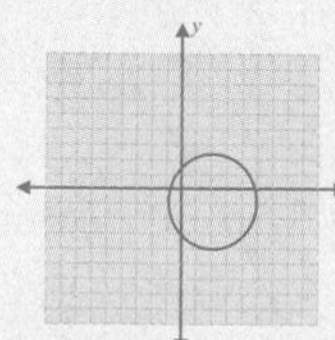

31. a. $a_n = 10 - 2n$

b. -10

32. a. $a_n = 16\left(\frac{1}{2}\right)^{n-1}$

b. $\frac{63}{2}$

Solve the systems of equations in Exercises 23 and 24.

23. Solve the following system by using Cramer's rule.

$$\begin{cases} 9x + 2y = 8 \\ 4x + 3y = -7 \end{cases}$$

24. Solve the following system of equations by using the Gaussian elimination method.

$$\begin{cases} 3x + y - 2z = 4 \\ x - 4y - 3z = -5 \\ 2x + 2y + z = 3 \end{cases}$$

25. If $f(x) = 2x - 7$ and $g(x) = x^2 + 1$, find:

a. $f^{-1}(x)$

b. $f[g(x)]$

c. $g(x+1) - g(x)$

26. The following is a graph of $y = f(x)$. Sketch the graph of $y = f(x-2) - 1$.

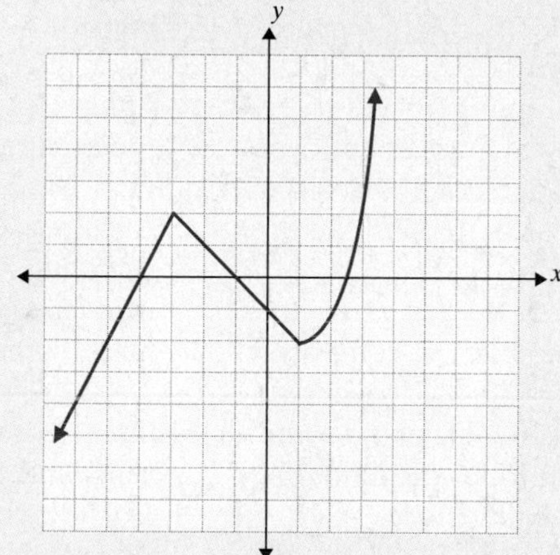

27. Solve the following system by graphing. $\begin{cases} 4x - 3y = 17 \\ 5x + 2y = 4 \end{cases}$

Graph each of the equations in Exercises 28 – 30.

28. $y = 4x^2 - 8x + 9$

29. $\frac{x^2}{4} + \frac{y^2}{16} = 1$

30. $x^2 - 4x + y^2 + 2y = 4$

31. If $\{a_n\}$ is an arithmetic sequence where $a_3 = 4$ and $a_8 = -6$,

a. find a_n.

b. find $\sum_{k=1}^{10} a_k$.

32. If $\{a_k\}$ is a geometric sequence where $a_1 = 16$ and $r = \frac{1}{2}$,

a. find a_n.

b. find $\sum_{k=1}^{6} a_k$.

33. $x^6+12x^5y+60x^4y^2+160x^3y^3+240x^2y^4+192xy^5+64y^6$

34. 210

35. 1680

36. 2002

37. 10 mph, 30 mph

38. 16 lbs. at $1.40, 4 lbs. at $2.60

39. 16 ft. by 26 ft.

40. 12.56 hrs.

41. $16,058.71

42. 443.7 liters

43. 12

44. 3024

45. 150

46. $x \approx -0.4746266, 1.395337$

47. $x \approx -1.9646356, 1.0580064$

48. $x \approx 0.24083054, 2.7336808$

49. $x = 1$

50. $x = -3, 1, 2$

33. Use the Binomial Theorem to expand $(x+2y)^6$.

Evaluate each expression in Exercises 34 – 36.

34. $\binom{10}{4}$ **35.** ${}_8P_4$ **36.** ${}_{14}C_5$

37. Two cars start together and travel in the same direction, one traveling 3 times as fast as the other. At the end of 3.5 hours, they are 140 miles apart. How fast is each traveling?

38. A grocer mixes two kinds of nuts. One costs $1.40 per pound and the other costs $2.60 per pound. If the mixture weighs 20 pounds and costs $1.64 per pound, how many pounds of each kind did he use?

39. A rectangular yard is 20 ft. by 30 ft. A rectangular swimming pool is to be built leaving a strip of grass of uniform width around the pool. If the area of the grass strip is 184 sq. ft., find the dimensions of the pool.

40. A radioactive isotope decomposes according to $A = A_0e^{-0.0552t}$, where t is measured in hours. Determine the half-life of the isotope. Round the solution to two decimal places.

41. Joan started her job exactly 5 years ago. Her original salary was $12,000 per year. Each year she received a raise of 6% of her current salary. What is her salary after this year's raise?

42. A tank holds 1000 liters of a liquid that readily mixes with water. After 150 liters are drained out, the tank is filled by adding water. Then 150 liters of the mixture are drained out and the tank is filled by adding water. If this process is continued 5 times, how much of the original liquid is left? Round the solution to one decimal place.

43. An architect is drawing house plans for a contractor. The contractor requested 4 different floor plans. Each floor plan has 3 different exterior designs. How many different plans must the architect draw?

44. A sailor has 9 different flags to use for signaling. A signal consists of displaying 4 flags in a specific order. How many signals can he send?

45. On a geometry test, there are six theorems and five constructions to choose from. If you are required to do 4 theorems and 2 constructions, in how many ways could you work the test?

Use a graphing calculator to solve (or estimate the solutions of) the equations in Exercises 46 – 50. Proceed by using the following steps:

1. *Get 0 on one side of the equation.*
2. *Press the* **Y=** *key and enter the nonzero side of the equation as the function* **Y1**.
3. *Press* **GRAPH**.
4. *Use the* **TRACE** *and* **ZOOM** *features of the calculator to estimate the zeros of the function. (Remember that the zeros are x-values where the graph intersects the x-axis.)*

46. $x^4 = 2x+1$ **47.** $e^x = -x^2+4$ **48.** $\ln x = x^2-2x-1$

49. $x^3 = 3x^2-3x+1$ **50.** $x^3 = 7x-6$

A.1 Absolute Value Inequalities

Objectives

After completing this section, you will be able to:

Solve absolute value inequalities.

Teaching Notes:
You might show your students that an inequality such as $-3 < x < 3$ can be interpreted graphically as the intersection of two graphs. Thus,

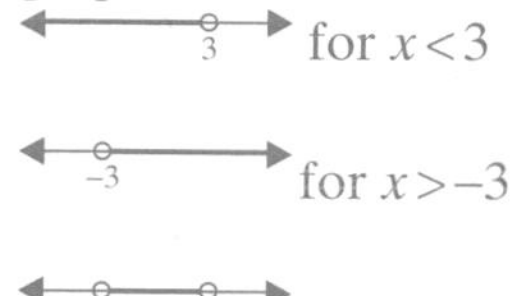

for the intersection $-3 < x < 3$.

Absolute Value Inequalities

Now consider an inequality with absolute value such as $|x| < 3$. For a number to have an absolute value less than 3, it must be within 3 units of 0. That is, the numbers between –3 and 3 have their absolute values less than 3 because they are within 3 units of 0. Thus, for $|x| < 3$,

Algebraic Notation	Graph	Interval Notation
$-3 < x < 3$	3 units 3 units; –3 0 3	x is in $(-3, 3)$

The inequality $|x-5| < 3$ means that the distance between x and 5 is less than 3. That is, we want all the values of x that are within 3 units of 5. The inequality is solved algebraically as follows:

$\|x-5\| < 3$	
$-3 < x-5 < 3$	$x-5$ is between –3 and 3.
$-3+5 < x-5+5 < 3+5$	Add +5 to each part of the expression, just as in solving linear inequalities.
$2 < x < 8$	Simplify each expression.
x is in $(2, 8)$	Using interval notation

2 5 8
3 units 3 units

The values for x are between 2 and 8 and are within 3 units of 5.

Solving Absolute Value Inequalities: $< c$

For $c > 0$;

a. *If* $|x| < c$, *then* $-c < x < c$.

b. *If* $|ax + b| < c$, *then* $-c < ax + b < c$.

Note: *These inequalities are true if* $<$ *is replaced by* $\leq$.

Example 1: Solving Absolute Value Inequalities

Solve the following absolute value inequalities and graph the solution sets.

a. $|x| \leq 6$

Solution: $|x| \leq 6$

$-6 \leq x \leq 6$

or x is in $[-6, 6]$

b. $|x + 3| < 2$

Solution: $|x + 3| < 2$

$-2 < x + 3 < 2$

$-2 - 3 < x + 3 - 3 < 2 - 3$

$-5 < x < -1$

or x is in $(-5, -1)$

c. $|2x - 7| < 1$

Solution: $|2x - 7| < 1$

$-1 < 2x - 7 < 1$

$6 < 2x < 8$

$3 < x < 4$

or x is in $(3, 4)$

We have been discussing inequalities in which the absolute value is less than some positive number. Now consider an inequality where the absolute value is greater than some positive number, such as $|x| > 3$. For a number to have an absolute value greater than 3, its distance from 0 must be greater than 3. That is, numbers that are greater than 3 **or** less than –3 will have absolute values greater than 3. Thus, for $|x| > 3$,

Algebraic Notation	Graph	Interval Notation
$x > 3$ or $x < -3$	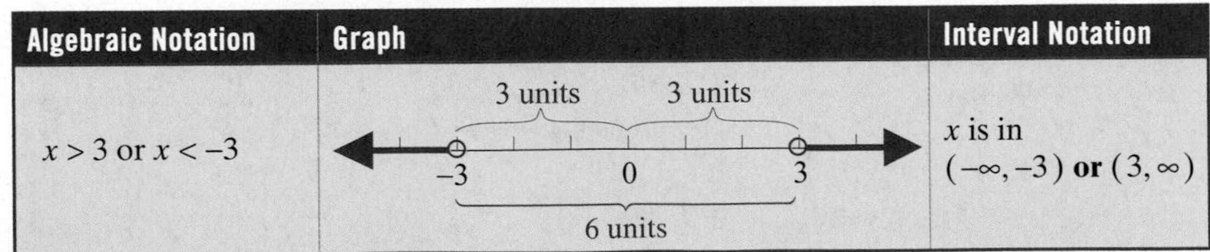	x is in $(-\infty, -3)$ **or** $(3, \infty)$

The expression $x > 3$ **or** $x < -3$ **cannot** be combined into one inequality expression. The word **or** must separate the inequalities since any number that satisfies one **or** the other is a solution to the absolute value inequality. There are **no** numbers that satisfy **both** inequalities.

The inequality $|x-5| > 6$ means that the distance between x and 5 is more than 6. That is, we want all values of x that are more than 6 units from 5. The inequality is solved algebraically as follows:

$|x-5| > 6$ indicates that

$x-5 < -6$ **or** $x-5 > 6$ — $x-5$ is less than -6 or greater than 6

Solving both inequalities gives,

$x-5+5 < -6+5$ **or** $x-5+5 > 6+5$, — Add 5 to each side, just as in solving linear inequalities.

$x < -1$ **or** $x > 11$. — Simplify.

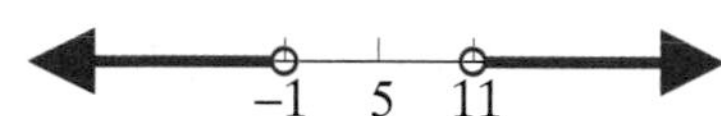

Note: The values for x less than -1 or greater than 11 are more than 6 units from 5. Thus, we can interpret the equality $|x-5| > 6$ to mean that the distance from x to 5 is greater than 6.

Solving Absolute Value Inequalities: $> c$

For $c > 0$;

a. *If $|x| > c$, then $x > c$ **or** $x < -c$.*

b. *If $|ax+b| > c$, then $ax+b < -c$ **or** $ax+b > c$.*

Note: *The inequalities in **a.** and **b.** are true if $>$ and $<$ are replaced by $\geq$ and $\leq$, respectively.*

Example 2: Solving Absolute Value Inequalities

Solve the following absolute value inequalities and graph the solution set.

a. $|x| \geq 5$

Solution: $|x| \geq 5$

$x \leq -5$ or $x \geq 5$

So, x is in $(-\infty, -5]$ or $[5, \infty)$.

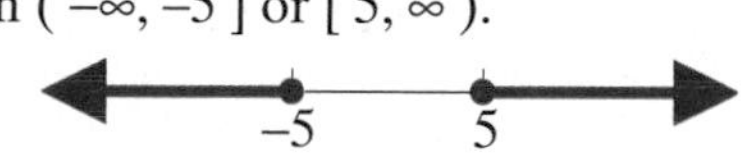

continued on next page ...

b. $|4x-3|>2$

Solution:

$$|4x-3|>2$$

$$4x-3<-2 \text{ or } 4x-3>2$$

$$4x<1 \quad \text{or} \quad 4x>5$$

$$x<\frac{1}{4} \quad \text{or} \quad x>\frac{5}{4}$$

So, x is in $\left(-\infty, \frac{1}{4}\right)$ or $\left(\frac{5}{4}, \infty\right)$.

c. $|3x-8|>-6$

Solution: There is nothing to do here except observe that no matter what is substituted for x, the absolute value will be greater than –6. Absolute value is always nonnegative (greater than or equal to 0). The solution to the inequality is all real numbers, so shade the entire number line. In interval notation, x is in $(-\infty, \infty)$.

d. $|x+9|<-\frac{1}{2}$

Solution: Since absolute value is always nonnegative (greater than or equal to 0), there is no solution to this inequality. No number has an absolute value less than $-\frac{1}{2}$.

e. $|2x+6|+4<9$

Solution:

$$|2x+6|+4<9$$

$$|2x+6|<5$$

$$-5<2x+6<5$$

$$-11<2x<-1$$

$$-\frac{11}{2}<x<-\frac{1}{2}$$

Add –4 to both sides in order to isolate the absolute value expression on one side. Then rewrite as a double inequality.

So, x is in $\left(-\frac{11}{2}, -\frac{1}{2}\right)$.

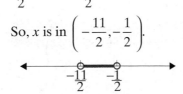

A.1 Exercises

1. 
-2
0
2
$(-\infty, \infty)$

2. 
-5 -3
3 5
$(-\infty, -3] \cup [3, \infty)$

3. 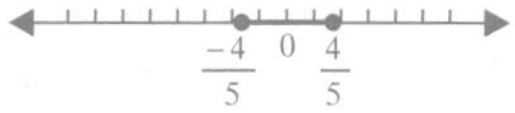
-4/5 0 4/5
$\left[-\frac{4}{5}, \frac{4}{5}\right]$

4. 
-9/2 -7/2
7/2 9/2
$\left(-\infty, -\frac{7}{2}\right] \cup \left[\frac{7}{2}, \infty\right)$

5. 
-1 1
5 7
$(-\infty, 1) \cup (5, \infty)$

6. 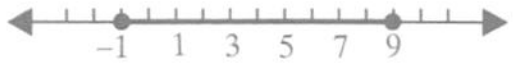
-1 1 3 5 7 9
$[-1, 9]$

7. 
-6 -4 -2 0 2
$[-6, 2]$

8. 
0
$(-\infty, \infty)$

9. 
-10 -8 -6 -4 -2
$[-10, -2]$

10. 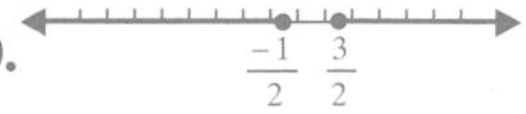
-1/2 3/2
$\left(-\infty, -\frac{1}{2}\right] \cup \left[\frac{3}{2}, \infty\right)$

11. No solution, $\varnothing$

12. 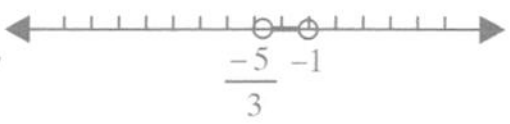
-5/3 -1
$\left(-\frac{5}{3}, -1\right)$

13. 
-2/3
6
$\left(-\infty, -\frac{2}{3}\right] \cup [6, \infty)$

14. 
-35/6
7/2
$\left(-\infty, -\frac{35}{6}\right) \cup \left(\frac{7}{2}, \infty\right)$

Solve the absolute value inequalities in Exercises 1 – 21 and graph the solution sets. Write each solution in interval notation.

1. $|y| \geq -2$ **2.** $|x| \geq 3$ **3.** $|t| \leq \frac{4}{5}$

4. $|x| \geq \frac{7}{2}$ **5.** $|y-3| > 2$ **6.** $|y-4| \leq 5$

7. $|t+2| \leq 4$ **8.** $|3x+4| > -8$ **9.** $|x+6| \leq 4$

10. $|2y-1| \geq 2$ **11.** $|3-2t| < -2$ **12.** $|3x+4| - 1 < 0$

13. $\left|\frac{3x}{2} - 4\right| \geq 5$ **14.** $\left|\frac{3}{7}y + \frac{1}{2}\right| > 2$ **15.** $|5t+2| + 3 < 4$

16. $|7x-3| + 4 \geq 6$ **17.** $|2x-9| - 7 \leq 4$ **18.** $5 > |4-2y| + 2$

19. $-4 < |6y-1| + 4$ **20.** $7 > |8-5x| + 3$ **21.** $|3x-7| + 4 \leq 4$

In Exercises 22 – 30 an interval is given. Represent this interval by using an absolute value inequality.

Example: Represent the closed interval [–2,4] as an absolute value inequality.

Solution: Note that the midpoint of the interval is $x = 1$ (the average of the two endpoints: $\frac{-2+4}{2} = 1$) and each endpoint is 3 units from 1. Thus, the inequality $|x-1| \leq 3$ represents the interval [–2, 4].

22. $[2, 6]$ **23.** $[-4, 8]$ **24.** $(-5, 3)$

25. $\left(-\frac{1}{2}, \frac{3}{4}\right)$ **26.** $(-1, 1)$ **27.** $(6, 10)$

28. $(-5.2, -1.4)$ **29.** $[-10, -2]$ **30.** $[1.9, 2.1]$

In Exercises 31 – 40, a graph of a set of real numbers is shown on a real number line. Represent this set of real numbers using an absolute value inequality.

31. 
-2 0 2

32. 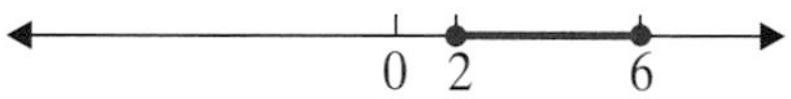
0 2
6

33. 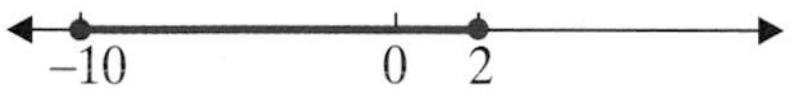
-10
0 2

34. 
-4 0 4

35. 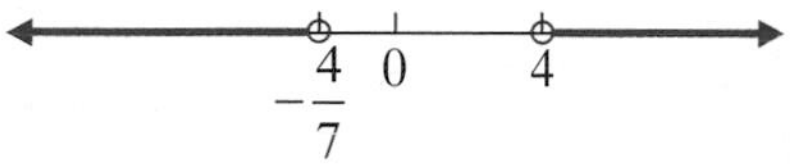
-4/7 0
4

36. 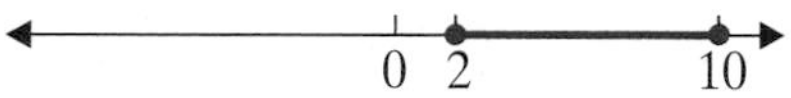
0 2
10

37. 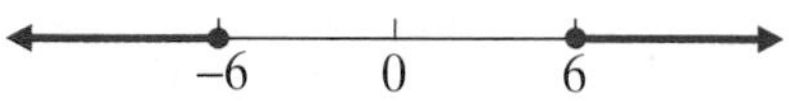
-6
0
6

38. 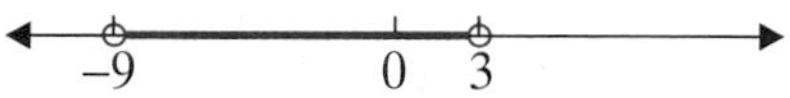
-9
0 3

39. 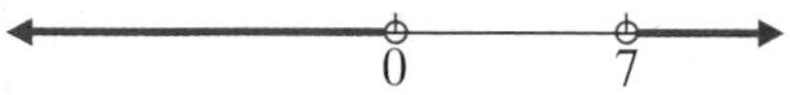
0
7

40. 
-4 0 4

15. $-\frac{3}{5}$ $-\frac{1}{5}$

$\left(-\frac{3}{5}, -\frac{1}{5}\right)$

16. 0 $\frac{1}{7}$ $\frac{5}{7}$ 1

$\left(-\infty, \frac{1}{7}\right] \cup \left[\frac{5}{7}, \infty\right)$

17. −1 1 3 5 7 10

$[-1,10]$

18. $\frac{1}{2}$ $\frac{7}{2}$

$\left(\frac{1}{2}, \frac{7}{2}\right)$

19. 0

$(-\infty, \infty)$

Writing and Thinking About Mathematics

In Exercises 41 – 45 a set of real numbers is described. ***(a)*** *Sketch a graph of the set on a real number line.* ***(b)*** *Represent each set by using absolute value notation.* ***(c)*** *If the set is one interval, state what type of interval it is.*

41. the set of real numbers between −10 and 10, inclusive

42. the set of real numbers within 7 units of 4

43. the set of real numbers more than 6 units from 8

44. the set of real numbers greater than or equal to 3 units from −1

45. the set of real numbers within 2 units of −5

Hawkes Learning Systems: Introductory & Intermediate Algebra

Solving Absolute Value Inequalities

20.

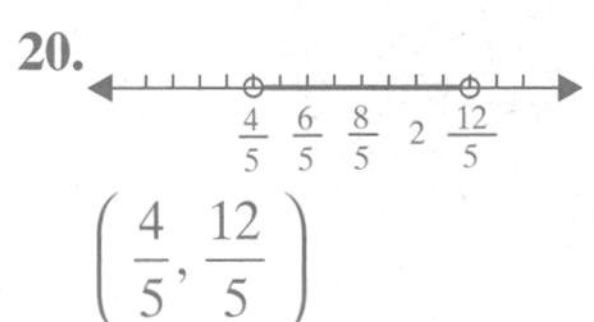

$\left(\frac{4}{5}, \frac{12}{5}\right)$

21.

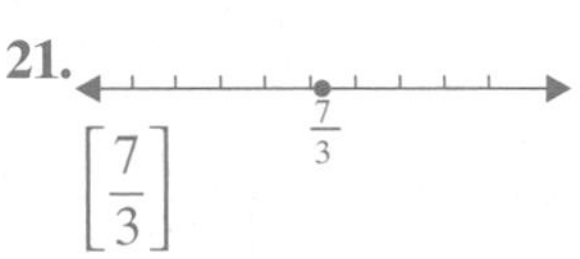

$\left[\frac{7}{3}\right]$

22. $|x-4| \le 2$

23. $|x-2| \le 6$

24. $|x+1| < 4$

25. $\left|x-\frac{1}{8}\right| < \frac{5}{8}$

26. $|x| < 1$

27. $|x-8| < 2$

28. $|x+3.3| < 1.9$

29. $|x+6| \le 4$

30. $|x-2| \le 0.1$

31. $|x| \le 2$

32. $|x-4| \le 2$

33. $|x+4| \le 6$

34. $|x| < 4$

35. $\left|x-\frac{12}{7}\right| > \frac{16}{7}$

36. $|x-6| \le 4$

37. $|x| \ge 6$

38. $|3+x| < 6$

39. $\left|x-\frac{7}{2}\right| > \frac{7}{2}$

40. $|x| \le 4$

41. a. −10 10

b. $|x| \le 10$

c. $[-10,10]$, closed interval

42. a. −3 4 11

b. $|x-4| \le 7$

c. $[-3,11]$, closed interval

43. a. 2 14

b. $|x-8| > 6$

c. $(-\infty,2) \cup (14,\infty)$, open interval

44. a.

b. $|x+1| \ge 3$

c. $(-\infty,-4] \cup [2,\infty)$, half-open interval

45. a. −7 −5 −3

b. $|x+5| \le 2$

c. $[-7,-3]$, closed interval

A.2 Synthetic Division

Objectives

After completing this section, you will be able to:

Divide polynomials by using synthetic division.

Synthetic Division

In the special case **when the divisor is first-degree with leading coefficient 1**, the division can be simplified by omitting the variables entirely and writing only certain coefficients. The procedure is called **synthetic division**. The following analysis describes how the procedure works for $\frac{5x^3+11x^2-3x+1}{x+3}$. (Note that $x + 3$ is first-degree with leading coefficient 1.)

a. With Variables

$$\begin{array}{r} 5x^2-4x+9 \\ x+3\overline{)5x^3+11x^2-3x+1} \\ \underline{5x^3+15x^2} \\ -4x^2-3x \\ \underline{-4x^2-12x} \\ 9x+1 \\ \underline{9x+27} \\ -26 \end{array}$$

b. Without Variables

$$\begin{array}{r} 5-4+9 \\ 1+3\overline{)5+11-3+1} \\ \underline{\boxed{5}+15} \\ -4\boxed{-3} \\ \underline{\boxed{-4}-12} \\ 9\boxed{+1} \\ \underline{\boxed{9}+27} \\ -26 \end{array}$$

The boxed numbers in step **b** can be omitted since they are repetitions of the numbers directly above them.

c. Boxed numbers omitted

$$\begin{array}{r} 5-4+9 \\ 1+3\overline{)5+11-3+1} \\ \underline{+15} \\ -4 \\ \underline{-12} \\ 9 \\ \underline{+27} \\ -26 \end{array}$$

d. Numbers moved up to fill in spaces

$$\begin{array}{r} 5-4+9 \\ 1+3\overline{)5+11-3+1} \\ \underline{+15-12+27} \\ -4+9-26 \end{array}$$

Next, we omit the 1 in the divisor, change +3 to –3, and write the opposites of the boxed numbers (because the quotient coefficient will now be multiplied by –3 instead of +3), as shown in steps **e** and **f**. This allows the numbers to be added instead of subtracted. The number 5 is written on the bottom line, and the top line is omitted. The quotient and remainder can now be read from the bottom line.

e.

$$\begin{array}{r|rrrr} & 5 & -4 & +9 & \\ 1+3 & 5 & +11 & -3 & +1 \\ & & \boxed{+15} & \boxed{-12} & \boxed{+27} \\ \hline & & -4 & +9 & -26 \end{array}$$

f.

$$\begin{array}{r|rrrr} -3 & 5 & +11 & -3 & +1 \\ & \downarrow & -15 & +12 & -27 \\ \hline & 5 & -4 & +9 & -26 \end{array}$$

Represents

$$5x^2 - 4x + 9 + \frac{-26}{x+3}$$

The numbers on the bottom now represent the coefficients of a polynomial of **one degree less than the dividend**, along with the remainder. The last number to the right is the remainder.

In summary, synthetic division can be accomplished as follows:

1. Write only the coefficients of the dividend and the opposite of the constant in the divisor.

$$\begin{array}{r|rrrr} -3 & 5 & 11 & -3 & 1 \\ \hline & & & & \end{array}$$

2. Rewrite the first coefficient as the first coefficient in the quotient.

$$\begin{array}{r|rrrr} -3 & 5 & 11 & -3 & 1 \\ & \downarrow & & & \\ \hline & 5 & & & \end{array}$$

3. Multiply the coefficient by the constant divisor and **add** this product to the second coefficient.

$$\begin{array}{r|rrrr} -3 & 5 & 11 & -3 & 1 \\ & \downarrow & -15 & & \\ \hline & 5 \nearrow & -4 & & \end{array}$$

4. Continue to multiply each new coefficient by the constant divisor and add this product to the next coefficient in the dividend.

$$\begin{array}{r|rrrr} -3 & 5 & 11 & -3 & 1 \\ & \downarrow & -15 & 12 & -27 \\ \hline & 5 \nearrow & -4 \nearrow & 9 \nearrow & -26 \end{array}$$

5. The constants on the bottom line are the coefficients of the quotient and the remainder.

$$\frac{5x^3 + 11x^2 - 3x + 1}{x+3} = 5x^2 - 4x + 9 + \frac{-26}{x+3}$$

$$= 5x^2 - 4x + 9 - \frac{26}{x+3}$$

Example 1: Synthetic Division

Use synthetic division to write each expression in the form $Q+\frac{R}{D}$.

a. $\dfrac{4x^3+10x^2+11}{x+5}$

Solution:

$$\begin{array}{r|rrrr} \underline{-5} & 4 & 10 & 0 & 11 \\ & \downarrow & -20 & 50 & -250 \\ \hline & 4 & -10 & 50 & -239 \end{array}$$

Since there is no x-term, 0 is the coefficient. The coefficient is 0 for any missing term.

$$\frac{4x^3+10x^2+11}{x+5}=4x^2-10x+50+\frac{-239}{x+5}$$

$$=4x^2-10x+50-\frac{239}{x+5}$$

b. $\dfrac{2x^4-x^3-5x^2-2x+7}{x-2}$

Solution:

$$\begin{array}{r|rrrrr} \underline{2} & 2 & -1 & -5 & -2 & 7 \\ & \downarrow & 4 & 6 & 2 & 0 \\ \hline & 2 & 3 & 1 & 0 & 7 \end{array}$$

$$\frac{2x^4-x^3-5x^2-2x+7}{x-2}=2x^3+3x^2+x+\frac{7}{x-2}$$

NOTES

Remember that synthetic division is used only when the divisor is first-degree of the form $(x+c)$ or $(x-c)$.

The Remainder Theorem

Synthetic division can be used for several purposes, one of which is to find the value of a polynomial for a particular value of x. For example, we know (from Section 5.3) that if

$$P(x)=x^3-5x^2+7x-10$$

then $$P(2)=2^3-5\cdot2^2+7\cdot2-10=-8.$$

With synthetic division of $x^3 - 5x^2 + 7x - 10$ by $x - 2$ we have

$$\begin{array}{r|rrrr} 2 & 1 & -5 & +7 & -10 \\ & & 2 & -6 & +2 \\ \hline & 1 & -3 & +1 & -8 \end{array} \longleftarrow \text{Remainder}$$

The fact that the remainder is the same as $P(2)$ is not an accident. In fact, as the following theorem states, the remainder when a polynomial is divided by a first-degree factor of the form $(x - c)$ will always be $P(c)$.

The Remainder Theorem

If a polynomial, $P(x)$, is divided by $(x - c)$, then the remainder will be $P(c)$.

Proof:

By the division algorithm we know that $\frac{P(x)}{x-c} = Q(x) + \frac{R}{x-c}$ where R is a constant.

(Remember that the degree of the remainder must be less than the degree of the divisor.)

Now, multiplying through by $(x - c)$, we have

$$P(x) = (x - c) \cdot Q(x) + R$$

and substituting $x = c$ gives

$$\begin{aligned} P(c) &= (c - c) \cdot Q(c) + R \\ &= 0 \cdot Q(c) + R \\ &= 0 + R \\ &= R \end{aligned}$$

The proof is complete.

Example 2: The Remainder Theorem and Synthetic Division

a. Use synthetic division to find $P(5)$ given $P(x) = -2x^2 + 15x - 50$.

Solution:

$$\begin{array}{r|rrr} 5 & -2 & 15 & -50 \\ & & -10 & 25 \\ \hline & -2 & 5 & -25 \end{array} \longleftarrow \text{Remainder} = P(5)$$

Thus, $P(5) = -25$

[Checking shows $P(5) = -2 \cdot 5^2 + 15 \cdot 5 - 50 = -50 + 75 - 50 = -25$.]

b. Use synthetic division to find $P(-3)$ given $P(x)=3x^4+10x^3-5x^2+125$.
Note: To evaluate $P(-3)$ we must think of the divisor of the form $(x+3)=(x-(-3))$. That is, in the form $(x-c)$, $c=-3$.

Solution:

$$\begin{array}{r|rrrrr} -3 & 3 & 10 & -5 & 0 & 125 \\ & & -9 & -3 & 24 & -72 \\ \hline & 3 & 1 & -8 & 24 & 53 \end{array} \longleftarrow \text{Remainder} = P(-3)$$

Thus, $P(-3)=53$.

c. Use synthetic division to show that $(x-6)$ is a factor of $P(x)=x^3-14x^2+53x-30$.

Solution:

$$\begin{array}{r|rrrr} 6 & 1 & -14 & 53 & -30 \\ & & 6 & -48 & 30 \\ \hline & 1 & -8 & 5 & 0 \end{array} \longleftarrow \text{Remainder} = P(6)$$

Thus, the remainder is $P(6)=0$ and $\boldsymbol{x-6}$ **is a factor of** $\boldsymbol{P(x)}$.

[**Note:** The coefficients in the quotient tell us that x^2-8x+5 is also a factor of $P(x)$.]

A.2 Exercises

In Exercises 1 – 30, divide by using synthetic division. ***a.*** *Write the answer in the form* $Q+\frac{R}{D}$ *where R is a constant.* ***b.*** *In each exercise, D = (x – c). State the value of c and the value of P(c).*

1. $\dfrac{x^2-12x+27}{x-3}$ **2.** $\dfrac{x^2-12x+35}{x-5}$ **3.** $\dfrac{x^2-5x+4}{x-1}$

4. $\dfrac{x^2+4x-21}{x+3}$ **5.** $\dfrac{4x^2+22x+10}{x+5}$ **6.** $\dfrac{3x^2-5x-12}{x-3}$

7. $\dfrac{2x^2+8x+15}{x+2}$ **8.** $\dfrac{4x^2-5x-19}{x-4}$ **9.** $\dfrac{x^3+4x^2+x-1}{x+8}$

10. $\dfrac{x^3-6x^2+8x-5}{x-2}$ **11.** $\dfrac{3x^3+x^2-x+5}{x-1}$ **12.** $\dfrac{4x^3-2x^2+3x+10}{x+1}$

13. $\dfrac{2x^3+5x^2-x-6}{x+2}$ **14.** $\dfrac{x^3-8x^2+27x-20}{x-5}$ **15.** $\dfrac{4x^3+2x^2-3x+1}{x+2}$

1. **a.** $x-9$ **b.** $c=3$, $P(3)=0$
2. **a.** $x-7$ **b.** $c=5$, $P(5)=0$
3. **a.** $x-4$ **b.** $c=1$, $P(1)=0$
4. **a.** $x+1+\frac{-24}{x+3}$ **b.** $c=-3$, $P(-3)=-24$
5. **a.** $4x+2$ **b.** $c=-5$, $P(-5)=0$
6. **a.** $3x+4$ **b.** $c=3$, $P(3)=0$
7. **a.** $2x+4+\frac{7}{x+2}$ **b.** $c=-2$, $P(-2)=7$
8. **a.** $4x+11+\frac{25}{x-4}$ **b.** $c=4$, $P(4)=25$
9. **a.** $x^2-4x+33+\frac{-265}{x+8}$ **b.** $c=-8$, $P(-8)=-265$
10. **a.** $x^2-4x+\frac{-5}{x-2}$ **b.** $c=2$, $P(2)=-5$
11. **a.** $3x^2+4x+3+\frac{8}{x-1}$ **b.** $c=1$, $P(1)=8$
12. **a.** $4x^2-6x+9+\frac{1}{x+1}$ **b.** $c=-1$, $P(-1)=1$
13. **a.** $2x^2+x-3$ **b.** $c=-2$, $P(-2)=0$
14. **a.** $x^2-3x+12+\frac{40}{x-5}$ **b.** $c=5$, $P(5)=40$

15. a. $4x^2 - 6x + 9 + \frac{-17}{x+2}$

b. $c=-2, P(-2)=-17$

16. a. $3x^2 + 3x + 5 + \frac{-10}{x+1}$

b. $c=-1, P(-1)=-10$

17. a. $x^2 + 7x + 55 + \frac{388}{x-7}$

b. $c=7, P(7)=388$

18. a. $2x^2 - 8x + 25 + \frac{-98}{x+4}$

b. $c=-4, P(-4)=-98$

19. a. $2x^2 - 2x + 6 + \frac{-27}{x+3}$

b. $c=-3, P(-3)=-27$

20. a. $4x^2 + 3x + 3 + \frac{16}{x-1}$

b. $c=1, P(1)=16$

21. a. $x^3 + 2x + 5 + \frac{17}{x-3}$

b. $c=3, P(3)=17$

22. a. $x^3 - 5x^2 + 26x - 155 + \frac{927}{x+6}$

b. $c=-6, P(-6)=927$

23. a. $x^3 + 2x^2 + 6x + 9 + \frac{23}{x-2}$

b. $c=2, P(2)=23$

24. a. $3x^3 - x^2 + 3x - 2 + \frac{1}{x+1}$

b. $c=-1, P(-1)=1$

16. $\frac{3x^3+6x^2+8x-5}{x+1}$ **17.** $\frac{x^3+6x+3}{x-7}$ **18.** $\frac{2x^3-7x+2}{x+4}$

19. $\frac{2x^3+4x^2-9}{x+3}$ **20.** $\frac{4x^3-x^2+13}{x-1}$ **21.** $\frac{x^4-3x^3+2x^2-x+2}{x-3}$

22. $\frac{x^4+x^3-4x^2+x-3}{x+6}$ **23.** $\frac{x^4+2x^2-3x+5}{x-2}$ **24.** $\frac{3x^4+2x^3+2x^2+x-1}{x+1}$

25. $\frac{x^4-x^2+3}{x-\frac{1}{2}}$ **26.** $\frac{x^3+2x^2+1}{x-\frac{2}{3}}$ **27.** $\frac{x^5-1}{x-1}$

28. $\frac{x^5-x^3+x}{x+\frac{1}{2}}$ **29.** $\frac{x^4-2x^3+4}{x+\frac{4}{5}}$ **30.** $\frac{x^6+1}{x+1}$

Writing and Thinking About Mathematics

31. State and prove the Remainder Theorem.

32. Suppose that a polynomial is divided by $(3x-2)$ and the answer is given as $x^2+2x+4+\frac{20}{3x-2}$. What is the polynomial? Explain how you arrived at this conclusion.

Collaborative Learning Exercise

33. The class should be divided into teams of 3 or 4 students. Each team should then develop answers to the following questions and be prepared to discuss these answers in class.

a. First use long division to divide the polynomial $P(x) = 2x^3 - 8x^2 + 10x + 15$ by $2x - 1$.

Then use synthetic division to divide the same polynomial by $x - \frac{1}{2}$.

Do the same process with two or three other polynomials and divisors. Next compare the corresponding long and synthetic division answers and explain how the answers are related.

b. Use the results from part (a) and explain algebraically the relationship of the answers when a polynomial is divided (using long division) by $ax - b$ and (using synthetic division) by $x - \frac{b}{a}$.

c. Show how the Remainder Theorem should be restated if $x - c$ is replaced by $ax - b$.

Hawkes Learning Systems: Introductory & Intermediate Algebra

Synthetic Division

25. a. $x^3+\frac{x^2}{2}-\frac{3}{4}x-\frac{3}{8}+\frac{45}{16\left(x-\frac{1}{2}\right)}$

b. $c=\frac{1}{2},\ P\left(\frac{1}{2}\right)=\frac{45}{16}$

26. a. $x^2+\frac{8}{3}x+\frac{16}{9}+\frac{59}{27\left(x-\frac{2}{3}\right)}$

b. $c=\frac{2}{3},\ P\left(\frac{2}{3}\right)=\frac{59}{27}$

27. a. $x^4+x^3+x^2+x+1$

b. $c=1,\ P(1)=0$

28. a. $x^4-\frac{x^3}{2}-\frac{3}{4}x^2+\frac{3}{8}x+\frac{13}{16}-\frac{13}{32\left(x+\frac{1}{2}\right)}$

b. $c=-\frac{1}{2},\ P\left(-\frac{1}{2}\right)=-\frac{13}{32}$

29. a. $x^3-\frac{14}{5}x^2+\frac{56}{25}x-\frac{224}{125}+\frac{3396}{625\left(x+\frac{4}{5}\right)}$

b. $c=-\frac{4}{5},\ P\left(-\frac{4}{5}\right)=\frac{3396}{625}$

30. a. $x^5-x^4+x^3-x^2+x-1+\frac{2}{x+1}$

b. $c=-1,\ P(-1)=2$

31. See page 1122.

32. $3x^3+4x^2+8x+12$

33. a. $x^2-\frac{7}{2}x+\frac{13}{4}+\frac{73}{4(2x-1)}$;

$2x^2-7x+\frac{13}{2}+\frac{73}{4\left(x-\frac{1}{2}\right)}$

b. Answers will vary.

c. Answers will vary.

A.3 Pi

As discussed in the text on page 6, π is an irrational number, and so the decimal form of π is an infinite nonrepeating decimal. Mathematicians even in ancient times realized that π is a constant value obtained from the ratio of a circle's circumference to its diameter, but they had no sense that it might be an irrational number. As early as about 1800 B.C., the Babylonians gave π a value of 3, and around 1600 B.C., the ancient Egyptians were using the approximation of 256/81, what would be a decimal value of about 3.1605. In the third century B.C., the Greek mathematician Archimedes used polygons approximating a circle to determine that the value of π must lie between 223/71(≈3.1408) and 22/7(≈3.1429). He was thus accurate to two decimal places. About seven hundred years later, in the fourth century A.D., Chinese mathematician Tsu Chung-Chi refined Archimedes' method and expressed the constant as 355/113, which was correct to six decimal places. By 1610, Ludolph van Ceulen of Germany had also used a polygon method to find π accurate to 35 decimal places.

Knowing that the decimal expression of π would not terminate, mathematicians still sought a repeating pattern in its digits. Such a pattern would mean that π was a rational number and that there would be some ratio of two whole numbers that would produce the correct decimal representation. Finally, in 1767, Johann Heinrich Lambert provided a proof to show that π is indeed irrational and thus is nonrepeating as well as nonterminating.

Since Lambert's proof, mathematicians have still made an exercise of calculating π to more and more decimal places. The advent of the computer age in this century has made that work immeasurably easier, and on occasion you will still see newspaper articles pronouncing that mathematics researchers have reached a new high in the number of decimal places in their approximations. In 1988 that number was 201,326,000 decimal places. Within 1 year that record was more than doubled, and most recent approximations of π now reach beyond 1.24 trillion decimal places! For your understanding, appreciation and interest, the value of π is given in the table on the next page to a mere 3742 decimal places as calculated by a computer program. To show π calculated to one billion decimal places would take every page of nearly 300 copies of this text!

The Value of π

π =
3.141592653589793238462643383279502884197169399375105820974944592307816406286
20899862803482534211706798214808651328230664709384460955058223172535940812848
11174502841027019385211055596446229489549303819644288109756659334461284756482
33786783165271201909145648566923460348610454326648213393607260249141273724587
00660631558817488152092096282925409171536436789259036001133053054882046652138
41469519415116094330572703657595919530921861173819326117931051185480744623799
62749567351885752724891227938183011949129833673362440656643086021394946395224
73719070217986094370277053921717629317675238467481846766940513200056812714526
35608277857713427577896091736371787214684409012249534301465495853710507922796
89258923542019956112129021960864034418159813629774771309960518707211349999998
37297804995105973173281609631859502445945534690830264252230825334468503526193
11881710100031378387528865875332083814206171776691473035982534904287554687311
59562863882353787593751957781857780532171226806613001927876611195909216420198
93809525720106548586327886593615338182796823030195203530185296899577362259941
38912497217752834791315155748572424541506959508295331168617278558890750983817
54637464939319255060400927701671139009848824012858361603563707660104710181942
95559619894676783744944825537977472684710404753464620804668425906949129331367
70289891521047521620569660240580381501935112533824300355876402474964732639141
99272604269922796782354781636009341721641219924586315030286182974555706749838
50549458858692699569092721079750930295532116534498720275596023648066549911988
18347977535663698074265425278625518184175746728909777727938000816470600161452
49192173217214772350141441973568548161361157352552133475741849468438523323907
39414333454776241686251898356948556209921922218427255025425688767179049460165
34668049886272327917860857843838279679766814541009538837863609506800642251252
05117392984896084128488626945604241965285022210661186306744278622039194945047
12371378696095636437191728746776465757396241389086583264599581339047802759009
94657640789512694683983525957098258226205224894077267194782684826014769909026
40136394437455305068203496252451749399651431429809190659250937221696461515709
85838741059788595977297549893016175392846813826868386894277415599185592524595
39594310499725246808459872736446958486538367362226260991246080512438843904512
44136549762780797715691435997700129616089441694868555848406353422072225828488
64815845602850601684273945226746767889525213852254995466672782398645659611635
48862305774564980355936345681743241125150760694794510965960940252288797108931
45669136867228748940560101503308617928680920874760917824938589009714909675985
26136554978189312978482168299894872265880485756401427047755513237964145152374
62343645428584447952658678210511413547357395231134271661021359695362314429524
84937187110145765403590279934403742007310578539062198387447808478489683321445
71386875194350643021845319104848100537061468067491927819119793995206141966342
87544406437451237181921799983910159195618146751426912397489409071864942319615
67945208095146550225231603881930142093762137855956638937787083039069792077346
72218256259966150142150306803844773454920260541466592520149744285073251866600
21324340881907104863317346496514539057962685610055081066587969981635747363840
52571459102897064140110971206280439039759515677157700420337869936007230558763
17635942187312514712053292819182618612586732157919841484882916447060957527069
57220917567116722910981690915280173506712748583222871835209353965725121083579
15136988209144421006751033467110314126711136990865851639831501970165151168517
14376576183515565088490998985998238734552833163550764791853589322618548963213
29330898570642046752590709154814165498594616371802709819943099244889575712828
905923233260972997120844335732654893823911932597...

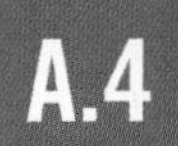

Powers, Roots, and Prime Factorizations

No.	Square	Square Root	Cube	Cube Root	Prime Factorization
1	1	1.0000	1	1.0000	
2	4	1.4142	8	1.2599	prime
3	9	1.7321	27	1.4422	prime
4	16	2.0000	64	1.5874	$2 \cdot 2$
5	25	2.2361	125	1.7100	prime
6	36	2.4495	216	1.8171	$2 \cdot 3$
7	49	2.6458	343	1.9129	prime
8	64	2.8284	512	2.0000	$2 \cdot 2 \cdot 2$
9	81	3.0000	729	2.0801	$3 \cdot 3$
10	100	3.1623	1000	2.1544	$2 \cdot 5$
11	121	3.3166	1331	2.2240	prime
12	144	3.4641	1728	2.2894	$2 \cdot 2 \cdot 3$
13	169	3.6056	2197	2.3513	prime
14	196	3.7417	2744	2.4101	$2 \cdot 7$
15	225	3.8730	3375	2.4662	$3 \cdot 5$
16	256	4.0000	4096	2.5198	$2 \cdot 2 \cdot 2 \cdot 2$
17	289	4.1231	4913	2.5713	prime
18	324	4.2426	5832	2.6207	$2 \cdot 3 \cdot 3$
19	361	4.3589	6859	2.6684	prime
20	400	4.4721	8000	2.7144	$2 \cdot 2 \cdot 5$
21	441	4.5826	9261	2.7589	$3 \cdot 7$
22	484	4.6904	10,648	2.8020	$2 \cdot 11$
23	529	4.7958	12,167	2.8439	prime
24	576	4.8990	13,824	2.8845	$2 \cdot 2 \cdot 2 \cdot 3$
25	625	5.0000	15,625	2.9240	$5 \cdot 5$
26	676	5.0990	17,576	2.9625	$2 \cdot 13$
27	729	5.1962	19,683	3.0000	$3 \cdot 3 \cdot 3$
28	784	5.2915	21,952	3.0366	$2 \cdot 2 \cdot 7$
29	841	5.3852	24,389	3.0723	prime
30	900	5.4772	27,000	3.1072	$2 \cdot 3 \cdot 5$
31	961	5.5678	29,791	3.1414	prime
32	1024	5.6569	32,768	3.1748	$2 \cdot 2 \cdot 2 \cdot 2 \cdot 2$
33	1089	5.7446	35,937	3.2075	$3 \cdot 11$
34	1156	5.8310	39,304	3.2396	$2 \cdot 17$

No.	Square	Square Root	Cube	Cube Root	Prime Factorization
35	1225	5.9161	42,875	3.2711	5 · 7
36	1296	6.0000	46,656	3.3019	2 · 2 · 3 · 3
37	1369	6.0828	50,653	3.3322	prime
38	1444	6.1644	54,872	3.3620	2 · 19
39	1521	6.2450	59,319	3.3912	3 · 13
40	1600	6.3246	64,000	3.4200	2 · 2 · 2 · 5
41	1681	6.4031	68,921	3.4482	prime
42	1764	6.4807	74,088	3.4760	2 · 3 · 7
43	1849	6.5574	79,507	3.5034	prime
44	1936	6.6332	85,184	3.5303	2 · 2 · 11
45	2025	6.7082	91,125	3.5569	3 · 3 · 5
46	2116	6.7823	97,336	3.5830	2 · 23
47	2209	6.8557	103,823	3.6088	prime
48	2304	6.9282	110,592	3.6342	2 · 2 · 2 · 2 · 3
49	2401	7.0000	117,649	3.6593	7 · 7
50	2500	7.0711	125,000	3.6840	2 · 5 · 5
51	2601	7.1414	132,651	3.7084	3 · 17
52	2704	7.2111	140,608	3.7325	2 · 2 · 13
53	2809	7.2801	148,877	3.7563	prime
54	2916	7.3485	157,464	3.7798	2 · 3 · 3 · 3
55	3025	7.4162	166,375	3.8030	5 · 11
56	3136	7.4833	175,616	3.8259	2 · 2 · 2 · 7
57	3249	7.5498	185,193	3.8485	3 · 19
58	3364	7.6158	195,112	3.8709	2 · 29
59	3481	7.6811	205,379	3.8930	prime
60	3600	7.7460	216,000	3.9149	2 · 2 · 3 · 5
61	3721	7.8102	226,981	3.9365	prime
62	3844	7.8740	238,328	3.9579	2 · 31
63	3969	7.9373	250,047	3.9791	3 · 3 · 7
64	4096	8.0000	262,144	4.0000	2 · 2 · 2 · 2 · 2 · 2
65	4225	8.0623	274,625	4.0207	5 · 13
66	4356	8.1240	287,496	4.0412	2 · 3 · 11
67	4489	8.1854	300,763	4.0615	prime
68	4624	8.2462	314,432	4.0817	2 · 2 · 17
69	4761	8.3066	328,509	4.1016	3 · 23
70	4900	8.3666	343,000	4.1213	2 · 5 · 7
71	5041	8.4261	357,911	4.1408	prime

No.	Square	Square Root	Cube	Cube Root	Prime Factorization
72	5184	8.4853	373,248	4.1602	2 · 2 · 2 · 3 · 3
73	5329	8.5440	389,017	4.1793	prime
74	5476	8.6023	405,224	4.1983	2 · 37
75	5625	8.6603	421,875	4.2172	3 · 5 · 5
76	5776	8.7178	438,976	4.2358	2 · 2 · 19
77	5929	8.7750	456,533	4.2543	7 · 11
78	6084	8.8318	474,552	4.2727	2 · 3 · 13
79	6241	8.8882	493,039	4.2908	prime
80	6400	8.9443	512,000	4.3089	2 · 2 · 2 · 2 · 5
81	6561	9.0000	531,441	4.3267	3 · 3 · 3 · 3
82	6724	9.0554	551,368	4.3445	2 · 41
83	6889	9.1104	571,787	4.3621	prime
84	7056	9.1652	592,704	4.3795	2 · 2 · 3 · 7
85	7225	9.2195	614,125	4.3968	5 · 17
86	7396	9.2736	636,056	4.4140	2 · 43
87	7569	9.3274	658,503	4.4310	3 · 29
88	7744	9.3808	681,472	4.4480	2 · 2 · 2 · 11
89	7921	9.4340	704,969	4.4647	prime
90	8100	9.4868	729,000	4.4814	2 · 3 · 3 · 5
91	8281	9.5394	753,571	4.4979	7 · 13
92	8464	9.5917	778,688	4.5144	2 · 2 · 23
93	8649	9.6437	804,357	4.5307	3 · 31
94	8836	9.6954	830,584	4.5468	2 · 47
95	9025	9.7468	857,375	4.5629	5 · 19
96	9216	9.7980	884,736	4.5789	2 · 2 · 2 · 2 · 2 · 3
97	9409	9.8489	912,673	4.5947	prime
98	9604	9.8995	941,192	4.6104	2 · 7 · 7
99	9801	9.9499	970,299	4.6261	3 · 3 · 11
100	10,000	10.0000	1,000,000	4.6416	2 · 2 · 5 · 5

Answers

Chapter 1

Exercises 1.1, pages 12 – 14

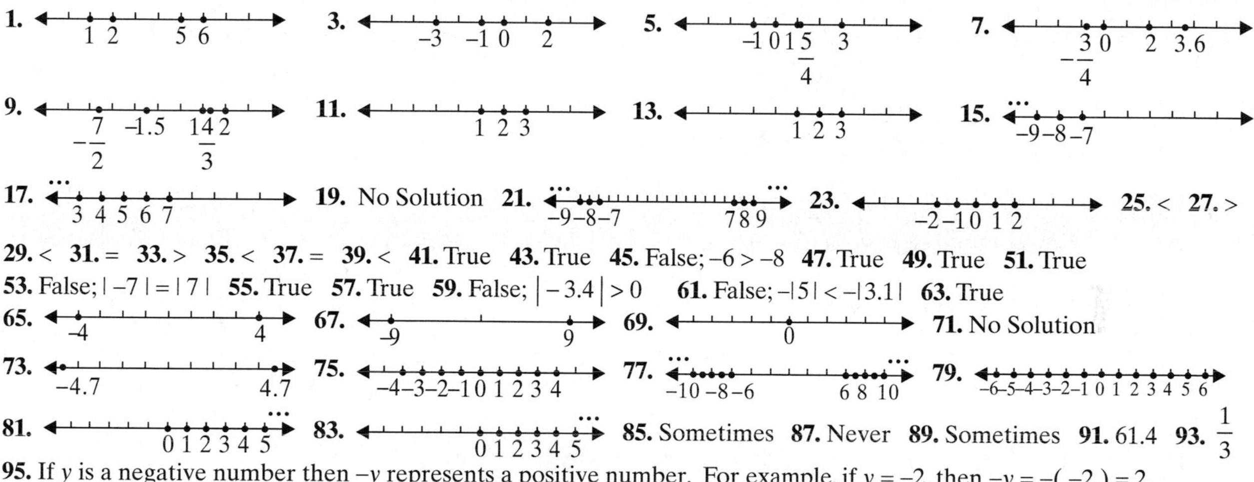

19. No Solution 25. < 27. > 29. < 31. = 33. > 35. < 37. = 39. < 41. True 43. True 45. False; −6 > −8 47. True 49. True 51. True 53. False; |−7| = |7| 55. True 57. True 59. False; $|-3.4| > 0$ 61. False; −|5| < −|3.1| 63. True 71. No Solution 85. Sometimes 87. Never 89. Sometimes 91. 61.4 93. $\frac{1}{3}$ 95. If y is a negative number then $-y$ represents a positive number. For example, if $y = -2$, then $-y = -(-2) = 2$.

Exercises 1.2, pages 18 – 19

1. 13 **3.** −4 **5.** 0 **7.** 5 **9.** −13 **11.** −8 **13.** −10 **15.** 0 **17.** 17 **19.** −29 **21.** −9 **23.** −3 **25.** 22 **27.** −54 **29.** −7 **31.** −16 **33.** −26 **35.** 0 **37.** −32 **39.** 5 **41.** −83 **43.** 12 **45.** −32 **47.** −2 is a solution **49.** −4 is a solution **51.** −6 is a solution **53.** 18 is a solution **55.** −10 is a solution **57.** −2 is not a solution **59.** −72 is not a solution **61.** Sometimes **63.** Never **65.** Never **67.** Sometimes **69.** Always **71.** 84 **73.** −97,714 **75.** −6143

Exercises 1.3, pages 23 – 25

1. −11 **3.** 6 **5.** −47 **7.** 0 **9.** 52 **11.** 5 **13.** −10 **15.** 12 **17.** 3 **19.** −16 **21.** 24 **23.** −15 **25.** −16 **27.** −57 **29.** −54 **31.** 1 **33.** 8 **35.** −26 **37.** 1 **39.** −8 **41.** −10 **43.** −6 **45.** −139 **47.** −1 > −7 **49.** 10 > −10 **51.** −6 < 6 **53.** −4 > −5 **55.** −37 < −34 **57.** −3 is a solution **59.** 3 is a solution **61.** 4 is not a solution **63.** −10 is a solution **65.** 12 is a solution **67.** −9 is a solution **69.** 16 is not a solution **71.** 1044 > −39 **73.** −15,254 > −35,090 **75.** 2316 points **77.** 6° below 0 (or −6°)

Exercises 1.4, pages 30 – 32

1. −12 **3.** 56 **5.** 57 **7.** 56 **9.** −30 **11.** 26 **13.** −60 **15.** −24 **17.** −288 **19.** 0 **21.** 4 **23.** −6 **25.** −3 **27.** 13 **29.** 0 **31.** Undefined **33.** −11 **35.** −4 **37.** Negative **39.** Negative **41.** Negative **43.** 0 **45.** Undefined **47.** True **49.** True **51.** True **53.** False; 17+(−3) > (−14)+(−4) **55.** True **57.** −12 is a solution **59.** −72 is a solution **61.** −8 is not a solution **63.** 5 is a solution **65.** −4 is a solution **67.** −34,459,110 **69.** −2671 **71.** Dividing 0 by any number other than 0 gives a quotient of 0.

Exercises 1.5, pages 39 – 40

1. 5^2 **3.** 7^2 **5.** 11^2 **7.** 8^2 or 4^3 or 2^6 **9.** 2^3 **11.** 5^3 **13.** 3^5 **15.** 10^2 **17.** 64 **19.** 100 **21.** 343 **23.** 10,000 **25.** 1 **27.** 225 **29.** 27,000 **31.** 10,000 **33.** 36 **35.** −64 **37.** 15,625 **39.** 262,144 **41.** −225 **43.** −8 **45.** 196 **47.** $2 \cdot 5 \cdot 7$ **49.** 43 is prime

51. $3^2 \cdot 5^2$ **53.** $3 \cdot 11^2$ **55.** $2 \cdot 3^2 \cdot 5$ **57.** $2^2 \cdot 5 \cdot 7$ **59.** $2^3 \cdot 5^3$ **61. a.** 36 **b.** 16 **63.** –25 **65.** –10 **67.** –45 **69.** –137 **71.** 152 **73.** –6 **75.** –2 **77.** 1270 **79.** 35 **81.** –100 **83.** $(3^2 - 9) = 0$ and division by 0 is undefined.

Exercises 1.6, pages 48 – 50

1. $\frac{5}{6} = \frac{5}{6} \cdot \frac{8}{8} = \frac{40}{48}$ **3.** $\frac{0}{9} = \frac{0}{9} \cdot \frac{7b}{7b} = \frac{0}{63b}$ **5.** $\frac{2}{5}$ **7.** $\frac{3x}{7}$ **9.** $\frac{3ab}{25}$ **11.** $\frac{-2}{3y}$ **13.** $\frac{-1}{2x^2}$ **15.** $\frac{-12y}{35}$ **17.** $\frac{3}{8}$ **19.** 35 **21.** $-\frac{1}{6}$ **23.** $\frac{81}{100}$ **25.** $\frac{-7}{4}$ **27.** $\frac{50}{9}$ **29.** $\frac{4}{3y}$ **31.** $\frac{-21}{8}$ **33.** Undefined **35.** $\frac{22}{a^2}$ **37.** $\frac{25}{98p^3q^2}$ **39.** $\frac{16}{5}$ **41.** $\frac{25}{32}$ **43.** 2 inches **45. a.** more than 60 **b.** less than 60 **c.** 72 **47.** $\frac{7}{20}$ and $\frac{13}{20}$ **49.** $\frac{135}{32}$ ft. or $4\frac{7}{32}$ ft. **51.** $\frac{-25}{24}$ **53.** Division by zero is undefined. For example we could write $0 = \frac{0}{1}$. Then the reciprocal would be $\frac{1}{0}$, but this reciprocal is undefined since division by zero is undefined. Thus, 0 does not have a reciprocal.

Exercises 1.7, pages 58 – 60

1. a. 1, 2, 3, 6 **b.** 6, 12, 18, 24, 30, 36 **3. a.** 1, 3, 5, 15 **b.** 15, 30, 45, 60, 75, 90 **5.** 120 **7.** $40xy$ **9.** $210x^2y^2$ **11.** $\frac{7}{9}$ **13.** $\frac{5}{23}$ **15.** $-\frac{1}{6}$ **17.** $\frac{4}{3}$ **19.** $-\frac{28}{25x}$ **21.** $\frac{83}{60}$ **23.** $\frac{5}{42}$ **25.** $\frac{33}{70}$ **27.** $\frac{3}{4x}$ **29.** $\frac{5}{24x}$ **31.** 0 **33.** $\frac{12+y}{2y}$ **35.** $\frac{x-2}{14}$ **37.** $\frac{6a-25}{30}$ **39.** $\frac{5b+22}{20}$ **41.** $\frac{11}{30}$ **43.** $\frac{31}{24}$ **45.** $\frac{7}{2}$ **47.** $\frac{-23}{21y}$ **49.** $\frac{7}{5}$ **51.** $\frac{1}{18009460}$ **53.** $\frac{92}{19}$ **55.** 344 yd. **57. a.** $\frac{13}{30}$ **b.** \$1170

Exercises 1.8, pages 71 – 73

1. 0.86; eighty-six hundredths **3.** 5.1; five and one tenth **5.** –18.06; negative eighteen and six hundredths **7.** 0.087 **9.** –5.14 **11.** 7.0021 **13.** 91.1 **15.** 0.50 **17.** 67.057 **19.** 317.23 **21.** 263.51 **23.** –55.58 **25.** –152.83 **27.** 108.72 **29.** –7.626 **31.** 15.1 **33.** 0.375 **35.** 0.05 **37.** –23 + 13 – 6; –16 lbs. **39.** 4 + 3 – 9; –2° **41.** 47 – 22 + 8 – 45; –\$12 **43.** 14 – 6 + 11 – 15; 4° **45.** 187 – 241 + 82 + 26; \$54 **47.** 4 yd. gain **49.** 8th floor **51.** \$150 **53.** \$760 **55.** 35° F, 5° F **57.** 54 years **59.** 30,280 ft. **61.** 0.4375 **63.** $-0.\overline{54}$ **65.** $1.\overline{185}$ **67.** $\frac{13}{10}$ **69.** $\frac{1323}{250}$ **71.** $\frac{69}{40}$ **73. a.** For all positive numbers, the product will be less than the other number. Ex: $\frac{1}{4} \cdot 2 = \frac{1}{2}$, $\frac{1}{2} < 2$ **b.** The product will equal the other number when the other number is 0. Ex: $\frac{1}{2} \cdot 0 = 0$, $0 = 0$ **c.** The product will be more than the other number when the other number is negative. Ex: $\frac{1}{4}(-1) = -\frac{1}{4}$, $-\frac{1}{4} > -1$

Exercises 1.9, pages 80 – 81

1. a. –1, 0, 4 **b.** –3.56, $-\frac{5}{8}$, –1, 0, 0.7, 4, $\frac{13}{2}$ **c.** $-\sqrt{8}$, π, $\sqrt[3]{25}$ **d.** All are real numbers **3.** 3 + 7 **5.** 4 · 19 **7.** 30 + 48 **9.** $(2 \cdot 3) \cdot x$ **11.** $(3 + x) + 7$ **13.** $0 \cdot 6 = 0$ **15.** $x + 7$ **17.** $2x - 24$ **19.** Commutative property of multiplication **21.** Additive identity **23.** Commutative property of addition **25.** Commutative property of multiplication **27.** Commutative property of multiplication **29.** Commutative property of addition; 19 + 3 = 3 + 19 = 22 **31.** Associative property of multiplication; $(2 \cdot 7) \cdot 4 = 2 \cdot (7 \cdot 4) = 56$ **33.** Associative property of addition; (2(3) + 14) + 3 = 2(3) + (14 + 3) = 23 **35.** Commutative property of multiplication; $11 \cdot 4 = 4 \cdot 11 = 44$ **37.** Distributive property; 3(–2 + 15) = –6 + 45 = 39

39. Commutative property of multiplication; $(-2+2)(-2-4)=(-2-4)(-2+2)=0$ **41.** Commutative property of addition; $3+(4+4)=(4+4)+3=11$ **43.** $6(11)=66$ and $6\cdot 3+6\cdot 8=66$ **45.** $10(-7)=-70$ and $10\cdot 2-10\cdot 9=-70$ **47.** $5(-2)=-10$ and $5(14)-5(16)=-10$.

Chapter 1 Test, pages 89 – 90

1. a. The integers are -10 and 0 **b.** The rational numbers are -10, $\frac{-3}{4}$, 0, $\frac{7}{6}$, $3\frac{4}{9}$ and 7.121212... **c.** The irrational numbers are $-\pi$ and $-\sqrt{5}$ **d.** All are real numbers **2. a.** True, since for any integer n, $n=\frac{n}{1}$ which is a rational number since the numerator and the denominator are both integers. **b.** False, because rational numbers also include non-integers, such as $\frac{2}{5}$ **3. a.** $<$ **b.** $>$ **c.** $=$ **4.** 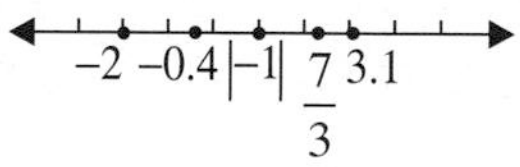

5. 

6. $y=7$ or $y=-7$

7. 2, 3, 5, 7, 11, 13, 17, 19, 23, 29, 31, 37, 41, 43 and 47 **8.** $\{-2,-1,0,1,2\}$ 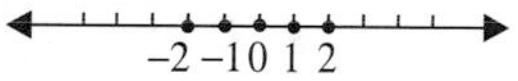

9. $\{\ldots,-10,-9,9,10,\ldots\}$ 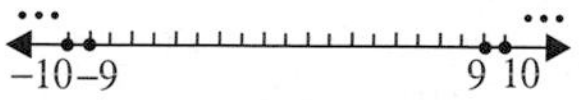

10. -18 **11.** 19 **12.** 1 **13.** 162 **14.** 7 **15.** 0 **16. a.** -25 **b.** -265 **c.** -44 **17. a.** $2^4\cdot 5$ **b.** $5^2\cdot 7$ **18. a.** Multiplicative identity **b.** Commutative property of addition **c.** Associative property of addition **d.** Zero factor property **e.** Distributive property **f.** Commutative property of multiplication **19. a.** $-19{,}487{,}171$ **b.** 0 **20.** 96 **21.** $90x^2y^2$ **22.** $\frac{-9}{2}$ **23.** $\frac{3}{5}$ **24.** $\frac{9}{10}$ **25.** $\frac{-2}{y}$ **26.** $\frac{29}{40}$ **27.** $\frac{15+n}{5n}$ **28.** Seventeen and three thousandths

29. 83.15 **30.** 36.53 **31.** -16.5 **32.** 17.952 **33.** 1.35 **34.** $\frac{-33}{16}$ **35. a.** -0.7 **b.** \$15.80 **36. a.** more **b.** less **c.** \$80 **37. a.** 15 gallons **b.** No, \$1 short. **38.** $-33°$ F

Chapter 2

Exercises 2.1, pages 98 – 99

1. -5, $\frac{1}{6}$ and 8 are like terms; $7x$ and $9x$ are like terms. **3.** $-x^2$ and $2x^2$ are like terms; $5xy$ and $-6xy$ are like terms; $3x^2y$ and $5x^2y$ are like terms. **5.** 24, 8.3 and -6 are like terms; $1.5xyz$, $-1.4xyz$ and xyz are like terms. **7.** 64 **9.** -121 **11.** $15x$ **13.** $3x$ **15.** $-2n$ **17.** $5y^2$ **19.** $12x^2$ **21.** $7x+2$ **23.** $x-3y$ **25.** $8x^2+3y$ **27.** $2n+3$ **29.** $7a-8b$ **31.** $8x+y$ **33.** $2x^2-x$ **35.** $-2n^2+2n$ **37.** $3x^2-xy+y^2$ **39.** $2x$ **41.** $-y$ **43. a.** $3x+4$ **b.** 16 **45. a.** $3.6x^2$ **b.** 57.6 **47. a.** $\frac{17x}{8}$ **b.** 8.5 **49. a.** 0 **b.** 0 **51. a.** $-2x-8$ **b.** -16 **53. a.** $9y+2$ **b.** 29 **55. a.** $10a+13$ **b.** -7 **57. a.** $3ab+b^2+b^3$ **b.** 6 **59. a.** $8a$ **b.** -16 **61. a.** $6a$ **b.** $9a^2$ **c.** $\frac{2}{3a}$ **d.** $\frac{1}{9a^2}$ **e.** 1 **63. a.** $-14ab$ **b.** $49a^2b^2$ **c.** $\frac{2}{7ab}$ **d.** $\frac{1}{49a^2b^2}$ **e.** 1 **65. a.** $3a$ **b.** $2.25a^2$ **c.** $\frac{4}{3a}$ **d.** $\frac{4}{9a^2}$ **e.** -1

Exercises 2.2, pages 104 – 105

1. 4 times a number **3.** 1 more than twice a number **5.** 5.3 less than 7 times a number **7.** -2 times the difference between a number and 8 **9.** 5 times the sum of twice a number and 3 **11.** 6 times the difference between a number and 1 **13.** 3 times a number plus 7; 3 times the sum of a number and 7 **15.** The product of 7 and a number minus 3; 7 times the difference between a number and 3 **17.** $x+6$ **19.** $x-4$ **21.** $3x-5$ **23.** $\frac{x-3}{7}$ **25.** $3(x-8)$ **27.** $3x-5$ **29.** $8-2x$ **31.** $8(x-6)+4$ **33.** $3x-5x$ **35.** $2(x-7)-6$ **37.** $2(17+x)+9$ **39. a.** $x-6$ **b.** $6-x$ **41.** $\$4.95x$ **43.** $7t+3$ **45.** $6t+3$ **47.** $x+8+2x=3x+8$ **49.** $c+0.2c=1.2c$

Exercises 2.3, pages 115 – 118

1. $y = 6$ **3.** $n = -10$ **5.** $y = 5.2$ **7.** $x = 3$ **9.** $x = 25$ **11.** $a = 0$ **13.** $x = \frac{1}{2}$ **15.** $x = 20$ **17.** $x = -4$ **19.** $y = -2$ **21.** $a = 1$ **23.** $x = -1$ **25.** $a = 6.2$ **27.** 25 m, 34 m, 12 m **29.** 16.125 m **31.** $x = 2$ in. **33.** 36 in. **35.** Yes

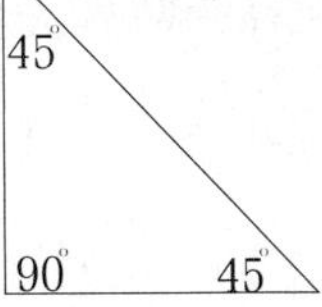

37. 0.25 years (or 3 months) **39.** 24% annual interest rate, \$4,400 **41.** \$855 **43.** \$1690 **45.** 3.5 hrs. **47.** $x = -17.214$ **49.** $x = 246$

Exercises 2.4, pages 124 – 128

1. Use the distributive property. Add 12 to both sides. Simplify. Add x to both sides. Simplify. Divide both sides by 4. Simplify. **3.** Multiply both sides by 30. Use the distributive property. Simplify. Add –5 to both sides. Simplify. Add $-12a$ to both sides. Simplify. Divide both sides by –2. Simplify. **5.** $x = -3$ **7.** $y = -4$ **9.** $x = -5$ **11.** $x = -0.12$ **13.** $n = 0$ **15.** $x = 0$ **17.** $z = -1$ **19.** $y = \frac{1}{5}$ **21.** $x = -4$ **23.** $x = -3$ **25.** $x = -21$ **27.** $y = 0$ **29.** $x = -5$ **31.** $n = \frac{1}{6}$ **33.** $n = 2$ **35.** $x = \frac{7}{60}$ **37.** $x = \frac{8}{5}$ **39.** $x = \frac{2}{3}$ **41.** $n = -2, \frac{2}{3}$ **43.** $x = -22, 26$ **45.** No solution **47.** $x = 2$ **49.** $x = -\frac{3}{2}, 2$ **51.** $l = 64$ ft. **53.** 15 ft. **55.** 10 cm **57.** 12 ft., 38 ft. **59.** \$490 **61.** $x = -9.293$ **63.** $x = 4.58$

Exercises 2.5, pages 134 – 137

1. $x - 5 = 13 - x$; 9 **3.** $36 = 2x + 4$; 16 **5.** $7x = 2x + 35$; 7 **7.** $3x + 14 = 6 - x$; –2 **9.** $\frac{2x}{5} = x + 6$; –10 **11.** $4(x - 5) = x + 4$; 8 **13.** $\frac{2x+5}{11} = 4 - x$; 3 **15.** $x - 21 = 8x$; –3 **17.** $x + x + 4 = 24$; 10 cm **19.** $x + x + 2x - 3 = 45$; 12 cm, 12 cm, 21 cm **21.** $3x + 1500 = 12{,}000$; \$3500 **23.** $n + n + 2 = 60$; 29, 31 **25.** $2n + 3(n + 1) = 83$; 16, 17 **27.** $n + 2 + n + 4 - n = 66$; 60, 62, 64 **29.** $n + n + 2 + n + 4 = n + 2 + 168$; 82, 84, 86 **31.** $2l + 2(75) = 410$; 130 yards **33.** $\frac{2l}{3} = 18$; 27 ft. **35.** $x + x + 2 + x + 6 = 29$; 7 ft., 9 ft., 13 ft. **37.** $c + c + 49.50 = 125.74$; calculator: \$38.12, textbook: \$87.62 **39.** $2n + 3(n + 2) = 2(n + 4) + 7$; 3, 5, 7

Note: Answers for 41 – 45 may vary. **41.** Find two consecutive integers whose sum is 33; 16, 17 **43.** Find 3 consecutive even integers such that the sum of the first and the third is 3 times the second; –2, 0, 2 **45.** The quotient of a number and 2 increased by $\frac{1}{3}$ is equal to 3 times the number divided by 4; $\frac{4}{3}$

Exercises 2.6, pages 144 – 148

1. 91% **3.** 137% **5.** 37.5% **7.** 150% **9.** 0.69 **11.** 0.113 **13.** 0.005 **15.** 0.82 **17.** $\frac{7}{20}$ **19.** $\frac{13}{10}$ **21. a.** 32% **b.** 28% **c.** 8% **23. a.** 59.7% **b.** 16.7% **25.** 61.56 **27.** 40 **29.** 2180 **31.** 125% **33.** 80 **35. a.** 8% **b.** 10% **c.** b **37. a.** \$990 **b.** 22% **39.** \$1952.90 **41.** 1.5% **43.** 40 **45. a.** \$9 **b.** \$3.75 **47. a.** \$7134.29 **b.** \$6609.23 **49. a.** 10% **b.** 11.11% **c.** The first percentage is a percentage of his original weight while the second percentage is a percentage of his weight after he lost weight. **51. a.** Because the 6% comission is against the selling price, not the \$141,000 the couple wanted. **b.** The selling price is 100%. Therefore, \$141,000 is 94% of the selling price (selling price(100%) – realtor fee (6%) = amount for couple(94%)). The selling price should be \$150,000.

Chapter 2 Test, pages 153 – 154

1. $7-5x$ **2.** $4.4y$ **3.** $7a^2-5a+10$ **4. a.** $\frac{19y}{24}$ **b.** $\frac{19}{8}$ **5. a.** $2x-16$ **b.** -20 **6. a.** $3x+5$ **b.** 17 **7. a.** y^3+y^2-y-2 **b.** -17 **8.** $6x-3$ **9.** $2(x+5)$ **10.** $2(x+5)-4$ **11.** $3-2x$ **12.** $3(x+5)$ **13.** $\frac{x}{10}+x$ **14.** The quotient of a number and 6 decreased by twice the number **15.** 9 less than twice a number decreased by 7 **16.** $-8.2y$ **17.** $\frac{2x}{3}$ **18.** $12x$ **19.** $36x^2$

20. $\frac{1}{36x^2}$ **21.** 1 **22.** $x=-3$ **23.** $x=5$ **24.** $x=0$ **25.** $x=-\frac{9}{8}$ **26.** $x=20$ **27.** $x=-2$ **28.** $x=-\frac{3}{2},2$ **29.** No solution

30. 111.6 **31.** 32% **32.** \$337.50 **33.** The \$6000 investment is the better investment since it has a higher percent of profit, 6%, than the \$10,000 investment (5%). **34.** 3 m, 4 m, 5 m **35.** $(2y+5)+y=-22$; $-9,-13$ **36.** $2n+3(n+1)=83$; 16, 17 **37. a.** $(.75)x=547.50$; \$730 **b.** $(0.06)547.50+547.50+25=x$; \$605.35 **38.** $3(n+2)=n+(n+4)+27$; $n=25,27,29$ **39.** $(10+c)+(c-45.50)+c=156.50$; book = \$74.00, manual = \$18.50, calculator = \$64.00

Chapter 2 Cumulative Review, pages 155 – 158

1. 20 **2.** -14 **3.** 9 **4.** -15 **5.** -126 **6.** 459 **7.** 39 **8.** -124 **9.** True **10.** True **11.** False; $\frac{3}{4}\le|-1|$

12. −6 0 6 **13.** 0 1 2 3 4 5 6 … **14.** −2 −1 0 1 2 **15.** Refer to discussion on "Division by 0 is Undefined" on page 28 **16.** 2, 3, 5, 7, 11, 13, 17, 19, 23, 29, 31, 37, 41, 43, 47 **17. a.** $2\cdot3\cdot5^2$ **b.** 5^3 **c.** $2^4\cdot3\cdot5$ **18. a.** 108 **b.** $72xy^2$ **19.** Associative property of multiplication **20.** Commutative property of multiplication **21.** Distributive property **22.** Commutative property of addition **23.** Associative property of addition **24.** Multiplicative identity

25. $\frac{43}{40}$ **26.** $-\frac{1}{36}$ **27.** $\frac{22}{15y}$ **28.** $\frac{1}{4a}$ **29.** $\frac{18+x}{6x}$ **30.** $\frac{5}{6}$ **31.** $\frac{7}{5}$ **32.** $\frac{2}{9}$ **33.** 17.27 **34.** -15.31 **35.** 14.6 **36.** 58.19

37. 5.292 **38.** 19 **39.** 18 **40.** 90 **41.** 12 **42.** 164 **43.** 13 **44.** 37 **45.** $\frac{49}{40}$ **46.** $-\frac{31}{5}$ **47.** $\frac{1}{6}$ **48.** $\frac{-7}{45}$ **49.** $\frac{263}{576}$ **50.** $\frac{3}{5}$

51. 0 **52.** $\frac{-7}{16}$ **53.** $-2y-12$; -18 **54.** $4x+7$; -1 **55.** x^2+13x; -22 **56.** $x-7$; -9 **57.** $9-2x$ **58.** $3(x+10)$ **59.** $24x+5$

60. $6T+E+3F$ **61.** $x=-1$ **62.** $x=3$ **63.** $x=-11$ **64.** $x=-4$ **65.** $x=\frac{31}{8}$ **66.** $y=\frac{24}{5}$ **67.** $y=\frac{15}{2}$ **68.** $y=0$ **69.** $-\frac{5}{51}$

70. -5 **71.** $\frac{10}{3}$ cups (or $3\frac{1}{3}$ cups) **72.** 8 lollipops **73. a.** 1000 **b.** 2500 **74.** \$80 **75.** \$506.94 **76.** $x=11$ **77.** 18, 20, 22

78. $-2,-1,0,1$ **79.** <33.33 miles **80.** The value of furniture sold was \$12,500, Jay's and Kay's salary was \$1250. **81.** \$10,000 **82.** These are equally good investments since the percent of profit, 8%, is the same for each investment. **83.** \$50,000 **84.** <100 balloon animals

Chapter 3

Exercises 3.1, pages 166 – 170

1. \$120 **3.** 72 days **5.** \$10,000 **7.** 176 ft./sec. **9.** 4 milliliters **11.** \$1120 **13.** 14 **15.** 336 in. or 28 ft. **17.** \$337.50 **19.** \$2400

21. $b=P-a-c$ **23.** $m=\frac{F}{a}$ **25.** $w=\frac{A}{l}$ **27.** $n=\frac{R}{p}$ **29.** $P=A-I$ **31.** $m=2A-n$ **33.** $t=\frac{I}{Pr}$ **35.** $b=\frac{P-a}{2}$

37. $\beta=180-\alpha-\gamma$ **39.** $h=\frac{V}{lw}$ **41.** $b=\frac{2A}{h}$ **43.** $\pi=\frac{A}{r^2}$ **45.** $g=\frac{mv^2}{2K}$ **47.** $y=\frac{6-2x}{3}$ **49.** $x=\frac{11-2y}{5}$

51. $b=\frac{2A-hc}{h}$ or $b=\frac{2A}{h}-c$ **53.** $x=\frac{8R+36}{3}$ **55.** $y=-x-12$ **57.** $h=\frac{3V}{\pi r^2}$ **59.** $g=\frac{V^2-v^2}{2h}$ **61.** $a=S(1-r)=S-Sr$

63. $R = \frac{nE}{I} - nr$ **65.** $C = nt + 9$ **67.** $C = 325n + 5400$ **69. a.** 0; No, because the numerator will be zero and thus the whole fraction will be equal to zero for all values of s. **b.** $x < 70$ **c.** Answers will vary.

Exercises 3.2, pages 178 – 181

1. e **3.** k **5.** c **7.** d **9.** g **11.** l **13.** m **15.** h **17.** 70 cm; 300 cm^2 **19.** 54 mm; 126 mm^2 **21.** 31.4 in.; 78.5 in.^{2} **23.** 1436.03 in.^{3} **25.** 729 m^3 **27.** 56 in. **29.** 63 cm^2 **31.** 6154.4 ft.^{3} **33.** 125 ft.^{3} **35.** 272 cm^2 **37.** 529.875 in.^{3} **39.** $\alpha = 180 - \beta - \gamma$, 43° **41.** $w = \frac{P - 2l}{2}$, 13 ft. **43.** $a = P - b - c$, 61 in. **45.** $A = \pi r^2$, 196π sq. ft. or 615.4 sq. ft. **47.** 28 cm; 48 cm^2 **49.** 212.04 m^2 **51.** 130.08 in.^{2} **53.** 6,334,233.21 cm^3

Exercises 3.3, pages 189 – 195

1. 71 **3.** 9 **5.** 12 **7.** 7 **9.** 18 **11.** 8, 30 **13.** 46 ft. by 84 ft. **15.** $1500 **17.** 78 min. **19.** 7 hrs. **21.** 3 hrs. **23.** $4\frac{4}{5}$ hrs. **25.** 36 mph, 60 mph **27.** $112.50 **29.** 2,000 baskets **31.** 62,500 pounds **33.** 20 at $300; 24 at $250 **35.** $14,000 at 5%; $11,000 at 6% **37.** 6.5% on $4,000; 6% on $3,000 **39.** $7,000 at 6%; $9,000 at 8% **41.** 5° **43.** 72.0 in. **45.** 31.5 hrs. **47.** 94 or more **49. a.** 53.25° F **b.** 44° F **c.** 16° F **51. a.** 3.2 in. **b.** 7.1 in. **c.** 6.9 in. **53. a.** 11.3 days **b.** 17 days **c.** Feb. 3 and Nov. 12

Exercises 3.4, pages 205 – 210

1. 1 **3.** $\frac{1}{3}$ **5.** $\frac{1}{2}$ **7.** 50 miles/hr. **9.** 1 **11.** $\frac{\$7 \text{ profit}}{\$100 \text{ invested}}$ **13.** True **15.** True **17.** True **19.** $x = 12$ **21.** $x = 12\frac{1}{2}$ **23.** $x = 156$ **25.** $A = 180$ **27.** $w = 6$ **29.** $y = 72$ **31.** $8.40 **33.** 5.5 in. **35.** $135 **37.** Investor B, $100 **39.** 259,200 revolutions **41.** width = 3 in., length = 7.5 in. **43. a.** 7.5 mph **b.** 55 mph **45.** $x = 3, y = 3$ **47.** $x = 6, y = 4$ **49.** $x = \frac{5}{2}, y = 2$ **51.** $x = 20°, y = 100°$ **53.** $a = 7, b = 5$ **55.** $x = 80°, y = 50°$ **57.** $a = 10, b = 8, x = 60°, y = 30°$ **59. a.** 55 lbs. **b.** 6 bags **c.** $72 **61.** 225, 100 **63.** 54 minutes **65. a.** The statement is misleading because the numbers 4 and 5 are not in the same units. **b.** The ratio of 4 quarters to 5 dollars is 1:5.

Exercises 3.5, pages 219 – 222

1. False; $3 > -3$ **3.** True **5.** True **7.** True **9.** True **11.** half-open interval

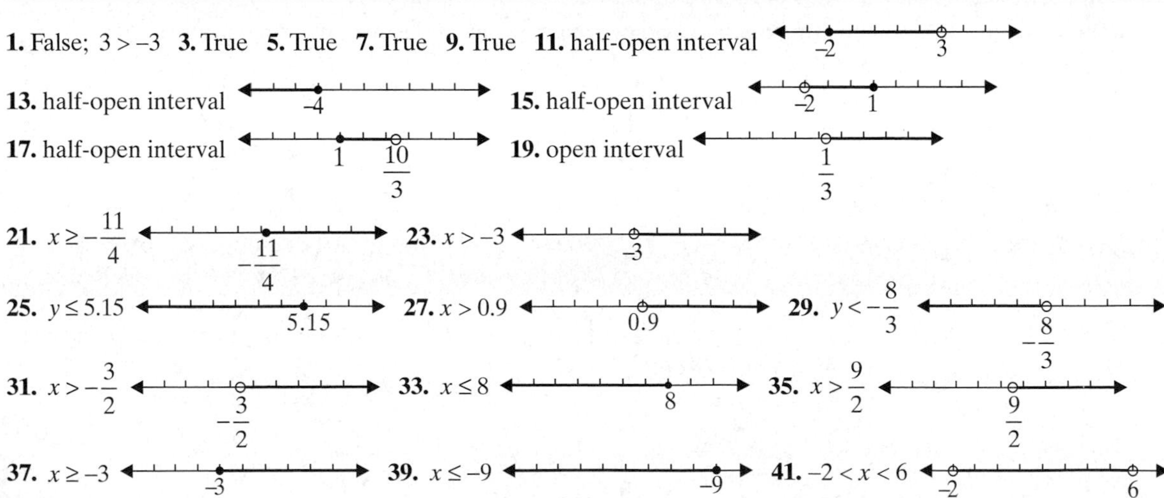

43. $1 < x < 2$ **45.** $-\frac{8}{5} \le x < \frac{6}{5}$ **47.** $-\frac{21}{2} \le x \le 3$

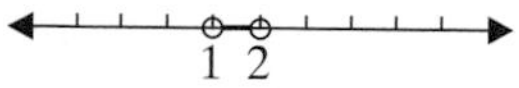

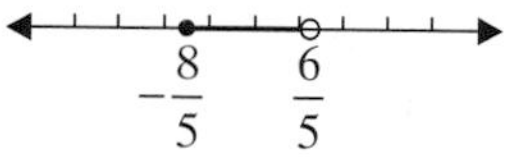

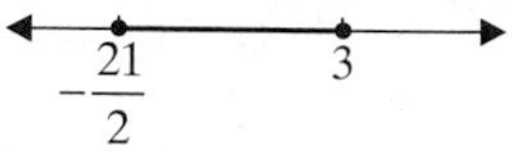

49. $\frac{20}{3} \le x < \frac{33}{4}$ **51.** $3 \le x \le 15$ **53.** $\frac{13}{3} \le x < \frac{22}{3}$

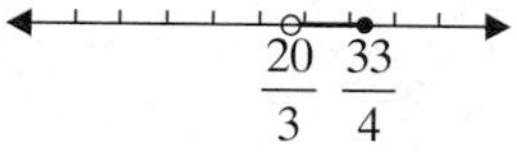

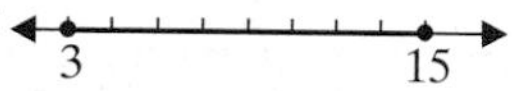

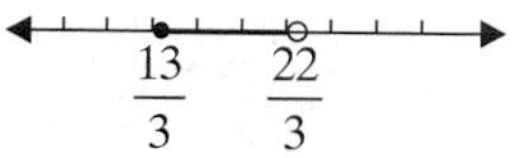

55. $44 \le x \le 100$ **57.** 59° to 149° F **59.** The second side is between 11 cm and 35 cm and the third side is between 6 cm and 18 cm. **61.** $86 \le x \le 100$ **63.** 171 adult tickets **65.** less than 100 miles **67. a.** The student cannot earn an A for the course. **b.** The student must score at least 192 to earn an A for the course. **69.** The second side must be more than 13 mm and less than 33 mm and the third more than 5 mm and less than 15 mm. **71.** You would have to drive more than 10 miles.

Chapter 3 Test, pages 227 – 228

1. $m = \frac{N-p}{rt}$ **2.** $y = \frac{7-5x}{3}$ **3.** $x = \frac{1}{96}$ **4.** $x = 14$ **5.** $x = 30$ **6.** $x \le -\frac{10}{3}$

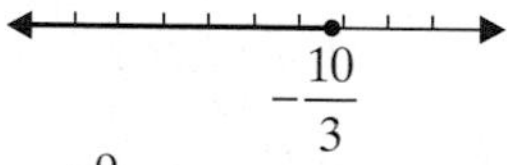

7. $x < -20$ **8.** $x > \frac{-13}{3}$ 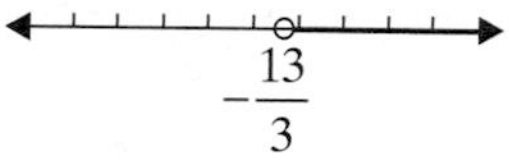 **9.** $\frac{-9}{4} \le x \le -1$

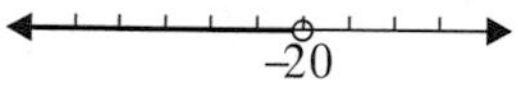

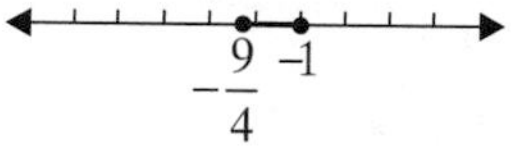

10. $x \ge 3$ 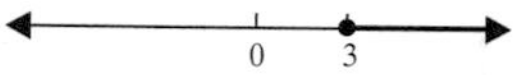 **11.** $0.1 < x < 13.2$ **12. a.** $-4 < x < 4$ **b.** $3 \le x \le 7$ **c.** $x < 0.75$ **13. a.** 3.14 cm **b.** 3.14 cm^2 **14.** 8 cm

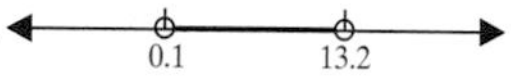

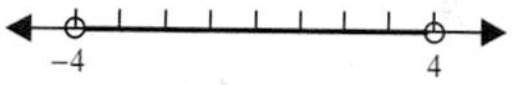

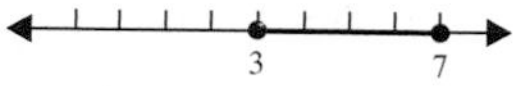

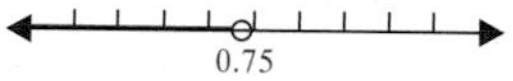

15. a. 27.42 ft. **b.** 50.13 ft.^{2} **16.** 4186.67 cm^3 **17.** –20° C **18.** $x = \frac{26}{3}$ **19. a.** 24 in. **b.** 888 in.^{2} **20. a.** 200 ft.^{3} **b.** 7.4 yd.^{3}

21. 100.48 in.^{3} **22.** $11.25 **23.** 3.2 in. **24.** $x = 3, y = 6.75$ **25.** $358.40 to $377.60 **26.** Between 9.1 and 18 gallons

Chapter 3 Cumulative Review, pages 229 – 232

1. $x + 15$ **2.** $3x + 8$ **3.** 540 **4.** $120a^2b^3$ **5.** 2, 16 **6.** –4,–9 **7.** –4, 14 **8.** 4, –12 **9.** False, $-15 < 5$ **10.** True

11. False, $\frac{7}{8} \ge \frac{7}{10}$ **12.** True **13.** $x = \frac{8}{5}$ **14.** $x = -\frac{1}{2}$ **15.** $x = \frac{3}{8}$ **16.** $x = -3$ **17.** 133.33 **18.** 22 **19.** 16.38

20. $v = \frac{h+16t^2}{t}$ **21.** $r = \frac{A-P}{Pt}$ **22.** $y = -\frac{14-5x}{3}$ **23.** $h = \frac{3V}{\pi r^2}$ **24.** $d = \frac{C}{\pi}$ **25.** $y = \frac{10-3x}{5}$ **26. a.** 17 **b.** –26

27. 21.7 **28.** $(-\infty, 2]$ 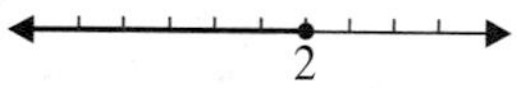 **29.** $\left[\frac{13}{3}, \infty\right)$ **30.** $\left(\frac{9}{28}, \infty\right)$

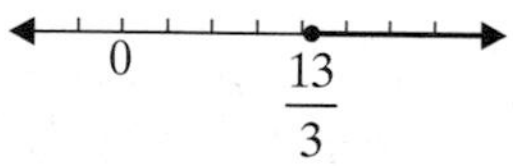

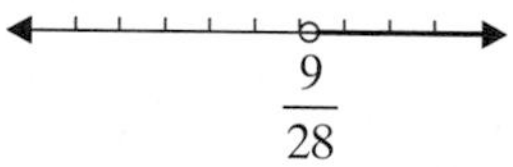

31. $[-1.1, 7.4]$ 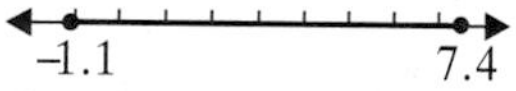 **32.** $\left[-\frac{5}{2}, \frac{3}{2}\right]$ 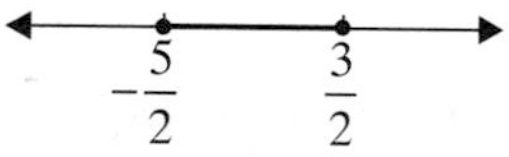 **33.** $(-6, 2)$

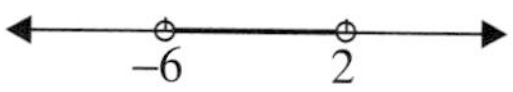

34. No solution **35.** No solution **36.** $a = -1.9, 0.9$ **37.** $x = -\frac{21}{2}, \frac{3}{2}$ **38.** $x = -\frac{8}{3}, 4$ **39.** $x = -30, 60$

40. a. $|x| \le 3$ **b.** $|x-7| < 1$ **c.** $|x-1.1| \le 0.2$ **41.** $x = 0$ **42.** $A = 100$ **43.** $x = \frac{1}{98}$ **44.** $x = 30$ **45.** $x = 3,\ y = 7$

46. $x = 9,\ y = 20°$ **47. a.** 45.7 cm **b.** 139.25 cm^2 **48.** 1177.5 in.^{3} **49.** 10.71 in. **50.** $86 \le x \le 100$ **51. a.** 5 in., 12 in., 13 in. **b.** 30 in.^{2} **52.** 20, 22, 24 **53.** length = 25 cm, width = 16 cm **54.** –19 **55.** $54 \le x \le 100$ **56.** less than 140 miles **57.** 15, 17, 19 **58.** 87.25

Chapter 4

Exercises 4.1, pages 241 – 246

1. $\{(-5,1),(-3,3),(-1,1),(1,2),(2,-2)\}$ **3.** $\{(-3,-2),(-1,-3),(-1,3),(0,0),(2,1)\}$ **5.** $\{(-4,4),(-3,-4),(0,-4),(0,3),(4,1)\}$
7. $\{(-6,2),(-1,6),(0,0),(1,-7),(6,3)\}$ **9.** $\{(-5,0),(-2,2),(-1,-4),(0,6),(2,0)\}$

11.
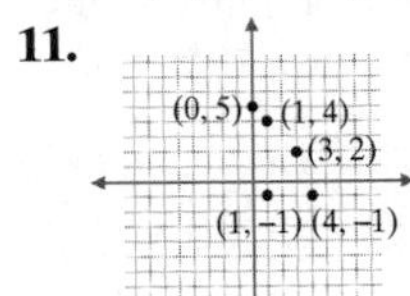

13.
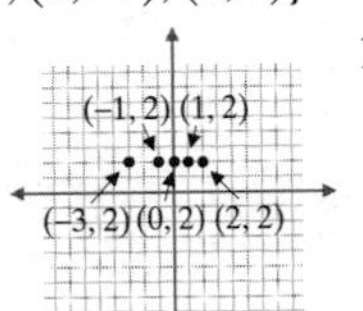

15.
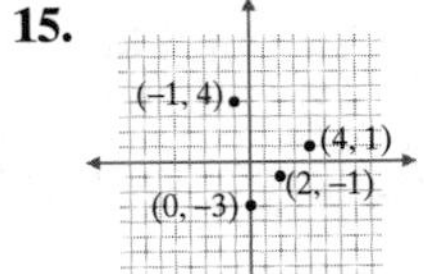

17.
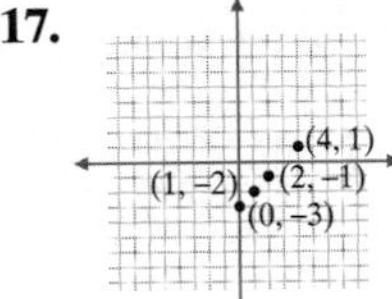

19.
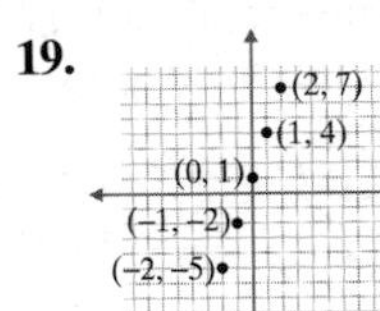

21.
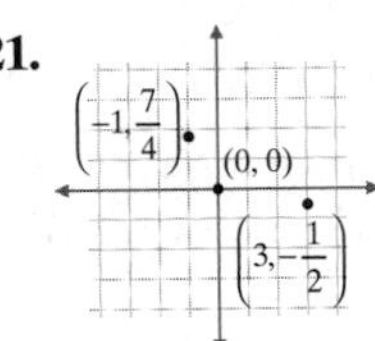

23.

25. b, c, d **27.** a, c **29.** a, c, d **31.** $(0,-4),(2,-2),(4,0),(1,-3)$ **33.** $(0,3),(2,2),(6,0),(-2,4)$

35. $(0,-8),(1,-4),(2,0),(1,-4)$ **37.** $(0,2),\left(-1,\frac{8}{3}\right),(3,0),(6,-2)$

39. $\left(0,-\frac{7}{4}\right),(1,-1),\left(\frac{7}{3},0\right),\left(3,\frac{1}{2}\right)$

41. $0,-1,-6,2$ **43.** $-3,1,-7,\frac{7}{4}$ **45.** $7,\frac{7}{3},10,6$ **47.** $2,4,-1,-1$ **49.** $-\frac{9}{5},3,-3,\frac{4}{3}$ **51.** $-5,2,-10,\frac{8}{5}$

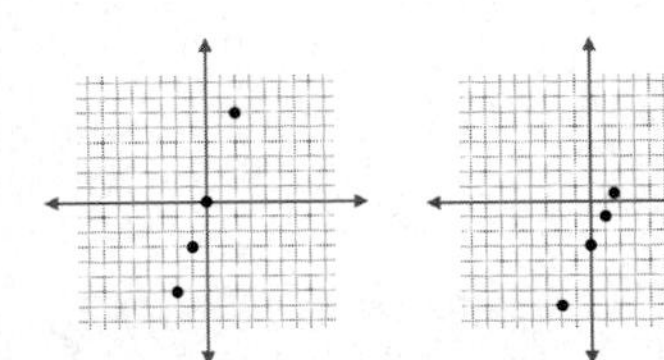
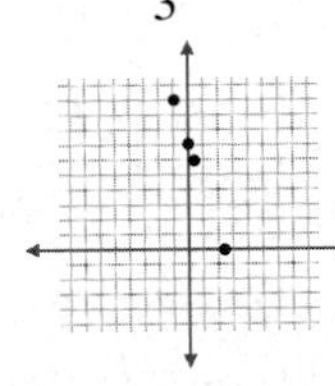
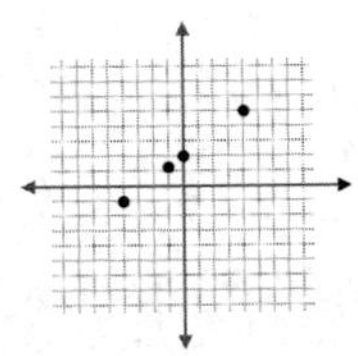
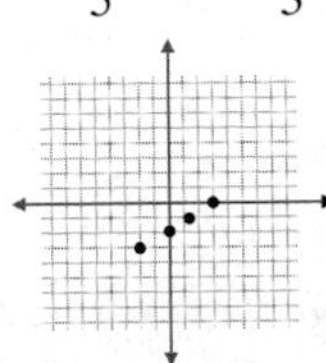
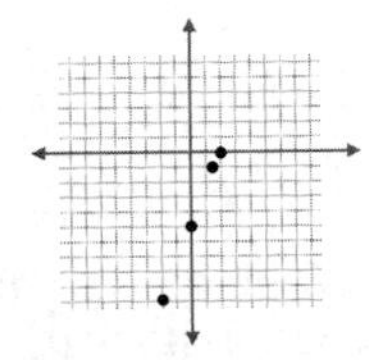

53. $-1.5,0,3.8,-1$ **55.** $-0.8,-2.4,-1,-7.2$ **57.** $-4,14,23,32,41,50,59$

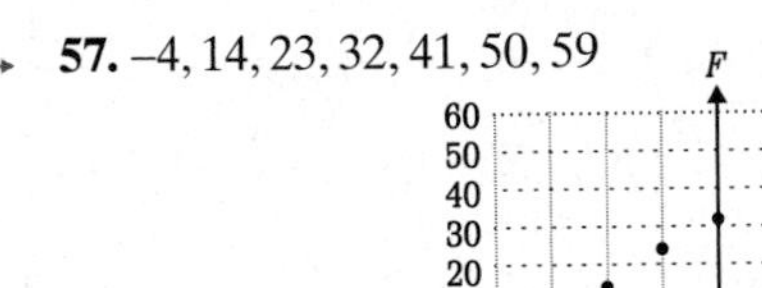

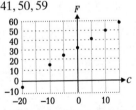

59. Answers will vary.

Exercises 4.2, pages 257 – 260

1. a. $(0,5)$ **b.** $\left(\frac{5}{2},0\right)$ **c.** $(-2,9)$ **d.** $(1,3)$ **3. a.** $(0,-4)$ **b.** $\left(\frac{4}{3},0\right)$ **c.** $(2,2)$ **d.** $(3,5)$ **5. a.** $(0,5)$ **b.** $\left(\frac{5}{2},0\right)$ **c.** $(2,1)$
d. $(-1,7)$ **7. a.** $(0,-3)$ **b.** $(2,0)$ **c.** $(-2,-6)$ **d.** $(4,3)$ **9.** a **11.** d **13.** f

15.

17.
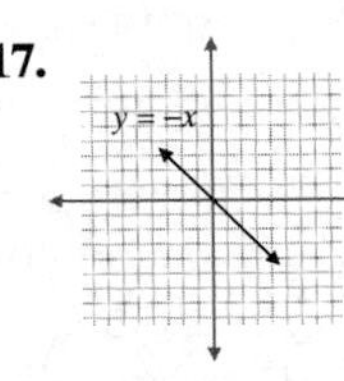

19.
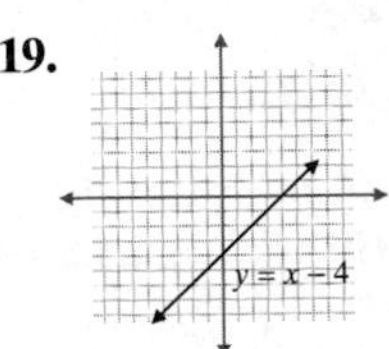

21.
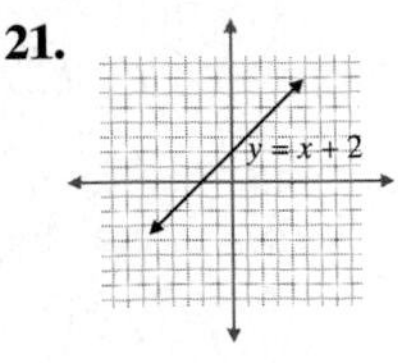

23.
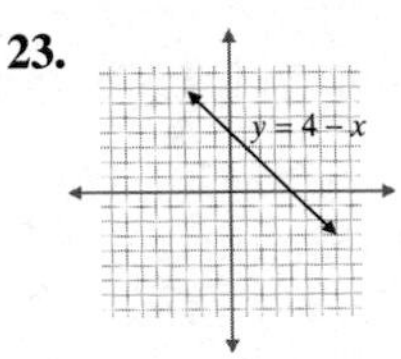

25.
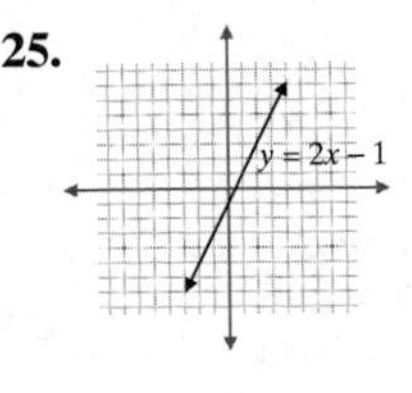

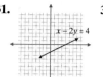

27. $3y = 12$ **29.** $2x - 8 = 0$ **31.** $x - 2y = 4$ **33.** $2x + y = 0$ **35.** $5y = 0$ **37.** $4x = 0$

39. $2x + 3y = 7$

41.

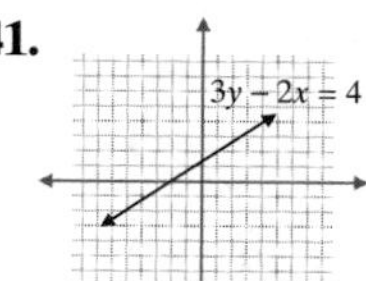

43. $5x + 2y = 9$ **45.** $4x + 2y = -10$ **47.** $y = \frac{1}{3}x - 3$

49.

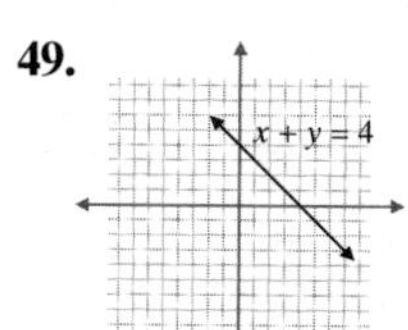

51. $3x - 2y = 6$

53.

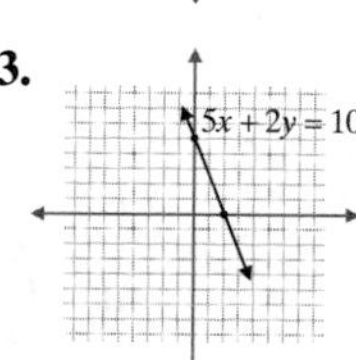

55.

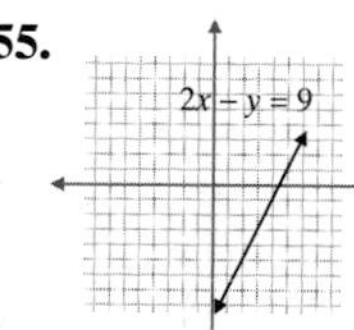

57.

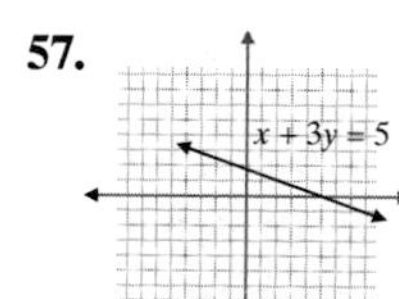

59.

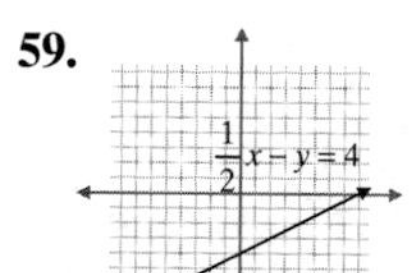

61.

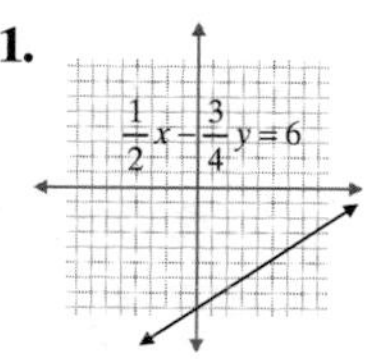

63. $2x + 3y = 5$

65.

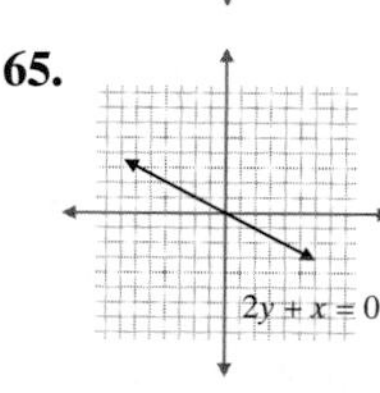

67.

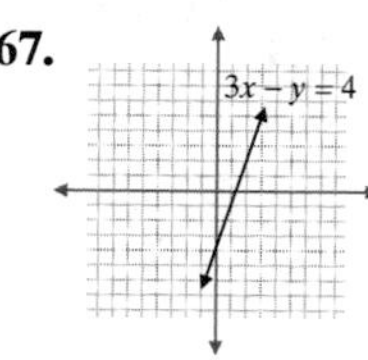

69.

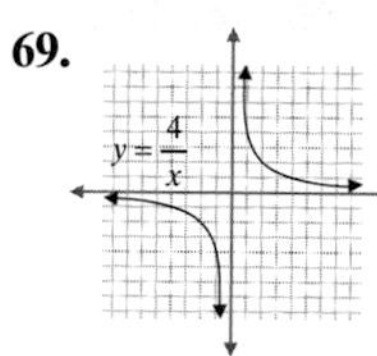

71.

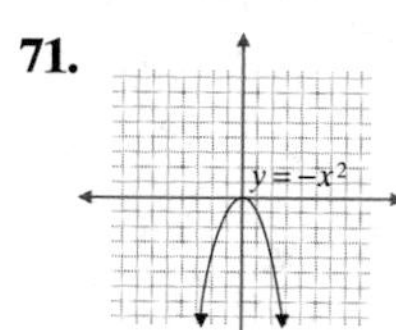

73.

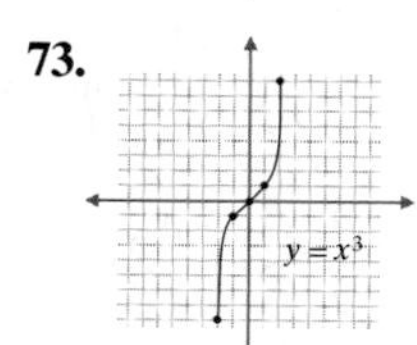

Exercises 4.3, pages 269 – 273

1. $m = 5$

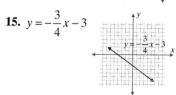

3. $m = -\frac{8}{7}$

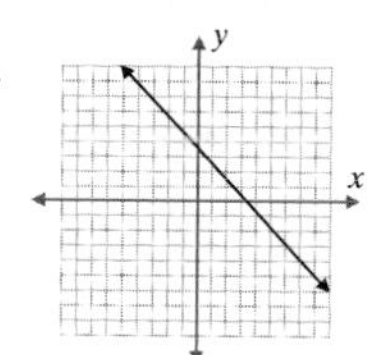

5. $m = 0$

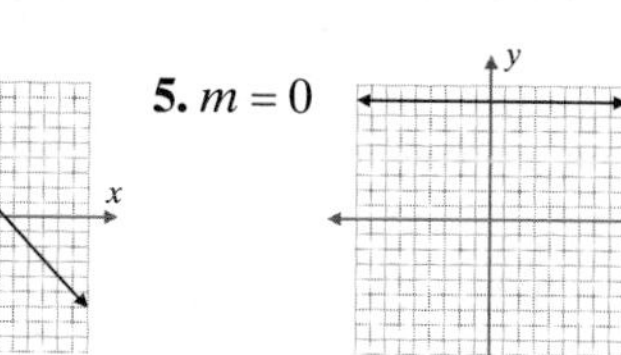

7. $m = -\frac{3}{10}$

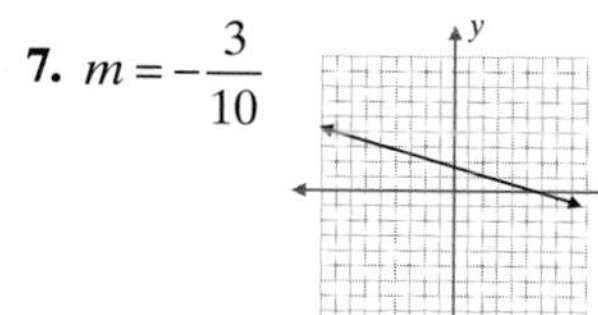

9. Slope is undefined.

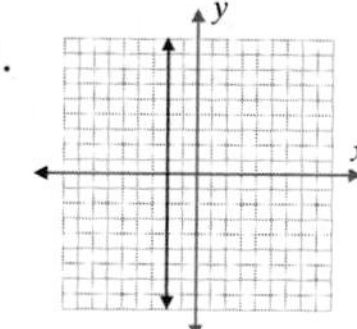

11. $m = \frac{1}{5}$

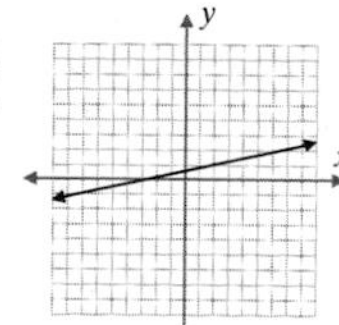

13. $y = \frac{2}{3}x$

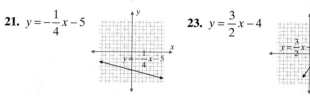

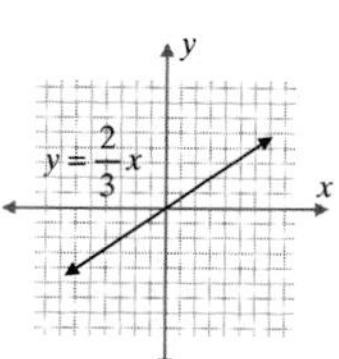

15. $y = -\frac{3}{4}x - 3$

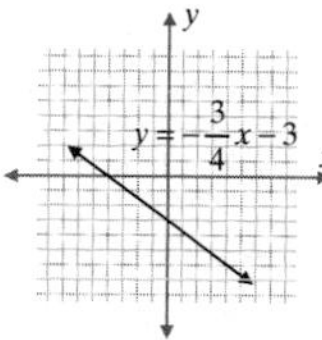

17. $y = -\frac{5}{3}x + 3$

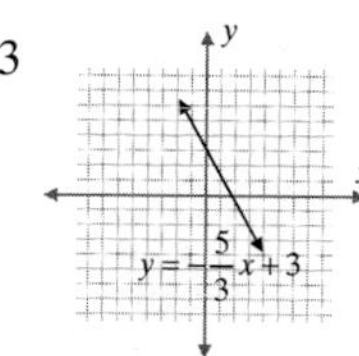

19. $y = 2x - 1$

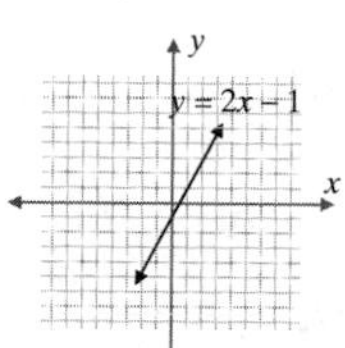

21. $y = -\frac{1}{4}x - 5$

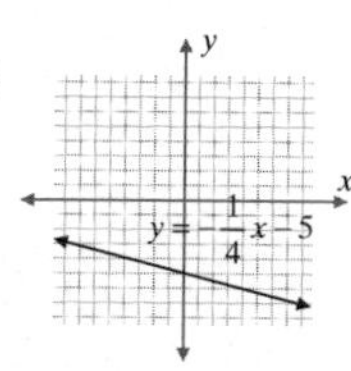

23. $y = \frac{3}{2}x - 4$

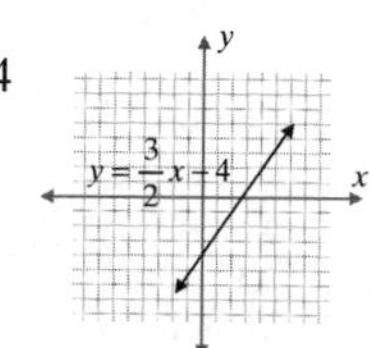

25. $y = 2x - 1$; $m = 2, b = -1$

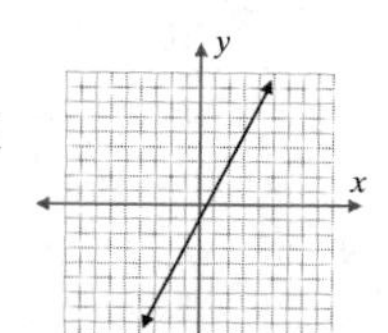

27. $y = -4x + 5;$ $m = -4, b = 5$

29. $y = \frac{2}{3}x + 2;\ m = \frac{2}{3},$ $b = 2$

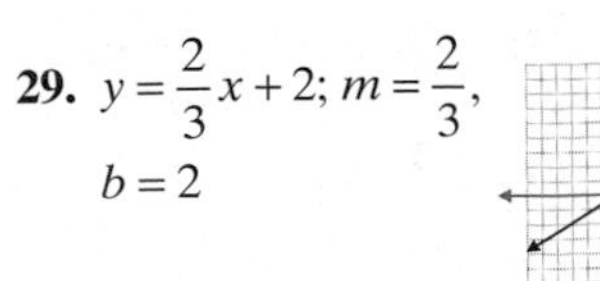

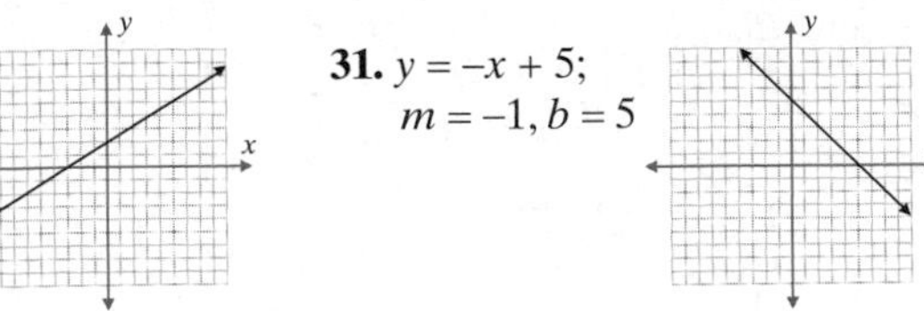

31. $y = -x + 5;$ $m = -1, b = 5$

33. $y = -\frac{1}{5}x + 2;$ $m = -\frac{1}{5},\ b = 2$

35. $y = 4;$ $m = 0,\ b = 4$

37. $y = -4x;$ $m = -4, b = 0$

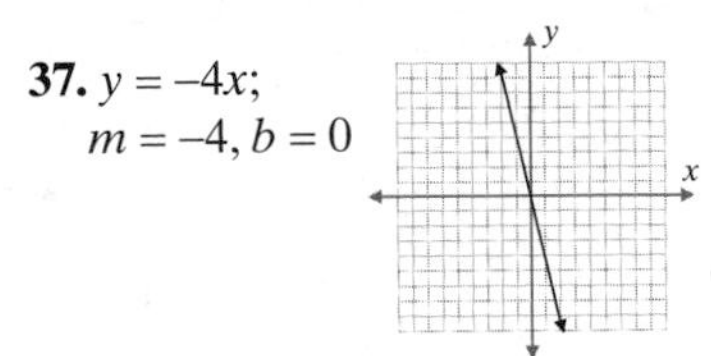

39. $y = \frac{2}{3}x - 2;\ m = \frac{2}{3},$ $b = -2$

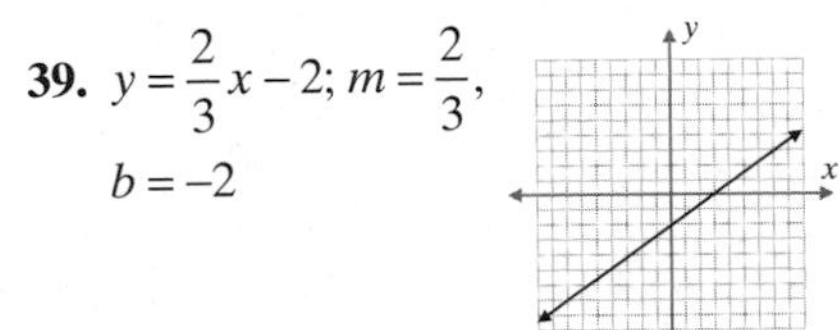

41. Cannot be written in slope intercept form, $x = -3$; slope is undefined, no y-intercept

43. $y = \frac{5}{6}x - \frac{5}{3};$ $m = \frac{5}{6},\ b = -\frac{5}{3}$

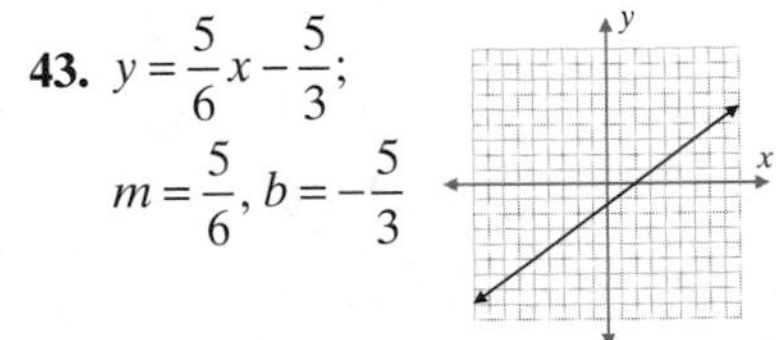

45. $y = -\frac{3}{4}x + \frac{5}{4};\ m = -\frac{3}{4},$ $b = \frac{5}{4}$

47. $y = -\frac{3}{2}x - \frac{7}{4};\ m = -\frac{3}{2},$ $b = -\frac{7}{4}$

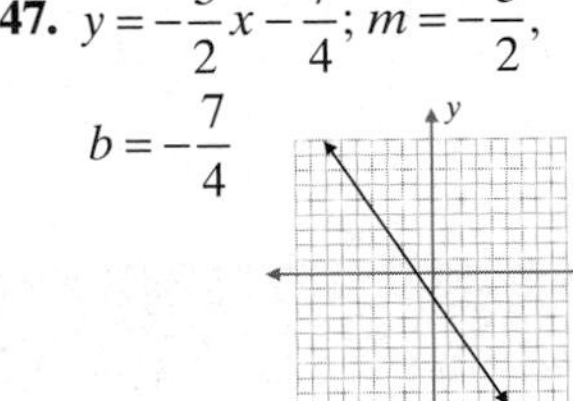

49. $y = \frac{1}{2}x + \frac{2}{3};\ m = \frac{1}{2},$ $b = \frac{2}{3}$

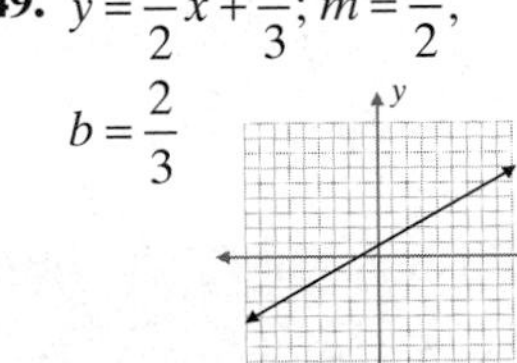

51. $y = \frac{5}{2}x + \frac{5}{2};\ m = \frac{5}{2},$ $b = \frac{5}{2}$

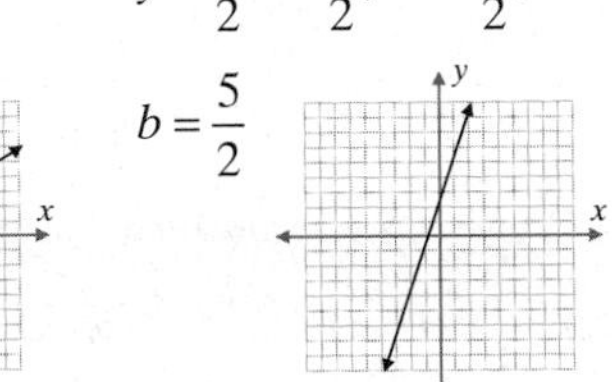

53. Answers will vary. **55. a.** undefined **b.** no y-intercept **c.** $x = 3$ **57. a.** -3 **b.** -3 **c.** $y = -3x - 3$ **59. a.** 2 **b.** $\frac{3}{4}$ **c.** $y = 2x + \frac{3}{4}$ **61. a.** 4 **b.** -7 **c.** $y = 4x - 7$ **63.** yes **65.** yes **67.** y-intercept $= -5$ (should be 5) **69.** y-intercept $= -9$ (should be -2) **71.** slope is reciprocal and y-intercept $= 2$ (should be 0) **73. a.** The x-axis **b.** The y-axis **75.** $y = mx + b$, m = slope of the line, b = y-intercept. **77.** Answers will vary.

Exercises 4.4, pages 283 – 287

1. $2x + y = -3$

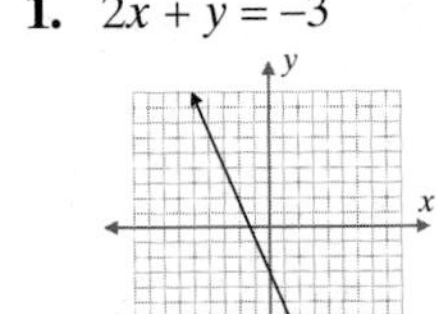

3. $x - 2y = -9$

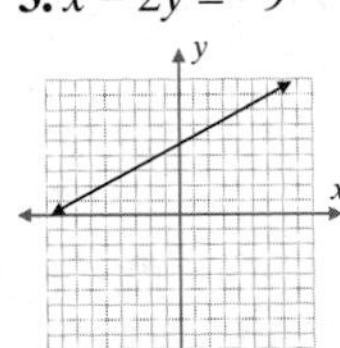

5. $x + 3y = 2$

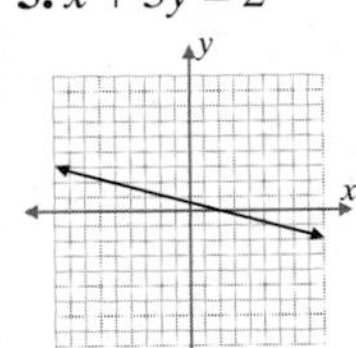

7. $x = 4$

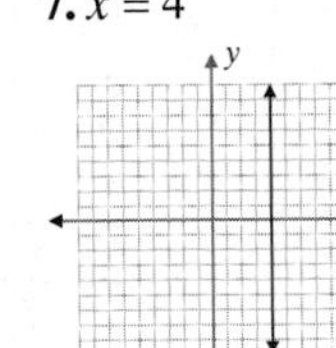

9. $y = 3$

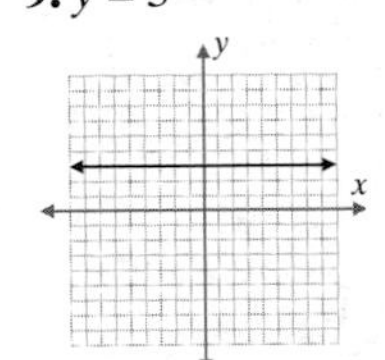

11. $6x + 5y = 23$

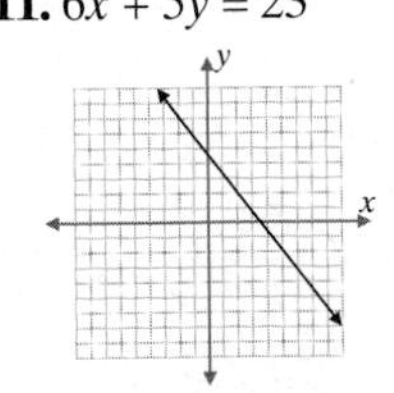

13. $2x + 27y = 5$

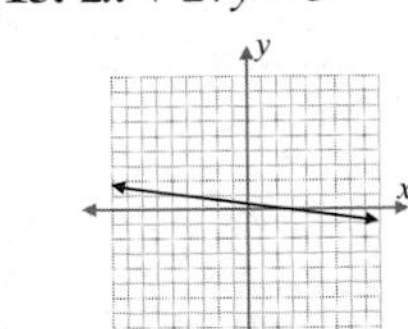

15. $12x + 9y = 17$

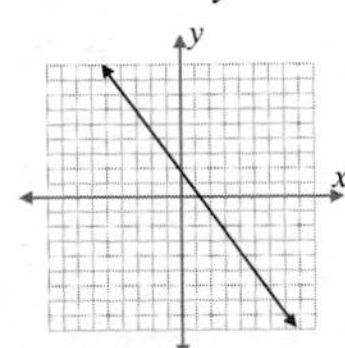

17.

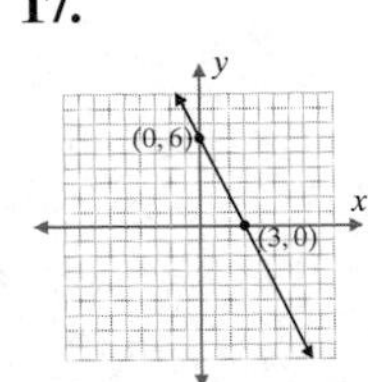

19.

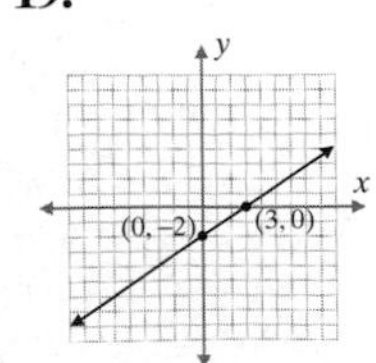

21.

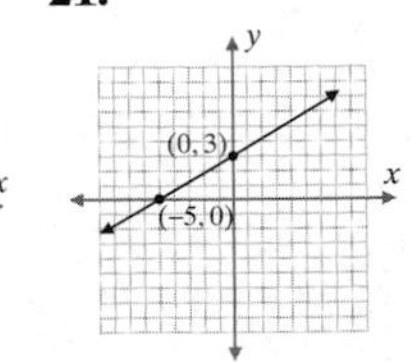

23.

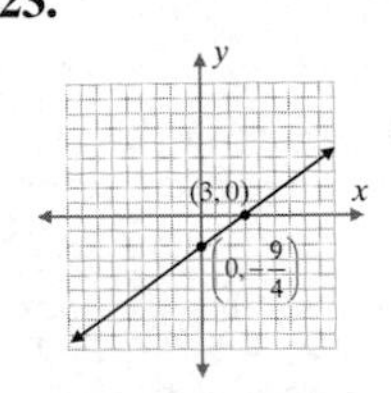

25. $x + 2y = 13$ **27.** $5x - y = -2$ **29.** $x = 2$ **31.** $3x + 5y = 7$ **33.** $2x - 3y = -2$ **35.** $x + 2y = -4$ **37.**

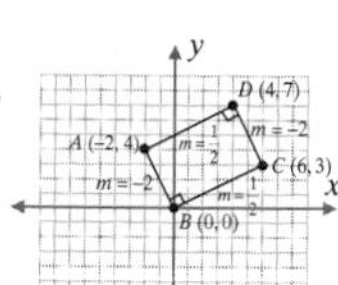
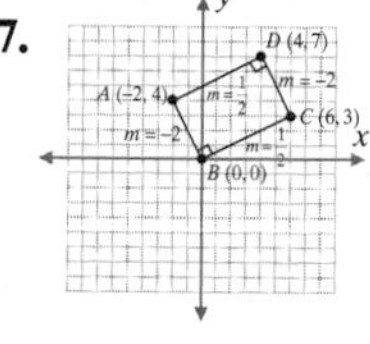

39. \$ 4000/year

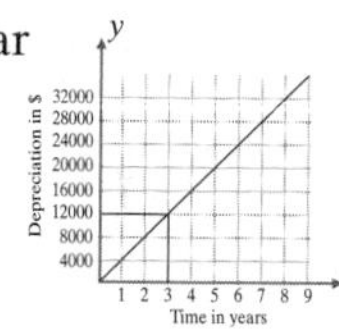

41. a. – b.

c. 29, 22, 19, 21
d. rate of change inc. 29 million people/year from 1998 - 1999, 22 million ppy from 1999 - 2000, 19 million ppy from 2000 - 2001, and 21 million ppy from 2001 - 2002

43. a. – b.

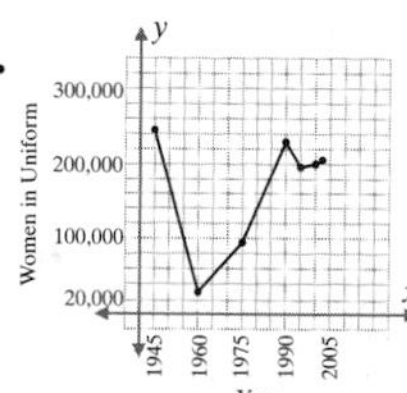

c. −15,647.067; 4354.53; 8676.67; −6180.4; 1297; 4587
d. rate of change dec. 15,647.067 women/year from 1945 - 1960; inc. 4354.53 wpy from 1960 - 1975; inc. 8676.67 wpy from 1975 - 1990; dec. 6180.4 wpy from 1990 - 1995; inc. 1297 wpy from 1995 - 2000; and inc. 4587 wpy from 2000 - 2001

45. a. 180 ft./min **b.** 0 ft./min **c.** 53.3 ft./min **d.** 73.3 ft./min. **47. a.** stg 11: 36.16 kph, stg 12: 33.19 kph, stg 19: 47.17 kph **b.** stg 11: 22.42 mph, stg 12: 20.58 mph, stg 19: 29.25 mph

49. Perpendicular

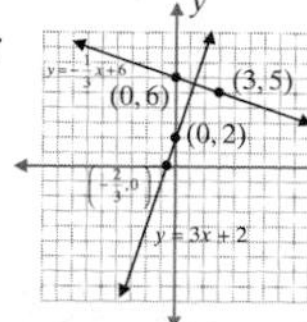

51. Neither

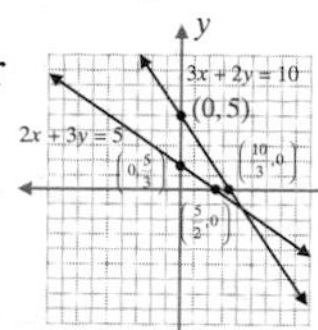

53. Perpendicular

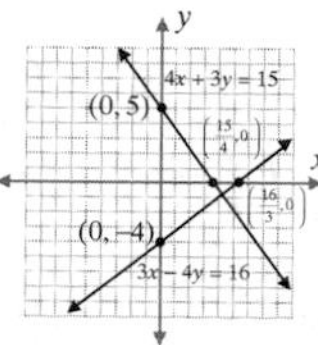

55. Use the two points to find the slope; then use the slope and one of the points in the point-slope form.

Exercises 4.5, pages 302 – 307

1. $\begin{Bmatrix}(-4,0),(-1,4),(1,2),\\(2,5),(6,-3)\end{Bmatrix}$; $D=\{-4,-1,1,2,6\}$; $R=\{0,4,2,5,-3\}$; function

3. $\begin{Bmatrix}(-5,-4),(-4,-2),\\(-2,-2),(1,-2),(2,1)\end{Bmatrix}$; $D=\{-5,-4,-2,1,2\}$; $R=\{-4,-2,1\}$; function

5. $\begin{Bmatrix}(-4,-3),(-4,1),\\(-1,-1),(-1,3),(3,-4)\end{Bmatrix}$; $D=\{-4,-1,3\}$; $R=\{-3,1,-1,3,-4\}$; not a function

7. $\begin{Bmatrix}(-5,-5),(-5,3),\\(0,5),(1,-2),(1,2)\end{Bmatrix}$; $D=\{-5,0,1\}$; $R=\{-5,3,5,-2,2\}$; not a function

9. $D=\{0,1,4,-3,2\}$ $R=\{0,6,-2,5,-1\}$ function

11. $D=\{-4,-3,1,2,3\}$ $R=\{4\}$ function

13. $D=\{0,-1,2,3,-3\}$ $R=\{2,1,4,5\}$ function

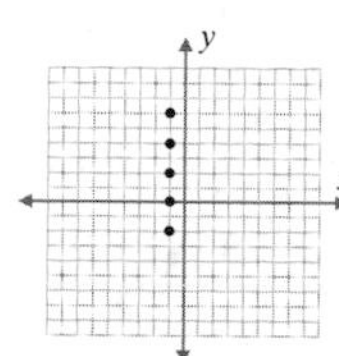

15. $D=\{-1\}$ $R=\{4,2,0,6,-2\}$ not a function

17. function, $D = (-\infty, \infty)$, $R = (0, \infty)$
19. function, $D = (-6, 6]$, $R\ \{-6, -4, -2, 0, 2, 4\}$
21. not a function
23. not a function

25. function, $D = (-\infty, \infty)$, $R = \left(-\frac{3}{2}, \frac{3}{2}\right)$ **27.** $\left\{(-9,-26),\left(-\frac{1}{3},0\right),(0,1),\left(\frac{4}{3},5\right),(2,7)\right\}$ **29.** $\left\{(-2,-11),(-1,-2),(0,1),(1,-2),(2,-11)\right\}$

31. $f(-2)=-3,\ f(-1)=-1, f(0)=1, f(1)=3, f(5)=11$ **33.** $g(-2)=10, g(-1)=6, g(0)=2, g(1)=-2, g(5)=-18$
35. $h(-2)=4, h(-1)=1, h(0)=0, h(1)=1, h(5)=25$ **37.** $F(-2)=16, F(-1)=9, F(0)=4, F(1)=1, F(5)=9$
39. $H(-2)=8, H(-1)=7, H(0)=0, H(1)=-7, H(5)=85$ **41.** $P(-2)=9.5, P(-1)=6, P(0)=3.5, P(1)=2, P(5)=6$
43. $f(-2)=-44, f(-1)=-15, f(0)=-2, f(1)=1, f(5)=33$ **45.** $D=\left\{x \mid x \neq -\frac{1}{2}\right\}$ **47.** $D=\left\{x \mid x \leq \frac{4}{3}\right\}$

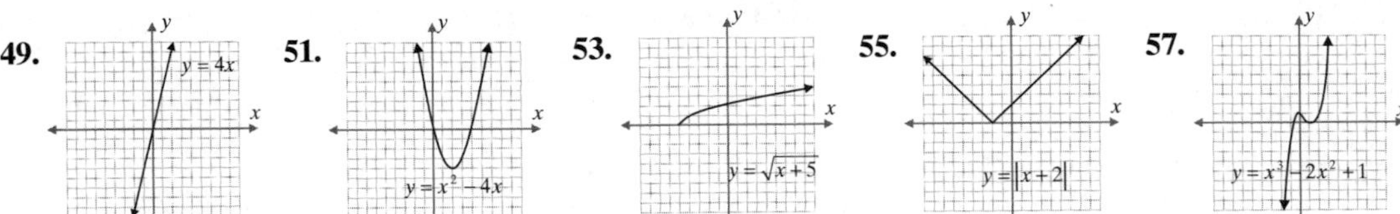

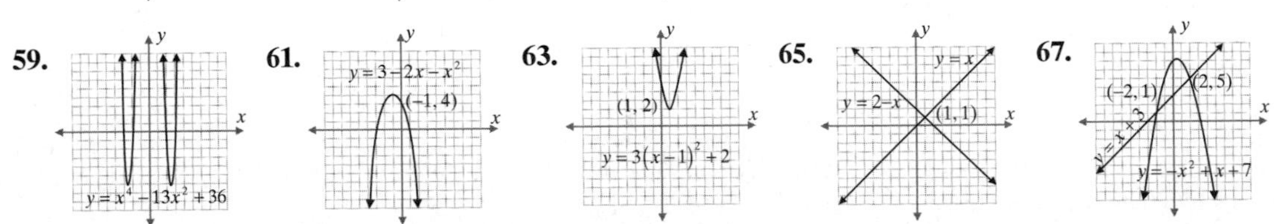

69. a. No restrictions **b.** $D=(-2,\infty)$ because the square root of a negative number does not exist.
c. $D=\left\{x \mid x \neq \frac{1}{3}\right\}$ because a zero in the denominator makes the function undefined.

Exercises 4.6, page 314

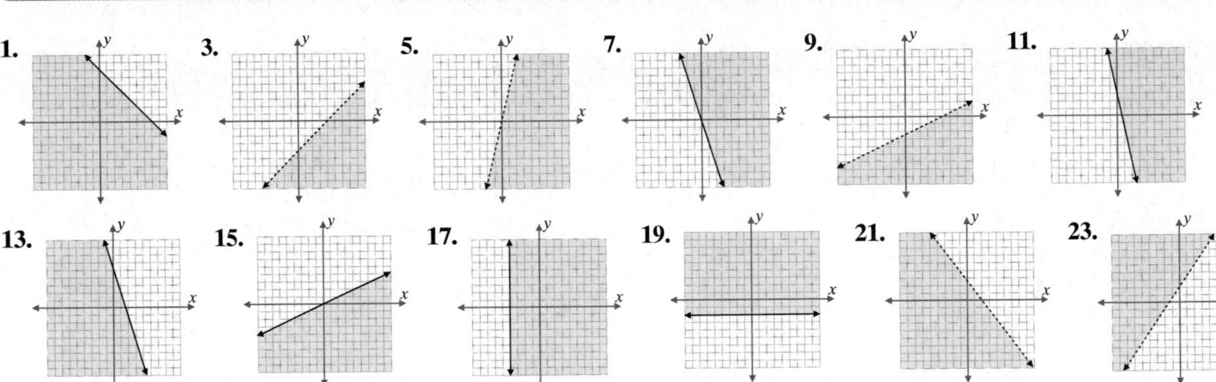

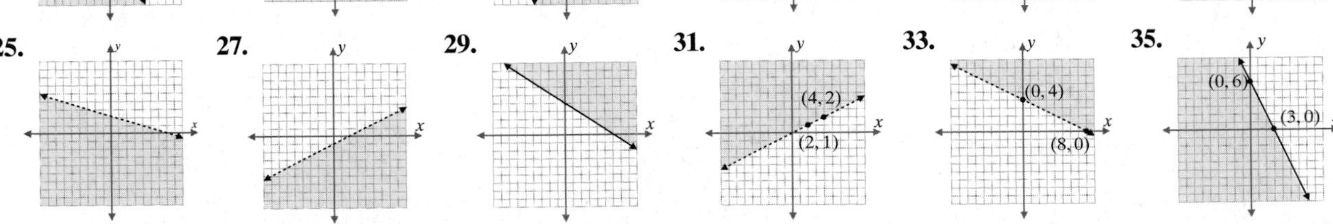

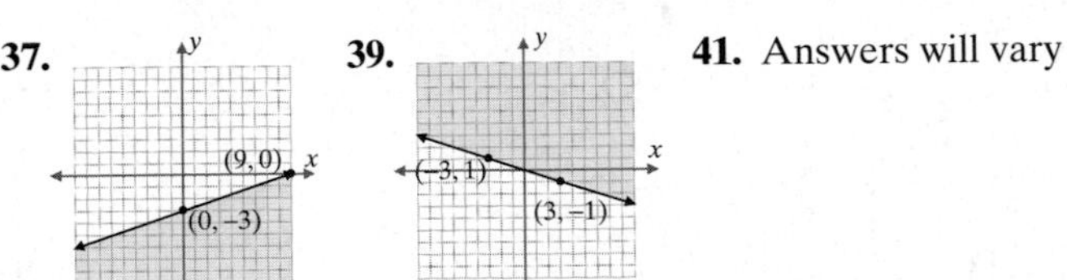

41. Answers will vary

Chapter 4 Test, pages 319 – 321

1. b and c **2.** (–3, 2), (–2, –1), (0, –3), (1, 2), (3, 4), (5, 0) **3.**

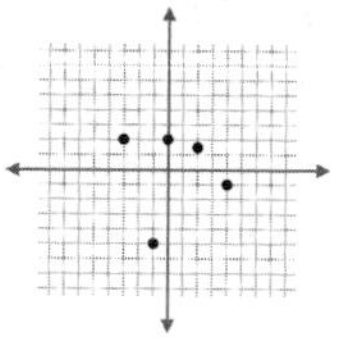

4.

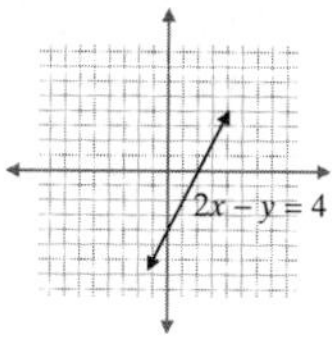

5. y-intercept $= \frac{9}{4}$
x-intercept = 3

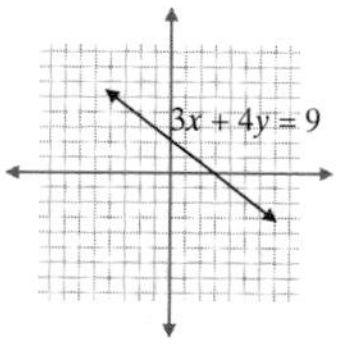

6.

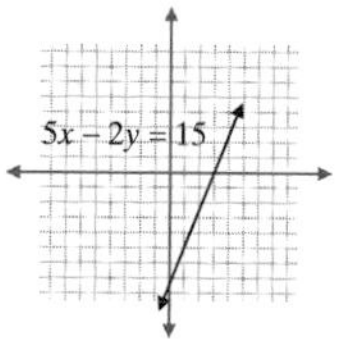

7. $m = -\frac{3}{5}, b = -3$

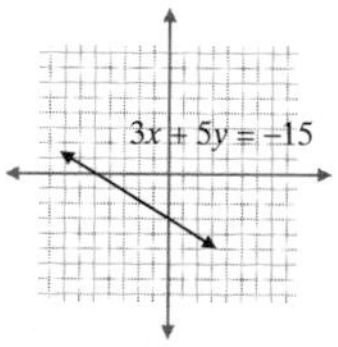

8. $m = -\frac{7}{4}$

9. $y - 2 = \frac{1}{4}(x+3)$ or $y = \frac{1}{4}x + \frac{11}{4}$ **10.** $y = -x + 4; m = -1$ **11.** $y = -2, m = 0$ **12. a.** $C = 4t + 9$ **b.** Because t represents the number of hours worked and one cannot work a negative number of hours. **c.** For every hour a person rents the carpet cleaning machine he will be charged \$4. **13.** These lines are parallel since they have the same slope, but different y-intercepts. **14.** $y = 2x - 8$ **15.** $D = \{-1, 0, 2, 5\}, R = \{-3, 4, 6\}$ **16.** It is not a function because the domain element, 2, has more than one range element, –5 and 0. **17. a.** –7 **b.** 14 **18. a.** 22 **b.** 7 **19.** It is a function. $D = \{-6, -4, -3, 3, 7\}, R = \{-4, -3, 0, 1, 3\}$ **20.** It is a function. $D: -2 \le x \le 2, R: 0 \le y \le 4$ **21.** Not a function **22.** It is a function. D: All real numbers, $R: y \ge 0$

23.

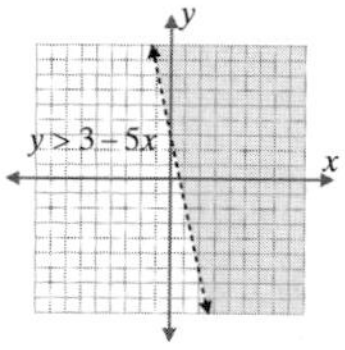

24.

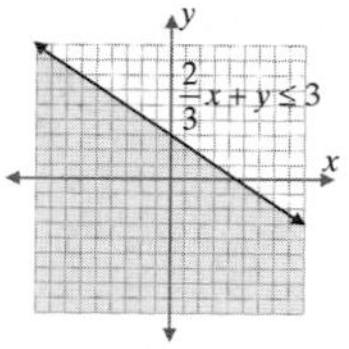

25.

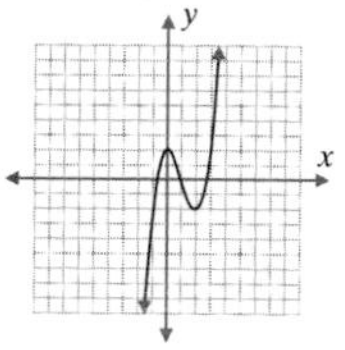

Chapter 4 Cumulative Review, pages 322 – 326

1. a. $\{\sqrt{9}\}$ **b.** $\{0, \sqrt{9}\}$ **c.** $\left\{ \begin{matrix} -10, -\sqrt{25}, \\ -1.6, 0, \frac{1}{5}, \sqrt{9} \end{matrix} \right\}$ **d.** $\{-10, -\sqrt{25}, 0, \sqrt{9}\}$ **e.** $\{-\sqrt{7}, \pi, \sqrt{12}\}$ **f.** $\left\{ \begin{matrix} -10, -\sqrt{25}, -1.6, \\ -\sqrt{7}, 0, \frac{1}{5}, \sqrt{9}, \\ \pi, \sqrt{12} \end{matrix} \right\}$

2.

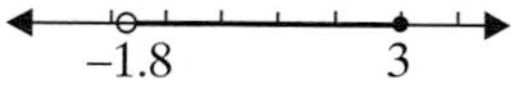

3.

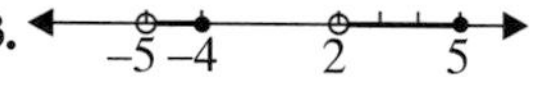

4. Associative property of addition **5.** Distributive property **6.** Inverse property of addition **7.** Identity property of multiplication **8.** Transitive property of order **9.** Trichotomy property of order **10.** –20 **11.** 11 **12.** 22 **13.** –1 **14.** –0.8 **15.** $-\frac{1}{12}$ **16.** 84 **17.** –40 **18.** –2 **19.** $\frac{2}{3}$ **20.** Undefined **21.** 0 **22.** –2 **23.** –35 **24.** 26 **25.** –59 **26.** $-2x - 12$ **27.** 0 **28.** $3x^3 - x^2 + 4x - 1$ **29.** $-10x + 4$ **30.** $x = 2$ **31.** $x = -4$ **32.** $x = -4$ **33.** $x = -8$ **34.** $x = 2.3$ or $x = -3.3$ **35.** $x = \frac{10}{3}$ or $x = 2$ **36.** $x = 62$ **37.** $x = 21$ **38.** $A = 2$ **39. a.** $n = 2A - m$ **b.** $w = \frac{P - 2l}{2}$ **40.** $-\frac{1}{3}$ **41.** 7 hrs **42.** 10 miles **43.** 81 **44.** 84 **45. a.** 5000 per hour **b.** 600,000 per week

46.

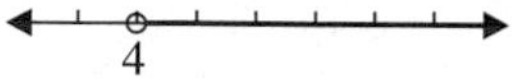

$(4, \infty)$ **47.**

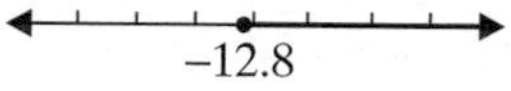

$[-12.8, \infty)$ **48.**

$(-\infty, 2]$

49.

$(-\infty, -5]$ **50.** No solution **51.** $x = -\frac{8}{3}, \frac{4}{3}$ **52.** 1.05×10^{-2} **53.** 9.0×10^{-4}

54. a. $(0,-4)$ **b.** $(2,0)$ **c.** $(1,-2)$ **d.** $(3,2)$ **55. a.** $(0,2)$ **b.** $(6,0)$ **c.** $\left(2,\frac{4}{3}\right)$ **d.** $(9,-1)$

56. $y=-\frac{1}{5}x+2;$ $m=-\frac{1}{5}, b=2$

57. $y=-3x+1;$ $m=-3, b=1$

58. $y=\frac{3}{7}x-1;$ $m=\frac{3}{7}, b=-1$

59. $y=-4x;$ $m=-4, b=0$

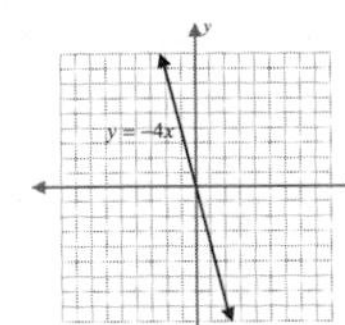

60. $y=-\frac{1}{2}x+2;$ $m=-\frac{1}{2}, b=2$

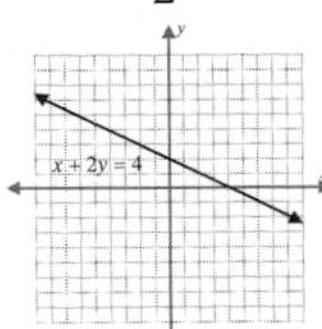

61. $x=-4;$ $m=$ undefined, no y-intercept

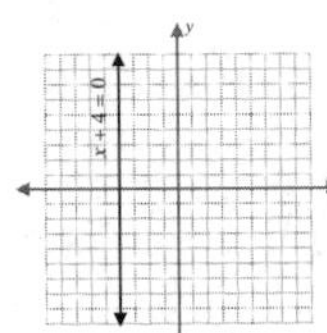

62. $2x-5y=17$ **63.** $4x-3y=-10$ **64.** $2x-y=0$ **65.** $x=5$ **66.** $4x+5y=15$ **67.** $2x+y=8$ **68.** $3x+2y=12$ **69.** $x=1$ **70.** $3x-4y=12$ **71.** $5x+3y=24$ **72.** $x-2y=-6$ **73. a.** -11 **b.** 22 **74. a.** -1 **b.** 76

75.

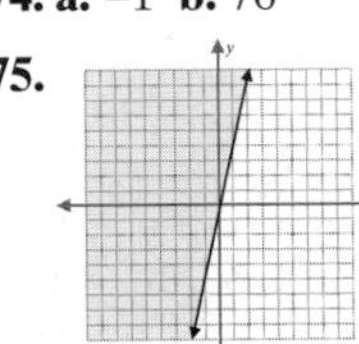

76.

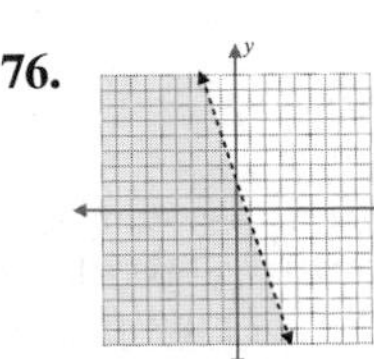

77. a. not a function **b.** it is a function
78. $D=\{-3,-2,-1,1,3\};$ $R=\{0,1,3\};$ It is a function
79. $D=\{-3,-1,1,4\};$ $R=\{-2,-1,1,3,4\};$ It is not a function
80. $D=\{-2,0,2,3,4\};$ $R=\{-1,0,1,2,3,4\};$ It is not a function

81. $D: -5<x\le-1$ and $0<x\le6;$ $R: \{-4,-2,2,4,5\};$ It is a function. **82. a.**

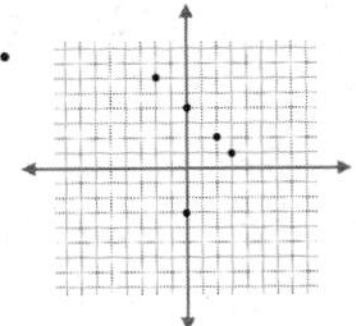

b. $D: \{-2,0,2,3\};$ $R: \{-3,1,2,4,6\}$ **c.** It is not a function

83. a.

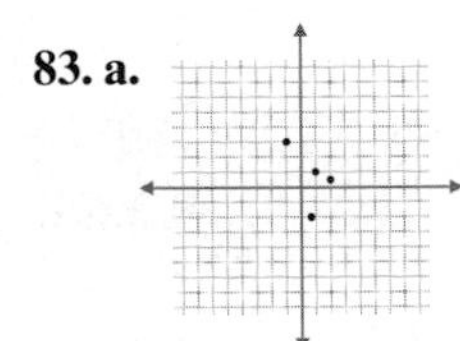

b. $D: \{-1, 1, \frac{2}{3}, 2\};$ $R: \{-2, 1, \frac{1}{2}, 3\}$ **c.** It is a function **84.** $D=\{x\,|\,x\ge-2\}$ **85. a.**

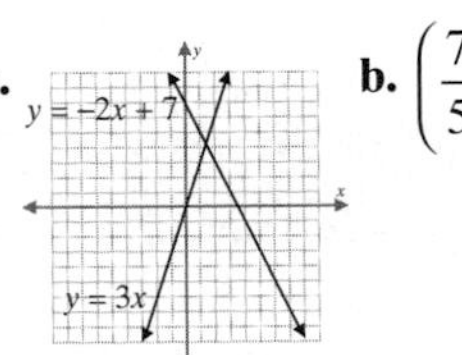

b. $\left(\frac{7}{5},\frac{21}{5}\right)$

86. a. $y=\frac{1}{2}x^2+x$ **b.** $\left(-1,-\frac{1}{2}\right)$ **87. a.**

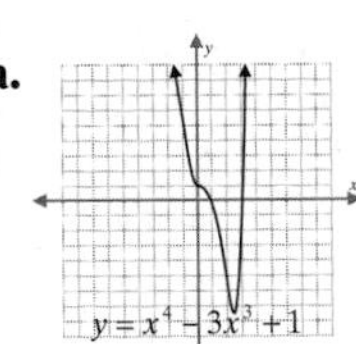

b. $(0.7648,0),\ (2.9615,0)$

Chapter 5

Exercises 5.1, pages 337 – 339

1. 27; product rule **3.** 512; product rule, 0 exponent rule **5.** $\frac{1}{16}$; negative exponent rule **7.** $\frac{1}{216}$; negative exponent rule

9. 24 **11.** 54 **13.** -54 **15.** $\frac{4}{9}$; negative exponent rule **17.** $\frac{-5}{4}$; negative exponent rule **19.** x^4; product rule

21. y^{11}; product rule **23.** $\frac{1}{y^2}$; negative exponent rule **25.** $\frac{5}{y^4}$; negative exponent rule **27.** $\frac{-10}{x^3}$; negative exponent rule **29.** $\frac{1}{y^2}$; negative exponent rule and 0 exponent rule **31.** 2; 0 exponent rule **33.** 729; quotient rule **35.** $\frac{1}{10000}$; quotient rule and negative exponent rule **37.** x^2; quotient rule **39.** x^2; quotient rule **41.** x^4; quotient rule **43.** $\frac{1}{x^4}$; quotient rule and negative exponent rule **45.** x^6; quotient rule **47.** x^2; quotient rule **49.** y^2; quotient rule **51.** –36 **53.** $\frac{1}{5}$ **55.** $-\frac{1}{64}$ **57.** x^{12} **59.** x^5 **61.** $\frac{1}{y^3}$ **63.** $\frac{1}{x^2}$ **65.** $\frac{1}{x}$ **67.** y^5 **69.** $\frac{1}{y^4}$ **71.** x^{10} **73.** $\frac{1}{x^3}$ **75.** x^8 **77.** 1 **79.** x^{k+1} **81.** x^{3k} **83.** x^{k-2} **85.** x^{k-2} **87.** $3y^5$; product rule **89.** $9x^3$; product rule **91.** $15x^6$; product rule **93.** $-18y^5$; product rule **95.** $30y^9$; product rule **97.** $4x^3$; quotient rule **99.** $-5x^4$; quotient rule **101.** $-4x^3$; quotient rule **103.** 10^7; product and quotient rules **105.** $\frac{1}{100}$; product, quotient, and negative exponent rules **107.** $-18x^5y^7$; product rule **109.** $\frac{-2y^2}{x}$; quotient and negative exponent rules **111.** 1; 0 exponent rule **113.** $-24ab^4$; product rule **115.** $\frac{5}{x}$; quotient and negative exponent rules **117.** 25.6036 **119.** 56,800.9424896 **121. a.** $a^m \cdot a^n = a^{m+n}$ **b.** $3^2 \cdot 2^2 = 3 \cdot 3 \cdot 2 \cdot 2 = 6^2$ **c.** $a^m \cdot a^n = a^{m+n}$

Exercises 5.2, pages 349 – 352

1. –81 **3.** 81 **5.** 1,000,000 **7.** $\frac{1}{x^4}$ **9.** $4x^8$ **11.** $36x^6$ **13.** $\frac{x^4}{y^6}$ **15.** $-108x^6$ **17.** –3 **19.** $\frac{a^4}{b^8}$ **21.** $\frac{27x^3}{y^3}$ **23.** 1 **25.** $4x^4y^4$ **27.** $\frac{b}{2a}$ **29.** $\frac{1}{3xy^2}$ **31.** $-\frac{x^3y^6}{27}$ **33.** m^2n^4 **35.** $-\frac{y^2}{49x^2}$ **37.** $\frac{25x^6}{y^2}$ **39.** $\frac{x^6}{y^6}$ **41.** $\frac{36y^{14}}{x^4}$ **43.** $\frac{24y^5}{x^7}$ **45.** $\frac{36b^4}{a^2}$ **47.** $\frac{x^2}{36y^6}$ **49.** $\frac{x^4}{25y^2}$ **51.** $x^{2k}y^k$ **53.** $x^{2k+1}y^{2+n}$ **55.** x^ny^{k+1} **57.** $\frac{x^4}{y}$ **59.** $\frac{9y^7}{x^6}$ **61.** $\frac{a^6}{b^3}$ **63.** $\frac{7}{2x^3y}$ **65.** $\frac{x^{13}}{y^{20}}$ **67.** $\frac{y^6}{x^6}$ **69.** 472,000 **71.** 0.000000128 **73.** 0.00923 **75.** 0.042 **77.** 7,560,000 **79.** 851,500,000 **81.** 8.6×10^4 **83.** 3.62×10^{-2} **85.** 1.83×10^7 **87.** 4.79×10^5 **89.** 8.71×10^{-7} **91.** 4.29×10 **93.** 1.44×10^5 **95.** 6×10^{-4} **97.** 1.2×10^{-1} **99.** 5×10^3 **101.** 5.6×10^{-2} **103.** 1.8×10^{12} cm/min.; 1.08×10^{14} cm/hr. **105.** 6×10^8 ounces **107.** 10,000,000,000; 19,025.87519 years **109.** $\approx 6.26\times10^{-4}$ light years **111.** 6.642×10^{-26}kg **113.** 1.15×10^7 sq. mi. **115.** 4.356×10^8 sq. ft. **117.** 3.9 **119.** 2.0×10^{-7} **121.** 1.0×10^{14}

Exercises 5.3, pages 357 – 358

1. Monomial **3.** Not a polynomial **5.** Trinomial **7.** Not a polynomial **9.** Binomial **11.** $4y$; first degree monomial; $a_1 = 4, a_0 = 0$ **13.** $x^3 + 3x^2 - 2x$; third degree trinomial; $a_3 = 1, a_2 = 3, a_1 = -2, a_0 = 0$ **15.** $-2x^2$; second degree monomial; $a_2 = -2, a_1 = 0, a_0 = 0$ **17.** 0; not a polynomial; $a_0 = 0$ **19.** $6a^5 - 7a^3 - a^2$; fifth degree trinomial; $a_5 = 6,\ a_4 = 0,\ a_3 = -7;\ a_2 = -1, a_1 = 0, a_0 = 0$ **21.** $2y^3 + 4y$; third degree binomial; $a_3 = 2, a_2 = 0,\ a_1 = 4, a_0 = 0$ **23.** 4; monomial of degree 0; $a_0 = 4$ **25.** $2x^3 + 3x^2 - x + 1$; third degree polynomial; $a_3 = 2, a_2 = 3,\ a_1 = -1, a_0 = 1$ **27.** $4x^4 - x^2 + 4x - 10$; fourth degree polynomial; $a_4 = 4, a_3 = 0, a_2 = -1,\ a_1 = 4, a_0 = -10$ **29.** $9x^3 + 4x^2 + 8x$; third degree trinomial; $a_3 = 9, a_2 = 4, a_1 = 8, a_0 = 0$ **31.** $p(-1) = -16$ **33.** $p(3) = -41$ **35.** $p(-2) = 81$ **37.** $p(-1) = 13$ **39.** $p(2) = 24$

41. a. **b.** **c.**

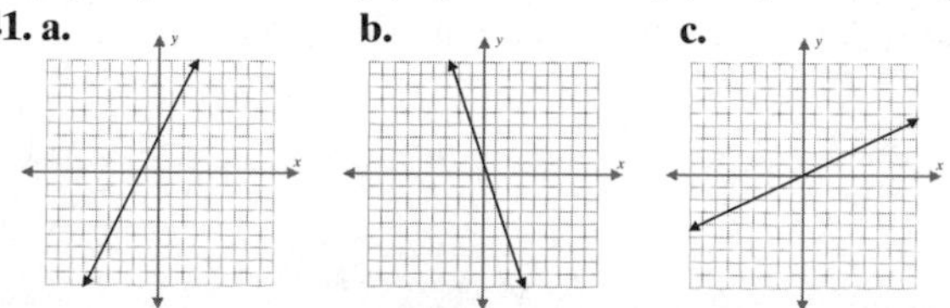

43. a. **b.** **c.**

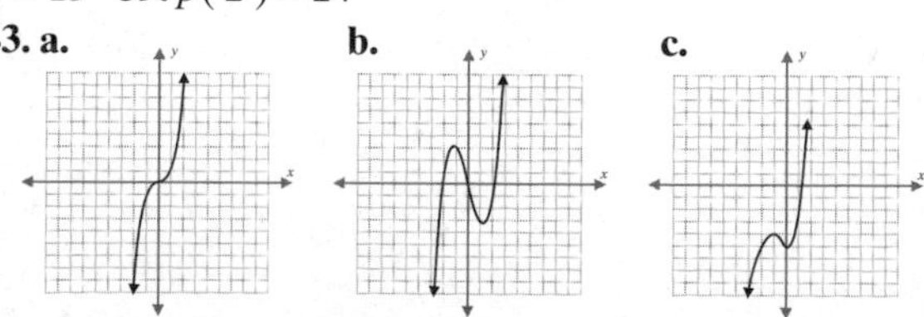

Exercises 5.4, pages 362 – 364

1. $3x^2+7x+2$ **3.** $2x^2+11x-7$ **5.** $3x^2$ **7.** $x^2-5x+17$ **9.** $3x^2+14x-7$ **11.** $-5x^2-3xy-8y^2$ **13.** $-x^2+2x-7$ **15.** $-x^2+2x$ **17.** $-x^2+7x-3$ **19.** $4x^3+7$ **21.** $-x^2-3x-7$ **23.** $2x^3-2x-3$ **25.** $10x^4+2x^3-4x^2+2x+1$ **27.** $2x^3+11x^2-4x-3$ **29.** $5x^3+11x^2+10x-14$ **31.** $2x^2+5x-11$ **33.** $-4x^4-7x^3-11x^2+5x+13$ **35.** $3x^2-4x-8$ **37.** $2x^2+13x+9$ **39.** $2x^4-5x^3-x-3$ **41.** $3x^4-7x^3+3x^2-5x+9$ **43.** $-4x^2+6x-11$ **45.** $-2x^3+3x^2-2x-8$ **47.** $2x^3+4x^2+3x-10$ **49.** $8x^4+6x^2+15$ **51.** $-5x^2+8x-12$ **53.** $4x^3-16x^2+10x$ **55.** $-5x^4+9x^3-8x^2-12x-12$ **57.** $4x-13$ **59.** $-7x+9$ **61.** $2x+17$ **63.** $3x^3+13x^2+9$ **65.** $8x^2-x-2$ **67.** $3x-19$ **69.** $7x-1$ **71.** $4x^2-x$ **73.** $x^2+11x-8$ **75.** $7x^3-x^2-7$ **77.** Any monomial or algebraic sum of monomials. **79.** The largest of the degrees of its terms after like terms have been combined. **81. a.** Answers will vary. **b.** Answers will vary.

Exercises 5.5, pages 370 – 372

1. $5x^3-10x^2+15x$ **3.** $x^3y^2+4xy^3$ **5.** $-6x^5-15x^3$ **7.** $-20x^4+30x^2$ **9.** $-y^5+8y-2$ **11.** $-4x^8+8x^7-12x^4$ **13.** $-35t^5+14t^4+7t^3$ **15.** $-x^4-5x^2+4x$ **17.** $6x^2-x-2$ **19.** $9a^2-25$ **21.** $-10x^2+39x-14$ **23.** x^3+5x^2+8x+4 **25.** x^2+x-12 **27.** $a^2-2a-48$ **29.** x^2-3x+2 **31.** $x^2-3x-18$ **33.** x^2-9x+8 **35.** $3t^2-3t-60$ **37.** x^3+11x^2+24x **39.** $2x^2-7x-4$ **41.** $6x^2+17x-3$ **43.** $4x^2-9$ **45.** $16x^2+8x+1$ **47.** $2y^2-11y-6$ **49.** $3x^2-19x+20$ **51.** $6y^2+13y+6$ **53.** $8x^2-37x-15$ **55.** $27x^2-15x-2$ **57.** y^3+2y^2+y+12 **59.** $x^{k+2}+3x^2$ **61.** $x^{2k}-2x^k-15$ **63.** $x^{2k}+5x^k+4$ **65.** $3x^{2k}+17x^k+10$ **67.** $5x^3+6x^2-22x-9$ **69.** $2x^4+7x^3+5x^2+x-15$ **71.** $6x^3-13x^2-23x-6$ **73.** $2x^4-3x^3-6x^2+17x-12$ **75.** $4x^4-6x^3-28x^2+26x+24$ **77. a.** $3x^2+2x-8$ **b.** 8 **79. a.** $2x^2+3x-5$ **b.** 9 **81. a.** $6x^2-13x-8$ **b.** -10 **83. a.** $7x^2-13x-2$ **b.** 0 **85. a.** $4x^2+12x+9$ **b.** 49 **87. a.** $x^3+3x^2-4x-12$ **b.** 0 **89. a.** $4x^2-49$ **b.** -33 **91. a.** x^3+1 **b.** 9 **93. a.** $49a^2-28a+4$ **b.** 144 **95. a.** $2x^3+x^2-5x-3$ **b.** 7 **97. a.** $x^3+6x^2+11x+6$ **b.** 60 **99. a.** a^4-a^2+2a-1 **b.** 15 **101. a.** $t^4+6t^3+13t^2+12t+4$ **b.** 144 **103.** $2y^2-6y+7$ **105.** $3t^2+16t+5$

Exercises 5.6, pages 379 – 382

1. x^2-9, difference of two squares **3.** $x^2-10x+25$, perfect square trinomial **5.** x^2-36, difference of two squares **7.** $x^2+16x+64$, perfect square trinomial **9.** $2x^2+x-3$, neither **11.** $9x^2-24x+16$, perfect square trinomial **13.** $4x^2-1$, difference of two squares **15.** $9x^2-12x+4$, perfect square trinomial **17.** x^2+6x+9, perfect square trinomial **19.** $x^2-10x+25$, perfect square trinomial **21.** $25x^2-81$, difference of two squares **23.** $4x^2+28x+49$, perfect square trinomial **25.** $81x^2-4$, difference of two squares **27.** $10x^4-11x^2-6$, neither **29.** $49x^2+14x+1$, perfect square trinomial

31. $5x^2+11x+2$ **33.** $4x^2+13x-12$ **35.** $3x^2-25x+42$ **37.** $x^2+10x+25$ **39.** x^4-1 **41.** x^4+6x^2+9 **43.** x^6-4x^3+4 **45.** x^4+3x^2-54 **47.** $x^2-\frac{4}{9}$ **49.** $x^2-\frac{9}{16}$ **51.** $x^2+\frac{6}{5}x+\frac{9}{25}$ **53.** $x^2-\frac{5}{3}x+\frac{25}{36}$ **55.** $x^2-\frac{1}{4}x-\frac{1}{8}$ **57.** $x^2+\frac{5}{6}x+\frac{1}{6}$ **59.** $9x^2+6x+1$ **61.** $25x^2-20xy+4y^2$ **63.** $16x^2-49$ **65.** $4x^2-9y^2$ **67.** $9x^5-16x$ **69.** x^3-1 **71.** $x^3+9x^2+27x+27$ **73.** x^6+4x^3+4 **75.** $4x^6-49$ **77.** x^3-27y^3 **79.** $4x^6y-144x^2y^5$ **81.** $10x^6y-13x^4y^3-3x^2y^5$ **83.** $x^2+2xy+y^2-4$ **85.** $25x^2+y^2+30x-6y-10xy+9$ **87.** $x^2+8x-4xy-16y+4y^2+16$ **89. a.** $A(x)=400-4x^2$ **b.** $P(x)=80$ **91.** $A(x)=8x+15$ **93. a.** $A(x)=150-4x^2$ **b.** $P(x)=50$ **c.** $V(x)=150x-50x^2+4x^3$ **95.** 1 **97.** factor: x^2-3x+9; sum: x^3+27

Exercises 5.7, pages 391 – 393

1. $x^2+2x+\frac{3}{4}$ **3.** $2x^2-3x-\frac{3}{5}$ **5.** $2x+5$ **7.** $x+6-\frac{3}{x}$ **9.** $2x+3-\frac{3}{2x}$ **11.** $2x-3-\frac{1}{x}$ **13.** $x+3y-\frac{11y}{7x}$
15. $\frac{3x^2}{4}-2y-\frac{y^2}{x}$ **17.** $\frac{5x}{8}-1+\frac{2}{y}$ **19.** $\frac{8x^2}{9}-xy+\frac{5}{9y}$ **21.** $12+\frac{2}{23}$ **23.** $7+\frac{24}{59}$ **25.** $x-2$ **27.** $a-\frac{15}{a-2}$
29. $y-5-\frac{22}{y+4}$ **31.** $5y-11+\frac{48}{y+5}$ **33.** $4c-5+\frac{1}{2c+3}$ **35.** $2m+1-\frac{3}{5m-3}$ **37.** $x-2+\frac{10}{x+5}$ **39.** $y^2-7y+12$
41. $3a^2+1+\frac{2}{4a-1}$ **43.** $x^2+7x+35+\frac{170}{x-5}$ **45.** x^2-3x+9 **47.** $2x+\frac{3x+3}{x^2-2}$ **49.** $4x^2+6x+9$ **51.** $4x+3+\frac{8}{x-1}$
53. $3x^2+2x+5$ **55.** $2a+3-\frac{4a}{a^2+2}$ **57.** $5x+2+\frac{-3}{3x-4}$ **59.** $2x^2-x+4$ **61.** $6x^2+8x+21+\frac{35}{x-2}$
63. $x^2-x+3+\frac{7}{2x-1}$ **65.** $x^2+3x-1+\frac{-6}{6x+1}$ **67.** $2x^2-2x+6+\frac{-27}{x+3}$ **69.** $4x^2-3x+4$ **71.** $2x^2-x-6+\frac{-18}{2x-3}$
73. $x^2+x+\frac{3x-1}{2x^2+3}$ **75.** $2x^2+x+2+\frac{8x+8}{x^2-5}$ **77.** $x-6+\frac{-3x-8}{x^2-3x+5}$ **79.** $2x^2+15x+42+\frac{117x-33}{x^2-3x+1}$
81. $x^2-x+2+\frac{-4x+5}{3x^2+x-1}$ **83.** $x^2-5x+25$ **85.** x^3+1

Chapter 5 Test, pages 397 – 398

1. $-10a^5$ **2.** 1 **3.** $\frac{4x^3}{y^7}$ **4.** $\frac{x}{3y^2}$ **5.** $\frac{x^2}{4y^2}$ **6.** $4x^2y^4$ **7. a.** 135,000 **b.** 0.0000027 **8. a.** 1.25×10^8 **b.** 5.2×10^{-4} **9.** $8x^2+3x$; second degree binomial **10.** $-x^3+3x^2+3x-1$; third degree polynomial **11.** $5x^5+2x^4-11x+3$; fifth degree polynomial **12. a.** 20 **b.** – 110 **13.** $-3x+2$ **14.** $20x-8$ **15.** $5x^3-x^2+6x+5$ **16.** $-2x^2+x-6$ **17.** $7x^4+14x^2+4$ **18.** $7x^3-2x^2-7x+1$ **19.** $15x^7-20x^6+15x^5-40x^4-10x^2$ **20.** $49x^2-9$; **21.** $16x^2+8x+1$ **22.** $36x^2-60x+25$ **23.** $10x^2+29x+10$ **24.** $6x^3-69x^2+189x$ **25.** $7x^2+7x+14$ **26.** $10x^4+4x^3-15x^2-41x-14$ **27.** x^3-64 **28.** $x^2+4x+4-y^2$ **29.** $2x+\frac{3}{2}-\frac{3}{x}$ **30.** $\frac{5}{3}+2b+\frac{b^2}{a}$ **31.** $x-6-\frac{2}{2x+3}$ **32.** $x-9+\frac{15x-12}{x^2+x-3}$ **33.** $x^2+4x+14+\frac{11}{x-3}$

34. a.
$$\begin{array}{r} x^3 - 8x + 5 \\ x^2+1\overline{)x^5 - 7x^3 + 5x^2 - 8x + 5} \\ \underline{-x^5 - x^3} \\ -8x^3 + 5x^2 - 8x \\ \underline{8x^3 + 8x} \\ 5x^2 + 5 \\ \underline{-5x^2 - 5} \\ 0 \end{array}$$
b. x^3-8x+5

Chapter 5 Cumulative Review, pages 398 – 401

1. $(x+15)$ **2.** $(3x+8)$ **3.** 540 **4.** $120a^2b^3$ **5.** 16, 2 **6.** –9, –4 **7.** 14, –4 **8.** 4, –12 **9.** $5x^2+16x$; second degree binomial; $a_2=5, a_1=16, a_0=0$ **10.** $5x-18$; first degree binomial; $a_1=5, a_0=-18$ **11.** $x-9$; first degree binomial; $a_1=1, a_0=-9$ **12.** $x^4-2x^3+4x^2-10x+40$; fourth degree polynomial; $a_4=1, a_3=-2, a_2=4, a_1=-10, a_0=40$ **13.** $x=\frac{8}{5}$ **14.** $x=-\frac{1}{2}$

15. $x=\frac{3}{8}$ **16.** $x=-3$ **17.** $x=\frac{y-b}{m}$ **18.** $y=\frac{10-3x}{5}$ **19.** $[-6,\infty)$ **20.** $(-7,4)$ **21.** $y=\frac{2}{3}x+\frac{8}{3}$

22. $5x+8y=-17$

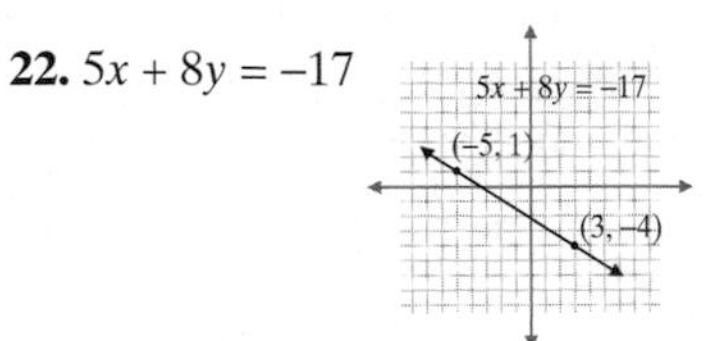

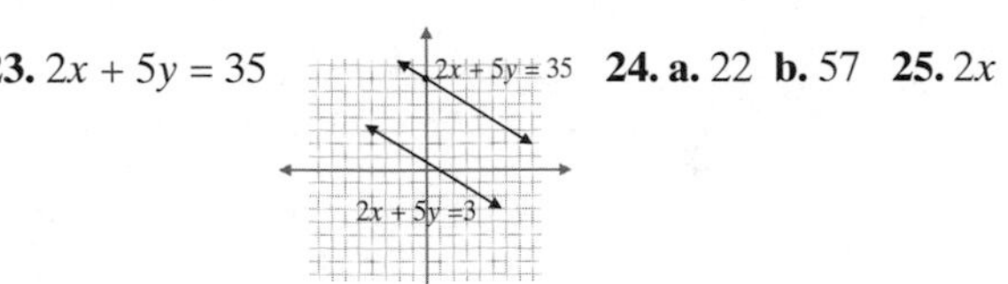

23. $2x+5y=35$ **24. a.** 22 **b.** 57 **25.** $2x$

26. $64x^6y^3$ **27.** $\frac{49x^{10}}{y^4}$ **28.** $\frac{36x^4}{y^{10}}$ **29.** $\frac{a^6}{b^4}$ **30.** $\frac{x}{3y^4}$ **31.** 1 **32.** $\frac{x^8}{9y^6}$ **33. a.** 0.00000028 **b.** 35,100

34. $1.5\times10^{-3}\cdot4.2\times10^{3};6.3\times10^{0}$ **35.** $\frac{8.4\times10^{2}}{2.1\times10^{-4}};4\times10^{6}$ **36.** $\frac{5\times10^{-3}\times7.7\times10}{1.1\times10^{-2}\times3.5\times10^{3}};1\times10^{-2}$ **37. a.** x^2+4x **b.** second degree binomial **c.** 21 **d.** 5 **38. a.** $3x^2+8x$ **b.** second degree binomial **c.** 51 **d.** 35 **39. a.** $-x^4+3x^3+x^2-2x$ **b.** fourth degree polynomial **c.** 3 **d.** −965 **40. a.** $-x^3-6x^2+2x-4$ **b.** third degree polynomial **c.** −79 **d.** −39 **41.** x^3-2x^2-6x-7 **42.** $-x^2+x-5$ **43.** $3x^3+4x^2-7x-3$ **44.** $-x^3+7x^2-6$ **45.** $6x^2-7x+1$ **46.** $-x-17$ **47.** $-3x^3+12x^2-3x$ **48.** $5x^5+10x^4$ **49.** x^2-36 **50.** x^2+x-12 **51.** $9x^2+42x+49$ **52.** $4x^2-1$ **53.** x^4-25 **54.** x^4-4x^2+4 **55.** $-x^2+5x-2$ **56.** $x^2-8x+16$ **57.** $2x^2-x-36$ **58.** $5x^2-27x-18$ **59.** $6x^2+x-12$ **60.** $9x^2-64$ **61.** $4x^3-3x^2-x$ **62.** y^3-125 **63.** x^3+27y^3 **64.** a^3+1 **65.** $x^6+2x^3y^3+y^6$ **66.** $9x^2-6x+1+6xy-2y+y^2$ **67.** $4x-7+\frac{3}{x}$ **68.** $\frac{13x}{5}+2y+\frac{y^2}{x}$ **69.** $\frac{1}{7}-3y+\frac{1}{xy}-4y^2$ **70.** $x-2$ **71.** $x+2+\frac{4}{4x-3}$ **72.** $2x^2-x+3-\frac{2}{x+3}$ **73.** $2x^3-x^2-5x+8-\frac{48}{x^2+2x+3}$ **74.** $2x^3-7x^2+14x-14+\frac{18}{x+2}$ **75. a.** $A(x)=900-4x^2$ **b.** $P(x)=120$ **76.** Function **77.** Function **78.** Not a function **79.** Function **80.** 20, 22, 24 **81.** 16 cm by 25 cm **82.** −19 **83.** 38, 38, 42 **84.** $35,000 at 8% and $65,000 at 6% **85.** length = 105 ft., width = 90 ft. **86.** 10 miles

Chapter 6

Exercises 6.1, pages 411 – 413

1. 5 **3.** 8 **5.** 1 **7.** $10x^3$ **9.** $13ab$ **11.** $-14cd^2$ **13.** $12xy$ **15.** x^4 **17.** $-4y$ **19.** $3x^3$ **21.** $2x^2y$ **23.** $m+9$ **25.** $x-6$ **27.** $b+1$ **29.** $3y+4x+1$ **31.** $11(x-11)$ **33.** $4y(4y^2+3)$ **35.** $-8(a+2b)$ **37.** $-3a(2x-3y)$ **39.** $2x^2y(8x^2-7)$ **41.** $-14x^2y(y^2+1)$ **43.** $5(x^2-3x-1)$ **45.** $4m^2(2x^3-3y+z)$ **47.** $51x^3y^5(x^2y-1)$ **49.** $x^4y^2(15+24x^2y^4-32x^3y)$ **51.** $(y+3)(7y^2+2)$ **53.** $(x-4)(3x+1)$ **55.** $(x-2)(4x^3-1)$ **57.** $(2y+3)(10y-7)$ **59.** $(x-2)(a-b)$ **61.** $(x^2+3)(15x-4)$ **63.** $(y^3-4)(13y^2+2y-1)$ **65.** $(a^3+2a+5)(a^2+1)$ **67.** $(b+c)(x+1)$ **69.** $(x^2+6)(x+3)$ **71.** $(x+6y)(x-4)$ **73.** $(y-4)(5x+z)$ **75.** $(2x-3y)(z-8)$ **77.** $(a+3)(x+5y)$ **79.** $(4y+3)(x-1)$ **81.** $(x-1)(y+1)$ **83.** Not factorable **85.** Not factorable

Exercises 6.2, pages 422 – 424

1. $5(x^2+3)$ **3.** $xy(x-2+y)$ **5.** $5x^2y(y+4)$ **7.** $3x^2y(1+7xy+y^2)$ **9.** Not factorable (or Prime) **11.** $(3x+5)(3x-5)$ **13.** $(x-5)^2$ **15.** $(3y+2)^2$ **17.** $(x+6)(x+3)$ **19.** $(x-9)(x+3)$ **21.** $-1(y-7)(y+2)$ **23.** $(x+7)^2$ **25.** $2(2x+3)(3x+2)$

27. $-1(2x-5)(x+7)$ **29.** $(4x-5)^2$ **31.** $(5x-3)(7x+6)$ **33.** $4x(x-4)(x+4)$ **35.** $x^2(7x+8)(3x-4)$

37. $2x^3y^2(3x^2-14xy-3y^2)$ **39.** $2x(3x^2+2)(2x^2+5)$ **41.** $x(3x^2-4)(3x^2+4)$ **43.** $(2x^2-5y)(x^2+3y)$

45. Not factorable **47.** $(2x^3+y^2)(x^3+4y^2)$ **49.** $(x-5)(x+7)(x-3)$ **51.** $(2x+1)(2x+3)(x-6)$ **53.** $(x+y+3)^2$

55. $(x+2y-5)(x+2y+5)$ **57.** $(x+5y+6)(x+5y+2)$ **59.** $(3x-y-1)(3x-y+4)$ **61.** $l=2x+30$

63. a. $x(36-2x)(12-2x)$ **b.** $V(2)=512$ in.^{3}, $V(4)=448$ in.^{3} **65.** 20, –20 **67.** 64 **69.** Answers will vary. **71.** Answers will vary.

Exercises 6.3, pages 429 – 432

1. $(x-5)(x^2+5x+25)$ **3.** $(x-2y)(x^2+2xy+4y^2)$ **5.** $(x+6)(x^2-6x+36)$ **7.** $(x+y)(x^2-xy+y^2)$

9. $(x-1)(x^2+x+1)$ **11.** $(3x+2)(9x^2-6x+4)$ **13.** $3(x+3)(x^2-3x+9)$ **15.** $(5x-4y)(25x^2+20xy+16y^2)$

17. $2(3x-y)(9x^2+3xy+y^2)$ **19.** $y(x+y)(x^2-xy+y^2)$ **21.** $x^2y^2(1-y)(1+y+y^2)$ **23.** $3xy(2x+3y)(4x^2-6xy+9y^2)$

25. $(x^2-y^3)(x^4+x^2y^3+y^6)$ **27.** $[3x+(y^2-1)][9x^2-3xy^2+3x+y^4-2y^2+1]$ **29.** $(x+y-1)(x^2+7x-xy+y^2-8y+19)$

31. $(x+y-4)(x^2+5xy+7y^2+4x+16y+16)$ **33.** $(x-3)(y+4)$ **35.** $(x+4)(y+6)$ **37.** Not factorable **39.** $(4x+3)(y-7)$

41. Not factorable **43.** $(x-y-6)(x-y+6)$ **45.** $(4x+1-y)(4x+1+y)$ **47.** $(x^2+6)(x-5)$ **49.** $(x+12)(x+2)(x-2)$

51. $(x+2)(x+2)(x-2)$ **53.** $(y+5)(x+3)(x-3)$ **55.** $(x^k+2)(x^{2k}-2x^k+4)$ **57.** $3x^{3k}(1-2y^k)(1+2y^k+4y^{2k})$

59. $4x^{-2}(x+2)$ **61.** $x^{-3}(2x+3y-6)$ **63.** $5x^{-4}(x+2)(x-2)$ **65.** $x^{-3}(x+3)(x+1)$ **67.** $2x^{-2}(x-5)(x+1)$

69. a. x^2-16 **b.**  $x-4$ / $x+4$ **71.** factor: $x-5$; difference: x^3-125 **73.** factor: $x+6$; sum: x^3+216

75. Sum; factor: $2x+1$; sum: $8x^3+1$ **77. a.** $xy+xy+x^2+y^2=x^2+2xy+y^2=(x+y)^2$ **b.** $(x+y)(x+y)=(x+y)^2$

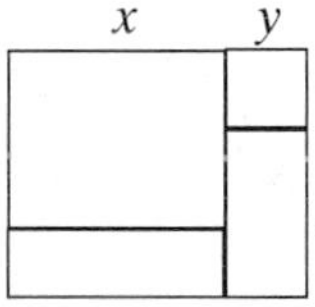

79. I. $abc=100a+10b+c$
$=(99+1)a+(9+1)b+c$
$=9(11a+b)+a+b+c$

So, if the sum $(a+b+c)$ is divisible by 3 (or 9), then the number abc will be divisible by 3 (or 9).

II. $abcd=1000a+100b+10c+d$
$=(999+1)a+(99+1)b+(9+1)c+d$
$=9(111a+11b+c)+a+b+c+d$

So, if the sum $(a+b+c+d)$ is divisible by 3 (or 9), then the number $abcd$ will be divisible by 3 (or 9).

Exercises 6.4, pages 438 – 440

1. $x=2$ or $x=3$ **3.** $x=-2$ or $x=\frac{9}{2}$ **5.** $x=-3$ **7.** $x=-5$ **9.** $x=0$ or $x=2$ **11.** $y^2-y-6=0$ **13.** $8x^2-10x+3=0$

15. $p^3-p^2-6p=0$ **17.** $x^2+8x+15=0$ **19.** $y^3-4y^2-3y+18=0$ **21.** $x=-9$ or $x=-4$ **23.** $x=6$ or $x=8$

25. $x=-5$ or $x=10$ **27.** $x=-7$ or $x=7$ **29.** $x=\frac{2}{3}$ or $x=5$ **31.** $x=\frac{4}{3}$ or $x=\frac{1}{2}$ **33.** $x=-6$ or $x=6$ **35.** $z=-\frac{7}{2}$ or $z=\frac{7}{2}$

37. $z=\frac{1}{2}$ or $z=3$ **39.** $x=-\frac{3}{4}$ or $x=\frac{2}{3}$ **41.** $x=-\frac{5}{2}$ or $x=\frac{7}{3}$ **43.** $y=-\frac{4}{5}$ or $y=\frac{4}{5}$ **45.** $x=-6$ or $x=8$ **47.** $x=2$ or $x=6$

49. $x=-11$ or $x=3$ **51.** $y=\frac{1}{6}$ or $y=\frac{2}{3}$ **53.** $x=-\frac{2}{9}$ or $x=\frac{6}{7}$ **55.** $x=-3$ or $x=-2$ or $x=0$ **57.** $x=-\frac{5}{2}$ or $x=0$ or $x=\frac{5}{2}$

59. $x=-\frac{5}{2}$ or $x=0$ or $x=\frac{7}{3}$ **61.** $x=-\frac{3}{2}$ or $x=0$ or $x=\frac{7}{5}$ **63.** $x=3$ **65.** $x=-5$ or $x=10$ **67.** $x=-2$ or $x=-6$ **69.** $x=\frac{1}{2}$

71. $x=0$ or $x=2$ or $x=4$ **73.** $x=0$ or $x=-\frac{1}{2}$ or $x=-\frac{2}{3}$ **75.** $x=\pm10$ **77.** $x=\pm5$ **79.** $x=-4$ **81.** $x=3$ **83.** $x=0$ or $x=\frac{1}{2}$ or $x=-\frac{1}{3}$

85. $x=-1$ or $x=2$ or $x=6$ **87.** $x=-\frac{1}{4}$ or $x=\frac{2}{3}$ or $x=8.5$ or $x=\frac{1}{6}$

Exercises 6.5, pages 446 – 451

1. $x^2=7x$; $x=0,7$ **3.** $x^2=x+12$; $x=4$ **5.** $x(x+7)=78$; $x=-13,6$; so the numbers are –13 and –6 or 13 and 6.

7. $x^2+3x=54$; $x=6$; so the number is 6. **9.** $x+(x+8)^2=124$; $x=3$; so the numbers are 3 and 11.

11. $x^2+(x-5)^2=97$; $x=-4,9$; so the numbers are –9 and –4 or 9 and 4. **13.** $x(x+1)=72$; $x=8$; so the numbers are 8 and 9. **15.** $x^2+(x+1)^2=85$; $x=6$; so the numbers are 6 and 7. **17.** $(x+1)^2-x^2=17$; $x=8$; so the numbers are 8 and 9. **19.** $x(x+2)=120$; $x=-12,10$; so the numbers are –12 and –10 or 12 and 10. **21.** $w(2w)=72$; width = 6 in. and length = 12 in. **23.** $w(w+12)=85$; $w=5$, width = 5 m and length = 17 m. **25.** $l(l-4)=117$; $l=13$, width = 9 ft. and length = 13 ft. **27.** $\frac{1}{2}(h+5)h=42$; $h=7$, height is 7 m and base is 12 m. **29.** $\frac{1}{2}(h-6)h=56$; $h=14$, height is 14 ft.

31. $2(l+w)=20$; $lw=24$; width = 4 cm, length = 6 cm. **33.** $x(x+13)=140$; $x=7$; so there are 7 trees in each row.

35. $x(x+7)=144$; $x=9$; so there are 9 rows. **37.** $(l+6)(l+1)=300$; $l=14$ m; so the rectangle is 14 m by 9 m.

39. $w(52-2w)=320$; $w=10$ yd., 16 yd.; so the corral is 10 yd. by 32 yd. or 16 yd. by 20 yd. **41.** $w(3w)=48$; $w=4$; width = 4 ft. and length = 12 ft. **43.** $x(x+8)=-16$; $x=-4$; so the numbers are –4 and 4. **45.** $x^2+(x+1)^2=113$; $x=7$; so the numbers are 7 and 8. **47.** $x(x+2)=420+3x$; $x=21$; so the numbers are 21 and 23.
49. $x^2+(x+2)^2+(x+4)^2=440$; $x=-14,10$; so the numbers are 10, 12, and 14 or –14, –12, and –10.

51. $6x(x+2)=(x+1)+(x+3)^2$; $x=-2,1$; so the numbers are 1, 2, 3, and 4 or –2, –1, 0, and 1. **53.** $x^2+(2x+2)^2=13^2$; $x=5$; length = 12 in. and width = 5 in. **55.** $x^2+(x+1)^2=29^2$; $x=20$; so the two mountain peaks are 20 miles and 21 miles away. **57. a.** 4 sec. **b.** 2 sec. **c.** 2 sec. **59.** 128 in.^2; use the Pythagorean Theorem

Exercises 6.6, pages 452 – 455

1. 2260.8 in.^3 **3.** 5 in. **5.** 800 lb. **7.** 6 in. **9.** 10 in. **11.** 200 ft. **13.** 4 seconds and 6 seconds **15.** 12 amps or 20 amps
17. \$16 or \$20 **19.** \$4

Exercises 6.7, pages 461 – 462

1. $x=\pm2$ **3.** $x=\pm\sqrt{2}=\pm1.41$ **5.** $x=-2,6$ **7.** $x=2.46,-4.46$ **9.** $x=4.7,-1.70$ **11.** $\varnothing$ **13.** $x=1.91,-1.57$
15. $x=-2.49,0.66,1.83$ **17.** $x=0,1,3$ **19.** $x=-3,-1,5$ **21.** $x=3.40$ **23.** $x=-0.91,1.71,3.20$ **25.** $x=-1.73,0,1.73$
27. $x=-4,7$ **29.** $x=-1.67,3$ **31.** $x=-2,0$ **33.** $x=-5,1$ **35.** $x=3.5$ **37.** $x=[-3,3]$ **39.** $x=(-\infty,3)$ or $(7,\infty)$
41. $x=(-\infty,1.33)$ or $(4,\infty)$ **43.** $x=(-\infty,-8)$ or $(16,\infty)$ **45.** $x=(-\infty,-16]$ or $[-8,\infty)$

Exercises 6.8, pages 464 – 465

1. $(m+6)(m+1)$ **3.** $(x+9)(x+2)$ **5.** $(x-10)(x+10)$ **7.** $(m-3)(m+2)$ **9.** Not Factorable **11.** $(8a-1)(8a+1)$
13. $(x+5)(x+5)$ **15.** $(x+12)(x-3)$ **17.** $(x+4)(x+9)$ **19.** $-5(x-6)(x-8)$ **21.** $-4(x-10)(x+5)$ **23.** $3(x-7)(x+7)$

25. $3n(n+3)(n+2)$ **27.** $4x(2x-5)(2x+5)$ **29.** $-(x-5)(3x-2)$ **31.** $2(2x-1)(x-3)$ **33.** $(2x-5)(6x-1)$
35. $(3t-7)(2t+5)$ **37.** $(4x-7)(2x+5)$ **39.** $(5x+6)(4x-9)$ **41.** $3\left(4n^2-20n-25\right)$ **43.** $-1(x-3)(3x+8)$
45. $-(2a-3)(4a-5)$ **47.** $(4y+5)(5y-4)$ **49.** $(6x-1)(3x-2)$ **51.** $-7x(5x-6)(5x+6)$ **53.** $a\left(21a^2-13a-2\right)$
55. $3x(3x-2)(4x+5)$ **57.** $2x(2x-1)(4x-11)$ **59.** $5\left(24m^2+2m+15\right)$ **61.** $(y-4)(x+3)$ **63.** $(x+2y)(x-6)$
65. $-\left(x^2-5\right)\left(x-8\right)$ **67.** $x^4(y^3-1)$ **69.** $\left(2a^2+3b^2\right)\left(4a^4-6a^2b^2+9b^4\right)$ **71.** $6a^{-1}(a+1)(a+1)$
73. $\left(5x-y^2\right)\left(25x^2+5xy^2+y^4\right)$ **75.** $18(x^2+3)$ **77.** $(x+5)(x^2-2)(x^2+2)$ **79.** $x^{-4}(x+4)(x+4)$
81. $(3x+y+6)(3x-y-6)$ **83.** $(y+10+7x)(y+10-7x)$ **85.** $4(2t^4-3s^4)(4t^8+6t^4s^4+9s^8)$ **87.** $(y+14)(y+6)$
89. $(2x-y+11)(2x-y+3)$

Chapter 6 Test, pages 470 – 471

1. 15 **2.** $8x^2y$ **3.** $2x^3$ **4.** $-7y^2$ **5.** $10x^3\left(2y+1\right)\left(y+1\right)$ **6.** $(x-5)(x-4)$ **7.** $-(x+7)(x+7)$ **8.** $6(x+1)(x-1)$
9. $2(6x-5)(x+1)$ **10.** $(x+3)(3x-8)$ **11.** $(4x-5y)(4x+5y)$ **12.** $x(x+1)(2x-3)$ **13.** $(2x-3)(3x-2)$ **14.** $(y+7)(2x-3)$
15. Not factorable **16.** $-3x\left(x^2-2x+2\right)$ **17.** $\left(x^k+5\right)\left(x^{2k}-5x^k+25\right)$ **18.** $3x^{-5}\left(x^2+5\right)$ **19.** $(y-2+5x)(y-2-5x)$
20. $\left(y-4x^2\right)\left(y^2+4x^2y+16x^4\right)$ **21.** $x=-2, \frac{5}{3}$ **22.** $x=8,-1$ **23.** $x=0,-6$ **24.** $x=5,-3$ **25.** $x=5, -\frac{3}{4}$ **26.** $x=-\frac{5}{4}, \frac{3}{2}$
27. $x=4, \frac{3}{2}$ **28.** $x=6,-4$; so the numbers are 6 and 20 or –4 and –30. **29.** Length = 15 cm, Width = 11 cm **30.** $n=18, 19$
31. 12, 3 **32. a.** $A(x)=240-(x^2+3x)=-(x^2+3x-240)$ **b.** $P(x)=64$ **33.** 10 meters and 24 meters
34. $x=-2.5$ and 3 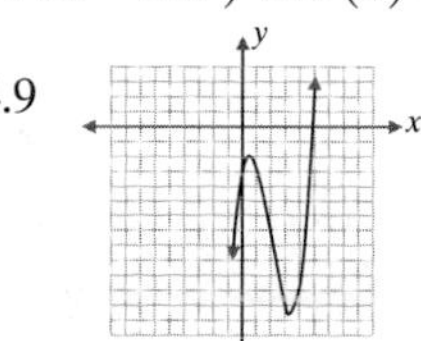 **35.** $x=4.9$

Chapter 6 Cumulative Review, pages 471 – 474

1. 120 **2.** $168x^2y$ **3.** $\frac{55}{48}$ **4.** $\frac{19}{60a}$ **5.** $\frac{3}{10}$ **6.** $\frac{75x}{23}$ **7.** $-8x^2y$ **8.** $4x^2y$ **9.** $-\frac{xy^4}{3}$ **10.** $7x^2y$ **11. a.** $D=\{2,3,5,7.1\}$
b. $R=\{-3,-2,0,3.2\}$ **c.** It is a function because each first coordinate (domain) has only one corresponding second coordinate (range). **12. a.** 58 **b.** 4 **c.** $\frac{11}{4}$ **13.** $x+y+2=0$ 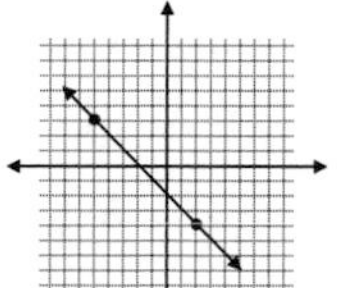 **14.** $2x-y+8=0$ 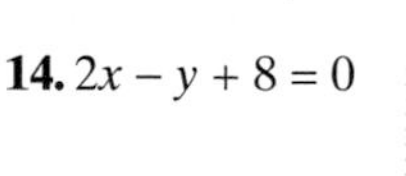 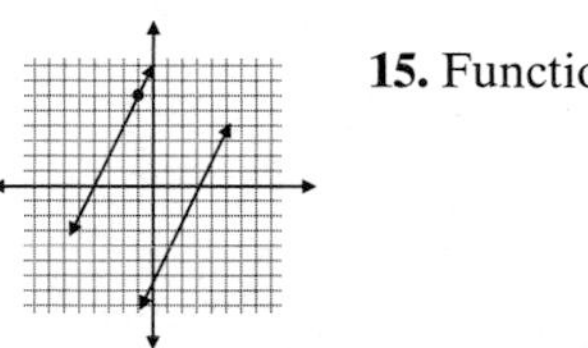 **15.** Function

16. Not a function **17.** (–2.076, –.692) and (2.409, .803) **18.** $13x+1$ **19.** $x^2+12x+3$ **20.** $3x^2-5x+3$
21. $4x^2+5x-8$ **22.** x^2-3x-5 **23.** $-4x^2-2x+3$ **24.** $2x^2+x-28$ **25.** $-3x^2-17x+6$ **26.** $x^2+12x+36$
27. $4x^2-28x+49$ **28.** $3x+1+\frac{22}{x-2}$ **29.** $x^2-8x+38-\frac{142}{x+4}$ **30.** $4x+4-\frac{11x+18}{x^2+2}$ **31.** $6x^4+9x^3+16x^2+24x+\frac{63}{2}+\frac{197}{2(2x-3)}$

32. $4(2x-5)$ **33.** $6(x-16)$ **34.** $(x-6)(2x-3)$ **35.** $(2x-3)(3x+4)$ **36.** $5(x-2)$ **37.** $-4x(3x+4)$ **38.** $8xy(2x-3)$ **39.** $5x^2(2x^2-5x+1)$ **40.** $(2x+1)(2x-1)$ **41.** $(y-10)(y-10)$ **42.** $3(x+4y)(x-4y)$ **43.** $(x-9)(x+2)$ **44.** $5(x+4)(x+4)$ **45.** Not Factorable **46.** $(5x+2)(5x+2)$ **47.** $(x+1)(3x+2)$ **48.** $2x(x-5)(x-5)$ **49.** $4x(x^2+25)$ **50.** $(x+2)(y+3)$ **51.** $(x-2)(a+b)$ **52.** $(2x+3y)(4x^2-6xy+9y^2)$ **53.** $(x^2-6)(x^4+6x^2+36)$ **54.** $3(x^k-2)(x^{2k}+2x^k+4)$ **55.** $x^{-4}(x^2+5x+1)$ **56.** $(x^k-5y)(x^k+5y)$ **57.** $(x+y^2+3)(x-y^2+3)$ **58.** $6x^{-4}(x^2-6)$ **59.** $(x^4-5)(x^8+5x^4+25)$ **60.** $x^2-3x-10$ **61.** $6y^2+y-1$ **62.** $4p^2-29p+7$ **63.** $x=-1,7$ **64.** $x=0,-\frac{5}{3}$ **65.** $x=-3,\frac{3}{4}$ **66.** $x=0,7$ **67.** $x=-2,-6$ **68.** $x=-4,7$ **69.** $x=0,1,-6$ **70.** $x=-10,6$ **71.** $x=-5,1$ **72.** $x=0,-7$ **73.** $x=0,-2$ **74.** $x=0,4$ **75.** $x=5,-\frac{5}{2}$ **76.** $x=1,-\frac{5}{4}$ **77.** $x=0,-5,2$ **78.** boat = 5 mph, current = 2 mph

79. \$7500 at 6%, \$2500 at 8% **80.** 13 and 8 **81.** 8 and 9 or −8 and −9 **82.** $t=2.5, 3$ seconds **83.** $p=\$8$ **84.** 11 cm **85.** 7, 24, 25 **86.** 10, 12, 14

87. No real solution **88.** $x=0.33, 3$ **89.** $x=-6, 3$ **90.** $x=-2.09$

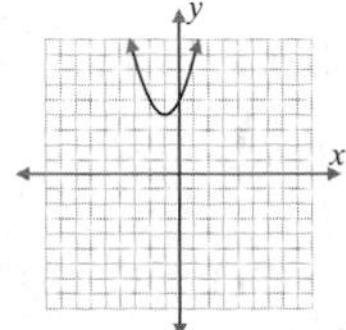

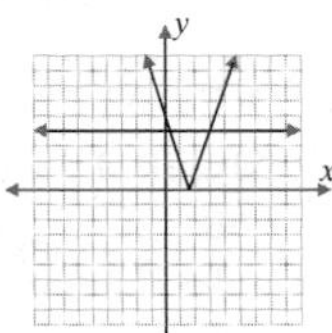

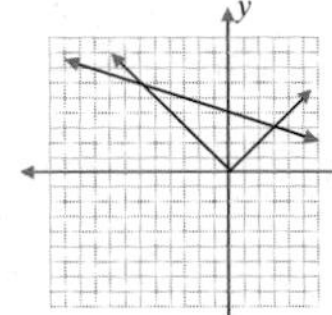

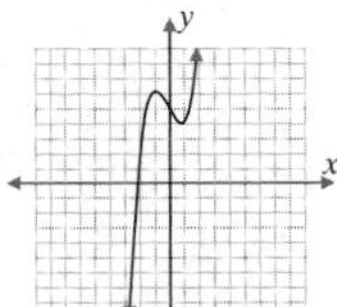

Chapter 7

Exercises 7.1, pages 484 – 486

1. $-12x^4y$ **3.** $4x^2+8x$ **5.** x^2+x-2 **7.** $-8(x+2)$ **9.** $7(x^2-4x+16)$ **11.** $\frac{3x}{4y}; x\neq 0; y\neq 0$ **13.** $\frac{2x^3}{3y^3}; x\neq 0; y\neq 0$

15. $\frac{1}{(x-3)}; x\neq 0,3$ **17.** $7; x\neq 2$ **19.** $-\frac{3}{4}; x\neq 3$ **21.** $-\frac{3}{4}; x\neq 3$ **23.** $\frac{x}{(x-1)}; x\neq -6,1$ **25.** $\frac{x-y}{3x}; x\neq 0,-y$

27. $\frac{x^2-4x+16}{(2x-7)}; x\neq -4,\frac{7}{2}$ **29.** $\frac{x^2+5}{x^2+2x+4}; x\neq 2$ **31.** $\frac{ab}{y}$ **33.** $\frac{8x^2y^3}{15}$ **35.** $\frac{x+3}{x}$ **37.** $\frac{x-1}{(x+1)}$ **39.** $-\frac{1}{x-8}$ **41.** $\frac{x-2}{x}$

43. $\frac{4x+20}{x(x+1)}$ **45.** $\frac{x}{(x+3)(x-1)}$ **47.** $-\frac{x+4}{x(x+1)}$ **49.** $\frac{x+2y}{(x-3y)(x-2y)}$ **51.** $\frac{x-1}{x(2x-1)}$ **53.** $\frac{1}{x+1}$ **55.** $\frac{2x-1}{2x+1}$

57. $\frac{49y^3}{6x^5}$ **59.** $\frac{1}{12x}$ **61.** $\frac{7x}{x+2}$ **63.** $\frac{x}{6}$ **65.** $\frac{x+3}{x+4}$ **67.** $\frac{x+5}{x}$ **69.** $-\frac{x+4}{x(2x-1)}$ **71.** $\frac{x-2}{x+2}$ **73.** $\frac{3x^3+8x^2-9x-18}{(6x^2+x-9)(x-2)}$

75. $\frac{8x^2-14x+5}{(x+4)(x+1)}$ **77.** $\frac{9x^3-6x^2+4x}{(9x^2-12x+4)(9x^2+6x+4)}$ **79.** $-\frac{3x-2}{x^2}$ **81.** $\frac{x+3}{2x-1}$ **83.** $\frac{x-1}{x-8}$ **85. a.** Answers will vary.

b. $\frac{(x-1)}{(x+2)(x-3)}$ **c.** $\frac{1}{(x+5)}$ **87. a.** $x=4$ **b.** $x=10, x=-10$

Exercises 7.2, pages 495 – 496

1. 3 **3.** 2 **5.** 1 **7.** $\frac{2}{x-1}$ **9.** $\frac{14}{7-x}$ **11.** 4 **13.** $\frac{x^2-x+1}{x^2+x-12}$ **15.** $\frac{x-2}{x+2}$ **17.** $\frac{4x+5}{2(7x-2)}$ **19.** $\frac{6x+15}{(x+3)(x-3)}$

21. $\frac{x^2-2x+4}{(x+2)(x-1)}$ **23.** $\frac{x^2+3x+6}{(x+3)(x-3)}$ **25.** $\frac{8x^2+13x-21}{6(x+3)(x-3)}$ **27.** $\frac{3x^2-20x}{(x+6)(x-6)}$ **29.** $\frac{-4x}{x-7}$ **31.** $\frac{4x^2-x-12}{(x+7)(x-4)(x-1)}$

33. $\frac{4}{(x+2)(x+2)(x-2)}$ **35.** $\frac{-3x^2+17x+15}{(x-3)(x-4)(x+3)}$ **37.** $\frac{4x-19}{(7x+4)(x-1)(x+2)}$ **39.** $\frac{-7x-9}{(4x+3)(x-2)}$ **41.** $\frac{4x^2-41x+3}{(x+4)(x-4)}$

43. $\frac{-2x^2-4x+15}{(x-4)(x-2)}$ **45.** $\frac{x^2-4x-6}{(x+2)(x-2)(x-1)}$ **47.** $\frac{6x+2}{(x-1)(x+3)}$ **49.** $\frac{5x+12}{(x+3)(x+1)}$

51. $\frac{9x^3-19x^2+22x+12}{(3x+1)(x+3)(5x+2)(x-1)}$ **53.** $\frac{5x+1}{(y-3)(x+1)}$ **55.** $\frac{5xy-8y+2}{(y-4)(y+1)(x-2)}$ **57.** $\frac{6x^2-12x+5}{2(2x+1)(4x^2-2x+1)}$

59. $\frac{14x-43}{(x^2+6)(x-5)(x-2)}$ **61.** Answers will vary.

Exercises 7.3, pages 500 – 502

1. $\frac{4}{5xy}$ **3.** $\frac{8}{7x^2y}$ **5.** $\frac{2x(x+3)}{2x-1}$ **7.** $\frac{7}{2(x+2)}$ **9.** $\frac{x}{x-1}$ **11.** $\frac{4x}{3(x+6)}$ **13.** $\frac{7x}{x+2}$ **15.** $\frac{x}{6}$ **17.** $\frac{24y+9x}{18y-20x}$ **19.** $\frac{xy}{x+y}$

21. $\frac{y+x}{y-x}$ **23.** $\frac{2(x+1)}{x+2}$ **25.** $\frac{x^2-3x}{x-4}$ **27.** $\frac{x+1}{2x-3}$ **29.** $-\frac{2x+h}{x^2(x+h)^2}$ **31.** $-(x-2y)(x-y)$ **33.** $\frac{2-x}{x+2}$ **35.** $\frac{-x^2y^2}{x+y}$

37. $\frac{-5}{x+1}$ **39.** $\frac{29}{4(4x+5)}$ **41.** $\frac{x^2-3x-6}{x(x-1)}$ **43.** $\frac{x^2-4x-2}{(x-4)(x+4)}$ **45.** $\frac{R_1R_2}{R_2+R_1}$ **47. a.** $\frac{8}{5}$ **b.** 1 **c.** $\frac{x^4+x^3+3x^2+2x+1}{x^3+x^2+2x+1}$

Exercises 7.4, pages 510 – 512

1. $x=7$ **3.** $x=-\frac{74}{9}$ **5.** $x=\frac{1}{4}$ **7.** $x=\frac{3}{2}$ **9.** $x=4$ **11.** $x=\frac{10}{3}$ **13.** $x=\frac{1}{4}$ **15.** $x=-\frac{3}{16}$ **17.** $x=-3$ **19.** $x=-39$

21. $x=\frac{62}{7}$ **23.** $x=-3$ **25.** $x=2$ **27.** $x=\frac{2}{3}$ **29.** $x=-2$ **31.** No solution **33.** $x=\frac{7}{11}$

35. $x \le -4$ or $x > 0$ 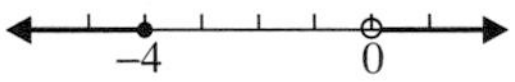 **37.** $x < -6$

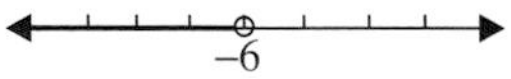

39. $x < -9$ or $x > -3$ 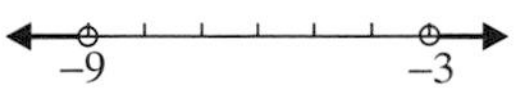 **41.** $2 < x < \frac{5}{2}$ **43.** $x > 7$

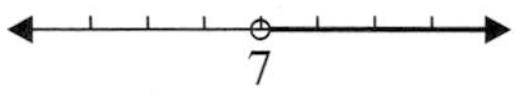

45. $\frac{7}{5} \le x < 4$ **47.** $-9 < x < -\frac{4}{3}$

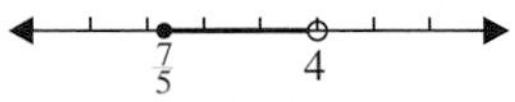

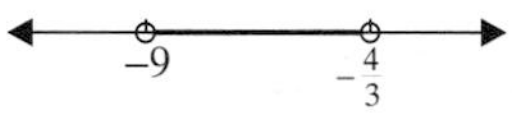

49. $x \le -4$ or $0 \le x < 3$ **51.** $r=\frac{S-a}{S}$ **53.** $s=\frac{x-\bar{x}}{z}$ **55.** $R_{\text{total}}=\frac{R_1R_2}{R_1+R_2}$ **57.** $P=\frac{A}{1+r}$

59. $x=\frac{b-yd}{yc-a}$ **61. a.** $\frac{4x^2+41x-10}{x(x-1)}$ **b.** $x=\frac{1}{4}$, 10 and $x \ne 0$, 1

Exercises 7.5, pages 520 – 523

1. 10 **3.** 14 **5.** $\frac{12}{7}$ **7.** 15, 9 **9.** 2, 5 **11.** \$1500 **13.** 1875 miles **15.** 52 mph, 48 mph **17.** 45 minutes **19.** 120 mph, 300 mph **21.** 250 mph, 500 mph **23.** 50 mph **25.** John: 11 hours, Ralph: 22 hours, Denny: 33 hours **27.** 1 min. and 20 sec. **29.** 1 mph **31.** $AC = 2, ST = 12$ **33.** $ST = 8, TU = 12, QR = 24$ **35. a.** 5 and 7 **b.** 2 and 4

Exercises 7.6, pages 529 – 534

1. $y = kx, \frac{7}{3}$ **3.** $y = \frac{k}{x}, 2$ **5.** $y = \frac{k}{x}, 10$ **7.** $y = k\sqrt{x}, 36$ **9.** $y = kx^3, 24$ **11.** $z = kxy, -\frac{27}{5}$ **13.** $z = \frac{kx}{y^2}, 40$

15. $z = \frac{k\sqrt{x}}{y}, 54$ **17.** $z = \frac{kx^2}{\sqrt{y}}, 32$ **19.** $L = \frac{kmn}{P}, 27$ **21.** $d = kt^2$, 256 feet **23.** $s = kw$, 6 in. **25.** $A = kr^2$, 254.34 feet

27. $C = kd$, 4.71 feet **29.** $E = \frac{kml}{A}$, 0.0073 cm **31.** $L = \frac{kwd^2}{l}$, 1890 lbs. **33.** $F = \frac{km_1m_2}{d^2}, 9\times10^{-11}\text{N}$ **35.** 15,000 lbs.

37. 25,200 lbs. **39.** 1.8 cu. ft. **41.** 15 ohms **43.** 5.2 ohms **45.** 35 teeth **47.** 900 lbs. **49.** $10\frac{2}{3}$ feet from the 120 lb. weight

Chapter 7 Test, pages 539 – 540

1. $\frac{x}{x+4}$ **2.** $-\frac{x^2+4x+16}{x+4}$ **3. a.** $5-x$ **b.** x^2+x-2 **4.** $\frac{x+3}{x+4}$ **5.** $\frac{3x-2}{3x+2}$ **6.** $\frac{-2x^2-13x}{(x+5)(x+2)(x-2)}$

7. $\frac{-2x^2-7x+18}{(3x+2)(x-4)(x+1)}$ **8.** $2x^2$ **9.** $\frac{x^2-7x+1}{(x+3)(x-3)}$ **10. a.** $\frac{3x}{2-x}$ **b.** $\frac{x-2}{x}$ **11. a.** $\frac{7x+11}{2x(x+1)}$ **b.** $x = \frac{1}{3}$ **12.** $x = -\frac{7}{2}$

13. $x = -1$ **14.** $\left(-\infty, -\frac{5}{2}\right] \cup (3, \infty)$ [number line: $-\frac{5}{2}$, 3] **15.** $\left(-\infty, -\frac{5}{3}\right) \cup \left(-\frac{1}{2}, \infty\right)$ [number line: $-\frac{5}{3}$, $-\frac{1}{2}$] **16.** $\frac{2}{7}$

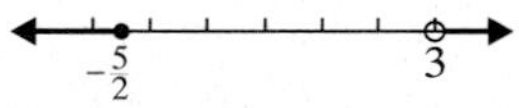

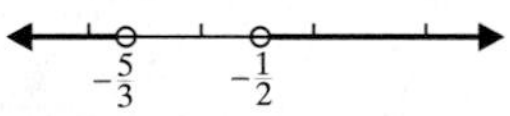

17. 4 hr. **18.** 42 mph: Carlos; 57 mph: Mario **19.** 11 mph **20.** $\frac{400}{9}$ **21.** $\frac{15}{2}$ cm **22.** 25,000 lbs. **23.** 3 ohms **24.** 6 hrs.

Chapter 7 Cumulative Review, pages 540 – 544

1. a. $\frac{19}{12}$ **b.** $-\frac{3}{14}$ **c.** $\frac{1}{4}$ **d.** $\frac{1}{12}$ **2.** 5 **3.** $3x^2-5x-28$ **4.** $4x^2-20x+25$ **5.** $6x-23$ **6.** $10x+4$ **7.** $t = \frac{A-P}{Pr}$ **8.** $r = \frac{C}{2\pi}$

9. $(2x+5)(2x+3)$ **10.** $2(x+5)(x-2)$ **11.** $(3x+2)(5x+4)$ **12.** $(x+4)(2x^2+3)$ **13.** $2(2x-3)(4x^2+6x+9)$

14. $x^{-3}(2x+1)(x-3)$ **15. a.** $D = \{-1, 0, 5, 6\}$ **b.** $R = \{0, 5, 2\}$ **c.** It is not a function because the x-coordinate -1 has more than one corresponding y-coordinate. **16. a.** 8 **b.** -1 **c.** $-\frac{59}{27}$ **17.** $y = \frac{3}{4}x + \frac{11}{2}$;

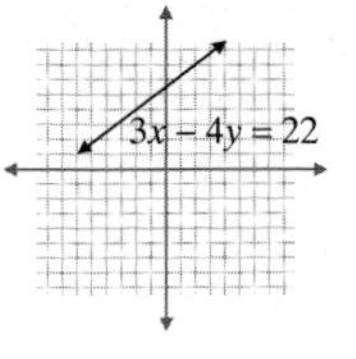

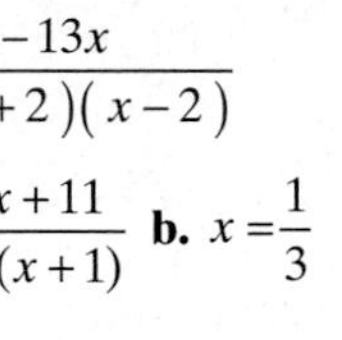

18. $y + 2x = -8$ **19. a.** $(-2, 6]$ **b.** $\left(\frac{1}{2}, 2\right)$

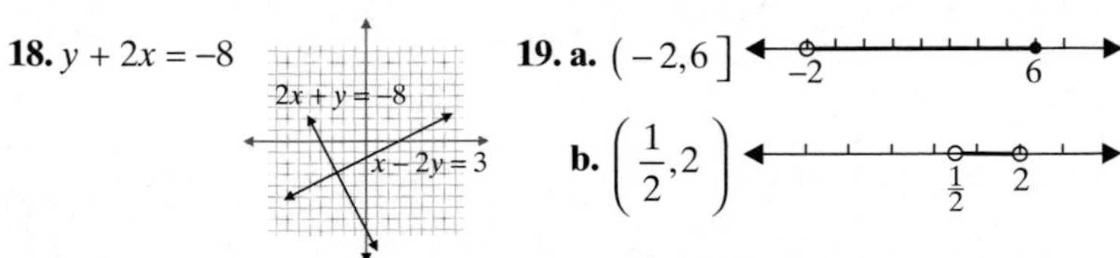

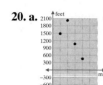

20. a. **b.** 528, –968, –616, –1144 **c.** Juan is accelerating at a rate of 528 feet per minute; decelerating 968 feet per minute; decelerating 616 feet per minute; decelerating 1144 feet per minute.

21. Function **22.** Function **23. a.** Not a function **b.** Function; $D = \{-7, -3, 0, 5, 6\}$; $R = \{-2, 1, 4\}$ **c.** Function; $D: (-\infty, \infty)$; $R: [0, \infty)$

24. (–1.68, .326) and (1.484, 1.594)

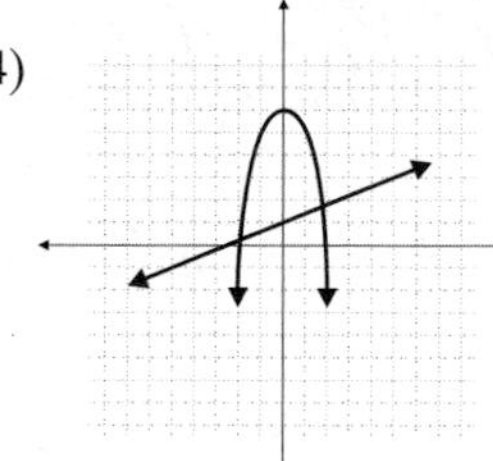

25. 6.8 **26.** 1.05×10^6 **27.** $8x^2$ **28.** $5x^3y^2$ **29.** $\frac{y^6}{9x^4}$ **30.** $\frac{x^6}{y^6}$ **31.** $\frac{3}{2x^4y}$

32. a. $x^2+9x+20=0$ **b.** $x^3-6x^2+11x-6=0$

33. $\frac{3}{5}x+2y+\frac{y^2}{x}$ **34.** $1-3y+\frac{8}{7}y^2$ **35.** $x-15-\frac{1}{x+1}$ **36.** $2x^2-x+3-\frac{2}{x+3}$ **37.** $4x^2+3x+8+\frac{15}{x-2}$

38. $2x^2+x-3+\frac{9}{x+2}$ **39.** $x-4+\frac{3(2x-5)}{x^2-6}$ **40.** $\frac{4x^2}{x-3}, x\neq 3, -\frac{1}{2}$ **41.** Does not reduce, $x\neq 2$ **42.** $x-3, x\neq -3$

43. $\frac{1}{x+1},\ x\neq 0, -1$ **44.** $-\frac{1}{3},\ x\neq 4$ **45.** $\frac{2}{3},\ x\neq -3$ **46.** $\frac{x}{x+4},\ x\neq -4, -3$ **47.** $\frac{x+5}{2(x-3)},\ x\neq 3$ **48.** $x-y$ **49.** $\frac{4(4x-7)}{x(x-4)}$

50. $\frac{4}{(y+2)(y+3)}$ **51.** $\frac{x}{3x+3}$ **52.** $\frac{x+4}{3x}$ **53.** $\frac{3x(x+2)^2}{x+3}$ **54.** $\frac{2x+1}{x-1}$ **55.** $\frac{2x^2+14x-8}{(x+3)(x-1)(x-2)}$ **56.** $\frac{2(x+2)}{(x-1)(x+4)}$

57. $\frac{x-4}{(x+2)(x-2)}$ **58.** $\frac{2x-1}{(x+1)(x-1)}$ **59.** $\frac{x^2+15x-26}{(x-4)(x+2)(2x+3)}$ **60.** $\frac{19}{14}$ **61.** $\frac{x-1}{x+1}$ **62.** $\frac{3x}{2-x}$ **63.** $-\frac{3}{5}$ **64.** $x=6,-5$

65. $x=2,5$ **66.** $x=2,-10$ **67.** $x=-45$ **68.** $x=-\frac{9}{8}$ **69.** $x=\frac{29}{11}$ **70.** $x=-\frac{4}{3}, x=4$

71. $\left(-\infty,-\frac{1}{4}\right]$ $-\frac{1}{4}$ **72.** $\left(-\infty,-\frac{50}{3}\right)$ $-\frac{50}{3}$ **73.** $(-\infty,12]$ 12

74. a. $\frac{9-2x}{x(x+3)}$ **b.** $x=\frac{9}{2}$ **75. a.** $\frac{6x^2-16x-20}{(x-4)(x+2)}$ **b.** $x=7$ **76.** The father takes $2\frac{2}{3}$ hrs. and the daughter takes 8 hrs.

77. 2.5 fc **78.** $7\frac{9}{13}$ in. **79.** 256 feet **80.** 24 feet **81.** 16, 18

Chapter 8

Exercises 8.1, pages 556 – 558

1. –5, –3.2, –0.3, $\frac{1}{4}$, $2\frac{1}{2}$ **3. a.** –7, 7 **b.** –7, 0, 7 **c.** –7, 0, 7

5. False, $|6|>-6$ **7.** False, $|-1.8|>-2$ **9.** –10 **11.** –27 **13.** 38 **15.** –1 **17.** –17.3 **19.** –2 **21.** 12.2 **23.** –160 **25.** –13

27. $\frac{11}{2a^2}$ **29. a.** $2^2\cdot3^2\cdot5$ **b.** $2\cdot3^3\cdot5$ **c.** $2^4\cdot5^4$ **31.** 720 women, 480 men **33.** $\frac{36}{25}$ **35.** $\frac{10}{11}$ **37.** $\frac{5}{42}$ **39.** $\frac{8x-18}{9x}$ **41.** $\frac{x-80}{8x}$

43. $\frac{18}{5}$ **45.** $\frac{63y}{160x}$ **47. a.** 17.85 **b.** 920.01 **c.** 0.58 **49.** 183 lbs. **51.** 316.06 **53.** 33.9 **55.** – 2.7 **57.** 85.33 **59.** 16.1

61. 1838.2656 **63.** 24.1456 **65.** 15.4847 **67.** 2.5 **69.** Commutative of Multiplication **71.** Identity of Addition **73.** Distributive Property **75.** Associative Property of Multiplication **77. a.** 0, 4 **b.** $-2.48, \frac{13}{5}, \frac{-1}{3}, 2.3, 0, 4$ **c.** $\pi, \sqrt{7}$ **d.** $-2.48, -\frac{1}{3}, 0, 2.3, \pi, 4, \sqrt{7}, \frac{13}{5}$ **79.** $\frac{37}{50}$

Exercises 8.2, pages 565 – 567

1. $12x - 15y$ **3.** $4x - 4$ **5.** $x^2 + 3x$ **7.** $2y - 20$ **9.** $6a^3 + 10ab - 8a$ **11.** 10 **13.** 99.9 **15.** $\frac{53}{8}$ **17.** $-\frac{25}{12}$ **19.** $\frac{145}{12}$ **21.** $5 + 2x$ **23.** $-6 + x$ **25.** $2(x + 5)$ **27.** $3x - 20$ **29.** $\frac{x}{5} + x$ **31.** 7 more than 8 times a number **33.** 5 times the sum of a number and 10 **35.** Sum of 6 times a number and 7 times the same number **37.** 5 less than the quotient of a number and 14 **39.** 11 more than the product of –9 and a number **41.** 0.21 (rounded to 2 decimals) **43.** 7 **45.** $-\frac{3}{2}$ **47.** –19 **49.** –4 **51.** 0 **53.** –4 **55.** 18.5 **57.** $a = -3, \frac{7}{5}$ **59.** $n = 75$ **61.** 12 **63.** 72 **65.** 150 **67.** 62.5% **69.** 48 **71.** $\frac{-15}{136}$ **73.** $5243.75 **75. a.** 7% **b.** 8%, b is better. 8% is better than 7%, so a profit of $1200 on an investment of $15,000 is better **77.** 31, 32, 33, 34 **79.** 8, 11, 15

Exercises 8.3, pages 573 – 576

1. Square, $P = 4a$ Rectangle, $P = 2(l + w)$ Circle, $P = 2\pi r$

a w l r

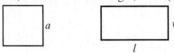

3. Rectangular solid, $V = lwh$ Right circular cylinder, $V = \pi r^2 h$ Right circular cone, $V = \frac{1}{3}\pi r^2 h$

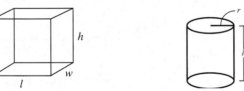

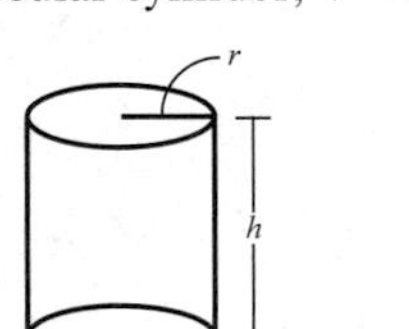

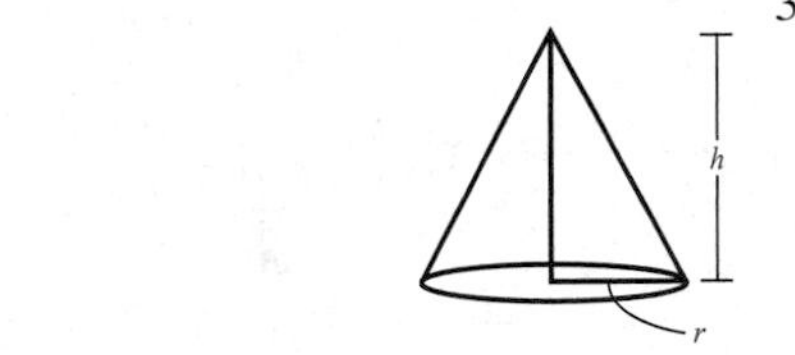

5. 70 **7.** 38.465 **9.** 5200 **11.** $r = \frac{d}{t}$ **13.** $\beta = 180 - \alpha - \gamma$ **15.** $\pi = \frac{3V}{4r^3}$ **17.** $y = 3x - 14$ **19.** $n = \frac{L - a}{d} + 1$ **21.** $a = \frac{84}{13}$ **23.** $w = 2$ **25.** $x = 25$ **27.** $x = 50$ **29.** $y = \frac{75}{14}$ **31.** $x = 9, y = 6$ **33.** $x = 17.5, y = 15$ **35.** 1962.5 **37.** 2200 ft.^{2}

39. $P = 56$ cm, $A = 192$ cm **41.** 3.75 in. **43.** $30,000 at 3.5% and $20,000 at 10% **45.** $48.9\overline{3}$ **47. a.** 64 feet per second, the object is going up since the velocity is positive. **b.** –160 feet per second, the object is falling down since the velocity is negative. **49.** $(-\infty, 2]$ **51.** $\left(\frac{1}{3}, \infty\right)$ **53.** $(-\infty, 0]$

55. $\left[-\frac{1}{3}, 3\right]$ **57.** $[-12, -7]$ **59.** $(-5, 7)$

61. $\left[-\frac{3}{5}, \frac{3}{10}\right]$ **63.** $\left(\frac{6}{5}, 3\right)$ **65.** $(0, 6]$

67. 4710 in.^{3} **69.** 7 cm **71.** 50.24 m^2 **73.** Area = 168.75 cm^2, width = 7.5 cm, length = 22.5 cm

Exercises 8.4, pages 586 – 588

1. **3.** **5.** $m=-\frac{2}{3}$ **7.** $m=0$ **9.** m = undefined or ∞ **11.** $m=-2$, y-intercept = 1

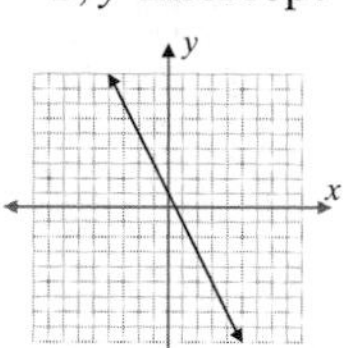

13. $m=-\frac{3}{2}$, y-intercept = 5 **15.** $m=\frac{1}{5}$, y-intercept = 3 **17.** $3x-4y=-1$ **19.** $x=-7$ **21.** Neither **23.** Parallel **25.** Perpendicular **27.** $5x-8y=-4$ **29.** $x+5y=-20$

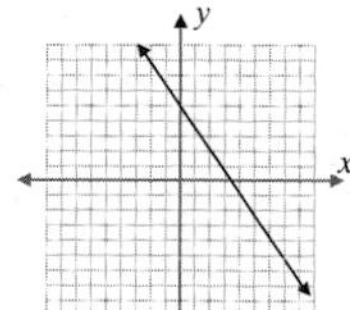

31. Function; Domain : $\begin{Bmatrix}-6,-7,-8,\\-9\end{Bmatrix}$, Range : $\{2,3\}$

33. Function; Domain : $\left\{\frac{5}{6},\frac{9}{10},\frac{11}{5},\frac{15}{2}\right\}$, Range : $\left\{\frac{1}{6},\frac{2}{3},\frac{3}{5}\right\}$

35. Function; Domain : $\begin{Bmatrix}-10,-8,-6,\\-2,0,5,7\end{Bmatrix}$, Range : $\{-2,1,2,3,5\}$ **37.** $h(-3)=29$, $h(4)=-13$ **39.** $f(-3)=33$, $f(1)=1$ **41.** Function

43. Not a function **45.** Function **47.**

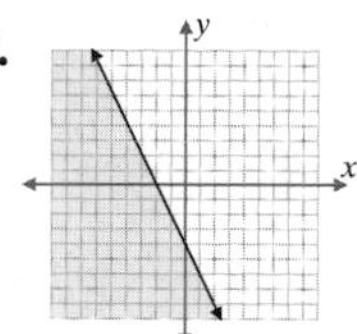

49. **51.** $x=1.3\overline{3}$

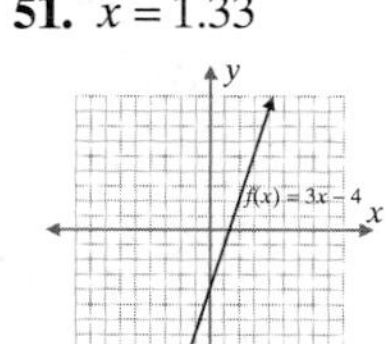

53. $x=5$

55. No values of x satisfy as this function does not cross or touch the x-axis.

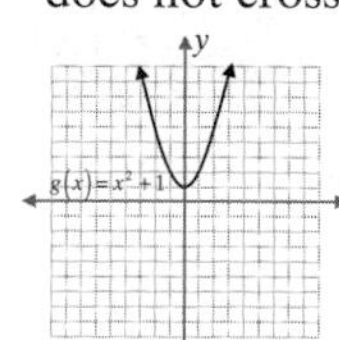

57. $x=-1.31$

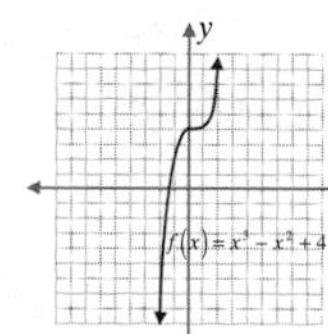

59. $x=-0.518, x=0.518,$ $x=-1.91, x=1.91$

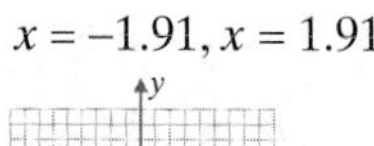

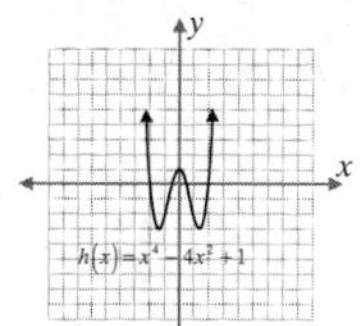

Exercises 8.5, pages 595 – 598

1. 1 **3.** $\frac{1}{4^2}=0.0625$ **5.** $-15a^6$ **7.** y^3 **9.** $\frac{x^3}{2}$ **11.** x^5 **13.** $9x^6$ **15.** $\frac{a^6}{216}$ **17.** $\frac{x^4}{25y^4}$ **19.** 1 **21.** $a^{k+5}b^{3k-2}$ **23.** $\frac{50}{a^7b^7}$

25. $20ab^{15}$ **27.** 94100000 **29.** 0.0000000003792 **31.** 5.37×10^7 **33.** 6×10^{-3} **35.** 2.24×10^{-2} **37.** 1.625×10^{-18} g

39. Fifth degree, binomial **41. a.** –10 **b.** 54 **43. a.** 50.24 **b.** 7.065 **45. a.** 100 **b.** 564 **47.** $-2x^2+6x-3$

49. $4y^3+y^2-18y+15$ **51.** $22x^3-4x^2-x-17$ **53.** $4y^3-8y-1$ **55.** $4x^3-5x^2-9x+26$ **57.** $x^4-6x^3-8x^2+2x$

59. $4x^2+3x+1$ **61.** $20x^5+28x^4+24x^3$ **63.** $x^4+x^3-6x^2+5x+15$ **65.** $2y^2+9y+10$ **67.** $49x^2-1$

69. $20x^3-29x^2-9x+18$ **71.** $25x^2+60x+36$ **73.** x^3-8y^3 **75. a.** $P(x)=10x^5-20x^4+10x^3$ **b.** $p=0.0321$ **c.** $p=0.3125$

77. $9y^2-3y+1$ **79.** $8ab+4-\frac{2b}{a}$ **81.** $4x-28+\frac{58}{x+2}$ **83.** $x^2+x-1+\frac{-5x+2}{x^2+1}$ **85.** $2x^2-3x-14+\frac{-30x+37}{x^2-3x+2}$

87. $a^2+4a+16$ **89.** y^3+2y^2+6y+9

Exercises 8.6, pages 607 – 609

1. 15 **3.** $6x^2y$ **5.** $10a^2bc$ **7.** $-2x^2(x-1)$ **9.** $6x^2y(5+8y^2+9y^3)$ **11.** $2(x^2+9x+10)$ **13.** $-(x+5)(x+5)$

15. $(y+5)(y+1)$ **17.** $(3x-2)(y+7)$ **19.** $(x^k+4)(x^{2k}-4x^k+16)$ **21.** $-(3x-2)(2x-3)$ **23.** $(6x^k+5)(6x^k-5)$

25. $(x-2+9y^2)(x-2-9y^2)$ **27.** $a^{-4}(11a+12)(11a-12)$ **29.** $y(y-6)^2$ **31.** $x=0, x=-6$ **33.** $x=\frac{3}{2}, x=-\frac{1}{4}$

35. $x=\frac{-1+\sqrt{61}}{6}, x=\frac{-1-\sqrt{61}}{6}$ **37.** $x=\frac{1}{4}, x=-\frac{3}{2}$ **39.** $x=0, x=2, x=5$ **41.** $x=0, x=4$ **43.** $x=4, x=\frac{2}{3}$

45. $x=\frac{5}{8}, x=\frac{-5}{9}, x=0$ **47.** 11, 12 **49.** 12, 18 **51. a.** 5 sec. **b.** 3.5 sec. **c.** 2 sec. **53.** 20 yards, 25 yards

55. length = 12 in., width = 10 in. **57.** $x=-2.24, x=2.24$ **59.** No values of x satisfy, as this function does not cross or touch the x-axis.

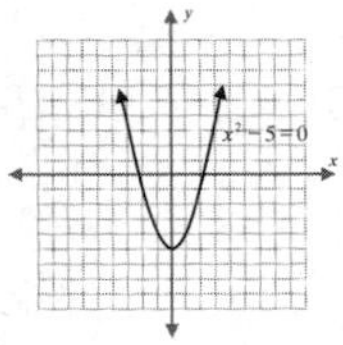

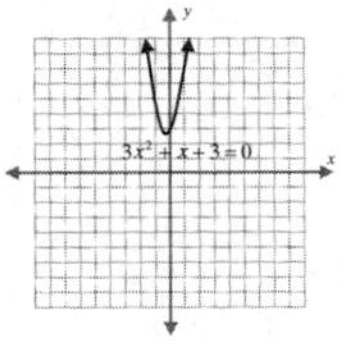

61. $x=.66, x=2.44, x=-3.1$ **63.** $x=-4, x=7$ **65.** $x=3.5$

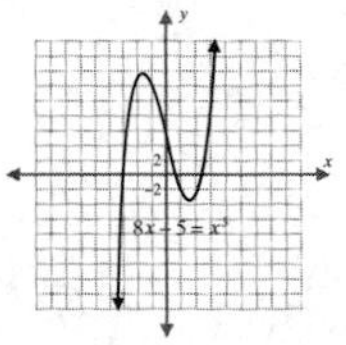

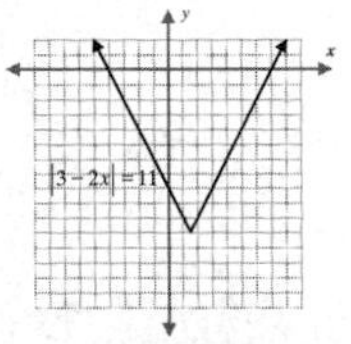

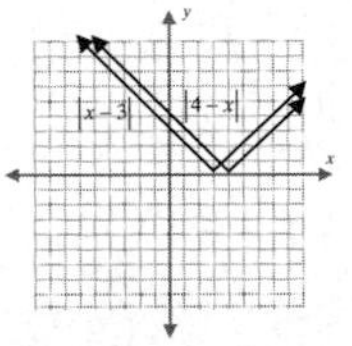

Exercises 8.7, pages 619 – 620

1. $x^4+7x^3+12x^2$ **3.** $32x^4-32x^3+10x^2-x$ **5.** $6a^4-36a^3-216a^2+1296a$ **7.** $\frac{3}{5(x-3)}, x\neq -3, 3$

9. $\frac{x^2+4x+16}{3(x+4)}, x\neq -4, 4$ **11.** $\frac{3}{2x}$ **13.** $\frac{3}{x-3}$ **15.** $\frac{15}{y+5}$ **17.** $\frac{x+5}{x^2}$ **19.** $\frac{7a+15}{a^2+5a+4}$ **21.** $\frac{2x-5}{3x+5}$ **23.** $\frac{x+1}{x-1}$

25. $\frac{xy^2-x^3-y^3+yx^2}{x^3}$ **27.** $t=3$ **29.** $a=5$ **31.** $x=-5$ **33.** $x=0, 8$ **35.** $x=0$ **37.** $\left[-\frac{1}{2},\frac{1}{2}\right)$ **39.** $(-\infty,-4]\cup(0,\infty)$

41. Woman = 4.5 hrs., Daughter = 9 hrs. **43.** 4.62 cm **45.** $\frac{3}{10}$ **47.** 2 mph **49.** Speed of private plane = 150 mph, Speed of commercial airline = 350 mph

Chapter 8 Test, pages 621 – 624

1. $x=-4$ **2.** $x=\frac{15}{2}$ **3.** $x=-16, 0$ **4.** $y=-\frac{1}{2}$ **5.** $[-1.1, 7.4]$ 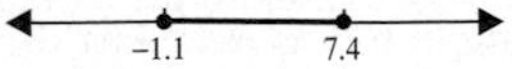 **6.** $r=\frac{A-P}{Pt}$ **7. a.** 36.56 cm **b.** 89.12 cm **8.** 18, 20, 22

9. a. **b.** Domain $= \{-2, -1, 0, 3, 4\}$

c. Range $= \left\{-2, \frac{1}{2}, 1, 2, 3\right\}$

d. Function, because it passes the vertical line test

10. a. $x + 2y = 4$ **b.**

11. a. $3x - 4y + 7 = 0$ **b.**

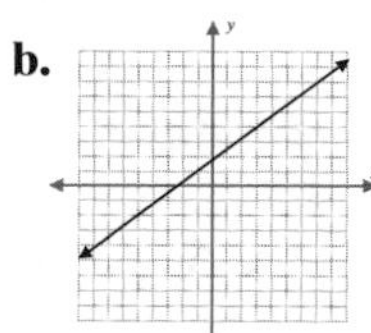

12. a. 52 **b.** 4 **13.** $x = -0.5, y = 3$ **14.** $AB = 6, EF = 15$
15. length = 14 m, width = 8 m **16.** Joe = 1.5 hours, Bruce = 3 hours
17. 15 ohms **18.** $1.844 \cdot 10^{-8}$ **19.** $-2x^2 - x - 17$ **20.** $5x^2 + 33x + 18$

21. $4x^2 + 20x + 25$ **22.** $2x^2 - x + 3 - \frac{2}{x+3}$ **23.** $(5x)(x-1)(x-2)$ **24.** $(2x-5)(2x+3)$ **25.** $x = -\frac{5}{3}, x = 2$

26. $x = -5, x = 5$ **27.** 5, 7 **28.** $\frac{5x}{2}$ **29.** 1 **30.** $\frac{5x^2 + 18x}{x^3 + 4x^2 - 9x - 36}$ **31.** $x = 2, x = -9$ **32.** $x = 0$

33. John = 60 mph, Bill = 45 mph **34.** 144 ft. **35. a.** $\frac{\sqrt{7} \cdot y^3}{4x^2}$ **b.** $\frac{1 + 3\sqrt{2}}{5}$ **c.** $12a^4$ **36.** $x = 2, x = 12$

37. a. $x^{\frac{13}{12}}$ **b.** $48x^{\frac{7}{6}}$ **38.** 6.7082 **39.** Answers will vary. **40. a. – d.** Answers will vary. **41.** 4 hours and 12 hours

42. $\frac{35}{12}$ hours, or 2 hours, 55 minutes **43.** 90 minutes **44.** factor: $x^2 + 5xy^2 + 25y^4$, difference: $x^3 - 125y^6$

45. factor: $x^k + 4$, sum: $x^{3k} + 64$ **46.** difference; factor: $9x^2 + 6xy^2 + 4y^4$, difference: $27x^3 - 8y^6$

Chapter 9

Exercises 9.1, pages 634 – 637

1. c **3.** a, c **5.** (0, 2) **7.** (4, 2) **9.** $m_1 = -2, b_1 = 3, m_2 = -2, b_2 = \frac{5}{2}$ **11.** $m_1 = 3, b_1 = -8, m_2 = 3, b_2 = -6$

13. Consistent (2, 0) **15.** Inconsistent **17.** Consistent (2, 3) **19.** Dependent **21.** Inconsistent **23.** Consistent $\left(\frac{1}{2}, 1\right)$

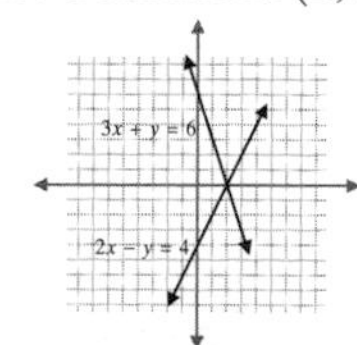

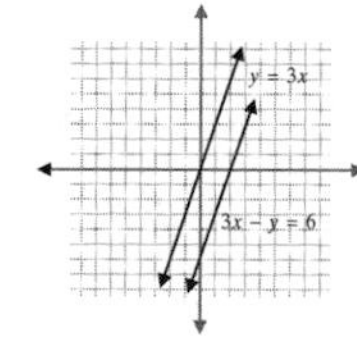

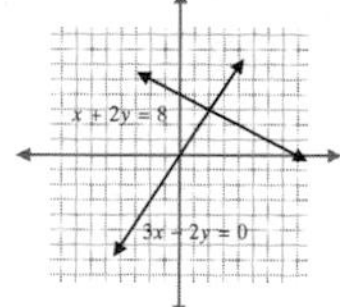

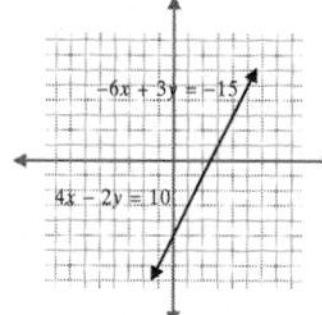

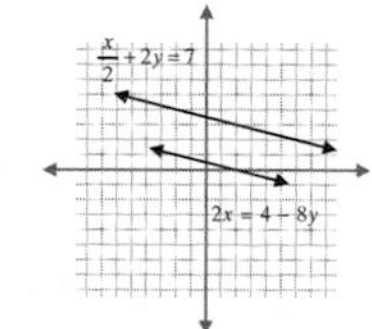

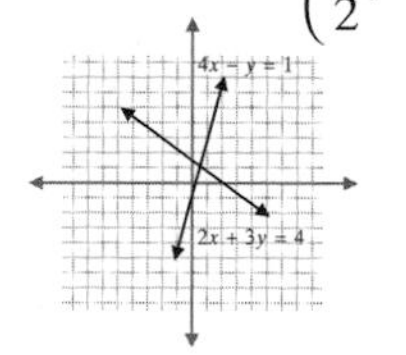

25. Inconsistent **27.** Dependent **29.** Consistent (3, 0) **31.** Consistent (1, 2) **33.** Consistent (–3,–2)

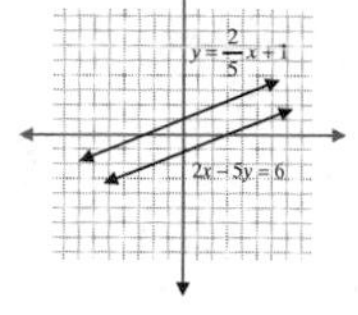

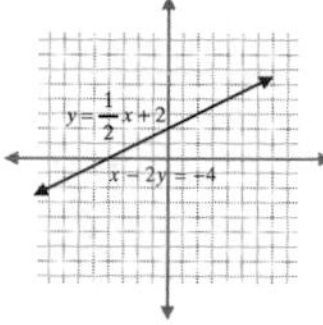

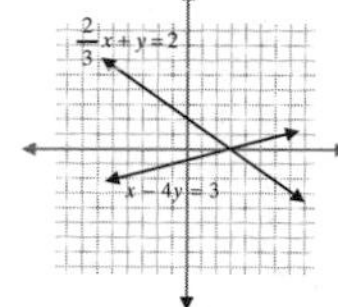

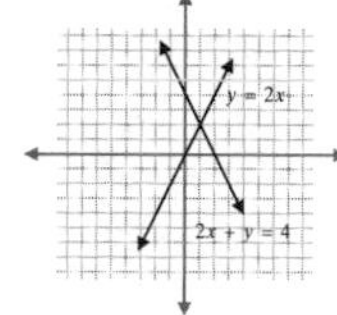

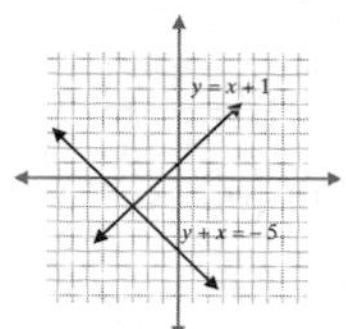

35. Consistent $(-1, 2)$ **37.** Consistent $(1, -1)$ **39.** $(5, 0)$ **41.** $\varnothing$ **43.** $(-6, -4)$ **45.** $(-1, 1)$

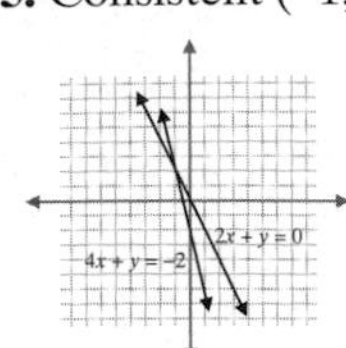

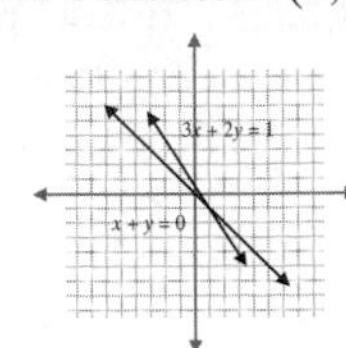

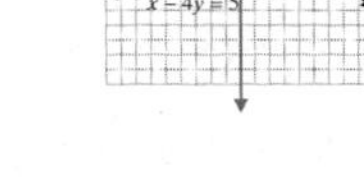

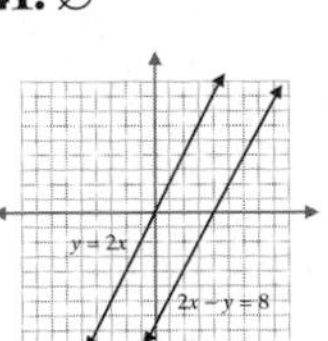

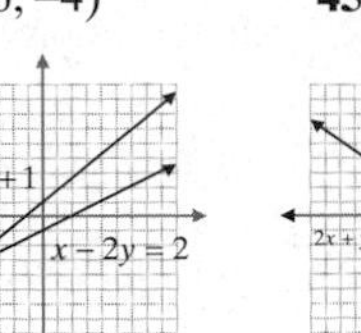

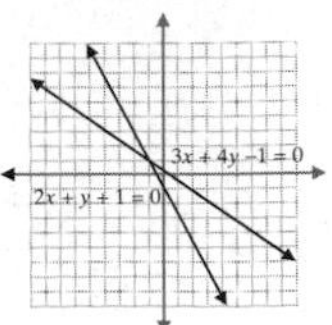

47. $(3, -4)$ **49.** $x = 20, y = 5$ **51.** $x = 2, y = 8$ **53.** $x = 1, y = 4$ **55.** $x = \frac{3}{2}, y = 2$

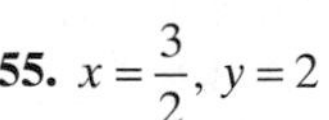

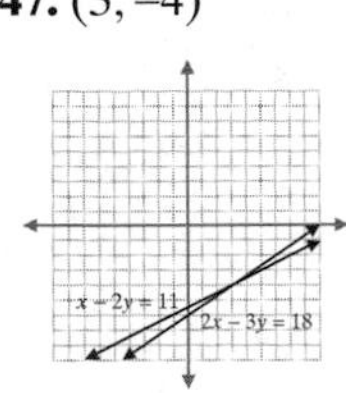

57. $x = \frac{8}{3}, y = -\frac{7}{3}$ **59.** $x = \frac{2}{3}, y = -\frac{4}{3}$ **61.** $x = -\frac{3}{5}, y = -\frac{11}{5}$ **63.** $x = 13, y = -8$

Exercises 9.2, pages 641 – 643

1. $(2, 4)$ **3.** $(1, -2)$ **5.** $(-6, -2)$ **7.** $(4, 1)$ **9.** Inconsistent **11.** $(3, 2)$ **13.** Dependent **15.** $(2, 2)$ **17.** $\left(\frac{1}{2}, -4\right)$ **19.** $\left(2, \frac{5}{2}\right)$ **21.** $\left(-2, \frac{1}{2}\right)$ **23.** $(2, -10)$ **25.** $\left(\frac{5}{2}, \frac{1}{8}\right)$ **27.** $(4, -1)$ **29.** $\left(\frac{13}{5}, \frac{-39}{5}\right)$ **31.** $(10, 4)$ **33.** $(9, -3)$ **35.** $(6, -4)$ **37.** $(-7.5, 0)$ **39.** $(20, 5)$ **41.** $(2, 8)$

Exercises 9.3, pages 650 – 653

1. $x = 3, y = -1$ **3.** $x = 1, y = -\frac{3}{2}$ **5.** Inconsistent **7.** $x = 1, y = -5$ **9.** Dependent **11.** $x = 4, y = 3$ **13.** $x = -3, y = -1$ **15.** $x = 2, y = -1$ **17.** $x = -2, y = -3$ **19.** Inconsistent **21.** $x = 10, y = 2$ **23.** $x = 7, y = 0$ **25.** $x = 5, y = 6$ **27.** $x = -3, y = 7$ **29.** $x = \frac{44}{17}, y = -\frac{2}{17}$ **31.** $x = 6, y = 4$ **33.** $x = 4, y = 7$ **35.** $x = \frac{1}{2}, y = \frac{2}{3}$ **37.** $x = -\frac{45}{7}, y = \frac{92}{7}$ **39.** $y = 5x - 7, m = 5, b = -7$

41. $y = \frac{1}{9}x + \frac{13}{9};\ m = \frac{1}{9}, b = \frac{13}{9}$ **43.** $y = \frac{11}{4}x - \frac{49}{4};\ m = \frac{11}{4}, b = -\frac{49}{4}$ **45.** $x = 3, y = -1$

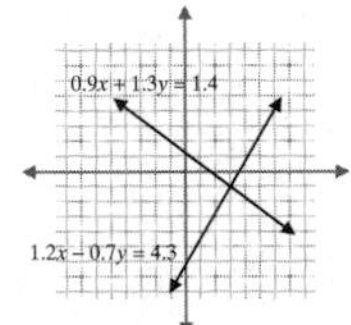

47. $x = 1.2, y = 1.6$ **49.** $x \approx 6.726, y \approx 0.134$

51. $x = \$4000$ at 10%, $y = \$6000$ at 6%

53. $x = \frac{7}{2}$ hrs., $y = \frac{7}{2}$ hrs.

55. 40 ounces of 12% solution (x), 20 ounces of 30% solution (y) **57.** 25 balcony seats (x), 15 standard seats (y)

Exercises 9.4, pages 658 – 662

1. $\begin{cases} x+y=56 \\ x-y=10 \end{cases}$; $x=33, y=23$ **3.** $\begin{cases} x+y=36 \\ 2x+3y=87 \end{cases}$; $x=21, y=15$ **5.** $\begin{cases} \frac{1}{3}(x+y)=4 \\ \frac{1}{2}(x-y)=4 \end{cases}$; Rate of boat $(x)=10$ mph, Rate of current $(y)=2$ mph

7. $\begin{cases} x+y=3\frac{1}{2} \\ 52x+56y=190 \end{cases}$; He traveled $1\frac{1}{2}$ hrs. at the first rate (x) and 2 hours at the second rate (y). **9.** $\begin{cases} x-y=10 \\ \frac{200}{x}+\frac{200}{y}=9 \end{cases}$; where x is his going rate and y is his return rate. His rate of speed to the city was 50 mph.

11. $\begin{cases} x=4y \\ 3(x+y)=105 \end{cases}$; Steve (x) traveled at 28 mph and Fred (y) traveled at 7 mph **13.** $\begin{cases} x=y+8 \\ 3(x+y)=324 \end{cases}$; Mary's speed (x) was 58 mph and Linda's speed (y) was 50 mph.

15. $\begin{cases} x+y=\frac{8}{5} \\ 10x=6y \end{cases}$; He jogged about 12 miles. **17.** $\begin{cases} x=\frac{7}{2}y \\ 2(x-y)=580 \end{cases}$; speed of airliner $(x)=406$ mph and speed of private plane $(y)=116$ mph.

19. $\begin{cases} q+d=27 \\ 0.25q+0.1d=5.4 \end{cases}$; 18 quarters and 9 dimes. **21.** $\begin{cases} A-2=\frac{1}{2}(B-2) \\ A+8=\frac{2}{3}(B+8) \end{cases}$; Anna is 12 years old and Beth is 22 years old.

23. $\begin{cases} l=2w-1 \\ 2(l+4)+2(w+4) \\ \quad =116 \end{cases}$; Length is 33 m and width is 17 m. **25.** $\begin{cases} -1=m(-2)+b \\ -7=m(6)+b \end{cases}$; $y=-\frac{3}{4}x-\frac{5}{2}$.

27. $\begin{cases} p+n=182 \\ 0.01p+0.05n=3.90 \end{cases}$; $n=52$ and $p=130$. **29.** $\begin{cases} a+s=3500 \\ 3.5a+2.5s=9550 \end{cases}$; 800 adults and 2700 students attended.

31. $\begin{cases} l=3w \\ 2(l+w)=260 \end{cases}$; Length is 97.5 m and width is 32.5 m. **33.** $\begin{cases} g+r=12{,}500 \\ 2g+3.5r=36{,}250 \end{cases}$ 5000 general admission and 7500 reserved tickets were sold.

35. $\begin{cases} 70c+160a=620 \\ 2c+a=7 \end{cases}$; 1 adult ticket is \$3 and 1 child's ticket is \$2. **37.** $\begin{cases} 2x+y=55 \\ x+2y=68 \end{cases}$; $x=\$14$ for shirts, $y=\$27$ for pair of slacks.

39. $\begin{cases} 4x+3y=58 \\ 8x+7y=126 \end{cases}$; They produced 7 of Model X and 10 of Model Y **41.** $\begin{cases} a+b=13 \\ 10b+a=10a+b+45 \end{cases}$; $a=4$; $b=9$; The number is 49.

Exercises 9.5, pages 667 – 669

1. \$5500 at 6%, \$3500 at 10% **3.** \$7400 at 5.5%, \$2600 at 6% **5.** \$450 at 8%, \$650 at 10% **7.** \$3500 in each or \$7000 total **9.** \$20,000 at 24%, \$11,000 at 18% **11.** \$800 at 5%, \$2100 at 7% **13.** \$8500 at 9%, \$3500 at 11% **15.** 20 pounds of 20%, 30 pounds of 70% **17.** 20 ounces of 30%, 30 ounces of 20% **19.** 450 pounds of 35%, 1350 pounds of 15% **21.** 20 pounds of 40%, 30 pounds of 15% **23.** 15 pounds of cashews, 5 pounds of peanuts

Chapter 9 Test, pages 672 – 673

1. c **2.** d **3.** Consistent, $\left(\frac{4}{3}, \frac{-14}{3}\right)$ **4.** Consistent, (4, 1) **5.** $x = 6, y = 4$ **6.** $x = 2,\ y = 1.\overline{11}$

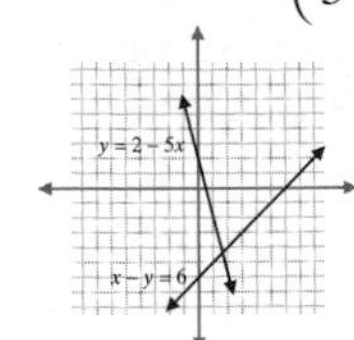

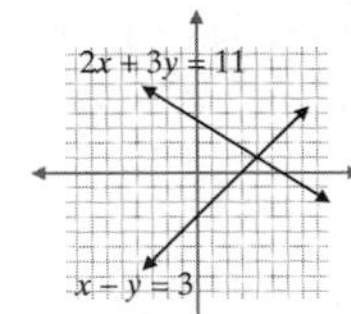

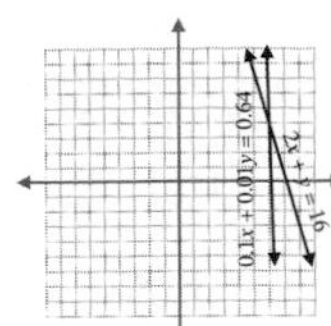

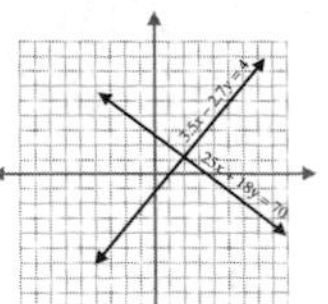

7. Consistent, $x = -8, y = -20$ **8.** Consistent, $x = -2, y = 6$ **9.** Consistent, $x = -\frac{3}{14}, y = \frac{13}{7}$ **10.** Consistent, $x = \frac{43}{12}, y = -\frac{11}{24}$

11. Consistent, $x = -3, y = 5$ **12.** Inconsistent **13.** Consistent, $x = 3, y = -6$ **14.** $a = 8, b = -1$ **15.** Speed of the boat = 14 mph; Speed of the current = 2 mph **16.** Pen price = \$0.79; Pencil price = \$0.08 **17.** \$1600 at 8% and \$960 at 6% **18.** 1600 lbs. of 83% and 400 lbs. of 68% **19.** 13 in. by 17 in. **20.** nickels = 45, quarters = 60 **21.** 80, 15 **22.** Scott is 9 years old and Clayton is 3 years old **23.** 4 gallons of 65%, 10 gallons of 30%

Chapter 9 Cumulative Review, pages 674 – 678

1. −165 **2.** 28 **3.** 55 **4.** 8750 **5.** $x = \frac{1}{5}$ **6.** $x = \frac{11}{4}$ **7.** $y = -7$ **8.** $a = \frac{25}{11}$ **9.** $x \geq 13$ (number line: 13)

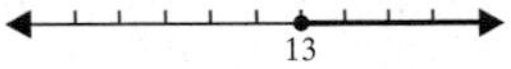

10. $-5 \leq x < -4.2$ (number line: −5, −4.2) **11.** $t = \frac{v-k}{g}$ **12.** $y = \frac{3x-6}{4}$ **13.** $m = -\frac{1}{2},\ b = \frac{11}{4}$

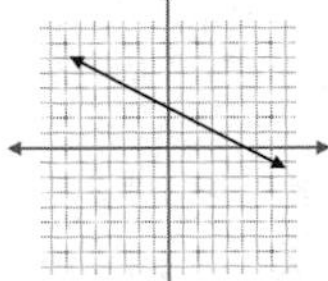

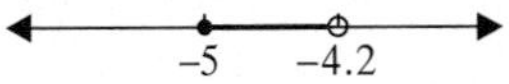

14. $x - 3y = -5$ **15.** $4x + 5y - 2 = 0$ **16.** $2x - y - 10 = 0$ **17.** $x - 6y = -17$

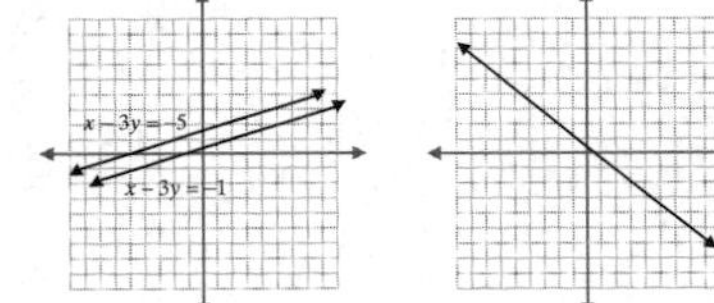

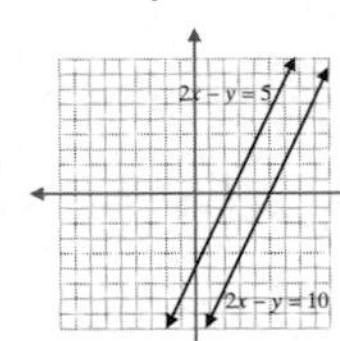

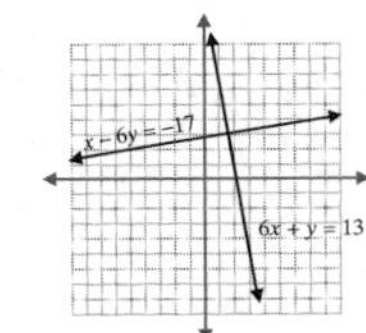

18. $y = 2x + 21$; original line is $y = -\frac{1}{2}x + 10$

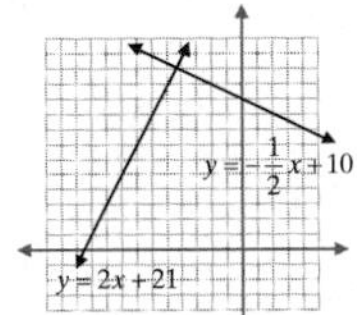

19. $a = 5, b = 1$ **20.** b, c, d **21.** $x = \frac{13}{7}, y = \frac{44}{7}$ **22.** $x = \frac{37}{30}, y = \frac{119}{30}$

23. $x = -\frac{5}{2}, y = \frac{7}{4}$ **24.** $x = 0, y = 6$ **25.** $x = -\frac{12}{7}, y = \frac{31}{7}$ **26.** $x = \frac{11}{5}, y = \frac{19}{10}$ **27.** $x = \frac{11}{40}, y = \frac{16}{5}$ **28.** Consistent, (2, 0)

29. Consistent, (−1, 4) **30.** Consistent, $\left(\frac{13}{10}, -\frac{9}{10}\right)$ **31.** Inconsistent **32.** Consistent, (−6, 2) **33.** Inconsistent

34. Dependent **35.** Consistent, (2, −4) **36.** Consistent, (2, 3) **37.** Consistent, (−1, −6) **38.** Inconsistent **39.** Dependent

40. Consistent, (8, 5) **41.** Consistent, $\left(\frac{22}{21}, \frac{13}{21}\right)$ **42.** Consistent, (1, 2) **43.** Consistent, $\left(\frac{54}{11}, \frac{-5}{11}\right)$ **44.** Consistent, (5, 0)

45. Dependent **46.** Consistent, $\left(\frac{9}{5}, \frac{39}{10}\right)$ **47.** $x = 1, y = 7$ **48.** $x = -\frac{94}{25}, y = -\frac{57}{25}$ **49.** $y = -7x + 20$ **50.** $y = -\frac{4}{3}x + \frac{7}{3}$

51. a. The measures of the angles are 20° and 70°. **b.** The measures of the angles are 40° and 140°. **52.** $a = 3, b = -4$

53. \$35,000 at 8% and \$65,000 at 6% **54.** 6 ounces of 10% mixture and 2 ounces of 30% mixture **55.** Ten 25¢ stamps and five 34¢ stamps. **56.** speed of the boat = 10 mph and speed of the current = 2 mph **57.** length = 14 yards; width = 11 yards. **58.** A = 4; B = 18 **59.** Eastbound train is traveling at 40 mph, Westbound train is traveling at 45 mph.

60. train's speed = 60 mph; speed of airplane = 230 mph **61.** 60, 24 **62.** 3 mph **63.** $x = 3, y = 11$ **64.** $\frac{2}{3}$ hr. or 40 minutes

65. 120 sq. yds. **66.** $N = 950t - 850$ **67.** $\{(-5,-2),(-3,-2),(-2,4),(1,-1),(2,4)\}$; $D = \{-5,-3,-2,1,2\}$; $R = \{-2,-1,4\}$; It is a function **68.** Not a function **69.** Is a function **70. a.** –22 **b.** 23 **c.** 3.5 **71. a.** –10 **b.** –1 **c.** 80

Chapter 10

Exercises 10.1, pages 691 – 693

1. 0.02 **3.** 32 **5.** 67.0820 **7.** 0.02 **9.** 2.3208 **11.** $2\sqrt{3}$ **13.** $7\sqrt{2}$ **15.** $-9\sqrt{2}$ **17.** $2\sqrt[3]{2}$ **19.** $3\sqrt[3]{4}$ **21.** Nonreal **23.** $2yx^5\sqrt{6x}$ **25.** $ac\sqrt[3]{a^2b^2}$ **27.** Nonreal **29.** $2x^3y^4$ **31.** $9ab^2\sqrt[3]{ab^2}$ **33.** $5xy^3\sqrt{5x}$ **35.** $2bc\sqrt{3ac}$ **37.** $-5x^2y^3z^4\sqrt{3}$ **39.** $2xy^2z^3\sqrt[3]{3x^2y}$ **41.** $5|y|$ **43.** $-8|a^3|$ **45.** $4x^2y^4\sqrt{2}$ **47.** $3b^3\sqrt[3]{4a}$ **49.** $3xy^2\sqrt[3]{3x^2y}$ **51.** $\frac{\sqrt{x}}{8x}$ **53.** $\frac{-\sqrt{6y}}{3y}$ **55.** $\frac{y\sqrt{10x}}{4x}$ **57.** $\frac{2\sqrt{2y}}{y}$ **59.** $\frac{y\sqrt[3]{2x}}{3x}$ **61.** $\frac{\sqrt[3]{30a^2b^2}}{5b}$ **63.** $\sqrt{x^2}$ represents the positive square root. **65.** A cube root has no restrictions.

Exercises 10.2, pages 702 – 704

1. 3 **3.** $\frac{1}{10}$ **5.** –512 **7.** Nonreal **9.** $\frac{3}{7}$ **11.** –4 **13.** –5 **15.** $\frac{1}{4}$ **17.** $-\frac{27}{8}$ **19.** $\frac{9}{16}$ **21.** $\frac{2}{5}$ **23.** $\frac{1}{81}$ **25.** $-\frac{1}{16{,}807}$ **27.** 2187 **29.** 4.6416 **31.** 0.0131 **33.** 158.7401 **35.** 7.7460 **37.** 1.0605 **39.** 1.6083 **41.** $8x$ **43.** $3a^2$ **45.** $8x^{\frac{5}{2}}$ **47.** $5a^{\frac{13}{6}}$ **49.** $a^{\frac{8}{5}}$ **51.** $x^{\frac{1}{2}}$ **53.** $a^{\frac{7}{6}}$ **55.** $a^{\frac{1}{4}}$ **57.** $\frac{y^{\frac{1}{2}}}{x^{\frac{4}{3}}}$ **59.** $\frac{1}{a^{\frac{3}{4}}b^{\frac{1}{2}}}$ **61.** $\frac{x^{\frac{3}{2}}}{16y^{\frac{2}{5}}}$ **63.** $\frac{x^2y^4}{z^4}$ **65.** $\frac{c^3}{3ab^2}$ **67.** $\frac{8b^{\frac{9}{4}}}{a^3c^3}$ **69.** $x^{\frac{1}{4}}y^{\frac{5}{4}}$ **71.** $\frac{y^{\frac{2}{3}}}{50x^{\frac{4}{3}}}$ **73.** $\frac{6a}{b^2}$ **75.** $\frac{b^{\frac{5}{12}}}{a^{\frac{11}{12}}}$ **77.** $\frac{3x^{\frac{1}{2}}}{20y^{\frac{2}{3}}}$ **79.** $x\sqrt[15]{x^4}$ **81.** $\frac{1}{x\sqrt[12]{x}}$ **83.** $\sqrt[12]{a^5}$ **85.** $\sqrt[10]{x}$ **87.** $\sqrt[4]{a}$ **89.** $a^9b^{18}c^3$ **91.** $a^{20}b^5c^{10}$

Exercises 10.3, pages 711 – 713

1. $-6\sqrt{2}$ **3.** $5\sqrt{x}$ **5.** $-3\sqrt[3]{7x^2}$ **7.** $10\sqrt{3}$ **9.** $-7\sqrt{2}$ **11.** $20\sqrt{3}$ **13.** $3\sqrt{6}+7\sqrt{3}$ **15.** $20x\sqrt{5x}$ **17.** $14\sqrt{5}-3\sqrt{7}$ **19.** $11\sqrt{2x}+14\sqrt{3x}$ **21.** $4\sqrt[3]{5}-13\sqrt[3]{2}$ **23.** $x^2\sqrt{2x}$ **25.** $4x^2y\sqrt{x}$ **27.** $3x-9\sqrt{3x}+8$ **29.** $24+10\sqrt{2x}+2x$ **31.** 2 **33.** $13+4\sqrt{10}$ **35.** –3 **37.** $6+\sqrt{30}-2\sqrt{3}-\sqrt{10}$ **39.** $5-\sqrt{33}$ **41.** $49x-2$ **43.** $9x+6\sqrt{xy}+y$ **45.** $12x+7\sqrt{2xy}+2y$ **47.** $-\frac{3\sqrt{3}+15}{22}$ **49.** $\frac{3\sqrt{6}-\sqrt{42}}{2}$ **51.** $4\sqrt{5}+4\sqrt{3}$ **53.** $\frac{-5\sqrt{2}-4\sqrt{5}}{3}$ **55.** $\frac{59+21\sqrt{5}}{44}$ **57.** $8+3\sqrt{6}$ **59.** $\frac{x^2+4x\sqrt{y}+4y}{x^2-4y}$ **61.** $\frac{1}{\sqrt{7}+2}$ **63.** $\frac{2}{\sqrt{15}-\sqrt{3}}$ **65.** $\frac{1}{\sqrt{x}-\sqrt{5}}$ **67.** $\frac{2y-x}{x\sqrt{2y}+x\sqrt{x}}$ **69.** $\frac{1}{\sqrt{2+h}+\sqrt{2}}$ **71.** 4.33975

73. 31.60000 **75.** –57.00000 **77.** 251.94353 **79.** 0.40863 **81. a.** $l^2 - l - 1 = 0$ **b.** 97.08 feet **c.** The yellow rectangle is "golden".

Exercises 10.4, pages 720 – 722

1. a. $\sqrt{5}$ and 2.2361 **b.** $\sqrt{9}$ and 3.0000 **c.** $\sqrt{50}$ and 7.0711 **d.** $\sqrt{4}$ and 2.0000 **3. a.** $\sqrt[3]{27}$ and 3.0000 **b.** $\sqrt[3]{-1}$ and –1.0000

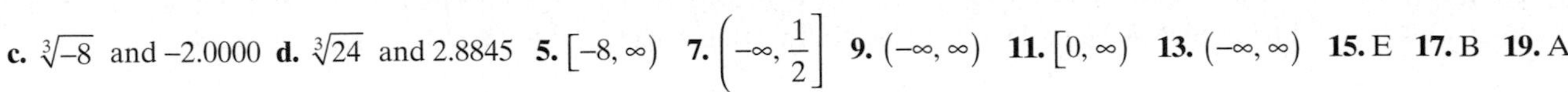

c. $\sqrt[3]{-8}$ and –2.0000 **d.** $\sqrt[3]{24}$ and 2.8845 **5.** $[-8, \infty)$ **7.** $\left(-\infty, \frac{1}{2}\right]$ **9.** $(-\infty, \infty)$ **11.** $[0, \infty)$ **13.** $(-\infty, \infty)$ **15.** E **17.** B **19.** A

21.

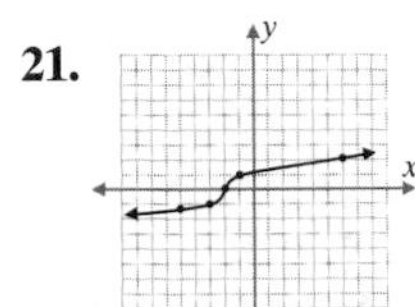

23.

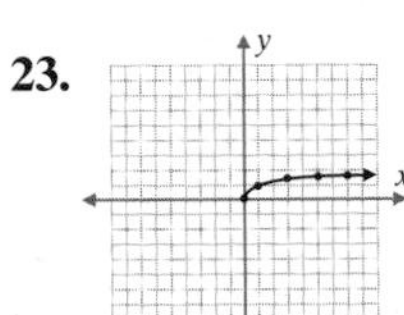

25.

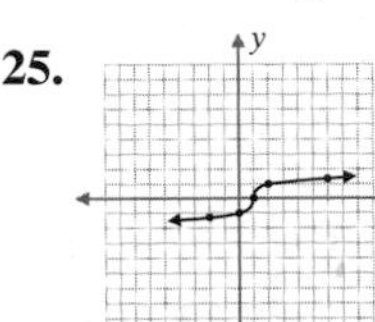

27.

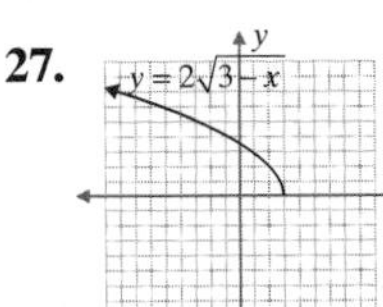

29.

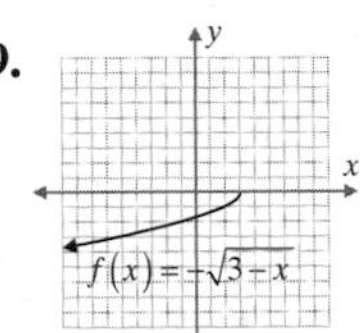

31.

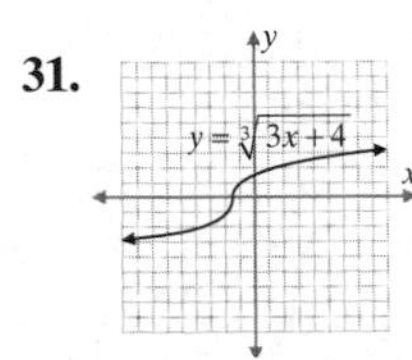

33.

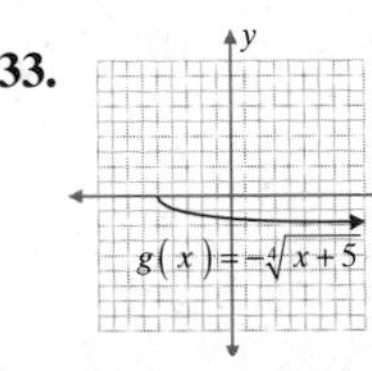

35.

$y = \sqrt[5]{x+7}$

Exercises 10.5, pages 728 – 729

1. Real part is 4, imaginary part is –3 **3.** Real part is –11, imaginary part is $\sqrt{2}$ **5.** Real part is $\frac{2}{3}$, imaginary part is $\sqrt{17}$ **7.** Real part is $\frac{4}{5}$, imaginary part is $\frac{7}{5}$ **9.** Real part is $\frac{3}{8}$, imaginary part is 0. **11.** $7i$ **13.** $-8i$ **15.** $21\sqrt{3}$ **17.** $10i\sqrt{6}$ **19.** $-12i\sqrt{3}$ **21.** $11\sqrt{2}$ **23.** $10i\sqrt{10}$ **25.** $6 + 2i$ **27.** $1 + 7i$ **29.** $2 - 6i$ **31.** $14i$ **33.** $\left(3+\sqrt{5}\right)-6i$ **35.** $5+\left(\sqrt{6}+1\right)i$ **37.** $\sqrt{3}-5$ **39.** $-2 - 5i$ **41.** $11 - 16i$ **43.** 2 **45.** $3 + 4i$ **47.** $x = 6, y = -3$ **49.** $x = -2, y = \sqrt{5}$ **51.** $x = \sqrt{2}-3, y = 1$ **53.** $x = 2, y = -8$ **55.** $x = 2, y = -6$ **57.** $x = 3, y = 10$ **59.** $x = -\frac{4}{3}, y = -3$ **61. a.** Yes **b.** No

Exercises 10.6, pages 734 – 735

1. $16 + 24i$ **3.** $-7\sqrt{2}+7i$ **5.** $3 + 12i$ **7.** $1 - i\sqrt{3}$ **9.** $5\sqrt{2}+10i$ **11.** $2 + 8i$ **13.** $-7 - 11i$ **15.** $13 + 0i$ **17.** $-3 - 7i$ **19.** $5 + 12i$ **21.** $5 - i\sqrt{3}$ **23.** $21 + 0i$ **25.** $23 - 10i\sqrt{2}$ **27.** $(2+\sqrt{10})+(2\sqrt{2}-\sqrt{5})i$ **29.** $(9-\sqrt{30})+(3\sqrt{5}+3\sqrt{6})i$ **31.** $0 + 3i$ **33.** $0 - \frac{5}{4}i$ **35.** $-\frac{1}{4}+\frac{1}{2}i$ **37.** $-\frac{4}{5}+\frac{8}{5}i$ **39.** $\frac{24}{25}+\frac{18}{25}i$ **41.** $-\frac{1}{13}+\frac{5}{13}i$ **43.** $-\frac{1}{29}-\frac{12}{29}i$ **45.** $-\frac{17}{26}-\frac{7}{26}i$ **47.** $\frac{4+\sqrt{3}}{4}+\left(\frac{4\sqrt{3}-1}{4}\right)i$ **49.** $-\frac{1}{7}+\frac{4\sqrt{3}}{7}i$ **51.** $0 + i$ **53.** $-1 + 0i$ **55.** $0 + i$ **57.** $1 + 0i$ **59.** $0 - i$ **61.** x^2+9 **63.** x^2+2 **65.** $5y^2+4$ **67.** $x^2+4x+40$ **69.** $y^2-6y+13$ **71.** Answers will vary **73.** $a^2 + b^2 = 1$

Chapter 10 Test, pages 742 – 743

1. 4 **2.** $\frac{1}{27}$ **3.** $4x^{\frac{7}{6}}$ **4.** $\frac{7x^{\frac{1}{4}}}{y^{\frac{1}{3}}}$ **5.** $\frac{8y^{\frac{3}{2}}}{x^3}$ **6.** $\sqrt[3]{4x^2}$ **7.** $2^{\frac{1}{2}}x^{\frac{1}{3}}y^{\frac{2}{3}}$ **8.** $\sqrt[12]{x^{11}}$ **9.** $4\sqrt{7}$ **10.** $2y\sqrt[3]{6x^2y^2}$ **11.** $\frac{y}{4x^2}\sqrt{10x}$

12. $17\sqrt{3}$ **13.** $(7x-6)\sqrt{x}$ **14.** $8\sqrt[3]{3}$ **15.** $5-2\sqrt{6}$ **16.** $-xy\sqrt{y}$ **17.** $30-7\sqrt{3x}-6x$ **18. a.** $-\left(\sqrt{3}+\sqrt{5}\right)$ **b.** $\frac{1+\sqrt{x^3}}{1+x+x^2}$

19. a. $\left[-\frac{4}{3},\infty\right)$ **b.** $(-\infty,\infty)$ **20. a.**

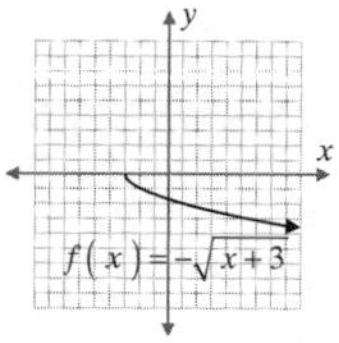

b.

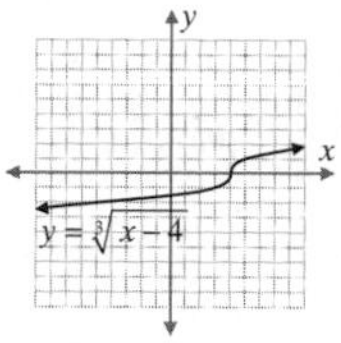

21. $16+4i$ **22.** $-5+16i$ **23.** $23-14i$ **24.** $\frac{8}{13}-\frac{1}{13}i$

25. $x=\frac{11}{2}, y=3$ **26.** $0-i$ **27.** x^2+4 **28.** $x^2+6x+12$ **29.** Answers will vary **30.** 0.1250 **31.** 3.0711 **32.** 0.9758

Chapter 10 Cumulative Review, pages 743 – 746

1. $2x^3-4x^2+8x-4$ **2.** $6x^2+19x-7$ **3.** $-5x^2+18x+8$ **4.** $n=\frac{s-a}{d}+1$ or $\frac{s-a+d}{d}$ **5.** $p=\frac{A}{1+rt}$ **6.** $(4x+3)(3x-4)$

7. $(7+2x)(4-x)$ **8.** $5(x-4)(x^2+4x+16)$ **9.** $(x+4)(x+1)(x-1)$ **10.** $\frac{4x^2-12x-1}{(x+3)(x-2)(x-5)}$

11. $\frac{4x^2+6x+17}{(2x+1)(2x-1)(x+3)}$ **12.** $\frac{x^2-1}{2x-4}$ **13.** $\frac{x+6}{x-1}$ **14.** $x=-\frac{5}{7}$ **15.** $x=\frac{83}{4}$ **16.** $\left(-2,\frac{3}{5}\right]$

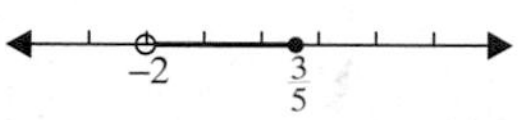

17. $\left[-2,\frac{1}{3}\right)$

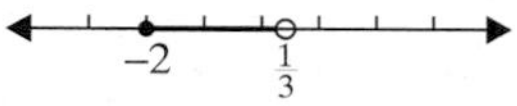

18. $\frac{50}{3}$ in.^{3} or $16.\overline{6}$ in.^{3} **19.** 8 ohms **20.** $(x-16)(x+3)=0$ $x=-3$ and 16

21. $(2x+3)(x-2)=0, x=-\frac{3}{2}$ and 2 **22.** $(5x-2)(3x-1)=0$, $x=\frac{1}{3}$ and $\frac{2}{5}$ **23.** $a=5, b=-4$ **24.** $(1,-2)$

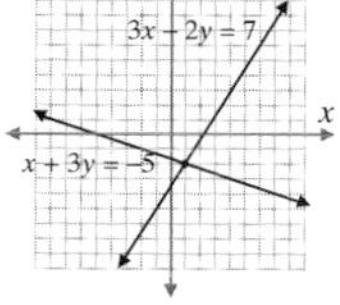

25. $x=-1, y=1$ **26.** $x=\frac{85}{13}, y=\frac{60}{13}$ **27. a.** $x^2-11x+28=0$ **b.** $x^3-11x^2-14x+24=0$

28. $(1.5, 2.5)$

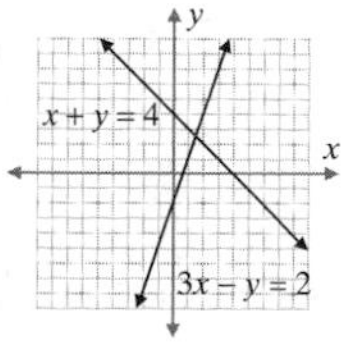

29. $x-9+\frac{21x-24}{x^2+2x-1}$ **30.** x^2-4x+3

31. a.

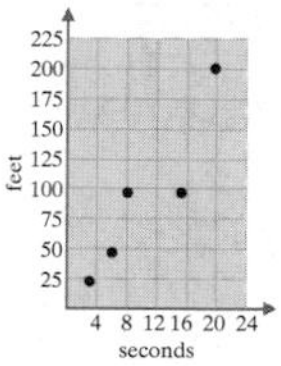

b. $m_1=8, m_2=24, m_3=0, m_4=\frac{104}{5}$

c. Rate was 8 ft./sec. (fps), then increased to 24 fps, then went to 0 fps, then increased to $\frac{104}{5}$ fps.

32. a. Function, $D=(-\infty,\infty)$, $R=(-\infty,0]$ **b.** Not a function **c.** Function, $D=(-\infty,\infty)$, $R=[-1,1]$ **33.** 15,17 **34.** $x=1$

35. $x=\pm\sqrt{12}\approx\pm3.46, \pm\sqrt{6}\approx\pm2.45$ **36.** $\frac{1}{16}$ **37.** $x^{\frac{7}{3}}$ **38.** $12\sqrt{2}$ **39.** $2x^2y^3\sqrt[3]{2y}$ **40.** $14-2i$ **41.** $-\frac{8}{17}-\frac{19}{17}i$ **42.** 10 mph

43. 7.5 hours **44.** $35,000 in 6% and $15,000 in 10% **45.** 60 pounds of first type ($1.25 candy) and 40 pounds of second type ($2.50 candy) **46.** $x\approx-1.372, x\approx4.372$ **47.** $x=1, x\approx-2.303, x\approx1.303$ **48.** $x=-4, x=1, x=5$ **49.** 279,936

50. 0.04000 **51.** 1.30268 **52.** 6.45522

Chapter 11

Exercises 11.1, pages 756 – 757

1. $x^2-12x+\underline{36}=(x-6)^2$ **3.** $x^2+6x+\underline{9}=(x+3)^2$ **5.** $x^2-5x+\underline{\frac{25}{4}}=\left(x-\frac{5}{2}\right)^2$ **7.** $y^2+y+\underline{\frac{1}{4}}=\left(y+\frac{1}{2}\right)^2$

9. $x^2+\frac{1}{3}x+\underline{\frac{1}{36}}=\left(x+\frac{1}{6}\right)^2$ **11.** $x=\pm12$ **13.** $x=\pm5i$ **15.** $x=\pm3i\sqrt{2}$ **17.** $x=9,\ x=-1$ **19.** $x=\pm2\sqrt{3}$

21. $x=1\pm\sqrt{5}$ **23.** $x=-8\pm3i$ **25.** $x=5\pm i\sqrt{10}$ **27.** $x=1,\ x=-7$ **29.** $x=4\pm\sqrt{13}$ **31.** $z=-2\pm\sqrt{6}$

33. $x=-1\pm i\sqrt{5}$ **35.** $y=5\pm\sqrt{21}$ **37.** $x=\frac{5\pm\sqrt{5}}{2}$ **39.** $x=\frac{-1\pm i\sqrt{7}}{2}$ **41.** $x=-2\pm\sqrt{7}$ **43.** $x=-1\pm i\sqrt{3}$ **45.** $y=1,\ y=-\frac{4}{3}$

47. $x=\frac{-7\pm\sqrt{17}}{8}$ **49.** $x=\frac{1\pm i\sqrt{11}}{4}$ **51.** $y=\frac{-3\pm i\sqrt{11}}{2}$ **53.** $x=2\pm\sqrt{2}$ **55.** $x=\frac{-5\pm i\sqrt{7}}{2}$ **57.** $x^2-5=0$

59. $z^2-4z+2=0$ **61.** $x^2-2x-11=0$ **63.** $x^2+49=0$ **65.** $y^2+5=0$ **67.** $x^2+6x+13=0$ **69.** $x^2-4x+7=0$
71. Answers will vary

Exercises 11.2, pages 763 – 765

1. 68, two real solutions **3.** 0, one real solution **5.** –44, two nonreal solutions **7.** 4, two real solutions

9. 19,600, two real solutions **11.** –11, two nonreal solutions **13.** $k<16$ **15.** $k=\frac{81}{4}$ **17.** $k>3$ **19.** $k>-\frac{1}{36}$ **21.** $k=\frac{49}{48}$

23. $k>\frac{4}{3}$ **25.** $x=\frac{-3\pm\sqrt{29}}{2}$ **27.** $x=\frac{5\pm\sqrt{17}}{2}$ **29.** $x=\frac{-7\pm\sqrt{33}}{4}$ **31.** $x=1,\ x=-\frac{1}{6}$ **33.** $x=\pm\sqrt{\frac{4}{3}}$ **35.** $x=\frac{9\pm\sqrt{65}}{2},\ x=0$

37. $x=\frac{-3\pm\sqrt{5}}{2},\ x=0$ **39.** $x=-1,\ x=4$ **41.** $x=\frac{-4\pm i\sqrt{2}}{2}$ **43.** $x=\pm\sqrt{7}$ **45.** $x=0,\ x=-1$ **47.** $x=\frac{7\pm i\sqrt{51}}{10}$

49. $x=-2,\ x=\frac{5}{3}$ **51.** $x=\pm\frac{3}{2}i$ **53.** $x=\frac{2\pm\sqrt{3}}{3}$ **55.** $x=\frac{7\pm i\sqrt{287}}{12}$ **57.** $\frac{2\pm\sqrt{2}}{2}$ **59.** $x=\frac{-7\pm\sqrt{17}}{4}$ **61.** x = 60.4007, 2.5993 **63.** x = 2.0110, –0.7862 **65.** x = –0.5806,–4.1334 **67.** x = –4.7693i, +4.7693i **69.** x^4-13x^2+36, answers will vary

Exercises 11.3, pages 770 – 774

1. 6, 13 **3.** 4, 14 **5.** –6,–11 **7.** $-5-4\sqrt{3}$ **9.** $\frac{1+\sqrt{33}}{4}$ **11.** $7\sqrt{2}$ cm, $7\sqrt{2}$ cm

13. Mel: 30 mph, John: 40 mph **15.** 4 feet × 10 feet **17.** 5 m × 8 m **19.** 70
21. 3 m × 9 m **23.** 7 cm × 9 cm **25.** 2 amperes or 8 amperes
27. a. \$307.20 **b.** 45 cents or 35 cents **29.** \$13 and \$22 **31.** 64 mph
33. Sam: 7.5 hours, Bob: 12.5 hours **35.** 150 mph **37.** 190 reserved,150 general
39. a. 6.75 sec. **b.** 3 sec., 3.75 sec. **41. a.** 3.5 sec. **b.** 144 ft. **43.** 14.1 cm
45. a. 94.20 ft., 706.50 ft.^{2} **b.** 84.84 ft., 449.86 ft.^{2} **47. a.** No **b.** Home plate **c.** No
49. 8.49 cm

51.

$$ax^2+bx+c=0$$
$$ax^2+bx=-c$$
$$x^2+\frac{b}{a}x=-\frac{c}{a}$$
$$\left(x^2+\frac{b}{a}x+\frac{b^2}{4a^2}\right)=-\frac{c}{a}+\frac{b^2}{4a^2}$$
$$\left(x+\frac{b}{2a}\right)^2=-\frac{c}{a}+\frac{b^2}{4a^2}$$
$$x+\frac{b}{2a}=\pm\sqrt{-\frac{c}{a}+\frac{b^2}{4a^2}}$$
$$x=-\frac{b}{2a}\pm\frac{\sqrt{-4ac+b^2}}{2a}$$
$$x=\frac{-b\pm\sqrt{b^2-4ac}}{2a}$$

Exercises 11.4, pages 779 – 780

1. $x=3$ **3.** $x=13$ **5.** $x=3$ **7.** $x=2$ **9.** $x=-4, x=1$ **11.** $x=-5, x=\frac{5}{2}$ **13.** $x=-2$ **15.** $x=\pm 5$ **17.** $x=2$ **19.** $x=4$
21. $x=0$ **23.** $x=5$ **25.** No solution **27.** $x=4$ **29.** $x=-1, x=3$ **31.** $x=5$ **33.** $x=2$ **35.** $x=6$ **37.** $x=-4$ **39.** $x=12$
41. No solution **43.** $x=40$ **45.** $(a+b)^2=(a+b)(a+b)$
$=a^2+2ab+b^2$
$\neq a^2+b^2$

Exercises 11.5, pages 784 – 785

1. $x=\pm 2, x=\pm 3$ **3.** $x=\pm 2, x=\pm\sqrt{5}$ **5.** $y=\pm\sqrt{7}, y=\pm 2i$ **7.** $y=\pm\sqrt{5}, y=\pm i\sqrt{5}$ **9.** $z=\frac{1}{6}, z=-\frac{1}{4}$ **11.** $x=4, x=\frac{25}{4}$
13. $x=1, x=4$ **15.** $x=\frac{1}{8}, x=-8$ **17.** $x=\frac{1}{25}$ **19.** $x=-\frac{1}{3}, x=\frac{3}{8}$ **21.** $x=0, x=-27, x=-8$ **23.** $x=1, x=2$
25. $x=-3, x=-\frac{7}{2}$ **27.** $x=0, x=8$ **29.** $x=-1, x=-\frac{1}{2}$ **31.** $x=\pm\sqrt{1+i}, x=\pm\sqrt{1-i}$ **33.** $x=\pm\sqrt{1+3i}, x=\pm\sqrt{1-3i}$
35. $x=\pm\sqrt{2\pm i\sqrt{3}}$ **37.** $x=\pm\frac{1}{5}\sqrt{5}, x=\pm 1$ **39.** $x=\pm\frac{1}{2}\sqrt{6}, x=\pm\frac{1}{3}i$ **41.** $x=-\frac{1}{4}$ **43.** $x=-4, x=1$ **45.** $x=-\frac{1}{2}$
47. $x=-7, x=-\frac{3}{2}$ **49.** $x=\frac{26}{5}$ **51.** Because of the square root, the solution must be positive.

Chapter 11 Test, pages 789 – 790

1. a. ± 4 **b.** $\pm 4i$ **2.** $x=-\frac{1}{2}, x=0$ **3. a.** $x^2-30x+\underline{225}=(x-15)^2$ **b.** $x^2+5x+\underline{\frac{25}{4}}=\left(x+\frac{5}{2}\right)^2$ **4. a.** $x^2+8=0$
b. $x^2-2x-4=0$ **5.** $x=-2\pm\sqrt{3}$ **6.** 73, two real solutions **7.** $k=\pm 2\sqrt{6}$ **8.** $x=\frac{-1\pm i\sqrt{7}}{4}$ **9.** $x=\frac{3\pm\sqrt{41}}{4}$ **10.** $x=\frac{2\pm i\sqrt{2}}{2}$
11. $x=1$ **12.** $x=0, x=36$ **13.** $x=\pm 1, x=\pm 3$ **14.** $x=\frac{3}{2}, x=-1$ **15.** $x=0, x=1$

16. Triangle is not a right triangle because $6^2+8^2\neq 11^2$.

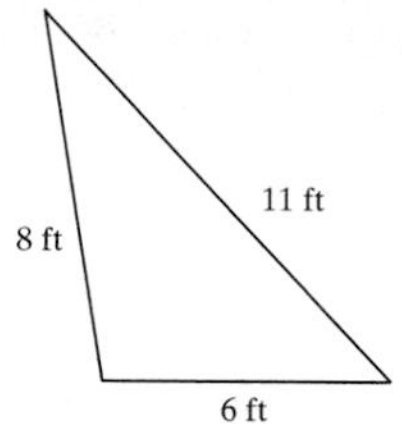

17. a. $t=7$ seconds **b.** $t=6.464$ seconds **18.** 12 inches, 16 inches
19. towards home 50 mph, away 40 mph **20.** 0.0714, −175.0714
21. height = 20 feet, base = 10 feet **22.** length = 8 feet, brace = 10 feet
23. 4.5 seconds **24.** 21 members

Chapter 11 Cumulative Review, pages 790 – 792

1. $x=6$ **2.** $x=\frac{7}{8}$ **3.** $\left[-\frac{19}{2},\infty\right)$

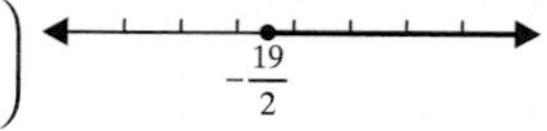

4. $\left(-\infty,\frac{7}{5}\right]$

5. $y=-\frac{3}{2}x+3$
6. $y=-\frac{3}{2}x+10$ **7.** $\frac{1}{x^5}$ **8.** x^3y^6 **9.** $\frac{9y^{\frac{1}{2}}}{x^2}$ **10.** $\frac{27x^2}{8y}$ **11.** $-3x^2y^2\sqrt[3]{y^2}$ **12.** $2x^2y^3\sqrt[4]{2xy^3}$ **13.** $\frac{9}{2}\sqrt{2}$ **14.** $\frac{4}{3}\sqrt{3}$

15. $4x^2+17x-15=0$ **16.** $x^2-2x-19=0$ **17.** $x=-1,\ y=-2$ **18.** $x=5,\ y=4$ **19.** Inconsistent **20.** $(-1.3,\ 3.1)$

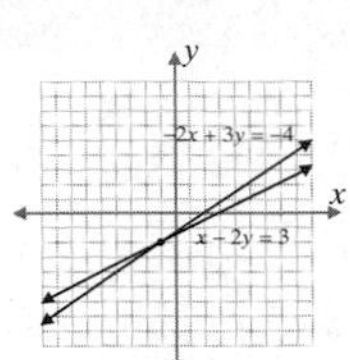

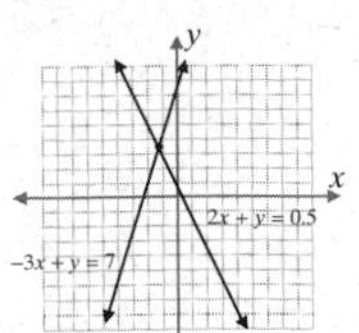

21. $x=-\frac{3}{2},\ x=\frac{2}{5}$ **22.** $x=\frac{-7\pm\sqrt{33}}{8}$ **23.** $x=-1$ **24.** $x=-2,\ x=\frac{4}{3}$ **25.** $x^2-12x+26-\frac{32}{x+2}$ **26.** $x^3-8x^2+4x-\frac{2}{x-2}$

27. a. 5.9136 **b.** 14.8306 **c.** –5.8284 **28.** $5\left(\sqrt{3}-\sqrt{2}\right)$ **29.** $(x+9)\left(\sqrt{x}-3\right)$ **30.** $\sqrt{5}\left(\sqrt{6}-1\right)$ **31.** $4+2i$ **32.** $19+4i$

33. $\frac{-11+16i}{13}$ **34.** x-int: (–1.58, 0) and (1.58, 0) **35.** x-int: (–0.62, 0), (1, 0), and (1.62, 0)

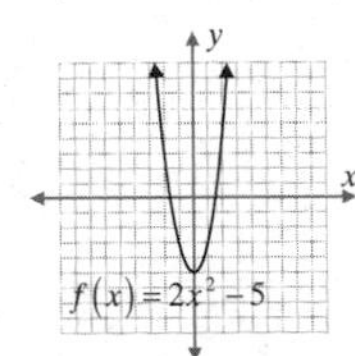

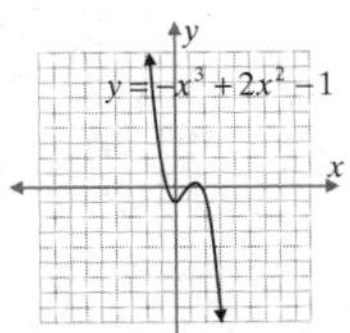

36. x-int: none **37.** x-int: (–0.60, 0) **38.** $D = [3, \infty)$, $R = [0, \infty)$ **39.** $D = (-\infty, 1]$, $R = [0, \infty)$ **40.** $D = [-1, \infty)$, $R = (-\infty, 0]$

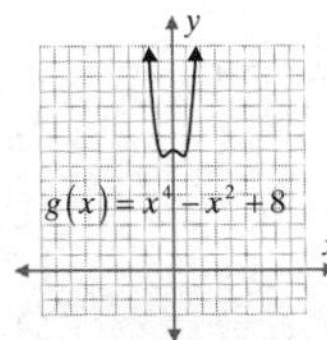

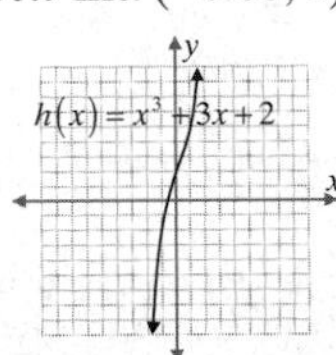

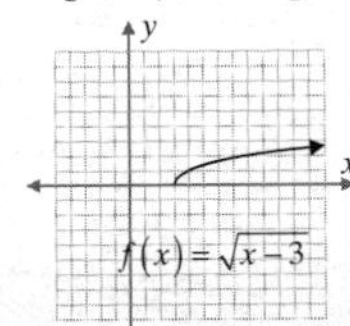

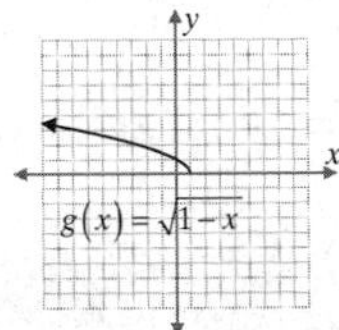

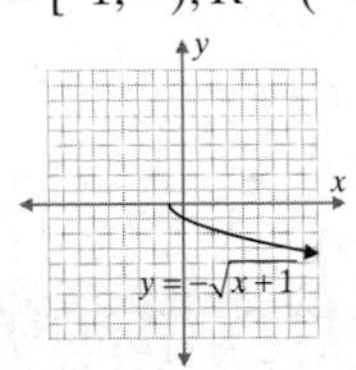

41. \$35 **42.** 20, 26 **43.** 12, 14, 16 **44.** length of the base =19 cm, altitude = 8 cm **45.** If any vertical line intersects the graph of a relation at more than one point, then the relation graphed is not a function. **46.** 30 seats **47.** 1.8 seconds

Chapter 12

Exercises 12.1, pages 805 – 808

1. $x=0,(0,-4),\{x|x \text{ is real}\},\{y|y\geq-4\}$ **3.** $x=0,(0,9),\{x|x \text{ is real}\},\{y|y\geq 9\}$ **5.** $x=0,(0,1),\{x|x \text{ is real}\},\{y|y\leq 1\}$

7. $x=0,\left(0,\frac{1}{5}\right),\{x|x \text{ is real}\},\left\{y|y\leq\frac{1}{5}\right\}$ **9.** $x=-1,(-1,0),\{x|x \text{ is real}\},\{y|y\geq 0\}$ **11.** $x=4,(4,0),\{x|x \text{ is real}\},\{y|y\leq 0\}$

13. $x=-3,(-3,-2),\{x|x \text{ is real}\},\{y|y\geq-2\}$ **15.** $x=-2,(-2,-6),\{x|x \text{ is real}\},\{y|y\geq-6\}$

17. $x=\frac{3}{2},\left(\frac{3}{2},\frac{7}{2}\right),\{x|x \text{ is real}\},\left\{y|y\leq\frac{7}{2}\right\}$ **19.** $x=\frac{4}{5},\left(\frac{4}{5},-\frac{11}{5}\right),\{x|x \text{ is real}\},\left\{y|\ y\geq-\frac{11}{5}\right\}$ **21.a.**

b. **c.** **d.** **23. a.** **b.** **c.**

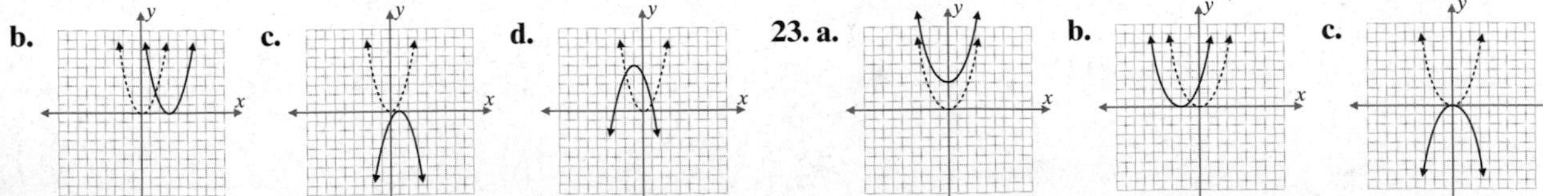

d. 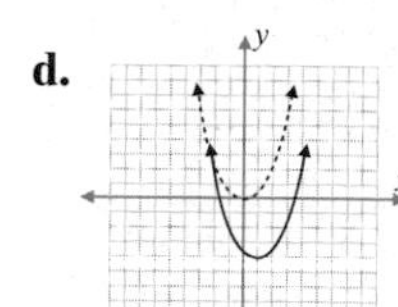

25. $y = 2(x-1)^2$, Vertex $= (1,0)$, $R_f = \{ y | y \geq 0 \}$, Zeros : $x = 1$

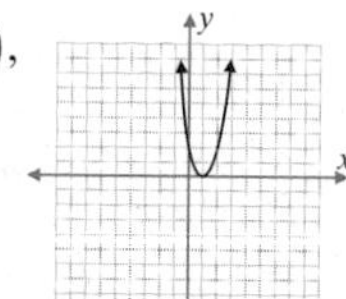

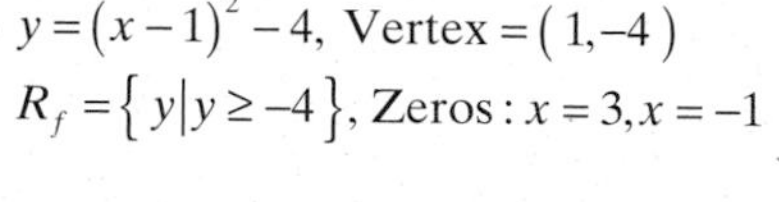

27. $y = (x-1)^2 - 4$, Vertex $= (1,-4)$ $R_f = \{ y | y \geq -4 \}$, Zeros : $x = 3, x = -1$

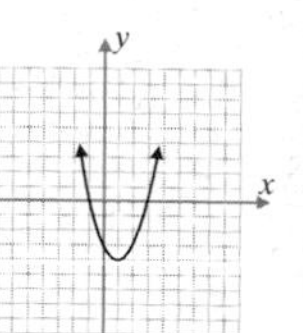

29. $y = (x+3)^2 - 4$, Vertex $= (-3,-4)$ $R_f = \{ y | y \geq -4 \}$, Zeros : $x = -5, x = -1$

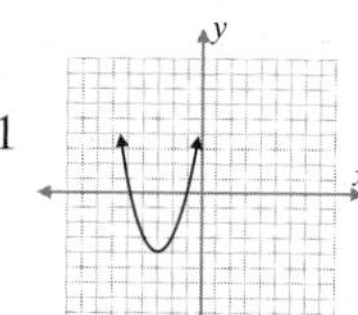

31. $y = 2(x-2)^2 - 3$, Vertex $= (2,-3)$ $R_f = \{ y | y \geq -3 \}$, Zeros : $x = \dfrac{4 \pm \sqrt{6}}{2}$

33. $y = -3(x+2)^2 + 3$, Vertex $= (-2,3)$ $R_f = \{ y | y \leq 3 \}$, Zeros : $x = -1, x = -3$

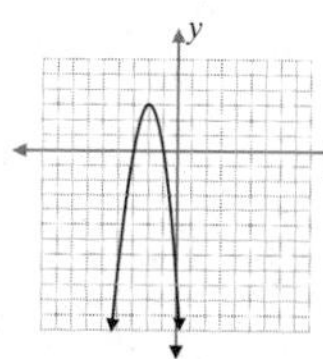

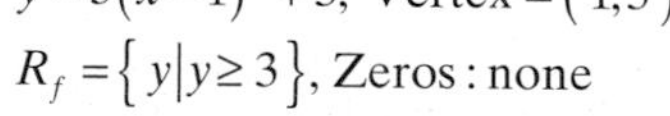

35. $y = 5(x-1)^2 + 3$, Vertex $= (1,3)$ $R_f = \{ y | y \geq 3 \}$, Zeros : none

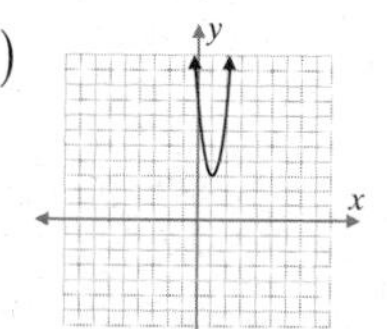

37. $y = -\left(x + \dfrac{5}{2}\right)^2 + \dfrac{17}{4}$, Vertex $= \left(-\dfrac{5}{2}, \dfrac{17}{4}\right)$ $R_f = \left\{ y \middle| y \leq \dfrac{17}{4} \right\}$, Zeros : $x = \dfrac{-5 \pm \sqrt{17}}{2}$

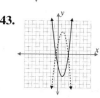

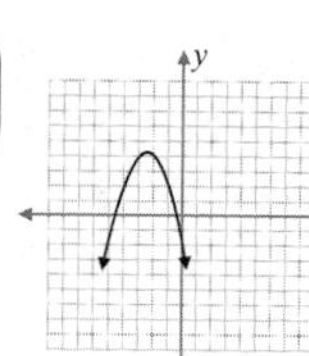

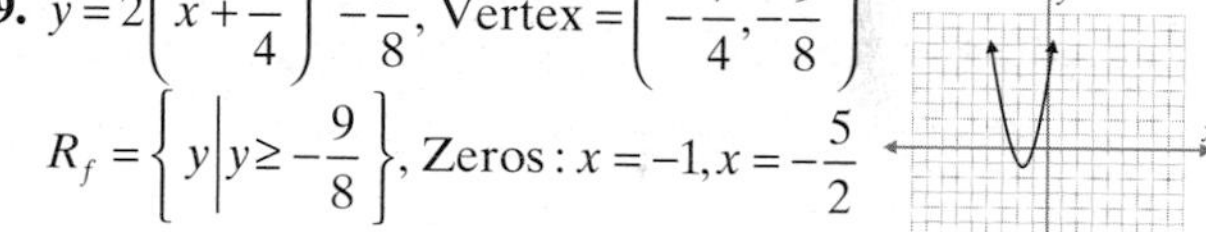

39. $y = 2\left(x + \dfrac{7}{4}\right)^2 - \dfrac{9}{8}$, Vertex $= \left(-\dfrac{7}{4}, -\dfrac{9}{8}\right)$ $R_f = \left\{ y \middle| y \geq -\dfrac{9}{8} \right\}$, Zeros : $x = -1, x = -\dfrac{5}{2}$

41. **a.** No **b.** Yes **c.** Answers will vary.

43. **a.** No **b.** Yes **c.** Answers will vary.

45. a. $3\frac{1}{2}$ sec. **b.** 196 ft. **47. a.** 4 sec. **b.** 288 ft. **49.** \$20 **51. a.** \$30 **b.** \$1800 **53.** zeros: $x \approx 2.732, x \approx -0.732$; vertex $(1, -3)$ **55.** zeros: $x \approx 2.158, x \approx -1.158$; vertex $\left(\dfrac{1}{2}, \dfrac{11}{2}\right)$ **57.** no real zeros; vertex $\left(-\dfrac{3}{2}, \dfrac{3}{4}\right)$ **59. a.** a parabola **b.** $x = -\dfrac{b}{2a}$ **c.** $x = h$ **d.** No. Answers will vary. **61.** Domain $= (-\infty, \infty)$. For $|a| > 0$, Range $= [k, \infty)$. For $|a| < 0$, Range $= (-\infty, k]$.

Exercises 12.2, pages 817 – 819

1. $(-2, 6)$

3. $\left(-\infty, \dfrac{2}{3}\right) \cup (5, \infty)$

5. $(-\infty, -7] \cup \left[\dfrac{5}{2}, \infty\right)$

7. $\left[-2, -\dfrac{1}{3}\right]$

9. $\left(-\infty, -\dfrac{4}{3}\right) \cup (0, 5)$

11. $x = -2$

13. $\left(-\infty, -\dfrac{5}{2}\right) \cup (3, \infty)$

15. $\left(-\dfrac{1}{4}, \dfrac{3}{2}\right)$

17. $\left(-\infty,\frac{1}{2}\right]\cup[2,\infty)$ **19.** $\left(-\frac{2}{3},-\frac{1}{2}\right)$ **21.** $\left(-\infty,\frac{5}{2}\right)\cup\left(\frac{5}{2},\infty\right)$ **23.** $\left[-\frac{5}{2},\frac{7}{4}\right]$

25. $(-1,0)\cup(3,\infty)$ **27.** $(0,1)\cup(4,\infty)$ **29.** $(-\infty,-1)\cup(4,\infty)$ **31.** $(-\infty,-2)\cup(-1,1)\cup(2,\infty)$

33. $[-3,-2]\cup[2,3]$ **35.** $(-\infty,-4]\cup[2,\infty)$ **37.** $\left(-\frac{2}{3},2\right)$ **39.** $\left(-\infty,-1-\sqrt{5}\right)\cup\left(-1+\sqrt{5},\infty\right)$

41. $\left(-\infty,-3-\sqrt{2}\right]\cup\left[-3+\sqrt{2},\infty\right)$ **43.** $\left(\frac{-5-\sqrt{13}}{6},\frac{-5+\sqrt{13}}{6}\right)$ **45.** $\left(-\infty,-\frac{1}{2}\right]\cup[0,4]$ **47.** $(-\infty,\infty)$ **49.** $\varnothing$

49. $\varnothing$ **51.** $(-\infty,-3.1623)\cup(3.1623,\infty)$ **53.** $\varnothing$ **55.** $(-\infty,-3)\cup(0,3)$ **57.** $[-1.8868,\infty)$

59. $(-2.3344,-0.7420)\cup(0.7420,2.3344)$ **61. a.** $x=\frac{7\pm\sqrt{89}}{2}$ **b.** $\left(-\infty,\frac{7-\sqrt{89}}{2}\right)\cup\left(\frac{7+\sqrt{89}}{2},\infty\right)$ **c.** $\left(\frac{7-\sqrt{89}}{2},\frac{7+\sqrt{89}}{2}\right)$

63. a. $x=-5,1$ **b.** $(-5,1)$ **c.** $(-\infty,-5)\cup(1,\infty)$ **65. a.** $(-4,0)\cup(1,\infty)$ **b.** $(-\infty,-4)\cup(0,1)$ **c.** The function is undefined at $x=0$.

Exercises 12.3, pages 830 – 834

1. a. 12 **b.** 4 **c.** $a+8$ **3. a.** 4 **b.** 31 **c.** x^2-4x-1 **5. a.** 1 **b.** $4a+5$ **c.** $4x+4h-3$ **d.** 4 **7. a.** 0 **b.** a^2-6a+5 **c.** $x^2+2xh+h^2-4$ **d.** $2x+h$ **9. a.** -3 **b.** $2a^2-8a+5$ **c.** $2x^2+4xh+2h^2-3$ **d.** $4x+2h$

11. (1, −2)

13. 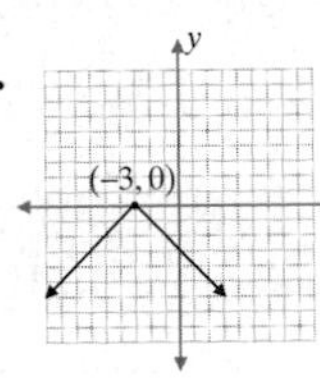

15. 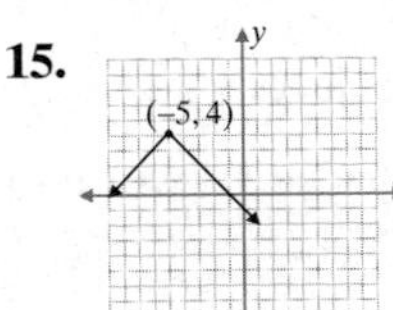

17. 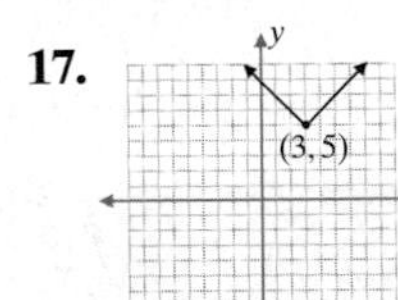

19. 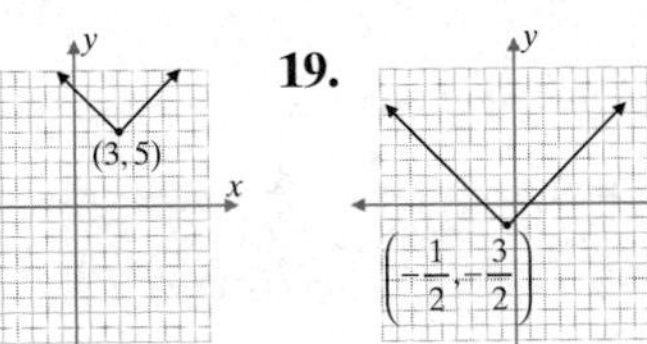

21. 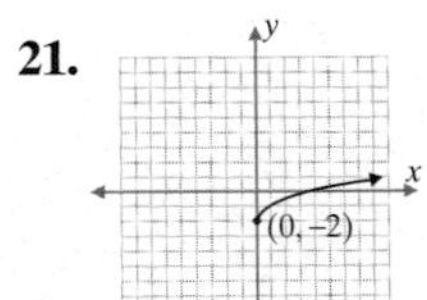

23. 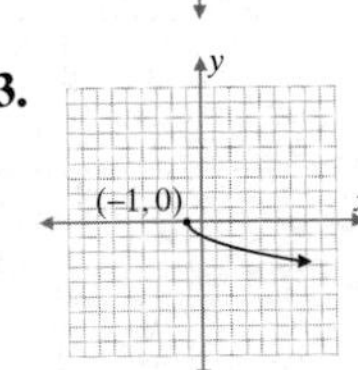

25. 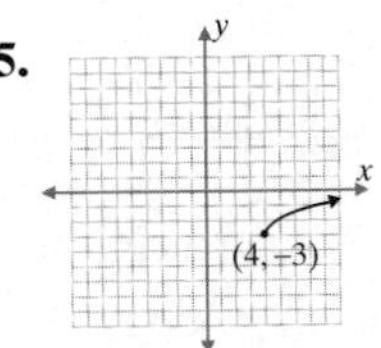

27. 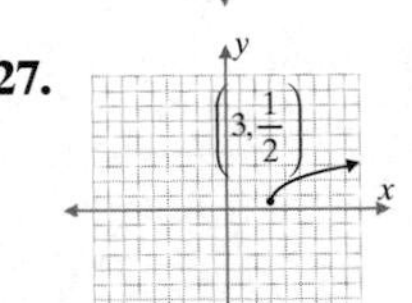

29. 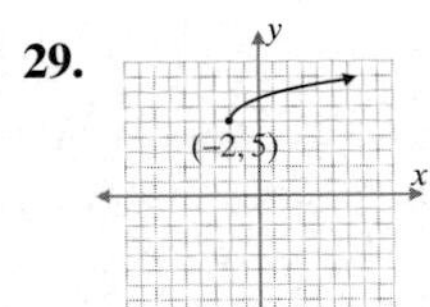

31. 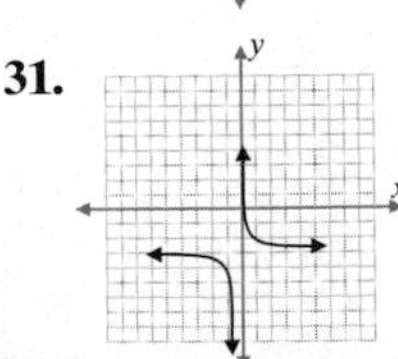

33. 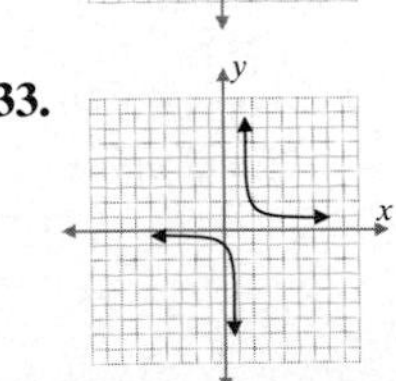

35. 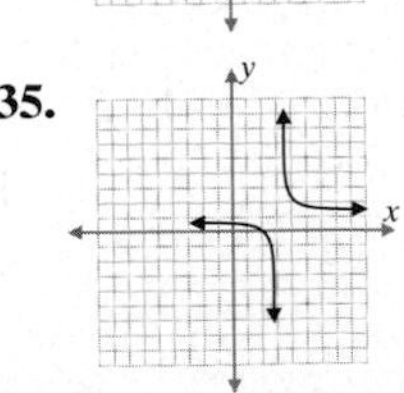

37. 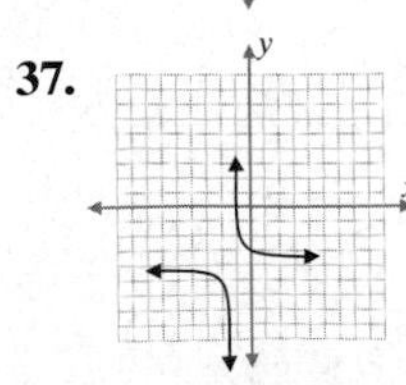

39. 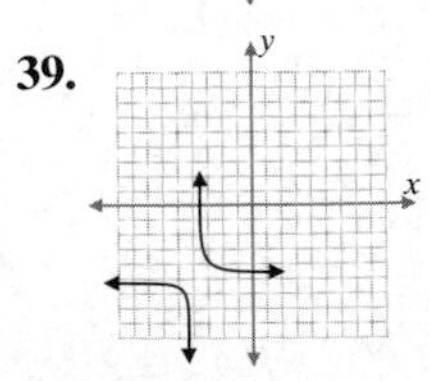

41. 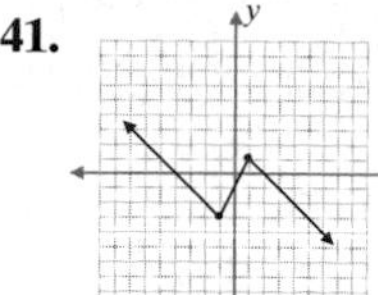 **43.** 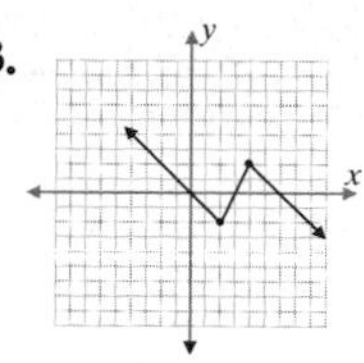 **45.** 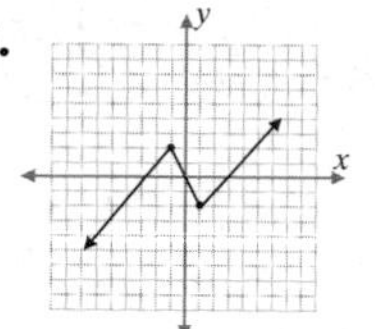 **47.** 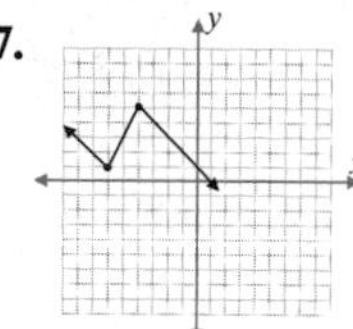 **49.** 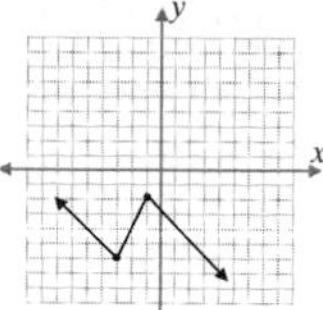

51. iii **53.** ii **55.**

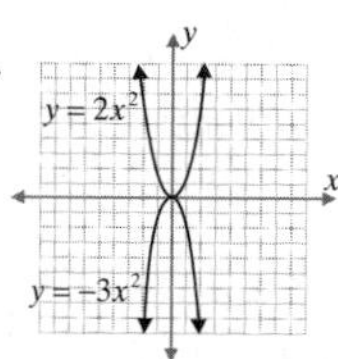

57.

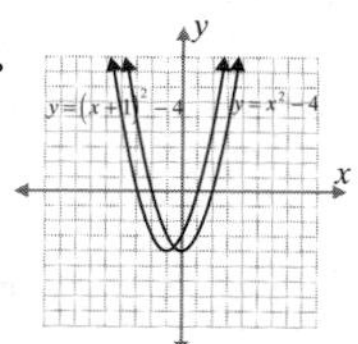

59.

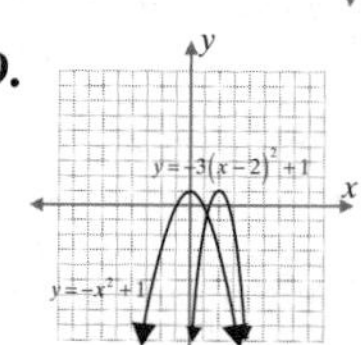

Exercises 12.4, pages 841 – 842

1. a. $(4,0)$ **b.** none **c.** $y=0$

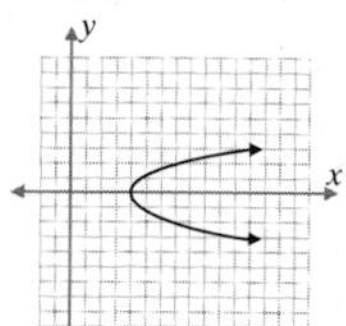

3. a. $(-3,0)$ **b.** $\left(0,\sqrt{3}\right)$, $\left(0,-\sqrt{3}\right)$ **c.** $y=0$

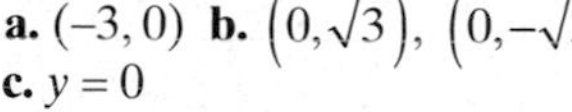

5. a. $(3,0)$ **b.** none **c.** $y=0$

7. a. $(0,3)$ **b.** $(0,3)$ **c.** $y=3$

9. a. $(4,-2)$ **b.** none **c.** $y=-2$

11. a. $(-1,1)$ **b.** $(0,0), (0,2)$ **c.** $y=1$

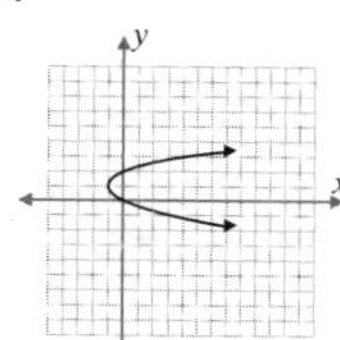

13. a. $(0,-2)$ **b.** $(0,-2)$ **c.** $y=-2$

15. a. $(0,4)$ **b.** $(0,4)$ **c.** $y=4$

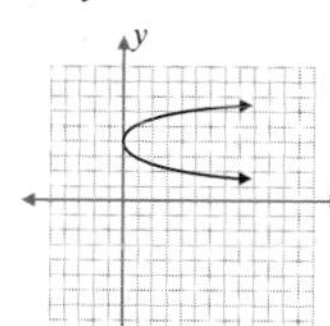

17. a. $(-2,9)$ **b.** $(0,5)$ **c.** $x=-2$

19. a. $(-3,-4)$ **b.** $(0,5)$ **c.** $x=-3$

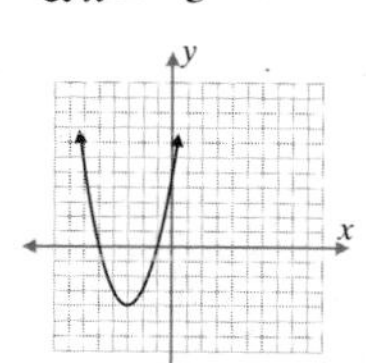

21. a. $(1,2)$ **b.** $(0,1), (0,3)$ **c.** $y=2$

23. a. $\left(-\frac{1}{4},-\frac{9}{8}\right)$ **b.** $(0,-1)$ **c.** $x=-\frac{1}{4}$

25. a. $(-8,-1)$ **b.** $\left(0,-1+\frac{2\sqrt{6}}{3}\right)$, $\left(0,-1-\frac{2\sqrt{6}}{3}\right)$ **c.** $y=-1$

27. a. $\left(\frac{9}{8},\frac{5}{4}\right)$ **b.** $\left(0,\frac{1}{2}\right)$, $(0,2)$ **c.** $y=\frac{5}{4}$

29. a. $\left(\frac{3}{2},0\right)$ **b.** $(0,9)$ **c.** $x=\frac{3}{2}$

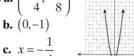

31. $(0, 1.225), (0, -1.225)$ **33.** $(0,2), (0,0)$

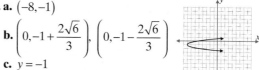

35. no y-intercept **37.** $(0, 0.658), (0, -2.658)$ **39.** $(0, 2.581), (0, -0.581)$ **41.** ii **43.** i **45.** The smaller $|a|$ is, the more open the graph will be.

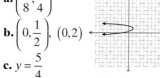

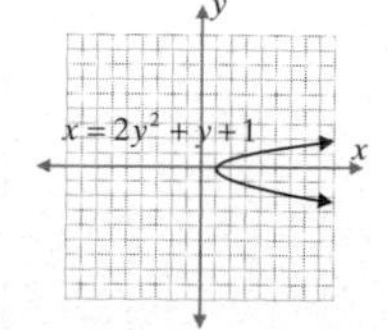

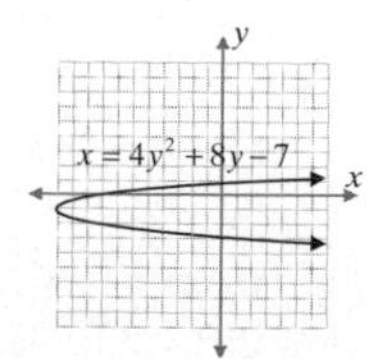

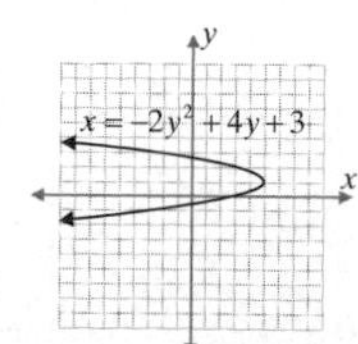

Exercises 12.5, pages 852 – 855

1. 5;(4, 5.5) **3.** 13;(3, 4.5) **5.** $\sqrt{29}$;(2, 4.5) **7.** 3;(5.5, −3) **9.** $\sqrt{13}$;(6, −3.5) **11.** 17;(−3, −4.5) **13.** $x^2+y^2=16$

15. $x^2+y^2=3$ **17.** $x^2+y^2=11$ **19.** $x^2+y^2=\frac{4}{9}$ **21.** $x^2+(y-2)^2=4$ **23.** $(x-4)^2+y^2=1$

25. $(x+2)^2+y^2=8$ **27.** $(x-3)^2+(y-1)^2=36$ **29.** $(x-3)^2+(y-5)^2=12$ **31.** $(x-7)^2+(y-4)^2=10$

33. $x^2+y^2=9$
Center: $(0,0)$, $r=3$

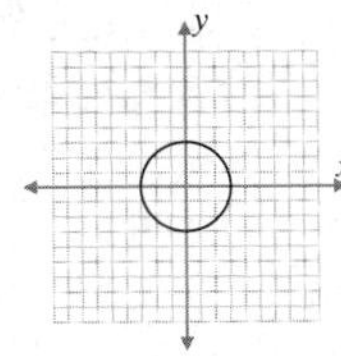

35. $x^2+y^2=49$
Center: $(0,0)$, $r=7$

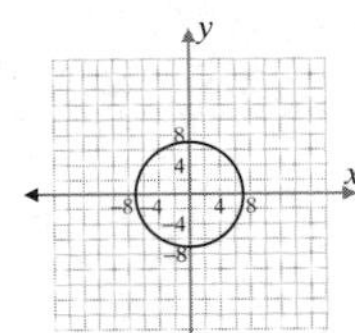

37. $x^2+y^2=18$
Center: $(0,0)$, $r=3\sqrt{2}$

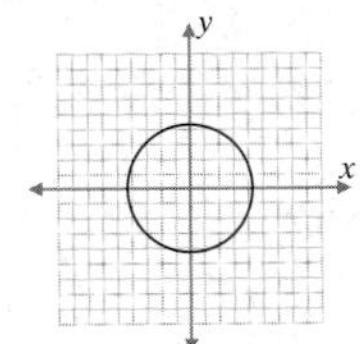

39. $(x+1)^2+y^2=9$
Center: $(-1,0)$, $r=3$

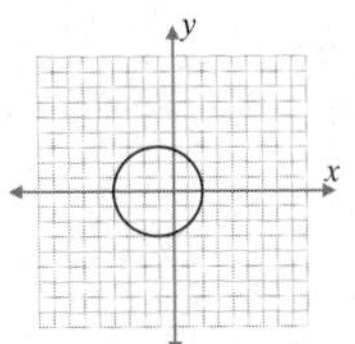

41. $x^2+(y-2)^2=4$
Center: $(0,2)$, $r=2$

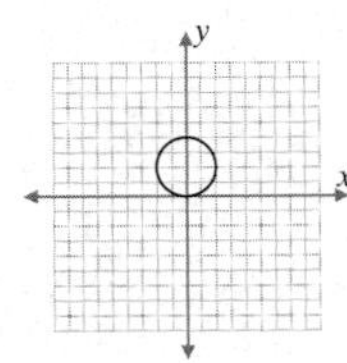

43. $(x+1)^2+(y+2)^2=16$
Center: $(-1,-2)$, $r=4$

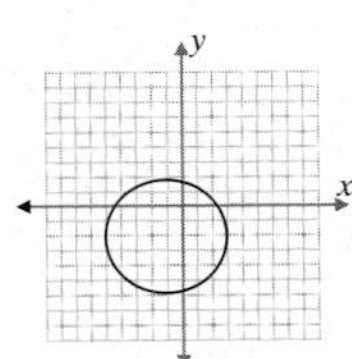

45. $(x+2)^2+(y+2)^2=16$
Center: $(-2,-2)$, $r=4$

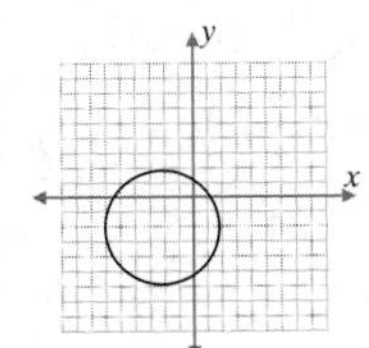

47. $(x-2)^2+(y-3)^2=8$
Center: $(2,3)$, $r=2\sqrt{2}$

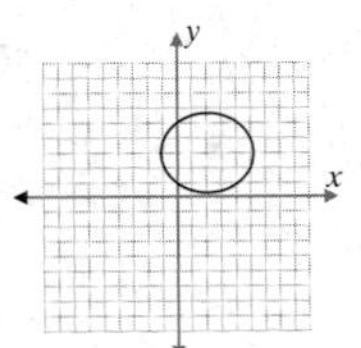

49. $|\overline{AB}|^2=45, |\overline{AC}|^2=65$
$|\overline{BC}|^2=20$
$|\overline{AB}|^2+|\overline{BC}|^2=|\overline{AC}|^2$
Right Triangle

51. $|\overline{AB}|=|\overline{AC}|=4\sqrt{5}$ **53.** $|\overline{AB}|=|\overline{AC}|=|\overline{BC}|=4$ **55.** $|\overline{AC}|=|\overline{BD}|=\sqrt{61}$ **57.** $10+4\sqrt{5}$

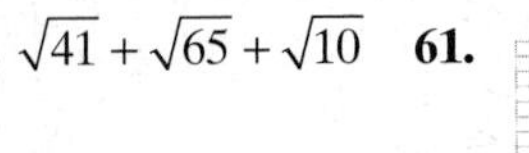

59. $\sqrt{41}+\sqrt{65}+\sqrt{10}$ **61.**

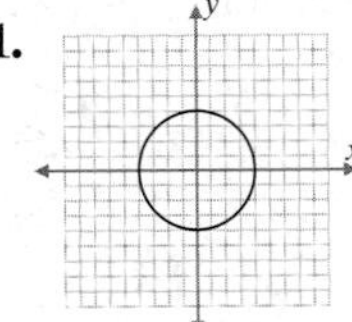

63.

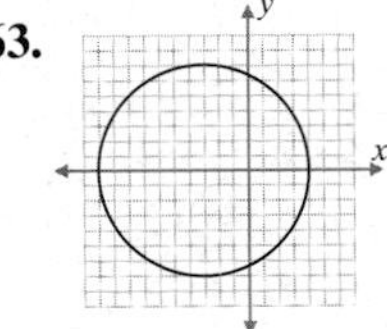

65. **a.** $d=\sqrt{x^2+(y-p)^2}$
b. $d=y+p$
c. $y+p=\sqrt{x^2+(y-p)^2}$
$(y+p)^2=x^2+(y-p)^2$
$y^2+2py+p^2=x^2+y^2-2py+p^2$
$x^2=4py$

67. **a.** $d=\sqrt{(x-p)^2+y^2}$
b. $d=x+p$
c. $x+p=\sqrt{(x-p)^2+y^2}$
$(x+p)^2=(x-p)^2+y^2$
$x^2+2px+p^2=x^2-2px+p^2+y^2$
$y^2=4px$

Exercises 12.6, pages 865 – 867

1. $\frac{x^2}{36}+\frac{y^2}{4}=1$ **3.** $\frac{x^2}{25}+\frac{y^2}{4}=1$ **5.** $\frac{x^2}{1}+\frac{y^2}{16}=1$

7. $\dfrac{x^2}{1}-\dfrac{y^2}{1}=1$

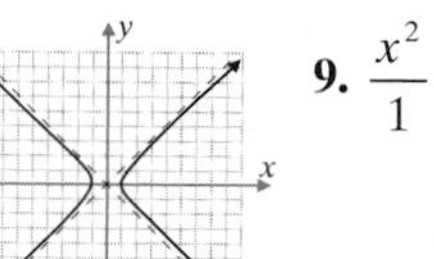

9. $\dfrac{x^2}{1}-\dfrac{y^2}{9}=1$

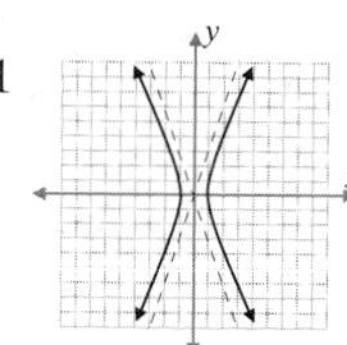

11. $\dfrac{x^2}{9}-\dfrac{y^2}{4}=1$

13. $\dfrac{x^2}{4}+\dfrac{y^2}{8}=1$

15. $\dfrac{x^2}{20}+\dfrac{y^2}{4}=1$

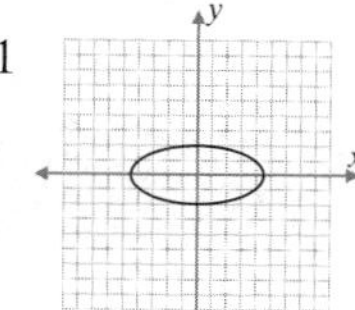

17. $\dfrac{y^2}{9}-\dfrac{x^2}{9}=1$

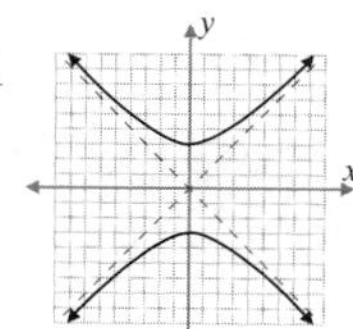

19. $\dfrac{y^2}{8}-\dfrac{x^2}{4}=1$

21. $\dfrac{y^2}{18}-\dfrac{x^2}{9}=1$

23. $\dfrac{x^2}{6}+\dfrac{y^2}{9}=1$

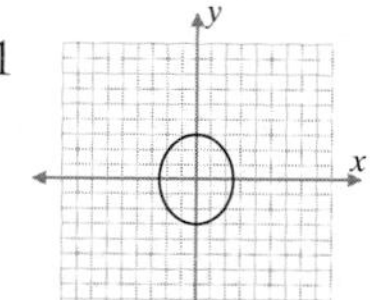

25. $\dfrac{x^2}{5}+\dfrac{y^2}{4}=1$

27. $\dfrac{x^2}{25}-\dfrac{y^2}{15}=1$

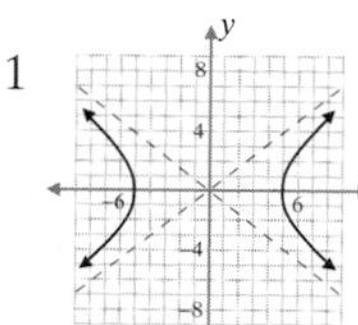

29. $\dfrac{y^2}{12}-\dfrac{x^2}{9}=1$

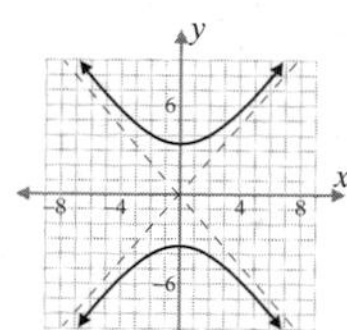

31. E **33.** D **35.** B

37.

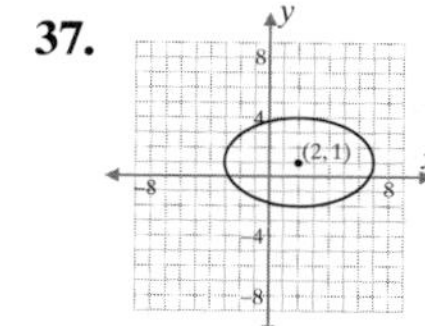

39.

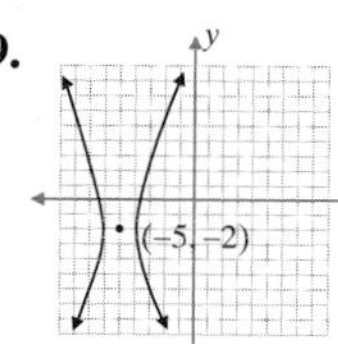

41.

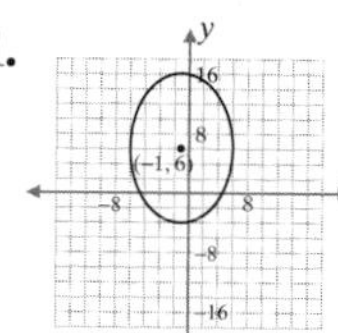

43. b. Set up the equation $\sqrt{(x+c)^2+(y-0)^2}+\sqrt{(x-c)^2+(y-0)^2}=2a$ and square both sides twice and simplify. **c.** At the y-intercept $(0,b)$ a right triangle is formed with hypotenuse a and sides b and c. The Pythagorean Theorem gives $a^2=b^2+c^2$.

Exercises 12.7, pages 873 – 874

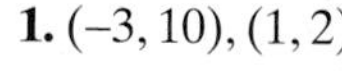

1. $(-3,10),(1,2)$ **3.** $(1,1),(2,0)$ **5.** $(-2,-4),(4,2)$ **7.** $(5,3)$ **9.** $(-3,0),(3,0)$ **11.** $(2,3),(2,-3)$

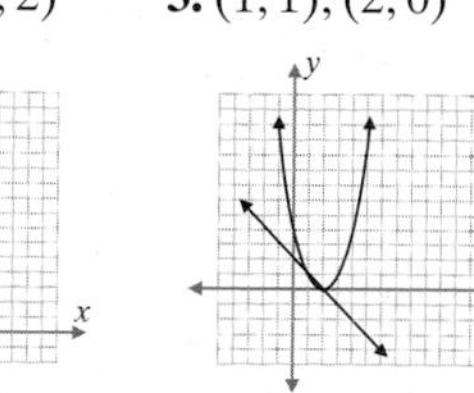

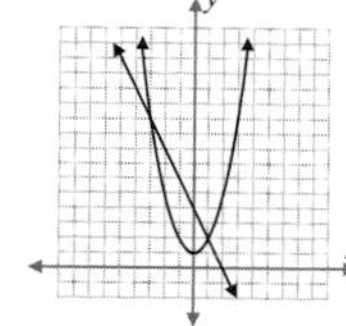

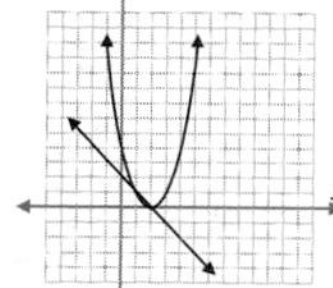

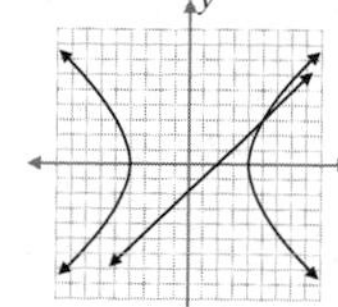

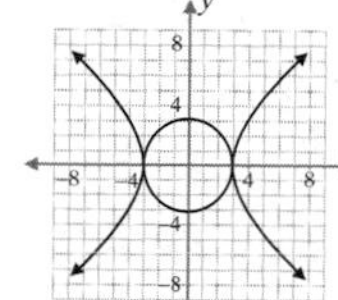

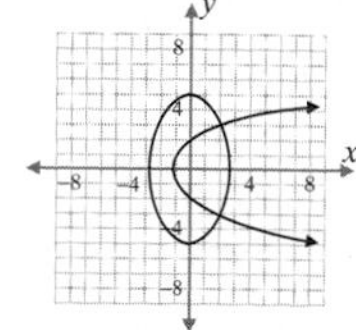

13. $(-2,4),(1,1)$

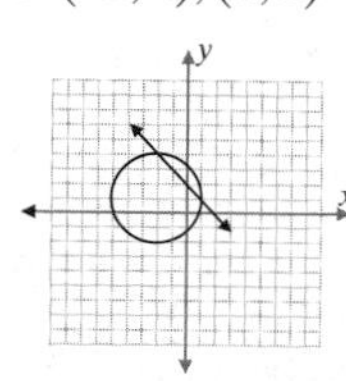

15. $(-3,-2),(-3,2),(3,-2),(3,2)$

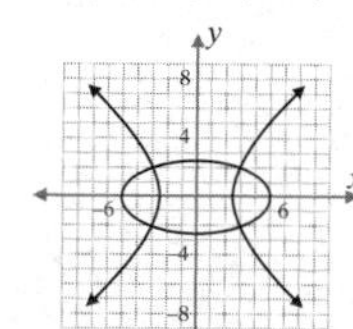

17. $\left(-3\sqrt{5},5\right),\left(-6,4\right),\left(3\sqrt{5},5\right),\left(6,4\right)$

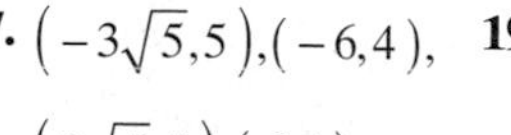

19. $(1,3),(-1,3)$ **21.** $\left(-3,\sqrt{11}\right),\left(-3,-\sqrt{11}\right),\left(3,\sqrt{11}\right),\left(3,-\sqrt{11}\right)$

23. $\left(-\dfrac{3}{2},-1\right)\left(\dfrac{1}{2},\dfrac{1}{2}\right)$

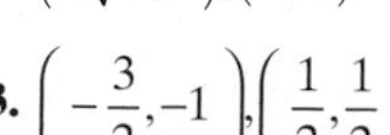

25. $(4,4),(-2,-2)$ **27.** $(5,-3),(-5,1)$ **29.** $(0,3),(2,3)$

31. (0, 3), (−1, 4)

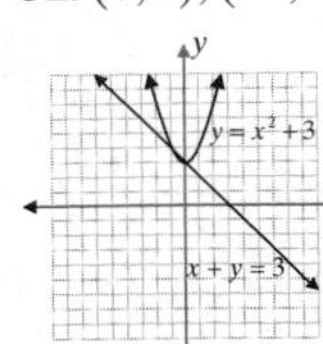

33. (−1.414, 5.828), (1.414, 0.172)

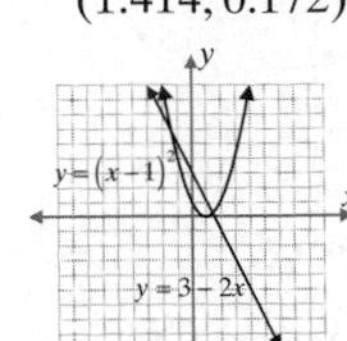

35. (2.679, 1.679), (−1.679, −2.679)

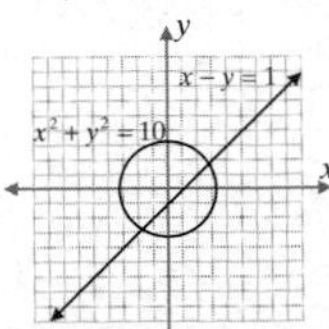

Chapter 12 Test, pages 880 – 881

1. $y = (x-3)^2 - 1$
Vertex: $(3, -1)$
Axis: $x = 3$
Domain: $(-\infty, \infty)$
Range: $y \geq -1$
Zeros: $x = 2, 4$

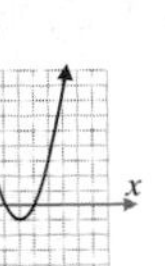

2. $y = -2\left(x - \frac{3}{2}\right)^2 + \frac{15}{2}$
Vertex: $\left(\frac{3}{2}, \frac{15}{2}\right)$
Axis: $x = \frac{3}{2}$
Domain: $(-\infty, \infty)$
Range: $y \leq \frac{15}{2}$
Zeros: $x = \frac{3 \pm \sqrt{15}}{2}$

3. $y = 2(x-3)^2 - 9$
Vertex: $(3, -9)$
Axis: $x = 3$
Domain: $(-\infty, \infty)$
Range: $y \geq -9$
Zeros: $x = \frac{6 \pm 3\sqrt{2}}{2}$

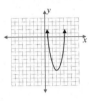

4. a. 13
b. $2x^2 + 6$
c. $2x^2 + 4x + 7$
d. $4x + 2h$

5. $(-\infty, -3] \cup [5, \infty)$

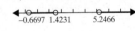

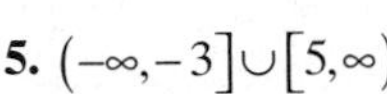

6. $\left(-4, -\frac{1}{2}\right)$

−4 $-\frac{1}{2}$

7. [1, 3]

1 3

8. (−0.6697, 1.4231) ∪ (5.2466, ∞)

−0.6697 1.4231 5.2466

9.

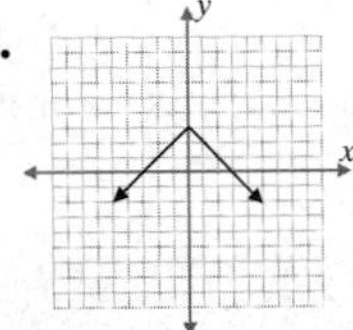

10.

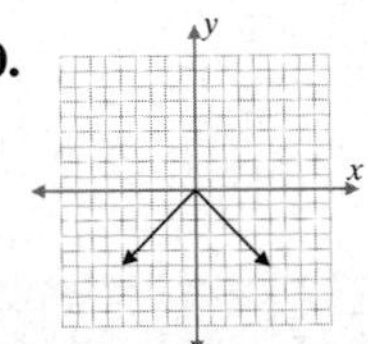

11.

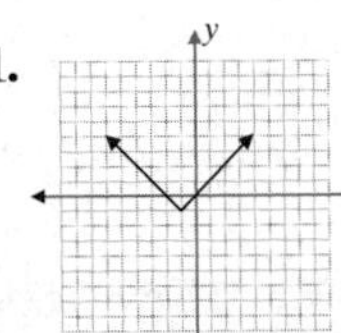

12. −25

13. $5\frac{1}{2}$ in. by $5\frac{1}{2}$ in.

14. vertex: (−5, 0)
y-intercept: $(0, \sqrt{5}), (0, -\sqrt{5})$
line of symmetry: $y = 0$

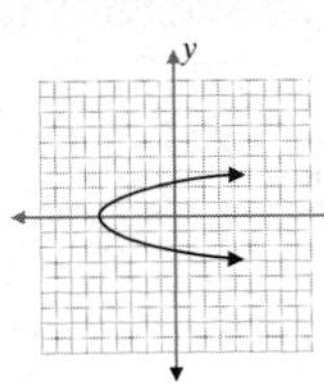

15. a. vertex: $\left(-\frac{25}{4}, -\frac{3}{2}\right)$
y-intercept: $(0, -4), (0, 1)$
line of symmetry: $y = -\frac{3}{2}$

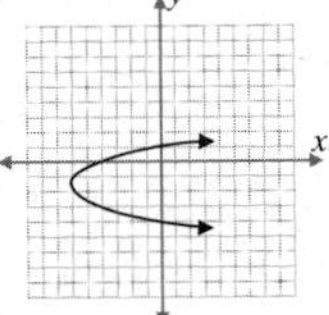

16. $3\sqrt{10}$

17. $\left|\overline{AB}\right|^2 = 52$,
$\left|\overline{AC}\right|^2 = 104$
$\left|\overline{BC}\right|^2 = 52$
$\left|\overline{AB}\right|^2 + \left|\overline{BC}\right|^2 = \left|\overline{AC}\right|^2$

18. $(x+3)^2 + (y+1)^2 = 25$

19. $x^2 + (y-1)^2 = 9$
Center: $(0, 1)$, $r = 3$

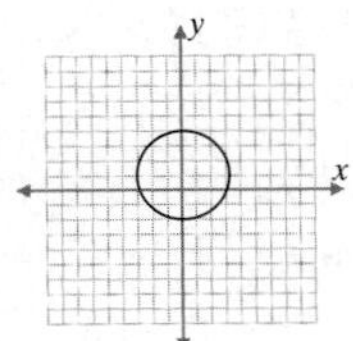

20. $\frac{x^2}{4} - \frac{y^2}{9} = 1$
$y = \frac{3}{2}x,\ y = -\frac{3}{2}x$

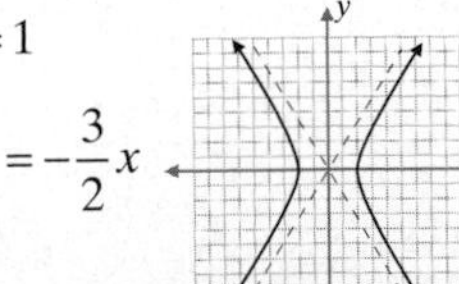

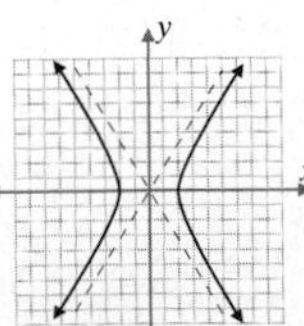

21. $\frac{x^2}{9} + \frac{y^2}{\frac{9}{4}} = 1$

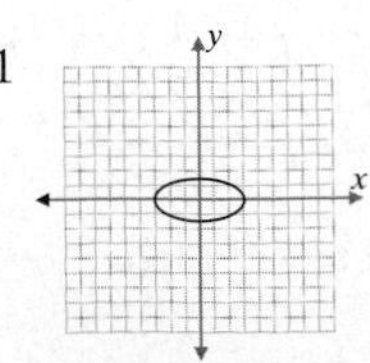

22. $\frac{y^2}{9} - \frac{x^2}{16} = 1$
$y = \frac{3}{4}x,\ y = -\frac{3}{4}x$

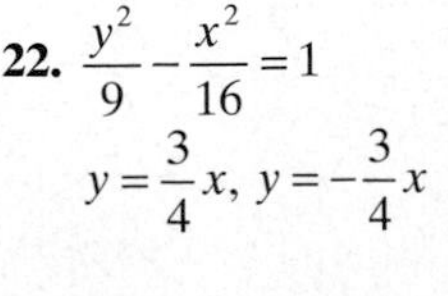

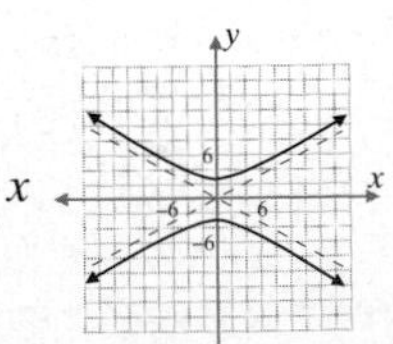

23. $\dfrac{x^2}{4}+\dfrac{y^2}{25}=1$ **24.** 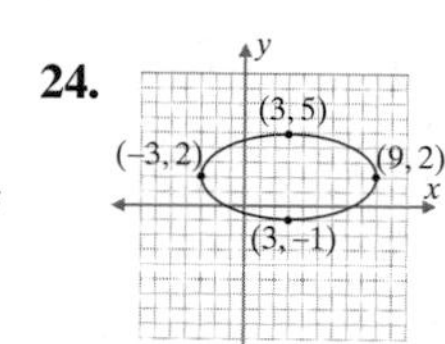 **25.** 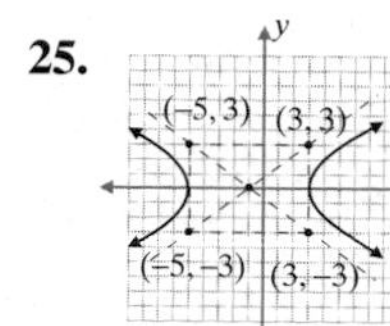 **26.** $(-2,-5),(5,2)$

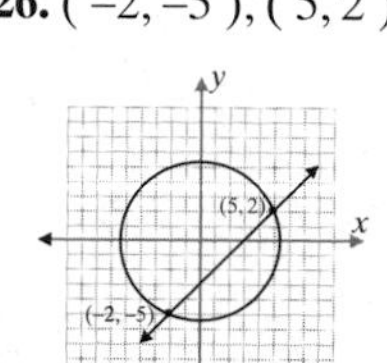

27. $(0,\sqrt{2}),\ (0,-\sqrt{2}),\ (-2,0)$ **28.** $(4,3),(4,-3),(-4,3),(-4,-3)$

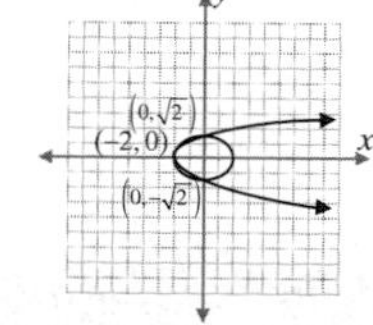

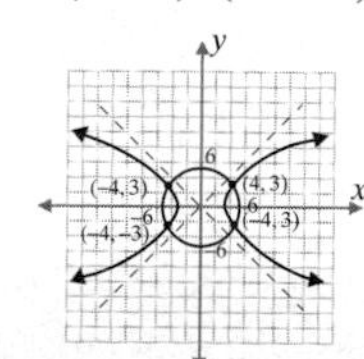

29. Maximum Area = 11,250 square feet, Dimensions = 75 × 150
30. Maximum Area = 6050 square feet, Dimensions = 55 × 110
31. Kim drives 60 mph and her sister drives 70 mph.

Chapter 12 Cumulative Review, pages 882 – 884

1. 1 **2.** $\dfrac{9x^8}{4y^8}$ **3.** $5x^{\frac{3}{4}}$ **4.** $\dfrac{8y^{\frac{3}{5}}}{x}$ **5.** $\sqrt[3]{(7x^3y)^2}=x^2\sqrt[3]{49y^2}$ **6.** $(32x^6y)^{\frac{1}{3}}=2x^2(4y)^{\frac{1}{3}}$ **7.** $2(x+3)(x^2-3x+9)$

8. $x^{-2}(2x-5)(x+7)$ **9.** $(x-4)(x^2+3)$ **10.** $13\sqrt{3}$ **11.** $2\sqrt{3x}+10\sqrt{6}$ **12.** $5x-2y=-16$

13. $3x-4y=16$ **14.** $\dfrac{2x^2+x+4}{(2x+3)(x-4)(3x-2)}$ **15.** -1 **16.** $x=\dfrac{-1\pm\sqrt{7}}{3}$ **17.** $x=-9$ **18.** $x=2$ **19.** $x=-2,\ y=-3$

20. $x=\dfrac{28}{13},\ y=\dfrac{10}{13}$ **21.** $y=\dfrac{1}{2}(x+0)^2-3$
vertex: $(0,-3)$
axis : $x=0$
$D_f=(-\infty,\infty)$
$R_f=\{y|y\geq-3\}$
Zeros: $x=\pm\sqrt{6}$

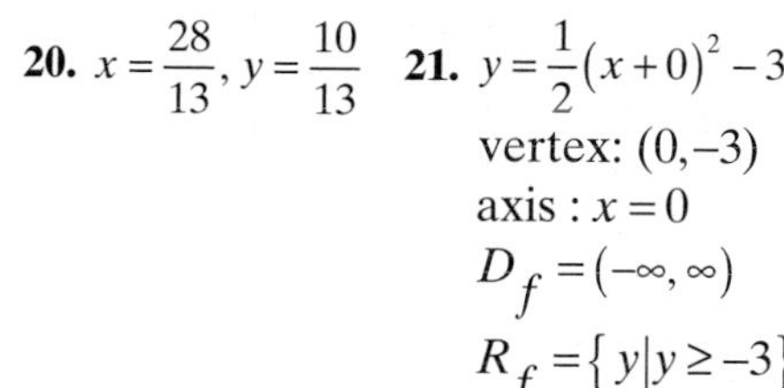
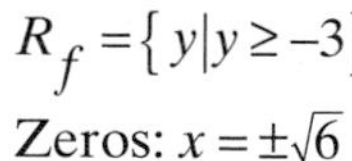
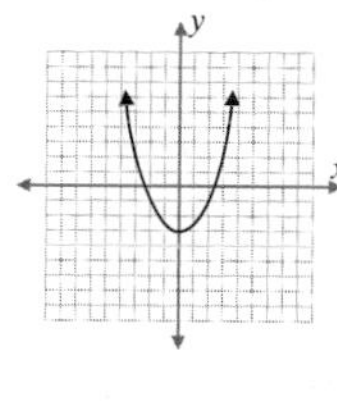

22. $y=-(x-2)^2+0$
vertex: $(2,0)$
axis : $x=2$
$D_f=(-\infty,\infty)$
$R_f=\{y|y\leq0\}$
Zeros: $x=2$

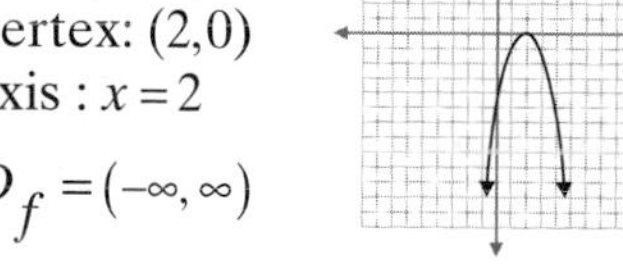
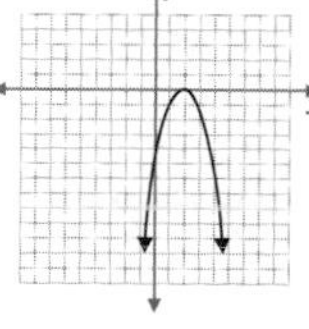

23.

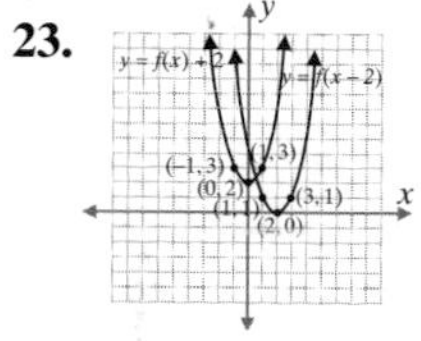

24. $x\geq5,\ -1\leq x<0$ **25.**

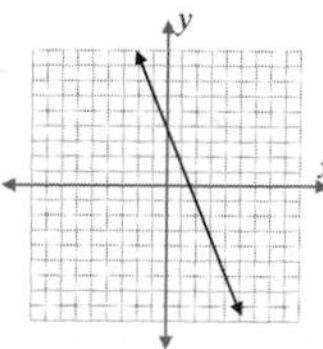

26.

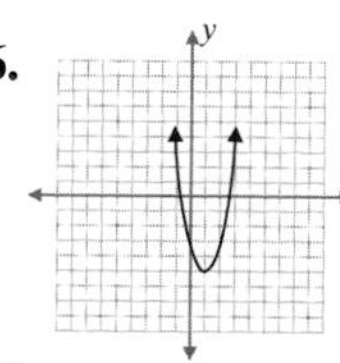

27.

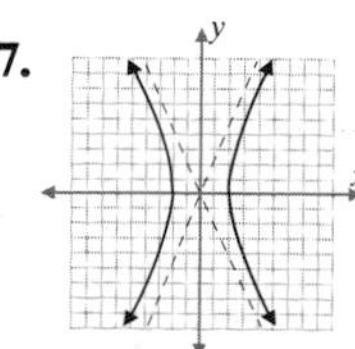

28.

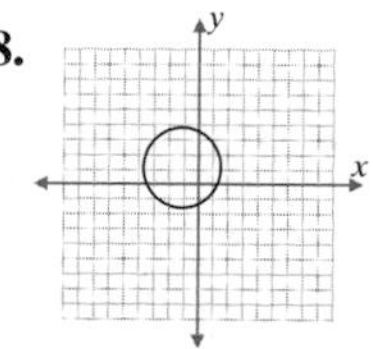

29. a. 5 **b.** $4x-11$ **c.** $4x-4$ **d.** 4 **30. a.** 13 **b.** $2x^2+3x-3$ **c.** $2x^2-5x+1$ **d.** $4x+2h+3$ **31.** \$3600 at 7%; \$5400 at 8%

32. 36 **33. a.** $t=2\frac{1}{2}$ sec. **b.** 148 ft. **34.** \$15 **35.**

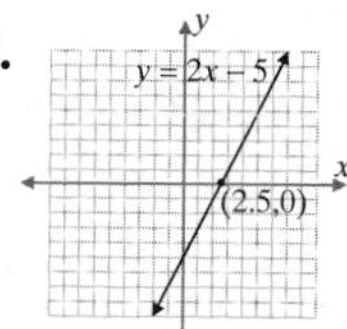

36.

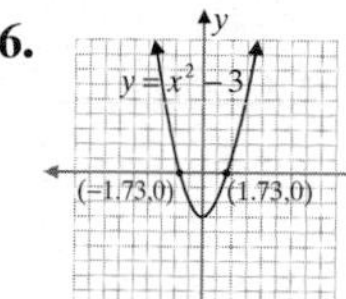

37.

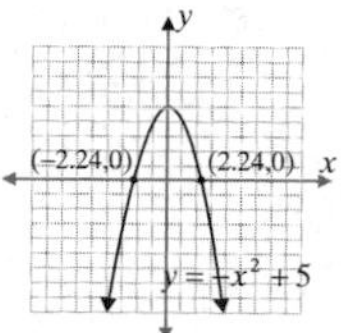

38. $y=(x+3)^2-6$; $(-5.45,0)$, $(-0.55,0)$

39. a. $(-2,0),(5,0)$ **b.** $(-\infty,-2)\cup(5,\infty)$ **c.** $(-2,5)$ **40. a.** $(-1,0),(2.5,0)$ **b.** $(-1,2.5)$ **c.** $(-\infty,-1)\cup(2.5,\infty)$

41. 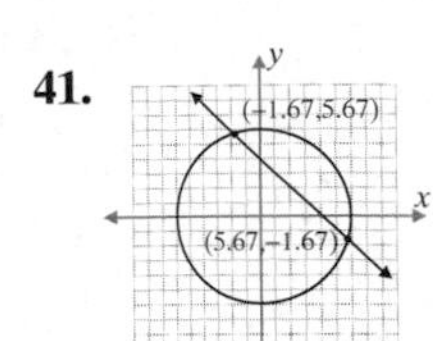

42. 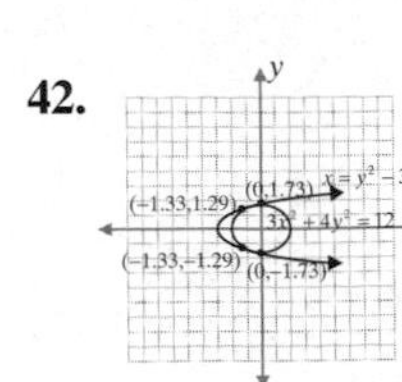

43. $2x^2+2x+20-\dfrac{50}{x-5}$ **44.** $a^2+b^2=c^2$ **45.** Domain: $[7,\infty)$, Range: $[0,\infty)$

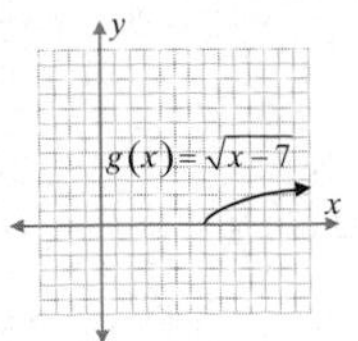

Chapter 13

Exercises 13.1, pages 892 – 895

1. a. $2x-3$ **b.** 7 **c.** $x^2-3x-10$ **d.** $\dfrac{x+2}{x-5}, x\neq 5$ **3. a.** x^2+3x-4 **b.** x^2-3x+4 **c.** $3x^3-4x^2$ **d.** $\dfrac{x^2}{3x-4}, x\neq\dfrac{4}{3}$

5. a. x^2+x-12 **b.** x^2-x-6 **c.** $x^3-3x^2-9x+27$ **d.** $x+3, x\neq 3$ **7. a.** $3x^2+x+2$ **b.** x^2+x-2 **c.** $2x^4+x^3+4x^2+2x$

d. $\dfrac{2x^2+x}{x^2+2}$ **9. a.** $2x^2+2$ **b.** $8x$ **c.** x^4-14x^2+1 **d.** $\dfrac{x^2+4x+1}{x^2-4x+1}, x\neq 2\pm\sqrt{3}$ **11.** 9 **13.** $-a^2-a-1$ **15.** 27 **17.** $\dfrac{8}{5}$ **19.** −31

21. $\sqrt{2x-6}+x+4$, Domain: $[3,\infty)$ **23.** $3x^2-19x-14$, Domain: $(-\infty,\infty)$ **25.** $\dfrac{x-5}{\sqrt{x+3}}$, Domain: $(-3,\infty)$

27. $-3x\sqrt{x-3}$, Domain: $[3,\infty)$ **29.** $\sqrt[3]{x+3}+\sqrt{5+x}$, Domain: $[-5,\infty)$ **31.** 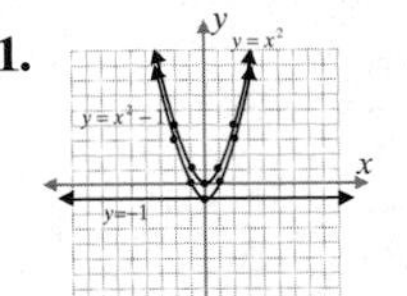

33. 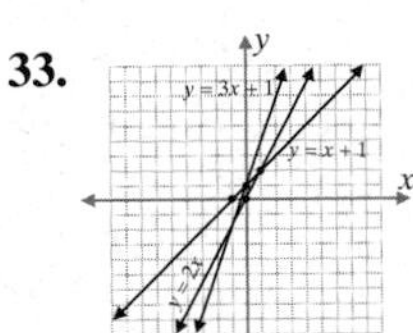

35. 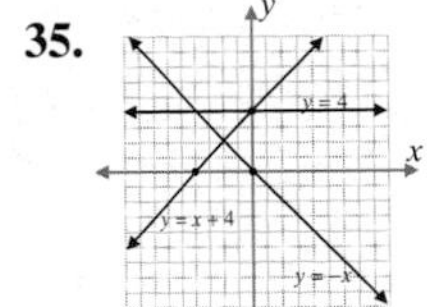

37. 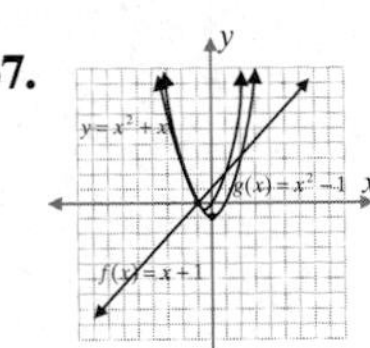

39. 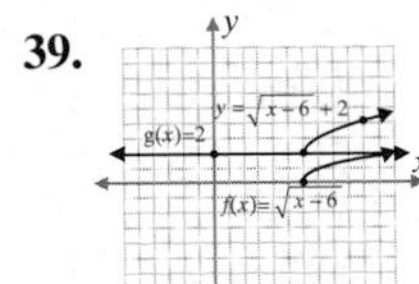

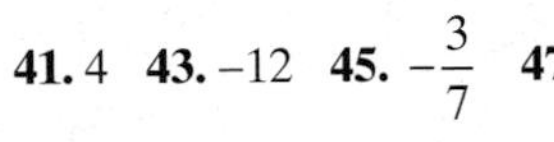
41. 4 **43.** −12 **45.** $-\dfrac{3}{7}$ **47.** 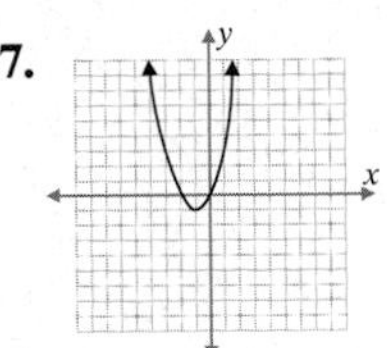

49. 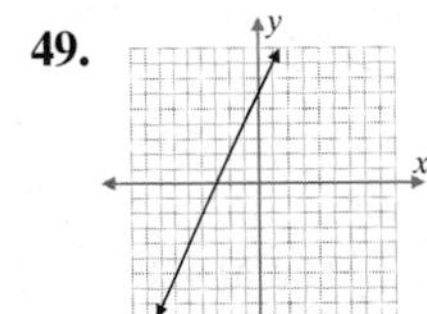

51. 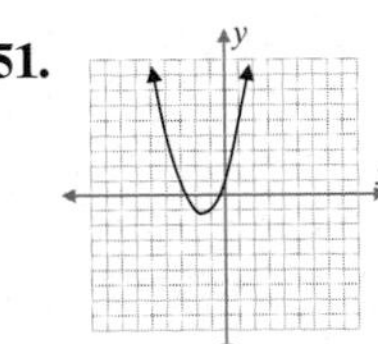

53. 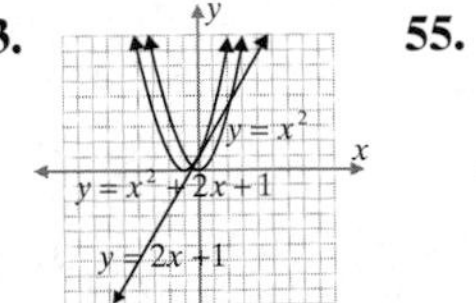

55. 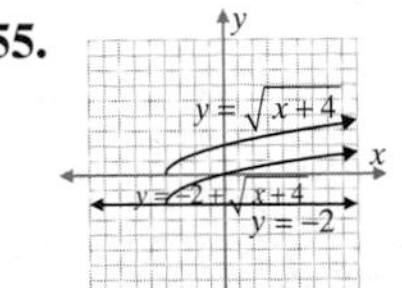

57. 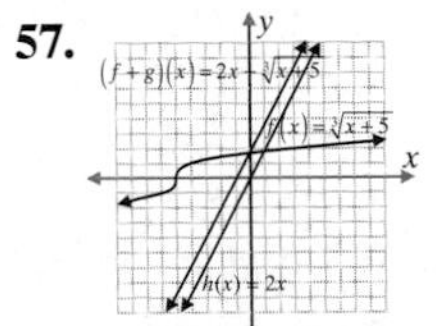

59. 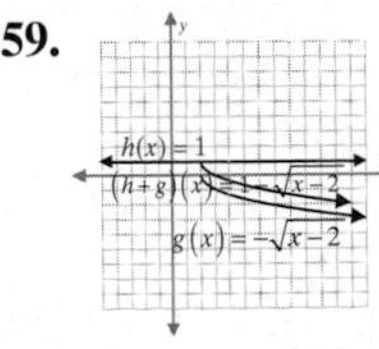

61. Answers will vary **63. a.** 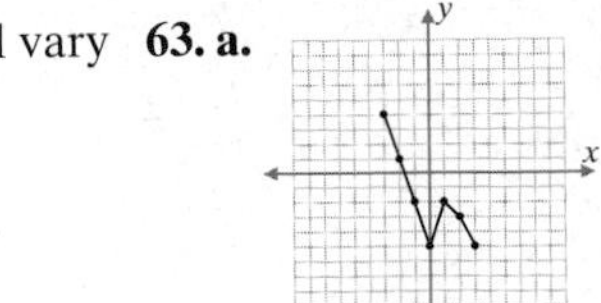

b. 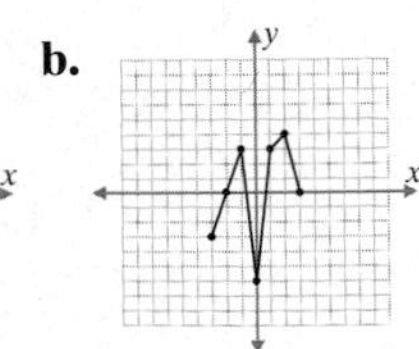

c. 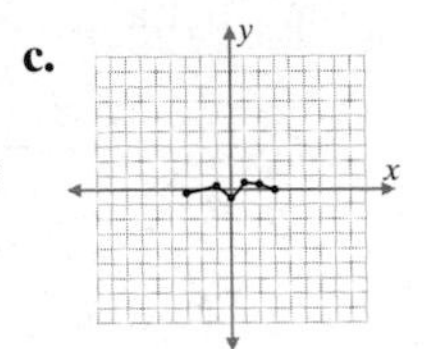

Exercises 13.2, pages 907 – 910

1. $f(g(x)) = \frac{3}{2}x + 11$
$g(f(x)) = \frac{3}{2}x + \frac{9}{2}$

3. $f(g(x)) = 4x^2 + 12x + 9$
$g(f(x)) = 2x^2 + 3$

5. $f(g(x)) = \frac{1}{5x - 8}$
$g(f(x)) = \frac{5}{x} - 8$

7. $f(g(x)) = \frac{1}{x^2} - 1$
$g(f(x)) = \frac{1}{(x-1)^2}$

9. $f(g(x)) = x^3 + 3x^2 + 4x + 3$
$g(f(x)) = x^3 + x + 2$

11. $f(g(x)) = \sqrt{x - 2}$
$g(f(x)) = \sqrt{x} - 2$

13. $f(g(x)) = \sqrt{x^2} = |x|$
$g(f(x)) = (\sqrt{x})^2 = x$

15. $f(g(x)) = \frac{1}{\sqrt{x^2 - 4}}$
$g(f(x)) = \left(\frac{1}{\sqrt{x}}\right)^2 - 4 = \frac{1}{x} - 4$

17. $f(g(x)) = x$
$g(f(x)) = x$

19. $f(g(x)) = (x-8)^{\frac{3}{2}}$
$g(f(x)) = \sqrt{x^3 - 8}$

21. **23.** **25.**

27. **29.**

31. $f^{-1}(x) = \frac{x+3}{2}$

33. $g^{-1}(x) = x$

35. $f^{-1}(x) = \frac{x-1}{5}$ **37.** $g^{-1}(x) = \frac{1-x}{3}$ **39.** $f^{-1}(x) = x^2$ **41.** $f^{-1}(x) = \sqrt{x - 1}$ **43.** $f^{-1}(x) = -x - 2$

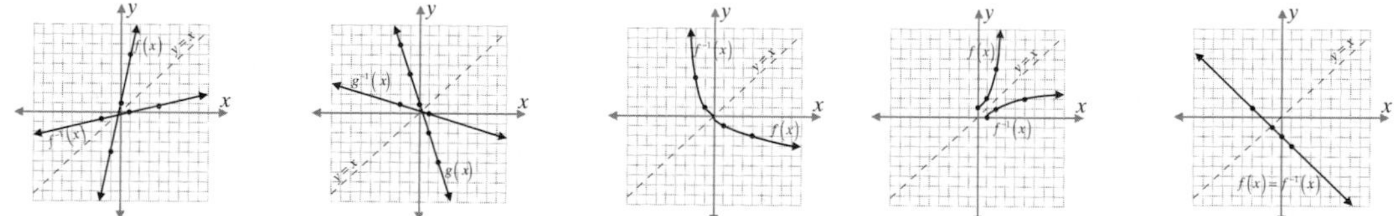

45. One-to-one

47. Not one-to-one

49. Not one-to-one

51. One-to-one

53. One-to-one

55. Is 1-1

57. Is not 1-1

59. Is 1-1

61. Is 1-1

63. Is not 1-1

65.

67. **69.** **71.** **73.** **75.**

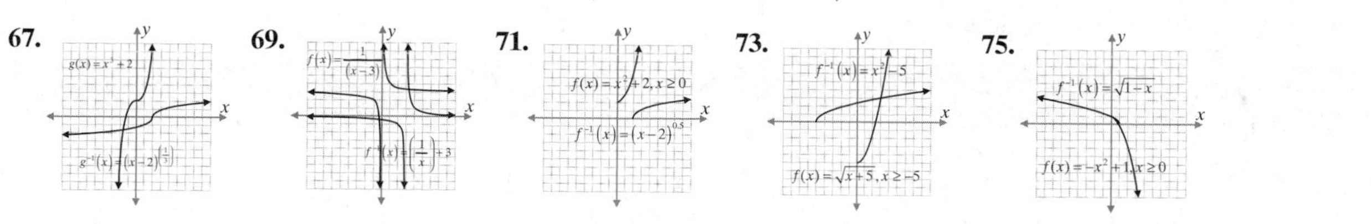

Exercises 13.3, pages 920 – 922

1. 3. 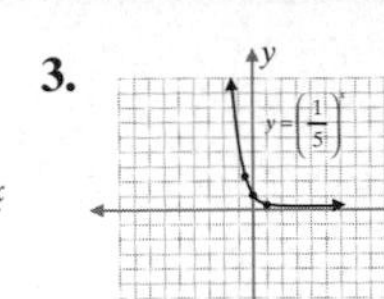

5. 7. 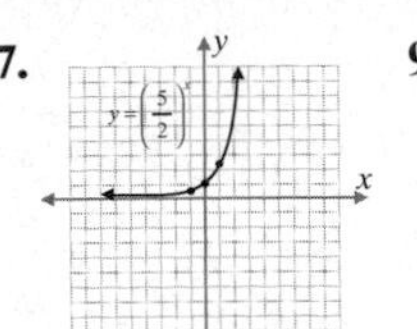

9. 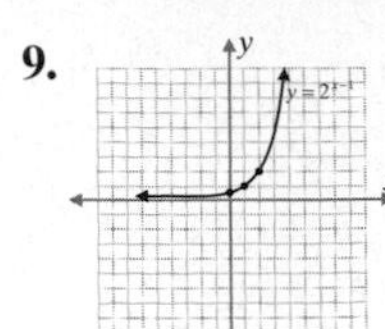

11. 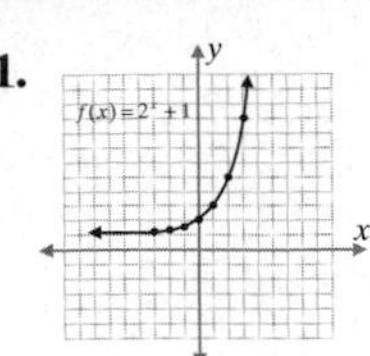

13. 15. 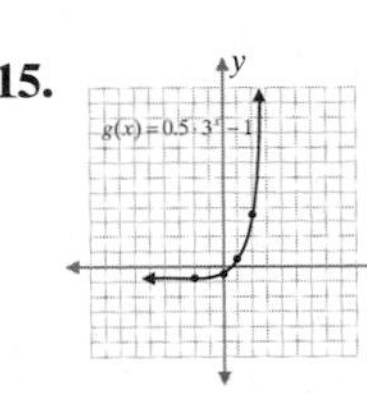

17. 19. 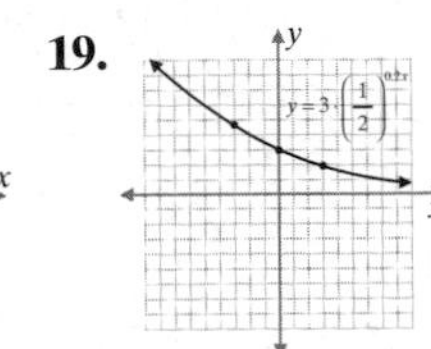

21. 48 **23.** 30 **25.** 6008.332 **27.** 20,000 bacteria **29. a.** \$2621.59 **b.** \$2633.62 **c.** \$2639.86 **d.** \$2646.19 **e.** \$2646.26 **31.** \$3210.06 **33.** \$53.33 **35.** $\frac{1}{64}$ **37.** \$17,182.82, see page 918

39. \$1228.25 **41.** Value after 3 years = \$5925.93, $t = 6$ years **43.** Population up to March = 288 bunnies, Population at the end of the year = 34, 035 bunnies **45.** Answers will vary

Exercises 13.4, pages 933 – 935

1. $\log_7 49 = 2$ **3.** $\log_5 \frac{1}{25} = -2$ **5.** $\log_2 \frac{1}{32} = -5$ **7.** $\log_{2/3} \frac{4}{9} = 2$ **9.** $\ln 17 = x$ **11.** $\log 10 = 1$ **13.** $3^2 = 9$ **15.** $9^{\frac{1}{2}} = 3$

17. $7^{-1} = \frac{1}{7}$ **19.** $e^{1.74} = N$ **21.** $b^4 = 18$ **23.** $n^x = y^2$ **25.** $x = 16$ **27.** $x = 2$ **29.** $x = \frac{1}{6}$ **31.** $x = 11$ **33.** $x = 32$ **35.** $x = -2$

37. $x = 1.52$ **39.** $x = 3$ **41.** 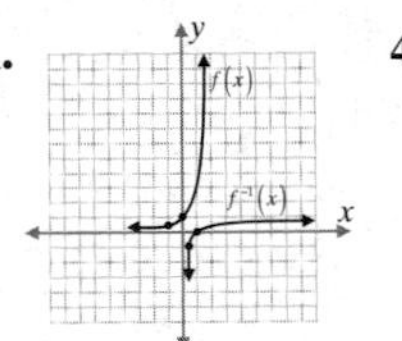

43. **45.** 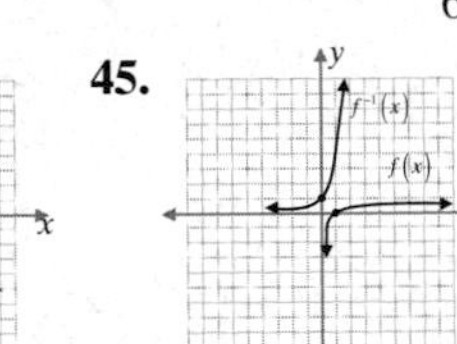

47. 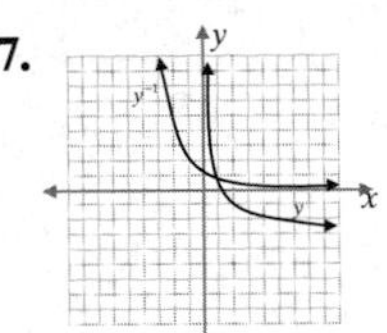

49.

51. 2.23805 **53.** 1.94645 **55.** −1.24185 **57.** 3.62434 **59.** Error **61.** −5.37953 **63.** 204.17379 **65.** 0.0199526 **67.** 0.95719 **69.** 175.91484 **71.** 0.00024852 **73.** 1.04029

75. $D = (-1, \infty)$ $R = (-\infty, \infty)$ **77.** 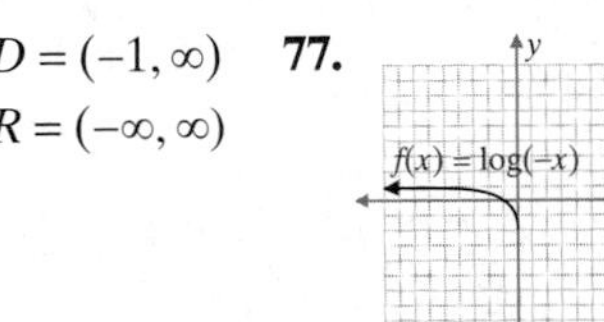

$D = (-\infty, 0)$ $R = (-\infty, \infty)$

79. 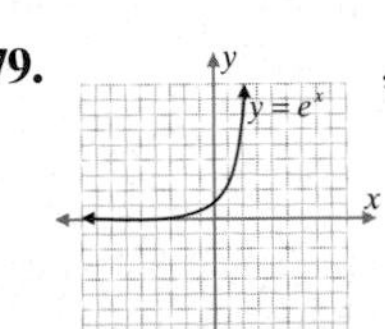

, Answers will vary.

81. Answers will vary. **83.** Answers will vary.

Exercises 13.5, pages 942 – 943

1. 5 **3.** −2 **5.** $\frac{1}{2}$ **7.** 10 **9.** $\sqrt{3}$ **11.** $\log 5 + 4 \log x$ **13.** $\ln 2 - 3 \ln x + \ln y$ **15.** $\log 2 + \log x - 3 \log y$

17. $2 \ln x - \ln y - \ln z$ **19.** $-2 \log x - 2 \log y$ **21.** $\frac{1}{3}\log x + \frac{2}{3}\log y$ **23.** $\frac{1}{2}\ln x + \frac{1}{2}\ln y - \frac{1}{2}\ln z$ **25.** $\log 21 + 2\log x + \frac{2}{3}\log y$

27. $-\frac{1}{2}\log x - \frac{5}{2}\log y$ **29.** $-9\ln x - 6\ln y + 3\ln z$ **31.** $\ln\frac{9x}{5}$ **33.** $\log\frac{7x^2}{8}$ **35.** $\log x^2 y$ **37.** $\ln\frac{x^3}{y^2}$ **39.** $\ln\sqrt{\frac{x}{y}}$

41. $\log\frac{xz}{y}$ **43.** $\log\frac{x}{y^2z^2}$ **45.** $\log(2x^2 + x)$ **47.** $\ln\left(x^2 + 2x - 3\right)$ **49.** $\log\frac{x^2 - 2x - 3}{x - 3}$ **51.** $\log\frac{x+6}{2x^2 + 9x - 18}$

53. Answers will vary.

Exercises 13.6, pages 951 – 953

1. $x = 11$ **3.** $x = 9$ **5.** $x = \frac{1}{6}$ **7.** $x = \frac{7}{12}$ **9.** $x = \frac{7}{2}$ **11.** $x = -2$ **13.** $x = -3$ **15.** $x = 4, x = -1$ **17.** $x = -1, x = \frac{3}{2}$

19. $x = -2, x = 3$ **21.** $x = 2$ **23.** $x = 0, x = -\frac{3}{2}$ **25.** $x = 3, x = -1$ **27.** $x \approx 0.7154$ **29.** $x \approx 7.5098$ **31.** $x \approx -3.648$

33. $x \approx 24.7312$ **35.** $t \approx -653.0008$ **37.** $t \approx -1.5193$ **39.** $x \approx 3.322$ **41.** $x \approx -1.4307$ **43.** $x = 1$ **45.** $x \approx 1.2203$

47. $x \approx -1.6467$ **49.** $x \approx 1.1292$ **51.** $x \approx 4.854$ **53.** $x \approx 1.252$ **55.** $x \approx 25.1189$ **57.** $x \approx 31.6228$ **59.** $x = 0.0001$

61. $x \approx 4.953$ **63.** $x \approx \pm 0.3329$ **65.** $x = 3$ **67.** $x = 100$ **69.** $x = 6$ **71.** $x = 20$ **73.** $x = \frac{25}{2}$ **75.** $x = 1001$ **77.** No solution

79. $x \approx 1.1353$ **81.** $x \approx 22.0855$ **83.** $x = -10, x = 8$ **85.** 2.2619 **87.** 0.3223 **89.** −0.6279 **91.** 2.4391 **93.** −1.2222
95. 2.3219 **97.** 3.1133 **99.** 0.839 **101.** Answers will vary. **103.** Answers will vary, $7^{1+x}, 7^{2x}$

Exercises 13.7, pages 957 – 959

1. $4027.51 **3.** 13.9 years **5.** $f = \frac{3}{10}$ **7.** 1.73 hours **9.** 8166 bees **11.** 12.28 lb. per sq. in. **13.** 2350 years **15.** 2.3 days

17. 8.75 years **19.** 9 years **21. a.** 13.86 years **b.** 6.93 years **23.** 294.41 days **25.** 100 **27.** 8.64 million **29.** 2083, 2437, 2744

Chapter 13 Test, pages 965 – 966

1. a. $\sqrt{x-3} + x^2 + 1$ **b.** $\sqrt{x-3} - x^2 - 1$ **c.** $\left(\sqrt{x-3}\right)\cdot\left(x^2+1\right)$ **d.** $\frac{\sqrt{x-3}}{x^2+1}$ **e.** a. $x \geq 3$ b. $x \geq 3$ c. $x \geq 3$ d. $x \geq 3$

2. a. $-4x^2 + 1$ **b.** $-8x^2 + 40x - 47$ **3. a.** Not inverse to each other **b.** Inverse to each other **c.** Inverse to each other

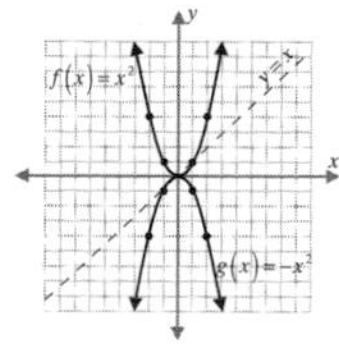

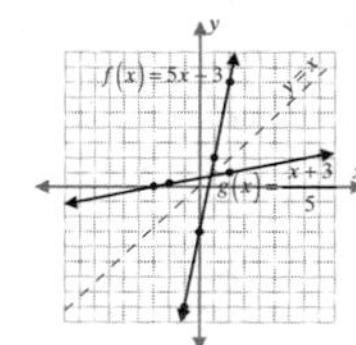

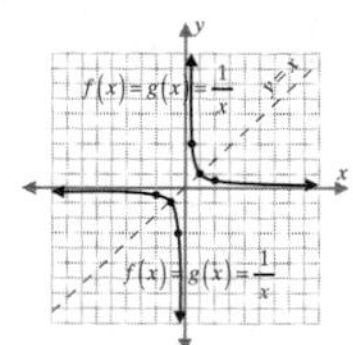

4. $f^{-1}(x) = \frac{1}{x} + 2$ **5.**

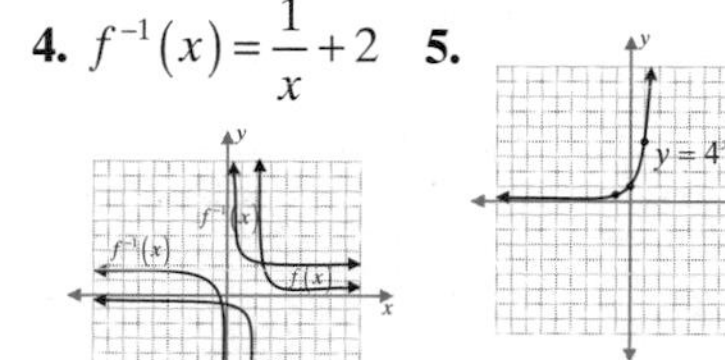

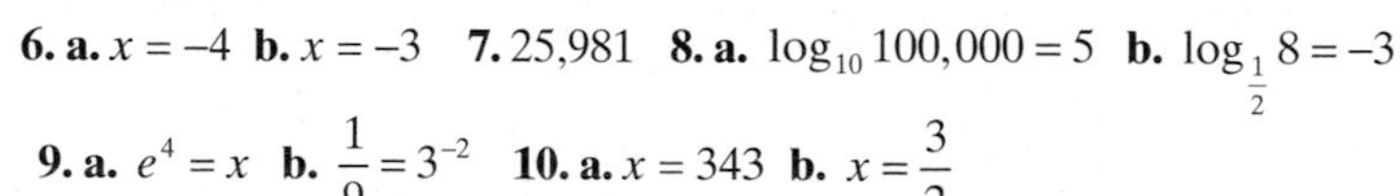

6. a. $x = -4$ **b.** $x = -3$ **7.** 25,981 **8. a.** $\log_{10} 100{,}000 = 5$ **b.** $\log_{\frac{1}{2}} 8 = -3$

9. a. $e^4 = x$ **b.** $\frac{1}{9} = 3^{-2}$ **10. a.** $x = 343$ **b.** $x = \frac{3}{2}$

11.

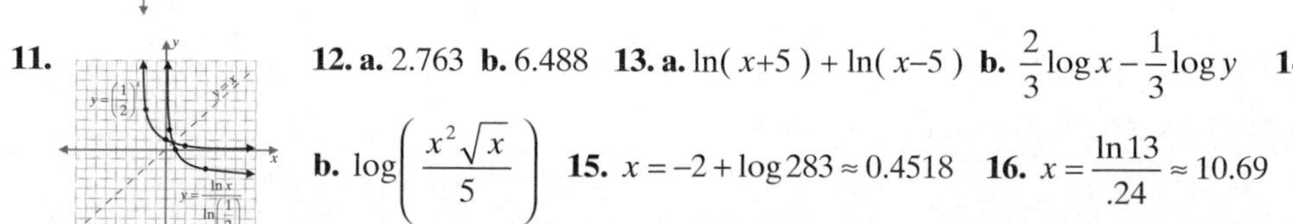

12. a. 2.763 **b.** 6.488 **13. a.** $\ln(x+5) + \ln(x-5)$ **b.** $\frac{2}{3}\log x - \frac{1}{3}\log y$ **14. a.** $\ln(x^2 + x - 20)$

b. $\log\left(\frac{x^2\sqrt{x}}{5}\right)$ **15.** $x = -2 + \log 283 \approx 0.4518$ **16.** $x = \frac{\ln 13}{.24} \approx 10.69$ **17.** $x = \frac{\ln 12}{\ln 4} \approx 1.79$

18. No solution **19.** $x = 1 + e^3 \approx 21.09$ **20.** 19.07 years **21. a.** 134.3 years **b.** 198 years **22.** about 73.26 minutes **23.** half-life = 5730 years **24.** $35.58

Chapter 13 Cumulative Review, pages 966 – 970

1. 1 **2.** $-\dfrac{x+3}{x+1}$ **3.** $\dfrac{2}{x+4}$ **4.** $\dfrac{x-4}{x+1}$ **5.** $\dfrac{x^{\frac{1}{6}}}{y^{\frac{1}{3}}}$ **6.** $\dfrac{x^{\frac{1}{2}}}{2y^2}$ **7.** $x^2y^{\frac{3}{2}}$ **8.** $x^{\frac{2}{3}}y$ **9.** Dependent, $x - 5y = 10$ **10.** $x = 2, y = 5$

11. Vertex: $(3, -11)$, Range: $y \geq -11$

Zeros: $x = 3 \pm \sqrt{11}$

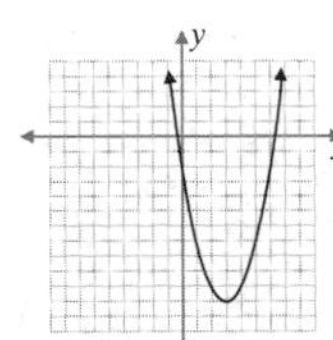

12. Vertex: $(-2, -5)$, Range: $(y \geq -5)$

Zeros: $x = -2 \pm \dfrac{\sqrt{10}}{2}$

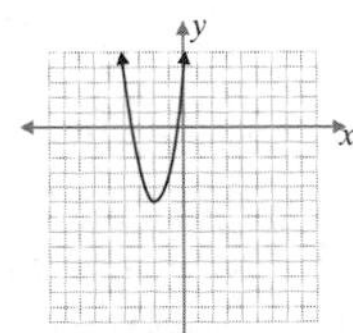

13. $x = \dfrac{-5 \pm \sqrt{33}}{2}$ **14.** $x = \pm 2, x = \pm 3$

15. $x = 6$ **16.** $x = 4 \pm \sqrt{6}$ **17.** $x \approx -0.9212$ **18.** $x = 5 + e^{2.5} \approx 17.18$ **19.** $x^2 + y^2 = 12$ **20.** $(x-1)^2 + (y-2)^2 = 9$

21.

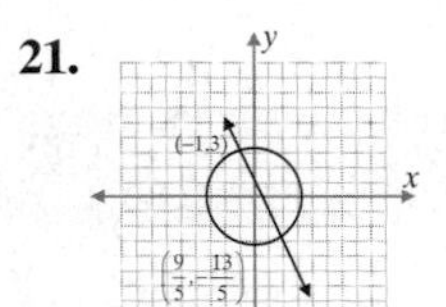

22.

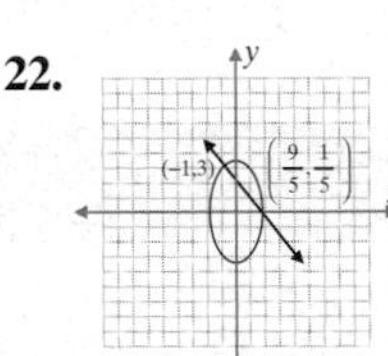

23.

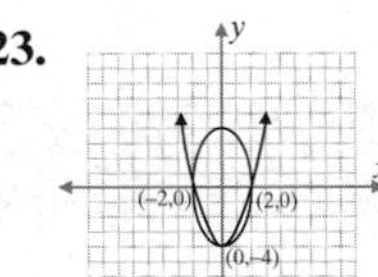

24. Is a function **25.** Is a function **26.** Is not a function **27.** Is a function **28.** $\{(-2, 13), (-1, 8), (0, 5), (1, 4), (2, 5)\}$ **29.** $\{(-2, -28), (-1, -6), (0, 0), (1, -4), (2, -12)\}$

30.

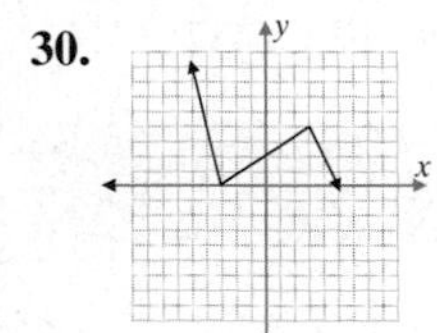

31.

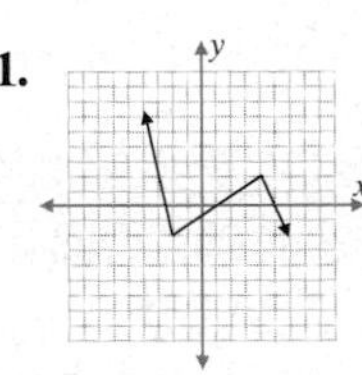

32.

33. Is not a 1-1 function **34.** Is not a 1-1 function

35. Is a 1-1 function **36.** Is a 1-1 function

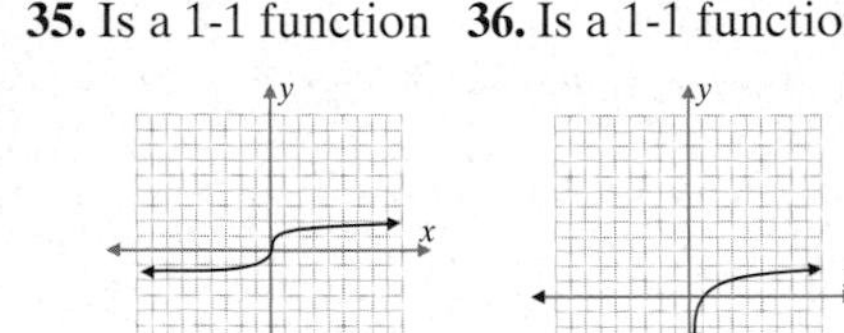

37. a. $x^2 + x + \sqrt{x+2}$ **b.** $x^2 + x - \sqrt{x+2}$ **c.** $\left(x^2 + x\right)\sqrt{x+2}$ **d.** $\dfrac{x^2 + x}{\sqrt{x+2}}$

e. a. $x \geq -2$ b. $x \geq -2$ c. $x \geq -2$ d. $x > -2$

38. a. $x^2 + 2x + 4$ **b.** $x^2 - 2x + 2$ **c.** $2x^3 + x^2 + 6x + 3$ **d.** $\dfrac{x^2+3}{2x+1}$ **e.** a. All real numbers b. All real numbers c. All real numbers d. All real numbers except $x = -\dfrac{1}{2}$

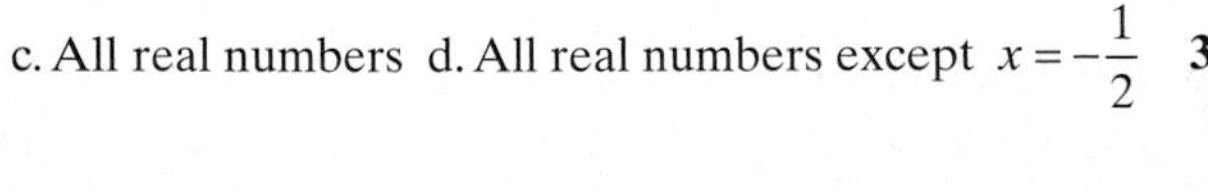

39.

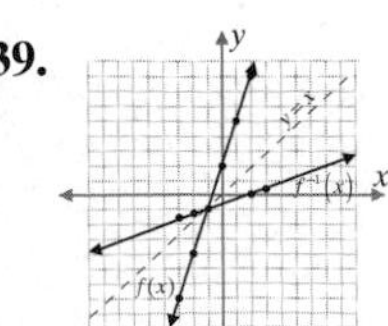

40.

41.

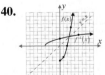

42.

43. 13 clothbound, 30 paperback **44.** 10 mph, 4 mph **45.** 14.616 **46.** $\left|\overline{AB}\right|^2 + \left|\overline{AC}\right|^2 = \left|\overline{BC}\right|^2 = 65$ **47.** $\dfrac{7}{8}$ **48.** $609.20 **49.** 17.33 centuries **50.** 4.6 sec.

Chapter 14

Exercises 14.1, pages 981 – 982

1. 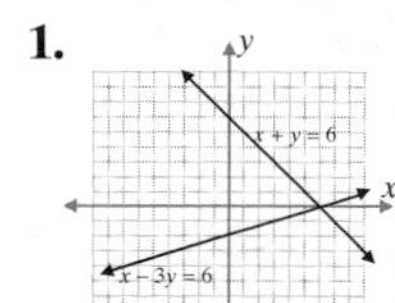

$(6,0)$ **3.** 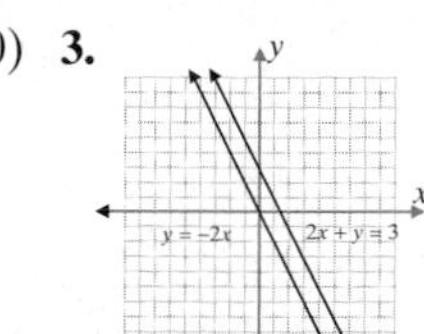

$\varnothing$ **5.** 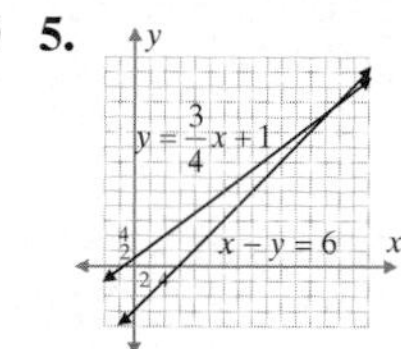

$(28,22)$ **7.** 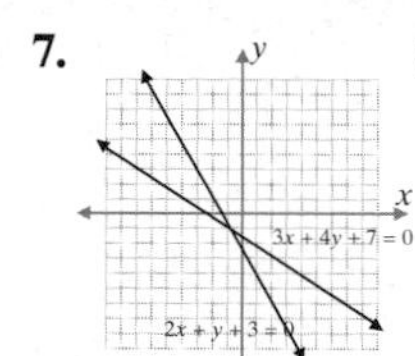

$(-1,-1)$ **9.** 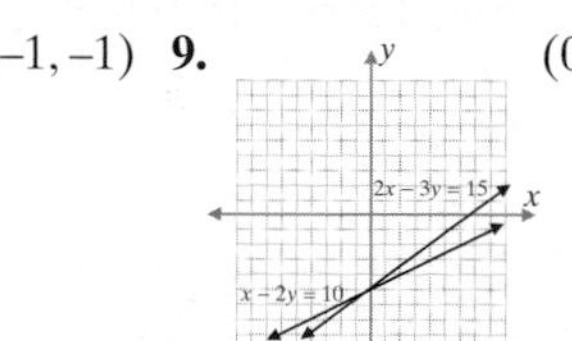

$(0,-5)$

11. $x=2,\ y=1$, consistent **13.** $x=-1,\ y=-3$, consistent **15.** No solution, inconsistent **17.** Infinite solutions, dependent **19.** $x=1,\ y=6$, consistent **21.** $x=\frac{3}{2},\ y=1$, consistent **23.** $x=3,\ y=-\frac{3}{4}$, consistent **25.** No solution, inconsistent **27.** $x=\frac{1}{2}, y=\frac{1}{4}$, consistent **29.** $x=5,\ y=2$, consistent **31.** $x=3, y=3$, consistent **33.** $x=7,\ y=5$, consistent **35.** $x=4,\ y=7$, consistent **37.** $x=\frac{10}{3},\ y=-\frac{8}{3}$ **39.** $x=\frac{4}{5},\ y=-\frac{4}{5}$ **41.** $x=-\frac{8}{5},\ y=-\frac{11}{5}$ **43.** $x=-10, y=8$

Exercises 14.2, pages 986 – 989

1. 23, 79 **3.** 80°, 100° **5.** 100 yards × 45 yards **7.** 40 liters of 12%, 50 liters of 30% **9.** 240 gal. of 5%, 120 gal. of 2% **11.** $87,000 in bonds, $37,000 in certificates **13.** 325 at $3.50/share, 175 at $6.00/share **15.** 40 lbs. at $3.90/lb., 30 lbs. at $2.50/lb. **17.** 16 lbs. at $.70/lb., 4 lbs. at $1.30/lb. **19.** $5.50 for paperback, $9.00 for hardback **21.** 150 legislators voted "for" the bill **23.** 6:00 P.M. **25.** Commercial jet: 300 mph, Private: 125 mph **27.** $a=-2,\ b=3$ **29.** $a=10,\ b=-2$ **31.** $11.20/hr. for labor, $4.80/lb. for materials **33.** 9 lbs. of Ration I, 2 lbs. of Ration II **35.** 65°, 65°, 50°

Exercises 14.3, pages 996 – 999

1. $x=1, y=0, z=1$ **3.** $x=1, y=2, z=-1$ **5.** Infinite solutions **7.** $x=-2, y=9, z=1$ **9.** $x=4, y=1, z=1$ **11.** $x=-2, y=3, z=1$ **13.** No solution **15.** Infinite solutions **17.** $x=\frac{1}{2}, y=\frac{1}{3}, z=-1$ **19.** $x=2, y=1, z=-3$ **21.** 34, 6, 27 **23.** 11 nickels, 7 dimes, 5 quarters **25.** $a=1, b=3, c=-2$ **27.** 19 cm, 24 cm, 30 cm **29.** home: $90,000, lot: $22,000, improve: $11,000 **31.** savings: $30,000, bonds: $55,000, stocks: $15,000 **33.** 100°, 30°, 50° **35.** 3 liters of 10%, 4.5 liters of 30%, 1.5 liters of 40% **37.** A = 2, B = −1, C = 3

Exercises 14.4, pages 1011 – 1013

1. $\begin{bmatrix} 2 & 2 \\ 5 & -1 \end{bmatrix}, \left[\begin{array}{cc|c} 2 & 2 & 13 \\ 5 & -1 & 10 \end{array}\right]$ **3.** $\begin{bmatrix} 7 & -2 & 7 \\ -5 & 3 & 0 \\ 0 & 4 & 11 \end{bmatrix}, \left[\begin{array}{ccc|c} 7 & -2 & 7 & 2 \\ -5 & 3 & 0 & 2 \\ 0 & 4 & 11 & 8 \end{array}\right]$ **5.** $\begin{bmatrix} 3 & 1 & -1 & 2 \\ 1 & -1 & 2 & -1 \\ 0 & 2 & 5 & 1 \\ 1 & 3 & 0 & 3 \end{bmatrix}, \left[\begin{array}{cccc|c} 3 & 1 & -1 & 2 & 6 \\ 1 & -1 & 2 & -1 & -8 \\ 0 & 2 & 5 & 1 & 2 \\ 1 & 3 & 0 & 3 & 14 \end{array}\right]$ **7.** $\begin{cases} -3x+5y=1 \\ -x+3y=2 \end{cases}$

9. $\begin{cases} x+3y+4z=1 \\ 2x-3y-2z=0 \\ x+y=-4 \end{cases}$ **11.** $x=-1, y=2$ **13.** $x=-1, y=-1$ **15.** $x=-1, y=-2, z=3$ **17.** $x=1, y=0, z=1$ **19.** $x=2, y=1, z=-1$ **21.** $x=-2, y=-1, z=5$ **23.** $x=1, y=2, z=1$ **25.** $x=1, y=-3, z=2$ **27.** 52, 40, 77 **29.** 20 small, 32 medium, 16 large **31.** $x=0, y=-4$ **33.** $x=2, y=1, z=7$ **35.** $x=\frac{13}{12}, y=\frac{5}{4}, z=\frac{8}{3}$ **37.** $x=-\frac{1}{3}, y=1, z=\frac{16}{3}$ **39.** Answers will vary.

Exercises 14.5, pages 1020 – 1022

1. −22 **3.** −212 **5.** 11 **7.** 3 **9.** 47 **11.** 36 **13.** −3 **15.** −4 **17.** −25 **19.** 20 **21.** $x = 7$ **23.** $x = -7$ **25.** $x = -3$

27. $2x - 7y = -11$ **29.** $A = 1$

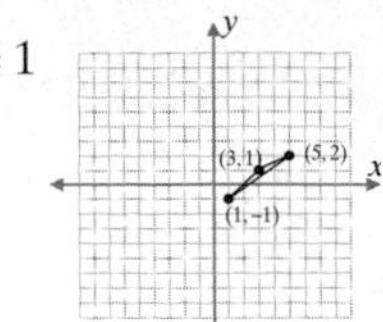

31. $A = \frac{31}{2}$

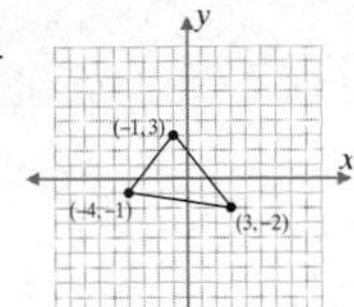

33. −25 **35.** −33.28 **37.** 0

Exercises 14.6, pages 1028 – 1029

1. $x = 4, y = 3$ **3.** $x = \frac{2}{3}, y = -\frac{1}{4}$ **5.** No solution **7.** $x = -\frac{1}{4}, y = \frac{3}{2}$ **9.** $x = \frac{31}{17}, y = \frac{2}{17}$ **11.** $x = \frac{39}{44}, y = \frac{41}{44}$

13. $x = \frac{18}{7}, y = -\frac{3}{7}$ **15.** $x = -\frac{7}{61}, y = \frac{266}{183}$ **17.** $x = \frac{210}{41}, y = -\frac{40}{123}$ **19.** $x = \frac{525}{124}, y = \frac{109}{93}$ **21.** $x = -2, y = 1, z = 3$

23. No solution **25.** $x = 4, y = 2, z = 6$ **27.** Infinite solutions **29.** $x = -\frac{2}{3}, y = \frac{11}{3}, z = 2$ **31.** 6 feet, 17 feet, 20 feet
33. \$3,500,000 in mutual funds; \$2,500,000 in stocks

Exercises 14.7, pages 1037 – 1039

1. **3.** **5.** **7.** **9.** **11.**

13. **15.** **17.** **19.** **21.**

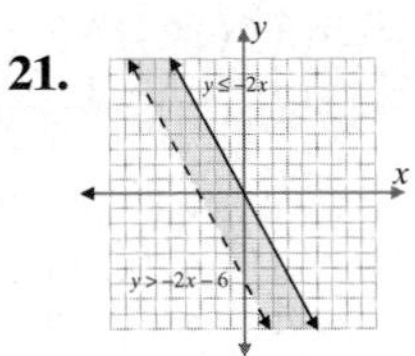

23. **25.** **27.** **29.** **31.**

33. **35.** **37.** **39.** **41.**

43. $F = -2$
45. $F = 26$
47. $F = 37$
49. \$225
51. \$546

Chapter 14 Test, pages 1046 – 1047

1. $x = 3, y = -1$

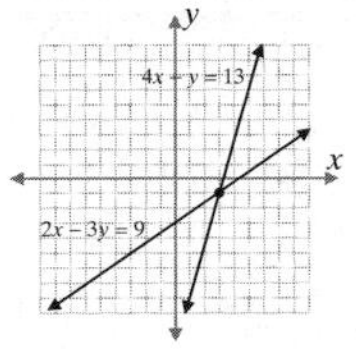

2. $x = 7, y = 2$ **3.** Inconsistent, No solution **4.** $x = \frac{1}{2}, y = \frac{1}{4}$ **5.** $x = 11, y = -7$ **6.** $a = -3, b = 4$

7. 8 ft. × 23 ft. **8.** $x = -1, y = 1, z = -2$ **9.** Inconsistent, No solution **10. a.** $\begin{bmatrix} 1 & 2 & -3 \\ 1 & -1 & -1 \\ 1 & 3 & 2 \end{bmatrix}_{3\times3}$, $\left[\begin{array}{ccc|c} 1 & 2 & -3 & -11 \\ 1 & -1 & -1 & 2 \\ 1 & 3 & 2 & -4 \end{array}\right]_{3\times4}$ **b.** $x = 1, y = -3, z = 2$

11. 60 33-cent stamps, 20 55-cent stamps, 10 78-cent stamps. **12.** 42 **13.** 5 **14.** $x = 6$ **15.** $x = \frac{17}{3}$ **16.** $x = -\frac{78}{5}, y = \frac{38}{5}$

17. $x = -\frac{5}{26}, y = \frac{33}{26}, z = \frac{9}{26}$ **18.** 50°, 60°, 70° **19.**

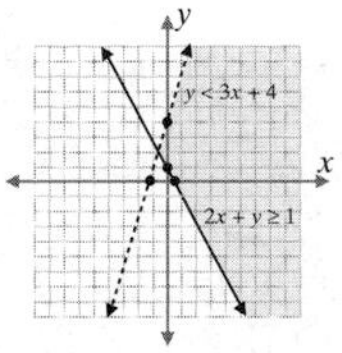

20. $x = -\frac{53}{11}, y = \frac{29}{11}, z = -\frac{2}{11}$ **21.** $D_x = -53$

22. $F = \frac{34}{3}$

Chapter 14 Cumulative Review, pages 1048 – 1050

1. a. $-3, 0, 1$ **b.** $-3, -\frac{1}{2}, 0, \frac{5}{8}, 1$ **c.** $-\sqrt{13}, \sqrt{2}, \pi$ **d.** $-\sqrt{13}, -3, -\frac{1}{2}, 0, \frac{5}{8}, 1, \sqrt{2}, \pi$ **2.** 0 **3.** 320 **4.** $x = 2$ **5.** $x = -6$

6. $x = -\frac{29}{2}$ or $x = \frac{27}{2}$ **7.** $x = -\frac{2}{3}$ **8.** $m = \frac{2gK}{v^2}$ **9.**

$(-\infty, 4)$ **10.**

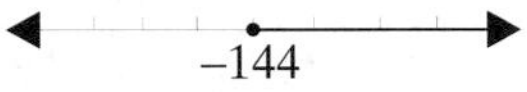

$[-144, \infty)$

11. $\frac{1}{x^8}$ **12.** $\frac{x^5}{y}$ **13.** $\frac{6}{x^2}$ **14.** 2.7×10^{-3} **15. a.** $m = -\frac{1}{2}$ **b.** $x + 2y = -1$ **c.**

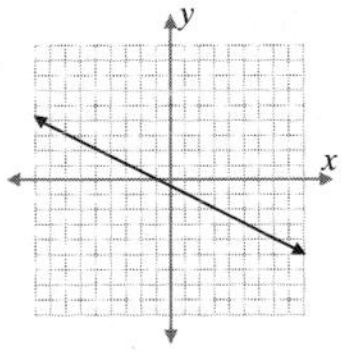

16. $y = \frac{2}{5}x$

17. x-intercepts: $(-3.45, 0)$ and $(1.45, 0)$ maximum: $(-1, 6)$

18. $(-1.65, -0.29)$ and $(3.65, 10.29)$ **19.** $x = 1, y = 4$

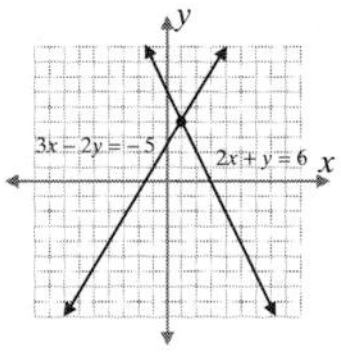

20. $x = 1, y = 3$ **21.** $x = 4, y = 3, z = 2$

22. $x = 2, y = \frac{2}{5}, z = -\frac{1}{2}$ **23.** 3 batches of Choc-O-Nut and 2 batches of Chocolate Krunch **24.** \$4200 at 7%, \$2800 at 8%

25. $a = 2, b = 3, c = 4$ **26.** solution: $(-1, 6)$

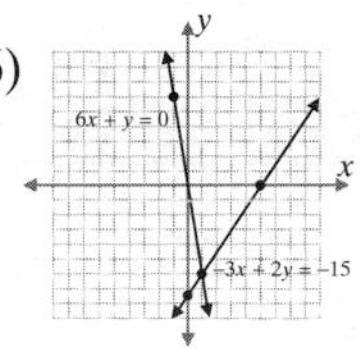

27. $x = \frac{8}{5}, y = \frac{27}{10}$ **28.** $x = -3, y = 0, z = 5$ **29.** $x = 1, y = 1$

30. $x = 1, y = 2, z = 3$ **31.**

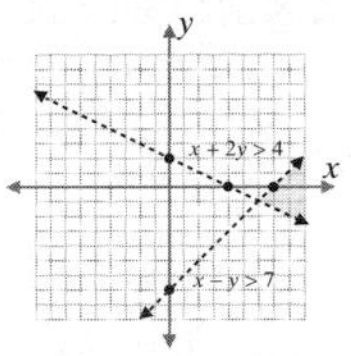

32.

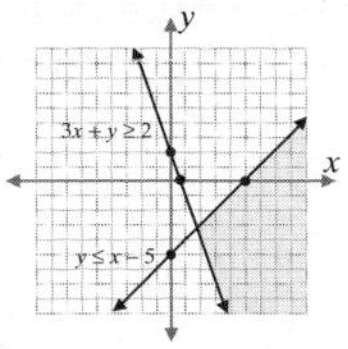

Chapter 15

Exercises 15.1, pages 1057 – 1058

1. $1, 3, 5, 7$ **3.** $2, \frac{3}{2}, \frac{4}{3}, \frac{5}{4}$ **5.** $2, 6, 12, 20$ **7.** $2, 4, 8, 16$ **9.** $-2, 5, -10, 17$ **11.** $-\frac{1}{5}, \frac{1}{7}, -\frac{1}{9}, \frac{1}{11}$ **13.** $1, 0, -1, 0$ **15.** $0, 1, 0, 1$

17. $\{4n+1\}$ **19.** $\{(-1)^{n+1}(2n-1)\}$ **21.** $\{n^2\}$ **23.** $\left\{\frac{1}{n+2}\right\}$ **25.** $\{n^2+1\}$ **27.** Decreasing **29.** Decreasing **31.** Increasing

33. $\left\{250\left(\frac{2}{5}\right)^n\right\}$; 6.4 cm when $n = 4$ **35.** $\left\{20{,}000 \cdot \left(\frac{97}{100}\right)^n\right\}$; 17,175 students when $n = 5$

Exercises 15.2, pages 1062 – 1063

1. $2, 7, 15, 26$ **3.** $\frac{1}{2}, \frac{7}{6}, \frac{23}{12}, \frac{163}{60}$ **5.** $1, -3, 6, -10$ **7.** $\frac{1}{2}, \frac{3}{4}, \frac{7}{8}, \frac{15}{16}$ **9.** $-\frac{2}{3}, -\frac{2}{9}, -\frac{14}{27}, -\frac{26}{81}$ **11.** $2+4+6+8+10=30$

13. $5+6+7+8+9=35$ **15.** $\frac{1}{2}+\frac{1}{3}+\frac{1}{4}=\frac{13}{12}$ **17.** $2+4+8=14$ **19.** $16+25+36+49+64=190$

21. $3+1+(-1)+(-3)=0$ **23.** $6+(-12)+20+(-30)=-16$ **25.** $\frac{1}{2}+\frac{2}{3}+\frac{3}{4}+\frac{4}{5}+\frac{5}{6}=\frac{71}{20}$ **27.** $\sum_{k=1}^{5}(2k-1)$ **29.** $\sum_{k=1}^{5}(-1)^k$

31. $\sum_{k=2}^{6}(-1)^k\left(\frac{1}{k^3}\right)$ **33.** $\sum_{k=4}^{15}\frac{k}{k+1}$ **35.** $\sum_{k=5}^{12}\frac{k+1}{k^2}$ **37.** 7 **39.** 48 **41.** 10 **43.** 295 **45.** -152

Exercises 15.3, pages 1071 – 1072

1. Arithmetic sequence, $d = 3, \{3n-1\}$ **3.** Arithmetic sequence, $d = -2, \{9-2n\}$ **5.** Not an arithmetic sequence

7. Arithmetic sequence, $d = -4, \{10-4n\}$ **9.** Arithmetic sequence, $d = \frac{1}{2}, \left\{\frac{n-1}{2}\right\}$ **11.** $1, 3, 5, 7, 9$; arithmetic sequence

13. $-1, 4, -7, 10, -13$; not an arithmetic sequence **15.** $-1, -7, -13, -19, -25$; arithmetic sequence

17. $\frac{20}{3}, \frac{19}{3}, 6, \frac{17}{3}, \frac{16}{3}$; arithmetic sequence **19.** $\frac{1}{2}, \frac{1}{4}, \frac{1}{6}, \frac{1}{8}, \frac{1}{10}$; not an arithmetic sequence **21.** $\left\{\frac{2n+1}{3}\right\}$ **23.** $\{9-2n\}$

25. $\left\{\frac{17+3n}{2}\right\}$ **27.** $\{63-5n\}$ **29.** $\left\{\frac{81+3n}{2}\right\}$ **31.** 232 **33.** 44 **35.** 7 **37.** 19 **39.** 154 **41.** 235 **43.** 126 **45.** 231 **47.** $\frac{143}{3}$

49. 132 **51.** 171 **53.** -100 **55.** 9 days **57.** \$64,250 **59.** Answers will vary.

Exercises 15.4, pages 1082 – 1084

1. Not a geometric sequence **3.** Geometric sequence, $r = -\frac{1}{2}$, $\left\{3\left(-\frac{1}{2}\right)^{n-1}\right\}$ **5.** Geometric sequence, $r = \frac{3}{8}$, $\left\{\frac{32}{27}\left(\frac{3}{8}\right)^{n-1}\right\}$

7. Not a geometric sequence **9.** Geometric sequence, $r = -\frac{1}{4}$, $\left\{48\left(-\frac{1}{4}\right)^{n-1}\right\}$ **11.** $9, -27, 81, -243$; geometric sequence

13. $\frac{2}{3},\frac{4}{3},2,\frac{8}{3}$; not a geometric sequence **15.** $-\frac{8}{5},\frac{32}{25},-\frac{128}{125},\frac{512}{625}$; geometric sequence **17.** $3\sqrt{2},6,6\sqrt{2},12$; geometric sequence **19.** 0.3, –0.09, 0.027, –0.0081; geometric sequence **21.** $\left\{3(2)^{n-1}\right\}$ **23.** $\left\{\frac{1}{3}\left(-\frac{1}{2}\right)^{n-1}\right\}$ **25.** $\left\{\left(\sqrt{2}\right)^{n-1}\right\}$

27. $\left\{\frac{1}{3}(3)^{n-1}\right\}$ **29.** $\left\{-5\left(-\frac{3}{4}\right)^{n-1}\right\}$ **31.** $\frac{1}{4}$ **33.** $\frac{32}{9}$ **35.** $\frac{1}{4}$ or $-\frac{1}{4}$ **37.** $n = 4$ **39.** 363 **41.** $\frac{1023}{64}$ **43.** $-\frac{333}{64}$

45. $\frac{422}{243}$ **47.** $\frac{75}{128}$ **49.** 4 **51.** $-\frac{1}{3}$ **53.** $\frac{2}{9}$ **55.** $\frac{4}{11}$ **57.** \$13,248.23 **59.** \$4352 **61.** 12.8 grams

63. a. 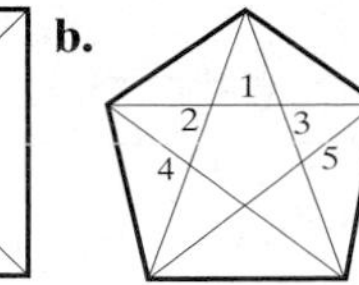

b. 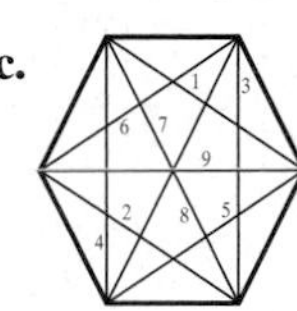

65. In the formula for S_n, $r^n \to \infty$ (or $r^n \to -\infty$) if $|r| > 1$

Exercises 15.5, pages 1092 – 1093

1. 56 **3.** $\frac{1}{15}$ **5.** 4 **7.** $(n-1)!$ **9.** $(k+3)(k+2)(k+1)$ **11.** 20 **13.** 35 **15.** 1 **17.** $x^7+7x^6y+21x^5y^2+35x^4y^3$
19. $x^9+9x^8+36x^7+84x^6$ **21.** $x^5+15x^4+90x^3+270x^2$ **23.** $x^6+12x^5y+60x^4y^2+160x^3y^3$
25. $2187x^7-5103x^6y+5103x^5y^2-2835x^4y^3$ **27.** $x^{18}-36x^{16}y+576x^{14}y^2-5376x^{12}y^3$
29. $x^6+6x^5y+15x^4y^2+20x^3y^3+15x^2y^4+6xy^5+y^6$ **31.** $x^7-7x^6+21x^5-35x^4+35x^3-21x^2+7x-1$
33. $243x^5+405x^4y+270x^3y^2+90x^2y^3+15xy^4+y^5$ **35.** $x^4+8x^3y+24x^2y^2+32xy^3+16y^4$
37. $81x^4-216x^3y+216x^2y^2-96xy^3+16y^4$ **39.** $243x^{10}-405x^8y+270x^6y^2-90x^4y^3+15x^2y^4-y^5$ **41.** $3360x^6y^4$
43. $1{,}140{,}480x^8$ **45.** $-27{,}500x^6y^{18}$ **47.** 1.062 **49.** 0.808 **51.** 64.363 **53.** 0.851

Exercises 15.6, pages 1098 – 1099

1. 840 **3.** 24 **5.** 120 **7.** 60,480 **9.** 1,814,400 **11.** 110 **13.** 336 **15.** 72 **17.** 360 **19.** 64 **21.** 576 **23.** 288 **25.** 69,300

Exercises 15.7, pages 1102 – 1103

1. 35 **3.** 1 **5.** 84 **7.** 6 **9.** 120 **11.** 816 **13.** 792 **15.** 190 **17.** 30 **19.** 108 **21.** 14,700 **23.** 66,528 **25.** 600

27. a. 1 2 **b.** 1 2 3 4 5 **c.** 1 2 3 4 5 6 7 8 9 **29.** 45

Chapter 15 Test, page 1109

1. $\frac{1}{4}, \frac{1}{7}, \frac{1}{10}, \frac{1}{13}, \ldots$; neither **2.** $\left\{\frac{n}{2n+1}\right\}$ **3.** 26 **4.** $a_n = 8 - 2n$ **5.** 243 **6.** $\frac{1}{8}$ **7.** $a_n = \left(\sqrt{3}\right)^{n-2}$ **8.** $\frac{21}{16}$ **9.** 68 **10.** $-\frac{40}{729}$

11. $\frac{15}{100} + \frac{15}{10{,}000} + \ldots = \frac{5}{33}$ **12.** 74 **13.** \$13,050.16 **14.** 462 **15.** $32x^5 - 80x^4y + 80x^3y^2 - 40x^2y^3 + 10xy^4 - y^5$ **16.** $5670x^4y^4$

17. 151,200 **18.** 792 **19.** 720 **20.** 168 **21.** Answers will vary.

Chapter 15 Cumulative Review, pages 1110 – 1112

1. $11x + 3$ **2.** $-x^2 + 3x + 5$ **3.** $(4x+3)(16x^2 - 12x + 9)$ **4.** $(3x-5)(2x+9)$ **5.** $(2x+1)(5x+7)(x-2)$ **6.** $\frac{11x+12}{15}$

7. $\frac{-6x-4}{(x+4)(x-4)(x-1)}$ **8.** $\frac{1}{x^2}$ **9.** $\frac{8^{1/2}y^{1/3}}{x^2}$ **10.** $\frac{2x^{\frac{4}{3}}}{3y}$ **11.** $3\sqrt{3x}$ **12.** $\frac{-13+11i}{29}$ **13.** $x > 3$ 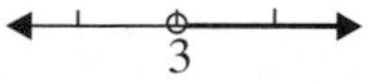

14. $-\frac{5}{2} < x < \frac{9}{4}$ 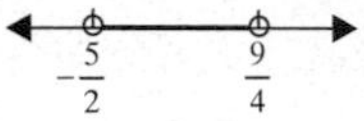

15. $x = \frac{26}{7}$ **16.** $x = \frac{-2 \pm i\sqrt{2}}{2}$ **17.** $x = 4$ **18.** $x = \frac{39}{17}$ **19.** $x = 5$ **20.** $x = e^{2.4} \approx 11.02$

21. No solution **22.** $n = \frac{A-P}{Pi}$ **23.** $x = 2, y = -5$ **24.** $x = 0, y = 2, z = -1$ **25. a.** $y = \frac{x+7}{2}$ **b.** $2x^2 - 5$ **c.** $2x + 1$

26. 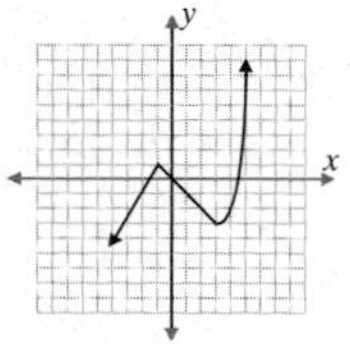

27. 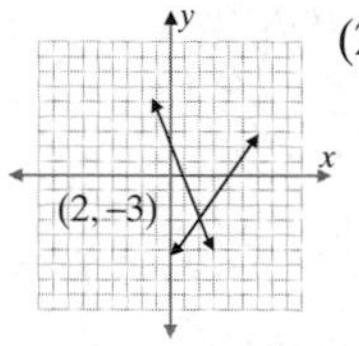

(2, −3) **28.** 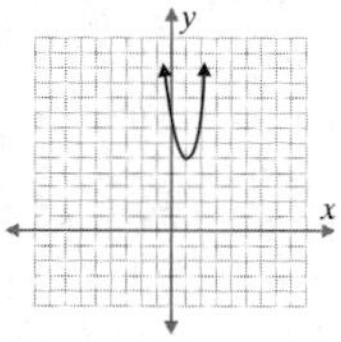

29. 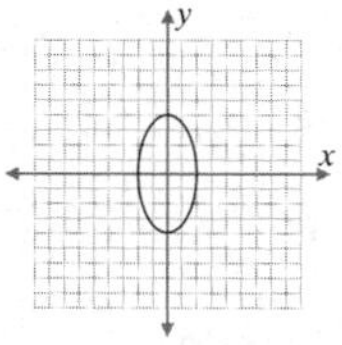

30. 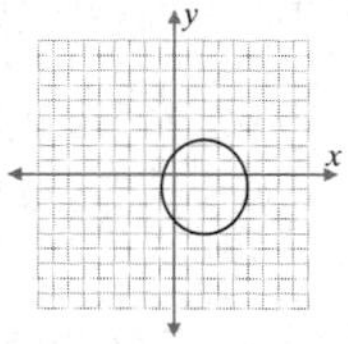

31. a. $a_n = 10 - 2n$ **b.** -10 **32. a.** $a_n = 16\left(\frac{1}{2}\right)^{n-1}$ **b.** $\frac{63}{2}$ **33.** $x^6 + 12x^5y + 60x^4y^2 + 160x^3y^3 + 240x^2y^4 + 192xy^5 + 64y^6$ **34.** 210

35. 1680 **36.** 2002 **37.** 10 mph, 30 mph **38.** 16 lbs. at \$1.40, 4 lbs. at \$2.60 **39.** 16 ft. by 26 ft. **40.** 12.56 hrs. **41.** \$16,058.71

42. 443.7 liters **43.** 12 **44.** 3024 **45.** 150 **46.** $x \approx -0.4746266, 1.395337$ **47.** $x \approx -1.9646356, 1.0580064$

48. $x \approx 0.24083054, 2.7336808$ **49.** $x = 1$ **50.** $x = -3, 1, 2$

Appendix

Appendix 1, pages 1117 – 1118

1. 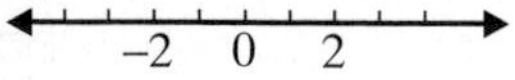

$(-\infty, \infty)$ **3.** 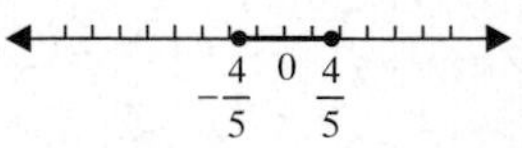

$\left[-\frac{4}{5}, \frac{4}{5}\right]$ **5.** 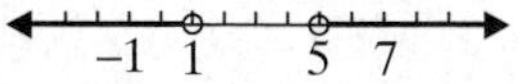

$(-\infty, 1) \cup (5, \infty)$

7. 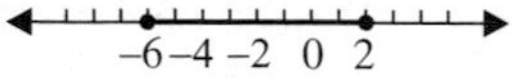

$[-6, 2]$ **9.** 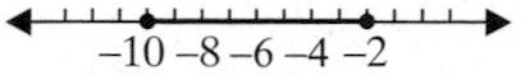

$[-10, -2]$ **11.** No solution, $\varnothing$

13. 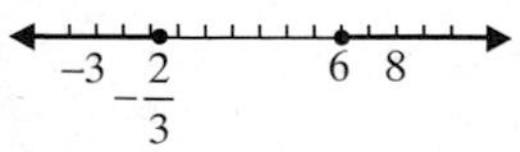

$\left(-\infty, -\frac{2}{3}\right] \cup [6, \infty)$ **15.** 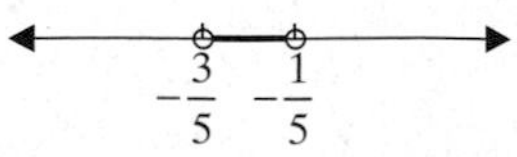

$\left(-\frac{3}{5}, -\frac{1}{5}\right)$

17. 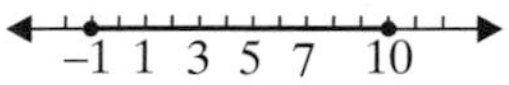

$[-1, 10]$ **19.** 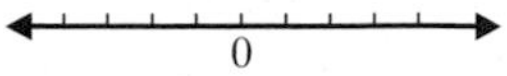

$(-\infty, \infty)$ **21.** 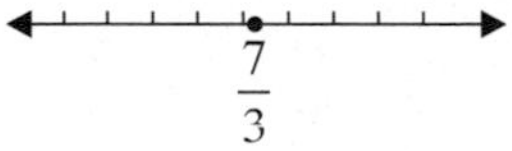

$\left[\frac{7}{3}\right]$

23. $|x-2| \le 6$ **25.** $\left|x-\frac{1}{8}\right| < \frac{5}{8}$ **27.** $|x-8| < 2$ **29.** $|x+6| \le 4$ **31.** $|x| \le 2$ **33.** $|x+4| \le 6$ **35.** $\left|x-\frac{12}{7}\right| > \frac{16}{7}$ **37.** $|x| \ge 6$

39. $\left|x-\frac{7}{2}\right| > \frac{7}{2}$ **41. a.** 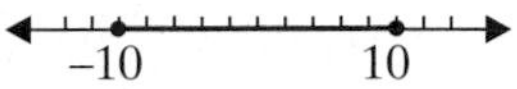

b. $|x| \le 10$ **c.** $[-10, 10]$, closed interval **43. a.** 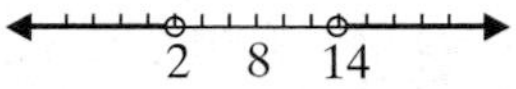

b. $|x-8| > 6$ **c.** $(-\infty, 2) \cup (14, \infty)$, open interval **45. a.** 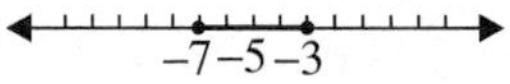

b. $|x+5| \le 2$ **c.** $[-7, -3]$, closed interval

Appendix 2, pages 1123 – 1124

1. a. $x-9$ **b.** $c=3, P(3)=0$ **3. a.** $x-4$ **b.** $c=1, P(1)=0$ **5. a.** $4x+2$ **b.** $c=-5, P(-5)=0$ **7. a.** $2x+4+\frac{7}{x+2}$ **b.** $c=-2, P(-2)=7$

9. a. $x^2-4x+33+\frac{-265}{x+8}$ **b.** $c=-8, P(-8)=-265$ **11. a.** $3x^2+4x+3+\frac{8}{x-1}$ **b.** $c=1, P(1)=8$ **13. a.** $2x^2+x-3$

b. $c=-2, P(-2)=0$ **15. a.** $4x^2-6x+9+\frac{-17}{x+2}$ **b.** $c=-2, P(-2)=-17$ **17. a.** $x^2+7x+55+\frac{388}{x-7}$ **b.** $c=7, P(7)=388$

19. a. $2x^2-2x+6+\frac{-27}{x+3}$ **b.** $c=-3, P(-3)=-27$ **21. a.** $x^3+2x+5+\frac{17}{x-3}$ **b.** $c=3, P(3)=17$

23. a. $x^3+2x^2+6x+9+\frac{23}{x-2}$ **b.** $c=2, P(2)=23$ **25. a.** $x^3+\frac{x^2}{2}-\frac{3}{4}x-\frac{3}{8}+\frac{45}{16\left(x-\frac{1}{2}\right)}$ **b.** $c=\frac{1}{2}, P\left(\frac{1}{2}\right)=\frac{45}{16}$

27. a. $x^4+x^3+x^2+x+1$ **b.** $c=1, P(1)=0$ **29. a.** $x^3-\frac{14}{5}x^2+\frac{56}{25}x-\frac{224}{125}+\frac{3396}{625\left(x+\frac{4}{5}\right)}$ **b.** $c=-\frac{4}{5}, P\left(-\frac{4}{5}\right)=\frac{3396}{625}$

31. See page 1122. **33. a.** $x^2-\frac{7}{2}x+\frac{13}{4}+\frac{73}{4(2x-1)}$; $2x^2-7x+\frac{13}{2}+\frac{73}{4\left(x-\frac{1}{2}\right)}$ **b.** Answers will vary. **c.** Answers will vary.

INDEX

A

B

C

D

E

F

G

H

O

P

Q

R

S

T

U

V

W

X

Y

Z

CHAPTER 9 Systems of Linear Equations I

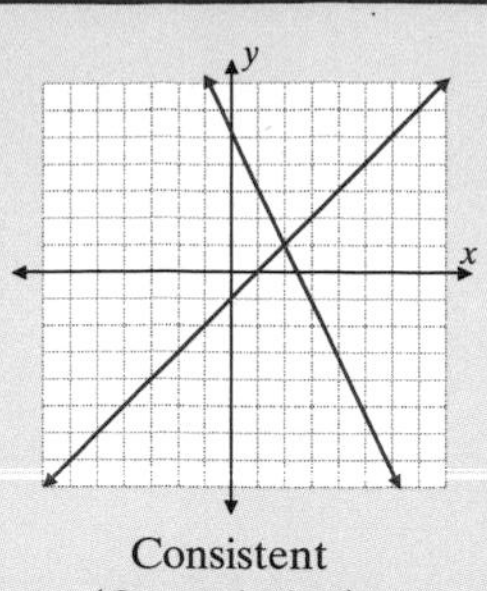

Consistent
(One solution)

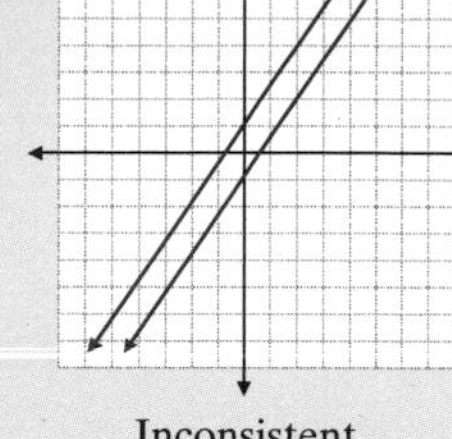

Inconsistent
(No solution)

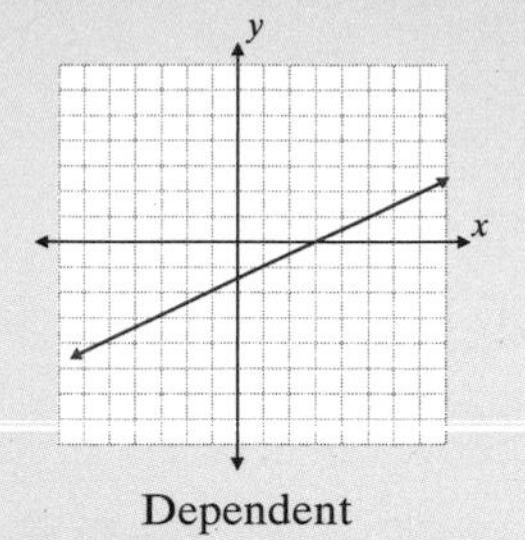

Dependent
(Infinite number of solutions)

CHAPTER 10 Roots, Radicals, and Complex Numbers

Properties of Square Roots:

1. If $b^2 = a$, then b is called the **square root** of a $(a \geq 0)$.
2. $\sqrt{ab} = \sqrt{a}\sqrt{b}$, where a and b are positive real numbers.
3. $\sqrt{\frac{a}{b}} = \frac{\sqrt{a}}{\sqrt{b}}$, where a and b are positive real numbers.
4. $\sqrt{x^{2m}} = |x^m|$, m is a positive integer.
5. $\sqrt{x^{2m+1}} = |x^m|\sqrt{x}$, m is a positive integer.

Properties of Radicals:

1. If $b^n = a$, then $b = \sqrt[n]{a} = a^{\frac{1}{n}}$, n is a positive integer.
2. $a^{\frac{m}{n}} = \left(a^{\frac{1}{n}}\right)^m = \left(a^m\right)^{\frac{1}{n}}$ or $a^{\frac{m}{n}} = \left(\sqrt[n]{a}\right)^m = \sqrt[n]{a^m}$, where n is a positive integer and m is any integer.

Complex Numbers:

Numbers of the form $a + bi$.

Powers of i:

$i = \sqrt{-1}$ and $i^2 = \left(\sqrt{-1}\right)^2 = -1$

CHAPTER 11 Quadratic Equations

Square Root Property:

If $x^2 = c$, then $x = \pm\sqrt{c}$.

If $(x - a)^2 = c$, then $x - a = \pm\sqrt{c}$ $\left(\text{or } x = a \pm \sqrt{c}\right)$.

Quadratic Formula:

$$x = \frac{-b \pm \sqrt{b^2 - 4ac}}{2a}, \quad a \neq 0$$

Discriminant: $b^2 - 4ac$

$b^2 - 4ac > 0 \rightarrow$ Two Real Solutions

$b^2 - 4ac = 0 \rightarrow$ One Real Solution

$b^2 - 4ac < 0 \rightarrow$ Two Nonreal Solutions

CHAPTER 12 Quadratic Functions and Conic Sections

Parabolas:

Vertical Parabolas:

Parabolas of the form $y = ax^2 + bx + c$:

If $a > 0$, the parabola opens upward.

If $a < 0$, the parabola opens downward.

Vertex: $\left(-\frac{b}{2a}, \frac{4ac - b^2}{4a}\right)$

Line of Symmetry: $x = -\frac{b}{2a}$

Parabolas of the form $y = a(x - h)^2 + k$:

Vertex: (h, k)

Line of Symmetry: $x = h$

y
Line of Symmetry
x
Vertex

Horizontal Parabolas:

Parabolas of the form $x = ay^2 + by + c$ or $x = a(y - k)^2 + h$

If $a > 0$, the parabola opens right.

If $a < 0$, the parabola opens left.

Vertex: (h, k)

Line of Symmetry: $y = k$

Vertical and Horizontal Translations:

Given a graph $y = f(x)$, the graph of $y = f(x - h) + k$ is:

1. a horizontal translation of $f(x)$ by h units and
2. a vertical translation of $f(x)$ by k units.

Distance Formula (distance between two points):

$$d = \sqrt{\left(x_2 - x_1\right)^2 + \left(y_2 - y_1\right)^2}$$

Midpoint Formula:

$$\left(\frac{x_1 + x_2}{2}, \frac{y_1 + y_2}{2}\right)$$

continued on next page ...

CHAPTER 12 Quadratic Functions and Conic Sections (continued)

Circles:

Standard form: $(x-h)^2+(y-k)^2=r^2$

Center: (h, k)

Radius: r

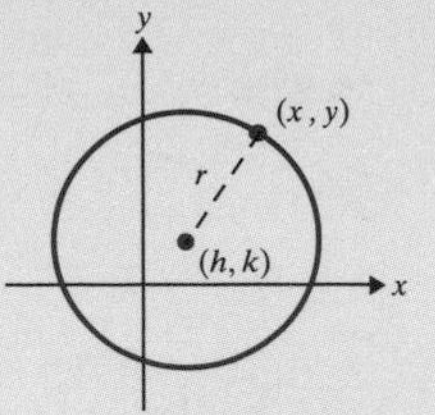

Ellipses:

Standard form: $\dfrac{(x-h)^2}{a^2}+\dfrac{(y-k)^2}{b^2}=1$ or $\dfrac{(x-h)^2}{b^2}+\dfrac{(y-k)^2}{a^2}=1;\ a^2 \ge b^2$

Center: (h, k)

Length of Major Axis: $2a$

Length of Minor Axis: $2b$

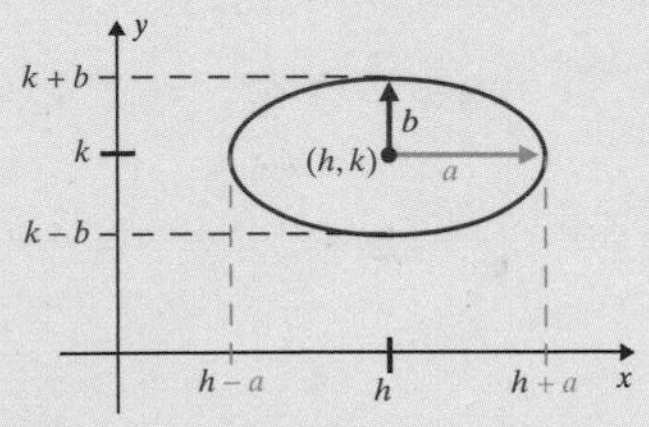

Hyperbolas:

Centered at the origin:

1. Standard form: $\dfrac{x^2}{a^2}-\dfrac{y^2}{b^2}=1$

   Curves open left and right.

   x-intercepts: $(a, 0)$ and $(-a, 0)$

   Asymptotes: $y=\dfrac{b}{a}x$ and $y=-\dfrac{b}{a}x$

2. Standard form: $\dfrac{y^2}{a^2}-\dfrac{x^2}{b^2}=1$

   Curves open up and down.

   y-intercepts: $(0, a)$ and $(0, -a)$

   Asymptotes: $y=\dfrac{a}{b}x$ and $y=-\dfrac{a}{b}x$

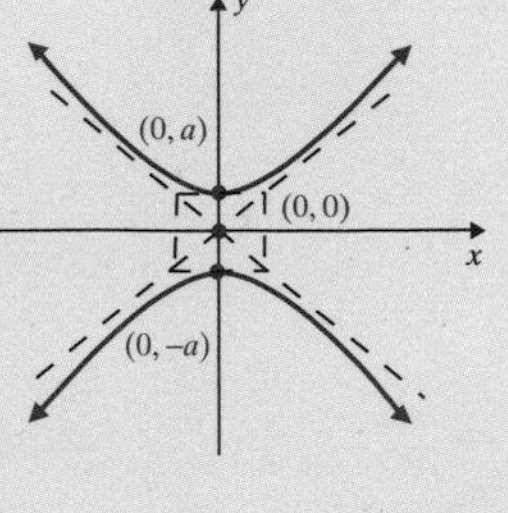

Centered at (h, k):

Standard form: $\dfrac{(x-h)^2}{a^2}-\dfrac{(y-k)^2}{b^2}=1$ or $\dfrac{(y-k)^2}{a^2}-\dfrac{(x-h)^2}{b^2}=1$

CHAPTER 13 Exponential and Logarithmic Functions

Algebraic Operations with Functions:

$(f+g)(x)=f(x)+g(x)$

$(f-g)(x)=f(x)-g(x)$

$(f\cdot g)(x)=f(x)\cdot g(x)$

$\left(\dfrac{f}{g}\right)(x)=\dfrac{f(x)}{g(x)},\quad g(x)\neq 0$

Composite Function:

$(f\circ g)(x)=f(g(x))$

One-to-One Function:

A function is a **1-1 function** if for each value of y in the range there is only one corresponding value of x in the domain.

Inverse Function:

1. If f is a 1-1 function with ordered pairs of the form (x, y), then its **inverse function**, denoted as f^{-1}, is also a 1-1 function with ordered pairs of the form (y, x).
2. If f and g are 1-1 functions and $f(g(x))=x$ for all x in D_g and $g(f(x))=x$ for all x in D_f, then f and g are **inverse functions**. That is, $g=f^{-1}$ and $f=g^{-1}$.

Logarithm:

$x=b^y$ if and only if $y=\log_b x$, $b>0$ and $b\neq 1$

Properties of Logarithms:

For $b, x, y>0$, $b\neq 1$, and any real number r:

1. $\log_b 1=0$
2. $\log_b b=1$
3. $x=b^{\log_b x}$
4. $\log_b xy=\log_b x+\log_b y$
5. $\log_b \dfrac{x}{y}=\log_b x-\log_b y$
6. $\log_b x^r=r\cdot\log_b x$

Properties of Equations with Exponents and Logarithms:

For $b>0$ and $b\neq 1$:

1. If $b^x=b^y$, then $x=y$.
2. If $x=y$, then $b^x=b^y$.
3. If $\log_b x=\log_b y$, then $x=y$ $(x>0$ and $y>0)$.
4. If $x=y$, then $\log_b x=\log_b y$ $(x>0$ and $y>0)$.

Change-of-Base:

$$\log_b x=\frac{\log_a x}{\log_a b}$$